Atomic numbers and atomic weights[a]

Actinium	Ac	89	227.0278
Aluminum	Al	13	26.98154
Americium	Am	95	(243)
Antimony	Sb	51	121.75
Argon	Ar	18	39.948
Arsenic	As	33	74.9216
Astatine	At	85	(210)
Barium	Ba	56	137.33
Berkelium	Bk	97	(247)
Beryllium	Be	4	9.01218
Bismuth	Bi	83	208.9804
Boron	B	5	10.81
Bromine	Br	35	79.904
Cadmium	Cd	48	112.41
Calcium	Ca	20	40.08
Californium	Cf	98	(251)
Carbon	C	6	12.011
Cerium	Ce	58	140.12
Cesium	Cs	55	132.9054
Chlorine	Cl	17	35.453
Chromium	Cr	24	51.996
Cobalt	Co	27	58.9332
Copper	Cu	29	63.546
Curium	Cm	96	(247)
Dysprosium	Dy	66	162.50
Einsteinium	Es	99	(254)
Erbium	Er	68	167.26
Europium	Eu	63	151.96
Fermium	Fm	100	(257)
Fluorine	F	9	18.99840
Francium	Fr	87	(223)
Gadolinium	Gd	64	157.25
Gallium	Ga	31	69.72
Germanium	Ge	32	72.59
Gold	Au	79	196.9665
Hafnium	Hf	72	178.49
Helium	He	2	4.00260
Holmium	Ho	67	164.9304
Hydrogen	H	1	1.0079
Indium	In	49	114.82
Iodine	I	53	126.9045
Iridium	Ir	77	192.22
Iron	Fe	26	55.847
Krypton	Kr	36	83.80
Lanthanum	La	57	138.9055
Lawrencium	Lr	103	(260)
Lead	Pb	82	207.2
Lithium	Li	3	6.941
Lutetium	Lu	71	174.97
Magnesium	Mg	12	24.305
Manganese	Mn	25	54.9380
Mendelevium	Md	101	(258)
Mercury	Hg	80	200.59
Molybdenum	Mo	42	95.94
Neodymium	Nd	60	144.24
Neon	Ne	10	20.179
Neptunium	Np	93	237.0482
Nickel	Ni	28	58.70
Niobium	Nb	41	92.9064
Nitrogen	N	7	14.0067
Nobelium	No	102	(259)
Osmium	Os	76	190.2
Oxygen	O	8	15.9994
Palladium	Pd	46	106.4
Phosphorous	P	15	30.97376
Platinum	Pt	78	195.09
Plutonium	Pu	94	(244)
Polonium	Po	84	(209)
Potassium	K	19	39.0983
Praseodymium	Pr	59	140.9077
Promethium	Pm	61	(145)
Protactinium	Pa	91	231.0359
Radium	Ra	88	226.0254
Radon	Rn	86	(222)
Rhenium	Re	75	186.207
Rhodium	Rh	45	102.9055
Rubidium	Rb	37	85.4678
Ruthenium	Ru	44	101.07
Samarium	Sm	62	150.4
Scandium	Sc	21	44.9559
Selenium	Se	34	78.96
Silicon	Si	14	28.0855
Silver	Ag	47	107.868
Sodium	Na	11	22.98977
Strontium	Sr	38	87.62
Sulfur	S	16	32.06
Tantalum	Ta	73	180.9479
Technetium	Tc	43	(97)
Tellurium	Te	52	127.60
Terbium	Tb	65	158.9254
Thallium	Tl	81	204.37
Thorium	Th	90	232.0381
Thulium	Tm	69	168.9342
Tin	Sn	50	118.69
Titanium	Ti	22	47.90
Tungsten	W	74	183.85
Uranium	U	92	238.029
Vanadium	V	23	50.9414
Xenon	Xe	54	131.30
Ytterbium	Yb	70	173.04
Yttrium	Y	39	88.9059
Zinc	Zn	30	65.38
Zirconium	Zr	40	91.22

[a] From *Pure Appl. Chem.*, **47,** 75 (1976). A value in parentheses is the mass number of the longest-lived isotope of the element.

PHYSICAL CHEMISTRY

PHYSICAL CHEMISTRY

IRA N. LEVINE

Chemistry Department
Brooklyn College
City University of New York
Brooklyn, New York

McGraw-Hill Book Company

New York St. Louis San Francisco Auckland Bogotá Düsseldorf Johannesburg
London Madrid Mexico Montreal New Delhi Panama Paris
São Paulo Singapore Sydney Tokyo Toronto

This book was set in Times Roman.
The editors were Donald C. Jackson and Michael Gardner;
the designer was Anne Canevari Green;
the production supervisor was Dominick Petrellese.
The drawings were done by J & R Services, Inc.
R. R. Donnelley & Sons Company was printer and binder.

PHYSICAL CHEMISTRY

1 2 3 4 5 6 7 8 9 0 DODO 7 8 3 2 1 0 9 8

Library of Congress Cataloging in Publication Data

Levine, Ira N date
Physical Chemistry.

Bibliography: p.
Includes index.
1. Chemistry, Physical and theoretical. I. Title.
QD453.2.L48 541'.3 77-12399
ISBN 0-07-037418-X

TO MY MOTHER,
FAN DAVIDOFF LEVINE
AND TO THE MEMORY OF MY FATHER,
LOUIS LEVINE

CONTENTS

PREFACE

This is a textbook for the standard undergraduate course in physical chemistry.

In writing this book, I have kept in mind the goals of clarity, accuracy, and depth. I have tried to make the presentation easy to follow by giving careful definitions and explanations of concepts, by giving full details of most derivations, by giving derivations that appear motivated rather than arbitrary, and by reviewing topics in mathematics and physics. At the same time, I have avoided a superficial, watered-down treatment, which would leave the student with little real understanding of physical chemistry. Instead, I have aimed at a treatment that is as accurate, as fundamental, and as up-to-date as can be readily presented at the undergraduate level.

Although the treatment is an in-depth one, the mathematics has been kept at a reasonable level and advanced mathematics unfamiliar to the student has been avoided. Since mathematics has proved to be a stumbling block for many students trying to master physical chemistry, I have included brief reviews of those aspects of calculus that are important to physical chemistry. Although it might be objected that students "ought" to know these things before they come to physical chemistry, the fact is that many students in physical-chemistry courses have made little use of calculus in previous science courses and hence have forgotten most of the calculus they once knew.

The book is organized so that the student can see the broad structure and logic of physical chemistry, rather than feel that he is being bombarded with a hodgepodge of formulas and ideas presented in random order. In line with this, the thermodynamics chapters are grouped together, as are those on quantum chemistry.

Statistical mechanics is not taken up until after thermodynamics and quantum chemistry. It is sometimes argued that thermodynamics and statistical mechanics ought to be interwoven so as to clarify the macroscopic thermodynamic concepts with the molecular statistical-mechanical ones. My own belief is that students find it easier to learn one thing at a time and that thermodynamics and statistical

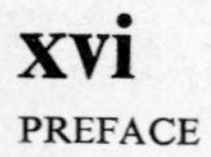

mechanics are best studied in succession. Their simultaneous presentation often leaves students confused about which results are thermodynamic and which statistical-mechanical. The logical structures of thermodynamics and of statistical mechanics are better grasped if they are studied in sequence, rather than together.

I have not, however, given a rigorously macroscopic presentation of thermodynamics. Instead, wherever possible, I have used molecular concepts in a qualitative way to illuminate the meaning of the thermodynamic concepts. For example, a discussion of entropy in terms of probability on a molecular level is given following the macroscopic presentation. I have tried to avoid the impression often given that thermodynamics is mainly the study of ideal gases by including thermodynamic discussions of nonideal systems. I have clearly indicated the conditions of applicability of the important thermodynamic equations; too often students are confused about when an equation applies.

The presentation of quantum chemistry steers a middle course between an excessively mathematical treatment that would obscure the physical ideas for most undergraduates and a purely qualitative treatment that does little beyond repeat what the student has already learned in freshman chemistry and organic chemistry.

The content of physical chemistry courses has expanded greatly in recent years. In a one-year course, there is not enough time available for a detailed presentation of all the required material. Because this book gives derivations in full detail, the instructor is freed from the necessity of presenting them in class; the class can be devoted to discussing concepts and answering questions, and more material can be covered.

Each chapter has a wide variety of problems, and answers to many of the numerical problems are given. The class time available for going over problems is usually limited, so a manual of solutions to the problems has been prepared and can be purchased by students upon authorization of the instructor.

A fair number of biological applications have been included. No justification of this is needed in an age where it has been said that "The only remaining scientific question worth investigating is: Are we machines?"

Material on polymers has been integrated into sections on osmotic pressure, transport properties, and solids. Nuclear chemistry is discussed at the end of Chap. 17 on kinetics.

The book uses both SI and non-SI units. Students are best served if they are made familiar with both the officially recommended SI units and widely used non-SI units. I have taken care to nearly always include the proper units with physical quantities. For the most part, the symbols recommended by the International Union of Pure and Applied Chemistry are used.

Professors Gene B. Carpenter, Howard D. Mettee, Roland R. Roskos, Theodore Sakano, and Peter E. Yankwich reviewed various portions of the manuscript, and I thank them for their many valuable comments and suggestions. Professors Fritz Steinhardt and Vicki Steinhardt provided helpful mathematical advice.

I welcome any suggestions for improvement that readers may have.

IRA N. LEVINE

1

THERMODYNAMICS

1.1 PHYSICAL CHEMISTRY

Physical chemistry applies the methods of physics to the study of chemical systems. A chemical system can be studied from either a microscopic or a macroscopic viewpoint. The *microscopic* viewpoint makes explicit use of the concept of molecules. The *macroscopic* viewpoint studies large-scale properties of matter (e.g., volume, pressure, composition) without explicit use of the molecule concept. The first half of this book uses mainly a macroscopic viewpoint; the second half uses mainly a microscopic viewpoint.

We can divide physical chemistry into four main areas: thermodynamics, quantum chemistry, statistical mechanics, and kinetics. *Thermodynamics* is a macroscopic science that studies the interrelationships between the various equilibrium properties of a system. Thermodynamics is treated in Chaps. 1 to 14.

Molecules and the particles (electrons and nuclei) that compose them do not obey classical (Newtonian) mechanics; instead their motions are governed by the laws of quantum mechanics (Chap. 18). Application of quantum mechanics to atomic structure, molecular bonding, and spectroscopy gives us *quantum chemistry* (Chaps. 19 to 21).

The macroscopic science of thermodynamics is a consequence of what is happening at a molecular (microscopic) level. The molecular and macroscopic levels are related to each other by the branch of science called *statistical mechanics.* Statistical mechanics gives insight into why the laws of thermodynamics hold and allows calculation of macroscopic thermodynamic properties from molecular properties. We shall study statistical mechanics in Chaps. 15, 16, 22, and 23.

Kinetics is the study of rate processes such as chemical reactions, diffusion, and the flow of charge in an electrochemical cell. The theory of rate processes is not as well developed as the theories of thermodynamics, quantum mechanics, and statistical mechanics. Kinetics uses relevant portions of thermodynamics, quantum chemistry, and statistical mechanics. Chapters 16, 17, and 23 deal with kinetics.

1.2 THERMODYNAMICS

We begin our study of physical chemistry with thermodynamics. *Thermodynamics* (from the Greek words for "heat" and "power") studies heat, work, energy, and the changes they produce in the states of systems; in a broader sense, thermodynamics studies the relationships between the macroscopic properties of a system. We shall be studying *classical* (or *equilibrium*) thermodynamics, which deals with systems in equilibrium. (*Irreversible* thermodynamics deals with nonequilibrium systems and rate processes.) Classical thermodynamics is a macroscopic science and is independent of any theories of molecular structure. Strictly speaking, the word "molecule" is not part of the vocabulary of thermodynamics; however, we won't adopt a purist attitude but will often refer to molecular concepts to aid in understanding thermodynamics. Thermodynamics is not applicable to systems of molecular size; a system must consist of a large number of molecules for it to be treated thermodynamically. When we use the term thermodynamics, we shall always mean equilibrium thermodynamics.

The part of the universe under study in thermodynamics is called the *system*; the parts of the universe that can interact with the system are called the *surroundings*. A thermodynamic system must be of macroscopic size.

If, for example, we are studying the vapor pressure of water as a function of temperature, we might put a sealed container of water (with any air evacuated) in a constant-temperature bath and connect a manometer to the container to measure the pressure (Fig. 1.1). Here, the system consists of the liquid water and the water vapor in the container, and the surroundings are the constant-temperature bath and the mercury in the manometer.

An *open* system is one where transfer of matter between system and surroundings can occur. A *closed* system is one where no transfer of matter can occur between system and surroundings. An *isolated* system is one that does not interact in any way at all with its surroundings. An isolated system is obviously a closed system, but not every closed system is isolated. For example, in Fig. 1.1, the system of liquid water plus water vapor in the sealed container is closed (since no matter can enter or leave), but it is not isolated (since it can be warmed or cooled by the surrounding bath). For an isolated system, neither matter nor energy can be transferred between system and surroundings. For a closed system, energy but not matter can be transferred between system and surroundings. For an open system, both matter and energy can be

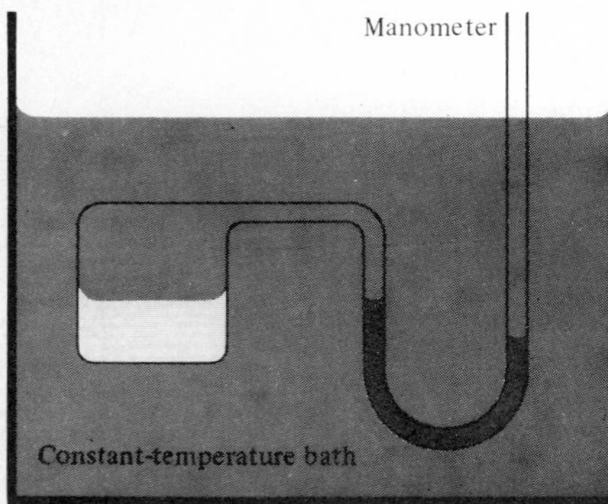

Figure 1.1
Measurement of the vapor pressure of a liquid.

transferred between system and surroundings. Most commonly, we shall deal with closed systems. It is important to note the kind of system under study, since thermodynamic statements valid for one kind of system may be invalid for other kinds.

A system may be separated from its surroundings by various kinds of walls. (Thus, in Fig. 1.1, the system is separated from the bath by the container walls.) A wall can be either *rigid* or *nonrigid* (i.e., movable). A wall may be *permeable* or *impermeable*, where by impermeable we mean that it allows no matter to pass through it. (Semipermeable walls will be discussed later.) Finally, a wall may be *adiabatic* or *diathermic*. In plain language, an adiabatic wall is one that does not conduct heat at all, whereas a diathermic wall does conduct heat. However, we have not yet defined heat, and hence to have a logically correct development of thermodynamics, adiabatic and diathermic walls must be defined without reference to heat. This is done as follows.

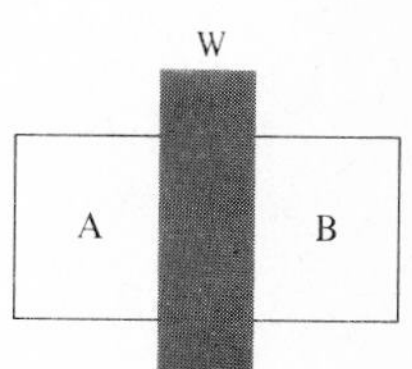

Figure 1.2 Systems A and B are separated by a wall W.

Suppose we have two separate systems A and B, each of whose properties are observed to be constant with time. We then bring A and B into contact via a rigid, impermeable wall (Fig. 1.2). If, no matter what the initial values of the properties of A and B are, we observe no change in the values of these properties (e.g., pressures, volumes) with time, then the wall separating A and B is said to be *adiabatic*. If we generally observe changes in the properties of A and B with time when they are brought in contact via a rigid, impermeable wall, then this wall is called *diathermic* (or *thermally conducting*). (As an aside, when two systems at different temperatures are brought in contact through a thermally conducting wall, heat flows from the hotter to the colder system, thereby changing the temperatures and other properties of the two systems; with an adiabatic wall, any temperature difference is maintained. Since heat and temperature are still undefined, these remarks are logically out of place, but they have been included to illuminate the definitions of adiabatic and thermally conducting walls.) An adiabatic wall is an idealization, but it can be approximated, e.g., by a very thick layer of asbestos.

In Fig. 1.1, the container walls are impermeable (so as to keep the system closed) and are thermally conducting (so as to allow the system's temperature to be adjusted to that of the surrounding bath); the container walls are essentially rigid, but if the interface between the water vapor and the mercury in the manometer is considered to be a "wall," then this wall is movable. Frequently, we shall deal with a system separated from its surroundings by a piston, which acts as a movable wall.

A system surrounded by a rigid, impermeable, and adiabatic wall cannot interact with the surroundings and is isolated.

Classical thermodynamics deals with systems in *equilibrium*. An isolated system is in equilibrium when its properties are observed to remain constant with time. A nonisolated system is in equilibrium when the following two conditions hold: (*a*) the system's properties remain constant with time; (*b*) removal of the system from contact with its surroundings causes no change in the properties of the system. [If condition (*a*) holds but (*b*) does not hold, the system is in a *steady state*. An example of a steady state is a metal rod in contact at one end with a large body at 50°C and in contact at the other end with a large body at 40°C. After sufficient time has elapsed, the metal rod satisfies condition (*a*); a uniform temperature gradient is set up along the rod. However, if we remove the rod from contact with its surroundings, the temperatures of its parts change until the whole rod is at 45°C.]

The equilibrium concept can be divided into the following three kinds of equilibrium. For *mechanical equilibrium*, there are no unbalanced forces acting on or within the system; hence the system undergoes no acceleration, and there is no turbulence within the system. For *material equilibrium*, no net chemical reactions are occurring in the system, nor is there any net transfer of matter from one part of the system to another; the concentrations of the chemical species in the various parts of the system are constant in time. For *thermal equilibrium* between a system and its surroundings, there must be no change in the properties of the system or surroundings when they are separated by a thermally conducting wall; likewise, we can insert a thermally conducting wall between two parts of a system to test whether the parts are in thermal equilibrium with each other. For thermodynamic equilibrium, all three kinds of equilibrium must be present.

What properties does thermodynamics use to characterize a system in equilibrium? Clearly, the *composition* must be specified; this can be done by stating the mass of each chemical species that is present in each phase. The *volume* V is a characteristic property of the system. The pressure P is another thermodynamic variable; *pressure* is defined as the magnitude of the normal (i.e., perpendicular) force per unit area exerted by the system on its surroundings:

$$P \equiv F/A \tag{1.1}^*$$

where F is the magnitude of the normal force exerted on a boundary wall of area A. The symbol $\equiv$ indicates a definition. (An equation with an asterisk after the number is best memorized.) Pressure is a scalar, not a vector. For a system in mechanical equilibrium, the pressure throughout the system is uniform and equal to the pressure of the surroundings. (We are ignoring the effect of the earth's gravitational field, which causes a slight increase in pressure as one goes from the top to the bottom of the system.) Still another thermodynamic property is the *temperature*, which will be defined in the next section. If external electric or magnetic fields act on the system, the field strengths are thermodynamic variables; we won't consider systems with such fields. Later, further thermodynamic properties (e.g., internal energy, entropy) will be defined.

Thermodynamic properties are classified as extensive or intensive. An *extensive* property is one whose value is equal to the sum of its values for the parts of the system. Thus, if we divide a system into parts, the mass of the system is the sum of the masses of the parts; mass is an extensive property. So is volume. Properties that do not depend on the amount of matter in the system are called *intensive*. Density and pressure are examples of intensive properties. We can take a drop of water or a swimming pool full of water and both systems will have the same density.

If each intensive property is constant throughout a system, the system is *homogeneous*. If a system is not homogeneous, it may consist of a number of homogeneous parts; each such homogeneous part is called a *phase*. For example, if the system consists of a crystal of AgBr in equilibrium with an aqueous solution of AgBr, the system has two phases, the solid AgBr and the solution. A phase can consist of several disconnected pieces. For example, in a system composed of several AgBr crystals in equilibrium with an aqueous solution, all the crystals are considered to be part of the same phase. A system composed of two or more phases is *heterogeneous*. The *density* ρ (rho) of a phase of mass m and volume V is

$$\rho \equiv m/V \tag{1.2}^*$$

A heterogeneous system shows a sudden, discontinuous change in some of its intensive properties as we go from one phase to another. There are systems that show a continuous change in one or more intensive properties; e.g., a long vertical column of gas in the earth's gravitational field shows a continuous variation of pressure and density with height. This is an example of a *continuous* system. Although most thermodynamic experiments are done in the earth's gravitational field, the effect of this field is negligible unless there are great differences in altitude between different parts of the system. We shall therefore neglect gravitational effects. Thus any system we consider will be either homogeneous (one phase) or heterogeneous (several phases).

Suppose that for two separate thermodynamic systems A and B the measured value of every thermodynamic property in system A equals the measured value of the corresponding property in system B. The systems are then said to be in the same thermodynamic *state*. The state of a thermodynamic system is thus defined by specifying the values of its thermodynamic properties. However, it is not necessary to specify all the properties to define the state; there is a minimum number of properties specification of which determines the values of the remaining properties. For example, suppose we take 6.66 g of pure H_2O at 1 atm (atmosphere) pressure and 24°C. It is found that (in the absence of external fields) all the remaining properties (volume, heat capacity, index of refraction, etc.) are fixed. (This statement ignores the possibility of surface effects, which are considered in Chap. 13.) Thus, two thermodynamic systems each consisting of 6.66 g of H_2O at 24°C and 1 atm are in the same thermodynamic state.

A thermodynamic system in a given equilibrium state then has a particular value for each thermodynamic property. These properties are therefore also called *state functions*, since their values are functions of the system's state. The terms thermodynamic variable, thermodynamic property, and state function are synonymous. Note especially that the value of a state function depends only on the present state of a system and not on its past history. (It doesn't matter whether we got the 6.66 g of water at 1 atm and 24°C by melting ice and warming the water or by condensing steam and cooling the water.)

1.3 TEMPERATURE

Suppose two systems are separated by a nonrigid, impermeable, adiabatic wall, and further suppose that the two systems are in mechanical equilibrium with each other. Because we have mechanical equilibrium, no unbalanced forces act and each system exerts an equal and opposite force on the separating wall; therefore each system exerts an equal pressure on this wall. Systems in mechanical equilibrium with each other have the same pressure. What about systems that are in thermal equilibrium with each other?

Just as systems in *mechanical* equilibrium have a common *pressure*, we expect there to be some thermodynamic property common to systems in *thermal* equilibrium; this property is what we *define* as the *temperature*, symbolized by θ (theta). By definition, two systems in thermal equilibrium with each other have the same temperature; two systems not in thermal equilibrium have different temperatures. Having defined temperature as a thermodynamic property that determines whether or not two systems are in thermal equilibrium, we now devise a way to measure the temperature θ of a system. As we shall see, our definition does not lead to a unique temperature scale. (The following reasoning is rather abstract; if you don't fully grasp it, don't worry too much; no one understands everything fully.)

We start by stating a law involving thermal equilibrium. Let systems A and B be in states such that when A and B are brought in contact via a thermally conducting wall, they are found to be in thermal equilibrium (as defined in Sec. 1.2). Similarly, let systems B and C be in states that are found to be in thermal equilibrium. It is then an experimental result that if A and C are brought in contact via a thermally conducting wall, they will be found to be in thermal equilibrium.

Two systems that are each found to be in thermal equilibrium with a third system will be found to be in thermal equilibrium with each other.

This unexciting generalization from experience is *the zeroth law of thermodynamics*. It is so called because only after the first, second, and third laws of thermodynamics had been formulated was it realized that the zeroth law is needed for the development of thermodynamics; moreover, a statement of the zeroth law logically precedes the other three.

To set up a temperature scale, we first pick some reference system r, which we call a *thermometer*. For simplicity, r is taken as homogeneous with a fixed composition; for example, r might be a fixed amount of liquid mercury. Experience shows that the thermodynamic state of r is determined once its pressure P_r and volume V_r are specified. (There are a few exceptions to this statement; see below.) We then make a duplicate of the thermometer to give two thermometers each containing the same amount of the same homogeneous substance. Each possible pair of pressure and volume values of one thermometer is then tested to see whether it is in thermal equilibrium with each possible pair of pressure and volume values of the duplicate thermometer. On a plot of P_r vs. V_r, states found to be in thermal equilibrium with one another are joined by a line. Figure 1.3 shows the typical appearance of such a plot. All states on a given curve are in thermal equilibrium and by the definition of temperature must be assigned the same temperature. A line that connects states of constant temperature is called an *isotherm*.

For a very few substances, some of the isotherms cross each other to give a plot like Fig. 1.4. The main example is liquid water (Prob. 1.24). For such a substance, two different states can have the same

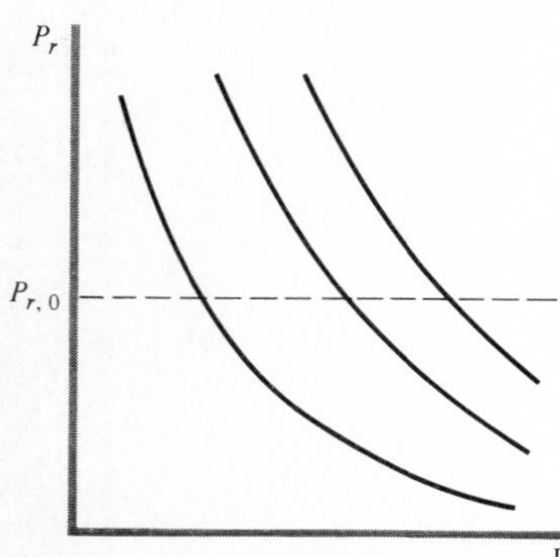

Figure 1.3
Isotherms (solid curves) of a system r to be used as a thermometer.

pair of values of P and V. For example, the intersection point of isotherms 1 and 2 corresponds to two thermodynamic states with different temperatures but the same P and V values; such states will also have different values for other thermodynamic properties (e.g., index of refraction) besides temperature. A substance that gives a P-V plot where isotherms cross cannot be used as a thermometer.

The next step in setting up a temperature scale is to hold the pressure of the thermometer r fixed at some convenient value $P_{r,0}$. Figure 1.3 shows that at fixed pressure, any two thermometer states that differ in volume are not in thermal equilibrium and so must be assigned different temperatures. Moreover, since isotherms do not cross one another, a given value of the thermometer volume at fixed pressure corresponds to a unique thermodynamic state and a unique temperature.

To each value of the thermometer volume V_r at the fixed pressure $P_{r,0}$ we now assign a number θ, which we call the *temperature* of the thermometer. The numbers θ can be assigned in an arbitrary manner, provided that two different values of V_r (at fixed P_r) do not get assigned the same temperature. (By definition, the temperature must differ for two states that are not in thermal equilibrium.) Although the θ values can be assigned arbitrarily, it is most convenient to choose θ values that are smooth, continuous functions of the thermometer's volume V_r. It is also conventional to assign θ values so that states with larger θ values are experienced physiologically as hotter than states with smaller values. Once the θ assignments are made, a measurement of the thermometer's volume V_r gives us its temperature θ.

Having armed ourselves with a thermometer, we can now determine the temperature of any system B. To do so, we put system B in contact with the thermometer through a thermally conducting wall, wait till thermal equilibrium is achieved, and then measure the thermometer volume V_r. We then know the thermometer's temperature, and since B is in thermal equilibrium with the thermometer, B's temperature equals that of the thermometer. (Of course, you've done this many times, sticking thermometers into beakers of liquid or into your mouth.)

Let systems A and B be found to have the same temperature ($\theta_A = \theta_B$), and let systems B and C have different temperatures ($\theta_B \neq \theta_C$). Suppose we set up a second temperature scale using a different fluid for our thermometer and assigning temperature values in a different manner. Although the numerical values of the temperatures of systems A, B, and C on the second scale will differ from those on the first temperature scale, it follows from the zeroth law that on the second scale systems A and B

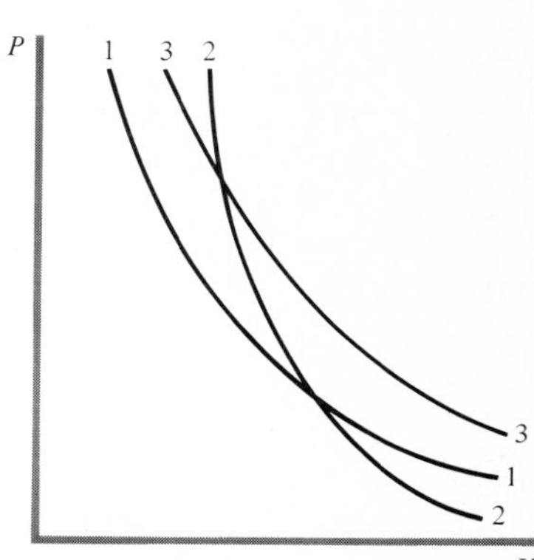

Figure 1.4
A system for which isotherms cross.

will still be found to have the same temperature and systems B and C to have different temperatures. Thus, although numerical values on any temperature scale are arbitrary, the zeroth law assures us that the temperature scale will fulfill its function of telling whether or not two systems are in thermal equilibrium.

Now consider a system D that is homogeneous with a fixed composition. Experience shows its thermodynamic state to be specified once its pressure P and volume V are specified. By putting D in thermal equilibrium with our thermometer, we can assign a temperature θ to every possible state of D. Thus, given values of P and V of system D, the value of θ for system D is determined. But this is exactly what is meant by the statement that θ is a function of P and V. We therefore have $\theta = g(P, V)$, where g is some function that depends both on the nature of system D and on how we made the numerical temperature assignments for the thermometer r. If the restriction that D have fixed composition is dropped, then the state of D will depend on its composition C as well as on P and V, and we shall then have

$$\theta = f(P, V, C) \tag{1.3}$$

where f is some function. The composition C of the homogeneous system can be specified by giving the masses of its components. [If the system is not homogeneous, we can separate each phase and end up with an equation like (1.3) for each phase.] The relation (1.3) between pressure, volume, composition, and temperature is called an *equation of state*.

Having discussed in an abstract way how a temperature scale is set up, we now set up a particular temperature scale. We choose as our thermometer r a fixed quantity of liquid mercury. Investigation shows that isotherms of liquid mercury on a P-V plot never cross, so this is a suitable thermometric fluid. We fix the mercury's pressure at some constant value and assign a different value of the temperature θ to each mercury volume V_r at the fixed pressure. The simplest way to make these assignments is to take θ as a linear function of V_r. We thus *define* the temperature θ to be the following linear function of mercury volume at fixed pressure: $\theta \equiv aV_r + b$, where a and b are constants. The mercury is placed in a glass container that consists of a bulb connected to a narrow tube. Let the cross-sectional area of the tube be A, and let the mercury rise to a length l in the tube. The mercury volume equals the sum of the mercury volumes in the bulb and the tube, so

$$\theta \equiv aV_r + b = a(V_{\text{bulb}} + Al) + b = aAl + (aV_{\text{bulb}} + b) \equiv cl + d$$

where c and d are constants defined as $c \equiv aA$ and $d \equiv aV_{\text{bulb}} + b$.

To fix c and d, we define the temperature of equilibrium between pure ice and air-saturated liquid water at 1 atm pressure as 0°C (for centigrade), and we define the temperature of equilibrium between pure liquid water and water vapor at 1 atm pressure (the normal boiling point of water) as 100°C. (These points are called the *ice point* and the *steam point.*) Since our scale is linear with length of the mercury column, we mark off 100 equal intervals between 0 and 100°C and extend the marks above and below these temperatures.

Of course, the reader will have realized the completely arbitrary way we defined our scale. This scale depends on the expansion properties of a particular substance, liquid mercury. If we had chosen, say, ethanol instead of mercury as the thermometric fluid, temperatures on the ethanol scale would show small but significant differences from those on the mercury scale. Moreover, there is at this point no reason (apart from simplicity) for choosing a linear relation between temperature and mercury volume. We could just as well have chosen θ to vary as $aV_r^2 + b$. Temperature is a fundamental concept of thermodynamics, and one naturally feels that it should be formulated less arbitrarily. Some of the arbitrariness will be removed in Sec. 1.5, where the ideal-gas temperature scale is defined. Finally, in Sec. 3.3 we shall define the most fundamental temperature scale, the thermodynamic scale. (As we shall see, even the thermodynamic scale has some arbitrariness.) The mercury centigrade scale defined in this section is not in current use, but we shall use it until we define a better scale in Sec. 1.5.

Thermodynamics is a macroscopic science and does not provide any explanation of the molecular meaning of temperature. We shall later use statistical mechanics to show that temperature is related to the average molecular kinetic energy; increasing temperature corresponds to increasing average molecular kinetic energy (provided the temperature scale is chosen to give higher temperatures to hotter states).

1.4 THE MOLE

We now review the concept of the mole, which is used throughout chemical thermodynamics.

The ratio of the mass of an atom of an element to the mass of some chosen standard is called the *atomic weight* of that element. The standard used since 1961 is $\frac{1}{12}$ times the mass of the isotope ^{12}C; the atomic weight of ^{12}C is thus exactly 12, by definition. The ratio of the mass of a molecule of a substance to $\frac{1}{12}$ times the mass of a ^{12}C atom is called the *molecular weight* of that substance. The statement that the molecular weight of H_2O is 18.015 means that a water molecule has on the average a mass that is 18.015/12 times the mass of a ^{12}C atom. (We say "on the average" to acknowledge the existence of naturally occurring isotopes of H and O.) Since atomic and molecular weights are *relative* masses, these "weights" are dimensionless numbers. A better name than the traditional term "molecular weight" is *relative molecular mass*. (For an ionic compound, the mass of one formula unit replaces the mass of one molecule in the definition of the molecular weight. For example, we say that the molecular weight of NaCl is 58.443, even though there are no individual NaCl molecules in an NaCl crystal.)

A quantity of an element whose mass in grams is numerically equal to its atomic weight is called 1 gram-atomic weight of that element. One gram-atomic weight of hydrogen is 1.0079 g of hydrogen; 1 gram-atomic weight of carbon is 12.011 g of carbon; this number is greater than 12 due to the presence of ^{13}C. The number of ^{12}C atoms in exactly 12 g of ^{12}C is called *Avogadro's number*. Experiment gives 6.02×10^{23} as the value of Avogadro's number. If we counted out Avogadro's number of ^{12}C atoms, they would have a mass of 12 g, exactly. Suppose we count out Avogadro's number of hydrogen atoms. What would their mass be? We have equal numbers of H and ^{12}C atoms, and each H atom has a mass 1.0079/12 times the mass of a ^{12}C atom. Hence the total mass of hydrogen is 1.0079/12 times the total ^{12}C mass, or $(1.0079/12)(12\text{ g}) = 1.0079$ g. But this is just 1 gram-atomic weight of hydrogen. This reasoning shows that 1 gram-atomic weight of any element contains Avogadro's number of atoms. Similarly, it follows that 1 gram-molecular weight of any substance contains Avogadro's number of molecules.

A *mole* of some substance is defined as an amount of that substance which contains Avogadro's number of elementary units. For example, a mole of hydrogen atoms contains 6.02×10^{23} H atoms; a mole of water molecules contains 6.02×10^{23} molecules of H_2O. The conclusion of the preceding paragraph shows that a mole of any type of atom is the same as 1 gram-atomic weight of the element, and a mole of some molecular species is the same as 1 gram-molecular weight of that substance. Therefore, if $M_{r,i}$ is the molecular weight of species i (the r subscript stands for relative), then the mass of 1 mole of species i equals $M_{r,i}$ grams. The mass per mole of substance i is called its *molar mass* M_i. (For example, for H_2O, $M_i = 18.015$ g/mole.) The molar mass is given by

$$M_i \equiv m_i/n_i \tag{1.4}^*$$

where m_i is the mass of substance i in a sample and n_i is the number of moles of i in the sample.

Frequently, the composition of a system is specified by stating the mole fraction of each species present. If the total number of moles of all species present is n_{tot}, and if n_i moles of component i is present, then the *mole fraction* x_i of species i in the system is defined as

$$x_i \equiv n_i/n_{tot} \tag{1.5}^*$$

The sum of the mole fractions of all components equals 1:

$$x_1 + x_2 + \cdots = \frac{n_1}{n_{tot}} + \frac{n_2}{n_{tot}} + \cdots = \frac{n_1 + n_2 + \cdots}{n_{tot}} = \frac{n_{tot}}{n_{tot}} = 1$$

A few sentences ago, n_i was called "the number of moles" of species i. Strictly speaking, this is incorrect. In the officially recommended SI units (Sec. 2.2), the *amount of substance* is taken as one of the fundamental physical quantities (along with mass, length, time, etc.), and the unit of this physical quantity is the mole (abbreviated mol). (Both the spelled-out form and the abbreviation will be used interchangeably in this book; they mean exactly the same.) Just as the SI unit of mass is the kilogram, the SI unit of amount of substance is the mole. Just as the symbol m_i stands for the mass of substance i, the symbol n_i stands for the amount of substance i. The quantity m_i is not a pure number but is a number times a unit of mass; for example, m_i might be 4.18 kg (4.18 kilograms). Likewise, n_i is not a pure number but is a number times a unit of amount of substance; for example, n_i might be 1.26 mol (1.26 moles). Thus the correct statement is that n_i is the amount of substance i. The number of moles of i is a pure number and equals n_i/mol, since n_i has a factor of 1 mol included in itself. The dimensionally incorrect statement that n_i is the number of moles of i occurs almost universally, and will be found in later chapters of this book.

If n_i/mol is the number of moles of species i present in a system, the number of molecules N_i of i present equals n_i/mol times Avogadro's number:

$$N_i = (n_i/\text{mol}) \cdot (\text{Avogadro's number})$$

The quantity (Avogadro's number)/mol is called the *Avogadro constant* N_0, and we have

$$N_i = n_i N_0 \qquad \text{where } N_0 \equiv 6.02 \times 10^{23}\ \text{mol}^{-1} \tag{1.6}^*$$

Avogadro's number is a pure number, whereas the Avogadro constant N_0 has units of moles^{-1}.

Equation (1.6) applies to any collection of elementary entities, whether they are atoms, molecules, ions, radicals, electrons, photons, or whatever. Written in the form $n_i \equiv N_i/N_0$, Eq. (1.6) gives the definition of the amount of substance n_i of species i; in this equation, N_i is the number of elementary entities of species i.

From the discussion preceding (1.4), the molar mass M_i and the molecular weight $M_{r,i}$ of i are related by $M_i = M_{r,i} \times 1$ g/mol, where mol stands for mole (not molecule), and where $M_{r,i}$ is dimensionless.

1.5 IDEAL GASES

The laws of thermodynamics are quite general and make no reference to the specific nature of the system under study. Before beginning the study of these laws, we find it convenient to describe the properties of a particular kind of system, namely, an ideal gas. We shall then be able to illustrate the application of thermodynamic laws to an ideal-gas system.

The chemist and physicist Robert Boyle investigated the relation between the pressure and volume of gases in 1662 and found that for a fixed amount of gas kept at a fixed temperature, P and V are inversely proportional:

$$PV = k \qquad \text{constant } \theta, m \tag{1.7}$$

where k is a constant and m is the gas mass. Careful investigation shows that Boyle's law holds only approximately for real gases, with deviations from the law approaching zero in the limit of zero pressure (and also in the limit of very high temperature). Figure 1.5*a* shows some observed P-vs.-V curves for 28 g of N_2 at two temperatures. Figure 1.5*b* shows plots of PV vs. P for 28 g of N_2; note the near constancy of PV at low pressures and the significant deviations from Boyle's law at high pressures. Note also how the axes in Fig. 1.5 are labeled. The quantity P equals a pure number times a unit; for example, P might be 4.0 atm $= 4.0 \times 1$ atm. Hence, P/atm is a pure number. Boyle's law is understandable from the picture of a gas as consisting of a very large number of molecules moving essentially independently of one another. The pressure exerted by the gas is due to the impacts of the molecules on the walls. A decrease in volume causes the molecules to hit the walls more frequently, thereby increasing the pressure. We shall derive Boyle's law from the molecular picture in Chap. 15.

Before continuing, we discuss the units for pressure. From the definition (1.1), pressure has units of force divided by area. In the cgs system (Sec. 2.2), its units are dynes per square centimeter (dyn/cm^2); in the mks system, its units are newtons per square meter (N/m^2), also called *pascals* (Pa). Equation (2.23) gives

$$1 \text{ Pa} \equiv 1 \text{ N/m}^2 = (10^5 \text{ dyn})/(10^2 \text{ cm})^2 = 10 \text{ dyn/cm}^2$$

Chemists customarily use other units. One *torr* (or 1 mmHg) is the pressure exerted at 0°C by a column of mercury one millimeter high when the gravitational field has the standard value $g = 980.665$ cm/s^2. The downward force exerted by the mercury column equals its mass m times the gravitational acceleration g. Thus a mercury column of height h, mass m, cross-sectional area A, volume V, and density ρ exerts a pressure P given by

$$P = mg/A = \rho Vg/A = \rho Ahg/A = \rho gh \tag{1.8}$$

The density of mercury at 0°C and 1 atm is 13.5951 g/cm^3; hence

$$1 \text{ torr} = (13.5951 \text{ g/cm}^3)(980.665 \text{ cm/s}^2)(10^{-1} \text{ cm})$$
$$1 \text{ torr} = 1333.22 \text{ dyn/cm}^2 = 133.322 \text{ N/m}^2$$

One *atmosphere* (atm) is defined as exactly 760 torr:

$$1 \text{ atm} \equiv 760 \text{ torr} = 1.01325 \times 10^6 \text{ dyn/cm}^2 = 101{,}325 \text{ N/m}^2$$

Still another unit of pressure is the *bar*:

$$1 \text{ bar} \equiv 10^6 \text{ dyn/cm}^2 = 10^5 \text{ N/m}^2 = 0.986923 \text{ atm}$$

Charles (1787) and Gay-Lussac (1802) investigated the thermal expansion of gases and found a linear increase of volume with temperature (measured on the mercury centigrade scale) at constant pressure and fixed amount of gas:

$$V = a_1 + a_2\theta \qquad \text{const. } P, m$$

where a_1 and a_2 are constants. For example, Fig. 1.6 shows the observed relation between V and θ for 28 g of N_2 at a few pressures. Note the near linearity of the curves, which are at low pressures. The content of Charles' law is simply that the thermal expansions of gases and of liquid mercury are quite similar. (The molecular explanation for Charles' law lies in the fact that the average kinetic energy of the gas molecules is related to the temperature. An increase in temperature means the molecules are moving faster and hitting the walls harder. Therefore, the volume must increase if the pressure is to remain constant.)

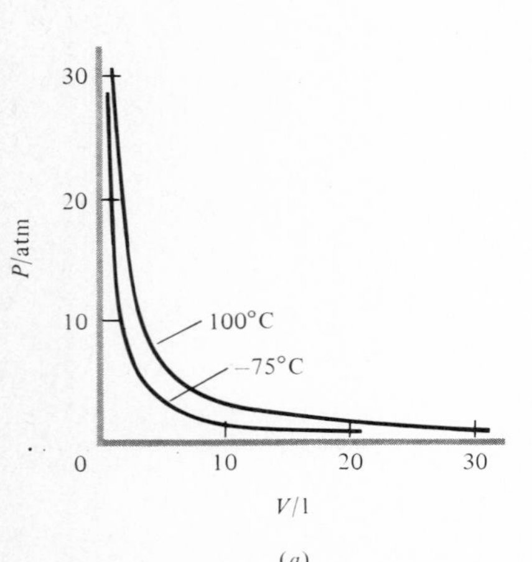

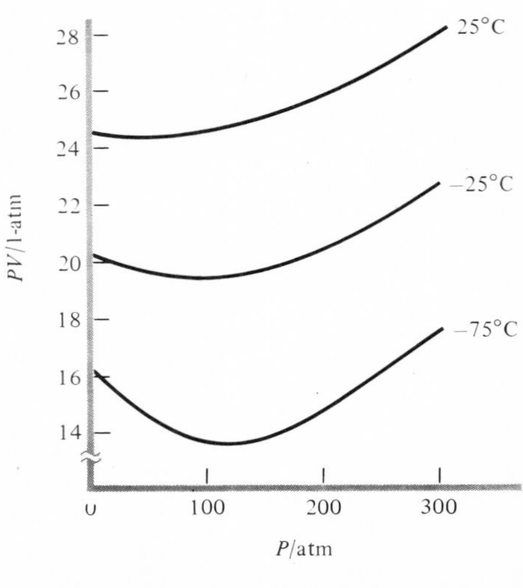

Figure 1.5
Plots of (*a*) P vs. V and (*b*) PV vs. P for 1 mole of N_2 at constant temperature.

At $\theta = 0°C$, we have $a_1 = V_0$, where V_0 is the gas volume at 0°C and the fixed pressure P. Hence $V = V_0 + a_2\theta$. If we define a new constant $A \equiv a_2/V_0$, Charles' law reads

$$V = V_0(1 + A\theta) \qquad \text{const. } P, m \tag{1.9}$$

For nonzero values of P, all gases show deviations from this law; moreover, the deviations are different for different gases. In the limit of zero pressure, gases still show small deviations from (1.9). [These deviations in the zero-pressure limit can be ascribed to the "nonideality" (whatever that might mean) of the expansion of liquid mercury, which is the basis for the θ temperature scale.] However, in the limit of zero pressure, the deviations are the *same* for different gases; moreover, the constant A has the same value for all gases in the limit of zero pressure. Thus, in the limit of zero pressure, all gases show the same temperature-vs.-volume behavior at constant pressure. Hence, to get a temperature scale that is independent of the properties of any one substance, we *define* the *ideal-gas temperature* t by the requirement that the temperature-vs.-volume behavior of a gas satisfy Charles' law exactly in the limit of zero pressure; i.e., we want t defined so that the equation

$$V = V_{\text{ref}}(1 + Bt) \qquad \text{const. } P, m \tag{1.10}$$

holds exactly in the limit of zero pressure, where V_{ref} and B are constants analogous to V_0 and A in (1.9). Solving for t, we have as the definition of t

$$t \equiv \frac{1}{B} \lim_{P \to 0} \left(\frac{V}{V_{\text{ref}}} - 1 \right) \qquad \text{const. } P, m \tag{1.11}$$

where V is the measured volume of gas at temperature t and V_{ref} is the measured volume of gas at a certain defined reference temperature t_{ref}.

To complete the definition, we must specify the two constants in (1.11). We first note from Eq. (1.10) that at $t = -1/B$, the volume is predicted to become zero. (Of course, in actuality the gas condenses to a liquid before V reaches zero.) Since it seems likely that the temperature at which an ideal gas is predicted to have zero volume might have fundamental importance, we choose to shift our scale by adding a constant to all temperatures to make the zero-volume temperature equal zero degrees on the shifted scale. Thus, we define the *absolute ideal-gas temperature* T as

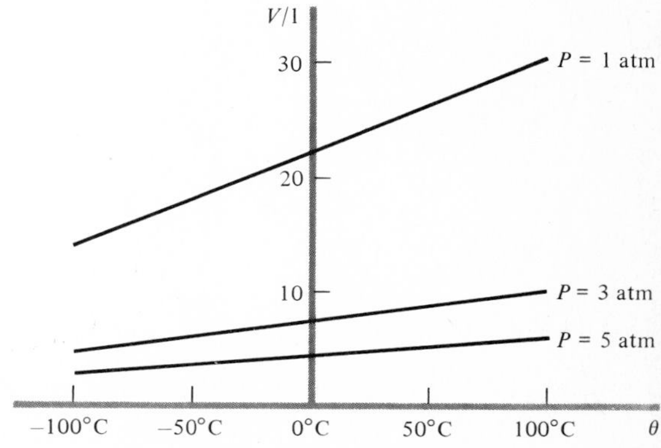

Figure 1.6
Plots of volume versus centigrade temperature for 1 mole of N_2 at constant pressure.

$T \equiv t + 1/B$. Using Eq. (1.11) and recalling that the limit of a difference equals the difference of the limits, we find

$$T \equiv t + 1/B = (1/B) \lim_{P \to 0} (V/V_{\text{ref}}) - 1/B + 1/B$$

$$T \equiv \frac{1}{B} \lim_{P \to 0} \frac{V}{V_{\text{ref}}} \qquad \text{const. } P, m \tag{1.12}$$

To complete the definition, we must specify the reference point and its value of T. In 1954 it was decided by international agreement to use the triple point (tr) of water as the reference point. The water triple point is the temperature at which pure liquid water, ice, and water vapor are in mutual equilibrium. It was further decided to define the absolute temperature at the triple point T_{tr} as 273.16 K exactly. The K stands for the unit of temperature, the *kelvin*, formerly called the degree Kelvin (°K). (The numerical value 273.16 was chosen to give close agreement with the old centigrade scale.) By definition, then, when $V = V_{\text{tr}} = V_{\text{ref}}$, we have $T = 273.16$ K, so that Eq. (1.12) gives 273.16 K $= (1/B) \lim (V_{\text{tr}}/V_{\text{tr}}) = 1/B$. Hence (1.12) becomes

$$T \equiv (273.16 \text{ K}) \lim_{P \to 0} \frac{V}{V_{\text{tr}}} \qquad \text{const. } P, m \tag{1.13}$$

This completes the definition of the ideal-gas absolute temperature scale.

The reader may be somewhat unclear about how the limit $P \to 0$ is taken in (1.13). What is done is to take a quantity of gas at some pressure P, say, 200 torr. This gas is put in thermal equilibrium with the body whose temperature T is to be measured, keeping P constant at 200 torr and measuring the volume of the gas V. The gas thermometer is then put in thermal equilibrium with a water triple-point cell at 273.16 K, maintaining P of the gas at 200 torr and measuring V_{tr}. The ratio V/V_{tr} is then calculated for $P = 200$ torr. Next, the gas pressure is reduced to, say, 150 torr, and the gas volume at this pressure is measured at temperature T and at 273.16 K; this gives the ratio V/V_{tr} at $P = 150$ torr. The operations are repeated at successively lower pressures to give further ratios V/V_{tr}. These ratios are then plotted against P, and the curve is extrapolated to $P = 0$ to give the limit of V/V_{tr}. Multiplication of this limit by 273.16 K then gives the ideal-gas absolute temperature T of the body. [In actual practice, a constant-volume gas thermometer is more convenient to use than a constant-pressure one; here, V/V_{tr} at constant P in (1.13) is replaced by P/P_{tr} at constant V.]

Accurate measurement of a body's temperature with an ideal-gas thermometer clearly requires much painstaking work, and this thermometer is not useful for day-to-day laboratory work. What is done is to use the ideal-gas thermometer to determine accurate values for several fixed points that cover a wide temperature range; the fixed points are the normal melting and boiling points of certain pure substances (e.g., oxygen, water, gold). The specified values for these fixed points, together with specified interpolation formulas that use platinum resistance thermometers for temperatures between the fixed points, constitute the International Practical Temperature Scale of 1968 (IPTS-68). The IPTS-68 scale is designed to reproduce the ideal-gas absolute scale within experimental error and is used to calibrate laboratory thermometers. [Details of IPTS-68 are given in F. D. Rossini, *J. Chem. Thermodyn.*, **2**, 447 (1970); see also W. T. Gray and D. I. Finch, *Phys. Today*, Sept. 1971, p. 32.]

Since the ideal-gas temperature scale is independent of the properties of any one substance, it is superior to the mercury centigrade scale defined in a previous section. However, the ideal-gas scale still depends on the limiting properties of *gases* and so is not as fundamental as one would like. The thermodynamic temperature scale, defined in Sec. 3.3, is independent of the properties of any particular kind of matter. For now we shall use the ideal-gas scale.

The present definition of the Celsius (centigrade) scale t is in terms of the ideal-gas absolute temperature scale T, as follows:

$$t \equiv T - 273.15° \qquad (1.14)^*$$

The triple-point of water is at 273.16 K − 273.15° = 0.01°C, exactly. On the present Celsius and Kelvin scales, the ice and steam points are not fixed but are determined by experiment. Hence there is no guarantee that these points will be at 0 and 100°C. However, the value 273.16 K for the water triple point and the number 273.15 in (1.14) were chosen to give good agreement with the old centigrade scale; hence, we expect the ice and steam points to be little changed from their old values. Experiment gives 0.0000 ± 0.0001 and 99.975°C for the ice and steam points.

One final point. From Eq. (1.13), at constant P and m we have $V/T = V_{tr}/T_{tr}$. This equation holds exactly only in the limit of zero pressure but is pretty accurate provided the pressure is not too high. Since V_{tr}/T_{tr} is a constant for a fixed amount of gas at fixed pressure, we have

$$V/T = K \qquad \text{const. } P, m \qquad (1.15)$$

where K is a constant. This is Charles' law. However, logically speaking, this equation is not a law of nature but simply embodies the *definition* of the ideal-gas absolute temperature scale T. Later, after defining the thermodynamic temperature scale, we can once again view (1.15) as a law of nature.

Boyle's and Charles' laws apply when two of the thermodynamic variables (T and m or P and m) are held fixed. Now consider a more general change of state of an ideal gas, in which the pressure, volume, and temperature all change, going from P_1, V_1, T_1 to P_2, V_2, T_2, with m unchanged. To apply Boyle's and Charles' laws, we imagine this process to be carried out in two steps, as follows:

$$P_1, V_1, T_1 \xrightarrow{(a)} P_2, V_a, T_1 \xrightarrow{(b)} P_2, V_2, T_2 \qquad (1.16)$$

Since T and m are constant in step (a), Boyle's law applies and $P_1 V_1 = k = P_2 V_a$; hence $V_a = P_1 V_1/P_2$. Application of Charles' law to step (b) gives $V_a/T_1 = V_2/T_2$; substitution of $V_a = P_1 V_1/P_2$ into this equation gives $P_1 V_1/P_2 T_1 = V_2/T_2$, which can be rearranged to read

$$P_1 V_1/T_1 = P_2 V_2/T_2 \qquad \text{const. } m, \text{ ideal gas} \qquad (1.17)$$

Now consider the effect of varying the mass m of ideal gas while keeping P and T constant. The volume V is an extensive quantity, so V is directly proportional to m for any one-phase, one-component system at constant T and P. Thus V/m is constant at constant T and P. Combining this fact with the constancy of PV/T at constant m, we readily find (Prob. 1.9) that PV/mT remains constant for any variation in P, V, T, and m of any pure ideal gas: $PV/mT = c$, where c is a constant. There is no reason for c to be the same for different ideal gases, and in fact it isn't. To obtain a form of the ideal-gas law that has the same constant for every ideal gas, we need another experimental observation.

In 1808 Gay-Lussac formulated his law of combining volumes, which states that the volumes of gases that react with one another stand in the ratios of small whole numbers when these volumes are measured at the same temperature and pressure. For example, one finds that two volumes of hydrogen gas react with one volume of oxygen gas to form water. The chemical equation for this reaction is $2H_2 + O_2 = 2H_2O$, so that twice as many hydrogen molecules react as oxygen molecules. The two volumes of hydrogen must then contain twice the number of molecules as the one volume of oxygen, and therefore one volume of hydrogen will have the same number of molecules as one volume of oxygen at the same temperature and pressure. The same result is obtained for other gas-phase reactions. We conclude that equal volumes of different gases at the same temperature and pressure contain equal numbers of molecules. This idea was first recognized by Avogadro in 1811. (Of course, Gay-Lussac's law of combining volumes and Avogadro's hypothesis are strictly true for real gases only in the limit as P goes to zero.) We can restate Avogadro's hypothesis in macroscopic terms by saying that equal volumes of different gases at the same temperature and pressure have equal numbers of moles. (The number of molecules and the number of moles are proportional to each other.)

Since the mass of gas and the number of moles of gas are proportional to each other, the ideal-gas law $PV/mT = c$ can be rewritten as $PV/nT = R$ or $n = PV/RT$, where n is the number of moles of gas and R is some other constant. Avogadro's hypothesis says that if P, V, and T are the same for two different gases, then n must be the same. But this can only hold true if R has the same value for every gas. R is therefore a universal constant, called the *ideal-gas constant*. Our final version of the ideal-gas law is

$$PV = nRT \qquad \text{ideal gas} \tag{1.18}*$$

Equation (1.18) incorporates Boyle's law, Charles' law (more accurately, the definition of T), and Avogadro's hypothesis.

Using Eq. (1.4) to introduce the molar mass M of the gas, we can write the ideal-gas law as $PV = mRT/M$. This form allows us to find the molecular weight of a gas by measuring the volume occupied by a known mass at a known T and P. For accurate results, one carries out a series of measurements at successively lower pressures and extrapolates the results to zero pressure (see Prob. 1.6). We can also write the ideal-gas law in terms of the density $\rho = m/V$ as $P = \rho RT/M$. The only form worth remembering is $PV = nRT$, since all other forms are easily derived from this one.

The gas constant R can be experimentally evaluated by taking a known number of moles of some gas held at some known temperature and carrying out a series of pressure–volume measurements at successively lower pressures. Evaluation of the zero-pressure limit of PV/nT then gives R. The experimental result is

$$R = 82.06 \text{ (cm}^3 \text{ atm)/(mole K)} \tag{1.19}*$$

Since $1 \text{ atm} = 101{,}325 \text{ N/m}^2$ (see above), we have $1 \text{ cm}^3 \text{ atm} = (10^{-2} \text{ m})^3 \times 101{,}325 \text{ N/m}^2 = 0.101325 \text{ m}^3 \text{ N/m}^2 = 0.101325 \text{ J}$. [One newton-meter = one joule (J); see Sec. 2.2.] Hence $R = 82.06 \times 0.101325$ J/(mole K), or

$$R = 8.314 \text{ J/(mole K)} = 8.314 \text{ (m}^3 \text{ Pa)/(mole K)} \tag{1.20}*$$

One calorie equals 4.184 J (Sec. 2.6), so that

$$R = 1.987 \text{ cal/(mole K)} \tag{1.21*}$$

So far, we have considered only a pure ideal gas. In 1810 Dalton found that the total pressure of a mixture of gases equals the sum of the pressures each gas would exert if placed alone in the container. (Again, this law is exact only in the limit of zero pressure.) Suppose we have n_1 moles of gas 1. If placed alone in the container, gas 1 would exert a pressure $n_1 RT/V$ (where we assume the pressure low enough for the gas to behave essentially ideally). Dalton's law asserts that the total pressure in the gas mixture is $P_{\text{tot}} = n_1 RT/V + n_2 RT/V + \cdots = (n_1 + n_2 + \cdots)RT/V = n_{\text{tot}} RT/V$, so

$$P_{\text{tot}} V = n_{\text{tot}} RT \qquad \text{ideal gas mixture} \tag{1.22*}$$

Dalton's law makes sense from the molecular picture of gases. Ideal-gas molecules do not interact with one another, so the presence of gases 2, 3, ... has no effect on gas 1 and its contribution to the total pressure is the same as if it alone were present. Each gas acts independently, and the total pressure is the sum of the individual pressures. For real gases, the intermolecular interactions in a mixture differ from those in a pure gas, and Dalton's law does not hold accurately. A mixture of ideal gases that satisfies Dalton's law is called an *ideal mixture of ideal gases.*

The *partial pressure* P_i of gas i in a gas mixture (ideal or nonideal) is defined as

$$P_i \equiv x_i P \qquad \text{any gas mixture} \tag{1.23*}$$

where x_i is the mole fraction of i in the mixture and P is the mixture's (total) pressure. Some people define the partial pressure of gas i to be the pressure gas i would exert if it alone were present in the container. This definition is not convenient for thermodynamics. Moreover, the International Union of Pure and Applied Chemistry (IUPAC) has recommended the definition (1.23). For an ideal gas mixture, the two definitions are equivalent. This is seen as follows. The pressure gas i would exert if it alone were present is $P(i \text{ alone}) = n_i RT/V$. The total pressure in the mixture is $P = n_{\text{tot}} RT/V$. Division gives $P(i \text{ alone})/P = n_i/n_{\text{tot}} = x_i$, so $P(i \text{ alone}) = x_i P = P_i$, and $P(i \text{ alone})$ is the same as the partial pressure (1.23). For a nonideal gas mixture, the two definitions are not equivalent.

Example Find the volume of one mole of ideal gas at 0°C and 1 atm pressure.

Since the gas is ideal, we have $PV = nRT$ so $V = nRT/P$. We now must decide which of the values (1.19) to (1.21) to use for R. The units of (1.19) involve atmospheres, which is the pressure unit used in the statement of the problem; hence it is most appropriate to use (1.19). Since the T in $PV = nRT$ is the absolute temperature, we must use (1.14) to convert 0°C to an absolute temperature: $T = 0°\text{C} + 273.15 \text{ K} = 273.15 \text{ K}$. We have

$$V = nRT/P = 1.0000 \text{ mole } (82.06 \text{ cm}^3 \text{ atm mole}^{-1} \text{ K}^{-1})(273.15 \text{ K})/1.0000 \text{ atm}$$

$$= 22.41 \times 10^3 \text{ cm}^3 = 22.41 \text{ l}$$

Note that the units of temperature, pressure, and amount of substance (moles) canceled out. The fact that we ended up with units of cubic centimeters, which is a correct unit for a volume, provides a check on our work. Although it is hard to go astray on a simple problem like this one, there will be many instances later on where fairly involved conversions between units are required. *It is strongly recommended that the units of every physical quantity be written down when doing calculations.*

One final point. The *liter* (l) was originally defined as the volume of 1000 g of water at 3.98°C and 1 atm pressure. This made the liter equal to 1000.028 cm^3. However, in 1964 the liter was redefined as exactly 1000 cm^3. Not everyone is aware of this redefinition. Use of the liter as a unit is best avoided in precise work because of the possible confusion between the two definitions.

1.6 DIFFERENTIAL CALCULUS

Thermodynamics and other branches of physical chemistry use calculus extensively. We therefore review some ideas of differential calculus without bothering to give mathematically rigorous definitions or proofs.

To say that the variable y is a *function* of the variable x means that for any given value of x there is specified a value of y; we write $y = f(x)$. For example, the area of a circle is a function of its radius r, since the area can be calculated from r by the expression πr^2. The variable x is called the *independent variable*, and y is the *dependent variable*. Of course, we can solve for x in terms of y to get $x = g(y)$, so that it is a matter of convenience which variable is considered to be the independent one. Instead of $y = f(x)$, it is often more convenient in physical problems to write $y = y(x)$.

To say that the *limit* of the function $f(x)$ as x approaches the value a is equal to c [which is written as $\lim_{x\to a} f(x) = c$] means that for all values of x sufficiently close to a (but *not* necessarily equal to a) the difference between $f(x)$ and c can be made as small as we please. For example, suppose we want the limit as x goes to zero of the function $(\sin x)/x$. We first note that $(\sin x)/x$ is undefined at $x = 0$ since 0/0 is undefined. However, this fact is irrelevant to determining the limit. To find the limit, we calculate the following values of $(\sin x)/x$, where x is in radians: 0.99833 for $x = \pm 0.1$, 0.99958 for $x = \pm 0.05$, 0.99998 for $x = \pm 0.01$, etc. It thus seems clear that

$$\lim_{x\to 0} \frac{\sin x}{x} = 1 \tag{1.24}$$

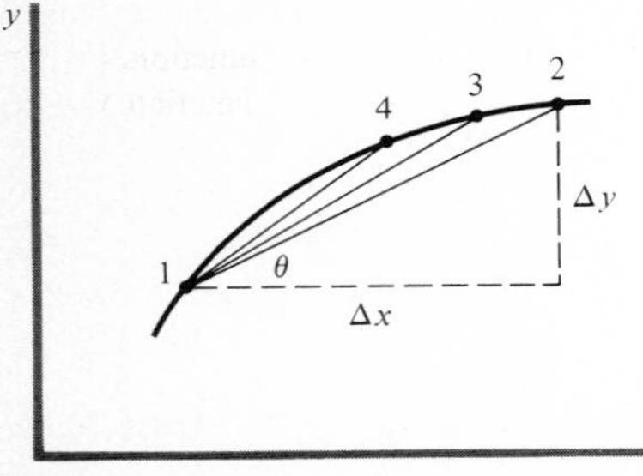

Figure 1.7
As point 2 approaches point 1, the quantity $\Delta y/\Delta x = \tan \theta$ approaches the slope of the tangent to the curve at point 1.

(Of course this isn't meant as a rigorous proof.) Note the resemblance to taking the limit as P goes to zero in Eq. (1.13); in this limit both V and V_{tr} become infinite as P goes to zero, but the limit has a well-defined value even though ∞/∞ is undefined.

Let $y = f(x)$. We shall consider the rate of change of y with x. Let the independent variable change its value from x to $x + h$; this will change y from $f(x)$ to $f(x + h)$. The average rate of change of y with x over this interval equals the change in y divided by the change in x. Using Δ to indicate a change in a variable, we have as the average rate of change

$$\frac{\Delta y}{\Delta x} = \frac{f(x+h) - f(x)}{(x+h) - x} = \frac{f(x+h) - f(x)}{h}$$

The *instantaneous* rate of change of y with x is the limit of this average rate of change taken as the change in x goes to zero. The instantaneous rate of change is called the *derivative* of the function f and is symbolized by f':

$$f'(x) \equiv \lim_{h \to 0} \frac{f(x+h) - f(x)}{h} = \lim_{\Delta x \to 0} \frac{\Delta y}{\Delta x} \qquad (1.25)^*$$

From Fig. 1.7 it is clear that the derivative of a function at a given point is equal to the slope of the line tangent to the curve of y vs. x at that point.

As a simple example, let $y = x^2$. Then

$$f'(x) = \lim_{h \to 0} \frac{(x+h)^2 - x^2}{h} = \lim_{h \to 0} \frac{2xh + h^2}{h} = \lim_{h \to 0} (2x + h) = 2x$$

Hence the derivative of x^2 is $2x$.

A function that makes a sudden jump in value at a certain point is said to be *discontinuous* at that point. An example is shown in Fig. 1.8*a*. Consider the function $y = |x|$, whose graph is shown in Fig. 1.8*b*. This function makes no jumps in value anywhere and so is everywhere continuous. However, the slope of the curve changes suddenly at $x = 0$; therefore, the derivative y' is discontinuous at this point; for negative x the function y equals $-x$ and y' equals -1, whereas for positive x the function y equals x and y' equals $+1$. [Strictly speaking, y' doesn't exist at $x = 0$, since we must get the same limit in (1.25) whether h approaches zero through positive values or through negative values in order for the derivative to exist at a point.]

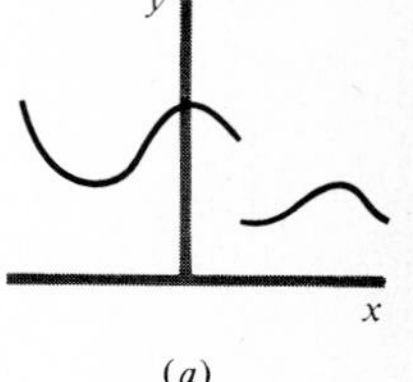

(*a*)

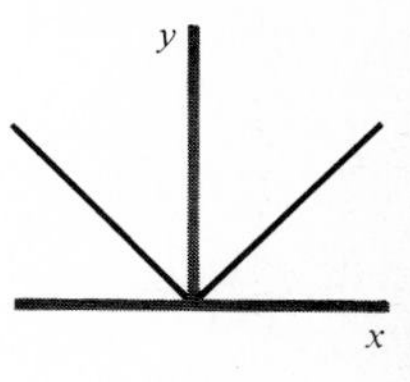

(*b*)

Figure 1.8
(*a*) A discontinuous function. (*b*) The function $y = |x|$.

Since $f'(x)$ is defined as the limit as Δx goes to zero of $\Delta y/\Delta x$, we know that for small changes in x and y, the derivative $f'(x)$ will be approximately equal to $\Delta y/\Delta x$. Thus $\Delta y \approx f'(x)\, \Delta x$ for Δx small. This equation becomes more and more accurate as Δx gets smaller. We can conceive of an infinitesimally small change in x, which we symbolize by dx; denoting the corresponding infinitesimally small change in y by dy, we have $dy = f'(x)\, dx$, or

$$dy = y'(x)\, dx \qquad (1.26)$$

The quantities dy and dx are called *differentials*. Equation (1.26) gives the alternative notation dy/dx for a derivative. Actually, the rigorous mathematical definition of dx and dy does not require these quantities to be infinitesimally small; instead they can be of any magnitude. (See *Thomas*, sec. 2-6; references with the author's name italicized are to books listed in the Bibliography.) However, in our applications of calculus to thermodynamics, we shall find it convenient always to conceive of dy and dx as infinitesimal changes.

Leibniz originally formulated calculus in terms of infinitesimal quantities, but since his formulation is gravely lacking in mathematical rigor, later mathematicians reformulated calculus to eliminate reference to infinitesimals. It was not until 1960 that a mathematically rigorous formulation of calculus that uses infinitesimal quantities was developed; see A. Robinson, *Non-standard Analysis*, North-Holland, 1966; R. A. Bonic et al., *Freshman Calculus*, Heath, 1971, pp. 416–417; *The New York Times*, Feb. 15, 1975, p. 22.

Let a and n be nonzero constants, and let u and v be functions of x: $u = u(x)$ and $v = v(x)$. Using the definition (1.25), one finds the following formulas for differentiation:

$$\frac{da}{dx} = 0, \qquad \frac{d(ax^n)}{dx} = nax^{n-1}, \qquad \frac{d(e^{ax})}{dx} = ae^{ax}$$

$$\frac{d \ln ax}{dx} = \frac{1}{x}, \qquad \frac{d \sin ax}{dx} = a \cos ax, \qquad \frac{d \cos ax}{dx} = -a \sin ax$$

$$\frac{d(u+v)}{dx} = \frac{du}{dx} + \frac{dv}{dx}, \qquad \frac{d(uv)}{dx} = u\frac{dv}{dx} + v\frac{du}{dx} \tag{1.27*}$$

$$\frac{d(u/v)}{dx} = \frac{d(uv^{-1})}{dx} = -uv^{-2}\frac{dv}{dx} + v^{-1}\frac{du}{dx}$$

The chain rule is used frequently to find derivatives. Let z be a function of x, where x is a function of r: $z = z(x)$, where $x = x(r)$. Then z can be expressed as a function of r: $z = z(x) = z[x(r)] = g(r)$, where g is some function. The *chain rule* states that $dz/dr = (dz/dx)(dx/dr)$. For example, suppose we want $(d/dr) \sin 3r^2$. Let $z = \sin x$ and $x = 3r^2$. Then $z = \sin 3r^2$, and the chain rule gives $dz/dr = (\cos x)(6r) = 6r \cos 3r^2$.

Equation (1.26) and the above list of derivatives give the following formulas for differentials:

$$d(ax^n) = nax^{n-1}\,dx, \qquad d(e^{ax}) = ae^{ax}\,dx$$

$$d(au) = a\,du, \qquad d(u+v) = du + dv, \qquad d(uv) = u\,dv + v\,du \tag{1.28}$$

We frequently want to find a maximum or minimum of some function. For a function with a continuous derivative, the slope of the tangent curve is zero at a maximum or minimum point (Fig. 1.9*a*). Hence to locate an extremum we look for those points where $dy/dx = 0$. Figure 1.9*a* shows that for a maximum point the tangent to the curve has positive slope to the left of the maximum and negative slope to its right; in other words, dy/dx is decreasing going through a maximum.

Let the notation $+0-$ indicate a point where dy/dx is positive to the immediate left of the point, zero at the point, and negative to the immediate right of the point. A point with the $+0-$ pattern of slopes is a maximum point. A point with the $-0+$ pattern is a minimum point. What about the pattern $+0+$? This is neither a maximum nor a minimum point but a *horizontal inflection point*. An example is $y = x^3$, which shows this pattern at $x = 0$ (Fig. 1.9*b*). The pattern $-0-$ is also a horizontal inflection point; an example is $y = -x^3$ at $x = 0$. For a horizontal inflection point, it is clear that dy/dx is either a minimum (in the $+0+$ case) or a maximum (in the $-0-$ case) at this point. Hence $(d/dx)(dy/dx) = d^2y/dx^2 = 0$ at a horizontal inflection point. (Any point where d^2y/dx^2 changes sign is called an *inflection point*. Clearly, $d^2y/dx^2 = 0$ at an inflection point; when $dy/dx = 0$ also, it is a horizontal inflection point.) (An example of a horizontal inflection point is the critical point of a fluid; see Chap. 8.)

In thermodynamics we usually deal with functions of two or more variables. Let z be a function of x and y: $z = f(x, y)$. We define the *partial derivative* of z with respect to x as follows:

$$\left(\frac{\partial z}{\partial x}\right)_y \equiv \lim_{\Delta x \to 0} \frac{f(x + \Delta x, y) - f(x, y)}{\Delta x} \qquad (1.29)$$

This definition is analogous to the definition (1.25) of the ordinary derivative, in that if y were a constant instead of a variable, the partial derivative $(\partial z/\partial x)_y$ would become just the ordinary derivative dz/dx. The variable being held constant in a partial derivative is often omitted and $(\partial z/\partial x)_y$ written simply as $\partial z/\partial x$. In thermodynamics there are many possible variables, and to avoid confusion it is usually imperative to indicate which variables are being held constant in a partial derivative. The partial derivative of z with respect to y at constant x is defined similarly to (1.29):

$$\left(\frac{\partial z}{\partial y}\right)_x \equiv \lim_{\Delta y \to 0} \frac{f(x, y + \Delta y) - f(x, y)}{\Delta y}$$

There may be more than two independent variables. For example, let $z = g(w, x, y)$. Then the partial derivative of z with respect to x at constant w and y is defined as

$$\left(\frac{\partial z}{\partial x}\right)_{w,y} \equiv \lim_{\Delta x \to 0} \frac{g(w, x + \Delta x, y) - g(w, x, y)}{\Delta x}$$

The evaluation of partial derivatives is quite simple. To find $(\partial z/\partial x)_y$ we take the ordinary derivative of z with respect to x while regarding y as a constant. For example, if $z = x^2y^3 + e^{yx}$, then $(\partial z/\partial x)_y = 2xy^3 + ye^{yx}$; also, $(\partial z/\partial y)_x = 3x^2y^2 + xe^{yx}$.

Let $z = f(x, y)$. Suppose x changes by an infinitesimal amount dx while y remains constant. What is the infinitesimal change dz in the variable z brought about by the infinitesimal change in x? If z were a function of x only, then [Eq. (1.26)] we would have $dz = (dz/dx)\,dx$; because z depends on y also, the infinitesimal change in z at constant y is given by the analogous equation $dz = (\partial z/\partial x)_y\,dx$. Similarly, if y were to undergo an infinitesimal change dy while x were held constant, we would

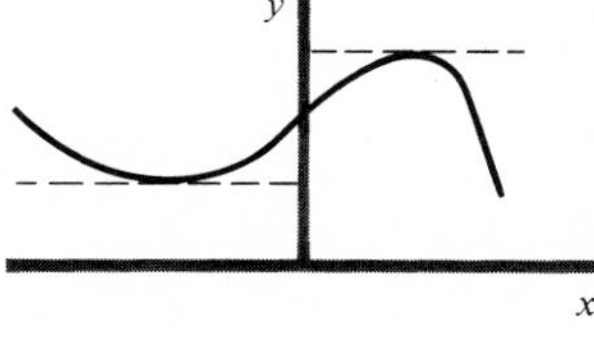

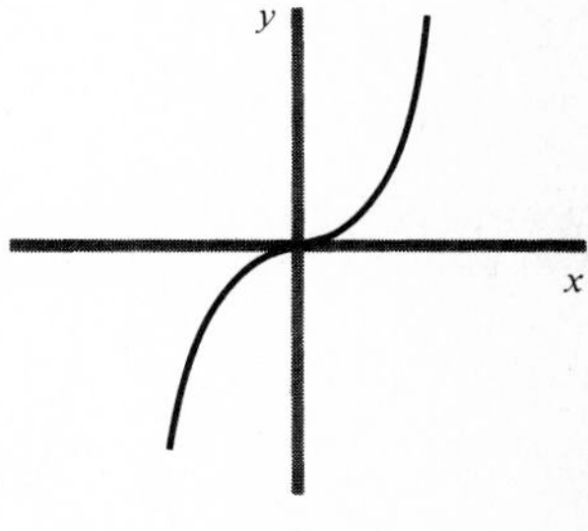

Figure 1.9
(*a*) Horizontal tangent at maximum and minimum points. (*b*) The function $y = x^3$.

have $dz = (\partial z/\partial y)_x\, dy$. If now both x and y undergo infinitesimal changes, the infinitesimal change in z is the sum of the infinitesimal changes due to dx and dy. Thus

$$dz = \left(\frac{\partial z}{\partial x}\right)_y dx + \left(\frac{\partial z}{\partial y}\right)_x dy \qquad (1.30)^*$$

In this equation dz is called the *total differential* of $z(x, y)$. Equation (1.30) is one of the most used equations in thermodynamics. An analogous equation holds for the total differential of a function of more than two variables. For example, if $z = z(r, s, t)$, then

$$dz = \left(\frac{\partial z}{\partial r}\right)_{s,t} dr + \left(\frac{\partial z}{\partial s}\right)_{r,t} ds + \left(\frac{\partial z}{\partial t}\right)_{r,s} dt$$

Three useful partial-derivative identities can be derived from (1.30). For an infinitesimal process in which y does not change, the infinitesimal change dy is 0, and (1.30) becomes

$$dz_y = \left(\frac{\partial z}{\partial x}\right)_y dx_y \qquad (1.31)$$

where the y subscripts on dz and dx indicate that these infinitesimal changes occur at constant y. Division by dz_y gives

$$1 = \left(\frac{\partial z}{\partial x}\right)_y \frac{dx_y}{dz_y} = \left(\frac{\partial z}{\partial x}\right)_y \left(\frac{\partial x}{\partial z}\right)_y$$

since from the definition of the partial derivative, the ratio of infinitesimals dx_y/dz_y equals $(\partial x/\partial z)_y$. Therefore

$$\left(\frac{\partial z}{\partial x}\right)_y = \frac{1}{(\partial x/\partial z)_y} \qquad (1.32)^*$$

Note that the same variable, y, is being held constant in both partial derivatives in (1.32); when y is held constant, there are only two variables, x and z, and you will probably recall that $dz/dx = 1/(dx/dz)$. [However, $d^2z/dx^2 \neq 1/(d^2x/dz^2)$.]

For an infinitesimal process in which z remains constant, Eq. (1.30) becomes

$$0 = \left(\frac{\partial z}{\partial x}\right)_y dx_z + \left(\frac{\partial z}{\partial y}\right)_x dy_z \qquad (1.33)$$

Dividing by dy_z and recognizing that dx_z/dy_z equals $(\partial x/\partial y)_z$, we get

$$0 = \left(\frac{\partial z}{\partial x}\right)_y \left(\frac{\partial x}{\partial y}\right)_z + \left(\frac{\partial z}{\partial y}\right)_x \quad \text{and} \quad \left(\frac{\partial z}{\partial x}\right)_y \left(\frac{\partial x}{\partial y}\right)_z = -\left(\frac{\partial z}{\partial y}\right)_x = -\frac{1}{(\partial y/\partial z)_x}$$

where (1.32) with x and y interchanged was used. Multiplication by $(\partial y/\partial z)_x$ gives

$$\left(\frac{\partial x}{\partial y}\right)_z \left(\frac{\partial y}{\partial z}\right)_x \left(\frac{\partial z}{\partial x}\right)_y = -1 \qquad (1.34)^*$$

Equation (1.34) looks intimidating but is actually easy to remember because of the simple pattern of variables: $\partial x/\partial y$, $\partial y/\partial z$, $\partial z/\partial x$; the variable held constant in each partial derivative is the one that doesn't appear in that derivative.

Sometimes students wonder why the ∂y's, ∂z's, and ∂x's in (1.34) don't cancel to give $+1$ (instead of -1). One can only cancel ∂y's, etc., when the same variable is held constant in each partial derivative. [Note that (1.32) can be written as $(\partial z/\partial x)_y(\partial x/\partial z)_y = 1$.]

Finally, let dy in (1.30) be zero, so that (1.31) holds. Let u be some other variable. Division of (1.31) by du_y gives

$$\frac{dz_y}{du_y} = \left(\frac{\partial z}{\partial x}\right)_y \frac{dx_y}{du_y}$$

$$\left(\frac{\partial z}{\partial u}\right)_y = \left(\frac{\partial z}{\partial x}\right)_y \left(\frac{\partial x}{\partial u}\right)_y \tag{1.35}$$

The ∂x's in (1.35) can be canceled because the same variable (y) is held constant in each partial derivative.

A function of two independent variables $z(x, y)$ has the following four second partial derivatives:

$$\left(\frac{\partial^2 z}{\partial x^2}\right)_y \equiv \left[\frac{\partial}{\partial x}\left(\frac{\partial z}{\partial x}\right)_y\right]_y, \qquad \left(\frac{\partial^2 z}{\partial y^2}\right)_x \equiv \left[\frac{\partial}{\partial y}\left(\frac{\partial z}{\partial y}\right)_x\right]_x$$

$$\frac{\partial^2 z}{\partial x\,\partial y} \equiv \left[\frac{\partial}{\partial x}\left(\frac{\partial z}{\partial y}\right)_x\right]_y, \qquad \frac{\partial^2 z}{\partial y\,\partial x} \equiv \left[\frac{\partial}{\partial y}\left(\frac{\partial z}{\partial x}\right)_y\right]_x$$

Provided the first partial derivatives are continuous (as is generally true in physical applications), one can show that the two mixed second partial derivatives are equal (see *Thomas*, sec. 14-12):

$$\frac{\partial^2 z}{\partial x\,\partial y} = \frac{\partial^2 z}{\partial y\,\partial x} \tag{1.36*}$$

Thus the order of partial differentiation is immaterial.

It is often convenient to write fractions with a slant line. The convention is that

$$a/bc + d \equiv \frac{a}{bc} + d$$

1.7 EQUATIONS OF STATE

The zeroth law of thermodynamics leads to the thermodynamic variable called the temperature; moreover, the temperature is expressible as a function of nonthermal variables of the system [see Eq. (1.3)]. Consider a system composed of a single phase and a single substance. Using the ideal-gas temperature scale T and using the number of moles n to specify the composition, we write (1.3) as

$$T = f(P, V, n) \tag{1.37}$$

where f is a function that depends on the nature of the system; f for liquid water differs from f for ice and from f for liquid benzene. An equation like (1.37) relating the thermodynamic variables that define the state of a system is an *equation of state*. Of course, we can solve (1.37) for P or for V and thereby get the alternative forms $P = g(V, T, n)$ or $V = h(P, T, n)$, where g and h are certain functions. The laws of thermodynamics are general and cannot be used to deduce equations of state for particular systems. Equations of state must be determined experimentally. (One can also use statistical mechanics to deduce an approximate equation of state starting from some assumed form for the intermolecular interactions in the system.)

We have already given an example of an equation of state, namely, $PV = nRT$, the equation of state of an ideal gas. The thermodynamic definition of an ideal gas is one that obeys $PV = nRT$. In reality, no gas obeys this equation of state.

The volume of a one-phase, one-component system is clearly proportional to the number of moles n of substance present at any given T and P. Hence the equation of state for any pure one-phase system can be written in the form

$$V = nk(T, P)$$

where the function k depends on what substance is being considered. Since the functional dependence of V on n is the same for any pure substance, and since we usually deal with closed systems (n fixed), it is convenient to eliminate n and write the equation of state using only intensive variables. To this end, we define the *molar volume* $\bar{V}$ of any pure, one-phase system as the volume per mole:

$$\bar{V} \equiv V/n \tag{1.38)*$$

Note that $\bar{V}$ is a function of T and P: $\bar{V} = k(T, P)$. (For an ideal gas, $\bar{V} = RT/P$.)

For any extensive property of a pure, one-phase system we can define a corresponding molar quantity. Thus the molar mass $\bar{m}$ of a substance is given by $\bar{m} \equiv m/n = M$, as in Eq. (1.4).

There are thus three variables (P, $\bar{V}$, and T) for a single-phase, pure, closed system. The equation of state of the system can be written in the form $P = G(\bar{V}, T)$. One can make a three-dimensional plot of the equation of state by plotting $\bar{V}$, T, and P on the x, y, and z axes. Each possible state of the system gives a point in space, and the locus of all such points gives a surface whose equation is the equation of state.

If we hold one of the three variables constant, we can make a two-dimensional plot. For example, holding T constant at the value T_1, we have as the equation of state of an ideal gas $P\bar{V} = RT_1$. An equation of the form $xy = \text{constant}$ gives a hyperbola when plotted. Choosing other values of T, we get a series of hyperbolas (Fig. 1.5*a*). The lines of constant temperature are called *isotherms*, and a constant-temperature process is called an *isothermal* process. We can also hold either P or $\bar{V}$ constant and plot *isobars* (P constant) or *isochores* ($\bar{V}$ constant).

We shall later find that thermodynamics allows us to express many thermodynamic properties of substances in terms of partial derivatives of P, $\bar{V}$, and T with respect to one another. This is useful because these partial derivatives can be readily measured. There are six such partial derivatives:

$$\left(\frac{\partial \bar{V}}{\partial T}\right)_P, \left(\frac{\partial \bar{V}}{\partial P}\right)_T, \left(\frac{\partial P}{\partial \bar{V}}\right)_T, \left(\frac{\partial P}{\partial T}\right)_{\bar{V}}, \left(\frac{\partial T}{\partial \bar{V}}\right)_P, \left(\frac{\partial T}{\partial P}\right)_{\bar{V}}$$

Equation (1.32) allows three of these six to be written as the reciprocals of the other three:

$$\left(\frac{\partial T}{\partial P}\right)_{\bar{V}} = \frac{1}{(\partial P/\partial T)_{\bar{V}}}, \qquad \left(\frac{\partial T}{\partial \bar{V}}\right)_P = \frac{1}{(\partial \bar{V}/\partial T)_P}, \qquad \left(\frac{\partial P}{\partial \bar{V}}\right)_T = \frac{1}{(\partial \bar{V}/\partial P)_T} \tag{1.39}$$

Furthermore, Eq. (1.34) with x, y, z replaced by P, $\bar{V}$, T gives

$$\left(\frac{\partial P}{\partial \bar{V}}\right)_T \left(\frac{\partial \bar{V}}{\partial T}\right)_P \left(\frac{\partial T}{\partial P}\right)_{\bar{V}} = -1$$

$$\left(\frac{\partial P}{\partial T}\right)_{\bar{V}} = -\left(\frac{\partial P}{\partial \bar{V}}\right)_T \left(\frac{\partial \bar{V}}{\partial T}\right)_P = -\frac{(\partial \bar{V}/\partial T)_P}{(\partial \bar{V}/\partial P)_T} \tag{1.40}$$

where (1.32) was used twice.

Hence there are only two independent partial derivatives: $(\partial \bar{V}/\partial T)_P$ and $(\partial \bar{V}/\partial P)_T$. The other four can be calculated from these two and need not be measured. We define the *thermal expansivity* α (alpha) and the *isothermal compressibility* κ (kappa) of a substance by

$$\alpha(T, P) \equiv \frac{1}{V}\left(\frac{\partial V}{\partial T}\right)_{P,n} \equiv \frac{1}{\bar{V}}\left(\frac{\partial \bar{V}}{\partial T}\right)_P \tag{1.41*}$$

$$\kappa(T, P) \equiv -\frac{1}{V}\left(\frac{\partial V}{\partial P}\right)_{T,n} \equiv -\frac{1}{\bar{V}}\left(\frac{\partial \bar{V}}{\partial P}\right)_T \tag{1.42*}$$

Usually, α is positive; however, liquid water decreases in volume with increasing temperature between 0 and 4°C at 1 atm. One can prove from the laws of thermodynamics that κ must always be positive (see *Zemansky*, sec. 16-9, for the proof). Equation (1.40) can be written as

$$\left(\frac{\partial P}{\partial T}\right)_{\bar{V}} = \frac{\alpha}{\kappa} \tag{1.43}$$

For an ideal gas, $\bar{V} = RT/P$, and one finds

$$\alpha = \frac{1}{\bar{V}}\left(\frac{\partial \bar{V}}{\partial T}\right)_P = \frac{1}{\bar{V}}\left[\frac{\partial}{\partial T}\left(\frac{RT}{P}\right)\right]_P = \frac{1}{\bar{V}}\left(\frac{R}{P}\right) = \frac{1}{T} \tag{1.44}$$

$$\kappa = -\frac{1}{\bar{V}}\left(\frac{\partial \bar{V}}{\partial P}\right)_T = -\frac{1}{\bar{V}}\left[\frac{\partial}{\partial P}\left(\frac{RT}{P}\right)\right]_T = \frac{1}{\bar{V}}\left(\frac{RT}{P^2}\right) = \frac{1}{P} \tag{1.45}$$

$$\left(\frac{\partial P}{\partial T}\right)_{\bar{V}} = \left[\frac{\partial}{\partial T}\left(\frac{RT}{\bar{V}}\right)\right]_{\bar{V}} = \frac{R}{\bar{V}} = \frac{P}{T} = \frac{\alpha}{\kappa} \tag{1.46}$$

which checks with (1.43).

For solids, α is typically 10^{-5} to 10^{-4} K^{-1}. For liquids, α is typically $10^{-3.5}$ to 10^{-3} K^{-1}. (A value $\alpha = 10^{-3}$ K^{-1} means a 1 percent increase in volume for a 10°C increase in temperature.) For gases, α can be estimated from the ideal-gas α, which is $1/T$; for typical temperatures of 100 to 1000 K, we thus have α in the range 10^{-2} to 10^{-3} K^{-1} for gases.

For solids, κ is typically 10^{-5} to 10^{-6} atm^{-1}. For liquids, κ is typically

10^{-4} atm^{-1}. Equation (1.45) for ideal gases gives κ as 1 and 0.1 atm^{-1} at P equal to 1 and 10 atm, respectively. Compared with gases, solids and liquids are quite incompressible; this is so because there isn't much space between molecules in liquids and solids.

The single most comprehensive collection of physical and chemical data is Landolt-Börnstein: *Zahlenwerte und Funktionen* (*Numerical Data and Functional Relationships*), published by Springer-Verlag. The sixth edition of Landolt-Börnstein consists of 27 books published from 1950 to 1976. A "New Series" of volumes was begun in 1961 and so far contains over 40 books, each with text in both English and German. A comprehensive list of sources of experimental data is given at the end of the *Handbook of Chemistry and Physics* (published annually by CRC Press).

So far in this section, we have been considering a pure substance. For a homogeneous mixture, the equation of state involves the variables describing the composition of the mixture (i.e., the mole fractions) in addition to P, T, and $\bar{V}$.

1.8 STUDY SUGGESTIONS

A common reaction to a physical-chemistry course is for a student to say to himself, "This looks like a tough course, so I'd better memorize all the equations, or I won't do well." Such a reaction is understandable, especially since many of us have had teachers who emphasized rote memory, rather than understanding, as the method of instruction. [John Holt's book *How Children Fail* (Dell, 1970) is a good account of this and other failings of educational systems.]

Actually, comparatively few equations need to be remembered (they have been marked), and most of these are simple enough to require little effort at conscious memorization. Being able to reproduce an equation is no guarantee of being able to apply that equation to solving problems. To use an equation properly, one must understand it. Understanding involves not only knowing what the symbols stand for but also knowing when the equation applies and when it does not apply. Everyone knows the ideal-gas equation $PV = nRT$, but it's amazing how often students will use this equation in problems involving liquids or solids. Another part of understanding an equation is knowing where the equation comes from. Is it simply a definition? Or is it a law that represents a generalization of experimental observations? Or is it a rough empirical rule with only approximate validity? Or is it a deduction from the four basic laws (the zeroth, first, second, and third laws) of thermodynamics made without approximations? Or is it a deduction from the laws of thermodynamics made using approximations and therefore of limited validity?

As well as understanding the important equations, you should also know the meanings of the various defined terms (closed system, ideal gas, etc.)

Working problems is an aid to learning physical chemistry. Suggestions for working problems are given in Sec. 2.15.

Make studying an active process. Read with a pencil at hand and use it to verify equations, to underline key ideas, to make notes in the margin, and to write down questions you want to ask your instructor. Try to sort out the basic principles from what is simply illustrative detail and digression. *After reading a section, make a written summary of the important points*; this is a far more effective way of learning than to keep reading and rereading the material.

A psychologist carried out a project in improving student study habits that raised student grades substantially. One of the techniques used was to have students close the textbook at the end of each section, sit on the book (to prevent looking at the book and slumping in the chair), and spend a few minutes outlining the material; the outline was then checked against the section in the book. [L. Fox in R. Ulrich et al. (eds.), *Control of Human Behavior*, Scott, Foresman, 1966, pp. 85–90.]

Before reading a chapter in detail, it is a good idea to browse through it first, reading only the section headings, the first paragraph of each section, and some of the problems at the end of the chapter. This gives an idea of the structure of the chapter and makes the reading of each section more meaningful. Reading the problems first lets you know what you are expected to learn from the chapter.

Since, as with all of us, your capabilities for learning and understanding are finite, it is best to accept the fact that there will probably be some material you may never fully understand.

If you lack motivation to study and work problems, read D. L. Watson and R. G. Tharp, *Self-Directed Behavior*, Brooks/Cole, 1972.

1.9 SUMMARY

Classical thermodynamics deals with the relationships between the macroscopic equilibrium properties of a system. Some important concepts in thermodynamics are *system* (*open, closed, isolated, homogeneous, heterogeneous*); *surroundings*; *walls* (*rigid* vs. *nonrigid*; *permeable* vs. *impermeable*; *adiabatic* vs. *thermally conducting*); *equilibrium* (*mechanical, material, thermal*); *state functions* (*extensive* vs. *intensive*); *phase*; and *equation of state.*

Temperature was defined as an intensive state function that has the same value for two systems in thermal equilibrium and a different value for two systems not in thermal equilibrium. The setting up of a temperature scale is arbitrary, but we chose to use the ideal-gas absolute scale defined by Eq. (1.13).

An ideal gas is one that obeys the equation of state $PV = nRT$. Real gases obey this equation only in the limit of zero density. At ordinary temperatures and pressures, the ideal-gas approximation will usually be adequate for our purposes.

Understanding, rather than rote memory, is the royal road to learning physical chemistry.

PROBLEMS

1.1 (*a*) It is said that a seventeenth-century physicist built a water barometer that projected through a hole in the roof of his house so that his neighbors could predict the weather by the height of the water. Suppose that at 25°C a mercury barometer reads 30.0 in. What would be the corresponding height of the column in a water barometer? (The densities of mercury and water at 25°C are 13.53 and 0.997 g/cm^3, respectively.) (*b*) What pressure in atmospheres corresponds to a 30.0-in. mercury-barometer reading at 25°C at a location where $g = 978$ cm/s^2?

1.2 Explain why the definition of an adiabatic wall in Sec. 1.2 specifies that the wall be rigid and impermeable.

1.3 Derive Eq. (1.17) from Eq. (1.18).

1.4 What is the pressure exerted by 85.0 g of carbon monoxide in a 5.00-l vessel at 0°C?

1.5 For a certain hydrocarbon gas, 20.0 mg exerts a pressure of 24.7 torr in a 500-cm^3 vessel at 25°C. Identify the gas.

1.6 The measured density of a certain gaseous amine at 0°C as a function of pressure is:

P/atm	0.2000	0.5000	0.8000
ρ/(g/l)	0.2796	0.7080	1.1476

Plot P/ρ vs. P and extrapolate to $P = 0$ to find an accurate molecular weight. Identify the gas.

1.7 After 1.60 mol of ammonia gas is placed in a 1600-cm^3 box at 25°C, the box is heated to 500 K. At this temperature the ammonia is partially decomposed to N_2 and H_2, and a pressure measurement gives 4.85×10^6 Pa. Find the number of moles of each component present at 500 K.

1.8 A student attempts to combine Boyle's law and Charles' law as follows. "We have $PV = K_1$ and $V/T = K_2$. Equals multiplied by equals are equal; multiplication of one equation by the other gives $PV^2/T = K_1K_2$. The product K_1K_2 of two constants is a constant, so PV^2/T is a constant for a fixed amount of ideal gas." What is the fallacy in this reasoning?

1.9 Prove that the equations $PV/T = C_1$ for m constant and $V/m = C_2$ for T and P constant lead to PV/mT = a constant.

1.10 Calculate the mass in grams of (*a*) one atom of carbon; (*b*) one molecule of water.

1.11 A 1.00-l bulb of methane at a pressure of 10.0 atm is connected to a 3.00-l bulb of hydrogen at 20.0 atm; both bulbs are at the same temperature. (*a*) After the gases mix, what is the total pressure? (*b*) What is the mole fraction of each component in the mixture?

1.12 A student decomposes $KClO_3$ and collects 36.5 cm^3 of O_2 over water at 23°C. The laboratory barometer reads 751 torr. The vapor pressure of water at 23°C is 21.1 torr. Find the volume the dry oxygen would occupy at 0°C and 1.000 atm.

1.13 Prove that $(d/dx)(\sin x) = \cos x$. You will need to use Eq. (1.24).

1.14 Find $\partial/\partial y$ of each of these functions: (*a*) $\sin axy$; (*b*) $\cos by^2z$; (*c*) $xe^{x/y}$; (*d*) $5x^3/\tan(3x + 1)$.

1.15 Let $z = ax^2y^3$. Find dz.

1.16 (*a*) Two evacuated bulbs of equal volume are connected by a tube of negligible volume. One bulb is placed in a 200-K constant-temperature bath and the other in a 300-K bath, and then 1.00 mole of an ideal gas is injected into the system. Find the final number of moles of gas in each bulb. (*b*) The same as (*a*), except that 1.00 g (instead of moles) of an ideal gas is injected and you are to find the mass of gas in each bulb.

1.17 Verify that $\alpha = -(\partial \ln \rho/\partial T)_P$.

1.18 Classify each property as intensive or extensive: (*a*) temperature; (*b*) mass; (*c*) density; (*d*) electric field strength; (*e*) α; (*f*) index of refraction.

1.19 For an ideal gas (*a*) sketch some isobars on a $\bar{V}$-T diagram; (*b*) sketch some isochores on a P-T diagram.

1.20 Let $z = x^5/y^3$. Evaluate the four second partial derivatives of z; check that $\partial^2 z/(\partial x\, \partial y) = \partial^2 z/(\partial y\, \partial x)$.

1.21 Use the following densities of water as a function of T and P to estimate α, κ, and $(\partial P/\partial T)_{\bar{V}}$ for water at 25°C and 1 atm: 0.997044 g/cm^3 at 25°C and 1 atm; 0.996783 g/cm^3 at 26°C and 1 atm; 0.997092 g/cm^3 at 25°C and 2 atm.

1.22 The definitions of the torr and the atmosphere given in Sec. 1.5 are slightly incorrect. The actual definitions are

$$1 \text{ atm} \equiv 1.01325 \times 10^5 \text{ N/m}^2 \qquad \text{exactly}$$

$$1 \text{ torr} \equiv \tfrac{1}{760} \text{ atm} \qquad \text{exactly}$$

One mmHg is defined as the pressure exerted by a 1-mm-high column of fluid of density 13.5951 g/cm^3 when the gravitational acceleration equals 980.665 cm/s^2. Use a calculator to verify that

$$1 \text{ mmHg} = 1.00000014 \text{ torr}$$

1.23 For O_2 gas in thermal equilibrium with boiling sulfur, the following values of $P\bar{V}$ vs. P are found:

P/torr	1000	500	250
$P\bar{V}$/(l atm mole^{-1})	59.03	58.97	58.93

(Since P has units of pressure, P/torr is dimensionless.) From a plot of these data, find the boiling point of sulfur.

1.24 At 1 atm, liquid water contracts when heated from 0 to 4.0°C and expands when heated above 4.0°C. At 50 atm, the temperature of maximum density of liquid water is 3.0°C; at 100 atm, it is 2.0°C. Let $\rho(P, T)$ denote the density of water at pressure P and temperature T. The following equalities hold quite well:

$$\rho(1 \text{ atm}, 0°\text{C}) = \rho(1 \text{ atm}, 8°\text{C})$$

$$\rho(1 \text{ atm}, 2°\text{C}) = \rho(1 \text{ atm}, 6°\text{C})$$

$$\rho(50 \text{ atm}, 0°\text{C}) = \rho(50 \text{ atm}, 6°\text{C})$$

$$\rho(50 \text{ atm}, 2°\text{C}) = \rho(50 \text{ atm}, 4°\text{C})$$

$$\rho(100 \text{ atm}, 0°\text{C}) = \rho(100 \text{ atm}, 4°\text{C})$$

Sketch the 0, 2, 4, 6, and 8°C isotherms of liquid water on a P-vs.-$\bar{V}$ plot for the pressure range 1 to 100 atm. [Recall that $(\partial P/\partial \bar{V})_T$ is always negative.] A quantitative plot is not required, just one that shows where isotherms cross.

1.25 For water at 17°C and 1 atm,

$$\alpha = 1.7 \times 10^{-4}\ \mathrm{K}^{-1}$$

and $\kappa = 4.7 \times 10^{-5}\ \mathrm{atm}^{-1}$. A closed, rigid container is completely filled with liquid water at 14°C and 1 atm. If the temperature is raised to 20°C, estimate the pressure in the container. (Neglect the pressure and temperature dependences of α and κ.)

1.26 An oil-diffusion pump aided by a mechanical forepump can readily produce a "vacuum" with pressure 10^{-6} torr. Various special vacuum pumps can reduce P to 10^{-11} torr. For 25°C, calculate the number of molecules per cm^3 at (*a*) 1 atm; (*b*) 10^{-6} torr; (*c*) 10^{-11} torr.

1.27 A certain mixture of He and Ne in a 356-cm^3 bulb weighs 0.1480 g and is at 20.0°C and 748 torr. Find the mass and mole fraction of He present.

1.28 The earth's radius is 6.37×10^6 m. Find the mass of the earth's atmosphere. (Neglect the dependence of g on altitude.)

1.29 (*a*) If $10^5 P/\mathrm{atm} = 6.4$, what is P? (*b*) If $10^{-2} T/\mathrm{K} = 4.60$, what is T?

1.30 A mixture of N_2 and O_2 has a density of 1.185 g/l at 25°C and 1.000 atm. Find the mole fraction of O_2 in the mixture.

THE FIRST LAW OF THERMODYNAMICS

2.1 INTEGRAL CALCULUS

Differential calculus was reviewed in Sec. 1.6. We now review integral calculus. (The reader impatient to get on to thermodynamics can skim Secs. 2.1 and 2.2 and come back to them later to pick up the details.)

Frequently, we want to find a function $y(x)$ whose derivative is known to be a certain function $f(x)$:

$$dy/dx = f(x) \tag{2.1}$$

The most general function y that satisfies (2.1) is called the *indefinite integral* of $f(x)$. The following notation is used for the indefinite integral of f:

$$y(x) = \int f(x)\, dx \tag{2.2}$$

Since the derivative of a constant is zero, the indefinite integral of any function contains an arbitrary additive constant. For example, if $f(x) = x$, its indefinite integral $y(x)$ is $\frac{1}{2}x^2 + C$, where C is an arbitrary constant. This result is readily verified by showing that y satisfies (2.1), that is, by showing that $(d/dx)(\frac{1}{2}x^2 + C) = x$. To save space, tables of indefinite integrals usually omit the arbitrary constant C.

From the derivatives given in Sec. 1.6, it follows that

$$\int dx = x + C, \qquad \int ax^n\, dx = \frac{ax^{n+1}}{n+1} + C \qquad \text{where } n \neq -1 \tag{2.3)*}$$

$$\int \frac{1}{x}\, dx = \ln x + C, \qquad \int e^{ax}\, dx = \frac{e^{ax}}{a} + C \tag{2.4)*}$$

$$\int \sin ax\, dx = -\frac{\cos ax}{a} + C, \qquad \int \cos ax\, dx = \frac{\sin ax}{a} + C \tag{2.5)*}$$

where a and n are nonzero constants and C is an arbitrary constant. This list of integrals and the derivatives given in Sec. 1.6 are best memorized. For more complicated integrals than the above, consult a table of integrals.

Particularly recommended is M. Klerer and F. Grossman, *A New Table of Indefinite Integrals*, Dover, 1971 (paperback). Klerer and Grossman used a computer to numerically check each integral in their table; thus their table is probably the most accurate available. (They did computer checks on eight well-known tables of indefinite integrals and found error rates ranging from 0.5 percent to an astonishing 27 percent.)

A second important concept in integral calculus is the definite integral. Let $f(x)$ be a continuous function, and let a and b be any two values of x. The *definite integral* of f between the limits a and b is denoted by the symbol

$$\int_a^b f(x)\,dx \tag{2.6}$$

(The reason for the resemblance to the notation for an indefinite integral will become clear shortly.) The definite integral (2.6) is a number whose value is found from the following definition. We divide the interval from a to b into n subintervals, each of width Δx, where $\Delta x = (b - a)/n$ (see Fig. 2.1). In each subinterval we pick any point we please, denoting the chosen points by $x_1, x_2, \ldots, x_n$. We evaluate $f(x)$ at each of the n chosen points and form the sum

$$\sum_{i=1}^{n} f(x_i)\,\Delta x = f(x_1)\,\Delta x + f(x_2)\,\Delta x + \cdots + f(x_n)\,\Delta x \tag{2.7}$$

[Recall that the definition of the summation notation is

$$\sum_{i=1}^{n} a_i \equiv a_1 + a_2 + \cdots + a_n \tag{2.8}$$

When the limits of a sum are clear, they are often omitted.] We now take the limit of the sum (2.7) as the number of subintervals n goes to infinity, and hence as the width Δx of each subinterval goes to zero. This limit is, by definition, the definite integral (2.6):

$$\int_a^b f(x)\,dx \equiv \lim_{\Delta x \to 0} \sum_{i=1}^{n} f(x_i)\,\Delta x \tag{2.9}$$

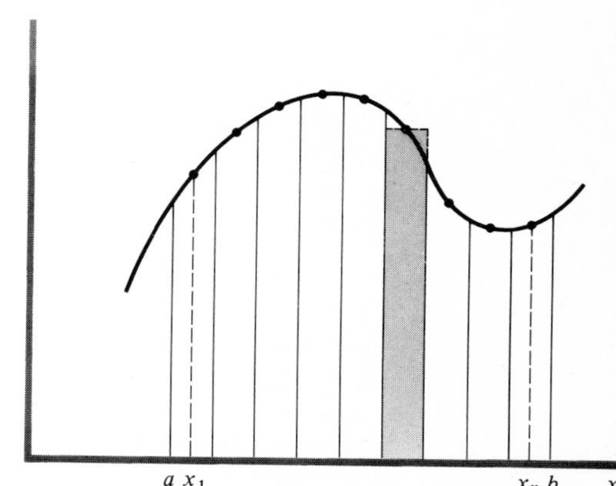

Figure 2.1
Definition of the definite integral.

The motivation for the definition (2.9) is that the quantity on the right side of (2.9) occurs very frequently in physical problems.

Each term in the sum (2.7) is the area of a rectangle of width Δx and height $f(x_i)$. A typical rectangle is indicated by the shading in Fig. 2.1. As the limit $\Delta x \to 0$ is taken, the total area of these n rectangles becomes equal to the area under the curve $f(x)$ between a and b. Thus we can interpret the definite integral as an area. Areas lying below the x axis, where $f(x)$ is negative, make negative contributions to the definite integral.

Use of the definition (2.9) to evaluate an indefinite integral would be tedious. The fundamental theorem of integral calculus (proved in any calculus text; see *Thomas*, sec. 4-9, for example) states that if $y(x)$ is an indefinite integral of $f(x)$ [that is, if y satisfies (2.1)], then

$$\int_a^b f(x)\,dx = y(b) - y(a) \tag{2.10*}$$

For example, if $f(x) = x$, $a = 2$, $b = 6$, we can take $y = \frac{1}{2}x^2$ (or $\frac{1}{2}x^2$ plus some constant) and (2.10) gives $\int_2^6 x\,dx = \frac{1}{2}(6^2) - \frac{1}{2}(2^2) = 16$.

The integration variable x in the definite integral on the left of (2.10) does not appear in the final result (the right side of this equation). It thus does not matter what symbol we use for this variable. If we evaluate $\int_2^6 z\,dz$, we still get 16. In general, $\int_a^b f(x)\,dx = \int_a^b f(z)\,dz$. For this reason the integration variable in a definite integral is called a *dummy variable*. (The integration variable in an indefinite integral is not a dummy variable.) Similarly it doesn't matter what symbol we use for the summation index in (2.8). Replacement of i by j gives exactly the same sum on the right; thus i in (2.8) is a dummy index.

Two identities that readily follow from (2.10) are $\int_a^b f(x)\,dx = -\int_b^a f(x)\,dx$ and $\int_a^b f(x)\,dx + \int_b^c f(x)\,dx = \int_a^c f(x)\,dx$.

An important method for evaluating integrals is a change of variables. For example, suppose we want $\int_2^3 x \exp(x^2)\,dx$. Let $z \equiv x^2$; then $dz = 2x\,dx$, and

$$\int_2^3 xe^{x^2}\,dx = \frac{1}{2}\int_4^9 e^z\,dz = \tfrac{1}{2}e^z\Big|_4^9 = \tfrac{1}{2}(e^9 - e^4) = 4024.2 \tag{2.11}$$

Note that the limits were changed in accord with the substitution $z = x^2$.

From (2.2) and (2.1), it follows that the derivative of an indefinite integral equals the integrand: $(d/dx)\int f(x)\,dx = f(x)$. Note, however, that a definite integral is simply a number and not a function; hence, $(d/dx)\int_a^b f(x)\,dx = 0$.

We can define integration with respect to x for a function of two variables in an analogous manner to (2.2) and (2.9). Thus if $y(x, z)$ is the most general function that satisfies

$$\left[\frac{\partial y(x, z)}{\partial x}\right]_z = f(x, z) \tag{2.12}$$

then the indefinite integral of $f(x, z)$ is

$$\int f(x, z)\,dx = y(x, z) \tag{2.13}$$

For example, if $f(x, z) = xz^3$, then $y(x, z) = \frac{1}{2}x^2z^3 + g(z)$, where g is an arbitrary function of z. If y satisfies (2.12), one can show [in analogy with (2.10)] that a definite integral of $f(x, z)$ is given by

$$\int_a^b f(x, z)\,dx = y(b, z) - y(a, z) \tag{2.14}$$

For example, $\int_2^6 xz^3\,dx = \frac{1}{2}(6^2)z^3 + g(z) - \frac{1}{2}(2^2)z^3 - g(z) = 16z^3$.

Note from (2.14) that the definite integral with respect to x of $f(x, z)$ is a function of z (but not of x). One can show (see *Sokolnikoff and Redheffer*, p. 348) that

$$\frac{d}{dz}\int_a^b f(x, z)\,dx = \int_a^b \frac{\partial f(x, z)}{\partial z}\,dx \tag{2.15}$$

The integrals (2.13) to (2.15) are similar to ordinary integrals of a function $f(x)$ of a single variable in that we regard the second independent variable z in these integrals as constant during the integration process; z acts as a parameter rather than a variable. However, in thermodynamics, we shall frequently have occasion to integrate a function of two (or more) variables in which all the variables are changing during the course of the integration. Such integrals are called line integrals and will be discussed in Sec. 2.4.

2.2 CLASSICAL MECHANICS

Two important concepts in thermodynamics are work and energy. Since these concepts originated in classical mechanics, it is useful to review this subject before we continue with thermodynamics. Some knowledge of classical mechanics is also needed for the study of quantum mechanics.

Classical mechanics (first formulated by the alchemist, theologian, physicist, and mathematician Isaac Newton) deals with the laws of motion of macroscopic bodies whose speeds are small compared with the speed of light c. For objects with speeds not small compared with c, one must use Einstein's *relativistic mechanics.* Since the thermodynamic systems we consider will not be moving at high speeds, we need not worry about relativistic effects. For nonmacroscopic objects (e.g., electrons), one must use *quantum mechanics.* Thermodynamic systems are of macroscopic size, so we shall not need quantum mechanics at this point.

The fundamental equation of classical mechanics is Newton's second law of motion:

$$\mathbf{F} = m\mathbf{a} \tag{2.16}^*$$

where m is the mass of a body, $\mathbf{F}$ is the vector sum of all forces acting on it at some instant of time, and $\mathbf{a}$ is the acceleration the body undergoes at that instant of time. Both $\mathbf{F}$ and $\mathbf{a}$ are vectors (as indicated by the boldface type). Vectors have both magnitude and direction. Scalars (for example, m) have only a magnitude. The meaning of acceleration is as follows. We set up a coordinate system with three mutually perpendicular axes x, y, and z. Let $\mathbf{r}$ be the vector from the coordinate origin to the particle. The *velocity* **v** is the instantaneous rate of change of the position vector **r** with respect to time:

$$\mathbf{v} \equiv d\mathbf{r}/dt \qquad (2.17)^*$$

The magnitude (length) of the vector **v** is the particle's speed v. The particle's *acceleration* **a** is the time rate of change of its velocity:

$$\mathbf{a} \equiv \frac{d\mathbf{v}}{dt} = \frac{d^2\mathbf{r}}{dt^2} \qquad (2.18)^*$$

A vector in three-dimensional space has three components, one along each of the coordinate axes. Equality of vectors means equality of their corresponding components, so a vector equation is equivalent to three scalar equations. Thus Newton's second law (2.16) is equivalent to the three equations

$$F_x = ma_x, \qquad F_y = ma_y, \qquad F_z = ma_z \qquad (2.19)$$

where F_x and a_x are the x components of the force and the acceleration. The x component of the position vector **r** is simply x, the value of the particle's x coordinate. Hence (2.18) gives $a_x = d^2x/dt^2$, and (2.19) becomes

$$F_x = m\frac{d^2x}{dt^2}, \qquad F_y = m\frac{d^2y}{dt^2}, \qquad F_z = m\frac{d^2z}{dt^2} \qquad (2.20)$$

There are two systems of units in common use in mechanics. In the mks system, the units of distance, time, and mass are meters (m), seconds (sec or s), and kilograms (kg), respectively. A force that produces an acceleration of one meter per second² when applied to a one-kilogram mass is defined as one *newton* (N):

$$1 \text{ N} \equiv 1 \text{ kg m/s}^2 \qquad (2.21)$$

The cgs units of length, time, and mass are centimeters (cm), seconds, and grams (g); the cgs unit of force is the *dyne* (dyn):

$$1 \text{ dyn} \equiv 1 \text{ g cm/s}^2 \qquad (2.22)$$

The reader can verify that

$$1 \text{ N} = 10^5 \text{ dyn} \qquad (2.23)$$

In 1960 the General Conference on Weights and Measures recommended a single system of units for worldwide scientific use. This system is called the International System of units, abbreviated SI. In mechanics, the SI uses mks units. If one were to adhere to SI units, pressures would always be given in newtons/meter² (pascals). However, it seems clear that many scientists will continue to use such units as atmospheres and torrs for many years to come. The current scientific literature shows a trend toward increased use of SI units, but since much of the literature continues to be in non-SI units, it is necessary to be familiar with both SI units and some of the commonly used non-SI units.

The *weight* W of a body is the gravitational force exerted on it by the earth. If g is the acceleration due to gravity, Newton's second law gives

$$W = mg \qquad (2.24)$$

If you were suddenly transported to the top of Mount Everest, your mass would be unchanged but your weight would be decreased due to the smaller value of g at the increased altitude.

We now define work. Suppose a force $\mathbf{F}$ acts on a body while the body undergoes an infinitesimal displacement. Let the displacement occur in the x direction. The infinitesimal amount of *work* dw done on the body by the force $\mathbf{F}$ is defined as

$$dw \equiv F_x \, dx \tag{2.25*}$$

where F_x is the component of the force in the direction of the displacement. If the infinitesimal displacement has components in all three directions, then

$$dw \equiv F_x \, dx + F_y \, dy + F_z \, dz \tag{2.26}$$

Consider now a finite (rather than infinitesimal) displacement. For simplicity, let the particle be moving in one dimension (rather than three). The particle is acted on by a force $F(x)$ whose magnitude depends on the particle's position. (Since we are in one dimension, F has only one component and need not be considered a vector.) The work w done by F during a displacement of the particle from x_1 to x_2 is the sum of the infinitesimal amounts of work (2.25) done during the displacement:

$$w = \sum F(x) \, dx \tag{2.27}$$

But the sum of the infinitesimal quantities in (2.27) is the definition of the definite integral [Eq. (2.9)], so

$$w = \int_{x_1}^{x_2} F(x) \, dx \tag{2.28}$$

In the special case that F is constant during the displacement, (2.28) becomes

$$w = F(x_2 - x_1) \tag{2.29}$$

From (2.25), the units of work are those of force times length. The mks and cgs units of work are the *joule* (J) and the *erg*, respectively:

$$1 \text{ J} \equiv 1 \text{ N m} \tag{2.30}$$

$$1 \text{ erg} \equiv 1 \text{ dyn cm} \tag{2.31}$$

Use of (2.23) gives

$$1 \text{ J} = 10^7 \text{ ergs} \tag{2.32}$$

Power P is defined as the rate at which work is done; if an agent does work dw in time dt, then

$$P \equiv dw/dt \tag{2.33}$$

The mks unit of power is the *watt* (W):

$$1 \text{ W} \equiv 1 \text{ J/s} \tag{2.34}$$

We now prove an important theorem in classical mechanics, the *work–energy theorem*. Let $\mathbf{F}$ be the total force acting on a particle, and let the particle undergo a displacement from point 1 with coordinates (x_1, y_1, z_1) to point 2 with coordinates (x_2, y_2, z_2). Integration of (2.26) gives as the total work done on the particle

$$w = \int_1^2 F_x\,dx + \int_1^2 F_y\,dy + \int_1^2 F_z\,dz \tag{2.35}$$

Newton's second law gives $F_x = ma_x = m(dv_x/dt)$. Also, $(dv_x/dt) = (dv_x/dx) \times (dx/dt) = (dv_x/dx)v_x$. Hence $F_x = mv_x(dv_x/dx)$, with similar equations for F_y and F_z. We thus have $F_x\,dx = mv_x\,dv_x$, and (2.35) becomes

$$w = \int_1^2 mv_x\,dv_x + \int_1^2 mv_y\,dv_y + \int_1^2 mv_z\,dv_z$$

$$w = \tfrac{1}{2}m(v_{x2}^2 + v_{y2}^2 + v_{z2}^2) - \tfrac{1}{2}m(v_{x1}^2 + v_{y1}^2 + v_{z1}^2) \tag{2.36}$$

We now define the *kinetic energy* T of the particle as

$$T \equiv \tfrac{1}{2}mv^2 = \tfrac{1}{2}m(v_x{}^2 + v_y{}^2 + v_z{}^2) \tag{2.37*}$$

The right side of (2.36) is then the final kinetic energy T_2 minus the initial kinetic energy T_1:

$$w = T_2 - T_1 = \Delta T \tag{2.38}$$

where ΔT is the change in kinetic energy. In words, the work done on the particle by the force acting on it equals the change in kinetic energy of the particle. This is the work–energy theorem. It is valid because we defined kinetic energy in such a manner as to make it valid.

Besides kinetic energy, there is another kind of energy in classical mechanics. Suppose we throw a body up into the air. As it rises, its kinetic energy decreases, reaching zero at the high point. What happens to the kinetic energy the body loses as it rises? It proves convenient to introduce the notion of a *field* (in this case, a gravitational field) and to say that the decrease in kinetic energy of the body is accompanied by a corresponding increase in the *potential energy* of the field. Likewise, as the body falls back to earth, it gains kinetic energy and the gravitational field loses a corresponding amount of potential energy. Usually, we don't refer explicitly to the field but simply ascribe a certain amount of potential energy to the body itself, the amount depending on the location of the body in the field. A similar situation occurs with a body oscillating on a frictionless spring. As the body oscillates, its kinetic energy undergoes changes, and we say that the amount of potential energy stored in the spring undergoes corresponding changes of opposite sign. Again, we commonly omit specific reference to the spring and simply say that the potential energy of the body changes as it oscillates, so that the sum of its kinetic energy and potential energy remains constant.

To put the concept of potential energy on a quantitative basis, we proceed as follows. Let the forces acting on the particle depend only on the particle's position (and not its velocity, or the time, or any other variable). Such a force **F** with $F_x = F_x(x, y, z)$, $F_y = F_y(x, y, z)$, $F_z = F_z(x, y, z)$ is called a *conservative force*, for a reason to be seen shortly. Examples of conservative forces are gravitational forces, electrical forces, and the Hooke's law force of a spring. Some nonconservative forces are air resistance, friction, and the force you exert when you kick a football. For a conservative force, we define the *potential energy* $V(x, y, z)$ as a function of x, y, and z whose partial derivatives satisfy

$$\frac{\partial V}{\partial x} \equiv -F_x, \qquad \frac{\partial V}{\partial y} \equiv -F_y, \qquad \frac{\partial V}{\partial z} \equiv -F_z \tag{2.39}$$

Since only the partial derivatives of V are defined, V itself has an arbitrary additive constant; we can set the zero level of potential energy anywhere we please. A common choice is to set $V = 0$ at the point where the force on the particle is zero.

From (2.35) and (2.39) it follows that

$$w = -\int_1^2 \frac{\partial V}{\partial x}\,dx - \int_1^2 \frac{\partial V}{\partial y}\,dy - \int_1^2 \frac{\partial V}{\partial z}\,dz \tag{2.40}$$

Since [Eq. (1.30)] $dV = (\partial V/\partial x)\,dx + (\partial V/\partial y)\,dy + (\partial V/\partial z)\,dz$, we have

$$w = -\int_1^2 dV = -(V_2 - V_1) = V_1 - V_2 \tag{2.41}$$

But Eq. (2.38) gives $w = T_2 - T_1$; hence $T_2 - T_1 = V_1 - V_2$ or

$$T_1 + V_1 = T_2 + V_2 \tag{2.42}$$

When only conservative forces act, the sum of the particle's kinetic energy and potential energy remains constant during the motion. This is the law of conservation of mechanical energy. Using E for the total energy, we have for a mechanical system

$$E = T + V \tag{2.43}$$

If only conservative forces act, E remains constant.

We have considered a one-particle system. Similar results hold for a many-particle system. (See H. Goldstein, *Classical Mechanics*, Addison-Wesley, 1950, sec. 1-2 for derivations.) The kinetic energy of an n-particle system is the sum of the kinetic energies of the individual particles:

$$T = T_1 + T_2 + \cdots + T_n = \frac{1}{2}\sum_{i=1}^{n} m_i v_i^2 \tag{2.44}$$

Let the particles exert forces (assumed conservative) on one another. The potential energy V of the system is not the sum of potential energies of the individual particles; instead V is a property of the system as a whole. V turns out to be the sum of contributions due to interactions between pairs of particles. Let V_{ij} be the contribution to V due to the forces acting between particles i and j. Then one finds

$$V = \sum_i \sum_{j>i} V_{ij} \tag{2.45}$$

The double sum indicates that we sum over all pairs of i and j values except those with i equal to or greater than j. Terms with $i = j$ are omitted because a particle does not exert a force on itself; also, only one of the terms V_{12} and V_{21} is included, to avoid counting the interaction between particles 1 and 2 twice. Thus, for a system of three particles, $V = V_{12} + V_{13} + V_{23}$. (If external forces act on the particles of the system, their contributions to V must also be included.)

V_{ij} is a function of the coordinates x_i, y_i, z_i and x_j, y_j, z_j of particles i and j and [cf. Eq. (2.39)] is defined to satisfy

$$\partial V_{ij}/\partial x_i = -F_{ij,x}, \qquad \partial V_{ij}/\partial y_i = -F_{ij,y}, \qquad \partial V_{ij}/\partial z_i = -F_{ij,z} \tag{2.46}$$

where $F_{ij,x}$, $F_{ij,y}$, $F_{ij,z}$ are the components of the force exerted on particle i by particle j. Similarly, $\partial V_{ij}/\partial x_j = -F_{ji,x}$, etc., where $F_{ji,x}$ is the x component of the force exerted on particle j by particle i. (Newton's third law gives $F_{ji,x} = -F_{ij,x}$.)

One finds that $T + V = E$ is constant for a many-particle system with only conservative forces acting.

Energy is a measure of the work the system can do. When a particle's kinetic energy decreases, the work–energy theorem (2.38) says that w, the work done on it is negative; i.e., the particle does work on the surroundings equal to its loss of kinetic energy. Since potential energy is convertible to kinetic energy, potential energy can also be converted ultimately to work done on the surroundings. The kinetic energy is energy the system possesses by virtue of its motion; the potential energy is energy the system possesses by virtue of its configuration (the positions of its particles).

Consider an example. Suppose someone lifts a 30.0-kg object to a height of 2.00 m. The force she exerted on the object is [from (2.24)] equal to $mg = (30.0 \text{ kg})(9.81 \text{ m/s}^2) = 294$ N. From (2.29), the work she did on the object is $(294 \text{ N})(2.00 \text{ m}) = 588$ J. The earth exerted an equal and opposite force on the object compared with the lifter, and the earth did -588 J of work on the object; this work is negative because the force and the displacement are in opposite directions. The total work done on the object by all forces is zero. The work–energy theorem (2.38) gives $w = \Delta T = 0$, in agreement with the fact that the object started at rest and ended at rest.

What is the potential energy of an object in the earth's gravitational field? Let the x axis point outward from the earth with origin at the earth's surface. We have $F_x = -mg$, $F_y = F_z = 0$. Equation (2.39) gives $\partial V/\partial x = mg$, $\partial V/\partial y = 0 = \partial V/\partial z$. Integration gives $V = mgx + C$, where C is a constant. (In doing the integration, we assumed the object's distance above the earth's surface was sufficiently small for g to be considered constant; for the general case, see Prob. 2.4.) Choosing the arbitrary constant as zero, we get

$$V = mgh \tag{2.47}$$

where h is the object's altitude above the earth's surface.

2.3 *P-V* WORK

In thermodynamics, the most common way system and surroundings do work on each other is by a change in the volume of the system.

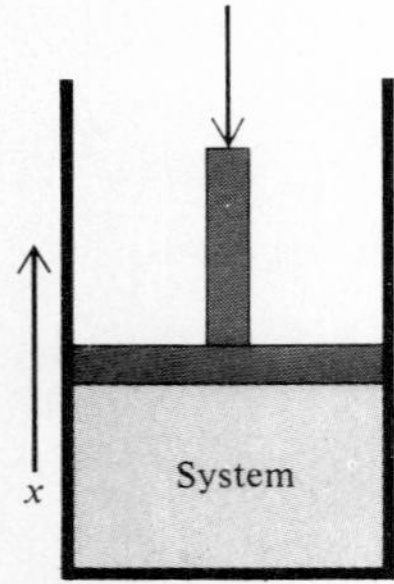

Figure 2.2
A system confined by a piston.

Consider the system of Fig. 2.2, where the piston has cross-sectional area A. Let the system (the matter contained within the piston and cylinder walls) be at pressure P, and let the external pressure on the piston also be P. Equal opposing forces act on the piston, and it is in mechanical equilibrium. Now suppose we decrease the external pressure on the piston by an infinitesimal amount. This causes an infinitesimal imbalance in pressure between system and surroundings, and the system will increase its volume by an infinitesimal amount $dV = A\,dx$, where dx is the distance the piston moves. During this infinitesimal expansion (which occurs at an infinitesimal rate), the system will be infinitesimally close to equilibrium; after the expansion, the system will be at equilibrium again, with its pressure reduced by an infinitesimal amount. The system exerted a force on the piston, and the piston moved. Hence the system did an infinitesimal amount of work on its surroundings (the piston). The work dw_{by} done *by* the system on its surroundings is [Eq. (2.25)]:

$$dw_{by} = F_{x,by}\,dx = PA\,dx = P\,dV \tag{2.48}$$

Note that dw_{by} is positive for the expansion because $F_{x,by}$, the force exerted by the system on its surroundings, is in the same direction as the displacement dx. During the expansion, the piston exerted a force $F_{x,on}$ on the system equal in magnitude and opposite in direction to the force exerted by the system on the piston (Newton's third law), and the system underwent a displacement dx at the point of application of this force; hence (2.25) gives for the work dw_{on} done *on* the system by its surroundings

$$dw_{on} = F_{x,on}\,dx = -F_{x,by}\,dx = -PA\,dx = -P\,dV \tag{2.49}$$

We shall use the symbol w without any subscript to denote work done *on the system* by its surroundings. (Many texts use w to mean work done by the system on its surroundings; their w is the negative of our w.) Hence we write (2.49) as

$$dw_{rev} = -P\,dV \qquad \text{closed system, reversible process} \tag{2.50)*}$$

To use (2.50) it is essential to have memorized the convention that w means work done *on* the system. (The meaning of "reversible" will be discussed shortly.) We have implicitly assumed a closed system in deriving (2.50); when matter is transported between system and surroundings, the meaning of work becomes ambiguous, and we shall not consider this case. We derived (2.50) for a particular shape of system, but it is readily shown to be valid for an arbitrary system shape (see *Zemansky*, p. 54).

Now suppose we carry out an infinite number of successive infinitesimal reductions of the external pressure. At each such reduction, the system expands by dV, and work $-P\,dV$ is done on the system, where P is the current value of the system's pressure. The total work w done on the system is the sum of the infinitesimal amounts of work, and this sum of infinitesimal quantities is the following definite integral:

$$w_{rev} = -\int_1^2 P\,dV \qquad \text{closed syst., rev. proc.} \tag{2.51}$$

where 1 and 2 are the initial and final states of the system, respectively.

We derived (2.50) and (2.51) by considering an expansion. A similar derivation shows that these equations also apply to a contraction of the system. Of course, dV is negative for a contraction; hence $dw > 0$ for contraction. In a contraction, positive work is done on the system and w is positive. In an expansion, positive work is done on the surroundings and w is negative.

The finite expansion process to which (2.51) applies consists of an infinite number of infinitesimal steps and takes an infinite amount of time to carry out. In this process, the difference between the pressures on the two sides of the piston is always infinitesimally small, so that finite unbalanced forces never act and the system remains infinitesimally close to equilibrium throughout the process. Moreover, the process can be reversed at any stage by an infinitesimal change in conditions, namely, by infinitesimally increasing the external pressure; reversal of the process will restore both system and surroundings to their initial conditions.

A *reversible process* is one where the system is always infinitesimally close to equilibrium, and an infinitesimal change in conditions can reverse the process to restore both system and surroundings to their initial states. A reversible process is obviously an idealization.

Equations (2.50) and (2.51) apply only to reversible expansions and contractions. More precisely, they apply to mechanically reversible volume changes; there could be a chemically irreversible process, such as a chemical reaction, occurring in the system during the expansion, but so long as the mechanical forces are only infinitesimally unbalanced, (2.50) and (2.51) apply.

For irreversible volume changes, we usually cannot calculate the work from thermodynamic considerations. For example, suppose the external pressure in Fig. 2.2 is suddenly decreased by a finite amount and is held fixed thereafter. The upward pressure on the piston is then greater than the downward pressure by a finite amount, and the piston is accelerated upward. The piston's acceleration causes a lack of pressure equilibrium within the enclosed gas, with the system's pressure being lower near the piston than farther away from it. Moreover, there is turbulence within the gas. Thus we cannot give a thermodynamic description of the state of the gas. At some point in the piston's rise, the pressure on the bottom face of the piston will be reduced to a value equal to the external pressure, so that the force on the piston is zero; however, the piston now possesses kinetic energy and will therefore continue to rise past this point, thus overshooting the equilibrium position. Eventually the piston will come to rest momentarily at some height, but now the downward pressure exceeds the upward pressure, so that the piston starts to fall. Thus the piston will oscillate. Even if we put a catch into the cylinder to stop the piston at some point in its rise, we still can't calculate w; because of the turbulence in the gas, we have no way (using only thermodynamics) of calculating the kinetic energy of the piston just before it hits the catch.

For an irreversible expansion, we can write $dw_{\text{irrev}} = -P_{\text{surf}}\, dV$, where P_{surf} is the pressure the system exerts on the surface where the expansion occurs (the inner face of the piston in the preceding example); however, we have no way of calculating the variation of P_{surf} during the course of the turbulent, irreversible expansion and hence cannot integrate this equation to find w_{irrev}. Note that by Newton's third law, P_{surf} is also the pressure the surroundings (the inner face of the piston in the example) exert on the system at the expansion surface. Further discussion of work in irreversible expansions is given in D. Kivelson and I. Oppenheim, *J. Chem. Educ.*, **43**, 233 (1966).

2.4 LINE INTEGRALS

The integral in (2.51) is not an ordinary integral. For a closed system of fixed composition, the equation of state has the form $P = P(T, V)$. Hence to calculate w, one must evaluate the negative of the integral

$$\int_1^2 P(T, V)\, dV \tag{2.52}$$

The integrand is a function of two variables T and V. In the integrals (2.13) to (2.15) the integrand depended on two variables, but we regarded the second variable (z) as constant during the integration process and so reduced these integrals (in effect) to ordinary integrals of a single variable. In (2.51), however, both the independent variables T and V will in general change during the expansion process, and to evaluate the integral we must know how T and V vary. Let us plot the variation of T and V during the work process as a line in the TV plane, where points on the line show the succession of equilibrium states through which the system passes during the process. Three possible processes are shown in Fig. 2.3*a*, all with the same initial state and the same final state. The *line integral* (2.52) is defined to be the sum of the infinitesimal quantities $P(T, V)\,dV$ for the process; thus, we divide the line corresponding to the process into small segments (Fig. 2.3*b*) and form the sum

$$\sum_i P(T_i, V_i)\,\Delta V_i \tag{2.53}$$

where ΔV_i is the change that V undergoes in the ith segment and $P(T_i, V_i)$ is any value of P in the ith segment; the line integral (2.52) is then defined as the limit of (2.53) when the lengths of all the segments go to zero. (Often the letter L is put under the integral sign of a line integral; we usually won't use this notation, simply recognizing line integrals from their context.)

As an example, suppose the system has the equation of state $PV = bT^2$, where b is a constant. Then (2.52) becomes

$$b\int_1^2 \frac{T^2}{V}\,dV \tag{2.54}$$

To evaluate this line integral, we must know the process used. For process B in Fig. 2.3*a*, T varies linearly with V, and we substitute $T = c_1 V + c_2$ (where c_1 and c_2 are certain known constants) into (2.54); this reduces (2.54) to an ordinary integral with a single variable V, and we can then evaluate (2.54). Now consider process A of Fig. 2.3*a*. We can break this process into two parts. In the first part, T is constant at T_1, and (2.54) gives for the first part

$$bT_1^2 \int_1^2 \frac{1}{V}\,dV = bT_1^2(\ln V_2 - \ln V_1) = bT_1^2 \ln \frac{V_2}{V_1}$$

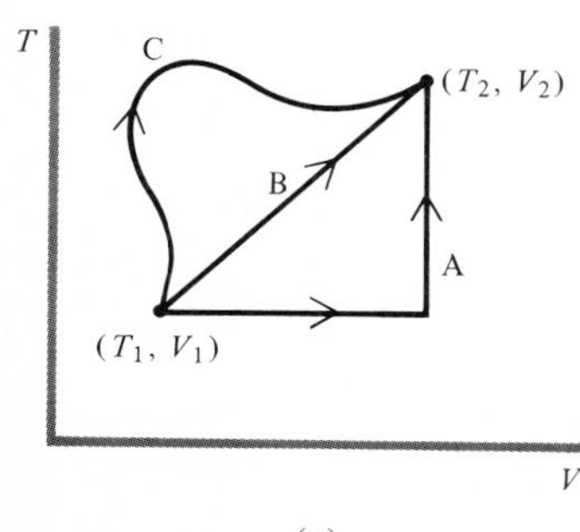

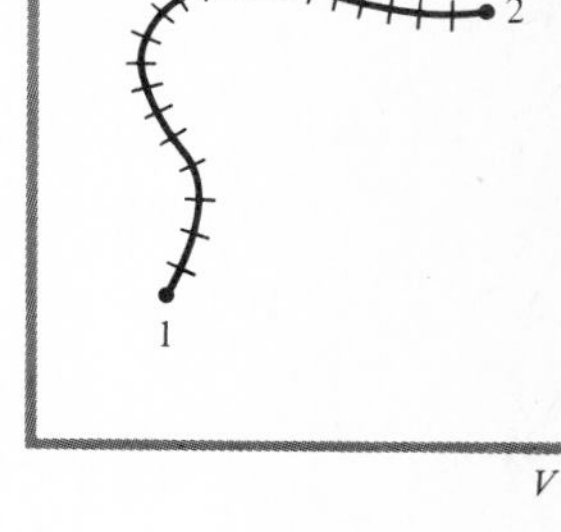

Figure 2.3
(*a*) Three different paths connecting states 1 and 2. (*b*) Definition of the line integral.

For the second part of process A, V is constant at V_2, no work is done, and we get zero contribution to the line integral.

Thus, the work w_{rev} depends on how P varies during the volume change from V_1 to V_2. We say that w_{rev} depends on the *path* from V_1 to V_2, meaning that it depends on the specific process used to go from state 1 to state 2. [Note that the work integral in (2.35) is a line integral since F_x, F_y, and F_z are each functions of all three variables x, y, and z.]

Since w for a reversible process is the negative of the definite integral in (2.52), we can represent w as the negative of the area under a P-V plot between V_1 and V_2 (Fig. 2.4). The work done *by* the system $(-w)$ equals the area under the curve. There are an infinite number of ways of going from the state (P_1, V_1) to the state (P_2, V_2), and w can have any positive or negative value for a given change of state.

The plot of Fig. 2.4 implies internal pressure equilibrium within the system during the process. In the irreversible expansion discussed in the previous section, the system has no single well-defined pressure, and we cannot plot this process on a P-V diagram.

2.5 HEAT

When two bodies at unequal temperatures are placed in contact via a thermally conducting wall, they eventually reach thermal equilibrium at a common intermediate temperature. We say that heat has flowed from the hotter body to the colder one. Let the two bodies have masses m_1 and m_2 and initial temperatures T_1 and T_2, with $T_2 > T_1$; let T_f be the final equilibrium temperature. Provided the two bodies are isolated from the rest of the universe and no phase change or chemical reaction occurs, one experimentally observes the following equation to be satisfied for all values of T_1 and T_2:

$$m_2 c_2(T_2 - T_f) = m_1 c_1(T_f - T_1) \equiv q \tag{2.55}$$

where c_1 and c_2 are constants (evaluated experimentally) that depend on the composition of bodies 1 and 2. We call c_1 the *specific heat capacity* (or *specific heat*) of body 1. We define q, the amount of *heat* that flowed from body 2 to 1, as equal to $m_2 c_2(T_2 - T_f)$.

The unit of heat commonly used in the nineteenth and early twentieth centuries was the *calorie*, defined as the quantity of heat needed to raise one gram of water from 14.5 to 15.5°C at one atmosphere pressure. (This definition is no longer used, as we shall see in Sec. 2.6.) Thus $c_{H_2O} = 1.00$ cal/(g °C) at 15°C and 1 atm.

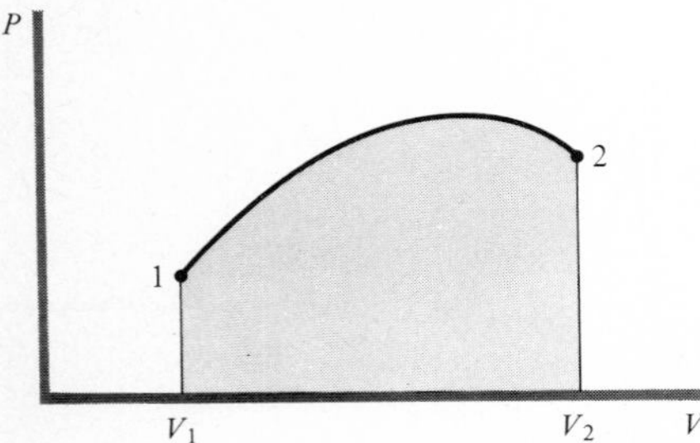

Figure 2.4
The work w for the process corresponding to the curved line is minus the shaded area.

Actually, (2.55) does not hold exactly, because experiment shows the specific heat capacities of substances to be functions of temperature and pressure. Thus when an infinitesimal amount of heat dq_P flows at constant pressure P into a body of mass m and specific heat capacity at constant pressure c_P, the body's temperature is raised by dT and

$$dq_P \equiv mc_P\, dT \tag{2.56}$$

where c_P is a function of T and P. Summing up the infinitesimal flows of heat, we get the total heat that flowed as a definite integral:

$$q_P = m \int_{T_1}^{T_2} c_P(T)\, dT \qquad \text{closed syst., } P \text{ const.} \tag{2.57}$$

(We omitted the pressure dependence of c_P because P is held fixed for the process.) The quantity mc_P is the *heat capacity at constant pressure* C_P of the body: $C_P \equiv mc_P$. From (2.56) we have

$$C_P = dq_P/dT \tag{2.58}$$

Equation (2.55) is more accurately written as

$$m_2 \int_{T_f}^{T_2} c_{P2}(T)\, dT = m_1 \int_{T_1}^{T_f} c_{P1}(T)\, dT = q_P \tag{2.59}$$

If the dependence of c_{P2} and c_{P1} on T is negligible, (2.59) reduces to (2.55).

We gave examples in Sec. 2.3 of reversible and irreversible ways of doing work on a system. Likewise, heat can be transferred reversibly or irreversibly. A reversible transfer of heat requires that the temperature difference between the two bodies be infinitesimal. When there is a finite temperature difference between the bodies, the heat flow is irreversible.

Two bodies need not be in direct physical contact for heat to flow from one to the other; the process of radiation transfers heat between two bodies at different temperatures (e.g., the sun and the earth). The transfer occurs by emission of electromagnetic waves by one body and absorption of these waves by the second body. (An adiabatic wall must be able to block radiation.)

Equation (2.57) was written with the implicit assumption that the system is closed (m fixed). As is true for work, the meaning of heat is ambiguous for open systems. (See R. Haase, *Thermodynamics of Irreversible Processes*, Addison-Wesley, 1969, pp. 17–21 for a discussion of open systems.)

2.6 THE FIRST LAW OF THERMODYNAMICS

As a rock falls toward the earth, its potential energy is transformed into kinetic energy. When it hits the earth and comes to rest, what has happened to its energy of motion? Or consider a billiard ball rolling on a billiard table. Eventually it comes to rest. Again, what happened to its energy of motion? Or imagine that we stir some water in a beaker. Eventually the water comes to rest, and we again ask: What happened to its energy of motion? (The reader may answer that in each case the kinetic energy was converted into heat. We shall soon see that this answer is thermodynamically incorrect.) Careful measurement will show very slight increases in the temperatures of the rock, the billiard ball, and the water (and in their immediate surroundings). Knowing that matter is composed of molecules, we find it easy to believe that the macroscopic kinetic energies of motion of the rock, the ball, and the water were converted into energy at the molecular level; the average molecular translational, rotational, and vibrational energies in the bodies were increased slightly, and these increases were reflected in the temperature rises.

We therefore ascribe an *internal energy* to a body, in addition to its macroscopic kinetic energy T and potential energy V, discussed in Sec. 2.2. This internal energy consists of: molecular translational, rotational, and vibrational energies; kinetic energy and potential energy of the electrons within the molecules (electronic energy); kinetic energy and potential energy of the nucleons (protons and neutrons) within the nucleus; the relativistic rest-mass energy ($m_{rest}c^2$) of the electrons, protons, and neutrons; and potential energy of interaction between the molecules.

The total energy of a body is therefore $E_{tot} = T + V + E_{int}$, where T and V are the macroscopic kinetic and potential energies and E_{int} is the internal energy. We discussed the nature of E_{int} in the preceding paragraph. However, since thermodynamics is a macroscopic science, the development of thermodynamics requires no knowledge of the nature of E_{int}; all that is needed is some means of measuring the change in E_{int} for a process. This will be provided by the first law of thermodynamics. We shall use the symbol U to denote E_{int} and the symbol E to denote the total energy: $U \equiv E_{int}$, $E \equiv E_{tot}$. Thus

$$E = T + V + U \tag{2.60}$$

(Many workers use E as the symbol for E_{int}.) In most of the applications of thermodynamics that we shall consider, the system will be at rest and external fields will not be present; hence, T and V will be zero and the total and internal energies will be equal.

With our present knowledge of the molecular structure of matter, we take it for granted that a flow of heat between two bodies involves a transfer of internal energy between them. However, in the eighteenth and nineteenth centuries the molecular theory of matter was a subject of great controversy. The nature of heat was not well understood until about 1850. In the late eighteenth century, most scientists accepted the caloric theory of heat. (Some students still do, unhappily.) Caloric was a hypothetical fluid substance present in matter and supposed to flow from a hot body to a cold one. (In 1789 Lavoisier listed caloric as one of the chemical elements.) The amount of caloric lost by the hot body equalled the amount gained by the cold body. The total amount of caloric was believed to be conserved in all processes. (Remnants of the caloric theory survive in some of the nomenclature involving heat; e.g., heat capacity. Heat capacity was conceived of as a measure of the ability of a body to store caloric.)

Strong evidence against the caloric theory was provided by Count Rumford in 1798. In charge of the army of Bavaria, he observed that in boring cannon, a virtually unlimited amount of heating was produced by friction, in contradiction to the caloric-theory notion of conservation of heat. Rumford did quantitative experiments, in one of which he found that a cannon borer driven by one horse for 2.5 hr heated 27 lb of ice-cold water to its boiling point. In a paper read before the Royal Society of London in 1798, Rumford argued that his experiments had proved the incorrectness of the caloric theory.

Rumford began life as Benjamin Thompson of Woburn, Mass. At nineteen he married a wealthy widow of thirty. He served the British during the American Revolution and settled in Europe after the war. He secured a position as Minister of War for Bavaria, where he earned extra money by spying for the British. In 1798 he traveled to London, where he founded the Royal Institution, which became one of Britain's leading scientific laboratories. In 1805 he married Lavoisier's widow, adding further to his wealth. He died in 1814. His will left some money to Harvard to establish the Rumford chair of physics, which still exists. It has been estimated that Rumford stole about $\$10^6$ in his lifetime.

Despite Rumford's work, the caloric theory held sway until the 1840s. In 1842 Julius Mayer, a German physician, noted that the food that organisms consume goes partly to produce heat to maintain body temperature and partly to produce mechanical work performed by the organism. He then speculated that work and heat were both forms of energy and that the total amount of energy was conserved. Mayer's arguments were not found convincing, and it remained for James Joule to deal the death blow to the caloric theory.

Joule was the son of a wealthy English brewer. Working in a laboratory adjacent to the brewery, Joule carried out a series of experiments in the 1840s showing that the same changes produced by heating a substance could also be produced by doing mechanical work on the substance, without transfer of heat. His most famous experiment used descending weights to turn paddlewheels in liquids (e.g., water, sperm-whale oil). The potential energy of the weights was converted to kinetic energy of the liquid; the viscosity of the liquid then converted the liquid's kinetic energy to internal energy, producing a rise in temperature. Besides stirring, Joule used several other means of producing temperature increases of liquids by mechanical work; e.g., he compressed a gas in a cylinder immersed in the liquid. Joule found that to increase the temperature of one pound of water by one degree Fahrenheit requires the expenditure of 772 foot-pounds of mechanical energy. Based on Joule's work, the first clear, convincing statement of the principle of conservation of energy was published by the German surgeon, physiologist, and physicist Hermann von Helmholtz in 1847.

The internal energy of a system can be changed in several ways. Internal energy is an extensive property and thus depends on the amount of matter in the system. The internal energy of 20 g of H_2O at a given T and P is twice the internal energy of 10 g of H_2O at that T and P. We usually deal with closed systems; here the amount of matter is held fixed. Besides changing the amount of matter, we can transfer energy to a system by doing work on it or by heating it. The *first law of thermodynamics* asserts that there exists an extensive state function E (called the *total energy* of the system) such that for any process in a closed system

$$\Delta E = q + w \qquad \text{closed syst.} \tag{2.61}$$

where ΔE is the energy change undergone by the system in the process, q is the heat added to the system during the process, and w is the work done on the system during the process. The change of energy ΔE of the system is accompanied by a change of energy of the surroundings equal to $-\Delta E$, so that the total energy of system plus surroundings remains constant (is conserved) for any process.

We shall restrict ourselves to systems at rest in the absence of external fields; hence $T = 0 = V$, and from (2.60) we have $E = U$. Thus (2.61) becomes

$$\Delta U = q + w \qquad \text{closed syst. at rest, no fields} \qquad (2.62)^*$$

where ΔU is the change in internal energy of the system. U is an extensive state function.

Note that when we write ΔU, we mean ΔU_{syst}; we always focus attention on the system, and *all thermodynamic state functions refer to the system* (unless otherwise specified). The conventions for the signs of q and w are set from the system's viewpoint: when heat flows into the system from the surroundings during a process, q is positive ($q > 0$); an outflow of heat from the system to the surroundings means q is negative. When work is done on the system by the surroundings (e.g., in a compression of the system), w is positive; when the system does work on its surroundings, w is negative. A positive q and a positive w each increase the internal energy of the system.

For an infinitesimal process, Eq. (2.62) becomes

$$dU = dq + dw \qquad \text{closed syst.} \tag{2.63}$$

where the other two conditions of (2.62) are implicitly understood. dU is the infinitesimal change in system energy in a process with infinitesimal heat dq flowing into the system and infinitesimal work dw done on the system.

The internal energy U is (just like P or V or T) a function of the state of the system. For any process, ΔU thus depends only on the final and initial states of the system and is independent of the path used to bring the system from the initial state to the final state. If the system goes from state 1 to state 2 by any process, then

$$\Delta U = U_2 - U_1$$

A process in which the final state of the system is the same as the initial state is called a *cyclic process*; here $U_2 = U_1$, and

$$\Delta U = 0 \qquad \text{cyclic proc.} \tag{2.64}$$

as must obviously be true for the change in any state function in a cyclic process.

In contrast to U, the quantities q and w are not state functions. Given only the initial and final states of the system, we cannot find q or w. The heat q and the work w depend on the path used to go from state 1 to 2.

Suppose, for example, that we take 1.00 mole of H_2O at 25.0°C and 1.00 atm and raise its temperature to 30.0°C, the final pressure being 1.00 atm. What is q? The answer is that we cannot calculate q because the process is not specified. We could, if we like, increase the temperature by heating at 1 atm. In this case, $q = mc_P\,\Delta T =$ 18.0 g × 1.00 cal/(g °C) × 5.0°C = 90 cal. However, we could instead emulate James Joule and increase the temperature solely by doing work on the water, stirring it with a paddle (made of an adiabatic substance) until the water reached 30.0°C. In this case, $q = 0$. Or we could heat the water to some temperature between 25 and 30°C and then do enough stirring to bring it up to 30°C. In this case, q is between 0 and 90 cal. (We could even carry out this change of state by a process for which q is negative.) Each of these processes also has a different value of w. However, no matter how we bring the water from 25°C and 1.00 atm to 30.0°C and 1.00 atm, ΔU is always the same, since the final and initial states are the same in each process. Let us find ΔU for the water for this change of state. We can use any process we like to calculate ΔU; it is convenient to consider the process in which the water is reversibly heated from 25 to 30°C at a fixed pressure of 1 atm. For this process, $q = 90$ cal. During the heating, the water expanded slightly, doing work on the surrounding atmosphere. P is constant, and Eq. (2.51) gives $w = -P\,\Delta V$. We have $\Delta V = m(1/\rho_2 - 1/\rho_1)$, where ρ_2 and ρ_1 are the final and initial densities of the water; a handbook gives these densities as 0.9956 and 0.9970 g/cm³. We find $\Delta V = 0.025$ cm³ and

$$w = -0.025 \text{ cm}^3 \text{ atm} = -(0.025 \text{ cm}^3 \text{ atm})\frac{1.987 \text{ cal mole}^{-1} \text{ K}^{-1}}{82.06 \text{ cm}^3 \text{ atm mole}^{-1} \text{ K}^{-1}}$$

$$= -0.0006 \text{ cal}$$

[Note the use of the two values of R (1.21) and (1.19) to convert w from cm³ atm to calories.] Thus w is negligible compared with q, and $\Delta U = q + w = 90$ cal.

Since q and w are not state functions, it is meaningless to ask how much heat a system contains (or how much work it contains). Although one often says that "heat and work are forms of energy," this language, unless properly understood, can mislead one into the error of regarding heat and work as state functions. Heat and work are defined only in terms of processes; before and after the process of energy transfer between system and surroundings, heat and work do not exist. (Recall the preceding example of water brought from 25 to 30°C.) Heat is an energy transfer between system and surroundings due to a temperature difference. Work is an energy transfer between system and surroundings due to a macroscopic force acting through a distance. Heat and work are forms of energy *transfer* rather than forms of energy. Work is energy transfer due to the action of macroscopically observable forces. Heat is energy transfer due to the action of forces at a molecular level; when bodies at different temperatures are placed in contact, collisions between molecules of the two bodies lead to energy transfer from the hotter to the colder body; heat is work done at the molecular level.

Heat and work are measures of energy transfer, and both have the same units as energy. The unit of heat can therefore be defined in terms of the joule. Thus the definition of the calorie given in Sec. 2.5 is no longer in use. The present definition of the calorie is

$$1 \text{ cal} \equiv 4.184 \text{ J} \qquad \text{exactly} \tag{2.65}^*$$

where the value 4.184 was chosen to give good agreement with the old definition of the calorie. The calorie defined by (2.65) is called the *thermochemical calorie*, often designated cal_{th}. (Over the years, about half a dozen slightly different calories have been used.)

It is not necessary to measure heat in calories: the joule can be used as the unit of heat. This is what is done in the officially recommended SI units (Sec. 2.2), but since most of the available thermochemical tabulations use calories, we shall use both joules and calories as the units of heat, work, and internal energy.

For a pure substance, the *molar internal energy* $\bar{U}$ is defined as

$$\bar{U} \equiv U/n \tag{2.66}$$

where n is the number of moles of pure substance.

Often the term *thermal energy* is used. Some people define thermal energy as $U - U_0$, where U is the internal energy of the system and U_0 is the system's internal energy at $T = 0$ K. The thermal energy then differs from the internal energy simply by a constant. However, some people use thermal energy to mean that part of U which is due to molecular kinetic energy.

Although we won't be considering systems with bulk (i.e., macroscopic) kinetic or potential energy, it is worthwhile to consider a possible source of confusion that can arise when dealing with such systems. Consider a rock falling in vacuum toward the earth's surface. Its total energy is $E = T + V + U$. Since the gravitational potential energy V is included as part of the system's energy, the gravitational field (in which the potential energy resides) must be considered to be part of the system. In the first-law equation $\Delta E = q + w$, we do not include work that one part of the system does on another part of the system. Hence w in the first law does not include the work done by the gravitational field on the falling body. Thus for the falling rock, w is zero; also q is zero. Hence $\Delta E = q + w$ is zero, and E remains constant as the body falls (although both T and V vary).

Sometimes people get the idea that Einstein's special relativity equation $E = mc^2$ invalidates the conservation of energy, the first law of thermodynamics. This is not so. All $E = mc^2$ says is that a mass m always has an energy mc^2 associated with it and an energy E always has a mass $m = E/c^2$ associated with it. The total energy of system plus surroundings is still conserved in special relativity; likewise, the total relativistic mass of system plus surroundings is conserved in special relativity. Energy cannot disappear; mass cannot disappear. The equation $\Delta E = q + w$ is still valid in special relativity. Consider, for example, nuclear fission. Although it is true that the sum of the *rest* masses of the nuclear fragments is less than the rest mass of the original nucleus, the fragments are moving at high speed. The relativistic mass of a body increases with increasing speed, and the total relativistic mass of the fragments exactly equals the relativistic mass of the original nucleus.

2.7 STATE FUNCTIONS AND LINE INTEGRALS

The state of a system at rest in the absence of applied fields is specified by its composition (the number of moles of each component present in each phase) and by any two of the three variables P, V, and T; commonly, P and T are used. For a closed system of fixed composition, the state is specified by P and T. Any state function has a definite value once the system's state is specified. Hence any state function of a closed system of fixed composition is a function of T and P of the system; e.g., for such a system we have $U = U(T, P)$, where the nature of the functional dependence on T and P depends on the composition of the system.

We now discuss two criteria that can be used to test whether some quantity is a state function. Let the system undergo a change of state from state 1 to state 2 by some process. We subdivide the process into infinitesimal steps. Let db be some infinitesimal quantity associated with each infinitesimal step. For example, db might be the infinitesimal amount of heat that flows into the system in an infinitesimal step ($db = dq$), or it might be the infinitesimal change in system pressure for an infinitesimal step ($db = dP$), or it might be the infinitesimal heat flow divided by the system's temperature ($db = dq/T$), etc. To determine whether db is the differential of a state function, we consider the line integral

$$\int_{1}^{2} db \tag{2.67}$$
$$L$$

where the L indicates that the integral's value depends in general on the process (path) used to go from state 1 to state 2 of the system. If b is a state function, then the quantity $\Delta b \equiv b_2 - b_1$ [which equals the line integral (2.67)] is independent of the path from state 1 to state 2 but depends only on the initial and final states. (Thus if b is temperature, then $\Delta b = T_2 - T_1$, the temperature difference between the final and initial states, and it doesn't matter how we carried out the change of state.) Therefore if the value of the line integral (2.67) is found to depend on the path used to go from state 1 to state 2, b cannot be a state function.

Conversely, if (2.67) has the same value for every possible path from state 1 to 2, b is a state function whose value for any state of the system can be defined as follows. We pick a reference state r and assign it some value of b, which we denote by b_r. The b value of an arbitrary state 2 is then defined by

$$b_2 - b_r = \int_r^2 db \qquad (2.68)$$

Since, by hypothesis, the integral in (2.68) is independent of the path, the value of b_2 depends only on state 2: $b_2 = b_2(T_2, P_2)$, and b is thus a state function.

As noted in the previous section, if A is any state function, ΔA must be zero for any cyclic process. To indicate a cyclic process, one adds a circle to the line integral (2.67). We now show that if

$$\oint db = 0 \qquad (2.69)$$

for every cyclic process, then the value of the line integral (2.67) is independent of the path and hence b is a state function. Figure 2.5 shows three processes I, II, and III connecting states 1 and 2. Processes I and II constitute a cycle; hence (2.69) gives

$$\underset{\text{I}}{\int_2^1} db + \underset{\text{II}}{\int_1^2} db = 0 \qquad (2.70)$$

Likewise, processes I and III constitute a cycle, and

$$\underset{\text{I}}{\int_2^1} db + \underset{\text{III}}{\int_1^2} db = 0 \qquad (2.71)$$

Subtraction of (2.71) from (2.70) gives

$$\underset{\text{II}}{\int_1^2} db = \underset{\text{III}}{\int_1^2} db \qquad (2.72)$$

Since processes II and III are arbitrary processes connecting states 1 and 2, Eq. (2.72) shows that the line integral (2.67) has the same value for any process between states 1 and 2. Therefore b must be a state function.

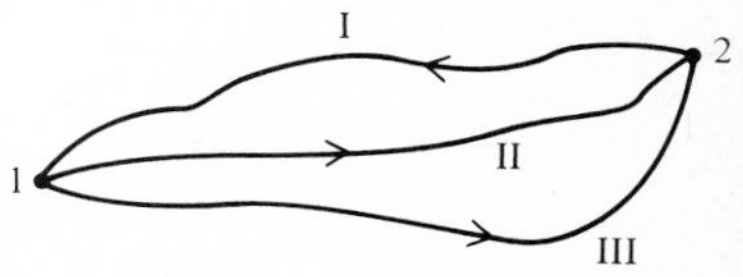

Figure 2.5
Three processes connecting states 1 and 2.

Summarizing, b is a state function if the value of the line integral (2.67) is independent of the path from 1 to 2. Likewise, b is a state function if (2.69) holds for every cyclic process.

2.8 ALTERNATIVE FORMULATION OF THE FIRST LAW

In 1909 the Greek–German mathematician Constantin Carathéodory gave a formulation of the first law of thermodynamics and a definition of heat that are more logical than the historical formulations discussed earlier in this chapter. This reformulation of the first law will not supply us with anything essentially new, but it does improve the logical foundations of thermodynamics and is therefore presented in this section. (This section may be skimmed.)

We begin with the observation that all efforts to construct a perpetual motion machine that will perform work with no input of energy have failed. A machine that is to operate continuously must work in cycles. Therefore, if we consider a machine that is enclosed in adiabatic walls (to prevent energy input via heat flow from the surroundings), it is impossible for the machine to undergo a cycle and give a net output of work for the cycle, reflected, say, in the lifting of a weight in the surroundings. Although mad scientists spend their time looking for a machine that produces net work *output* with no energy input, experience also indicates that an adiabatically enclosed machine cannot undergo a cycle with a net *input* of work.

Thus, experience leads us to postulate that: No closed system can undergo an adiabatic cyclic process that has $w \neq 0$. (As before, w is the work done on the system; in general, w can be positive, negative, or zero.) Hence for any cycle that an adiabatically enclosed closed system undergoes, we must have $w_{\text{ad, cyc}} = 0$, where ad stands for adiabatic and cyc for cycle. The total work $w_{\text{ad, cyc}}$ for the cycle is the sum of the dw's for each infinitesimal part of the cycle, and this sum is the line integral around the cycle. Our postulate gives for a closed system

$$w_{\text{ad, cyc}} = \oint dw_{\text{ad}} = 0 \tag{2.73}$$

Therefore (Sec. 2.7), w_{ad} must define a state function. We shall call this state function the total energy E. Let us restrict ourselves to the case of no bulk motion of the system and no applied fields; then E becomes the *internal energy* U. By definition, the change dU in the internal energy for an infinitesimal process equals dw_{ad} for the process: $dU \equiv dw_{\text{ad}}$. For a noninfinitesimal change from state 1 to state 2, the line integral $\int_1^2 dw_{\text{ad}}$ is independent of the path from 1 to 2 and defines the change $\Delta U = U_2 - U_1$ in internal energy for this process: $\Delta U = \int_1^2 dw_{\text{ad}}$, or

$$U_2 - U_1 = w_{\text{ad}}(1 \rightarrow 2) \qquad \text{closed syst.} \tag{2.74}$$

where $w_{\text{ad}}(1 \rightarrow 2)$ is the work done on the system in any adiabatic process from state 1 to state 2. [Equation (2.74) gives only the energy difference $U_2 - U_1$. To assign a value to U_2, the internal energy of state 2, we pick some convenient reference state r and arbitrarily assign a value U_r as its internal energy. Then (2.74) with state 1 replaced by state r gives $U_2 = U_r + w_{\text{ad}}(r \rightarrow 2)$, and measurement of $w_{\text{ad}}(r \rightarrow 2)$ gives U_2.]

The statement that

For a closed system, the work $w_{ad}(1 \rightarrow 2)$ ***is the same for all adiabatic paths between states 1 and 2***

is the Carathéodory formulation of the first law of thermodynamics.

So far in this section, the term *heat* has been undefined. (Recall that the definition of an adiabatic wall in Sec. 1.2 makes no reference to heat.) For a closed system that is not surrounded by adiabatic walls, experiment shows that the work w done on the system in a process is not necessarily equal to the change in internal energy ΔU, where ΔU is defined by (2.74). For example, using a falling weight to turn paddle wheels in water, one finds that to change 18 g of water from 25°C and 1 atm (state 1) to 26°C and 1 atm (state 2) requires 75 J of adiabatic work; by (2.74), $U_2 - U_1$ is 75 J; however, by use of a bunsen burner, the same change of state can be carried out in a process with $w = 0$. Therefore, if a closed system is not adiabatically enclosed, its internal energy can be changed by means other than work. We define the *heat* q added to the system in any process as the contribution to the change in internal energy of the closed system (assumed at rest in the absence of fields) from effects other than mechanical work. Subtraction of the work contribution w from the change in internal energy gives as the definition of the heat q for a process:

$$q \equiv \Delta U - w = (U_2 - U_1) - w \qquad \text{closed syst.} \tag{2.75}$$

which gives Eq. (2.62) again. Since $U_2 - U_1$ is [from (2.74)] the work done on going from state 1 to 2 adiabatically, Eq. (2.75) gives us a purely mechanical definition of heat: $q = w_{ad}(1 \rightarrow 2) - w$. The heat q for any process is found by subtracting the work w done on the system in that process from the work done on the system when the same change of state is carried out by an adiabatic process.

2.9 ENTHALPY

The *enthalpy* H of a system is defined as

$$H \equiv U + PV \tag{2.76}^*$$

Since U, P, and V are each state functions, H is a state function. Note from (2.50) that the product of P and V has the dimensions of work and hence of energy; therefore it is legitimate to add U and PV. Naturally, H has units of energy.

Of course, we could take any dimensionally correct combination of state functions to define a new state function. Thus, we might define $(3U - 5PV)/T^3$ as the state function "enwhoopee." The motivation for giving a special name to the state function $U + PV$ is that this combination of U, P, and V is of frequent occurrence in thermodynamics. For example, let q_P be the heat absorbed in a constant-pressure process in a closed system. The first law (2.62) gives

$$U_2 - U_1 = q + w = q - \int_{V_1}^{V_2} P\,dV = q_P - P\int_{V_1}^{V_2} dV = q_P - P(V_2 - V_1)$$

$$q_P = (U_2 + PV_2) - (U_1 + PV_1) = H_2 - H_1$$

$$\Delta H = q_P \qquad \text{const. } P\text{, closed syst., } P\text{-}V \text{ work only} \tag{2.77}^*$$

In the derivation of (2.77), we used (2.51) for the work w. Equation (2.51) gives the work associated with a volume change of the system. Besides a volume change, there are other ways that system and surroundings can exchange work, but we won't consider these possibilities until Chaps. 13 and 14. Thus (2.77) is valid only when no kind of work other than volume-change work is done. Note also that (2.51) is for a mechanically reversible process. A constant-pressure process is mechanically reversible since if there were unbalanced mechanical forces acting, the system's pressure P would not remain constant. Equation (2.77) says that for a closed system that can do only P-V work the heat q_P absorbed in a constant-pressure process equals the system's enthalpy change.

In (2.77), $P_1 = P_2 = P$. For a general change of state, the enthalpy change is

$$\Delta H = H_2 - H_1 = U_2 + P_2V_2 - (U_1 + P_1V_1) \tag{2.78}$$

Since U and V are extensive, H is extensive. The molar enthalpy of a pure substance is $\bar{H} = H/n = (U + PV)/n = \bar{U} + P\bar{V}$.

Consider now a constant-volume process. If the closed system can do only P-V work, then $dw = -P\,dV = 0$, since $dV = 0$. The sum of the dw's is zero, so $w = 0$. The first law $\Delta U = q + w$ then becomes for a constant-volume process

$$\Delta U = q_V \qquad \text{closed syst., } P\text{-}V \text{ work only, } V \text{ const.} \tag{2.79}$$

where q_V is the heat absorbed at constant volume. Comparison of (2.79) and (2.77) shows that in a constant-pressure process H plays a role analogous to that played by U in a constant-volume process.

2.10 HEAT CAPACITIES

The *heat capacity* C_{pr} of a closed system for an infinitesimal process pr is defined as

$$C_{\text{pr}} \equiv \frac{dq_{\text{pr}}}{dT} \tag{2.80}^*$$

where dq_{pr} and dT are the heat added to the system and the temperature change of the system in the process. The subscript on C indicates that the heat capacity depends on the nature of the process. For example, for a constant-pressure process we get the *heat capacity at constant pressure* C_P [Eq. (2.58)]:

$$C_P \equiv dq_P/dT \tag{2.81}^*$$

Similarly, the *heat capacity at constant volume* C_V of a closed system is

$$C_V \equiv dq_V/dT \tag{2.82}^*$$

where dq_V and dT are the heat added to the system and the system's temperature change in an infinitesimal constant-volume process. Strictly speaking, Eqs. (2.80) to (2.82) apply only to reversible processes. In an irreversible heating, the system may develop temperature gradients, and there will then be no single temperature assignable to the system. If T is undefined, the infinitesimal change in temperature dT is undefined.

Equations (2.77) and (2.79) written for an infinitesimal process give $dq_P = dH$

at constant pressure and $dq_V = dU$ at constant volume. Hence (2.81) and (2.82) can be written as

$$C_P = \left(\frac{\partial H}{\partial T}\right)_P, \qquad C_V = \left(\frac{\partial U}{\partial T}\right)_V \tag{2.83)*}$$

where these equations apply to closed systems capable of doing only P-V work.

For a closed system in equilibrium, the state function H can be taken as a function of T and P; the partial derivative $[\partial H(T, P)/\partial T]_P$ is also a function of T and P. Hence C_P is a function of T and P and is thus a state function. Similarly, U can be taken as a function of T and V, and C_V is a state function.

For a pure substance, the molar heat capacities at constant pressure and constant volume are $\bar{C}_P = C_P/n$ and $\bar{C}_V = C_V/n$. Some values of $\bar{C}_P$ in cal mole^{-1} K^{-1} at 25°C and 1 atm are (see also the Appendix):

KCl	$CaCO_3$	sucrose	H_2O	CCl_4	C_8H_{18}	CH_4	C_2H_6	H_2
12.3	19.6	102	18.0	31.5	60.7	8.5	12.6	6.9

where the first three substances are solids, the next three are liquids, and the last three are gases. Clearly, $\bar{C}_P$ increases with increasing size of the molecules.

One can prove from the laws of thermodynamics that C_P and C_V must each be positive. (See *Zemansky*, pp. 583, 293.)

What is the relation between C_P and C_V? We have

$$C_P - C_V = \left(\frac{\partial H}{\partial T}\right)_P - \left(\frac{\partial U}{\partial T}\right)_V = \left[\frac{\partial(U + PV)}{\partial T}\right]_P - \left(\frac{\partial U}{\partial T}\right)_V \tag{2.84}$$

$$C_P - C_V = \left(\frac{\partial U}{\partial T}\right)_P + P\left(\frac{\partial V}{\partial T}\right)_P - \left(\frac{\partial U}{\partial T}\right)_V \tag{2.85}$$

We expect $(\partial U/\partial T)_P$ and $(\partial U/\partial T)_V$ in (2.85) to be related to each other. In $(\partial U/\partial T)_V$, the internal energy is taken as a function of T and V: $U = U(T, V)$. From (1.30), the total differential of $U(T, V)$ is

$$dU = \left(\frac{\partial U}{\partial T}\right)_V dT + \left(\frac{\partial U}{\partial V}\right)_T dV \tag{2.86}$$

Equation (2.86) is valid for any infinitesimal process, but since we want to relate $(\partial U/\partial T)_V$ to $(\partial U/\partial T)_P$, we impose the restriction of constant P on (2.86) to give

$$dU_P = \left(\frac{\partial U}{\partial T}\right)_V dT_P + \left(\frac{\partial U}{\partial V}\right)_T dV_P$$

where the P subscripts indicate that the infinitesimal changes dU, dT, and dV occur at constant P. Division by dT_P gives

$$\frac{dU_P}{dT_P} = \left(\frac{\partial U}{\partial T}\right)_V + \left(\frac{\partial U}{\partial V}\right)_T \frac{dV_P}{dT_P}$$

The ratio of infinitesimals dU_P/dT_P is the partial derivative $(\partial U/\partial T)_P$, so

$$\left(\frac{\partial U}{\partial T}\right)_P = \left(\frac{\partial U}{\partial T}\right)_V + \left(\frac{\partial U}{\partial V}\right)_T \left(\frac{\partial V}{\partial T}\right)_P \tag{2.87}$$

Substitution of (2.87) into (2.85) gives the desired relation:

$$C_P - C_V = \left[\left(\frac{\partial U}{\partial V}\right)_T + P\right]\left(\frac{\partial V}{\partial T}\right)_P \tag{2.88}$$

What is the physical reason for the fact that $C_P \neq C_V$? The definitions $C_P = dq_P/dT$ and $C_V = dq_V/dT$ show that the origin of the difference lies in the difference between dq_P and dq_V, the heats added at constant pressure and at constant volume. The first law (2.63) gives $dq = dU - dw = dU + P\,dV$ for a closed system with only P-V work. It follows that $dq_P = dU_P + P\,dV_P$ and $dq_V = dU_V$, where the subscripts indicate constant P or V. Therefore,

$$dq_P - dq_V = dU_P - dU_V + P\,dV_P \tag{2.89}$$

There are thus two reasons for the difference between dq_P and dq_V. In a constant-pressure process part of the added heat goes into the work of expansion (the $P\,dV_P$ term), whereas in a constant-volume process the system does no work on the surroundings. Also, the change in internal energy that occurs at constant pressure differs from the change that occurs at constant volume; $dU_P \neq dU_V$.

The state function $(\partial U/\partial V)_T$ in (2.88) has dimensions of pressure and is sometimes called the internal pressure. Clearly, $(\partial U/\partial V)_T$ is related to that part of the internal energy U which is due to intermolecular potential energy; a change in V will change the average intermolecular distance and hence the average intermolecular potential energy. Where the intermolecular potential energy is small (as in a gas at low or moderate pressures), we expect $(\partial U/\partial V)_T$ to be small; where the intermolecular potential energy is large (as in a liquid or solid), we expect $(\partial U/\partial V)_T$ to be large. Measurement of $(\partial U/\partial V)_T$ in gases is discussed in the next section.

2.11 THE JOULE AND JOULE–THOMSON EXPERIMENTS

In 1843 Joule did an experiment that allows measurement of $(\partial U/\partial V)_T$ for a gas; Joule measured the temperature change after a free expansion of a gas into vacuum. This experiment has been repeated several times since with improved setups. A sketch of the setup used by Keyes and Sears in 1924 is shown in Fig. 2.6. Initially,

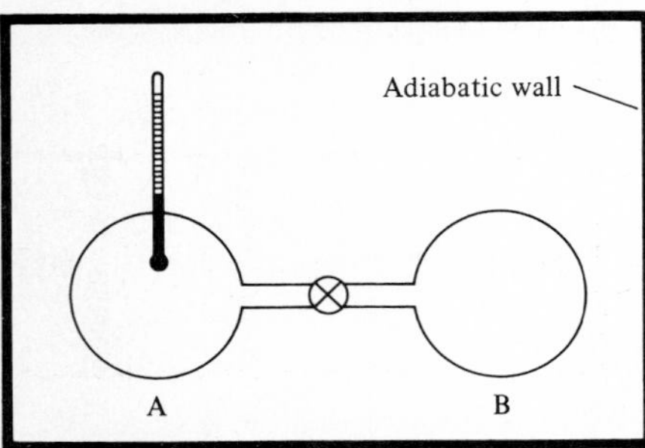

Figure 2.6
Sketch of the Keyes–Sears modification of the Joule experiment.

chamber A is filled with a gas, and chamber B is evacuated. The valve between the chambers is then opened. After equilibrium is reached, the temperature change in the system is measured by the thermometer. Because the system is surrounded by adiabatic walls, q is 0; no heat flows into or out of the system. The expansion into vacuum is highly irreversible; finite unbalanced forces act within the system, and as the gas rushes into B, there is turbulence and lack of pressure equilibrium. Hence (2.51) does not apply; however, we can readily calculate the work $-w$ done by the system. The only motion that occurs is within the system itself; therefore the gas does no work on its surroundings, and vice versa. Hence $w = 0$ for expansion into a vacuum. Since $\Delta U = q + w$ for a closed system, we have $\Delta U = 0 + 0 = 0$; this is a constant-energy process. The experiment therefore measures the temperature change with change in volume at constant internal energy $(\partial T/\partial V)_U$. [More precisely, the experiment measures $\Delta T/\Delta V$ at constant U; the method used to get $(\partial T/\partial V)_U$ from $\Delta T/\Delta V$ measurements is similar to that described later in this section for $(\partial T/\partial P)_H$.]

We define the *Joule coefficient* μ_J (mu jay) as

$$\mu_J \equiv (\partial T/\partial V)_U \tag{2.90}$$

How is the measured quantity $(\partial T/\partial V)_U = \mu_J$ related to $(\partial U/\partial V)_T$? The three variables in these two partial derivatives are the same (namely, T, U, and V). Hence we can use (1.34) to relate the partial derivatives. Replacement of x, y, z in (1.34) with T, U, V gives

$$\left(\frac{\partial T}{\partial U}\right)_V \left(\frac{\partial U}{\partial V}\right)_T \left(\frac{\partial V}{\partial T}\right)_U = -1$$

$$\left(\frac{\partial U}{\partial V}\right)_T = -\left[\left(\frac{\partial V}{\partial T}\right)_U\right]^{-1} \left[\left(\frac{\partial T}{\partial U}\right)_V\right]^{-1} = -\left(\frac{\partial T}{\partial V}\right)_U \left(\frac{\partial U}{\partial T}\right)_V$$

$$\left(\frac{\partial U}{\partial V}\right)_T = -C_V \mu_J \tag{2.91}$$

where (1.32), (2.90), and (2.83) were used.

Joule's 1843 experiment gave zero for μ_J and hence zero for $(\partial U/\partial V)_T$. However, his experimental setup was so poor that his result was meaningless. The 1924 Keyes–Sears experiment showed that $(\partial U/\partial V)_T$ is small but definitely nonzero for gases; because of experimental difficulties, only a few rough measurements were made.

In 1853 Joule and William Thomson (later Lord Kelvin) did an experiment similar to the Joule experiment but allowing far more accurate results to be obtained. The *Joule–Thomson experiment* involves the slow throttling of a gas through a rigid porous plug. An idealized sketch of the experiment is shown in Fig. 2.7. The system is

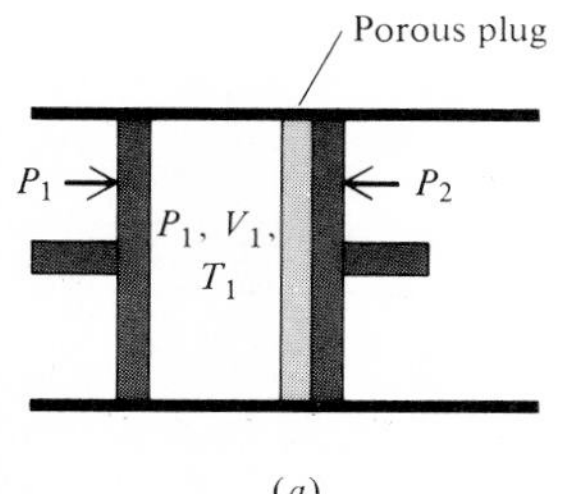

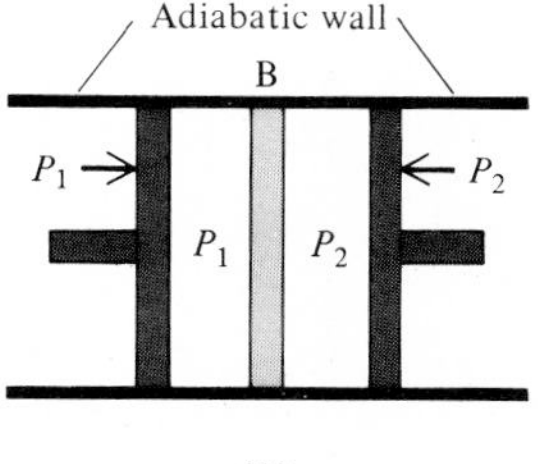

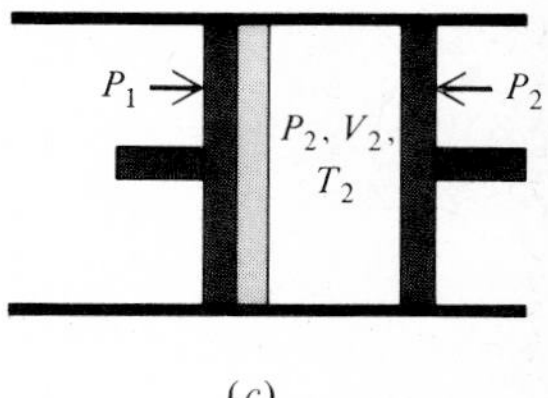

(a) (b) (c)

Figure 2.7 The Joule–Thomson experiment.

enclosed in adiabatic walls. The left piston is held at a fixed pressure P_1; the right piston is held at a fixed pressure $P_2 < P_1$. The partition B is porous but not greatly so; this allows the gas to be slowly forced from one chamber to the other. Because the throttling process is slow, pressure equilibrium is maintained in each chamber. Essentially all the pressure drop from P_1 to P_2 occurs in the porous plug.

We want to calculate w, the work done on the gas in throttling it through the plug. The overall process is irreversible since P_1 exceeds P_2 by a finite amount, and an infinitesimal change in pressures cannot reverse the process. However, the pressure drop occurs almost completely in the plug; the plug is rigid, and the gas does no work on the plug, or vice versa. The exchange of work between system and surroundings occurs solely at the two pistons. Pressure equilibrium is maintained at each piston, and hence we can use $dw_{\text{rev}} = -P\,dV$ to calculate the work at each piston. The left piston does work w_L on the gas; we have $dw_L = -P_L\,dV = -P_1\,dV$, where we use subscripts L and R for left and right. Let all the gas be throttled through; the initial and final volumes of the left chamber are V_1 and 0, so $w_L = -\int_{V_1}^{0} P_1\,dV = -P_1 \int_{V_1}^{0} dV = -P_1(0 - V_1) = P_1 V_1$. The right piston does work dw_R on the gas. (Of course, w_R is negative, since the gas in the right chamber does positive work on the piston.) We have $w_R = -\int_0^{V_2} P_2\,dV = -P_2 V_2$. The work done on the gas is $w = w_L + w_R$ or $w = P_1 V_1 - P_2 V_2$.

The first law for this adiabatic process ($q = 0$) gives $U_2 - U_1 = q + w = w$, so that $U_2 - U_1 = P_1 V_1 - P_2 V_2$ or $U_2 + P_2 V_2 = U_1 + P_1 V_1$. Since $H \equiv U + PV$, we have

$$H_2 = H_1 \qquad \text{or} \qquad \Delta H = 0 \tag{2.92}$$

The initial and final enthalpies are equal in a Joule–Thomson expansion.

The experiment involves measurement of the temperature change $\Delta T = T_2 - T_1$. Hence the Joule–Thomson experiment measures $\Delta T/\Delta P$ at constant H. This may be compared to the Joule experiment, which measures $\Delta T/\Delta V$ at constant U.

We define the *Joule–Thomson coefficient* μ_{JT} by

$$\mu_{JT} \equiv \left(\frac{\partial T}{\partial P}\right)_H \tag{2.93*}$$

μ_{JT} is the ratio of infinitesimal changes in two intensive properties and hence is an intensive property. Like any intensive property, it is a function of T and P (and the nature of the gas).

A single Joule–Thomson experiment yields only $(\Delta T/\Delta P)_H$. To find $(\partial T/\partial P)_H$ values for a gas, we proceed as follows. Starting with some initial P_1 and T_1, we pick a value of P_2 less than P_1 and do the throttling experiment, measuring T_2. We then plot the two points (T_1, P_1) and (T_2, P_2) on a T-P diagram; these are points 1 and 2 in Fig. 2.8. From (2.92), the states 1 and 2 have equal enthalpies. A repetition of the experiment with the same initial P_1 and T_1 but with the pressure on the right piston set at a new value P_3 gives point 3 on the diagram. Several repetitions, each with a different final pressure, yield a series of points that correspond to states of equal enthalpy. We join these points with a smooth curve (called an *isenthalpic curve*). The slope of this curve at any point gives $(\partial T/\partial P)_H$ for the temperature and pressure at that point. Values of T and P for which μ_{JT} is negative (points to the right of point 4) correspond to warming on Joule–Thomson expansion. At point 4, μ_{JT} is zero. To the left of point 4, μ_{JT} is positive, and the gas is cooled by throttling. To generate further isenthalpic curves and get more values of $\mu_{JT}(T, P)$, we use different initial temperatures T_1.

Values of μ_{JT} for gases range from $+3$ to -0.1 °C/atm, depending on the gas and on its temperature and pressure.

Joule–Thomson throttling is frequently used to liquefy gases. For a gas to be cooled by a Joule–Thomson expansion, its μ_{JT} must be positive over the range of T and P involved. In Joule–Thomson liquefaction of gases, the porous plug is replaced by a narrow opening (a needle valve). (Another method of gas liquefaction is an approximately reversible adiabatic expansion against a piston.)

A procedure similar to that used to derive (2.91) yields (Prob. 2.18*a*)

$$\left(\frac{\partial H}{\partial P}\right)_T = -C_P\mu_{JT} \tag{2.94}$$

We can use thermodynamic identities to relate the Joule and Joule–Thomson coefficients. In fact (Prob. 2.18*b*),

$$\mu_{JT} = -(V/C_P)(\kappa C_V \mu_J - \kappa P + 1) \tag{2.95}$$

where κ is defined by (1.42).

2.12 PERFECT GASES AND THE FIRST LAW

In Chap. 1 we defined an ideal gas as one that obeys the equation of state $PV = nRT$. Our molecular picture of an ideal gas is that the intermolecular interactions are negligible. It therefore seems likely that $(\partial U/\partial V)_T$ would vanish for an ideal gas; however, at this point we are not yet able to prove this thermodynamically. To maintain the logical development of thermodynamics, we therefore now define a *perfect gas* as one that obeys both the following equations:

$$PV = nRT \quad \text{perfect gas} \tag{2.96*}$$

$$(\partial U/\partial V)_T = 0 \quad \text{perfect gas} \tag{2.97*}$$

An ideal gas is required to obey only (2.96). Once we have postulated the second law of thermodynamics, we shall prove that (2.97) follows from (2.96), so that there is in fact no distinction between an ideal gas and a perfect gas. Until then, we shall maintain the distinction between the two.

For a closed system in equilibrium, the internal energy (and any other state function) can be expressed as a function of temperature and volume: $U = U(T, V)$.

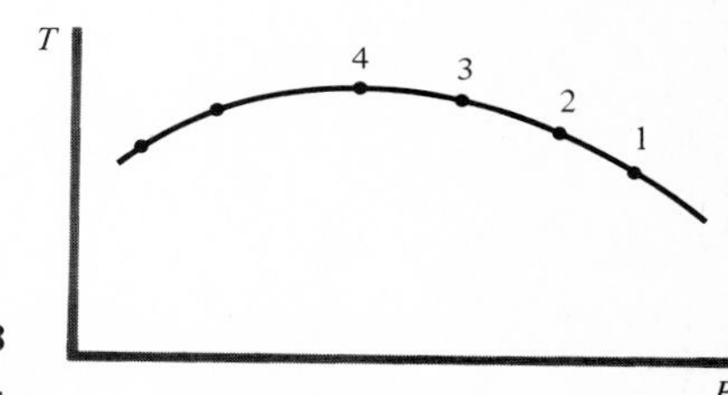

Figure 2.8
An isenthalpic curve.

However, (2.97) states that for a perfect gas U is independent of volume; hence U of a perfect gas depends only on temperature:

$$U = U(T) \qquad \text{perf. gas} \tag{2.98*}$$

Since U is independent of V for a perfect gas, the partial derivative $(\partial U/\partial T)_V$ in Eq. (2.83) for C_V becomes an ordinary derivative:

$$C_V = dU/dT \qquad \text{perf. gas} \tag{2.99*}$$

It follows from (2.98) and (2.99) that C_V of a perfect gas depends only on T:

$$C_V = C_V(T) \qquad \text{perf. gas} \tag{2.100*}$$

What about H? We have for a perfect gas $H = U + PV = U + nRT$; hence (2.98) shows that H depends only on T for a perfect gas. Using (2.83), we then have

$$H = H(T), \qquad C_P = dH/dT, \qquad C_P = C_P(T) \qquad \text{perf. gas} \tag{2.101*}$$

From (2.88) and (2.97) we have

$$C_P - C_V = P(\partial V/\partial T)_P \qquad \text{perf. gas} \tag{2.102}$$

Equation (2.96) gives $(\partial V/\partial T)_P = nR/P$; hence for a perfect gas $C_P - C_V = nR$ or

$$\bar{C}_P - \bar{C}_V = R \qquad \text{perf. gas} \tag{2.103*}$$

It is an experimental fact that at moderate temperatures and pressures $\bar{C}_V$ for *monatomic* gases equals $\frac{3}{2}R$. The reason for this will become clear after we study statistical mechanics.

Mayer used (2.102) in 1842 to calculate the equivalence between heat in calories and work in mechanical units. Using known values for air of C_P and C_V in cal/°C and of the expansion coefficient $P(\partial V/\partial T)_P$ in mechanical units per degree, he found that "the warming of a given weight of water from 0 to 1 degree centigrade corresponds to the fall of an equal weight from a height of about 365 meters."

It follows from (2.97) and (2.91) that $\mu_J = 0$ for a perfect gas. Since H depends only on T for a perfect gas, we have $(\partial H/\partial P)_T = 0$ for such a gas. Hence (2.94) becomes

$$\mu_{JT} = 0 \qquad \text{perf. gas} \tag{2.104}$$

Let us apply the first law $dU = dq + dw$ to a perfect gas. For a reversible process with P-V work only, we have [Eq. (2.50)] $dw = -P\,dV$. Also, (2.99) gives $dU = C_V\,dT$ for a perfect gas. Hence

$$dU = C_V\,dT = dq - P\,dV \qquad \text{perf. gas, rev. proc., } P\text{-}V \text{ work only} \tag{2.105}$$

where the system is assumed closed.

Consider the special case of a reversible *isothermal process* (T constant) in a perfect gas. We have $\Delta U = q + w$. However, since T is constant, U must be constant [Eq. (2.98)] and $\Delta U = 0$. [Equivalently, setting $dT = 0$ in (2.105), we get $dU = 0$, and ΔU is zero.] Hence $q + w = 0$, or $q = -w$. Integration of $dw = -P\,dV$ and use of $PV = nRT$ gives

$$w = -\int_1^2 P\,dV = -\int_1^2 \frac{nRT}{V}\,dV = -nRT\int_1^2 \frac{dV}{V} = -nRT(\ln V_2 - \ln V_1)$$

$$w = -q = nRT \ln \frac{V_1}{V_2} = nRT \ln \frac{P_2}{P_1} \qquad \text{rev. isothermal proc., perf. gas} \tag{2.106}$$

where Boyle's law was used. If the process is an expansion ($V_2 > V_1$), then w (the work done *on* the gas) is negative and q (the heat added to the gas) is positive; all the added heat appears as work done by the gas, maintaining U as constant for the perfect gas. [It is best not to memorize an equation like (2.106), since it can be quickly derived from $dw = -P\,dV$.]

Example A cylinder fitted with a piston contains 3.00 moles of He gas at $P = 1.00$ atm and is in a large constant-temperature bath at 400 K. The pressure is reversibly increased to 5.00 atm. Find w, q, and ΔU for this process.

It is an excellent approximation to consider the helium as a perfect gas. Since T is constant, ΔU is zero. Equation (2.106) gives

$$w = (3.00 \text{ moles})(8.314 \text{ J mole}^{-1}\text{ K}^{-1})(400 \text{ K}) \ln (5.00/1.00)$$
$$= (9980 \text{ J}) \ln 5.00$$

The logarithm in this equation is to the base e. Logarithms are reviewed in Sec. 2.14, where it is shown that $\ln x = 2.303 \log x$, where $\log x$ signifies $\log_{10} x$. Therefore

$$w = (9980 \text{ J})(2.303 \log 5.00) = (9980 \text{ J})(2.303)(0.699) = 1.61 \times 10^4 \text{ J}$$

Also, $q = -w = -1.61 \times 10^4$ J. (If desired, q and w can be calculated in calories by using $R = 1.987$ cal mole^{-1} K^{-1}; the result is $w = 3.84 \times 10^3$ cal.) Of course, w (the work done on the gas) is positive for the compression. The heat q is negative because heat must flow from the gas to the surrounding constant-temperature bath to maintain the gas at 400 K as it is compressed.

Now consider a reversible *adiabatic process* ($q = 0$) in a perfect gas. Setting $dq = 0$ in (2.105) and using $PV = nRT$, we have $C_V\,dT = -P\,dV = -(nRT/V)\,dV$ or $C_V\,dT/T = -nR\,dV/V$. Use of (2.103) gives $C_V\,dT/T = -(C_P - C_V)\,dV/V$. Defining the *heat-capacity ratio* γ (gamma) as

$$\gamma \equiv C_P/C_V \qquad (2.107)^*$$

we have: $dT/T = -(\gamma - 1)\,dV/V$. Integration between the initial and final states 1 and 2 gives

$$\int_1^2 \frac{1}{\gamma - 1}\frac{dT}{T} = -\int_1^2 \frac{dV}{V}$$

Both C_P and C_V of a perfect gas depend on T [Eqs. (2.100) and (2.101)]. Hence $\gamma = \gamma(T)$ for a perfect gas. However, it is an experimental fact that the heat-capacity ratio is only weakly dependent on T for most gases. (For example, for CO_2 at moderate pressures, γ changes from 1.4 to 1.3 when T goes from 0 to 2000°C.) Hence it is usually a good approximation to take γ as independent of T. We then have

$$\frac{1}{\gamma - 1}\ln\frac{T_2}{T_1} = \ln\frac{V_1}{V_2}$$
$$\ln (T_2/T_1) = (\gamma - 1)\ln (V_1/V_2) = \ln [(V_1/V_2)^{\gamma - 1}]$$

If ln x = ln y, then $x = y$. Hence

$$\frac{T_2}{T_1} = \left(\frac{V_1}{V_2}\right)^{\gamma-1} \qquad \text{perf. gas, rev. adiabatic proc., } \gamma \text{ const.} \qquad (2.108)$$

From (2.103) we have $C_P > C_V$; hence $C_P/C_V = \gamma > 1$ for a perfect gas. Hence (2.108) says that when $V_2 > V_1$, we have $T_2 < T_1$. A perfect gas is cooled by a reversible adiabatic expansion. In expanding adiabatically, the gas does work on its surroundings, and since q is zero, U must decrease; hence T decreases. (A near-reversible, near-adiabatic expansion is one method used in refrigeration.)

A more convenient form than (2.108) is obtained by using (1.17). We have $T_2/T_1 = P_2V_2/P_1V_1$; Eq. (2.108) then becomes $P_2V_2/P_1V_1 = (V_1/V_2)^{\gamma-1}$ or

$$P_1V_1^{\gamma} = P_2V_2^{\gamma} \qquad \text{perf. gas, rev. ad. proc., } \gamma \text{ const.} \qquad (2.109)$$

We might compare a reversible isothermal expansion of a perfect gas with a reversible adiabatic expansion of the gas. Let the gas start from the same initial P_1 and V_1 and go to the same V_2. For the isothermal process, $T_2 = T_1$. For the adiabatic expansion, we showed that $T_2 < T_1$. Hence the final pressure P_2 for the adiabatic expansion must be less than P_2 for the isothermal expansion (Fig. 3.4*b*).

In summary, a perfect gas obeys $PV = nRT$, has $(\partial U/\partial V)_T = (\partial H/\partial P)_T = \mu_J = \mu_{JT} = 0$, has U, H, C_V, and C_P depending on T only, has $C_P - C_V = nR$, and has $dU = C_V\,dT$ and $dH = C_P\,dT$. These equations are valid only for a perfect gas. One of the most common errors students make in thermodynamics is to use one of these equations where it is not applicable.

2.13 CALCULATION OF FIRST-LAW QUANTITIES

We now summarize the available thermodynamic methods for the calculation of q, w, ΔU, and ΔH in a process.

Calculation of q. The simplest case is an adiabatic process, where $q = 0$ by definition. For a general reversible process pr, the heat-capacity definition (2.80) gives $dq_{\text{pr}} = C_{\text{pr}}\,dT$ and $q_{\text{pr}} = \int_1^2 C_{\text{pr}}\,dT$, where we must use the heat capacity appropriate to the process. q is not a state function but depends on the path used to go from state 1 to state 2. The most common case (especially for liquids and solids) is a constant-pressure heating, where we use C_P. For a constant-volume process, we use C_V. For other kinds of processes, C_{pr} is usually not known, so we can't readily calculate q from this equation.

A reversible phase change (e.g., the melting of ice or the freezing of water at 0°C and 1 atm) has $dq \neq 0$ and $dT = 0$. Hence C_{pr} is infinite for a reversible phase change and cannot be used to calculate q for a phase change. For a reversible phase change, the experimentally measured latent heat is used to find q.

Sometimes it is easiest to calculate q from the first law $\Delta U = q + w$. For example, for an isothermal process in a perfect gas, ΔU is zero, and $q = -w$, where w is given by (2.106).

Thermodynamics cannot furnish us with values of the heat capacities C_P of substances or of the latent heats of phase changes. These quantities must be measured experimentally. (One can use statistical mechanics to theoretically calculate the heat capacities of certain systems, as we shall later see.)

Calculation of w. For now, we shall deal only with cases where w is P-V work of compression or expansion. For a mechanically reversible volume change, w is given by Eq. (2.51), which involves a line integral. Evaluation of w then depends on knowledge of how P varies during the process. The simplest case is a constant-pressure process, where $w = -P\,\Delta V$.

For an ideal gas, $P = nRT/V$; if T is constant (isothermal process), integration gives $w = -nRT \ln (V_2/V_1)$, as in (2.106); if T of the ideal gas is not constant, we must know how it varies with volume to evaluate the line integral and find w.

For a constant-volume process, $w = 0$.

If ΔU and q are known, w can be calculated from the first law as $w = \Delta U - q$. For example, for an adiabatic process in a perfect gas, Eq. (2.99) gives $dw = dU = C_V(T)\,dT$, and if $C_V(T)$ is known, we can integrate to get w.

In general, thermodynamics cannot calculate w for mechanically irreversible processes, but there are two kinds of irreversible processes where w is readily calculated. For the (irreversible) expansion of a gas into vacuum (Joule experiment), $w = 0$ since the only motion is within the system. For a Joule–Thomson (irreversible) throttling, we can calculate w (Sec. 2.11) since all the irreversibility is confined to the rigid porous plug, where no work is done.

Calculation of ΔU. Once q and w have been calculated, the first law gives ΔU as $q + w$ for a closed system. (Any equation with q or w in it applies only to closed systems, since q and w are undefined for open systems.)

Another approach is to note that for a closed equilibrium system, we have $U = U(T, V)$; the total differential of U is then

$$dU = \left(\frac{\partial U}{\partial T}\right)_V dT + \left(\frac{\partial U}{\partial V}\right)_T dV = C_V\,dT + \left(\frac{\partial U}{\partial V}\right)_T dV \tag{2.110}$$

where (2.83) was used. To make (2.110) generally useful, we must relate $(\partial U/\partial V)_T$ to easily measured properties of a system. This will be done in Sec. 4.5. Until then, (2.110) will only be of use in a constant-volume process ($dV = 0$), where it becomes

$$dU = C_V\,dT, \qquad \Delta U = \int_1^2 C_V\,dT = q_V, \qquad V \text{ const.} \tag{2.111}$$

Since $w = 0$ for a constant-volume process, ΔU must equal q for such a process and (2.111) doesn't give us anything we didn't already know.

In certain cases (see below), it may be easier to find ΔH than to find ΔU. Once ΔH is known, it is easy to find ΔU. We have $\Delta H = H_2 - H_1 = (U_2 + P_2 V_2) - (U_1 + P_1 V_1)$; thus $H_2 - H_1 = U_2 - U_1 + P_2 V_2 - P_1 V_1$ and

$$\Delta H = \Delta U + \Delta(PV) \tag{2.112}$$

An error students sometimes make is to equate $\Delta(PV)$ with $P\,\Delta V + V\,\Delta P$. The quantity $\Delta(PV)$ is by definition $(PV)_2 - (PV)_1 = P_2 V_2 - P_1 V_1$. We have

$$\Delta(PV) = P_2 V_2 - P_1 V_1 = (P_1 + \Delta P)(V_1 + \Delta V) - P_1 V_1$$

$$= P_1\,\Delta V + V_1\,\Delta P + \Delta P\,\Delta V$$

$$\Delta(PV) = P\,\Delta V + V\,\Delta P + \Delta P\,\Delta V \tag{2.113}$$

where P and V on the right side of (2.113) are the initial pressure and initial volume. Because of the $\Delta P\ \Delta V$ term, $\Delta(PV)$ is not equal to $P\ \Delta V + V\ \Delta P$.

If we take the limit of (2.113) as the changes become infinitesimally small, we get

$$d(PV) = P\ dV + V\ dP \tag{2.114}$$

since the product $dP\ dV$ of two infinitesimals can be neglected in comparison with the other terms, which each involve only one infinitesimal. Equation (2.114) is a special case of the differential identity $d(uv) = u\ dv + v\ du$, given in Sec. 1.6.

For a perfect gas, $dU = C_V\ dT$ [Eq. (2.99)], and integration gives

$$\Delta U = U_2 - U_1 = \int_{T_1}^{T_2} C_V(T)\ dT \qquad \text{perf. gas} \tag{2.115}$$

If $C_V(T)$ is known, the integration can be performed. If ΔT is reasonably small, it may be a good approximation to take C_V independent of T; then (2.115) gives $\Delta U \approx C_V\ \Delta T$.

Most commonly, ΔU is calculated from $\Delta U = q + w$, or from (2.115) *if* the system is a perfect gas.

Calculation of ΔH. At this point we have three methods for the calculation of ΔH. We can calculate ΔH for an arbitrary process by using (2.112), provided we are able to first calculate ΔU.

The second method is restricted to constant-pressure processes, for which we showed $\Delta H = q_P$. For a constant-pressure process, it is often easier to calculate ΔH first and then find ΔU from ΔH. Two common constant-pressure processes are: (*a*) A reversible phase change at constant pressure. Here ΔH is calculated from the latent heat of the phase transition. For example, the measured latent heat of fusion of H_2O at 0°C and 1 atm is 79.7 cal/g. Hence $\Delta \bar{H} = 1436$ cal/mole for fusion (melting) of H_2O at 0°C and 1 atm; for the freezing of liquid water at 0°C and 1 atm, $\Delta \bar{H} = -1436$ cal/mole, since q_P is negative for freezing (heat flows from system to surroundings during the freezing of the liquid system). (*b*) Constant-pressure reversible heating of a substance without a phase change. Here the definition of C_P as dq_P/dT gives

$$\Delta H = q_P = \int_{T_1}^{T_2} C_P(T)\ dT \qquad \text{const. } P \tag{2.116}$$

Since P is constant, we didn't bother to indicate that C_P depends on P as well as on T. (The dependence of C_P and C_V on pressure is rather weak; hence, unless one deals with high pressures, a value of C_P measured at 1 atm can be used at other pressures.)

Since H is a state function, we can use the integral in (2.116) to calculate ΔH for any process whose initial and final states have the same pressure, whether or not the entire process occurs at constant pressure and whether or not the process is reversible. Since ΔH is independent of the path, we can imagine a reversible constant-pressure path between the initial and final states and use (2.116). (Of course, q will be different for different paths between the same pair of states.)

The third method for calculating ΔH is restricted to perfect gases. Equation (2.101) gives

$$\Delta H = \int_{T_1}^{T_2} C_P(T)\ dT \qquad \text{perf. gas} \tag{2.117}$$

Alternatively, we can calculate ΔU from (2.115) and then use $\Delta H = \Delta U + \Delta(PV) = \Delta U + \Delta(nRT) = \Delta U + nR\,\Delta T$ for a perfect gas.

A word about units. Most heat-capacity and latent-heat data are tabulated in calories, so q is usually calculated in calories. Pressures are usually given in atmospheres, so P-V work is usually calculated in cm^3 atm. The SI-recommended unit for q, w, ΔU, and ΔH is the joule. Hence we often want to interconvert between joules, calories, and cm^3 atm; this is done using the three values of R in (1.19) to (1.21); an example was given in Sec. 2.6.

2.14 LOGARITHMS

Because logarithms are used so frequently in physical chemistry, we shall review their properties in this section. If $x = a^s$, then s is said to be the *logarithm* (log) of x to the base a: if $a^s = x$, then $\log_a x = s$. The most important base is the irrational number $e = 2.71828 \cdots$, defined as the limit of $(1 + b)^{1/b}$ as $b \to 0$. Logs to the base e are called *natural* logarithms and are written as $\ln x$. The importance of natural logs is due to the fact that $\int x^{-1}\,dx = \ln x$. For practical calculations, one often uses logs to the base 10, called *common* logarithms and written as $\log x$. Thus

$$\ln x \equiv \log_e x, \qquad \log x \equiv \log_{10} x \qquad (2.118)^*$$

$$\text{If } 10^t = x \qquad \text{then } \log x = t \qquad (2.119)$$

$$\text{If } e^s = x \qquad \text{then } \ln x = s \qquad (2.120)$$

These last two equations can be written as

$$e^{\ln x} = x \qquad \text{and} \qquad 10^{\log x} = x \qquad (2.121)$$

From (2.120) it follows that $\ln e^s = s$. (Since $e^{\ln x} = x = \ln e^x$, we say that the exponential and natural logarithmic functions are inverses of each other.) The function e^x is often written as $\exp x$. Since $e^1 = e$, $e^0 = 1$, and $e^{-\infty} = 0$, Eq. (2.120) gives

$$\ln e = 1, \qquad \ln 1 = 0, \qquad \ln 0 = -\infty \qquad (2.122)$$

An important identity is $\ln xy = \ln x + \ln y$. To prove this, let $e^s = x$ and $e^t = y$; hence, $s = \ln x$ and $t = \ln y$. We have $e^s e^t = xy$, or $e^{s+t} = xy$. This last equation shows that $\ln xy = s + t$, which becomes $\ln xy = \ln x + \ln y$. Q.E.D. Similarly, $\ln (x/y) = \ln x - \ln y$. Thus

$$\ln xy = \ln x + \ln y, \qquad \ln (x/y) = \ln x - \ln y \qquad (2.123)^*$$

Let $\ln x = s$, so that $e^s = x$. Raising this last equation to the kth power, we get $e^{ks} = x^k$; hence, $\ln x^k = ks$, which becomes

$$\ln x^k = k \ln x \qquad (2.124)^*$$

To relate base 10 and base e logarithms, we use (2.121) to write $e^{\ln x} = 10^{\log x}$. Taking base 10 logs of this equation and using $\log y^k = k \log y$, we get $\ln x \log e = \log x \log 10 = \log x$. Hence $\ln x = \log x/\log e = \log x/0.43429$, and

$$\ln x = 2.3026 \log x \qquad (2.125)^*$$

To find the common log of a number, we first write it in scientific notation with one figure to the left of the decimal point and then use $\log ab = \log a + \log b$. Since log 1 = 0 and log 10 = 1, the log of a number between 1 and 10 lies between 0 and 1. For example,

$$\log 0.00275 = \log (2.75 \times 10^{-3}) = \log 2.75 + \log 10^{-3} = 0.4393 - 3 = -2.5607$$

To find the antilog of a number, we reverse the above process, starting by writing the log as a number between 0 and 1 plus (or minus) an integer. For example, to find the number whose common log is -2.585, we have $10^{-2.585} = \text{antilog}\,(-2.585) = \text{antilog}\,(0.415 - 3) = 2.60 \times 10^{-3}$.

One can only take the logarithm or the exponential of a dimensionless quantity.

2.15 PROBLEM SOLVING

Trying to learn physical chemistry solely by reading a textbook without working problems is about as effective as trying to improve your physique by reading a book on body conditioning without doing the recommended physical exercises.

If you don't see how to work a problem, it often helps to carry out the following steps: (1) List all the relevant information that is given. (2) List what quantities are to be calculated. (3) Ask yourself what equations, laws, or theorems connect what is known to what is unknown. (4) Apply the relevant equations to calculate what is unknown from what is given.

Although these steps are just common sense, they can be quite useful. The point is that problem solving is an active process. Listing the given information and the unknown quantities and actively searching for relationships that connect them gets your mind working on the problem, whereas simply sitting and reading the problem over and over may not get you anywhere.

In steps 1 and 2, sketches of the system and the processes involved may be helpful. In working a problem in thermodynamics, one must have clearly in mind just which portion of the universe is the system and which is the surroundings. Likewise, be aware of the nature of the process involved—whether it is adiabatic ($q = 0$), isothermal (T constant), isobaric (P constant), isochoric (V constant), or whatever.

Of course, the main hurdle is step 3. Because of the many equations in physical chemistry, it might seem a complex task to find the right equation to use in a problem. However, as noted earlier, there are relatively few equations that are best committed to memory; these are usually the most fundamental equations, and usually they have fairly simple forms. For example, we have several equations for mechanically reversible volume-change work in a closed system: Eq. (2.50) gives the work in an infinitesimal process; Eq. (2.51) gives the work in a finite process; the work in a constant-pressure process is $-P\,\Delta V$; Eq. (2.106) gives the work in an isothermal process in a perfect gas. The only one of these equations worth memorizing is $dw = -P\,dV$, since the others can be quickly derived from $dw = -P\,dV$ whenever needed. Moreover, rederiving an equation from a fundamental equation reminds you of the conditions under which the equation is valid. Many of the errors students make in thermodynamics arise from using an equation where it does not apply. (To help prevent this, many of the equations have the conditions of validity stated next to them.) Readers who have invested their time mainly in achieving an understanding of the ideas and equations of physical chemistry will do better than those who have spent their time memorizing formulas.

As to step 4, performing the actual calculations, errors can be minimized by carrying units of all quantities as part of the calculation. In doing calculations it is usually best to avoid substitution of numerical values until the final formula has been obtained; frequently, intermediate quantities will cancel. For numerical calculations use an electronic calculator. Exponentials and logs are used so frequently in physical chemistry that it is worthwhile to buy a calculator that has keys for these functions. (Now that inexpensive pocket calculators are available, it may not be long before arithmetic becomes a lost art. See Isaac Asimov's short story "The Feeling of Power" in *Nine Tomorrows*, Doubleday, 1959; Fawcett, 1969.) After the calculation is completed, it is a good idea to check the entire solution. If you are like most of us, you are probably too lazy to do a complete check, but it only takes a few seconds to check that the sign and the magnitude of the answer are physically reasonable. Sign errors are especially common in thermodynamics, since most quantities can be either positive or negative.

A solutions manual for problems in this textbook is available from the publisher on authorization of your instructor.

PROBLEMS

2.1 Evaluate (*a*) $\int \sin ax\, dx$; (*b*) $\int_0^\pi \sin ax\, dx$; (*c*) $(d/da) \int_0^\pi \sin ax\, dx$.

2.2 Express each of these units as a combination of meters, kilograms, and seconds: (*a*) joule; (*b*) pascal; (*c*) liter; (*d*) newton; (*e*) watt.

2.3 Suppose an object is released at a height h_1 above the earth's surface with initial speed zero. (*a*) Use conservation of mechanical energy to show that its speed v at a height h_2 is $v = [2g(h_1 - h_2)]^{1/2}$. Neglect air resistance. (*b*) Derive the same result for v by solving $F = ma = m\, dv/dt$.

2.4 Provided the particle is not below the earth's surface, Newton's law of gravitation gives the gravitational force on a particle as $F = -GmM/r^2$, where G is the gravitational constant, m is the particle's mass, M is the earth's mass, and r is the particle's distance from the earth's center. F is negative because the particle is attracted toward the earth's center. (When $r = R$, the earth's radius, then $GM/R^2 = g$, the acceleration due to gravity at the earth's surface.) (*a*) Use $F = -dV/dr$ to show that the potential energy of the particle is $V = -GmM/r$, where we choose the zero of potential energy to coincide with the location that has $F = 0$. (*b*) Show that the work done in removing a particle from the earth's surface to infinity is $w = GmM/R$. (*c*) Show that the speed needed for an object launched from the earth's surface to escape the earth is $v_{esc} = (2GM/R)^{1/2}$. (*d*) Calculate v_{esc} in m/s and in mi/hr. (See Prob. 1.28.)

2.5 Verify Eqs. (2.23) and (2.32).

2.6 Consider the line integral $\int_1^2 (5x^2 + 3xy + y^2)\, dx$, where point 1 is (0, 0) and point 2 is (1, 2). Evaluate this integral for each of the following two paths: (*a*) a straight line from point 1 to point 2; (*b*) a straight line from (0, 0) to (1, 0), followed by a straight line from (1, 0) to (1, 2).

2.7 (*a*) Use Rumford's data given in Sec. 2.6 to estimate the relation between the "old" calorie (as defined in Sec. 2.5) and the joule. (Use 1 horsepower = 746 watts.) (*b*) The same as (*a*) using Joule's data given in Sec. 2.6. (*c*) The same as (*a*) using Mayer's statement given in Sec. 2.12.

2.8 One food Calorie = 10^3 cal = 1 kcal. A typical person ingests 2500 kcal/day. Show that a person uses energy at about the same rate as a 100-W lightbulb.

2.9 A hypothetical substance has the equation of state $P\bar{V}^2 = BT$, where B is a constant. The substance undergoes a change of state from (T_1, V_1) to (T_2, V_2) along a path where T varies linearly with V, according to $T = c_1 V + c_2$. (*a*) Find an expression for the work w in terms of B, c_1, c_2, V_1, and V_2. (*b*) Express c_1 and c_2 in terms of T_1, T_2, V_1, and V_2.

2.10 Reynolds tells the following apocryphal story (W. C. Reynolds, *Thermodynamics*, 2d ed., McGraw-Hill, 1968, p. 43). One Friday afternoon in December 1843, Prof. J. Yule left his laboratory in charge of his assistant Dr. B. T. Ewe. On returning the following Monday, Yule found that a beaker of water that on Friday had been at 25°C was now at 40°C. Yule's student Cal

Orick said that Ewe must have heated the water. Ewe denied this, stating that in fact the only heat transfer involved was a cooling of the beaker by placing it on a block of ice for a while. Shortly thereafter, Cal Orick went mad, and J. Yule became a member of the Royal Society. Explain how B. T. Ewe increased the water's temperature by a process with $q < 0$.

2.11 For which of these systems is the system's energy conserved in any process (*a*) a closed system; (*b*) an open system; (*c*) an isolated system?

2.12 William Thomson tells of running into Joule in 1847 at Mont Blanc; Joule had with him his bride and a long thermometer with which he was going to "try for elevation of temperature in waterfalls." Niagara Falls is 190 ft high. Calculate the temperature difference between the water at the top and bottom of the falls.

2.13 Imagine an isolated system divided into two parts 1 and 2 by a rigid, impermeable, thermally conducting wall. Let heat q_1 flow into part 1. Use the first law to show that the heat flow for part 2 must be $q_2 = -q_1$.

2.14 A mole of water vapor initially at 200°C and 1 atm undergoes a cyclic process for which $w = 145$ J. Find q for this process.

2.15 One mole of liquid water at 20°C is adiabatically compressed, P increasing from 1.00 to 10.00 atm. Since liquids and solids are rather incompressible, it is a fairly good approximation to take V as unchanged for this process. With this approximation, calculate q, ΔU, and ΔH for this process.

2.16 Sometimes one sees the notation Δq and Δw for the heat flow into a system and the work done during a process. Explain why this notation is misleading.

2.17 Rossini and Frandsen found that for air at 28°C and pressures in the range 1 to 40 atm, $(\partial \bar{U}/\partial P)_T = -6.08$ J mole^{-1} atm^{-1}. Calculate $(\partial \bar{U}/\partial \bar{V})_T$ for air at (*a*) 28°C and 1.00 atm; (*b*) 28°C and 2.00 atm. *Hint:* Use (1.35).

2.18 (*a*) Derive Eq. (2.94). (*b*) Derive Eq. (2.95). *Hint:* Start by taking $(\partial/\partial P)_T$ of $H = U + PV$.

2.19 (This problem is especially instructive.) For each of the following processes deduce whether each of the quantities q, w, ΔU, ΔH is positive, zero, or negative. (*a*) Reversible melting of solid benzene at 1 atm and the normal melting point. (*b*) Reversible melting of ice at 1 atm and 0°C. (*c*) Reversible adiabatic expansion of a perfect gas. (*d*) Reversible isothermal expansion of a perfect gas. (*e*) Adiabatic expansion of a perfect gas into vacuum (Joule experiment). (*f*) Joule–Thomson adiabatic throttling of a perfect gas. (*g*) Reversible heating of a perfect gas at constant P. (*h*) Reversible cooling of a perfect gas at constant V.

2.20 For each process state whether each of q, w, ΔU is positive, zero, or negative. (*a*) Combustion of benzene in a sealed container with rigid, adiabatic walls. (*b*) Combustion of benzene in a sealed container that is immersed in a water bath at 25°C and that has rigid, thermally conducting walls.

2.21 The molar heat capacity at constant pressure of oxygen for temperatures in the range 300 to 400 K and for low or moderate pressures can be approximated as $\bar{C}_P = a + bT$, where $a = 6.15$ cal mole^{-1} K^{-1} and $b = 0.00310$ cal mole^{-1} K^{-2}. (*a*) Calculate q, w, ΔU, and ΔH when 2.00 moles of O_2 is reversibly heated from 27 to 127°C with P held fixed at 1.00 atm. (Assume perfect-gas behavior.) (*b*) Calculate q, w, ΔU, and ΔH when 2.00 moles of O_2 initially at 1.00 atm is reversibly heated from 27 to 127°C with V held fixed.

2.22 Give the value of C_{pr} for (*a*) the melting of ice at 0°C and 1 atm; (*b*) the freezing of water at 0°C and 1 atm; (*c*) the reversible isothermal expansion of a perfect gas; and (*d*) the reversible adiabatic expansion of a perfect gas.

2.23 Show that for the reversible adiabatic expansion of a perfect gas with γ constant

$$w = \frac{nRT_1}{\gamma - 1}\left[\left(\frac{V_1}{V_2}\right)^{\gamma-1} - 1\right] = \frac{P_2V_2 - P_1V_1}{\gamma - 1}$$

Hint: One approach is to note that (2.108) holds when T_2 and V_2 are replaced by T and V, where these quantities refer to any intermediate state of the process.

2.24 For this problem use 79.7 and 539.4 cal/g as the latent heats of fusion and vaporization of water at the normal melting and boiling points, $c_P = 1.00$ cal/(g K) for liquid water, $\rho = 0.917$ g/cm^3 for ice at 0°C, $\rho = 1.000$ g/cm^3 and 0.958 g/cm^3 for water at 1 atm and 0 and 100°C, respectively. Calculate q, w, ΔU, and ΔH for (*a*) the melting of 1 mole of ice at 0°C and 1 atm; (*b*) the reversible constant-pressure heating of 1 mole of liquid water from 0 to 100°C at 1 atm; (*c*) the vaporization of 1 mole of water at 100°C and 1 atm.

2.25 Calculate ΔU and ΔH for each of the following changes of state of 2.50 moles of a perfect monatomic gas with $\bar{C}_V = 1.5R$ for all temperatures: (*a*) (1.50 atm, 400 K) → (3.00 atm, 600 K); (*b*) (2.50 atm, 20.0 l) → (2.00 atm, 30.0 l); (*c*) (28.5 l, 400 K) → (42.0 l, 400 K).

2.26 Can q and w be calculated for the processes of Prob. 2.25? If the answer is yes, calculate them for each process.

2.27 Calculate q, w, ΔU, and ΔH for the isothermal expansion at 300 K of 5.00 moles of a perfect gas from 500 to 1500 cm^3.

2.28 A student attempting to remember a certain formula comes up with $C_P - C_V = TV\alpha^m/\kappa^n$, where m and n are certain integers whose values the student has forgotten, and where the remaining symbols have their usual meanings. Use dimensional considerations to find m and n.

2.29 A certain perfect gas has $\bar{C}_P = 3.5R$ at all temperatures. Calculate q, w, ΔU, and ΔH for the reversible adiabatic compression of 0.200 mole of this gas from (400 torr, 1000 cm^3) to a final volume of 250 cm^3.

2.30 [For this problem recall that 1 decimeter (dm) = 0.1 m.] A certain perfect gas has $\bar{C}_V = 2.5R$ at all temperatures. Calculate q, w, ΔU, and ΔH when 1.00 mole of this gas undergoes each of the following processes: (*a*) a reversible isobaric expansion from (1.00 atm, 20.0 dm^3) to (1.00 atm, 40.0 dm^3); (*b*) a reversible isochoric change of state from (1.00 atm, 40.0 dm^3) to (0.500 atm, 40.0 dm^3); (*c*) a reversible isothermal compression from (0.500 atm, 40.0 dm^3) to (1.00 atm, 20.0 dm^3). Sketch each process on the same P-V diagram and calculate q, w, ΔU and ΔH for a cycle that consists of steps (*a*), (*b*), and (*c*).

2.31 For air at temperatures near 25°C and pressures in the range 0 to 50 atm, the μ_{JT} values are all reasonably close to 0.2° C/atm. Estimate the final temperature of the gas if 58 g of air at 25°C and 50 atm undergoes a Joule–Thomson throttling to a final pressure of 1 atm.

2.32 The potential energy stored in a spring is $\frac{1}{2}kx^2$, where k is the force constant of the spring and x is the distance the spring is stretched from equilibrium. Suppose a spring with force constant 125 N/m is stretched by 10.0 cm, placed in 112 g of water in an adiabatic container, and released. The mass of the spring is 20 g, and its specific heat capacity is 0.30 cal/(g °C). The initial temperature of the water is 18.000°C, and its specific heat capacity is 1.00 cal/(g °C). Find the final temperature of the water.

2.33 Why do you suppose the symbol $\int$ was chosen to represent integration?

3

THE SECOND LAW OF THERMODYNAMICS

3.1 THE SECOND LAW OF THERMODYNAMICS

A major application of thermodynamics to chemistry is to provide information about equilibrium in chemical systems. If we mix nitrogen and hydrogen gases (together with a catalyst), portions of each gas react to form ammonia. The first law assures us that the total energy of system plus surroundings remains constant during the reaction, but the first law cannot say what the final equilibrium concentrations will be. We shall see that the second law provides such information.

In 1824 a French engineer named Sadi Carnot published a study of the theoretical efficiency of steam engines. This book (*Reflections on the Motive Power of Fire*) pointed out that for a heat engine to produce continuous mechanical work, it must exchange heat with two bodies at different temperatures, absorbing heat from the hot body and discarding heat to the cold body; without a cold body for the discard of heat, the engine cannot function continuously. This is the essential idea of one form of the second law of thermodynamics. Carnot's work had almost no influence on scientists at the time of its publication. Carnot worked when the caloric theory of heat held sway, and his book used this theory, incorrectly setting the heat discarded to the cold body equal to the heat absorbed from the hot body. When Carnot's book was rediscovered in the 1840s by William Thomson (Lord Kelvin) and others, it caused confusion for a while, since Joule's work had led to the overthrow of the caloric theory. Finally, around 1850, Thomson and Rudolph Clausius corrected Carnot's work to conform with the first law of thermodynamics.

Carnot died in 1832 of cholera at age 36. Notes of his that were not published until 1878 showed that in fact he believed the caloric theory to be false and planned experiments to demonstrate this. These planned experiments included the vigorous agitation of water, mercury, and alcohol " in a small cask " and measurement of " the motive power consumed and the heat produced." Carnot's notes stated that " Heat is simply motive power, or rather motion, which has changed its form.... [Motive] power is, in quantity, invariable in nature; it is ... never either produced or destroyed...." Thus, in a sense, we could consider Carnot the discoverer of both the first and second laws, although his conclusions about heat and work remained unpublished until others had discovered the first law.

There are several equivalent ways of stating the second law. We shall use the following statement (the *Kelvin–Planck statement*), due originally to William Thomson and later rephrased by Planck:

It is impossible for a system to undergo a cyclic process whose sole effects are the flow of heat into the system from a heat reservoir and the performance of an equivalent amount of work by the system on the surroundings.

By a *heat reservoir* (or *heat bath*) we mean a body that is in internal equilibrium at a constant temperature and that is sufficiently large for flow of heat between it and the system to cause no significant change in the temperature of the reservoir.

The second law says that it is impossible to build a cyclic machine that converts heat into work with 100 percent efficiency (Fig. 3.1). Such a machine is called a perpetual motion machine of the second kind. Note that its existence would not violate the first law, since energy is conserved in the operation of the machine.

Like the first law, the second law is a generalization from experience. There are three kinds of evidence for the second law. First is the failure of anyone to construct a machine like that shown in Fig. 3.1. If such a machine were available, it could use the atmosphere as a heat reservoir, continuously withdrawing energy from the atmosphere and converting it completely to useful work; it would be nice to have such a machine, but no one has been able to build one. Second, and more convincing, is the fact that the second law leads to many conclusions about equilibrium in chemical systems and these conclusions have been verified. [For example, we shall see that the second law shows that the equilibrium vapor pressure of a pure substance varies with temperature according to $dP/dT = \Delta H/(T\,\Delta V)$, where ΔH and ΔV are the heat of vaporization and the volume change in vaporization, and this equation has been experimentally verified.] Third, statistical mechanics shows that the second law follows as a consequence of certain assumptions about the molecular level.

The first law tells us that work output cannot be produced without an equivalent amount of energy input. The second law tells us that it is impossible to have a cyclic machine that completely converts the random molecular energy of heat flow into the ordered motion of mechanical work. As some wit has put it: The first law says you can't win; the second law says you can't even break even.

Note that the second law does not forbid the complete conversion of heat to work in a *noncyclic* process. Thus, if we reversibly and isothermally heat a perfect gas, the gas expands and since $\Delta U = 0$, the work done by the gas equals the heat input [Eq. (2.106)]. Such an expansion, however, cannot be made the basis of a continuously operating machine; eventually, the piston will fall out of the cylinder. A continuously operating machine must be based on a cyclic process.

An alternative statement of the second law is the *Clausius statement*:

It is impossible for a system to undergo a cyclic process whose sole effects are the flow of heat into the system from a cold reservoir and the flow of an equal amount of heat out of the system into a hot reservoir.

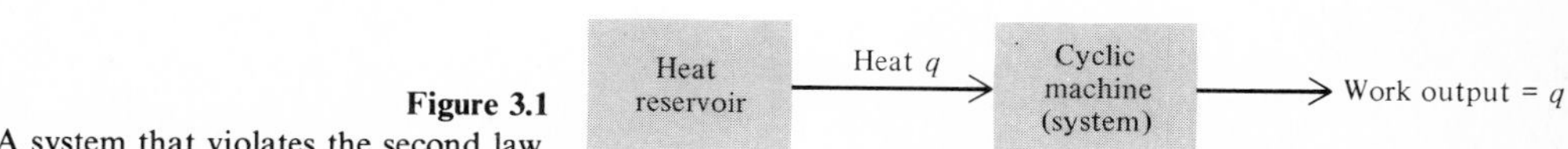

Figure 3.1
A system that violates the second law.

Proof of the equivalence of the Clausius and Kelvin–Planck statements is outlined in Prob. 3.3.

3.2 HEAT ENGINES

We shall use the second law to deduce certain theorems about the efficiency of heat engines. As chemists rather than engineers, we have little interest in heat engines, but our study of heat engines is part of a chain of reasoning that will lead to the criterion for determining the position of chemical equilibrium in a system. Moreover, study of the efficiency of heat engines is related to the basic question of what limitations exist on the conversion of heat to work.

A *heat engine* converts some of the random molecular energy of heat flow into macroscopic mechanical energy (work). The working substance (e.g., steam in a steam engine) is heated in a cylinder, and its expansion causes a piston to move, thereby doing mechanical work. If the engine is to operate continuously, the working substance has to be cooled back to its original state and the piston has to return to its original location before we can heat the working substance again and get another work-producing expansion. Hence the working substance goes through a cyclic process. The essentials of the cycle are the absorption of heat q_H by the working substance from a hot body (e.g., the boiler), the performance of work $-w$ by the working substance on the surroundings, and the emission of heat $-q_C$ by the working substance to a cold body (e.g., the condenser), with the working substance returning to its original state at the end of the cycle. The system is the working substance.

Our convention is that w is work done *on* the system; hence work done *by* the system is $-w$. Likewise, q means heat added to the system; hence $-q_C$ is the heat that flows from the system to the cold body. For a heat engine, $q_H > 0$, $-w > 0$, and $-q_C > 0$, so $w < 0$ and $q_C < 0$. The quantity w is negative for a heat engine because the engine does positive work on its surroundings; q_C is negative for a heat engine because positive heat flows out of the system to the cold body.

The preceding is an idealization of how real heat engines work but contains the essential features of a real heat engine.

The *efficiency* e of a heat engine is the fraction of energy input that appears as useful energy output, i.e., that appears as work. The energy input per cycle is the heat input q_H to the engine. (The source of this energy might be the burning of oil or coal to heat the boiler.) Hence

$$e = \frac{\text{work output}}{\text{energy input}} = \frac{-w}{q_H} = \frac{|w|}{q_H} \tag{3.1}$$

For a cycle of operation, ΔU is zero by the first law. Hence $0 = q + w = q_H + q_C + w$, so that

$$-w = q_H + q_C \tag{3.2}$$

where the quantities in (3.2) are for one cycle of operation. Use of (3.2) in (3.1) gives

$$e = \frac{q_H + q_C}{q_H} = 1 + \frac{q_C}{q_H} \tag{3.3}$$

Since q_C is negative and q_H positive, the efficiency is less than 1.

To further simplify the analysis, we assume that the heat q_H is absorbed from a hot reservoir and that $-q_C$ is emitted to a cold reservoir, each reservoir being large enough to ensure that its temperature is unchanged by interaction with the engine. Figure 3.2 is a schematic diagram of the heat engine.

Since our analysis (at this point) will not require specification of the temperature scale, instead of denoting temperatures with the symbol T (which indicates use of the ideal-gas scale; Sec. 1.5), we shall use the symbol τ (tau); thus we call the temperatures of the hot and cold reservoirs τ_H and τ_C. The τ scale might be the ideal-gas scale, or it might be a scale based on the expansion of liquid mercury, or it might be some other scale. The only restriction we set is that the τ scale always give readings such that the temperature of the hot reservoir is greater than that of the cold reservoir: $\tau_H > \tau_C$. The motivation for leaving the temperature scale unspecified will become clear in Sec. 3.3.

We now use the second law to prove Carnot's principle: *No heat engine can be more efficient than a reversible heat engine when both engines work between the same pair of temperatures τ_H and τ_C.* (Equivalently, the maximum amount of work from a given supply of heat is obtained with a reversible engine.)

To prove Carnot's principle, we assume it to be false and show that this assumption leads to a violation of the second law. Thus, let there exist a superengine whose efficiency e_{super} exceeds the efficiency e_{rev} of some reversible engine working between the same pair of temperatures as the superengine:

$$e_{\text{super}} > e_{\text{rev}} \tag{3.4}$$

where, from (3.1),

$$e_{\text{super}} = \frac{-w_{\text{super}}}{q_{H,\,\text{super}}}, \qquad e_{\text{rev}} = \frac{-w_{\text{rev}}}{q_{H,\,\text{rev}}} \tag{3.5}$$

Let us run the reversible engine in reverse, doing positive work w_{rev} *on* it, thereby causing it to absorb heat $q_{C,\,\text{rev}}$ from the cold reservoir and emit heat $-q_{H,\,\text{rev}}$ to the hot reservoir, where these quantities are for one cycle. It thereby functions as a heat pump, or refrigerator. Because this engine is reversible, the magnitudes of the two heats and the work are the *same* for a cycle of operation as a heat pump as for a cycle of operation as a heat engine, except that all signs are changed. We couple the reversible heat pump with the superengine, so that the two systems use the same pair of reservoirs (Fig. 3.3).

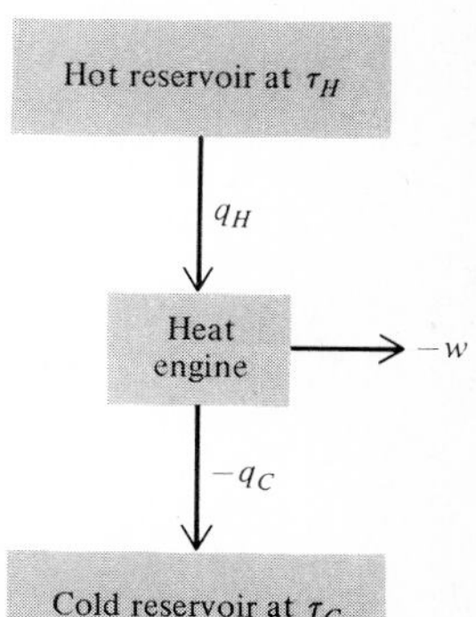

Figure 3.2
A heat engine operating between two temperatures.

We shall run the superengine at such a rate that it withdraws heat from the hot reservoir at the same rate that the reversible heat pump deposits heat into this reservoir. (Thus, suppose 1 cycle of the superengine absorbs 1.3 times as much heat from the hot reservoir as 1 cycle of the reversible heat pump deposits in the hot reservoir. We would then have the superengine complete 10 cycles in the time that the reversible heat pump completes 13 cycles. After each 10 cycles of the superengine, both the devices are back in their original states, so the combined device is a cyclic one.) Since the magnitude of the heat exchange with the hot reservoir is the same for the two engines, and since the superengine is by assumption more efficient than the reversible engine, the superengine will deliver more work output than the work put into the reversible heat pump. Hence we can use *part* of the mechanical work output of the superengine to supply *all* the work needed to run the reversible heat pump and still have some net work output from the superengine left over. This work output must (by the first law) have come from some input of energy to the system of reversible heat pump + superengine; since there is no net absorption or emission of heat to the hot reservoir, this energy input must have come from a net absorption of heat from the cold reservoir. The net result is an absorption of heat from the cold reservoir and its complete conversion to work by a cyclic process.

However, this cyclic process violates the second law of thermodynamics and is therefore impossible. We were led to this impossible conclusion by our initial assumption of the existence of a superengine with $e_{\text{super}} > e_{\text{rev}}$. We therefore conclude that this assumption is false. We have proved that

$$e(\text{any engine}) \leq e(\text{a reversible engine}) \qquad (3.6)$$

for heat engines that operate between the same pair of temperatures.

Now consider two reversible engines A and B that work between the same pair of temperatures with efficiencies $e_{\text{rev, A}}$ and $e_{\text{rev, B}}$. If we replace the superengine in the above reasoning with engine A (running forward) and use engine B running backward as the heat pump, the same reasoning that led to (3.6) gives $e_{\text{rev, A}} \leq e_{\text{rev, B}}$. If we now interchange A and B, running B forward and A backward, the same reasoning gives $e_{\text{rev, B}} \leq e_{\text{rev, A}}$. These two relations can be satisfied only if $e_{\text{rev, A}} = e_{\text{rev, B}}$.

We have shown that (1) all reversible heat engines operating between the same pair of temperatures must have the same efficiency e_{rev} and (2) this reversible

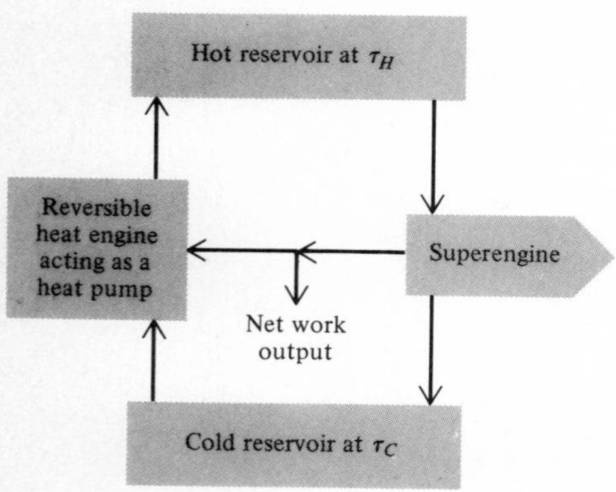

Figure 3.3
Reversible heat pump coupled with a superengine.

efficiency is the maximum possible for any heat engine that operates between this pair of temperatures, i.e.,

$$e_{\text{irrev}} \leq e_{\text{rev}} \tag{3.7}$$

These conclusions are independent of the nature of the working substance used in the engines and of the kind of work, holding also for non-P-V work. The only assumption in the derivation of (3.7) is the validity of the second law of thermodynamics.

Since the efficiency of any reversible engine working between the temperatures τ_H and τ_C is the same, this efficiency e_{rev} can depend only on τ_H and τ_C:

$$e_{\text{rev}} = f(\tau_H, \tau_C) \tag{3.8}$$

The function f naturally depends on the temperature scale used. We now find f for the ideal-gas scale, taking $\tau = T$. Since e_{rev} is independent of the nature of the working substance, we can use any working substance we please to find f. We know the most about a perfect gas, so we choose this as the working substance in the reversible engine.

Consider first the nature of the cycle we used to derive (3.8). The first step involves absorption of heat q_H from a reservoir whose temperature remains at T_H. Since we are considering a reversible engine, the gas also must remain at temperature T_H throughout the heat absorption from the reservoir. (Heat flow between two bodies with a finite difference in temperature is an irreversible process.) Thus the first step of the cycle is an *isothermal* process. Moreover, since $\Delta U = 0$ for an isothermal process in a perfect gas [Eq. (2.98)], it follows that to maintain U as constant, the gas must expand and do work on the surroundings equal to the heat absorbed in the first step. The first step of the cycle is thus a reversible isothermal expansion, as shown by the line from state 1 to state 2 in Fig. 3.4*a*. Similarly, when the gas gives up heat at T_C, we have a reversible isothermal compression at temperature T_C. The T_C isotherm lies below the T_H isotherm and is the line from state 3 to state 4 in Fig. 3.4*a*. To have a complete cycle there must be steps that connect states 2 and 3 and states 4 and 1. We assumed that heat is transferred only at T_H and T_C; hence the two isotherms in Fig. 3.4*a* must be connected by two steps with no heat transfer, i.e., by two reversible *adiabats*.

This reversible cycle is called a *Carnot cycle* (Fig. 3.4*b*). The working substance need not be a perfect gas. A Carnot cycle is defined as a reversible cycle that consists of two isothermal steps at different temperatures and two adiabatic steps. Although

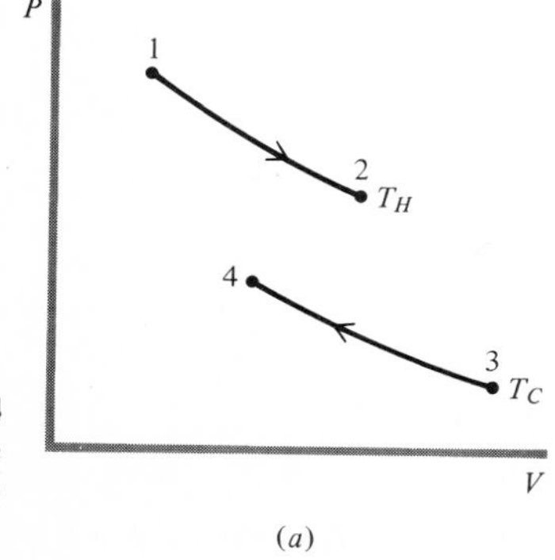

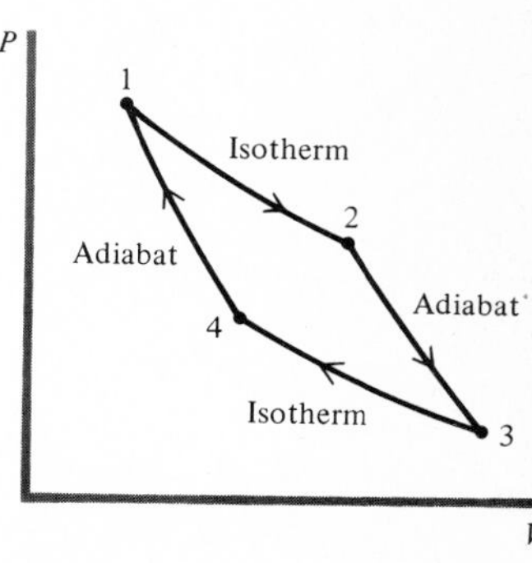

Figure 3.4
(*a*) Isothermal steps of the reversible heat-engine cycle. (*b*) The complete Carnot cycle.

we thought in terms of a perfect gas when we obtained the appearance of the Carnot cycle in Fig. 3.4*b*, one can use the second law to prove that a Carnot cycle for any working substance must have this general appearance. (See *Denbigh*, p. 27.)

We now calculate the Carnot-cycle efficiency e_{rev} on the ideal-gas temperature scale T. We use a perfect gas as working substance and restrict ourselves to P-V work. The first law gives $dU = dq + dw = dq - P\,dV$ for a reversible volume change. For a perfect gas, $P = nRT/V$ and $dU = C_V(T)\,dT$. Hence the first law becomes $C_V\,dT = dq - nRT\,dV/V$ for a perfect gas. Dividing by T and integrating over the Carnot cycle, we get

$$\oint C_V(T)\frac{dT}{T} = \oint \frac{dq}{T} - nR\oint \frac{dV}{V} \tag{3.9}$$

Each integral in (3.9) is the sum of four line integrals, one for each step of the Carnot cycle in Fig. 3.4*b*. We have

$$\oint C_V(T)\frac{dT}{T} = \int_{T_1}^{T_2} C_V(T)\frac{dT}{T} + \int_{T_2}^{T_3} C_V(T)\frac{dT}{T} + \int_{T_3}^{T_4} C_V(T)\frac{dT}{T} + \int_{T_4}^{T_1} C_V(T)\frac{dT}{T} \tag{3.10}$$

But $T_1 = T_2$ $(= T_H)$, so that the first integral on the right of (3.10) equals zero; likewise, $T_3 = T_4$, and the third integral on the right is zero. Furthermore

$$\int_{T_2}^{T_3} C_V(T)\frac{dT}{T} = \int_{T_H}^{T_C} C_V(T)\frac{dT}{T} = -\int_{T_C}^{T_H} C_V(T)\frac{dT}{T} \tag{3.11}$$

$$\int_{T_4}^{T_1} C_V(T)\frac{dT}{T} = \int_{T_C}^{T_H} C_V(T)\frac{dT}{T} \tag{3.12}$$

so that the sum of the second and fourth integrals on the right of (3.10) is zero. Hence

$$\oint C_V(T)\frac{dT}{T} = 0 \tag{3.13}$$

The alert reader may realize that the cyclic integral in (3.13) must vanish because $[C_V(T)/T]\,dT$ is the differential of a state function, namely, a certain function of T whose derivative is $C_V(T)/T$. (Recall Sec. 2.7.) Note, however, that the integral of $P\,dV$ does not vanish for a cycle since $P\,dV$ is not the differential of a state function.

The second integral on the right of (3.9) must also vanish; this is because dV/V is the differential of a state function (namely, $\ln V$), and its line integral is therefore zero for a cyclic process.

Hence (3.9) becomes

$$\oint \frac{dq}{T} = 0 \qquad \text{Carnot cycle, perf. gas} \tag{3.14}$$

We have

$$\oint \frac{dq}{T} = \int_1^2 \frac{dq}{T} + \int_2^3 \frac{dq}{T} + \int_3^4 \frac{dq}{T} + \int_4^1 \frac{dq}{T} \tag{3.15}$$

Since the processes $2 \rightarrow 3$ and $4 \rightarrow 1$ are adiabatic with $dq = 0$, the second and fourth integrals on the right of (3.15) are zero. For the isothermal process $1 \rightarrow 2$, we have

$T = T_H$; since T is constant, it can be taken outside the integral: $\int_1^2 T^{-1}\,dq = T_H^{-1}\int_1^2 dq = q_H/T_H$. Similarly, $\int_3^4 dq/T = q_C/T_C$. Equation (3.15) becomes

$$\oint \frac{dq}{T} = \frac{q_H}{T_H} + \frac{q_C}{T_C} = 0 \qquad \text{Carnot cycle, perf. gas} \tag{3.16}$$

We now find e_{rev}, the maximum possible efficiency for the conversion of heat to work. Equations (3.3) and (3.16) give $e = 1 + q_C/q_H$ and $q_C/q_H = -T_C/T_H$; hence

$$e_{\text{rev}} = 1 - \frac{T_C}{T_H} = \frac{T_H - T_C}{T_H} \qquad \text{Carnot cycle} \tag{3.17*}$$

We derived (3.17) using a perfect gas as working substance, but since we earlier proved that e_{rev} is independent of the working substance, Eq. (3.17) must hold for any working substance undergoing a Carnot cycle. Moreover, since the equations $e_{\text{rev}} = 1 + q_C/q_H$ and $e_{\text{rev}} = 1 - T_C/T_H$ hold for any working substance, we must have $q_C/q_H = -T_C/T_H$ or $q_C/T_C + q_H/T_H = 0$ for any working substance; hence

$$\oint \frac{dq}{T} = \frac{q_C}{T_C} + \frac{q_H}{T_H} = 0 \qquad \text{Carnot cycle} \tag{3.18}$$

Equation (3.18) holds for any closed system undergoing a Carnot cycle. We shall use (3.18) to derive the state function entropy in Sec. 3.4.

Note from (3.17) that the smaller T_C is and the larger T_H is, the closer the reversible efficiency approaches the limit of 1, which represents complete conversion of the heat input into work output. Of course, a reversible heat engine is an idealization of real heat engines, which involve some degree of irreversibility in their operation. The efficiency (3.17) is an upper limit to the efficiency of real heat engines [Eq. (3.7)].

A modern steam power plant might have the boiler at 600°C (with the pressure correspondingly high) and the condenser at 40°C. If it operated on a Carnot cycle, then $e_{\text{rev}} = 1 - (313\text{ K})/(873\text{ K}) = 64$ percent. The actual cycle of a steam engine is not a Carnot cycle because of irreversibility and because heat is transferred at temperatures between T_H and T_C, as well as at T_H and T_C. Both these factors make the actual efficiency less than 64 percent. Typical efficiencies of currently available heat engines run 10 to 40 percent. An automobile engine has an efficiency around 20 percent.

The second-law requirement that some heat be discarded to a cold reservoir in the operation of a heat engine leads to environmental problems. Most of our electric power is produced by steam engines that drive conducting wires through magnetic fields, thereby generating electric currents. Typically, the condenser of the steam plant is cooled by river water at, say, 25°C. The second law requires that the fraction of input energy q_H to be discarded as q_C be at least $|q_C/q_H| = T_C/T_H$, which is substantial for typical values of T_C and T_H. This discarded heat raises the river temperature; the thermal pollution disrupts the ecological balance of the river.

The analysis of this section applies only to heat engines, which are engines that convert heat to work. Not all engines are heat engines. For example, in an engine that consists of a battery used to drive a motor, the energy of a chemical reaction is converted in the battery to electrical energy, which in turn is converted to mechanical energy; thus, chemical energy is converted to work, and this is not a heat engine. Likewise, the human body converts chemical energy to work and is not a heat engine.

3.3 THE THERMODYNAMIC TEMPERATURE SCALE

In developing thermodynamics, we have so far used the ideal-gas temperature scale, which is based on the properties of a particular kind of substance, an ideal gas. The state functions P, V, U, and H are not defined in terms of any particular kind of substance, and it would seem desirable that a fundamental property like temperature be defined in a more general way than in terms of ideal gases. Lord Kelvin pointed out that the second law of thermodynamics provides the basis for defining a temperature scale that is independent of the properties of any kind of substance. We shall call this scale the *thermodynamic temperature scale.*

We showed that for a Carnot cycle between temperatures τ_C and τ_H, the efficiency e_{rev} is independent of the nature of the system (the working substance) but depends on only the temperatures: $e_{\text{rev}} = 1 + q_C/q_H = f(\tau_C, \tau_H)$, where τ symbolizes any temperature scale whatever. It follows that the heat ratio $-q_C/q_H$ (which equals $1 - e_{\text{rev}}$) is independent of the nature of the system that undergoes the Carnot cycle. We have

$$-q_C/q_H = 1 - f(\tau_C, \tau_H) \equiv g(\tau_C, \tau_H) \tag{3.19}$$

where the function g (defined as $1 - f$) depends on the choice of temperature scale but is independent of the nature of the system. We shall use the equation $-q_C/q_H = g(\tau_C, \tau_H)$ to define the thermodynamic temperature scale.

Before doing so, we examine the requirements the function g must satisfy. By considering two Carnot engines working with one reservoir in common, one can show that Carnot's principle (3.6) (which is a consequence of the second law) requires that the function g have the form

$$g(\tau_C, \tau_H) = \phi(\tau_C)/\phi(\tau_H) \tag{3.20}$$

where ϕ (phi) is some function. [The proof of (3.20) is outlined in Prob. 3.5.] Hence (3.19) becomes

$$-q_C/q_H = \phi(\tau_C)/\phi(\tau_H) \tag{3.21}$$

For convenience we shall also require that the temperature scale be such that the temperature of the hotter reservoir will always be numerically greater than the temperature of the colder reservoir.

Within the framework of these two requirements, we are free to choose a scale defined in terms of the ratio $-q_C/q_H$. For example, we might choose ϕ as the function "take the square root." Then the τ scale would be defined by $-q_C/q_H = \tau_C^{1/2}/\tau_H^{1/2}$; the temperature ratio is then defined as $\tau_C/\tau_H = (-q_C)^2/q_H^2$, where q_C and q_H are experimentally measured heats for a Carnot cycle. Since we have only fixed the temperature ratio τ_C/τ_H, we would complete the definition of the scale by choosing some reference temperature and assigning it a numerical value.

The simplest choice for the function ϕ is "take the first power." This choice gives the *thermodynamic temperature scale* Θ. Temperature ratios on the thermodynamic scale are thus defined by

$$\frac{\Theta_C}{\Theta_H} \equiv \frac{-q_C}{q_H} \tag{3.22}$$

We complete the definition of the Θ scale by choosing the temperature of the triple point of water as $\Theta_{tr} = 273.16°$.

To measure the thermodynamic temperature Θ of an arbitrary body, we use it as one of the heat reservoirs in a Carnot cycle and use a body composed of water at its triple point as the second reservoir. We then put any system through a Carnot cycle between these two reservoirs and measure the heat q exchanged with the reservoir at Θ and the heat q_{tr} exchanged with the reservoir at 273.16°. The thermodynamic temperature Θ is then calculated from (3.22) as

$$\Theta = 273.16° \frac{|q|}{|q_{tr}|} \tag{3.23}$$

Since the heat ratio in (3.23) is independent of the nature of the system put through the Carnot cycle, the Θ scale does not depend on the properties of any kind of substance.

How is the thermodynamic scale Θ related to the ideal-gas scale T? We proved in Sec. 3.2 that, on the ideal-gas temperature scale, $T_C/T_H = -q_C/q_H$ for any system that undergoes a Carnot cycle; see Eq. (3.18). Moreover, we chose the ideal-gas temperature at the water triple point as 273.16 K. Hence for a Carnot cycle between an arbitrary temperature T and the triple-point temperature, we have

$$T = 273.16 \text{ K} \frac{|q|}{|q_{tr}|} \tag{3.24}$$

where q is the heat exchanged with the reservoir at T. Comparison of (3.23) and (3.24) shows that *the ideal-gas temperature scale and the thermodynamic temperature scale are numerically identical.* We need not distinguish between them and will henceforth use the same symbol T for each scale. The thermodynamic scale is the fundamental scale of science, but as a matter of practical convenience, extrapolated measurements on gases (rather than Carnot-cycle measurements) are used to measure temperatures accurately.

3.4 ENTROPY

For any closed system that undergoes a Carnot cycle, Eq. (3.18) shows that the integral of dq_{rev}/T around the cycle is zero. (The subscript rev reminds us of the reversible nature of a Carnot cycle.)

We now extend this result to an *arbitrary* reversible cycle, removing the constraint that heat be exchanged with the surroundings only at T_H and T_C. To make it easier we first establish a preliminary result. Let the smooth solid curve ab in Fig. 3.5 be an *arbitrary* reversible process from state a to state b. We draw in the curves for adiabatic reversible processes that pass through state a and through state b; these are shown as dashed lines. We now choose the isotherm mn from one adiabat to the other, such that the area under the zigzag curve $amnb$ equals the area under the smooth curve ab. Since these areas give the reversible work done by the system in each process, we have $w_{ab} = w_{amnb}$, where ab is the process along the smooth curve and $amnb$ is the zigzag process along the two adiabats and the isotherm. Also, since ΔU is path-independent, ΔU is the same for processes ab and $amnb$. Hence it follows from the first law that $q_{ab} = q_{amnb}$. Since am and nb are adiabats, we have $q_{amnb} = q_{mn}$. Hence $q_{ab} = q_{mn}$. We have shown that given an arbitrary reversible process from a to

b with heat transfer q_{ab}, there exists a reversible process from a to b consisting of two adiabats and an isotherm such that the heat transfer along the isotherm equals q_{ab}.

Now consider an arbitrary reversible cyclic process, e.g., the curve in Fig. 3.6*a*. We draw adiabats that divide the cycle into adjacent strips (Fig. 3.6*b*). Consider one such strip, bounded by curves *ab* and *cd* at the top and bottom. We use the result just proved to draw the zigzag path *amnb* with $q_{mn} = q_{ab}$; likewise, we draw *crsd* such that $q_{sr} = q_{dc}$, where *mn* and *rs* are isotherms. Since *mn* and *rs* are reversible isotherms and *ns* and *rm* are reversible adiabats, we could use these four curves to carry out a Carnot cycle; use of (3.18) then gives $q_{mn}/T_{mn} + q_{sr}/T_{sr} = 0$, and

$$\frac{q_{ab}}{T_{mn}} + \frac{q_{dc}}{T_{sr}} = 0 \tag{3.25}$$

We can do exactly the same with every other strip in Fig. 3.6*b* to get an equation similar to (3.25) for each strip.

Now consider the limit as we draw the adiabats closer and closer together in Fig. 3.6*b*, ultimately dividing the cycle into an infinite number of infinitesimally narrow strips, in each of which we draw the zigzags at the top and bottom. Thus, imagine the adiabat *bd* coming closer and closer to the adiabat *ac*. As this happens, point b on the smooth curve comes closer to point a and the temperature variation along *ab* diminishes; finally, in the limit, the temperature T_b at b differs only infinitesimally from that at a; let T_{ab} denote this essentially constant temperature. Moreover, T_{mn} in (3.25) (which lies between T_a and T_b) becomes essentially the same as T_{ab}. The same thing happens at the bottom of the strip. Also, the heats transferred become infinitesimal quantities in the limit. Thus (3.25) becomes in this limit

$$\frac{dq_{ab}}{T_{ab}} + \frac{dq_{dc}}{T_{dc}} = 0 \tag{3.26}$$

Exactly the same thing happens in every other strip when we take the limit, and an equation similar to (3.26) holds for each infinitesimal strip. We now add all the equations like (3.26) for each strip. Each term in the sum will be infinitesimal and will be of the form dq/T, where dq is the heat transfer along an infinitesimal portion of the arbitrary reversible cycle and T is the temperature at which this heat transfer occurs. The sum of the infinitesimals is thus a line integral around the cycle, and we get

$$\oint \frac{dq_{\text{rev}}}{T} = 0 \tag{3.27}$$

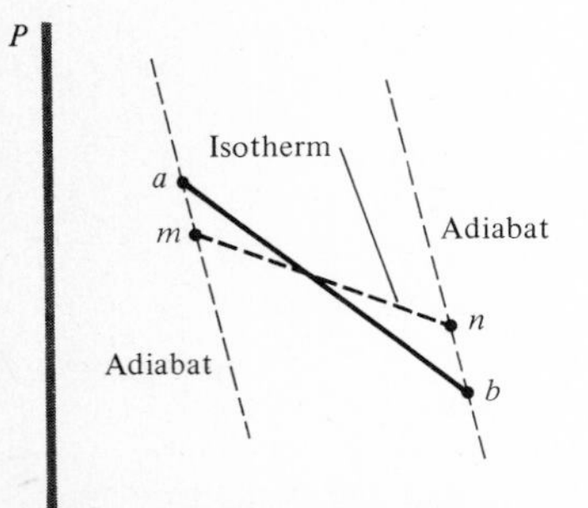

Figure 3.5
Two paths between a and b having equal heat transfers.

The subscript rev reminds us that the cycle under consideration is reversible. [If it isn't reversible, we can't relate it to Carnot cycles and (3.27) need not hold.] Apart from the reversibility requirement, the cycle in (3.27) is arbitrary, and (3.27) is the desired generalization of (3.18).

Since the integral of dq_{rev}/T around any reversible cycle is zero, it follows (see Sec. 2.7) that the value of the line integral $\int_1^2 dq_{rev}/T$ is independent of the path between states 1 and 2, depending only on the initial and final states. Hence dq_{rev}/T is the differential of a state function. We shall call this state function the *entropy* S:

$$dS \equiv \frac{dq_{rev}}{T} \qquad \text{closed syst., rev. proc.} \qquad (3.28)^*$$

The entropy change on going from state 1 to state 2 is given by the integral of (3.28):

$$\Delta S = S_2 - S_1 = \int_1^2 \frac{dq_{rev}}{T} \qquad \text{closed syst., rev. proc.} \qquad (3.29)$$

(Throughout this chapter we have been considering only closed systems; q is undefined for an open system.)

If a system goes from state 1 to state 2 by an irreversible process, the intermediate states it passes through may well not be states of thermodynamic equilibrium and the entropies, temperatures, etc., of intermediate states may be undefined. However, since S is a state function, it doesn't matter how the system went from state 1 to state 2; ΔS is the same for any process (reversible or irreversible) that connects states 1 and 2. But it is only for a reversible process that the integral of dq/T gives the entropy change. Calculation of ΔS in irreversible processes is considered in the next section.

Clausius discovered the state function S in 1854 and called it the transformation content (*Verwandlunginhalt*). Later, he renamed it entropy, from the Greek word *trope*, meaning "transformation" (since the entropy is related to the transformation of heat to work).

Entropy is an extensive state function. To see this, imagine a system in equilibrium to be divided into two parts; each part, of course, is at the same temperature T. Let parts 1 and 2 receive heats dq_1 and dq_2, respectively, in a reversible process. From (3.28), the entropy changes of the parts are $dS_1 = dq_1/T$ and $dS_2 = dq_2/T$. But the entropy change dS for the whole system is

$$dS = dq/T = (dq_1 + dq_2)/T = dq_1/T + dq_2/T = dS_1 + dS_2 \qquad (3.30)$$

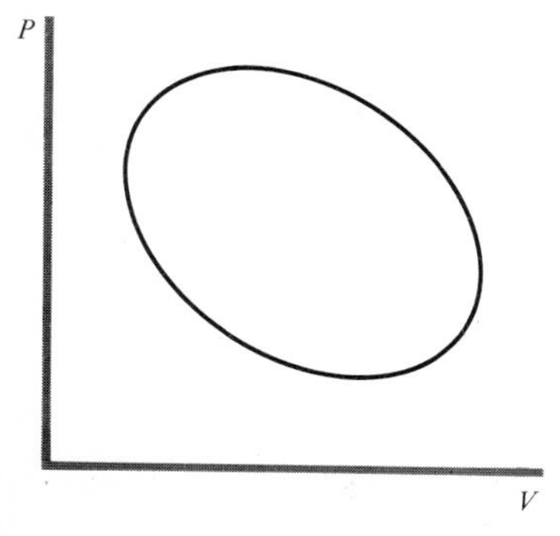

(*a*)

(*b*)

Figure 3.6
An arbitrary reversible cycle and its relation to Carnot cycles.

Integration gives $\Delta S = \Delta S_1 + \Delta S_2$; hence $S = S_1 + S_2$, and S is extensive.

For a pure substance, the molar entropy is $\bar{S} = S/n$.

The units of S in (3.28) are J/K or cal/K. The units of $\bar{S}$ are J/(mole K) or cal/(mole K).

The path from the postulation of the second law to the existence of S has been a long one, so let us review the chain of reasoning that led to entropy.

1. Experience shows that complete conversion of heat to work in a cyclic process is impossible. (This assertion is the Kelvin–Planck statement of the second law.)

2. From statement 1, we proved that the efficiency of any heat engine that operates on a (reversible) Carnot cycle is independent of the nature of the working substance but depends only on the temperatures of the reservoirs: $e_{\text{rev}} = -w/q_H = 1 + q_C/q_H = f(\tau_C, \tau_H)$.

3. We used a perfect gas as the working substance in a Carnot cycle and used the ideal-gas temperature scale to find that $e_{\text{rev}} = 1 - T_C/T_H$; from statement 2, this equation holds for any system as working substance. Equating this expression for e_{rev} with that in statement 2, we get $q_C/T_C + q_H/T_H = 0$ for any system that undergoes a Carnot cycle.

4. We showed that an arbitrary reversible cycle can be divided into an infinite number of infinitesimal strips, each strip being a Carnot cycle; hence for each strip $dq_C/T_C + dq_H/T_H = 0$. Summing the dq/T's from each strip, we proved that $\oint dq_{\text{rev}}/T = 0$ for any reversible cycle that any system undergoes. It follows that the integral of dq_{rev}/T is independent of the path; hence dq_{rev}/T is the differential of a state function, which we call the entropy S: $dS \equiv dq_{\text{rev}}/T$.

3.5 CALCULATION OF ENTROPY CHANGES

The entropy change on going from state 1 to state 2 is given by (3.29) as $\Delta S = S_2 - S_1 = \int_1^2 dq_{\text{rev}}/T$. For a reversible process, we can apply (3.29) directly to calculate ΔS. For an irreversible process pr, we cannot integrate dq_{pr}/T to obtain ΔS because dS equals dq/T only for reversible processes; for an irreversible process, dS is not necessarily equal to dq_{irrev}/T. However, S is a state function, and therefore ΔS depends only on the initial and final states. We can therefore find ΔS for an irreversible process that goes from state 1 to state 2 if we can conceive of a reversible process that goes from 1 to 2. We then calculate ΔS for this reversible change from 1 to 2, and this is the same as ΔS for the irreversible change from 1 to 2.

In summary, to calculate ΔS for any process: (1) identify the initial and final states 1 and 2; (2) devise a convenient *reversible* path from 1 to 2; (3) calculate ΔS from Eq. (3.29).

Let us calculate ΔS for some common processes. [Note that, as before, all state functions refer to the *system* and ΔS means ΔS_{syst}. Equation (3.29) gives ΔS_{syst} and does not include any entropy changes that may occur in the surroundings.]

1. Reversible adiabatic process. Here $dq_{\text{rev}} = 0$; hence

$$\Delta S = 0 \qquad \text{rev. ad. proc.} \tag{3.31}$$

Two of the four steps of a Carnot cycle are reversible adiabatic processes.

2. Reversible phase change at constant T and P. Here T is constant, and (3.29) gives

$$\Delta S = \int_1^2 \frac{dq_{\text{rev}}}{T} = \frac{1}{T}\int_1^2 dq_{\text{rev}} = \frac{q_{\text{rev}}}{T} \tag{3.32}$$

q_{rev} is the latent heat of the transition. Since P is constant, Eq. (2.77) gives $q_{\text{rev}} = \Delta H$; hence

$$\Delta S = \Delta H/T \qquad \text{rev. phase change} \tag{3.33}$$

An example might be the melting of 5.0 g of ice at 0°C and 1 atm. The latent heat of fusion is 79.7 cal/g; hence

$$\Delta S = (79.7 \text{ cal/g})(5.0 \text{ g})/(273 \text{ K}) = 1.46 \text{ cal/K} = 6.1 \text{ J/K}$$

For the freezing of 5.0 g of liquid water at 0°C and 1 atm, q_{rev} is negative, and $\Delta S = -6.1$ J/K.

3. Reversible isothermal process. Here T is constant, and $\Delta S = \int_1^2 dq_{\text{rev}}/T = q_{\text{rev}}/T$. Thus

$$\Delta S = q_{\text{rev}}/T \qquad \text{isotherm. proc.} \tag{3.34}$$

Examples include a reversible phase change (case 2 above) and two of the four steps of a Carnot cycle.

4. Reversible change of state of a perfect gas. From the first law and Sec. 2.12, we have for a reversible process in a perfect gas

$$dq_{\text{rev}} = dU - dw_{\text{rev}} = C_V\,dT + P\,dV = C_V\,dT + nRT\,dV/V$$

$$dS = dq_{\text{rev}}/T = C_V\,dT/T + nR\,dV/V$$

$$\Delta S = \int_1^2 C_V(T)\frac{dT}{T} + nR\int_1^2 \frac{dV}{V}$$

$$\Delta S = \int_1^2 \frac{C_V(T)}{T}\,dT + nR\ln\frac{V_2}{V_1} \qquad \text{perf. gas} \tag{3.35}$$

If the temperature change is not large, it may be a good approximation to take C_V constant, in which case $\Delta S \approx C_V \ln(T_2/T_1) + nR\ln(V_2/V_1)$.

5. Irreversible change of state of a perfect gas. Let n moles of a perfect gas at P_1, V_1, T_1 irreversibly change its state to P_2, V_2, T_2. We can readily conceive of a reversible process to carry out this same change of state. For example, we might (a) put the gas (enclosed in a cylinder fitted with a frictionless piston) in a large constant-temperature bath at temperature T_1 and infinitely slowly change the pressure on the piston until the gas reaches volume V_2; (b) then remove the gas from contact with the bath, hold the volume fixed at V_2, and reversibly heat (or cool) the gas until its temperature reaches T_2. Since S is a state function, ΔS for this reversible change from state 1 to 2 is the same as ΔS for the irreversible change from state 1 to 2 (even though q is not necessarily the same for the two processes). Therefore Eq. (3.35) gives ΔS for the irreversible change; note that the value of the right side of (3.35) depends only on T_2, V_2, and T_1, V_1, the state functions of the final and initial states.

An example of an irreversible change of state in a perfect gas is the adiabatic free expansion of a perfect gas into vacuum (the Joule experiment). The initial state is T_1, V_1, and the final state is T_1, V_2, where $V_2 > V_1$. T is constant because $\mu_J = (\partial T/\partial V)_U$ is zero for a perfect gas. Although the process is adiabatic ($q = 0$), ΔS is not zero because the process is irreversible. Equation (3.35) gives $\Delta S = nR \ln (V_2/V_1)$, since the temperature integral in (3.35) is zero when $T_2 = T_1$. If the original container and the evacuated container are of equal volume, then $V_2 = 2V_1$ and $\Delta S = nR \ln 2$; per mole of gas we have

$$\Delta S/n = \Delta \bar{S} = R \ln 2 = [8.314 \text{ J/(mol K)}](2.303)(0.3010) = 5.76 \text{ J/(mol K)}$$

6. Constant-pressure heating. First, suppose the heating is done *reversibly*. At constant pressure (provided no phase change occurs), Eq. (2.81) gives $dq_{\text{rev}} = C_P\, dT$; hence (3.29) becomes

$$\Delta S = \int_{T_1}^{T_2} \frac{C_P}{T}\, dT \qquad \text{const. } P\text{, no phase change} \tag{3.36}$$

If C_P is essentially constant over the temperature range, then $\Delta S = C_P \ln T_2/T_1$. For example, suppose 100 g of water is reversibly warmed from 25 to 50°C at 1 atm. The heat capacity C_P for water is nearly constant over this temperature range, and we shall take the specific heat capacity as $c_P = 1.00$ cal/(g °C). [Accurate values for c_P of liquid water at 0, 15, 25, 50, and 100°C in cal/(g °C) are 1.008, 1.000, 0.9989, 0.9992, and 1.0076, respectively.] For the 100 g of water, $C_P = 100 \text{ g} \times 1.00 \text{ cal/(g °C)} = 100 \text{ cal/K}$. (One Celsius degree equals one kelvin, by definition.) For the heating process,

$$\Delta S = (100 \text{ cal/K}) \ln (323 \text{ K}/298 \text{ K}) = 8.06 \text{ cal/K} = 33.7 \text{ J/K}$$

Now suppose we heat the water *irreversibly* from 25 to 50°C at 1 atm (say, by using a bunsen-burner flame). The initial and final states are the same as for the reversible heating. Hence the integral on the right of (3.36) gives ΔS for the irreversible heating; note that ΔS in (3.36) depends only on T_1, T_2, and the value of P (since C_P will depend somewhat on P); that is, ΔS depends only on the initial and final states. Thus ΔS for heating 100 g of water from 25 to 50°C at 1 atm is 8.1 cal/K, whether the heating is done reversibly or irreversibly. Even though the heating may be done irreversibly, with temperature gradients existing during the process, we can imagine doing it reversibly and apply (3.36) to find ΔS (provided the initial and final states are equilibrium states). Likewise, if we carry out the change of state by stirring at constant pressure (à la Joule), rather than heating, we can still use (3.36).

To reversibly heat a system, we start with the system at temperature T_1 and put it in contact with a heat reservoir at temperature $T_1 + dT$. Heat flows into the system until its temperature reaches $T_1 + dT$, and this flow is reversible since system and surroundings have temperatures that differ only infinitesimally. Next we put the system into contact with a reservoir at temperature $T_1 + 2dT$, etc. We thus use an infinite number of reservoirs that span the temperature range from T_1 to T_2 in infinitesimal intervals. If we can't afford an infinite number of heat reservoirs, we can do the reversible heating another way (see paragraph 8).

7. General change of state from (P_1, T_1) to (P_2, T_2). In paragraph 6 we considered the change of entropy with change in temperature at constant pressure. Here we also need to know how S varies with pressure. This case will be discussed in Sec. 4.5; see Eqs. (4.61) and (4.64).

8. **Irreversible, infinitely slow heat transfer.** Consider an isolated system that consists of two parts: part 1 is at temperature T_1, and part 2 is at T_2, with $T_2 > T_1$. Further, let the parts be separated by a wall that is a very poor conductor of heat, so that the rate of heat flow from part 2 to 1 is infinitesimal; still further, let each part by itself be in internal equilibrium. Let heat $dq > 0$ flow from 2 to 1; what is dS for the composite system? Because the heat flows infinitely slowly, each separate part remains in internal equilibrium (even though the two parts are not in thermal equilibrium with each other). However, because there is a finite temperature difference between the two parts, the heat flow is irreversible. To find ΔS, we carry out the same change of state in a reversible manner, as follows.

We put part 2 at temperature T_2 in thermal contact with a perfect gas also at temperature T_2; we then carry out a reversible isothermal expansion of the gas, in the course of which heat $-dq$ flows from part 2 of the system into the gas [Eq. (2.106)]. The entropy change of part 2 in this reversible process is $-dq/T_2$. Removing the gas from contact with part 2, we carry out a reversible adiabatic expansion of the gas until its temperature falls to T_1. We then put it in contact with part 1 and reversibly and isothermally compress the gas until heat dq flows into part 1. The entropy change of part 1 in this reversible process is dq/T_1. The total entropy change for parts 1 and 2 is thus

$$dS = \frac{dq}{T_1} - \frac{dq}{T_2} \tag{3.37}$$

Note that dS is positive since $T_2 > T_1$. (Of course, the perfect gas suffers an entropy change in the reversible process, but this change is not relevant to finding dS for the system composed of parts 1 and 2.)

9. **Irreversible phase change.** Suppose we want ΔS for the transformation of 1 mole of supercooled liquid water at $-10°C$ and 1 atm to 1 mole of ice at $-10°C$ and 1 atm. This transformation is irreversible; intermediate states consist of mixtures of water and ice at $-10°C$, and these are not equilibrium states; moreover, withdrawal of an infinitesimal amount of heat from the ice at $-10°C$ will not cause any of the ice to go back to supercooled water at $-10°C$. To find ΔS, we use the following reversible path for this transformation. We first reversibly warm the supercooled liquid to 0°C and 1 atm (paragraph 6); we then reversibly freeze it at 0°C and 1 atm (paragraph 2); finally, we reversibly cool the ice to $-10°C$ and 1 atm (paragraph 6). ΔS for the irreversible transformation at $-10°C$ equals the sum of the entropy changes for the three reversible steps since the irreversible process and the reversible process each start from the same state and each end at the same state. Figure 3.7 shows the two paths. Numerical calculations are left as a problem (Prob. 3.15).

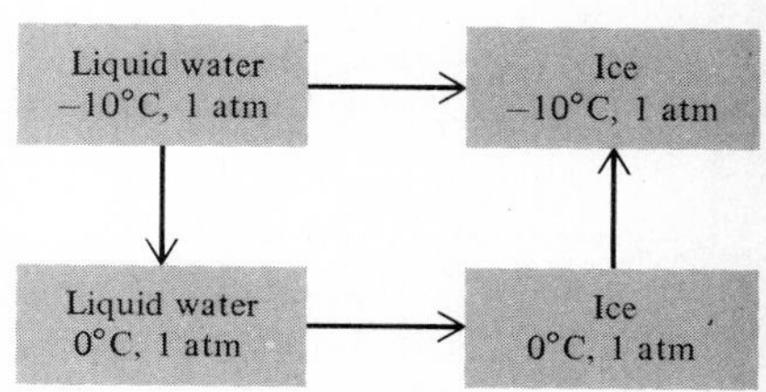

Figure 3.7
Irreversible and reversible paths connecting liquid water and ice at $-10°C$.

10. Mixing of different inert perfect gases at constant P and T. Let n_a and n_b moles of the inert perfect gases a and b, each at the same initial P and T, mix (Fig. 3.8). By inert gases, we mean that no chemical reaction occurs on mixing. Since the gases are perfect, there are no intermolecular interactions either before or after the partition is removed; hence the total internal energy is unchanged on mixing, and T is unchanged on mixing.

The mixing is irreversible. To find ΔS, we must find a way to carry out this change of state reversibly. This can be done in two steps. In step 1, we put each gas in a constant-temperature bath and reversibly and isothermally expand each gas separately to a volume equal to the final volume V. Note that step 1 is not adiabatic; instead heat flows into each gas to balance the work done by each gas. Since S is extensive, ΔS for step 1 is the sum of ΔS for each gas, and Eq. (3.35) gives

$$\Delta S_1 = \Delta S_a + \Delta S_b = n_a R \ln (V/V_a) + n_b R \ln (V/V_b)$$

Step 2 is a reversible isothermal mixing of the expanded gases. This can be done as follows. We suppose it possible to obtain two *semipermeable* membranes, one of which is permeable to gas a only and one of which is permeable to gas b only. (For example, heated palladium is permeable to hydrogen but not to oxygen or nitrogen.) We set up the unmixed state of the two gases as shown in Fig. 3.9*a*. (We assume the absence of friction.) We then move the two coupled membranes very slowly to the left. Figure 3.9*b* shows an intermediate state of the system. Since the membranes move slowly, membrane equilibrium exists, meaning that the partial pressures of gas a on each side of the membrane permeable to a are equal, and similarly for gas b. The gas pressure in region I of Fig. 3.9*b* is P_a and in region III is P_b. Because of membrane equilibrium for each of the semipermeable membranes, the partial pressure of gas a in region II is P_a, and that of gas b in region II is P_b; the total pressure in region II is thus $P_a + P_b$. The total force to the right on the two movable coupled membranes is due to gas pressure in regions I and III and equals $(P_a + P_b)A$, where A is the area of each membrane. The total force to the left on these membranes is due to gas pressure in region II and equals $(P_a + P_b)A$. These two forces are equal. Hence any intermediate state is an equilibrium state, and only an infinitesimal force is needed to move the membranes. Since we pass through equilibrium states and exert only infinitesimal forces, step 2 is reversible. The final state (Fig. 3.9*c*) is the desired mixture. The internal energy of a perfect gas depends only on T. Since T is constant for step 2, ΔU is zero for step 2. Since only an infinitesimal force was exerted on the membranes, $w = 0$ for step 2. Hence $q = \Delta U - w = 0$ for step 2. Step 2 is adiabatic as well as reversible; hence Eq. (3.31) gives $\Delta S_2 = 0$ for the reversible mixing of two perfect gases.

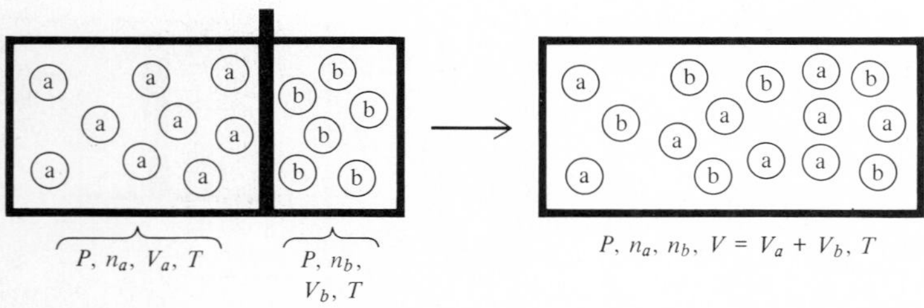

Figure 3.8
Mixing of perfect gases at constant T and P.

ΔS for the irreversible mixing of Fig. 3.8 equals $\Delta S_1 + \Delta S_2$, so

$$\Delta S = n_a R \ln (V/V_a) + n_b R \ln (V/V_b) \tag{3.38}$$

We can rewrite (3.38) as follows. Boyle's law gives $PV_a = P_a V$; hence $V/V_a = P/P_a$. From Eq. (1.23), $P_a/P = x_a = n_a/(n_a + n_b)$. Hence $V/V_a = (n_a + n_b)/n_a$, with a similar equation for V/V_b. Therefore

$$\Delta S = n_a R \ln [(n_a + n_b)/n_a] + n_b R \ln [(n_a + n_b)/n_b] \tag{3.39}$$

Note that ΔS is positive.

The term *entropy of mixing* for ΔS in (3.39) is perhaps misleading, in that the entropy change comes entirely from the volume change of each gas (step 1) and is zero for the reversible mixing (step 2).

Of course, (3.39) applies only when a and b are different gases. If they are identical gases, then the "mixing" at constant T and P corresponds to no change in state and $\Delta S = 0$.

3.6 ENTROPY, REVERSIBILITY, AND IRREVERSIBILITY

In the previous section, we calculated ΔS for the *system* in various processes. In this section we shall consider the total entropy change that occurs in a process; i.e., we shall examine the sum of the entropy changes in the system and the surroundings: $\Delta S_{\text{syst}} + \Delta S_{\text{surr}}$. We call this sum the entropy change of the universe:

$$\Delta S_{\text{univ}} = \Delta S_{\text{syst}} + \Delta S_{\text{surr}} \tag{3.40}$$

where the subscript univ stands for universe. Here, "universe" refers to the system plus those parts of the world which can interact with the system. Whether the conclusions of this section about ΔS_{univ} apply to the entire universe in a cosmic sense will be considered in Sec. 3.8. We shall give separate consideration to ΔS_{univ} for reversible processes and irreversible processes.

Reversible processes. In a reversible process, any heat flow between system and surroundings must occur with no finite temperature difference between system and surroundings; otherwise the heat flow would be irreversible. Let dq_{rev} be the heat

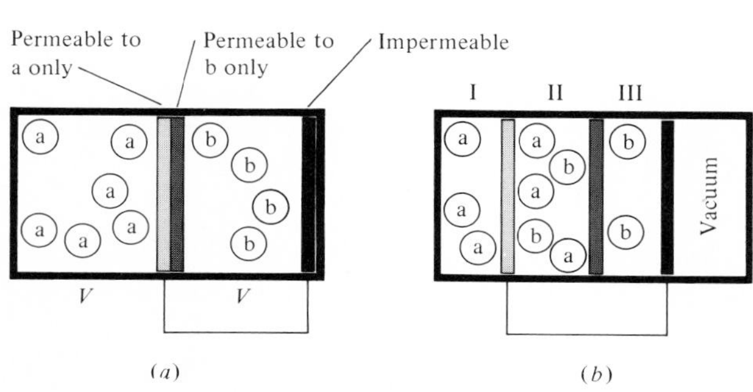

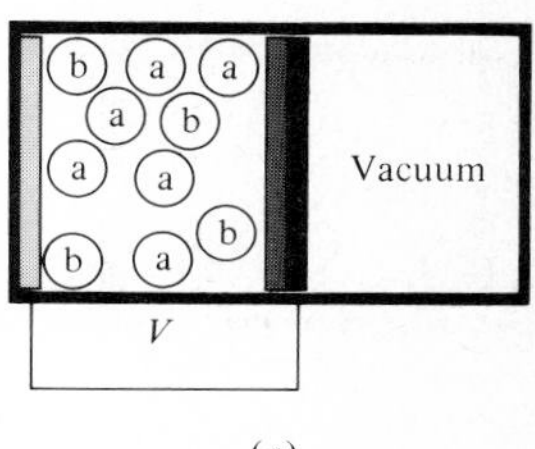

Figure 3.9 Reversible isothermal mixing of gases. The system is in a constant-temperature bath (not shown).

flow into the system from the surroundings during an infinitesimal part of the reversible process; the corresponding heat flow into the surroundings is $-dq_{\text{rev}}$. We have

$$dS_{\text{univ}} = dS_{\text{syst}} + dS_{\text{surr}} = dq_{\text{rev}}/T_{\text{syst}} + (-dq_{\text{rev}}/T_{\text{surr}})$$

$$= dq_{\text{rev}}/T_{\text{syst}} - dq_{\text{rev}}/T_{\text{syst}} = 0$$

Integration gives

$$\Delta S_{\text{univ}} = 0 \qquad \text{rev. proc.} \tag{3.41}$$

Although S_{syst} and S_{surr} may both change in a reversible process, $S_{\text{syst}} + S_{\text{surr}} = S_{\text{univ}}$ is unchanged in a reversible process.

Irreversible processes. We first consider the special case of an *adiabatic* irreversible process in a closed system. This special case will lead us to the desired general result. Let the system go from state 1 to state 2 in an irreversible adiabatic process; the disconnected arrowheads from 1 to 2 in Fig. 3.10 indicate the irreversibility and the fact that an irreversible process cannot in general be plotted on a P-V diagram since it usually involves nonequilibrium states. To evaluate $S_2 - S_1 = \Delta S_{\text{syst}}$ we connect states 1 and 2 by the following reversible path. From state 2, we do work adiabatically and reversibly on the system to increase its temperature to T_{hr}, the temperature of a certain heat reservoir; this brings the system to state 3. From Eq. (3.31), ΔS is zero for a reversible adiabatic process; hence $S_3 = S_2$. (As always, state functions refer to the system unless otherwise specified; thus S_3 and S_2 are the system's entropies in states 3 and 2.) We next either add or withdraw sufficient heat $q_{3\to 4}$ isothermally and reversibly at temperature T_{hr} to make the entropy of the system equal to S_1; this brings the system to state 4 with $S_4 = S_1$. ($q_{3\to 4}$ is positive if heat flows into the system from the reservoir during the process $3 \to 4$ and negative if heat flows out of the system into the reservoir during $3 \to 4$.) We have

$$S_4 - S_3 = \int_3^4 \frac{dq_{\text{rev}}}{T} = \frac{1}{T_{hr}} \int_3^4 dq_{\text{rev}} = \frac{q_{3\to 4}}{T_{hr}}$$

Since states 4 and 1 have the same entropy, they lie on a line of constant S, an *isentrop*. What is an isentrop? For an isentrop, $dS = 0 = dq_{\text{rev}}/T$, so that $dq_{\text{rev}} = 0$; that is, an isentrop is a reversible adiabat. Hence to go from 4 to 1, we carry out a

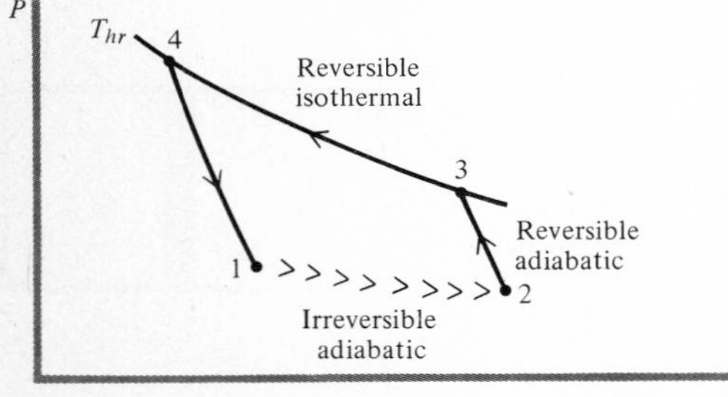

Figure 3.10
Irreversible and reversible paths between states 1 and 2.

reversible adiabatic process (with the system doing work on the surroundings). Since S is a state function, we have for the cycle $1 \to 2 \to 3 \to 4 \to 1$:

$$0 = \oint dS_{\text{syst}} = (S_2 - S_1) + (S_3 - S_2) + (S_4 - S_3) + (S_1 - S_4)$$

$$\oint dS_{\text{syst}} = (S_2 - S_1) + 0 + q_{3\to4}/T_{hr} + 0 = 0$$

$$S_2 - S_1 = -q_{3\to4}/T_{hr}$$

The sign of $S_2 - S_1$ is thus determined by the sign of $-q_{3\to4}$. We have for the cycle

$$\oint dU = 0 = \oint (dq + dw) = q_{3\to4} + w$$

The work done on the system in the cycle is thus $w = -q_{3\to4}$; the work done by the system on the surroundings is $-w = q_{3\to4}$. Suppose $q_{3\to4}$ were positive. Then the work $-w$ done on the surroundings would be positive, and we would have a cycle $(1 \to 2 \to 3 \to 4 \to 1)$ whose sole effect is extraction of heat $q_{3\to4}$ from a reservoir and its complete conversion to work $-w = q_{3\to4} > 0$. Such a cycle is impossible, since it violates the second law. Hence $q_{3\to4}$ cannot be positive; we have $q_{3\to4} \le 0$. Therefore

$$S_2 - S_1 = -q_{3\to4}/T_{hr} \ge 0 \tag{3.42}$$

We now strengthen this result by showing that $S_2 - S_1 = 0$ can be ruled out. To do this, consider the nature of reversible and irreversible processes. In a reversible process, we can make things go the other way by an infinitesimal change in circumstances; when the process is reversed, both system *and* surroundings are restored to their original states; i.e., the universe is restored to its original state. In an irreversible process, the universe cannot be restored to its original state. Now suppose $S_2 - S_1 = 0$. Then $q_{3\to4}$, which equals $-T_{hr}(S_2 - S_1)$, would be zero; also, w, which equals $-q_{3\to4}$, would be zero. (Points 3 and 4 would coincide.) After the irreversible process $1 \to 2$, the path $2 \to 3 \to 4 \to 1$ restores the system to state 1. Moreover, since $q = 0 = w$ for the cycle $1 \to 2 \to 3 \to 4 \to 1$, this cycle would have no net effect on the surroundings, and at the end of the cycle, the surroundings would be restored to their original state. Thus we would be able to restore the universe (system + surroundings) to its original state. But by hypothesis, the process $1 \to 2$ is irreversible, and so the universe cannot be restored to its original state after this process has occurred. Hence $S_2 - S_1$ cannot be zero. Equation (3.42) now tells us that $S_2 - S_1$ must be positive.

We have proved that the entropy of a closed system must increase in an irreversible adiabatic process:

$$\Delta S_{\text{syst}} > 0 \qquad \text{irrev. ad. proc., closed syst.} \tag{3.43}$$

A special case of this result is of great importance. If we have an isolated system, the system is necessarily closed and, moreover, any process must be adiabatic (since no heat can flow between the isolated system and its surroundings). Hence (3.43) applies to any irreversible process in an isolated system:

$$\Delta S_{\text{syst}} > 0 \qquad \text{irrev. proc., isolated syst.} \tag{3.44}$$

Now consider $\Delta S_{\text{univ}} = \Delta S_{\text{syst}} + \Delta S_{\text{surr}}$ for an irreversible process. Since we only want to examine the effect on S_{univ} of the interaction between the system and its surroundings, we must consider that during the irreversible process the surroundings interact only with the system and not with any other part of the world. Hence, for the duration of the irreversible process, we can regard the system plus its surroundings (syst + surr) as forming an isolated system. Equation (3.44) then gives $\Delta S_{\text{syst+surr}} \equiv \Delta S_{\text{univ}} > 0$ for an irreversible process. We have shown that S_{univ} increases in an irreversible process:

$$\Delta S_{\text{univ}} > 0 \qquad \text{irrev. proc.} \tag{3.45}$$

where ΔS_{univ} is the sum of the entropy changes for the system and surroundings.

We previously showed $\Delta S_{\text{univ}} = 0$ for a reversible process. Hence

$$\Delta S_{\text{univ}} \geq 0 \tag{3.46*}$$

depending on whether the process is reversible or irreversible. Energy cannot be created or destroyed. Entropy can be created but not destroyed.

The statement that

dq_{rev}/T is the differential of a state function S that has the property $\Delta S_{\text{univ}} \geq 0$ for any process

can be taken as a third formulation of the second law of thermodynamics, equivalent to the Kelvin–Planck and the Clausius statements. (See Prob. 3.20.)

We have shown (as a deduction from the Kelvin–Planck statement of the second law) that S_{univ} increases for an irreversible process and remains the same for a reversible process. A reversible process is an idealization that generally cannot be precisely attained in real processes. In virtually all real processes there will be some effect (e.g., friction, lack of precise thermal equilibrium, small amounts of turbulence, irreversible mixing, etc.; see *Zemansky*, chap. 8, for a full discussion) which makes the process irreversible. Since (virtually) all real processes are irreversible, we can say as a deduction from the second law that S_{univ} is continually increasing with time. (See Sec. 3.8 for comment on this statement.)

3.7 WHAT IS ENTROPY?

Each of the first three laws of thermodynamics leads to the existence of a state function. The zeroth law leads to temperature; the first law leads to internal energy; the second law leads to entropy. It is not the business of thermodynamics (which is a macroscopic science) to explain the microscopic nature of these state functions. All thermodynamics needs to do is to tell us how to measure T, ΔU, and ΔS. Nevertheless it is nice to have a molecular picture of the macroscopic thermodynamic state functions.

Temperature is readily interpreted as some sort of measure of the average molecular energy. Internal energy is interpreted as the total molecular energy. Although we have shown how ΔS can be calculated for various processes, the reader may feel frustrated at not having any clear picture of the physical nature of entropy. Although entropy is not as easy a concept to grasp as temperature or internal energy, we can get some understanding of its physical nature. The following discussion is based on the molecular picture of matter; hence the ideas and results of this section are not part of thermodynamics but belong to statistical mechanics.

Equation (3.44) shows that for any irreversible process that occurs in the isolated system, ΔS is positive. Since all real processes are irreversible, when processes are occurring in an isolated system, its entropy is increasing. Irreversible processes (mixing, chemical reactions, flow of heat from hot to cold bodies, etc.) accompanied by an increase in S will continue to occur in the isolated system until it has reached its maximum possible value of S. (For example, paragraph 8 of Sec. 3.5 shows that heat flow from a hot to a cold body is accompanied by an increase in entropy; hence, if two parts of an isolated system are at different temperatures, heat will flow from the hot part to the cold part until the temperatures of the parts are equalized, and this equalization of temperatures maximizes the system's entropy.) When the entropy of the isolated system is maximized, things cease happening (on a macroscopic scale), because any further processes can only decrease S, which would violate the second law. By definition, the isolated system has reached equilibrium when processes cease occurring. Hence:

Thermodynamic equilibrium in an isolated system is reached when the system's entropy is maximized.

(Thermodynamic equilibrium in nonisolated systems will be discussed in Chap. 4.)

Thermodynamics says nothing about the rate at which equilibrium is attained. An isolated mixture of H_2 and O_2 at room temperature will remain unchanged in the absence of a catalyst; however, the system is not in a state of true thermodynamic equilibrium; when a catalyst is introduced, the gases react to produce H_2O, with an increase in entropy. Likewise, diamond is thermodynamically unstable with respect to conversion to graphite at room temperature, but the rate of conversion is zero, so no one need worry about loss of her engagement ring. ("Diamonds are forever.") It can even be said that pure hydrogen is in a sense thermodynamically unstable at room temperature, since fusion of the hydrogen nuclei to helium nuclei is accompanied by an increase in S_{univ}; of course, the rate of nuclear fusion is zero at room temperature, and we can completely ignore the possibility of this process.

Since S of an isolated system is maximized at equilibrium, we ask ourselves: What else is maximized at equilibrium? In other words, what really determines the equilibrium position of an isolated thermodynamic system? To answer this, consider a simple example, the mixing at constant temperature and pressure of equal volumes of two different inert perfect gases in an isolated system (Fig. 3.11). The motion of the gas molecules is completely random, and the molecules do not interact with one another. What then makes 2 in Fig. 3.11 the equilibrium state and 1 a nonequilibrium state? Why is the passage from the unmixed state 1 to the mixed state 2 irreversible? (From 2, an isolated system will never go back to 1.)

Clearly the answer is *probability*. If the molecules move at random, any A molecule has a 50 percent chance of being in the left half of the container. The probability that all the A molecules will be in the left half and all the B molecules in

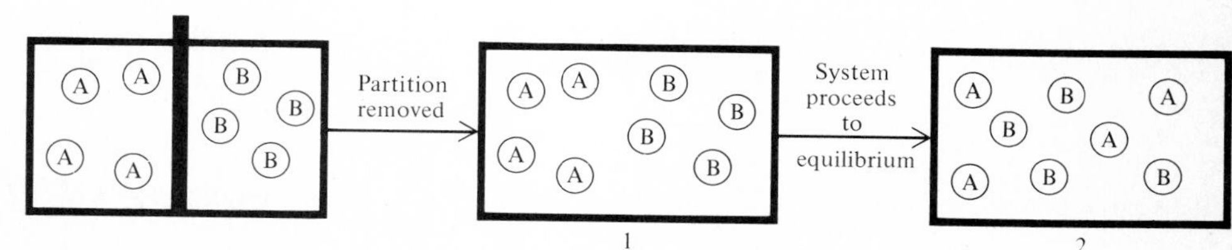

Figure 3.11 Irreversible mixing of gases.

the right half (state 1) is extremely tiny. The most probable distribution has A and B molecules each equally distributed between the two halves of the container (state 2). An analogy to the spatial distribution of 1 mole of A molecules would be tossing a coin 6×10^{23} times; the chance of getting 6×10^{23} heads is extremely minute; the most probable outcome is 3×10^{23} heads and 3×10^{23} tails, and only outcomes with a very nearly equal ratio of heads to tails have significant probabilities; the probability maximum is very sharply peaked at 50 percent heads. Likewise, any spatial distribution of the A molecules that differs significantly from 50 percent A in each container has an extremely tiny probability, due to the large number of A molecules. Similarly for the B molecules.

It seems clear that the equilibrium thermodynamic state of an isolated system is the most probable state. The increase in S as an isolated system proceeds toward equilibrium is directly related to the system's going from a state of low probability to one of high probability. We shall therefore postulate that the entropy S of a system is a function of the probability p of the system's thermodynamic state:

$$S = f(p) \tag{3.47}$$

It may seem amazing, but use of the single fact that entropy is an extensive state function allows us to determine the function f in our postulate (3.47). To do this, we consider a system composed of two independent, noninteracting parts, 1 and 2. (Imagine the two parts to be separated by a rigid, impermeable, adiabatic wall that prevents flow of heat, work, and matter between them.) Entropy is an extensive property, so the entropy of the composite system 1 + 2 is $S_{1+2} = S_1 + S_2$, where S_1 and S_2 are the entropies of parts 1 and 2. Substitution of (3.47) gives

$$f(p_{1+2}) = f(p_1) + f(p_2) \tag{3.48}$$

What is the relation between the probability p_{1+2} of the composite system's thermodynamic state and the probabilities p_1 and p_2 of the states of parts 1 and 2? The probability of two independent events' both happening is shown in probability theory to be the product of the probabilities for each event. (For example, the probability of getting two heads when two separate coins are tossed is $\frac{1}{2} \times \frac{1}{2} = \frac{1}{4}$.) Since parts 1 and 2 behave independently of each other, we have $p_{1+2} = p_1 p_2$. Equation (3.48) becomes

$$f(p_1 p_2) = f(p_1) + f(p_2) \tag{3.49}$$

Our task is thus to find the function that satisfies

$$f(xy) = f(x) + f(y) \tag{3.50}$$

[Before reading ahead, you might try and guess a solution for f in (3.50).]

It isn't hard to prove that the only function that satisfies (3.50) is a logarithmic function. Problem 15.36 shows that f in (3.50) must have the form

$$f(x) = k \ln x + C \tag{3.51}$$

where k and C are constants.

Since we postulated $S = f(p)$ in Eq. (3.47), we have from (3.51)

$$S = k \ln p + C \tag{3.52}$$

where k and C are constants and p is the probability of the system's thermodynamic state. Since the second law only allows us to calculate *changes* in entropy, we cannot use thermodynamics to find C. We can, however, evaluate k, as follows.

Consider again the spontaneous mixing of equal volumes of two different perfect gases (Fig. 3.11). State 1 is the unmixed state of the middle drawing of Fig. 3.11, and state 2 is the mixed state. Equation (3.52) gives for the process $1 \rightarrow 2$:

$$\Delta S = S_2 - S_1 = k \ln p_2 + C - k \ln p_1 - C$$

$$S_2 - S_1 = k \ln (p_2/p_1) \tag{3.53}$$

(Don't confuse the probabilities p_1 and p_2 with pressures.) We want p_2/p_1. The probability that any particular A molecule is in the left half of the container is $\frac{1}{2}$. Since the perfect-gas molecules move independently of one another, the probability that every A molecule is in the left half of the container is the product of the independent probabilities for each A molecule, namely, $(\frac{1}{2})^{n_A N_0}$, where n_A is the number of moles of A, N_0 is the Avogadro constant, and $n_A N_0$ is the number of A molecules. Likewise, the probability that all the B molecules are in the right half of the container is $(\frac{1}{2})^{n_B N_0}$. Since A and B molecules move independently, the simultaneous probability that all A molecules are in the left half of the box and all B molecules are in the right half is the product of the two separate probabilities, namely,

$$p_1 = (\tfrac{1}{2})^{n_A N_0}(\tfrac{1}{2})^{n_B N_0} = (\tfrac{1}{2})^{(n_A + n_B)N_0} = (\tfrac{1}{2})^{2n_A N_0} \tag{3.54}$$

since $n_A = n_B$. (We took equal volumes of A and B at the same T and P.)

State 2 is the thermodynamic state in which to within the limits of macroscopic measurement the gases A and B are uniformly distributed through the container. As noted a few paragraphs ago, the probability of any departure from a uniform distribution that is sufficiently large to be directly detectable is vanishingly small, due to the large number of molecules composing the system. Hence the probability of the final state 2 is only "infinitesimally" less than one and can be taken as one: $p_2 = 1$. Therefore (3.53) and (3.54) give for the mixing

$$S_2 - S_1 = k \ln 2^{2n_A N_0} = 2n_A N_0 k \ln 2 \tag{3.55}$$

However, in Sec. 3.5 we used thermodynamics to calculate ΔS for the isothermal, isobaric irreversible mixing of two perfect gases; Eq. (3.39) with $n_A = n_B$ gives

$$S_2 - S_1 = 2n_A R \ln 2 \tag{3.56}$$

Equating the thermodynamic ΔS of (3.56) to the statistical-mechanical ΔS of (3.55), we get $N_0 k = R$, or

$$k = \frac{R}{N_0} = \frac{8.314 \text{ J mole}^{-1} \text{ K}^{-1}}{6.022 \times 10^{23} \text{ mole}^{-1}} = 1.38 \times 10^{-23} \text{ J/K} \tag{3.57}$$

We have evaluated k in the statistical-mechanical formula $S = k \ln p + C$. The fundamental physical constant k, called *Boltzmann's constant*, plays a key role in statistical mechanics. The connection between entropy and probability was first recognized in the late nineteenth century by the physicist Ludwig Boltzmann. The application of $S = k \ln p + C$ to situations more complicated than the mixing of perfect gases requires knowledge of quantum and statistical mechanics; in Chap. 22

we shall obtain an equation that expresses the entropy of a system in terms of its quantum-mechanical energy levels. Our main conclusion for now is that entropy is a measure of the *probability* of a state; apart from an additive constant, the entropy is proportional to the log of the probability of the thermodynamic state.

Disordered states generally have higher probabilities than ordered states; e.g., in the mixing of two gases, the disordered, mixed state is far more probable than the ordered, unmixed state. Hence it is often said that entropy is a measure of the *disorder* of a state; increasing entropy means increasing disorder. However, order and disorder are subjective concepts, whereas probability is a precise quantitative concept. Hence it is preferable to relate S to probability rather than to disorder. Moreover, the idea of disorder can sometimes lead one astray. For example, an isolated system that consists of a supersaturated solution will spontaneously and irreversibly go to a state with the excess solute crystallized out of solution; this process is accompanied by an increase in entropy (and an increase in probability), but one would probably say that the disorder has decreased with the separation of some of the solute from the solution.

For the mixing of two different gases, the connection between probability and entropy is clear. Let us consider some other kinds of processes. Suppose two parts of a system are at different temperatures. Heat then flows spontaneously and irreversibly between the parts, accompanied by an increase in entropy. How is probability involved here? The heat flow occurs via collisions between molecules of the hot part with molecules of the cold part; in such collisions, it is more probable for the high-energy molecules of the hot part to lose some of their energy to the low-energy molecules of the cold part than for the reverse to happen. Thus, internal energy is transferred from the hot to the cold body until thermal equilibrium is attained, at which point it is equally probable for molecular collisions to transfer energy from one part to the second part as to do the opposite. It is thus more probable for the internal molecular translational, vibrational, and rotational energy to be spread out among the parts of the system than for there to be an excess of such energy in one part.

Now consider an isolated mixture of H_2, Br_2, and HBr at a temperature high enough for chemical reaction to occur at a nonzero rate. During molecular collisions, energy transfers can occur that break bonds and allow the formation of new chemical species. There will be a probability for each possible outcome of each possible kind of collision, and these probabilities, together with the numbers of molecules of each species present, determine whether there is a net reaction to give more HBr or more H_2 and Br_2. When equilibrium is reached, the system has attained the most probable distribution of the species present over the available energy levels of H_2, Br_2, and HBr.

The mixing of two different gases results in a loss of *information*. In the unmixed state, we knew which half of the box each molecule was in; this information is lost in the mixed state. Entropy is often discussed in terms of information theory, but we shall not go into this relationship.

What light does this discussion throw on the second law of thermodynamics, which can be formulated as $\Delta S \geq 0$ for an isolated system (where $dS = dq_{\text{rev}}/T$)? The reason S increases is because an isolated system tends to go to a state of higher probability. *However*, it is not absolutely impossible for a macroscopic isolated system to go spontaneously to a state of lower probability, but such an occurrence is highly unlikely. Hence the second law is only a law of probability; there is an extremely small, but nonzero, chance that it might be violated. For example, there is a possibility of observing the spontaneous unmixing of two mixed gases, but because of the huge numbers of molecules present, the probability of its happening is fantastically small. There is an extremely tiny probability that the random motions of oxygen molecules in the air around you might carry them all to one corner of the room, causing you to die for lack of oxygen, but this possibility is nothing to lose any sleep over. The mixing of gases is irreversible because the mixed state is far, far more probable than any state with significant unmixing.

To show the extremely small probability of significant macroscopic deviations from the second law, consider the mixed state of Fig. 3.11; let there be $N_A = 0.6 \times 10^{24}$ molecules of the perfect gas A distributed between the two equal volumes. The most likely distribution is one with 0.3×10^{24} molecules of A in each half of the container, and similarly for the B molecules. (For simplicity we shall consider only the distribution of the A molecules, but the same considerations apply to the B molecules.) The probability for each A molecule to be in the left half of the container is $\frac{1}{2}$; probability theory (*Sokolnikoff and Redheffer*, p. 645) shows that the standard deviation of the number of A molecules in the left volume equals $\frac{1}{2}N_A^{1/2} = 0.4 \times 10^{12}$. *The standard deviation* is a measure of the typical deviation that is observed from the most probable value, 0.3×10^{24} in this case. Probability theory shows that when a large number of observations are made, 68 percent of them will give a result that lies within 1 standard deviation from the most probable value. (This statement applies whenever the distribution of probabilities is a normal, or gaussian, distribution; the gaussian distribution is the familiar bell-shaped curve.) In our example, we can expect that 68 percent of the time, the number of A molecules in the left volume will lie in the range $0.3 \times 10^{24} \pm 0.4 \times 10^{12}$. Although the standard deviation 0.4×10^{12} molecules is a very large number of molecules, compared with the total number of A molecules in the left volume, 0.3×10^{24}, it is negligible. A deviation of 0.4×10^{12} out of 0.3×10^{24} would mean a fluctuation in gas density of 1 part in 10^{12}, which is much too small to be directly detectable experimentally. A directly detectable density fluctuation might be 1 part in 10^6, or 0.3×10^{18} molecules out of 0.3×10^{24}. This is a fluctuation of about 10^6 standard deviations; the probability of a fluctuation this large or larger is found (Prob. 3.24) to be on the order of

$$1/10^{200,000,000,000} \tag{3.58}$$

The age of the universe is perhaps 10^{10} years. If we measured the density of the gas sample once every second, it would take (Prob. 3.25) on the order of

$$\frac{.7 \times 10^{200,000,000,000}}{3 \times 10^7} \approx 10^{200,000,000,000} \tag{3.59}$$

years of measurements for the probability of finding a detectable density fluctuation of 1 part in 10^6 to reach 50 percent. Thus, for all practical purposes, such a fluctuation in a macroscopic system is "impossible."

Probability theory shows that we can expect fluctuations about the equilibrium density to be of the order of $\sqrt{N}$, where N is the number of molecules per unit volume. These fluctuations correspond to continual fluctuations of the entropy about its equilibrium value. Such fluctuations are generally unobservable for systems of macroscopic size but can be detected in special situations (see below). If a system had

100 molecules, we would get fluctuations of about 10 molecules, which is an easily detectable 10 percent fluctuation; a system of 10^6 molecules would show fluctuations of 0.1 percent, which is still significant; for 10^{12} molecules ($\approx 10^{-12}$ mole), fluctuations are 1 part per million, which is perhaps the borderline of detectability. The validity of the second law is limited to systems where N is large enough to make fluctuations essentially undetectable.

In certain situations, fluctuations about equilibrium are experimentally observable. One case is that of tiny (but still macroscopic) dust particles or colloidal particles suspended in a fluid. Observation of such particles through a microscope shows them to undergo a ceaseless random motion (Fig. 3.12), known as *Brownian motion* (after the botanist Robert Brown, who first observed the phenomenon). These motions are due to collisions with the molecules of the fluid. If the fluid pressure on all parts of the colloidal particle were always the same, the particle would remain at rest. (More accurately, it would sink to the bottom of the container due to gravity.) However, tiny fluctuations in fluid pressures on the colloidal particle cause the random motion. (Such motion can be regarded as a small-scale violation of the second law.)

Similarly, random fluctuations in electron densities in an electrical resistor produce tiny internal currents, which, when amplified, give the "noise" that is always present in an electronic circuit. This noise limits the size of an experimentally detectable electronic signal, since amplification of the signal also amplifies the noise.

The realization that the second law is not an absolute law but only one for which observation of macroscopic violations is in general overwhelmingly improbable need not disconcert us. Most laws dealing with the macroscopic behavior of matter are really statistical laws whose validity follows from the random behavior of very large numbers of molecules. For example, throughout thermodynamics, we refer to the pressure P of a system. The pressure a gas exerts on the container walls results from the collisions of molecules with the walls. There is a possibility that at some instant, the gas molecules might all be moving inward toward the interior of the container, so that the gas would exert zero pressure on the container; likewise, there is a possibility that the molecular motion at a given instant might make the pressure on some walls differ significantly from that on other walls. However, such situations are so overwhelmingly improbable that we can with complete confidence ascribe a single uniform pressure to the gas.

3.8 ENTROPY, TIME, AND COSMOLOGY

Consider yet again the spontaneous mixing of two different gases. In the mixing process, the molecules move according to Newton's second law, $\mathbf{F} = m\, d^2\mathbf{r}/dt^2 =$

Figure 3.12
Path of a particle undergoing Brownian motion.

$m\,d\mathbf{v}/dt$. This law is symmetric with respect to time, meaning that if t is replaced by $-t$ and $\mathbf{v}$ by $-\mathbf{v}$, the law is unchanged. Thus, a reversal of all particle motions gives a set of motions that is also a valid solution of Newton's equation. Hence it is possible for the molecules to become spontaneously unmixed, and this unmixing does not violate the law of motion $\mathbf{F} = m\mathbf{a}$. However, as noted in the previous section, motions that correspond to a detectable degree of unmixing are extremely improbable (even though not absolutely impossible). Although Newton's laws of motion (which govern the motion of individual molecules) do not single out a direction of time, when the behavior of a very large number of molecules is considered, the second law of thermodynamics (which is a statistical law) tells us that states of an isolated system with lower entropy must precede in time states with higher entropy. The second law is not time-symmetric but singles out the direction of increasing time; we have $dS/dt > 0$ for an isolated system, so that the signs of dS and dt are the same. If someone showed us a film of two gases mixing spontaneously and then ran the film backward, we would not see any violation of $\mathbf{F} = m\mathbf{a}$ in the unmixing process, but the second law would tell us which showing of the film corresponded to how things actually happened. Likewise, if we saw a film of someone being spontaneously propelled out of a swimming pool of water, with the concurrent subsidence of waves in the pool, we would know that we were watching a film run backward; although tiny pressure fluctuations in a fluid can propel colloidal particles about, the Brownian motion of an object the size of a person is too improbable to occur.

Our statement that the laws of molecular motion are time-symmetric might be objected to on the grounds that molecules actually obey quantum mechanics and not classical mechanics. However, the laws of quantum mechanics are also symmetric with respect to time reversal, so quantum mechanics does not single out a direction of time.

The second law of thermodynamics singles out the direction of increasing time. The astrophysicist Eddington put things nicely with his statement that "entropy is time's arrow." The fact that $dS/dt > 0$ for an isolated system gives us the *thermodynamic arrow* of time. Besides the thermodynamic arrow, there is a *cosmological arrow* of time. Spectral lines in light reaching us from other galaxies show wavelengths that are longer than the corresponding wavelengths of light from objects at rest (the famous red shift). This red shift indicates that all galaxies are moving away from us. (The frequency shift results from the Doppler effect.) Thus the universe is expanding with increasing time, and this expansion gives the cosmological arrow. Many physicists believe that the thermodynamic and the cosmological arrows are directly related, but this question is still undecided. [See T. Gold, *Am. J. Phys.*, **30**, 403 (1962).]

There is strong (but not conclusive) evidence that the decay of one of the elementary particles (the neutral K meson) follows a law that is not symmetric with respect to time reversal. Thus there may also be a microscopic arrow of time, in addition to the thermodynamic and cosmological arrows. [See R. S. Casella, *Phys. Rev. Lett.*, **21**, 1128 (1968); **22**, 554 (1969); Y. Ne'eman, *Int. J. Theoret. Phys.*, **3**, 1 (1970).]

The second law of thermodynamics shows that S increases with time for an isolated system. Can this statement be applied to the entire physical universe? In Sec. 3.6 we used universe to mean the system plus those parts of the world which interact with the system. In this section, universe shall mean everything that exists—the entire cosmos of galaxies, intergalactic matter, electromagnetic radiation, etc. Physicists in the late nineteenth century generally believed that the second law is valid for the entire universe, but nowadays people are not so sure. Most of our experimental thermodynamic observations are on systems that are not of astronomic size, and hence we must be cautious about extrapolating thermodynamic results to encompass the entire universe. There is no guarantee that laws that hold on a terrestrial scale must also hold on a cosmic scale. Although there is no evidence for a cosmic violation of the second law, our experience is insufficient to rule out such a violation.

Likewise, although there is no evidence for any cosmic violation of the first law—conservation of energy—such a violation cannot be absolutely ruled out. In the 1960s, the astrophysicists Gold, Bondi, and Hoyle advocated a steady-state model of the universe in which there is continuous creation of hydrogen atoms in space, the atoms being created out of nothing; this continuous creation compensates for the expansion of the universe and maintains the density of matter in the universe constant with time. This steady-state model has E of the universe increasing with time and violates the first law. However, the model has only three hydrogen atoms created per year in a volume of 1 km^3. On a small scale, such a first-law violation would be undetectable; because of the vast volume of intergalactic space, such a violation would have significant consequences on a cosmic scale. The evidence of astronomical observations has been heavily against the steady-state model, and it has been largely abandoned.

At present the most widely accepted cosmological model is the "big bang" model, in which it is hypothesized that some 10 or 20 billion (10 or 20×10^9) years ago all the matter of the universe was gathered together. Explosion of this ball of matter gave rise to the presently observed expanding universe.

Astronomical observations are insufficient at present to decide whether this expansion will continue forever or whether there is sufficient matter in the universe to allow gravitational attractions in time to overcome the force of the initial explosion. If the latter is true, the universe will eventually reach a maximum expansion, from which it will begin to contract, ultimately bringing all matter together again. Perhaps a new big bang would then initiate a new cycle of expansion and contraction. Thus we could have a cyclic or oscillatory cosmological model. [The philosopher Nietzsche (1844–1900), who favored the idea of "eternal recurrence," spent much time studying physics to find evidence that would support his philosophical speculations.]

If the cyclic expansion–contraction cosmological model is the correct one, what would happen in the contraction phase of the universe? If the universe returns to a state essentially the same as the initial state that preceded the big bang (and if we accept the applicability of the concept of entropy to the entire universe), then the entropy of the universe would be decreasing during the contraction phase; this expectation is further supported by the arguments for a direct connection between the thermodynamic and cosmological arrows of time. But what would a universe with decreasing entropy be like? Would time run backward in a contracting universe? What is the meaning of the statement that "time runs backward"? One speculation is that since we can only experience time running forward, and since time would actually run backward in a contracting universe, a contracting universe would be unobservable; even though S would actually decrease during the contraction, people would experience S as increasing. [See P. T. Landsberg, *Pure Appl. Chem.*, **20**, 543 (1970).]

A concept that follows from the second law of thermodynamics is the *principle of the degradation of energy*. One can prove (see *Zemansky*, sec. 9-10) that any process that occurs makes an amount of energy $T_C \, \Delta S_{\text{univ}}$ unavailable for conversion to work, where T_C is the temperature of the coldest reservoir at hand and ΔS_{univ} is the entropy change in the process. Since S_{univ} is continually increasing, more and more energy is continually being made unavailable for conversion to work. If the second law applies to the entire universe, ultimately no energy will be available for doing work; the entropy of the universe will be maximized, and all processes (including life) will cease. This gloomy prospect has been christened the *heat death* of the universe. In the past, there has been much philosophical speculation based on this supposed heat death. It seems clear at present that we need to know a lot more about cosmology before we can be certain what the fate of the universe will be. (Entropy and heat death are favorite themes of science-fiction writers; see, for example, Isaac Asimov's story "The Last Question," in *Nine Tomorrows*, Doubleday, 1959, Fawcett, 1969.)

PROBLEMS

3.1 Consider a heat engine that uses reservoirs at 800 and 0°C. (*a*) Calculate the maximum possible efficiency. (*b*) If q_H is 1000 J, find the maximum value of $-w$ and the minimum value of $-q_C$.

3.2 Suppose the coldest reservoir we have at hand is at 10°C. If we want a heat engine that is at least 90 percent efficient, what is the minimum temperature of the required hot reservoir?

3.3 Prove that the Clausius statement of the second law is equivalent to the Kelvin–Planck statement. [To prove that statements A and B are logically equivalent, we must show that (*a*) if we assume A to be true, then B must be true; (*b*) if we assume B to be true, then A must be true. Thus we must be able to deduce the Clausius statement from the Kelvin–Planck statement and vice versa. Here are some hints on how to proceed. First, assume the Kelvin–Planck statement to be true. Temporarily, suppose the Clausius statement to be false. Let an anti-Clausius device (a cyclic device that absorbs heat from a cold reservoir and delivers an equal amount of heat to a hot reservoir with no other effects) be coupled with a heat engine that uses the same pair of reservoirs. Show that if the heat engine is run so that it discards heat to the cold reservoir at the same rate the anti-Clausius device removes heat from this reservoir, we have a device that violates the Kelvin–Planck statement. Hence the existence of an anti-Clausius device is incompatible with the truth of the Kelvin–Planck statement and the Clausius statement has been deduced from the Kelvin–Planck statement. To deduce the Kelvin–Planck statement from the Clausius statement, assume the Clausius statement to be true, and couple an anti-Kelvin–Planck heat engine with a heat pump.]

3.4 The slopes of the lines of the Carnot cycle of Fig. 3.4*b* are appropriate for a gas as working substance. Suppose we use as working substance a system that consists of a pure liquid (e.g., water) in equilibrium with its pure vapor. Further suppose that all pressures and temperatures of the cycle are such as to maintain a two-phase system. Sketch on a *P-V* diagram the appearance of a Carnot cycle of this two-phase system. (You may want to review the concept of vapor pressure by skimming Sec. 7.2.)

3.5 Let the Carnot-cycle reversible heat engine A absorb heat q_3 per cycle from a reservoir at τ_3 and discard heat $-q_{2A}$ per cycle to a reservoir at τ_2. Let Carnot engine B absorb heat q_{2B} per cycle from the reservoir at τ_2 and discard heat $-q_1$ per cycle to a reservoir at τ_1. Further, let $-q_{2A} = q_{2B}$, so that engine B absorbs an amount of heat from the τ_2 reservoir equal to the heat deposited in this reservoir by engine A. Show that

$$g(\tau_2, \tau_3)g(\tau_1, \tau_2) = -q_1/q_3$$

where the function g is defined as $1 - e_{rev}$. The heat reservoir at τ_2 can be omitted and the combination of engines A and B can be viewed as a single Carnot engine operating between τ_3 and τ_1; hence $g(\tau_1, \tau_3) = -q_1/q_3$. Therefore

$$g(\tau_1, \tau_2) = \frac{g(\tau_1, \tau_3)}{g(\tau_2, \tau_3)} \tag{3.60}$$

Since τ_3 does not appear on the left side of (3.60), it must cancel out of the numerator and denominator on the right side; after τ_3 is canceled, the numerator takes the form $\phi(\tau_1)$ and the denominator takes the form $\phi(\tau_2)$, where ϕ is some function; we then have

$$g(\tau_1, \tau_2) = \frac{\phi(\tau_1)}{\phi(\tau_2)} \tag{3.61}$$

which is the desired result, Eq. (3.20) [A more rigorous derivation of (3.61) from (3.60) is given in *Denbigh*, p. 30.]

3.6 Willard Rumpson (in later life Baron Melvin, K.C.B.) defined a temperature scale with the function ϕ in (3.21) as "take the square root" and with the water triple-point temperature defined as 200.00°M. (*a*) What is the temperature of the steam point on the Melvin scale? (*b*) What is the temperature of the ice point on the Melvin scale?

3.7 Which of these cyclic integrals must vanish? (*a*) $\oint P\,dV$; (*b*) $\oint (P\,dV + V\,dP)$; (*c*) $\oint V\,dV$; (*d*) $\oint dq_{\text{rev}}/T$; (*e*) $\oint H\,dT$; (*f*) $\oint dU$.

3.8 For each of the processes of Probs. 2.19 and 2.20*a*, state whether ΔS is negative, zero, or positive.

3.9 Find ΔS when 2.00 moles of O_2 is heated from 27 to 127°C with P held fixed at 1.00 atm. Use $\bar{C}_P$ from Prob. 2.21.

3.10 Find ΔS for the conversion of 1.00 mole of ice at 0°C and 1.00 atm to 1.00 mole of water vapor at 100°C and 0.50 atm. Use data from Prob. 2.24.

3.11 Find ΔS for each process of Prob. 2.25.

3.12 Find ΔS for the process of Prob. 2.14.

3.13 Find ΔS for the process of Prob. 2.29.

3.14 Find ΔS for each process of Prob. 2.30 and for the entire cycle.

3.15 Find ΔS for the conversion of 10.0 g of supercooled water at -10°C and 1.00 atm to ice at -10°C and 1.00 atm. Average c_P values for ice and supercooled water in the range 0 to -10°C are 0.50 and 1.01 cal/(g °C), respectively. See also Prob. 2.24.

3.16 State whether each of q, w, ΔU, and ΔS is negative, zero, or positive for each step of a Carnot cycle of a perfect gas.

3.17 After 200 g of gold [$c_P = 0.0313$ cal/(g °C)] at 120.0°C is dropped into 25.0 g of water at 10.0°C, the system is allowed to reach equilibrium in an adiabatic container. Find (*a*) the final temperature; (*b*) ΔS_{Au}; (*c*) ΔS_{H_2O}; (*d*) ΔS_{univ}.

3.18 Calculate ΔS for the mixing of 10.0 g of He at 120°C and 1.50 bars with 10.0 g of O_2 at 120°C and 1.50 bars.

3.19 (*a*) What is ΔS for each step of a Carnot cycle? (*b*) What is ΔS_{univ} for each step of a Carnot cycle?

3.20 Prove the equivalence of the Kelvin–Planck statement and the entropy statement [the set-off statement after Eq. (3.46)] of the second law. *Hint:* Since the entropy statement was derived from the Kelvin–Planck statement, all we need do to show the equivalence is to assume the truth of the entropy statement and derive the Kelvin–Planck statement (or the Clausius statement, which we proved equivalent to the Kelvin–Planck statement in Prob. 3.3) from the entropy statement.

3.21 The Carnot cycle of Fig. 3.4*b* is plotted on a *P-V* diagram. Plot a Carnot cycle on a *T-S* diagram, showing the correspondence with the steps on a *P-V* diagram.

3.22 For each of the processes of Probs. 2.19 and 2.20, state whether ΔS_{univ} is negative, zero, or positive.

3.23 Consider the mixing of different perfect gases—(paragraph 10 of Sec. 3.5). Explain why we expanded each gas to the final volume V before using the reversible mixing device of Fig. 3.9.

3.24 For the gaussian probability distribution, the probability of observing a value that deviates from the mean value by at least x standard deviations is given by the following infinite series (M. L. Abramowitz and I. A. Stegun, Handbook of Mathematical Functions, *Natl. Bur. Stand. Appl. Math. Ser.* 55, 1964, pp. 931–932):

$$\frac{2}{\sqrt{2\pi}}\,e^{-x^2/2}\left(\frac{1}{x}-\frac{1}{x^3}+\frac{3}{x^5}-\cdots\right)$$

where the series is useful for reasonably large values of x. (*a*) Show that 99.7 percent of observations lie within ± 3 standard deviations from the mean. (*b*) Calculate the probability of a deviation $\geq 10^6$ standard deviations.

3.25 If the probability of observing a certain event in a single trial is p, then clearly the probability of not observing it in one trial is $1 - p$. The probability of not observing it in n independent trials is then $(1 - p)^n$; the probability of observing it at least once in n independent trials is $1 - (1 - p)^n$. (*a*) Use these ideas to verify the calculation of Eq. (3.59). (*b*) How many times must a coin be tossed to reach a 99 percent probability of observing at least one head?

3.26 What is the relevance to thermodynamics of the following refrain from the Gilbert and Sullivan operetta *H.M.S. Pinafore*? "What, never? No, never! What, *never*? Hardly ever!"

3.27 Prove that two reversible adiabats cannot intersect on a *P-V* diagram. *Hint:* Assume they do intersect and show this leads to a violation of the second law.

3.28 In the tropics, water at the surface of the ocean is warmer than water well below the surface. Someone proposes to draw heat from the warm surface water, convert part of it to work, and discard the remainder to cooler water below the surface. Does this proposal violate the second law?

3.29 Express $\Delta S/(n_a + n_b)$ in Eq. (3.39) as a function of the mole fractions x_a and x_b.

3.30 A system consists of 1.00 mg of ClF gas. A mass spectrometer separates the gas into the species ^{35}ClF and ^{37}ClF. Calculate ΔS. (Isotopic abundances: $^{19}F = 100$ percent; $^{35}Cl = 75.5$ percent; $^{37}Cl = 24.5$ percent.)

3.31 Use (3.17) to show that it is impossible to attain the absolute zero of temperature.

3.32 Use sketches of w_{by} for each step of a Carnot cycle to show that w_{by} for the cycle equals the area enclosed by the curve of the cycle on a P-V plot.

MATERIAL EQUILIBRIUM

4.1 MATERIAL EQUILIBRIUM

The second law of thermodynamics allows us to determine whether a given process is possible. A process that decreases S_{univ} is impossible; one that increases S_{univ} is possible (and irreversible). Reversible processes have $\Delta S_{\text{univ}} = 0$; such processes are possible in principle but difficult to achieve in practice. Our aim is to use this entropy criterion to derive specific criteria for material equilibrium.

We subdivide *material equilibrium* into (*a*) *reaction equilibrium*, which is equilibrium with respect to conversion of one set of chemical species to another set, and (*b*) *phase equilibrium*, which is equilibrium with respect to transport of matter between phases of the system (without conversion of one species to another). The general condition for material equilibrium will be derived in Sec. 4.6. This condition will be applied to phase equilibrium in Sec. 4.7 and to reaction equilibrium in Sec. 4.8. Later chapters will go into the details of reaction equilibrium and phase equilibrium.

The initial application of the laws of thermodynamics to material equilibrium is largely the work of Josiah Willard Gibbs (1839–1903). Gibbs received his doctorate in engineering from Yale in 1863 with a thesis on gear design. (Gibbs's doctorate was the first awarded in engineering in the United States. The first American Ph.D. in any field was awarded in 1861.) From 1866 to 1869 Gibbs studied mathematics and physics in Europe. In 1871 he was appointed Professor of Mathematical Physics, without salary, at Yale; at that time his only published work was a railway brake patent. In 1876–1878 he published in the *Transactions of the Connecticut Academy of Arts and Sciences* a 300-page monograph titled "On the Equilibrium of Heterogeneous Substances." This work used the first and second laws of thermodynamics to deduce the conditions of material equilibrium. Gibbs's second major contribution was his book *Elementary Principles in Statistical Mechanics* (1902), which laid much of the foundations of statistical mechanics. Gibbs also developed vector analysis. Gibbs's life was rather uneventful; he never married and lived in his family's house until his death. Ostwald wrote of Gibbs: "To physical chemistry he gave form and content for a hundred years." Planck wrote that Gibbs' "... name ... will ever be reckoned among the most renowned theoretical physicists of all times ...."

4.2 THERMODYNAMIC PROPERTIES OF A NONEQUILIBRIUM SYSTEM

In this chapter we shall consider systems in which chemical reactions or transport of matter from one phase to another are occurring. Since such systems are not in thermodynamic equilibrium, we first examine to what extent we can ascribe definite values of thermodynamic properties to nonequilibrium systems.

Consider a system that is not in material equilibrium. We shall assume that the system is in mechanical and thermal equilibrium, so that P and T are uniform throughout the system. We further assume that within each phase of the system the composition is uniform. We thus assume that the rate of diffusion within a phase is rapid compared with the rate of transport of components from one phase to another. We also assume that any chemical reactions do not occur at an explosive rate, which would destroy thermal and mechanical equilibrium. Our ultimate interest is the *position* of equilibrium in the system; thermodynamics can give no information on the *rate* of a process. Since the final equilibrium position is independent of the rates of the various processes (provided these rates are nonzero), we are free to make convenient assumptions about the relative rates of processes.

We first consider systems where there is a lack of phase equilibrium. To be concrete, consider a system that initially consists of a very large crystal of NaCl separated by a partition from an unsaturated solution of NaCl in water (Fig. 4.1), with P and T held fixed. Since U and S are extensive, we have

$$U = U_{\text{soln}} + U_{\text{NaCl}}, \qquad S = S_{\text{soln}} + S_{\text{NaCl}} \tag{4.1}$$

Now the frictionless partition is removed. It requires only an infinitesimal force to do this, and the removal is done reversibly and adiabatically. q and w for the removal of the partition are zero; therefore ΔU and ΔS are zero for its removal. Thus, immediately after the removal of the partition, Eq. (4.1) still holds. The instant after the partition is removed, we no longer have phase equilibrium, since the solid NaCl starts to dissolve in the unsaturated solution. Despite this lack of phase equilibrium, we have shown that it is still meaningful to ascribe a value to U and a value to S for the system, namely, the values (4.1). Of course, as NaCl dissolves in the solution, the values of U and S change, but at any concentration of dissolved NaCl, we can imagine replacing the partition without changing U and S, and then we can apply (4.1) at this concentration. Hence (4.1) is valid at any stage of the solution process. Even though the system is not in phase equilibrium when the partition is absent, we can still ascribe values of U and S to it, namely, the values we would assign if the partition were present.

Now consider systems not in reaction equilibrium. Let us imagine mixing arbitrary amounts of H_2, O_2, and H_2O gases at some temperature and pressure. Provided there is no catalyst present (and the temperature is moderate), the gases

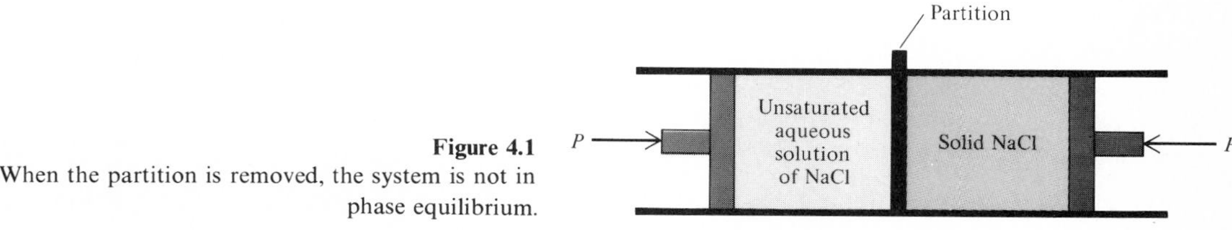

Figure 4.1
When the partition is removed, the system is not in phase equilibrium.

will not undergo any chemical reaction when mixed. We can use the first law to measure the ΔU of the mixing; also, with the aid of semipermeable membranes (Sec. 3.5), it is possible (at least in principle) to do the mixing reversibly and hence measure ΔS for the mixing process. Hence it makes sense to ascribe definite values of U and S to the mixture for any composition whatever. However, the mixture is not necessarily at reaction equilibrium; if we add the appropriate catalyst, we shall find that reaction ensues, changing the mixture's composition. At any point during the reaction, we can withdraw the catalyst, stopping the reaction. At this new composition, we can ascribe new values to U and S of the mixture. (These values can be determined by measurements done in a reversible separation process.)

We conclude that we can ascribe values of U and S to a system that is in mechanical and thermal equilibrium and has a uniform composition in each phase, even though the system is not in material equilibrium. Of course, such systems also have well-defined values of P, V, and T.

We can go even further. Thus suppose a system lacks thermal equilibrium and has a temperature gradient from one end to the other. We can imagine the system cut into "infinitesimal" slices such that the temperature within each slice is essentially constant. We can then assign values of thermodynamic variables (T, P, V, U, S, composition) to each slice. The total S and U of the system is the sum of the values for the slices. (Since thermodynamics is a macroscopic science, each infinitesimal slice must contain enough molecules to make it meaningful to assign it a macroscopic property like temperature. The number of molecules in each slice should be much, much greater than 1 but much, much less than 10^{23}.)

Suppose a system lacks a uniform composition in each phase. (Thus we might have a concentration gradient in a solution as NaCl dissolves in water.) We can imagine the system divided into infinitesimal parts such that the concentrations are essentially constant within each part. Each tiny part can then be assigned values of thermodynamic variables, and the system's U and S is the sum of the internal energies and entropies of the parts.

In this discussion, we have perhaps crossed the border between classical (equilibrium) thermodynamics and irreversible (nonequilibrium) thermodynamics.

4.3 ENTROPY AND EQUILIBRIUM

Consider an *isolated* system that is not at material equilibrium. The spontaneous chemical reactions and/or transport of matter between phases that are occurring in this nonequilibrium system are irreversible processes that increase the entropy. These processes continue until the system's entropy is maximized. Once S is maximized, any further processes can only decrease S, which would violate the second law. Thus, the criterion for equilibrium in an *isolated* system is the maximization of the system's entropy S.

When we deal with material equilibrium in a closed system, the system is ordinarily not isolated; instead it can exchange heat (and work) with its surroundings. Under these conditions, we can take the system itself *plus* the surroundings with which it interacts to constitute an isolated system, and *the condition for chemical equilibrium in the system is then the maximization of the total entropy of the system plus its surroundings:*

$$S_{\text{syst}} + S_{\text{surr}} \text{ a maximum at equilib.} \tag{4.2}$$

Thus changes (chemical reactions, transport of matter between phases) continue in a system until $S_{syst} + S_{surr}$ has been maximized.

It is usually most convenient to deal with properties of the system and not have to worry about changes in the thermodynamic properties of the surroundings as well. Thus, although the criterion (4.2) for material equilibrium is perfectly valid and general, it will be more useful to have a criterion for material equilibrium that refers only to thermodynamic properties of the system itself. Since S_{syst} is a maximum at equilibrium only for an isolated system, consideration of the entropy of a system does not furnish us with an equilibrium criterion. We must look for another system state function to find the equilibrium criterion.

Reaction equilibrium is ordinarily studied under one of two conditions. For reactions that involve gases, the chemicals are put in a container of fixed volume, and the system is allowed to reach equilibrium at constant T and V in a constant-temperature bath. For reactions in liquid solutions, the system is held at atmospheric pressure and allowed to reach equilibrium at constant T and P.

To find equilibrium criteria for these conditions, we consider the situation of Fig. 4.2. The system at temperature T is placed in a bath also at T. The system and surroundings are isolated from the rest of the world. The system is not in material equilibrium but is in mechanical and thermal equilibrium. The surroundings are in material, mechanical, and thermal equilibrium. System and surroundings can exchange energy (as heat and work) but not matter. Let chemical reaction or transport of matter between phases (or both) be occurring in the system (at rates small enough to maintain thermal and mechanical equilibrium). Let heat dq_{syst} flow into the system as a result of the changes that occur in the system during an infinitesimal time period. (For example, if an endothermic chemical reaction is occurring, dq_{syst} is positive.) Since system plus surroundings are isolated from the rest of the world, we have $dq_{surr} = -dq_{syst}$ or

$$dq_{syst} + dq_{surr} = 0 \tag{4.3}$$

The chemical reaction or matter transport within the nonequilibrium system is irreversible. Hence (3.45) gives

$$dS_{univ} = dS_{syst} + dS_{surr} > 0 \tag{4.4}$$

for the process. (Recall from Sec. 4.2 that one can meaningfully assign an entropy to a system that is not in material equilibrium.) The surroundings are in thermodynam-

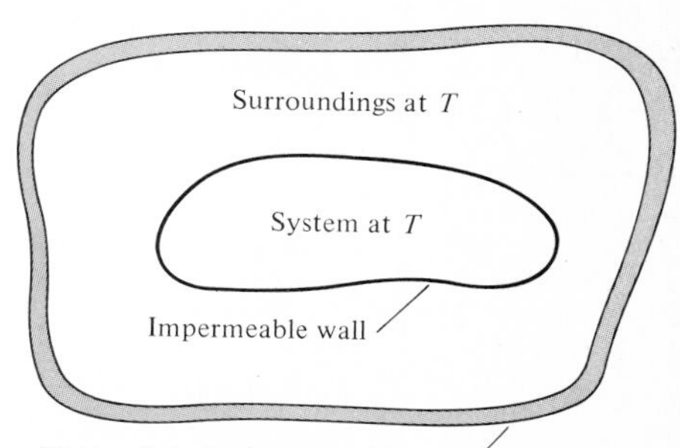

Figure 4.2
A system that is in mechanical and thermal equilibrium but not in material equilibrium.

ic equilibrium throughout the process; hence, as far as the surroundings are concerned, the heat transfer is reversible; therefore

$$dS_{\text{surr}} = dq_{\text{surr}}/T \tag{4.5}$$

However, the system is not in thermodynamic equilibrium, and the process involves an irreversible change in the system; therefore $dS_{\text{syst}} \neq dq_{\text{syst}}/T$. Equations (4.3) to (4.5) give $dS_{\text{syst}} > -dS_{\text{surr}} = -dq_{\text{surr}}/T = dq_{\text{syst}}/T$. Therefore

$$dS_{\text{syst}} > dq_{\text{syst}}/T \tag{4.6}$$

$$dS > dq_{\text{irrev}}/T \qquad \text{closed syst. in therm. and mech. equilib.} \tag{4.7}$$

where we dropped the subscript syst from S and q since, by convention, unsubscripted symbols refer to the system.

When the system has reached material equilibrium, any infinitesimal process represents a change from a system at equilibrium to one infinitesimally close to equilibrium and hence is a reversible process. Thus, at material equilibrium we have

$$dS = dq_{\text{rev}}/T \tag{4.8}$$

Combining (4.7) and (4.8), we have

$$dS \geq \frac{dq}{T} \qquad \text{material change, closed syst. in mech. and therm. equilib.} \tag{4.9}$$

where the equality sign holds only when the system is in material equilibrium.

The first law for a closed system is $dq = dU - dw$. But (4.9) gives $dq \leq T\,dS$. Hence for a closed system in mechanical and thermal equilibrium, we have $dU - dw \leq T\,dS$, or

$$dU \leq T\,dS + dw \qquad \text{material change, closed syst. in mech. and therm. equilib.} \tag{4.10}$$

where the equality sign applies only at material equilibrium.

4.4 THE GIBBS AND HELMHOLTZ FUNCTIONS

We now use (4.10) to deduce criteria for material equilibrium in terms of state functions of the system. We first consider material equilibrium in a system at constant T and V. Here $dV = 0$ and $dT = 0$ throughout the irreversible approach toward equilibrium. The inequality (4.10) involves dS and dV (since $dw = -P\,dV$ for P-V work only). To introduce dT into (4.10), we add and subtract $S\,dT$ on the right. [Note that $S\,dT$ has the dimensions of entropy times temperature, the same dimensions as the term $T\,dS$ that appears in (4.10).] We have

$$dU \leq T\,dS + S\,dT - S\,dT + dw \tag{4.11}$$

The differential relation $d(uv) = u\,dv + v\,du$ of Sec. 1.6 gives $d(TS) = T\,dS + S\,dT$, and Eq. (4.11) becomes

$$dU \leq d(TS) - S\,dT + dw \tag{4.12}$$

The differential relation $d(u + v) = du + dv$ of Sec. 1.6 gives $dU - d(TS) = d(U - TS)$, and (4.12) becomes

$$d(U - TS) \leq -S\,dT + dw \tag{4.13}$$

If the system can do only *P-V* work, then $dw = -P\,dV$ (we use dw_{rev} since we are assuming mechanical equilibrium); we have

$$d(U - TS) \leq -S\,dT - P\,dV \tag{4.14}$$

At constant T and V, (4.14) becomes

$$d(U - TS) \leq 0 \qquad \text{const. } T \text{ and } V\text{, closed syst. in therm. and mech. eq., } P\text{-}V \text{ work only} \tag{4.15}$$

where the equality sign holds at material equilibrium.

Thus, for a closed system held at constant T and V, the state function $U - TS$ continually decreases during the spontaneous, irreversible processes of chemical reaction and matter transport between phases until material equilibrium is reached. At material equilibrium, $d(U - TS) = 0$, and $U - TS$ has reached a minimum. Any spontaneous change (at constant T and V) away from equilibrium (in either direction) would mean an increase in $U - TS$, which, working back through the preceding equations, would mean a decrease in $S_{univ} = S_{syst} + S_{surr}$; this decrease would violate the second law. The approach to and achievement of material equilibrium is a consequence of the second law.

Thus the criterion for material equilibrium in a closed system capable of doing only *P-V* work and held at constant T and V is the minimization of the system's state function $U - TS$. This state function is called the *Helmholtz free energy*, the *Helmholtz energy*, the *Helmholtz function*, or the *work function* and is symbolized by A:

$$A \equiv U - TS \tag{4.16}$$

Now consider material equilibrium for constant T and P conditions, $dP = 0$, $dT = 0$. To introduce dP and dT into (4.10) with $dw = -P\,dV$, we add and subtract $S\,dT$ and $V\,dP$:

$$dU \leq T\,dS + S\,dT - S\,dT - P\,dV + V\,dP - V\,dP$$

$$dU \leq d(TS) - S\,dT - d(PV) + V\,dP$$

$$d(U + PV - TS) \leq -S\,dT + V\,dP$$

$$d(H - TS) \leq -S\,dT + V\,dP \tag{4.17}$$

Therefore, for a material change at constant T and P in a closed system in mechanical and thermal equilibrium and capable of doing only *P-V* work, we have

$$d(H - TS) \leq 0 \tag{4.18}$$

where the equality sign holds at material equilibrium.

Thus, the state function $H - TS$ continually decreases during material changes at constant T and P until equilibrium is reached. The criterion for material equilibrium in a closed system doing *P-V* work only at constant T and P is the minimization of the system's state function $H - TS$. This state function is called the *Gibbs function*, the *Gibbs energy*, the *Gibbs free energy*, or the *free enthalpy* and is symbolized by G:

$$G \equiv H - TS \equiv U + PV - TS \tag{4.19}$$

G decreases during the approach to equilibrium at constant T and P, reaching a minimum at equilibrium.

Both A and G have units of energy. However, they are not energies in the sense of being conserved. $G_{syst} + G_{surr}$ need not be constant in a process, nor need $A_{syst} + A_{surr}$ remain constant. Note that A and G are defined for any system to which meaningful values of U, T, S, P, V can be assigned, not just for systems held at constant T and V or constant T and P.

Summarizing, we have shown that in a closed system capable of doing only P-V work, the constant-T-and-V equilibrium condition is the minimization of the Helmholtz function A:

$$dA = 0 \tag{4.20}$$

whereas the constant-T-and-P equilibrium condition is the minimization of the Gibbs function G:

$$dG = 0 \tag{4.21}$$

The names "work function" and "Gibbs free energy" arise for the following reasons. Let us drop the restriction that only P-V work be performed. From (4.13) we have for a closed system in thermal and mechanical equilibrium

$$dA \leq -S\,dT + dw \tag{4.22}$$

For a constant-temperature process in such a system,

$$dA \leq dw \qquad \text{const. } T \tag{4.23}$$

For a finite isothermal process, $\Delta A \leq w$. Our convention is that w is the work done *on* the system. The work w_{by} done *by* the system on its surroundings is $w_{by} = -w$, and $\Delta A \leq -w_{by}$ for an isothermal process. Multiplication of an inequality by -1 reverses the direction of the inequality; hence

$$w_{by} \leq -\Delta A \qquad \text{const. } T \tag{4.24}$$

The name "work function" (*Arbeitsfunktion*) for A arises from (4.24). The work done by the system in an isothermal process is less than or equal to the negative of the change in the state function A. The equality sign in (4.24) holds for a reversible process; moreover, $-\Delta A$ is a fixed quantity for a given change of state. Hence the maximum work output by a closed system for a given *isothermal* change of state is obtained when the change is carried out reversibly.

From $G = A + PV$, we have $dG = dA + P\,dV + V\,dP$, and use of (4.22) gives for a closed system in thermal and mechanical equilibrium

$$dG \leq -S\,dT + dw + P\,dV + V\,dP \tag{4.25}$$

For a process at constant T and P in such a system

$$dG \leq dw + P\,dV \qquad \text{const. } T, P \tag{4.26}$$

Let us divide the work w into P-V work and non-P-V work, designating the latter by w_{other}. (The most common kind of w_{other} is electrical work.) If the P-V work is done in a mechanically reversible manner, then

$$dw = -P\,dV + dw_{\text{other}} \tag{4.27}$$

and (4.26) becomes

$$dG \le dw_{\text{other}} \qquad \text{or} \qquad \Delta G \le w_{\text{other}} \qquad \text{const. } T, P \tag{4.28}$$

For a reversible change, the equality sign holds, and $\Delta G = w_{\text{other}}$. The non-$P$-$V$ work done *by* the system in a constant-T-and-P reversible process is $w_{\text{by, other}} = -\Delta G$. In many cases (e.g., a battery) the P-V expansion work is not useful work, but $w_{\text{by, other}}$ is the useful work output. In a reversible change at constant T and P, the system's G decreases by an amount equal to the "useful" work output $w_{\text{by, other}}$. Hence the name "free energy." (Of course, for a system with P-V work only, $dw_{\text{other}} = 0$ and $dG = 0$ for a reversible, isothermal, isobaric process.)

4.5 THERMODYNAMIC RELATIONS FOR A SYSTEM OF CONSTANT COMPOSITION

In the last section we introduced two new thermodynamic state functions A and G. We shall apply the conditions (4.20) and (4.21) for material equilibrium in Sec. 4.6. Before doing so, we investigate the properties of A and G. In fact, in this section we shall consider the broader question of the thermodynamic relations between all state functions in systems of constant composition. This will allow us to consolidate and summarize our thermodynamic development. This section is rather long, and can best be studied in small portions. Although the many equations may look complicated, once the underlying pattern is grasped, things will appear relatively straightforward.

All thermodynamic state-function relations can be derived from six fundamental equations. The first law for a closed system is $dU = dq + dw$. If only P-V work is possible, and if the work is done reversibly, then $dw = dw_{\text{rev}} = -P\,dV$. For a reversible process, $dq = dq_{\text{rev}} = T\,dS$. Hence, under these conditions, $dU = T\,dS - P\,dV$. This is the first fundamental equation; it combines the first and second laws. The next three fundamental equations are the definitions (2.76), (4.16), and (4.19) of H, A, and G. Finally, we have the relations (2.81) to (2.83) for C_V and C_P: $C_V = dq_V/dT = (\partial U/\partial T)_V$ and $C_P = dq_P/dT = (\partial H/\partial T)_P$. The six fundamental equations are thus

$$dU = T\,dS - P\,dV \qquad \text{closed syst., rev. proc., } P\text{-}V \text{ work only} \tag{4.29*}$$

$$H \equiv U + PV \tag{4.30*}$$

$$A \equiv U - TS \tag{4.31*}$$

$$G \equiv H - TS \tag{4.32*}$$

$$C_V = \left(\frac{\partial U}{\partial T}\right)_V \qquad \text{closed syst., rev. proc., } P\text{-}V \text{ work only} \tag{4.33*}$$

$$C_P = \left(\frac{\partial H}{\partial T}\right)_P \qquad \text{closed syst., rev. proc., } P\text{-}V \text{ work only} \tag{4.34*}$$

The heat capacities C_V and C_P have alternative expressions that are also fundamental. Consider a reversible flow of heat accompanied by a temperature change dT. By definition, $C_X = dq_X/dT$, where X is the variable (P or V) held constant. But $dq_{\text{rev}} = T\,dS$, and we have $C_X = T\,dS/dT$, where dS/dT is for constant X. Putting X equal to V and P, we have

$$C_V = T\left(\frac{\partial S}{\partial T}\right)_V, \qquad C_P = T\left(\frac{\partial S}{\partial T}\right)_P \tag{4.35*}$$

Equation (4.29) applies to a reversible process in a closed system. Let us consider processes that change the system's composition. There are two ways the composition can change. First, one can add or remove one or more substances. However, the requirement of a closed system ($dU \neq dq + dw$ for an open system) rules out addition or removal of matter. Second, the composition can change by chemical reactions or by transport of matter from one phase to another in the system. The usual way of carrying out a chemical reaction is to mix the chemicals and allow them to reach equilibrium. This spontaneous chemical reaction is irreversible (since the system passes through nonequilibrium states). The requirement of reversibility ($dq \neq T\,dS$ for an irreversible chemical change) rules out a chemical reaction as ordinarily conducted. Likewise, if we put several phases together and allow them to reach equilibrium, we have an irreversible composition change. For example, if we throw a handful of salt into water, the solution process consists of nonequilibrium states (except for the final state) and is irreversible. The equation $dU = T\,dS - P\,dV$ does not apply to such irreversible composition changes in a closed system.

We can, if we like, carry out a composition change reversibly in a closed system. If we start with a system that is initially in material equilibrium and reversibly vary the temperature or pressure of the system, we generally get a shift in the equilibrium position, and this shift is reversible. For example, if we have a closed system that consists of an NaCl crystal in equilibrium with an aqueous solution of NaCl at 22°C and 1 atm and we slowly and reversibly warm the system to 30°C, the number of moles of NaCl in the liquid phase increases at the expense of the solid phase and this composition change is reversible since the system passed through equilibrium states only: at every stage of the process, solid–solution equilibrium was maintained. Likewise, if we have a mixture of N_2, H_2, and NH_3 (together with a catalyst) and we reversibly vary T or P, the position of chemical-reaction equilibrium shifts reversibly. For such reversible composition changes, the equation $dU = T\,dS - P\,dV$ does apply.

In the next section, we shall find the equation for dU that applies to an irreversible composition change, but for now we shall not consider irreversible changes. Equations in this section will be restricted in applicability to *reversible* processes in *closed* systems. Most commonly, we shall consider reversible changes in a closed system of *constant composition* (but the equations of this section also apply to processes in which the composition of the closed system changes *reversibly*).

We now derive an expression for dH that corresponds to (4.29) for dU. Equations (4.30) and (4.29) give

$$dH = d(U + PV) = dU + d(PV) = dU + P\,dV + V\,dP$$

$$= (T\,dS - P\,dV) + P\,dV + V\,dP$$

$$dH = T\,dS + V\,dP \tag{4.36}$$

Similarly, $dA = d(U - TS) = dU - T\,dS - S\,dT = T\,dS - P\,dV - T\,dS - S\,dT$
$= -S\,dT - P\,dV;$

also, $dG = d(H - TS) = dH - T\,dS - S\,dT = T\,dS + V\,dP - T\,dS - S\,dT$
$= -S\,dT + V\,dP,$

where (4.36) was used.

Collecting the expressions for dU, dH, dA, and dG, we have

$$dU = T\,dS - P\,dV \qquad (4.37)^*$$

$$dH = T\,dS + V\,dP \qquad (4.38)$$

$$dA = -S\,dT - P\,dV \qquad (4.39)$$

$$dG = -S\,dT + V\,dP \qquad (4.40)^*$$

closed syst., rev. proc., P-V work only

These four equations are the *Gibbs equations.* (As we shall see, many equations in chemical thermodynamics bear Gibbs' name.) The first of these can be written down from the first law $dU = dq + dw$ and knowledge of the expressions for dw_{rev} and dq_{rev}. The other three can be quickly derived from the first by use of the definitions of H, A, and G. Thus they need not be memorized. The expression for dG is used so frequently, however, that it saves time to commit it to memory.

Our aim is to be able to express any thermodynamic property of a system of fixed composition in terms of easily measured physical quantities. The power of thermodynamics is that it gives expressions for quantities difficult to measure in terms of easily measured properties. The three easily measured properties most commonly used for this purpose are

$$C_P(T, P), \qquad \alpha(T, P) \equiv \frac{1}{V}\left(\frac{\partial V}{\partial T}\right)_P, \qquad \kappa(T, P) \equiv -\frac{1}{V}\left(\frac{\partial V}{\partial P}\right)_T \qquad (4.41)^*$$

Since these are state functions, they are functions of T, P, and composition; we are considering mainly constant-composition systems, so we omit the composition dependence. [Note that α and κ can be found from the equation of state $V = V(T, P)$ if this is known.]

To relate a desired property to C_P, α, and κ, we use the fundamental equations (4.29) to (4.35) and the Gibbs equations (4.37) to (4.40), plus various mathematical partial-derivative identities. Before proceeding, there is one further partial-derivative identity we shall need. Let z be a function of x and y: $z = z(x, y)$. Equation (1.30) gives

$$dz = \left(\frac{\partial z}{\partial x}\right)_y dx + \left(\frac{\partial z}{\partial y}\right)_x dy \equiv M\,dx + N\,dy \qquad (4.42)$$

where we defined M and N as

$$M(x, y) \equiv (\partial z/\partial x)_y, \qquad N(x, y) \equiv (\partial z/\partial y)_x \qquad (4.43)$$

From Eq. (1.36), the order of partial differentiation is immaterial:

$$\frac{\partial}{\partial y}\left(\frac{\partial z}{\partial x}\right) = \frac{\partial}{\partial x}\left(\frac{\partial z}{\partial y}\right) \qquad (4.44)$$

Hence if $dz = M\,dx + N\,dy$, Eqs. (4.42) to (4.44) give

$$\left(\frac{\partial M}{\partial y}\right)_x = \left(\frac{\partial N}{\partial x}\right)_y \tag{4.45*}$$

Equation (4.45) is the *Euler reciprocity relation.*

The Gibbs equation $dU = T\,dS - P\,dV$ implies that U is being considered as a function of the variables S and V. From $U = U(S, V)$, we have

$$dU = \left(\frac{\partial U}{\partial S}\right)_V dS + \left(\frac{\partial U}{\partial V}\right)_S dV \tag{4.46}$$

Since dS and dV are arbitrary and independent of each other, comparison of (4.46) with (4.37) gives

$$\left(\frac{\partial U}{\partial S}\right)_V = T, \qquad \left(\frac{\partial U}{\partial V}\right)_S = -P \tag{4.47}$$

A quick way to get these two equations is to first put $dV = 0$ in $dU = T\,dS - P\,dV$ to give $(\partial U/\partial S)_V = T$ and then put $dS = 0$ in $dU = T\,dS - P\,dV$ to give $(\partial U/\partial V)_S = -P$. The other three Gibbs equations (4.38) to (4.40) give in a similar manner

$$\left(\frac{\partial H}{\partial S}\right)_P = T, \qquad \left(\frac{\partial H}{\partial P}\right)_S = V \tag{4.48}$$

$$\left(\frac{\partial A}{\partial T}\right)_V = -S, \qquad \left(\frac{\partial A}{\partial V}\right)_T = -P \tag{4.49}$$

$$\left(\frac{\partial G}{\partial T}\right)_P = -S, \qquad \left(\frac{\partial G}{\partial P}\right)_T = V \tag{4.50}$$

We now apply the Euler reciprocity relation (4.45). Consider the first Gibbs equation (4.37): $dU = T\,dS - P\,dV = M\,dx + N\,dy$, where we set $M = T$, $N = -P$, $x = S$, and $y = V$. The Euler relation $(\partial M/\partial y)_x = (\partial N/\partial x)_y$ gives $(\partial T/\partial V)_S = [\partial(-P)/\partial S]_V = -(\partial P/\partial S)_V$.

Application of the Euler relation to the other three Gibbs equations yields three more thermodynamic relations. We find (Prob. 4.1)

$$\left(\frac{\partial T}{\partial V}\right)_S = -\left(\frac{\partial P}{\partial S}\right)_V, \qquad \left(\frac{\partial T}{\partial P}\right)_S = \left(\frac{\partial V}{\partial S}\right)_P \tag{4.51}$$

$$\left(\frac{\partial S}{\partial V}\right)_T = \left(\frac{\partial P}{\partial T}\right)_V, \qquad \left(\frac{\partial S}{\partial P}\right)_T = -\left(\frac{\partial V}{\partial T}\right)_P \tag{4.52}$$

These four equations are the *Maxwell relations* (after James Clerk Maxwell, one of the greatest of nineteenth-century physicists). The first two Maxwell relations do not involve α, κ, or C_P and are of little practical value. However, the last two are quite useful since they relate the isothermal pressure and volume variations of entropy to measurable properties.

We now find the dependence of U, H, S, and G on the variables of the system. The most common independent variables are T and P. We shall relate the temperature and pressure variations of U, H, S, and G to the directly measurable properties C_P, α, and κ.

Pressure dependence of U. We want $(\partial U/\partial P)_T$. The Gibbs equation (4.37) gives the general expression for dU as $dU = T\,dS - P\,dV$. The partial derivative $(\partial U/\partial P)_T$ corresponds to an isothermal process; for such a process (4.37) reads

$$dU_T = T\,dS_T - P\,dV_T \tag{4.53}$$

where the T subscripts indicate that the infinitesimal changes dU, dS, and dV are for a constant-T process. Since $(\partial U/\partial P)_T$ is wanted, we divide (4.53) by dP_T, the infinitesimal pressure change at constant T, to give

$$\frac{dU_T}{dP_T} = T\frac{dS_T}{dP_T} - P\frac{dV_T}{dP_T} \tag{4.54}$$

From the definition of a partial derivative, the quantity dU_T/dP_T is the partial derivative $(\partial U/\partial P)_T$, and (4.54) reads

$$\left(\frac{\partial U}{\partial P}\right)_T = T\left(\frac{\partial S}{\partial P}\right)_T - P\left(\frac{\partial V}{\partial P}\right)_T \tag{4.55}$$

The second Maxwell relation in (4.52) gives $(\partial S/\partial P)_T = -(\partial V/\partial T)_P$, so

$$\left(\frac{\partial U}{\partial P}\right)_T = -T\left(\frac{\partial V}{\partial T}\right)_P - P\left(\frac{\partial V}{\partial P}\right)_T = -TV\alpha + PV\kappa \tag{4.56}$$

which is the desired expression for $(\partial U/\partial P)_T$ in terms of easily measured properties.

Temperature dependence of U. We want $(\partial U/\partial T)_P$. Starting with the Gibbs equation (4.37) for dU, imposing the condition of constant P, and dividing by dT_P, we get

$$\left(\frac{\partial U}{\partial T}\right)_P = T\left(\frac{\partial S}{\partial T}\right)_P - P\left(\frac{\partial V}{\partial T}\right)_P \tag{4.57}$$

Use of the fundamental relation (4.35) gives the desired result:

$$\left(\frac{\partial U}{\partial T}\right)_P = C_P - P\left(\frac{\partial V}{\partial T}\right)_P = C_P - PV\alpha \tag{4.58}$$

Temperature dependence of H. The fundamental equation (4.34) is the desired relation: $(\partial H/\partial T)_P = C_P$. [Alternatively, we can start with the Gibbs equation (4.38) and impose the condition of constant P, which eliminates the $V\,dP$ term to give $dH_P = T\,dS_P$; division by dT_P gives $(\partial H/\partial T)_P = T(\partial S/\partial T)_P$; use of (4.35) then gives $(\partial H/\partial T)_P = C_P$.]

Pressure dependence of H. Imposing the condition of constant T on the Gibbs equation (4.38) and dividing by dP_T, we get $(\partial H/\partial P)_T = T(\partial S/\partial P)_T + V$. Use of the second Maxwell relation in (4.52) then gives the desired result:

$$\left(\frac{\partial H}{\partial P}\right)_T = -T\left(\frac{\partial V}{\partial T}\right)_P + V = -TV\alpha + V \tag{4.59}$$

Temperature dependence of S. The fundamental equation (4.35) for C_P is the desired relation:

$$\left(\frac{\partial S}{\partial T}\right)_P = \frac{C_P}{T} \tag{4.60}$$

Pressure dependence of S. The Euler reciprocity relation applied to $dG = -S\,dT + V\,dP$ gives

$$\left(\frac{\partial S}{\partial P}\right)_T = -\left(\frac{\partial V}{\partial T}\right)_P = -\alpha V \tag{4.61}$$

as already noted in Eq. (4.52).

Temperature and pressure dependences of G. In $dG = -S\,dT + V\,dP$, we set $dP = 0$ to get $(\partial G/\partial T)_P = -S$. In $dG = -S\,dT + V\,dP$, we set $dT = 0$ to get $(\partial G/\partial P)_T = V$. Thus [Eq. (4.50)]

$$\left(\frac{\partial G}{\partial T}\right)_P = -S, \qquad \left(\frac{\partial G}{\partial P}\right)_T = V \tag{4.62}$$

Having found how U, S, and H vary with T and P, we can use these equations to find ΔU, ΔS, and ΔH for an arbitrary process in a closed system of constant composition. Thus suppose the system goes from state (P_1, T_1) to state (P_2, T_2) by any path, including, possibly, an irreversible path. We have $S = S(T, P)$, and

$$dS = \left(\frac{\partial S}{\partial T}\right)_P dT + \left(\frac{\partial S}{\partial P}\right)_T dP = \frac{C_P}{T}\,dT - \alpha V\,dP \tag{4.63}$$

where (4.60) and (4.61) were used. Integration gives

$$\Delta S = S_2 - S_1 = \int_1^2 \frac{C_P}{T}\,dT - \int_1^2 \alpha V\,dP \tag{4.64}$$

Since C_P, α, and V each depend on both T and P, these integrals are line integrals.

Since S is a state function, ΔS is independent of the path used to connect states 1 and 2. A convenient path (Fig. 4.3) is first to hold P constant and change T from T_1 to T_2; then T is held constant and P is changed from P_1 to P_2. For step (a), $dP = 0$, and (4.64) gives

$$\Delta S_a = \int_{T_1}^{T_2} \frac{C_P}{T}\,dT \qquad \text{const. } P \tag{4.65}$$

With P held constant, C_P in (4.65) depends only on T, and we have an ordinary integral, which is easily evaluated (assuming we know how C_P varies with T). For step (b), $dT = 0$, and (4.64) gives

$$\Delta S_b = -\int_{P_1}^{P_2} \alpha V\,dP \qquad \text{const. } T \tag{4.66}$$

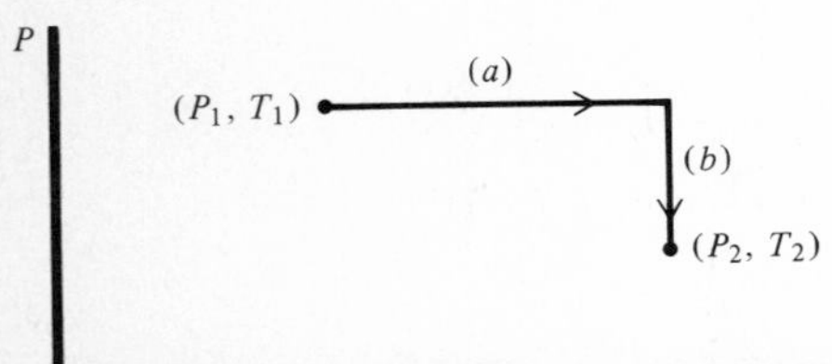

Figure 4.3
Path to calculate ΔS.

With T held constant, α and V in (4.66) are functions of P only, and the integral is an ordinary integral. ΔS for the process $(P_1, T_1) \to (P_2, T_2)$ is then the sum of ΔS_a and ΔS_b.

If the system undergoes a phase transition in a process, we must make separate allowance for this change. For example, to calculate ΔS for the isobaric heating of ice at -5°C and 1 atm to liquid water at 5°C and 1 atm, we use (4.65) to calculate the entropy change for warming the ice to 0°C and for warming the water from 0 to 5°C, but we must also add in the entropy change [Eq. (3.33)] that accompanies the melting process; during the melting process, $C_P = dq_P/dT$ is infinite, and Eq. (4.65) is not applicable.

Similar to (4.64) for ΔS, we have

$$\Delta U = \int_1^2 (C_P - PV\alpha)\, dT - \int_1^2 (TV\alpha - PV\kappa)\, dP \qquad (4.67)$$

$$\Delta H = \int_1^2 C_P\, dT + \int_1^2 (V - TV\alpha)\, dP \qquad (4.68)$$

where, again, separate allowance must be made for phase changes.

A word about calculation of ΔG. From the definition $G = H - TS$ and Eq. (2.113), we have $G_2 - G_1 = \Delta G = \Delta H - \Delta(TS) = \Delta H - T_1\, \Delta S - S_1\, \Delta T - \Delta S\, \Delta T$. However, thermodynamics does not define entropies but only gives entropy *changes*; thus S_1 is undefined in the expression for ΔG. Hence ΔG is undefined unless $\Delta T = 0$. For an *isothermal* process, $G = H - TS$ gives

$$\Delta G = \Delta H - T\, \Delta S \qquad \text{const. } T \qquad (4.69)$$

Thus ΔG is defined for an isothermal process. To calculate ΔG for an isothermal process, we first calculate ΔH and ΔS [Secs. 2.13 and 3.5, and Eqs. (4.64) and (4.68)] and then use (4.69). Alternatively, ΔG for an isothermal process that does not involve an irreversible composition change can be calculated from (4.62) as

$$\Delta G = \int_{P_1}^{P_2} V\, dP \qquad \text{const. } T \qquad (4.70)$$

Similar considerations apply for ΔA.

Consider the relative magnitudes of the temperature and pressure dependences of S and G.

We have $(\partial S/\partial T)_P = C_P/T$. Since C_P is substantial, we expect S to vary substantially with temperature.

We have $(\partial \bar{S}/\partial P)_T = -\alpha \bar{V}$. As noted in Sec. 1.7, α is somewhat larger for gases than for condensed phases. Moreover, $\bar{V}$ at usual temperatures and pressures is about 10^3 times as great for gases as for liquids and solids. Thus, the variation of entropy with pressure is small for liquids and solids but is substantial for gases.

For G, we have $(\partial \bar{G}/\partial P)_T = \bar{V}$. For solids and liquids, the molar volume is relatively small, and hence $\bar{G}$ for condensed phases is rather insensitive to moderate changes in pressure, a fact we shall use frequently. For gases, $\bar{V}$ is large, and $\bar{G}$ depends strongly on pressure.

We also have $(\partial G/\partial T)_P = -S$. However, thermodynamics does not define absolute entropies, only entropy differences; the entropy S has an arbitrary additive constant. Thus the function $(\partial G/\partial T)_P$ is without physical meaning in thermodynamics, and it is impossible to measure $(\partial G/\partial T)_P$ of a system. However, from Eq. (4.62), we can derive $(\partial \Delta G/\partial T)_P = -\Delta S$. This equation has direct physical meaning and will be useful.

Having found the temperature and pressure variations of U, H, S, and G, we now derive further thermodynamic identities.

Volume dependence of U. We want $(\partial U/\partial V)_T$. The Gibbs equation $dU = T\,dS - P\,dV$ becomes $dU_T = T\,dS_T - P\,dV_T$ for a constant-T process. Division by dV_T gives $(\partial U/\partial V)_T = T(\partial S/\partial V)_T - P$. Use of the first Maxwell equation in (4.52) and of Eq. (1.46) gives the desired result:

$$\left(\frac{\partial U}{\partial V}\right)_T = T\left(\frac{\partial P}{\partial T}\right)_V - P = \frac{\alpha T}{\kappa} - P \qquad (4.71)$$

For a liquid at room temperature, Eq. (4.71) indicates that $(\partial U/\partial V)_T$ is typically 3000 atm $\approx$ 100 cal/cm^3 (Prob. 4.8).

Ideal-gas $(\partial U/\partial V)_T$. An ideal gas was defined as one that obeys the equation of state $PV = nRT$, whereas a perfect gas obeys both $PV = nRT$ *and* $(\partial U/\partial V)_T = 0$. For an ideal gas, $(\partial P/\partial T)_V = nR/V$, and Eq. (4.71) gives $(\partial U/\partial V)_T = nRT/V - P = P - P = 0$; thus

$$(\partial U/\partial V)_T = 0 \qquad \text{ideal gas} \qquad (4.72)$$

We have proved that all ideal gases are perfect, so there is no distinction between an ideal gas and a perfect gas. (It was necessary to maintain the fiction of a distinction between the two for the logical development of thermodynamics.) From now on, we shall drop the term perfect gas.

Joule–Thomson coefficient. Equations (2.93) and (2.94) give $\mu_{JT} = -(\partial H/\partial P)_T/C_P$. Use of (4.59) gives

$$\mu_{JT} = (1/C_P)[T(\partial V/\partial T)_P - V] = (V/C_P)(\alpha T - 1) \qquad (4.73)$$

which relates μ_{JT} to α and C_P.

Ideal-gas Joule–Thomson coefficient. For an ideal gas, $\mu_{JT} = 0$. This follows from (4.73), since $(\partial V/\partial T)_P = nR/P = V/T$ for an ideal gas. Alternatively, this follows from (2.104) and the identity of ideal and perfect gases.

Heat-capacity difference. Use of (4.71) in (2.88) gives $C_P - C_V = (\alpha T/\kappa)(\partial V/\partial T)_P$. Using (4.41), we get

$$C_P - C_V = TV\alpha^2/\kappa \qquad (4.74)$$

For a condensed phase (liquid or solid), C_P is readily measured, but C_V would be difficult to measure. Equation (4.74) gives a way to calculate C_V from the measured C_P.

Note the following: (1) As $T \to 0$, $C_P \to C_V$. (2) The compressibility κ can be proved to be always positive (*Zemansky*, sec. 16-9). Hence $C_P \ge C_V$. (3) If $\alpha = 0$, then $C_P = C_V$. For liquid water at 1 atm, the molar volume reaches a maximum at 3.98°C. Hence $(\partial V/\partial T)_P = 0$ and $\alpha = 0$ for water at this temperature. Thus $C_P = C_V$ for water at 1 atm and 3.98°C.

Just for fun, let us do a calculation of C_V from C_P. For water at 30°C and 1 atm: $\alpha = 3.04 \times 10^{-4}$ K^{-1}, $\kappa = 4.52 \times 10^{-5}$ atm^{-1} $= 4.46 \times 10^{-10}$ m^2/N, $\bar{C}_P = 17.99$ cal/(mol K), $\bar{V} = 18.1$ cm^3/mol. We have

$$\frac{T\bar{V}\alpha^2}{\kappa} = \frac{(303 \text{ K})(18.1 \times 10^{-6} \text{ m}^3 \text{ mol}^{-1})(3.04 \times 10^{-4} \text{ K}^{-1})^2}{4.46 \times 10^{-10} \text{ m}^2/\text{N}}$$

$$T\bar{V}\alpha^2/\kappa = 1.14 \text{ J mol}^{-1} \text{ K}^{-1} = 0.27 \text{ cal mol}^{-1} \text{ K}^{-1}$$

$$\bar{C}_V = 17.72 \text{ cal/(mole K)} \tag{4.75}$$

Some values of $\bar{C}_P$ and $\bar{C}_V$ in cal mol^{-1} K^{-1} for condensed phases at 25°C and 1 atm are: Hg, 6.6 and 5.6; $CHCl_3$, 27.8 and 17.4; Cu, 5.9 and 5.7; C_6H_6, 32.1 and 21.9. Thus $\bar{C}_P$ and $\bar{C}_V$ can differ substantially for a condensed phase, a fact that is not always appreciated.

As a reminder, we again note that the equations of this section apply to a closed system of fixed composition (and also to closed systems where the composition is changed reversibly).

4.6 CHEMICAL POTENTIALS

The fundamental equation $dU = T\,dS - P\,dV$ and the related equations (4.38) to (4.40) for dH, dA, and dG are not applicable when the composition is changing due to interchange of matter with the surroundings or to irreversible chemical reaction or irreversible interphase transport of matter within the system. We now develop equations that will hold during such processes.

Consider a system that consists of a single homogeneous phase whose composition can be varied. Let the system be in thermal and mechanical equilibrium but not necessarily in material equilibrium. Since thermal and mechanical equilibrium exist, the temperature and pressure have well-defined values and the system's thermodynamic state is defined by the values of $T, P, n_1, n_2, \ldots, n_k$, where the n_i's ($i = 1, 2, \ldots, k$) are the mole numbers of the k components of the one-phase system. As discussed in Sec. 4.2, even though the system is not in material equilibrium, we can still assign meaningful values to U and S of the system (relative to the values of U and S in some chosen reference state). Since T, P, V, U, and S have values, the functions H, A, and G also have values. The state functions U, H, A, and G can each be expressed as functions of T, P, and the n_i's.

At any instant during a chemical process in the system, the Gibb's energy is

$$G = G(T, P, n_1, \ldots, n_k) \tag{4.76}$$

Let T, P, and the n_i's change by the infinitesimal amounts $dT, dP, dn_1, \ldots, dn_k$ due to an irreversible chemical reaction or irreversible transport of matter into the system. We want dG for this infinitesimal process. Since G is a state function, we shall replace the actual irreversible change by a reversible change and calculate dG for the reversible change. We imagine using an anticatalyst to "freeze out" any chemical reactions in the system. We then (1) reversibly add dn_1 moles of component 1, dn_2 moles of 2, etc., and (2) reversibly change T and P by dT and dP.

To reversibly add component 1 to a system, we would use a rigid, semipermeable membrane (Sec. 3.5) that is permeable to component 1 only; if pure component 1 is on one side of the membrane and the system is on the other side, we can adjust the pressure of pure 1 so that there is no tendency for component 1 to flow between system and surroundings. An infinitesimal increase or decrease of the pressure of pure 1 then reversibly varies the value of n_1 in the system. Similarly for the other components. (Of course such a procedure is impractical, but it is possible in principle to perform it.)

The total differential of (4.76) is

$$dG = \left(\frac{\partial G}{\partial T}\right)_{P,n_i} dT + \left(\frac{\partial G}{\partial P}\right)_{T,n_i} dP + \left(\frac{\partial G}{\partial n_1}\right)_{T,P,n_{j \neq 1}} dn_1 + \cdots + \left(\frac{\partial G}{\partial n_k}\right)_{T,P,n_{j \neq k}} dn_k \tag{4.77}$$

where the following conventions are used: the subscript n_i on a partial derivative means that all mole numbers are held constant; the subscript $n_{j \neq i}$ on a partial derivative means that all mole numbers except n_i are held fixed. For a reversible process where no change in composition occurs, Eq. (4.40) gives

$$dG = -S\,dT + V\,dP \qquad \text{rev. proc., } n_i \text{ fixed, } P\text{-}V \text{ work only} \tag{4.78}$$

It follows from (4.78) that

$$\left(\frac{\partial G}{\partial T}\right)_{P,n_i} = -S, \qquad \left(\frac{\partial G}{\partial P}\right)_{T,n_i} = V \tag{4.79}$$

where we added the subscripts n_i to (4.50) to emphasize the constant composition. Substitution of (4.79) in (4.77) gives for dG in a reversible process in a system with only P-V work

$$dG = -S\,dT + V\,dP + \sum_{i=1}^{k} \left(\frac{\partial G}{\partial n_i}\right)_{T,P,n_{j \neq i}} dn_i \tag{4.80}$$

Now consider a process in which the state variables change due to an irreversible material change. Since G is a state function, dG is independent of the process that connects states $(T, P, n_1, n_2, \ldots)$ and $(T + dT, P + dP, n_1 + dn_1, n_2 + dn_2, \ldots)$; therefore dG for the irreversible change is the same as dG for a reversible change that connects these two states. Hence Eq. (4.80) gives dG for the irreversible material change. Note also that all the state functions in (4.80) are defined for the system during the irreversible composition change (Sec. 4.2). Thus (4.80) is the desired relation for dG.

To save time in writing, the *chemical potential* μ_i (mu eye) of component i in the one-phase system is defined as

$$\mu_i \equiv \left(\frac{\partial G}{\partial n_i}\right)_{T,P,n_{j \neq i}} \tag{4.81*}$$

Equation (4.80) then becomes

$$dG = -S\,dT + V\,dP + \sum_i \mu_i\,dn_i \qquad \begin{array}{l}\text{one-phase syst. in therm.}\\ \text{and mech. equilib., } P\text{-}V \text{ work only}\end{array} \tag{4.82*}$$

Equation (4.82) is the fundamental equation of chemical thermodynamics. It applies to a process in which the single-phase system is in thermal and mechanical equilibrium but is not necessarily in material equilibrium. Thus (4.82) holds during an irreversible chemical reaction and during transport of matter into or out of the system. (Our previous equations were for closed systems, but we now have an equation applicable to open systems.)

Let us obtain the equation for dU that corresponds to (4.82). From $G \equiv U + PV - TS$, we have $dU = dG - P\,dV - V\,dP + T\,dS + S\,dT$. Use of (4.82) gives

$$dU = T\,dS - P\,dV + \sum_i \mu_i\,dn_i \tag{4.83}$$

where the same restrictions apply as for (4.82). Equation (4.83) may be compared with $dU = T\,dS - P\,dV$ for a reversible process in a closed system.

Setting $dS = 0$, $dV = 0$, and $dn_{j \neq i} = 0$ in (4.83), we get the following alternative expression for the chemical potential:

$$\mu_i = \left(\frac{\partial U}{\partial n_i}\right)_{S,V,n_{j \neq i}} \tag{4.84}$$

It also follows from (4.83) that $(\partial U/\partial V)_{S,n_i} = -P$ and $(\partial U/\partial S)_{V,n_i} = T$ [cf. (4.47)]. Note that the variables held constant in (4.84) and (4.81) differ. Therefore, $\mu_i \neq (\partial U/\partial n_i)_{T,P,n_{j \neq i}}$.

From $H = U + PV$ and $A = U - TS$, together with (4.83), we can obtain expressions for dH and dA for irreversible chemical changes. Collecting together the expressions for dU, dH, dA, and dG, we have

$$dU = T\,dS - P\,dV + \sum_i \mu_i\,dn_i \tag{4.85*}$$

$$dH = T\,dS + V\,dP + \sum_i \mu_i\,dn_i \tag{4.86}$$

$$dA = -S\,dT - P\,dV + \sum_i \mu_i\,dn_i \tag{4.87}$$

$$dG = -S\,dT + V\,dP + \sum_i \mu_i\,dn_i \tag{4.88*}$$

one-phase syst. in mech. and therm. equilib., P-V work only

These equations are the extensions of the Gibbs equations (4.37) to (4.40) to processes involving exchange of matter with the surroundings or irreversible composition changes; the extra terms in (4.85) to (4.88) allow for the effect of the composition changes on the state functions U, H, A, and G. Equations (4.85) to (4.88) are also called the Gibbs equations.

Consider the chemical potentials μ_i. The state function G is a function of T, P, $n_1, n_2, \ldots$. Hence its partial derivative $\partial G/\partial n_i$ in (4.81) is also a function of these variables:

$$\mu_i = \mu_i(T, P, n_1, n_2, \ldots) \qquad \text{one-phase syst.} \tag{4.89}$$

The chemical potential of component i in the phase is a state function that depends on the temperature, pressure, and composition of the phase. Since μ_i is the ratio of infinitesimal changes in two extensive properties, it is an intensive property. The state function μ_i was introduced into thermodynamics by Gibbs.

We can use (4.83) to derive an expression for the heat during a chemical reaction. Consider a closed, one-phase system capable of only P-V work; let the system be in thermal and mechanical equilibrium but not necessarily in material equilibrium. For any process (reversible or irreversible) in a closed system, we have $dU = dq + dw$; also, $dw = -P\,dV$ (since we have mechanical equilibrium); hence $dU = dq - P\,dV$. Comparison of this equation with (4.85) gives

$$dq = T\,dS + \sum_i \mu_i\,dn_i \tag{4.90}$$

for a one-phase closed system in mechanical and thermal equilibrium. This expression gives dq during a chemical reaction in a closed system. Since the reaction is irreversible, $dq \neq T\,dS$.

Equations (4.85) to (4.88) are for a one-phase system. The state function $G = U + PV - TS$ is extensive; hence, if the system has several phases, we just add the Gibbs energy of each phase to get G for the system. Let G^α be the Gibbs free energy of phase α, and let G be the Gibbs energy of the several-phase system. Then $G = \sum_\alpha G^\alpha$. The differential of this equation is $dG = \sum_\alpha dG^\alpha$. Use of Eq. (4.88) for dG^α gives

$$dG = -\sum_\alpha S^\alpha\, dT + \sum_\alpha V^\alpha\, dP + \sum_\alpha \sum_i \mu_i^\alpha\, dn_i^\alpha \tag{4.91}$$

where S^α and V^α are the entropy and volume of phase α, μ_i^α is the chemical potential of component i in phase α, and n_i^α is the number of moles of i in phase α. (We have taken the temperature of each phase to be the same and the pressure of each phase to be the same; this will be true for a system in mechanical and thermal equilibrium provided no rigid or adiabatic walls separate the phases.) Since S and V are extensive, the sums over the entropies and volumes of the phases equal the total entropy S of the system and the total volume V of the system. Hence

$$dG = -S\, dT + V\, dP + \sum_\alpha \sum_i \mu_i^\alpha\, dn_i^\alpha \tag{4.92*}$$

syst. in mech. and therm. equilib., P-V work only

Equation (4.92) is the extension of (4.88) to a several-phase system. Equations similar to (4.92) hold for dU, dH, and dA in a several-phase system.

We now use (4.92) to derive the condition for material equilibrium (including both phase equilibrium and reaction equilibrium). Consider a closed system in mechanical and thermal equilibrium and held at constant T and P. We showed in Sec. 4.4 that during an irreversible material process (chemical reaction or interphase transport of matter) in a system at constant T and P, the Gibbs function G is decreasing ($dG < 0$); at equilibrium, G has reached a minimum, and $dG = 0$ for any infinitesimal change at constant T and P [Eq. (4.21)]. At constant T and P, $dT = 0 = dP$, and from (4.92) the equilibrium condition $dG = 0$ becomes

$$\sum_\alpha \sum_i \mu_i^\alpha\, dn_i^\alpha = 0 \qquad \text{material equilib., closed syst., } P\text{-}V \text{ work only, const. } T,\ P \tag{4.93}$$

This is the desired relation.

Now consider material equilibrium in a closed system at constant T and V. Generalizing (4.87) to a several-phase system, we have

$$dA = -S\, dT - P\, dV + \sum_\alpha \sum_i \mu_i^\alpha\, dn_i^\alpha \tag{4.94}$$

syst. in mech. and therm. equilib., P-V work only

The Helmholtz energy A is a minimum for chemical equilibrium at constant T and V. Hence $dA = 0$ for constant-T-and-V equilibrium, and (4.94) gives

$$\sum_\alpha \sum_i \mu_i^\alpha\, dn_i^\alpha = 0 \qquad \text{material equilib., closed syst., } P\text{-}V \text{ work only, const. } T,\ V \tag{4.95}$$

which is the same as (4.93) for material equilibrium at constant T and P.

The material-equilibrium condition (4.93) not only is valid for equilibrium reached under conditions of constant T and P or constant T and V but in fact holds no matter how the closed system reaches equilibrium. To show this, consider an infinitesimal reversible process in a closed system with P-V work only. Equation (4.92) applies; also, Eq. (4.40) applies. Subtraction of (4.40) from (4.92) gives

$$\sum_{\alpha} \sum_{i} \mu_i^{\alpha} \, dn_i^{\alpha} = 0 \qquad \text{rev. proc., closed syst., } P\text{-}V \text{ work only} \tag{4.96}$$

Equation (4.96) must hold for any reversible process in a closed system with P-V work only. An infinitesimal process in a system that is in equilibrium is a reversible process (since it connects an equilibrium state with one infinitesimally close to equilibrium); hence (4.96) must hold for any infinitesimal change in a system that has reached material equilibrium. Thus (4.96) holds for any closed system in material equilibrium. If the system reaches material equilibrium under the conditions of constant T and P, then G is minimized at equilibrium; if equilibrium is reached under conditions of constant T and V, then A is minimized at equilibrium; if equilibrium is reached under some other conditions, then neither A nor G is necessarily minimized at equilibrium, but in all cases, Eq. (4.96) holds at equilibrium. Thus (4.96) is the desired general condition for material equilibrium.

We shall apply (4.96) to phase and reaction equilibria in the next two sections. Before doing so, we consider some further points about the all-important chemical potentials μ_i^{α}. As noted in Eq. (4.89), the chemical potential of component i in phase α depends on the temperature, pressure, and composition of the phase:

$$\mu_i^{\alpha} = \mu_i^{\alpha}(T^{\alpha}, P^{\alpha}, x_1^{\alpha}, x_2^{\alpha}, \ldots) \tag{4.97}$$

where, since μ_i^{α} is an intensive property, we can use mole fractions instead of moles as variables. Note that even if component i is absent from phase α ($n_i^{\alpha} = 0$), its chemical potential μ_i^{α} in phase α is not zero; there is always the possibility of introducing component i into the phase; when dn_i^{α} moles of i is introduced at constant T, P, and $n_{j \neq i}$, the Gibbs energy of the phase changes by dG^{α} and μ_i^{α} is given by $dG^{\alpha}/dn_i^{\alpha}$.

The simplest possible system is a one-phase, one-component system—a single phase of pure substance i, for example, solid gold or liquid water. Let $\bar{G}_i(T, P)$ be the molar Gibbs energy of pure i at the temperature and pressure of the system. Since G is extensive, the Gibbs energy G of the system is $G = n_i \bar{G}_i(T, P)$. Partial differentiation of this equation gives

$$\mu_i \equiv \left(\frac{\partial G}{\partial n_i}\right)_{T,P} = \bar{G}_i \qquad \text{one-phase, one-comp. syst.} \tag{4.98*}$$

For a pure substance, μ_i is simply the molar Gibbs free energy. (It might be thought that in a one-phase mixture, μ_i would also equal the molar Gibbs energy of pure component i, but this isn't so, due to different intermolecular interactions in the mixture compared with the intermolecular interactions in pure substance i.)

4.7 PHASE EQUILIBRIUM

As noted in Sec. 4.1, there are two kinds of material equilibrium, phase equilibrium and reaction equilibrium. Phase equilibrium is considered in this section and reaction equilibrium in the next. The condition for material equilibrium in a closed system with P-V work only is given by Eq. (4.96) [see also (4.93) and (4.95)], which holds for any possible infinitesimal change of the mole numbers n_i^{α}. Consider a several-phase system that is in equilibrium, and suppose that dn_j moles of substance j were to flow from phase β (beta) to phase δ (delta). For this process, Eq. (4.96) becomes $\mu_j^\beta\, dn_j^\beta + \mu_j^\delta\, dn_j^\delta = 0$. However, conservation of matter gives $dn_j^\beta = -dn_j^\delta$. Hence $-\mu_j^\beta\, dn_j^\delta + \mu_j^\delta\, dn_j^\delta = 0$, and

$$(\mu_j^\delta - \mu_j^\beta)\, dn_j^\delta = 0 \qquad (4.99)$$

Since $dn_j^\delta \neq 0$, we must have $\mu_j^\delta - \mu_j^\beta = 0$, or

$$\mu_j^\beta = \mu_j^\delta \qquad \text{phase equilib. in closed syst., } P\text{-}V \text{ work only} \qquad (4.100)^*$$

We have shown that for phase equilibrium in a closed system with P-V work only, the chemical potential of a given component is the same in every phase of the system.

Now consider a closed system that has not yet reached phase equilibrium. For definiteness, we suppose that the system is held at constant T and P, but our conclusions will also hold for other conditions. Let dn_j moles of component j flow spontaneously from phase β to phase δ. For this irreversible process, dG for the system must be negative. Hence (4.92) gives

$$(\mu_j^\delta - \mu_j^\beta)\, dn_j^\delta < 0 \qquad (4.101)$$

where we used $dn_j^\beta = -dn_j^\delta$. By assumption, the flow of j is from phase β to phase δ; hence dn_j^δ is positive; therefore (4.101) requires that $\mu_j^\delta - \mu_j^\beta$ be negative: $\mu_j^\delta < \mu_j^\beta$. Hence:

Substance j flows spontaneously from a phase with higher chemical potential μ_j to a phase with lower chemical potential μ_j.

This flow will continue until the chemical potential of component j has been equalized in all the phases of the system. Similarly for the other components. (As a substance flows from one phase to another, the compositions of the phases are changed and hence the chemical potentials of the phases are changed.)

If $T^\beta > T^\delta$, heat flows spontaneously from phase β to phase δ until $T^\beta = T^\delta$. If $P^\beta > P^\delta$, work "flows" from phase β to phase δ until $P^\beta = P^\delta$. If $\mu_j^\beta > \mu_j^\delta$, substance j flows spontaneously from phase β to phase δ until $\mu_j^\beta = \mu_j^\delta$. The state function T determines whether or not there is thermal equilibrium between phases. The state function P determines whether or not there is mechanical equilibrium between phases. The state functions μ_i determine whether or not there is material equilibrium between phases.

We can use (4.100) to simplify the material equilibrium condition (4.96). Since at equilibrium the chemical potential of component i is the same in every phase, we can drop the superscript α from μ_i^α; thus

$$\sum_i \sum_\alpha \mu_i^\alpha\, dn_i^\alpha = \sum_i \left[\mu_i \left(\sum_\alpha dn_i^\alpha\right)\right] = \sum_i \mu_i\, dn_i$$

where dn_i is the change in the total number of moles of component i in the closed system and μ_i is the chemical potential of i in any phase. Equation (4.96) becomes

$$\sum_i \mu_i\, dn_i = 0 \qquad \text{material equilib. in closed syst., } P\text{-}V \text{ work only} \qquad (4.102)$$

There is one exception to the phase-equilibrium condition $\mu_j^\beta = \mu_j^\delta$, which we now examine. We found above that a substance flows from a phase where its chemical potential is higher to a phase where its chemical potential is lower. Suppose that substance j is initially absent from phase δ. Although there is no j in phase δ, the chemical potential μ_j^δ is a defined quantity, since we could, in principle, introduce dn_j moles of j into δ and measure $(\partial G^\delta/\partial n_j^\delta)_{T,P,n_{i\neq j}} = \mu_j^\delta$ (or use statistical mechanics to calculate μ_j^δ). If initially $\mu_j^\beta > \mu_j^\delta$, then j flows from phase β to phase δ until phase equilibrium is reached. However, if initially $\mu_j^\delta > \mu_j^\beta$, then j cannot flow out of δ (since it is absent from δ); the system will therefore remain unchanged with time and hence is in equilibrium. Thus, when a component is absent from a phase, the equilibrium condition becomes

$$\mu_j^\delta \geq \mu_j^\beta \qquad \text{phase equilib., } j \text{ absent from } \delta \tag{4.103}$$

for all phases β in equilibrium with δ.

The principal conclusion of this section is that:

In a closed system in thermodynamic equilibrium, the chemical potential of any given component is the same in every phase in which that component is present.

4.8 REACTION EQUILIBRIUM

We now apply the material-equilibrium condition (4.102) to reaction equilibrium. Let the reaction be

$$aA_1 + bA_2 + \cdots \rightarrow eA_m + fA_{m+1} + \cdots \tag{4.104}$$

where $A_1, A_2, \ldots$ are the reactants, $A_m, A_{m+1}, \ldots$ are the products, and $a, b, \ldots, e, f, \ldots$ are the coefficients. Thus for

$$2C_6H_6 + 15O_2 \rightarrow 12CO_2 + 6H_2O \tag{4.105}$$

we have $A_1 = C_6H_6$, $A_2 = O_2$, $A_3 = CO_2$, $A_4 = H_2O$ and $a = 2$, $b = 15$, $e = 12$, $f = 6$. The substances in the reaction (4.104) need not all occur in the same phase since Eq. (4.102) is applicable to several-phase systems.

We adopt the convention of transposing the reactants in (4.104) to the right side of the equation to get

$$0 = -aA_1 - bA_2 - \cdots + eA_m + fA_{m+1} + \cdots \tag{4.106}$$

We now let

$$\nu_1 \equiv -a,\ \nu_2 \equiv -b,\ \ldots,\ \nu_m \equiv e,\ \nu_{m+1} \equiv f,\ \ldots$$

and write the reaction (4.106) in the compact form

$$0 = \sum_i \nu_i A_i \tag{4.107}$$

where the *stoichiometric coefficients* ν_i (nu i) are negative for reactants and positive for products. Thus for the reaction (4.105), we have $\nu_1 = -2$, $\nu_2 = -15$, $\nu_3 = 12$, $\nu_4 = 6$. The stoichiometric coefficients are pure numbers with no units.

Suppose we start with a nonequilibrium system in which the amount of substance A_1 is $n_{1,0}$, the amount of A_2 is $n_{2,0}$, etc., where the subscript zero indicates the initial composition. ($n_{1,0}$ is the initial number of moles of A_1 multiplied by the unit 1 mol.) Let certain amounts of A_1, A_2, etc., react. As a measure of how much reaction has occurred, we define the *extent of reaction* ξ (xi) as follows. The number of moles of A_1, A_2, etc., that react are proportional to the coefficients a, b, ... in (4.104). Suppose that a moles of A_1 in the mixture were to react with b moles of A_2, etc., to produce e moles of A_m, f moles of A_{m+1}, etc. We would then say that ξ was 1 mol. If 0.2a moles of A_1 were to react with 0.2b moles of A_2, etc., to produce 0.2e moles of A_m, etc., then ξ would be 0.2 mol. In general, ξa of A_1 will react with ξb of A_2, etc., to produce ξe of A_m, ξf of A_{m+1}, etc. (The quantity ξ is defined to have units of moles; hence ξa is the amount of substance A_1 that reacts; ξa/mol is the number of moles of A_1 that react.) The amounts of each component present after this much reaction will be

$$n_1 = n_{1,0} - a\xi = n_{1,0} + \nu_1 \xi, \; n_2 = n_{2,0} + \nu_2 \xi, \; \ldots$$

$$n_m = n_{m,0} + e\xi = n_{m,0} + \nu_m \xi, \; n_{m+1} = n_{m+1,0} + \nu_{m+1} \xi, \; \ldots$$

where the equation following (4.106) was used. Thus at any time during the reaction the amount of each species present is

$$n_i = n_{i,0} + \nu_i \xi \tag{4.108}$$

Equation (4.108) gives the formal definition of ξ as

$$\xi \equiv \Delta n_i/\nu_i = (n_i - n_{i,0})/\nu_i \tag{4.109}$$

The variable ξ is positive if the reaction has proceeded left to right and negative if the reaction has proceeded right to left.

Since $n_{i,0}$ and ν_i are constants, the differential of (4.108) is

$$dn_i = \nu_i \, d\xi \tag{4.110}$$

Substitution of (4.110) in the equilibrium condition (4.102) gives $(\sum_i \nu_i \mu_i)\, d\xi = 0$ at equilibrium. This equation must hold for arbitrary infinitesimal values of $d\xi$. Hence

$$\sum_i \nu_i \mu_i = 0 \qquad \text{reaction equilib. in closed syst., } P\text{-}V \text{ work only} \tag{4.111*}$$

This is the fundamental condition for chemical-reaction equilibrium. Its relation to the more familiar concept of the equilibrium constant will become clear in later chapters. Note that (4.111) is valid no matter how the closed system reaches equilibrium; e.g., it holds for equilibrium reached at constant T and P, or at constant T and V, or in an isolated system.

The equilibrium condition (4.111) is easily remembered by noting that it is obtained by simply replacing each substance in the reaction equation (4.104) by its chemical potential. Thus, for the reaction (4.105), the equilibrium condition (4.111) is $2\mu_{C_6H_6} + 15\mu_{O_2} = 12\mu_{CO_2} + 6\mu_{H_2O}$.

If the reaction occurs under conditions of constant T and P, the Gibbs function G is minimized at equilibrium. Note from (4.82) that the sum on the left of (4.102) equals dG at constant T and P: $dG_{T,P} = \sum_i \mu_i \, dn_i$. Use of (4.110) gives $dG_{T,P} = \sum_i \nu_i \mu_i \, d\xi$. Hence

$$\frac{dG}{d\xi} = \sum_i \nu_i \mu_i \qquad \text{const. } T, P \tag{4.112}$$

At equilibrium, $dG/d\xi = 0$, and G is minimized. The μ_i's in (4.112) are the chemical potentials of the substances in the reaction mixture, and they depend on the composition of the mixture (the n_i's). Hence the chemical potentials vary during the course of the reaction. This variation continues until G (which depends on the μ_i's and the n_i's at constant T and P) is minimized and (4.111) is satisfied. Figure 4.4 sketches G vs. ξ for a reaction run at constant T and P. (For constant T and V, G is replaced by A in the preceding discussion.)

Note the resemblance of the reaction-equilibrium condition (4.111) to the phase-equilibrium condition (4.100).

4.9 ENTROPY AND LIFE

The second law of thermodynamics is the law of increase of entropy. Increasing entropy means increasing disorder. Living organisms maintain a high degree of internal order. Hence one might ask whether life processes violate the second law.

We first ask whether we can meaningfully define the entropy of a living organism. Processes are continually occurring in a living organism, and the organism is not in an equilibrium state. However, the discussion of Sec. 4.2 shows that the entropy of nonequilibrium systems can be defined. We therefore can consider whether living systems obey the second law.

The first thing to note is that the statement $\Delta S \geq 0$ applies only to systems that are both closed and thermally isolated from their surroundings; see Eq. (3.43). Living organisms are open systems since they both take in and expel matter; further, they exchange heat with their surroundings. According to the second law, we must have $\Delta S_{\text{syst}} + \Delta S_{\text{surr}} \geq 0$ for an organism, but ΔS_{syst} (ΔS of the organism) can be positive, negative, or zero. Any decrease in S_{syst} must, according to the second law, be compensated for by an increase in S_{surr} that is at least as great as the magnitude of the decrease in S_{syst}. (For example, during the freezing of water to the more ordered state of ice, S_{syst} decreases, but the heat flow from system to surroundings increases S_{surr}.)

We can analyze entropy changes in open systems as follows. Let dS be the change in the entropy of any system (open or closed) during an infinitesimal time interval dt. Let d_iS be the system's entropy change due to processes that occur inside the system during dt; let d_eS be the system's entropy change due to exchanges of energy and matter between system and surroundings during dt. Then

$$dS = d_iS + d_eS \tag{4.113}$$

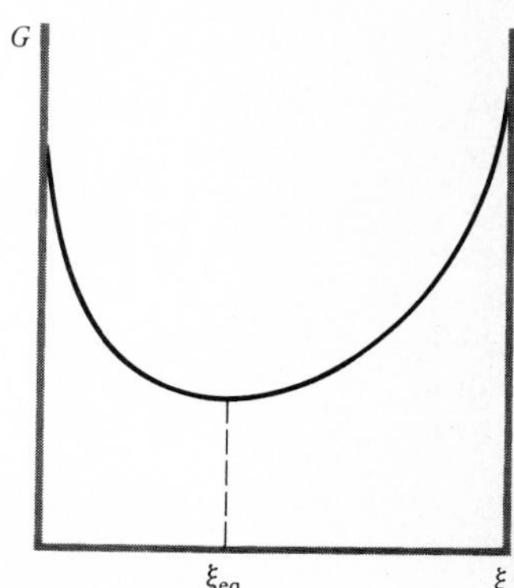

Figure 4.4
Gibbs energy vs. extent of reaction in a system held at constant T and P.

As far as internal changes in the system are concerned, we can consider the system to be isolated from its surroundings; hence Eqs. (3.43) and (3.41) give

$$d_i S \geq 0 \tag{4.114}$$

where the inequality sign holds for irreversible internal processes. However, $d_e S$ can be positive, negative, or zero, and dS can be positive, negative, or zero.

The state of a fully grown living organism remains about the same from day to day. (The organism is not in an equilibrium state, but it is, to a good approximation, in a steady state.) Thus over a 24-hr period, ΔS of a fully grown organism is about zero: $\Delta S \approx 0$. The internal processes of chemical reaction, diffusion, blood flow, etc., are irreversible; hence $\Delta_i S > 0$ for the organism. Thus $\Delta_e S$ must be negative to compensate for the positive $\Delta_i S$. We can break $\Delta_e S$ into a term due to heat exchange with the surroundings and a term due to matter exchange with the surroundings. The sign of q, and hence the sign of that part of $\Delta_e S$ due to heat exchange, can be positive or negative, depending on whether the surroundings are hotter or colder than the organism. We shall concentrate on that part of $\Delta_e S$ which is due to matter exchange. The organism takes in foodstuffs that contain highly ordered, low-entropy polymeric molecules such as proteins and starch and excretes waste products that contain smaller, less ordered molecules. Thus the entropy of the food intake is less than the entropy of the excretion products returned to the surroundings; it is this which keeps $\Delta_e S$ negative. The organism discards matter with a greater entropy content than the matter it takes in, thereby losing entropy to the environment to compensate for the entropy produced in internal irreversible processes.

In his book *What Is Life?* (Doubleday, 1956, p. 71), Schrödinger wrote, "... a living organism continually ... produces positive entropy [$\Delta_i S > 0$] ... and thus tends to approach the dangerous state of maximum entropy, which is death. It can only keep ... alive by continually drawing from its environment negative entropy .... [The] essential thing in metabolism is that the organism succeeds in freeing itself from all the entropy it cannot help producing while alive."

As the preceding analysis shows, there is no reason to believe that living organisms violate the second law. Of course, a quantitative demonstration of this (by measurement of $\Delta S_{\text{syst}} + \Delta S_{\text{surr}}$ for an organism and its surroundings) would be a difficult task. In the absence of such a quantitative demonstration, there have been occasional speculations that organisms violate the second law of thermodynamics.

PROBLEMS

4.1 Verify Eqs. (4.51) and (4.52).

4.2 Derive (4.58) by substitution of (4.30) into (4.34).

4.3 For the processes of Probs. 2.19*a*, *b*, *d*, *e*, and *f*, state whether each of ΔA and ΔG is positive, zero, or negative.

4.4 Calculate ΔA and ΔG for the process of Prob. 2.25*c*. (Save time by using the results of Probs. 2.25 and 3.11.)

4.5 Calculate ΔA and ΔG for the process of Prob. 2.14.

4.6 (*a*) Find ΔG for the fusion of 50.0 g of ice at 0°C and 1 atm. (*b*) Find ΔG for the process of Prob. 3.15.

4.7 Use the values of α and κ given preceding Eq. (4.75) to calculate $(\partial U/\partial V)_T$ for water at 30°C and 1 atm.

4.8 For a liquid with the (typical) values $\alpha = 10^{-3}\ \text{K}^{-1}$, $\kappa = 10^{-4}\ \text{atm}^{-1}$, $\bar{V} = 50\ \text{cm}^3\ \text{mol}^{-1}$, $\bar{C}_P = 40\ \text{cal mol}^{-1}\ \text{K}^{-1}$, calculate at 25°C and 1 atm (*a*) $(\partial \bar{U}/\partial T)_P$; (*b*) $(\partial \bar{U}/\partial P)_T$; (*c*) $(\partial U/\partial V)_T$; (*d*) $(\partial \bar{S}/\partial T)_P$; (*e*) $(\partial \bar{S}/\partial P)_T$; (*f*) $\bar{C}_V$. (Use calories as the energy unit in all answers.)

4.9 A certain gas obeys the equation of state $P\bar{V} = RT(1 + bP)$, where b is a constant. Prove that for this gas (a) $(\partial U/\partial V)_T = bP^2$; (b) $\bar{C}_P - \bar{C}_V = R(1 + bP)^2$; (c) $\mu_{JT} = 0$.

4.10 Find μ_{JT} for water at 30°C and 1 atm, using data given preceding Eq. (4.75).

4.11 Show that $(\partial H/\partial V)_T = \alpha T/\kappa - 1/\kappa$.

4.12 Use Eqs. (4.34), (4.44), and (4.59) to show that $(\partial C_P/\partial P)_T = -T(\partial^2 V/\partial T^2)_P$. The volumes of substances increase approximately linearly with T, so $\partial^2 V/\partial T^2$ is usually quite small. Consequently, the pressure dependence of C_P can usually be neglected unless one is dealing with high pressures.

4.13 The volume of Hg in the temperature range 0 to 100°C at 1 atm is given by

$$V = V_0(1 + at + bt^2)$$

$$a = 0.18182 \times 10^{-3}\ °\text{C}^{-1}$$

$$b = 0.78 \times 10^{-8}\ °\text{C}^{-2}$$

where V_0 is the volume at 0°C and t is the Celsius temperature. The density of mercury at 1 atm and 0°C is 13.595 g/cm^3. (a) Use the result of Prob. 4.12 to calculate $(\partial \bar{C}_P/\partial P)_T$ for Hg at 25°C and 1 atm. (b) Given that $\bar{C}_P = 6.66$ cal mol^{-1} K^{-1} for Hg at 1 atm and 25°C, estimate $\bar{C}_P$ of Hg at 25°C and 10^4 atm.

4.14 Suppose 1 mole of water initially at 27°C and 1 atm undergoes a process whose final state is 100°C and 50 atm. Use data given preceding Eq. (4.75) and the approximation that the temperature and pressure variations of α, κ, and C_P can be neglected to calculate: (a) ΔU; (b) ΔH; (c) ΔS.

4.15 Calculate ΔG for the isothermal compression of 30.0 g of water from 1.0 atm to 100.0 atm at 25°C; neglect the variation of V with P.

4.16 A reversible adiabatic process is an isentropic process. (a) Let $\alpha_S \equiv V^{-1}(\partial V/\partial T)_S$. Show that $\alpha_S = -C_V\kappa/TV\alpha$. *Hint:* Start by applying (1.34) to $(\partial V/\partial T)_S$. (b) The adiabatic compressibility κ_S is defined as $\kappa_S \equiv -V^{-1}(\partial V/\partial P)_S$. Show that $\kappa_S = (C_V/C_P)\kappa$. *Hint:* Form the ratio κ_S/κ and apply (1.34) to both numerator and denominator; then use (1.32) and (1.35).

4.17 Show that $\mu_i = (\partial H/\partial n_i)_{S,P,n_{j\neq i}} = (\partial A/\partial n_i)_{T,V,n_{j\neq i}}$.

4.18 Verify that (4.68) reduces to (2.117) for an ideal gas.

4.19 Consider a two-phase system that consists of liquid water in equilibrium with water vapor; the system is kept in a constant-temperature bath. (a) Suppose we reversibly increase the system's volume, holding T and P constant, causing some of the liquid to vaporize. State whether each of ΔH, ΔS, ΔS_{univ}, and ΔG is positive, zero, or negative. (b) Suppose we suddenly remove some of the water vapor, holding T and V constant. This reduces the pressure below the equilibrium vapor pressure of water, and liquid water will evaporate at constant T and V until the equilibrium vapor pressure is restored. For this evaporation process state whether each of ΔU, ΔS, ΔS_{univ}, and ΔA is positive, zero, or negative.

4.20 For each of the following processes, state which of ΔU, ΔH, ΔS, ΔA, and ΔG must be zero. (a) A nonideal gas is taken around a Carnot cycle; (b) hydrogen is burned in an adiabatic calorimeter of fixed volume; (c) a nonideal gas undergoes a Joule–Thomson expansion; (d) ice is melted at 0°C and 1 atm.

4.21 For the freezing of 1 mole of water at 0°C and 1 atm, calculate $\Delta_i S$ and $\Delta_e S$. (Heat of fusion = 79.7 cal/g.) *Hint:* Use a generalization of (4.90) to prove that for a closed system at constant T and P, $dS = dq/T - dG/T$; compare with (4.113).

4.22 Use the result of Prob. 4.6b to calculate $\Delta_i S$ and $\Delta_e S$ for the freezing of 1 mole of supercooled water at 1 atm and −10°C. (See Prob. 4.21.)

4.23 If 1.00 mole of water at 30.00°C is reversibly and adiabatically compressed from 1.00 to 10.00 atm, calculate the final volume by using expressions from Prob. 4.16 and neglecting the temperature and pressure variation of κ_S. Next calculate the final temperature. Then (assuming $\bar{C}_V$ to be constant) calculate ΔU and compare with the approximate answer of Prob. 2.15. What is w for this process? [See Eq. (4.75) and data preceding it.]

4.24 Verify that $[\partial(G/T)/\partial T]_P = -H/T^2$. (This is the *Gibbs–Helmholtz equation.*)

4.25 Show that $\mu_J = (P - \alpha T\kappa^{-1})/C_V$, where μ_J is the Joule coefficient.

4.26 Derive Eq. (4.60) by using (1.35) with z, x, u, and y replaced by S, H, T, and P, respectively.

4.27 The speed of sound in a phase can be shown to equal $(\rho\kappa C_V/C_P)^{-1/2}$, where ρ is the density. (For the proof, see *Zemansky*, sec. 5-9.) Use data given preceding Eq. (4.75) to calculate the speed of sound in water at 30°C and 1 atm. Compare with the speed of sound in air, which is 10^3 ft/s.

4.28 A certain gas obeys the equation of state $P\bar{V} = RT(1 + bP + cP^2)$, where b and c are constants. Find expressions for $\Delta\bar{U}$ and $\Delta\bar{S}$ for a change of state of this gas from (P_1, T_1) to (P_2, T_2); neglect the temperature and pressure dependence of $\bar{C}_P$.

5

CONVENTIONAL VALUES OF THERMODYNAMIC STATE FUNCTIONS

5.1 STANDARD STATES

Thermodynamics tells us how to measure changes in U and S but does not provide absolute values of U and S. In applying thermodynamics to chemistry, one chooses reference states for substances and assigns values of U, S, and related thermodynamic functions to these states; one can then find values of U, S, etc., relative to these chosen reference states. Before discussing the choice of reference states, we first consider the related idea of standard states.

The *standard state* of a pure substance is defined as follows. For a pure solid or a pure liquid, the standard state is defined as the state with pressure $P = 1$ atm and temperature T, where T is some temperature of interest. Thus for each value of T, there is a single standard state for a pure substance. The symbol for a standard state is a degree superscript, with the temperature written as a subscript. For example, the molar volume of a pure solid or liquid at 1 atm and 300 K is symbolized by $\bar{V}^{\circ}_{300}$, where the degree superscript indicates the standard pressure of 1 atm and 300 stands for 300 K. For a pure gas, the standard state at temperature T is chosen as the state with $P = 1$ atm and the gas behaving as an ideal gas. Since real gases do not behave ideally at 1 atm, the standard state of a pure gas is a fictitious state. The method used to relate properties of the gas in the fictitious standard state to properties of the real gas will be discussed in Sec. 5.2. Summarizing, the standard states for pure substances are:

$$\begin{aligned} &\text{Solid or liquid:} \qquad && P = 1 \text{ atm},\ T \\ &\text{Gas:} && P = 1 \text{ atm},\ T, \text{ gas ideal} \end{aligned} \qquad (5.1)^*$$

Standard states for components of solutions will be discussed in later chapters.

5.2 CONVENTIONAL ENTHALPIES

The first law of thermodynamics allows us to measure changes in enthalpies and internal energies (ΔH and ΔU) but does not provide absolute values of H and U, only relative values. Thus U and H of a system each contain an arbitrary additive constant. For calculations in chemical thermodynamics it is convenient to set up tables of molar enthalpies $\bar{H}$ and entropies $\bar{S}$ of pure substances. How can we construct such a table?

It is possible in principle to assign an absolute value for the internal energy (and hence the enthalpy) of a system by appealing to the special theory of relativity. This theory (developed by Einstein in 1905) shows that energy and mass are always associated with each other according to the formula

$$E = mc^2 \tag{5.2}$$

where m is the relativistic mass of the system of energy E and c is the speed of light in vacuum. For a system at rest in the absence of external fields, E is the internal energy U [Eq. (2.60)]. Thus, relativity allows us to calculate the absolute internal energy of a mole of water at a given T and P simply by multiplying the molar mass by c^2. However, such a calculation is useless for a chemist. Chemical energy changes are so small that the Δm corresponding to a change ΔU is too small to be detected; $\Delta m = \Delta U/c^2$, and c^2 is very large. Thus, for the melting of 1 mole of ice,

$$\Delta U = (79.7 \text{ cal/g})(18.0 \text{ g}) = 1440 \text{ cal} = 6000 \text{ J} = 6.0 \times 10^{10} \text{ ergs}$$

since w is negligible compared with q. The corresponding mass change is

$$\Delta m = (6.0 \times 10^{10} \text{ ergs})/(3.0 \times 10^{10} \text{ cm/s})^2 = 7 \times 10^{-11} \text{ g}$$

One mole of liquid water at 0°C and 1 atm is 7×10^{-11} g heavier than 1 mole of ice at these conditions, but the mass difference is undetectable. Hence, in practice, we have no way of finding absolute values of U of sufficient accuracy to be useful to chemists.

Since we cannot get useful values of absolute internal energies and enthalpies, we construct a table of *conventional* (or *relative*) *enthalpies*, as follows. We (1) pick sufficient reference substances and arbitrarily assign definite values to their enthalpies in certain specified reference states; (2) use thermodynamic equations and experimental data to calculate ΔH for the change from reference substances in their reference states to the substance of interest in the state of interest.

The most convenient choice of reference substances for a chemist are the pure elements. The enthalpy reference state chosen is the standard state (Sec. 5.1) at 25°C = 298.15 K. The arbitrary value chosen for the molar enthalpy $\bar{H}$ of each element in this state is zero. Thus

$$\bar{H}^\circ_{298} = 0 \qquad \text{for each element in its stable form} \tag{5.3}*$$

where 298 stands for 298.15 K. Recall that the standard state (degree superscript) for a pure solid or liquid is the substance at 1 atm but for a gas is the fictitious ideal gas at 1 atm. Certain elements have two or more forms (allotropes). Thus carbon occurs as diamond and as graphite. For such elements, the form that is stable at 25°C and 1 atm is chosen as the reference substance. Thus $\bar{H}^\circ_{298} = 0$ for graphite but not for diamond. Note that the convention (5.3) applies only to *elements* and not to compounds.

Suppose we want $\bar{H}$ for a solid or liquid element at an arbitrary temperature and pressure T' and P'. Equation (5.3) gives $\bar{H}(T', P') = \bar{H}(T', P') - \bar{H}^\circ_{298}$, and use of (4.68) for ΔH gives

$$\bar{H}(T', P') = \int_{298}^{T'} \bar{C}_P(T, 1 \text{ atm})\, dT + \int_{1 \text{ atm}}^{P'} [\bar{V}(T', P) - T'\bar{V}\alpha]\, dP \qquad (5.4)$$

where we carried out the change of state from (298.15 K, 1 atm) to (T', P') in two steps: a reversible isobaric temperature change at 1 atm followed by a reversible isothermal pressure change at T'. We use experimental values of heat capacity, density, and thermal expansivity in (5.4) to find $\bar{H}$ for the element at the desired temperature and pressure. If any phase changes occur, we make separate allowance for them.

For a gaseous element we must allow for the fact that $\bar{H}^\circ_{298}$ refers to a fictitious state. Let $\bar{H}_{\text{re}}(298.15 \text{ K}, 1 \text{ atm})$ be the conventional molar enthalpy of the real gaseous element at 298.15 K and 1 atm. The conventional $\bar{H}$ of the corresponding fictitious ideal gas is $\bar{H}_{\text{id}}(298.15 \text{ K}, 1 \text{ atm}) = \bar{H}^\circ_{298} = 0$. (The subscripts re and id stand for real and ideal.) To relate $\bar{H}_{\text{re}}$ and $\bar{H}_{\text{id}}$, we use the following hypothetical isothermal process:

$$\text{Real gas at 1 atm} \xrightarrow{(a)} \text{real gas at 0 atm} \xrightarrow{(b)} \text{ideal gas at 0 atm} \xrightarrow{(c)} \text{ideal gas at 1 atm} \qquad (5.5)$$

In step (a), we isothermally reduce the pressure of 1 mole of the real gas from 1 atm to zero. In step (b), we wave a magic wand that eliminates intermolecular interactions, thereby changing the real gas into an ideal gas at zero pressure. In step (c), we isothermally increase the pressure of the ideal gas from 0 to 1 atm. The overall process converts 1 mole of the real gas at 1 atm and T into 1 mole of ideal gas at 1 atm and T. For this process,

$$\Delta H = \bar{H}_{\text{id}}(T, 1 \text{ atm}) - \bar{H}_{\text{re}}(T, 1 \text{ atm}) = \Delta H_a + \Delta H_b + \Delta H_c \qquad (5.6)$$

The enthalpy change ΔH_a for step (a) is calculated from the integrated form of (4.59) [Eq. (4.68) with $dT = 0$]:

$$\Delta H_a = \bar{H}_{\text{re}}(T, 0 \text{ atm}) - \bar{H}_{\text{re}}(T, 1 \text{ atm}) = \int_{1 \text{ atm}}^{0} (\bar{V} - T\bar{V}\alpha)\, dP$$

For step (b), $\Delta H_b = \bar{H}_{\text{id}}(T, 0 \text{ atm}) - \bar{H}_{\text{re}}(T, 0 \text{ atm})$. In the limit $P \to 0$, intermolecular interactions become negligible for the real gas; therefore

$$\lim_{P \to 0} \bar{U}_{\text{re}}(T, P) = \bar{U}_{\text{id}}(T)$$

Also, as P goes to zero, the equation of state of the real gas approaches that of the ideal gas; therefore $(PV)_{\text{re}}$ equals $(PV)_{\text{id}}$ in the zero-pressure limit. Hence $H_{\text{re}} \equiv U_{\text{re}} + (PV)_{\text{re}}$ equals H_{id} in the zero-pressure limit:

$$\bar{H}_{\text{re}}(T, 0 \text{ atm}) = \bar{H}_{\text{id}}(T, 0 \text{ atm}) \qquad (5.7)$$

$$\Delta H_b = 0$$

For step (c), ΔH_c is zero, since H of an ideal gas is independent of pressure. Equation (5.6) becomes

$$\bar{H}_{\text{id}}(T, 1 \text{ atm}) - \bar{H}_{\text{re}}(T, 1 \text{ atm}) = \int_0^{1 \text{ atm}} \left[T\left(\frac{\partial \bar{V}}{\partial T}\right)_P - \bar{V} \right] dP \tag{5.8}$$

where (1.41) was used. The integral in (5.8) is evaluated using P-V-T data of the real gas. (Since T is constant, this is an ordinary integral.) At $T = 298.15$ K and $P = 1$ atm, the conventional $\bar{H}_{\text{id}}$ of the element is zero by definition and (5.8) then gives the conventional enthalpy of the real gaseous element as the negative of the integral in (5.8). The difference between $\bar{H}_{\text{re}}$ and $\bar{H}_{\text{id}}$ is small but is included in precise work.

So far, only elements have been considered. How do we find the conventional enthalpy of a compound at some T and P? We define the *standard enthalpy of formation* (or *standard heat of formation*) $\Delta H^\circ_{f,T}$ of a pure substance at temperature T as ΔH for the process in which 1 mole of the substance in its standard state at T is formed from the corresponding separated elements, each element being in its standard state at T. For example, for liquid water, $\Delta H^\circ_{f,T}$ is the enthalpy change for the process

$$H_2(\text{ideal gas}, T, 1 \text{ atm}) + \tfrac{1}{2}O_2(\text{ideal gas}, T, 1 \text{ atm}) \rightarrow H_2O(\text{liq.}, T, 1 \text{ atm})$$

The gases on the left are in their standard states, which means that they are unmixed, each in its pure state at $P = 1$ atm and temperature T. (We shall later find that for ideal gases, ΔH is zero for isothermal mixing.) From the definition of $\Delta H^\circ_{f,T}$, it follows that for water $\Delta H^\circ_{f,T}(H_2O) = \bar{H}^\circ_T(H_2O) - \bar{H}^\circ_T(H_2) - \tfrac{1}{2}\bar{H}^\circ_T(O_2)$.

At 25°C, the conventional enthalpies of the elements in their standard states are, by definition, zero. Hence at $T = 298.15$ K, this last equation becomes

$$\Delta H^\circ_{f,298}(H_2O) = \bar{H}^\circ_{298}(H_2O)$$

A similar equation holds for any compound. Obviously, $\Delta H^\circ_{f,T}$ for any element is zero; "formation" of an element from itself is a "process" in which nothing happens. Hence for any substance, element or compound, the conventional enthalpy at 25°C equals the standard enthalpy of formation at 25°C:

$$\bar{H}^\circ_{298} = \Delta H^\circ_{f,298} \tag{5.9}*$$

To find $\Delta H^\circ_{f,298}$ and hence the conventional enthalpy of a compound at 25°C and 1 atm, we carry out the following steps: (1) If any of the elements are gases at 25°C and 1 atm, we calculate (as described above) ΔH for the "transformation" of each such element as an ideal gas at 25°C and 1 atm to a real gas at 25°C and 1 atm. (2) We measure ΔH for mixing the pure elements at 25°C and 1 atm. (3) We use (4.68) to find ΔH for the process of bringing the mixture from 25°C and 1 atm to the conditions under which we plan to carry out the reaction to form the compound. (For example, in the combustion of an element with oxygen in a constant-volume container, we may want to have the system initially at, say, 30 atm pressure.) (4) We measure ΔH for the reaction in which the compound is formed from the mixed elements. (5) We use (4.68) to find ΔH for the process of bringing the compound from the state in which it is formed in step 4 to its standard state at 25°C. The standard heat of formation $\Delta H^\circ_{f,298}$ for the compound is then the sum of these five ΔH's. Once we have the compound's conventional enthalpy at 25°C and 1 atm, its conventional enthalpy in any other state is calculable by use of (4.68). (Measurement of ΔH for a chemical reaction as in step 4 is described in the next section.) The main contribution to $\Delta H^\circ_{f,298}$ comes from step 4, but in precise work one includes all the steps.

Once conventional enthalpies have been found, we can calculate conventional internal energies from $U = H - PV$. The enthalpies are more useful in most calculations, so tabulations usually list conventional enthalpies rather than internal energies. (The conventional U of an element at 1 atm and 25°C is negative, since the conventional H is zero for these conditions.)

The convention of assigning $\bar{H}^\circ_{298} = 0$ for every element is sometimes objected to on the grounds that the enthalpies of different elements in their standard states at 25°C are not *really* equal. Although the absolute enthalpies of different elements certainly do differ, the convention (5.3) cannot lead to any contradictions in ordinary chemical changes since elements are never interconverted in chemical reactions and enthalpy differences between elements are thus not determinable in chemical reactions. The actual (absolute) enthalpies of the elements are irrelevant to finding ΔH for any chemical process.

If, however, one were to apply thermodynamics to nuclear reactions, the convention (5.3) would have to be dropped. For nuclear reactions, experimentally measured nuclear mass changes together with $\Delta U = c^2\,\Delta m$ are used to find the differences between the internal energies of different elements.

5.3 ENTHALPY OF REACTION

For any chemical reaction, we define the *standard enthalpy of reaction* ΔH°_T as the enthalpy change for the process of transforming stoichiometric numbers of moles of the pure, separated reactants, each in its standard state at temperature T, to stoichiometric numbers of moles of the pure separated products, each in its standard state at temperature T. (Often ΔH°_T is called the *heat of reaction*.) We define ΔU°_T in a similar manner. For the reaction [Eqs. (4.104) and (4.107)]

$$aA_1 + bA_2 + \cdots \rightarrow eA_m + fA_{m+1} + \cdots \qquad \text{or} \qquad 0 = \sum_i \nu_i A_i$$

the standard enthalpy change ΔH°_T is

$$\Delta H^\circ_T = e\bar{H}^\circ_T(A_m) + f\bar{H}^\circ_T(A_{m+1}) + \cdots - a\bar{H}^\circ_T(A_1) - b\bar{H}^\circ_T(A_2) - \cdots$$

$$\Delta H^\circ_T = \sum_i \nu_i \bar{H}^\circ_{T,i} \qquad (5.10)^*$$

where the ν_i's are the stoichiometric coefficients, and $\bar{H}^\circ_{T,i}$ is the standard molar enthalpy of substance A_i at temperature T. For example, for the reaction (4.105), we have

$$\Delta H^\circ_T = 12\bar{H}^\circ_T(CO_2) + 6\bar{H}^\circ_T(H_2O) - 2\bar{H}^\circ_T(C_6H_6) - 15\bar{H}^\circ_T(O_2)$$

Since the stoichiometric coefficients ν_i in (5.10) are dimensionless, the units of ΔH°_T are the same as those of $\bar{H}^\circ_{T,i}$, namely, cal/mol or J/mol. The subscript T in ΔH°_T is often omitted.

Note that ΔH° depends on how the reaction is written. For

$$2H_2(g) + O_2(g) \rightarrow 2H_2O(l) \qquad (5.11)$$

the standard enthalpy of reaction ΔH_T° is twice that for

$$H_2(g) + \tfrac{1}{2}O_2(g) \rightarrow H_2O(l) \tag{5.12}$$

(The coefficients give the mole numbers of reactants and products.) For (5.11) one finds $\Delta H_{298}^\circ = -136.6$ kcal/mol, whereas for (5.12) $\Delta H_{298}^\circ = -68.3$ kcal/mol (which is the standard heat of formation $\Delta H_{f,298}^\circ$ for water at 25°C). The factor mol^{-1} in ΔH° indicates that we are giving the enthalpy change for $\xi = 1$ mole for the specified reaction. When $\xi = 1$ mole for (5.11), 2 moles of water is formed, whereas when $\xi = 1$ mole for (5.12), 1 mole of water is formed [Eq. (4.109)].

In (5.12), the symbols l and g denote the liquid and gaseous states. A crystalline solid is denoted by either s or c.

The steps needed to find ΔH_T° (and ΔU_T°) are the same as outlined in the last section for finding $\Delta H_{f,T}^\circ$, except that the reactants are not necessarily elements. To carry out step 4 of the procedure of the last section, we must measure ΔH for the relevant chemical reaction. This is done in a calorimeter.

To be suitable for calorimetric study, the reaction should be rather fast, so that the heat exchange between the calorimeter and the surroundings can be made negligible during the reaction. The reaction preferably should go essentially to completion, to avoid having to analyze the mixture after reaction to determine how much material reacted. The reaction should be "clean," with no significant amount of other reactions occurring. The most common type of reaction studied calorimetrically is combustion. One also measures heats of hydrogenation, halogenation, neutralization, solution, dilution, mixing, phase transitions, etc. Heat capacities are also determined in a calorimeter. Reactions where some of the species are gases (e.g., combustion reactions) are studied in a constant-volume calorimeter; reactions not involving gases are studied in a constant-pressure calorimeter.

Some heats of combustion at 25°C (ΔH_{298}° per mole of compound burned) in kcal/mole are:

H_2	C(*gr*)	CH_4	C_2H_6	C_2H_4	C_3H_8	$C_2H_5OH(l)$	sucrose(*s*)
−68	−94	−213	−373	−337	−531	−327	−1350

where *gr* stands for graphite.

Of the many types of calorimeters, we shall consider only the adiabatic bomb calorimeter (Fig. 5.1), which is used to measure heats of combustion. Let R stand for the mixture of reactants, P for the product mixture, and K for the calorimeter walls plus the surrounding water bath. Suppose we start with the reactants at 25°C. Let the measured temperature rise due to the reaction be ΔT. Since the closed system of bomb plus water bath has a constant volume and is thermally insulated, we have $q = 0 = w$. Hence $\Delta U = 0$ for the reaction, as noted in step (a) of Fig. 5.2. After ΔT is measured to high accuracy, we cool the system back to 25°C. Then we accurately measure the amount of electrical energy U_{el} that must be supplied to raise the system's temperature from 25°C to 25°C + ΔT; this is step (b) in Fig. 5.2. (We have $U_{\text{el}} = VIt$, where V, I, and t are the voltage, current, and time.) The desired quantity, $\Delta U_{\text{reaction, }298}$, is shown as step ($c$) in Fig. 5.2. The change in the state function U must be the same for path (a) as for path (c) + (b), since both paths start and end at the same states. Hence, $0 = \Delta U_{\text{reaction, 298}} + U_{\text{el}}$, and $\Delta U_{\text{reaction, 298}} = -U_{\text{el}}$. Thus, the measured U_{el} allows $\Delta U_{\text{reaction, 298}}$ to be found.

To find the standard internal energy change ΔU°_{298} for the reaction, we must allow for the changes in U_R and U_P that occur when the reactants and products are brought from the states that occur in the calorimeter to their standard states.

For reactions that do not involve gases, one can use an adiabatic constant-pressure calorimeter. The discussion is similar to that for the adiabatic bomb calorimeter, except that P is held fixed instead of V, and ΔH of reaction is measured instead of ΔU.

Thus one determines either ΔU° or ΔH°. Use of $H = U + PV$ allows interconversion between ΔU° and ΔH°. Equation (5.10) gives $\Delta H^\circ = \sum_i \nu_i(\bar{U}_i^\circ + P_i^\circ \bar{V}_i^\circ)$. Since the standard pressure is the same (1 atm) for all species, we have

$$\Delta H^\circ = \sum_i \nu_i \bar{U}_i^\circ + P^\circ \sum_i \nu_i \bar{V}_i^\circ = \Delta U^\circ + P^\circ \,\Delta V^\circ \tag{5.13}$$

where ΔV° is the change in standard-state volumes for the reaction.

Knowledge of the standard-state molar volumes of reactants and products is all that is needed to find ΔH° from ΔU° or vice versa. Since the molar volumes of gases are many times greater than the molar volumes of condensed phases, we can, to an excellent approximation, consider only the gaseous reactants and products in applying (5.13). The standard states of gases are the hypothetical ideal gases at 1 atm. Hence $\bar{V}^\circ_{i(g)} = RT/P^\circ$, where g indicates a gas. Using a prime to indicate summation over gases only, we have

$$P^\circ \sum_i{}' \nu_{i(g)} \bar{V}^\circ_{i(g)} = P^\circ \sum_i{}' \nu_{i(g)}(RT/P^\circ) = RT \sum_i{}' \nu_{i(g)}$$

With the volumes of liquids and solids neglected, Eq. (5.13) becomes

$$\Delta H^\circ = \Delta U^\circ + RT \sum_i{}' \nu_{i(g)}$$

The sum in this last equation is the change in number of moles of gases for the reaction as written. For example, for $C_3H_8(g) + 5O_2(g) \rightarrow 3CO_2(g) + 4H_2O(l)$, this sum is $3 - 1 - 5 = -3$, where the stoichiometric coefficient of the liquid water is omitted. We shall symbolize this sum by $\Delta|\nu_g|$:

$$\Delta|\nu_g| \equiv \sum_i{}' \nu_{i(g)} \tag{5.14}$$

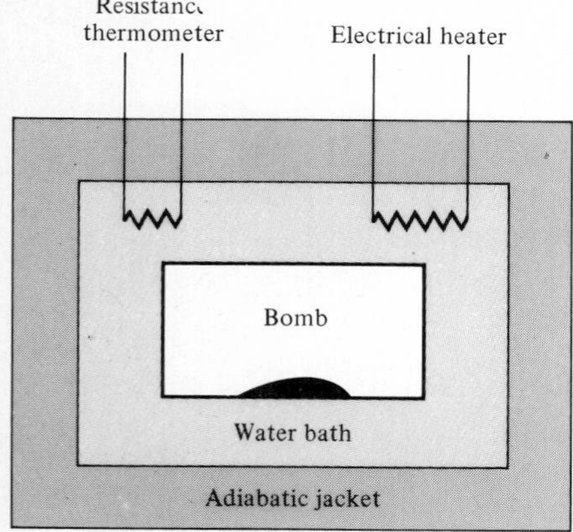

Figure 5.1
Schematic diagram of an adiabatic bomb calorimeter. (Not shown are the electrical wires used to ignite the sample.)

[We use the absolute value because the stoichiometric coefficients for reactants are negative; thus for the above combustion of C_3H_8, we have $\Delta|\nu_g| = |\nu_{g,\text{final}}| - |\nu_{g,\text{initial}}| = 3 - (1 + 5) = -3$.] Our final equation relating the standard enthalpy and internal-energy changes is

$$\Delta H_T^\circ = \Delta U_T^\circ + \Delta|\nu_g|\, RT \tag{5.15}$$

For example, the reaction (5.12) has $\Delta|\nu_g| = 0 - 1.5 = -1.5$; Eq. (5.15) gives

$$\Delta H_{298}^\circ - \Delta U_{298}^\circ = (-1.5)(1.987 \text{ cal mol}^{-1}\text{ K}^{-1})(298.15 \text{ K}) = -888.6 \text{ cal/mol}$$

Knowledge of one of ΔH_{298}° and ΔU_{298}° then allows calculation of the other. Without the approximation of neglecting the volume of $H_2O(l)$, one finds (Prob. 5.23):

$$\Delta(PV)^\circ = -888.2 \text{ cal mol}^{-1} \tag{5.16}$$

so that neglect of the liquid volume is justified.

Suppose we want the standard enthalpy of formation $\Delta H_{f,298}^\circ$ of ethane at 25°C. This is the enthalpy change for the transformation $2C + 3H_2 \rightarrow C_2H_6$, where all species are in their standard states at 25°C. Unfortunately, we cannot react graphite with hydrogen and expect to get ethane. Hence the heat of formation of ethane cannot be measured directly. (This is true for most compounds.) Instead, we determine the heats of combustion of ethane, hydrogen, and graphite, these heats being readily measured. The following values are found at 25°C:

$$C_2H_6 + \tfrac{7}{2}O_2 \rightarrow 2CO_2 + 3H_2O(l) \qquad \Delta H_{298}^\circ = -1560 \text{ kJ/mol} \tag{1}$$

$$C(gr) + O_2 \rightarrow CO_2 \qquad \Delta H_{298}^\circ = -393\tfrac{1}{2} \text{ kJ/mol} \tag{2}$$

$$H_2 + \tfrac{1}{2}O_2 \rightarrow H_2O(l) \qquad \Delta H_{298}^\circ = -286 \text{ kJ/mol} \tag{3}$$

Therefore

$$-(-1560 \text{ kJ/mol}) = -2\bar{H}^\circ(CO_2) - 3\bar{H}^\circ(H_2O) + \bar{H}^\circ(C_2H_6) + 3.5\bar{H}^\circ(O_2)$$

$$2(-393\tfrac{1}{2} \text{ kJ/mol}) = 2\bar{H}^\circ(CO_2) - 2\bar{H}^\circ(O_2) - 2\bar{H}^\circ(C)$$

$$3(-286 \text{ kJ/mol}) = 3\bar{H}^\circ(H_2O) - 3\bar{H}^\circ(H_2) - 1.5\bar{H}^\circ(O_2)$$

where the subscript 298 on the $\bar{H}^\circ$'s is understood. Addition of these three equations gives

$$-85 \text{ kJ/mol} = \bar{H}^\circ(C_2H_6) - 2\bar{H}^\circ(C) - 3\bar{H}^\circ(H_2) \tag{5.17}$$

But the quantity on the right of (5.17) is ΔH° for the desired formation reaction

$$2C + 3H_2 \rightarrow C_2H_6 \tag{5.18}$$

Hence $\Delta H_{f,298}^\circ = -85 \text{ kJ/mol} = -20$ kcal/mol for ethane.

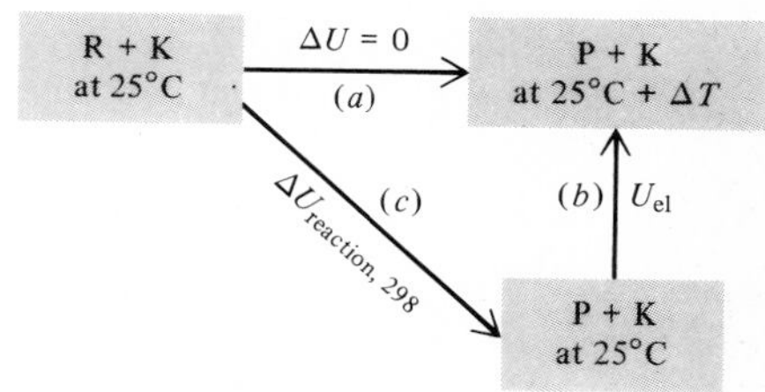

Figure 5.2
Energy relations for an adiabatic bomb calorimeter

We can save time in writing if we just look at the chemical reactions (1) to (3), figure out what factors are needed to multiply each reaction so that they add up to the desired reaction (5.18), and apply these factors to the $\Delta H°$ values. For example, multiplication of reactions (1) to (3) by the factors -1, $+2$, and $+3$, followed by addition, gives reaction (5.18). Hence $\Delta H°_{298}$ for (5.18) is $[-(-1560) + 2(-393\frac{1}{2}) + 3(-286)]$ kJ/mol. The validity of dealing directly with the chemical reactions rather than with equations for the $\Delta H°$'s follows from the similar form of (5.10) and (4.107); replacement of chemical species in a reaction equation by their standard enthalpies gives $\Delta H°$.

The procedure of combining heats of several reactions to obtain the heat of a desired reaction is *Hess's* law.

Since the reactants and products are not ordinarily in their standard states when we carry out a reaction, the actual enthalpy change ΔH for a reaction differs somewhat from $\Delta H°$. However, this difference is generally small, and it would be unlikely that ΔH and $\Delta H°$ would have different signs. For the discussion of this paragraph, we shall assume that ΔH and $\Delta H°$ have the same sign. If this sign is positive, the reaction is said to be *endothermic*; if this sign is negative, the reaction is *exothermic*. For a reaction run at constant pressure, ΔH equals q_P, the heat absorbed by the system. For an endothermic reaction, ΔH is positive and q_P is positive, meaning that the system absorbs heat from the surroundings. An exothermic reaction gives off heat to the surroundings. The combustion reactions (1) to (3) above are all exothermic.

Some values of $\Delta H°_{f,298} = \bar{H}°_{298}$ in kcal/mole are:

$H_2O(l)$	$CO_2(g)$	$NO(g)$	$NaCl(s)$	$C_2H_2(g)$	$C_2H_6(g)$	sucrose(s)
−68	−94	22	−98	54	−20	−531

The positive value for NO indicates that the bonds in $N_2(g)$ and $O_2(g)$ are stronger than those in $2NO(g)$. A more extensive table is given in the Appendix. Once we have built up such a table, we can use Eq. (5.10) to find $\Delta H°_{298}$ for any reaction whose species are listed. For example, for the combustion of 4 moles of ammonia,

$$4NH_3(g) + 3O_2(g) \rightarrow 2N_2(g) + 6H_2O(l) \tag{5.19}$$

the Appendix enthalpy values give for $\Delta H°_{298}$

$$[2(0) + 6(-68.315) - 3(0) - 4(-11.02)] \text{ kcal/mol} = -365.8 \text{ kcal/mol} \tag{5.20}$$

5.4 TEMPERATURE DEPENDENCE OF REACTION HEATS

Suppose we have determined $\Delta H°$ for a reaction at temperature T_1, and we want $\Delta H°$ at T_2. Differentiation of Eq. (5.10) with respect to T gives $d\,\Delta H°/dT = \sum_i \nu_i\, d\bar{H}_i°/dT$, where we used the fact that the derivative of a sum equals the sum of the derivatives. (The derivatives are not partial derivatives; since P is fixed at the standard-state value 1 atm, $\bar{H}_i°$ and $\Delta H°$ depend only on T.) Use of $(\partial \bar{H}_i/\partial T) = \bar{C}_{P,i}$ [Eq. (4.34)] gives

$$\frac{d\,\Delta H°}{dT} = \sum_i \nu_i \bar{C}°_{P,i} \equiv \Delta C°_P \tag{5.21}$$

where $\bar{C}^\circ_{P,i}$ is the molar heat capacity of substance i in its standard state at the temperature of interest, and where we defined the *standard heat-capacity change* ΔC°_P for the reaction as equal to the sum in (5.21).

Integration of (5.21) between the limits T_1 and T_2 gives

$$\Delta H^\circ_{T_2} - \Delta H^\circ_{T_1} = \int_{T_1}^{T_2} \Delta C^\circ_P \, dT \tag{5.22}$$

which is the desired relation (*Kirchhoff's law*).

An easy way to see the validity of (5.22) is from the following diagram:

$$\begin{array}{ccc} \text{Standard-state reactants at } T_2 & \overset{(a)}{\rightarrow} & \text{standard-state products at } T_2 \\ \downarrow (b) & & \uparrow (d) \\ \text{Standard-state reactants at } T_1 & \overset{(c)}{\rightarrow} & \text{standard-state products at } T_1 \end{array} \tag{5.23}$$

We can go from reactants to products at T_2 by a path consisting of step (a) or by a path consisting of steps $(b) + (c) + (d)$. Since enthalpy is a state function, ΔH is independent of path and $\Delta H_a = \Delta H_b + \Delta H_c + \Delta H_d$. Use of (2.116) for ΔH_b and ΔH_d then gives Eq. (5.22).

The quantity $\bar{C}^\circ_{P,i}$ $(= d\bar{H}^\circ_i/dT)$ depends on T only. (Over a short temperature range, we can often neglect the temperature dependence of ΔC°_P.) The temperature dependence of $\bar{C}^\circ_P$ of a substance is commonly expressed by a power series of the form

$$\bar{C}^\circ_P = a + bT + cT^2 \tag{5.24}$$

where the coefficients a, b, c are determined by a least-squares fit of the experimental $\bar{C}^\circ_P(T)$ data. Sometimes a series of the form $\bar{C}^\circ_P = d + eT + fT^{-2}$ is used. Such power series are valid only over the temperature range of the data used to determine the coefficients and have no theoretical significance. Figure 5.3 shows $\bar{C}^\circ_P$ vs. T for some substances.

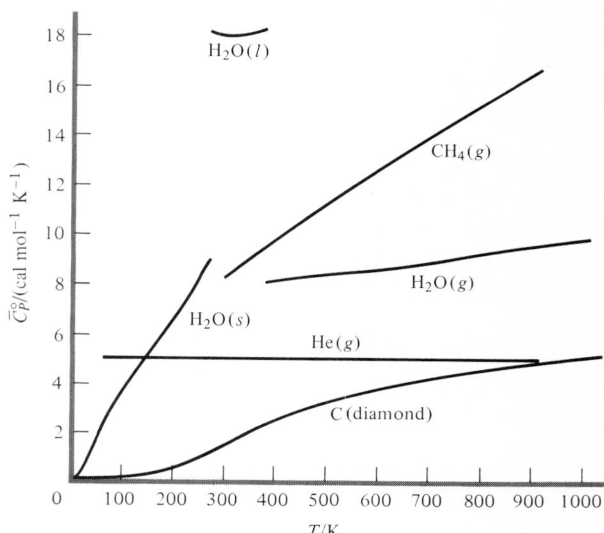

Figure 5.3
$\bar{C}^\circ_P$ vs. T for sundry substances.

As an example, the standard molar heat capacities at constant pressure of O_2, CO, and CO_2 in the range 300 to 1500 K can each be represented by a power series of the form (5.24) with the following coefficients:

	$a/(\text{cal mol}^{-1}\ \text{K}^{-1})$	$10^3 b/(\text{cal mol}^{-1}\ \text{K}^{-2})$	$10^7 c/(\text{cal mol}^{-1}\ \text{K}^{-3})$
O_2	6.0954	3.2533	−10.171
CO	6.3424	1.8363	−2.801
CO_2	6.3957	10.1933	−35.333

(Note that a divided by cal mol^{-1} K^{-1} is a dimensionless number; hence a has units of cal mol^{-1} K^{-1}, as it should.) The Appendix yields $\Delta H^\circ_{f,298}$ values for O_2, CO, and CO_2 of 0, −26,416, and −94,051 cal/mol, respectively. Hence, ΔH°_{298} for the reaction

$$2CO + O_2 \rightarrow 2CO_2 \tag{5.25}$$

is

$$\Delta H^\circ_{298}/(\text{cal mol}^{-1}) = 2(-94{,}051) - 2(-26{,}416) - 0 = -135{,}270$$

The $\bar{C}^\circ_P$'s have the form of Eq. (5.24). Hence, $\Delta C^\circ_P = \Delta a + T\,\Delta b + T^2\,\Delta c$, where $\Delta a = 2a(CO_2) - 2a(CO) - a(O_2)$, with similar expressions for Δb and Δc. Substitution of ΔC°_P into (5.22) and integration gives

$$\Delta H^\circ_{T_2} - \Delta H^\circ_{T_1} = \Delta a(T_2 - T_1) + \tfrac{1}{2}\Delta b(T_2^2 - T_1^2) + \tfrac{1}{3}\Delta c(T_2^3 - T_1^3)$$

The above-listed a, b, and c values give

$$\Delta a = [2(6.3957) - 2(6.3424) - 6.0954]\ \text{cal mol}^{-1}\ \text{K}^{-1} = -5.9888\ \text{cal mol}^{-1}\ \text{K}^{-1}$$

$$\Delta b = 13.4607 \times 10^{-3}\ \text{cal mol}^{-1}\ \text{K}^{-2} \qquad \Delta c = -54.893 \times 10^{-7}\ \text{cal mol}^{-1}\ \text{K}^{-3}$$

Letting $T_1 = 298.15$ K, we therefore have

$$\begin{aligned}\Delta H_{T_2}/(\text{cal/mol}) = {} & -135{,}270 - 5.9888(T_2/\text{K} - 298.15) \\ & + 6.7304 \times 10^{-3}[(T_2/\text{K})^2 - 298.15^2] \\ & - 18.298 \times 10^{-7}[(T_2/\text{K})^3 - 298.15^3]\end{aligned}$$

For example, substitution of $T_2 = 1000$ K gives $\Delta H^\circ_{1000} = -135{,}122$ cal/mol.

5.5 CONVENTIONAL ENTROPIES AND THE THIRD LAW

The second law of thermodynamics tells us how to measure changes in entropy but does not provide absolute entropies. To set up a table of conventional entropy values, we proceed as with enthalpy to (1) assign an arbitrary S value to each element in a chosen reference state and (2) find ΔS for the transformation from elements in their reference states to the substance of interest in the state of interest.

The choice of reference state for entropy differs from that for enthalpy. The choice is the pure element in its stable condensed form (solid or liquid) at 1 atm in the limit $T \to 0$ K. We arbitrarily set $\bar{S}$ for each *element* in this state equal to zero:

$$\bar{S}_0^\circ \equiv \lim_{T\to 0} \bar{S}_T^\circ = 0 \qquad \text{element in stable condensed form} \tag{5.26*}$$

The degree superscript in (5.26) indicates the standard pressure of 1 atm; the subscript zero indicates a temperature of absolute zero. As we shall see, absolute zero is unattainable, so we use the limit in (5.26). (Helium remains a liquid as T goes to zero at 1 atm; all other elements are solids in this limit.)

To find the conventional entropy $\bar{S}(T, P)$ for an element at any T and P, we use (4.64) and (5.26), including also the ΔS of any phase changes that occur between absolute zero and T.

How do we find the conventional entropy of a *compound*? We saw that ΔU or ΔH values for reactions are readily measured as q_V or q_P for the reactions, and these ΔH values then allow us to set up a table of conventional enthalpies for compounds. However, ΔS for a chemical reaction is not so easily measured. We have $\Delta S = q_{\text{rev}}/T$ for constant temperature. However, a chemical reaction is an irreversible process, and measurement of the isothermal irreversible heat of a reaction does not give ΔS for the reaction. As we shall see in Chap. 14, one can carry out a chemical reaction reversibly in an electrochemical cell and use measurements on such cells to find ΔS values for reactions. However, the number of reactions that can be carried out in an electrochemical cell is too limited to allow us to set up a complete table of conventional entropies of compounds, so we have a problem.

The solution to our problem is provided by the third law of thermodynamics, which we now discuss. Around 1900, Theodore William Richards measured ΔG° as a function of temperature for several chemical reactions carried out reversibly in electrochemical cells. Walther Nernst pointed out that Richards's data indicated that the slope of the ΔG°-vs.-T curve for a reaction goes to zero as T goes to absolute zero. Therefore in 1907 Nernst postulated that for any change

$$\lim_{T\to 0} (\partial \Delta G/\partial T)_P = 0 \tag{5.27}$$

From (4.62), we have $(\partial G/\partial T)_P = -S$; hence $(\partial \Delta G/\partial T)_P = \partial(G_2 - G_1)/\partial T = \partial G_2/\partial T - \partial G_1/\partial T = -S_2 + S_1 = -\Delta S$. Thus (5.27) implies that

$$\lim_{T\to 0} \Delta S = 0 \tag{5.28}$$

Nernst believed (5.28) to be valid for any process. However, later experimental work by Simon and others showed (5.28) to hold only for changes in which only pure substances each in internal equilibrium are involved. Thus (5.28) does not hold for a transition involving a supercooled liquid (which is not in internal equilibrium) or for mixing (which involves impure substances).

We thus adopt as the *Nernst–Simon statement of the third law of thermodynamics*:

For any isothermal process that involves only pure substances, each in internal equilibrium, the entropy change satisfies

$$\lim_{T\to 0} \Delta S = 0 \tag{5.29*}$$

To see how (5.29) is used to determine conventional entropies of compounds, consider the process

$$H_2(s) + \tfrac{1}{2}O_2(s) \rightarrow H_2O(s) \tag{5.30}$$

where the pure, separated elements at 1 atm and T are converted to the compound H_2O at 1 atm and T. For this process,

$$\Delta S = \bar{S}^\circ(H_2O) - \bar{S}^\circ(H_2) - \tfrac{1}{2}\bar{S}^\circ(O_2) \tag{5.31}$$

Our arbitrary choice of the entropy of each element as zero at 0 K and 1 atm [Eq. (5.26)] gives

$$\lim_{T\to 0} \bar{S}^\circ(H_2) = 0 = \lim_{T\to 0} \bar{S}^\circ(O_2) \tag{5.32}$$

The third law, Eq. (5.29), gives for the process (5.30): $\lim_{T\to 0} \Delta S = 0$. Hence in the limit $T \to 0$, Eq. (5.31) becomes

$$\lim_{T\to 0} \bar{S}^\circ(H_2O) = 0 \tag{5.33}$$

which we write more succinctly as $\bar{S}_0^\circ(H_2O) = 0$.

Exactly the same argument applies for any compound. Hence $\bar{S}_0^\circ = 0$ for any pure substance (element or compound) in internal equilibrium. The third law (5.29) shows that an isothermal pressure change of a pure substance in internal equilibrium in the limit of absolute zero has $\Delta S = 0$. Hence we can drop the superscript degree (which indicates $P = 1$ atm). Also, if $\bar{S}_0 = 0$, then $S_0 = 0$ for any amount of pure substance. Our conclusion is that the conventional entropy of any pure substance (element or compound) in internal equilibrium is zero in the limit $T \to 0$:

$$S_0 = 0 \qquad \text{pure substance in int. equilib.} \tag{5.34*}$$

Now that we have the conventional entropies of substances at $T = 0$, their conventional entropies at any T and P are readily calculated from Eq. (4.64). We are most interested in $\bar{S}_T^\circ$, the conventional molar entropy in the standard state at temperature T. Let state 1 in (4.64) be 0 K and 1 atm; then $S_1 = 0$. Let state 2 in (4.64) be the standard state at temperature T_2. Since P is the same for states 1 and 2, we need only calculate the entropy change due to the temperature increase; of course, we must also include contributions to ΔS due to phase changes.

Consider, for example, a substance which is a liquid at T_2 and 1 atm. From Eqs. (4.64) [or (3.36)], (3.33), and (5.34), the conventional standard-state molar entropy of a pure liquid at an arbitrary temperature T_2 is

$$\bar{S}_{T_2}^\circ = \int_0^{T_m} \frac{\bar{C}_P^\circ(s)}{T}\,dT + \frac{\Delta \bar{H}_m^\circ}{T_m} + \int_{T_m}^{T_2} \frac{\bar{C}_P^\circ(l)}{T}\,dT \tag{5.35}$$

where T_m is the normal melting point of the substance, $\bar{C}_P^\circ(s)$ and $\bar{C}_P^\circ(l)$ are the molar heat capacities at 1 atm of the solid and liquid forms of the substance, $\Delta \bar{H}_m^\circ$ is the molar enthalpy change on melting at 1 atm. Frequently a solid undergoes one or more phase transitions from one crystalline form to another before the melting point is reached. For example, the stable low-temperature form of sulfur is rhombic sulfur; at 95°C, solid rhombic sulfur is transformed to solid monoclinic sulfur (whose melting point is 119°C). The entropy contribution of each such solid–solid phase transition must be included in (5.35) as an additional term $\Delta\bar{H}_{tr}/T_{tr}$, where $\Delta\bar{H}_{tr}$ is the molar enthalpy change of the phase transition at temperature T_{tr}.

For a substance that is a gas at 1 atm and T_2, we include the ΔS of vaporization at the boiling point T_b plus the ΔS of heating the gas from T_b to T_2. In addition, since the standard state is the *ideal* gas at 1 atm, we include the (small) correction for the difference between the ideal-gas and real-gas entropies. The quantity $\bar{S}_{id}(T, 1 \text{ atm}) - \bar{S}_{re}(T, 1 \text{ atm})$ is calculated from the hypothetical isothermal three-step process (5.5). For step (*a*) of (5.5), Eq. (4.61) [or (4.64)] gives

$$\Delta S_a = -\int_{1 \text{ atm}}^{0} \left(\frac{\partial \bar{V}}{\partial T}\right)_P dP = \int_{0}^{1 \text{ atm}} \left(\frac{\partial \bar{V}}{\partial T}\right)_P dP \tag{5.36}$$

For step (*b*), we use a result of statistical mechanics which shows that the entropy of a real gas and the entropy of the corresponding ideal gas (no intermolecular interactions) become equal in the limit of zero density (see Prob. 22.54); hence $\Delta S_b = 0$. For step (*c*), Eq. (4.61) and $P\bar{V} = RT$ give

$$\Delta S_c = -\int_{0}^{1 \text{ atm}} \frac{R}{P} dP \tag{5.37}$$

The desired ΔS is the sum $\Delta S_a + \Delta S_b + \Delta S_c$:

$$\bar{S}_{id}(T, 1 \text{ atm}) - \bar{S}_{re}(T, 1 \text{ atm}) = \int_{0}^{1 \text{ atm}} \left[\left(\frac{\partial \bar{V}}{\partial T}\right)_P - \frac{R}{P}\right] dP \tag{5.38}$$

Knowledge of the P-V-T behavior of the real gas allows calculation of the contribution (5.38) to $\bar{S}^\circ$, the standard-state molar entropy of the gas.

The first integral in (5.35) presents a problem in that $T = 0$ is unattainable (Sec. 5.8); moreover, it is impractical to measure $\bar{C}_P^\circ(s)$ below a few degrees Kelvin. Debye's statistical-mechanical theory of solids (Sec. 24.10) and experimental data show that specific heats of nonmetallic solids at very low temperatures have the form

$$\bar{C}_P^\circ \approx \bar{C}_V^\circ = aT^3 \tag{5.39}$$

where a is a constant characteristic of the substance. At the very low temperatures to which (5.39) applies, the difference $TV\alpha^2/\kappa$ between C_P and C_V is negligible [Eq. (4.74)]; besides T's vanishing, α also vanishes in the limit of absolute zero (see Prob. 5.26). For metals, a statistical-mechanical treatment [*Kestin and Dorfman*, sec. 9.5.2] and experimental data show that at very low temperatures

$$\bar{C}_P^\circ \approx \bar{C}_V^\circ = aT^3 + bT \tag{5.40}$$

where a and b are constants. (The term bT arises from the behavior of the conduction electrons.) One uses measured values of $\bar{C}_P^\circ$ at very low temperatures to determine the constant(s) in (5.39) or (5.40). Then one uses (5.39) or (5.40) to extrapolate $\bar{C}_P^\circ$ to $T = 0$ K. (Note that C_P vanishes as T goes to zero.) For example, let $\bar{C}_P^\circ(T_{low})$ be the observed value of $\bar{C}_P^\circ$ of a nonconductor at the lowest temperature for which $\bar{C}_P^\circ$ is conveniently measurable (typically about 10 K). Provided T_{low} is low enough for (5.39) to apply, we have

$$aT_{low}^3 = \bar{C}_P^\circ(T_{low}) \tag{5.41}$$

We write the first integral in (5.35) as

$$\int_0^{T_m} \frac{\bar{C}_P^\circ}{T}\,dT = \int_0^{T_{\text{low}}} \frac{\bar{C}_P^\circ}{T}\,dT + \int_{T_{\text{low}}}^{T_m} \frac{\bar{C}_P^\circ}{T}\,dT \tag{5.42}$$

The first integral on the right of (5.42) is evaluated by use of (5.39) and (5.41)

$$\int_0^{T_{\text{low}}} \frac{\bar{C}_P^\circ}{T}\,dT = \int_0^{T_{\text{low}}} \frac{aT^3}{T}\,dT = \frac{aT^3}{3}\bigg|_0^{T_{\text{low}}} = \frac{aT_{\text{low}}^3}{3} = \frac{\bar{C}_P^\circ(T_{\text{low}})}{3} \tag{5.43}$$

To evaluate the second integral on the right of (5.42) and the integral from T_m to T_2 in (5.35), we can fit polynomials like (5.24) to the measured $\bar{C}_P^\circ(T)$ data and then integrate the resulting expressions for $\bar{C}_P^\circ/T$. Alternatively, we can use graphical integration: we plot the measured values of $\bar{C}_P^\circ(T)/T$ against T between the relevant temperature limits, draw a smooth curve joining the points, and measure the area under the curve to evaluate the integral. [Equivalently, since $(C_P/T)\,dT = C_P\,d\ln T$, we can plot C_P vs. $\ln T$ and measure the area under the curve.]

An example is the calculation of the conventional entropy $\bar{S}_{298}^\circ$ of SO_2. The measured $\bar{C}_P^\circ$ at 15 K is 0.83 cal K^{-1} mol^{-1}; hence from (5.43), integration from 0 to 15 K gives an entropy contribution of 0.3 cal mol^{-1} K^{-1}. Integration of $\bar{C}_P^\circ/T$ from 15 K to the normal melting point 197.6 K gives a contribution of 20.1 cal mol^{-1} K^{-1} (Fig. 5.4). The heat of fusion is 1769 cal/mole, giving an entropy contribution of (1769/197.6) cal mol^{-1} K^{-1} = 8.95 cal mol^{-1} K^{-1}. Integration of $\bar{C}_P^\circ/T$ from 197.6 K to the normal boiling point 263.1 K contributes 6.0 cal mol^{-1} K^{-1}. The latent heat of vaporization is 5960 cal/mole, giving an entropy contribution of 22.65 cal mol^{-1} K^{-1}. Integration of $\bar{C}_P^\circ$ of the gas from 263.1 to 298.15 K contributes 1.2 cal mol^{-1} K^{-1}. Use of (5.38) to calculate $\bar{S}_{\text{id}} - \bar{S}_{\text{re}}$ at 298.15 K gives 0.1 cal mol^{-1} K^{-1}. Thus the conventional molar standard-state entropy of SO_2 gas at 298.15 K is

$$\begin{aligned}\bar{S}_{298}^\circ &= (0.3 + 20.1 + 8.95 + 6.0 + 22.65 + 1.2 + 0.1)\text{ cal mol}^{-1}\text{ K}^{-1}\\ &= 59.3\text{ cal mol}^{-1}\text{ K}^{-1}\end{aligned} \tag{5.44}$$

Some conventional entropies $\bar{S}_{298}^\circ$ in cal mole^{-1} K^{-1} are:

NaCl(s)	Sucrose(s)	$H_2O(l)$	$N_2(g)$	$CH_4(g)$	$C_2H_6(g)$	n-$C_4H_{10}(g)$
17.2	86.1	16.7	45.8	44.5	54.8	74.1

A more complete table is in the Appendix.

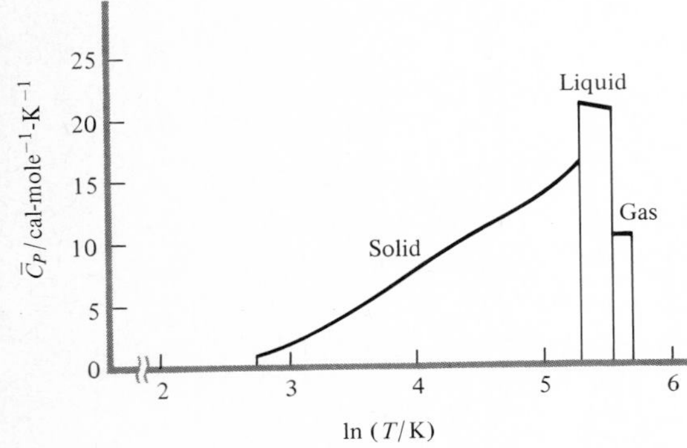

Figure 5.4
Integration of $\bar{C}_P\,d\ln T$ for SO_2.

Conventional entropies are often called absolute entropies. However, this name is inappropriate in that these entropies are not absolute entropies but relative (conventional) entropies. Since full consideration of this question requires statistical mechanics, we postpone discussion until Sec. 22.9.

From the listed $\bar{S}^\circ_{298}$ values, one can use (4.64) to find the conventional entropy $\bar{S}$ of a substance at any T and P.

One can calculate ΔS°_{298} (the *standard entropy change* at 298.15 K) for a chemical reaction from the tabulated entropy values and [compare Eq. (5.10)]

$$\Delta S^\circ_T = \sum_i \nu_i \bar{S}^\circ_{T,i} \tag{5.45}$$

Thus for the reaction (5.19), the Appendix entropy values give

$$\Delta S^\circ_{298}/(\text{cal mol}^{-1}\ \text{K}^{-1}) = 2(45.77) + 6(16.71) - 3(49.003) - 4(45.97)$$

$$\Delta S^\circ_{298} = -139.09\ \text{cal/(mol K)} \tag{5.46}$$

From the entropy values $\bar{S}^\circ_{298}$ listed in the Appendix, we note the following. Molar entropies of gases tend to be higher than those of liquids; molar entropies of liquids tend to be higher than those of solids. Molar entropies tend to increase with increasing number of atoms in a molecule.

Differentiation of (5.45) with respect to T, use of Eqs. (4.60) and (5.21), followed by integration, gives (Prob. 5.18)

$$\Delta S^\circ_{T_2} - \Delta S^\circ_{298} = \int_{298}^{T_2} \frac{\Delta C^\circ_P}{T}\, dT \tag{5.47}$$

so that ΔS° at any temperature can be calculated from ΔS°_{298}. Note that (5.47) and (5.22) are applicable only if no species undergoes a phase change in the temperature interval. An easy way to get (5.47) is to equate ΔS for the two paths connecting reactants and products at T_2 in (5.23).

With the statement of the third law of thermodynamics, we have completed the laws of thermodynamics.

5.6 CONVENTIONAL GIBBS ENERGIES

The Gibbs function G is defined as $H - TS$. Once conventional values of enthalpy and entropy are assigned (as discussed in preceding sections), the conventional value of G is fixed. For a pure substance in its standard state at temperature T, the conventional molar Gibbs energy is

$$\bar{G}^\circ_T = \bar{H}^\circ_T - T\bar{S}^\circ_T \tag{5.48}$$

In particular, at 298.15 K, $\bar{G}^\circ_{298} = \bar{H}^\circ_{298} - (298.15\ \text{K})\bar{S}^\circ_{298}$. For an element, the conventional $\bar{H}^\circ_{298}$ is zero, but $\bar{S}^\circ_{298}$ is not zero. ($\bar{S}^\circ_0$ is zero.) Hence the conventional $\bar{G}^\circ_{298}$ for an element is not zero.

The *standard change in Gibbs free energy* ΔG°_T for a chemical reaction is defined as the change in G for the process of converting stoichiometric numbers of moles of the separated pure reactants, each in its standard state at T, into the separated pure products in their standard states at T. Similar to Eqs. (5.45) and (5.10), we have

$$\Delta G^\circ_T = \sum_i \nu_i \bar{G}^\circ_{T,i} \tag{5.49}$$

If the reaction is one of formation of a compound from its elements, then ΔG_T° is the standard Gibbs free energy of formation $\Delta G_{f,T}^\circ$. (Of course, for any element, $\Delta G_{f,T}^\circ$ is zero, since "formation" of an element from itself is no change at all.) Recall from Sec. 4.5 that ΔG is physically meaningful only for isothermal processes.

Consider some examples. For $H_2O(l)$, substitution of the Appendix values of $\bar{H}_{298}^\circ$ and $\bar{S}_{298}^\circ$ in (5.48) gives

$$\bar{G}_{298}^\circ(H_2O,\ l) = -68{,}315 \text{ cal mol}^{-1} - (298.15 \text{ K})(16.71 \text{ cal mol}^{-1} \text{ K}^{-1})$$

$$= -73{,}297 \text{ cal/mole} \tag{5.50}$$

For $H_2(g)$, the Appendix gives $\bar{H}_{298}^\circ = 0$, $\bar{S}_{298}^\circ = 31.208$ cal mol^{-1} K^{-1}. Use of (5.48) gives $\bar{G}_{298}^\circ(H_2, g) = -9305$ cal/mol. For O_2, we find $\bar{G}_{298}^\circ(O_2, g) = -14{,}610$ cal/mol. Hence from (5.49), the standard Gibbs free energy of formation of water [Eq. (5.12)] at 298.15 K is

$$\Delta G_{f,298}^\circ(H_2O,\ l) = [-73{,}297 - (-9305) - \tfrac{1}{2}(-14{,}610)] \text{ cal mol}^{-1}$$

$$= -56{,}687 \text{ cal/mole}$$

which checks with the value listed in the Appendix.

Now consider the calculation of ΔG_{298}° for a general chemical reaction. Once we have used $\bar{H}_{298}^\circ$ and $\bar{S}_{298}^\circ$ values to calculate $\bar{G}_{298}^\circ$ values for the reactants and products, we can find ΔG_{298}° for the reaction from Eq. (5.49). Thus if conventional Gibbs energies $\bar{G}_{298}^\circ$ of substances were tabulated, we could readily use them to calculate ΔG_{298}° for desired reactions. However, tables do not list $\bar{G}_{298}^\circ$ values of substances; instead, they list $\Delta G_{f,298}^\circ$ values (as in the Appendix). We shall show that

$$\Delta G_T^\circ = \sum_i \nu_i \, \Delta G_{f,T,i}^\circ \tag{5.51*}$$

where $\Delta G_{f,T,i}^\circ$ is the standard Gibbs energy of formation of substance i at temperature T. Thus, instead of using $\bar{G}^\circ$ values as in (5.49), we can use ΔG_f° values.

To prove (5.51), consider the reaction $a\text{A} + b\text{B} \rightarrow c\text{C} + d\text{D}$, where a, b, c, d are the unsigned stoichiometric coefficients and A, B, C, D are compounds. Figure 5.5 shows two different isothermal paths from reactants to products in their standard states. Step 1 is a direct conversion of reactants to products. Step 2 is a conversion of reactants to elements; step 3 is a conversion of elements to products. (Of course, the same elements produced by the decomposition of the reactants will form the products.) Since G is a state function, ΔG is independent of path and $\Delta G_1 = \Delta G_2 + \Delta G_3$. We have $\Delta G_1 = \Delta G_T^\circ$ for the reaction. The reverse of process 2 would form $a\text{A} + b\text{B}$ from their elements; hence, $-\Delta G_2 = a\,\Delta G_{f,T}^\circ(\text{A}) + b\,\Delta G_{f,T}^\circ(\text{B})$, where $\Delta G_{f,T}^\circ(\text{A})$ is the Gibbs energy of formation of compound A at temperature T. Step 3 is the

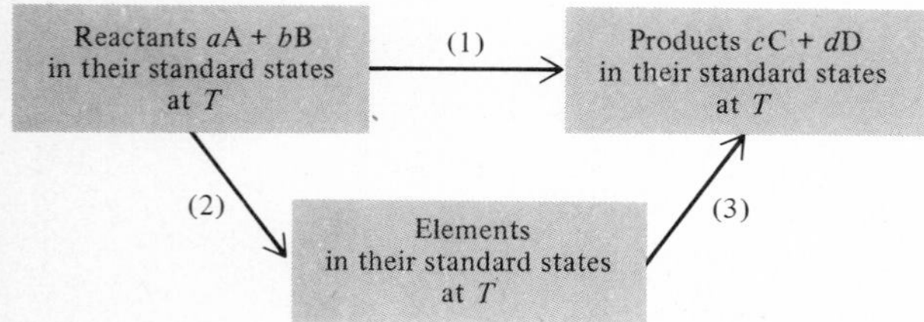

Figure 5.5
Steps used to relate ΔG° of a reaction to ΔG_f° of reactants and products.

formation of cC + dD from their elements, so $\Delta G_3 = c\ \Delta G^\circ_{f,T}(\text{C}) + d\ \Delta G^\circ_{f,T}(\text{D})$. The relation $\Delta G_1 = \Delta G_2 + \Delta G_3$ becomes

$$\Delta G^\circ_T = -a\ \Delta G^\circ_{f,T}(\text{A}) - b\ \Delta G^\circ_{f,T}(\text{B}) + c\ \Delta G^\circ_{f,T}(\text{C}) + d\ \Delta G^\circ_{f,T}(\text{D})$$

which is Eq. (5.51). Q.E.D.

The same argument shows that an equation like (5.51) holds for changes in other state functions for a reaction. For example,

$$\Delta H^\circ_T = \sum_i \nu_i\ \Delta H^\circ_{f,T,i} \tag{5.52}$$

which may be compared with Eq. (5.10). Note from Eq. (5.9) that at 298.15 K the conventional enthalpy $\bar{H}^\circ_{298,i}$ equals the enthalpy of formation $\Delta H^\circ_{f,298,i}$, so that at this temperature (5.10) and (5.52) are identical equations. However, at temperatures other than 298.15 K, the conventional enthalpies and the enthalpies of formation differ (since the conventional enthalpies of elements at temperatures other than 298.15 K are not zero).

Some $\Delta G^\circ_{f,298}$ values in kcal/mole are:

$H_2O(l)$	NaCl(s)	K(s)	NO(g)	$N_2O_5(g)$	$CH_4(g)$	$C_2H_6(g)$	sucrose(s)
−57	−92	0	21	28	−12	−8	−369

The Appendix lists further data.

As an example, Eq. (5.51) and the Appendix $\Delta G^\circ_{f,298}$ values give ΔG°_{298} for the reaction (5.19) as

$$[2(0) + 6(-56.687) - 3(0) - 4(-3.94)]\ \text{kcal/mol} = -324.4\ \text{kcal/mol} \tag{5.53}$$

Suppose we want ΔG° for a reaction at a temperature other than 298.15 K. We previously showed how to find ΔS° and ΔH° at temperatures other than 298.15 K. Use of $\Delta G^\circ_T = \Delta H^\circ_T - T\ \Delta S^\circ_T$ then gives ΔG° at any temperature T.

Tabulations of thermodynamic data most commonly list $\Delta H^\circ_{f,298} = \bar{H}^\circ_{298}$, $\bar{S}^\circ_{298}$, $\Delta G^\circ_{f,298}$, and $\bar{C}^\circ_{P,298}$. Values for other temperatures are often listed also. Instead of $\Delta H^\circ_{f,298}$, one often finds either $\bar{H}^\circ_T - \bar{H}^\circ_0$ or $\bar{H}^\circ_T - \bar{H}^\circ_{298}$ tabulated. Sometimes $(\bar{H}^\circ_T - \bar{H}^\circ_0)/T$ or $(\bar{H}^\circ_T - \bar{H}^\circ_{298})/T$ is tabulated; the reason for dividing by T is to make the function vary slowly with T, which allows accurate interpolation at temperatures not listed.

The standard tabulation of thermodynamic data for inorganic compounds, one- and two-carbon organic compounds, and ions, is D. D. Wagman et al., Selected Values of Chemical Thermodynamic Properties, *Natl. Bur. Stand. Tech. Note* 270-3, 270-4, 270-5, 270-6, 270-7, . . . , Washington, 1968–. (Eventually, these *Technical Notes* will be collected in a single volume.) Thermodynamic data for organic compounds are tabulated in *Selected Values of Properties of Hydrocarbons and Related Compounds*, 1966–, and in *Selected Values of Properties of Chemical Compounds*, 1966–, both published by the Thermodynamics Research Center, Texas A&M University. Data for biochemical compounds are tabulated by R. C. Wilhoit in chap. 2 of H. D. Brown (ed.), *Biochemical Microcalorimetry*, Academic, 1969. Other sources of thermodynamic data are D. R. Stull and H. Prophet, JANAF Thermochemical Tables, 2d ed., *Natl. Bur. Stand. Publ.* NSRDS-NBS 37, Washington, 1971; D. R. Stull, E. F. Westrum, and G. C. Sinke, *The Chemical Thermodynamics of Organic Compounds*, Wiley, 1969.

Almost all existing tabulations use the thermochemical calorie (= 4.184 J) as the unit of energy. (Many physicists and engineers use the international-steam-table calorie, defined as 4.1868 J.) The calorie is a historical relic that would best be discarded in favor of the joule, but this has not yet happened.

We have discussed calculation of thermodynamic properties from calorimetric data. We shall later see that statistical mechanics allows thermodynamic properties of an ideal gas to be calculated from molecular data (bond distances and angles, vibrational frequencies, etc.).

5.7 ESTIMATION OF THERMODYNAMIC PROPERTIES

Frequently one wants values of thermodynamic properties for a compound for which data do not exist. Several methods have been proposed for estimating thermodynamic quantities.

Many molecular properties (thermodynamic and nonthermodynamic) can be estimated as the sum of contributions from the bonds. One uses experimental data on compounds for which data exist to arrive at typical values for the bond contributions to the property in question. These bond contributions are then used to estimate the property in molecules for which data are unavailable. It should be emphasized that this approach is only an approximation.

Benson and Buss constructed a table of bond contributions to $\bar{C}^\circ_{P,298}$, $\bar{S}^\circ_{298}$, and $\Delta H^\circ_{f,298}$ for compounds *in the ideal-gas state*. Addition of these contributions allows one to estimate ideal-gas thermodynamic properties of a compound. [For a compound that is a liquid (or solid) at 25°C and 1 atm, the ideal-gas state (like a supercooled liquid state) is not stable. Let P' be the liquid's equilibrium vapor pressure at 25°C. To relate observed thermodynamic properties of the liquid at 25°C and 1 atm to ideal-gas properties at 25°C and 1 atm, we consider the following four-step process: (*a*) reversibly vaporize the liquid at 25°C and P'; (*b*) isothermally reduce the gas pressure to zero; (*c*) wave a magic wand that transforms the real gas to an ideal gas; (*d*) isothermally compress the ideal gas to $P = 1$ atm. Since the difference between real and ideal gas properties at 1 atm is usually quite small, we can usually replace steps (*b*), (*c*), and (*d*) by a compression of the real gas from pressure P' to 1 atm.]

For details of the method (including a contribution to $\bar{S}^\circ_{298}$ related to the symmetry of the molecule), see S. W. Benson and J. H. Buss, *J. Chem. Phys.*, **29**, 546 (1958).

The Benson–Buss method of bond contributions gives $\bar{S}^\circ_{298}$ and $\bar{C}^\circ_{P,298}$ values with typical errors of 1 to 2 cal mol^{-1} K^{-1} and $\Delta H^\circ_{f,298}$ values with typical errors of 3 to 6 kcal mol^{-1}.

An improvement is the method of group contributions. Here one estimates thermodynamic quantities as the sum of contributions from groups in the molecule. In this method, a *group* is defined as a polyligated atom together with the atoms bonded to it. (A polyligated atom is an atom that is bonded to more than one other atom.) For example, CH_3CH_2OH has the following three groups: **C**$(H_3)(C)$, **C**$(C)(H_2)(O)$, and **O**$(C)(H)$, where the central atom of each group is in boldface type and the ligands are in parentheses. For the group-contribution method one obviously needs tables with more entries than for the bond-contribution method, but greater accuracy is attained.

Extensive tables of group contributions to ΔH°_f, $\bar{C}^\circ_P$, and $\bar{S}^\circ$ for temperatures from 300 to 1500 K are given in S. W. Benson et al., *Chem. Rev.*, **69**, 279 (1969); S. W. Benson, *Thermochemical Kinetics*, 2d ed., Wiley-Interscience, 1976. These tables give $\bar{C}^\circ_P$ and $\bar{S}^\circ$ ideal-gas values with typical errors of 1 cal mol^{-1} K^{-1} and give ΔH°_f ideal-gas values with typical errors of 1 or 2 kcal/mol.

A useful rule of thumb for relating entropies of liquids and gases is *Trouton's rule*, which states that $\Delta\bar{S}$ for the vaporization of a liquid at its normal boiling point is roughly 21 cal mol^{-1} K^{-1}. The rule fails for highly polar liquids and for liquids that boil below 150 K. Some values of $\Delta\bar{H}$ and $\Delta\bar{S}$ of vaporization at the normal boiling point T_{nbp} follow.

	N_2	NH_3	CCl_4	H_2O	C_2H_6	SO_2
$\Delta\bar{H}_{vap}$/cal-mol^{-1}	1333	5580	7170	9720	3520	5960
T_{nbp}/K	77.6	240	350	373	185	263
$\Delta\bar{S}_{vap}$/cal-mol^{-1}-K^{-1}	17.2	23.3	20.5	26.0	19.0	22.7

$\Delta\bar{H}_{vap}$ is a rough measure of the strength of intermolecular attractions in the liquid.

One finds that $\Delta\bar{S}$ of fusion varies widely from compound to compound. Some values in cal mol^{-1} K^{-1} are H_2O, 5.3; C_2H_6, 7.6; CCl_4, 2.4; SO_2, 8.9.

Closely related to the concept of bond contributions to ΔH_f° is the concept of *average bond energy*. Consider the gas-phase atomization process

$$CH_4(g) \rightarrow C(g) + 4H(g) \tag{5.54}$$

We define the average C–H bond energy in methane as one-fourth of ΔH_{298}° for the reaction (5.54). From the Appendix, $\Delta H_{f,298}^\circ$ of CH_4 is -17.9 kcal mol^{-1}. The standard heat of sublimation of graphite is 171.3 kcal mol^{-1} at 25°C; hence $\Delta H_{f,298}^\circ$ of $C(g)$ is 171.3 kcal mol^{-1}, as listed in the Appendix. (Recall that ΔH_f° is zero for the *stable* form of an element; at 25°C the stable form of carbon is graphite and not gaseous carbon atoms.) The standard heat of formation at 25°C of $H(g)$ is listed as 52.1 kcal mol^{-1}. [This is ΔH_{298}° for the process $\frac{1}{2}H_2 \rightarrow H(g)$.] Using Eq. (5.52) to calculate ΔH_{298}° for the process (5.54), we have

$$\Delta H_{298}^\circ = [171.3 + 4(52.1) - (-17.9)] \text{ kcal/mol} = 397.6 \text{ kcal/mol}$$

Hence the average C–H bond energy in CH_4 is 99.4 kcal mol^{-1}. (In principle, one ought to use ΔU_0° values rather than ΔH_{298}° values to calculate bond energies. However, the average-bond-energy concept is only an approximation, so use of ΔH_{298}° values is good enough.)

To arrive at an average carbon–carbon single-bond energy, consider the process

$$C_2H_6(g) \rightarrow 2C(g) + 6H(g) \tag{5.55}$$

The $\Delta H_{f,298}^\circ$ values in the Appendix give $\Delta H_{298}^\circ = 675.4$ kcal/mol for (5.55). This ΔH_{298}° is then taken as the sum of contributions from six C–H bonds and one C–C bond. Using the CH_4 value of 99.4 kcal mol^{-1} for the C–H bond, we get as the C–C bond energy

$$[675.4 - 6(99.4)] \text{ kcal/mol} = 79.0 \text{ kcal/mol}$$

The average-bond-energy method would then estimate the heat of atomization for propane $CH_3CH_2CH_3$ at 25°C as $[8(99.4) + 2(79.0)]$ kcal/mol = 953 kcal/mol. We break the formation of propane into two steps:

$$3C(gr) + 4H_2(g) \rightarrow 3C(g) + 8H(g) \rightarrow C_3H_8(g)$$

From the Appendix, ΔH°_{298} for the first step is

$$[3(171.3) + 8(52.1) - 3(0) - 4(0)] \text{ kcal/mol} = 931 \text{ kcal/mol}$$

We have estimated ΔH°_{298} for the second step as -953 kcal/mol. Hence the average-bond-energy estimate of $\Delta H^\circ_{f,298}$ for propane is -22 kcal/mol. The experimental value is -24.8 kcal/mol, so we are off by a couple of kilocalories.

Some values for average bond energies are listed in Table 20.1. The C–H and C–C values listed differ somewhat from the ones we calculated using methane and ethane data, so as to give better overall agreement with experiment.

Note that for the process $CH_4(g) \to CH_3(g) + H(g)$, we cannot expect ΔH°_{298} to be the same as the average C–H bond energy, since the product CH_3 differs from the products in (5.54).

5.8 THE UNATTAINABILITY OF ABSOLUTE ZERO

Besides the Nernst–Simon formulation of the third law, another formulation of this law is often given, the *unattainability formulation.* In 1912, Nernst gave a "derivation" of the unattainability of absolute zero from the second law of thermodynamics (see Prob. 3.31). However, Einstein showed that Nernst's argument was fallacious, so the unattainability statement cannot be derived from the second law. [For details, see M. L. Boas, *Am. J. Phys.*, **28**, 675 (1960).]

The unattainability of absolute zero is usually regarded as a formulation of the third law of thermodynamics, equivalent to the entropy formulation (5.29). Supposed proofs of this equivalence are given in several texts. However, in a careful study of the question [P. T. Landsberg, *Rev. Mod. Phys.*, **28**, 363 (1956)] Landsberg found that the unattainability and the entropy formulations of the third law are not equivalent. R. Haase (pp. 86–96 in *Eyring, Henderson, and Jost*, vol. I) also concludes that the entropy and unattainability formulations are not equivalent. Haase's analysis shows that the unattainability of absolute zero follows as a consequence of the first and second laws plus the Nernst–Simon statement of the third law.

In summary, the unattainability of absolute zero (*a*) cannot be derived from the second law of thermodynamics, (*b*) is not equivalent to the Nernst–Simon statement of the third law, and (*c*) is best regarded as a consequence of the laws of thermodynamics and not as a statement of the third law. The third law of thermodynamics is Eq. (5.29), the Nernst–Simon entropy formulation.

Although absolute zero is unattainable, temperatures as low as 0.000001 K have been reached. One can use the Joule–Thomson effect to liquefy helium gas. By pumping away the helium vapor above the liquid, thereby causing the liquid helium to evaporate rapidly, one can attain temperatures of around 1 K. To reach lower temperatures, adiabatic demagnetization is used. Here, liquid helium cools a paramagnetic salt to a low temperature; then a magnetic field is applied isothermally, magnetizing the salt; finally, the magnetic field is removed adiabatically, causing a drop in the salt's temperature. Temperatures of 10^{-3} K have been reached in this manner. Application of adiabatic demagnetization to nuclear magnetic moments (instead of the electron magnetic moments of paramagnetic salts) allows temperatures of 10^{-6} K to be reached. (See chap. 14 of *Zemansky* for details.)

PROBLEMS

5.1 For H_2(ideal gas), use data in the Appendix to find the conventional molar enthalpy $\bar{H}$ at (*a*) 298.15 K and 1 atm; (*b*) 398.15 K and 1 atm (neglect the temperature dependence of C_P); (*c*) 298.15 K and 100 atm.

5.2 For $H_2O(l)$, use data in the Appendix and preceding Eq. (4.75) to find the conventional molar enthalpy at (*a*) 298.15 K and 1 atm; (*b*) 398.15 K and 1 atm; (*c*) 298.15 K and 100 atm. (Neglect the temperature variation of C_P and the pressure variation of α and $\bar{V}$.)

5.3 (*a*) A gas obeys the equation of state $P(\bar{V} - b) = RT$, where b is a constant. Show that for this gas, $\bar{H}_{id}(T, P) - \bar{H}_{re}(T, P) = -bP$. (*b*) A gas obeys $P(\bar{V} - b) = RT$ with $b = 45\ \text{cm}^3/\text{mole}$. Calculate $\bar{H}_{id} - \bar{H}_{re}$ for this gas at 25°C and 1 atm.

5.4 The following reaction is carried out in an adiabatic bomb calorimeter:

$$2A_1(g) + 3A_2(l) \rightarrow 5A_3(g) + A_4(g)$$

An excess of A_1 is added to 1.450 g of A_2; the molecular weight of A_2 is 168.1. The reaction goes essentially to completion. The initial temperature is 25.000°C; after the reaction, the temperature is 27.913°C. A direct current of 12.62 mA (milliamperes) flowing through the calorimeter heater for 812 s is needed to bring the product mixture from 25.000 to 27.913°C, the potential drop across the heater being 8.412 V. Neglecting the changes in thermodynamic functions that occur when the reactants and products are brought from their standard states to the states that occur in the calorimeter, estimate ΔU°_{298} and ΔH°_{298} for this reaction. Express each answer in J/mol and in cal/mol. (One watt = one volt × one ampere = 1 J/s.)

5.5 The standard enthalpy of combustion at 25°C of solid naphthalene ($C_{10}H_8$) to $CO_2(g)$ and $H_2O(l)$ is $-1232\ \text{kcal mol}^{-1}$. (By definition, ΔH° of combustion refers to 1 mole of compound being burned.) Find (*a*) $\Delta H^\circ_{f,298}$ of $C_{10}H_8(s)$; (*b*) $\Delta U^\circ_{f,298}$ of $C_{10}H_8(s)$.

5.6 The standard enthalpy of combustion at 25°C of $CS_2(l)$ to $CO_2(g)$ and $SO_2(g)$ is $-257.4\ \text{kcal mol}^{-1}$. Find $\Delta H^\circ_{f,298}$ and $\Delta U^\circ_{f,298}$ of $CS_2(l)$. (Use data in the Appendix.)

5.7 For $Na(s) + HCl(g) \rightarrow NaCl(s) + \frac{1}{2}H_2(g)$, ΔH°_{298} is $-76\ \text{kcal mol}^{-1}$. Find ΔH°_{298} for:

(*a*) $2Na(s) + 2HCl(g) \rightarrow 2NaCl(s) + H_2(g)$

(*b*) $4Na(s) + 4HCl(g) \rightarrow 4NaCl(s) + 2H_2(g)$

(*c*) $NaCl(s) + \frac{1}{2}H_2(g) \rightarrow Na(s) + HCl(g)$

5.8 Given the following ΔH°_{298} values in kcal/mol, where, as before, *gr* stands for graphite,

$Fe_2O_3(s) + 3C(gr) \rightarrow 2Fe(s) + 3CO(g)$	117
$FeO(s) + C(gr) \rightarrow Fe(s) + CO(g)$	37
$2CO(g) + O_2(g) \rightarrow 2CO_2(g)$	−135
$C(gr) + O_2(g) \rightarrow CO_2(g)$	−94

find $\Delta H^\circ_{f,298}$ of $FeO(s)$ and of $Fe_2O_3(s)$.

5.9 Use data in the Appendix to find ΔH°_{298} for:

(*a*) $2H_2S(g) + 3O_2(g) \rightarrow 2H_2O(l) + 2SO_2(g)$

(*b*) $2H_2S(g) + 3O_2(g) \rightarrow 2H_2O(g) + 2SO_2(g)$

(*c*) $2HN_3(g) + 2NO(g) \rightarrow H_2O_2(l) + 4N_2(g)$

5.10 Use data in the Appendix and the approximation of neglecting the temperature dependence of $\bar{C}^\circ_P$ to find ΔH°_{370} for the reactions of Prob. 5.9.

5.11 For $O_2(g)$, $\bar{C}^\circ_{P,298} = 7.016\ \text{cal mol}^{-1}\ \text{K}^{-1}$, and $\bar{C}^\circ_{P,600} = 7.670\ \text{cal mol}^{-1}\ \text{K}^{-1}$. For $O_3(g)$, $\bar{C}^\circ_{P,298} = 9.37\ \text{cal mol}^{-1}\ \text{K}^{-1}$, $\bar{C}^\circ_{P,600} = 11.92\ \text{cal mol}^{-1}\ \text{K}^{-1}$, and $\Delta H^\circ_{f,298} = 34.10\ \text{kcal mol}^{-1}$. Make the approximation of using average $\bar{C}^\circ_P$ values over the temperature range of interest to calculate $\Delta H^\circ_{f,600}$ for $O_3(g)$. Compare with the correct value 34.05 kcal mol^{-1}.

5.12 Compute $\Delta H^\circ_{f,1000}$ of $HCl(g)$ from data in the Appendix and the following expressions, which hold from 300 to 1500 K:

$$\bar{C}^\circ_P(H_2) = (6.9469 - 0.1999 \times 10^{-3}T/\text{K} + 4.808 \times 10^{-7}T^2/\text{K}^2)\ \text{cal mol}^{-1}\ \text{K}^{-1}$$

$$\bar{C}^\circ_P(Cl_2) = (7.5755 + 2.4244 \times 10^{-3}T/\text{K} - 9.650 \times 10^{-7}T^2/\text{K}^2)\ \text{cal mol}^{-1}\ \text{K}^{-1}$$

$$\bar{C}^\circ_P(HCl) = (6.7319 + 0.4325 \times 10^{-3}T/\text{K} + 3.697 \times 10^{-7}T^2/\text{K}^2)\ \text{cal mol}^{-1}\ \text{K}^{-1}$$

5.13 Use data in the Appendix and data preceding Eq. (4.75) and make certain approximations to calculate the conventional molar entropy of $H_2O(l)$ at (*a*) 298.15 K and 1 atm; (*b*) 348.15 K and 1 atm; (*c*) 298.15 K and 100 atm; (*d*) 348.15 K and 100 atm.

5.14 Andrews and Westrum [*J. Chem. Thermodyn.*, **2**, 245 (1970)] measured $\bar{C}^\circ_P(T)$ of triptycene ($C_{20}H_{14}$) with the following results [where the first number in each pair is T/K and the second number is $\bar{C}^\circ_P/(\text{cal mol}^{-1}\ \text{K}^{-1})$]. *Solid:* 10, 0.863; 20, 4.303; 30, 7.731; 40, 10.649; 50, 13.17; 60, 15.40; 70, 17.43; 80, 19.33; 90, 21.16; 100, 22.98; 120, 26.67; 140, 30.55; 160, 34.63; 180, 38.91; 200, 43.37; 220, 48.01; 240, 52.83; 260, 57.79; 280, 62.88; 298.15, 67.56; 320, 73.16; 350,

80.67; 400, 92.53; 450, 103.85; 500, 113.98; 527.18, 119.38. *Liquid*: 527.18, 130.86; 530, 130.90; 540, 131.56; 550, 133.45. This compound melts at 527.18 K, and $\Delta H_m^\circ = 7236$ cal mol^{-1}. Find (*a*) $\bar{S}_{298}^\circ$ for $C_{20}H_{14}(s)$; (*b*) $\bar{S}_{550}^\circ$ for $C_{20}H_{14}(l)$; (*c*) $\bar{H}_{298}^\circ - \bar{H}_0^\circ$ for $C_{20}H_{14}(s)$. Graphical integration can be done by counting the number of squares under the curve (estimating the fractions of squares partly under the curve) or by cutting out the area under the curve, weighing it, and comparing with the weight of a known number of squares.

5.15 For the reactions of Prob. 5.9, find ΔS_{298}° from data in the Appendix.

5.16 For the reactions in Prob. 5.9, find ΔG_{298}° using (*a*) the results of Probs. 5.9 and 5.15; (*b*) $\Delta G_{f,298}^\circ$ values in the Appendix.

5.17 For the reactions in Prob. 5.9, find ΔS_{370}°; neglect the temperature variation of ΔC_P°.

5.18 Derive Eq. (5.47).

5.19 For the reactions of Prob. 5.9, use the results of Probs. 5.10 and 5.17 to find ΔG_{370}°.

5.20 A certain gas obeys the equation of state $P\bar{V}/RT = 1 + f(T)P$, where $f(T)$ is a certain function of T. Show that for this gas

$$\bar{S}_{\text{id}}(T, P) - \bar{S}_{\text{re}}(T, P) = RP[f(T) + Tf'(T)]$$

5.21 Look up in one of the references cited near the end of Sec. 5.6 ΔG_f° data at 1000 K to find ΔG_{1000}° for $2CH_4(g) \rightarrow C_2H_6(g) + H_2(g)$.

5.22 A certain liquid compound A has a vapor pressure of 23.8 torr at 25°C; its molar enthalpy of vaporization at 25°C and 23.8 torr is 10.5 kcal mol^{-1}. Assume the vapor behaves ideally, and calculate (*a*) $\bar{H}_{298}^\circ(\text{A}, g) - \bar{H}_{298}^\circ(\text{A}, l)$; (*b*) $\bar{S}_{298}^\circ(\text{A}, g) - \bar{S}_{298}^\circ(\text{A}, l)$; (*c*) $\bar{G}_{298}^\circ(\text{A}, g) - \bar{G}_{298}^\circ(\text{A}, l)$; (*d*)

$$\Delta G_{f,298}^\circ(\text{A}, g) - \Delta G_{f,298}^\circ(\text{A}, l).$$

(*e*) Use data in the Appendix to compare your answer to (*d*) with the value for H_2O.

5.23 Verify Eq. (5.16).

5.24 Use bond energies listed in Sec. 20.1 to estimate ΔH_{298}° for $CH_3CH_2OH(g) \rightarrow CH_3OCH_3(g)$. Compare with the true value 12.2 kcal/mol.

5.25 Use data of the Appendix and bond energies in Sec. 20.1 to estimate $\Delta H_{f,298}^\circ$ of $CH_3CH_2CN(g)$.

5.26 Show that for any pure substance $\lim_{T\rightarrow 0} \alpha = 0$. *Hint*: Use one of the Maxwell relations.

5.27 A certain nonmetallic solid has $\bar{C}_P^\circ = 0.200$ cal mol^{-1} K^{-1} at 10 K. Find $\bar{S}_{10}^\circ$ for this substance.

5.28 Use the relation between entropy and disorder to explain why the normal-boiling-point ΔS of vaporization of hydrogen-bonded liquids usually exceeds the Trouton's rule value.

5.29 Below 368 K, the stable form of S(*s*) is rhombic sulfur; above 368 K, the stable form is monoclinic sulfur. Monoclinic sulfur can be supercooled to extremely low temperatures without changing to rhombic sulfur. Explain how measurements on these two forms of sulfur can be used to check the third law of thermodynamics.

REACTION EQUILIBRIUM IN IDEAL GAS MIXTURES

6.1 CHEMICAL POTENTIALS IN AN IDEAL GAS MIXTURE

Before applying the chemical-reaction equilibrium condition to ideal gas mixtures, we develop the thermodynamic properties of such mixtures. The molecular picture of an ideal gas mixture is the same as for a pure ideal gas, namely, molecules that have negligible volume and negligible intermolecular interaction energy. For thermodynamic purposes, we want a macroscopic definition of an ideal gas mixture.

We define an *ideal gas mixture* as a gas mixture having the following three (macroscopic) properties: (1) The equation of state

$$PV = \left(\sum_i n_i\right) RT = n_{\text{tot}} RT \tag{6.1*}$$

is obeyed for all temperatures, pressures, and compositions; in this equation, n_i is the number of moles of gas i in the mixture, and n_{tot} is the total number of moles in the mixture. (2) If the mixture and a system consisting of pure gas i (where i is any one of the mixture's components) are separated from each other by a rigid membrane permeable to gas i only, then at equilibrium the partial pressure [defined by (1.23)] of gas i in the mixture is equal to the pressure of the pure-gas-i system. (3) When the gas mixture is formed by isothermally mixing the pure components, the heat of mixing is zero (provided no chemical reaction occurs on mixing). (Note that we implicitly used conditions 2 and 3 in paragraph 10 of Sec. 3.5.)

Conditions 1 to 3 make sense from a molecular viewpoint. Since there are no intermolecular interactions either in the pure gases or in the mixture, we expect the mixture to obey the same equation of state obeyed by each pure gas; hence the first condition holds. If two samples of pure gas i were separated by a membrane permeable to i, equilibrium (equal rates of passage of i through the membrane from each side) would be reached with equal pressures of i on each side. Because there are no intermolecular interactions, the presence of other gases on one side of the membrane has no influence on the rate of passage of i from the mixture to the pure gas. Hence the second condition is obeyed. On mixing the pure components, the volume occupied by each component changes; however, the internal energy of each ideal-gas component depends only on T, and not on the volume; moreover, there are no intermolecular interactions in the mixture; hence we expect no change in internal energy for isothermal mixing of ideal gases; no work is done during the mixing; hence we expect that q $(= \Delta U - w)$ is zero for mixing ideal gases.

Dalton's law can be derived from condition 1, as follows. Equation (6.1) gives

$$P = n_1 RT/V + n_2 RT/V + \cdots \tag{6.2}$$

When the composition of the mixture is pure gas 1 (so that $n_2 = n_3 = \cdots = 0$), Eq. (6.1) gives $P = n_1 RT/V$, so that $n_1 RT/V$ is the pressure if gas 1 alone is present. Similarly for gases 2, 3, ... Therefore, (6.2) says that the pressure of the gas mixture is the sum of the pressures each gas would exert if it alone were present.

Reaction equilibrium is determined by the chemical potentials (4.81); it is therefore important to find the expression for the chemical potential μ_i of component i in an ideal gas mixture. To do so, we proceed as follows.

Let system A be an ideal gas mixture at temperature T and pressure P, and let system B consist of pure gas i. Further, let systems A and B be separated by a rigid, thermally conducting membrane permeable to gas i only. When phase equilibrium between systems A and B is reached, we must have $\mu_i^{\text{A}} = \mu_i^{\text{B}}$, as in Eq. (4.100). Further, condition 2 of the definition of an ideal gas mixture states that at equilibrium the pressure of pure gas i in system B equals the partial pressure P_i of gas i in the mixture A. We therefore write μ_i^{B} as $\mu_i^*(T, P_i)$ or [using (1.23)] as $\mu_i^*(T, x_i P)$, where the star indicates pure gas i, and where x_i is the mole fraction of gas i in the mixture. (We shall frequently use a star superscript to distinguish thermodynamic properties of a pure substance from those of a component of a mixture.) If μ_i denotes the chemical potential of gas i in the mixture, the equilibrium condition $\mu_i^{\text{A}} = \mu_i^{\text{B}}$ becomes

$$\mu_i(T, P, n_1, n_2, \ldots) = \mu_i^*(T, x_i P)$$

Further, the chemical potential μ_i^* of pure gas i equals the molar Gibbs free energy $\bar{G}_i^*$ of pure gas i [Eq. (4.98)]. Hence the chemical potential μ_i of component i of an ideal gas mixture at temperature T and (total) pressure P is

$$\mu_i = \mu_i^*(T, x_i P) = \bar{G}_i^*(T, x_i P) \qquad \text{ideal gas mixture} \tag{6.3}$$

where $x_i P$ is the partial pressure of gas i in the ideal gas mixture. The chemical potential μ_i of gas i in an ideal gas mixture at T and P equals the molar Gibbs energy of pure gas i at temperature T and pressure P_i (its partial pressure in the mixture). Equation (6.3) makes sense from a molecular viewpoint; the absence of intermolecular interactions makes the presence of other gases in the mixture have no effect on the chemical potential of gas i.

The thermodynamic properties U, H, S, G, and C_P of an ideal gas mixture can readily be expressed in terms of the corresponding properties of the pure components. The result (Prob. 6.22) is that each of U, H, S, G, and C_P for an ideal gas mixture is the sum of the corresponding thermodynamic functions for the pure gases calculated for each pure gas occupying a volume equal to the mixture's volume at a pressure equal to its partial pressure in the mixture and at a temperature equal to its temperature in the mixture. These results make sense from the molecular picture of each gas's having no interaction with the other gases in the mixture.

6.2 IDEAL-GAS REACTION EQUILIBRIUM

The condition for reaction equilibrium is given by Eq. (4.111) as $\sum_i \nu_i \mu_i = 0$, where the ν_i's are the stoichiometric coefficients. We now specialize to the case where all reactants and products are ideal gases. (Other cases will be considered in Chaps. 9 and 11.) For an ideal gas mixture, each μ_i is given by (6.3) as equal to $\bar{G}_i^*(T, P_i)$, where P_i is the partial pressure of gas i in the mixture. The reacting system is usually held at constant T and either constant total pressure or constant V. During the reaction, the amounts of the reacting gases vary, and hence their partial pressures vary, until the equilibrium partial pressures are reached. We therefore investigate how μ_i varies with P_i, so that we can apply the equilibrium condition.

Consider 1 mole of pure ideal gas i at temperature T and pressure P_i. Its molar Gibbs energy depends on T and P_i. The fundamental equation (4.40) gives for a fixed quantity of a pure substance $dG_i^* = -S_i^*\,dT + V_i^*\,dP_i$. Division by n_i and use of (4.98) give $d\mu_i^* = d\bar{G}_i^* = -\bar{S}_i^*\,dT + \bar{V}_i^*\,dP_i$. At constant temperature,

$$d\mu_i^* = d\bar{G}_i^* = \bar{V}_i^*\,dP_i = (RT/P_i)\,dP_i \qquad \text{const. } T\text{, ideal gas} \tag{6.4}$$

Now consider the isothermal change of state of pure gas i from 1 to 2, where state 1 is the standard state (Sec. 5.1) with pressure $P_{i,1} = 1$ atm and temperature T, and state 2 has pressure $P_{i,2}$ and temperature T. Integration of (6.4) gives

$$\int_1^2 d\mu_i^* = RT \int_{1\text{ atm}}^{P_{i,2}} \frac{1}{P_i}\,dP_i \tag{6.5}$$

$$\mu_i^*(T, P_{i,2}) - \mu_i^{*\circ}(T) = RT \ln\,(P_{i,2}/1\text{ atm}) \tag{6.6}$$

where, as always, the degree superscript indicates the standard state of i at 1 atm pressure.

The subscript 2 in (6.6) is superfluous, and we drop it. From Eq. (6.3), the chemical potential $\mu_i^*(T, P_i)$ of pure gas i is equal to the chemical potential μ_i of gas i in an ideal gas mixture that is at temperature T and has gas i present at partial pressure P_i. The notation $\mu_i^{*\circ}$ is overdone, in that the standard state of gas i of the mixture is defined to be the pure gas i at 1 atm and T; hence the degree superscript alone indicates we are talking about pure gas i, and we drop the star from $\mu_i^{*\circ}$. With these three changes and with the introduction of the symbol P° for the standard pressure of 1 atm, Eq. (6.6) becomes

$$\mu_i = \mu_i^\circ(T) + RT \ln\,(P_i/P^\circ) \qquad \text{ideal gas mixture, } P^\circ \equiv 1\text{ atm} \tag{6.7}^*$$

Equation (6.7) is the fundamental thermodynamic equation for an ideal gas mixture. In (6.7), μ_i is the chemical potential of component i in an ideal gas mixture at temperature T, P_i is the partial pressure of gas i in the mixture, and μ_i° $(= \bar{G}_i^\circ)$ is the chemical potential of pure ideal gas i at temperature T and pressure 1 atm (the standard pressure). Note that μ_i° depends only on T because the pressure is fixed at 1 atm for the standard state.

Use of (6.7) in the equilibrium condition (4.111) gives

$$\sum_i \nu_i \mu_i = \sum_i \nu_i \mu_i^\circ(T) + RT \sum_i \nu_i \ln\,(P_{i,\text{eq}}/P^\circ) = 0 \tag{6.8}$$

where the subscript eq emphasizes that this equation holds only at reaction equilibrium. Recall [Eqs. (5.49) and (4.98)] that

$$\Delta G_T^\circ \equiv \sum_i \nu_i \bar{G}_i^\circ(T) = \sum_i \nu_i \mu_i^\circ(T) \tag{6.9}$$

where ΔG_T° is the standard Gibbs free-energy change for the reaction at temperature T. Using the definition (6.9) in (6.8), we get

$$\Delta G_T^\circ = -RT \sum_i \nu_i \ln (P_{i,\text{eq}}/P^\circ) \tag{6.10}$$

Use of the logarithmic identity $r \ln b = \ln b^r$ transforms (6.10) to

$$\Delta G_T^\circ = -RT \sum_i \ln (P_{i,\text{eq}}/P^\circ)^{\nu_i} \tag{6.11}$$

The sum of logs equals the log of the product:

$$\ln a_1 + \ln a_2 + \ln a_3 + \cdots = \ln a_1 a_2 a_3 \cdots \tag{6.12}$$

A convenient notation for a product (analogous to the sigma notation for sums) is a large capital pi:

$$\prod_{i=1}^{n} a_i \equiv a_1 a_2 \cdots a_n \tag{6.13}$$

(As with sums, the limits are often omitted when they are clear from the context.) Using the product notation (6.13), we write (6.12) as

$$\sum_i \ln a_i = \ln \left(\prod_i a_i \right) \tag{6.14}$$

Using (6.14), we have for Eq. (6.11)

$$\Delta G_T^\circ = -RT \ln \left[\prod_i (P_{i,\text{eq}}/P^\circ)^{\nu_i} \right] \tag{6.15}$$

We define K_P° as the product that occurs in (6.15):

$$K_P^\circ \equiv \prod_i (P_{i,\text{eq}}/P^\circ)^{\nu_i} \qquad \text{ideal-gas reaction equilib.} \tag{6.16*}$$

Equation (6.15) becomes

$$\Delta G^\circ = -RT \ln K_P^\circ \qquad \text{ideal-gas reaction equilib.} \tag{6.17*}$$

Recall that if $y = \ln_e x$, then $x = e^y$ [Eq. (2.120)]. Thus (6.17) can be written as

$$K_P^\circ = e^{-\Delta G^\circ/RT} \tag{6.18}$$

From (6.9), it is clear that ΔG° depends only on T. It therefore follows from (6.18) that K_P° for a given reaction is a function of T only:

$$K_P^\circ = K_P^\circ(T) \tag{6.19}$$

K_P° is independent of the pressure, the volume, and the number of moles of the reaction species present in the mixture. At a given temperature, K_P° is a constant for a given reaction. K_P° is the *standard equilibrium constant* (or the *standard pressure equilibrium constant*) for the ideal-gas reaction.

Despite the seeming complexity of all the subscripts, superscripts, sums, and products in the derivation, we have reached a quite simple result. Consider, for example, the gas-phase reaction

$$N_2 + 3H_2 = 2NH_3 \tag{6.20}$$

Let the density be reasonably low, so that the gases behave essentially ideally. Recall that the stoichiometric coefficients are negative for reactants and positive for products. For the reaction (6.20): $\nu(N_2) = -1$; $\nu(H_2) = -3$; $\nu(NH_3) = 2$. The standard equilibrium constant K_P° in (6.16) becomes

$$K_P^\circ = [P(NH_3)_{eq}/P^\circ]^2[P(N_2)_{eq}/P^\circ]^{-1}[P(H_2)_{eq}/P^\circ]^{-3}$$

$$K_P^\circ = \frac{[P(NH_3)_{eq}/P^\circ]^2}{[P(N_2)_{eq}/P^\circ][P(H_2)_{eq}/P^\circ]^3} \tag{6.21}$$

where the pressures are the equilibrium partial pressures of the gases in the reaction mixture. Since the stoichiometric coefficients of products are positive and those of reactants are negative, the pressures of the products end up in the numerator of K_P° and those of the reactants end up in the denominator. At any given temperature, the equilibrium partial pressures must be such as to satisfy (6.21). If the partial pressures do not satisfy (6.21), the system is not in reaction equilibrium.

For the ideal-gas reaction $aA + bB = cC + dD$, the standard (pressure) equilibrium constant is

$$K_P^\circ = \frac{(P_{C,eq}/P^\circ)^c(P_{D,eq}/P^\circ)^d}{(P_{A,eq}/P^\circ)^a(P_{B,eq}/P^\circ)^b} \tag{6.22}$$

Since P_i/P° in (6.16) is dimensionless, the standard equilibrium constant K_P° is dimensionless. In (6.17), the log of K_P° is taken; one can only take the log of a dimensionless number. It is sometimes convenient to work with an equilibrium constant that omits the P° in (6.16). We define the *equilibrium constant* (or *pressure equilibrium constant*) K_P as

$$K_P \equiv \prod_i (P_{i,eq})^{\nu_i} \tag{6.23}$$

K_P has dimensions of pressure raised to the change in mole numbers for the reaction as written. For example, for the reaction (6.20), K_P has dimensions of pressure^{-2}.

The existence of a standard equilibrium constant K_P° that depends only on T is a rigorous deduction from the laws of thermodynamics. The only assumption is that we have an ideal gas mixture. Our results are a good approximation for real gas mixtures at relatively low densities.

Example For the ideal-gas reaction $A(g) + 3B(g) = 2C(g) + 4D(g)$, analysis of an equilibrium mixture at 500 K yields the following partial pressures: A, 300 torr; B, 500 torr; C, 400 torr; D, 100 torr. Find K_P° and ΔG° at 500 K.

Substitution into (6.16) gives

$$K_P^\circ = \frac{(P_C/P^\circ)^2(P_D/P^\circ)^4}{(P_A/P^\circ)(P_B/P^\circ)^3} = \frac{(400\text{ torr}/760\text{ torr})^2(100\text{ torr}/760\text{ torr})^4}{(300\text{ torr}/760\text{ torr})(500\text{ torr}/760\text{ torr})^3}$$

$$= 7.4 \times 10^{-4} \tag{6.24}$$

where we didn't bother to use the eq subscript on the equilibrium partial pressures. From (6.17), we have

$$\begin{aligned}\Delta G^\circ_{500} &= -(1.987 \text{ cal mol}^{-1} \text{ K}^{-1})(500 \text{ K})[2.303 \log (7.4 \times 10^{-4})] \\ &= -(993 \text{ cal/mol})(2.303)(-3.131) = 7160 \text{ cal/mole}\end{aligned} \tag{6.25}$$

Let us make some qualitative remarks about equilibrium. (These remarks apply in a general way to all kinds of equilibria, not just to ideal-gas reactions.) The standard equilibrium constant K_P° is the product and quotient of positive numbers and must therefore be positive: $0 < K_P^\circ < \infty$. If K_P° is very large ($K_P^\circ \gg 1$), its numerator must be considerably greater than its denominator, and this means that the equilibrium pressures of the products are usually greater than those of the reactants. Conversely, if K_P° is very small ($K_P^\circ \ll 1$), its denominator is large compared with its numerator and the reactant equilibrium pressures are usually larger than the product equilibrium pressures. A moderate value of K_P° usually means substantial equilibrium pressures of both products and reactants. (The word "usually" has been used because it is not the pressures that appear in the equilibrium constant but the pressures raised to the stoichiometric coefficients.) A large value of the equilibrium constant favors products; a small value of the equilibrium constant favors reactants.

From (6.18) we have $K_P^\circ = 1/e^{\Delta G^\circ/RT}$. If ΔG° is a large positive number ($\Delta G^\circ \gg 0$), then $e^{\Delta G^\circ/RT}$ is very large and K_P° is very small. If ΔG° is a large negative number ($\Delta G^\circ \ll 0$), then $K_P^\circ = e^{-\Delta G^\circ/RT}$ is very large. If $\Delta G^\circ \approx 0$, then $K_P^\circ \approx 1$. A large negative value of ΔG° favors products, while a large positive value of ΔG° favors reactants.

Since $G^\circ = H^\circ - TS^\circ$, we have for an isothermal process

$$\Delta G^\circ = \Delta H^\circ - T\,\Delta S^\circ \qquad \text{const. } T \tag{6.26}$$

so that ΔG° is determined by ΔH°, ΔS°, and T. If T is rather low, the factor T in (6.26) is rather small and the first term on the right of (6.26) is dominant; the fact that ΔS° goes to zero as T goes to zero (the third law) adds to the dominance of ΔH° over $T\,\Delta S^\circ$ at low temperatures. Thus in the limit $T \to 0$, ΔG° approaches ΔH°. For low temperatures, we have the following rough relation:

$$\Delta G^\circ \approx \Delta H^\circ \qquad \text{low } T \tag{6.27}$$

For an exothermic reaction, ΔH° is negative, and hence from (6.27) ΔG° is negative at low temperatures. Thus at low temperatures, products of an exothermic reaction are favored over reactants. It turns out that for a majority of reactions, the values of ΔH° and $T\,\Delta S^\circ$ are such that at room temperature (and below) the first term on the right of (6.26) is the dominant one. Thus, for most exothermic reactions, products are favored at room temperature. However, ΔH° alone does not determine the equilibrium constant, and there are many *endo*thermic reactions with ΔG° negative and products favored at room temperature, due to the influence of the $-T\,\Delta S^\circ$ term.

For very high temperatures, the factor T makes the second term on the right of (6.26) the dominant one, and we have the following rough relation:

$$\Delta G^\circ \approx -T\,\Delta S^\circ \qquad \text{high } T \tag{6.28}$$

At high temperatures, a reaction with a positive ΔS° has a negative ΔG°, and products are favored.

A simple example is the breaking of a chemical bond, for example, $N_2(g) \rightleftharpoons 2N(g)$. Since a bond is broken, the reaction is highly endothermic ($\Delta H^\circ \gg 0$). Hence at reasonably low temperatures, ΔG° is highly positive, and N_2 is not significantly dissociated at low temperatures (including room temperature). The breaking of the bond leads to gaseous atoms, which is a more disordered state than gaseous molecules; hence we expect ΔS° to be positive for $N_2 = 2N$. (In fact, from the Appendix we find $\Delta S^\circ_{298} = 27.5$ cal mol^{-1} K^{-1} for this reaction.) Thus for high temperatures, we expect from (6.28) that ΔG° for $N_2 = 2N$ will be negative, favoring dissociation to atoms.

Another example is the denaturation of a protein. A protein molecule is a long-chain polymer of amino acids. Enzymes are globular proteins. In a globular protein, certain portions of the chain are coiled into helical segments that are stabilized by hydrogen bonds between one turn of a helix and the next, while other portions have no fixed conformation and are called *random-coil* portions. In addition, the partly coiled protein folds on itself to give a roughly ellipsoidal overall shape; the folding is not random, but is determined by van der Waals forces (Sec. 22.10) and by S–S covalent bonds between sulfur-containing amino acids. In the denaturation reaction, the protein unfolds into a completely random-coil structure. Because of the breaking of hydrogen bonds, denaturation has ΔH° positive. Since denaturation gives a more disordered structure, denaturation has ΔS° positive. Thus as the temperature is raised, $T\,\Delta S^\circ$ will eventually exceed ΔH° and denaturation will occur. An example of protein denaturation is the change in egg white (albumen) that occurs on hard-boiling an egg.

In freshman chemistry you probably saw gas-phase equilibrium constants expressed in terms of concentrations rather than partial pressures. Equations (1.23) and (6.1) give $P_i \equiv x_i P = (n_i/n_{\text{tot}})(n_{\text{tot}}RT/V)$, so

$$P_i = n_i RT/V \qquad \text{ideal gas mixture} \tag{6.29}$$

where n_i is the number of moles of ideal gas i present at partial pressure P_i in the vessel of volume V. The *concentration* c_i of species i in the mixture is defined as

$$c_i \equiv n_i/V \qquad (6.30)^*$$

The most common units of c_i are moles per liter. Equation (6.29) becomes $P_i = c_i RT$. Substitution into (6.22) gives for the ideal-gas reaction $a\text{A} + b\text{B} = f\text{F} + d\text{D}$

$$K_P^\circ = \frac{(c_{\text{F,eq}}RT/P^\circ)^f(c_{\text{D,eq}}RT/P^\circ)^d}{(c_{\text{A,eq}}RT/P^\circ)^a(c_{\text{B,eq}}RT/P^\circ)^b} = \frac{(c_{\text{F,eq}}/c^\circ)^f(c_{\text{D,eq}}/c^\circ)^d}{(c_{\text{A,eq}}/c^\circ)^a(c_{\text{B,eq}}/c^\circ)^b}\left(\frac{c^\circ RT}{P^\circ}\right)^{f+d-a-b} \tag{6.31}$$

where c°, defined as $c^\circ \equiv 1$ mole/liter, was introduced to make all fractions on the right of (6.31) dimensionless. Note that $c^\circ RT$ has the same dimensions as P°. Generalizing (6.31) to the reaction $0 = \sum_i \nu_i A_i$, we have

$$K_P^\circ = (c^\circ RT/P^\circ)^{\Delta|\nu|}\prod_i (c_{i,\text{eq}}/c^\circ)^{\nu_i} \tag{6.32}$$

where $\Delta|\nu|$ is defined as

$$\Delta|\nu| \equiv \sum_i \nu_i \tag{6.33}$$

$\Delta|\nu|$ is the net change in mole numbers for the reaction as written. [See the discussion preceding Eq. (5.15).] Defining the *standard concentration equilibrium constant* K_c° as

$$K_c^\circ \equiv \prod_i (c_{i,\text{eq}}/c^\circ)^{\nu_i} \qquad \text{where } c^\circ \equiv 1 \text{ mole/liter} \tag{6.34*}$$

we have for (6.32)

$$K_P^\circ = K_c^\circ[RT/(1 \text{ atm mol}^{-1})]^{\Delta|\nu|} \tag{6.35}$$

Knowing K_P°, we can use (6.35) to find K_c°. The constant K_c° is (like K_P°) dimensionless. Likewise, the quantity in brackets in (6.35) is dimensionless. Since K_P° depends only on T, it follows from (6.35) that K_c° [which, apart from units, equals $K_P^\circ(RT)^{-\Delta|\nu|}$] is also a function of T only.

One can also define a *mole-fraction equilibrium constant* K_x:

$$K_x \equiv \prod_i (x_{i,\text{eq}})^{\nu_i} \tag{6.36}$$

It is readily shown (Prob. 6.4) that

$$K_P^\circ = K_x(P/\text{atm})^{\Delta|\nu|} \tag{6.37}$$

Note that (except for reactions with $\Delta|\nu| = 0$) the equilibrium constant K_x depends on P as well as on T and hence is not as useful as K_P°.

Introduction of K_c° and K_x is simply a convenience, and we can solve any ideal-gas equilibrium problem using only K_P°. Since the standard state of an ideal gas is defined as having 1 atm pressure, ΔG° is directly related to K_P° by Eq. (6.17) as $\Delta G^\circ = -RT \ln K_P^\circ$ but is only indirectly related to K_c° and K_x through (6.35) and (6.37).

6.3 TEMPERATURE DEPENDENCE OF THE EQUILIBRIUM CONSTANT

The ideal-gas standard equilibrium constant K_P° is a function of temperature only. Let us derive its temperature dependence. Equation (6.17) gives $\ln K_P^\circ = -\Delta G^\circ/RT$. Differentiation of this equation with respect to T gives

$$\frac{d \ln K_P^\circ}{dT} = \frac{\Delta G^\circ}{RT^2} - \frac{1}{RT}\frac{d(\Delta G^\circ)}{dT} \tag{6.40}$$

Use of (5.49) gives

$$\frac{d}{dT}\Delta G^\circ = \frac{d}{dT}\sum_i \nu_i \bar{G}_i^\circ = \sum_i \nu_i \frac{d\bar{G}_i^\circ}{dT} \tag{6.41}$$

From $d\bar{G} = -\bar{S}\,dT + \bar{V}\,dP$, we have for a pure substance $\partial\bar{G}/\partial T = -\bar{S}$; hence

$$d\bar{G}_i^\circ/dT = -\bar{S}_i^\circ \tag{6.42}$$

(The degree superscript indicates the pressure of pure ideal gas i is fixed at the standard value 1 atm; hence $\bar{G}_i^\circ$ depends only on T, and the partial derivative becomes an ordinary derivative.) Using (6.42) in (6.41), we have

$$\frac{d\,\Delta G^\circ}{dT} = -\sum_i \nu_i \bar{S}_i^\circ = -\Delta S^\circ \tag{6.43}$$

where ΔS° is the standard entropy change for the reaction, Eq. (5.45). Hence (6.40) becomes

$$\frac{d \ln K_P^\circ}{dT} = \frac{\Delta G^\circ}{RT^2} + \frac{\Delta S^\circ}{RT} = \frac{\Delta G^\circ + T\,\Delta S^\circ}{RT^2} \tag{6.44}$$

Since $\Delta G^\circ = \Delta H^\circ - T\,\Delta S^\circ$, we end up with

$$\frac{d \ln K_P^\circ}{dT} = \frac{\Delta H^\circ}{RT^2} \tag{6.45}*$$

This is the *van't Hoff equation.* In (6.45), $\Delta H^\circ = \Delta H_T^\circ$ is the standard enthalpy change for the ideal-gas reaction at temperature T [Eq. (5.10)]. [The degree superscript in (6.45) is actually superfluous, since H of an ideal gas is independent of pressure and the presence of other ideal gases.]

Multiplication of (6.45) by dT and integration between the limits T_1 and T_2 gives

$$\ln \frac{K_P^\circ(T_2)}{K_P^\circ(T_1)} = \int_{T_1}^{T_2} \frac{\Delta H^\circ(T)}{RT^2}\, dT \tag{6.46}$$

To evaluate the integral in (6.46), we need ΔH° as a function of temperature; $\Delta H^\circ(T)$ can be obtained by integration of ΔC_P°, as discussed in Sec. 5.4. Thus, in Eq. (5.22), let T_1 be some fixed temperature (for example, 298.15 K), let T_2 be replaced by an arbitrary temperature T, and let the dummy integration variable be changed to T' (to avoid double use of the symbol T); Eq. (5.22) then reads

$$\Delta H^\circ(T) = \Delta H^\circ(T_1) + \int_{T_1}^{T} \Delta C_P^\circ(T')\, dT' \tag{6.47}$$

Substitution of (6.47) into (6.46) then allows evaluation of K_P° at an arbitrary temperature T_2 from its known value at T_1.

If the temperature interval between T_1 and T_2 is reasonably small, we can take ΔH° as approximately constant. Moving ΔH° outside the integral sign in (6.46) and integrating, we get

$$\ln \frac{K_P^\circ(T_2)}{K_P^\circ(T_1)} \approx \frac{\Delta H^\circ}{R}\left(\frac{1}{T_1} - \frac{1}{T_2}\right) \tag{6.48}$$

Since $d(T^{-1}) = -T^{-2}\, dT$, Eq. (6.45) can be written as

$$\frac{d \ln K_P^\circ}{d(1/T)} = -\frac{\Delta H^\circ}{R} \tag{6.49}$$

The slope of a graph of $\ln K_P^\circ$ vs. $1/T$ at a given temperature multiplied by $-R$ equals ΔH° at that temperature; if ΔH° is essentially constant over the temperature range plotted, the graph is a straight line.

6.4 IDEAL-GAS EQUILIBRIUM CALCULATIONS

For an ideal-gas reaction, once we know the value of K_P° at a given temperature, we can find the equilibrium composition of any given reaction mixture at that temperature. K_P° can be determined by chemical analysis of a single mixture that has reached equilibrium at the temperature of interest. However, it is generally simpler to determine K_P° from ΔG°, according to Eq. (6.17): $\Delta G^\circ = -RT \ln K_P^\circ$. In Chap. 5, we showed how calorimetric measurements (heat capacities and latent heats of phase transitions of pure substances, and heats of reaction) allow one to determine $\Delta G_{f,T}^\circ$ values for a great many compounds. Once these values are known, we can calculate ΔG_T° for any chemical reaction between these compounds, and from ΔG° we get K_P°. Thus thermodynamics allows us to find K_P° for a reaction without making any measurements on an equilibrium mixture. This knowledge is of obvious value in assessing the maximum possible yield of product in a chemical reaction. If ΔG_T° is found to be highly positive for a reaction under consideration, we know that this reaction will not be useful for producing the desired product. If ΔG_T° is negative (or only slightly positive), the reaction *may* be useful. Even though the equilibrium position yields substantial amounts of products, we must still consider the *rate* of the reaction (a subject outside the purview of thermodynamics). Frequently, a reaction with a negative ΔG° is found to proceed extremely slowly. Hence we may have to search for a suitable catalyst for the reaction, to speed up attainment of equilibrium. Frequently, several different reactions can occur between a given set of reactants, and we must then consider the rates and the equilibrium constants of several simultaneous reactions.

We now give some examples of equilibrium calculations for ideal-gas reactions. We shall use K_P° in all these examples; K_c° could also have been used, but consistent use of K_P° avoids having to learn any formulas with K_c°. In these examples, the density is assumed low enough to permit treating the gas mixtures as ideal.

Example For the reaction $H_2 + D_2 \rightleftharpoons 2HD$ (where $D \equiv {}^2H$ is the isotope deuterium), data in the Appendix give $\Delta G_{298}^\circ = -0.700$ kcal mol^{-1}. (The reasons ΔG° is not zero will be discussed in Sec. 22.8.) Suppose 0.300 mol of H_2 and 0.100 mol of D_2 are placed in a 2.00-l vessel at 25°C and a catalyst for the isotope exchange reaction is added. Find the equilibrium amounts of H_2, D_2, and HD.

Equations (6.17) and (2.125) give 2.303 log $K_P^\circ = -\Delta G^\circ/RT = 1.18_2$, and

$$K_P^\circ = 3.26 = \frac{[P(\text{HD})/\text{atm}]^2}{[P(\text{H}_2)/\text{atm}][P(\text{D}_2)/\text{atm}]} = \frac{[P(\text{HD})]^2}{P(\text{H}_2)P(\text{D}_2)}$$

Use of $P_i = n_i RT/V$ for the partial pressures gives

$$K_P^\circ = \frac{[n(\text{HD})RT/V]^2}{[n(\text{H}_2)RT/V][n(\text{D}_2)RT/V]} = \frac{[n(\text{HD})]^2}{n(\text{H}_2)n(\text{D}_2)} = 3.26 \tag{6.50}$$

(To save time, we omit the eq subscripts in these examples.) Equation (6.50) is valid only because $\Delta|\nu| = 0$ here.

Let x be the number of moles of H_2 that have reacted when equilibrium is reached. The equilibrium amounts are then

$$n(\text{H}_2) = (0.300 - x)\text{ mol}, \qquad n(\text{D}_2) = (0.100 - x)\text{ mol}, \qquad n(\text{HD}) = 2x\text{ mol}$$

Substitution in (6.50) gives after simplification: $0.227x^2 + 0.400x - 0.0300 = 0$. The solutions to $ax^2 + bx + c = 0$ are $x = [-b \pm (b^2 - 4ac)^{1/2}]/2a$. We find $x = 0.072$ and -1.83. Since we started with 0 mol of HD and nonzero amounts of H_2 and D_2,

x must be positive. Hence $x = 0.072$ is the correct solution, and the equilibrium amounts are $n(H_2) = 0.228$ mol, $n(D_2) = 0.028$ mol, and $n(HD) = 0.144$ mol.

Example For the ideal-gas reaction $2A + B = C + D$, we are given that $\Delta G^\circ_{800} = -3.000$ kcal/mol. If 3.000 mol of A, 1.000 mol of B, and 4.000 mol of C are placed in an 8000-cm³ vessel at 800 K, find the equilibrium amounts of all species.

Using (6.17), we find $K^\circ_P = 6.60$. Also

$$K^\circ_P = \frac{(P_C/\text{atm})(P_D/\text{atm})}{(P_A/\text{atm})^2(P_B/\text{atm})} = \frac{(n_C RT/V)(n_D RT/V)\text{ atm}}{(n_A RT/V)^2(n_B RT/V)} = \frac{n_C n_D}{n_A{}^2 n_B}\frac{V\text{ atm}}{RT}$$

where (6.29) was used. Letting x be the number of moles of B that react, we have

$$6.6 = \frac{(4+x)x\text{ mol}^2}{(3-2x)^2(1-x)\text{ mol}^3}\frac{8000\text{ cm}^3\text{ atm}}{(82.06\text{ cm}^3\text{ atm mol}^{-1}\text{ K}^{-1})(800\text{ K})}$$

$$x^3 - 3.995x^2 + 5.269x - 2.250 = 0 \tag{6.51}$$

where we divided by the coefficient of x^3. We have a cubic equation to solve. The formula for the roots of a cubic equation is quite complicated. Moreover, equations of degree higher than quartic often arise in equilibrium calculations, and there is no formula for the roots of such equations. Hence we shall solve (6.51) by trial and error. Since we started with 1 mol of B, we know that x must be between 0 and 1. For $x = 0$, the left side of (6.51) equals -2.250; for $x = 1$, the left side equals 0.024. Hence x is much closer to 1 than to 0. Guessing $x = 0.9$, we get -0.015 for the left side. Hence the root is between 0.9 and 1.0; interpolation gives an estimate of $x = 0.94$. For $x = 0.94$, the left side equals 0.003, so we are still a bit high. Trying $x = 0.93$, we get -0.001 for the left side. Hence the root is 0.93 (to two places). The equilibrium amounts are then $n_A = 1.14$ mol, $n_B = 0.07$ mol, $n_C = 4.93$ mol, and $n_D = 0.93$ mol.

Example Ideal gases A and B are in equilibrium according to $A \rightleftharpoons 2B$. Let T and P be the temperature and pressure of the equilibrium mixture. Find expressions for the equilibrium mole fractions in terms of $K^\circ_P(T)$ and P.

We have at equilibrium

$$K^\circ_P = \frac{(P_B/P^\circ)^2}{P_A/P^\circ} = \frac{P_B{}^2}{P_A P^\circ} = \frac{(x_B P)^2}{x_A P P^\circ} = \frac{x_B{}^2}{x_A}\frac{P}{P^\circ} = \frac{x_B{}^2}{1-x_B}\frac{P}{\text{atm}} \tag{6.52}$$

where we used $P_i = x_i P$, Eq. (1.23), and the fact that the mole fractions add to 1: $x_A + x_B = 1$. Defining $z \equiv K^\circ_P/(P/\text{atm})$, we have from (6.52): $x_B{}^2 + zx_B - z = 0$, and $x_B = \frac{1}{2}[-z \pm (z^2 + 4z)^{1/2}]$. The quantity z is always positive; the mole fraction x_B cannot be negative; we therefore discard the negative root to get

$$x_B = \tfrac{1}{2}[(z^2 + 4z)^{1/2} - z] \qquad \text{where } z \equiv K^\circ_P/(P/\text{atm}) \tag{6.53}$$

Knowing P and K°_P, we can calculate z; from z we calculate x_B; from x_B, we calculate x_A as $1 - x_B$.

Example Suppose we start with an amount of substance n_0 of a pure species, and it dissociates to some extent. We define the *degree of dissociation* α as the fraction of n_0 that has dissociated when equilibrium is reached. For the ideal-gas reaction $A = 2B$, find an expression for α as a function of K°_P and P.

We start with n_0 of A, of which αn_0 dissociates, leaving $n_0 - \alpha n_0$ of A at equilibrium, and producing $2\alpha n_0$ of B; the equilibrium amounts are $n_A = n_0(1 - \alpha)$ and $n_B = 2\alpha n_0$. The equilibrium mole fractions are $n_A/(n_A + n_B)$ and $n_B/(n_A + n_B)$. Since $n_A + n_B = n_0(1 + \alpha)$, we have $x_A = (1 - \alpha)/(1 + \alpha)$ and $x_B = 2\alpha/(1 + \alpha)$. Substituting for x_B in (6.52) gives $K_P^\circ = [4\alpha^2/(1 - \alpha^2)]P/\text{atm}$. Solving for α, we get

$$\alpha = \left[\frac{K_P^\circ}{K_P^\circ + 4(P/\text{atm})}\right]^{1/2} \tag{6.54}$$

Consider the effect of P on α. The equilibrium constant K_P° depends only on T and is independent of pressure. As P goes to infinity in (6.54), the degree of dissociation of A to 2B goes to zero at any given temperature. As P goes to zero, the degree of dissociation goes to 1. High pressure shifts the equilibrium to the undissociated gas; low pressure shifts it to the dissociated gas.

Example Dinitrogen tetroxide at room temperature is partially dissociated according to $N_2O_4(g) = 2NO_2(g)$. Find the composition of an N_2O_4–NO_2 equilibrium mixture at 25°C and 0.500 atm; assume ideal-gas behavior.

The Appendix $\Delta G_{f,298}^\circ$ values give for the reaction $\Delta G_{298}^\circ = 1140$ cal/mol. From $\Delta G^\circ = -RT \ln K_P^\circ$, we find $K_P^\circ = 0.146$. Therefore z in (6.53) is: $z = 0.146/0.500 = 0.292$. Use of (6.53) gives $x(NO_2) = 0.414$. Hence, $x(N_2O_4) = 0.586$.

Example Find the equilibrium composition of an NO_2–N_2O_4 mixture at 400 K and 0.500 atm.

To find K_P° at 400 K, we can use Eq. (6.45). For an accurate calculation, we need $\Delta H^\circ(T)$, which is found from $\Delta C_P^\circ(T)$. We shall make the approximation that ΔH° is independent of T over the range 298 to 400 K and use the approximate equation (6.48). The Appendix data give $\Delta H_{298}^\circ = 13.67$ kcal/mol for $N_2O_4 = 2NO_2$. From the last example, we have $K_{P,298}^\circ = 0.146$. Substitution in (6.48) gives $\ln (K_{P,400}^\circ/0.146) = 5.87_5$ and $K_{P,400}^\circ = 52.0$. Use of (6.53) gives $x(NO_2) = 0.99$ and $x(N_2O_4) = 0.01$.

For improved accuracy, we can look up $\Delta G_{f,400}^\circ$ of NO_2 and N_2O_4 in the JANAF tables (Sec. 5.6); we find $\Delta G_{400}^\circ = -3117$ cal/mol. This accurate value of ΔG_{400}° yields $K_{P,400}^\circ = 50.5$. This improved K_P° value also gives $x(NO_2) = 0.99$.

Example For the ideal-gas reaction (reaction I) $2A + B = 2C$, we are given that $\Delta G_{500}^\circ = -1000$ cal mol^{-1}. Find K_P° at 500 K for the reaction (reaction II) $A + \frac{1}{2}B = C$.

From $\Delta G^\circ = -RT \ln K_P^\circ$, we find $K_P^\circ = 2.74$ for reaction I. The equilibrium-constant expressions for reactions I and II are

$$K_{P,\text{I}}^\circ = \frac{(P_C/\text{atm})^2}{(P_A/\text{atm})^2(P_B/\text{atm})} \quad \text{and} \quad K_{P,\text{II}}^\circ = \frac{P_C/\text{atm}}{(P_A/\text{atm})(P_B/\text{atm})^{1/2}}$$

We see that $K_{P,\text{I}}^\circ$ is the square of $K_{P,\text{II}}^\circ$; hence $2.74 = (K_{P,\text{II}}^\circ)^2$, and $K_{P,\text{II}}^\circ = 1.66$.

Alternatively, ΔG° for II is one-half ΔG° for I and is thus -500 cal mol^{-1}; this ΔG_{500}° yields $K_{P,\text{II}}^\circ = 1.66$.

Two final remarks about equilibrium calculations. When $|\Delta G^\circ|$ is large, one finds a very large (or very small) value for K_P°. For example, if $\Delta G_{298}^\circ = 33{,}000$ cal mol^{-1}, then $K_{P,298}^\circ = 10^{-24}$. From this value of K_P°, we might well calculate that at equilibrium only a few molecules (or even only a fraction of one molecule) of a product are present. Of course, when the number of molecules of a species is small, thermodynamics is not rigorously applicable and the system shows continual fluctuations about the thermodynamically predicted number of molecules (Sec. 3.7).

Tables of thermodynamic data often list ΔH_f° and ΔG_f° values to 0.01 kcal/mole. However, experimental errors in measured ΔH_f° values typically run about $\frac{1}{2}$ kcal mol^{-1} (although they may be as little as 0.05 kcal mol^{-1} or as much as 5 kcal mol^{-1}). An error in ΔG_{298}° of $\frac{1}{2}$ kcal mol^{-1} corresponds to a factor of 2 in K_P°. Thus, the reader should take equilibrium constants calculated from thermodynamic data with a grain of NaCl(*s*).

6.5 SHIFTS IN IDEAL-GAS REACTION EQUILIBRIA

In this section, we consider the effects of changes in various conditions on the equilibrium position of an ideal-gas reaction. We suppose that after the mixture has reached equilibrium, we change one of the thermodynamic variables, and we examine the effect of this change.

Isobaric temperature change. Suppose we change T, keeping P constant. Since $d \ln y = (1/y)\,dy$, Eq. (6.45) gives $dK_P^\circ/dT = K_P^\circ\,\Delta H^\circ/RT^2$. Since K_P° and RT^2 are positive, the sign of dK_P°/dT is the same as the sign of ΔH°.

If ΔH° is positive (an endothermic reaction), then dK_P°/dT is positive; for a temperature increase ($dT > 0$), dK_P° is then positive, and K_P° increases. Since product partial pressures are in the numerator of K_P°, an increase in K_P° means an increase in the equilibrium values of the product partial pressures and a decrease in reactant partial pressures. Since $P_i = x_i P$, and P is held fixed, the mole fractions undergo changes proportional to the changes in the partial pressures. Thus for an endothermic reaction, an increase in temperature at constant pressure will shift the equilibrium to the right.

If ΔH° is negative (an exothermic reaction), then dK_P°/dT is negative and a positive dT gives a negative dK_P°. Thus an isobaric temperature increase shifts the equilibrium to the left for an exothermic reaction.

These results can be summarized in the easily remembered rule that a change in temperature at constant pressure leads to an equilibrium shift tending to counteract the effect of the temperature change. Consider, for example, an endothermic reaction. If T is increased at constant P, the equilibrium shifts so as to counteract the temperature increase, i.e., to cool the system. Since the forward reaction is endothermic, a shift to the right will cool the system.

Isothermal pressure change. Suppose we isothermally change the volume of the system, thereby changing the total pressure $P\ (= n_{\text{tot}}RT/V)$ and changing the partial pressure $P_i\ (= n_i RT/V)$ of each gas. Since K_P° is independent of P, this change has no effect on K_P°. However, K_x in (6.37) is pressure dependent (unless $\Delta|\nu| = 0$). Equation (6.37) gives $K_x = K_P^\circ(P/\text{atm})^{-\Delta|\nu|}$. If $\Delta|\nu|$ is positive, an isothermal increase in P will decrease K_x and hence will shift the equilibrium to the left; an isothermal decrease in P will increase K_x and shift the equilibrium to the right. If $\Delta|\nu|$ is negative, an isothermal increase in P will shift the equilibrium to the right. If $\Delta|\nu|$ is zero, an isothermal pressure change has no effect on the equilibrium amounts. These results can be summarized in the easily remembered rule that an isothermal pressure change shifts the equilibrium in a direction tending to counteract the pressure change. (This rule and the corresponding rule for an isobaric temperature change constitute *Le Châtelier's principle.*) Thus, an isothermal increase in pressure will shift the equilibrium to the side that has a smaller number of moles, thereby decreasing the pressure; if $\Delta|\nu| > 0$, there are fewer moles of reactants than of products and a pressure increase shifts the equilibrium to the left; if $\Delta|\nu| < 0$, there are fewer moles of products than of reactants and a pressure increase shifts the equilibrium to the right.

An alternative way to look at the effect of a pressure change is this. Consider a reaction with $\Delta|\nu|$ positive; for example, $A \rightleftharpoons 2B$. We define Q_P as $Q_P \equiv P_B{}^2/P_A$, where P_A and P_B are the partial pressures of the gases A and B in the system at some instant. When the system is in equilibrium, we have $Q_P = K_P$; when the system is not in equilibrium, then $Q_P \neq K_P$. Let equilibrium be established, and suppose we then double the pressure at constant T (by isothermally compressing the mixture) and thereafter hold P constant at its new value. K_P in (6.23) is unchanged (since T is unchanged). Since $P_i = x_i P$, this doubling of P doubles P_A and doubles P_B. This quadruples the numerator of Q_P and doubles its denominator, so that Q_P is doubled. Before the pressure increase, Q_P was equal to K_P, but after the pressure increase, Q_P has been increased and is greater than K_P. The system is no longer in equilibrium, and Q_P will have to decrease to restore equilibrium. Q_P decreases when the equilibrium shifts to the left, thereby decreasing P_B and increasing P_A. Thus a pressure increase shifts the equilibrium $A = 2B$ to the left, the side with fewer moles, as we concluded above.

Isochoric addition of inert gas. Suppose we add some inert gas to an equilibrium mixture, holding V and T constant. Since $P_i = n_i RT/V$, the partial pressure of each gas taking part in the reaction is unaffected by the addition of an inert gas. Hence the quotient

$$Q_P \equiv \prod_i (P_i)^{\nu_i} \tag{6.55}$$

is unaffected and remains equal to K_P. Thus, there is no shift in equilibrium. Isochoric, isothermal addition of an inert gas does not affect the equilibrium position of an ideal-gas reaction. This makes sense, in that because of the absence of intermolecular interactions, the reacting ideal gases have no way of knowing whether there is any inert gas present or not.

Isobaric addition of inert gas. Suppose we add an inert gas to an equilibrium mixture, holding P and T constant. To keep P constant with gas being added, the volume V must be increased. Since $P_i = n_i RT/V$, this increase in volume decreases each partial pressure P_i by the same percentage. If $\Delta|\nu| \neq 0$, the quotient (6.55) will be affected and Q_P will no longer equal K_P. Hence the equilibrium will shift. If $\Delta|\nu|$ is positive, the volume increase will decrease the numerator of Q_P more than it will decrease the denominator; hence the equilibrium will shift to the right, making the quotient Q_P again equal to K_P. If $\Delta|\nu|$ is negative, the equilibrium will shift to the left. Note that the shift is not due directly to the added inert gas but is a consequence of the increase in volume that accompanies addition of the inert gas at constant P and T.

Addition of a reactant gas. Suppose that for the reaction A + B = 2C + D we add some A to an equilibrium mixture of A, B, C, and D while holding T and V constant. Since $P_i = n_i RT/V$, this addition increases P_A and does not change the other partial pressures. Since P_A appears in the denominator of the quotient (6.55) (ν_A is negative), addition of A at constant T and V makes the quotient Q_P less than K_P. The equilibrium must then shift to the right, so as to increase the numerator of Q_P and make Q_P equal to K_P again. Thus addition of A at constant T and V shifts the equilibrium to the right, thereby consuming some of the added A. Similarly, addition of a reaction product at constant T and V shifts the equilibrium to the left, thereby consuming some of the added substance. Removal of some of a reaction product from a mixture held at constant T and V shifts the equilibrium to the right, producing more product.

It might be thought that the same conclusions apply to addition of a reactant while holding T and P constant. Surprisingly, however, there are certain circumstances where constant-T-and-P addition of a reactant will shift the equilibrium so as to produce more of the added species. For example, consider the ideal-gas equilibrium $N_2 + 3H_2 = 2NH_3$. Suppose equilibrium is established at a temperature and pressure for which K_x is 8.33: $K_x = 8.33 = [x(NH_3)]^2/x(N_2)[x(H_2)]^3$. Let the amounts $n(N_2) = 3.00$ mol, $n(H_2) = 1.00$ mol, and $n(NH_3) = 1.00$ mol be present at this T and P. Defining Q_x as $Q_x \equiv \prod_i (x_i)^{\nu_i}$, we find that for these amounts,. $Q_x = (0.2)^2/0.6(0.2)^3 = 8.33$. Since $Q_x = K_x$, the system is in equilibrium. Now, holding T and P constant, we add 0.1 mol of N_2. Because T and P are constant, K_x is still 8.33. After the N_2 is added but before any shift in equilibrium occurs, we have

$$Q_x = \frac{(1/5.1)^2}{(3.1/5.1)(1/5.1)^3} = 8.39$$

Q_x now exceeds K_x, and the equilibrium must therefore shift to the left, so as to reduce Q_x to 8.33; this shift produces more N_2. Addition of N_2 under these conditions shifts the equilibrium to produce more N_2. Although the addition of N_2 increases x_{N_2}, it also decreases x_{H_2} (and x_{NH_3}), and the fact that x_{H_2} is cubed in the denominator of Q_x outweighs the increase in x_{N_2} and the decrease in x_{NH_3}; hence, in this case, Q_x increases on addition of N_2. For the general conditions under which addition of a reagent at constant T and P shifts the equilibrium to produce more of the added species, see Prob. 6.18.

Le Châtelier's principle is often stated as follows: in a system at equilibrium, a change in one of the variables that determines the equilibrium will shift the equilibrium in the direction counteracting the change in that variable. The example just given shows this statement is *false*. A change in a variable may or may not shift the equilibrium in a direction that counteracts the change. Thus, use of Le Châtelier's principle should be restricted to changes in T at constant P and changes in P at constant T. [Le Châtelier's principle can be formulated in a more general way, which, however, bears little resemblance to the statement at the beginning of this paragraph; for a discussion, see J. de Heer, *J. Chem. Educ.*, **34**, 375 (1957); **35**, 133 (1958).]

PROBLEMS

6.1 Find the conventional value of μ for $NH_3(g)$ at 25°C and 2.000 atm. Use data in the Appendix.

6.2 For the gas-phase reaction $2NO + Br_2 = 2NOBr$, observed equilibrium partial pressures for a certain mixture at 324 K are $P(NO) = 96.5$ torr; $P(Br_2) = 33.9$ torr; $P(NOBr) = 100.8$ torr. Find K_P° and ΔG° at 324 K. (Assume ideal gases.)

6.3 For the ideal-gas reaction $A + B = 2C + 3D$, one finds the following equilibrium amounts at 1000 K and 2.00 atm: $n_A = 1.00$ mol; $n_B = 1.00$ mol; $n_C = 1.00$ mol; $n_D = 2.00$ mol. Find K_P° and ΔG° at 1000 K.

6.4 Derive Eq. (6.37).

6.5 Prove that for an ideal-gas reaction

$$\frac{d \ln K_c^\circ}{dT} = \frac{\Delta U^\circ}{RT^2}$$

6.6 Prove that for an ideal-gas reaction

$$\left(\frac{\partial \ln K_x}{\partial T}\right)_P = \frac{\Delta H^\circ}{RT^2}, \qquad \left(\frac{\partial \ln K_x}{\partial P}\right)_T = -\frac{\Delta|\nu|}{P}$$

6.7 At high temperatures, I_2 vapor is partially dissociated to I atoms. Let P_0 be the expected pressure of I_2 calculated ignoring dissociation, and let P be the observed pressure. Some values of P_0 and P for I_2 samples are:

T/K	973	1073	1173	1274
P_0/atm	0.0576	0.0631	0.0684	0.0736
P/atm	0.0624	0.0750	0.0918	0.1122

(*a*) Show that the degree of dissociation is $\alpha = (P - P_0)/P_0$. (*b*) Use the equation immediately preceding (6.54) to calculate K_P° values. (*c*) Find the average ΔH° for $I_2 = 2I$ for temperatures in the range of the above data.

6.8 Replacing T_2 by T and considering T_1 as a fixed temperature, we can write the approximate equation (6.48) in the form $\ln K_P^\circ(T) \approx -\Delta H^\circ/RT + C$, where the constant C equals $\ln K_P^\circ(T_1) + \Delta H^\circ/RT_1$. Derive the following exact equation:

$$\ln K_P^\circ(T) = -\Delta H_T^\circ/RT + \Delta S_T^\circ/R.$$

(The derivation is very short.)

6.9 After 0.600 mol of A is put in a cylinder fitted with a piston, the cylinder is heated to 150°C; with the volume held fixed at 5000 cm^3, a catalyst is introduced that establishes the ideal-gas equilibrium $A = 2B + C$. A pressure measurement gives $P = 8.000$ atm at equilibrium. Find K_P° at 150°C.

6.10 For the ideal-gas reaction $A + B = C$, a mixture with $n_A = 1.000$ mol, $n_B = 3.000$ mol, and $n_C = 2.000$ mol is at equilibrium at 300 K and 1.000 atm. Suppose the pressure is isothermally increased to 2.000 atm; find the new equilibrium amounts.

6.11 For the reaction $PCl_5(g) = PCl_3(g) + Cl_2(g)$, use data in the Appendix to find K_P° at 25°C and at 500 K. Assume ideal-gas behavior, and neglect the temperature variation of ΔH°. Calculate the degree of dissociation α at 25°C and 1.000 atm and at 500 K and 1.000 atm.

6.12 (*a*) For the reaction (5.25) assume ideal-gas behavior and use data in the Appendix and the expression for ΔH° found in Sec. 5.4 to find an expression for $\ln K_P^\circ(T)$ valid from 300 K to 1500 K. (*b*) Calculate K_P° at 1000 K for this reaction.

6.13 Consider the ideal-gas dissociation reaction $A = 2B$. For A and B, we have $\bar{C}_{P,A}^\circ = a + bT + cT^2$ and $\bar{C}_{P,B}^\circ = e + fT + gT^2$, where a, b, c, e, f, g are known constants, and these equations are valid over the temperature range from T_1 to T_2. Further, suppose that $\Delta H_{T_1}^\circ$ and $K_P^\circ(T_1)$ are known. Find an expression for $\ln K_P^\circ(T)$ valid between T_1 and T_2.

6.14 The synthesis of ammonia from N_2 and H_2 is an exothermic reaction. Hence the equilibrium yield of ammonia decreases as T increases. Explain why the synthesis of ammonia from its elements (Haber process) is run at the high temperature of 800 K rather than at a lower temperature.

6.15 Suppose 1.00 mol of CO_2 and 1.00 mol of COF_2 are placed in a very large vessel at 25°C and a catalyst for the gas-phase reaction $2COF_2 = CO_2 + CF_4$ is added. Use data in the Appendix to find the equilibrium amounts.

6.16 For the ideal-gas reaction $A + B = 2C + 2D$, it is given that $\Delta G_{500}^\circ = 1250$ cal mol^{-1}. (*a*) If 1.000 mol of A and 1.000 mol of B are placed in a vessel at 500 K and P is held fixed at 1200 torr, find the equilibrium amounts. (*b*) If 1.000 mol of A and 2.000 mol of B are placed in a vessel at 500 K and P is held fixed at 1200 torr, find the equilibrium amounts.

6.17 Suppose that for a certain ideal-gas reaction, the error in ΔG° is 750 cal mol^{-1}. What error in K_P° does this cause?

6.18 (*a*) Show that

$$\frac{\partial \ln Q_x}{\partial n_j} = \frac{1}{Q_x}\frac{\partial Q_x}{\partial n_j} = \frac{\nu_j - x_j\,\Delta|\nu|}{n_j}$$

where $Q_x \equiv \prod_i (x_i)^{\nu_i}$. (*b*) Use the result of part (*a*) to show that addition at constant T and P of a small amount of reacting species j to an ideal-gas equilibrium mixture will shift the equilibrium to produce more j when the following two conditions are both satisfied: (1) the species j appears on the side of the reaction equation that has the greater sum of the coefficients; (2) the equilibrium mole fraction x_j is greater than $\nu_j/\Delta|\nu|$. (*c*) For the reaction $N_2 + 3H_2 = 2NH_3$, when will addition of N_2 to an equilibrium mixture held at constant T and P shift the equilibrium to produce more N_2? Answer the same question for H_2 and for NH_3. (Assume ideal-gas behavior.)

6.19 A certain ideal-gas dissociation reaction A = 2B has $\Delta G^\circ_{1000} = 1000$ cal mol^{-1}, which gives $K_P^\circ = 0.605$ at 1000 K. If pure A is put in a vessel at 1000 K and 1 atm and held at constant T and P, then A will partially dissociate to give some B. Someone presents the following chain of reasoning: "The second law of thermodynamics tells us that a process in a closed system at constant T and P that corresponds to $\Delta G > 0$ is forbidden [Eq. (4.18)]. The standard Gibbs free-energy change for the reaction A = 2B is positive. Therefore, any amount of dissociation of A to B at constant T and P corresponds to an increase in G and is forbidden. Hence gas A held at 1000 K and 1 atm will not give any B at all." Point out the fallacy in this argument.

6.20 A certain gas mixture held at 395°C has the following initial partial pressures: $P(Cl_2) = 351.4$ torr; $P(CO) = 342.0$ torr; $P(COCl_2) = 0$. At equilibrium, the total pressure is 439.5 torr. Find K_P° at 395°C for $CO + Cl_2 = COCl_2$. [$COCl_2$ (phosgene) was used as a poison gas in World War I.] (V is held constant.)

6.21 For an ideal-gas reaction with $\Delta|\nu| \neq 0$, suppose we decrease V at constant T. Does the equilibrium shift in a direction that tends to counteract the volume decrease or tends to augment it?

6.22 (*a*) Consider the isothermal mixing of several pure ideal gases to form an ideal gas mixture; use the first law and the definition of an ideal gas mixture to show that the internal energy U and enthalpy H of the mixture are

$$U = \sum_i U_i^*(T, n_i)$$

$$= \sum_i n_i \bar{U}_i^*(T) \qquad \text{ideal gas mixture}$$

$$H = \sum_i H_i^*(T, n_i)$$

$$= \sum_i n_i \bar{H}_i^*(T) \qquad \text{ideal gas mixture}$$

where U_i^* is the internal energy of n_i moles of pure gas i and n_i is the number of moles of i in the mixture. (*b*) Use results in paragraph 10 of Sec. 3.5 to show that the entropy S of an ideal gas mixture is

$$S = \sum_i S_i^*(T, P_i, n_i)$$

$$= \sum_i n_i \bar{S}_i^*(T, P_i) \qquad \text{ideal gas mixture}$$

[This last equation does not contradict the entropy-of-mixing formula (3.39); Eq. (3.39) refers to a process in which each gas is initially at a pressure equal to the total pressure of the final mixture.] (*c*) Show that

$$G = \sum_i G_i^*(T, P_i, n_i)$$

$$= \sum_i n_i \bar{G}_i^*(T, P_i) \qquad \text{ideal gas mixture}$$

$$C_P = \sum_i n_i \bar{C}_{P,i}^*(T) \qquad \text{ideal gas mixture}$$

6.23 Use the results of Prob. 6.22 to answer the following. (*a*) If 2.000 mol of inert ideal gas A at 300 K and 1.000 atm and 3.000 mol of inert ideal gas B at 300 K and 1.000 atm are mixed to give a mixture with $T = 300$ K and $P = 1.000$ atm, find ΔU, ΔH, ΔS, and ΔG for this process. (*b*) If 2.000 mol of inert ideal gas A at 300 K and 2.000 atm and 3.000 mol of inert ideal gas B at 300 K and 3.000 atm are mixed and the mixture is then compressed to give a final state with $T = 300$ K and $P = 5.000$ atm, find ΔU, ΔH, ΔS, and ΔG for the overall process.

6.24 For the gas-phase reaction

$$I_2 + \text{cyclopentene} = \text{cyclopentadiene} + 2HI$$

measured K_P° values in the range 450 to 700 K are fitted by $\log K_P^\circ = 7.55 - (4.83 \times 10^3)(\text{K}/T)$. Calculate ΔG°, ΔH°, ΔS°, and ΔC_P° for this reaction at 500 K. (Assume ideal gases.)

7

ONE-COMPONENT PHASE EQUILIBRIUM

7.1 THE PHASE RULE

The general condition for material equilibrium, Eq. (4.96), was derived from the second law of thermodynamics in Sec. 4.6. From (4.96), we found in Sec. 4.7 the conditions for phase equilibrium in a closed system: the chemical potential of any given chemical species i must have the same value in every phase in which i is present. In Secs. 7.2 to 7.5 we shall consider phase equilibrium in systems that have only one component. Before specializing to one-component systems, we want to answer the general question of how many independent variables are needed to define the equilibrium state of a multiphase, multicomponent system.

To describe the equilibrium state of a system with several phases and several chemical species, we can specify the mole numbers of each species in each phase and the temperature and pressure, T and P. (Provided no rigid or adiabatic walls separate phases, T and P are the same in all phases at equilibrium.) Specifying mole numbers is not what we shall do, however, since the mass of each phase of the system is of no real interest. The mass (or size) of each phase doesn't affect the phase-equilibrium position, since the equilibrium position is determined by equality of chemical potentials, which are intensive variables. (For example, in a two-phase system consisting of an aqueous solution of NaCl and solid NaCl at fixed T and P, the equilibrium concentration of dissolved NaCl in the saturated solution is independent of the mass of each phase; it doesn't matter whether we have a small or large crystal of solid NaCl or a small or large volume of solution; provided that both phases are present at equilibrium, the equilibrium concentration in the solution has a unique value at a given T and P.) We shall therefore deal with the mole fractions of each species in each phase, rather than with the mole numbers. Let n_j^α be the number of moles of species j in phase α; then the mole fraction of j in phase α is

$$x_j^\alpha \equiv n_j^\alpha \Big/ \sum_i n_i^\alpha \tag{7.1}$$

where the sum goes over all chemical species of phase α.

We define the number of *degrees of freedom* (or the *variance*) f of an equilibrium system as the number of independent intensive variables needed to specify its state. Let the system have c different chemical species and p phases. We describe the equilibrium state by specifying the intensive variables P, T, and all the mole fractions. As we shall see, these variables are not all independent. (In addition, complete specification of the state requires that the mass of each phase be given, but these extensive variables are of no interest to us.)

We shall initially make two assumptions, which will later be eliminated: (1) No chemical reactions occur. (2) Every chemical species is present in every phase.

From assumption 2, there are c chemical species in each phase and hence a total of pc mole fractions. Adding in T and P, we have

$$pc + 2 \tag{7.2}$$

intensive variables to describe the state of the equilibrium system. However, these $pc + 2$ variables are not all independent: there are relations between them. First of all, the sum of the mole fractions in each phase must be 1:

$$x_1^\alpha + x_2^\alpha + \cdots + x_c^\alpha = 1 \tag{7.3}$$

There is an equation like (7.3) for each phase, and hence there are p such equations. We can solve these equations for $x_1^\alpha, x_1^\beta, \ldots$, thereby eliminating p of the intensive variables.

In addition to the relations (7.3), we have the following equalities between chemical potentials:

$$\mu_1^\alpha = \mu_1^\beta = \mu_1^\gamma = \cdots \tag{7.4}$$

$$\mu_2^\alpha = \mu_2^\beta = \mu_2^\gamma = \cdots \tag{7.5}$$

$$\cdots\cdots\cdots\cdots\cdots\cdots$$

$$\mu_c^\alpha = \mu_c^\beta = \mu_c^\gamma = \cdots \tag{7.6}$$

Since there are p phases, Eq. (7.4) contains $p - 1$ equality signs and $p - 1$ independent equations. Since there are c different chemical species, there are a total of $c(p - 1)$ equality signs in the set of equations (7.4) to (7.6); we thus have $c(p - 1)$ independent relations between chemical potentials. Each chemical potential is a function of T, P, and the composition of the phase [Eq. (4.97)]; for example, $\mu_1^\alpha = \mu_1^\alpha(T, P, x_1^\alpha, \ldots, x_c^\alpha)$. Hence the $c(p - 1)$ equations (7.4) to (7.6) provide $c(p - 1)$ simultaneous relations between T, P, and the mole fractions, which we can solve for $c(p - 1)$ of these variables, thereby eliminating $c(p - 1)$ intensive variables.

We started out with $pc + 2$ intensive variables in (7.2). We eliminated p of them using (7.3) and $c(p - 1)$ of them using (7.4) to (7.6). Hence, the number of independent intensive variables (which, by definition, is the number of degrees of freedom f) is $f = pc + 2 - p - c(p - 1) = c - p + 2$. Thus

$$f = c - p + 2 \tag{7.7}^*$$

Equation (7.7) is the *phase rule*, first derived by Gibbs.

Now let us drop assumption 2 and allow for the possibility that one or more chemical species might be absent from one or more phases. (An example is a saturated aqueous salt solution in contact with pure solid salt.) If species i is absent from phase δ, the number of intensive variables is reduced by 1 since x_i^δ is identically zero and is not a variable. However, the number of relations between the intensive variables is also reduced by 1, since we drop μ_i^δ from the set of equations (7.4) to (7.6); recall from Eq. (4.103) that when substance i is absent from phase δ, μ_i^δ need not equal the chemical potential of i in the other phases. Thus, the phase rule (7.7) still holds when every species does not appear in every phase.

Now we drop assumption 1 and suppose that chemical reactions can occur. For each independent chemical reaction that occurs, we have an equilibrium condition given by Eq. (4.111) as $\sum_i \nu_i \mu_i = 0$, where μ_i is the chemical potential of substance i in any phase in which it is present. Each independent chemical reaction provides one relation between the chemical potentials, and [like the relations (7.4) to (7.6)] each such relation can be used to eliminate one variable from T, P, and the mole fractions. If there are r independent chemical reactions, then the number of independent intensive variables is reduced by r and the phase rule (7.7) becomes

$$f = c - p + 2 - r \tag{7.8}$$

(By independent chemical reactions, we mean that no reaction can be written as a combination of the others; see *Denbigh*, secs. 4.16 and 4.17 for a discussion.)

In addition to reaction equilibrium relations, there may be other restrictions on the intensive variables of the system. For example, suppose we have a gas-phase system in which we introduce some NH_3 but no H_2 or N_2; we then add a catalyst to establish the equilibrium $2NH_3 = N_2 + 3H_2$; further, we refrain from introducing any N_2 or H_2 from outside. Since all the N_2 and H_2 come from the dissociation of NH_3, we must have $n(H_2) = 3n(N_2)$ and $x(H_2) = 3x(N_2)$. This provides an additional relation between the intensive variables besides the equilibrium relation $2\mu(NH_3) = \mu(N_2) + 3\mu(H_2)$. In ionic solutions, the condition of electrical neutrality provides such an additional relation. For example, an aqueous solution of HCN has the five species: H_2O, HCN, H^+, OH^-, CN^-; there are two equilibrium conditions (from the reactions $H_2O = H^+ + OH^-$ and $HCN = H^+ + CN^-$) and one additional condition that arises from electroneutrality: $\mu(H_2O) = \mu(H^+) + \mu(OH^-)$, $\mu(HCN) = \mu(H^+) + \mu(CN^-)$, and $x(H^+) = x(CN^-) + x(OH^-)$. If, besides the r reaction equilibrium conditions of the form (4.111), there are a additional restrictions on the intensive variables (arising from stoichiometric and electroneutrality conditions), then the number of degrees of freedom is reduced by a and the phase rule (7.8) becomes

$$f = c - p + 2 - r - a \tag{7.9}^*$$

We can preserve the simple form (7.7) for the phase rule by defining the number of *independent components* c_{ind} as

$$c_{\text{ind}} \equiv c - r - a \tag{7.10}$$

Equation (7.9) then reads

$$f = c_{\text{ind}} - p + 2 \tag{7.11}$$

(Many books call c_{ind} simply the number of components, but this terminology can be misleading.)

As an example, for the aqueous solution of HCN mentioned above, $r = 2$ and $a = 1$, so $c_{\text{ind}} = 5 - 2 - 1 = 2$. Since $p = 1$, we have $f = 3$. This makes sense, since once the three intensive variables T, P, and the HCN mole fraction are specified, all the remaining mole fractions can be calculated using the H_2O and HCN dissociation equilibrium constants.

Another example is a system containing $CaCO_3(s)$, $CaO(s)$, and $CO_2(g)$ where all the CaO and CO_2 come from dissociation of the $CaCO_3$, according to $CaCO_3(s) = CaO(s) + CO_2(g)$. There are three chemical species ($c = 3$). There is one reaction equilibrium condition, $\mu[CaCO_3(s)] = \mu[CaO(s)] + \mu[CO_2(g)]$, so $r = 1$. Are there any additional restrictions on the intensive variables? It is true that the number of moles of $CaO(s)$ must equal the number of moles of $CO_2(g)$: $n[CaO(s)] = n[CO_2(g)]$; however, this equation cannot be converted into a relation between the mole fractions in each phase and does not provide an additional relation between intensive variables. Hence $c_{ind} = 3 - 1 = 2$, and $f = 2 - 3 + 2 = 1$. The value $f = 1$ makes sense, since once T is fixed, the pressure of CO_2 gas in equilibrium with the $CaCO_3$ is fixed by the reaction equilibrium condition, and thus P of the system is fixed.

In doubtful cases, rather than applying (7.9) or (7.11), it is often best to first list the intensive variables and then list all the independent restrictive relations between them. Subtraction gives f. For example, for the $CaCO_3$–CaO–CO_2 example just given, the intensive variables are T, P, and the mole fractions in each phase. Since each phase is pure, we know that in each phase the mole fraction of each of $CaCO_3$, CaO, and CO_2 is either 0 or 1; hence the mole fractions are fixed and are not variables. There is one independent relation between intensive variables (namely, the above-stated reaction equilibrium condition). Hence $f = 2 - 1 = 1$. [Knowing f, we can then calculate c_{ind} from (7.11) if c_{ind} is wanted.]

Note the following restrictions on the applicability of the phase rule (7.9). There must be no walls between phases. (We equated the temperatures of the phases, the pressures of the phases, and the chemical potentials of a given component in the phases. These equalities need not hold if adiabatic, rigid, or impermeable walls separate phases.) The system must be capable of P-V work only. (If, for example, we can do electrical work on the system by applying an electric field, then the electric field strength is an additional intensive variable that must be specified to define the system's state.)

[Gibbs's phase rule $f = c_{ind} - p + 2$ has the same appearance as Euler's formula relating the number of vertices V, edges E, and faces F of a simple polyhedron: $V = E - F + 2$. The relationship between the two formulas is discussed in J. Mindel, *J. Chem. Educ.*, **39**, 512 (1962).]

7.2 ONE-COMPONENT PHASE EQUILIBRIUM

In the rest of this chapter, we specialize to phase equilibrium in systems with one independent component. (Chapter 12 deals with multicomponent phase equilibrium.) We shall be concerned in this chapter with pure substances. As an example, consider a one-phase system of pure liquid water. If we ignore the dissociation of H_2O, we would say that there is only one species present ($c = 1$), and there are no reactions or additional restrictions ($r = 0$, $a = 0$); hence $c_{ind} = 1$ and $f = 2$. If we take account of the dissociation $H_2O = H^+ + OH^-$, we have three chemical species ($c = 3$), one reaction equilibrium condition $[\mu(H_2O) = \mu(H^+) + \mu(OH^-)]$, and one electroneutrality or stoichiometry condition $[x(H^+) = x(OH^-)]$; hence $c_{ind} = 3 - 1 - 1 = 1$, and $f = 2$. The dissociation of OH^- in water according to

$OH^- = H^+ + O^{2-}$ occurs to a negligible extent, but if we allow for this dissociation also, we now have four chemical species (H_2O, H^+, OH^-, O^{2-}), two reaction equilibrium conditions, and one electroneutrality condition [$x(H^+) = x(OH^-) + 2x(O^{2-})$]; hence $c_{ind} = 4 - 2 - 1 = 1$, and $f = 2$. Thus, whether or not we take dissociation into account, we have one independent component and two degrees of freedom.

With $c_{ind} = 1$, the phase rule (7.11) becomes

$$f = 3 - p \tag{7.12}$$

If $p = 1$, then $f = 2$; if $p = 2$, then $f = 1$; if $p = 3$, then $f = 0$. The maximum f is 2. Hence for a one-component system, specification of at most two intensive variables specifies the state (apart from specification of the sizes of the phases). Thus we can represent any state (apart from specification of the sizes of the phases) of a one-component system by a point on a two-dimensional P vs. T diagram, where each point corresponds to a definite T and P. Such a diagram is an example of a *phase diagram.*

A P-T phase diagram for pure water is shown in Fig. 7.1. The one-phase regions are the open areas; here $p = 1$ and there are 2 degrees of freedom, in that both P and T must be specified to specify the state (apart from specification of the mass of the phase).

Along the lines (except for point A), there are two phases present in equilibrium; hence $f = 1$ along a line. Thus, with liquid and vapor in equilibrium, we can vary T anywhere along the line AC, but once T is fixed, then P, the (*equilibrium*) *vapor pressure* of liquid water at temperature T, is fixed. The *boiling point* of a liquid at a given pressure P is the temperature at which its equilibrium vapor pressure equals P; the *normal boiling point* is the temperature at which the liquid's vapor pressure is 1 atm. Line AC gives the boiling point of water as a function of pressure. (The H_2O normal boiling point is not precisely 100°C; Sec. 1.5.)

Point A is the *triple point*. Here solid, liquid, and vapor are in mutual equilibrium, and $f = 0$. Since there are no degrees of freedom, the triple point occurs at a definite T and P. Recall that the water triple point is used as the reference temperature for the thermodynamic temperature scale. By definition, the water triple-point temperature is exactly 273.16 K. The water triple-point pressure is found to be 4.585 torr. The present definition of the Celsius scale is Eq. (1.14); hence the water triple-point temperature is 0.01°C exactly.

The *melting point* of a solid at a given pressure P is the temperature at which solid and liquid are in equilibrium for pressure P. Line AD in Fig. 7.1 is the solid–liquid equilibrium line for H_2O and gives the melting point of ice as a function of pressure. We see that the melting point of ice decreases (slowly) with increasing pressure. The *normal melting point* of a solid is the melting point at $P = 1$ atm. For water, the normal melting point is 0.0024°C. The ice point (Secs. 1.3 and 1.5), which occurs at 0.0000°C, is the equilibrium temperature of ice and *air-saturated* liquid water at 1 atm pressure; the equilibrium temperature of ice and *pure* liquid water at 1 atm pressure is 0.0024°C. (The dissolved N_2 and O_2 lower the freezing point compared with that of pure water; see Sec. 12.3.) Note that the triple point is the melting point of ice at the triple-point temperature. For a pure substance, the *freezing point* of the liquid at a given pressure equals the melting point of the solid.

The phase diagram of Fig. 7.1 is for pure water. Ordinarily, one determines melting points in the laboratory in a container open to the air. The air dissolved in the liquid affects the solid–liquid equilibrium temperature slightly. (In reporting precise work on properties of liquids, it is important to state whether the liquid is air-free or air-saturated.) The melting points on line AD corresponding to equilibrium between ice and air-free liquid water might be called *true* melting points to distinguish them from melting points as ordinarily determined in the laboratory. [For further discussion, see R. C. Parker and D. S. Kristol, *J. Chem. Educ.*, **51**, 658 (1974); R. C. Wilhoit, ibid., **52**, 276 (1975).]

Along line BA, there is equilibrium between solid and vapor. Ice heated at a pressure below 4.58 torr will sublime to vapor rather than melt to liquid.

Suppose we put some H_2O in a closed container fitted with a piston. We place the system in a constant-temperature bath at 200°C and set the piston pressure at 0.5 atm. These values correspond to point F in Fig. 7.1; at F, the equilibrium state is vapor. Hence, whether we originally put solid, liquid, or gaseous H_2O in the container, at equilibrium we end up with vapor. Now let us slowly increase the piston pressure, holding T constant. When we reach the pressure of point G, the vapor

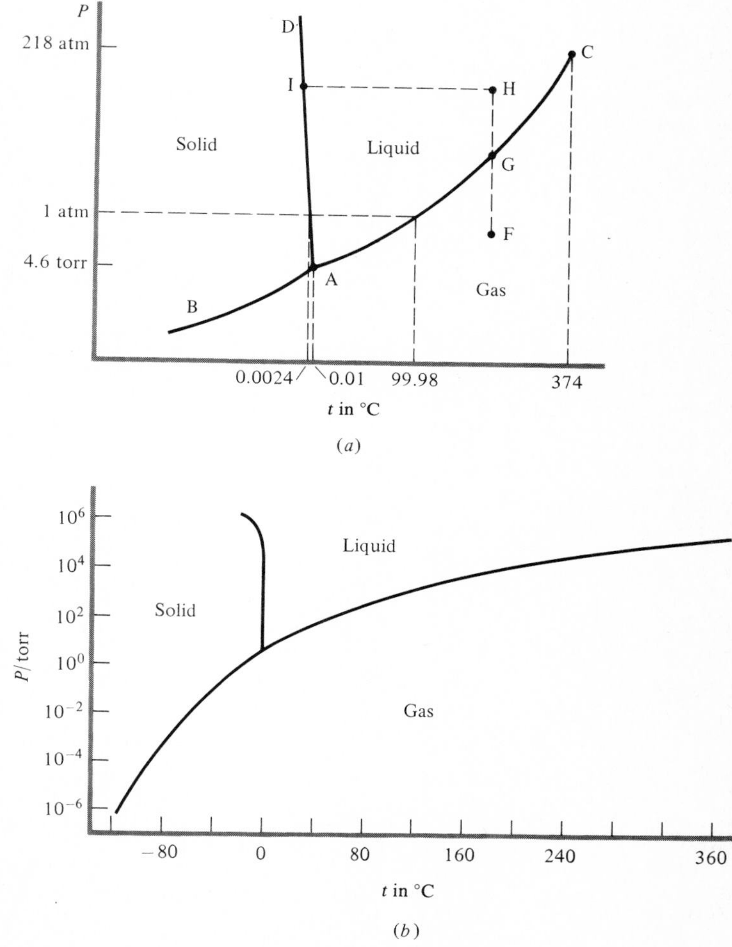

Figure 7.1
The H_2O phase diagram at low and moderate pressures. (*a*) Caricature of the diagram. (*b*) The diagram drawn accurately. (The vertical scale is logarithmic.)

starts to condense to liquid; this condensation continues at constant pressure until all the vapor has condensed. During condensation, the volume of the system decreases; the amounts of liquid and vapor present at point G can be varied by varying the total volume (see Fig. 8.3). (Note that the masses of the liquid and vapor phases and the volume are extensive variables.) Once the vapor has all condensed, we might isothermally increase the pressure further, say, to point H; between G and H there is only one phase present, liquid water. If at point H we remove the system from the bath and cool the system at constant pressure, we eventually reach point I, where the liquid begins to freeze, and two phases (solid and liquid) are in equilibrium. The temperature will remain fixed until all the liquid has frozen. Further isobaric cooling just lowers the temperature of the one-phase system.

Suppose we now start at point G with liquid and vapor in equilibrium and very slowly heat the closed system, holding V fixed. The temperature will continually increase. Since we are proceeding reversibly, liquid–vapor equilibrium is maintained and the system moves from point G along the liquid–vapor line toward point C, with both T and P increasing. During this process, the liquid-phase density is observed to decrease, and the vapor-phase density increases. Eventually we reach point C, at which the liquid and vapor densities (and all other intensive properties) become equal to each other. At point C, the two-phase system becomes a one-phase system, and the liquid–vapor line ends. Point C is the *critical point*; the temperature and pressure at this point are called the *critical temperature* and the *critical pressure*, T_c and P_c. For water, $T_c = 647$ K and $P_c = 218$ atm. At any temperature above T_c, liquid and vapor phases cannot coexist in equilibrium. At any temperature above T_c, isothermal compression of the vapor will not cause condensation, in contrast to compression below T_c. Note that it is possible to go from point F (vapor) to point H (liquid) without condensation occurring by varying T and P in such a manner as to go around the critical point C without crossing the liquid–vapor line AC. In such a process, the density increases smoothly and continuously, so that there is a smooth transition from vapor to liquid, rather than a sudden transition as in condensation. (See also Sec. 8.3.)

The phase diagram for CO_2 is shown in Fig. 7.2. For CO_2, an increase in pressure increases the melting point. The triple-point pressure of CO_2 is 5.1 atm; hence at 1 atm, solid CO_2 will sublime to vapor when warmed rather than melt to liquid. Hence the name "dry ice."

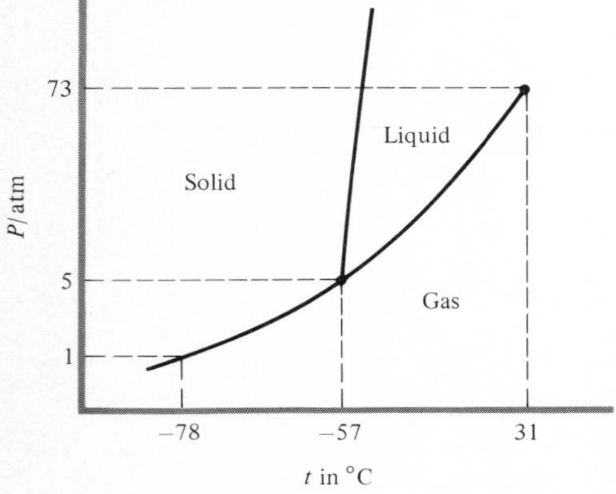

Figure 7.2
Caricature of the CO_2 phase diagram.

The liquid–vapor line on a P-T phase diagram ends in a critical point at a certain T and P; above T_c, there is no distinction between liquid and vapor. One might ask whether the solid–liquid line ends in a critical point at high pressure. The answer to this question is not known, but no solid–liquid critical point has ever been found.

7.3 THE CLAPEYRON EQUATION

The Clapeyron equation gives the slope dP/dT of a two-phase equilibrium line on a P-T phase diagram of a one-component system. To derive it, we consider two infinitesimally close points 1 and 2 on such a line (Fig. 7.3). The line in Fig. 7.3 might involve solid–liquid, solid–vapor, or liquid–vapor equilibrium or even solid–solid equilibrium (Sec. 7.4). We shall call the two phases involved α and β. From Eq. (4.100), the condition for phase equilibrium is $\mu^\alpha = \mu^\beta$. (No subscript is needed because we have only one component.) For a pure substance, μ equals $\bar{G}$ [Eq. (4.98)]. Hence $\mu^\alpha = \mu^\beta$ becomes $\bar{G}^\alpha = \bar{G}^\beta$ for any point on the α-β equilibrium line. The molar Gibbs energies of two one-component phases in equilibrium are equal. At point 1 in Fig. 7.3, we therefore have $\bar{G}_1{}^\alpha = \bar{G}_1{}^\beta$. Likewise, at point 2, $\bar{G}_2{}^\alpha = \bar{G}_2{}^\beta$, or $\bar{G}_1{}^\alpha + d\bar{G}^\alpha = \bar{G}_1{}^\beta + d\bar{G}^\beta$, where $d\bar{G}^\alpha$ and $d\bar{G}^\beta$ are the infinitesimal changes in molar Gibbs energies of phases α and β as we go from point 1 to point 2. Use of $\bar{G}_1{}^\alpha = \bar{G}_1{}^\beta$ in the last equation gives

$$d\bar{G}^\alpha = d\bar{G}^\beta \tag{7.13}$$

To apply (7.13), we proceed as follows. For a pure (one-component) phase, Eq. (4.82) reads

$$dG = -S\,dT + V\,dP + \mu\,dn \qquad \text{pure phase} \tag{7.14}$$

We have $\bar{G} \equiv G/n$, or $G = n\bar{G}$. Hence $dG = n\,d\bar{G} + \bar{G}\,dn$, and (7.14) becomes $n\,d\bar{G} + \bar{G}\,dn = -S\,dT + V\,dP + \mu\,dn$. Since $\mu = \bar{G}$ for a pure phase, we get $n\,d\bar{G} = -S\,dT + V\,dP$; division by n gives

$$d\bar{G} = -\bar{S}\,dT + \bar{V}\,dP \qquad \text{one-phase, one-comp. syst.} \tag{7.15}$$

Note that (7.15) is applicable to open systems, as well as closed systems. A quick way to obtain (7.15) is to divide (4.40) by n; although (4.40) applies to a closed system, $\bar{G}$ is an intensive property and $d\bar{G}$ is unaffected by a change in size of the system.

Use of (7.15) in (7.13) gives

$$-\bar{S}^\alpha\,dT + \bar{V}^\alpha\,dP = -\bar{S}^\beta\,dT + \bar{V}^\beta\,dP \tag{7.16}$$

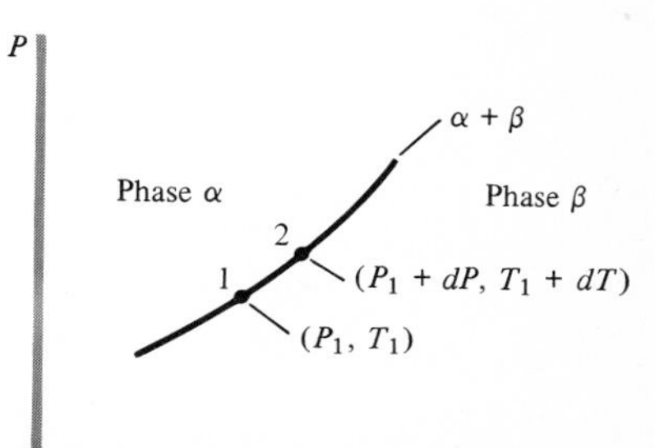

Figure 7.3
Two neighboring points on a two-phase line of a one-component system.

where dT and dP are the infinitesimal changes in T and P on going from point 1 to point 2 along the α-β equilibrium line. Rewriting (7.16), we have $(\bar{V}^\alpha - \bar{V}^\beta)\, dP = (\bar{S}^\alpha - \bar{S}^\beta)\, dT$, or

$$\frac{dP}{dT} = \frac{\bar{S}^\alpha - \bar{S}^\beta}{\bar{V}^\alpha - \bar{V}^\beta} = \frac{\Delta \bar{S}}{\Delta \bar{V}} = \frac{\Delta S}{\Delta V} \tag{7.17}$$

where ΔS and ΔV are the entropy and volume changes for the phase transition $\beta \to \alpha$. (Of course, for the transition $\alpha \to \beta$, ΔS and ΔV are each reversed in sign, and their quotient is unchanged; hence it doesn't matter which phase we call α.)

For a phase change at equilibrium (a reversible phase change), we have $\Delta S = \Delta H/T$, Eq. (3.33). Hence (7.17) becomes

$$\frac{dP}{dT} = \frac{\Delta \bar{H}}{T\, \Delta \bar{V}} = \frac{\Delta H}{T\, \Delta V} \tag{7.18}^*$$

Equation (7.18) is the *Clapeyron equation* (also called the Clausius–Clapeyron equation). Its derivation involved no approximations; (7.18) is an exact result for a one-component system.

Consider some examples. For a liquid-to-vapor transition, both ΔH and ΔV are positive; hence dP/dT is positive; the liquid–vapor line on a one-component P-T phase diagram has positive slope. The same is true of the solid–vapor line. For a solid-to-liquid transition, ΔH is always positive; ΔV is usually positive but is negative in a few cases, for example, H_2O, Ga, Bi. Because of the volume decrease for the melting of ice, the solid–liquid equilibrium line slopes to the left in the water P-T diagram (Fig. 7.1). For nearly all other substances, the solid–liquid line has positive slope.

For liquid–vapor (and solid–vapor) equilibrium $\bar{V}_{\text{gas}}$ is much greater than $\bar{V}_{\text{liq}}$ (or $\bar{V}_{\text{solid}}$) unless T is near the critical temperature, in which case the vapor and liquid densities are rather close. Hence when one of the phases is a gas, we have $\Delta \bar{V} = \bar{V}_{\text{gas}} - \bar{V}_{\text{liq}}$ (or $\bar{V}_{\text{gas}} - \bar{V}_{\text{solid}}$) $\approx \bar{V}_{\text{gas}}$. If we assume the vapor to behave approximately ideally, then $\bar{V}_{\text{gas}} \approx RT/P$. With these two approximations, the Clapeyron equation (7.18) becomes $dP/dT \approx P\, \Delta \bar{H}/RT^2$, and

$$\frac{d \ln P}{dT} \approx \frac{\Delta \bar{H}}{RT^2} \qquad \text{solid–vapor or liquid–vapor equilib.} \tag{7.19}$$

[Note the resemblance to Eq. (6.45), which, however, is an exact equation for ideal-gas reaction equilibrium.] Equation (7.19) does not hold at high vapor densities, where the gas is far from ideal and the liquid volume is not negligible compared to the gas volume. In (7.19), the quantity $\Delta \bar{H} = \bar{H}_{\text{gas}} - \bar{H}_{\text{liq}}$ (or $\bar{H}_{\text{gas}} - \bar{H}_{\text{solid}}$) depends on the temperature of the phase transition. (Once T of the transition is fixed, the transition pressure is fixed, so P is not an independent variable along the equilibrium line.) If we make a third approximation and take $\Delta \bar{H}$ to be constant along the equilibrium line, integration of (7.19) gives $\int_1^2 d \ln P \approx \Delta \bar{H} \int_1^2 dT/RT^2$, and

$$\ln \frac{P_2}{P_1} \approx -\frac{\Delta \bar{H}}{R}\left(\frac{1}{T_2} - \frac{1}{T_1}\right) \qquad \text{solid–gas or liquid–gas equilib.} \tag{7.20}$$

If P_1 is 1 atm, then T_1 is the normal boiling point, T_b. Dropping the superfluous subscript 2, we have

$$\ln (P/1 \text{ atm}) \approx -(\Delta\bar{H}/R)(1/T) + \Delta\bar{H}/RT_b \tag{7.21}$$

To the extent that our approximations are accurate, a plot of ln $(P/1 \text{ atm})$ vs. $1/T$ yields a straight line of slope $-\Delta\bar{H}/R$. This is one method of determining heats of vaporization and sublimation. The name Clausius–Clapeyron equation is often applied to (7.20) or (7.19).

Equation (7.21) gives $P/\text{atm} \approx Be^{-\Delta\bar{H}/RT}$, where (for liquids) $B \equiv e^{\Delta\bar{H}/RT_b}$. The exponential form of this equation gives a very rapid increase of vapor pressure with temperature for solids and liquids. Some vapor-pressure data for ice and liquid water are:

t	−90°C	−60°C	−40°C	−20°C	0°C
P_{ice}/torr	0.00007	0.008	0.097	0.78	4.6

t	0°C	20°C	40°C	60°C	80°C	100°C	120°C	374°C
P_{liq}/torr	4.6	17.5	55.3	149	355	760	1489	165,000

Actually, $\Delta\bar{H}$ of vaporization is reasonably constant over only a short temperature range, and (7.20) and (7.21) must not be applied over a large temperature range. For example, some values of $\Delta\bar{H}$ of vaporization (in kcal/mole) for liquid water at various temperatures are: 10.8 at 0°C, 9.7 at 100°C, 8.4 at 200°C, 6.0 at 300°C, 3.8 at 350°C, 1.9 at 370°C, and 0 at 374°C. Note the rapid drop in $\Delta\bar{H}$ as the critical temperature 374°C is approached.

The integration of (7.19) taking the temperature variation of $\Delta\bar{H}$ into account is discussed in *Denbigh*, secs. 6.3 and 6.4; one finds that a more accurate (but still approximate) version of (7.21) has the form

$$\ln (P/\text{atm}) \approx a/T + b \ln T + c \tag{7.22}$$

where a, b, and c are constants. An equation of the form (7.22) is often used to represent the vapor pressure of a liquid over a temperature range.

Example Find the melting point of ice at 100 atm.

The first impulse of someone who has blindly memorized equations might be to use (7.19). However, (7.19) applies only to solid–vapor or liquid–vapor equilibria. For this solid–liquid equilibrium, we must use (7.18), which is valid for any one-component phase equilibrium. We have for fusion $dP = (\Delta\bar{H}_{\text{fus}}/T\,\Delta\bar{V}_{\text{fus}})\,dT$; integration gives

$$\int_1^2 dP = \int_1^2 \frac{\Delta\bar{H}_{\text{fus}}}{T\,\Delta\bar{V}_{\text{fus}}}\,dT \tag{7.23}$$

It is a reasonable approximation to take $\Delta\bar{H}_{\text{fus}}$ as constant, for the following reasons. $\Delta\bar{H}_{\text{fus}}$ equals $\bar{H}_l - \bar{H}_s$. The molar enthalpies of liquid and solid each vary along the equilibrium line, due to the variations in temperature and pressure along this line. (Of course, the pressure and temperature variations are not independent.) Since $\bar{V}$ is small for condensed phases, Eq. (4.59) shows that $(\partial \bar{H}/\partial P)_T$ is small for condensed phases, so we expect $\bar{H}_l$ and $\bar{H}_s$ to be little affected by the pressure variation and their difference will be little affected by the change in pressure. $(\partial \bar{H}/\partial T)_P$ is large for condensed phases (and for gases), but the temperature change along the solid–liquid equilibrium line will be small for a change of 99 atm in P: the Clapeyron equation (7.18) gives $dT = T(\Delta \bar{V}/\Delta \bar{H})\, dP$; because $\Delta \bar{V}$ is quite small for a solid–liquid transition (in contrast to a solid–vapor or liquid–vapor transition), a large change in pressure produces only a small change in melting point. Hence we expect $\Delta \bar{H}_{\text{fus}}$ to be little changed along the solid–liquid line (unless P changes by a huge amount).

Moreover, because solids and liquids are rather incompressible, and because the melting point only changes by a small amount, we can take $\Delta \bar{V}_{\text{fus}}$ as essentially constant for this problem.

Hence (7.23) becomes

$$P_2 - P_1 = (\Delta H_{\text{fus}}/\Delta V_{\text{fus}}) \ln (T_2/T_1) \tag{7.24}$$

The problem is completed by use of data of Prob. 2.24. For 1 g of ice, we have $\Delta H_{\text{fus}} = 79.7$ cal and $\Delta V_{\text{fus}} = 1.000\ \text{cm}^3 - 1.091\ \text{cm}^3 = -0.091\ \text{cm}^3$. Also, $\Delta P = 99$ atm. Equation (7.24) becomes

$$\ln \frac{T_2}{273.15\ \text{K}} = \frac{(99\ \text{atm})(-0.091\ \text{cm}^3)}{79.7\ \text{cal}} \frac{1.987\ \text{cal mol}^{-1}\ \text{K}^{-1}}{82.06\ \text{cm}^3\ \text{atm mol}^{-1}\ \text{K}^{-1}}$$

where the two values of R were introduced to convert calories to cm^3 atm, so as to give a dimensionless number for the logarithm. We find $\ln (T_2/273.15\ \text{K}) = -0.00274$ and $T_2/273.15\ \text{K} = 0.99727$. Hence, $T_2 = 272.40$ K. The pressure increase of 99 atm has lowered the melting point by only 0.75 K.

7.4 SOLID–SOLID PHASE TRANSITIONS

Many substances have more than one solid form, each such form being thermodynamically stable over certain ranges of P and T. This phenomenon is called *polymorphism*. Polymorphism in elements is called *allotropy*. (Recall from Sec. 5.5 that in finding the conventional entropy of a substance, we must take any solid–solid phase transitions into account.) Polymorphism makes for a more complicated phase diagram.

The phase diagram for sulfur is shown in Fig. 7.4. At 1 atm, slow heating of (solid) rhombic sulfur transforms it at 95°C to (solid) monoclinic sulfur. The normal melting point of monoclinic sulfur is 119°C. The stability of monoclinic sulfur is confined to a closed region of the P-T diagram. Note the existence of three *triple points* (three-phase points) on the sulfur phase diagram: rhombic–monoclinic–vapor equilibrium at 95°C, monoclinic–liquid–vapor equilibrium at 119°C, and rhombic–monoclinic–liquid equilibrium at 151°C.

If rhombic sulfur is heated very rapidly at 1 atm, it will melt at 114°C to liquid sulfur, without first being transformed to monoclinic sulfur. Although rhombic sulfur is thermodynamically unstable between 95 and 114°C at 1 atm, it can exist for short periods of time under these conditions, where its $\bar{G}$ is greater than that of monoclinic sulfur. (The situation is similar to that of a supercooled liquid.)

The phase diagram for water is actually far more complex than shown in Fig. 7.1. At high pressures, the familiar form of ice is not stable, and other forms of ice exist (Fig. 7.5). Note the existence of several triple points at high pressure and the high freezing point of water at high pressure.

Ordinary ice is ice Ih (where the h is for hexagonal, and describes the crystal structure). Ice Ic (cubic) is a metastable form (not shown on the phase diagram) that is obtained by condensation of water vapor below $-80°C$. (Phase α is *metastable* with respect to phase β at a given T and P if $\bar{G}^{\alpha} > \bar{G}^{\beta}$ at that T and P and if the rate of conversion of α to β is slow enough to allow α to exist for a finite period of time at that T and P. An example is rhombic sulfur between 95 and 114°C at 1 atm. Data in the Appendix show that diamond is metastable with respect to graphite at 25°C and 1 atm.) Ice IV (not shown on the phase diagram) is a metastable form which exists in the same region as ice V. The very-high-pressure forms ice VII and ice VIII have the same molar volume and the same crystal structure except for the positions of the hydrogen atoms in the hydrogen bonds; because $\Delta V = 0$ and $\Delta H \neq 0$ here, the ice VII–ice VIII equilibrium line is vertical on a P-T diagram. The ice VII liquid equilibrium line has been followed out to 240,000 atm and 440°C. The structures of the

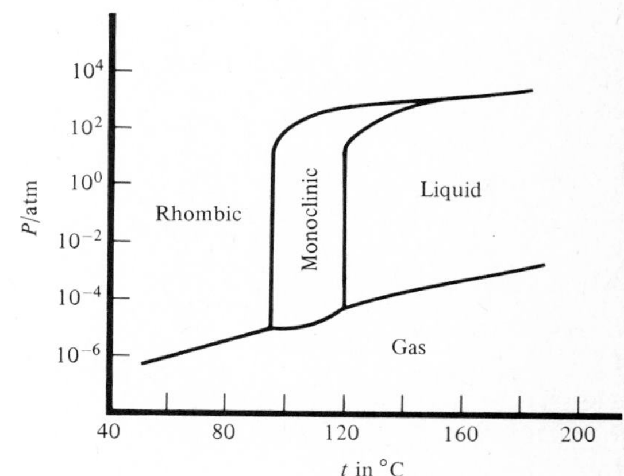

Figure 7.4
The sulfur phase diagram. (The vertical scale is logarithmic.)

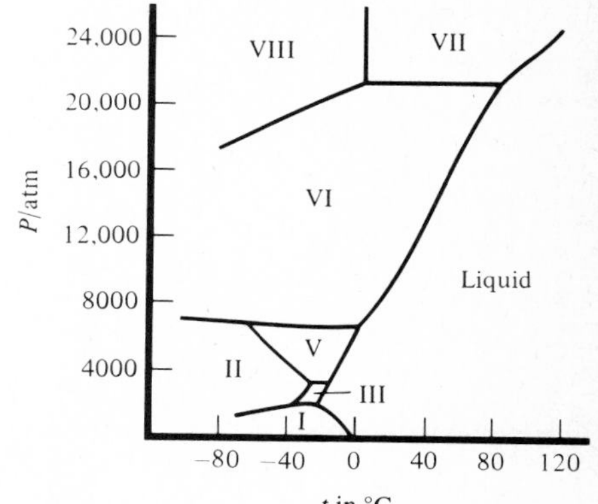

Figure 7.5
The H_2O phase diagram at high pressures. Not shown are the metastable forms ice IV and ice Ic and the form ice IX (which exists below $-100°C$ at pressures around 3000 atm).

various forms of ice are discussed in F. Franks (ed.), *Water: A Comprehensive Treatise*, vol. 1, Plenum, 1972, pp. 116–129.

[The plot of Kurt Vonnegut's novel *Cat's Cradle* (Dell, 1963), written when only ices I to VIII were known, involves the discovery of ice IX, a form supposed to exist at 1 atm with a melting point of 114°F, relative to which liquid water is unstable; ice IX brings about the destruction of life on earth.]

7.5 HIGHER-ORDER PHASE TRANSITIONS

For the common, garden-variety equilibrium phase transitions at constant T and P discussed in Secs. 7.2 to 7.4, the transition is accompanied by a transfer of heat $q_P \neq 0$ between system and surroundings; also, the system is generally observed to undergo a volume change. Such transitions are called *first order*.

For a first-order transition, $C_P = (\partial H/\partial T)_P$ of the two phases is observed to differ. C_P may either increase (as in the transition ice to water) or decrease (as in the transition water to steam) on going from the low-T to high-T phase (see Fig. 5.3). Right at the transition temperature, $C_P = dq_P/dT$ is infinite, since the nonzero latent heat is absorbed by the system with no temperature change (Fig. 7.6*a*).

Certain rather special phase transitions are observed to occur with $q_P = \Delta H = T\,\Delta S = 0$ and with $\Delta V = 0$. These are called *higher-order* (phase) transitions. For such a transition, the Clapeyron equation $dP/dT = \Delta H/T\,\Delta V$ is, of course, meaningless. For a higher-order transition, $\Delta U = \Delta(H - PV) = \Delta H - P\,\Delta V = 0$. The known higher-order transitions are either second-order transitions or lambda transitions.

A *second-order* transition is defined as one where $\Delta H = T\,\Delta S = 0$, $\Delta V = 0$, and C_P does not become infinite at the transition temperature but does change by a finite amount (Fig. 7.6*b*). The only firmly established example of a second-order transition is that between normal conductivity and superconductivity in certain metals; some metals, for example, Hg, Sn, Pb, Al, on being cooled to characteristic temperatures (4.2 K for Hg at 1 atm; 7.2 K for Pb at 1 atm) become superconductors with zero electrical resistance.

A *lambda* transition is one where $\Delta H = T\,\Delta S = 0$, $\Delta V = 0$, and C_P goes to infinity at the transition temperature (Fig. 7.6*c*); the shape of the C_P-vs.-T curve resembles the Greek letter λ, lambda. Examples of lambda transitions include the transition between liquid helium I and liquid helium II in ^{4}He; the transition between ferromagnetism and paramagnetism in metals like Fe or Ni; order–disorder transitions in certain alloys, for example, β-brass, and certain compounds, for example, NH_4Cl, HF, and CH_4. It was formerly believed that these transitions were

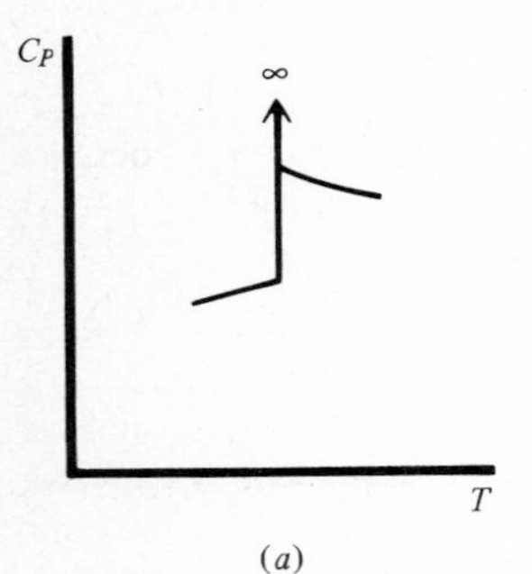

(*a*)

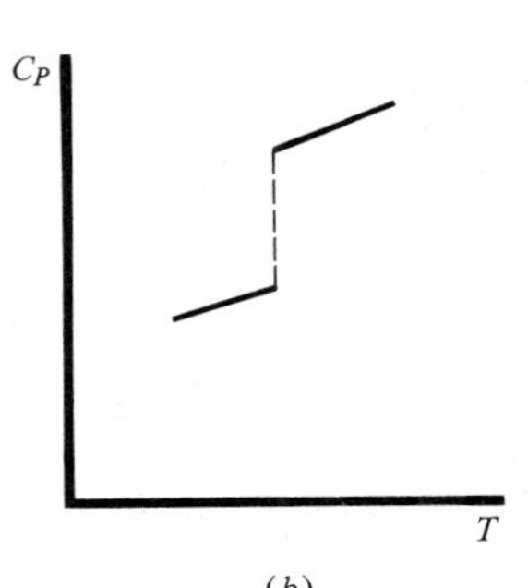

(*b*)

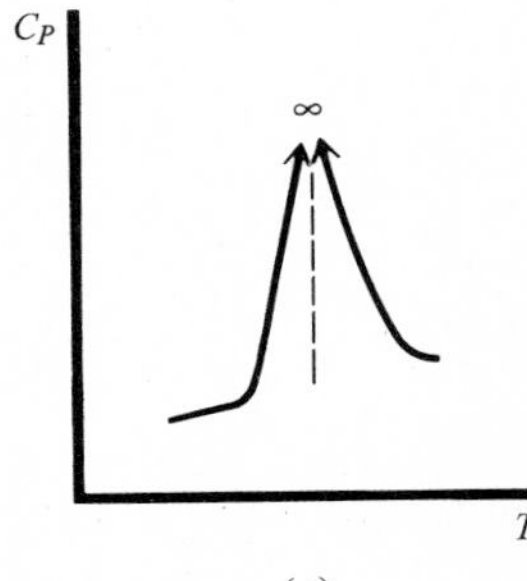

(*c*)

Figure 7.6
C_P vs. T in the region of (*a*) a first-order transition; (*b*) a second-order transition; (*c*) a lambda transition.

second-order transitions; however, experimental work and theoretical statistical-mechanical calculations by Onsager and others on model systems both indicate that C_P increases without limit as the transition temperature is approached from above and from below. Hence these transitions are now classified as lambda rather than second-order transitions.

β-brass is a nearly equimolar mixture of Zn and Cu; for simplicity, let us assume an exactly equimolar mixture. The crystal structure (body-centered cubic) has each atom surrounded by eight nearest neighbors which lie at the corners of a cube. Interatomic forces are such that the lowest-energy arrangement of atoms in the crystal is a completely ordered structure with each Zn atom surrounded by eight Cu atoms and each Cu atom surrounded by eight Zn atoms. (Imagine two interpenetrating cubic arrays, one of Cu atoms and one of Zn atoms.) In the limit of absolute zero, this is the crystal structure. As the alloy is warmed from T near zero, part of the added heat is used to randomly interchange Cu and Zn atoms; the degree of disorder gradually increases as T increases. Eventually the state of maximum disorder is reached in which there is a 50-50 chance for any nearest neighbor of a Cu atom to be Cu or Zn; the temperature (742 K) at which this maximum disorder is reached is the temperature of the lambda transition.

Lambda transitions in compounds like NH_4Cl and HF were originally thought to occur at a temperature at which free rotation of molecules or ions in the crystal began. However, this explanation is now known to be incorrect. These transitions are also order–disorder transitions and correspond to achievement of disorder in the spatial orientation of the axes about which vibrations occur in the crystal. For example, in NH_4Cl, each NH_4^+ ion is surrounded by eight Cl^- ions at the corners of a cube; the four protons of the NH_4^+ ion lie on lines going from N to four of the eight Cl^- ions. There are two equivalent orientations of an NH_4^+ ion with respect to the surrounding chloride ions. At very low T, all the NH_4^+ ions have the same orientation. As T is increased, the NH_4^+ orientations become more and more random; at the lambda point (and above), any two NH_4^+ ions have a 50-50 chance of having the same or different orientations; the NH_4^+ orientations have become completely random.

The transition from ferromagnetism to paramagnetism involves a transition from ordered to disordered orientations of electron spins.

For more on higher-order transitions, see *Münster*, secs. 48–51, 67; *Zemansky*, sec. 12-10.

PROBLEMS

7.1 (*a*) If a system has c_{ind} independent components, what is the maximum number of phases that can exist in equilibrium? (*b*) In the book *Regular Solutions* by J. H. Hildebrand and R. L. Scott (Prentice-Hall, 1962), there is a photograph of a system with 10 liquid phases in equilibrium. What must be true about the number of independent components in this system?

7.2 (*a*) For an aqueous solution of H_3PO_4, write down the reaction equilibrium conditions and the electroneutrality condition. What is f? (*b*) For an aqueous solution of KBr and NaCl, write down the stoichiometric relations between ion mole fractions. Does the electroneutrality condition give an independent relation? What is f?

7.3 Find f for the following systems: (*a*) a gaseous mixture of N_2, H_2, and NH_3 with no catalyst present (so that the rate of reaction is zero); (*b*) a gaseous mixture of N_2, H_2, and NH_3 with a catalyst present to establish reaction equilibrium; (*c*) a system formed by adding a catalyst to pure $NH_3(g)$ so as to establish reaction equilibrium between NH_3, N_2, and H_2; (*d*) a system formed by heating pure $CaCO_3(s)$ to give $CaCO_3(s)$, $CaO(s)$, $CO_2(g)$, $CaCO_3(g)$, and $CaO(g)$.

7.4 Find the relation between f, c_{ind}, and p if (*a*) rigid, permeable, thermally conducting walls separate all the phases of a system; (*b*) movable, permeable, adiabatic walls separate all the phases of a system; (*c*) movable, impermeable, thermally conducting walls separate all the phases of a system.

7.5 The vapor pressure of water at 25°C is 23.76 torr. (*a*) If 0.360 g of H_2O is placed in a container at 25°C with $V = 10.0$ l, state what phase(s) are present at equilibrium and the mass of H_2O in each phase. (*b*) The same as (*a*), except that $V = 20.0$ l. (State any approximations you make.)

7.6 Why do solid–liquid equilibrium lines have much steeper slopes than solid–vapor or liquid–vapor equilibrium lines on a *P-T* diagram?

7.7 The vapor pressure of water at 25°C is 23.76 torr. Calculate the average value of $\Delta \bar{H}$ of vaporization of water over the temperature range 25 to 100°C.

7.8 ΔH of vaporization of water is 539.4 cal/g at the normal boiling point. (*a*) Many bacteria can survive at 100°C by forming spores. Most bacterial spores die at 120°C. Hence, autoclaves used to sterilize medical and laboratory instruments are pressurized to raise the boiling point of water to 120°C. At what pressure does water boil at 120°C? (*b*) What is the boiling point of water on top of Pike's Peak (altitude 14,100 ft), where the atmospheric pressure is typically 446 torr?

7.9 Some vapor pressures of liquid Hg are:

t	80°C	100°C	120°C	140°C
P/torr	0.08880	0.2729	0.7457	1.845

(*a*) Find the average $\Delta \bar{H}$ of vaporization over this temperature range. (*b*) Find the vapor pressure at 160°C. (*c*) Estimate the normal boiling point of Hg.

7.10 Some vapor pressures of solid CO_2 are:

t	−120°C	−110°C	−100°C	−90°C
P/torr	9.81	34.63	104.81	279.5

(*a*) Find the average $\Delta \bar{H}$ of vaporization over this temperature range. (*b*) Find the vapor pressure at −75°C.

7.11 Show that Eq. (7.19) can be written as

$$\frac{d \ln P}{d(1/T)} \approx -\frac{\Delta \bar{H}}{R} \tag{7.25}$$

Hence a plot of ln P vs. $1/T$ has slope $-\Delta \bar{H}_T/R$ at temperature T. Measurement of the slope at various temperatures allows determination of the temperature variation of $\Delta \bar{H}$.

7.12 The heat of fusion of Hg at its normal melting point −38.9°C is 2.82 cal/g. The densities of Hg(*s*) and Hg(*l*) at −38.9°C and 1 atm are 14.193 and 13.690 g/cm³, respectively. Find the melting point of Hg at (*a*) 100 atm; (*b*) 800 atm.

7.13 Use Trouton's rule to show that the increase ΔT in normal boiling point T_b of a liquid due to a small change ΔP in pressure is roughly

$$\Delta T \approx T_b \, \Delta P/(11 \text{ atm})$$

7.14 In the application of (7.23) to solid–liquid equilibrium, if P does not change by a huge amount, it is a good approximation to take T in the integrand as constant (in addition to taking ΔH and ΔV as constant). Thus (7.23) becomes

$$P_2 - P_1 \approx (\Delta \bar{H}_{fus}/T_1 \, \Delta \bar{V}_{fus})(T_2 - T_1) \tag{7.26}$$

Use this equation to find the melting point of water at 100 atm, and compare the result with that found using (7.24).

7.15 The densities of diamond and graphite at 25°C and 1 atm are 3.52 and 2.25 g/cm³, respectively. Use data in the Appendix to find the minimum pressure needed at 25°C to convert graphite to diamond. (State any approximations you make.)

7.16 The vapor pressure of liquid water at 0.01°C is 4.585 torr. Find the vapor pressure of ice at 0.01°C.

7.17 (*a*) Consider a two-phase system, where one phase is pure liquid A and the second phase is an ideal gas mixture of A vapor with inert gas B (assumed insoluble in liquid A). The presence of gas B changes μ_A^l, the chemical potential of liquid A, because B increases the total pressure on the liquid phase. However, since the vapor is assumed ideal, the presence of B does not affect μ_A^g, the chemical potential of A in the vapor phase [see Eq. (6.7)]. Because of its effect on μ_A^l, gas B affects the liquid–vapor equilibrium position, and its presence changes the equilibrium vapor pressure of A. Imagine an isothermal infinitesimal change dP in the total pressure P of the system. Show that this causes a change dP_A in the vapor pressure of A given by

$$\frac{dP_A}{dP} = \frac{\bar{V}_A{}^l}{\bar{V}_A{}^g} = \frac{\bar{V}_A{}^l P_A}{RT} \qquad \text{const. } T \qquad (7.27)$$

Equation (7.27) is often called the Gibbs equation. Because $\bar{V}_A{}^l$ is much less than $\bar{V}_A{}^g$, the presence of gas B at low or moderate pressures has only a small effect on the vapor pressure of A. (*b*) The vapor pressure of water at 25°C is 23.76 torr. Calculate the vapor pressure of water at 25°C in the presence of 1 atm of inert ideal gas insoluble in water.

7.18 The vapor pressure of water at 25°C is 23.766 torr. Calculate ΔG°_{298} for the process $H_2O(l) \rightarrow H_2O(g)$. (Assume the vapor is ideal.) Compare with the value found from data in the Appendix.

7.19 Benzene obeys Trouton's rule, and its normal boiling point is 80.1°C. (*a*) Derive an equation for the vapor pressure of benzene as a function of T. (*b*) Find the vapor pressure of C_6H_6 at 25°C. (*c*) Find the boiling point of C_6H_6 at 620 torr.

7.20 The vapor pressure of $SO_2(s)$ is 1.00 torr at 177.0 K and 10.0 torr at 195.8 K. The vapor pressure of $SO_2(l)$ is 33.4 torr at 209.6 K and 100.0 torr at 225.3 K. (*a*) Find the temperature and pressure of the SO_2 triple point. (State any approximations made.) (*b*) Find $\Delta\bar{H}$ of fusion of SO_2 at the triple point.

7.21 The normal melting point of Ni is 1452°C. The vapor pressure of liquid Ni is 0.100 torr at 1606°C and 1.00 torr at 1805°C. The molar heat of fusion of Ni is 4.2_5 kcal/mol. Making reasonable approximations, estimate the vapor pressure of solid Ni at 1200°C.

8

REAL GASES

8.1 COMPRESSIBILITY FACTORS

An ideal gas obeys the equation of state $P\bar{V} = RT$. In this chapter, we shall discuss the P-V-T behavior of real gases.

As a measure of the deviation from ideality of the behavior of a real gas, we define the *compressibility factor* (or *compression factor*) Z of a gas as

$$Z(P, T) \equiv P\bar{V}/RT \tag{8.1}$$

(Do not confuse the compressibility factor Z with the isothermal compressibility κ.) Since $\bar{V}$ in (8.1) is a function of T and P (Sec. 1.7), Z is a function of T and P. For an ideal gas, $Z = 1$ for all temperatures and pressures. Figure 8.1*a* shows the variation of Z with P at 0°C for several gases. Figure 8.1*b* shows the variation of Z with P for methane at several temperatures.

These curves show that ideal behavior ($Z = 1$) is approached in the limit $P \to 0$ and also in the limit $T \to \infty$. For each of these limits, the gas volume goes to infinity for a fixed quantity of gas, and the density goes to zero. Deviations from ideality are due to intermolecular forces and to the nonzero volume of the molecules themselves. At zero density, the molecules are infinitely far apart, and intermolecular forces are zero; at infinite volume, the volume of the molecules themselves is negligible compared with the infinite volume the gas occupies. Hence the ideal-gas equation of state is obeyed in the limit of zero gas density.

A real gas then obeys $PV = Z(P, T)nRT$. Numerical tables of $Z(P, T)$ are available for many gases.

8.2 REAL-GAS EQUATIONS OF STATE

Instead of using numerical tables, it is often convenient to have an analytic expression (by this is meant an algebraic formula) for the equation of state of a real gas. Perhaps the most famous such equation is the *van der Waals equation*:

$$\left(P + \frac{a}{\bar{V}^2}\right)(\bar{V} - b) = RT \tag{8.2}$$

Dividing by $\bar{V} - b$ and solving for P, we have the alternate form

$$P = \frac{RT}{\bar{V} - b} - \frac{a}{\bar{V}^2} \tag{8.3}$$

In addition to the universal gas constant R, the van der Waals equation contains two other constants, a and b, whose values differ for different gases. A method for determining a and b values is given in Sec. 8.4. The term $a/\bar{V}^2$ in (8.2) is meant to correct for the effect of intermolecular attractive forces on the gas pressure. The nonzero volume of the molecules themselves makes the volume available for the molecules to move in less than V, so some volume b is subtracted from $\bar{V}$. The van der Waals equation gives a very substantial improvement compared with the ideal-gas equation; however, at very high pressures, the van der Waals equation is unsatisfactory.

Many other equations of state have been proposed for gases. Berthelot's equation and Dieterici's equation are each more complicated than the van der Waals equation, but since they are less accurate, we won't give their forms. [The Dieterici equation is not only inaccurate but also violates the laws of thermodynamics; see R. J. Tykodi and E. P. Hummel, *Amer. J. Phys.*, **41**, 340 (1973).]

A quite accurate two-parameter equation of state for gases is the *Redlich–Kwong equation* [O. Redlich and J. N. S. Kwong, *Chem. Rev.*, **44**, 233 (1949)]:

$$\left[P + \frac{a}{\bar{V}(\bar{V} + b)T^{1/2}}\right](\bar{V} - b) = RT \tag{8.4}$$

which is useful over very wide ranges of T and P. The Redlich–Kwong parameters a and b differ in value for any given gas from the van der Waals a and b.

The *virial equation of state* involves a power series in $1/\bar{V}$:

$$P\bar{V} = RT\left[1 + \frac{B_2(T)}{\bar{V}} + \frac{B_3(T)}{\bar{V}^2} + \frac{B_4(T)}{\bar{V}^3} + \cdots\right] \tag{8.5}$$

The functions of temperature B_2, B_3, ... are called the *second, third, ... virial coefficients.* They can be determined from experimental P-V-T data of gases. The virial equation of state is of great theoretical interest because one can use statistical

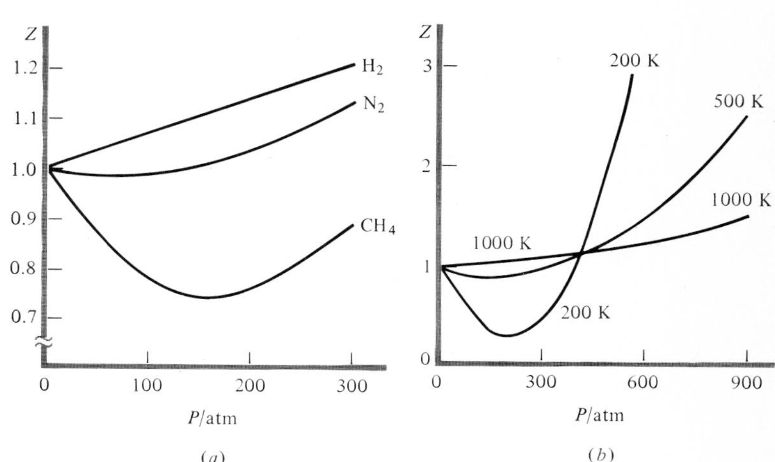

Figure 8.1
(*a*) Compressibility factors of gases at 0°C. (*b*) Methane compressibility factors at various temperatures.

mechanics to deduce the virial coefficients from the intermolecular forces (see Sec. 22.11). Unfortunately, not enough is known about these forces to yield accurate theoretical expressions for the virial coefficients of most real gases.

A form of the virial equation equivalent to (8.5) uses a power series in P rather than $1/\bar{V}$:

$$P\bar{V} = RT[1 + A_2(T)P + A_3(T)P^2 + A_4(T)P^3 + \cdots] \tag{8.6}$$

To relate the virial coefficients in (8.6) to those in (8.5), we solve (8.5) for P, substitute this expression into the right side of (8.6), and compare the coefficient of each power of $1/\bar{V}$ with that in (8.5); the results for the first few coefficients are (Prob. 8.2)

$$B_2 = A_2RT, \qquad B_3 = (A_2{}^2 + A_3)R^2T^2, \qquad B_4 = (A_2{}^3 + 3A_2A_3 + A_4)R^3T^3 \tag{8.7}$$

$$A_2 = B_2/RT, \qquad A_3 = (B_3 - B_2{}^2)/R^2T^2, \qquad A_4 = (B_4 + 2B_2{}^3 - 3B_2B_3)/R^3T^3$$

At ordinary pressures, terms beyond A_3P^2 in (8.6) are usually negligible, and even the A_3P^2 term is small. Thus one cuts off the series after two or three terms. At high pressures, the higher terms become important. At very high pressures, the virial equation fails.

A comparison of equations of state for gases [K. K. Shah and G. Thodos, *Ind. Eng. Chem.*, **57**(3), 30 (1965)] concluded that the Redlich–Kwong equation is the best two-parameter equation; in fact, the two-parameter Redlich–Kwong equation is "at least as good as" the five-parameter Beattie–Bridgeman equation and the eight-parameter Benedict–Webb–Rubin equation. Because of its simplicity and accuracy, the Redlich–Kwong equation is widely used.

So far we have considered pure real gases. For a real gas mixture, V depends on the mole fractions, as well as on T and P. One approach to the P-V-T behavior of real gas mixtures is to use a two-parameter equation of state like the van der Waals or Redlich–Kwong with the parameters a and b taken as functions of the mixture's composition. For a mixture of two gases, 1 and 2, it is usual to take

$$a = x_1{}^2a_1 + 2x_1x_2(a_1a_2)^{1/2} + x_2{}^2a_2 \qquad \text{and} \qquad b = x_1b_1 + x_2b_2$$

where x_1 and x_2 are the mole fractions of the two components. The parameter b is related to the molecular size, so b is taken as a weighted average of b_1 and b_2. The parameter a is related to intermolecular attractions; the quantity $(a_1a_2)^{1/2}$ is an estimate of what the intermolecular interaction between gas 1 and gas 2 molecules might be. (In applying an equation of state to a mixture, $\bar{V}$ is replaced by V/n_{tot}.)

8.3 CONDENSATION

A very significant deviation from ideal-gas behavior is the fact that, provided T is below the critical temperature, any real gas condenses to a liquid when the pressure is increased sufficiently. Figure 8.2 shows a plot of several isotherms for H_2O on a P-V diagram. (Figures 8.2 and 7.1 are each cross sections of a three-dimensional P-V-T phase diagram.) For temperatures below 374°C, we observe condensation of the gas to a liquid when P is increased. Consider the points ABCDEF on the 300°C isotherm. The physical situation is shown in Fig. 8.3. To go from A to B, we slowly push in the piston, decreasing V and $\bar{V}$ and increasing P (while keeping the gas in a constant-

Table 8.1. Some values of T_c, P_c, and $\bar{V}_c$

Species	T_c/K	P_c/atm	$\bar{V}_c$/(l/mol)
Neon (Ne)	44.4	27.2	0.0417
Nitrogen (N_2)	126.2	33.5	0.0895
Water (H_2O)	647.1	217.6	0.056
Carbon dioxide (CO_2)	304.2	72.8	0.0940
Hydrogen chloride (HCl)	324.6	82.0	0.081
Methanol (CH_3OH)	512.6	79.9	0.118
n-Octane (C_8H_{18})	568.8	24.5	0.492

temperature bath). Having reached B, we now observe that pushing the piston further in causes some of the gas to liquefy. As the volume is further decreased, more of the gas liquefies until at E we have all liquid. For all points between B and E on the isotherm, there are two phases present; moreover, the gas pressure above the liquid (its vapor pressure) remains constant for all points between B and E. (The term *saturated vapor* is sometimes used to refer to a gas in equilibrium with its liquid; from B to E, the vapor is saturated.) Going from E to F by pushing in the piston still further, we observe a steep increase in pressure with a small decrease in volume; liquids are relatively incompressible (small value of κ).

Above the critical temperature (374°C for water) no amount of compression will cause the separation out of a liquid phase in equilibrium with the gas. Note that as we approach the critical isotherm from below, the length of the horizontal portion of an isotherm where liquid and gas coexist decreases until it reaches zero at the critical point. The molar volumes of liquid and gas at 300°C are given by the points E and B. As T is increased, the difference between molar volumes of liquid and gas decreases, becoming zero at the critical point.

The pressure, temperature, and molar volume at the critical point are the *critical pressure* P_c, the *critical temperature* T_c, and the *critical (molar) volume* $\bar{V}_c$. At the critical point, the gas is in the *critical state*. Table 8.1 lists some data.

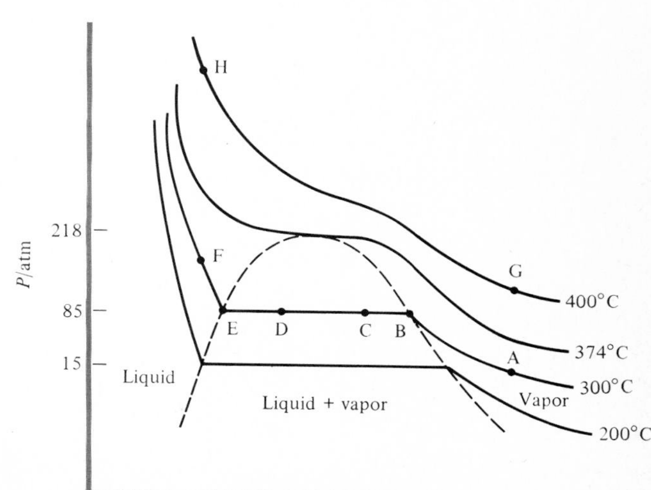

Figure 8.2
Isotherms of H_2O. (Not drawn to scale.) The critical point is at the top of the dashed curve and has $\bar{V} = 0.056$ l/mol.

For most substances, T_c is roughly 1.6 times the absolute temperature of the normal boiling point. For example, for H_2O, T_c equals 1.73 × 373 K. Also, $\bar{V}_c$ is usually close to 2.7 times the molar volume of the liquid at the normal boiling point. The stronger the intermolecular forces, the higher the critical temperature.

Usually one thinks of converting a gas to a liquid by a process that involves a sudden (discontinuous) change in density between gas and liquid, so that we go through a two-phase region in the liquefaction process. For example, in the preceding discussion of going along the isotherm ABCDEF, there are two phases present for points between B and E: a gas phase of molar volume $\bar{V}_B$ and a liquid phase of molar volume $\bar{V}_E$. (Because T and P are constant along BE, the gas and liquid molar volumes each remain constant along BE; of course, the actual amounts of gas and liquid are changing going from B to E, so the actual volumes of gas and liquid vary along BE.) Since $\bar{V}_B > \bar{V}_E$, the gas density is less than the liquid density. One can, however, change a gas into a liquid by a process in which there is always present only a single phase whose density varies smoothly, without undergoing any discontinuous changes.

One way to do this is as follows. We start with 1 mole of gas at point A of Fig. 8.2. Keeping V constant, we heat the gas to some temperature above T_c, say to point G at 400°C. We then put the system in a constant-temperature bath at 400°C and slowly push in the piston until the volume decreases to $\bar{V}_F$ at point H. Finally, we again hold V fixed and cool the system back to 300°C, point F. We end up with a liquid at F, but during the process AGHF, the properties of the system varied in a continuous manner and there is no point at which we can say that the system changed from gas to liquid. Thus we have demonstrated the continuity between the gaseous and the liquid states. In recognition of this continuity, the term *fluid* is used to mean either a liquid or a gas. What is ordinarily called a liquid can be viewed as a very dense gas. Only when both phases are present in the system (as in Fig. 8.3C) is there a clear-cut distinction between liquid and gaseous states. However, for a single-phase fluid system it is customary to define as a *liquid* a fluid whose temperature is below the critical temperature T_c and whose molar volume is less than $\bar{V}_c$ (so that its density is greater than the critical density); if these two conditions are not met, the fluid is called a *gas*. (Some people make a further distinction between *gas* and *vapor*, but we shall use these words interchangeably.)

8.4 CRITICAL DATA AND EQUATIONS OF STATE

Critical-point data can be used to calculate values for parameters in analytic equations of state such as the van der Waals equation. Along a horizontal two-phase line

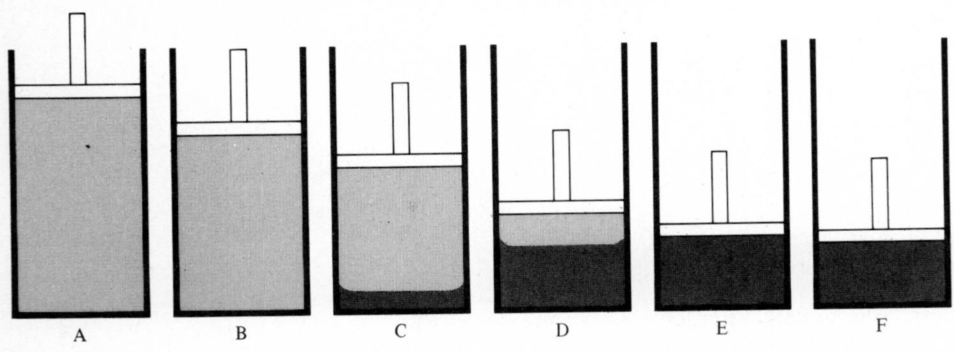

Figure 8.3 Condensation of a gas. The system is surrounded by a constant-T bath (not shown).

like EB in Fig. 8.2, the isotherm has zero slope; hence $(\partial P/\partial \bar{V})_T = 0$ along this part of an isotherm. The critical point is the limiting point of a series of such horizontal two-phase lines, so that $(\partial P/\partial \bar{V})_T = 0$ holds at the critical point. Any point with zero slope must be a maximum, a minimum, or a horizontal inflection point (Sec. 1.6). Clearly, the critical point is not a maximum or minimum on the critical isotherm, and so it must be an inflection point. For any inflection point, the second derivative is zero. Hence the critical point is a horizontal inflection point with both first and second derivatives vanishing:

$$(\partial P/\partial \bar{V})_T = 0 \qquad \text{and} \qquad (\partial^2 P/\partial \bar{V}^2)_T = 0 \tag{8.8}$$

These conditions allow us to determine parameters in equations of state.

For example, differentiating the van der Waals equation (8.3), we get

$$\left(\frac{\partial P}{\partial \bar{V}}\right)_T = -\frac{RT}{(\bar{V}-b)^2} + \frac{2a}{\bar{V}^3} \qquad \text{and} \qquad \left(\frac{\partial^2 P}{\partial \bar{V}^2}\right)_T = \frac{2RT}{(\bar{V}-b)^3} - \frac{6a}{\bar{V}^4}$$

Application of the conditions (8.8), which hold at the critical point, then gives

$$\frac{RT_c}{(\bar{V}_c-b)^2} = \frac{2a}{\bar{V}_c^{\,3}} \qquad \text{and} \qquad \frac{RT_c}{(\bar{V}_c-b)^3} = \frac{3a}{\bar{V}_c^{\,4}} \tag{8.9}$$

Moreover, the van der Waals equation itself gives at the critical point

$$P_c = \frac{RT_c}{\bar{V}_c - b} - \frac{a}{\bar{V}_c^{\,2}} \tag{8.10}$$

Division of the first equation in (8.9) by the second yields $\bar{V}_c - b = 2\bar{V}_c/3$, or

$$\bar{V}_c = 3b \tag{8.11}$$

Substitution of $\bar{V}_c = 3b$ into the first equation in (8.9) gives $RT_c/4b^2 = 2a/27b^3$, or

$$T_c = 8a/27Rb \tag{8.12}$$

Substitution of (8.11) and (8.12) into (8.10) gives $P_c = (8a/27b)/2b - a/9b^2$, or

$$P_c = a/27b^2 \tag{8.13}$$

We thus have *three* equations [(8.11) to (8.13)] relating the three critical constants P_c, $\bar{V}_c$, T_c to the *two* parameters to be determined, a and b. If the van der Waals equation were accurately obeyed in the critical region, it would not matter which two of the three equations were used to solve for a and b; however, this is not the case, and the values of a and b obtained are significantly dependent on which two of the three critical constants are used. It is customary to choose P_c and T_c. Solving (8.12) and (8.13) for a and b, we get

$$b = RT_c/8P_c, \qquad a = 27R^2T_c^{\,2}/64P_c \tag{8.14}$$

Thus the van der Waals parameters of a gas can be found from the experimentally measured values of P_c and T_c.

For a given gas, the calculated b value is of the order of magnitude of the molar volume of the corresponding liquid. This is to be expected if b is to correct for the volume of the molecules themselves. For example, for water, Eq. (8.14) gives

$$b = (82.1 \text{ cm}^3 \text{ atm mol}^{-1} \text{ K}^{-1})(647 \text{ K})/8(218 \text{ atm}) = 30.5 \text{ cm}^3/\text{mol}$$

compared with a molar volume of 18 cm^3/mol for liquid water.

Some van der Waals a and b values calculated from the critical data of the last section are:

Gas	Ne	N_2	H_2O	HCl	CH_3OH	n-C_8H_{18}
$10^{-6}a/(\text{cm}^6 \text{ atm mol}^{-2})$	0.21	1.35	5.47	3.65	9.34	37.5
$b/(\text{cm}^3 \text{ mol}^{-1})$	16.7	38.6	30.5	40.6	65.8	238

Combination of (8.11) to (8.13) shows that the van der Waals equation predicts for the compressibility factor at the critical point

$$Z_c \equiv P_c\bar{V}_c/RT_c = \tfrac{3}{8} = 0.375 \tag{8.15}$$

This may be compared to the ideal-gas prediction $P_c\bar{V}_c/RT_c = 1$. Of the known Z_c values, 80 percent lie between 0.25 and 0.30, significantly less than predicted by the van der Waals equation. The smallest known Z_c is 0.12 for HF; the largest is 0.47 for N_2O_4. [Critical data are tabulated in A. P. Kudchadker, G. H. Alani, and B. J. Zwolinski, *Chem. Rev.*, **68**, 659 (1968); J. F. Mathews, ibid., **72**, 71 (1972).]

For the Redlich–Kwong equation, a similar treatment gives (the algebra is complicated, so we omit the derivation)

$$a = R^2T_c^{5/2}/9(2^{1/3} - 1)P_c = 0.4275R^2T_c^{5/2}/P_c \tag{8.16}$$

$$b = (2^{1/3} - 1)RT_c/3P_c = 0.08664RT_c/P_c \tag{8.17}$$

$$P_c\bar{V}_c/RT_c = \tfrac{1}{3} = 0.333 \tag{8.18}$$

Since there is a continuity between the liquid and gaseous states, it should be possible to develop an equation of state that would apply to liquids as well as gases. The van der Waals equation completely fails to reproduce the isotherms in the liquid region of Fig. 8.2. However, the Redlich–Kwong equation does work fairly well in the liquid region for some (but not all) liquids. Of course this equation does not reproduce the horizontal portion of isotherms in the two-phase region of Fig. 8.2. The slope $(\partial P/\partial \bar{V})_T$ is discontinuous at points B and E in the figure. A simple analytic expression like the Redlich–Kwong equation will not have such discontinuities in $(\partial P/\partial \bar{V})_T$. What happens is that a Redlich–Kwong isotherm oscillates in the two-phase region (Fig. 8.4). This is because the Redlich–Kwong equation is a cubic equation in $\bar{V}$, so that for a given T and P, there may be three real values of $\bar{V}$ that satisfy the equation. The Peng–Robinson equation of state is an improvement on the Redlich–Kwong equation and works well for liquids as well as gases; see D.-Y. Peng and D. B. Robinson, *Ind. Eng. Chem. Fundam.*, **15**, 59 (1976).

8.5 THE LAW OF CORRESPONDING STATES

The (dimensionless) *reduced pressure* P_r, *reduced temperature* T_r, and *reduced volume* V_r of a gas in the state $(P, \bar{V}, T)$ are defined as

$$P_r \equiv P/P_c, \qquad V_r \equiv \bar{V}/\bar{V}_c, \qquad T_r \equiv T/T_c \tag{8.19}$$

where P_c, $\bar{V}_c$, T_c are the critical constants of the gas. Van der Waals pointed out in 1881 that if one uses reduced variables to express the states of gases, then, to a pretty good approximation, all gases show the same P-$\bar{V}$-T behavior; in other words, if two different gases are each at the same P_r and T_r, they have (nearly) the same $\bar{V}_r$ values. This observation is called the *law of corresponding states.* Mathematically,

$$V_r = f(P_r, T_r) \tag{8.20}$$

where approximately the same function f applies to any gas.

Equation (8.20) contains three parameters, the three critical constants. However, the condition that the equation of state approach $P\bar{V} = RT$ as $V_r \to \infty$ reduces the number of independent parameters to two. Thus the law of corresponding states implies that there exists a universal two-parameter equation of state for gases. Conversely, a two-parameter equation of state like the van der Waals or Redlich–Kwong can be expressed as an equation of the form (8.20) relating P_r, V_r, and T_r, with the constants a and b eliminated. For example, use of (8.14) and (8.15) allows one to put the van der Waals equation (8.2) in the form (Prob. 8.6)

$$\left(P_r + \frac{3}{V_r^2}\right)(V_r - \tfrac{1}{3}) = \tfrac{8}{3}T_r \tag{8.21}$$

The law of corresponding states can be explained by the fact that the interaction between two gas molecules can be crudely approximated by a potential-energy function containing only two parameters. (An example is the Lennard-Jones potential discussed in Sec. 22.10.) Starting from a two-parameter intermolecular potential-energy function, one can then use statistical mechanics to deduce an equation of state that contains only two parameters; this equation of state can then be put in reduced form to give V_r as some universal function of P_r and T_r.

If we multiply the law of corresponding states (8.20) by P_r/T_r, we get $P_r V_r/T_r = P_r f(P_r, T_r)/T_r$. The right side of this equation is some function of P_r and T_r, which we shall call $g(P_r, T_r)$. Thus

$$P_r V_r/T_r = g(P_r, T_r) \tag{8.22}$$

where (to the extent that the law of corresponding states holds) the function g is the same for all gases.

The law of corresponding states makes a prediction about Z_c, which we deduce as follows. Since every gas obeys $P\bar{V} = RT$ in the limit of zero density, we have for any gas

$$\lim_{V \to \infty} \frac{P\bar{V}}{RT} = 1 \tag{8.23}$$

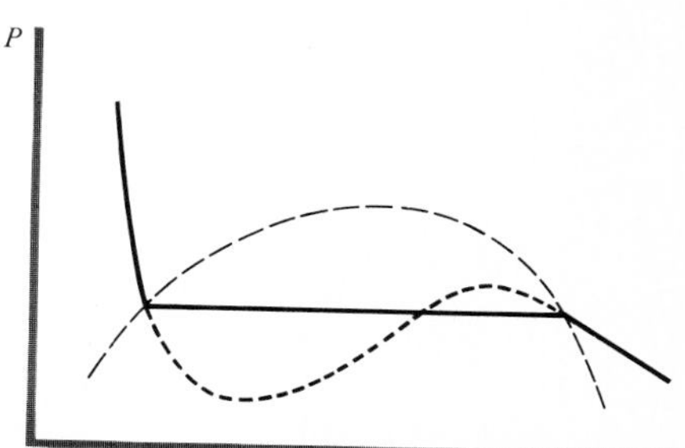

Figure 8.4
Behavior of a van der Waals or Redlich–Kwong isotherm (heavy dashed line) in the two-phase region.

If both sides of (8.23) are multiplied by $RT_c/P_c\bar{V}_c$ (which is a constant for a given gas) and (8.19) and (8.22) are used, we get $\lim (P_r V_r/T_r) = RT_c/P_c\bar{V}_c$ and

$$\lim_{V \to \infty} g(P_r, T_r) = 1/Z_c \tag{8.24}$$

Since g is the same function for every gas, its limiting value as V goes to infinity must be the same constant for every gas. Calling this constant K, we have the prediction that $Z_c = 1/K$ for every gas. The law of corresponding states predicts that the critical compressibility factor is the same for every gas. As mentioned above, Z_c varies from 0.12 to 0.47, so this prediction is far from true.

If Eq. (8.22) is multiplied by $Z_c = P_c\bar{V}_c/RT_c$, it becomes

$$P\bar{V}/RT = Z_c g(P_r, T_r) \equiv G(P_r, T_r)$$

$$Z = G(P_r, T_r) \tag{8.25}$$

Since the law of corresponding states predicts Z_c to be the same constant for all gases and g to be the same function for all gases, the function G, defined as $Z_c g$, is the same for all gases. Thus the law of corresponding states predicts that the compressibility factor Z is a universal function of P_r and T_r. Since in reality Z_c varies from one gas to another, the predictions (8.22) and (8.25) are not actually equivalent to each other. Experimental data show that (8.25) holds more accurately than (8.22). To apply (8.25), a graphical approach is often used. One takes data for a representative sample of gases and calculates average Z values at various values of P_r and T_r; these average values are then plotted, with the result shown in Fig. 8.5. Such graphs can be used to predict P-V-T data for gases to within an error of a few percent.

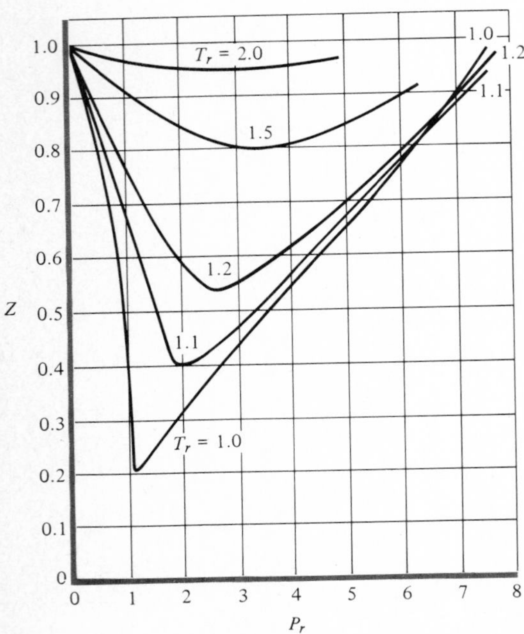

Figure 8.5
Average compressibility factor as a function of reduced variables. [*Adapted from Gouq-Jen Su, Ind. Eng. Chem.*, **38,** *803* (*1946*).]

8.6 STANDARD-STATE THERMODYNAMIC PROPERTIES OF GASES

The standard state of a gas at a given temperature is defined to be the hypothetical ideal gas at 1 atm pressure (Secs. 5.1 and 5.2). To find the values of conventional thermodynamic properties (enthalpy, entropy, etc.) of gases in their standard states, we must know the P-V-T behavior of the real gas. We can use an empirical equation of state like the van der Waals to approximate this P-V-T behavior in calculating conventional thermodynamic properties.

Equations (5.8) and (5.38) allow calculation of conventional H° and S° values. Having calculated these quantities, we can find the standard-state Gibbs energy G° as $H^\circ - TS^\circ$. An equivalent approach that is more unified is first to find G° and then use G° to find both H° and S°. The conventional standard-state molar Gibbs energy $\bar{G}^\circ$ of a gas is defined as $\bar{G}_{\text{id}}(T, P^\circ)$, where the standard pressure P° equals 1 atm. We want to calculate $\bar{G}_{\text{id}}(T, P^\circ) - \bar{G}_{\text{re}}(T, P^\circ)$. We shall understand unsubscripted variables to refer to the real gas: $\bar{G}(T, P^\circ) \equiv \bar{G}_{\text{re}}(T, P^\circ)$, $\bar{V} \equiv \bar{V}_{\text{re}}$. As before, id stands for ideal and re for real.

Since $\Delta G = \Delta H - T\,\Delta S$ for an isothermal process, we have

$$(\bar{G}_{\text{id}} - \bar{G}) = (\bar{H}_{\text{id}} - \bar{H}) - T(\bar{S}_{\text{id}} - \bar{S}) \tag{8.26}$$

where all the functions are evaluated at the same temperature T and at pressure P°. Substitution of (5.8) for $\bar{H}_{\text{id}} - \bar{H}$ and (5.38) for $\bar{S}_{\text{id}} - \bar{S}$ in (8.26) gives

$$\bar{G}_{\text{id}}(T, P^\circ) - \bar{G}(T, P^\circ) = \int_0^{P^\circ} \left(\frac{RT}{P} - \bar{V}\right) dP \qquad \text{const. } T \tag{8.27}$$

Once we have found $\bar{G}_{\text{id}} - \bar{G}$, we can get $\bar{S}_{\text{id}} - \bar{S}$ and $\bar{H}_{\text{id}} - \bar{H}$ as follows. From $d\bar{G} = -\bar{S}\,dT + \bar{V}\,dP$, we have $(\partial\bar{G}/\partial T)_P = -\bar{S}$; hence $\bar{S} = -(\partial\bar{G}/\partial T)_P$, $\bar{S}_{\text{id}} = -(\partial\bar{G}_{\text{id}}/\partial T)_P$, and

$$\bar{S}_{\text{id}}(T, P^\circ) - \bar{S}(T, P^\circ) = -\left\{\frac{\partial[\bar{G}_{\text{id}}(T, P^\circ) - \bar{G}(T, P^\circ)]}{\partial T}\right\}_P \tag{8.28}$$

Having found $\bar{S}_{\text{id}} - \bar{S}$ and $\bar{G}_{\text{id}} - \bar{G}$, we use (8.26) to find $\bar{H}_{\text{id}} - \bar{H}$.

To use (8.27), we need $\bar{V}$ of the real gas as a function of T and P. An equation of state that gives $\bar{V}(T, P)$ directly is the virial equation in the form (8.6). Equation (8.6) gives $RT/P - \bar{V} = RT/P - RT/P - RTA_2 - RTA_3P - \cdots$. Substitution in (8.27) gives after integration

$$\bar{G}_{\text{id}}(T, P^\circ) - \bar{G}(T, P^\circ) = -RT[A_2(T)P^\circ + \tfrac{1}{2}A_3(T)(P^\circ)^2 + \tfrac{1}{3}A_4(T)(P^\circ)^3 + \cdots] \tag{8.29}$$

Substitution of (8.29) in (8.28) gives $\bar{S}_{\text{id}} - \bar{S}$. Use of (8.26) then gives $\bar{H}_{\text{id}} - \bar{H}$. The tedious details are left as an exercise for the reader (Prob. 8.16). Thus, if the virial coefficients are known, we can use them to find standard-state (ideal-gas) thermodynamic properties.

The van der Waals and Redlich–Kwong equations are cubics in $\bar{V}$, so direct application of (8.27) with these equations of state is troublesome. The simplest procedure is to first put the equation of state in the virial form (8.5) and then use (8.7) and (8.29) followed by (8.28) and (8.26). To put the van der Waals equation $P = RT/(\bar{V} - b) - a/\bar{V}^2$ into an expansion in powers of $1/\bar{V}$, we start with the identity

$$\frac{1}{\bar{V} - b} = \frac{1}{\bar{V}} \frac{1}{1 - b/\bar{V}} \tag{8.30}$$

The parameter b is of the order of magnitude of the volume of the gas molecules; hence (except at very high pressures) b is much less than the gas molar volume $\bar{V}$; therefore, $b/\bar{V} \ll 1$. For $|x| < 1$, the following expansion holds:

$$\frac{1}{1 - x} = 1 + x + x^2 + x^3 + \cdots \qquad \text{for } |x| < 1 \tag{8.31}$$

The reader may recall (8.31) from high-school algebra study of geometric series. Equation (8.31) can also be derived as a Taylor series (Sec. 8.7). Setting $x = b/\bar{V}$ in (8.31) and multiplying by $1/\bar{V}$, we get

$$\frac{1}{\bar{V}} \frac{1}{1 - b/\bar{V}} = \frac{1}{\bar{V}}\left(1 + \frac{b}{\bar{V}} + \frac{b^2}{\bar{V}^2} + \cdots\right)$$

The van der Waals equation (8.3) becomes

$$P = (RT/\bar{V})(1 + b/\bar{V} + b^2/\bar{V}^2 + \cdots) - a/\bar{V}^2$$

$$P\bar{V} = RT\left[1 + \left(b - \frac{a}{RT}\right)\frac{1}{\bar{V}} + \frac{b^2}{\bar{V}^2} + \frac{b^3}{\bar{V}^3} + \cdots\right] \tag{8.32}$$

which is the desired virial form of the van der Waals equation. Comparison of (8.32) with (8.5) gives

$$B_2 = b - a/RT,\ B_3 = b^2,\ B_4 = b^3,\ \ldots \tag{8.33}$$

From (8.33) and (8.7) we get

$$A_2 = (bRT - a)/R^2T^2,\ A_3 = (2abRT - a^2)/R^4T^4,\ \ldots \tag{8.34}$$

Substitution into (8.29) gives for a van der Waals gas

$$\bar{G}_{\text{id}}(T, P^\circ) - \bar{G}(T, P^\circ) = (a/RT - b)P^\circ + \cdots \tag{8.35}$$

Since P° is only 1 atm, higher terms are negligible. Use of (8.28) and (8.26) gives for a van der Waals gas at T and P°

$$\bar{S}_{\text{id}} - \bar{S} = (a/RT^2)P^\circ + \cdots, \qquad \bar{H}_{\text{id}} - \bar{H} = (2a/RT - b)P^\circ + \cdots \tag{8.36}$$

The differences between real-gas and ideal-gas thermodynamic properties at 25°C and 1 atm are very slight. For example, Eq. (8.36) gives for HCl at 25°C and 1 atm $\bar{H}_{\text{id}} - \bar{H} = 0.006$ kcal mol^{-1} and $\bar{S}_{\text{id}} - \bar{S} = 0.01$ cal mol^{-1} K^{-1}. Note from (8.36) that at lower temperatures these differences are larger.

8.7 TAYLOR SERIES

In the last section, we used the Taylor-series expansion of the function $1/(1 - x)$. We now discuss Taylor series.

Let $f(x)$ be a function of the real variable x, and let f and all its derivatives exist at the point $x = a$ and in some neighborhood of a. It may then be possible to represent $f(x)$ as the following expansion (called a *Taylor series*) in powers of $(x - a)$:

$$f(x) = f(a) + \frac{f'(a)(x-a)}{1!} + \frac{f''(a)(x-a)^2}{2!} + \frac{f'''(a)(x-a)^3}{3!} + \cdots$$

$$f(x) = \sum_{n=0}^{\infty} \frac{f^{(n)}(a)}{n!}(x-a)^n \qquad (8.37)^*$$

In (8.37), $f^{(n)}(a)$ is the nth derivative $d^n f(x)/dx^n$ evaluated at $x = a$. The zeroth derivative of f is defined to be f itself; 0! is defined to equal 1. The derivation of (8.37) is given in most calculus texts.

To use (8.37), we must know for what range of values of x the infinite series represents $f(x)$. The infinite series in (8.37) will converge to $f(x)$ for all values of x within some interval centered at $x = a$:

$$a - b < x < a + b \qquad (8.38)$$

where b is some positive number. The value of b can frequently be determined by taking the distance between the point a and the real singularity of $f(x)$ nearest to a. (A *singularity* of f is a point where f or one of its derivatives doesn't exist.) For example, the function $1/(1 - x)$ expanded about $a = 0$ gives the Taylor series (8.31); the nearest real singularity to $x = 0$ is at $x = 1$, since $1/(1 - x)$ becomes infinite at $x = 1$. For this function, $b = 1$, and the Taylor series (8.31) converges to $1/(1 - x)$ for all x in the range $-1 < x < 1$. In some cases, b is less than the distance to the nearest real singularity. The general method of finding b is given in Prob. 8.19.

As an example, let us find the Taylor series for $\sin x$ with $a = 0$. We have

$$\begin{aligned} f(x) &= \sin x & f(a) &= \sin 0 = 0 \\ f'(x) &= \cos x & f'(a) &= \cos 0 = 1 \\ f''(x) &= -\sin x & f''(a) &= -\sin 0 = 0 \\ f'''(x) &= -\cos x & f'''(a) &= -\cos 0 = -1 \\ f^{(iv)}(x) &= \sin x & f^{(iv)}(a) &= \sin 0 = 0 \\ &\cdots\cdots\cdots\cdots\cdots\cdots \end{aligned} \qquad (8.39)$$

so that the values of $f^{(n)}(a)$ are the set of numbers 0, 1, 0, -1 repeated again and again. The Taylor series (8.37) is

$$\sin x = 0 + \frac{1(x-0)}{1!} + \frac{0(x-0)^2}{2!} + \frac{(-1)(x-0)^3}{3!} + \frac{0(x-0)^4}{4!} + \cdots$$

$$\sin x = x - x^3/3! + x^5/5! - x^7/7! + \cdots \qquad (8.40)$$

The function $\sin x$ has no singularities for real values of x. A full mathematical investigation shows that (8.40) is valid for all values of x.

Another example is $\ln x$. Since $\ln 0$ doesn't exist, we can't take $a = 0$ in (8.37). A convenient choice is $a = 1$. We find (Prob. 8.9)

$$\ln x = (x-1) - (x-1)^2/2 + (x-1)^3/3 - \cdots \qquad \text{for } 0 < x < 2 \qquad (8.41)$$

The nearest singularity to $a = 1$ is at $x = 0$ (where f doesn't exist), and the series (8.41) converges to ln x for $0 < x < 2$. Two other useful Taylor series (Probs. 8.10 and 8.11) are

$$e^x = 1 + x + \frac{x^2}{2!} + \frac{x^3}{3!} + \cdots = \sum_{n=0}^{\infty} \frac{x^n}{n!} \tag{8.42}$$

$$\cos x = 1 - x^2/2! + x^4/4! - x^6/6! + \cdots \tag{8.43}$$

Both these series are valid for all values of x.

PROBLEMS

8.1 Verify that the van der Waals, the virial, and the Redlich–Kwong equations all reduce to $PV = nRT$ in the limit of zero density.

8.2 Verify (8.7).

8.3 For ethane, $P_c = 48.2$ atm and $T_c = 305.4$ K. Calculate the pressure exerted by 74.8 g of ethane in a 200-cm^3 vessel at 37.5°C using (*a*) the ideal-gas law; (*b*) the van der Waals equation; (*c*) the Redlich–Kwong equation. (The observed value is 135 atm.)

8.4 The Berthelot equation of state for gases is

$$(P + a/T\bar{V}^2)(\bar{V} - b) = RT$$

(*a*) Express a and b in terms of P_c and T_c. (*b*) What value of Z_c is predicted? (*c*) Write the Berthelot equation in reduced form.

8.5 Show that if all terms after $B_3/\bar{V}^2$ are omitted from the virial equation (8.5), this equation predicts $Z_c = \frac{1}{3}$.

8.6 Verify the reduced van der Waals equation (8.21) by substituting (8.14) for a and b and (8.15) for R in (8.2).

8.7 (*a*) Evaluate the van der Waals a and b for CO_2 from the critical data in Sec. 8.3. (*b*) Use (8.35) to evaluate $\bar{G}_{id} - \bar{G}$ for CO_2 at 1 atm and 25°C. (*c*) Use (8.36) to find $\bar{S}_{id} - \bar{S}$ and $\bar{H}_{id} - \bar{H}$ for CO_2 at 25°C and 1 atm.

8.8 Use (8.37) to verify the Taylor series (8.31) for $1/(1 - x)$.

8.9 Verify the Taylor series (8.41) for ln x.

8.10 Verify the Taylor series (8.42) for e^x.

8.11 Derive the Taylor series (8.43) for cos x (*a*) by using (8.37); (*b*) by differentiating (8.40).

8.12 Use (8.40) to calculate the sine of 1° to four significant figures. [Before beginning, decide whether x in (8.40) is in degrees or in radians.]

8.13 Starting from the Taylor series (8.41) for ln x, derive the Taylor series (8.31) for $1/(1 - x)$. *Hint:* Use the substitution $y = 1 - x$.

8.14 (*a*) Use the virial equation (8.6) to show that

$$\mu_{JT} = \frac{RT^2}{\bar{C}_P}\left(\frac{dA_2}{dT} + \frac{dA_3}{dT}P + \frac{dA_4}{dT}P^2 + \cdots\right) \tag{8.44}$$

$$\lim_{P\to 0} \mu_{JT} = (RT^2/\bar{C}_P)(dA_2/dT) \neq 0 \tag{8.45}$$

Thus, even though the Joule–Thomson coefficient of an ideal gas is zero, the Joule–Thomson coefficient of a real gas does not become zero in the limit of zero pressure. (*b*) Use (8.5) to show that for a real gas, $(\partial U/\partial V)_T \to 0$ as $P \to 0$. (*c*) For N_2, $T_c = 126$ K, $P_c = 33.5$ atm, $\bar{C}_P$(298 K, 1 atm) = 6.96 cal mol^{-1} K^{-1}. Use (8.44) and (8.34) to estimate μ_{JT} for N_2 at 298 K and 1 atm. (The experimental value is 0.222 °C/atm.)

8.15 Use the virial equation of state to show that for a real gas

$$\lim_{P\to 0} (\bar{V} - \bar{V}_{id}) = B_2(T)$$

8.16 Use (8.29), (8.28), and (8.26) to show that for a nonideal gas

$$\bar{S}_{id}(T, P^\circ) - \bar{S}(T, P^\circ)$$
$$= [A_2R + RT(dA_2/dT)]P^\circ + \cdots \tag{8.46}$$
$$\bar{H}_{id}(T, P^\circ) - \bar{H}(T, P^\circ) = RT^2(dA_2/dT)P^\circ + \cdots$$

8.17 The vapor pressure of water at 25°C is 23.766 torr. Calculate ΔG°_{298} for the process $H_2O(l) \to H_2O(g)$; do not assume ideal vapor; instead use (8.35) and data in Sec. 8.4 to correct for nonideality. Compare your answer with that to Prob. 7.18 and with the value found from $\Delta G^\circ_{f,298}$ values in the Appendix.

8.18 The normal boiling point of benzene is 80°C. The density of liquid benzene at 80°C is 0.81 g/cm^3. Estimate P_c, T_c, and $\bar{V}_c$ for benzene.

8.19 This problem is only for those familiar with the notion of the complex plane (in which the real and imaginary parts of a number are plotted on the horizontal and vertical axes). The *radius of convergence b* in (8.38) for the Taylor series (8.37) can be shown to equal the distance between point a and the singularity in the complex plane that is nearest to a (see *Sokolnikoff and Redheffer*, sec. 8.10). Find the radius of convergence for the Taylor-series expansion of $1/(x^2 + 4)$ about $a = 0$.

9

SOLUTIONS

9.1 SOLUTIONS

A *solution* is a homogeneous mixture; i.e., a solution is a one-phase system with more than one component. The phase may be solid, liquid, or gas. (We considered solutions of ideal gases in Sec. 6.1.) Much of this chapter deals with liquid solutions, but most of the equations of Secs. 9.1 to 9.3 apply to all solutions.

There are several ways to specify the composition of a solution. Let the solution be composed of the species 1, 2, ..., r. The *mole fraction* x_i of species i is defined by Eq. (1.5) as

$$x_i \equiv n_i/n_{\text{tot}} \tag{9.1)*$$

where n_i is the number of moles of component i and n_{tot} is the total number of moles of all species in the solution. The *concentration* c_i of species i is defined by (6.30) as

$$c_i \equiv n_i/V \tag{9.2)*$$

where V is the solution's volume. For liquid solutions, the concentration of a species in moles per liter is called the molarity, a term which is no longer recommended. One liter equals $10^3 \text{ cm}^3 = 10^3(10^{-2} \text{ m})^3 = 10^{-3} \text{ m}^3 = (10^{-1} \text{ m})^3 = (1 \text{ dm})^3 = 1 \text{ dm}^3$, where one decimeter (dm) equals 0.1 m. Thus 1 liter $= 1 \text{ dm}^3$. From here on, we shall use the cubic decimeter instead of the liter.

For liquid (and solid) solutions, it is often convenient to treat one substance (called the *solvent*) differently from the others (called the *solutes*). Usually, the solvent mole fraction is greater than the mole fraction of each solute. The *molality* m_i of species i in a solution is defined as the amount of substance of i divided by the mass of the solvent. Thus, let a solution contain n_B moles of solute B (plus certain amounts of other solutes) and n_A moles of solvent A. Let M_A be the solvent molar mass. From Eq. (1.4), the solvent mass w_A equals $n_A M_A$. (We use w for mass, to avoid confusion with molality.) The solute molality m_B is

$$m_B \equiv \frac{n_B}{w_A} = \frac{n_B}{n_A M_A} \tag{9.3)*$$

Most commonly, the units of n_B are moles (although millimoles, micromoles, etc., could also be used). The units of M_A are commonly either grams per mole or kilograms per mole. Chemists almost always use moles per kilogram as the unit of molality. Hence it is desirable (but not essential) that M_A in (9.3) be in kg/mol. Valid units for the molality m_B include mol/kg, mmol/g, mol/g, mmol/kg, etc., but we shall generally use mol/kg. (The international body that recommended the adoption of SI units also recommended the use of decimal multiples and submultiples of SI units. Hence, even if one is committed to the SI system, there is no objection to such units as grams, centimeters, millimeters, etc.)

The *weight percent* of species B in a solution equals $(w_B/w) \times 100\%$, where w_B is the mass of B and w is the mass of the solution.

We adopt the convention that the solvent is denoted by the letter A.

Since the volume of a solution depends on T and P, the concentrations c_i change with changing T and P. Mole fractions and molalities, being independent of T and P, are preferable to concentrations for specifying solution compositions.

Example An aqueous solution of $AgNO_3$ that is 12.000 percent $AgNO_3$ by weight has a density 1.1080 g/cm^3 at 20°C and 1 atm. Find the mole fraction, the concentration at 20°C and 1 atm, and the molality of the solute $AgNO_3$.

The table of atomic weights gives $M_A = 18.015$ g/mol and $M_B = 169.87$ g/mol. In 100.00 g of solution there are 12.00 g of $AgNO_3$ and 88.00 g of H_2O. Equation (1.4) gives as the amounts of $AgNO_3$ and H_2O in 100.00 g of solution: $n(AgNO_3) = 0.07064$ mol and $n(H_2O) = 4.885$ mol. Equation (9.1) gives $x(AgNO_3) = 0.07064/4.955_6 = 0.01425$. The volume of 100.0 g of solution is $V = (100.0 \text{ g})/(1.1080 \text{ g/cm}^3) = 90.25 \text{ cm}^3$. Equations (9.2) and (9.3) give

$$c(AgNO_3) = (7.064 \times 10^{-2} \text{ mol})/(90.25 \text{ cm}^3) = 7.827 \times 10^{-4} \text{ mol/cm}^3$$

$$c(AgNO_3) = (7.827 \times 10^{-4} \text{ mol/cm}^3)(10^3 \text{ cm}^3/1 \text{ dm}^3) = 0.7827 \text{ mol/dm}^3$$

$$m(AgNO_3) = (7.064 \times 10^{-2} \text{ mol})/(88.00 \text{ g}) = 0.8027 \times 10^{-3} \text{ mol/g}$$

$$m(AgNO_3) = (0.8027 \times 10^{-3} \text{ mol/g})(10^3 \text{ g/kg}) = 0.8027 \text{ mol/kg}$$

Note especially that M_A in (9.3) is the molar mass (*not* the molecular weight) of A and must have the proper dimensions.

In expressing concentrations and molalities, chemists traditionally use the symbols M and m to stand for mol/l and mol/kg, respectively. Neither of these notations is currently recommended by the IUPAC.

9.2 PARTIAL MOLAR QUANTITIES

Suppose we form a solution by mixing at constant temperature and pressure n_1, $n_2, \ldots, n_r$ moles of substances $1, 2, \ldots, r$. The total volume of the unmixed components at T and P is

$$V_{\text{unmix}} = n_1 \bar{V}_1^* + n_2 \bar{V}_2^* + \cdots + n_r \bar{V}_r^* = \sum_i n_i \bar{V}_i^* \qquad (9.4)$$

where we use a star to indicate a property of a pure substance and unmix stands for unmixed. $\bar{V}_i^*$ is the molar volume of pure i at T and P. After mixing, one finds that the volume V of the solution is not in general equal to V_{unmix}: $V \neq V_{\text{unmix}}$. (This is because intermolecular interactions differ in the solution compared with those in the pure, separated components.) The same situation holds for other extensive properties, for example, U, H, S, A, G. For each of these, we can write an equation like (9.4), and for each of these, the property generally changes on mixing.

We want to find expressions for the volume V of the solution and for its other extensive properties. Each such property is a function of the solution's state, which can be specified by the variables T, P, n_1, n_2, ..., n_r. Hence

$$V = V(T, P, n_1, n_2, \ldots, n_r), \qquad U = U(T, P, n_1, n_2, \ldots, n_r) \tag{9.5}$$

with similar equations for H, S, etc. The total differential of V in (9.5) is

$$dV = \left(\frac{\partial V}{\partial T}\right)_{P,n_i} dT + \left(\frac{\partial V}{\partial P}\right)_{T,n_i} dP + \left(\frac{\partial V}{\partial n_1}\right)_{T,P,n_{i\neq 1}} dn_1 + \left(\frac{\partial V}{\partial n_2}\right)_{T,P,n_{i\neq 2}} dn_2 + \cdots + \left(\frac{\partial V}{\partial n_r}\right)_{T,P,n_{i\neq r}} dn_r \tag{9.6}$$

The subscript n_i in the first two partial derivatives indicates that all mole numbers are held constant; the subscript $n_{i\neq 1}$ indicates that all mole numbers except n_1 are held constant. We now define the *partial molar volume* $\bar{V}_j$ of substance j in the solution as

$$\bar{V}_j \equiv \left(\frac{\partial V}{\partial n_j}\right)_{T,P,n_{i\neq j}} \tag{9.7*}$$

where V is the solution volume, and where the partial derivative is taken with T, P, and all mole numbers except n_j held constant. Equation (9.6) becomes

$$dV = \left(\frac{\partial V}{\partial T}\right)_{P,n_i} dT + \left(\frac{\partial V}{\partial P}\right)_{T,n_i} dP + \sum_i \bar{V}_i\, dn_i \tag{9.8}$$

The partial molar volumes are partial derivatives of V and hence depend on the same variables as V. Thus,

$$\bar{V}_i = \bar{V}_i(T, P, n_1, n_2, \ldots, n_r) \tag{9.9}$$

From (9.7), if dV is the infinitesimal change in solution volume that occurs when dn_j moles of substance j is added to the solution with T, P, and all mole numbers except n_j held constant, then $\bar{V}_j$ equals dV/dn_j. (Often the bar is omitted and V_j used as the symbol for partial molar volume.)

We previously used a bar to denote a molar property of a pure substance. The volume of a pure substance is given by $V = n\bar{V}(T, P)$, where $\bar{V}(T, P)$ is the molar volume of the substance and n is the number of moles. Therefore $(\partial V/\partial n)_{T,P} = \bar{V}$ for a pure substance, which is consistent with (9.7) for a one-component "solution." Hence the notation (9.7) is a generalization of the previous usage of the bar and is consistent with it. The partial molar volume of a pure substance is simply its molar volume. However, the partial molar volume of component i of a solution is not necessarily equal to the molar volume of pure i. When we want to refer to both these quantities, we denote the molar volume of pure i by $\bar{V}_i^*$, to avoid confusion with the partial molar volume $\bar{V}_i$ of i in solution.

As a simple example, let us find the partial molar volume of a component of an ideal gas mixture. We have

$$V = (n_1 + n_2 + \cdots + n_i + \cdots + n_r)RT/P \tag{9.10}$$

$$\bar{V}_i \equiv (\partial V/\partial n_i)_{T,P,n_{j \neq i}} = RT/P \qquad \text{ideal gas mixture} \tag{9.11}$$

Of course, RT/P is the molar volume of pure gas i, so $\bar{V}_i = \bar{V}_i^*$ for an ideal gas mixture, a result not true for nonideal gas mixtures.

We now find an expression for the volume V of a solution. V depends on temperature, pressure, and the mole numbers. We know, however, that for fixed values of T, P, and solution mole fractions x_i, the volume (which is an extensive property) is directly proportional to the total number of moles n in the solution. (If we double all the mole numbers at constant T and P, then V is doubled; if we triple the mole numbers, then V is tripled; etc.) Thus, we have $V \propto n$ for fixed T, P, x_1, $x_2, \ldots, x_r$, and V must have the form

$$V = nf(T, P, x_1, x_2, \ldots) \tag{9.12}$$

where $n \equiv \sum_i n_i$, and where f is some function of T, P, and the mole fractions. Differentiation of (9.12) at constant T, P, $x_1, \ldots, x_r$ gives

$$dV = f(T, P, x_1, x_2, \ldots)\, dn \qquad \text{const. } T, P, x_i \tag{9.13}$$

Equation (9.8) becomes for constant T and P

$$dV = \sum_i \bar{V}_i\, dn_i \qquad \text{const. } T, P \tag{9.14}$$

We have $x_i = n_i/n$ or $n_i = x_i n$. Hence $dn_i = x_i\, dn + n\, dx_i$. At fixed x_i, we have $dx_i = 0$, and $dn_i = x_i\, dn$. Substitution into (9.14) gives

$$dV = \sum_i x_i \bar{V}_i\, dn \qquad \text{const. } T, P, x_i \tag{9.15}$$

Comparison of the two expressions (9.13) and (9.15) for dV gives (after division by dn): $f = \sum_i x_i \bar{V}_i$. Equation (9.12) becomes $V = n \sum_i x_i \bar{V}_i$ or (since $x_i = n_i/n$)

$$V = \sum_i n_i \bar{V}_i \tag{9.16}^*$$

which is the desired expression for the solution's volume V.

The change in volume on mixing the solution from its pure components at constant T and P is given by the difference of (9.16) and (9.4):

$$\Delta V_{\text{mxg}} \equiv V - V_{\text{unmix}} = \sum_i n_i(\bar{V}_i - \bar{V}_i^*) \tag{9.17}$$

where the subscript mxg stands for mixing.

The same ideas just developed for the volume V apply to any extensive property of the solution. For example, the solution's internal energy U is a function of T, P, $n_1, \ldots, n_r$ [Eq. (9.5)], and by analogy with (9.7) we define the *partial molar internal energy* $\bar{U}_i$ of component i in the solution by

$$\bar{U}_i \equiv (\partial U/\partial n_i)_{T,P,n_{j \neq i}} \tag{9.18}$$

By analogy with (9.8) we have

$$dU = \left(\frac{\partial U}{\partial T}\right)_{P,n_i} dT + \left(\frac{\partial U}{\partial P}\right)_{T,n_i} dP + \sum_i \bar{U}_i \, dn_i \tag{9.19}$$

where we obtained (9.19) by taking the total differential of U in (9.5) and using (9.18). The same arguments that gave (9.16) give (simply replace the symbol V by U in all the equations of the derivation)

$$U = \sum_i n_i \bar{U}_i \tag{9.20}$$

Note that $\bar{U}_i$ is not the same as the chemical potential μ_i. Equation (4.84) gives $\mu_i = (\partial U/\partial n_i)_{S,V,n_{j\neq i}}$; here S and V are held constant, compared with T and P in (9.18). All partial molar quantities are defined with T, P, and $n_{j\neq i}$ held constant.

We also have partial molar enthalpies $\bar{H}_i$, partial molar entropies $\bar{S}_i$, partial molar Helmholtz functions $\bar{A}_i$, partial molar Gibbs functions $\bar{G}_i$, and partial molar heat capacities $\bar{C}_{P,i}$:

$$\bar{H}_i \equiv (\partial H/\partial n_i)_{T,P,n_{j\neq i}}, \qquad \bar{S}_i \equiv (\partial S/\partial n_i)_{T,P,n_{j\neq i}} \tag{9.21}$$

$$\bar{A}_i \equiv (\partial A/\partial n_i)_{T,P,n_{j\neq i}}, \qquad \bar{G}_i \equiv (\partial G/\partial n_i)_{T,P,n_{j\neq i}} \tag{9.22}$$

$$\bar{C}_{P,i} \equiv (\partial C_P/\partial n_i)_{T,P,n_{j\neq i}} \tag{9.23}$$

The partial molar Gibbs function is of special significance since it is identical to the chemical potential [Eq. (4.81)]:

$$\bar{G}_i \equiv (\partial G/\partial n_i)_{T,P,n_{j\neq i}} \equiv \mu_i \tag{9.24*}$$

Analogous to (9.16) and (9.20), the Gibbs energy G of a solution is

$$G = \sum_i n_i \bar{G}_i \equiv \sum_i n_i \mu_i \tag{9.25}$$

Since $G \equiv U + PV - TS$, Eq. (9.25) gives the following expression for the internal energy of a phase:

$$U = -PV + TS + \sum_i n_i \mu_i \tag{9.26}$$

If Y is any extensive property of a solution, the corresponding partial molar property of component i of the solution is defined by

$$\bar{Y}_i \equiv (\partial Y/\partial n_i)_{T,P,n_{j\neq i}} \tag{9.27}$$

Analogous to (9.8) and (9.19), dY is

$$dY = \left(\frac{\partial Y}{\partial T}\right)_{P,n_i} dT + \left(\frac{\partial Y}{\partial P}\right)_{T,n_i} dP + \sum_i \bar{Y}_i \, dn_i \tag{9.28}$$

The same reasoning that led to (9.16) gives for the Y value of the solution

$$Y = \sum_i n_i \bar{Y}_i \tag{9.29*}$$

Partial molar quantities are the ratio of two infinitesimal extensive quantities and hence are intensive properties.

Equation (9.29) suggests that we view $n_i \bar{Y}_i$ as the contribution of solution component i to the extensive property Y of the phase. However, such a view is oversimplified. The partial molar quantity $\bar{Y}_i$ is a function of T, P, and the solution mole fractions. Hence, $\bar{Y}_i$ is a property of the solution as a whole, and not a property of component i alone.

For many of the thermodynamic relations between extensive properties of a homogeneous system, there are corresponding relations with the extensive variables replaced by partial molar quantities. For example, we have

$$G = H - TS \tag{9.30}$$

If we differentiate (9.30) partially with respect to n_i at constant T, P, and $n_{j\neq i}$ and use (9.21), (9.22), and (9.24), we get

$$(\partial G/\partial n_i)_{T,P,n_{j\neq i}} = (\partial H/\partial n_i)_{T,P,n_{j\neq i}} - T(\partial S/\partial n_i)_{T,P,n_{j\neq i}}$$

$$\mu_i \equiv \bar{G}_i = \bar{H}_i - T\bar{S}_i \tag{9.31}$$

which corresponds to (9.30).

Another example is the first equation of (4.79):

$$(\partial G/\partial T)_{P,n_j} = -S \tag{9.32}$$

Partial differentiation of (9.32) with respect to n_i gives

$$-\left(\frac{\partial S}{\partial n_i}\right)_{T,P,n_{j\neq i}} = \left[\frac{\partial}{\partial n_i}\left(\frac{\partial G}{\partial T}\right)_{P,n_j}\right]_{T,P,n_{j\neq i}} = \left[\frac{\partial}{\partial T}\left(\frac{\partial G}{\partial n_i}\right)_{T,P,n_{j\neq i}}\right]_{P,n_j} \tag{9.33}$$

where the order of partial differentiation was reversed. Use of (9.21) and (9.24) gives

$$\left(\frac{\partial \mu_i}{\partial T}\right)_{P,n_j} \equiv \left(\frac{\partial \bar{G}_i}{\partial T}\right)_{P,n_j} = -\bar{S}_i \tag{9.34}$$

which corresponds to (9.32) with extensive variables replaced by partial molar quantities. Similarly, partial differentiation with respect to n_i of $(\partial G/\partial P)_{T,n_j} = V$ leads to

$$\left(\frac{\partial \mu_i}{\partial P}\right)_{T,n_j} \equiv \left(\frac{\partial \bar{G}_i}{\partial P}\right)_{T,n_j} = \bar{V}_i \tag{9.35}$$

The subscript n_j in (9.34) and (9.35) indicates that all mole numbers are held constant.

We now prove two useful relations involving mixing. The Gibbs energy G of a solution is given by (9.25); the Gibbs energy of the unmixed components is $\sum_i n_i \bar{G}_i^*$, where $\bar{G}_i^*$ is the molar Gibbs energy of pure i. Hence, ΔG for the process of forming the solution from its pure components at constant T and P is

$$\Delta G_{\text{mxg}} = \sum_i n_i(\bar{G}_i - \bar{G}_i^*) \tag{9.36}$$

which is analogous to (9.17). Partial differentiation of (9.36) with respect to P gives

$$(\partial \Delta G_{\text{mxg}}/\partial P)_{T,n_j} = \sum_i n_i[(\partial \bar{G}_i/\partial P)_{T,n_j} - (\partial \bar{G}_i^*/\partial P)_T] \tag{9.37}$$

(We omit the n_j subscript on $\partial \bar{G}_i^*/\partial P$ because $\bar{G}_i^*$ is a function only of T and P.) In the notation of the present chapter, Eq. (7.15) reads

$$d\bar{G}_i^* = -\bar{S}_i^*\, dT + \bar{V}_i^*\, dP \qquad \text{one-phase, one-comp. syst.} \tag{9.38}$$

Hence $(\partial \bar{G}_i^*/\partial P)_T = \bar{V}_i^*$. Use of this equation and of (9.35) gives the right side of (9.37) as $\sum_i n_i(\bar{V}_i - \bar{V}_i^*)$; therefore

$$\left(\frac{\partial \Delta G_{\text{mxg}}}{\partial P}\right)_{T,n_j} = \Delta V_{\text{mxg}} \tag{9.39}$$

where (9.17) was used. Similarly, taking $\partial/\partial T$ of (9.36), we get

$$\left(\frac{\partial \Delta G_{\text{mxg}}}{\partial T}\right)_{P,n_j} = -\Delta S_{\text{mxg}} \tag{9.40}$$

where we used $(\partial \bar{G}_i^*/\partial T)_P = -\bar{S}_i^*$, Eq. (9.34), and Eq. (9.17) with V replaced by S. Note that ΔG_{mxg}, ΔV_{mxg}, and ΔS_{mxg} in (9.39) and (9.40) all refer to the same temperature and pressure. Equations (9.39) and (9.40) are easily remembered since they have the same structure as (4.79).

9.3 THE GIBBS–DUHEM EQUATION

Taking the total differential of (9.25), we find as the change in G of the solution in any infinitesimal process (including processes that change the amounts of the components of the solution)

$$dG = d\sum_i n_i\mu_i = \sum_i d(n_i\mu_i) = \sum_i (n_i\, d\mu_i + \mu_i\, dn_i)$$

$$dG = \sum_i n_i\, d\mu_i + \sum_i \mu_i\, dn_i \tag{9.41}$$

Substitution of (4.82) for dG gives

$$-S\, dT + V\, dP + \sum_i \mu_i\, dn_i = \sum_i n_i\, d\mu_i + \sum_i \mu_i\, dn_i$$

$$\sum_i n_i\, d\mu_i + S\, dT - V\, dP = 0 \tag{9.42}$$

Equation (9.42) is the *Gibbs–Duhem equation*, and applies to any infinitesimal process. Its most common application is to a process at constant T and P ($dT = 0 = dP$), where it becomes

$$\sum_i n_i\, d\mu_i \equiv \sum_i n_i\, d\bar{G}_i = 0 \qquad \text{const. } T, P \tag{9.43}$$

Equation (9.43) can be generalized to any partial molar quantity, as follows. From (9.29), we get $dY = \sum_i n_i\, d\bar{Y}_i + \sum_i \bar{Y}_i\, dn_i$, and use of (9.28) with $dT = 0 = dP$ gives

$$\sum_i n_i\, d\bar{Y}_i = 0 \qquad \text{const. } T, P \tag{9.44}$$

The Gibbs–Duhem equation (9.44) shows that the $\bar{Y}_i$'s are not all independent. Knowing the values of $r - 1$ of the $\bar{Y}_i$'s as a function of composition for a solution of r components, we can integrate (9.44) to obtain $\bar{Y}_r$. By dividing (9.44) by the total number of moles in the solution, we get an alternative form that uses mole fractions:

$$\sum_i x_i\, d\bar{Y}_i = 0 \qquad \text{const. } T, P \tag{9.45}$$

Equations (9.44) and (9.45) apply to isothermal, isobaric changes in solution composition. For example, suppose we take a solution of substances A and B with $x_A = 0.2$, $x_B = 0.8$, and add infinitesimal amounts of A and/or B while holding T and P fixed. If $d\bar{V}_A$ and $d\bar{V}_B$ are the changes in the partial molar volumes of A and B produced by this infinitesimal composition change, then (9.45) with $Y = V$ gives $x_A\, d\bar{V}_A + x_B\, d\bar{V}_B = 0$, or

$$d\bar{V}_A = -(x_B/x_A)\, d\bar{V}_B \qquad \text{const. } T, P \tag{9.46}$$

Hence, $d\bar{V}_A$ and $d\bar{V}_B$ are related to each other. In particular, for the solution with $x_A = 0.2$ and $x_B = 0.8$, we have $d\bar{V}_A = -4\, d\bar{V}_B$.

Pierre Duhem (1861–1916) made major contributions to the philosophy of science, the history of science, thermodynamics, hydrodynamics, and physical chemistry. The first Ph.D. thesis he submitted was rejected (through Berthelot's influence) because (following Gibbs and Helmholtz) it contradicted Berthelot's erroneous contention that the heat of reaction (rather than the free-energy change) provides the criterion of spontaneity for a chemical reaction. Duhem's rejected thesis was published as a book and contains the Gibbs–Duhem equation. Duhem's reactionary political and religious views and the polemical character of some of his scientific writings led to much of his work being ignored or forgotten.

9.4 DETERMINATION OF PARTIAL MOLAR QUANTITIES

We start by discussing the determination of partial molar volumes from experimental data. For simplicity, we consider a two-component solution. Let A be the solvent, and B the solute. To measure $\bar{V}_B \equiv (\partial V/\partial n_B)_{T,P,n_A}$, we make up solutions at the desired T and P all of which contain a fixed number of moles of component A but varying values of n_B. We then plot the measured solution volumes V vs. n_B. The slope of the V-vs.-n_B curve at any composition is then $\bar{V}_B$ for that composition.

The slope at any point on a curve is found by drawing the tangent line at that point and measuring its slope. [Methods for drawing the tangent line accurately are discussed in *Shoemaker, Garland, and Steinfeld*, p. 49; C. K. Patel and R. D. Patel, *J. Chem. Educ.*, **51**, 230 (1974); B. J. Luberoff, ibid., **51**, 556 (1974).]

Figure 9.1 plots V vs. $n(MgSO_4)$ for aqueous solutions of this salt that contain a fixed amount (1000 g or 55.5 mol) of the solvent (H_2O) at 20°C and 1 atm. (For 1000 g of solvent, n_B is *numerically* equal to the solute molality in mol/kg.) Note that (due to attractions between the solute ions and the solvent molecules) the solution volume V initially decreases with increasing $n(MgSO_4)$ at fixed $n(H_2O)$; the negative slope means that the partial molar volume $\bar{V}(MgSO_4)$ is *negative* for molalities less than 0.07 mol/kg.

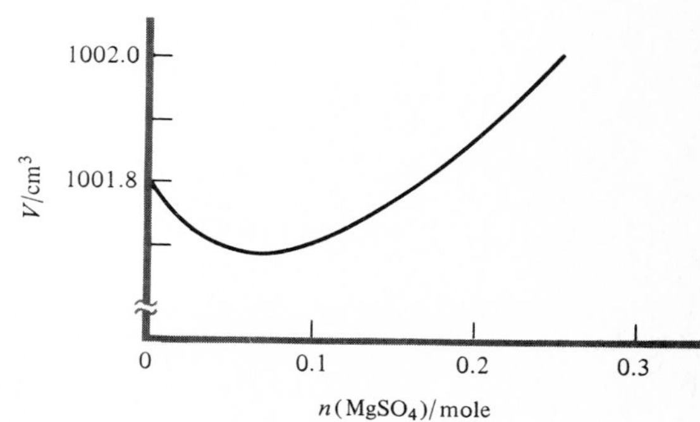

Figure 9.1
Volumes at 20°C of solutions containing 1000 g of water and n moles of $MgSO_4$.

Once $\bar{V}_B$ has been determined by the slope method, we can find $\bar{V}_A$ from (9.16). For a two-component solution, Eq. (9.16) gives $V = n_A\bar{V}_A + n_B\bar{V}_B$ or

$$\bar{V}_A = (V - n_B\bar{V}_B)/n_A \tag{9.47}$$

Figure 9.2 plots $(V - V_{unmix})/n$ (where $n = n_A + n_B$ is the total number of moles in the solution) as a function of composition for water–ethanol solutions at 20°C and 1 atm. From Eq. (9.4), the unmixed volume is

$$V_{unmix}/n = (n_A\bar{V}_A^* + n_B\bar{V}_B^*)/n = x_A\bar{V}_A^* + x_B\bar{V}_B^* \tag{9.48}$$

For all water–ethanol solution compositions, there is a shrinkage in total volume on mixing: $\Delta V_{mxg} < 0$. The data in Fig. 9.2 allow calculation of the partial molar volumes. Figure 9.3 shows $\bar{V}_A$ and $\bar{V}_B$ as a function of composition for water–ethanol solutions. Note that when $\bar{V}(H_2O)$ is decreasing, $\bar{V}(C_2H_5OH)$ is increasing, and vice versa. The opposite signs of $d\bar{V}_A$ and $d\bar{V}_B$ follow from the Gibbs–Duhem equation in the form (9.46).

Besides the above-mentioned *slope method*, another way to determine partial molar quantities is the *intercept method*. In the intercept method for a two-component solution, one plots V/n vs. x_B, where V is the solution volume, $n = n_A + n_B$, and x_B is the B mole fraction. One draws the tangent line to the curve at some particular composition x_B'; the intercept of the tangent line with the V/n axis gives $\bar{V}_A$ at the composition x_B'; the intercept of the tangent line with the vertical line $x_B = 1$ gives $\bar{V}_B$ at x_B' (see Fig. 9.4). Proof of these statements is outlined in Prob. 9.17.

Now consider determination of partial molar enthalpies. Whereas we can measure absolute (i.e., actual) volumes V of a solution, and hence determine absolute partial molar volumes $\partial V/\partial n_i$, we cannot determine the absolute enthalpy of a solution or of any other system (Sec. 5.2). Instead, we must deal with the enthalpy of a solution relative to the enthalpy of some reference system, which we may take to be the unmixed components. (Other choices are also used, as we shall later see.) We shall thus be determining relative partial molar enthalpies. Instead of measuring the enthalpy of a solution, we shall measure the enthalpy change for mixing the solution from its components. Aside from the replacement of absolute with relative values, the determination of partial molar enthalpies is similar to the determination of partial molar volumes.

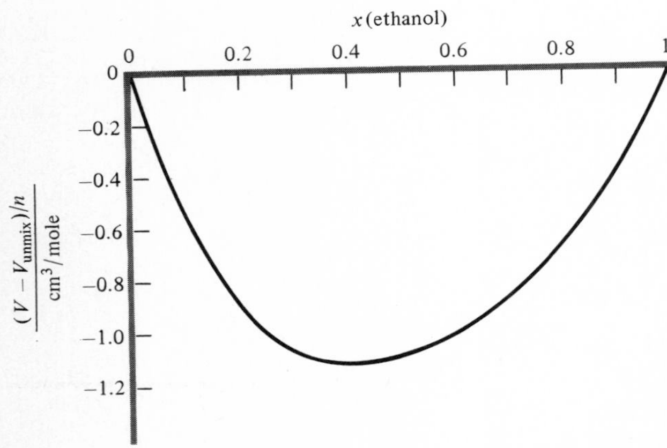

Figure 9.2
$\Delta V_{mxg}/n$ for water–ethanol solutions at 20°C.

The enthalpy H of a solution is given by (9.29) with Y replaced by H: $H = \sum_i n_i \bar{H}_i$, where the partial molar enthalpies $\bar{H}_i$ are defined by (9.21). Analogous to (9.17) and (9.36), the enthalpy of mixing at constant T and P is

$$\Delta H_{\text{mxg}} = H - H_{\text{unmix}} = \sum_i n_i(\bar{H}_i - \bar{H}_i^*) \tag{9.49}$$

ΔH_{mxg} is measured in a constant-pressure calorimeter, the same way that heats of reaction are measured (Sec. 5.3). For a two-component solution,

$$\Delta H_{\text{mxg}} = n_{\text{A}}(\bar{H}_{\text{A}} - \bar{H}_{\text{A}}^*) + n_{\text{B}}(\bar{H}_{\text{B}} - \bar{H}_{\text{B}}^*) \tag{9.50}$$

Analogous to plotting V vs. n_{B} at fixed n_{A}, T, and P, ΔH_{mxg} vs. n_{B} is plotted at fixed n_{A}, T, and P. The slope of the tangent line to the curve at any composition is $(\partial \Delta H_{\text{mxg}}/\partial n_{\text{B}})_{T,P,n_{\text{A}}}$ for that composition. We assert that

$$\left(\frac{\partial \Delta H_{\text{mxg}}}{\partial n_{\text{B}}}\right)_{T,P,n_{\text{A}}} = \bar{H}_{\text{B}} - \bar{H}_{\text{B}}^* \tag{9.51}$$

A proof of (9.51) is as follows. From (9.49) we have

$$\left(\frac{\partial \Delta H_{\text{mxg}}}{\partial n_{\text{B}}}\right)_{T,P,n_{\text{A}}} = \left(\frac{\partial H}{\partial n_{\text{B}}}\right)_{T,P,n_{\text{A}}} - \left(\frac{\partial H_{\text{unmix}}}{\partial n_{\text{B}}}\right)_{T,P,n_{\text{A}}} \tag{9.52}$$

From (9.21), $(\partial H/\partial n_{\text{B}})_{T,P,n_{\text{A}}} = \bar{H}_{\text{B}}$. Since $H_{\text{unmix}} = n_{\text{A}}\bar{H}_{\text{A}}^*(T, P) + n_{\text{B}}\bar{H}_{\text{B}}^*(T, P)$, the second partial derivative on the right of (9.52) is $\bar{H}_{\text{B}}^*$. Equation (9.52) therefore reduces to (9.51).

Equation (9.51) shows that the slope of the ΔH_{mxg}-vs.-n_{B} curve gives $\bar{H}_{\text{B}} - \bar{H}_{\text{B}}^*$. Usually we know a relative value for $\bar{H}_{\text{B}}^*$ (determined as discussed in Sec. 5.2), so that (9.51) allows relative values for the partial molar enthalpy $\bar{H}_{\text{B}}$ to be found.

The following terminology is used in connection with solution enthalpies. The

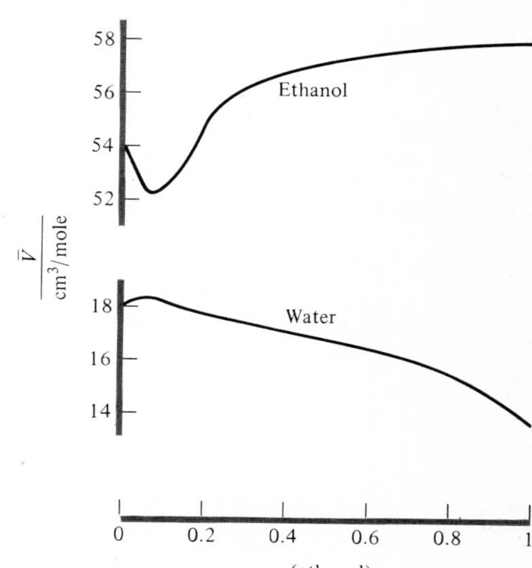

Figure 9.3
Partial molar volumes in water–ethanol solutions.

quantity $\Delta H_{mxg}/n_B$ is called the *integral heat of solution* $\Delta H_{int,B}$ per mole of B in the solvent A:

$$\Delta H_{int,B} \equiv \Delta H_{mxg}/n_B = (n_A/n_B)(\bar{H}_A - \bar{H}_A^*) + (\bar{H}_B - \bar{H}_B^*) \tag{9.53}$$

where ΔH_{mxg} is given by (9.50). The limit of $\Delta H_{int,B}$ as the solvent mole fraction x_A goes to 1 is the *integral heat of solution at infinite dilution* $\Delta H_{int,B}^{\infty}$ of B in A. The quantity $\Delta H_{int,B}^{\infty}$ equals the heat absorbed by the system when 1 mole of solute B is dissolved in an infinite amount of solvent A at constant T and P. Figure 9.5 plots ΔH_{int} for H_2SO_4 in water at 25°C and 1 atm vs. $n(H_2SO_4)$ for $n(H_2O)$ fixed at 1 mole. The curve intersects the vertical axis [which corresponds to $n(H_2SO_4) = 0$] at -23 kcal/mol, and this is ΔH_{int}^{∞} of the solute H_2SO_4 in water at 25°C.

Suppose we add an infinitesimal amount dn_B of solute B to a solution of n_B moles of B in n_A moles of solvent A while holding T and P constant. The infinitesimal enthalpy change for this isobaric process equals the heat absorbed by the system. Before mixing, the enthalpy of the solution is given by Eq. (9.29) with $Y = H$ as $\bar{H}_A n_A + \bar{H}_B n_B$, and the enthalpy of the added pure B is $\bar{H}_B^* \, dn_B$; hence the initial enthalpy is $\bar{H}_A n_A + \bar{H}_B n_B + \bar{H}_B^* \, dn_B$. From (9.29) with $Y = H$, the enthalpy after mixing is $(\bar{H}_A + d\bar{H}_A)n_A + (\bar{H}_B + d\bar{H}_B)(n_B + dn_B) = \bar{H}_A n_A + \bar{H}_B(n_B + dn_B)$, where we used (9.44) with $Y = H$ and neglected the product of two infinitesimals. Hence, the enthalpy change for the mixing process is $(\bar{H}_B - \bar{H}_B^*) \, dn_B$. The *differential heat of solution* $\Delta H_{diff,B}$ of B in A at the specified temperature, pressure, and solution composition is defined as the enthalpy change divided by dn_B for this infinitesimal mixing process:

$$\Delta H_{diff,B} = \bar{H}_B - \bar{H}_B^* = (\partial \Delta H_{mxg}/\partial n_B)_{T,P,n_A} \tag{9.54}$$

where (9.51) was used. Determination of $\bar{H}_B - \bar{H}_B^*$ was discussed above.

Example Use Fig. 9.5 to find $\Delta H_{int}(H_2SO_4)$, $\Delta H_{diff}(H_2SO_4)$, and $\Delta H_{diff}(H_2O)$ for an aqueous H_2SO_4 solution at 25°C with $x(H_2SO_4) = 0.50$.

Figure 9.5 is plotted for $n(H_2O) = 1.00$ mol, so the composition $x(H_2SO_4) = 0.50$ corresponds to $n(H_2SO_4) = 1.00$ mol. At this point on the curve, we read off the value $\Delta H_{int}(H_2SO_4) = -7$ kcal/mol.

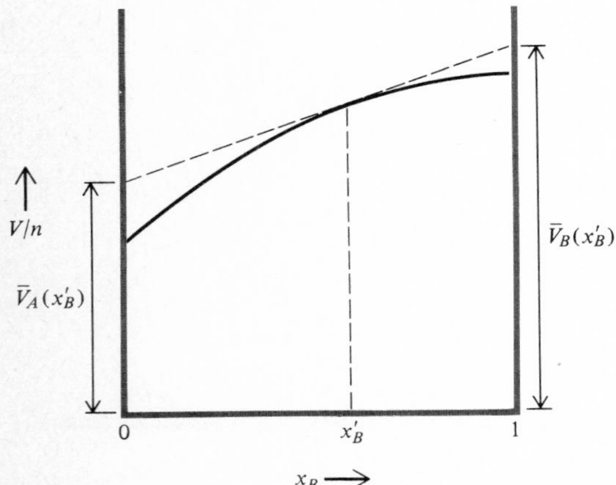

Figure 9.4
The intercept method for determining partial molar quantities.

Equation (9.54) says that $\Delta H_{\text{diff}}(H_2SO_4)$ equals $(\partial/\partial n_B)_{n_A}$ of ΔH_{mxg}. Figure 9.5 plots $\Delta H_{\text{int,B}} = \Delta H_{\text{mxg}}/n_B$ vs. n_B at constant n_A, and the slope in Fig. 9.5 equals $(\partial/\partial n_B)_{n_A}$ of $\Delta H_{\text{int,B}}$. To relate these two partial derivatives, we take $(\partial/\partial n_B)_{n_A}$ of the first equation in (9.53); we get

$$\frac{\partial \Delta H_{\text{int,B}}}{\partial n_B} = \frac{1}{n_B}\frac{\partial \Delta H_{\text{mxg}}}{\partial n_B} - \frac{\Delta H_{\text{mxg}}}{n_B{}^2} = \frac{\Delta H_{\text{diff,B}} - \Delta H_{\text{int,B}}}{n_B}$$

where (9.54) and the first equation of (9.53) were used. Drawing the tangent line in Fig. 9.5 at $n(H_2SO_4) = 1.00$ mol, we measure its slope to be 5 kcal/mol^2. Substitution in the above equation gives

$$5 \text{ kcal/mol}^2 = [\Delta H_{\text{diff}}(H_2SO_4) - (-7 \text{ kcal/mol})]/1 \text{ mol}$$

$$\Delta H_{\text{diff}}(H_2SO_4) = -2 \text{ kcal/mol}$$

The quantity $\Delta H_{\text{diff}}(H_2O)$ is given by (9.54) as $\bar{H}_A - \bar{H}_A^*$, where A is H_2O. Substitution in (9.53) gives

$$-7 \text{ kcal/mol} = (1 \text{ mol}/1 \text{ mol})[\Delta H_{\text{diff}}(H_2O)] - 2 \text{ kcal/mol}$$

$$\Delta H_{\text{diff}}(H_2O) = -5 \text{ kcal/mol}$$

Determination of partial molar Gibbs energies (chemical potentials) and partial molar entropies is discussed in Sec. 10.8.

9.5 IDEAL SOLUTIONS

The discussion in Secs. 9.1 to 9.4 applies to all solutions. In the rest of this chapter, we deal with special kinds of solutions. This section considers ideal solutions.

The molecular picture of an ideal gas mixture is one with no intermolecular interactions. For a condensed phase (solid or liquid), the molecules are close together, and we could never legitimately assume no intermolecular interactions. Our molecular picture of an *ideal solution* (liquid or solid) will be a solution where all intermolecular forces are the same and where the molecular volume of each species is

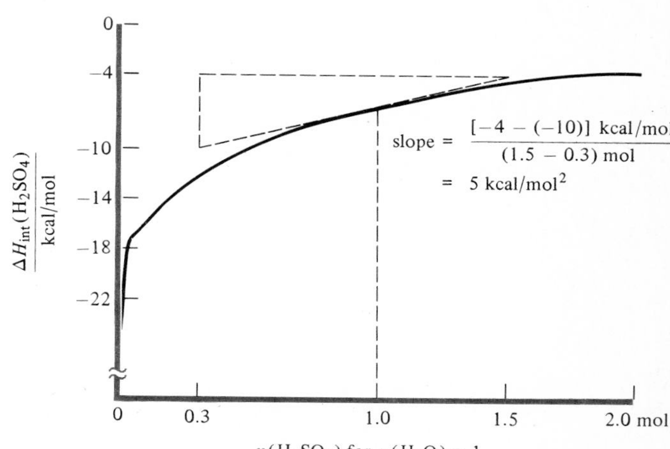

Figure 9.5
Integral heat of solution for H_2SO_4 in water at 25°C.

the same. Thus, if the solution is a mixture of A, B, and C molecules, the intermolecular forces between A and A, A and B, A and C, B and B, B and C, and C and C are all the same and the A, B, and C molecules have equal volumes. On this basis, we would expect no volume change on mixing the components A, B, and C and would expect no heat to be evolved on mixing. (A molecules have no preference for being with other A molecules or with B or C molecules.) Obviously, ionic species would not satisfy these conditions, so ideal solutions must involve nonelectrolytes. (Since the various intermolecular forces are identical, we expect that at the temperature and pressure of the solution, all pure components of an ideal liquid solution will be liquids.)

To have an ideal solution, we need identical forces between all pairs of molecules in the solution. The only truly ideal solutions would thus involve isotopic species, e.g., a mixture of $^{12}CH_3I$ and $^{13}CH_3I$. [Strictly speaking, even here there would be very slight departures from ideal behavior. The difference in isotopic masses leads to a difference in the magnitudes of molecular zero-point vibrations (Chap. 21), which causes the bond lengths and the dipole moments of the two isotopic species to differ very slightly. Hence the molecular sizes and intermolecular forces will differ *very* slightly for the isotopic species.] Apart from isotopic species, there are some pairs of liquids for which we would expect quite similar A-A, A-B, and B-B intermolecular interactions and quite similar A and B molecular volumes and hence would expect nearly ideal behavior. Examples include benzene–toluene, *m*-xylene–*p*-xylene, and C_2H_5Cl–C_2H_5Br.

Thermodynamics is a macroscopic science, and the above molecular definition of an ideal solution is not acceptable in thermodynamics. To arrive at a macroscopic, thermodynamically valid definition of an ideal solution, we use the following rough, kinetic–molecular argument.

Consider first the equilibrium between pure liquid i and its vapor. Let P_i^* be the (equilibrium) vapor pressure of pure liquid i at temperature T. At equilibrium, the rate of condensation of i vapor equals the rate of evaporation of liquid i. The rate of condensation of i vapor is proportional to the number of collisions per second of gas i molecules with the liquid's surface, and this number is proportional to the vapor pressure P_i^*. Therefore,

$$r^* = bP_i^* \tag{9.55}$$

where r^* is the rate of condensation of i vapor (defined as the number of vapor molecules condensing on unit area of the liquid in unit time) and b is a constant. The (equilibrium) rate of evaporation of pure liquid i also equals (9.55).

Now consider an ideal solution of i with other liquids. Let P_i be the equilibrium partial pressure of i vapor above the solution at temperature T. The rate of condensation of i is proportional to P_i, and we assert that we can use the same proportionality constant b as in (9.55) for the vapor above pure liquid i. There are two reasons for this: (1) We assume the vapors above pure liquid i and above the solution to be ideal gases. Hence, there are no intermolecular interactions in the gas phase to affect the rate of condensation. (2) Since the intermolecular forces between a gas i molecule and each species of molecule of the solution are the same, a collision of a gas i molecule with any molecule of the solution is just as likely to lead to condensation as a collision of a gas i molecule with a molecule of pure liquid i. Hence, we may use the same condensation proportionality constant for the solution as for the pure liquid. The rate of condensation of i for the solution is thus

$$r_{\text{con}} = bP_i \tag{9.56}$$

We now compare the rates of evaporation of i molecules from unit surface area of pure liquid i and from unit surface area of the ideal solution, both systems being at temperature T. If x_i is the mole fraction of i in the solution, then (since the molecules of the ideal solution all have the same size), the number of i molecules in unit surface area of the solution will be x_i times the number of i molecules in unit surface area of pure i. With fewer i molecules in the surface, the rate of evaporation of i from the solution's surface will be reduced by a factor x_i compared with that from the surface of pure i and will equal x_i times the rate of (9.55):

$$r_{\text{evap}} = x_i bP_i^* \tag{9.57}$$

Equating the i evaporation and condensation rates (9.56) and (9.57), we get $bP_i = x_i bP_i^*$, or $P_i = x_i P_i^*$. For each component i of the ideal solution, we therefore have

$$P_i = x_i P_i^* \tag{9.58}$$

where P_i is the equilibrium partial pressure of i in the vapor above an ideal solution with i mole fraction x_i, with temperature T, and with pressure P equal to the sum of the partial pressures of the components of the vapor above it, and where P_i^* is the equilibrium vapor pressure of pure liquid i at temperature T. (Note that at $x_i = 1$, we get $P_i = P_i^*$, and at $x_i = 0$, we get $P_i = 0$, as we should.) Equation (9.58) is *Raoult's law*. (The molecular argument just given for Raoult's law should not be taken too seriously. A more rigorous, statistical-mechanical derivation of Raoult's law is given in *Denbigh*, pp. 246–248.)

Just as the ideal-gas law $PV = nRT$ is approached by real gases in the limit of zero density, the ideal-solution law $P_i = x_i P_i^*$ is experimentally observed to be approached by liquid and solid solutions in the limit as the solution components resemble one another more and more closely (without, however, becoming identical).

Since Raoult's law is a macroscopic law, we might take the thermodynamic definition of an ideal solution as a solution whose components all obey Raoult's law for all concentrations. However, it is more convenient to use a different definition (given below), which (provided certain very good approximations are made) is equivalent to the Raoult's law definition.

Let us see what Raoult's law implies about the chemical potential $\mu_i \equiv \bar{G}_i$ of any component i of an ideal solution. At equilibrium between the liquid (or solid) solution and its vapor, we have from (4.100) that $\mu_i = \mu_{i,\text{vap}}$, where $\mu_{i,\text{vap}}$ is the chemical potential of i in the vapor above the solution and μ_i is its chemical potential in the solution. We shall assume the vapor to be an ideal gas mixture; hence, (6.7) gives $\mu_{i,\text{vap}} = \mu_{i,\text{vap}}^\circ + RT \ln (P_i/\text{atm})$, where $\mu_{i,\text{vap}}^\circ$ is the chemical potential of pure ideal gas i at temperature T and pressure 1 atm and P_i is the partial pressure of i in the vapor above the solution. Substitution into $\mu_i = \mu_{i,\text{vap}}$ and use of Raoult's law (9.58) gives $\mu_i = \mu_{i,\text{vap}}^\circ + RT \ln (x_i P_i^*/\text{atm})$, or

$$\mu_i = \mu_{i,\text{vap}}^\circ + RT \ln (P_i^*/\text{atm}) + RT \ln x_i \tag{9.59}$$

where x_i is the mole fraction of i in the solution. Suppose the solution is pure i; then $x_i = 1$, and μ_i becomes equal to $\mu_i^*(T, P_i^*)$, the chemical potential of pure liquid i at temperature T and pressure P_i^*, the equilibrium vapor pressure of pure i at T. For $x_i = 1$, (9.59) reduces to

$$\mu_i^*(T, P_i^*) = \mu_{i,\text{vap}}^\circ + RT \ln (P_i^*/\text{atm}) \tag{9.60}$$

Use of (9.60) for the first two terms on the right of (9.59) gives

$$\mu_i = \mu_i^*(T, P_i^*) + RT \ln x_i \tag{9.61}$$

We shall take the standard state of the ideal-solution component i to be pure liquid i at the temperature T and pressure P of the solution; hence $\mu_i^\circ \equiv \mu_i^*(T, P)$, where, as always, the degree superscript denotes the standard state and the star superscript indicates a pure substance. We noted following Eq. (4.70) that $\bar{G}$ for solids and liquids is insensitive to moderate changes in pressure. Therefore, $\mu_i^*(T, P_i^*) \approx \mu_i^*(T, P) \equiv \mu_i^\circ$, and (to an excellent approximation) Eq. (9.61) becomes

$$\mu_i = \mu_i^\circ + RT \ln x_i \qquad \text{where } \mu_i^\circ \equiv \mu_i^*(T, P) \tag{9.62}$$

Summarizing, we used molecular arguments to guess Raoult's law for a solution whose components are very similar; we then used Raoult's law together with two approximations (ideal gas mixture above the solution, insensitivity of μ^* to pressure changes) to reach Eq. (9.62). Since we used approximations to reach (9.62), it is not precisely equivalent to Raoult's law. It proves most convenient to regard (9.62) as the *definition* of an ideal solution. Thus, everything that preceded (9.62) in this section is simply motivation for the definition (9.62) and is not part of our logical development of the thermodynamics of solutions. We therefore discard all the equations preceding (9.62) in this section, since they are based on molecular arguments.

9.6 THERMODYNAMIC PROPERTIES OF IDEAL SOLUTIONS

Consider a solid or liquid solution with mole fractions $x_1, x_2, \ldots, x_i, \ldots$ at temperature T and pressure P (Fig. 9.6*a*). As a result of the discussion of the last section, we define an *ideal solution* as a solution for which the chemical potential μ_i of each component in the solution satisfies the following equation for all values of the solution mole fractions:

$$\mu_i = \mu_i^\circ + RT \ln x_i \qquad \text{ideal soln.} \tag{9.63)*}$$

$$\mu_i^\circ \equiv \mu_i^*(T, P) \tag{9.64)*}$$

where the standard state (degree superscript) of ideal-solution component i is defined as pure i (star superscript) at the temperature T and pressure P of the solution. Note that in the limit $x_i \to 1$ at constant T and P, $\ln x_i$ goes to zero and (9.63) gives $\lim_{x_i \to 1} \mu_i = \mu_i^\circ$, which is consistent with (9.64). Let us now derive some properties of ideal solutions.

1. Vapor pressure. Suppose we isothermally reduce the pressure in Fig. 9.6*a* until the ideal solution begins to vaporize. This yields a two-phase system consisting of solution in equilibrium with its vapor. The vapor mole fractions will in general differ from the solution mole fractions since the vapor will be richer in the more volatile components. As we reduce the pressure still further, more of the solution will vaporize. (See Sec. 12.6 for details.) We shall consider a system (Fig. 9.6*b*) consisting of an ideal solution with mole fractions $x_1, x_2, \ldots, x_i, \ldots$ in equilibrium with vapor with mole fractions $x_{1,\text{vap}}, x_{2,\text{vap}}, \ldots, x_{i,\text{vap}}, \ldots$; the system's temperature and pressure are T and P. (Of course, the value of P in Fig. 9.6*b* is less than P in Fig. 9.6*a*; also, the solution mole fractions in Fig. 9.6*b* differ from those in Fig. 9.6*a* due to evaporation of differing amounts of the solution components.) We want to derive the relation between the composition of the ideal solution and the composition of the vapor above it.

For equilibrium between the ideal solution and its vapor, Eq. (4.100) gives $\mu_i = \mu_{i,\text{vap}}$, where μ_i and $\mu_{i,\text{vap}}$ are the chemical potentials of substance i in the solution and in the vapor, respectively. (In this chapter, the symbols μ_i, μ_i°, x_i, μ_i^* without a phase subscript refer to the solution or to pure liquid i; the partial-pressure symbols P_i and P_i^* refer to the vapor above the solution or above pure liquid i.) Equation (9.63) gives μ_i in the ideal solution. We shall assume that the vapor is an ideal gas mixture (which is a very good assumption at the low or moderate pressures at which solutions are usually studied); hence, (6.7) gives $\mu_{i,\text{vap}} = \mu_{i,\text{vap}}^\circ + RT \ln (P_i/\text{atm})$, where $\mu_{i,\text{vap}}^\circ$ is the chemical potential of pure ideal gas i at T and 1 atm and $P_i = x_{i,\text{vap}} P$ is the partial pressure of gas i in the vapor with total pressure P above the solution. Thus, $\mu_i = \mu_{i,\text{vap}}$ becomes

$$\mu_i^\circ(T, P) + RT \ln x_i = \mu_{i,\text{vap}}^\circ(T) + RT \ln (P_i/\text{atm}) \qquad (9.65)$$

[The pressure P on the left of (9.65) is the pressure of the solution, which equals the (total) pressure of the vapor in equilibrium with it.]

Let P_i^* be the vapor pressure of pure liquid i at temperature T. For equilibrium between pure liquid i and its vapor, we have $\mu_i^*(T, P_i^*) = \mu_{i,\text{vap}}^*(T, P_i^*)$ or

$$\mu_i^*(T, P_i^*) = \mu_{i,\text{vap}}^\circ(T) + RT \ln (P_i^*/\text{atm}) \qquad (9.66)$$

Subtraction of (9.66) from (9.65) gives

$$\mu_i^\circ(T, P) - \mu_i^*(T, P_i^*) + RT \ln x_i = RT \ln (P_i/P_i^*) \qquad (9.67)$$

From (9.64), we have $\mu_i^\circ(T, P) \equiv \mu_i^*(T, P) \equiv \bar{G}_i^*(T, P)$. Since $\bar{G}_i^*$ is relatively insensitive to pressure (Sec. 4.5), we have $\bar{G}_i^*(T, P) \approx \bar{G}_i^*(T, P_i^*) \equiv \mu_i^*(T, P_i^*)$. Hence, $\mu_i^\circ(T, P)$ is very nearly equal to $\mu_i^*(T, P_i^*)$. Taking them as equal, we have for (9.67) $RT \ln x_i = RT \ln (P_i/P_i^*)$; hence, $x_i = P_i/P_i^*$, or

$$P_i = x_i P_i^* \qquad \text{ideal soln., ideal vapor, } \mu_i^* \text{ ind. of } P \qquad (9.68)^*$$

In *Raoult's law* (9.68), P_i is the partial pressure of substance i in the vapor above an ideal solution at temperature T, x_i is the mole fraction of i in the ideal solution, and

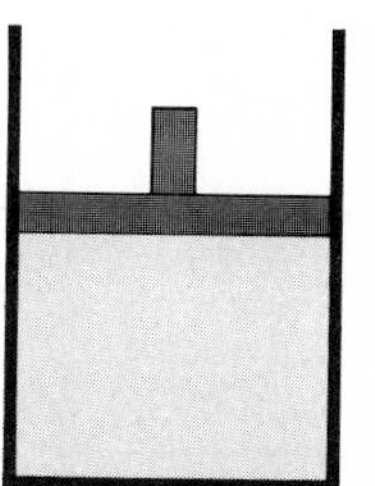
(*a*)

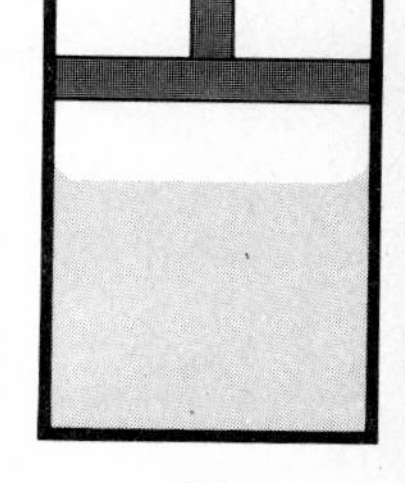
(*b*)

Figure 9.6
A solution held at fixed T.

P_i^* is the vapor pressure of pure liquid i at temperature T. Raoult's law can also be written as

$$x_{i,\text{vap}} P = x_i P_i^* \tag{9.69}$$

where P is the total vapor pressure of the ideal solution and $x_{i,\text{vap}}$ is the mole fraction of i in the vapor above the solution.

The total vapor pressure P in equilibrium with the ideal solution is the sum of the partial pressures. For a two-component ideal solution, Raoult's law gives

$$P = P_\text{A} + P_\text{B} = x_\text{A} P_\text{A}^* + x_\text{B} P_\text{B}^* = x_\text{A} P_\text{A}^* + (1 - x_\text{A}) P_\text{B}^*$$

$$P = (P_\text{A}^* - P_\text{B}^*) x_\text{A} + P_\text{B}^* \tag{9.70}$$

At fixed temperature, P_A^* and P_B^* are constants, and the two-component ideal-solution vapor pressure P varies linearly with x_A. For $x_\text{A} = 0$, we have pure B, and $P = P_\text{B}^*$; for $x_\text{A} = 1$, the solution is pure A, and $P = P_\text{A}^*$. Figure 9.7 shows the Raoult's law partial pressures P_A and P_B [Eq. (9.68)] and the total vapor pressure P for an ideal solution as a function of composition at fixed temperature. A nearly ideal solution like benzene–toluene shows a vapor-pressure curve that conforms very closely to Fig. 9.7.

2. ΔG_mxg. We want to find ΔG for the mixing of the amounts $n_1, n_2, \ldots, n_i, \ldots$ of the pure, separated liquids, each at T and P, to form the ideal solution (Fig. 9.6*a*) at T and P. For the Gibbs energy G of an ideal solution, Eqs. (9.25), (9.63), and (9.64) give

$$G = \sum_i n_i \mu_i = \sum_i n_i \mu_i^*(T, P) + RT \sum_i n_i \ln x_i$$

The Gibbs energy of the unmixed components at T and P is

$$G_\text{unmix} = \sum_i n_i \bar{G}_i^* \equiv \sum_i n_i \mu_i^*(T, P)$$

Subtraction gives for $G - G_\text{unmix} \equiv \Delta G_\text{mxg}$:

$$\Delta G_\text{mxg} = RT \sum_i n_i \ln x_i \qquad \text{ideal soln., const. } T, P \tag{9.71}$$

Since $0 < x_i < 1$, we have $\ln x_i < 0$ and $\Delta G_\text{mxg} < 0$, as it must be for an isothermal, isobaric spontaneous process. We shall use (9.71) to obtain ΔS_mxg, ΔV_mxg, and ΔH_mxg.

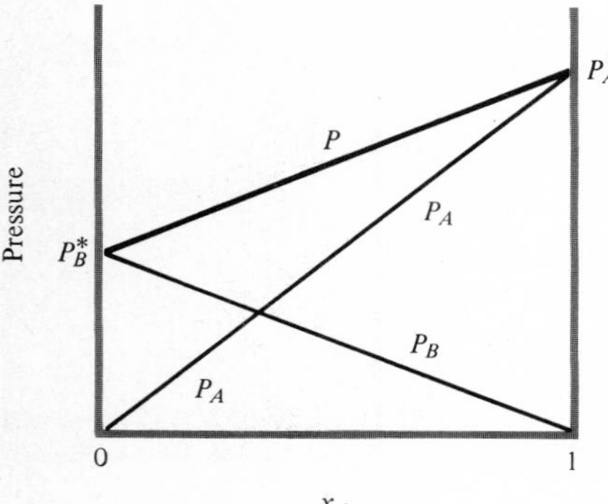

Figure 9.7
Partial vapor pressures P_A and P_B and total vapor pressure P above an ideal solution as a function of composition at fixed T.

3. ΔV_{mxg}. From Eq. (9.39), ΔV_{mxg} equals $(\partial/\partial P)_{T,n_i}$ of ΔG_{mxg}. For an ideal solution, Eq. (9.71) shows that ΔG_{mxg} depends on T and the mole fractions but not on P. Hence, $(\partial/\partial P)_{T,n_i}$ of ΔG_{mxg} is zero for an ideal solution, and

$$\Delta V_{\text{mxg}} = 0 \qquad \text{ideal soln., const. } T, P \tag{9.72}$$

There is no volume change on forming an ideal solution from its components at constant T and P (as we anticipated in Sec. 9.5).

4. ΔS_{mxg}. From (9.40), ΔS_{mxg} equals $-(\partial/\partial T)_{P,n_i}$ of ΔG_{mxg}. Taking $-\partial/\partial T$ of (9.71), we get

$$\Delta S_{\text{mxg}} = -R \sum_i n_i \ln x_i \qquad \text{ideal soln., const. } T, P \tag{9.73}$$

Since $\ln x_i$ is negative, ΔS_{mxg} is positive for an ideal solution. Note that (9.73) is the same as Eq. (3.39) for the mixing of ideal gases; see Prob. 9.37 for an explanation of this fact.

5. ΔH_{mxg}. Use of $\Delta G = \Delta H - T\,\Delta S$ at constant T and Eqs. (9.71) and (9.73) gives

$$\Delta H_{\text{mxg}} = \Delta G_{\text{mxg}} + T\,\Delta S_{\text{mxg}} = RT \sum_i n_i \ln x_i - RT \sum_i n_i \ln x_i$$

$$\Delta H_{\text{mxg}} = 0 \qquad \text{ideal soln., const. } T, P \tag{9.74}$$

There is no heat of mixing on formation of an ideal solution at constant T and P.

6. ΔU_{mxg}. From $\Delta H_{\text{mxg}} = \Delta U_{\text{mxg}} + P\,\Delta V_{\text{mxg}}$ at constant P and Eqs. (9.72) and (9.74), we have $\Delta U_{\text{mxg}} = 0$ for forming an ideal solution at constant T and P, as would be expected from the molecular picture of an ideal solution (Sec. 9.5).

Figure 9.8 plots $\Delta G_{\text{mxg}}/n$, $\Delta H_{\text{mxg}}/n$, and $T\,\Delta S_{\text{mxg}}/n$ for an ideal two-component solution vs. x_B at 25°C, where $n \equiv n_A + n_B$.

We derived ΔV_{mxg}, ΔS_{mxg}, and ΔH_{mxg} for an ideal solution from ΔG_{mxg}. An alternative approach is first to find the partial molar quantities $\bar{V}_i$, $\bar{H}_i$, and $\bar{S}_i$ from

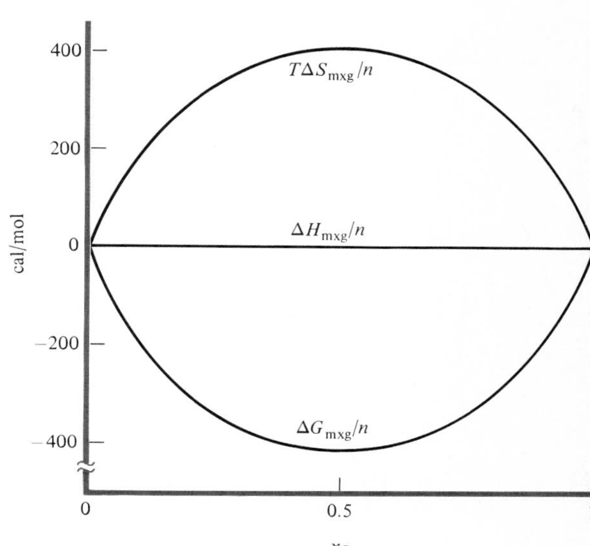

Figure 9.8
Mixing quantities for a two-component ideal solution as a function of composition at 25°C.

(9.63) and (9.64) and then use equations like (9.17) to find ΔV, ΔS, and ΔH of mixing. For an ideal solution, one finds (Prob. 9.25)

$$\bar{S}_i = \bar{S}_i^* - R \ln x_i, \qquad \bar{V}_i = \bar{V}_i^*, \qquad \bar{H}_i = \bar{H}_i^* \qquad \text{ideal soln.} \tag{9.75}$$

7. Miscibility. Two liquids that form an ideal solution are miscible in all proportions; see Prob. 9.38.

9.7 IDEALLY DILUTE SOLUTIONS

An ideal solution occurs in the limit where the molecules of the different components resemble one another very closely. A different kind of limit is where the solvent mole fraction approaches 1, so that all solutes are present in very low concentrations. We call such a solution an *ideally dilute* solution. In an ideally dilute solution, solute molecules interact essentially only with solvent molecules, due to the high dilution of the solutes.

Consider such a very dilute solution of nonelectrolytes. (We specify nonelectrolytes because the strong interionic forces in electrolyte solutions give substantial solute–solute interactions even at very high dilutions; hence, the ideally dilute solution model is not useful for electrolyte solutions. Also, each electrolyte component gives two or more ions in solution, and this has to be taken into account. Solutions of electrolytes are treated in Chap. 10.) We shall use A to denote the solvent and i to signify any one of the solutes. The condition of high dilution is expressed as

$$1 - \varepsilon \le x_A \le 1 \qquad \text{where } 0 \le \varepsilon \ll 1 \tag{9.76}$$

The A mole fraction differs from 1 by at most ε, where ε (epsilon) is a certain very small number. For such a very dilute solution, solute molecules are generally surrounded by only solvent molecules, so that all solute molecules are in an essentially uniform environment for all concentrations that satisfy (9.76).

To obtain an expression for the chemical potentials in an ideally dilute solution, we modify the kinetic–molecular arguments of Sec. 9.5. Equation (9.55) gives as the rate of evaporation and of condensation in pure solute i

$$r^* = bP_i^* \tag{9.77}$$

For the ideally dilute solution, the rate of condensation of i is proportional to the equilibrium partial vapor pressure P_i of i, but because i vapor molecules interact essentially only with solvent (A) molecules, we can't use the same proportionality constant as in (9.77) for pure i. Hence, for the solution

$$r_{\text{con}} = cP_i \tag{9.78}$$

where c is a constant that differs from b.

We now compare the rates of evaporation of i molecules from unit surface areas of pure i and of the ideally dilute solution. The number of i molecules in unit surface area of the solution will be reduced by a factor ax_i compared with the number in unit surface area of pure i, where a is a constant that corrects for the different sizes of the solute and solvent molecules and x_i is the mole fraction of i in the solution. The smaller number of i molecules in the surface reduces the rate of evaporation of i from the solution by the factor ax_i compared with pure i; moreover, in the ideally dilute solution, i molecules interact only with solvent A molecules, whereas, in pure i, they interact with other i molecules; this also affects the i evaporation rate, and we include a factor of d to allow for this effect, where d is a constant related to the difference between i-i and i-A interactions. The evaporation rate of solute i from the ideally dilute solution is thus dax_i times the evaporation rate from pure i [Eq. (9.77)] at the same temperature; hence, the rate of evaporation of i from the solution is

$$r_{\text{evap}} = dax_i bP_i^* \tag{9.79}$$

Equating (9.78) and (9.79), we get $P_i = (dabP_i^*/c)x_i$, where the symbols in parentheses are independent of x_i. Defining $k_i \equiv dabP_i^*/c$, we finally get

$$P_i = k_i x_i \qquad \text{solute in ideally dilute soln.} \tag{9.80}$$

Equation (9.80) is *Henry's law* and is experimentally observed to be obeyed by solutes in very dilute solutions of nonelectrolytes. The Henry's law constant k_i might be more fully written as $k_{i,\text{A}}$, since its value depends on the solvent A as well as on the solute i. Since the solute–solvent (i-A) interaction differs from the solute–solute (i-i) interaction, k_i is not equal to P_i^* (as in Raoult's law) but must be determined experimentally by vapor-pressure measurements. k_i is constant with respect to variations in solution composition over the range (9.76). k_i has dimensions of pressure.

To deal thermodynamically with ideally dilute solutions, we want expressions for the chemical potentials of the solutes and the solvent. Equating the liquid-phase and vapor-phase chemical potentials of i, we have for the chemical potential μ_i of each *solute* in the ideally dilute solution

$$\mu_i = \mu_{i,\text{vap}} = \mu_{i,\text{vap}}^\circ + RT \ln (P_i/\text{atm}) = \mu_{i,\text{vap}}^\circ + RT \ln (k_i x_i/\text{atm})$$

$$\mu_i = \mu_{i,\text{vap}}^\circ + RT \ln (k_i/\text{atm}) + RT \ln x_i \tag{9.81}$$

where we assumed the vapor is an ideal gas mixture [Eq. (6.7)] and used Henry's law (9.80). We now choose the standard state of the solute i such that the standard-state chemical potential μ_i° equals the sum of the first two terms in (9.81):

$$\mu_i^\circ \equiv \mu_{i,\text{vap}}^\circ + RT \ln (k_i/\text{atm}) \tag{9.82}$$

Equation (9.81) then becomes

$$\mu_i = \mu_i^\circ + RT \ln x_i \qquad \text{for } i \neq \text{A and } 1 - \varepsilon \leq x_\text{A} \leq 1 \tag{9.83}$$

where μ_i° is independent of composition. Despite the resemblance of (9.83) to (9.63), an ideally dilute solution is not the same as an ideal solution, as will be discussed in detail in the next section.

The "derivation" of Henry's law used molecular arguments and is not part of our logical development of thermodynamics. Following the procedure used for ideal solutions, we now *define* an ideally dilute solution as one where the chemical potential of each solute is given by (9.83), and we discard all equations and arguments that precede (9.83) in this section. The purpose of the molecular arguments is to give us plausible forms for μ_i in ideally dilute solutions and in ideal solutions. The laws of thermodynamics are general and cannot supply us with the explicit forms of equations of state or chemical potentials for specific systems. Such information must be obtained by appeal to molecular (statistical-mechanical) arguments and to experimental data (as in the use of $PV = nRT$ for low-density gases).

9.8 THERMODYNAMIC PROPERTIES OF IDEALLY DILUTE SOLUTIONS

As a result of the discussion of the last section, we define an *ideally dilute* solution (also called an *ideal dilute* solution) as one in which each solute chemical potential μ_i satisfies

$$\mu_i = \mu_i^\circ(T, P) + RT \ln x_i \qquad \text{for } i \neq \text{A} \tag{9.84}$$

for the following range of solvent mole fraction:

$$1 - \varepsilon \leq x_\text{A} \leq 1 \tag{9.85}$$

Outside the range (9.85), the solution is not ideally dilute. The standard-state solute chemical potential μ_i° is a function of T and P (but not of the mole fractions).

What about μ_A, the chemical potential of the solvent in an ideally dilute solution? Dividing the Gibbs–Duhem equation (9.43) by the total number of moles in the solution, we have

$$x_\text{A}\, d\mu_\text{A} + \sum_{i \neq \text{A}} x_i\, d\mu_i = 0 \qquad \text{const. } T, P \tag{9.86}$$

where the sum goes over all the solutes i but not the solvent. Taking the differential of (9.84) at constant T and P, we have $d\mu_i = RT\, d \ln x_i = (RT/x_i)\, dx_i$. Substitution in (9.86) gives

$$x_\text{A}\, d\mu_\text{A} + RT \sum_{i \neq \text{A}} dx_i = 0 \qquad \text{const. } T, P \tag{9.87}$$

The sum of all the mole fractions is one: $x_\text{A} + \sum_{i \neq \text{A}} x_i = 1$. The differential of this equation is

$$dx_\text{A} + \sum_{i \neq \text{A}} dx_i = 0 \qquad \text{or} \qquad \sum_{i \neq \text{A}} dx_i = -dx_\text{A} \tag{9.88}$$

Use of (9.88) in (9.87) gives $x_\text{A}\, d\mu_\text{A} - RT\, dx_\text{A} = 0$, or

$$d\mu_\text{A} = (RT/x_\text{A})\, dx_\text{A} \qquad \text{const. } T, P \tag{9.89}$$

We now integrate (9.89) at constant T and P between state 1, defined as $x_\text{A} = 1$ (pure solvent A), and state 2, defined as having mole fraction x_A, where x_A lies in the range (9.85); we get $\int_1^2 d\mu_\text{A} = RT \int_1^2 (1/x_\text{A})\, dx_\text{A}$ and

$$\mu_{\text{A},2} - \mu_{\text{A},1} = RT(\ln x_{\text{A},2} - \ln x_{\text{A},1}) = RT \ln (x_{\text{A},2}/x_{\text{A},1}) \tag{9.90}$$

Since state 1 is pure A, we have $\mu_{\text{A},1} = \mu_\text{A}^*$ and $x_{\text{A},1} = 1$. Dropping the superfluous subscript 2, we get

$$\mu_\text{A} = \mu_\text{A}^* + RT \ln x_\text{A} \tag{9.91}$$

We define the standard state of the solvent as pure A at the temperature and pressure T and P of the solution; hence, the solvent standard-state chemical potential is $\mu_\text{A}^\circ \equiv \mu_\text{A}^*(T, P)$. Equation (9.91) becomes

$$\mu_\text{A} = \mu_\text{A}^\circ + RT \ln x_\text{A} \qquad \text{solvent in ideally dil. soln.} \tag{9.92}$$

$$\mu_\text{A}^\circ \equiv \mu_\text{A}^*(T, P) \tag{9.93}$$

The solvent equation (9.92) is the same as (9.63) and (9.64) for each component of an ideal solution, except that (9.92) holds only for the very limited concentration range (9.85).

The solvent standard state is the pure solvent at the temperature and pressure of the solution. What choice of standard state for each solute is implied by Eq. (9.84)? When x_i in (9.84) is 1, we see that μ_i then equals the standard-state chemical potential μ_i°. Hence, it might be thought that the standard state of solute i is pure i at the temperature and pressure of the solution. This supposition is wrong. When x_i becomes large and approaches 1, the solution is no longer ideally dilute; hence, Eq. (9.84), whose validity is restricted to compositions for which $1 - \varepsilon \leq x_A \leq 1$, does not hold for x_i approaching 1. However, one could conceive of the *hypothetical* case where (9.84) holds for all values of x_i. In this hypothetical case, μ_i would become equal to μ_i° in the limit $x_i \to 1$. The choice of solute standard state uses this hypothetical situation. The standard state for solute i in an ideally dilute solution is defined to be the fictitious state at the temperature and pressure of the solution that arises by supposing that $\mu_i = \mu_i^\circ + RT \ln x_i$ holds for all values of x_i and setting $x_i = 1$. This fictitious state is an extrapolation of the properties of solute i in the very dilute solution to the limiting case $x_i \to 1$. Since the properties of i in the dilute solution depend very strongly on the solvent (which provides the environment for the i molecules), the fictitious standard state of solute i depends on what the solvent is. Of course, the properties of the standard state obtained also depend on T and P; hence, μ_i° is a function of T and P (but not of the mole fractions): $\mu_i^\circ = \mu_i^\circ(T, P)$. We might write $\mu_i^\circ = \mu_i^{\circ,A}(T, P)$ to indicate that the standard state depends on what solvent i is dissolved in, but we won't do so unless we are dealing with solutions of i in two different solvents.

We previously used a fictitious standard state for gases (Secs. 5.1 and 5.2), so there is nothing shocking about fictitious standard states. We shall soon see how the properties of solute i in this fictitious state can be found from experimental data. Note from (9.84) that μ_i° equals $\mu_i - RT \ln x_i$ for all values of x_i such that $1 - \varepsilon \leq x_A \leq 1$; in particular, this equality holds in the limit of infinite dilution ($x_A \to 1$); hence

$$\mu_i^\circ = \lim_{x_A \to 1} (\mu_i - RT \ln x_i) \equiv (\mu_i - RT \ln x_i)^\infty \tag{9.94}$$

where the superscript ∞ denotes infinite dilution.

Summarizing, we note that in an ideally dilute solution, the solute chemical potentials μ_i and the solvent chemical potential μ_A are

$$\mu_i = \mu_i^\circ + RT \ln x_i \qquad \text{for } i \neq A \qquad (9.95)^*$$

$$\mu_i^\circ(T, P) = (\mu_i - RT \ln x_i)^\infty \tag{9.96}$$

$$\mu_A = \mu_A^\circ + RT \ln x_A \qquad (9.97)^*$$

$$\mu_A^\circ \equiv \mu_A^*(T, P) \qquad (9.98)^*$$

where (9.95) and (9.97) hold only for the concentration range (9.85). The solvent standard state is pure liquid A at the temperature and pressure T and P of the solution. The standard state of solute i is the fictitious state at T and P obtained by taking the limit $x_i \to 1$ while pretending that (9.95) holds for all concentrations.

Although Eqs. (9.95) and (9.97) have the same appearance as (9.63) for an ideal solution, ideally dilute solutions and ideal solutions are not the same. Equations (9.95) and (9.97) hold only for the concentration range (9.85), whereas (9.63) holds for all solution compositions. Moreover, the standard state for all components of an ideal solution is the actual state of the pure component at T and P of the solution, whereas the standard state of each solute in an ideally dilute solution is the fictitious state defined above.

We now derive some thermodynamic properties of ideally dilute solutions.

1. Vapor pressure. Let P_i be the (equilibrium) vapor partial pressure of solute i of an ideally dilute solution. Equating the chemical potential μ_i of i in the solution to the chemical potential $\mu_{i,\text{vap}}$ of i in the vapor (assumed to be an ideal gas mixture) in equilibrium with the solution, we have $\mu_i = \mu_{i,\text{vap}}$, or

$$\mu_i^\circ + RT \ln x_i = \mu_{i,\text{vap}}^\circ + RT \ln (P_i/\text{atm})$$

$$(\mu_i^\circ - \mu_{i,\text{vap}}^\circ)/RT = \ln [P_i/(x_i \text{ atm})] \tag{9.99}$$

$$P_i/(x_i \text{ atm}) = \exp [(\mu_i^\circ - \mu_{i,\text{vap}}^\circ)/RT] \tag{9.100}$$

(where $\exp z \equiv e^z$). From Eqs. (9.96) and (6.7), μ_i° depends on T and P, and $\mu_{i,\text{vap}}^\circ$ depends on T. The right side of (9.100) is thus a function of T and P. We now define k_i as

$$k_i(T, P) \equiv 1 \text{ atm} \cdot \exp [(\mu_i^\circ - \mu_{i,\text{vap}}^\circ)/RT] \tag{9.101}$$

Since the standard-state chemical potential μ_i° of solute i in the solution depends on the nature of the solvent (as well as the solute), k_i differs for the same solute in different solvents. Equation (9.100) becomes

$$P_i = k_i x_i \qquad \text{solute in ideally dil. soln., ideal vapor} \tag{9.102*}$$

which is *Henry's law*. The Henry's law constant k_i is independent of the mole fractions.

The pressure dependence of k_i arises from the dependence of μ_i° on pressure. As noted several times previously, the chemical potentials in condensed phases vary only slowly with pressure. Hence, k_i depends only weakly on pressure, and its pressure dependence can be neglected (except at quite high pressures). We thus take k_i to depend only on T. This approximation corresponds to a similar approximation made in deriving Raoult's law (9.68).

Now consider the solvent vapor pressure. Equations (9.97) and (9.98) for the solvent chemical potential μ_A in an ideally dilute solution are the same as Eqs. (9.63) and (9.64) for the chemical potential of a component of an ideal solution. Therefore, the same derivation that gave (9.68) for the vapor pressure of an ideal-solution component gives as the vapor pressure of the solvent in an ideally dilute solution

$$P_A = x_A P_A^* \qquad \text{solvent in ideally dil. soln., ideal vapor} \tag{9.103*}$$

Of course, (9.103) [and (9.102)] hold only for the concentration range (9.85). In an ideally dilute solution, the solvent obeys Raoult's law and the solutes obey Henry's law.

At sufficiently high dilutions, all nonelectrolyte solutions become ideally dilute. For less dilute solutions, the solution is no longer ideally dilute, and we generally find deviations from Raoult's and Henry's laws. Two systems that show large deviations are graphed in Fig. 9.9.

The solid lines in Fig. 9.9*a* show the observed partial and total vapor pressures above acetone–chloroform solutions at 35°C. The three upper dashed lines show the partial and total vapor pressures that would occur for an ideal solution, where Raoult's law is obeyed by both species (Fig. 9.7). In the limit $x(\text{CHCl}_3) \rightarrow 1$, the solution becomes ideally dilute with $CHCl_3$ as the solvent and acetone as the solute. For $x(\text{CHCl}_3) \rightarrow 0$, the solution becomes ideally dilute with acetone as the solvent and $CHCl_3$ as the solute. Hence, near $x(\text{CHCl}_3) = 1$, the observed $CHCl_3$ partial pressure approaches the Raoult's law line very closely, whereas near $x(\text{CHCl}_3) = 0$, the observed acetone partial pressure approaches the Raoult's law line very closely. Near $x(\text{CHCl}_3) = 1$, the observed partial pressure of the solute acetone varies nearly linearly with mole fraction (Henry's law), whereas near $x(\text{CHCl}_3) = 0$, the observed partial pressure of the solute $CHCl_3$ varies nearly linearly with mole fraction. The two lower dashed lines show the Henry's law lines extrapolated from the observed limiting slopes of $P(\text{CHCl}_3)$ near $x(\text{CHCl}_3) = 0$ and $P(\text{CH}_3\text{COCH}_3)$ near $x(\text{CHCl}_3) = 1$.

For all compositions, the partial and total vapor pressures in Fig. 9.9*a* are below those predicted by Raoult's law; the solution is said to exhibit *negative deviations* from Raoult's law. The acetone–CS_2 system in Fig. 9.9*b* exhibits *positive deviations* from Raoult's law at all compositions. For certain systems, one component exhibits a positive deviation while the second component exhibits a negative deviation at the same composition; see M. L. McGlashan, *J. Chem. Educ.*, **40**, 516 (1963) for examples. A rough molecular interpretation of deviations from Raoult's law is as follows. If the A-B intermolecular attractions are weaker than the A-A and B-B attractions, it is easier for A and B molecules to vaporize from the solution than from the pure liquids and the vapor partial pressures will be higher than Raoult's law pressures (which are based on the assumption of equal A-A, A-B, and B-B interactions). If A-B attractions are stronger than both A-A and B-B attractions, we get

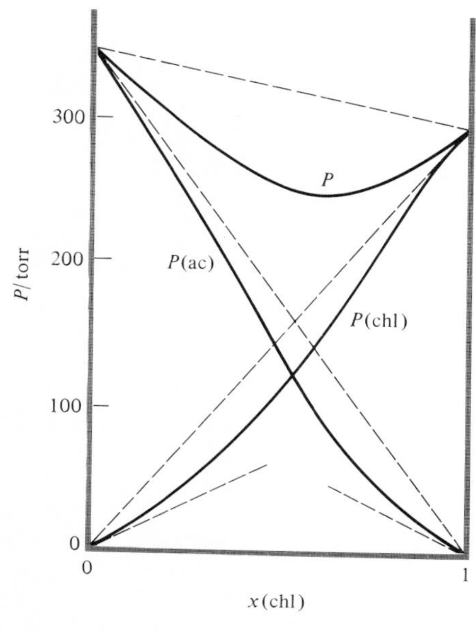

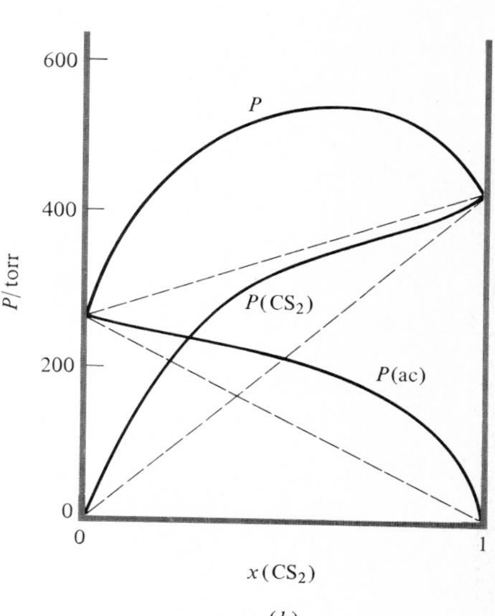

Figure 9.9 Partial and total vapor pressures for (*a*) acetone–chloroform solutions at 35°C; (*b*) acetone–CS_2 solutions at 29°C.

negative deviations from Raoult's law. If A-B attractions are stronger than A-A attractions but weaker than B-B attractions, then A shows negative deviations from Raoult's law and B shows positive deviations.

2. Partial molar quantities. For the solvent in an ideally dilute solution, use of (9.35), (9.97), (9.98), and (9.24) gives for the partial molar volume

$$\bar{V}_A = (\partial \mu_A/\partial P)_{T,n_j} = (\partial \mu_A^\circ/\partial P)_T = (\partial \mu_A^*/\partial P)_T = (\partial \bar{G}_A^*/\partial P)_T$$

From (7.15), we have $(\partial \bar{G}_A^*/\partial P)_T = \bar{V}_A^*$; hence

$$\bar{V}_A = \bar{V}_A^* \qquad \text{solvent in ideally dil. soln.} \tag{9.104}$$

The solvent partial molar volume equals the molar volume of the pure solvent at the temperature and pressure of the solution (as is also true for each component of an ideal solution).

For each solute, use of (9.35), (9.95), and (9.24) gives

$$\bar{V}_i = (\partial \mu_i/\partial P)_{T,n_j} = (\partial \mu_i^\circ/\partial P)_T = (\partial \bar{G}_i^\circ/\partial P)_T$$

Although the solute standard state is not experimentally realizable, all thermodynamic identities that hold for real states hold for the fictitious standard state; in particular, Eq. (7.15) gives $(\partial \bar{G}_i^\circ/\partial P)_T = \bar{V}_i^\circ$, where $\bar{V}_i^\circ$ is the molar volume of solute i in its (fictitious) standard state. Therefore $\bar{V}_i = \bar{V}_i^\circ$. Since μ_i° is a function of T and P only, its derivative $(\partial \mu_i^\circ/\partial P)_T = \bar{V}_i^\circ$ is a function of T and P only. Hence, $\bar{V}_i$ is independent of concentration for all concentrations in the range (9.85). In particular, $\bar{V}_i$ equals its limiting (infinite-dilution) value for $x_A \to 1$ and $x_i \to 0$; we have $\bar{V}_i = \lim_{x_A \to 1} \bar{V}_i \equiv \bar{V}_i^\infty$. Since $\bar{V}_i^\circ = \bar{V}_i$, we conclude that the solute standard-state partial molar volume $\bar{V}_i^\circ$ equals the solute partial molar volume $\bar{V}_i^\infty$ at infinite dilution. Thus, $\bar{V}_i^\circ$ can be found from measurements on very dilute solutions. In summary,

$$\bar{V}_i = \bar{V}_i^\circ = \bar{V}_i^\infty \qquad \text{solute in ideally dil. soln.} \tag{9.105}$$

Using (9.34) and then (9.31), we can find the partial molar entropies and enthalpies for the solvent A and the solutes i. Doing so is rather boring, so we shall leave it to the reader to prove (Prob. 9.23) that for an ideally dilute solution

$$\bar{S}_A = \bar{S}_A^* - R \ln x_A, \qquad \bar{H}_A = \bar{H}_A^* \tag{9.106}$$

$$\bar{S}_i = \bar{S}_i^\circ - R \ln x_i, \qquad \bar{S}_i^\circ = (\bar{S}_i + R \ln x_i)^\infty \tag{9.107}$$

$$\bar{H}_i = \bar{H}_i^\circ = \bar{H}_i^\infty \tag{9.108}$$

Of course, Eqs. (9.104) to (9.108) hold only for the composition range (9.85).

Although $\bar{V}_i^\circ = \bar{V}_i^\infty$ and $\bar{H}_i^\circ = \bar{H}_i^\infty$, the standard state of a solute in an ideally dilute solution is not the state of infinite dilution. The standard state (degree superscript) is a single state in which the substance has the properties μ_i°, $\bar{S}_i^\circ$, $\bar{H}_i^\circ$, $\bar{V}_i^\circ$, etc. Equation (9.96) shows that $\mu_i^\circ \neq \mu_i^\infty$. [In fact, taking the limit of (9.95) as $x_i \to 0$, we have $\mu_i^\infty = -\infty \neq \mu_i^\circ$.] Since $\mu_i^\circ \neq \mu_i^\infty$, the state of infinite dilution cannot be the standard state. The solute standard state is the (fictitious) state defined after Eq. (9.98). [The reason $\bar{V}_i^\circ$ equals $\bar{V}_i^\infty$ and $\bar{H}_i^\circ$ equals $\bar{H}_i^\infty$ is that the solute partial molar volume and partial molar enthalpy are independent of concentration in the range (9.85), and since the standard state is obtained by extrapolation of the properties of the ideally dilute solution to $x_i = 1$, we get $\bar{V}_i^\circ = \bar{V}_i^\infty$ and $\bar{H}_i^\circ = \bar{H}_i^\infty$.]

Note from Prob. 9.24 that ΔV_{mxg} and ΔH_{mxg} are not zero for an ideally dilute solution.

3. Solubility of gases in liquids. For gases that are sparingly soluble in liquids, the concentration of the dissolved gas is usually low enough for the solution to be approximately ideally dilute, and Henry's law (9.102) holds well. Therefore,

$$x_i = k_i^{-1} P_i \tag{9.109}$$

where x_i is the mole fraction of dissolved gas in the solution at a given temperature and P_i is the partial pressure of the gas above the solution. The gas solubility (as measured by x_i) is proportional to the partial pressure of the gas above the solution, provided the solution is ideally dilute. For He, H_2, and N_2 dissolved in water, an essentially linear relation between x_i and P_i is observed for pressures up to 300 atm; for O_2 in water, moderate deviations from Henry's law occur even at pressures below 100 atm.

For the low solute concentrations at which Henry's law applies, the solute molality m_i and the solute concentration c_i are each proportional to the solute mole fraction x_i (Prob. 9.1*c*); hence, we can use molalities or concentrations instead of mole fractions in Henry's law:

$$P_i \approx k_{i,m} m_i \qquad \text{or} \qquad P_i \approx k_{i,c} c_i \tag{9.110}$$

where $k_{i,m}$ and $k_{i,c}$ are constants related to k_i in (9.109).

Some values of k_i in (9.109) for gases in water at 25°C are:

Gas	H_2	N_2	O_2	CH_4	C_2H_6
$10^{-7} k_i$/torr	5.3	6.5	3.3	3.1	2.3

For the majority of gases in water, the solubility decreases with increasing temperature. For each of the noble gases (He, Ne, etc.) in water, the solubility goes through a minimum at a certain temperature (35°C for He at 1 atm), being larger both above and below this temperature. In organic solvents, many gases show increasing solubility with increasing temperature; this is especially true of gases with very low solubilities in the solvent.

Sometimes a dissolved gas reacts chemically with the solvent, for example, HCl, NH_3, and CO_2 in water. Equation (9.109) was derived by equating μ_i in the solution to $\mu_{i,\mathrm{vap}}$. Hence x_i in (9.109) must be the actual mole fraction in solution of the same chemical species *i* as occurs in the vapor. For example, to apply Henry's law to NH_3 in water, we must take account of the ionization equilibrium $NH_3 + H_2O = NH_4^+ + OH^-$; the true value of $x(NH_3)$ in solution is less than would be calculated from the number of moles of dissolved NH_3 since some of the dissolved NH_3 is present as NH_4^+. Moreover, the very high solubility of NH_3 in water (and the presence of ions) makes the assumption of an ideally dilute solution poor; hence, Henry's law will hold only at quite low NH_3 pressures.

The solubility of O_2 in water is of obvious interest to fish. The solubility of O_2 in blood is greatly increased by compound formation with hemoglobin.

Note from (9.101) that if we know k_i in a solvent A, we can find $\mu_i^\circ - \mu_{i,\mathrm{vap}}^\circ = \bar{G}_i^\circ - \bar{G}_{i,\mathrm{vap}}^\circ \equiv \Delta G_i^\circ$, the change in standard-state Gibbs energy of gas *i* when it dissolves in liquid A. If we know k_i as a function of T, we can find $\bar{H}_i^\circ - \bar{H}_{i,\text{vap}}^\circ \equiv \Delta H_i^\circ$ (Prob. 9.21); knowing ΔG_i° and ΔH_i°, we can find $\Delta S_i^\circ \equiv \bar{S}_i^\circ - \bar{S}_{i,\text{vap}}^\circ$.

The solubility of gases in liquids is relevant to anesthesia. There are two conflicting explanations of anesthesia. The Meyer–Overton hypothesis attributes anesthesia to gas dissolved in the lipid (fatty) portion of nerve-cell membranes, thereby blocking nerve conduction. The Pauling–Miller hypothesis attributes the anesthetic action to gas dissolved in the aqueous portion of nerve cells, which forms regions of high order and thus decreases the conductivity of nerve cells. Since entropy is related to the degree of order, Haberfield and Kivuls examined ΔS_i° (defined above) for anesthetic gases in water. [P. Haberfield and J. Kivuls, *J. Med. Chem.*, **16**, 942 (1973).] They found a good linear correlation between ΔS_i° and ln $x_{i,\text{anes}}$ (where $x_{i,\text{anes}}$ is the mole fraction of dissolved anesthetic gas in water at the minimum pressure that causes anesthesia) and argued that this correlation supports the Pauling–Miller hypothesis.

A comprehensive review of gas solubilities in liquids is R. Battino and H. L. Clever, *Chem. Rev.*, **66**, 395 (1966).

9.9 REACTION EQUILIBRIUM IN IDEAL AND IDEALLY DILUTE SOLUTIONS

Suppose a chemical reaction between some of the solution components is occurring in an ideal solution or in an ideally dilute solution. The equilibrium condition (4.111) is $\sum_i \nu_i \mu_i = 0$. For an ideal solution, each μ_i is given by (9.63) and (9.64). For an ideally dilute solution, the solvent chemical potential is given by (9.97) and (9.98), and each solute chemical potential is given by (9.95) and (9.96). These equations show that the chemical potential of any species in an ideal solution or an ideally dilute solution has the form

$$\mu_i = \mu_i^\circ + RT \ln x_i \tag{9.111}$$

Of course, the standard-state chemical potential μ_i° has a different meaning for solutes in an ideally dilute solution than for the solvent in an ideally dilute solution or the species of an ideal solution.

We now substitute (9.111) into $\sum_i \nu_i \mu_i = 0$. There is no need to go through the algebraic manipulations because (9.111) has the same form as μ_i in (6.7) for an ideal gas mixture, except that $P_i/(1 \text{ atm})$ is replaced by x_i and μ_i° depends on P as well as on T. Hence, the manipulations are the same as those leading to (6.16) and (6.17), except that we replace $P_i/(1 \text{ atm})$ by x_i in all equations. Thus, (6.9), (6.16), and (6.17) become

$$\Delta G^\circ \equiv \sum_i \nu_i \mu_i^\circ \tag{9.112}$$

$$K_x \equiv \prod_i (x_{i,\text{eq}})^{\nu_i} \qquad \text{ideal soln. or ideally dil. soln.} \tag{9.113}$$

$$\Delta G^\circ = -RT \ln K_x \tag{9.114}$$

where $x_{i,\text{eq}}$ is the equilibrium mole fraction of species i in solution. K_x is the *mole-fraction equilibrium constant*. The standard-state chemical potentials μ_i° are functions of both T and P. Hence, ΔG° depends on T and P, and K_x [which equals $\exp(-\Delta G^\circ/RT)$] also depends on both T and P: $K_x = K_x(T, P)$. However, the pressure dependence is generally small, since μ_i° varies slowly with pressure for condensed phases.

Differentiation of (9.114) with respect to T [the derivation is essentially the same as for (6.45)] gives

$$\left(\frac{\partial \ln K_x}{\partial T}\right)_P = \frac{\Delta H^\circ}{RT^2} \qquad \text{where } \Delta H^\circ \equiv \sum_i \nu_i \bar{H}_i^\circ \tag{9.115}$$

Differentiation of (9.114) with respect to P (Prob. 9.35) gives

$$\left(\frac{\partial \ln K_x}{\partial P}\right)_T = -\frac{\Delta V^\circ}{RT} \qquad \text{where } \Delta V^\circ \equiv \sum_i \nu_i \bar{V}_i^\circ \tag{9.116}$$

If the total of the product standard-state volumes is about the same as the total of the reactant standard-state volumes, K_x is essentially independent of pressure.

Further discussion of reaction equilibrium in solution is given in Chap. 11.

PROBLEMS

9.1 Let a solution have density ρ and volume V, and let n_j be the number of moles of species j with molar mass M_j in the solution. (*a*) Show that $\rho V = \sum_j n_j M_j$, where the sum goes over all components of the solution. (*b*) Let x_i, c_i, and m_i be the mole fraction, concentration, and molality of species i in the solution. Show that

$$c_i = \frac{\rho x_i}{\sum_j x_j M_j}, \qquad m_i = \frac{x_i}{x_A M_A} \tag{9.117}$$

(*c*) Consider a very dilute solution with solvent mole fraction near 1: $x_A = 1 - \varepsilon$, where ε is very small; thus, each solute mole fraction cannot exceed ε. Show that in such a very dilute solution, the equations of (*b*) become

$$c_i \approx \rho x_i / M_A$$

and

$$m_i \approx x_i / M_A \qquad \text{for } i \neq A \tag{9.118}$$

9.2 Find the solute mole fractions and molalities in a solution prepared by dissolving 145 mg of CH_3OH and 1.20 g of C_2H_5OH in 16.0 g of H_2O (the solvent).

9.3 Show that $m_B = (1000 n_B / n_A M_{r,A})$ mol/kg, where m_B is the molality of solute B and $M_{r,A}$ is the molecular weight (relative molecular mass) of the solvent.

9.4 The partial molar mass $\bar{m}_i$ of component i of a solution of mass m is defined by an equation analogous to (9.8) for the partial molar volume $\bar{V}_i$. Show that $\bar{m}_i = M_i$.

9.5 Show that $\bar{H}_i = \bar{U}_i + P\bar{V}_i$.

9.6 Use Fig. 9.3 to calculate the volume at 20°C and 1 atm of a solution formed from 20.0 g of H_2O and 45.0 g of C_2H_5OH.

9.7 The density of a methanol–water solution that is 12.000 weight percent methanol is 0.97942 g/cm³ at 15°C and 1 atm. For a solution that is 13.000 weight percent methanol, the density is 0.97799 g/cm³ at this T and P. Since the change in solution composition is small, we can estimate $\bar{V}_A$ by

$$\bar{V}_A \equiv (\partial V/\partial n_A)_{T,P,n_B} \approx (\Delta V/\Delta n_A)_{T,P,n_B}$$

Calculate $\bar{V}(CH_3OH)$ for a methanol–water solution at 15°C and 1 atm that is $12\frac{1}{2}$ percent CH_3OH by weight. Then calculate $\bar{V}(H_2O)$ for this solution.

9.8 Let V be the volume of an aqueous solution of NaCl at 25°C and 1 atm that contains 1000 g of water and n_B moles of NaCl. One finds that the following empirical formula accurately reproduces the experimental data:

$$V = a + bn_B + cn_B^{3/2} + kn_B^2 \qquad \text{for } n_A M_A = 1 \text{ kg}$$

$$a = 1001.38 \text{ cm}^3, \qquad b = 16.6253 \text{ cm}^3/\text{mol}$$

$$c = 1.7738 \text{ cm}^3/\text{mol}^{3/2}, \qquad k = 0.1194 \text{ cm}^3/\text{mol}^2$$

(*a*) Show that the NaCl partial molar volume $\bar{V}_B$ is

$$\bar{V}_B = b + (3c/2)n_B^{1/2} + 2kn_B \qquad \text{for } n_A M_A = 1 \text{ kg}$$

(*b*) Find $\bar{V}(\text{NaCl})$ for a solution with NaCl molality $m_B = 1.0000$ mol/kg. (*c*) Use (9.16) to show that the partial molar volume of the water in the solution is

$$\bar{V}_A = (M_A/1000 \text{ g})(a - \tfrac{1}{2}cn_B^{3/2} - kn_B^2) \qquad \text{for } n_A M_A = 1 \text{ kg}$$

(*d*) Show that the results of (*a*) and (*c*) can be written as

$$\bar{V}_B = b + (3c/2)(m_B \text{ kg})^{1/2} + 2km_B \text{ kg}$$

$$\bar{V}_A = (M_A/1000 \text{ g})(a - \tfrac{1}{2}cm_B^{3/2} \text{ kg}^{3/2} - km_B{}^2 \text{ kg}^2)$$

Since $\bar{V}_A$, $\bar{V}_B$, and m_B are all intensive quantities, we need not specify n_A in these equations. (*e*) Find $\bar{V}(H_2O)$ for a solution with $m_B = 1.0000$ mol/kg.

9.9 Some densities of aqueous solutions of $ZnCl_2$ at 20°C and 1 atm as a function of the weight percent of $ZnCl_2$ are:

Wt. %	8.0000	10.000	12.000	14.000
$\rho/(g/cm^3)$	1.0715	1.0891	1.1085	1.1275

Use the intercept method to find $\bar{V}(ZnCl_2)$ and $\bar{V}(H_2O)$ for a solution with $ZnCl_2$ molality 0.8000 mol/kg.

9.10 A certain two-component solution formed by mixing 2.000 mol of A with 1.500 mol of B has a volume of 425 cm^3. It is known that $\bar{V}_B = 250$ cm^3 mol^{-1} in this solution. Find $\bar{V}_A$.

9.11 Show that for a one-component "solution," the Gibbs–Duhem equation (9.42) reduces to (9.38).

9.12 Values of ΔH_{mxg} for dissolving 1 mole of H_2SO_4 (B) in water (A) at 25°C and 1 atm as a function of the amount of water are as follows:

n_A/mol	ΔH_{mxg}/kJ
0	0
0.5000	−15.73
1.0000	−28.07
2.0000	−41.92
5.0000	−58.03
10.000	−67.03
20.000	−71.50

(*a*) Find $\Delta H_{int,B}$ for the formation at 25°C and 1 atm of a solution with $x_B = 0.200$. (*b*) Find $\Delta H_{diff,A}$ and $\Delta H_{diff,B}$ for a solution with $x_B = 0.200$.

9.13 Benzene and toluene form nearly ideal solutions. At 20°C the vapor pressure of benzene is 74.7 torr, and that of toluene is 22.3 torr. (*a*) Find the equilibrium partial vapor pressures above a 20°C solution of 100.0 g of benzene plus 100.0 g of toluene. (*b*) Find the mole fractions in the vapor phase that is in equilibrium with the solution of part (*a*).

9.14 Find ΔG_{mxg}, ΔV_{mxg}, ΔS_{mxg}, and ΔH_{mxg} for mixing 100.0 g of benzene with 100.0 g of toluene at 20°C and 1 atm. Assume an ideal solution.

9.15 Find ΔS for mixing 1.000 mol of H_2O with 1.000 mol of D_2O at 4°C and 1 atm.

9.16 Verify that the expressions for the ideal-solution chemical potentials in (9.63) are consistent with the Gibbs–Duhem equation (9.43).

9.17 Prove the validity of the intercept method of determining partial molar quantities, as follows. Consider a two-component solution. (*a*) Show that

$$V/n = (\bar{V}_B - \bar{V}_A)x_B + \bar{V}_A \qquad \text{where } n \equiv n_A + n_B$$

$$\left[\frac{\partial(V/n)}{\partial x_B}\right]_{T,P} = \bar{V}_B - \bar{V}_A + \left(\frac{\partial \bar{V}_A}{\partial x_B}\right)_{T,P} + \left[\left(\frac{\partial \bar{V}_B}{\partial x_B}\right)_{T,P} - \left(\frac{\partial \bar{V}_A}{\partial x_B}\right)_{T,P}\right] x_B$$

(*b*) Divide the Gibbs–Duhem equation (9.45) (with $Y = V$) by dx_B at constant T and P, and use the result to show that

$$\left[\frac{\partial(V/n)}{\partial x_B}\right]_{T,P} = \bar{V}_B - \bar{V}_A$$

Hence, the slope of the V/n-vs.-x_B curve at any point is $\bar{V}_B - \bar{V}_A$. (*c*) Let $y = mx_B + b$ be the equation of the tangent line to the V/n-vs.-x_B curve at the point with $x_B = x_B'$ and $V/n = V'/n$; let $\bar{V}_B'$ and $\bar{V}_A'$ be the partial molar volumes at this composition. From (*b*), we have $m = \bar{V}_B' - \bar{V}_A'$. Also, since the curve passes through the point $(x_B', V'/n)$, we have

$$V'/n = mx_B' + b = (\bar{V}_B' - \bar{V}_A')x_B' + b.$$

Show that $b = \bar{V}_A'$. Hence the equation of the tangent line is $y = (\bar{V}_B' - \bar{V}_A')x_B + \bar{V}_A'$. At $x_B = 0$, we have $y = \bar{V}_A'$; at $x_B = 1$, we have $y = \bar{V}_B'$. Q.E.D.

9.18 At 20°C, 0.164 mg of H_2 dissolves in 100.0 g of water when the H_2 pressure above the water is 1.000 atm. (*a*) Find the Henry's law constant k_i for H_2 in water at 20°C. (*b*) Find the mass of H_2 that will dissolve in 100.0 g of water at 20°C when the H_2 pressure is 10.00 atm. (Neglect the pressure variation of k_i.)

9.19 Air is 21 percent O_2 and 78 percent N_2 by volume. Find the masses of O_2 and N_2 dissolved in 100.0 g of water at 20°C that is in equilibrium with air at 1.000 atm pressure. At 20°C, k_i for O_2 and N_2 are 2.95×10^7 and 5.75×10^7 torr, respectively.

9.20 The solubility of a gas in a liquid is often expressed in terms of the *Bunsen absorption coefficient* α_i, defined as the volume of dissolved gas i (reduced to

0°C and 1 atm) divided by the volume of the liquid when the gas partial pressure is 1 atm. Let ρ_A and M_A be the density and the molar mass of liquid A. Show that

$$\alpha_i k_i \approx R(273.15\text{ K})\rho_A/M_A \qquad \text{ideal gas}$$

(*b*) Calculate α_i for N_2 in water at 20°C, given that $k_i = 5.75 \times 10^7$ torr.

9.21 Show that the temperature and pressure variations of the Henry's law constant are

$$\left(\frac{\partial \ln k_i}{\partial T}\right)_P = \frac{\bar{H}^\circ_{i,\text{vap}} - \bar{H}^\circ_i}{RT^2} = \frac{\bar{H}^\circ_{i,\text{vap}} - \bar{H}^\infty_i}{RT^2} \tag{9.119}$$

$$\left(\frac{\partial \ln k_i}{\partial P}\right)_T = \frac{\bar{V}^\circ_i}{RT} = \frac{\bar{V}^\infty_i}{RT} \tag{9.120}$$

9.22 For O_2 in water, the Henry's law constant is 2.95×10^7 torr at 20°C and 3.52×10^7 torr at 30°C. (*a*) Does the solubility of O_2 in water increase or decrease from 20 to 30°C? (*b*) Use (9.119) to estimate $\bar{H}^\circ_i - \bar{H}^\circ_{i,\text{vap}}$ for O_2 in water for the range 20 to 30°C. (*c*) Use $k_i = 3.24 \times 10^7$ to calculate $\bar{G}^\circ_i - \bar{G}^\circ_{i,\text{vap}}$ for O_2 in water at 25°C. (*d*) Calculate $\bar{S}^\circ_i - \bar{S}^\circ_{i,\text{vap}}$ for O_2 in water at 25°C.

9.23 Derive (9.106) to (9.108).

9.24 Show that for an ideally dilute solution

$$\Delta V_{\text{mxg}} = \sum_{i \neq A} n_i(\bar{V}^\circ_i - \bar{V}^*_i)$$

and

$$\Delta H_{\text{mxg}} = \sum_{i \neq A} n_i(\bar{H}^\circ_i - \bar{H}^*_i)$$

9.25 Derive the equations of (9.75) from (9.63) and (9.64).

9.26 Consider a hypothetical solution where the chemical potential of each component is given by $\mu_i = \mu_i^*(T, P) + Tf(x_1, x_2, \ldots)$, where μ_i^* is the chemical potential of pure i and f is a certain function of the mole fractions. (*a*) Find ΔG_{mxg}. (*b*) Show that $\Delta H_{\text{mxg}} = 0$ and $\Delta V_{\text{mxg}} = 0$. (Note that since f is not necessarily $R \ln x_i$, this solution is not ideal; however, it still has ΔH and ΔV of mixing equal to zero.) (*c*) Assume ideal vapor in equilibrium with the solution, and show that the partial pressure of i vapor above the solution is $P_i = e^{f/R}P_i^*$.

9.27 From Fig. 9.9*b*, estimate k_i for acetone in CS_2 and for CS_2 in acetone at 29°C.

9.28 For the partial molar quantities $\bar{G}_i$, $\bar{H}_i$, and $\bar{S}_i$, we can find only relative, not absolute, values. Can we find absolute values for partial molar specific heats $\bar{C}_{P,i}$?

9.29 For a certain two-component solution, the Gibbs energy is known to be

$$G = n_A\mu_A^*(T, P) + n_B\mu_B^*(T, P) + RT(n_A \ln x_A + n_B \ln x_B) + C(T, P)\frac{n_A n_B^2 - n_A^2 n_B}{(n_A + n_B)^2}$$

where C is a certain function of T and P. (*a*) Derive an expression for μ_A. *Hint*: Use the definition of μ_A; don't overlook the fact that both x_A and x_B are functions of n_A. (*b*) Derive an expression for μ_B. (*c*) Derive an expression for μ_B using a different method from that used in part (*b*).

9.30 Would a liquid mixture of the two optical isomers of CHFClBr be an ideal solution? Explain.

9.31 Show that $\Delta C_{P,\text{mxg}} = 0$ for an ideal solution.

9.32 Let solvents α and β form two separate phases when brought in contact. Let a small amount of solute i be dissolved in these phases to form two ideally dilute solutions in equilibrium with each other. (*a*) Show that the mole-fraction ratio x_i^α/x_i^β of i in the two phases is

$$x_i^\alpha/x_i^\beta = \exp\left[(\mu_i^{\circ,\beta} - \mu_i^{\circ,\alpha})/RT\right] \equiv N(T, P) \tag{9.121}$$

This is the Nernst distribution law. The quantity $N(T, P)$ is the *partition coefficient* for solute i in the solvents α and β. (Recall happy times spent shaking separatory funnels in organic chemistry lab.) (*b*) Show that $N(T, P) = k_i^\beta/k_i^\alpha$, where the k's are the Henry's law constants for i in the solvents α and β.

9.33 For the dissociation of N_2O_4 to NO_2 in dilute chloroform solutions, K_x is approximately 5×10^{-7} at 0°C and 1 atm. If 1.00 g of N_2O_4 is dissolved in 200 g of $CHCl_3$ at 0°C and 1 atm, find the equilibrium mole fractions of N_2O_4 and NO_2.

9.34 At 100°C the vapor pressures of hexane and octane are 1836 and 354 torr, respectively. A certain liquid mixture of these two compounds has a vapor pressure of 666 torr at 100°C. Find the mole fractions in the liquid mixture. (Assume an ideal solution.)

9.35 Derive (9.116).

9.36 (*a*) Show that

$$x_i = \frac{m_i M_A}{1 + M_A \sum_{j \neq A} m_j} \tag{9.122}$$

where A is the solvent, the sum goes over all species except the solvent, and the remaining symbols are defined in Prob. 9.1. *Hint:* Note that the total number of moles is $n = n_A + \sum_{j \neq A} n_j$. (*b*) Show that

$$x_i = \frac{c_i}{\left(\rho - \sum\limits_{j \neq A} c_j M_j\right) \Big/ M_A + \sum\limits_{j \neq A} c_j} \tag{9.123}$$

where the symbols are defined in Prob. 9.1. *Hint:* First show that $x_i = c_i/(c_A + \sum_{j \neq A} c_j)$. Then write down an expression for ρ and solve it for c_A.

9.37 Consider an ideal gas mixture at T and P; show that for component i, $\mu_i = \mu_i^*(T, P) + RT \ln x_i$. Therefore [Eqs. (9.63) and (9.64)] an ideal gas mixture is an ideal solution, and Eqs. (9.71) to (9.74) hold for an ideal gas mixture. (Of course, an ideal solution is not necessarily an ideal gas mixture. Note also the different choice of standard state for an ideal-solution component and an ideal-gas-mixture component.)

9.38 Let phases α and β, each composed of liquids 1 and 2, be in equilibrium with each other. Show that if substances 1 and 2 form ideal solutions, then $x_1^\alpha = x_1^\beta$ and $x_2^\alpha = x_2^\beta$. Therefore, the two phases have the same composition and are actually one phase. Hence, liquids that form ideal solutions are miscible in all proportions.

9.39 The normal boiling points of benzene and toluene are 80.1 and 110.6°C, respectively. Both liquids obey Trouton's rule well. For a benzene–toluene liquid solution at 120°C with $x(C_6H_6) = 0.68$, estimate the vapor pressure and $x_{vap}(C_6H_6)$. State any approximations made. (The experimental values are 2.38 atm and 0.79.)

9.40 A solution of hexane and heptane at 30°C with hexane mole fraction 0.305 has a vapor pressure of 95.0 torr and a vapor-phase hexane mole fraction of 0.555. Find the vapor pressures of pure hexane and heptane at 30°C. State any approximations made.

10

NONIDEAL MIXTURES

10.1 ACTIVITIES AND ACTIVITY COEFFICIENTS

In Secs. 10.1 to 10.3 we consider nonideal liquid and solid solutions of nonelectrolytes. Electrolyte solutions are treated in Secs. 10.4 to 10.8. Nonideal gas mixtures are considered in Sec. 10.9.

Once we know G of a phase as a function of temperature, pressure, and composition, we know everything. From $(\partial G/\partial T)_{P,n_i} = -S$, we get the entropy. From $G = H - TS$, we get the enthalpy once G and S are known. From $(\partial H/\partial T)_{P,n_i} = C_P$, we get the heat capacity once H is known. From $(\partial G/\partial P)_{T,n_i} = V$, we get V as a function of T, P, and composition: $V = V(T, P, n_1, n_2, \ldots)$, which is the equation of state of the phase. From $\mu_i = (\partial G/\partial n_i)_{T,P,n_{j\neq i}}$, we get μ_i for each component; partial differentiation of μ_i gives $\bar{V}_i$ and $\bar{S}_i$; from μ_i and $\bar{S}_i$, we get $\bar{H}_i$. Knowing G, we know everything.

Equation (9.25) gives $G = \sum_i n_i \mu_i$ for a solution. To know G of the solution, we therefore want expressions for the chemical potentials μ_i as functions of T, P, and the mole fractions.

For an ideal solution or an ideally dilute solution, we have from Eq. (9.111)

$$\mu_i = \mu_i^\circ + RT \ln x_i \qquad \text{ideal or ideally dil. soln.} \qquad (10.1)^*$$

where μ_i° is the chemical potential in the appropriately defined standard state. Equation (10.1) gives $\ln x_i = (\mu_i - \mu_i^\circ)/RT$, or

$$x_i = \exp\,[(\mu_i - \mu_i^\circ)/RT] \qquad \text{ideal or ideally dil. soln.} \qquad (10.2)$$

A *nonideal solution* is defined as one that is neither ideal nor ideally dilute. For each component i of a nonideal solution, we choose a standard state and symbolize the chemical potential of i in the standard state by μ_i°. (Possible choices of standard states are discussed below.) We then define the *activity* a_i of component i in any solution by

$$a_i \equiv \exp\,[(\mu_i - \mu_i^\circ)/RT] \qquad \text{any soln.} \qquad (10.3)$$

Note the resemblance of (10.3) to (10.2). Taking logs of (10.3), we get $\ln a_i = (\mu_i - \mu_i^\circ)/RT$, or

$$\mu_i = \mu_i^\circ + RT \ln a_i \qquad \text{any soln.} \qquad (10.4)^*$$

Thus, the activity a_i replaces the mole fraction x_i in the expression for μ_i in a nonideal solution. From (10.1) and (10.4) [or from (10.2) and (10.3)] we have

$$a_i = x_i \qquad \text{ideal or ideally dil. soln.} \tag{10.5}$$

For any solution, we define the *activity coefficient* γ_i (gamma i) of component i as

$$\gamma_i \equiv a_i/x_i \qquad \text{any soln.} \tag{10.6}$$

$$a_i = \gamma_i x_i \tag{10.7}^*$$

The activity coefficient γ_i measures the degree of departure of component i's behavior from ideal or ideally dilute behavior. The activity a_i can be viewed as being obtained from the mole fraction x_i by correcting for nonideality. Equations (10.5) and (10.7) give

$$\gamma_i = 1 \qquad \text{ideal or ideally dil. soln.} \tag{10.8}$$

From (10.4) and (10.7), the chemical potentials in a nonideal solution are

$$\mu_i = \mu_i^\circ + RT \ln \gamma_i x_i \tag{10.9}$$

Since μ_i generally depends on T, P, and all the mole fractions, the activity a_i in (10.3) and the activity coefficient γ_i in (10.6) depend on these variables:

$$a_i = a_i(T, P, x_1, x_2, \ldots), \qquad \gamma_i = \gamma_i(T, P, x_1, x_2, \ldots)$$

The task of thermodynamics is to show how a_i and γ_i can be found from experimental data; see Sec. 10.2. (The task of statistical mechanics is to find a_i and γ_i from the intermolecular interactions in the solution.)

We now consider the choice of standard states for components of nonideal solutions. There are two different standard-state conventions used with Eq. (10.9).

Convention I. For a solution where the mole fractions of all components vary over a considerable range, one usually uses Convention I. The most common case is a solution of two or more liquids (e.g., ethanol plus water). The Convention I standard state of every solution component i is taken as pure liquid i at the temperature and pressure of the solution:

$$\mu_{\mathrm{I},i}^\circ \equiv \mu_i^*(T, P) \qquad \text{for all components} \tag{10.10}$$

where we added the subscript I to indicate the Convention I choice of standard states. Equation (10.10) is the same convention as used for ideal solutions in Sec. 9.6.

What does the Convention I choice of standard states imply about the activity coefficients? From (10.9) and (10.10), we have $\mu_i = \mu_i^* + RT \ln \gamma_i x_i$. As $x_i \to 1$ at constant T and P, the chemical potential μ_i goes to μ_i^*, since the solution becomes pure i. Hence, the $x_i \to 1$ limit of this last equation is $\mu_i^* = \mu_i^* + RT \ln \gamma_i$ or $\ln \gamma_i = 0$ and $\gamma_i = 1$:

$$\lim_{x_i \to 1} \gamma_{\mathrm{I},i} = 1 \qquad \text{for all } i \tag{10.11}$$

where the subscript I indicates a Convention I activity coefficient. We should have expected (10.11) to hold, since in the limit as $x_i \to 1$, the solution becomes an ideally dilute solution with i as solvent; moreover, the choice of solvent standard state for an ideally dilute solution [Eq. (9.98)] is the same as (10.10) for i in the nonideal solution. Hence, in the limit as $x_i \to 1$, the ideally dilute equation $\mu_i = \mu_i^\circ + RT \ln x_i$ should hold for component i. This means that γ_i in (10.9) must become 1 as x_i goes to 1.

Since the Convention I standard state of i is pure i, we have the following expressions for thermodynamic properties in the Convention I standard state:

$$\mu_{\mathrm{I},i}^\circ = \mu_i^*(T, P) = \bar{G}_i^*(T, P) \tag{10.12}$$

$$\bar{S}_{\mathrm{I},i}^\circ = \bar{S}_i^*(T, P), \qquad \bar{H}_{\mathrm{I},i}^\circ = \bar{H}_i^*(T, P), \qquad \bar{V}_{\mathrm{I},i}^\circ = \bar{V}_i^*(T, P) \tag{10.13}$$

for each component i. Note that Convention I puts all the components on the same footing and does not single out one component as the solvent.

Convention II. One uses this convention when one wants to treat one solution component (called the *solvent*) differently from the other components (the solutes). Common cases are solutions of solids or gases in a liquid solvent. The Convention II standard state of the solvent A is pure liquid A at the temperature and pressure of the solution:

$$\mu_{\mathrm{II,A}}^\circ = \mu_{\mathrm{A}}^*(T, P) \tag{10.14}$$

Substituting (10.14) in (10.9) and taking the limit as $x_{\mathrm{A}} \to 1$, we get

$$\lim_{x_{\mathrm{A}} \to 1} \gamma_{\mathrm{II,A}} = 1 \tag{10.15}$$

From (10.14) it follows that

$$\bar{S}_{\mathrm{II,A}}^\circ = \bar{S}_{\mathrm{A}}^*(T, P), \qquad \bar{H}_{\mathrm{II,A}}^\circ = \bar{H}_{\mathrm{A}}^*(T, P), \qquad \bar{V}_{\mathrm{II,A}}^\circ = \bar{V}_{\mathrm{A}}^*(T, P)$$

For each solute $i \neq \mathrm{A}$, Convention II chooses the standard state so that

$$\lim_{x_{\mathrm{A}} \to 1} \gamma_{\mathrm{II},i} = 1 \qquad \text{for } i \neq \mathrm{A} \tag{10.16}$$

In the limit of infinite dilution ($x_{\mathrm{A}} \to 1$), each Convention II solute activity coefficient approaches 1. Note that the limit in (10.16) is taken as the solvent mole fraction x_{A} goes to 1 (and hence $x_i \to 0$), which is quite different from (10.11), where the limit is taken as $x_i \to 1$. To arrive at a Convention II choice of standard state for each solute that is consistent with (10.16), we use the following considerations. Setting μ_i in (10.9) equal to μ_i°, we get $0 = RT \ln \gamma_i x_i$, so that $\gamma_{\mathrm{II},i} x_i$ must equal 1 for the standard state. When x_{A} is near 1 and the solute mole fractions are small, then by (10.16) the activity coefficient $\gamma_{\mathrm{II},i}$ is close to 1. We choose the standard state of each solute i as the *fictitious* state obtained as follows. We pretend that the behavior of μ_i that holds in the limit of infinite dilution (namely, $\mu_i = \mu_i^\circ + RT \ln x_i$) holds for all values of x_i, and we take the limit as $x_i \to 1$; this gives a fictitious standard state with $\gamma_i = 1$ and $x_i = 1$. (This is the same standard state as used in Sec. 9.8 for solutes in an ideally dilute solution.)

To get an explicit expression for $\mu_{\mathrm{II},i}^\circ$, we take the limit of (10.9) as $x_{\mathrm{A}} \to 1$ and $x_i \to 0$. Using (10.16), we have

$$\mu_{\mathrm{II},i}^\circ(T, P) = \lim_{x_{\mathrm{A}} \to 1} (\mu_i - RT \ln \gamma_{\mathrm{II},i} x_i) = \lim_{x_{\mathrm{A}} \to 1} (\mu_i - RT \ln x_i)$$

$$\mu_{\mathrm{II},i}^\circ = (\mu_i - RT \ln x_i)^\infty \qquad \text{for } i \neq \mathrm{A} \tag{10.17}$$

where the ∞ superscript indicates infinite dilution. Partial differentiation of (10.17) gives expressions for the Convention II solute standard-state thermodynamic properties as follows:

$$\bar{V}^\circ_{\mathrm{II},i} = (\partial \mu_i^\circ/\partial P)_T = (\partial \mu_i^\infty/\partial P)_T = \bar{V}_i^\infty$$

$$\bar{S}^\circ_{\mathrm{II},i} = -(\partial \mu_i^\circ/\partial T)_P = -[(\partial \mu_i/\partial T)_{P,x_j} - R \ln x_i]^\infty = (\bar{S}_i + R \ln x_i)^\infty$$

$$\bar{H}^\circ_{\mathrm{II},i} = \mu_i^\circ + T\bar{S}_i^\circ = (\mu_i - RT \ln x_i + T\bar{S}_i + RT \ln x_i)^\infty = (\mu_i + T\bar{S}_i)^\infty = \bar{H}_i^\infty$$

$$\bar{V}^\circ_{\mathrm{II},i} = \bar{V}_i^\infty, \qquad \bar{H}^\circ_{\mathrm{II},i} = \bar{H}_i^\infty, \qquad \bar{S}^\circ_{\mathrm{II},i} = (\bar{S}_i + R \ln x_i)^\infty \qquad \text{for } i \neq \mathrm{A} \quad (10.18)$$

The Convention II equations (10.17) and (10.18) for nonideal solutions are the same as equations in Sec. 9.8 for ideally dilute solutions.

Convention I uses the same choice of standard states as for an ideal solution. Hence, the departures of $\gamma_{\mathrm{I},i}$ from 1 measure the departure of the solution's behavior from ideal-solution behavior. Convention II uses the same choice of standard states as for an ideally dilute solution. Hence the departures of $\gamma_{\mathrm{II},i}$ and $\gamma_{\mathrm{II,A}}$ from 1 measure the departure of the solution's behavior from ideally dilute behavior.

Note that a_i and γ_i are dimensionless.

When solution component i is in its standard state, μ_i equals μ_i° and, from (10.3), its activity a_i equals 1 ($a_i^\circ = 1$). The difference between a_i and 1 is a measure of the difference between μ_i and μ_i°.

The thermodynamic properties of a solution of liquids are often expressed in terms of excess functions. The *excess Gibbs energy* G^E of a mixture of liquids is defined as the difference between the actual Gibbs energy G of the solution and the Gibbs energy G_{id} of a hypothetical ideal solution at the same T, P, and composition: $G^E \equiv G - G_{\mathrm{id}}$. Similar definitions hold for the excess entropy S^E, excess enthalpy H^E, and excess volume V^E. See Prob. 10.6.

The concepts of activity and activity coefficient were introduced by the American chemist G. N. Lewis in the early 1900s. (Recall Lewis dot structures, the Lewis octet rule, Lewis acids and bases.) Lewis spent the early part of his career at Harvard and M.I.T. In 1912, he became head of the chemistry department at the University of California at Berkeley. In 1916 he proposed that a chemical bond consists of a shared pair of electrons, a novel idea at the time. He measured ΔG_f° for many compounds and cataloged the available free-energy data, drawing the attention of chemists to the usefulness of such data. The concept of partial molar quantities is due to Lewis. His 1923 book *Thermodynamics* (written with Merle Randall) made thermodynamics accessible to chemists. Lewis was resentful that his early speculative ideas on chemical bonding and the nature of light were not appreciated by Harvard chemists, and in 1929 he refused an honorary degree from Harvard. His later years were spent working on relativity and photochemistry.

10.2 DETERMINATION OF ACTIVITIES AND ACTIVITY COEFFICIENTS

The formalism of Sec. 10.1 leads nowhere unless we can determine activity coefficients. One way to do this is by vapor-pressure measurements. Consider, for example, a solution of several liquids. We adopt Convention I. We shall assume the vapor in equilibrium with the solution to be an ideal mixture. Departures from ideality in gases are ordinarily much less than in liquids. (In precise work, gas nonideality is allowed for by replacing the partial pressures P_i in the equations of this section by fugacities f_i; fugacities are defined in Sec. 10.9.) Recall that for an ideal solution we started from $\mu_i = \mu_i^\circ + RT \ln x_i$ and derived Raoult's law $P_i = x_i P_i^*$ (Sec. 9.6). For a real solution, the activity replaces the mole fraction in μ_i, and $\mu_i = \mu^\circ_{\mathrm{I},i} + RT \ln a_{\mathrm{I},i}$; also, the Convention I standard states are the same as the ideal-solution standard states. Hence, exactly the same steps that gave Raoult's law in Sec. 9.6 will give for a nonideal solution

$$P_i = a_{\text{I},i} P_i^* \qquad \text{ideal vapor, } \mu_i^* \text{ ind. of } P \tag{10.19*}$$

where the subscript I on the activity indicates the Convention I choice of standard states. [From (10.3) it is clear that a_i depends on the choice of standard states.] Thus $a_{\text{I},i} = P_i/P_i^*$, where P_i is the equilibrium partial vapor pressure of i above the solution and P_i^* is the vapor pressure of pure i at the temperature of the solution. Since $a_{\text{I},i} = \gamma_{\text{I},i} x_i$, we have

$$P_i = \gamma_{\text{I},i} x_i P_i^* \qquad \text{or} \qquad x_{i,\text{vap}} P = \gamma_{\text{I},i} x_i P_i^* \tag{10.20}$$

where x_i is the mole fraction of i in the liquid (or solid) solution, $x_{i,\text{vap}}$ is its mole fraction in the vapor above the solution, and P is the total vapor pressure of the solution. To find a_i and γ_i, we just measure the solution vapor pressure and chemically analyze the vapor to find $x_{i,\text{vap}}$. Equation (10.20) is Raoult's law modified to allow for solution nonideality.

Example For acetone–chloroform solutions at 35.2°C, acetone vapor-phase mole fractions $x_{\text{vap}}(C_3H_6O)$ and total vapor pressures P as functions of the liquid-phase acetone mole fraction $x(C_3H_6O)$ are given in Table 10.1. (See also Fig. 9.9*a*.) Find the Convention I activity coefficients at 35.2°C. (For liquid solutions, the pressure dependence of the activity coefficients is weak, and it is not important to specify the pressure unless one is working at high pressures.)

Let Ac and Chl stand for acetone and chloroform. For $x(\text{Ac}) = 0.0821$, use of (10.20) gives

$$\gamma_{\text{I}}(\text{Ac}) = \frac{x_{\text{vap}}(\text{Ac})P}{x(\text{Ac})P^*(\text{Ac})} = \frac{0.0500(279.5 \text{ torr})}{0.0821(344.5 \text{ torr})} = 0.494$$

$$\gamma_{\text{I}}(\text{Chl}) = \frac{x_{\text{vap}}(\text{Chl})P}{x(\text{Chl})P^*(\text{Chl})} = \frac{0.9500(279.5 \text{ torr})}{0.9179(293 \text{ torr})} = 0.987$$

Similar treatment of the other data and use of (10.11) give

$x(\text{Ac})$	0	0.082	0.200	0.336	0.506	0.709	0.815	0.940	1
$\gamma_{\text{I}}(\text{Ac})$		0.494	0.544	0.682	0.824	0.943	0.981	0.997	1
$\gamma_{\text{I}}(\text{Chl})$	1	0.987	0.957	0.875	0.772	0.649	0.588	0.536	
$x(\text{Chl})$	1	0.918	0.800	0.664	0.494	0.291	0.185	0.060	0

Table 10.1. Vapor pressures and vapor compositions for acetone–chloroform solutions at 35.2°C

$x(C_3H_6O)$	$x_{\text{vap}}(C_3H_6O)$	P/torr	$x(C_3H_6O)$	$x_{\text{vap}}(C_3H_6O)$	P/torr
0.0000	0.0000	293	0.6034	0.6868	267
0.0821	0.0500	$279._5$	0.7090	0.8062	286
0.2003	0.1434	262	0.8147	0.8961	307
0.3365	0.3171	249	0.9397	0.9715	332
0.4188	0.4368	248	1.0000	1.0000	$344._5$
0.5061	0.5625	255			

The limiting values of $\gamma_I(\text{Ac})$ at $x(\text{Ac}) = 0$ and $\gamma_I(\text{Chl})$ at $x(\text{Chl}) = 0$ can be obtained by plotting the calculated γ's vs. mole fraction and extrapolating to $x(\text{Ac}) = 0$ and to $x(\text{Chl}) = 0$. Figure 10.1 shows the activity coefficients as functions of solution composition.

The large negative deviations of γ_I from 1 (and the corresponding large negative deviations in Fig. 9.9*a* from Raoult's law) indicate large deviations from ideal-solution behavior. This deviation can perhaps be attributed to hydrogen bonding between acetone and chloroform according to

$$\text{Cl}_3\text{C—H} \cdots \text{O=C(CH}_3)_2$$

The hydrogen bonding makes the acetone–chloroform intermolecular interaction stronger than the acetone–acetone and chloroform–chloroform interactions. The hydrogen bonding thus makes ΔH_{mxg} negative, compared with zero for formation of an ideal solution. Also, the hydrogen bonding gives a significant degree of order in the mixture, making ΔS_{mxg} less than for formation of an ideal solution. The enthalpy effect turns out to outweigh the entropy effect, and $\Delta G_{\text{mxg}} = \Delta H_{\text{mxg}} - T\,\Delta S_{\text{mxg}}$ is less than ΔG_{mxg} for formation of an ideal solution.

Now suppose we want to use Convention II. The standard states in Convention II are the same as for an ideally dilute solution. Whereas $\mu_i = \mu_i^\circ + RT \ln x_i$ for an ideally dilute solution, we have $\mu_i = \mu_{\text{II},i}^\circ + RT \ln a_{\text{II},i}$ in a nonideal solution. Hence, exactly the same steps that led to Henry's law (9.102) for the solutes and Raoult's law (9.103) for the solvent lead to modified forms of these laws with x_i replaced by $a_{\text{II},i}$ for nonideal solutions and Convention II. Therefore

$$P_i = k_i a_{\text{II},i} = k_i \gamma_{\text{II},i} x_i \qquad \text{for } i \neq \text{A, ideal vapor} \tag{10.21}$$

$$P_{\text{A}} = a_{\text{II,A}} P_{\text{A}}^* = \gamma_{\text{II,A}} x_{\text{A}} P_{\text{A}}^* \qquad \text{ideal vapor, } \mu_{\text{A}}^* \text{ ind. of } P \tag{10.22}$$

To apply (10.21), we need the Henry's law constant k_i; this can be determined by measurements on very dilute solutions where $\gamma_{\text{II},i} = 1$. Thus, vapor-pressure measurements give the Convention II activities and activity coefficients.

Example Find the Convention II activity coefficients at 35.2°C for acetone–chloroform solutions, taking acetone as the solvent.

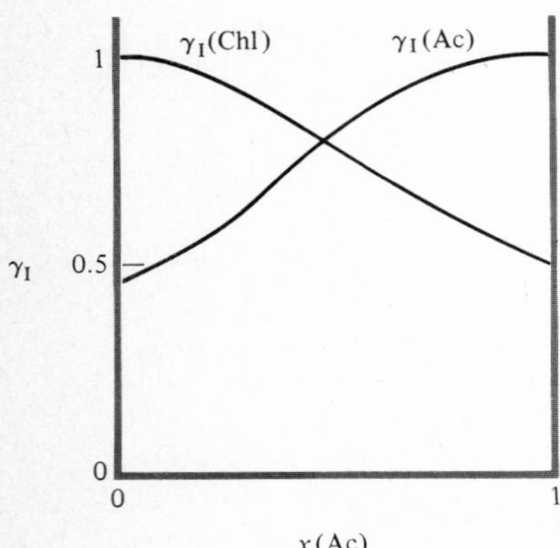

Figure 10.1
Activity coefficients in acetone–chloroform solutions at 35°C.

Ordinarily, one would use Convention I for acetone–chloroform solutions, but for illustrative purposes we use Convention II. Comparison of (10.21) and (10.22) with (10.20) gives

$$\gamma_{\text{II,A}} = \gamma_{\text{I,A}} \quad \text{and} \quad \gamma_{\text{II},i} = (P_i^*/k_i)\gamma_{\text{I},i}, \quad \text{for } i \neq \text{A, ideal vap.} \tag{10.23}$$

Since we are taking acetone as the solvent, we have $\gamma_{\text{II}}(\text{Ac}) = \gamma_{\text{I}}(\text{Ac})$, where the $\gamma_{\text{I}}(\text{Ac})$ values were calculated in the previous example.

To find γ_{II} for the solute chloroform from (10.23), we need the Henry's law constant k_i for chloroform. In the limit of infinite dilution ($x_{\text{A}} \to 1$ and $x_i \to 0$), Eq. (10.21) becomes $k_i = (P_i/x_i)^\infty$ (Henry's law); the quantity $(P_i/x_i)^\infty$ for chloroform can be found by plotting $P(\text{Chl})/x(\text{Chl})$ and extrapolating to $x(\text{Chl}) = 0$. Alternatively, k_i can be found from the plot in Fig. 10.1, as follows. At infinite dilution Eq. (10.23) becomes $k_i = (\gamma_{\text{I},i})^\infty P_i^*$, where (10.16) was used. Substitution into (10.23) gives $\gamma_{\text{II},i} = \gamma_{\text{I},i}/(\gamma_{\text{I},i})^\infty$. Figure 10.1 gives $\gamma_{\text{I}}(\text{Chl}) = 0.50$ in the limit $x(\text{Ac}) = 1$. Hence, $\gamma_{\text{II}}(\text{Chl}) = 2.0_0\, \gamma_{\text{I}}(\text{Chl})$. Use of $\gamma_{\text{I}}(\text{Chl})$ values found in the previous example gives:

$\gamma_{\text{II}}(\text{Chl})$	2.00	1.97	1.91	1.75	1.54	1.30	1.18	1.07	1
$x(\text{Chl})$	1	0.918	0.800	0.664	0.494	0.291	0.185	0.060	0

For a solution of a solid in a liquid solvent, the vapor partial pressure of the solute over the solution is generally immeasurably small and cannot be used to find the solute's activity coefficient. Measurement of the vapor pressure as a function of solution composition gives P_{A}, the solvent partial pressure, and hence allows calculation of the solvent activity coefficients γ_{A} as a function of composition. We then use the integrated Gibbs–Duhem equation to find solute activity coefficients γ_{B}. After division by $n_{\text{A}} + n_{\text{B}}$ the Gibbs–Duhem equation (9.43) gives

$$x_{\text{A}}\, d\mu_{\text{A}} + x_{\text{B}}\, d\mu_{\text{B}} = 0 \qquad \text{const. } T, P \tag{10.24}$$

From (10.9) we have $\mu_{\text{A}} = \mu_{\text{A}}^\circ(T, P) + RT \ln \gamma_{\text{A}} + RT \ln x_{\text{A}}$ and

$$d\mu_{\text{A}} = RT\, d \ln \gamma_{\text{A}} + (RT/x_{\text{A}})\, dx_{\text{A}} \qquad \text{const. } T, P \tag{10.25}$$

Similarly, $d\mu_{\text{B}} = RT\, d \ln \gamma_{\text{B}} + (RT/x_{\text{B}})\, dx_{\text{B}}$ at constant T and P. Substitution for $d\mu_{\text{A}}$ and $d\mu_{\text{B}}$ in (10.24) gives after division by RT:

$$x_{\text{A}}\, d \ln \gamma_{\text{A}} + dx_{\text{A}} + x_{\text{B}}\, d \ln \gamma_{\text{B}} + dx_{\text{B}} = 0 \qquad \text{const. } T, P \tag{10.26}$$

Since $x_{\text{A}} + x_{\text{B}} = 1$, we have $dx_{\text{A}} + dx_{\text{B}} = 0$, and

$$x_{\text{A}}\, d \ln \gamma_{\text{A}} + x_{\text{B}}\, d \ln \gamma_{\text{B}} = 0 \qquad \text{const. } T, P \tag{10.27}$$

$$d \ln \gamma_{\text{B}} = -(x_{\text{A}}/x_{\text{B}})\, d \ln \gamma_{\text{A}}$$

Integrating between states 1 and 2, and choosing Convention II, we get

$$\ln \gamma_{\text{II,B},2} - \ln \gamma_{\text{II,B},1} = -\int_1^2 \frac{x_{\text{A}}}{1 - x_{\text{A}}}\, d \ln \gamma_{\text{II,A}} \qquad \text{const. } T, P \tag{10.28}$$

We choose state 1 as pure solvent A. Therefore from (10.15) and (10.16), $\gamma_{\text{II,B},1} = 1$, $\gamma_{\text{II,A},1} = 1$, $\ln \gamma_{\text{II,B},1} = 0$, $\ln \gamma_{\text{II,A},1} = 0$. We plot $x_{\text{A}}/(1 - x_{\text{A}})$ vs. $\ln \gamma_{\text{II,A}}$; the area under the curve from $x_{\text{A}} = 1$ to $x_{\text{A}} = x_{\text{A},2}$ gives $-\ln \gamma_{\text{II,B},2}$. Note that the integrand

$x_A/(1 - x_A)$ becomes infinite as $x_A \to 1$; even though the curve goes to infinity as $x_A \to 1$, the area under the curve is finite. To avoid the waste of graph paper entailed with a curve that goes to infinity, one starts the integration not at $x_A = 1$ but at $x_A = 1 - \varepsilon$, where ε is sufficiently small for $\gamma_{II,B}$ to be essentially 1 at $x_A = 1 - \varepsilon$.

Some activity coefficients for aqueous solutions of sucrose at 25°C (calculated from vapor-pressure measurements) are:

$x(H_2O)$	0.999	0.995	0.980	0.960	0.930	0.900
$\gamma_{II}(H_2O)$	1.0000	0.9999	0.998	0.990	0.968	0.939
$\gamma_{II}(C_{12}H_{22}O_{11})$	1.009	1.047	1.231	1.58	2.31	3.23

Note from the equation following (10.27) that the sucrose activity coefficient must increase when the water activity coefficient decreases (at constant T and P). Because of the large size of a sucrose molecule (molecular weight 342) compared with a water molecule, the mole-fraction values can mislead one into thinking that a solution is more dilute than it actually is. For example, an aqueous sucrose solution with $x(\text{sucrose}) = 0.10$ is 68 percent sucrose by weight and is extremely concentrated; even though only 1 molecule in 10 is sucrose, the large size of sucrose molecules makes it highly likely for a given sucrose molecule to be close to several other sucrose molecules. Hence the large deviation of $\gamma_{II}(\text{sucrose})$ from 1 at this composition.

From the measured activity coefficients, one can calculate relative values of partial molar Gibbs energies μ_i and entropies $\bar{S}_i$; see Sec. 10.8.

Another method of finding activity coefficients will be discussed in Sec. 14.11.

10.3 ACTIVITY COEFFICIENTS ON THE MOLALITY AND CONCENTRATION SCALES

So far in this chapter, we have expressed solution compositions using mole fractions and written the chemical potential of each solution component as $\mu_i = \mu_i^\circ + RT \ln \gamma_i x_i$, where the standard state is chosen according to Convention I or II of Sec. 10.1. Although mole fractions have the greatest theoretical significance, it is common practice with solutions of solids or gases in a liquid to express solute chemical potentials in terms of solute molalities m_i.

To express μ_i in terms of m_i, we start with Eq. (10.9) for μ_i. Since we are singling out one component (A) as the solvent, we adopt Convention II and write (10.9) as

$$\mu_i = \mu_{II,i}^\circ + RT \ln \gamma_{II,i} x_i \tag{10.29}$$

Equation (9.3) gives $m_i = n_i/n_A M_A$; division of numerator and denominator by the total number of moles gives $m_i = x_i/x_A M_A$ or $x_i = m_i x_A M_A$. Therefore, (10.29) becomes

$$\mu_i = \mu_{II,i}^\circ + RT \ln (\gamma_{II,i} m_i x_A M_A m^\circ/m^\circ)$$

where, to keep subsequent equations dimensionally correct, we multiplied and divided the argument of the logarithm by m°, where m° is defined by

$$m^\circ \equiv 1 \text{ mol/kg} \tag{10.30}$$

We have

$$\mu_i = \mu^\circ_{\text{II},i} + RT \ln M_\text{A} m^\circ + RT \ln (x_\text{A} \gamma_{\text{II},i} m_i/m^\circ)$$

We now define $\mu^\circ_{m,i}$ and $\gamma_{m,i}$ as

$$\mu^\circ_{m,i} \equiv \mu^\circ_{\text{II},i} + RT \ln M_\text{A} m^\circ \tag{10.31}$$

$$\gamma_{m,i} \equiv x_\text{A} \gamma_{\text{II},i} \tag{10.32}$$

[Note that $M_\text{A} m^\circ$ is dimensionless. For example, for $\text{A} = H_2O$, we have $M_\text{A} m^\circ = (18 \text{ g/mol}) \times (1 \text{ mol/kg}) = 18/1000 = 0.018$.] With these definitions, μ_i becomes

$$\mu_i = \mu^\circ_{m,i} + RT \ln (\gamma_{m,i} m_i/m^\circ) \qquad \text{for } i \neq \text{A} \tag{10.33*}$$

$$\gamma_{m,i} \to 1 \text{ as } x_\text{A} \to 1 \tag{10.34}$$

where the limiting behavior of $\gamma_{m,i}$ follows from (10.32) and (10.16).

The motivation for the definitions (10.31) and (10.32) is to produce an expression for μ_i in terms of m_i that resembles the expression for μ_i in terms of x_i as closely as possible. Note the similarity between (10.33) and (10.29) and between (10.34) and (10.16). The definitions (10.31) and (10.32) are the only possible ones that give equations of the form (10.33) and (10.34). (See Prob. 10.9*b* for the proof of this statement.) We call $\gamma_{m,i}$ the *molality-scale activity coefficient* of solute i and $\mu^\circ_{m,i}$ the molality-scale standard-state chemical potential of i.

Although we use (10.33) for each solute, we retain the mole-fraction scale for the solvent:

$$\mu_\text{A} = \mu^\circ_\text{A} + RT \ln \gamma_\text{A} x_\text{A} \tag{10.35}$$

$$\mu^\circ_\text{A} = \mu^*_\text{A}(T, P), \qquad \gamma_\text{A} \to 1 \text{ as } x_\text{A} \to 1 \tag{10.36}$$

Note from (10.31) that $\mu^\circ_{m,i}$ is a function of T and P only: $\mu^\circ_{m,i} = \mu^\circ_{m,i}(T, P)$. Taking the limit of (10.33) as $x_\text{A} \to 1$ and using (10.34), we get

$$\mu^\circ_{m,i} = [\mu_i - RT \ln (m_i/m^\circ)]^\infty \tag{10.37}$$

Using either (10.37) or (10.31), we find (Prob. 10.8) the following standard-state values on the molality scale:

$$\bar{V}^\circ_{m,i} = \bar{V}^\infty_i, \qquad \bar{H}^\circ_{m,i} = \bar{H}^\infty_i, \qquad \bar{S}^\circ_{m,i} = (\bar{S}_i + R \ln m_i/m^\circ)^\infty \tag{10.38}$$

The activity coefficient γ_m in (10.32) has a peculiarity. For an ideally dilute solution, the Convention II mole-fraction activity coefficient γ_II equals 1 (Sec. 10.1). However, from (10.32), when γ_II equals 1, the molality activity coefficient γ_m is less than 1, due to the factor x_A, the solvent mole fraction. Thus, deviations of γ_m in (10.32) from 1 are due to two causes: (*a*) deviation of the solution from ideally dilute behavior; (*b*) deviation of x_A from 1. For this reason, it is preferable to use γ_II rather than γ_m in theoretical discussions. As an example, in an aqueous solution with the solute molality m_i equal to 1 mol/kg, we have

$$x_\text{A} = \frac{n_\text{A}}{n_\text{A} + n_i} = \frac{1}{1 + n_i/n_\text{A}} = \frac{1}{1 + m_i M_\text{A}} = \frac{1}{1.018} = 0.982$$

For reasonably dilute solutions (total solute molality less than, say, 1 mol/kg), the effect of x_A on γ_m is small.

Consider the molality-scale solute standard state. From (10.33), when μ_i equals $\mu^\circ_{m,i}$, we have $\gamma_{m,i} m_i/m^\circ = 1$. We shall take $m_i = m^\circ = 1$ mol/kg for the standard state (as is already implied in the notation m° for 1 mol/kg), and we must then have $\gamma_{m,i} = 1$ in the standard state. The molality-scale solute standard state will thus be the fictitious state with $m_i = 1$ mol/kg and $\gamma_{m,i} = 1$. Note that this standard state is not an ideally dilute solution since $\gamma_{\mathrm{II},i}$ $(= \gamma_{m,i}/x_\mathrm{A})$ exceeds 1 in the standard state.

People sometimes express the solute chemical potentials μ_i in terms of concentrations c_i rather than molalities. Instead of starting with (10.29) to find the expression for μ_i in terms of c_i, it is easier to work backward; we start with the desired expressions

$$\mu_i = \mu^\circ_{c,i} + RT \ln (\gamma_{c,i} c_i/c^\circ) \qquad \text{for } i \neq \mathrm{A} \tag{10.39}$$

$$\gamma_{c,i} \to 1 \text{ as } x_\mathrm{A} \to 1 \tag{10.40}$$

$$c^\circ \equiv 1 \text{ mol/dm}^3 \tag{10.41}$$

which have the same forms as (10.33), (10.34), and (10.30). To find the expressions for the activity coefficient $\gamma_{c,i}$ and the standard-state chemical potential $\mu^\circ_{c,i}$ on the concentration scale, we equate (10.39) and (10.29). The details of the derivation are left as a problem (Prob. 10.9). One finds

$$\mu^\circ_{c,i} = \mu^\circ_{\mathrm{II},i} + RT \ln \bar{V}^*_\mathrm{A} c^\circ \qquad \text{and} \qquad \gamma_{c,i} = (x_i/\bar{V}^*_\mathrm{A} c_i)\gamma_{\mathrm{II},i} \tag{10.42}$$

where $\bar{V}^*_\mathrm{A}$ is the solvent's molar volume. For the solvent, the mole-fraction scale is used [Eqs. (10.35) and (10.36)].

Because of the temperature dependence of c_i and $\bar{V}^*_\mathrm{A}$ in (10.42), the concentration-scale activity coefficients γ_c have even less theoretical significance than the molality-scale activity coefficients γ_m.

Equations (10.4), (10.33), and (10.39) give as the activities on the molality and concentration scales

$$a_{m,i} = \gamma_{m,i} m_i/m^\circ, \qquad a_{c,i} = \gamma_{c,i} c_i/c^\circ \tag{10.43}$$

which may be compared with (10.7).

In line with the notations γ_m and γ_c, the activity coefficient γ_{II} is often denoted by γ_x. The names *rational activity coefficient* and *practical activity coefficient* are sometimes used for γ_{II} and γ_m, respectively.

Some values of γ_{II}, γ_m, and γ_c for sucrose dissolved in water at 25°C and 1 atm are:

$x(H_2O)$	γ_{II}	γ_m	γ_c
0.995	1.047	1.042	1.103
0.980	1.23	1.21	1.50
0.960	1.58	1.51	2.26
0.930	2.31	2.15	4.07
0.900	3.23	2.91	6.78

Table 11.1 in Sec. 11.7 may help you keep straight the various conventions used for solution components.

10.4 SOLUTIONS OF ELECTROLYTES

An *electrolyte* is a substance that yields ions in solution, as evidenced by the solution's showing electrical conductivity. (A *polyelectrolyte* is an electrolyte that is a polymer.) For a given solvent, an electrolyte as classified as *weak* or *strong*, according to whether its solution is a poor or good conductor of electricity at moderate concentrations. For water as the solvent, some weak electrolytes are NH_3, CO_2, and CH_3COOH, and some strong electrolytes are NaCl, HCl, and $MgSO_4$.

An alternative classification, based on structure, is into true electrolytes and potential electrolytes. A *true electrolyte* consists of ions in the pure state. Most salts are true electrolytes. A crystal of NaCl, $CuSO_4$, or MgS consists of positive and negative ions. When an ionic crystal dissolves in a solvent, the ions break off from the crystal and go into solution as solvated ions. The term *solvated* indicates that each ion in solution is surrounded by a sheath of a few solvent molecules bound to the ion by electrostatic forces and traveling through the solution with the ion. Figure 10.2 shows solvated positive and negative ions.

Some salts of certain transition metals and Group IIIA and IVA metals have largely covalent bonding and are not true electrolytes. Thus, for $HgCl_2$, the interatomic distances in the crystal show the presence of $HgCl_2$ molecules (rather than Hg^{2+} and Cl^- ions), and $HgCl_2$ is largely molecular in aqueous solution (as evidenced by the low electrical conductivity). $HgCl_2$ is far more soluble in ethanol than in water. In contrast, $HgSO_4$, $Hg(NO_3)_2$, and HgF_2 are ionic. Aluminum bromide exists as covalent Al_2Br_6 molecules in the crystal and readily dissolves in benzene. $SnCl_4$ is covalent.

Figure 10.2
Hydration of ions in solution.

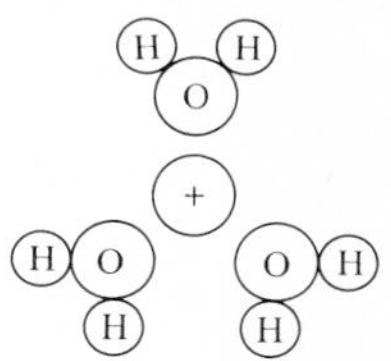

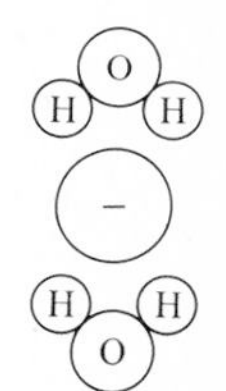

A *potential* electrolyte consists of uncharged molecules in the pure state, but when it is dissolved in a solvent, it reacts chemically with the solvent to some extent to yield ions. Thus, acetic acid reacts with water according to $HC_2H_3O_2 + H_2O \rightleftharpoons H_3O^+ + C_2H_3O_2^-$, yielding hydronium and acetate ions. Hydrogen chloride reacts with water according to $HCl + H_2O \rightleftharpoons H_3O^+ + Cl^-$. For the strong electrolyte HCl, the equilibrium lies far to the right. For the weak electrolyte acetic acid, the equilibrium lies far to the left (except in very dilute solutions).

In the pure liquid state, a true electrolyte is a good conductor of electricity. In contrast, a potential electrolyte is a poor conductor in the pure state.

Because of the strong, long-range forces between ions in solution, the use of activity coefficients in dealing with electrolyte solutions is essential, even for quite dilute solutions. Positive and negative ions occur together in solutions, and we cannot readily make observations on the positive ions alone to determine their activity. Hence, a special development of electrolyte activity coefficients is necessary. Our aim is to derive an expression for the chemical potential of an electrolyte in solution.

For simplicity, we consider a solution composed of a solvent A that is a nonelectrolyte, for example, H_2O or CH_3OH, and a single electrolyte that yields only two kinds of ions in solution, for example, Na_2SO_4, $MgCl_2$, or HNO_3 but not $KAl(SO_4)_2$. Let the electrolyte i have the formula $M_{\nu_+}X_{\nu_-}$, and let i yield the ions M^{z+} and X^{z-} in solution:

$$M_{\nu_+}X_{\nu_-} \rightarrow \nu_+ M^{z+} + \nu_- X^{z-} \tag{10.45}$$

For example, for $Ba(NO_3)_2$ and $BaSO_4$, we have:

$Ba(NO_3)_2$: $\quad M = Ba,\ X = NO_3;\quad \nu_+ = 1,\ \nu_- = 2;\quad z_+ = 2,\ z_- = -1$

$BaSO_4$: $\quad M = Ba,\ X = SO_4;\quad \nu_+ = 1,\ \nu_- = 1;\quad z_+ = 2,\ z_- = -2$

When $z_+ = 1$ and $|z_-| = 1$, we have a 1 : 1 electrolyte. $Ba(NO_3)_2$ is a 2 : 1 electrolyte; Na_2SO_4 is a 1 : 2 electrolyte; $MgSO_4$ is a 2 : 2 electrolyte. (Don't be intimidated by the notation. In the following discussion, the z's are charges, the ν's are numbers of ions in the chemical formula, the μ's are chemical potentials, and the γ's are activity coefficients.)

You probably learned in freshman chemistry that salts like $CuSO_4$, $MgCl_2$, and NaCl exist in aqueous solution only in the form of ions. Actually, this picture is inaccurate, and we shall see in Sec. 10.7 that (except for 1 : 1 electrolytes) there is a significant amount of association between oppositely charged ions in solution to yield ion pairs. For a true electrolyte, we start out with ions in the crystal, get solvated ions in solution as the crystal dissolves, and then get some degree of association of solvated ions to form ion pairs in solution. The equilibrium for ion-pair formation in solution is

$$M^{z_+} + X^{z_-} \rightleftharpoons MX^{z_+ + z_-} \tag{10.46}$$

For example, in a $Ca(NO_3)_2$ solution Eq. (10.46) reads $Ca^{2+} + NO_3^- \rightleftharpoons Ca(NO_3)^+$.

We shall restrict the treatment in this section to strong electrolytes. Let the solution be prepared by dissolving n_i moles of electrolyte i in n_A moles of solvent A. The species present in solution are A molecules, M^{z_+} ions, X^{z_-} ions, and $MX^{z_+ + z_-}$ ions. Let n_A, n_+, n_-, and n_{IP} be the numbers of moles of A, M^{z_+}, X^{z_-}, and $MX^{z_+ + z_-}$, respectively. Let μ_A, μ_+, μ_-, and μ_{IP} be the chemical potentials of these species.

The quantity μ_+ is by definition [Eq. (4.81)]

$$\mu_+ \equiv (\partial G/\partial n_+)_{T,P,n_{j \neq +}} \tag{10.47}$$

where G is the Gibbs energy of the solution. In (10.47), we must vary n_+ while holding fixed the amounts of all other species, including n_-. However, the requirement of electrical neutrality of the solution prevents varying n_+ while n_- is held fixed. [We can't readily vary $n(Na^+)$ in an NaCl solution while holding $n(Cl^-)$ fixed.] The same situation holds for μ_-. There is thus no simple way to determine μ_+ and μ_- experimentally. (Chemical potentials of single ions in solution can, however, be estimated theoretically using statistical mechanics; see Sec. 10.6.)

Since μ_+ and μ_- are not measurable, we define μ_i, the *chemical potential of the electrolyte as a whole*, by

$$\mu_i \equiv (\partial G/\partial n_i)_{T,P,n_A} \tag{10.48}$$

The number of moles n_i of dissolved electrolyte can readily be varied at constant n_A, so μ_i can be experimentally measured. [Definitions similar to (10.48) hold for other partial molar properties of the electrolyte as a whole. For example, $\bar{V}_i \equiv (\partial V/\partial n_i)_{T,P,n_A}$, where V is the solution volume. $\bar{V}_i$ for $MgSO_4$ in water was mentioned in Sec. 9.4.]

To relate μ_+ and μ_- to μ_i, we use the Gibbs equation (4.82), which reads for the electrolyte solution

$$dG = -S\,dT + V\,dP + \mu_A\,dn_A + \mu_+\,dn_+ + \mu_-\,dn_- + \mu_{IP}\,dn_{IP} \tag{10.49}$$

If no ion pairs were formed, then from (10.45), n_+ and n_- would be given by $n_+ = \nu_+ n_i$ and $n_- = \nu_- n_i$. With ion-pair formation, we have from (10.46) and (10.49):

$$n_+ = \nu_+ n_i - n_{IP}, \qquad n_- = \nu_- n_i - n_{IP} \tag{10.50}$$

$$dG = -S\,dT + V\,dP + \mu_A\,dn_A + \mu_+(\nu_+\,dn_i - dn_{IP})$$
$$+ \mu_-(\nu_-\,dn_i - dn_{IP}) + \mu_{IP}\,dn_{IP}$$

From (4.111), the equilibrium condition for the ion-pair formation reaction (10.46) is $\mu_{IP} = \mu_+ + \mu_-$. Use of this equation gives

$$dG = -S\,dT + V\,dP + \mu_A\,dn_A + (\nu_+\mu_+ + \nu_-\mu_-)\,dn_i \tag{10.51}$$

Setting $dT = dP = dn_A = 0$ in (10.51) and using (10.48), we get

$$\mu_i = \nu_+\mu_+ + \nu_-\mu_- \tag{10.52*}$$

which relates μ_i to μ_+ and μ_-. For example, the chemical potential of $CaCl_2$ in solution is $\mu(CaCl_2) = \mu(Ca^{2+}) + 2\mu(Cl^-)$.

We now consider the explicit expressions for μ_A and μ_i. The chemical potential μ_A of the solvent can be expressed on the mole-fraction scale [Eqs. (10.35) and (10.36)] as

$$\mu_A = \mu_A^*(T, P) + RT \ln \gamma_{x,A} x_A, \qquad (\gamma_{x,A})^\infty = 1 \tag{10.53}$$

where $\gamma_{x,A}$ is the mole-fraction activity coefficient.

The electrolyte chemical potentials μ_i, μ_+, and μ_- are usually expressed on the molality scale. Let m_+ and m_- be the molalities of the ions M^{z+} and X^{z-}, and let γ_+ and γ_- be the molality-scale activity coefficients of these ions. (The m subscript is omitted from the γ's, since in this section we shall use only the molality scale for solute species.) Equations (10.33), (10.34), and (10.30) give as the chemical potentials of the ions

$$\mu_+ = \mu_+^\circ + RT \ln(\gamma_+ m_+/m^\circ), \qquad \mu_- = \mu_-^\circ + RT \ln(\gamma_- m_-/m^\circ) \tag{10.54}$$

$$m^\circ \equiv 1 \text{ mol/kg}, \qquad \gamma_+^\infty = \gamma_-^\infty = 1 \tag{10.55}$$

where μ_+° and μ_-° are the molality-scale standard-state chemical potentials of the ions.

Substitution of (10.54) into (10.52) gives

$$\mu_i = \nu_+\mu_+^\circ + \nu_-\mu_-^\circ + \nu_+ RT \ln(\gamma_+ m_+/m^\circ) + \nu_- RT \ln(\gamma_- m_-/m^\circ) \tag{10.56}$$

$$\mu_i = \nu_+\mu_+^\circ + \nu_-\mu_-^\circ + RT \ln[(\gamma_+)^{\nu_+}(\gamma_-)^{\nu_-}(m_+/m^\circ)^{\nu_+}(m_-/m^\circ)^{\nu_-}] \tag{10.57}$$

Equations (10.57) and (10.55) give

$$\nu_+\mu_+^\circ + \nu_-\mu_-^\circ = \{\mu_i - RT \ln[(m_+/m^\circ)^{\nu_+}(m_-/m^\circ)^{\nu_-}]\}^\infty \tag{10.58}$$

where the infinity superscript denotes the limit of infinite dilution ($x_A \to 1$). Since μ_i and the ionic molalities are experimentally determinable, the linear combination $\nu_+\mu_+^\circ + \nu_-\mu_-^\circ$ can be determined experimentally from (10.58). Since μ_i, $\nu_+\mu_+^\circ + \nu_-\mu_-^\circ$, m_+, and m_- in (10.57) are experimentally measurable, the combination $(\gamma_+)^{\nu_+}(\gamma_-)^{\nu_-}$ of activity coefficients is experimentally determinable (even though

γ_+ and γ_- are not readily measurable individually). We therefore define the *molality-scale mean ionic activity coefficient* $\gamma_\pm$ of the electrolyte $M_{\nu_+}X_{\nu_-}$ by

$$(\gamma_\pm)^{\nu_+ + \nu_-} \equiv (\gamma_+)^{\nu_+}(\gamma_-)^{\nu_-} \tag{10.59*}$$

For example, for $BaCl_2$, $(\gamma_\pm)^3 = (\gamma_+)(\gamma_-)^2$ and $\gamma_\pm = (\gamma_+)^{1/3}(\gamma_-)^{2/3}$. The definition (10.59) applies also to solutions of several electrolytes. For example, for a solution of NaCl and KCl, there is a $\gamma_\pm$ for the ions K^+ and Cl^- and a different $\gamma_\pm$ for the ions Na^+ and Cl^-.

Rewriting (10.57) using (10.59) and defining $\mu_i^\circ(T, P)$ and ν as

$$\mu_i^\circ \equiv \nu_+ \mu_+^\circ + \nu_- \mu_-^\circ \tag{10.60}$$

$$\nu \equiv \nu_+ + \nu_- \tag{10.61*}$$

we have for the chemical potential of the electrolyte as a whole

$$\mu_i = \mu_i^\circ + RT \ln [(\gamma_\pm)^\nu (m_+/m^\circ)^{\nu_+}(m_-/m^\circ)^{\nu_-}] \tag{10.62}$$

$$\gamma_\pm^\infty = 1 \tag{10.63}$$

where the infinite-dilution behavior of $\gamma_\pm$ follows from (10.59) and (10.55).

The *stoichiometric molality* m_i of electrolyte i is defined by

$$m_i \equiv n_i/w_A \tag{10.64}$$

where the solution is prepared by dissolving n_i moles of electrolyte in a mass w_A of solvent. (Recall the example worked in Sec. 9.1.) To express μ_i in (10.62) as a function of m_i, we shall relate m_+ and m_- to m_i. Let α be the fraction of the ions M^{z+} that do not associate with X^{z-} ions to form ion pairs. From (10.45), if no ion pairing occurred, the number of moles of M^{z+} would equal $\nu_+ n_i$. With ion pairing, the number of moles of M^{z+} in solution is given by $n_+ = \alpha\nu_+ n_i$. There are a total of $\nu_+ n_i$ moles of M in the solution, present partly as M^{z+} ions and partly in $MX^{z_+ + z_-}$ ion pairs [Eq. (10.50)]. Hence the number of moles of the ion pair is $n_{IP} = \nu_+ n_i - n_+ = \nu_+ n_i - \alpha\nu_+ n_i = (1 - \alpha)\nu_+ n_i$. There are a total of $\nu_- n_i$ moles of X present, partly as X^{z-} and partly in the ion pairs; hence the number of moles of X^{z-} is $n_- = \nu_- n_i - n_{IP} = [\nu_- - (1 - \alpha)\nu_+]n_i$. Dividing the equations for n_+ and n_- by the solvent mass, we get as the molalities

$$m_+ = \alpha\nu_+ m_i, \qquad m_- = [\nu_- - (1 - \alpha)\nu_+]m_i$$

We now substitute these equations into (10.62). The quantity in brackets in (10.62) becomes

$$[\cdots] = (\gamma_\pm)^\nu (\alpha\nu_+)^{\nu_+}[\nu_- - (1 - \alpha)\nu_+]^{\nu_-}(m_i/m^\circ)^\nu \tag{10.65}$$

Before continuing, suppose that no ion pairing occurs. Then none of the M^{z+} would associate with X^{z-}, and α would be 1. For $\alpha = 1$, Eq. (10.65) becomes

$$[\cdots] = (\gamma_\pm)^\nu (\nu_+)^{\nu_+}(\nu_-)^{\nu_-}(m_i/m^\circ)^\nu \qquad \text{no ion pairs}$$

It is traditional to define $\nu_\pm$ [analogous to $\gamma_\pm$ in (10.59)] as

$$(\nu_\pm)^\nu \equiv (\nu_+)^{\nu_+}(\nu_-)^{\nu_-} \tag{10.66}$$

[For example, for $Mg_3(PO_4)_2$, $\nu_\pm = (3^3 \times 2^2)^{1/5} = 108^{1/5} = 2.551$.] Hence

$$[\cdots] = (\nu_\pm \gamma_\pm m_i/m^\circ)^\nu \qquad \text{no ion pairs} \tag{10.67}$$

Substitution of (10.67) for the argument of the logarithm in (10.62) gives

$$\mu_i = \mu_i^\circ + \nu RT \ln (\nu_\pm \gamma_\pm m_i/m^\circ) \qquad \text{no ion pairs} \tag{10.68}$$

where $\ln x^y = y \ln x$ was used.

Coming back to the case of ion-pair formation, we eliminate the unsightly mess in (10.65) by defining the electrolyte's *molality-scale stoichiometric activity coefficient* γ_i as

$$\gamma_i \equiv (\nu_\pm)^{-1}(\alpha\nu_+)^{\nu_+/\nu}[\nu_- - (1-\alpha)\nu_+]^{\nu_-/\nu}\gamma_\pm \tag{10.69}$$

Solving (10.69) for $\gamma_\pm$ and substituting in (10.65), we get

$$[\cdots] = (\nu_\pm \gamma_i m_i/m^\circ)^\nu$$

Equation (10.62) becomes

$$\mu_i = \mu_i^\circ + \nu RT \ln (\nu_\pm \gamma_i m_i/m^\circ) \tag{10.70}^*$$

which is the desired expression for the chemical potential of a strong electrolyte in solution. Note that $\gamma_i = 1$ in the limit of infinite dilution; that is, $\gamma_i^\infty = 1$. This follows from (10.69), (10.63), (10.66), and the fact that the extent of ion pairing goes to zero at infinite dilution ($\alpha^\infty = 1$).

Two special cases of (10.69) are of interest. For no ion-pair formation, α is 1 and

$$\gamma_i = \gamma_\pm \qquad \text{for } \alpha = 1 \tag{10.71}$$

where (10.66) was used. [Substitution of (10.71) in (10.70) gives (10.68).] For $\nu_+ = \nu_- = 1$ (a 1 : 1, 2 : 2, or 3 : 3 electrolyte), Eq. (10.61) gives $\nu = 2$, Eq. (10.66) gives $\nu_\pm = 1$, and (10.69) becomes

$$\gamma_i = \alpha\gamma_\pm \qquad \text{for } \nu_+ = \nu_- = 1$$

It is frequently difficult to make a reliable determination of the degree of association to form ion pairs, so α is often not known. One therefore measures and tabulates γ_i for strong electrolytes. The stoichiometric activity coefficient γ_i deviates from 1 due to (*a*) deviations of the solution from ideally dilute behavior; (*b*) deviations of the solvent mole fraction from 1 [see Eqs. (10.32) and (10.59)]; and (*c*) ion-pair formation, which makes α less than 1 [see Eqs. (10.69) and (10.71)]. Although it is actually γ_i that is tabulated for strong electrolytes, tables sometimes use the symbol $\gamma_\pm$ for γ_i. Strictly speaking, $\gamma_\pm = \gamma_i$ only for no ion-pair formation.

Concentrations instead of molalities are sometimes used for electrolyte solutions. The *stoichiometric concentration* c_i of electrolyte i is defined as $c_i \equiv n_i/V$, where V is the solution volume. Reasoning similar to that which led to (10.70) but with molalities replaced by concentrations gives

$$\mu_i = \mu_{c,i}^\circ + \nu RT \ln (\nu_\pm \gamma_{c,i} c_i/c^\circ), \qquad c^\circ \equiv 1 \text{ mol/dm}^3, \qquad \gamma_{c,i}^\infty = 1$$

where the subscript c on μ_i° and γ_i indicates use of the concentration scale. One finds (Prob. 10.10) the following relation between the molality-scale and concentration-scale activity coefficients:

$$\gamma_{c,i} = \frac{m_i \rho_A}{c_i}\gamma_{m,i} = \frac{V/n_A}{\bar{V}_A^*}\gamma_{m,i} \tag{10.72}$$

where $\gamma_{m,i}$ is the same as γ_i in (10.70); ρ_A and $\bar{V}_A^*$ are the density and molar volume of the pure solvent, and n_A is the number of moles of A in the solution.

We want an expression for the Gibbs energy G of an electrolyte solution. Equations (10.51) and (10.52) give

$$dG = -S\,dT + V\,dP + \mu_A\,dn_A + \mu_i\,dn_i \tag{10.73}$$

which has the same form as (4.82). Hence, the same reasoning that gave (9.25) and (9.43) gives for an electrolyte solution

$$G = n_A \mu_A + n_i \mu_i \tag{10.74}$$

$$n_A\, d\mu_A + n_i\, d\mu_i = 0 \qquad \text{const. } T, P \tag{10.75}$$

(See also Prob. 10.11.) Equation (10.75) is the Gibbs–Duhem equation for an electrolyte solution.

In principle, Eq. (10.70) can be applied to a solution of a weak electrolyte such as $HC_2H_3O_2$. For $HC_2H_3O_2$, the ions formed have $z_+ = 1$ and $z_- = -1$, so ion-pair formation is negligible in aqueous solution (Sec. 10.7). Equation (10.46) is written in reverse and becomes the dissociation equilibrium of the weak acid, rather than ion-pair formation. However, it is relatively simple to find the degree of dissociation α of a weak acid, so instead of using γ_i one uses $\gamma_\pm$. The dissociation equilibrium of a weak acid is treated in Sec. 11.3.

The main results of this section are the expressions (10.74) and (10.73) for G and dG of an electrolyte solution, Eq. (10.70) for μ_i of an electrolyte in terms of measurable quantities, and Eq. (10.52) relating μ_+ and μ_- to μ_i. From the expression for G, all other thermodynamic properties of the electrolyte solution can be derived, as noted in Sec. 10.1.

10.5 DETERMINATION OF ELECTROLYTE ACTIVITY COEFFICIENTS

The Gibbs–Duhem equation was used in Sec. 10.2 to find the activity coefficient of an involatile nonelectrolyte solute from known values of the solvent activity coefficient as a function of solution composition; see Eqs. (10.24) to (10.28). A similar procedure applies to a solution of an involatile electrolyte. We restrict the discussion to a solution of a single strong involatile electrolyte i with the formula $M_{\nu_+}X_{\nu_-}$.

Instead of expressing the solvent chemical potential μ_A according to (10.53), it is customary to express μ_A in an electrolyte solution in terms of the (*solvent*) *practical osmotic coefficient* ϕ (phi); the definition of ϕ for an electrolyte solution is

$$\phi \equiv -\frac{\ln a_A}{M_A \nu m_i} \equiv \frac{\mu_A^* - \mu_A}{RTM_A \nu m_i} \tag{10.76}$$

where ν and m_i are defined by (10.61) and (10.64), and where (10.4) and (10.36) were used. Note that νm_i would equal $m_+ + m_-$, the total molality of the ions, if there were no association of ions to ion pairs. M_A is the solvent molar mass (not molecular weight).

We can use vapor-pressure measurements on the electrolyte solution to find ϕ as a function of solution composition. We then use the Gibbs–Duhem equation to find the electrolyte activity coefficient from ϕ. The chemical potential of the solvent can be written as $\mu_A = \mu_A^* + RT \ln a_A$, where Convention I is used [Eq. (10.53)]. This expression for μ_A is the same as (10.4) and (10.10). Therefore, Eq. (10.19), which follows from (10.4) and (10.10), holds for the solvent in an electrolyte solution:

$$P_A = a_A P_A^* \qquad \text{ideal vap., } \mu_A^* \text{ ind. of } P \tag{10.77}$$

Substitution of (10.77) into (10.76) gives

$$\phi = \frac{1}{M_A \nu m_i} \ln \frac{P_A^*}{P_A} \qquad \text{ideal vap., } \mu_A^* \text{ ind. of } P \tag{10.78}$$

Since the electrolyte solute is assumed nonvolatile, P_A equals the vapor pressure of the solution and (10.78) allows ϕ to be found from vapor-pressure measurements.

Rewriting Eq. (10.76), we have

$$\mu_A = \mu_A^* - \phi RTM_A \nu m_i \qquad (10.79)$$

Use of (10.79) for μ_A and (10.70) for μ_i in the Gibbs–Duhem equation (10.75) allows the electrolyte activity coefficient γ_i to be related to the solvent osmotic coefficient ϕ. The derivation is similar to that of (10.28) and is left as an exercise (Prob. 10.15). One finds

$$\ln \gamma_i(m) = \phi(m) - 1 + \int_0^m \frac{\phi(m_i) - 1}{m_i}\, dm_i \qquad \text{const. } T, P \qquad (10.80)$$

Values of ϕ are available from (10.78). By plotting $(\phi - 1)/m_i$ vs. m_i and taking the area under the curve, we can evaluate the integral in (10.80) and obtain γ_i at the electrolyte molality m.

The reason for using ϕ instead of the solvent activity coefficient is that in dilute electrolyte solutions, the solvent activity coefficient may be very close to 1 even though the solute activity coefficient deviates substantially from 1 and the solution is far from ideally dilute. It is inconvenient to work with activity coefficients with values like 1.0001.

[For a solution of a single nonelectrolyte i, the practical osmotic coefficient ϕ is defined by an equation like (10.76), except that ν is omitted. A nonelectrolyte solute can be viewed as a special case of an electrolyte solute for which $\nu = 1$ and $\nu_\pm = 1$; note that (10.70) with $\nu = 1 = \nu_\pm$ reduces to (10.33). Therefore, Eq. (10.78) with ν taken as 1 is valid for a nonelectrolyte solution; also Eq. (10.80) is valid for a nonelectrolyte solution.]

Some experimental values of γ_i for aqueous electrolyte solutions at 25°C and 1 atm ($m^\circ \equiv 1$ mol/kg) are:

m_i/m°	$CaCl_2$	$CuSO_4$	LiBr	HCl
0.001	0.89	0.74	0.97	0.96
0.01	0.73	0.44	0.91	0.90
0.1	0.52	0.15	0.80	0.80
1	0.50	0.04	0.80	0.81
5	5.9		2.7	2.4
10	43		20	10

Figure 10.3 plots γ_i for several electrolytes as a function of molality. Extensive tabulations of γ_i are given in *Robinson and Stokes.*

Note that even at a molality of 0.001 mol/kg, electrolyte activity coefficients deviate substantially from 1, due to the long-range interionic forces. In concentrated solutions, both very large and very small values of γ_i can occur. For example, in aqueous solution at 25°C and 1 atm, $\gamma[UO_2(ClO_4)_2] = 1510$ at $m_i = 5.5$ mol/kg and $\gamma(CdI_2) = 0.017$ for $m_i = 2.5$ mol/kg.

10.6 THE DEBYE–HÜCKEL THEORY OF ELECTROLYTE SOLUTIONS

In 1923, Debye and Hückel started from a highly simplified model of an electrolyte solution and used statistical mechanics to derive theoretical expressions for the molality-scale ionic activity coefficients γ_+ and γ_-. In their model, the ions are taken to be uniformly charged hard spheres of diameter a. (The difference in size between the positive and negative ions is ignored, and a is interpreted as the mean ionic diameter.) The solvent A is treated as a structureless medium with dielectric constant $\varepsilon_{r,\text{A}}$. (If $\mathbf{F}$ is the force between two charges in vacuum and $\mathbf{F}_\text{A}$ is the force between the same charges immersed in the dielectric medium A, then $\mathbf{F}_\text{A}/\mathbf{F} = 1/\varepsilon_{r,\text{A}}$.)

The Debye–Hückel treatment assumes that the solution is quite dilute. This restriction allows several simplifying mathematical and physical approximations to be made. At high dilution, the main deviation from ideally dilute behavior comes from the long-range Coulomb's law attractions and repulsions between the ions. Debye and Hückel assumed that all the deviation from ideally dilute behavior is due to interionic forces.

The Debye–Hückel derivation of γ_+ and γ_- requires more knowledge of electrostatics than most chemists possess. Therefore, we shall just describe their derivation in words, omitting the mathematics. (For a full derivation, see *Bockris and Reddy*, sec. 3.3.) An ion in solution is surrounded by an atmosphere of solvent molecules and other ions. On the average, the positive ion M^{z+} will have more negative ions than positive ions in its immediate vicinity. Debye and Hückel used the Boltzmann distribution law of statistical mechanics (Sec. 22.5) to find the average distribution of charges in the neighborhood of an ion. They then calculated the average potential energy of electrostatic interaction of the ions with one another and equated this interaction energy to $G - G^\text{id}$, where G is the actual Gibbs energy of the solution and G^id is the Gibbs energy the solution would have if it were ideally dilute ($\gamma_+ = \gamma_- = 1$). This gave an expression for G. Partial differentiation of G with respect to n_+ and with respect to n_- then gives the ionic chemical potentials μ_+ and μ_- and hence gives the ionic activity coefficients γ_+ and γ_-.

Debye and Hückel's final result is

$$\ln \gamma_+ = -\frac{z_+^{\,2} C I_m^{1/2}}{1 + BaI_m^{1/2}}, \qquad \ln \gamma_- = -\frac{z_-^{\,2} C I_m^{1/2}}{1 + BaI_m^{1/2}} \tag{10.81}$$

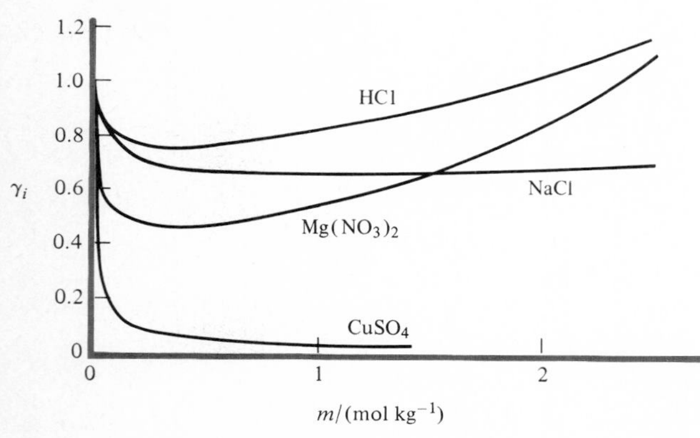

Figure 10.3
Stoichiometric activity coefficients of electrolytes in aqueous solutions at 25°C.

where C, B, and I_m are defined as

$$C \equiv (2\pi N_0 \rho_A)^{1/2}\left(\frac{e^2}{4\pi\varepsilon_0 \varepsilon_{r,A} kT}\right)^{3/2}, \qquad B \equiv e\left(\frac{2N_0 \rho_A}{\varepsilon_0 \varepsilon_{r,A} kT}\right)^{1/2} \tag{10.82}$$

$$I_m \equiv \frac{1}{2}\sum_j z_j^2 m_j \tag{10.83*}$$

In these equations (which are written in SI units), a is the mean ionic diameter, γ_+ and γ_- are the molality-scale activity coefficients for the ions M^{z+} and X^{z-}, respectively, N_0 is the Avogadro constant, k is Boltzmann's constant [Eq. (3.57)], e is the proton charge, ε_0 is the permittivity of vacuum (ε_0 occurs as a proportionality constant in Coulomb's law; see Sec. 14.1), ρ_A is the solvent density, $\varepsilon_{r,A}$ is the solvent dielectric constant, and T is the absolute temperature. I_m is called the (*molality-scale*) *ionic strength*; the sum in (10.83) goes over all ions in solution, m_j being the molality of ion j with charge z_j.

Although the Debye–Hückel theory gives γ of each ion, we cannot measure γ_+ or γ_- individually. Hence, we express the Debye–Hückel result in terms of the mean ionic activity coefficient $\gamma_\pm$. Taking the log of (10.59), we have

$$\ln \gamma_\pm = \frac{\nu_+ \ln \gamma_+ + \nu_- \ln \gamma_-}{\nu_+ + \nu_-} \tag{10.84}$$

Since the electrolyte $M_{\nu_+}X_{\nu_-}$ is electrically neutral, we have

$$\nu_+ z_+ + \nu_- z_- = 0 \tag{10.85}$$

Multiplication of (10.85) by z_+ yields $\nu_+ z_+{}^2 = -\nu_- z_+ z_-$; multiplication of (10.85) by z_- yields $\nu_- z_-{}^2 = -\nu_+ z_+ z_-$. Addition of these two equations gives

$$\nu_+ z_+{}^2 + \nu_- z_-{}^2 = -z_+ z_-(\nu_+ + \nu_-) = z_+|z_-|(\nu_+ + \nu_-) \tag{10.86}$$

since z_- is negative. Substitution of the Debye–Hückel equations (10.81) into (10.84) followed by use of (10.86) gives

$$\ln \gamma_\pm = -z_+|z_-|\frac{CI_m^{1/2}}{1 + BaI_m^{1/2}} \tag{10.87}$$

Using the SI values of N_0, k, e, and ε_0, and the values $\varepsilon_r = 78.40$, $\rho = 0.99705 \times 10^3$ kg/m³ for H_2O at 25°C and 1 atm, we have for (10.82)

$$C = 1.174\ (\text{kg/mol})^{1/2}, \qquad B = 3.284 \times 10^9\ (\text{kg/mol})^{1/2}\ \text{m}^{-1}$$

Molecular distances are of the order of 10^{-8} cm. We define the *angstrom* (Å) as

$$1\ \text{Å} \equiv 10^{-8}\ \text{cm} \equiv 10^{-10}\ \text{m} \tag{10.88*}$$

Substituting the numerical values for B and C into (10.87), using (10.88), and dividing C by 2.3026 to convert to base 10 logs, we get

$$\log \gamma_\pm = -0.510 z_+|z_-|\frac{(I_m/m^\circ)^{1/2}}{1 + 0.328(a/\text{Å})(I_m/m^\circ)^{1/2}} \qquad \text{dil. aq. soln. at 25°C} \tag{10.89}$$

where $m^\circ \equiv 1$ mol/kg. The ionic strength I_m has units of mol/kg, and the ionic diameter a has units of length, so log $\gamma_\pm$ is dimensionless, as it must be.

For very dilute solutions, I_m is very small, and the second term in the denominator in (10.89) can be neglected compared with 1. Therefore, for very dilute solutions,

$$\ln \gamma_\pm = -z_+|z_-|CI_m^{1/2} \tag{10.90}$$

$$\log \gamma_\pm = -0.510z_+|z_-|(I_m/m^\circ)^{1/2} \qquad \text{very dil. aq. soln., 25°C} \tag{10.91}$$

Equation (10.90) is called the *Debye–Hückel limiting law*, since it is valid only in the limit of infinite dilution. (Actually, most of the laws of science are limiting laws.)

How well does the Debye–Hückel theory work? Experimental data show that Eq. (10.91) does give the correct limiting behavior for electrolyte solutions as $I_m \to 0$ (Fig. 10.4). Equation (10.91) is found to be generally accurate for solutions with $I_m \le 0.01$ mol/kg. For a 2 : 2 electrolyte this corresponds to a molality $0.01/4 \approx 0.002$. (It is sometimes unkindly said that the Debye–Hückel theory applies to slightly contaminated distilled water.) The more complete equation (10.89) is reasonably accurate for aqueous solutions with $I_m \le 0.1$ mol/kg if we choose the ionic diameter a to give a good fit to the data. Values of a so found typically range from 3 to 8 Å for common inorganic salts, which are reasonable values for hydrated ions. At a given ionic strength, the theory works better the lower the value of $z_+|z_-|$; for example, at $I_m = 0.1$ mol/kg, the Debye–Hückel theory is considerably more reliable for 1 : 1 electrolytes than for 2 : 2 electrolytes. Part of the reason for this lies in ionic association (Sec. 10.7).

To eliminate the empirically determined ionic diameter a from (10.89), we note that for $a \approx 3$ Å, we have $0.328(a/\text{Å}) \approx 1$. Hence it is customary to simplify (10.89) to

$$\log \gamma_\pm = -0.510z_+|z_-|\frac{(I_m/m^\circ)^{1/2}}{1 + (I_m/m^\circ)^{1/2}} \qquad \text{dil. aq. soln., 25°C} \tag{10.92}$$

Example Use the Debye–Hückel equation to estimate $\gamma_\pm$ for aqueous $CaCl_2$ solutions at 25°C with molalities 0.001, 0.01, and 0.1 mol/kg.

For a 2 : 1 electrolyte, Eq. (10.83) gives $I_m = \frac{1}{2}(4m_+ + m_-)$. For the dilute solutions of this problem, it is a reasonably good approximation to neglect ion pairing for

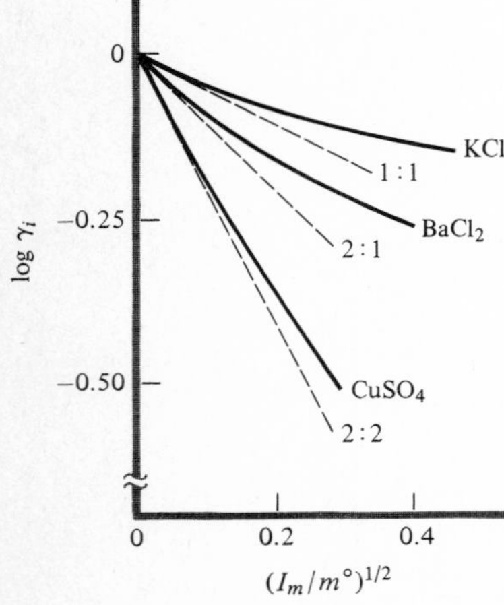

Figure 10.4
Plots of log γ_i vs. square root of ionic strength for aqueous electrolytes at 25°C. The dashed lines show the predictions of the Debye-Hückel limiting law (10.90).

a 2 : 1 electrolyte. Therefore, $m_+ = m_i$ and $m_- = 2m_i$. The ionic strength is $I_m = \frac{1}{2}(4m_i + 2m_i) = 3m_i$. Equation (10.92) with $z_+ = 2$, $|z_-| = 1$, and $I_m = 3m_i$ becomes

$$\log \gamma_\pm = -\frac{1.767(m_i/m^\circ)^{1/2}}{1 + 1.732(m_i/m^\circ)^{1/2}}$$

Substitution of $m_i/m^\circ = 0.001$ gives $\gamma_\pm = 0.885$. We find $\gamma_\pm = 0.707$ for $m_i = 0.01$ mol/kg, and $\gamma_\pm = 0.435$ for $m_i = 0.1$ mol/kg. With the assumption of no ion pairing, the activity coefficients $\gamma_\pm$ and γ_i are equal [Eq. (10.71)], so these calculated $\gamma_\pm$ values can be compared with the experimental γ_i values (0.89, 0.73, 0.52) listed in the last section.

The properties of very dilute electrolyte solutions frequently cannot be measured with the required accuracy. Hence, even though the range of validity of the Debye–Hückel theory is limited to quite dilute solutions, the theory is of great practical importance since it allows measured properties of electrolyte solutions to be reliably extrapolated into the region of very low concentrations.

Figure 10.3 shows that as the electrolyte's molality m_i increases from zero, its activity coefficient γ_i first decreases from the ideally dilute value 1 and then increases. The electrolyte's activity can be viewed roughly as its "effective" molality. The initial decrease in γ_i can be attributed to the long-range Coulomb attractions between oppositely charged ions, which substantially reduce the effective ionic molalities (even in quite dilute solutions), thereby reducing the electrolyte's activity and activity coefficient γ_i. The increase in γ_i at higher molalities can be rationalized as follows. Ions in solution are hydrated; hydration reduces the amount of "free" water molecules, thereby reducing the effective concentration of water in the solution and increasing the effective molality of the electrolyte. For example, for NaCl, experimental evidence (*Bockris and Reddy*, sec. 2.4.5) indicates the Na^+ ion carries about four H_2O molecules along with itself as it moves through the solution and the Cl^- ion carries about two H_2O molecules as it moves. Thus, each mole of NaCl in solution ties up about 6 moles of H_2O. One kilogram of water contains 55.5 moles. In a 0.1 mol/kg aqueous NaCl solution, there are $55.5 - 6(0.1) = 54.9$ moles of "free" water per kilogram of solvent, so the effect of hydration is slight in this dilute solution. However, in a 3 mol/kg aqueous NaCl solution, there are only $55 - 18 = 37$ moles of "free" water per kilogram of solvent, which is a very substantial reduction.

Stokes and Robinson modified the Debye–Hückel theory to take hydration into account. (See *Bockris and Reddy*, secs. 3.6 and 3.11.) Their equation for the activity coefficient contains the hydration number as an adjustable parameter and allows experimental activity-coefficient data to be fitted up to much higher molalities than with the Debye–Hückel equation. For NaCl, the best fit to the data is obtained with a hydration number of 3.5, substantially less than the value of 6 obtained by other methods. Of course, introduction of an adjustable parameter is bound to improve agreement with experiment. The Stokes–Robinson treatment is open to theoretical objections, since it uses the Debye–Hückel model in concentrated solutions, where this model is invalid.

It has been found empirically that addition of a term linear in I_m to the Debye–Hückel equation (10.92) improves agreement with experiment in less dilute solutions. Davies proposed the following expression containing no parameters (*Davies*, pp. 39–43):

$$\log \gamma_\pm = -0.51 z_+ |z_-| \left[\frac{(I_m/m^\circ)^{1/2}}{1 + (I_m/m^\circ)^{1/2}} - 0.30(I_m/m^\circ) \right] \qquad \text{in } H_2O \text{ at } 25°C \tag{10.93}$$

The Davies equation for $\log \gamma_+$ (or $\log \gamma_-$) is obtained by replacement of $z_+|z_-|$ in (10.93) with z_+^2 (or z_-^2). The Davies modification of the Debye–Hückel equation is typically in error by $1\frac{1}{2}$ percent at $I_m/m^\circ = 0.1$. The linear term in (10.93) causes $\gamma_\pm$ to go through a minimum and then increase as I_m increases, in agreement with the behavior in Fig. 10.3. As I_m/m° increases above 0.1, agreement of the Davies equation with experiment decreases; at $I_m/m^\circ = 0.5$, the error is typically 5 to 10 percent. It is best to use experimental values of $\gamma_\pm$, especially for ionic strengths above 0.1 mol/kg, but in the absence of experimental data, the Davies equation can serve to estimate $\gamma_\pm$. Note that since the Davies equation contains no parameters, it predicts that $\gamma_\pm$ will have the same value at a given I_m for any 1 : 1 electrolyte. In reality, $\gamma_\pm$ values for 1 : 1 electrolytes are equal only in the limit of high dilution.

10.7 IONIC ASSOCIATION

An important phenomenon in many electrolyte solutions is the association of ions to form ion pairs. Consider the salt $M_{\nu_+}X_{\nu_-}$, which yields the ions M^{z+} and X^{z-} in solution. For the Debye–Hückel model of spherical ions in a medium of dielectric constant ε_r, the electrostatic potential energy of interaction between oppositely charged ions is

$$(z_+ e)(z_- e)/4\pi\varepsilon_0 \varepsilon_r r \tag{10.94}$$

where r is the distance between the centers of the ions; $z_+ e$ and $z_- e$ are the charges of the ions in coulombs, e being the proton charge. Since z_- is negative, the energy (10.94) is negative. If the magnitude of the electrostatic attraction energy (10.94) is substantially greater than the thermal kinetic energy of species in the solution, then the pair of oppositely charged ions will form an entity with sufficient stability to exist over the course of several collisions with solvent molecules. We call such an entity an *ion pair*. Statistical mechanics (Chap. 22) shows that the average translational kinetic energy of a species is $\frac{3}{2}kT$, where k is Boltzmann's constant and T is the absolute temperature. Hence, when the interionic distance r is small enough to make (10.94) substantially greater than $\frac{3}{2}kT$, we have an ion pair.

The concept of ion pairs was introduced in 1926 by Bjerrum. He proposed that all pairs of ions in the solution for which the absolute value of (10.94) exceeds $2kT$ be considered as an ion pair. Bjerrum thus took as associated all oppositely charged ions for which $z_+|z_-|e^2/4\pi\varepsilon_0\varepsilon_r r > 2kT$, that is, all ions for which

$$r < z_+|z_-|e^2/8\pi\varepsilon_0\varepsilon_r kT \tag{10.95}$$

The quantity on the right of (10.95) is called the *critical distance* in the Bjerrum theory. For H_2O at 25°C and 1 atm, (10.95) becomes

$$r < 3.57 z_+|z_-| \text{ Å} \qquad \text{aq. soln. at } 25°C \tag{10.96}$$

For a 1 : 1 electrolyte in water at 25°C, Bjerrum considers as ion pairs all pairs of oppositely charged ions whose center-to-center distance is less than 3.57 Å. Two ions cannot approach each other more closely than the mean ionic diameter a (Sec. 10.6). As noted earlier, a values for common salts run from 3 to 8 Å. In the Bjerrum theory, 1 : 1 salts with a greater than 3.6 Å cannot form ion pairs in aqueous solution. (Their ions cannot approach each other closely enough to have the electrostatic attraction energy exceed $2kT$.) Even for 1 : 1 salts with a between 3 and $3\frac{1}{2}$ Å, we expect ion-pair formation in water to be slight, since there is only a narrow range of interionic distances for which the electrostatic attraction energy exceeds $2kT$.

The ion-pair association equilibrium is Eq. (10.46). Bjerrum used a model similar to that of Debye and Hückel to find a theoretical expression for the degree of association to ion pairs as a function of the electrolyte concentration, z_+, z_-, T, ε_r, and a. His theory indicated that association to ion pairs is usually negligible for 1 : 1 electrolytes in water but that ionic association in water can be quite substantial for electrolytes with higher $z_+|z_-|$ values, even at low concentrations.

The solvent H_2O has a high dielectric constant (due to the polarity of the water molecule). In solvents with lower values of ε_r, the magnitude of the electrostatic attraction energy (10.94) is greater than in aqueous solutions. Hence, ion-pair formation in these solvents is greater than in water. Even for 1 : 1 electrolytes, ion-pair formation is important in solvents with low dielectric constants.

Ionic association reduces the number of ions in solution and hence reduces the electrical conductivity of the solution. (For example, in $CaSO_4$ solutions, Ca^{2+} and SO_4^{2-} associate to form neutral $CaSO_4$ ion pairs; in MgF_2 solutions, Mg^{2+} and F^- form MgF^+ ion pairs.) The degree of association in an electrolyte solution can best be determined from conductivity measurements on the solution (Sec. 16.6). For example, one finds that $MgSO_4$ in water at 25°C is 10 percent associated at the molality 0.001 mol/kg; $CuSO_4$ in water at 25°C is 35 percent associated at 0.01 mol/kg and is 57 percent associated at 0.1 mol/kg. From conductivity data, one can calculate the equilibrium constant for the reaction (10.46). Conductivity data indicate that for 1 : 1 electrolytes ionic association is unimportant in dilute aqueous solutions but can be significant in concentrated aqueous solutions; for most electrolytes with higher $z_+|z_-|$ values, ionic association is important in both dilute and concentrated aqueous solutions. (In the limit of infinite dilution, the degree of association goes to zero.)

Conductivity measurements have shown that the qualitative conclusions of the Bjerrum theory are generally correct, but quantitative agreement with experiment is sometimes lacking. The Bjerrum theory is open to several objections. The choice of $2kT$ is rather arbitrary. In a solvent of low dielectric constant, this choice makes the critical distance abnormally large; e.g., for 1,4-dioxane at 25°C, $\varepsilon_r = 2.21$, and (10.95) becomes $r < 127$ Å for a 1 : 1 electrolyte.

Many improved theories of ionic association have been proposed since the Bjerrum theory. Fuoss published a treatment in which only ions in contact are regarded as forming an ion pair. (See Prob. 11.13.) For discussions of the Fuoss theory and other theories of ionic association, see *Bockris and Reddy*, sec. 3.8.6; B. E. Conway, chap. 2 of *Eyring, Henderson, and Jost*, vol. IXA.

Taking ionic association into account improves the accuracy of the Debye–Hückel equation (10.87). Formation of ion pairs reduces the number of ions in the solution. In calculating the ionic strength (10.83), one does not include those ions which are associated to form neutral ion pairs. Also, one uses (10.69) to relate the mean activity coefficient $\gamma_\pm$ calculated by the Debye–Hückel theory to the experimentally observed stoichiometric mean activity coefficient γ_i.

Formation of ion pairs in salt solutions should be distinguished from the dissociation equilibrium of an acid in solution. For CH_3COOH (or HCl), the ions H_3O^+ and CH_3COO^- (or H_3O^+ and Cl^-) react to give a compound with a polar covalent bond; the CH_3COOH and HCl molecules do not have ionic bonds. In contrast, an ion pair consists of two associated ions.

Ion pairs should also be distinguished from complex ions. Complex-ion formation is common in aqueous solutions of the halides of the transition metals (Zn, Cd, Hg, Ag, Cu, etc.). For example, an aqueous solution of $CdBr_2$ that is not highly dilute contains significant amounts of the species Cd^{2+}, $CdBr^+$, $CdBr_2$, $CdBr_3{}^-$, and $CdBr_4^{2-}$. In these complex ions, the Br^- ions are in direct contact with the central Cd^{2+} ion, and each Cd—Br bond has a substantial amount of covalent character. In contrast, the positive and negative ions of an ion pair retain their solvent sheaths and are held together by ionic (electrostatic) forces. The absorption spectrum of the solution can frequently be used to distinguish between ion-pair and complex-ion formation. In some solutions, both ion pairs and complex ions are present. The Bjerrum and Fuoss theories are of course inapplicable to complex-ion formation.

10.8 CONVENTIONAL THERMODYNAMIC PROPERTIES OF SOLUTION COMPONENTS

Determination of conventional values of H, S, G, etc., for pure substances was discussed in Chap. 5. We now consider conventional values for partial molar properties of components of a solution.

Consider first nonelectrolyte solutions where Convention I (Sec. 10.1) is used. The standard states are the pure components. We described a method for finding activity coefficients in Sec. 10.2. Knowledge of $\gamma_{\mathrm{I},i}$ in Eq. (10.9) gives $\mu_i - \mu^\circ_{\mathrm{I},i} = \mu_i - \mu_i^*$. If we have determined a conventional value of μ_i^* (which is the molar Gibbs energy for pure i), we then know the conventional value of $\mu_i = \bar{G}_i$ in the solution. Knowing $\gamma_{\mathrm{I},i}$ as a function of T, we can use $\bar{S}_i = -(\partial \bar{G}_i/\partial T)_{P,n_i}$ to find the conventional partial molar entropy $\bar{S}_i$ of i in the solution. From $\bar{G}_i$ and $\bar{S}_i$, we can calculate the conventional $\bar{H}_i$. Alternatively, $\bar{H}_i - \bar{H}_i^*$ can be measured as described in Sec. 9.4 and $\bar{S}_i$ found from $\bar{H}_i$ and $\bar{G}_i$.

For a nonelectrolyte solution where Convention II is used, measurement of activity coefficients gives $\mu_i - \mu^\circ_{\mathrm{II},i}$ for each solute. To find the conventional value of $\mu^\circ_{\mathrm{II},i}$, we use solubility data. Let $x_{i,\mathrm{sat}}$ and $\gamma_{\mathrm{II},i,\mathrm{sat}}$ be the mole fraction of i and the Convention II activity coefficient of i in a saturated solution of i in the solvent A at the desired temperature and pressure. Using (4.100), we equate the chemical potential of pure i to the chemical potential of i in the saturated solution:

$$\mu_i^*(T, P) \equiv \bar{G}_i^*(T, P) = \mu^\circ_{\mathrm{II},i}(T, P) + RT \ln \gamma_{\mathrm{II},i,\mathrm{sat}} x_{i,\mathrm{sat}} \tag{10.97}$$

We can readily measure $x_{i,\mathrm{sat}}$, and $\gamma_{\mathrm{II},i,\mathrm{sat}}$ is determined as discussed in Sec. 10.2. The conventional value for $\bar{G}$ of pure i is found as discussed in Chap. 5. Hence, we can use (10.97) to find the conventional value of $\mu^\circ_{\mathrm{II},i}$ for i in the solvent A at T and P. Since we know both $\mu_i - \mu^\circ_{\mathrm{II},i}$ and $\mu^\circ_{\mathrm{II},i}$, we then have the conventional value of $\mu_i \equiv \bar{G}_i$ in the solution. Conventional values of $\bar{H}_i$ and $\bar{S}_i$ are then found as discussed above for Convention I.

To deal with chemical equilibria in solutions (Chap. 11), it is desirable to have tables of standard-state thermodynamic properties of substances in solution. For aqueous solutions, the molality scale is most commonly used for solutes, and the solute standard state is the hypothetical state with $m_i = 1$ mol/kg and $\gamma_{m,i} = 1$ (Sec. 10.3). The standard-state partial molar Gibbs energy of formation $\Delta G^\circ_{f,T}$ for solute i in solution is defined as the molality-scale standard-state chemical potential μ°_m for i in solvent A at T and 1 atm minus the standard-state Gibbs energy at T of the pure, separated elements needed to form 1 mole of compound i:

$$\Delta G^\circ_{f,T}(i \text{ in soln.}) \equiv \mu^\circ_{m,i}(T, 1 \text{ atm}) - G^\circ_{\text{elem}}(T) \tag{10.98}$$

where the subscript elem stands for elements. Similarly,

$$\Delta H^\circ_{f,T}(i \text{ in soln.}) \equiv \bar{H}^\circ_{m,i}(T, 1 \text{ atm}) - H^\circ_{\text{elem}}(T) \tag{10.99}$$

Tables of $\Delta G^\circ_{f,298}$, $\Delta H^\circ_{f,298}$, and the standard-state conventional entropies $\bar{S}^\circ_{298}$ for biologically important solutes in aqueous solution are given in H. D. Brown (ed.), *Biochemical Microcalorimetry*, Academic, 1969, pp. 305–317. Standard-state thermodynamic properties of various solutes in aqueous solution are given in the *NBS Technical Notes* cited in Sec. 5.6.

Example The molality of a saturated solution of sucrose in water at 25°C and 1 atm is 6.05 mol/kg. Vapor-pressure measurements and the Gibbs–Duhem equation give $\gamma_m(C_{12}H_{22}O_{11}) = 2.87$ in the saturated solution. For pure sucrose at 25°C, $\Delta G^\circ_f = -369.1$ kcal/mol, $\Delta H^\circ_f = -531.0$ kcal/mol, and $\bar{S}^\circ = 86.1$ cal/(mol K). At 25°C and 1 atm, the differential heat of solution of sucrose in water at infinite dilution is 1.4 kcal/mol. Find $\Delta G^\circ_{f,298}$, $\Delta H^\circ_{f,298}$, and $\bar{S}^\circ_{298}$ for $C_{12}H_{22}O_{11}(aq)$.

For $\Delta G^\circ_{f,T}$ of pure substance i, we have

$$\bar{G}^*_i(T, P^\circ) - G_{\text{elem}}(T, P^\circ) \equiv \Delta G^\circ_{f,T}(\text{pure } i)$$

where $P^\circ \equiv 1$ atm. Equation (10.97) with the mole-fraction scale replaced by the molality scale and with $P = 1$ atm gives

$$\bar{G}^*_i(T, P^\circ) = \mu^\circ_{m,i}(T, P^\circ) + RT \ln (\gamma_{m,i,\text{sat}}\, m_{i,\text{sat}}/m^\circ) \tag{10.100}$$

where $m^\circ \equiv 1$ mol/kg. Subtracting (10.100) from the equation that precedes it and using (10.98), we get

$$\Delta G^\circ_{f,T}(i \text{ in soln.}) = \Delta G^\circ_{f,T}(\text{pure } i) - RT \ln (\gamma_{m,i,\text{sat}}\, m_{i,\text{sat}}/m^\circ) \tag{10.101}$$

Substitution into (10.101) gives $\Delta G^\circ_{f,298}[\text{sucrose}(aq)]$ as -370.8 kcal/mol.

Use of (10.99), (10.38), and (9.54) gives

$$\Delta H^\circ_{f,T}(i \text{ in soln.}) = \bar{H}^\infty_{m,i}(T, P^\circ) - H^\circ_{\text{elem}}(T)$$

$$= \Delta H^\infty_{\text{diff},i}(T, P^\circ) + \bar{H}^*_i(T, P^\circ) - H^\circ_{\text{elem}}(T)$$

$$\Delta H^\circ_{f,T}(i \text{ in soln.}) = \Delta H^\infty_{\text{diff},i}(T, P^\circ) + \Delta H^\circ_{f,T}(\text{pure } i)$$

Hence $\Delta H^\circ_{f,298}[\text{sucrose}(aq)] = (1.4 - 531.0)$ kcal/mol $= -529.6$ kcal/mol.

From (9.31) it follows that

$$\Delta G^\circ_{f,T}(i \text{ in soln.}) = \Delta H^\circ_{f,T}(i \text{ in soln.}) - T\, \Delta S^\circ_{f,T}(i \text{ in soln.}) \tag{10.102}$$

Using the above-calculated values of ΔH_f° and ΔG_f°, we get

$$\Delta S_{f,298}^\circ[\text{sucrose}(aq)] = [(-529.6 + 370.8)\ \text{kcal/mol}]/298.15\ \text{K}$$

$$\bar{S}_{298}^\circ[\text{sucrose}(aq)] - S_{\text{elem}}^\circ = -532.6\ \text{cal/(mol K)}$$

Using entropy data from the Appendix to find S_{elem}° for $C_{12}H_{22}O_{11}$, we get

$$\bar{S}_{298}^\circ[\text{sucrose}(aq)] = 96.6\ \text{cal mol}^{-1}\ \text{K}^{-1}$$

Example The solubility of $O_2(g)$ in water at 25°C and 1 atm pressure of O_2 above the solution is 1.15 mmol per kilogram of water. Find $\Delta G_{f,298}^\circ$ for O_2 in water.

Note from the derivation of (10.101) that $\gamma_{m,i,\text{sat}}$ and $m_{i,\text{sat}}$ are for $P = 1$ atm. Since the solution is very dilute, it is safe to set $\gamma_{m,i,\text{sat}}$ equal to 1 in (10.101). Therefore $\Delta G_{f,298}^\circ[O_2(aq)] = 0 - RT \ln 0.00115 = 4.0$ kcal/mol.

Now consider electrolyte solutions. Measurement of the stoichiometric mean activity coefficient γ_i and use of Eq. (10.70) give us $\mu_i - \mu_i^\circ$. To find μ_i°, we use solubility data. Corresponding to (10.100), we have

$$\bar{G}_i^* = \mu_i^\circ + \nu RT \ln (\nu_\pm \gamma_{i,\text{sat}} m_{i,\text{sat}}/m^\circ) \tag{10.103}$$

which allows determination of μ_i°. Knowing both μ_i° and $\mu_i - \mu_i^\circ$, we then know μ_i. Other thermodynamic properties of the electrolyte as a whole are then found as described above for nonelectrolyte solutions.

For electrolyte solutions, we can work with thermodynamic properties (μ_i, $\bar{H}_i$, $\bar{S}_i$, etc.) of the electrolyte as a whole, and these properties are experimentally determinable. Suppose we have 30 common cations and 30 common anions. This means we must measure thermodynamic properties for 900 electrolytes in water. If we could determine single-ion chemical potentials μ_+ and μ_- (and single-ion partial molar entropies and enthalpies), we would then need to measure values for only 60 ions, since μ_i for an electrolyte can be determined from (10.52) as $\nu_+\mu_+ + \nu_-\mu_-$. Unfortunately, as noted after Eq. (10.47), single-ion chemical potentials cannot readily be measured. What is traditionally done is to assign arbitrary values for thermodynamic properties to the aqueous H^+ ion. Thermodynamic properties of other aqueous ions are then tabulated relative to $H^+(aq)$.

The convention adopted is that ΔG_f° of the aqueous H^+ ion is zero at every temperature:

$$\Delta G_{f,T}^\circ[H^+(aq)] = 0 \qquad \text{by convention} \tag{10.104}$$

The reaction of formation of $H^+(aq)$ in its standard state at temperature T and pressure 1 atm $\equiv P^\circ$ from H_2 gas in its standard state [Eq. (5.1)] is

$$\tfrac{1}{2}H_2(\text{ideal gas}, P^\circ) \rightarrow H^+(aq, m = m^\circ, \gamma_m = 1) + e^-(ss) \tag{10.105}$$

where $e^-(ss)$ indicates 1 mole of electrons in some particular standard state (which we shall leave unspecified). Of course, ΔG° for (10.105) is not actually zero but has a value that is well defined once we define the standard state for the electrons in (10.105) and specify the temperature. Setting $\Delta G^\circ = 0$ for (10.105) is done simply to facilitate the thermodynamic discussion of ionic reactions in aqueous solutions. Whatever the value of ΔG° for (10.105) actually is, this value will cancel out in calculating thermodynamic property changes for ionic reactions in aqueous solutions. The value of ΔG° for (10.105) will not cancel out in calculations on reactions that involve transport of ions from one phase to another, for example, the reaction $H^+(g) \to H^+(aq)$, or on half-reactions, for example, (10.105). Hence, the convention (10.104) cannot be used to calculate thermodynamic quantities for ion-transport reactions or half-reactions. (Such reactions are not readily studied experimentally but can be discussed theoretically using statistical mechanics.)

We have $d\Delta G^\circ/dT = -\Delta S^\circ$. Since ΔG° for (10.105) is taken as zero at every temperature, $d\Delta G^\circ/dT$ for (10.105) equals zero and ΔS° for the $H^+(aq)$ formation reaction (10.105) is zero at every temperature:

$$\Delta S^\circ_{f,T}[H^+(aq)] = 0 \qquad \text{by convention} \tag{10.106}$$

We also have for the reaction (10.105), $\Delta H^\circ = \Delta G^\circ + T\,\Delta S^\circ = 0 + 0 = 0$. Hence,

$$\Delta H^\circ_{f,T}[H^+(aq)] = 0 \qquad \text{by convention} \tag{10.107}$$

Thermodynamic tables list ΔH°_f and ΔG°_f, which by convention are zero for $H^+(aq)$. Instead of listing ΔS°_f values, thermodynamic tables list $\bar{S}^\circ$ values (based on the convention of zero entropy for each element at absolute zero). Hence we need to assign a value for $\bar{S}^\circ[H^+(aq)]$. From (10.106) and (10.105) it follows that $\bar{S}^\circ_T[e^-(ss)] + \bar{S}^\circ_T[H^+(aq)] = \frac{1}{2}\bar{S}^\circ_T[H_2(g)]$. In the existing tables of thermodynamic properties of ions, the entropy of $H^+(aq)$ at any temperature is set equal to zero:

$$\bar{S}^\circ_T[H^+(aq)] = 0 \qquad \text{by convention} \tag{10.108}$$

Therefore, by convention, $\bar{S}^\circ_T[e^-(ss)] = \frac{1}{2}\bar{S}^\circ_T[H_2(g)]$.

Also, $\bar{C}^\circ_P$ of $H^+(aq)$ is taken as zero by convention:

$$\bar{C}^\circ_{P,T}[H^+(aq)] = 0 \qquad \text{by convention} \tag{10.109}$$

Having adopted conventions for $H^+(aq)$, we can determine thermodynamic properties of aqueous ions relative to those of $H^+(aq)$, as follows. Equation (10.60) gives for the electrolyte i with formula $M_{\nu_+}X_{\nu_-}$ in solution: $\mu^\circ_i = \nu_+\mu^\circ_+ + \nu_-\mu^\circ_-$, where the molality scale is used. Subtraction of G_{elem} from each side of this equation and use of (10.98) and corresponding equations for the ions gives

$$\Delta G^\circ_{f,T}[i(aq)] = \nu_+\ \Delta G^\circ_{f,T}[M^{z+}(aq)] + \nu_-\ \Delta G^\circ_{f,T}[X^{z-}(aq)] \tag{10.110}$$

For example, $\Delta G^\circ_{f,T}[BaCl_2(aq)] = \Delta G^\circ_{f,T}[Ba^{2+}(aq)] + 2\Delta G^\circ_{f,T}[Cl^-(aq)]$. Differentiation of (10.60) with respect to T and use of $(\partial\bar{G}/\partial T)_P = -\bar{S}$ gives

$$\bar{S}^\circ_T[i(aq)] = \nu_+\ \bar{S}^\circ_T[M^{z+}(aq)] + \nu_-\ \bar{S}^\circ_T[X^{z-}(aq)]$$

Subtraction of S_{elem} from each side of this equation gives

$$\Delta S^\circ_{f,T}[i(aq)] = \nu_+\ \Delta S^\circ_{f,T}[M^{z+}(aq)] + \nu_-\ \Delta S^\circ_{f,T}[X^{z-}(aq)] \tag{10.111}$$

From $\Delta H = \Delta G + T\,\Delta S$ and Eqs. (10.110) and (10.111), we get

$$\Delta H^\circ_{f,T}[i(aq)] = \nu_+\ \Delta H^\circ_{f,T}[M^{z+}(aq)] + \nu_-\ \Delta H^\circ_{f,T}[X^{z-}(aq)] \tag{10.112}$$

In these equations, the standard state for an ion is the hypothetical state with molality of that ion equal to 1 mol/kg and molality-scale activity coefficient of that ion equal to 1. From Eq. (10.70), the standard state of the electrolyte i as a whole has $\nu_\pm\gamma_i m_i = m^\circ$. We take the standard state of the electrolyte as a whole to have $\gamma_i = 1$

and $\nu_\pm m_i = m^\circ$. The quantity $\nu_\pm m_i$ is often called the *mean ionic molality* $m_\pm$. The properties of the electrolyte as a whole on the left sides of Eqs. (10.110) to (10.112) can be determined experimentally. Observations on the electrolyte $H_{\nu_+}X_{\nu_-}$ in solution together with the $H^+(aq)$ conventions thus allow the determination of thermodynamic properties of the anion $X^{z-}(aq)$ [relative to those of $H^+(aq)$]. Observations on the electrolyte $M_{\nu_+}X_{\nu_-}$ in solution then give thermodynamic properties of the cation $M^{z+}(aq)$ [relative to $H^+(aq)$].

Example Measurements on electrochemical cells (Sec. 14.11) give $\Delta G^\circ_{f,298}[HCl(aq)] = -31.38$ kcal/mol and $\Delta H^\circ_{f,298}[HCl(aq)] = -39.93$ kcal/mol. Also $\bar{S}^\circ_{298}[H_2(g)] = 31.21$ cal mol^{-1} K^{-1} and $\bar{S}^\circ_{298}[Cl_2(g)] = 53.29$ cal mol^{-1} K^{-1}. Find $\Delta G^\circ_{f,298}$, $\Delta H^\circ_{f,298}$, and $\bar{S}^\circ_{298}$ of $Cl^-(aq)$.

Equation (10.110) with $i =$ HCl, $M^{z+} = H^+$, and $X^{z-} = Cl^-$ gives

$$\Delta G^\circ_{f,298}[Cl^-(aq)] = -31.38 \text{ kcal/mol} \tag{10.113}$$

where the convention (10.104) was used. Similarly, (10.112) and (10.107) give

$$\Delta H^\circ_{f,298}[Cl^-(aq)] = -39.93 \text{ kcal/mol} \tag{10.114}$$

Also, $T\,\Delta S^\circ_f[HCl(aq)] = \Delta H^\circ_f[HCl(aq)] - \Delta G^\circ_f[HCl(aq)]$, so

$$\Delta S^\circ_{f,298}[HCl(aq)] = -28.68 \text{ cal mol}^{-1}\text{ K}^{-1}$$

$$\Delta S^\circ_f[HCl(aq)] = \bar{S}^\circ[HCl(aq)] - \tfrac{1}{2}\bar{S}^\circ[H_2(g)] - \tfrac{1}{2}\bar{S}^\circ[Cl_2(g)]$$

$$\bar{S}^\circ_{298}[HCl(aq)] = 13.57 \text{ cal mol}^{-1}\text{ K}^{-1} = \bar{S}^\circ[H^+(aq)] + \bar{S}^\circ[Cl^-(aq)]$$

$$\bar{S}^\circ_{298}[Cl^-(aq)] = 13.57 \text{ cal mol}^{-1}\text{ K}^{-1} \tag{10.115}$$

where (10.108) was used.

Example At 25°C and 1 atm, the differential heat of solution of KCl in water at infinite dilution is 4.11 kcal/mol. A saturated solution of KCl in water has a molality 4.82 mol/kg and a stoichiometric mean activity coefficient $\gamma_i = 0.588$. For pure KCl(*s*) at 25°C, $\Delta G^\circ_f = -97.59$ kcal/mol, $\Delta H^\circ_f = -104.18$ kcal/mol, and $\bar{S}^\circ = 19.8$ cal mol^{-1} K^{-1}. Find $\Delta G^\circ_{f,298}$, $\Delta H^\circ_{f,298}$, and $\bar{S}^\circ_{298}$ for $K^+(aq)$.

Subtraction of (10.103) from the equation preceding (10.100) gives for electrolyte i [analogous to (10.101) for a nonelectrolyte]:

$$\Delta G^\circ_{f,T}(i \text{ in soln.}) = \Delta G^\circ_{f,T}(\text{pure } i) - \nu RT \ln\left(\nu_\pm \gamma_{m,i,\text{sat}} m_{i,\text{sat}}/m^\circ\right) \tag{10.116}$$

For KCl, we have $\nu_+ = 1$, $\nu_- = 1$; hence $\nu = 2$ and $\nu_\pm = 1$. Substitution in (10.116) gives $\Delta G^\circ_{f,298}[KCl(aq)] = -98.82$ kcal/mol. Equations (10.110) and (10.113) give

$$\Delta G^\circ_{f,298}[K^+(aq)] = -67.44 \text{ kcal/mol}$$

For the heat of solution at infinite dilution, we have

$$4.11 \text{ kcal/mol} = \bar{H}^\infty[KCl(aq)] - \bar{H}^\circ[KCl(s)] = \bar{H}^\circ[KCl(aq)] - \bar{H}^\circ[KCl(s)]$$
$$= \bar{H}^\circ[K^+(aq)] + \bar{H}^\circ[Cl^-(aq)] - \bar{H}^\circ[KCl(s)]$$

Subtraction and addition of the standard enthalpies of K, $\frac{1}{2}Cl_2$, and e^- on the right side of this last equation gives

$$4.11 \text{ kcal/mol} = \Delta H^\circ_{f,298}[K^+(aq)] + \Delta H^\circ_{f,298}[Cl^-(aq)] - \Delta H^\circ_{f,298}[KCl(s)]$$

Use of (10.114) then gives $\Delta H^\circ_{f,298}[K^+(aq)] = -60.1$ kcal/mol.

To find the entropy of $K^+(aq)$, we note that ΔG°_{298} for dissolving KCl in water equals $\Delta G^\circ_{f,298}[KCl(aq)] - \Delta G^\circ_{f,298}[KCl(s)]$, which, from the above calculations, is -1.23 kcal/mol. Since ΔH°_{298} for the solution process is 4.11 kcal/mol, we find $\Delta S^\circ_{298} = 17.9$ cal mol^{-1} K^{-1} for dissolving KCl. Hence, using (10.115), we get

$$17.9 \text{ cal/(mol K)} = \bar{S}^\circ[KCl(aq)] - \bar{S}^\circ[KCl(s)]$$

$$= \bar{S}^\circ[K^+(aq)] + \bar{S}^\circ[Cl^-(aq)] - \bar{S}^\circ[KCl(s)]$$

$$\bar{S}^\circ_{298}[K^+(aq)] = 24.1 \text{ cal/(mol K)}$$

Values for standard-state conventional thermodynamic properties of some ions in aqueous solution at 25°C are listed in the Appendix.

10.9 NONIDEAL GAS MIXTURES

We now consider nonideal gas mixtures. The standard state of component i of a nonideal gas mixture is taken as pure gas i at the temperature T of the mixture, at 1 atm pressure, and such that i exhibits ideal-gas behavior. (This is the same choice of standard state as made in Sec. 5.1 for a pure nonideal gas and in Sec. 6.2 for a component of an ideal gas mixture.) The standard state of a real gas in a mixture is thus a fictitious state, but, as discussed earlier, thermodynamic properties of the fictitious standard state can be calculated once the behavior of the real gas is known.

The activity a_i of a component of a nonideal gas mixture is defined as in (10.3) as

$$a_i \equiv \exp\,[(\mu_i - \mu_i^\circ)/RT] \qquad (10.117)$$

where μ_i is the chemical potential of gas i in the mixture and μ_i° is the chemical potential of i in its standard state. We have, similar to (10.4),

$$\mu_i = \mu_i^\circ(T) + RT \ln a_i \qquad (10.118)$$

The choice of standard state (with $P = 1$ atm) makes μ_i° depend only on T for a component of a nonideal gas mixture. (For liquid or solid solutions, μ_i° depends on both T and P.)

The *fugacity* f_i of a component of any gas mixture is defined as $f_i \equiv a_i \times 1$ atm, so

$$f_i/P^\circ = a_i \qquad \text{where } P^\circ \equiv 1 \text{ atm} \qquad (10.119)$$

Since a_i is dimensionless, f_i has units of pressure. Since μ_i in (10.117) is an intensive property that depends on T, P, and the mixture's mole fractions, f_i is an intensive property that depends on these variables: $f_i = f_i(T, P, x_1, x_2, \ldots)$. Equation (10.118) becomes

$$\mu_i = \mu_i^\circ(T) + RT \ln (f_i/P^\circ) \qquad (10.120)^*$$

For an ideal gas mixture, (6.7) reads

$$\mu_{i,\text{id}} = \mu_i^\circ + RT \ln (P_i/P^\circ) \qquad \text{ideal gas mixt.} \tag{10.121}$$

Comparison with (10.120) shows that the fugacity f_i plays the same role in a nonideal gas mixture as the partial pressure P_i in an ideal gas mixture. Statistical mechanics shows that in the limit of zero pressure, μ_i approaches $\mu_{i,\text{id}}$. Moreover, μ_i° in (10.120) is the same as μ_i° in (10.121). Therefore f_i in (10.120) must approach P_i in the limit as the mixture's pressure P goes to zero:

$$P_i = \lim_{P\to 0} f_i \qquad \text{or} \qquad \lim_{P\to 0} (f_i/P_i) = 1 \tag{10.122}$$

For an ideal gas mixture, comparison of (10.120) and (10.121) gives $f_i = P_i$.

We now derive an expression for f_i in terms of measurable properties of the gas mixture. Equation (9.35) reads $(\partial\mu_i/\partial P)_{T,n_j} = \bar{V}_i$; hence $d\mu_i = \bar{V}_i\, dP$ at constant T and n_j, where constant n_j means that all mole numbers including n_i are held fixed. Equation (10.120) gives $d\mu_i = RT\, d \ln f_i$ at constant T and n_j. Equating these two expressions for $d\mu_i$, and integrating between states 1 and 2, we have $RT\, d \ln f_i = \bar{V}_i\, dP$ and

$$\ln \frac{f_{i,2}}{f_{i,1}} = \int_{P_1}^{P_2} \frac{\bar{V}_i}{RT}\, dP \qquad \text{const. } T,\ n_j \tag{10.123}$$

There are two fugacities in (10.123). To eliminate $f_{i,1}$, we take state 1 as the zero-pressure limit, since we know that $f_i \to P_i$ as $P \to 0$. However, taking state 1 as P and P_i equal to zero gives an infinity in (10.123), since $f_i \to 0$ as $P \to 0$. To avoid this infinity, we add $\ln (P_{i,1}/P_{i,2})$ to each side of (10.123). From (1.23), the partial pressure P_i of gas i in the mixture whose (total) pressure is P is defined as $P_i \equiv x_i P$. Hence, $\ln (P_{i,1}/P_{i,2}) = \ln (x_{i,1} P_1/x_{i,2} P_2) = \ln (P_1/P_2)$, since we are holding the composition constant ($x_{i,2} = x_{i,1}$). Addition of $\ln (P_{i,1}/P_{i,2}) = \ln (P_1/P_2)$ to (10.123) gives

$$\ln f_{i,2} - \ln f_{i,1} + \ln P_{i,1} - \ln P_{i,2} = \frac{1}{RT}\int_{P_1}^{P_2} \bar{V}_i\, dP + \ln P_1 - \ln P_2$$

$$\ln \frac{f_{i,2}}{P_{i,2}} - \ln \frac{f_{i,1}}{P_{i,1}} = \frac{1}{RT}\int_{P_1}^{P_2} \bar{V}_i\, dP - \int_{P_1}^{P_2} \frac{1}{P}\, dP \tag{10.124}$$

Letting P_1 go to zero and using (10.122), we get

$$\ln \frac{f_{i,2}}{P_{i,2}} = \int_0^{P_2} \left(\frac{\bar{V}_i}{RT} - \frac{1}{P}\right) dP \qquad \text{const. } T,\ n_j \tag{10.125}$$

where $P_{i,2} = x_i P_2$. To determine the fugacity of gas i in a mixture at temperature T, pressure P_2, and a certain composition, we must measure the partial molar volume $\bar{V}_i$ as a function of pressure at the desired temperature and composition; we then plot $\bar{V}_i/RT - 1/P$ vs. P and measure the area under the curve from $P = 0$ to $P = P_2$.

The deviation of the fugacity f_i from the partial pressure P_i in a gas mixture is measured by the *fugacity coefficient* χ_i, defined by

$$\chi_i \equiv f_i/x_i P \equiv f_i/P_i \qquad \text{or} \qquad f_i = \chi_i P_i \tag{10.126*}$$

Each χ_i equals 1 for an ideal gas mixture. For a nonideal mixture, χ_i (chi i) can be less than, equal to, or greater than 1, depending on T, P, and the composition.

For the special case of a one-component nonideal gas mixture, i.e., a pure nonideal gas, the partial molar volume $\bar{V}_i$ becomes the molar volume $\bar{V}$ of the gas, and Eqs. (10.117), (10.119), (10.120), (10.125), and (10.126) become

$$a \equiv \exp [(\mu - \mu^\circ)/RT], \qquad f \equiv aP^\circ, \qquad P^\circ \equiv 1 \text{ atm} \tag{10.127}$$

$$\mu = \mu^\circ(T) + RT \ln (f/P^\circ) \tag{10.128}$$

$$\ln \frac{f_2}{P_2} = \int_0^{P_2} \left(\frac{\bar{V}}{RT} - \frac{1}{P} \right) dP \qquad \text{const. } T \tag{10.129}$$

$$\chi \equiv f/P \qquad \text{or} \qquad f = \chi P \tag{10.130}$$

where f is a function of T and P; that is, $f = f(T, P)$.

The integral in (10.129) can be evaluated from measured values of $\bar{V}$ vs. P or from an analytic equation of state. In accord with the law of corresponding states (Sec. 8.5), different gases at the same reduced temperature and reduced pressure have approximately the same fugacity coefficient. Plots of χ (averaged over several gases) vs. P_R at various T_R values are available. [See R. H. Newton, *Ind. Eng. Chem.*, **27**, 302 (1935); R. H. Perry and C. H. Chilton, *Chemical Engineers' Handbook*, 5th ed., McGraw-Hill, 1973, p. **4**-52.]

Some experimental values of χ as a function of P for N_2 at 0°C and CO_2 at 75°C are:

P/atm	1	25	100	300	600	1000
$\chi(N_2)$	0.9996	0.99	0.97	1.00	1.24	1.84
$\chi(CO_2)$	0.9969	0.92	0.72	0.43	0.37	0.42

Equation (10.120) gives μ_i for each component of a nonideal gas mixture in terms of the fugacity f_i of i. Since all thermodynamic properties follow from μ_i (Sec. 10.1), we have in principle solved the problem of the thermodynamics of a nonideal gas mixture. The only hitch is that experimental evaluation of the fugacities from (10.125) requires a tremendous amount of work since the partial molar volume $\bar{V}_i$ of each component must be determined as a function of P; moreover, the fugacities so obtained apply to only one particular mixture composition. Usually, we want the f_i's for various mixture compositions, and for each such composition, the $\bar{V}_i$'s must be measured as a function of P and the integrations performed.

As a practical matter, one almost always estimates the mixture f_i's from fugacities of *pure* gases (which are comparatively easy to measure) using the approximation

$$f_i \approx x_i f_i^*(T, P) \tag{10.131}$$

where f_i is the fugacity of gas i in the mixture, x_i is its mole fraction in the mixture, and f_i^* is the fugacity of pure gas i at the temperature T and (total) pressure P of the mixture. Equation (10.131), the *Lewis fugacity rule*, works reasonably well in many cases. The Lewis fugacity rule yields a simple expression for the fugacity coefficient χ_i of a component of a nonideal gas mixture. For pure gas i Eq. (10.130) gives

$f_i^*(T, P) = \chi_i^*(T, P)P$. For gas i in the mixture Eq. (10.126) gives $f_i = \chi_i x_i P$. Substitution of these two expressions into (10.131) gives

$$\chi_i \approx \chi_i^*(T, P) \tag{10.132}$$

where χ_i is the fugacity coefficient of gas i in the mixture and $\chi_i^*(T, P)$ is the fugacity of pure gas i at the temperature T and (total) pressure P of the mixture.

A rough justification of (10.132) is as follows. The crudest approximation to the behavior of a gas mixture of species A, B, C, ... would be to assume zero intermolecular forces, in which case the mixture is ideal. A somewhat less crude picture would be to admit the existence of intermolecular forces but to ignore the differences between A-A, A-B, B-B, ... forces and take all intermolecular forces as the same. This picture is the molecular model of an ideal solution (Sec. 9.5) and leads to (9.62) for μ_i of each component. Thus we would have $\mu_i \approx \mu_i^*(T, P) + RT \ln x_i$. Subtraction of this equation from (10.120) gives $\ln (f_i/P^\circ x_i) \approx [\mu_i^*(T, P) - \mu_i^\circ(T)]/RT$. The right side of this last equation is a function of T and P only and is independent of x_i. Hence, the left side must remain constant for all values of x_i. If we let x_i go to 1 at constant T and P, the left side becomes $\ln (f_i^*/P^\circ)$, where f_i^* is the fugacity of pure i at T and P. Hence, $\ln (f_i/P^\circ x_i) \approx \ln (f_i^*/P^\circ)$, which gives (10.131).

For a nonideal gas mixture, f_i in (10.120) replaces P_i in (10.121) for an ideal gas mixture. Hence, to take gas nonideality into account, all the pressures and partial pressures in the liquid–vapor equilibrium equations of Sec. 10.2 are replaced by fugacities. For example, Eq. (10.20) becomes $f_i = \gamma_{\mathrm{I},i} x_i f_i^*$, where f_i^* is the fugacity of the vapor above pure liquid i at the temperature of the solution, x_i is the mole fraction of i in the liquid solution, $\gamma_{\mathrm{I},i}$ is its Convention I activity coefficient in the solution, and f_i is the fugacity of i in the nonideal gas mixture above the solution.

PROBLEMS

10.1 At 35°C, the vapor pressure of chloroform is 295.1 torr, and that of ethanol is 102.8 torr. A chloroform–ethanol solution at 35°C with $x(C_2H_5OH) = 0.200$ has a vapor pressure of 304.2 torr and has $x_{\mathrm{vap}}(C_2H_5OH) = 0.138$. (*a*) Calculate γ_I of chloroform and of ethanol in this solution. (*b*) Calculate ΔG for the mixing of 0.200 mol C_2H_5OH and 0.800 mol $CHCl_3$ at 35°C. (*c*) Calculate ΔG_{mxg} for the corresponding ideal solution.

10.2 For benzene–toluene solutions at 120°C, some liquid and vapor compositions and total vapor pressures are:

$x_{\mathrm{liq}}(C_6H_6)$	$x_{\mathrm{vap}}(C_6H_6)$	P/atm
0.00	0.00	1.34
0.136	0.200	1.51
0.262	0.363	1.68
0.587	0.711	2.22
0.850	0.912	2.70
0.927	0.961	2.85
1.00	1.00	2.98

For a solution at 120°C with $x_{\mathrm{liq}}(C_6H_6) = 0.262$ (*a*) calculate γ_I of benzene and of toluene; (*b*) calculate γ_{II} of benzene and of toluene if the solvent is taken as toluene; (*c*) calculate ΔG_{mxg} to form 125 g of this solution from the pure components at 120°C.

10.3 Activity coefficients of Zn in solutions of Zn in Hg at 25°C were determined by measurements on electrochemical cells. Hg is taken as the solvent. The data are fitted by $\gamma_{\mathrm{II,Zn}} = 1 - 3.92x_{\mathrm{Zn}}$ for solutions up to saturation. (*a*) Show that

$$\ln \gamma_{\mathrm{Hg}} = (2.92)^{-1}[3.92 \ln (1 - x_{\mathrm{Zn}}) - \ln (1 - 3.92x_{\mathrm{Zn}})]$$

for this composition range. (If you have forgotten the method of partial fractions, use a table of integrals.) (*b*) Calculate $\gamma_{\mathrm{II,Zn}}$, $a_{\mathrm{II,Zn}}$, γ_{Hg}, and a_{Hg} for $x_{\mathrm{Zn}} = 0.0400$.

10.4 For a fair number of solutions of two liquids, the Convention I activity coefficient of component 1 is well approximated by $\gamma_{\mathrm{I},1} = \exp bx_2^2$, where b is a function of T and P, and x_2 is the mole fraction of component 2. Such a solution is called a *simple* solution. Show that for a simple solution: (*a*) $\mu_1 = \mu_1^* + RT \ln x_1 + bRTx_2^2$; (*b*) $\gamma_{\mathrm{I},2} = \exp bx_1^2$. (*c*) Assume that acetone and chloroform form simple solutions at 35.2°C. Use the value of $\gamma_{\text{I}}(\text{CHCl}_3)$ at $x(\text{CHCl}_3) = 0$ in Fig. 10.1 to evaluate b. Then calculate γ_{I} for each component at $x(\text{CHCl}_3) = 0.494$ and compare with the experimental values.

10.5 For a dilute solution of a nonelectrolyte solute i, statistical mechanics and experimental data show that $\ln \gamma_{m,i} = Bm_i$, where B is a function of T and P. [Note from (10.90) and (10.83) that in a very dilute solution of an electrolyte, $\ln \gamma_{m,i}$ is proportional to $m_i^{1/2}$.] Use (10.27) and (10.32) to show that the dilute-solution behavior of the solvent activity coefficient (mole-fraction scale, of course) is

$$\ln \gamma_{\text{A}} = -B(\tfrac{1}{2}m_i^2 M_{\text{A}}^2 + \tfrac{1}{3}m_i^3 M_{\text{A}}^3)$$

10.6 (*a*) Use (9.25), (10.9), and (10.1) to show that the excess Gibbs energy $G^E \equiv G - G_{\text{id}}$ of a solution is $G^E = RT \sum_i n_i \ln \gamma_{\text{I},i}$. Hence, G^E can be calculated from the activity coefficients. (*b*) Show that $G^E = \Delta G_{\text{mxg}} - \Delta G_{\text{mxg,id}}$, where $\Delta G_{\text{mxg,id}}$ is given by (9.71). Hence, G^E is sometimes called the excess Gibbs energy of mixing. (*c*) Show that $G^E = H^E - TS^E$, where the excess enthalpy and entropy are defined by $H^E \equiv H - H_{\text{id}}$ and $S^E \equiv S - S_{\text{id}}$. (*d*) Show that $H^E = \Delta H_{\text{mxg}}$.

10.7 Use data in Sec. 10.2 and the result of Prob. 10.6*a* to calculate and plot G^E/n_{tot} for acetone–chloroform solutions at 35.2°C, where n_{tot} is the total number of moles in the solution.

10.8 Derive (10.38).

10.9 (*a*) Derive the first equation in (10.42) by equating (10.39) to (10.29) and taking the limit of the resulting equation as $x_{\text{A}} \to 1$. Then use this result to derive the second equation in (10.42). (*b*) Start with Eqs. (10.33) and (10.34) and derive Eqs. (10.31) and (10.32) by equating (10.33) to (10.29) and taking the limit as $x_{\text{A}} \to 1$.

10.10 Derive (10.72).

10.11 Starting from

$$G = n_{\text{A}}\mu_{\text{A}} + n_+\mu_+ + n_-\mu_- + n_{IP}\mu_{IP},$$

derive Eq. (10.74) for an electrolyte solution.

10.12 At 25°C, the vapor pressure of a 4.800 mol/kg solution of KCl in water is 20.02 torr. The vapor pressure of pure water at 25°C is 23.76 torr. For the solvent in this KCl solution, find (*a*) ϕ; (*b*) γ_{A}; (*c*) a_{A}.

10.13 Robinson and Stokes used vapor-pressure data to determine practical osmotic coefficients in aqueous solutions of sucrose (supplied by the Colonial Sugar Refining Company) at 25°C. They found the following expression to reproduce their results:

$$\phi = 1 + 0.07028(m/m^\circ) + 0.01847(m/m^\circ)^2 - 0.004045(m/m^\circ)^3 + 0.000228(m/m^\circ)^4$$

where m is the sucrose molality and $m^\circ \equiv 1$ mol/kg. (*a*) Use (10.80) to show that the sucrose molality-scale activity coefficients are

$$\log \gamma_m = 0.06104(m/m^\circ) + 0.01203(m/m^\circ)^2 - 0.002342(m/m^\circ)^3 + 0.000124(m/m^\circ)^4$$

(*b*) Calculate γ_m and γ_{II} of sucrose in a 6.00 mol/kg aqueous solution at 25°C.

10.14 Use the Debye–Hückel equation (10.92) to estimate (*a*) $\gamma_\pm$ for a 0.01 mol/kg aqueous solution of CuSO_4 at 25°C; (*b*) $\gamma_\pm$ of CuSO_4 in an aqueous solution at 25°C that has CuSO_4 molality 0.01 mol/kg, MgCl_2 molality 0.01 mol/kg, and $\text{Al(NO}_3)_3$ molality 0.01 mol/kg. (Neglect ion pairing.)

10.15 (*a*) Use (10.79) and (10.70) in the Gibbs–Duhem equation (10.75) to show that

$$d \ln \gamma_i = d\phi + [(\phi - 1)/m_i]\, dm_i \qquad \text{const. } T, P$$

(*b*) Use (10.76), (10.53), (8.41), and (10.64) to show that $\phi \to 1$ as $x_{\text{A}} \to 1$. (*c*) Show that integration of the result of (*a*) gives Eq. (10.80).

10.16 (*a*) Show that the result of Prob. 10.15*a* can be written as $d(m_i\phi) = dm_i + m_i\, d \ln \gamma_i$ at constant T and P. Then integrate this equation to show that

$$\phi(m) = 1 + \frac{1}{m}\int_0^m m_i\, d \ln \gamma_i \qquad \text{const. } T, P$$

(*b*) Use the Debye–Hückel limiting law (10.90) to show that in an electrolyte solution

$$\phi = 1 - \tfrac{1}{3}z_+|z_-|CI_m^{1/2} \qquad \text{very dil. soln.}$$

10.17 Use the result of Prob. 10.16*b* to calculate ϕ in 0.00100 and 0.0100 mol/kg aqueous solutions of HBr at 25°C.

10.18 For a solution of a single strong electrolyte with no ion pairing, show that $I_m = \frac{1}{2}z_+|z_-|\nu m_i$.

10.19 For $\text{Pb(NO}_3)_2$, the fraction of Pb^{2+} ions that associate with NO_3^- ions to form ion pairs is known to be $1 - \alpha = 0.43$ in a 0.100 mol/kg aqueous solution at 25°C. (*a*) Calculate I_m of this solution. (Note that the ion pair is charged.) (*b*) Use the Davies equation to calculate $\gamma_\pm$ for this solution. Then calculate γ_i. (The experimental γ_i is 0.395.)

10.20 Derive the following equations for partial molar properties of a solute in a nonelectrolyte solution:

$$\bar{S}_i = \bar{S}^\circ_{m,i} - R \ln (\gamma_{m,i} m_i/m^\circ) - RT(\partial \ln \gamma_{m,i}/\partial T)_{P,n_j}$$

$$\bar{V}_i = \bar{V}^\circ_{m,i} + RT(\partial \ln \gamma_{m,i}/\partial P)_{T,n_j}$$

$$\bar{H}_i = \bar{H}^\circ_{m,i} - RT^2(\partial \ln \gamma_{m,i}/\partial T)_{P,n_j}$$

10.21 Use data in the Appendix to find ΔG°_{298}, ΔH°_{298}, and ΔS°_{298} for (*a*) $H^+(aq) + OH^-(aq) \rightarrow H_2O(l)$; (*b*) $CO_3{}^{2-}(aq) + 2H^+(aq) \rightarrow H_2O(l) + CO_2(g)$.

10.22 (*a*) For a pure gas that obeys the virial equation (8.6), show that

$$\ln \chi = A_2(T)P + \tfrac{1}{2}A_3(T)P^2 + \tfrac{1}{3}A_4(T)P^3 + \cdots$$

(*b*) Use (8.34) to show that for a van der Waals gas

$$\ln \chi = \frac{bRT - a}{R^2T^2}P + \frac{2abRT - a^2}{2R^4T^4}P^2 + \cdots$$

10.23 For CO_2, the critical temperature and pressure are 304.2 K and 72.8 atm. Assume CO_2 obeys the van der Waals equation and use the result of Prob. 10.22*b* to estimate χ for CO_2 at 1.00 atm and 25°C and at 50.0 atm and 25°C. (*b*) Use the Lewis fugacity rule to estimate the fugacity and fugacity coefficient of CO_2 in a mixture of 1.00 mol CO_2 and 9.00 mol O_2 at 25°C and 50.0 atm.

10.24 (*a*) Calculate ΔG when 1.000 mol of an ideal gas at 0°C is isothermally compressed from 1.000 to 1000 atm. (*b*) Use data in Sec. 10.9 to calculate ΔG when 1.000 mol of N_2 is isothermally compressed from 1.000 to 1000 atm at 0°C.

10.25 For CH_4 at -50°C, measured values of $\bar{V}$ as a function of P are:

P/atm	$\bar{V}$/(cm^3/mol)
1	18,224
10	1,743
20	828
40	366
60	207
80	128.7
100	91.4
120	76.3

(*a*) Find the fugacity and fugacity coefficient of CH_4 at -50°C and 120 atm. (Note from Prob. 8.15 that $\bar{V} - RT/P$ does *not* go to zero as the gas pressure goes to zero.) (*b*) Give the value of the second virial coefficient B_2 for CH_4 at -50°C.

11

REACTION EQUILIBRIUM IN NONIDEAL SYSTEMS

11.1 THE ACTIVITY EQUILIBRIUM CONSTANT

This chapter deals with reaction equilibrium in nonideal systems. For the chemical reaction $\sum_i \nu_i A_i = 0$ of Eq. (4.107), the reaction equilibrium condition is given by (4.111) as $\sum_i \nu_i \mu_{i,\text{eq}} = 0$, where $\mu_{i,\text{eq}}$ is the equilibrium value of the chemical potential of the ith species.

To obtain a convenient expression for μ_i of each species A_i we proceed as follows. (1) We choose a convenient standard state for each species i and designate its chemical potential in that standard state by μ_i°. (2) We define the *activity* a_i of species i by

$$a_i \equiv e^{(\mu_i - \mu_i^\circ)/RT} \tag{11.1}$$

where $\mu_i(T, P, x_1, x_2, \ldots)$ is the chemical potential of i in the reaction mixture and x_1, x_2, ... are the mole fractions in the phase in which i occurs. (Species i can occur in more than one phase, but at equilibrium its chemical potential in any phase in which it occurs is the same.) Note that a_i depends on the choice of standard state and is meaningless unless the standard state has been specified. From (11.1), a_i depends on the same variables as μ_i; the activity a_i is a dimensionless intensive property. The reader should realize the purely formal nature of what we are doing. We are simply choosing a convenient definition of a_i and seeing where it leads. [Comparison of (11.1) with (10.3) for liquid and solid solutions and with (10.117) for gas mixtures shows that a_i in (11.1) is what we previously defined to be the activity of a component of a solid, liquid, or gaseous mixture.] Table 11.1 in Sec. 11.7 summarizes the various choices of standard states.

Taking logs of (11.1), we get

$$\mu_i = \mu_i^\circ + RT \ln a_i \tag{11.2}*$$

We now substitute (11.2) into the equilibrium condition $\sum_i \nu_i \mu_{i,\text{eq}} = 0$ and go through the same manipulations used to derive (6.17). We have

$$\sum_i \nu_i \mu_i^\circ + RT \sum_i \nu_i \ln a_{i,\text{eq}} = 0$$

where $a_{i,\text{eq}}$ is the equilibrium value of the activity a_i. The first sum in this equation is defined to be ΔG°, the standard Gibbs energy change for the reaction (reactants and products each in standard states). We have $\nu_i \ln a_{i,\text{eq}} = \ln (a_{i,\text{eq}})^{\nu_i}$; also, the sum of logs equals the log of the product. Hence

$$\Delta G^\circ + RT \ln \prod_i (a_{i,\text{eq}})^{\nu_i} = 0$$

Defining K_a to be the product in this equation, we have

$$\Delta G^\circ = -RT \ln K_a \tag{11.3*}$$

$$\Delta G^\circ \equiv \sum_i \nu_i \mu_i^\circ \tag{11.4*}$$

$$K_a \equiv \prod_i (a_{i,\text{eq}})^{\nu_i} \tag{11.5*}$$

K_a is the *activity equilibrium constant* or the *standard equilibrium constant*. We always choose the standard states so that μ_i° depends at most on T and P. (For gases, μ_i° depends only on T.) Hence ΔG° depends at most on T and P, and K_a, which equals $e^{-\Delta G^\circ/RT}$, depends at most on T and P (and not on any mole fractions). We have thus "solved" the problem of chemical equilibrium in an arbitrary system. The equilibrium position occurs when the activities are such that $\prod_i (a_i)^{\nu_i}$ equals the equilibrium constant K_a, where K_a is found from (11.3) as $e^{-\Delta G^\circ/RT}$. To solve the problem in a practical sense, we must be able to obtain expressions for the activities of the reacting species.

11.2 REACTION EQUILIBRIUM IN NONELECTROLYTE SOLUTIONS

To apply the results of the last section to solutions of nonelectrolytes we choose one of the conventions of Chap. 10 and introduce the appropriate expressions for the activities a_i into the equilibrium constant K_a of (11.5).

Most commonly, one component of the solution is designated as the solvent. For the solvent, we always use the mole-fraction scale [Eqs. (10.35) and (10.36)]. For the solutes, one can use the mole-fraction scale, the molality scale, or the concentration scale.

If the mole-fraction scale is used for the solutes, then for the activity $a_{x,i}$ of species i Eq. (10.7) gives $a_{x,i} = \gamma_{\text{II},i} x_i$, where γ_{II} denotes the Convention II activity coefficient (Sec. 10.1) and the subscript x on a reminds us that the activity depends on which scale is used. The equilibrium constant K_a in (11.5) becomes

$$K_x = \prod_i (\gamma_{\text{II},i} x_i)^{\nu_i} \tag{11.6}$$

where the subscript on K denotes use of the mole-fraction scale, and where the eq subscript has been omitted for simplicity. Equations (11.3) and (11.4) become

$$\Delta G_x^\circ \equiv \sum_i \nu_i \mu_{\text{II},i}^\circ = -RT \ln K_x \tag{11.7}$$

where the subscript x on ΔG° indicates that the mole-fraction scale is used.

Although the mole-fraction scale for solutes is to be preferred from a theoretical viewpoint, most thermodynamic data for species in aqueous solution are tabulated based on the molality scale. Hence, one most commonly uses the molality scale for solutes. From (10.33) and (11.2), the activity $a_{m,i}$ of solute i on the molality scale is $a_{m,i} = \gamma_{m,i} m_i/m^\circ$ $(i \neq \mathrm{A})$, where the standard molality m° equals 1 mol/kg. Hence the equilibrium constant (11.5) becomes

$$K_m^\circ = (\gamma_{x,\mathrm{A}} x_\mathrm{A})^{\nu_\mathrm{A}} \prod_{i \neq \mathrm{A}} (\gamma_{m,i} m_i/m^\circ)^{\nu_i} \tag{11.9}$$

As in Sec. 6.2, the degree superscript on K indicates a dimensionless equilibrium constant; if the standard molality m° were omitted from (11.9), we would have an equilibrium constant with dimensions of molality raised to the $\Delta|\nu|$, where $\Delta|\nu|$ is given by (6.33). Note that the mole-fraction scale is retained for the solvent (species A), so that the solvent activity in (11.9) is different in form from the solute activities. The solvent stoichiometric coefficient ν_A equals zero if the solvent does not appear in the net chemical reaction. If the solution is quite dilute, both x_A and $\gamma_{x,\mathrm{A}}$ are quite close to 1 and it is a good approximation to omit the factor $(\gamma_{x,\mathrm{A}} x_\mathrm{A})^{\nu_\mathrm{A}}$ from K_m°. For the molality scale, Eqs. (11.3) and (11.4) become

$$\Delta G_m^\circ = -RT \ln K_m^\circ \tag{11.10}$$

$$\Delta G_m^\circ \equiv \nu_\mathrm{A} \mu_{\mathrm{I,A}}^\circ + \sum_{i \neq \mathrm{A}} \nu_i \mu_{m,i}^\circ \tag{11.11}$$

Note that μ° of the solvent cannot be omitted from ΔG_m° (unless $\nu_\mathrm{A} = 0$) even if the solution is quite dilute.

Sometimes the concentration scale is used for activities. From (10.39) and (11.2) we have for the concentration-scale activity of a solute: $a_{c,i} = \gamma_{c,i} c_i/c^\circ$. The equations for K_c° and ΔG° are the same as (11.9) to (11.11) except that the letter c is replaced by m throughout. For example,

$$\Delta G_c^\circ = -RT \ln K_c^\circ \tag{11.12}$$

Note that K_x, K_c°, and K_m° have different values for the same reaction. Likewise, ΔG_x°, ΔG_c°, and ΔG_m° differ for the same reaction, since the value of the standard-state quantity μ_i° depends on the choice of standard state for species i. Thus, in using Gibbs free-energy data to calculate equilibrium compositions, one must be clear what the choice of standard state is for the tabulated data.

To apply the above expressions to calculate equilibrium compositions, we use the procedures discussed in Chap. 10 to determine activity coefficients. If the solution is dilute, it is a good approximation to set the activity coefficients equal to 1; the equilibrium constant K_x in (11.6) then reduces to the expression (9.113) for ideally dilute solutions.

11.3 REACTION EQUILIBRIUM IN ELECTROLYTE SOLUTIONS

The most commonly studied solution equilibria are ionic equilibria in aqueous solutions. As well as being important in inorganic chemistry, ionic equilibria are significant in biochemistry; for the majority of biologically important reactions at least some of the species involved are ions. Examples include the organic phosphates (such as adenosine triphosphate, ATP) and the anions of certain acids (such as citric acid) involved in metabolic energy transformations; inorganic ions like H_3O^+ and Mg^{2+} participate in many biochemical reactions.

Since thermodynamic data for ionic species are usually tabulated for the molality-scale standard state (Sec. 10.3), we shall use the molality-scale equilibrium constant K_m° of Eq. (11.9) for electrolytes.

Many ionic reactions in solution are acid–base reactions. We adopt the Brønsted definition of an *acid* as a proton donor and a *base* as a proton acceptor.

The water molecule is *amphoteric*; i.e., water can act as either an acid or a base. In pure liquid water and in aqueous solutions, the following ionization reaction occurs to a slight extent:

$$H_2O + H_2O = H_3O^+ + OH^- \tag{11.13}$$

where one H_2O molecule donates a proton to the other. For (11.13), the activity equilibrium constant (11.5) is

$$K_w = \frac{a(H_3O^+)a(OH^-)}{[a(H_2O)]^2} \tag{11.14}$$

where the subscript w (for water) is traditional. Since the standard state of the solvent H_2O is pure H_2O, we have $a(H_2O) = 1$ for pure H_2O [Eq. (10.3)]. In aqueous solutions, $a(H_2O) = \gamma_x(H_2O)x(H_2O)$. For *dilute* aqueous solutions, the mole fraction $x(H_2O)$ is close to 1, and (since H_2O is an uncharged species) $\gamma_x(H_2O)$ is close to 1. Thus we usually approximate $a(H_2O)$ as 1 in dilute aqueous solutions. Hence K_w becomes

$$a(H_3O^+)a(OH^-) = [\gamma(H_3O^+)m(H_3O^+)/m^\circ][\gamma(OH^-)m(OH^-)/m^\circ]$$

where $m^\circ \equiv 1$ mol/kg, and where we omitted the subscript m from γ; all unsubscripted activity coefficients in this section will be understood to be on the molality scale. The above expression differs from Eq. (11.9) by the omission of the solvent's activity. Introduction of the mean molal ionic activity coefficient $\gamma_\pm$ of Eq. (10.59) gives

$$K_w^\circ = \gamma_\pm{}^2 m(H_3O^+)m(OH^-)/(m^\circ)^2 \qquad \text{dil. aq. soln.} \tag{11.15*}$$

Experiment (Chap. 14) gives the value 1.00×10^{-14} for K_w° at 25°C. Approximating $\gamma_\pm$ as 1 in pure water, we get $m(H_3O^+) = m(OH^-) = 1.00 \times 10^{-7}$ mol/kg in pure water at 25°C. This gives an ionic strength $I_m = 1.00 \times 10^{-7}$ mol/kg; the Davies equation (10.93) then gives $\gamma_\pm = 0.9996$ in pure water, which is essentially equal to 1. Hence the H_3O^+ and OH^- molalities are accurately equal to 1.00×10^{-7} mol/kg in pure water at 25°C. In an aqueous solution that is not extremely dilute, $\gamma_\pm$ in (11.15) will probably not be close to 1. (Since μ° for each species in solution depends on pressure, the quantity ΔG° for the reaction depends on pressure and the equilibrium constant for a reaction in solution depends on pressure. However, this dependence is weak. Ordinarily, equilibrium constants in solution are determined for P at or near 1 atm, and this value of P is assumed throughout this section.)

Next, consider the ionization of the weak acid HX in aqueous solution. The ionization reaction and the equilibrium constant (11.9) are

$$HX + H_2O = H_3O^+ + X^- \tag{11.16}$$

$$K_a^\circ = \frac{[\gamma(H_3O^+)m(H_3O^+)/m^\circ][\gamma(X^-)m(X^-)/m^\circ]}{\gamma(HX)m(HX)/m^\circ} \tag{11.17}$$

where the subscript a (for acid) is traditional, and where the activity of the solvent (H_2O) is approximated by 1 in dilute solutions. In most applications, the HX molality is rather low, and it is a good approximation to take $\gamma = 1$ for the uncharged species HX. However, even though the X^- and H_3O^+ molalities are usually much less than the HX molality, we cannot set $\gamma = 1$ for these ions; as noted at the end of Sec. 10.5, γ for an ion deviates significantly from 1 even in quite dilute solutions. Using (10.59), we have

$$K_a = \frac{\gamma_\pm{}^2 m(H_3O^+)m(X^-)}{m(HX)} \qquad \text{dil. soln.} \qquad (11.18)$$

where $\gamma_\pm$ is for the pair of ions H_3O^+ and X^- and differs from $\gamma_\pm$ in (11.15). In (11.18) we have omitted dividing each molality by the standard molality m° (= 1 mol/kg), so that K_a has the dimensions of molality (mol/kg); correspondingly, the degree superscript on K_a is omitted. [Note that the subscript a stands for different things in (11.17) and (11.5).]

As an example, consider a 0.100 mol/kg aqueous solution of acetic acid ($HC_2H_3O_2$). Experiment (Chap. 14) gives $K_a = 1.75 \times 10^{-5}$ mol/kg for acetic acid at 25°C. To solve (11.18) for $m(H_3O^+)$ we need $\gamma_\pm$. To use the Davies equation (10.93) to estimate $\gamma_\pm$ we need the ionic strength I_m, which can't be calculated until $m(H_3O^+)$ is known. The solution to this dilemma is to first estimate $m(H_3O^+)$ and $m(X^-)$ by setting $\gamma_\pm = 1$ in (11.18) and solving for the ionic molalities. With these approximate molalities, we calculate an approximate I_m and then use the Davies equation to find an approximate $\gamma_\pm$, which we use in (11.18) to find a more accurate value for the molalities. If necessary, we can then use these more accurate molalities to find a more accurate I_m, and so on. Let $m(X^-) = x$. Equation (11.16) gives $m(HX) = 0.100$ mol/kg $- x$ and $m(H_3O^+) = x$ [since the H_3O^+ formed by (11.13) is negligible compared with that from the acetic acid]. Setting $\gamma_\pm = 1$ in (11.18), we have

$$1.75 \times 10^{-5}\ \text{mol/kg} \approx \frac{x^2}{0.100\ \text{mol/kg} - x} \qquad (11.19)$$

We can solve (11.19) using the quadratic formula, but a faster method is an iterative solution, as follows. Since K_a is much less than the acid's stoichiometric molality (0.100 mol/kg), the degree of ionization will be slight and 0.100 mol/kg $- x$ can be well approximated by 0.100 mol/kg. (In extremely dilute solutions the degree of ionization is substantial, and this approximation cannot be made.) Therefore $x^2/(0.100\ \text{mol/kg}) \approx 1.75 \times 10^{-5}$ mol/kg, and $x \approx 1.32 \times 10^{-3}$ mol/kg. With this value of x, the denominator in (11.19) becomes 0.100 mol/kg $-$ 0.001 mol/kg $=$ 0.099 mol/kg. Hence, $x^2/(0.099\ \text{mol/kg}) \approx 1.75 \times 10^{-5}$ mol/kg, and we again get $x \approx 1.32 \times 10^{-3}$ mol/kg.

Thus, with $\gamma_\pm$ taken as 1, we find $m(H_3O^+) = m(X^-) \approx 1.32 \times 10^{-3}$ mol/kg and $I_m \approx 1.32 \times 10^{-3}$ mol/kg [Eq. (10.83)]. The Davies equation (10.93) then gives $\gamma_\pm = 0.960$. Equation (11.18) becomes

$$1.75 \times 10^{-5}\ \text{mol/kg} = \frac{(0.960)^2 x^2}{0.100\ \text{mol/kg} - x}$$

Solving iteratively, as above, we get $x = 1.37 \times 10^{-3}$ mol/kg $= m(H_3O^+)$. This value is not greatly changed from our initial estimate of 1.32×10^{-3} mol/kg, so there is little point in redoing the calculation with further improved I_m and $\gamma_\pm$ values.

In the example just given, I_m is quite low, so it doesn't make much difference whether we include $\gamma_\pm$ or not. Consider, however, the calculation of $m(H_3O^+)$ in a 25°C solution that has the stoichiometric molalities $m(HC_2H_3O_2) = 0.100$ mol/kg and $m(NaC_2H_3O_2) = 0.100$ mol/kg (a buffer solution). The salt $NaC_2H_3O_2$ (which is a 1 : 1 electrolyte) exists virtually entirely in the form of positive and negative ions in solution; also, the ionization of $HC_2H_3O_2$ will contribute little to I_m in comparison with the contribution from the $NaC_2H_3O_2$. Hence, $I_m = \frac{1}{2}(0.100 + 0.100)$ mol/kg $= 0.100$ mol/kg. The Davies equation then gives $\gamma_\pm = 0.781$. Substitution into (11.18) gives

$$1.75 \times 10^{-5} \text{ mol/kg} = \frac{(0.781)^2 m(H_3O^+)(0.100 \text{ mol/kg})}{0.100 \text{ mol/kg}}$$

where $m(C_2H_3O_2^-)$ is well approximated by considering only the acetate ion from the sodium acetate, and where we set $m(HC_2H_3O_2)$ equal to 0.100 mol/kg since the degree of ionization of acetic acid is far less than in the previous example, due to the added sodium acetate (common-ion effect). Solving, we find $m(H_3O^+) = 2.87 \times 10^{-5}$ mol/kg. Note that if $\gamma_\pm$ were omitted in this example, we would get 1.75×10^{-5} mol/kg for $m(H_3O^+)$, which is in error by a whopping 39 percent. Except for solutions of quite low ionic strength, ionic equilibrium calculations that omit activity coefficients are likely to give only qualitatively correct answers. Many of the ionic equilibrium calculations you did in freshman chemistry are correct only in the exponent of 10, due to neglect of activity coefficients (and neglect of ionic association in salt solutions).

As another example, consider a 1.0×10^{-4} mol/kg aqueous solution of the weak acid HIO, which has $K_a = 2.3 \times 10^{-11}$ mol/kg at 25°C. We have

$$HIO + H_2O = H_3O^+ + IO^- \tag{11.20}$$

$$K_a = \gamma_\pm\, m(H_3O^+) m(IO^-)/m(HIO) \tag{11.21}$$

where $\gamma(HIO)$ and $a(H_2O)$ have each been taken as 1. Because of the extremely low value of K_a, the ionic strength is extremely low, and we can take $\gamma_\pm = 1$. If we proceed as we did with $HC_2H_3O_2$, we have $m(H_3O^+) = m(IO^-) = x$ and $m(HIO) =$ 0.00010 mol/kg $- x$. Since K_a is far less than the stoichiometric molality, we can set 0.00010 mol/kg $- x$ equal to 0.00010 mol/kg. We have

$$2.3 \times 10^{-11} \text{ mol/kg} = x^2/(0.00010 \text{ mol/kg})$$

which gives $x = 4.8 \times 10^{-8}$ mol/kg $= m(H_3O^+)$. However, this answer cannot possibly be correct. We know that $m(H_3O^+)$ equals 1.0×10^{-7} mol/kg in pure water. A solution with $m(H_3O^+) = 4.8 \times 10^{-8}$ mol/kg would have a lower H_3O^+ molality than pure water and would therefore be basic; however, HIO is an acid. The error in this calculation is failure to consider the contribution to $m(H_3O^+)$ from the ionization (11.13) of water. In previous examples, the H_3O^+ from the weak-acid ionization far exceeded that due to the ionization of water, but this is not true here. We must thus consider the two simultaneous equilibria (11.20) and (11.13).

When two or more simultaneous ionic equilibria occur, the following systematic procedure can be used:

1. Write down the equilibrium-constant expression for each reaction.
2. Write down the condition for electrical neutrality of the solution.
3. Write down relations that express the conservation of matter for substances added to the solution.
4. Solve the resulting set of simultaneous equations, making judicious approximations where possible.

We shall apply this procedure to the above problem of finding $m(H_3O^+)$ in a 1.0×10^{-4} mol/kg aqueous solution of HIO. The species in solution are H_2O, H_3O^+, OH^-, HIO, and IO^-. The species H_2O does not enter into the equilibrium expressions in this dilute solution. The equilibrium-constant expressions for the equilibria (11.13) and (11.20) are

$$K_w = m(H_3O^+)m(OH^-) \tag{11.22}$$

$$K_a = m(H_3O^+)m(IO^-)/m(HIO) \tag{11.23}$$

(Since the ionic strength is extremely low, all activity coefficients are taken as 1.) For electrical neutrality, the total charge on the positive ions must equal that on the negative ions; therefore

$$m(H_3O^+) = m(OH^-) + m(IO^-) \tag{11.24}$$

Let m be the stoichiometric HIO molality, where $m = 1.0 \times 10^{-4}$ mol/kg in this problem. The IO group of atoms exists in the solution in the two forms HIO and IO^-. Conservation of matter requires that the sum of the HIO and IO^- molalities must equal the HIO molality before ionization:

$$m = m(HIO) + m(IO^-) \tag{11.25}$$

We have the four equations (11.22) to (11.25) in the four unknowns $m(H_3O^+)$, $m(OH^-)$, $m(HIO)$, and $m(IO^-)$. Using (11.22) to eliminate $m(OH^-)$ and (11.25) to eliminate $m(HIO)$, we find that (11.23) and (11.24) become

$$K_a = \frac{m(H_3O^+)m(IO^-)}{m - m(IO^-)} \qquad \text{and} \qquad m(H_3O^+) = \frac{K_w}{m(H_3O^+)} + m(IO^-) \tag{11.26}$$

Solving the second equation of (11.26) for $m(IO^-)$ and substituting into the first equation of (11.26), we get

$$[m(H_3O^+)]^3 + [m(H_3O^+)]^2 K_a - (K_w + mK_a)m(H_3O^+) - K_a K_w = 0 \tag{11.27}$$

which is a cubic equation that can be solved to give the H_3O^+ molality.

To avoid the tedium of solving cubic equations, we solve (11.27) for m as a function of $m(H_3O^+)$ to obtain

$$m = [m(H_3O^+)]^2/K_a + m(H_3O^+) - K_w/K_a - K_w/m(H_3O^+) \tag{11.28}$$

We then calculate m for a range of assumed values of $m(H_3O^+)$ and plot $m(H_3O^+)$ vs. the stoichiometric molality m. (See Prob. 11.7.) This graph can be used to estimate $m(H_3O^+)$ at any given stoichiometric molality.

As is evident, the drawback of the systematic procedure is that it leads to quite complicated equations if approximations are not made. To get a simpler, but still accurate, solution, we recognize that since $K_a \ll m$ in this example, nearly all the IO will be in the form of un-ionized HIO, so that $m(\text{IO}^-) \ll m$ and the first equation of (11.26) becomes $K_a = m(\text{H}_3\text{O}^+)m(\text{IO}^-)/m$. Substitution for $m(\text{IO}^-)$ from the second equation of (11.26) gives

$$m(\text{H}_3\text{O}^+) = (K_w + mK_a)^{1/2}$$

Substitution of numerical values gives $m(\text{H}_3\text{O}^+) = 1.1 \times 10^{-7}$ mol/kg, and the solution is slightly acidic, as expected.

For an aqueous solution of the weak acid HX with stoichiometric molality m, the *degree of dissociation* α is defined as

$$\alpha \equiv \frac{m(\text{X}^-)}{m} = \frac{m(\text{X}^-)}{m(\text{X}^-) + m(\text{HX})} = \frac{1}{1 + m(\text{HX})/m(\text{X}^-)} = \frac{1}{1 + \gamma_\pm{}^2 m(\text{H}_3\text{O}^+)/K_a} \tag{11.29}$$

where (11.18) was used. As m goes to zero, $\gamma_\pm$ goes to 1. Also, as m goes to zero, the contribution of the ionization of HX to $m(\text{H}_3\text{O}^+)$ becomes negligible and all the H_3O^+ comes from the ionization of water. Hence in the limit of infinite dilution, $m(\text{H}_3\text{O}^+)$ becomes $K_w^{1/2}$, and the degree of dissociation of HX approaches

$$\alpha^\infty = \frac{1}{1 + K_w^{1/2}/K_a} = \frac{1}{1 + (10^{-7}\ \text{mol kg}^{-1})/K_a} \qquad \text{at } 25^\circ\text{C}$$

At 25°C, an acid with $K_a = 10^{-5}$ mol/kg is 99 percent dissociated at infinite dilution; however, an acid with $K_a = 10^{-7}$ mol/kg is only 50 percent dissociated at infinite dilution. The H_3O^+ from water partially suppresses the ionization of the weak acid at infinite dilution. The acid HIO with $K_a = 2.3 \times 10^{-11}$ mol/kg is only 0.02 percent dissociated at infinite dilution.

Other types of aqueous ionic equilibria include reactions of cationic and anionic acids and bases (for example, $\text{NH}_4{}^+$, $\text{C}_2\text{H}_3\text{O}_2{}^-$, CO_3^{2-}) with water (Prob. 11.8); solubility equilibria (Sec. 11.4); association equilibria involving complex ions [the equilibrium constants for the reactions $\text{Ag}^+ + \text{NH}_3 = \text{Ag(NH}_3)^+$ and $\text{Ag(NH}_3)^+ + \text{NH}_3 = \text{Ag(NH}_3)_2{}^+$ are called *association* or *stability* constants, while the equilibrium constants for the reverse reactions are called *dissociation* constants]; association equilibria to form ion pairs (for example, $\text{Sr}^{2+} + \text{IO}_3{}^- = \text{SrIO}_3{}^+$), as discussed in Sec. 10.7.

Experimental values of ionic equilibrium constants (including ionization constants of acids and bases, solubility-product constants, stability constants of complex ions, and constants for ion-pair formation) are tabulated in L. G. Sillen and A. E. Martell, *Stability Constants of Metal-Ion Complexes*, The Chemical Society, London, 1964, and *Suppl.* 1, 1971.

Instead of listing the equilibrium constant K, tables often give pK, where p$K \equiv -\log K$.

Perhaps the reader has been a bit disturbed at the use of molalities, since it is traditional to use concentrations in freshman-chemistry equilibrium calculations. We now show the relation between these two approaches.

The concentration-scale equilibrium constant for the ionization of the acid HX is

$$K_{c,a} = \frac{[\gamma_c(H_3O^+)c(H_3O^+)][\gamma_c(X^-)c(X^-)]}{\gamma_c(HX)c(HX)a(H_2O)}$$

The ratio of the concentration-scale equilibrium constant to the molality-scale equilibrium constant is

$$\frac{K_{c,a}}{K_{m,a}} = \frac{[\gamma_c(H_3O^+)c(H_3O^+)/\gamma_m(H_3O^+)m(H_3O^+)][\gamma_c(X^-)c(X^-)/\gamma_m(X^-)m(X^-)]}{\gamma_c(HX)c(HX)/\gamma_m(HX)m(HX)}$$

Equation (10.72) gives for species i

$$\gamma_{c,i}c_i = \rho_A \gamma_{m,i} m_i \qquad (11.30)$$

Using (11.30), we get $K_{c,a}/K_{m,a} = \rho_A$. For water at 25°C, the density is 0.997 g/cm^3 = 0.997 kg/dm^3. Hence

$$K_{c,a} = (0.997 \text{ kg/dm}^3)K_{m,a} \qquad \text{in } H_2O \text{ at } 25°C \qquad (11.31)$$

To the accuracy with which most K_a values are known, we can consider $K_{c,a}$ and $K_{m,a}$ as numerically equal in aqueous solutions. (Of course, these constants may differ considerably for nonaqueous solvents.)

Furthermore, in dilute aqueous solutions, the molality (in mol/kg) and the concentration (in mol/dm^3) are numerically not very different (Prob. 11.10). Also, since ρ_A in (11.30) is numerically very close to 1 and c_i and m_i are numerically nearly equal in dilute aqueous solutions, it follows that $\gamma_{c,i}$ and $\gamma_{m,i}$ are nearly equal in dilute aqueous solutions.

11.4 REACTION EQUILIBRIA INVOLVING PURE SOLIDS OR PURE LIQUIDS

So far in this chapter, we have considered only reactions that occur in a single phase. However, many reactions involve one or more pure solids or pure liquids. An example is $CaCO_3(s) = CaO(s) + CO_2(g)$. The equilibrium condition $\sum_i \nu_i \mu_{i,eq} = 0$ applies whether or not all species are in the same phase. Equation (11.1) defines the activity a_i of a pure solid or liquid, and Eq. (11.2) then gives

$$\mu_i = \mu_i^\circ + RT \ln a_i \qquad \text{pure sol. or liq.} \qquad (11.32)$$

As in Sec. 5.1, we choose the standard state of the pure solid or liquid to be the state with $P = 1$ atm and T equal to the temperature of the reaction mixture. Hence μ_i° in (11.32) is a function of T only.

Equations (11.3) to (11.5) apply for the equilibrium constant, but to use them we need an expression for the activity a_i of the pure solid or liquid. From $d\bar{G} = d\mu = -\bar{S}\,dT + \bar{V}\,dP$ for a pure substance, we have $d\mu_i = \bar{V}_i\,dP$ at constant T. Integration from the standard state $P = 1$ atm to an arbitrary pressure gives

$$\mu_i(T, P) = \mu_i^\circ(T) + \int_{1 \text{ atm}}^{P} \bar{V}_i\,dP' \qquad \text{const. } T, \text{ pure sol. or liq.} \qquad (11.33)$$

where the prime was added to the dummy integration variable to avoid the use of P with two different meanings in the same equation. Equating the right sides of (11.32) and (11.33), we have

$$\ln a_i = \frac{1}{RT}\int_{1\text{ atm}}^{P} \bar{V}_i \, dP' \qquad \text{pure sol. or liq.} \tag{11.34}$$

where $\bar{V}_i$ is the molar volume of pure i. Since solids and liquids are rather incompressible, it is a good approximation to take $\bar{V}_i$ independent of P and remove it from the integral to give

$$\ln a_i \approx (P - 1\text{ atm})\bar{V}_i/RT \qquad \text{pure sol. or liq.} \tag{11.35}$$

At the standard pressure of 1 atm, the activity of a pure solid or liquid is 1 (since the substance is in its standard state). As noted in earlier chapters, G is relatively insensitive to pressure for condensed phases; hence we expect a_i to be rather insensitive to pressure for solids and liquids. For example, a solid with molecular weight 200 and density 2.00 g/cm^3 has $\bar{V}_i = 100$ cm^3. From (11.35), we find at $P = 20$ atm and $T = 300$ K that $\ln a_i = 0.077$ and $a_i = 1.08$. This value is pretty close to 1. Provided P remains moderate (below, say, 20 atm), we can reasonably approximate the activity of most pure solids and liquids as 1. (For a substance with a large $\bar{V}$, for example, a polymer, this approximation is invalid.)

As an example, consider the equilibrium

$$CaCO_3(s) = CaO(s) + CO_2(g) \tag{11.36}$$

Equation (11.5) gives $K_a = a[CaO(s)]a[CO_2(g)]/a[CaCO_3(s)]$. Provided P is not too high, we can take the activity of each solid as 1 and the activity of the gas (assumed ideal) as $P(CO_2)/1$ atm. Hence

$$K_a \approx a[CO_2(g)] \approx P(CO_2)/P^\circ \qquad \text{where } P^\circ \equiv 1\text{ atm} \tag{11.37}$$

Thus, at a given T, the CO_2 pressure above $CaCO_3(s)$ is constant. Note, however, that in the calculation of K_a from ΔG° using (11.3) it would be wrong to omit $\bar{G}$ of the solids $CaCO_3$ and CaO. The fact that a of each solid is nearly 1 means that $\mu - \mu^\circ$ $(= RT \ln a)$ of each solid is nearly 0; however, $\mu^\circ = \bar{G}^\circ$ of each solid is nowhere near zero and must be included in calculating ΔG°.

Now consider the equilibrium between a solid salt $M_{\nu_+}X_{\nu_-}$ and a saturated aqueous solution of the salt. The reaction is

$$M_{\nu_+}X_{\nu_-}(s) = \nu_+ M^{z_+}(aq) + \nu_- X^{z_-}(aq) \tag{11.38}$$

where z_+ and z_- are the charges on the ions and ν_+ and ν_- are the numbers of positive and negative ions. Choosing the molality scale for the solute species, we have as the activity equilibrium constant for (11.38)

$$K_a = \frac{(a_+)^{\nu_+}(a_-)^{\nu_-}}{a[M_{\nu_+}X_{\nu_-}(s)]} = \frac{(\gamma_+ m_+/m^\circ)^{\nu_+}(\gamma_- m_-/m^\circ)^{\nu_-}}{a[M_{\nu_+}X_{\nu_-}(s)]}$$

where a_+, γ_+, and m_+ are the activity, molality-scale activity coefficient, and molality of the ion $M^{z_+}(aq)$. Provided the system is not at high pressure, we can take $a = 1$ for the pure solid salt. Dropping m° from K_a and using (10.59), we have as the *solubility-product* equilibrium constant

$$K_{sp} = (\gamma_\pm)^{\nu_+ + \nu_-}(m_+)^{\nu_+}(m_-)^{\nu_-} \tag{11.39*}$$

Equation (11.39) is valid for any salt, but its main application is to salts only slightly soluble in water. (For a highly soluble salt, the ionic strength of a saturated solution is quite high, the mean ionic activity coefficient $\gamma_\pm$ differs substantially from 1, and we have no good way of estimating $\gamma_\pm$. Moreover, ion-pair formation may be substantial in concentrated solutions of a salt.)

Example K_{sp} for AgCl in water has been determined to be 1.78×10^{-10} mol²/kg² at 25°C. Find the solubility of AgCl at 25°C in (*a*) pure water; (*b*) a 0.100 mol/kg KNO_3 solution; (*c*) a 0.100 mol/kg KCl solution.

(*a*) For $AgCl(s) = Ag^+(aq) + Cl^-(aq)$, we have

$$K_{sp} = \gamma_\pm^2 m(Ag^+)m(Cl^-)$$

Because of the very small value of K_{sp}, the ionic strength of a saturated AgCl solution is extremely low, and $\gamma_\pm$ can be taken as 1. Since $m(Ag^+) = m(Cl^-)$ in a solution containing only dissolved AgCl, we have 1.78×10^{-10} mol²/kg² $= [m(Ag^+)]^2$; hence, $m(Ag^+) = 1.33 \times 10^{-5}$ mol/kg. The solubility of AgCl in pure water at 25°C is 1.33×10^{-5} mole per kilogram of solvent.

(*b*) The ionic strength of a 0.100 mol/kg KNO_3 solution is 0.100 mol/kg. The Davies equation (10.93) gives $\gamma_\pm = 0.78$. Setting $m(Ag^+) = m(Cl^-)$, we have 1.78×10^{-10} mol²/kg² $= (0.78)^2[m(Ag^+)]^2$; hence $m(Ag^+) = 1.71 \times 10^{-5}$ mol/kg. Note the 29 percent increase in solubility compared with that in pure water. The added KNO_3 reduces $\gamma_\pm$ and increases the solubility, a phenomenon called the *salt effect*.

(*c*) In 0.100 mol/kg KCl, the ionic strength is 0.100 mol/kg, and the Davies equation gives $\gamma_\pm = 0.78$. The Cl^- from AgCl is negligible compared with that from the KCl. Setting $m(Cl^-) = 0.100$ mol/kg, we have

$$1.78 \times 10^{-10} \text{ mol}^2/\text{kg}^2 = (0.78)^2 m(Ag^+)(0.100 \text{ mol/kg})$$

Hence $m(Ag^+) = 2.9 \times 10^{-9}$ mol/kg. Note the sharp decrease in solubility compared with either pure water or the KNO_3 solution (common-ion effect).

In the example just given, we ignored the possibility of ion-pair formation and assumed that all the silver chloride in solution exists as Ag^+ and Cl^- ions. This is a reasonable assumption for dilute solutions of a 1 : 1 electrolyte. However, in working with K_{sp} for other than 1 : 1 electrolytes, substantial error can frequently result if ion-pair formation is not taken into account. Calculations that allow for ion-pair formation are complicated; see Prob. 11.18 and L. Meites, J. S. F. Pode, and H. C. Thomas, *J. Chem. Educ.*, **43**, 667 (1966).

Although ion-pair formation can be neglected in the above example, there is another phenomenon that often cannot be neglected in AgCl solutions. The ions Ag^+ and Cl^- react in aqueous solution to form a series of four complex ions, as follows: $Ag^+ + Cl^- \rightleftharpoons AgCl(aq)$, $AgCl(aq) + Cl^- \rightleftharpoons AgCl_2^-$, $AgCl_2^- + Cl^- \rightleftharpoons AgCl_3^{2-}$, $AgCl_3^{2-} + Cl^- \rightleftharpoons AgCl_4^{3-}$. Consideration of complex-ion formation shows that although the results of (*a*) and (*b*) in the above example are correct, the result of (*c*) is in error; see Prob. 11.23.

Example The $\Delta G^\circ_{f,298}$ values for $Ag_2SO_4(s)$, $Ag^+(aq)$, and $SO_4^{2-}(aq)$ are -147.2, 18.4, and -177.3 kcal/mol, respectively. Find K_{sp} for Ag_2SO_4 in water at 25°C.

The reaction is $Ag_2SO_4(s) = 2Ag^+(aq) + SO_4^{2-}(aq)$. We calculate $\Delta G^\circ_{298} = 6700$ cal mol^{-1}. Substitution in (11.3) gives

$$K^\circ_{sp} = 1.2 \times 10^{-5} \qquad \text{and} \qquad K_{sp} = 1.2 \times 10^{-5}\ \text{mol}^3/\text{kg}^3$$

For a homogeneous reaction like $N_2(g) + 3H_2(g) = 2NH_3(g)$ or $HCN(aq) + H_2O = H_3O^+(aq) + CN^-(aq)$, there will always be some of each species present at equilibrium. In contrast, reactions involving pure solids have the possibility of going to completion. For example, the equilibrium constant for $CaCO_3(s) = CaO(s) + CO_2(g)$ is given by (11.37) as $K_a = P(CO_2)/P^\circ$. At 800°C, experiment gives $K_a = 0.22$ for this reaction. Suppose we have $CaO(s)$ in an evacuated container at 800°C and we introduce some CO_2. If the initial pressure of the CO_2 is below 0.22 atm, then no $CaCO_3(s)$ whatever will be formed. If the initial pressure of the CO_2 is above 0.22 atm, then CaO will combine with CO_2 until the CO_2 pressure drops to 0.22 atm; if sufficient CO_2 is present, all the CaO may be used up before the CO_2 pressure falls to 0.22 atm. Conversely, if we start with $CaCO_3$ at 800°C in a closed container, the $CaCO_3$ will decompose until $P(CO_2)$ reaches 0.22 atm; if the container volume is large enough, all the $CaCO_3$ may decompose before this equilibrium pressure can be attained. Similarly, if a crystal of AgCl is added to a sufficiently large volume of water, all the AgCl can dissolve without having $\gamma_\pm{}^2 m(Ag^+)m(Cl^-)$ reach K_{sp}.

11.5 REACTION EQUILIBRIUM IN NONIDEAL GAS MIXTURES

From Eqs. (10.119) and (10.126), the activity a_i of component i of a nonideal gas mixture is

$$a_i = f_i/P^\circ = \chi_i P_i/P^\circ = \chi_i x_i P/P^\circ \qquad \text{where } P^\circ \equiv 1 \text{ atm} \qquad (11.40)$$

where f_i, χ_i, P_i, and x_i are the fugacity, fugacity coefficient, partial pressure, and mole fraction of gas i and P is the (total) pressure of the mixture. Substitution into (11.5) gives for the standard equilibrium constant of a gas-phase reaction

$$K^\circ = \prod_i \left(\frac{f_i}{P^\circ}\right)^{\nu_i} = \prod_i \left(\frac{\chi_i x_i P}{P^\circ}\right)^{\nu_i} \qquad (11.41)$$

The standard state for each gas has P fixed at 1 atm, so ΔG° depends only on T. Hence the equilibrium constant K°, which is calculated from (11.3), depends only on T. Rewriting (11.41), we have

$$K^\circ \Big/ \prod_i (\chi_i)^{\nu_i} = \prod_i \left(\frac{x_i P}{P^\circ}\right)^{\nu_i} \qquad (11.42)$$

To calculate the equilibrium composition at a given T and P of a reacting nonideal gas mixture, the following approximate procedure is generally used. Tables of $\Delta G^\circ_{f,T}$ of the reacting gases are used to calculate ΔG°_T for the reaction. The equilibrium constant K° is then calculated from ΔG°_T. The fugacity coefficients $\chi_i^*(T, P)$ of the pure gases are found using either law-of-corresponding-states charts of χ_i^* as a function of reduced temperature and pressure (Sec. 10.9) or tabulations of $\chi_i^*(T, P)$

for the individual gases. The Lewis fugacity rule (10.132) is then used to estimate χ_i for each gas in the mixture. The quantity on the left side of (11.42) is calculated, and (11.42) is then used to find the equilibrium composition by the procedures of Sec. 6.4.

11.6 TEMPERATURE AND PRESSURE DEPENDENCE OF THE EQUILIBRIUM CONSTANT

Equation (11.3) gives

$$\ln K_a = -\Delta G^\circ/RT \tag{11.43}$$

where ΔG° is the standard change in Gibbs energy (all species in their standard states). For gases and for pure liquids and solids, we chose a fixed-pressure standard state ($P^\circ = 1$ atm) so that ΔG° and hence K_a are independent of pressure and depend only on T. For liquid solutions (and solid solutions) we chose a variable-pressure standard state, with standard-state pressure equal to the actual pressure of the solution, so that here ΔG° and K_a are functions of both T and P.

Differentiating (11.43) with respect to T and going through the same steps used to derive (6.45), we have

$$\left(\frac{\partial \ln K_a}{\partial T}\right)_P = \frac{\Delta G^\circ}{RT^2} - \frac{(\partial \Delta G^\circ/\partial T)_P}{RT} = \frac{\Delta G^\circ}{RT^2} + \frac{\Delta S^\circ}{RT} = \frac{\Delta G^\circ + T\,\Delta S^\circ}{RT^2}$$

$$\left(\frac{\partial \ln K_a}{\partial T}\right)_P = \frac{\Delta H^\circ}{RT^2} \tag{11.44}^*$$

where the partial derivative becomes an ordinary derivative when liquid or solid solutions are not involved in the reaction. ΔH° is equal to $\sum_i \nu_i \bar{H}_i^\circ$, where the ν_i's are the stoichiometric coefficients and the $\bar{H}_i^\circ$'s are the standard-state molar (or partial molar) enthalpies. For the application of (11.44) to reactions in solution, see Prob. 11.20.

Consider a reaction in which all the reactants and products are in a liquid or solid solution. Going through the same steps used to derive (9.116), we have

$$\left(\frac{\partial \ln K_a}{\partial P}\right)_T = -\frac{1}{RT}\left(\frac{\partial \Delta G^\circ}{\partial P}\right)_T = -\frac{\Delta V^\circ}{RT}$$

If a reaction involves both species in a liquid or solid solution and species not in a liquid or solid solution (e.g., a solubility product), then in calculating ΔV° we consider only those species which are in the solution. This is so because species not in solution have pressure-independent standard states and make no contribution to $\partial \Delta G^\circ/\partial P$. Therefore for any reaction

$$\left(\frac{\partial \ln K_a}{\partial P}\right)_T = -\frac{\Delta V^\circ_{\text{soln}}}{RT} \tag{11.45}$$

where the subscript is a reminder to include only species in solution in calculating $\Delta V^\circ_{\text{soln}}$. Usually $\Delta V^\circ_{\text{soln}}$ is small, and the pressure dependence of K_a can be ignored unless very high pressures are involved.

11.7 SUMMARY OF STANDARD STATES

The activity equilibrium constant K_a for a reaction is given by (11.5) as $\prod_i (a_{i,\mathrm{eq}})^{\nu_i}$. The activity a_i of species i is defined by (11.1) as $\exp(\mu_i - \mu_i^\circ)/RT$, where μ_i° is the chemical potential of i in its standard state. The choice of standard state therefore determines a_i and determines the form of the equilibrium constant.

Table 11.1 summarizes the choices of standard states made in earlier sections. Also listed are the forms of a_i and the numbers of the equations for μ_i. In all cases, μ_i equals $\mu_i^\circ + RT \ln a_i$. (Some workers take the standard state of each component of a liquid solution to have pressure equal to 1 atm, rather than equal to the pressure of the solution as we have done.)

11.8 COUPLED REACTIONS

Suppose two chemical reactions are occurring in a system, and further suppose that there is a chemical species that takes part in both reactions. The reactions are then said to be *coupled*, in that one reaction will influence the equilibrium position of the second reaction. Thus, suppose species M is a product of reaction 1 and is a reactant in reaction 2:

$$\begin{aligned} &\text{Reaction 1:} \qquad && A + B = M + D \\ &\text{Reaction 2:} \qquad && M + R = S + T \end{aligned} \tag{11.46}$$

Table 11.1 Summary of standard states and activities

Substance	Standard state	a_i	μ_i
Gas (pure or in gas mixture)	Pure ideal gas at 1 atm and T	$f_i/P^\circ = \chi_i x_i P/P^\circ$	(10.120)
Pure liquid or pure solid	Pure substance at 1 atm and T	(11.35) (≈ 1 at low P)	(11.33)
Solution component, convention I	Pure i at T and P of solution	$\gamma_{\mathrm{I},i} x_i$ ($\gamma_{\mathrm{I},i} = 1$ at $x_i = 1$)	(10.9), (10.10)
Solvent A	Pure A at T and P of solution	$\gamma_A x_A$ ($\gamma_A^\infty = 1$)	(10.9), (10.14)
Nonelectrolyte solute:			
convention II	Fictitious state with $x_i = 1 = \gamma_{\mathrm{II},i}$	$\gamma_{\mathrm{II},i} x_i$ ($\gamma_{\mathrm{II},i}^\infty = 1$)	(10.29), (10.17)
molality scale	Fictitious state with $m_i/m^\circ = 1 = \gamma_{m,i}$	$\gamma_{m,i} m_i/m^\circ$ ($\gamma_{m,i}^\infty = 1$)	(10.33)
concentration scale	Fictitious state with $c_i/c^\circ = 1 = \gamma_{c,i}$	$\gamma_{c,i} c_i/c^\circ$ ($\gamma_{c,i}^\infty = 1$)	(10.39)
Electrolyte solute:			
molality scale	Fictitious state with $\gamma_i = 1 = \nu_\pm m_i/m^\circ$	$(\nu_\pm \gamma_i m_i/m^\circ)^\nu$ ($\gamma_i^\infty = 1$)	(10.70)

If reaction 1 has $\Delta G^\circ \gg 0$ and equilibrium constant $K \ll 1$, then at ordinary concentrations very little A and B will react to produce M and D in the absence of reaction 2. If reaction 2 has $\Delta G^\circ \ll 0$ and $K \gg 1$, then if both reactions occur, reaction 2 will use up large amounts of M, thereby allowing reaction 1 to proceed to a substantial degree.

A simple example of coupled reactions is the observation that most phosphate salts are virtually insoluble in water but quite soluble in aqueous solutions of strong acids. In acidic solutions, the strong Brønsted base PO_4^{3-} reacts to a very great extent with H_3O^+ to yield the extremely weak acid HPO_4^{2-} (which reacts further with H_3O^+ to yield $H_2PO_4^-$), thereby greatly reducing the PO_4^{3-} concentration in solution and shifting the solubility-product equilibrium (11.38) of the phosphate salt to the right.

Coupled reactions are of considerable biological importance. In biological systems, coupling is often brought about by the action of enzymes. For example, a metabolic reaction common to many living organisms is the reduction of pyruvate ion CH_3COCOO^- to lactate ion $CH_3CH(OH)COO^-$. For this reaction, ΔG° in aqueous solution is highly positive, and the reaction is not favored. However, the enzyme lactate dehydrogenase catalyzes the conversion of pyruvate to lactate by coupling this reaction with the oxidation of the reduced form of nicotinamide adenine dinucleotide (NADH) to the oxidized form (NAD^+) of this molecule, a reaction with a highly negative ΔG° in aqueous solution. The enzyme binds NADH to itself at one site and binds pyruvate at another site; a certain group on the enzyme then supplies pyruvate with hydrogen atoms to reduce it to lactate, and then hydrogens from the oxidation of NADH to NAD^+ restore this group of the enzyme to its original state. Thus the enzyme transfers hydrogen from NADH to pyruvate, thereby coupling the oxidation of NADH with the reduction of pyruvate. The overall reaction is $CH_3COCOO^- + NADH + H^+ = CH_3CH(OH)COO^- + NAD^+$.

One must use caution in applying thermodynamics to living organisms. Organisms and the cells that compose them are open (rather than closed) systems and are not at equilibrium. The rates of the chemical reactions may thus be more relevant than the values of the equilibrium constants. For discussion on these and related points, see B. E. C. Banks, *Chem. Brit.*, **5**, 514 (1969); L. Pauling, ibid., **6**, 468 (1970); D. Wilkie, ibid., **6**, 472; A. F. Huxley, ibid., **6**, 477; R. A. Ross and C. A. Vernon, ibid., **6**, 541.

11.9 GIBBS ENERGY CHANGE FOR A REACTION

The term *Gibbs energy change for a reaction* has at least three different meanings, which we now discuss.

1. The standard molar Gibbs energy change ΔG° for a reaction is defined by (11.4) as $\Delta G^\circ \equiv \sum_i \nu_i \mu_i^\circ$, where μ_i° is the value of the chemical potential of substance i in its standard state. (Since μ_i° is an intensive quantity and ν_i is a dimensionless number, ΔG° is an intensive quantity with units cal/mol or J/mol.) For a gas-phase reaction, the standard state of each gas is the hypothetical pure ideal gas at 1 atm; for a reaction in liquid solution with use of the molality scale, the standard state of each solute is the hypothetical state with $m_i = 1$ mol/kg and $\gamma_{m,i} = 1$. These standard states do not correspond to the states of the reactants in the reaction mixture; hence, ΔG° (and ΔH°, ΔS°, etc.) refer not to the actual change in the reaction mixture but to a hypothetical change from standard states of the separated reactants to standard states of the separated products.

2. Equation (4.112) reads $dG/d\xi = \sum_i \nu_i \mu_i$ at constant T and P, where ξ is the extent of reaction, the μ_i's are the actual chemical potentials in the reaction mixture at some particular value of ξ, and dG is the infinitesimal change in Gibbs energy of the reaction mixture due to a change in the extent of reaction from ξ to $\xi + d\xi$. Therefore

$$\left(\frac{\partial G}{\partial \xi}\right)_{T,P} = \sum_i \nu_i \mu_i \qquad (11.47)$$

The sum on the right of (11.47) is frequently denoted by ΔG, but this notation is misleading in that $\sum_i \nu_i \mu_i$ is not the change in G of the system as the reaction occurs but is the instantaneous rate of change of G with respect to ξ. (If the reaction mixture were of infinite mass, so that a finite change in ξ would not change the μ_i's in the mixture, then $\sum_i \nu_i \mu_i \times 1$ mol would be ΔG for a change $\Delta\xi = 1$ mol.) Note that $(\partial G/\partial \xi)_{T,P}$ is the slope of the G-vs.-ξ curve (Fig. 4.4).

3. From Eq. (9.25), the Gibbs energy G of the reaction mixture at a given instant is equal to $\sum_i n_i \mu_i$, where n_i (not to be confused with the stoichiometric coefficient ν_i) is the number of moles of i in the mixture and μ_i its chemical potential in the mixture. If at times t_1 and t_2 these quantities are $n_{i,1}$, $\mu_{i,1}$ and $n_{i,2}$, $\mu_{i,2}$, respectively, then the actual change ΔG in Gibbs energy of the reacting system from time t_1 to t_2 is $\Delta G = \sum_i n_{i,2}\mu_{i,2} - \sum_i n_{i,1}\mu_{i,1}$.

One final point. Instead of using ΔG°, biologists frequently use a quantity $\Delta G^{\circ\prime}$, defined by

$$\Delta G^{\circ\prime} \equiv \sum_{i \neq \mathrm{H}^+} \nu_i \mu_i^\circ + \nu(\mathrm{H}^+)\mu[\mathrm{H}^+ \text{ at } a(\mathrm{H}^+) = 10^{-7}] \qquad (11.48)$$

Biological fluids have hydrogen-ion molalities close to 10^{-7} mol/kg, so that $\Delta G^{\circ\prime}$ values are more relevant to reactions in living organisms than ΔG° values (which involve the standard state with molality 1 mol/kg for H^+).

PROBLEMS

Where appropriate, use the Davies equation to estimate activity coefficients.

11.1 Calculate $m(H_3O^+)$ in a 0.20 mol/kg aqueous solution of NaCl at 25°C.

11.2 The human body is typically at 98.6°F = 37.0°C. Use the expression given in Prob. 11.6 to calculate $m(H_3O^+)$ in pure water at 37°C.

11.3 Find $m(H_3O^+)$ in a 1.00×10^{-8} mol/kg aqueous HCl solution at 25°C.

11.4 A 0.200 mol/kg solution of the acid HX is found to have $m(H_3O^+) = 1.00 \times 10^{-2}$ mol/kg. Find K_a for this acid.

11.5 Estimate $a(H_2O)$ in a 0.50 mol/kg aqueous NaCl solution; take $\gamma(H_2O) = 1$.

11.6 Ackermann found that the molality-scale ionization constant of water can be represented as the following function of temperature:

$$\log K_w^\circ = 948.8760 - 24746.26(\mathrm{K}/T) - 405.8639 \log (T/\mathrm{K}) + 0.48796(T/\mathrm{K}) - 0.0002371(T/\mathrm{K})^2$$

[See H. L. Clever, *J. Chem. Educ.*, **45**, 231 (1968) for a review of experimental work on K_w.] Calculate ΔG°, ΔS°, and ΔH° for the ionization of water at 25°C.

11.7 Use (11.28) to graph $-\log [m(H_3O^+)/m^\circ]$ vs. $\log m/m^\circ$ for an aqueous solution of the acid HX at 25°C for K_a values of 10^4, 10^{-2}, 10^{-4}, 10^{-6}, 10^{-8}, and 10^{-10} mol/kg. Plot all the graphs on the same sheet; have the plots cover the range of $\log (m/m^\circ)$ from -9 to 0. Comment on the limiting behavior for high K_a and on the limiting behavior for low m.

11.8 Find $m(H_3O^+)$ in a 0.10 mol/kg solution of $NaC_2H_3O_2$ in water at 25°C, given that $K_a = 1.75 \times 10^{-5}$ mol/kg for $HC_2H_3O_2$ at 25°C. *Hint:* The acetate ion is a base and reacts with water as follows: $C_2H_3O_2^- + H_2O = HC_2H_3O_2 + OH^-$. Show that the equilibrium constant for this reaction is $K_b = K_w/K_a$. Neglect the OH^- coming from the ionization of water.

11.9 Let MX be the salt of a weak acid and a strong base. Show that if activity coefficients are neglected but the ionization of water is not neglected, for an aqueous solution of MX with stoichiometric molality m and H_3O^+ molality y we have: $y^3 + (m + K_a)y^2 - K_w y - K_a K_w = 0$, where K_a is the ionization constant of the acid HX.

11.10 Show that in a dilute solution the molality and the concentration of a solute are related by $c_i \approx \rho_A m_i$. Since ρ of water is close to 1.00 kg/dm^3, the molality and concentration are numerically nearly equal in dilute aqueous solutions.

11.11 For H_2S, the ionization constant is 1.0×10^{-7} mol/kg in water at 25°C; for HS^-, the ionization constant is 1.3×10^{-13} mol/kg in water at 25°C. (The erroneous value 10^{-14} mol/kg is often given for this constant.) (*a*) Ignoring activity coefficients, calculate the H_3O^+, HS^-, and S^{2-} molalities in a 0.100 mol/kg aqueous H_2S solution at 25°C, making reasonable approximations to simplify the calculation. (*b*) The same as (*a*), except that activity coefficients are to be included in the calculations. For the ionization of HS^-, use the form of the Davies equation that corresponds to (10.81).

11.12 For $CuSO_4$, the equilibrium constant for association to form $CuSO_4$ ion pairs has been found from conductivity measurements to be 230 kg/mol in aqueous solution at 25°C. Use the Davies equation (10.93) to calculate the Cu^{2+} molality, $\gamma_\pm$, and γ_i [Eq. (10.69)] in a 0.0500 mol/kg aqueous $CuSO_4$ solution at 25°C. *Hint:* First estimate $\gamma_\pm$ by neglecting ion association; then use this estimated $\gamma_\pm$ to calculate an approximate Cu^{2+} molality; then calculate improved I_m and $\gamma_\pm$ values; and then recalculate the Cu^{2+} molality. Repeat as many times as necessary to obtain convergence.

11.13 Fuoss' theory of ion-pair formation (Sec. 10.7) gives the following expression (*in SI units*) for the concentration-scale equilibrium constant for the ion-association reaction $M^{z_+} + X^{z_-} = MX^{z_+ + z_-}$ in solution:

$$K_c = \tfrac{4}{3}\pi a^3 N_0 \exp b \tag{11.50}$$

where N_0 is the Avogadro constant, a is the mean ionic diameter (as in the Debye–Hückel theory), and b is defined as

$$b \equiv z_+ |z_-| e^2/4\pi\varepsilon_0 \varepsilon_{r,A} akT \tag{11.51}$$

where the symbols in (11.51) are defined following (10.83). [For the derivation of (11.50), see R. M. Fuoss, *J. Am. Chem. Soc.*, **80**, 5059 (1958) or *Bockris and Reddy*, sec. 3.8.6. Note that $\frac{4}{3}\pi a^3 N_0$ is essentially the mean molar volume of the ions and $-b$ is the potential energy of interaction between a positive and a negative ion in contact (in the absence of other ions) divided by kT.] For the value $a = 4.5$ Å, use the Fuoss equation to calculate the ion-association equilibrium constant K_c in aqueous solution at 25°C for (*a*) 1 : 1 electrolytes; (*b*) 2 : 1 electrolytes; (*c*) 2 : 2 electrolytes; (*d*) 3 : 2 electrolytes. *Hint:* Be careful with the units of a. Note that the traditional units of K_c are dm^3/mol. Conductivity measurements show that in aqueous solutions at 25°C, the ion-association equilibrium constant is typically of the order of magnitude 0.5 dm^3/mol for 1 : 1 electrolytes, 5 dm^3/mol for 2 : 1 electrolytes, 200 dm^3/mol for 2 : 2 electrolytes, and 4000 dm^3/mol for 3 : 2 electrolytes. How well does the Fuoss equation agree with these experimental values?

11.14 Use the Fuoss equation in Prob. 11.13 to calculate the ion-association equilibrium constant for a 1 : 1 electrolyte with $a = 4.5$ Å in the solvent ethanol at 25°C. The dielectric constant of ethanol at 25°C is 24.3. Compare with the value in water at 25°C (Prob. 11.13).

11.15 (*a*) Use ΔG_f° data in the Appendix to calculate K_{sp} for KCl in water at 25°C. (*b*) A saturated solution of KCl in water at 25°C has a molality 4.82 mol/kg. Calculate γ_i [Eq. (10.69)] of KCl in a saturated aqueous solution at 25°C.

11.16 Use data in the Appendix to calculate the equilibrium pressure of CO_2 above $CaCO_3$ (calcite) at 25°C.

11.17 The equilibrium constant for the reaction $Fe_3O_4(s) + CO(g) = 3FeO(s) + CO_2(g)$ was found to be 1.15 at 600°C. If a mixture of 2.00 mol Fe_3O_4, 3.00 mol CO, 4.00 mol FeO, and 5.00 mol CO_2 is brought to equilibrium at 600°C, find the equilibrium composition. (Assume the pressure is low enough for the gases to behave ideally.)

11.18 For $CaSO_4$ in water at 25°C, the equilibrium constant for the formation of ion pairs is 190 kg/mol. The solubility of $CaSO_4$ in water at 25°C is 2.08 g per kilogram of water. Calculate K_{sp} for $CaSO_4$ in water at

25°C. *Hint:* Get an initial estimate of the ion-pair molality and the ion molalities by ignoring activity coefficients. Get an initial estimate of I_m and use this to get an initial estimate of $\gamma_\pm$. Then recalculate the ionic molalities. Then calculate an improved $\gamma_\pm$ value and recalculate the ionic molalities. Keep repeating the calculations until convergence is obtained. Then calculate K_{sp}.

11.19 For NH_3, N_2, and H_2, the critical temperatures are 405.6, 126.2, and 33.3 K, respectively, and the critical pressures are 111.3, 33.5, and 12.8 atm, respectively. $\Delta G^\circ_{f,700}$ for NH_3 is 6.49 kcal/mol. Use the Lewis fugacity rule and law-of-corresponding-states graphs of fugacity coefficients (Sec. 10.9) to calculate the equilibrium composition at 700 K of a system that initially consists of 1.00 mol of NH_3 if the system's pressure is held fixed at 500 atm. *Note:* For H_2, to improve the fit of the observed fugacity coefficients to the law-of-corresponding-states graphs, one uses $T/(T_c + 8\text{ K})$ and $P/(P_c + 8\text{ atm})$ in place of the usual expressions for reduced temperature and pressure. *Hint:* The quartic equation which is obtained in solving this problem can be reduced to a quadratic equation by the substitution $z \equiv x^2$.

11.20 (*a*) Use (11.44) to show that

$$\left(\frac{\partial \ln K_m^\circ}{\partial T}\right)_P = \frac{\nu_A \bar{H}_A^* + \sum_{i \neq A} \nu_i \bar{H}_i^\infty}{RT^2} = \frac{\Delta H^\infty}{RT^2}$$

where A is the solvent. (*b*) Show that $K_c/K_m = \rho_A{}^b$, where $b \equiv \sum_{i \neq A} \nu_i$. [Use (11.30).] (*c*) Use the results of (*a*), (*b*), and Prob. 1.17 to show that

$$\left(\frac{\partial \ln K_c^\circ}{\partial T}\right)_P = \frac{\Delta H^\infty}{RT^2} - \alpha_A \sum_{i \neq A} \nu_i$$

(*d*) Use (11.44) and the result of (*c*) to show $\bar{H}^\circ_{c,i} = \bar{H}_i^\infty - RT^2\alpha_A$ for $i \neq A$.

11.21 Show that $\Delta G^{\circ\prime} = \Delta G^\circ - 16.118\nu(H^+)RT$, where $\Delta G^{\circ\prime}$ is defined by (11.48).

11.22 Consider the ideal-gas reaction A + 2B = C + D. Let $\bar{G}^\circ_{300}$ of A, B, C, and D be 1.00, 2.00, 4.00, and 0.00 kcal/mol, respectively. Suppose 3.00 mol of A at 1.00 atm and 300 K and 6.00 mol of B at 1.00 atm and 300 K are mixed and react to form C and D, with T and P being held constant at 300 K and 1.00 atm. Calculate and plot G of the reacting system vs. ξ for the full range of ξ.

11.23 The equilibrium constant for the formation of the complex ion AgCl(*aq*) from Ag^+ and Cl^- is $K_1 = 1.10 \times 10^3$ kg/mol, and the equilibrium constant for the formation of $AgCl_2{}^-$ from AgCl(*aq*) and Cl^- is $K_2 = 1.00 \times 10^2$ kg/mol, in aqueous solution at 25°C. The equilibrium constants K_3 and K_4 for the formation of $AgCl_3{}^{2-}$ and $AgCl_4{}^{3-}$ are negligible compared with K_1 and K_2; hence we shall neglect formation of these ions. Since silver can exist in the solution in the forms Ag^+, AgCl(*aq*), and $AgCl_2{}^-$, the solubility s of AgCl in a solution not originally containing any silver species is $s = m(Ag^+) + m[AgCl(aq)] + m(AgCl_2{}^-)$. (*a*) Show that

$$s = K_{sp}\left[\frac{1}{\gamma_\pm{}^2 m(Cl^-)} + K_1 + K_1 K_2 m(Cl^-)\right]$$

where K_{sp} is for AgCl(*c*), where $\gamma_\pm$ is for Ag^+ and Cl^-, and where we made the reasonable approximation $\gamma(AgCl_2{}^-) = \gamma(Cl^-)$. (*b*) Calculate the solubility of AgCl in a 25°C aqueous 0.1 mol/kg KCl solution. Compare with the incorrect result of Sec. 11.4. ($K^\circ_{sp} = 1.78 \times 10^{-10}$ for AgCl in water at 25°C.) (*c*) Calculate the solubility of AgCl in water at 25°C, and compare with the result of Sec. 11.4. *Hint:* Use the result of Sec. 11.4 to get initial estimates of $m(Cl^-)$ and $\gamma_\pm$. Use these estimates to calculate s. Then calculate improved $m(Cl^-)$ and $\gamma_\pm$ values. Keep repeating until convergence is obtained. (If the answer oscillates, average the results of two successive calculations.)

11.24 Measured CO_2 pressures above mixtures of $CaCO_3(s)$ and CaO(*s*) at various temperatures are:

P/torr	23.0	70	183	381	716
T/K	974	1021	1073	1125	1167

(*a*) At 800°C (1073 K), find ΔG°, ΔH°, and ΔS° for $CaCO_3(s) = CaO(s) + CO_2(g)$. (*b*) Estimate the CO_2 pressure above a $CaCO_3$–CaO mixture at 1000°C.

11.25 (*a*) If 5.0 g of $CaCO_3(s)$ is placed in a 4000-cm³ container at 1073 K, give the final amounts of $CaCO_3(s)$, CaO(*s*), and $CO_2(g)$ present. (See Prob. 11.24 for K_a.) (*b*) The same as (*a*), except that 0.50 g of $CaCO_3$ is placed in the container.

11.26 Show that $(\partial G/\partial \xi)_{T,P} = \Delta G^\circ + RT \ln Q$ for a reacting system, where the *reaction quotient* Q is defined by $Q \equiv \prod_i a_i^{\nu_i}$. When the system has reached equilibrium, the activities are equal to the equilibrium activities, Q equals K_a, and G is minimized with $(\partial G/\partial \xi)_{T,P} = 0$; hence, we get $\Delta G^\circ = -RT \ln K_a$.

12

MULTICOMPONENT PHASE EQUILIBRIUM

12.1 COLLIGATIVE PROPERTIES

One-component phase equilibrium was discussed in Chap. 7. We now consider multicomponent phase equilibrium.

We begin with a group of interrelated properties of solutions that historically have been called *colligative properties* (from the Latin *colligatus*, meaning "bound together"). In a solution at T and P, the partial molar Gibbs energy (chemical potential) of the solvent A is $\mu_A = \mu_A^*(T, P) + RT \ln a_A$, where $\mu_A^*(T, P)$ is the molar Gibbs energy of the pure liquid solvent at T and P. Thus μ_A in solution differs from μ_A^* of pure A. This change in solvent chemical potential leads to a change in the vapor pressure, the normal boiling point, and the normal freezing point and causes the phenomenon of osmotic pressure. These four properties are the colligative properties. Each involves an equilibrium between two phases.

12.2 VAPOR-PRESSURE LOWERING

In this section, we consider a solution of an involatile solute in a solvent. By "involatile" we mean that the contribution of the solute to the vapor pressure above the solution is negligible. This condition will hold for nearly all solid solutes but not for liquid or gaseous solutes. Even though a solid solute might have a nonnegligible vapor pressure at the temperature in question (e.g., naphthalene has a vapor pressure of 1 torr at 53°C), its mole fraction in the solution will generally be small enough to allow us to ignore its contribution to the solution's vapor pressure. The solution's total vapor pressure P is then due to the solvent A alone. (For simplicity, we shall assume pressures are low enough to treat all gases as ideal. If this is not so, pressures are to be replaced by fugacities.)

From Eqs. (10.22) and (10.77), the solution's vapor pressure P is

$$P = P_A = \gamma_A x_A P_A^* \qquad \text{involatile solute} \qquad (12.1)$$

where we use the mole-fraction scale for the solvent activity coefficient. The change in vapor pressure ΔP compared with pure A is $\Delta P = P - P_A^*$. Use of (12.1) gives

$$\Delta P = (\gamma_A x_A - 1)P_A^* \qquad \text{invol. solute} \tag{12.2}$$

As noted in Sec. 10.2, accurate measurement of solution vapor pressures allows determination of γ_A; use of the Gibbs–Duhem equation then gives the solute activity coefficient.

If the solution is dilute enough to be considered essentially ideally dilute, then γ_A can be set equal to 1 and

$$\Delta P = (x_A - 1)P_A^* \qquad \text{ideally dil. soln., invol. solute} \tag{12.3}$$

For a single nondissociating solute, $1 - x_A$ equals the solute mole fraction x_B, and $\Delta P = -x_B P_A^*$; note that under these conditions, ΔP is independent of the nature of B and depends only on its mole fraction in solution.

12.3 FREEZING-POINT DEPRESSION AND BOILING-POINT ELEVATION

The normal boiling point (Chap. 7) of a liquid or solution is the temperature at which its total vapor pressure equals 1 atm. An involatile solute lowers the vapor pressure (Sec. 12.2); hence it requires a higher temperature for the solution's vapor pressure to reach 1 atm; the normal boiling point of the solution is elevated compared with that of the pure solvent.

Addition of a solute to a solvent A lowers the freezing point. A rough, molecular explanation is as follows. Imagine pure liquid A to be in equilibrium with pure solid A at the normal freezing point (1 atm pressure). If we now dissolve some solute B in the liquid, the rate at which A freezes out of the solution is reduced, since the number of liquid A molecules colliding with solid A is reduced; however, the rate at which A molecules go from solid to liquid is unaffected. Hence there is no longer an equilibrium, and the solid A will melt. To restore equilibrium, the temperature must be lowered so as to favor formation of solid. Speaking thermodynamically, at the normal freezing point of pure A, the chemical potentials of solid A and liquid A are equal. Addition of a solute to liquid A changes the chemical potential of A in the liquid phase but not in the solid phase. To restore equality of μ_A in the two phases, the temperature must be lowered.

We now calculate the freezing-point depression for solute B in solvent A. We shall assume that only pure solid A freezes out of the solution when it is cooled to its freezing point. (This is the most common situation; for other cases, see Sec. 12.8.) The equilibrium condition at the normal freezing point is that the chemical potentials of the pure solid A and A in the solution must be equal. Using (10.9) and (10.14) [or (10.35) and (10.36)], we have

$$\mu_{A(s)}^*(T_f, P^\circ) = \mu_{A(l)}^*(T_f, P^\circ) + RT_f \ln \gamma_A x_A \tag{12.4}$$

where $P^\circ \equiv 1$ atm, s and l stand for solid and liquid, T_f is the solution's normal freezing point, γ_A is the solvent's mole-fraction-scale activity coefficient, and the star indicates a property of a pure substance. Since μ for a pure substance is just $\bar{G}$, we can write (12.4) as

$$\Delta\bar{G}_{\text{fus},A}(T_f) = -RT_f \ln \gamma_A x_A \tag{12.5}$$

where $\Delta\bar{G}$ of fusion of pure A at temperature T_f (and pressure P°) equals $\bar{G}^*$ of pure liquid A at T_f (and P°) minus $\bar{G}^*$ of pure solid A at T_f (and P°). Since P is fixed at 1 atm throughout this discussion, we shall often not indicate the pressure explicitly.

We have

$$\Delta\bar{G}_{\text{fus,A}}(T_f) = \Delta\bar{H}_{\text{fus,A}}(T_f) - T_f\,\Delta\bar{S}_{\text{fus,A}}(T_f) \tag{12.6}$$

Let T_f^* be the normal freezing point of pure A. At temperature T_f (which is below T_f^*), A(l) is a supercooled liquid. Thus the isothermal transformation A(l, T_f)$\rightarrow$ A(s, T_f) is irreversible; for this process, $\Delta S \neq \Delta H/T$ and $\Delta G \neq 0$. To find $\Delta\bar{G}_{\text{fus,A}}(T_f)$, we use the following reversible process to convert pure solid A at T_f to pure liquid A at T_f:

$$\text{A}(s,\, T_f) \xrightarrow{1} \text{A}(s,\, T_f^*) \xrightarrow{2} \text{A}(l,\, T_f^*) \xrightarrow{3} \text{A}(l,\, T_f)$$

Step 1 is the reversible warming of 1 mole of pure solid A from T_f to T_f^*. Step 2 is the reversible melting of solid A to liquid A at T_f^* (the normal melting point of pure A). Step 3 is the reversible cooling of liquid A from T_f^* to T_f (in the absence of dust particles, etc., to prevent the supercooled liquid from freezing). All steps are at constant pressure P°. For step 1, Eqs. (2.116) and (3.36) give

$$\Delta\bar{H}_1 = \int_{T_f}^{T_f*} \bar{C}_{P,\text{A}(s)}\,dT = -\int_{T_f*}^{T_f} \bar{C}_{P,\text{A}(s)}\,dT \tag{12.7}$$

$$\Delta\bar{S}_1 = \int_{T_f}^{T_f*} \bar{C}_{P,\text{A}(s)}\,\frac{dT}{T} = -\int_{T_f*}^{T_f} \bar{C}_{P,\text{A}(s)}\,\frac{dT}{T}$$

For steps 2 and 3

$$\Delta\bar{H}_2 = \Delta\bar{H}_{\text{fus,A}}(T_f^*), \qquad \Delta\bar{S}_2 = \Delta\bar{H}_{\text{fus,A}}(T_f^*)/T_f^*$$

$$\Delta\bar{H}_3 = \int_{T_f*}^{T_f} \bar{C}_{P,\text{A}(l)}\,dT, \qquad \Delta\bar{S}_3 = \int_{T_f*}^{T_f} \bar{C}_{P,\text{A}(l)}\,\frac{dT}{T}$$

Adding up the changes for the three steps, we obtain as $\Delta\bar{H}$ and $\Delta\bar{S}$ for the transformation of solid A at T_f to liquid A at T_f:

$$\Delta\bar{H}_{\text{fus,A}}(T_f) = \Delta\bar{H}_{\text{fus,A}}(T_f^*) + \int_{T_f*}^{T_f} \Delta\bar{C}_{P,\text{A}}\,dT \tag{12.8}$$

$$\Delta\bar{S}_{\text{fus,A}}(T_f) = \frac{\Delta\bar{H}_{\text{fus,A}}(T_f^*)}{T_f^*} + \int_{T_f*}^{T_f} \Delta\bar{C}_{P,\text{A}}\,\frac{dT}{T} \tag{12.9}$$

where

$$\Delta\bar{C}_{P,\text{A}} \equiv \bar{C}_{P,\text{A}(l)}(T,\, P^\circ) - \bar{C}_{P,\text{A}(s)}(T,\, P^\circ) \tag{12.10}$$

Substitution of (12.8) and (12.9) into (12.6) and use of (12.5) give

$$-RT_f \ln \gamma_\text{A} x_\text{A} = \Delta\bar{H}_{\text{fus,A}}(T_f^*)\left(1 - \frac{T_f}{T_f^*}\right) + \int_{T_f*}^{T_f} \Delta\bar{C}_{P,\text{A}}\left(1 - \frac{T_f}{T}\right) dT \tag{12.11}$$

The freezing-point depressions one ordinarily deals with are not extremely large, so the interval from T_f to T_f^* will not be very large. Also, we shall see below that the $\Delta\bar{C}_P$ term is an order of magnitude smaller than the $\Delta\bar{H}_{\text{fus}}$ term. For these two reasons it is an excellent approximation to take $\Delta\bar{C}_P$ as independent of temperature and evaluate it at the freezing point of pure A. Taking $\Delta \bar{C}_{P,A}$ outside the integral in (12.11), we have after division by $-T_f$

$$R \ln \gamma_A x_A = \Delta \bar{H}_{\text{fus,A}}(T_f^*)\left(\frac{1}{T_f^*} - \frac{1}{T_f}\right) - \Delta \bar{C}_{P,\text{fus,A}}(T_f^*)\left(\frac{T_f - T_f^*}{T_f} - \ln \frac{T_f}{T_f^*}\right) \tag{12.12}$$

Defining the freezing-point depression ΔT_f as

$$\Delta T_f \equiv T_f - T_f^* \tag{12.13}$$

we have as the coefficient of the $\Delta \bar{H}_{\text{fus}}$ term in (12.12)

$$\frac{1}{T_f^*} - \frac{1}{T_f} = \frac{T_f - T_f^*}{T_f^* T_f} = \frac{\Delta T_f}{T_f^*(T_f^* + \Delta T_f)} = \frac{\Delta T_f}{(T_f^*)^2} \frac{1}{1 + \Delta T_f/T_f^*}$$

Typically, T_f^* will be a couple of hundred kelvins, whereas ΔT_f will be several kelvins. Thus, $\Delta T_f/T_f^*$ will be far less than 1, and we use the Taylor series (8.31) with $x = -\Delta T_f/T_f^*$ to obtain

$$\frac{1}{T_f^*} - \frac{1}{T_f} = \frac{\Delta T_f}{(T_f^*)^2}\left(1 - \frac{\Delta T_f}{T_f^*} + \cdots\right) = \frac{1}{T_f^*}\left[\frac{\Delta T_f}{T_f^*} - \left(\frac{\Delta T_f}{T_f^*}\right)^2 + \cdots\right] \tag{12.14}$$

Since $\Delta T_f/T_f^* \ll 1$, each term in the series is an order of magnitude smaller than the preceding term.

Similarly, use of the Taylor series (8.31) and (8.41) to expand the coefficient of the $\Delta \bar{C}_{P,\text{fus}}$ term in (12.12) gives (see Prob. 12.39)

$$\frac{T_f - T_f^*}{T_f} - \ln \frac{T_f}{T_f^*} = -\frac{1}{2}\left(\frac{\Delta T_f}{T_f^*}\right)^2 + \cdots \tag{12.15}$$

Substitution of (12.15) and (12.14) into (12.12) gives

$$\ln \gamma_A x_A = \frac{\Delta \bar{H}_{\text{fus,A}}}{R T_f^*} \frac{\Delta T_f}{T_f^*} + \frac{T_f^* \, \Delta \bar{C}_{P,\text{fus,A}} - 2\Delta \bar{H}_{\text{fus,A}}}{2 R T_f^*}\left(\frac{\Delta T_f}{T_f^*}\right)^2 + \cdots \tag{12.16}$$

$$\ln \gamma_A x_A = c_1 \, \Delta T_f + c_2 (\Delta T_f)^2 + \cdots \tag{12.17}$$

where $\Delta \bar{H}_{\text{fus,A}}$ and $\Delta \bar{C}_{P,\text{fus,A}}$ are evaluated at the normal freezing point T_f^* of pure A, and where the definitions of c_1 and c_2 are obvious. For the solvent H_2O, one finds (Prob. 12.3)

$$c_1 = 0.00968 \text{ K}^{-1}, \qquad c_2 = -5 \times 10^{-6} \text{ K}^{-2} \tag{12.18}$$

The assumptions used to derive (12.12) and (12.16) are that (*a*) only pure solvent freezes out and (*b*) ΔT_f is a lot less than T_f^* (so that $\Delta \bar{C}_{P,\text{fus}}$ could be taken as constant and a series expansion used). Nowhere did we say anything about the solute's volatility. Whether the solute is volatile or involatile is irrelevant to the freezing-point depression.

Equation (12.16) allows the solvent activity coefficient γ_A to be determined from the measured freezing-point depression. Another application of freezing-point-depression measurements is to determine molecular weights of nonelectrolytes. For this purpose, we shall make further approximations to simplify (12.16).

If there is only one solute B in the solution, and if B is neither associated nor dissociated, then $x_A = 1 - x_B$ and

$$\ln \gamma_A x_A = \ln \gamma_A + \ln x_A = \ln \gamma_A + \ln (1 - x_B)$$

For the pure solvent, $\gamma_A = 1$. For the dilute solutions of nonelectrolytes used in molecular-weight determinations, it is generally a good approximation to take $\gamma_A = 1$ and $\ln \gamma_A = 0$; that is, we assume an ideally dilute solution. Using (8.41) with $x = 1 - x_B$, we then have

$$\ln \gamma_A x_A \approx -x_B - x_B^2/2 - \cdots \approx -x_B \tag{12.19}$$

where, since x_B is much less than 1, we can neglect x_B^2 and higher powers. Since the solution is very dilute and the freezing-point depression small, the $(\Delta T_f / T_f^*)^2$ term in (12.16) can be neglected. Use of (12.19) in (12.16) then gives

$$\Delta T_f = -x_B R (T_f^*)^2 / \Delta \bar{H}_{\text{fus,A}} \tag{12.20}$$

We have $x_B = n_B/(n_A + n_B) \approx n_B/n_A$, since $n_B \ll n_A$. The solute molality is $m_B = n_B/n_A M_A$, where M_A is the solvent molar mass. Hence for this very dilute solution, we have $x_B = M_A m_B$, and (12.20) becomes

$$\Delta T_f = -\frac{M_A R (T_f^*)^2}{\Delta \bar{H}_{\text{fus,A}}} m_B$$

$$\Delta T_f = -k_f m_B \qquad \text{ideally dil. soln., pure A freezes out} \tag{12.21*}$$

where the solvent's *molal freezing-point-depression constant* k_f is defined by

$$k_f \equiv M_A R (T_f^*)^2 / \Delta \bar{H}_{\text{fus,A}} \tag{12.22}$$

For water,

$$k_f = \frac{(18.015 \times 10^{-3}\ \text{kg/mol})(1.987\ \text{cal mol}^{-1}\ \text{K}^{-1})(273.15\ \text{K})^2}{(79.7\ \text{cal/g})(18.015\ \text{g/mol})}$$

$$= 1.86_0\ \text{K kg/mol} \tag{12.23}$$

Some other k_f values in K kg/mol are: benzene, 5.1; acetic acid, 3.8; camphor, 40. The large k_f of camphor makes it especially useful in molecular-weight determinations.

To determine a molecular weight, one measures ΔT_f for a *dilute* solution of B in solvent A and calculates m_B from (12.21). We then have [Eqs. (9.3) and (1.4)] $m_B = n_B/w_A = w_B/M_B w_A$, where w_B and w_A are the masses of B and A in the solution. Thus $M_B = w_B/w_A m_B = -k_f w_B/(w_A \Delta T_f)$:

$$M_B = -k_f w_B/(\Delta T_f w_A) \qquad \text{ideally dil. soln., pure A freezes out} \tag{12.24}$$

Since (12.21) applies only in an ideally dilute solution, a precise determination of a molecular weight requires that ΔT_f be found for a few different molalities; one then plots M_B [as calculated from (12.24)] vs. m_B and extrapolates to zero molality.

As a pure substance freezes at fixed pressure, the temperature of the system remains constant until all the liquid has frozen. As a dilute solution of B in solvent A freezes at fixed pressure, the freezing point keeps dropping, since as pure A freezes out, the molality of B in the liquid phase keeps increasing. To determine the freezing point of a solution, one can use the method of cooling curves (Sec. 12.8).

Freezing points are usually measured with the system open to the air. The dissolved air slightly lowers the freezing points of both pure A and the solution, but the depression due to the dissolved air will be virtually the same for pure A and for the solution and will cancel in the calculation of ΔT_f.

If there are several species in solution, then x_A in (12.17) equals $1 - \sum_{i \neq A} x_i$, where the sum goes over all solute species. Equation (12.19) becomes $\ln \gamma_A x_A \approx -\sum_{i \neq A} x_i$. For a dilute solution, we have $x_i \approx M_A m_i$, and Eq. (12.21) becomes for several solute species

$$\Delta T_f = -k_f \sum_{i \neq A} m_i \qquad \text{ideally dil. soln., pure A freezes out} \qquad (12.25)$$

Note that ΔT_f is independent of the nature of the species in solution and depends only on the total molality, provided the solution is dilute enough to be considered ideally dilute.

For electrolyte solutions, one cannot use (12.25), since an electrolyte solution only becomes ideally dilute at molalities too low to produce a measurable ΔT_f. One must retain γ_A in (12.17) for electrolyte solutions. (See Prob. 12.8.) [In the crudest approximation with $\gamma_A = 1$, we would expect from (12.25) that an electrolyte like NaCl that yields two ions in solution would give roughly twice the freezing-point depression as a nonelectrolyte at the same molality.]

Calculation of the boiling-point elevation proceeds in exactly the same manner as for the freezing-point depression. We start with an equation like (12.4) except that $\mu^*_{A(s)}$ is replaced by $\mu^*_{A(v)}$ (where v is for vapor) and T_f is replaced by T_b, the solution's boiling point. The analog of (12.5) reads $\Delta \bar{G}_{\text{vap},A}(T_b) = RT_b \ln \gamma_A x_A$. (Note the absence of the minus sign.) Going through the same steps as above, we derive equations that correspond to (12.21) and (12.22):

$$\Delta T_b = k_b m_B \qquad \text{ideally dil. soln., invol. solute} \qquad (12.26)^*$$

$$k_b \equiv M_A R (T_b^*)^2 / \Delta \bar{H}_{\text{vap},A} \qquad (12.27)$$

where $\Delta T_b \equiv T_b - T_b^*$ is the boiling-point elevation for the ideally dilute solution and T_b^* is the boiling point of pure solvent A. The assumption in (12.21) that only pure A freezes out of the solution corresponds to the assumption in (12.26) that only pure A vaporizes out of the solution, i.e., that the solute is involatile. For water, $k_b = 0.513$ °C kg/mol.

Boiling-point elevations can be used for molecular-weight determinations, but the method is subject to greater experimental errors than the freezing-point-depression method.

12.4 OSMOTIC PRESSURE

As noted earlier, there exist semipermeable membranes which allow only certain chemical species to pass through them. Imagine a box divided into two chambers by a rigid, thermally conducting, semipermeable membrane that allows solvent A to pass through it but does not allow the passage of solute B. In the left chamber we put pure A, and in the right, a solution of B in A (Fig. 12.1). We restrict A to be a nonelectrolyte.

Suppose that we initially fill the chambers so that the heights of the liquids in the two capillary tubes are equal. The chambers are thus initially at equal pressures:

$P_L = P_R$ where the subscripts stand for left and right. Since the membrane is thermally conducting, thermal equilibrium is always maintained: $T_L = T_R = T$. The chemical potential of A on the left is μ_A^*. If the solution on the right is dilute enough to be considered ideally dilute, then $\mu_{A,R} = \mu_A^* + RT \ln x_A$, which is less than $\mu_{A,L} = \mu_A^*$, since x_A is less than 1. Since $\mu_{A,L} > \mu_{A,R}$, substance A will flow through the membrane from left to right, slightly diluting the solution on the right (Sec. 4.7). The liquid in the right tube rises, thereby increasing the pressure in the right chamber. Equations (9.35) and (9.104) give $(\partial\mu_A/\partial P)_T = \bar{V}_A = \bar{V}_A^*$ for an ideally dilute solution. Since $\bar{V}_A^*$ is positive, the increase in pressure increases $\mu_{A,R}$ until eventually equilibrium is reached with $\mu_{A,R} = \mu_{A,L}$. (Since the membrane is impermeable to B, there is no equilibrium relation for μ_B. If the membrane were permeable to both A and B, the equilibrium condition would have equal concentrations of B and equal pressures in the two chambers.)

Let the equilibrium pressures in the left and right chambers be P and $P + \Pi$, respectively. We call Π the *osmotic pressure*. It is the extra pressure that must be applied to the solution to make μ_A in the solution equal to μ_A^*, so as to achieve membrane equilibrium for species A between the solution and pure A. At equilibrium, Eqs. (10.9) and (10.14) give

$$\mu_{A,L} = \mu_{A,R}$$

$$\mu_A^*(P, T) = \mu_A^*(P + \Pi, T) + RT \ln \gamma_A x_A \tag{12.28}$$

where we do not assume an ideally dilute solution. [Note that γ_A in (12.28) is the value at the pressure $P + \Pi$ of the solution.] From $d\mu_A^* = d\bar{G}_A^* = -\bar{S}_A^*\, dT + \bar{V}_A^*\, dP$, we have $d\mu_A^* = \bar{V}_A^*\, dP$ at constant T. Integration from P to $P + \Pi$ gives

$$\mu_A^*(P + \Pi, T) - \mu_A^*(P, T) = \int_P^{P+\Pi} \bar{V}_A^*\, dP' \qquad \text{const. } T \tag{12.29}$$

where a prime was added to the dummy integration variable to avoid the use of the symbol P with two different meanings. Substitution of (12.29) into (12.28) gives

$$RT \ln \gamma_A x_A = -\int_P^{P+\Pi} \bar{V}_A^*\, dP' \qquad \text{const. } T \tag{12.30}$$

Liquids are rather incompressible, so $\bar{V}_A^*$ (the molar volume of pure A at T and P') varies little with pressure and we can take $\bar{V}_A^*$ as essentially constant. The integral in (12.30) then becomes $\bar{V}_A^*(P + \Pi - P) = \bar{V}_A^*\Pi$, where $\bar{V}_A^*$ is best taken as the value

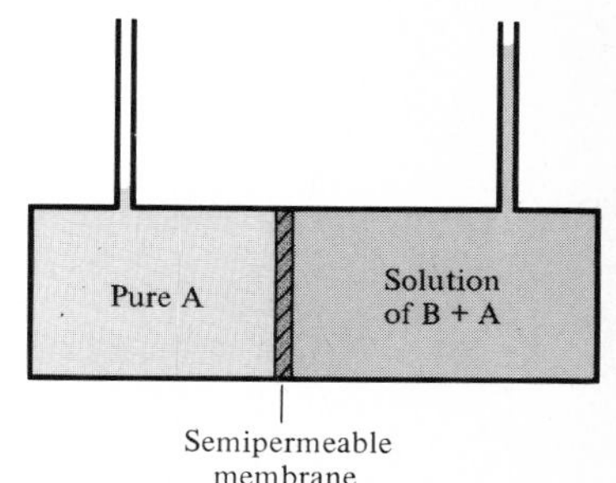

Figure 12.1
Setup for measurement of osmotic pressure.

at the middle of the range P to $P + \Pi$; that is, $\bar{V}_A^* = \bar{V}_A^*(P + \frac{1}{2}\Pi, T)$. Hence (12.30) becomes $RT \ln \gamma_A x_A = -\Pi \bar{V}_A^*$, or

$$\Pi = -(RT/\bar{V}_A^*) \ln \gamma_A x_A \tag{12.31}$$

For an ideally dilute solution, γ_A is 1 and $\ln \gamma_A x_A \approx -x_B$ [Eq. (12.19)]. Therefore

$$\Pi = (RT/\bar{V}_A^*)x_B \qquad \text{ideally dil. soln.} \tag{12.32}$$

Since the solution is quite dilute, we have $x_B = n_B/(n_A + n_B) \approx n_B/n_A$ and

$$\Pi = \frac{RT}{\bar{V}_A^*}\frac{n_B}{n_A} \qquad \text{ideally dil. soln.} \tag{12.33}$$

where n_A and n_B are the numbers of moles of solvent and solute in the solution that is in membrane equilibrium with pure solvent A.

Although (12.33) is best left as it is, a further approximation is sometimes made. Since the solution is quite dilute, the solution volume V is approximately equal to $n_A\bar{V}_A^*$ and (12.33) becomes $\Pi V = n_B RT$, or

$$\Pi = c_B RT \qquad \text{ideally dil. soln.} \tag{12.34}^*$$

where the concentration c_B equals n_B/V. (Note the formal resemblance to the equation of state of an ideal gas, $P = cRT$, where $c = n/V$.) Equation (12.34), which is called *van't Hoff's law*, is a limiting law valid in the limit of infinite dilution. Equation (12.33) also holds only in the limit of infinite dilution, but (12.33) is a better approximation than (12.34) in real dilute solutions.

For solutions that are not ideally dilute, Eq. (12.31) holds. However, a different (but equivalent) expression for Π in nonideally dilute solutions is often more convenient than (12.31). In 1945, McMillan and Mayer developed a statistical-mechanical theory for nonelectrolyte solutions. (For the McMillan–Mayer theory, see T. L. Hill, *Statistical Mechanics*, McGraw-Hill, 1956, chap. 6.) One result of their theory was a proof that the osmotic pressure in a nonideally dilute two-component solution is given by

$$\Pi = RT(M_B{}^{-1}\rho_B + A_2\rho_B{}^2 + A_3\rho_B{}^3 + \cdots) \tag{12.35}$$

where M_B is the solute molar mass, and where the solute *mass concentration* ρ_B is defined as

$$\rho_B \equiv w_B/V \tag{12.36}$$

where w_B is the mass of solute B in the solution with volume V. The quantities A_2, A_3, ... are related to the solute–solute intermolecular forces in the given solvent and are functions of temperature. [Note the formal resemblance of (12.35) to the virial equation (8.5) for gases.]

In the limit of infinite dilution, ρ_B goes to zero, and (12.35) becomes

$$\Pi = RT\rho_B/M_B = RTw_B/M_B V = RTn_B/V = c_B RT \tag{12.37}$$

which is the van't Hoff law.

The McMillan–Mayer theory provides a statistical-mechanical basis for the validity of the van't Hoff law (12.34) in an infinitely dilute solution. By reversing the steps we used to derive (12.34), we can derive the limiting law $\mu_A = \mu_A^* + RT \ln x_A$ for an ideally dilute solution from the van't Hoff equation.

Hence, the McMillan–Mayer theory provides a statistical-mechanical derivation for the chemical-potential expression in an ideally dilute solution.

The concept of osmotic pressure is sometimes misunderstood. Consider a 0.01 mol/kg solution of glucose in water at 25°C and 1 atm. When we say the freezing point of this solution is -0.02°C, we do not imply that the solution's temperature is actually -0.02°C. The freezing point is that temperature at which the solution would be in equilibrium with pure solid water at 1 atm. Likewise, when we say that the osmotic pressure of this solution is 0.24 atm (see the example below), we do not imply that the pressure in the solution is 0.24 atm (or 1.24 atm). Instead the osmotic pressure is the extra pressure that would have to be applied to the solution so that if it were placed in contact with a membrane permeable to water but not glucose, it would be in membrane equilibrium with pure water, as in Fig. 12.2.

Example Find the osmotic pressure at 25°C and 1 atm of a solution of 1.8016 g of glucose ($C_6H_{12}O_6$) in 1000.0 g of water.

The molecular weights are 180.16 and 18.015, so $n_B = 0.010000$ mol and $n_A = 55.509$ mol. The density of water at 25°C and 1 atm is 0.99704 g/cm³, so $\bar{V}^*_A = 18.068$ cm³/mol. Substitution in (12.33) gives

$$\Pi = \frac{(82.06 \text{ cm}^3 \text{ atm mol}^{-1} \text{ K}^{-1})(298.15 \text{ K})}{18.068 \text{ cm}^3 \text{ mol}^{-1}} \frac{0.010000 \text{ mol}}{55.509 \text{ mol}}$$

$$= 0.2439 \text{ atm} = 185.4 \text{ torr}$$

where we assumed the solution to be ideally dilute. In this calculation we used $\bar{V}^*_A$ at 1 atm. As noted above, it is best to use $\bar{V}^*_A$ at $P + \frac{1}{2}\Pi = 1.12$ atm. For water at 25°C and 1 atm, the compressibility is $\kappa = 4.6 \times 10^{-5}$ atm^{-1}. We have $\kappa = -\bar{V}^{-1}(\partial \bar{V}/\partial P)_T$, so $\Delta \bar{V} \approx -\kappa \bar{V}\, \Delta P = -10^{-4}$ cm³ mol^{-1} for $\Delta P = 0.12$ atm. Since this $\Delta \bar{V}$ is negligible compared with $\bar{V}^*_A = 18.068$ cm³/mol, the value for Π is unaffected by this correction.

Note especially the rather large value of Π for the very dilute 0.01 mol/kg glucose solution in this example. Since the density of water is 1/13.6 times that of mercury, an osmotic pressure of 185 torr (185 mmHg) corresponds to a height of 18.5 cm × 13.6 = 250 cm = 2.5 m = 8.2 ft of liquid in the right-hand tube in Fig. 12.1. In contrast, an aqueous 0.01 mol/kg solution will show a freezing-point depression of only 0.02°C. The large value of Π results from the fact (noted many times previously) that the chemical potential of a component of a condensed phase is rather insensitive to pressure. Hence it takes a large value of Π to change the chemical potential of A in the solution so that it equals the chemical potential of pure A at pressure P.

The large values of Π given by dilute solutions make osmotic-pressure measure-

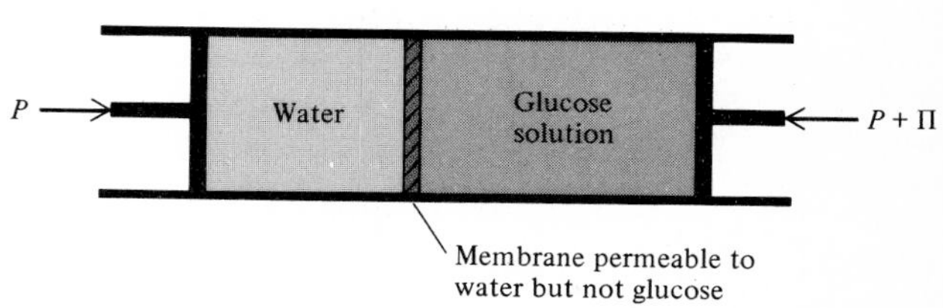

Figure 12.2
Pure water in equilibrium with water in a glucose solution.

ments valuable in determining molecular weights of substances with high molecular weights (e.g., polymers). For such substances, the freezing-point depression is too small to be of value. (For example, if $M_B = 4000$ g/mol, a solution of 1.0 g of B in 100 g of water has $\Delta T_f = -0.005°C$ and $\Pi = 46$ torr at 25°C.) Polymer solutions generally show quite large deviations from ideally dilute behavior even at quite low molalities. The large size of the molecules causes substantial solute–solute interactions in dilute polymer solutions. Hence for a polymer solution, it is essential to measure Π at several dilute concentrations and extrapolate to infinite dilution to obtain the true molecular weight. Π is given by the McMillan–Mayer expression (12.35). In dilute solutions it is usually adequate to terminate the series after the A_2 term. Thus, $\Pi/RT = \rho_B/M_B + A_2\rho_B^2$, or

$$\Pi/\rho_B = RT/M_B + RTA_2\rho_B \qquad \text{dil. soln.} \tag{12.38}$$

A plot of Π/ρ_B vs. ρ_B gives a straight line with intercept RT/M_B at $\rho_B = 0$.

A synthetic polymer usually consists of molecules of varying chain length, since chain termination in a polymerization reaction is a random process. We now find the expression for the apparent molecular weight of such a solute as determined by osmotic-pressure measurements. If there are several solute species in the solution, the solvent mole fraction x_A equals $1 - \sum_{i \neq A} x_i$, where the sum goes over the various solute species. Use of (8.41) gives

$$\ln x_A = \ln\left(1 - \sum_{i \neq A} x_i\right) \approx -\sum_{i \neq A} x_i \approx -\frac{1}{n_A}\sum_{i \neq A} n_i$$

Hence in place of (12.34) or (12.37), we get

$$\Pi = RT\sum_{i \neq A} c_i = \frac{RT}{V}\sum_{i \neq A} n_i \qquad \text{ideally dil. soln.} \tag{12.39}$$

If we pretended that there is only one solute species B, with molar mass M_B, we would use data extrapolated to infinite dilution to calculate M_B from (12.37) as $M_B = RTw_B/\Pi V$. Use of (12.39) for Π gives

$$M_B = \frac{w_B}{\sum_{i \neq A} n_i} = \frac{\sum_{i \neq A} w_i}{\sum_{i \neq A} n_i} = \frac{\sum_{i \neq A} n_i M_i}{\sum_{i \neq A} n_i} \tag{12.40}$$

where w_i, n_i, and M_i are the mass, the number of moles, and the molar mass of solute i. The quantity on the right of (12.40) is the *number average molar mass*. The number of moles n_i is proportional to the number of molecules of species i; hence, each value of M_i in (12.40) is weighted according to the number of molecules having that molecular weight. The same result is found for the molecular weight calculated from the other colligative properties.

If we consider the collection of solute molecules only, the denominator of the right side of (12.40) is n_{tot}, the total number of moles of solute, and n_i/n_{tot} is the mole fraction x_i of solute species i in the collection of solute molecules. (Of course, x_i is not the mole fraction of species i in the solution. We are now considering the solute species apart from the solvent.) Introducing the symbol M_n for the number average molar mass, we rewrite (12.40) as

$$M_n = \sum_i x_i M_i = \frac{w_{tot}}{n_{tot}} \tag{12.41}$$

where the sum goes over all solute species and where w_{tot} and n_{tot} are the total mass and total number of moles of solute species.

The mechanism of osmotic flow is not the business of thermodynamics, but we shall mention the three commonly cited mechanisms: (1) The pores of the membrane may be of such size as to allow small solvent molecules to pass through but not allow large solute molecules (such as polymers) to pass. (2) The volatile solvent may vaporize into the pores of the membrane and condense out on the other side, while the involatile solute does not do so. (3) The solvent may dissolve in the membrane.

A commonly used membrane for osmotic studies of polymers is cellulose in the form of cellophane film mounted on a rigid support. The pore diameter in cellophane has been estimated to be 30 to 70 Å.

In Fig. 12.2, the additional externally applied pressure Π causes membrane equilibrium to exist between the solution and the pure solvent. If the pressure on the solution were less than $P + \Pi$, then μ_A would be less in the solution than in the pure solvent and there would be a net flow of solvent from the pure solvent on the left to the solution on the right (a process called *osmosis*). If, however, the pressure on the solution is increased above $P + \Pi$, then μ_A in the solution becomes greater than μ_A in the pure solvent and there is a net flow of solvent from the solution to the pure solvent, a phenomenon called *reverse osmosis*. Reverse osmosis is used to desalinate seawater. Here, one requires a membrane that is substantially impermeable to salt ions, strong enough to withstand the pressure difference, and permeable enough to water to give a reasonably rapid flow of water. Membranes of cellulose acetate (prepared by treating cellulose with acetic acid and acetic anhydride) or hollow nylon fibers are used in desalination plants. It has been proposed that desalination be carried out by simply lowering into the ocean a long hollow pipe capped at the lower end by a semipermeable membrane. Below 230 m, the pressure difference on the membrane exceeds the osmotic pressure of salt water, and fresh water flows into the pipe; see O. Levenspiel and N. de Nevers, *Science*, **183**, 157 (1974).

Osmosis is of fundamental importance in biology. Cell membranes are permeable to water, to CO_2, O_2, and N_2, and to small organic molecules (e.g., amino acids, glucose) and are impermeable to large polymer molecules (e.g., proteins, polysaccharides). Inorganic ions and disaccharides (e.g., sucrose) generally pass quite slowly through cell membranes. The cells of an organism are bathed by body fluids (e.g., blood, lymph, sap) containing various solutes. The situation is more complex than in Fig. 12.1 in that there are solutes present on both sides of the membrane, which is permeable to water and some solutes (which we symbolize by B, C, ...) but impermeable to other solutes (which we symbolize by L, M, ...). In the absence of active transport (discussed below), water and solutes B, C, ... will move through the cell membrane until the chemical potentials of H_2O, of B, of C, ... are equalized on each side of the membrane. If the fluid surrounding a cell is more concentrated in solutes L, M, ... than the cell fluid is, the cell will lose water by osmosis; the surrounding fluid is said to be *hypertonic* with respect to the cell. If the surrounding fluid is less concentrated in L, M, ... than the cell, the cell gains water from the surrounding *hypotonic* fluid. When there is no net transfer of water between cell and surroundings, the two are said to be *isotonic*. Blood and lymph are approximately isotonic to the cells of an organism. Intravenous feeding and injections are performed using a salt solution that is isotonic with blood. (If water were injected, the red blood cells would gain water by osmosis and might ultimately burst.) Plant roots absorb water from the surrounding hypotonic soil fluids by osmosis.

Living cells are able to transport a chemical species through a membrane from a region of low chemical potential of that solute to a region of high chemical potential, a direction opposite that of spontaneous flow. Such transport (called *active transport*) is accomplished by coupling the transport with a chemical reaction that has a negative ΔG (Sec. 11.8). The details of such coupling processes are not known but are under investigation. Some examples of active transport are the transport of K^+ ions into cells from surrounding fluids with lower K^+ concentrations, the formation of the HCl solution used in digestion, and the movement of an amino acid from blood to cells with a higher concentration of that amino acid.

There is a logical gap in our derivation of the osmotic pressure. In Sec. 4.7, we derived $\mu_A^\alpha = \mu_A^\beta$ (the equality of chemical potentials in different phases in equilibrium) under the suppositions that $T^\alpha = T^\beta$ and $P^\alpha = P^\beta$. However, in osmotic equilibrium, $P^\alpha \neq P^\beta$ (where α and β are the phases separated by the semipermeable membrane). Hence we must show that $\mu_A^\alpha = \mu_A^\beta$ at equilibrium even when the phases are at different pressures. The proof is outlined in Prob. 12.18.

12.5 TWO-COMPONENT PHASE DIAGRAMS

Phase diagrams for one-component systems were discussed in Chap. 7. We discuss phase diagrams for two-component systems in Secs. 12.5 to 12.10 and for three-component systems in Sec. 12.11.

With $c_{\text{ind}} = 2$, the phase rule $f = c_{\text{ind}} - p + 2$ becomes $f = 4 - p$. For a one-phase, two-component system, $f = 3$, and there are three independent intensive variables. For convenience, we usually keep one of these variables constant and plot a two-dimensional phase diagram. This amounts to taking a cross section of a three-dimensional plot. The three independent variables are P, T, and one mole fraction. Usually, either P or T is held fixed. A two-component system is called a *binary* system.

Multicomponent phase equilibria have important applications in chemistry, geology, and materials science. *Materials science* studies the structure, properties, and applications of scientific and industrial materials. The main classes of materials are *metals*, *polymers*, *ceramics*, and *composites*. Traditionally, the term "ceramic" referred to materials produced by baking moist clay to form hard solids. Nowadays, the term is applied to a much broader class of materials. Most ceramics are compounds of one or more metals with a nonmetal (commonly oxygen) and are mechanically strong and resistant to heat and chemicals. Some examples of ceramics are rocks, most minerals, sand, porcelain, concrete, glass, ruby, asbestos, clay, graphite, diamond, MgO, SiC, and Mg_2SiO_4; many ceramics are silicates. A composite material is made of two or more materials and may possess properties not present in any one component. Bone is a composite of the soft, strong, polymeric protein collagen and the hard, brittle mineral hydroxyapatite [approximate formula $3Ca_3(PO_4)_2 \cdot Ca(OH)_2$]. Fiber glass is a composite containing a plastic strengthened by the addition of glass fibers.

12.6 TWO-COMPONENT LIQUID–VAPOR EQUILIBRIUM

Instead of plotting complete phase diagrams, we shall usually consider only one portion of the phase diagram at a time. This section deals with the liquid–vapor part of the phase diagram of a two-component system.

Ideal solution at fixed temperature. Consider two liquids A and B that form an ideal solution. We hold the temperature fixed at some value T that is above the freezing points of A and B. We shall plot the system's pressure P against x_A, the mole fraction of one component. Note that x_A will be the *overall* mole fraction of A in the system:

$$x_A \equiv \frac{n_{A,l} + n_{A,v}}{n_{A,l} + n_{A,v} + n_{B,l} + n_{B,v}} \tag{12.43}$$

where $n_{A,l}$ and $n_{A,v}$ are the number of moles of A in the liquid and vapor phases, respectively. For a closed system, x_A is fixed, although $n_{A,l}$ and $n_{A,v}$ may vary.

Imagine the system to be enclosed in a cylinder fitted with a piston and immersed in a constant-temperature bath (Fig. 12.3*a*). To see what the P-vs.-x_A phase diagram looks like, let us initially set the external pressure on the piston high enough for the system to be entirely liquid (point C in Fig. 12.3*b*). As the pressure is lowered below C, we eventually reach a pressure where the liquid just begins to vaporize (point D). At point D, the liquid has composition $x_{A,l}$ (where $x_{A,l}$ at D is equal to the overall mole fraction x_A since only an infinitesimal amount of liquid has vaporized). What is the composition of the first vapor that comes off? Raoult's law, Eq. (9.69), relates the vapor-phase mole fractions to the liquid composition, as follows:

$$x_{A,v} = \frac{P_A^*}{P} x_{A,l} \qquad \text{and} \qquad x_{B,v} = \frac{P_B^*}{P} x_{B,l} \tag{12.44}$$

where P_A^* and P_B^* are the vapor pressures of pure A and pure B at T, where the system's pressure P equals $P_A + P_B$, and where the vapor is assumed ideal.

From (12.44) we have

$$x_{A,v}/x_{B,v} = (x_{A,l}/x_{B,l})(P_A^*/P_B^*) \tag{12.45}$$

Let A be the more volatile component; i.e., let A have the higher vapor pressure at temperature T: $P_A^* > P_B^*$. Equation (12.45) then shows that $x_{A,v}/x_{B,v} > x_{A,l}/x_{B,l}$. Thus, the vapor above an ideal solution is richer than the liquid in the more volatile component. Of course, Eqs. (12.44) and (12.45) apply at any pressure where liquid–vapor equilibrium exists, not just at point D.

Now let us isothermally lower the pressure below point D, causing more liquid to vaporize. Eventually, we reach point F, where the last drop of liquid vaporizes. Below F, we have only vapor. For any point on the line between D and F, liquid and vapor phases coëxist in equilibrium.

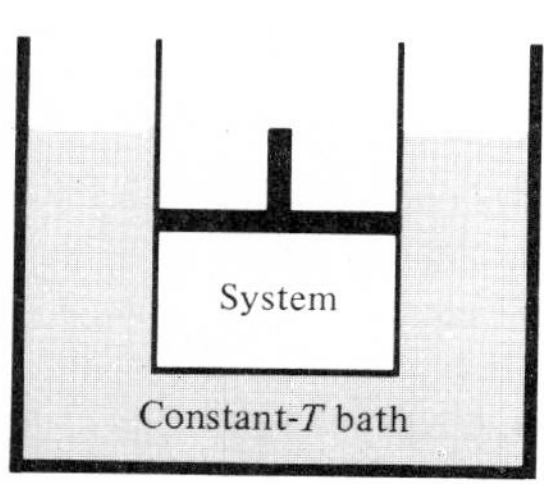

(*a*)

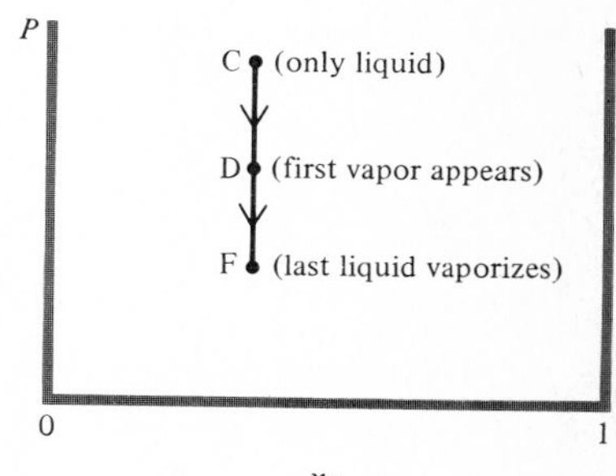

(*b*)

Figure 12.3
(*a*) A system held at constant T. (*b*) Points on the P-x_A phase diagram of the system in (*a*).

We can repeat this experiment many times, each time starting with a different composition for the closed system. For composition x'_A, we get points D′ and F′; for composition x''_A, we get points D″ and F″; and so on. We then plot the points D, D′, D″, ... and join them, and do the same for F, F′, F″, ... (Fig. 12.4).

What is the equation of the curve DD′D″? For each of these points, liquid of composition $x_{A,l}$ (or $x'_{A,l}$, etc.) is just beginning to vaporize. The equilibrium vapor pressure of this liquid is $P = P_A + P_B = x_{A,l}P_A^* + x_{B,l}P_B^* = x_{A,l}P_A^* + (1 - x_{A,l})P_B^*$, or

$$P = P_B^* + (P_A^* - P_B^*)x_{A,l} \qquad \text{ideal soln.} \tag{12.46}$$

This is the same as Eq. (9.70) and is the equation of a straight line that starts at P_B^* for $x_{A,l} = 0$ and ends at P_A^* for $x_{A,l} = 1$. All along the line DD′D″, the liquid is just beginning to vaporize, so that the overall mole fraction x_A is equal to the mole fraction of A in the liquid, $x_{A,l}$; DD′D″ is then simply a plot of the total vapor pressure P vs. $x_{A,l}$.

What is the equation of the curve FF′F″? Along this curve, the last drop of liquid is vaporizing, so that the overall x_A (which is what is plotted on the abscissa) will now equal $x_{A,v}$, the mole fraction of A in the vapor. FF′F″ is then a plot of the total vapor pressure P vs. $x_{A,v}$. To obtain P as a function of $x_{A,v}$ we must express $x_{A,l}$ in (12.46) as a function of $x_{A,v}$. Equation (12.45) reads

$$\frac{x_{A,v}}{1 - x_{A,v}} = \frac{x_{A,l}}{1 - x_{A,l}} \frac{P_A^*}{P_B^*} \tag{12.47}$$

Solving (12.47) for $x_{A,l}$, we find after a bit of algebra

$$x_{A,l} = \frac{x_{A,v}P_B^*}{x_{A,v}(P_B^* - P_A^*) + P_A^*} \tag{12.48}$$

Substitution of (12.48) into (12.46) gives after simplification

$$P = \frac{P_A^* P_B^*}{x_{A,v}(P_B^* - P_A^*) + P_A^*} \qquad \text{ideal soln.} \tag{12.49}$$

This is the desired equation for P vs. $x_{A,v}$ and is the FF′F″ curve.

We now redraw the phase diagram in Fig. 12.5. From the preceding discussion, the upper line is the P-vs.-$x_{A,l}$ curve, and the lower line is the P-vs.-$x_{A,v}$ curve.

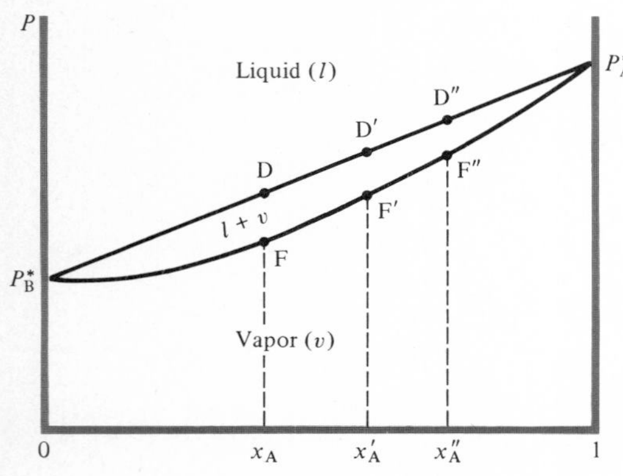

Figure 12.4
Pressure-vs.-composition liquid–vapor phase diagram for an ideal solution at fixed T.

Consider again the process of starting at point C (where P is high enough for only liquid to be present) and isothermally lowering the pressure. The system is closed, so that (even though the composition of the liquid and vapor phases may vary) the overall mole fraction of A remains fixed at x_A throughout the process; hence the process is represented by a vertical line on the P-vs.-x_A diagram. At point D with system pressure P_D, the liquid just begins to vaporize. What is the composition of the first vapor that comes off? What we want is the value of $x_{A,v}$ when liquid–vapor equilibrium exists and when the system's pressure P (which is also the total vapor pressure) equals P_D. The lower curve on the phase diagram is a plot of Eq. (12.49) and gives P as a function of $x_{A,v}$; alternatively, we can view the lower curve as giving $x_{A,v}$ as a function of P. Hence, to find the value of $x_{A,v}$ when P equals P_D, we simply find the point on the lower curve that corresponds to pressure P_D. This is point G and gives the composition (labeled $x_{A,1}$) of the first vapor that comes off.

As the pressure is lowered further, we reach point E. What are the liquid and vapor compositions at E? The upper curve relates P to $x_{A,l}$, and the lower curve relates P to $x_{A,v}$. Hence at point E with pressure P_E, we have $x_{A,v} = x_{A,2}$ (point I) and $x_{A,l} = x_{A,3}$ (point H). Finally, at point F with pressure P_F, the last liquid vaporizes; here, $x_{A,v} = x_A$ and $x_{A,l} = x_{A,4}$ (point J). Below F, we have vapor of composition x_A. Hence, as the pressure is lowered and the liquid vaporizes in the closed system, $x_{A,l}$ falls from D to J, that is, from x_A to $x_{A,4}$. This is because compound A is more volatile than B. Also, as the liquid vaporizes, $x_{A,v}$ falls from G to F, that is, from $x_{A,1}$ to x_A. This is because liquid vaporized later is richer in component B.

The P-vs.-x_A liquid–vapor phase diagram at constant T of two liquids that form an ideal solution thus has three regions. At any point above both curves in Fig. 12.5, only liquid is present. At any point below both curves, only vapor is present. At a typical point E between the two curves, there are two phases present, a liquid of composition H and a vapor of composition I; the overall composition is given by the x_A value at E. The horizontal line HEI is called a *tie line*. The two-phase region between the liquid and vapor curves is a gap in the phase diagram in which a single (homogeneous) phase cannot exist. A point in such a two-phase region gives the overall composition, and the compositions of the two phases are given by the points at the ends of the tie line.

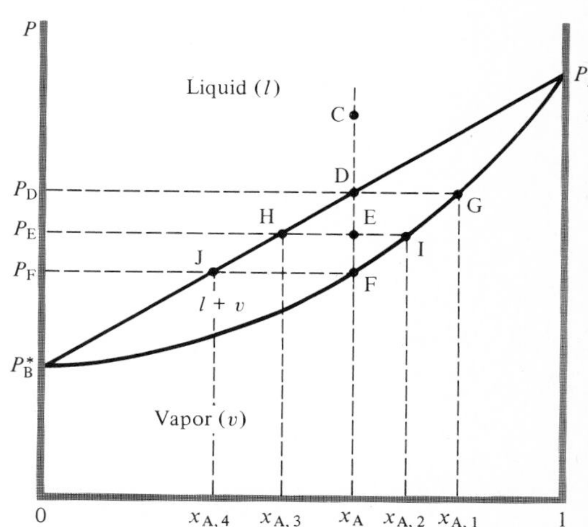

Figure 12.5
Pressure-vs.-composition liquid–vapor phase diagram for an ideal solution.

In the two-phase region, our two-component system has $f = 2 - 2 + 2 = 2$ degrees of freedom. The fact that T is fixed reduces f to 1 for a point in the two-phase region. Hence, once P is fixed, f is 0 in the two-phase region. For a fixed P, both $x_{A,v}$ and $x_{A,l}$ are thus fixed. For example, at pressure P_E in Fig. 12.5, $x_{A,l}$ is fixed as $x_{A,3}$ and $x_{A,v}$ is fixed as $x_{A,2}$. Of course, the overall x_A depends on the relative amounts of the liquid and vapor phases that are present in equilibrium. (Recall that the masses of the phases, which are extensive variables, are not considered in calculating f.)

Different relative amounts of liquid and vapor phases at the pressure P_E correspond to different points along the tie line HEI with different values of the overall mole fraction x_A but the same value of $x_{A,l}$ and the same value of $x_{A,v}$. We now derive a relation (called the *lever rule*) between the location of point E on the tie line and the relative amounts of the liquid and vapor phases. For the two-phase, two-component system, let n_A, n_l, and n_v be the total number of moles of A, the total number of moles in the liquid phase, and the total number of moles in the vapor phase, respectively. The overall mole fraction of A is $x_A = n_A/(n_l + n_v)$, so

$$n_A = x_A(n_l + n_v) \tag{12.50}$$

Also, $n_A = n_{A,l} + n_{A,v} = x_{A,l}n_l + x_{A,v}n_v$, so

$$n_A = x_{A,l}n_l + x_{A,v}n_v \tag{12.51}$$

Equating these two expressions for n_A, we get

$$x_A n_l + x_A n_v = x_{A,l}n_l + x_{A,v}n_v$$

$$n_l(x_A - x_{A,l}) = n_v(x_{A,v} - x_A) \tag{12.52}$$

$$n_l\overline{\text{EH}} = n_v\overline{\text{EI}} \tag{12.53}$$

where $\overline{\text{EH}}$ and $\overline{\text{EI}}$ are the lengths of the lines from E to the liquid and vapor curves in Fig. 12.5. Equation (12.53) is the lever rule. Note its resemblance to the lever law of physics: $m_1 l_1 = m_2 l_2$, where m_1 and m_2 are masses that balance each other on a seesaw with fulcrum a distance l_1 from mass m_1 and l_2 from m_2. When point E in Fig. 12.5 is close to point H on the liquid line, $\overline{\text{EH}}$ is less than $\overline{\text{EI}}$ and (12.53) tells us that n_l is greater than n_v. When E coincides with H, then $\overline{\text{EH}}$ is zero and n_v must be zero; there is only liquid present.

The above derivation of the lever rule clearly applies to any two-phase, two-component system, not just to liquid–vapor equilibrium. Therefore, if α and β are the two phases present, n_α and n_β are the total numbers of moles in phase α and in phase β, respectively, and l_α and l_β are the lengths of the lines from a point in a two-phase region of the phase diagram to the phase α and phase β lines, then, by analogy to (12.53) we have

$$n_\alpha l_\alpha = n_\beta l_\beta \tag{12.54*}$$

Frequently, the overall weight fraction of A or the overall weight percent of A (instead of x_A) is used as the abscissa of the phase diagram. In this case, the masses replace the numbers of moles in the above derivation, and the lever rule becomes

$$m_\alpha l_\alpha = m_\beta l_\beta \tag{12.55}$$

where m_α and m_β are the masses of phases α and β, respectively.

Example Let the two-component system of Fig. 12.5 contain 10 moles of A. Find the number of moles of A in the vapor phase at the pressure and composition represented by point E.

Using a ruler, we find $\overline{EI}/\overline{EH} = 0.8$. Hence (12.53) or (12.54) gives $n_l/n_v = 0.8$. Again using a ruler, we find that for point E, the overall mole fraction $x_A = 0.6$. Hence $0.6 = n_A/(n_l + n_v) = 10/(n_l + n_v)$, and $n_l + n_v = 17$. Therefore, $n_l/n_v = 0.8 = (17 - n_v)/n_v$. Solving, we find $n_v = 9.4$. The value of $x_{A,v}$ is $x_{A,2}$, and use of a ruler gives $x_{A,v} = 0.7$. Hence, $n_{A,v} = x_{A,v}n_v = 6.6$.

Ideal solution at fixed pressure. Now consider the fixed-pressure liquid–vapor phase diagram of two liquids that form an ideal solution. The explanation is quite similar to the fixed-temperature case just discussed in great (and somewhat repetitious) detail, so we can be brief here. We plot T vs. x_A, the overall mole fraction of one component. The phase diagram is Fig. 12.6.

T_B^* and T_A^* are the normal boiling points of pure B and pure A (assuming the fixed pressure is 1 atm). The lower curve gives T as a function of $x_{A,l}$ (or vice versa) for a system with liquid and vapor phases in equilibrium and is the boiling-point curve of the ideal solution. The upper curve gives T as a function of $x_{A,v}$ (or vice versa) for a system with liquid–vapor equilibrium. The vapor curve lies above the liquid curve on a T-vs.-x_A diagram but lies below the liquid curve on a P-vs.-x_A diagram (Fig. 12.5). This is obvious from the fact that the vapor phase is favored by high T and by low P.

If we isobarically heat a closed system of composition x_A, vapor will first appear at point L. As we raise the temperature and vaporize more of the liquid, the liquid will become richer in the less volatile, higher-boiling component B. Eventually, we reach point N, where the last drop of liquid vaporizes.

The first vapor that comes off when a solution of composition x_A is boiled has a value of $x_{A,v}$ given by point Q. If we remove this vapor from the system and condense it, we get liquid of composition $x_{A,1}$. Vaporization of this liquid gives vapor of initial composition $x_{A,2}$ (point R). Thus by successively condensing and revaporizing the mixture, we can ultimately separate A from B. This procedure is called *fractional distillation*. We get the maximum enrichment in A by taking just the first bit of vapor that comes off. This maximum degree of enrichment for any one distillation step is said to represent one *theoretical plate*. By packing the distillation column, we in effect get many successive condensations and revaporizations, giving a column with several theoretical plates.

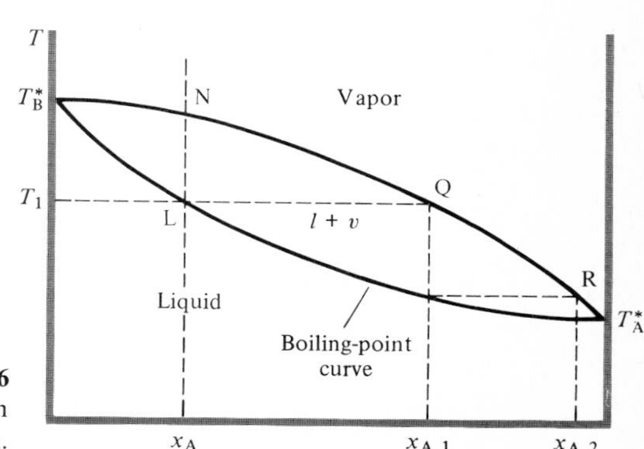

Figure 12.6
Temperature-vs.-composition liquid–vapor phase diagram for an ideal solution.

How do we plot the two curves in Fig. 12.6? We start with $P_A^*(T)$ and $P_B^*(T)$, the known vapor pressures of pure A and of pure B as functions of temperature. (These are obtained by direct experimental measurement or from the Clausius–Clapeyron equation.) Let the fixed pressure be $P^\#$. We have $P^\# = P_A + P_B$, where P_A and P_B are the partial pressures of A and B in the vapor. Raoult's law gives $P^\# = x_{A,l}P_A^*(T) + (1 - x_{A,l})P_B^*(T)$, or

$$x_{A,l} = \frac{P^\# - P_B^*(T)}{P_A^*(T) - P_B^*(T)} \qquad \text{ideal soln.} \tag{12.56}$$

Since P_A^* and P_B^* are known functions of temperature, we can use (12.56) to find $x_{A,l}$ at any given T and thereby plot the lower (liquid) curve. To plot the vapor curve, we use $x_{A,v} = P_A/P^\# = x_{A,l}P_A^*/P^\#$; substitution of (12.56) gives

$$x_{A,v} = \frac{P_A^*(T)}{P^\#}\,\frac{P^\# - P_B^*(T)}{P_A^*(T) - P_B^*(T)} \qquad \text{ideal soln.} \tag{12.57}$$

which is the desired equation for $x_{A,v}$ as a function of T. Note that (12.56) and (12.57) are the same equations as (12.46) and (12.49), except that P is now fixed at $P^\#$ and T is regarded as a variable.

Nonideal solutions. Having dealt with liquid–vapor equilibrium for ideal solutions, we now consider nonideal solutions. Liquid–vapor phase diagrams for nonideal systems are obtained by measurement of the pressure and composition of the vapor in equilibrium with liquid of known composition. If the solution is only slightly nonideal, the curves resemble those for ideal solutions and nothing is significantly changed. If, however, the solution has a great enough deviation from ideality to give a maximum or minimum in the P-vs.-$x_{A,l}$ curve (as in Fig. 9.9), a new phenomenon appears.

Suppose we have a positive deviation from Raoult's law of sufficient magnitude to give a maximum in the P-vs.-$x_{A,l}$ curve, which is the upper curve on the P-vs.-x_A phase diagram. What does the lower curve (the vapor curve) look like? Suppose we imagine the phase diagram to look like Fig. 12.7. Let point D in Fig. 12.7 be the maximum on the liquid curve. If we start at point C in a closed system and isothermally reduce the pressure, we shall reach point D, where the liquid just begins to

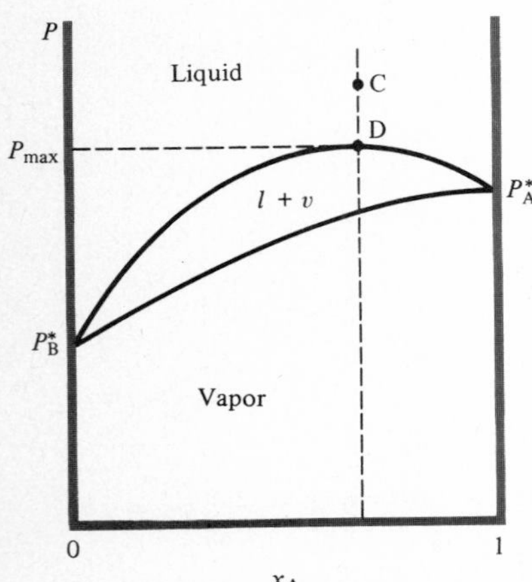

Figure 12.7
Erroneous pressure-vs.-composition liquid–vapor phase diagram with a maximum.

vaporize. What is the composition of the first vapor that comes off? To answer this, we want the value of $x_{A,v}$ that corresponds to the pressure (designated P_{max} in Fig. 12.7) at point D. However, there is no point on the vapor curve (the lower curve) in Fig. 12.7 that has pressure P_{max}. Hence, the phase diagram cannot look like Fig. 12.7. The only way (consistent with the requirement that the vapor phase be favored by low pressure and therefore always lie below the liquid phase) we can draw the phase diagram so that there is a point on the vapor curve with pressure P_{max} is to have the vapor curve touch the liquid curve at P_{max}, as in Fig. 12.8*a*.

What does the fixed-pressure T-vs.-x_A phase diagram that corresponds to Fig. 12.8*a* look like? Let T' be the temperature for which Fig. 12.8*a* is drawn, and let $x_{A,1}$ be the value of x_A that corresponds to P_{max}. If P is fixed at P_{max}, liquid with $x_{A,l}$ equal to $x_{A,1}$ will boil at temperature T'; however, liquid with $x_{A,l}$ less than or greater than $x_{A,1}$ will not have sufficient vapor pressure to boil at T' and will boil at higher temperatures. Therefore, a maximum on the P-x_A phase diagram will correspond to a minimum in the T-x_A diagram. The T-vs.-x_A phase diagram will then look like Fig. 12.8*b*.

Let the minimum in Fig. 12.8*b* occur at composition x'_A. (If the fixed value of P for Fig. 12.8*b* equals P_{max} in Fig. 12.8*a*, then x'_A in Fig. 12.8*b* equals $x_{A,1}$ in Fig. 12.8*a*. Usually, P is fixed at 1 atm, so x'_A and $x_{A,1}$ usually differ.) A liquid of composition x'_A when boiled will yield vapor with the same composition as the liquid. Since vaporization does not change the liquid's composition, the entire sample of liquid will boil at a constant temperature. Such a constant-boiling solution is called an *azeotrope*. The boiling behavior of an azeotropic solution resembles that of a pure compound and contrasts with that of most solutions of two liquids, which boil over a temperature range. However, since the azeotrope's composition depends on the pressure, a mixture which exhibits azeotropic behavior at one pressure will boil over a temperature range if the pressure is changed. Thus, an azeotrope can readily be distinguished from a compound.

Drawing lines in Fig. 12.8*b* similar to those in Fig. 12.6, we see that fractional distillation of a solution of two substances that form an azeotrope leads to separation into either pure A and azeotrope (if $x_{A,l} > x'_A$) or pure B and azeotrope (if $x_{A,l} < x'_A$). A liquid–vapor phase diagram with an azeotrope resembles two nonazeotropic liquid–vapor diagrams placed side by side.

The most famous azeotrope is that formed by water and ethanol. At 1 atm, the azeotropic composition is 96 percent C_2H_5OH by weight (192 proof); the boiling point is 78.2°C, which is below the boiling points of water and ethanol. Absolute

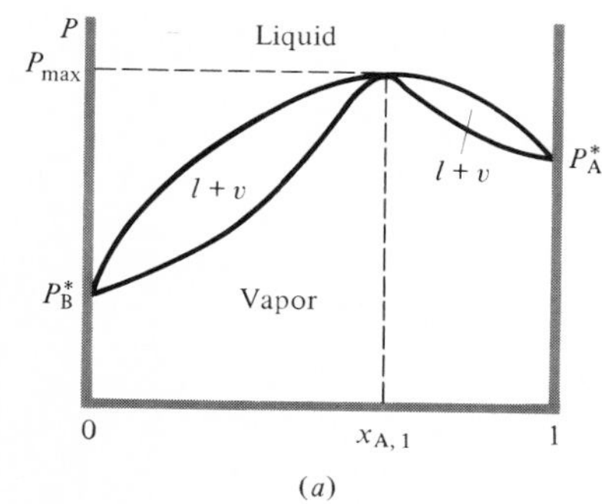

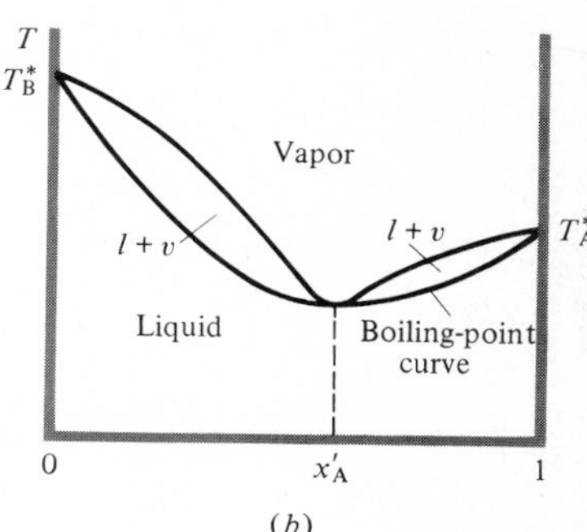

Figure 12.8
(*a*) Pressure-vs.-composition liquid–vapor phase diagram with a maximum. (*b*) The corresponding temperature-vs.-composition diagram.

(100%) ethanol cannot be prepared by distillation at 1 atm of a dilute aqueous solution of ethanol.

A tabulation of known azeotropes is L. H. Horsley, Azeotropic Data III, *Adv. Chem. Ser.* 116, American Chemical Society, 1973. About half of the binary systems examined show azeotropes.

Figure 12.6 shows that when no azeotrope is formed, the vapor in equilibrium with a liquid is always richer in the lower-boiling (more volatile) component than the liquid. When, however, a minimum-boiling azeotrope is formed, we see from Fig. 12.8*b* that for some liquid compositions, the vapor is richer in the higher-boiling component.

For a negative deviation from Raoult's law of sufficient magnitude to give a minimum in the P-vs.-$x_{A,l}$ curve, we get a maximum in the T-x_A phase diagram and a maximum-boiling azeotrope.

If the positive deviation from ideality is large enough, the two liquids may become only partially miscible with each other. Liquid–liquid equilibrium for partially miscible liquids is discussed in Sec. 12.7; liquid–vapor equilibrium for this case is considered in Prob. 12.29.

12.7 TWO-COMPONENT LIQUID–LIQUID EQUILIBRIUM

Let liquids A and B be partially miscible, meaning that A is soluble in B to a limited extent and vice versa. With P held fixed (typically at 1 atm), the most common form of the T-x_A liquid–liquid phase diagram is shown in Fig. 12.9.

To understand the diagram, imagine that we start with pure B and gradually add more and more A while keeping the temperature fixed at T_1. The system starts at point F (pure B) and moves horizontally to the right. Along CF we have one phase, a dilute solution of solute A in solvent B. At point C we have reached the maximum solubility of liquid A in liquid B (at T_1). Addition of more A then produces a two-phase system (all points between C and E): phase 1 is a dilute saturated solution of A in B and has composition $x_{A,1}$; phase 2 is a dilute saturated solution of B in A and has composition $x_{A,2}$. The overall composition of the two-phase system at a typical point D is $x_{A,3}$. The relative amounts of the two phases present in equilibrium are given by the lever rule. At D, there is more of phase 1 than phase 2. As we

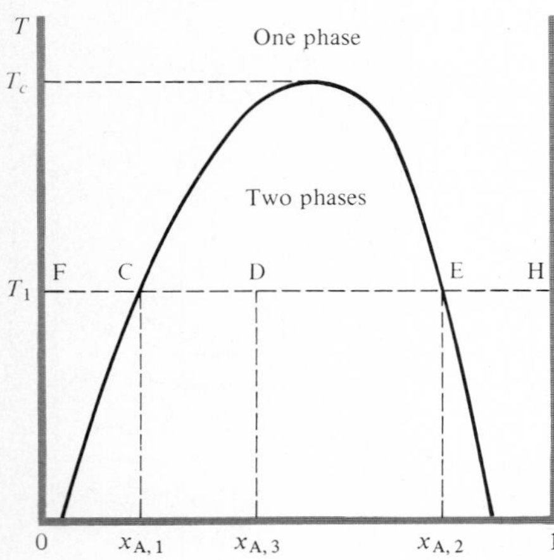

Figure 12.9
Temperature-vs.-composition liquid–liquid phase diagram for two partially miscible liquids.

continue to add more A, the overall composition eventually reaches point E. At E, there is just enough A present to allow all the B to dissolve in A to form a saturated solution of B in A. The system therefore again becomes a single phase at E. From E to H we are just diluting the solution of B in A. (To actually reach H requires the addition of an infinite amount of A.)

With two components and two phases present in equilibrium, the number of degrees of freedom is 2. However, since both P and T are fixed along line CE, f is 0 on CE. Two points on CE have the same value for each of the intensive variables P, T, $x_{A,1}$, $x_{B,1}$, $x_{A,2}$, $x_{B,2}$.

As the temperature is raised, the region of liquid–liquid immiscibility decreases, until at T_c (the *critical solution temperature*) it shrinks to zero. Above T_c, the liquids are completely miscible. (The critical point at the top of the two-phase region in Fig. 12.9 is similar to the liquid–vapor critical point of a pure substance, discussed in Sec. 8.3. In both cases, as the critical point is approached, the properties of two phases in equilibrium become more and more alike, until at the critical point the two phases become identical, yielding a one-phase system.)

For certain pairs of liquids, decreasing temperature leads to greater miscibility, and the liquid–liquid diagram resembles Fig. 12.10*a*. An example is water–triethylamine. Very occasionally, a system shows a combination of the behaviors in Figs. 12.9 and 12.10*a*, and the phase diagram resembles Fig. 12.10*b*. Such systems have lower and upper critical solution temperatures. Examples are nicotine–water and *m*-toluidine–glycerol.

The two-phase regions in Figs. 12.9 and 12.10 are called *miscibility gaps*.

Although it is often stated that gases are miscible in all proportions, in fact several cases of gas–gas miscibility gaps are known. Examples include CO_2–H_2O, NH_3–CH_4, and He–Xe. These gaps occur at temperatures above the critical temperatures of both components and hence by the conventional terminology of Sec. 8.3 involve two gases. Most such gaps occur at rather high pressures; however, *n*-butane–helium shows a miscibility gap at pressures as low as 40 atm. [See R. P. Gordon, *J. Chem. Educ.*, **49**, 249 (1972).]

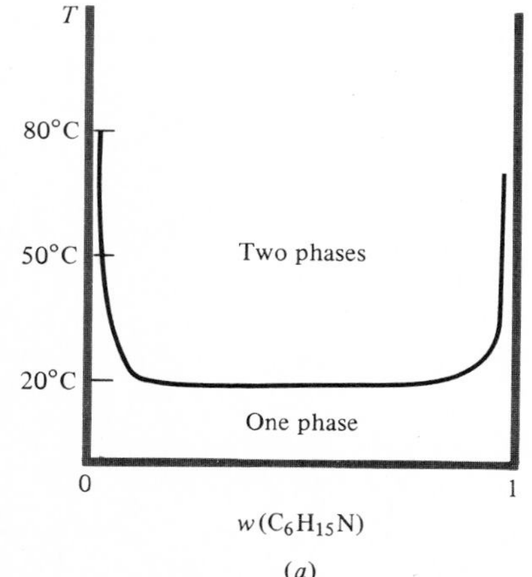

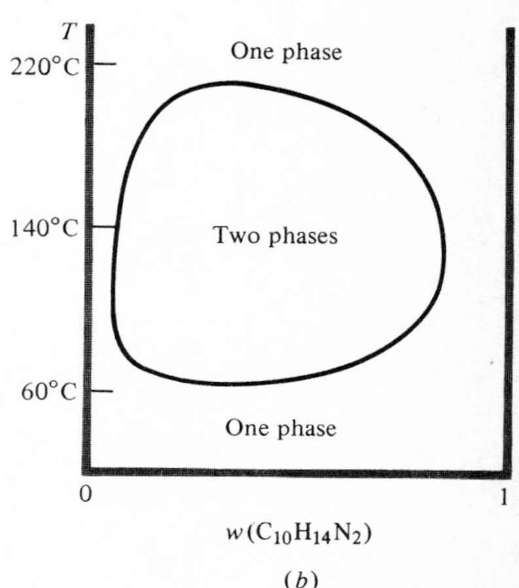

Figure 12.10
Temperature-vs.-composition liquid–liquid phase diagrams for (*a*) water–triethylamine; (*b*) water–nicotine. The horizontal axis is the weight fraction of the organic liquid. In (*b*), the pressure on the system equals the vapor pressure of the solution(s) and so is not fixed.

12.8 TWO-COMPONENT SOLID–LIQUID EQUILIBRIUM

We now discuss binary solid–liquid diagrams. The effect of pressure on condensed phases (i.e., solids and liquids) is slight, and unless one is interested in high-pressure phenomena, one generally holds P fixed at 1 atm and examines the T-x_A solid–liquid phase diagram.

Liquid-phase miscibility and solid-phase immiscibility. Let substances A and B be miscible in all proportions in the liquid phase and completely immiscible in the solid phase. Mixing any amounts of liquids A and B will produce a single-phase system that is a solution of A plus B. Since solids A and B are completely insoluble in each other, cooling a liquid solution of A and B will cause either pure A or pure B to freeze out of the solution.

The typical appearance of the solid–liquid phase diagram for this case is shown in Fig. 12.11. T_A^* and T_B^* are the freezing points of pure A and pure B.

The origin of the various regions on this diagram is as follows. In the low-temperature limit, we must have a two-phase mixture of pure solid A plus pure solid B (since the solids are immiscible). In the high-temperature limit, we must have a one-phase liquid solution of A plus B (since the liquids are miscible). Now consider cooling a solution of A and B that has $x_{A,l}$ close to 1 (the right side of the diagram). Eventually, we shall reach a temperature where the solvent A begins to freeze out, giving a two-phase region with solid A in equilibrium with a liquid solution of A and B. The curve DE thus gives the depression of the freezing point of A due to solute B. Likewise, if we cool a liquid solution of A plus B that has $x_{B,l}$ close to 1 (the left side of the diagram), we eventually get pure B freezing out and CFGE is the freezing-point-depression curve of B due to solute A. If we cool a two-phase mixture of solution plus either solid, the solution will eventually all freeze, giving a mixture of solid A plus solid B. The two freezing-point curves intersect at point E. For a solution with $x_{A,l}$ to the left of E, solid B will freeze out as T is lowered; for $x_{A,l}$ to the right of E, solid A will freeze out. At the values of T and $x_{A,l}$ corresponding to point E, the chemical potentials of solid A, solid B, and the solution are all equal, and both A and B freeze out. Point E is the *eutectic point* (from the Greek *eutektos*, "easily melted").

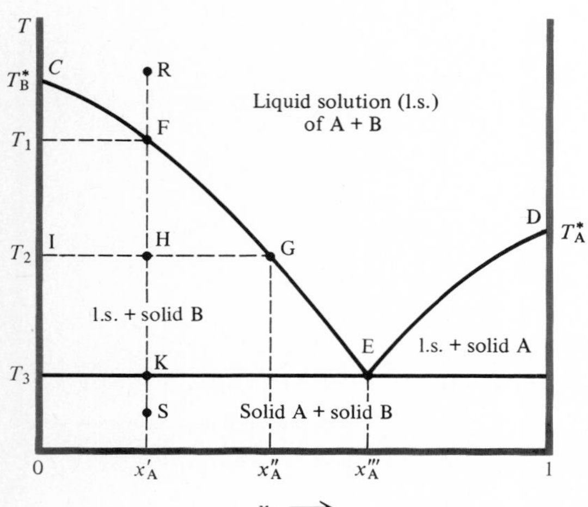

Figure 12.11
Solid–liquid phase diagram for complete liquid miscibility and solid immiscibility.

The portion of line DE with x_A very close to 1 can be calculated from Eq. (12.20), which is valid for a very dilute solution. Likewise, the portion of line CFGE with x_A very close to 0 (and hence x_B very close to 1) can be calculated from (12.20) with A and B interchanged. Away from the ends of these lines, Eq. (12.11) or (12.11) with A replaced by B applies. This exact equation is difficult to use to find the freezing point T_f as a function of x_A. Hence, we go to (12.12), which involves the approximation of taking $\Delta\bar{C}_{P,\text{fus}}$ as constant. To get a rough idea of the shape of the curves DE and CE, we make two further approximations. We set $\gamma_A = 1$ and $\gamma_B = 1$ over the entire range of solution composition; this amounts to assuming an ideal solution (an assumption that is usually quite poor). Further, we neglect the temperature dependence of $\Delta\bar{H}_{\text{fus},A}$ and $\Delta\bar{H}_{\text{fus},B}$. Since $\bar{C}_P = (\partial\bar{H}/\partial T)_P$, this amounts to setting $\Delta\bar{C}_{P,\text{fus},A} = 0$ and $\Delta\bar{C}_{P,\text{fus},B} = 0$. Thus, the very approximate equations for CE and DE are

CE: $$R \ln x_B \approx \Delta\bar{H}_{\text{fus},B}\left(\frac{1}{T_B^*} - \frac{1}{T}\right) \tag{12.58}$$

DE: $$R \ln x_A \approx \Delta\bar{H}_{\text{fus},A}\left(\frac{1}{T_A^*} - \frac{1}{T}\right) \tag{12.59}$$

Plotting these equations, we get curves with the general appearance of those in Fig. 12.11.

Now consider in detail what happens when we start at point R in Fig. 12.11 and isobarically cool a solution of A and B with composition x'_A. As usual, the system is closed, so the overall composition remains constant at x'_A, and we proceed vertically down from R. When T reaches T_1, solid B starts to freeze out. As B freezes out, the value of $x_{A,l}$ increases, and (since A is the solute here) the freezing point is depressed further. To cause more of the solvent (B) to freeze out, we therefore must lower the temperature further. At a typical temperature T_2, there is an equilibrium between a solution whose composition is given by point G as x''_A and solid B (whose composition is given by point I as $x_A = 0$). As usual, the points at the ends of the tie line (GHI) give the compositions of the two phases in equilibrium. The lever rule gives $n_{B,s}\overline{\text{HI}} = (n_{A,l} + n_{B,l})\overline{\text{HG}}$, where $n_{B,s}$ is the number of moles of solid B in equilibrium with a solution of $n_{A,l}$ moles of A plus $n_{B,l}$ moles of B. At point F, the lever rule gives $n_{B,s} = 0$. As the temperature drops along the line FHK, the horizontal distance to the line CFGE increases, indicating an increase in $n_{B,s}$.

As T is lowered further, we finally reach the eutectic temperature T_3 at point K. Here, the solution has composition x'''_A (point E), and now both solid B and solid A freeze out. The relative amounts of A and B that freeze out at E correspond to the eutectic composition x'''_A of the solution, and the entire remaining solution freezes out at T_3 with no further change in composition. At K, there are three phases in equilibrium (solution, solid A, and solid B), so the lever rule (12.54) does not apply. With three phases, we have $f = 2 - 3 + 2 = 1$ degree of freedom; this degree of freedom has been eliminated by the specification that P is fixed at 1 atm. Hence there are no degrees of freedom for the three-phase system, and the temperature must remain constant at T_3 until all the solution has frozen and the number of phases has dropped to 2. Below T_3, we are simply cooling a mixture of solid A plus solid B.

If we reverse the process and start at point S with solid A plus solid B, the first liquid formed will have composition x'''_A. The solid mixture will melt over the temperature range from T_3 to T_1. (Sharpness of melting point is one test organic chemists use for the purity of a compound.) A solid mixture that has the eutectic composition will melt entirely at one temperature (T_3). A solution of A and B that has the eutectic composition will freeze entirely at temperature T_3 to produce a eutectic mixture of solids A and B. However, a eutectic mixture is not a compound. Microscopic examination will show the eutectic solid to be an intimate mixture of crystals of A and crystals of B.

Systems with the solid–liquid phase diagram of Fig. 12.11 are called *simple eutectic* systems. Examples include Pb–Sb, benzene–naphthalene, Si–Al, KCl–AgCl, Bi–Cd, C_6H_6–CH_3Cl, and chloroform–aniline.

Solid solutions. Certain pairs of substances will form solid solutions. In a solid solution of A and B, there are no individual crystals of A or B. Instead the molecules (or atoms or ions) are mixed together at the molecular level, and the composition of the solution can be varied continuously over a certain range. Solid solutions can be prepared by condensing a vapor of A plus B or by cooling a liquid solution of A and B. We can have complete miscibility, partial miscibility, or complete immiscibility of solids A and B.

Solid solutions are classified as interstitial or substitutional. In an *interstitial* solid solution, the substance B molecules or atoms (which must be small) occupy interstices (i.e., holes) in the crystal structure of substance A. For example, steel is a solution in which carbon atoms occupy interstices in the Fe crystal structure. In a *substitutional* solid solution, molecules (or atoms or ions) of B substitute for those of A at random locations in the crystal structure. Examples include Cu–Ni, Cu–Zn, *p*-dichlorobenzene–*p*-dibromobenzene, and Na_2CO_3–K_2CO_3. Substitutional solids are formed by substances with atoms, molecules, or ions of similar size and structure.

Analysis of a transition-metal oxide or sulfide frequently shows an apparent violation of the law of definite proportions. For example, ZnO usually has a Zn/O mole ratio slightly greater than 1. The explanation is that the "zinc oxide" is actually an interstitial solid solution of Zn in ZnO.

Liquid-phase miscibility and solid-phase miscibility. Some pairs of substances are completely miscible in the solid state. Examples include Cu–Ni, Sb–Bi, Pd–Ni, KNO_3–$NaNO_3$, and *d*-carvoxime–*l*-carvoxime. With complete miscibility in both the liquid and the solid phases, the T-x_A binary phase diagram may look like Fig. 12.12, which is for Cu–Ni.

If a melt of Cu and Ni with any composition is cooled, a solid solution begins to freeze out; this solid solution is richer in Ni than the liquid solution. As the two-phase system of solid plus melt is cooled further, the mole fraction of Ni decreases in both the solid solution and the liquid melt. Eventually, a solid solution is formed that has the same composition as the liquid melt we started with.

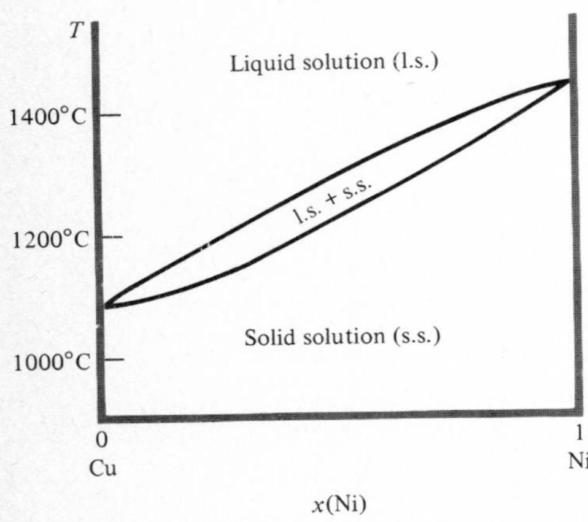

Figure 12.12
The Cu–Ni solid–liquid phase diagram.

Note that the freezing point of Cu is *raised* by the presence of a small amount of Ni. In discussing freezing-point depression in Sec. 12.3, we assumed solid-phase immiscibility, so that only pure solid solvent froze out. When the solids are miscible, the freezing point of the lower-melting component may be raised by the presence of the second component. An analogous situation is boiling-point elevation. When the solute is involatile, the solvent boiling point is elevated; however, if the solute is more volatile than the solvent, the solvent boiling point may be depressed. Note the resemblance between Figs. 12.12 and 12.6.

When the two miscible solids form an approximately ideal solid solution, the solid–liquid phase diagram resembles Fig. 12.12. However, when there are large deviations from ideality, the solid–liquid phase diagram may show a minimum or (less commonly) a maximum. Figure 12.13*a* for Cu–Au shows a minimum. Figure 12.13*b* for the optical isomers *d*-carvoxime–*l*-carvoxime ($C_{10}H_{14}NOH$) shows a maximum; here, the freezing point of each compound is elevated by presence of the other. The strong negative deviation from ideality indicates that in the solid state, *d*-carvoxime molecules prefer to associate with *l*-carvoxime molecules rather than with their own kind.

Liquid-phase miscibility and solid-phase partial miscibility. When A and B are completely miscible in the liquid phase and partly miscible in the solid phase, the T-x_A diagram looks like Fig. 12.14, which is for Cu–Ag.

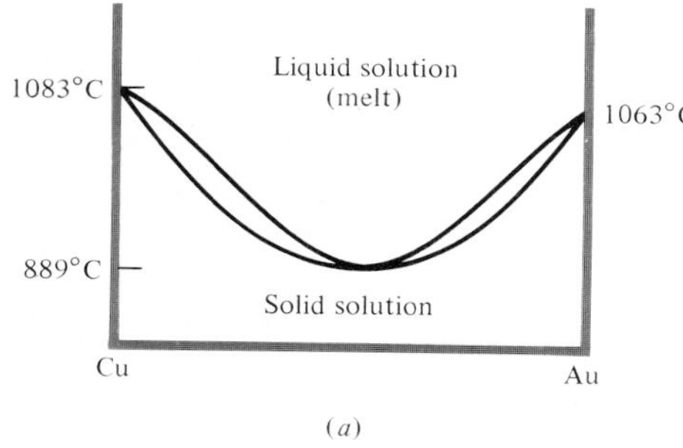

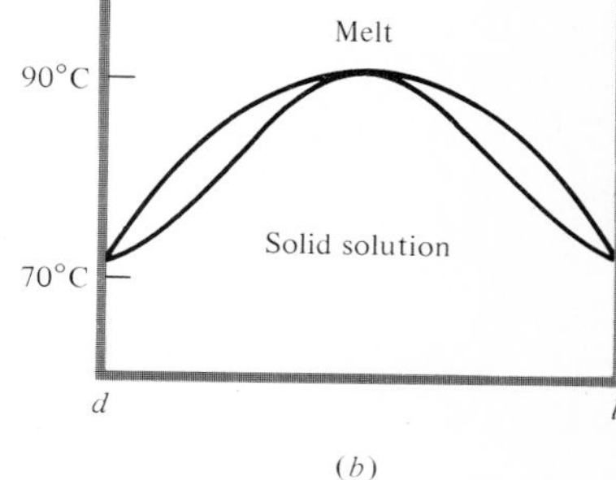

Figure 12.13 Solid–liquid T-x_A phase diagrams for (*a*) Cu–Au; (*b*) *d*-carvoxime–*l*-carvoxime.

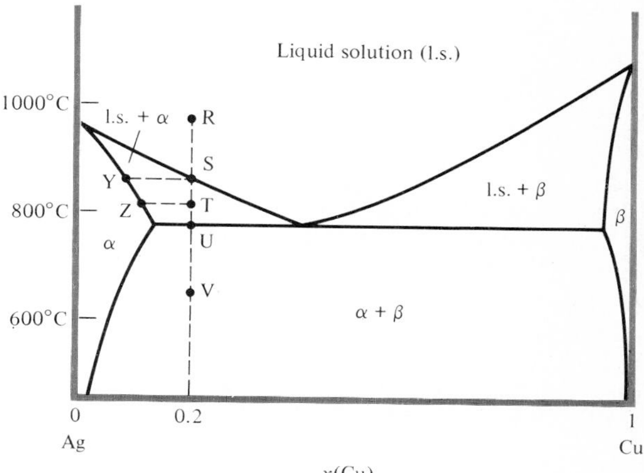

Figure 12.14 Solid–liquid phase diagram for Cu–Ag.

If a liquid melt (solution) of Cu and Ag with $x_{Cu} = 0.2$ is cooled, at point S a solid phase (called the α phase) that is a saturated solution of Cu in Ag begins to separate out. The initial composition of this solid solution is given by point Y at the end of the SY tie line. As the two-phase mixture of liquid solution plus solid solution is cooled further, the percentage of Cu in the solid solution that is in equilibrium with the melt increases. At point U, the melt has the eutectic composition, and two solid phases now freeze out—the α phase (solid Ag saturated with Cu) and the β phase (solid Cu saturated with Ag). Examination of the solid at V will show large crystals of phase α (which formed before point U was reached) and tiny crystals of phases α and β (which formed at U). A line of constant overall composition, for example, RSTU, is called an *isopleth*.

One complication is that diffusion of molecules, atoms, and ions through solids is quite slow, and it takes a long time for equilibrium to be reached in a solid phase. At point T, the solid in equilibrium with the melt has a composition given by point Z, whereas the first solid frozen out had a composition given by Y. It may be necessary to hold the system at point T for a long time before the solid phase becomes homogeneous with composition Z throughout.

The rate of diffusion in solids depends on the temperature. At elevated temperatures not greatly below the melting points of the solids, solid-state diffusion is generally rapid enough to allow equilibrium to be attained in a few days. At room temperature, diffusion is so slow that many years may be required to reach equilibrium in a solid. In a Manhattan park there stands the sculpture *3000* A.D. *Aluminum/Magnesium* by Terry Fugate-Wilcox. This artwork is a spire of alternating pieces of Al and Mg. Supposedly by A.D. 3000, solid-state diffusion will have transformed the sculpture into a homogeneous alloy of the two metals.

The two-phase region labeled $\alpha + \beta$ in Fig. 12.14 is a miscibility gap (Sec. 12.7). The two-phase regions α + liquid solution and β + liquid solution make up what is called a *phase-transition loop*. The two-phase regions in Figs. 12.5 and 12.12 illustrate the simplest kind of phase-transition loop. Figures 12.14 and 12.8*b* each show a phase-transition loop with a minimum. Figure 12.14 shows the intersection of a miscibility gap with a phase-transition loop that has a minimum. Figure 12.15 shows how we can imagine a phase diagram like Fig. 12.14 to arise by having a solid-phase miscibility gap approach and ultimately intersect a solid–liquid phase-transition loop that has a minimum. (The condensed-phase diagram of Ni–Au resembles Fig. 12.15*b*.)

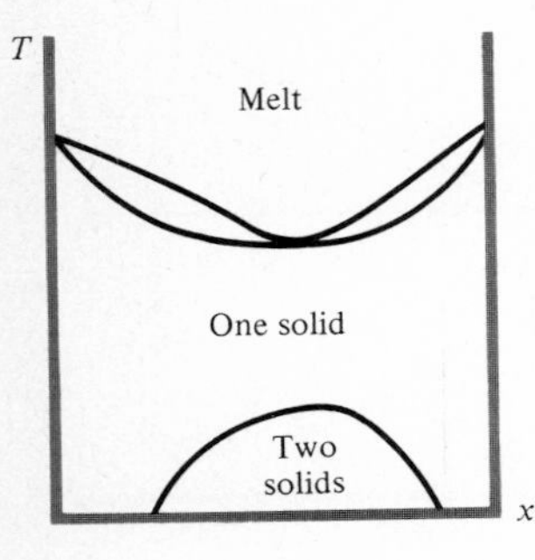

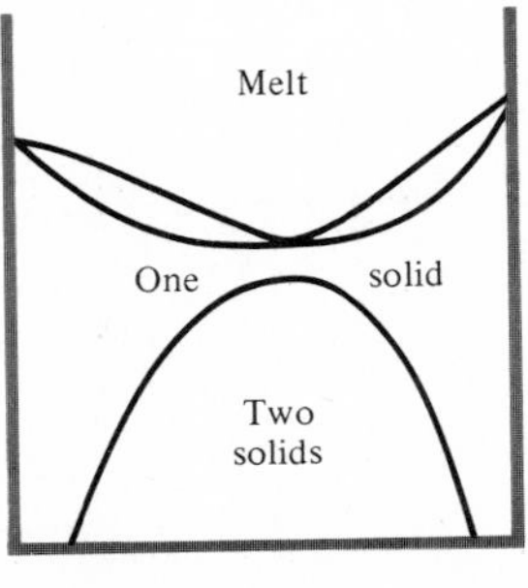

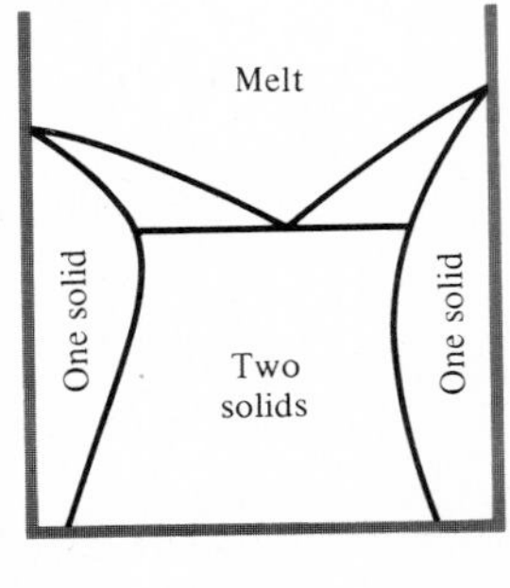

Figure 12.15 A solid-phase miscibility gap approaches and in (*c*) intersects a solid–liquid phase-transition loop.

Certain solid–liquid phase diagrams result from the intersection of a solid-phase miscibility gap with a simple solid–liquid phase-transition loop like the one in Fig. 12.12. This gives a phase diagram like Fig. 12.16. The α phase is a solid solution of A in the B crystal structure; the β phase is a solid solution of B in A. If solid α of composition F is heated, it starts to melt at point G, forming a two-phase mixture of α and liquid solution of initial composition N. However, when point H is reached, the remaining portion of phase α "melts" to form liquid of composition M plus solid phase β of composition R: $\alpha(s) \rightarrow \beta(s)$ + liquid solution. During this transition, the three phases α, β, and liquid are present, and the number of degrees of freedom is $f = 2 - 3 + 2 = 1$; however, since P is held fixed at 1 atm, the system has 0 degrees of freedom, and the transition from α to β + liquid must occur at a fixed temperature (called the peritectic temperature; see Sec. 12.9). Further heating after the transition at H brings us first into a two-phase region of β plus liquid solution and finally into a one-phase region of liquid solution.

Compound formation—liquid-phase miscibility and solid-phase immiscibility. A fairly common occurrence is for substances A and B to form a solid compound that can exist in equilibrium with the liquid. Figure 12.17 shows the solid–liquid phase diagram for phenol (P) plus aniline (A), which form the compound $C_6H_5OH \cdot C_6H_5NH_2$ (PA). The aniline mole fraction x_A on the abscissa is calculated pretending that only aniline and phenol (and no addition compound) are present. Although the system has $c = 3$ (instead of 2), the number of degrees of freedom is unchanged by compound formation, since we now have the equilibrium restriction $\mu_P + \mu_A = \mu_{PA}$; thus, $c - r - a = c_{ind}$ in Eq. (7.10) is still 2, and the system is binary.

Figure 12.17 can be understood qualitatively by imagining it to consist of a simple eutectic diagram for phenol–PA adjacent to a simple eutectic diagram for PA–aniline. The liquid solution at the top of the diagram is an equilibrium mixture of P, A, and PA. Depending on the solution's composition, solid phenol, solid PA, or solid aniline will separate out on cooling, until one of the two eutectic temperatures is reached, at which time a second solid also freezes out. If a solution with $x_A = 0.5$ is

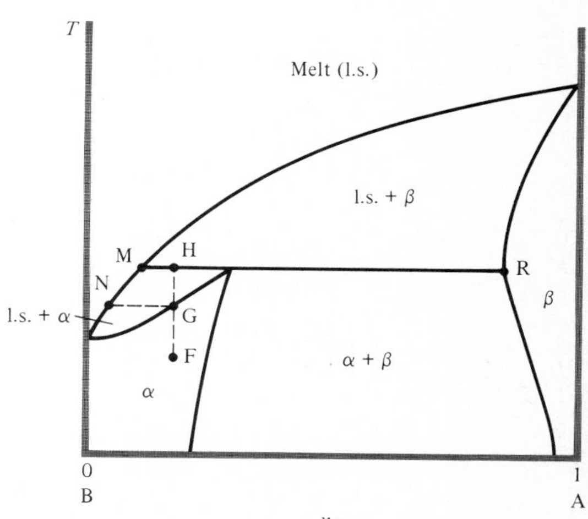

Figure 12.16
A solid–liquid phase diagram with a peritectic temperature.

cooled, only pure solid PA separates out and the solution freezes entirely at one temperature (31°C), the melting point of PA. Although the freezing-point-depression curves for P and for A each start off with nonzero slope, the freezing-point-depression curve of PA has zero slope at the PA melting point. [The thermodynamic proof of this is given in A. F. Berndt and D. J. Diestler, *J. Phys. Chem.*, **72**, 2263 (1968); see also A. F. Berndt, *J. Chem. Educ.*, **46**, 594 (1969).]

Some systems exhibit formation of several compounds. If n compounds are formed, the solid–liquid phase diagram can be viewed as consisting of $n + 1$ adjacent simple-eutectic phase diagrams (provided there are no peritectic points—see below). For an example, see Prob. 12.41.

Compound formation with incongruent melting—liquid-phase miscibility and solid-phase immiscibility. Figure 12.18*a* shows a phase diagram with formation of the solid compound A_2B. Now let the melting point of A be increased to give Fig. 12.18*b*. A further increase in the A melting point will yield Fig. 12.18*c*. In Fig. 12.18*c*, the

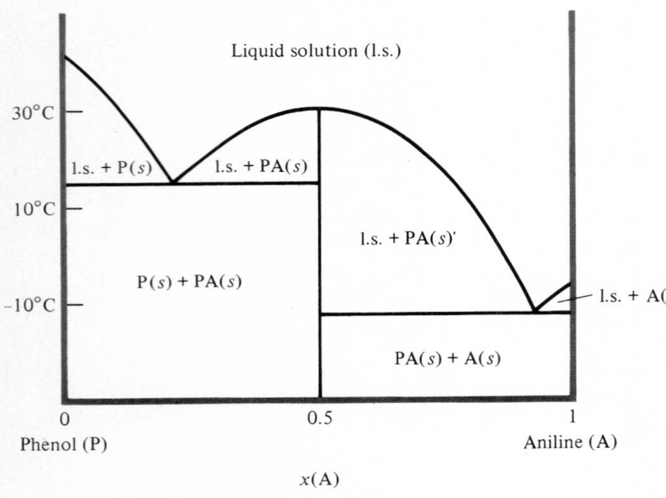

Figure 12.17
Phenol–aniline solid–liquid phase diagram. The symbols P(*s*), PA(*s*), and A(*s*) denote solid phenol, solid addition compound, and solid aniline, respectively.

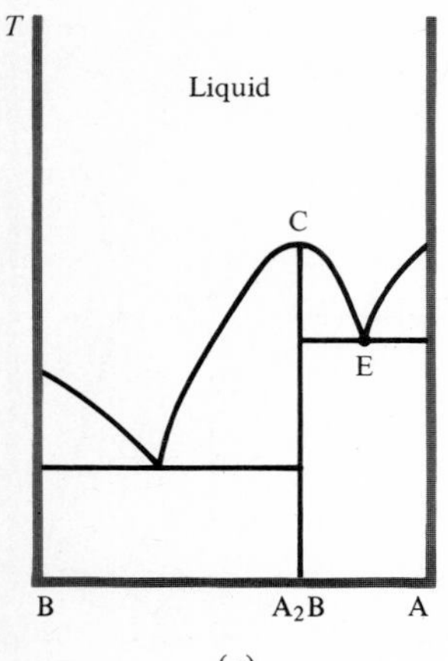

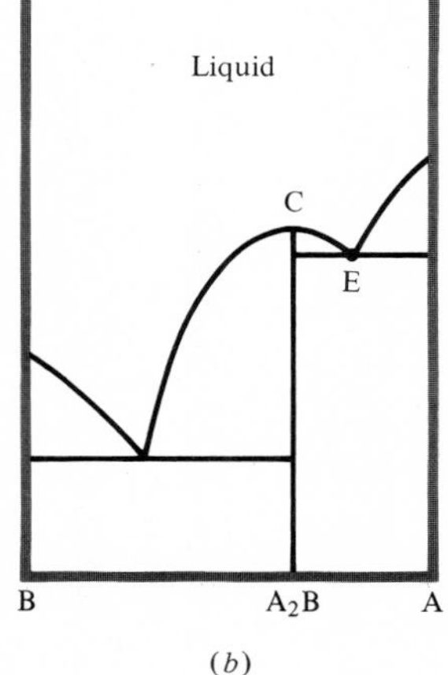

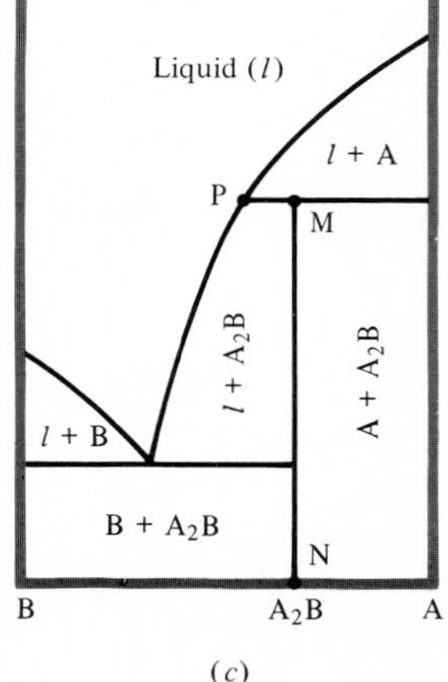

Figure 12.18
Origin of a peritectic point.

freezing-point-depression curve of A no longer intersects the right-hand freezing-point-depression curve of A_2B (curve CE in Fig. 12.18*a* and *b*), so the eutectic point between the compound A_2B and A is eliminated. Instead, the intersection at point P produces the phase diagram of Fig. 12.18*c*. (The system K–Na has the phase diagram of Fig. 12.18*c*; the compound formed is Na_2K.) Line MN is pure solid A_2B. If solid A_2B is heated, it melts sharply at temperature T_P to give a liquid solution (whose composition is given by point P) in equilibrium with pure solid A: $A_2B(s) \rightarrow A(s) +$ solution. Thus, at least some decomposition of the compound occurs on melting. Since the liquid solution formed has a different x_A value than the compound, the compound is said to melt *incongruently*. (The compound in Fig. 12.17 melts *congruently* to give a liquid with the same x_A as the solid compound.) Point P is called a *peritectic point*. When several compounds are formed, there is the possibility of more than one peritectic point. In the system Cu–La, the compounds LaCu and $LaCu_4$ melt incongruently, and the compounds $LaCu_2$ and $LaCu_6$ melt congruently.

Experimental methods. One way to determine a solid–liquid phase diagram experimentally is by *thermal analysis*. Here, one allows a liquid solution (melt) of the two components to cool and measures the system's temperature as a function of time; this is repeated for several different liquid compositions to give a set of cooling curves. The time variable t is approximately proportional to the amount of heat q lost from the system, so the slope dT/dt of a cooling curve is approximately proportional to the reciprocal of the system's heat capacity $C_P = dq_P/dT$. Typical cooling curves for the simple eutectic system of Fig. 12.11 are shown in Fig. 12.19. When pure B is cooled (curve 1), the temperature remains constant at the freezing point T_B^* while the entire sample freezes. The heat capacity of the system $B(s) + B(l)$ at T_B^* is infinite (Sec. 7.5). The slight dip below the freezing point is due to supercooling. After the sample is frozen, the temperature drops as solid B is cooled. Curve 2 is for a liquid mixture with the composition of point R in Fig. 12.11. Here, when solid B begins to freeze out at T_1, the cooling curve shows a decrease in the magnitude of its slope; this slope change is called a *break*; the heat capacity of the system $B(s)$ + liq. soln. is greater than that

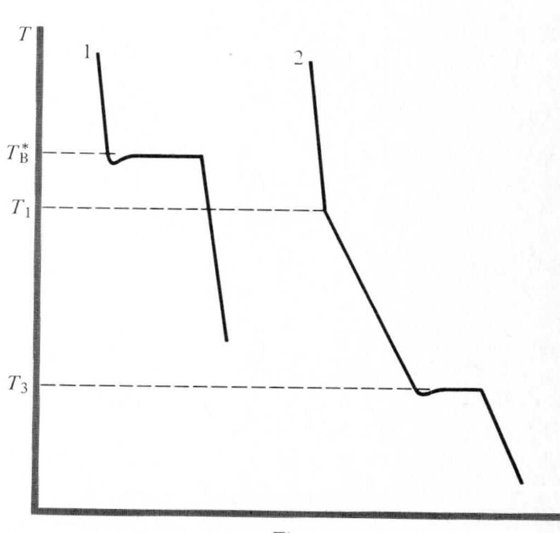

Figure 12.19
Two cooling curves for Fig. 12.11.

of the system consisting of liquid solution only, since a large fraction of the heat removed from the former system serves to convert liquid B to solid B rather than to decrease the system's temperature. When the system reaches the eutectic temperature T_3, the entire remaining liquid freezes at a constant temperature and the cooling curve becomes horizontal, exhibiting what is called a eutectic *halt*. By plotting the temperatures of the observed cooling-curve breaks against x_A, we generate the freezing-point-depression curves CE and DE of Fig. 12.11.

Another method for the determination of phase diagrams is simply to hold a system of known overall composition at a fixed temperature long enough for equilibrium to be attained; the phases present are then separated and analyzed chemically. This is repeated for many different compositions and temperatures to generate the phase diagram.

Solid–liquid equilibria are commonly studied with the system open to the atmosphere. The solubility of air in the solid and liquid phases is generally slight enough to be ignored, and the atmosphere simply acts as a piston that provides a constant external pressure of 1 atm. The atmosphere is not part of the system, and in applying the phase rule to the system, one therefore does not count the atmosphere as one of the phases or include O_2 and N_2 as components of the system; also, since the pressure is fixed, the number of degrees of freedom is reduced by 1.

12.9 STRUCTURE OF PHASE DIAGRAMS

The one-component (unary) P-T phase diagrams of Chap. 7 contain *one-phase areas* separated by *two-phase lines*; the two-phase lines intersect in *three-phase points* (triple points). A phase transition (e.g., solid to liquid) corresponds to moving from one one-phase area to another. During the course of the transition, the system contains two phases in equilibrium, and its state is represented by a point on a two-phase line. At a triple point, three phases coexist in equilibrium.

The binary T-x_A phase diagrams of Secs. 12.6 to 12.8 consist of *one-phase areas, two-phase areas, one-phase vertical lines*, and *three-phase horizontal lines.*

The two-phase areas are either miscibility gaps or phase-transition loops. Examples of miscibility gaps are the two-phase areas in Figs. 12.9 and 12.10, the $\alpha + \beta$ regions in Figs. 12.14 and 12.16, the solid A + solid B region of Fig. 12.11, and the regions at the bottom left and bottom right of Fig. 12.17. Examples of phase-transition loops are the $l + v$ regions in Figs. 12.6 and 12.8*b*, the l. s. + solid B and l. s. + solid A regions of Fig. 12.11, and the l. s. + α and l. s. + β regions of Fig. 12.14.

When the system consists of one phase, this phase is either a pure substance or a solution. If the phase is a pure substance, then it must have a fixed value of x_A and so must correspond to a vertical line on the T-x_A diagram. Examples are the lines at $x_A = 0$ and $x_A = 1$ in Figs. 12.11 and 12.17 and the line at $x_A = 0.5$ in Fig. 12.17. If the phase is a solution (solid, liquid, or gaseous), x_A can be varied continuously over a certain range, as can the temperature, so we get a one-phase area. Examples are the liquid regions at the tops of Figs. 12.11, 12.12, and 12.14, the vapor region at the top of Fig. 12.6, the solid regions at the bottoms of Figs. 12.12 and 12.13, and the α phase and the β phase at the left and right in Fig. 12.14. To repeat, a vertical one-phase line corresponds to a pure substance; a one-phase area corresponds to a solution.

When the system has three phases in equilibrium, $f = 2 - 3 + 2 = 1$, but since

P is fixed, f is reduced to 0. Hence T is fixed, and the existence of three phases in equilibrium in a binary system must correspond to a horizontal (i.e., isothermal) line on the T-x_A diagram. Examples are the horizontal lines in Figs. 12.11, 12.14, and 12.17.

The curved lines CE and DE in Fig. 12.11 are boundary lines marking the end of one region and the beginning of another, rather than lines corresponding to states of the system. On the isopleth RFHKS, the instant before the first crystal of B forms, the system is in the one-phase area of liquid; when the first tiny crystal of B has formed, the system is in the two-phase area of l. s. + B(s). In contrast, when point K on the horizontal line is reached, the three phases l. s., A(s), and B(s) are present in equilibrium and remain present for a considerable time (the eutectic halt) as the remaining liquid is gradually converted to A(s) + B(s).

The compositions of the three phases in equilibrium on a three-phase line are given by the two points at the ends of the line and a third point that occurs at the intersection of a vertical line or phase boundary lines with the three-phase line.

When the system crosses the boundary between a one-phase area and a two-phase area, the cooling curve shows a break. The cooling curve shows a halt at a three-phase line and at the transition from one one-phase region (line or area) to another one-phase region (e.g., at the freezing points of pure substances such as P, A, and PA in Fig. 12.17).

12.10 SOLUBILITY

At 25°C, naphthalene is a solid and benzene is a liquid. A saturated solution of naphthalene in benzene is found to contain 0.4 mole of naphthalene per mole of benzene at 25°C. (Throughout, we assume a constant pressure of 1 atm.) Suppose we add an infinitesimal bit more than 0.4 mole of naphthalene to 1.0 mole of benzene at 25°C. We get a system at 25°C that consists of an infinitesimal amount of solid naphthalene in equilibrium with a saturated solution of 0.4 mole of naphthalene in 1.0 mole of benzene. We call this system I.

The melting points of naphthalene and benzene are 80 and 5°C, and the naphthalene–benzene T-x_A phase diagram is shown in Fig. 12.20.

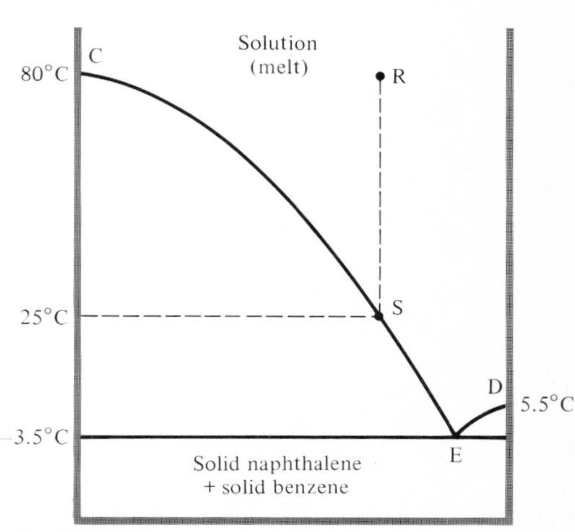

Figure 12.20
Solid–liquid phase diagram for benzene–naphthalene.

Suppose we take an infinitesimal amount more than 0.4 mole of naphthalene, heat it to 80°C to melt it, and to the molten naphthalene add 1.0 mole of benzene. The system is then at point R on the phase diagram. We now cool the liquid melt (solution) to 25°C. When 25°C is reached (point S), naphthalene will start to freeze out and the system will consist of an infinitesimal amount of solid naphthalene in equilibrium with a solution (melt) containing 0.4 mole of naphthalene and 1.0 mole of benzene. We call this system II.

How does system II compare with system I? Despite their different methods of preparation, the two systems are identical in all respects (temperature, pressure, and composition of each phase). The method of preparation of system I suggests that we have determined that the solubility of naphthalene in benzene at 25°C is 0.4 mole of naphthalene per mole of benzene. Since CE in Fig. 12.20 is the freezing-point-depression curve of naphthalene due to added benzene, the method of preparation of system II suggests that we have determined that the freezing point of naphthalene is lowered to 25°C by the addition of 1.0 mole of benzene to 0.4 mole of naphthalene.

Thus, solubility and freezing-point depression are really different words for the same thing. In both cases, we have solid X in equilibrium with a liquid solution of X + Y. The solubility of X (considered as the solute) in Y can just as well be interpreted as the freezing-point depression of X (considered as the solvent) due to added Y. The curve CE in Fig. 12.20 can be interpreted as giving the freezing-point depression of naphthalene due to added benzene or as giving the solubility of naphthalene in benzene as a function of temperature. (For example, CE tells us that a saturated solution of naphthalene in benzene at 60°C has naphthalene mole fraction equal to 0.7.) Likewise, curve DE can be interpreted as the freezing-point depression of benzene due to added naphthalene or as the solubility of benzene in supercooled liquid naphthalene. (The second interpretation sounds strange since naphthalene is a solid at room temperature.) Note from curve CE that the solubility of naphthalene in benzene goes to infinity as the solution's temperature approaches the melting point (80°C) of naphthalene. (The liquids are miscible in all proportions.)

System II above was prepared by starting with liquid naphthalene, and the addition of benzene kept the naphthalene from freezing out as the temperature was lowered to 25°C. Thus, in a solution of naphthalene in benzene at 25°C, the solute naphthalene is best considered to be in the liquid state. Note also that in the expression $\mu_i = \mu_i^\circ + RT \ln \gamma_i x_i$ for naphthalene in the solution (where Convention I is used), $\mu_i^\circ = \mu_i^*$ is the chemical potential of pure (supercooled) *liquid* naphthalene at the T and P of the solution.

What about the solubility of a salt in water? Figure 12.21 shows the T-x_A diagram for $NaNO_3$–H_2O. The melting point of $NaNO_3$ is 307°C. Figure 12.21 cheats a bit, since above 100°C the system's pressure is not held constant but is set at the vapor pressure of the solution by keeping the system in a sealed container. (If the system were kept open to the atmosphere, it would boil away above 100°C.) Since the effect of pressure on the chemical potentials of species in condensed phases is slight, the fact that P varies in part of the phase diagram is of no great consequence.

Curve CE in Fig. 12.21 gives the freezing-point depression of water due to added $NaNO_3$. Alternatively, we can interpret CE as giving the solubility of solid water in supercooled liquid $NaNO_3$. This second interpretation sounds bizarre but is quite valid. Curve DE gives the freezing-point depression of liquid $NaNO_3$ due to added water. Alternatively, we can interpret DE as giving the solubility of solid

$NaNO_3$ in water at various temperatures. The second interpretation sounds more natural, since we ordinarily prepare a liquid solution by adding solid $NaNO_3$ to liquid water (instead of cooling molten $NaNO_3 + H_2O$); also, the curve DE is ordinarily not investigated above 100°C.

Suppose we start with an equilibrium mixture of liquid water and a large amount of ice at 0°C (point C in Fig. 12.21) in a Dewar flask (a vessel with a vacuum jacket that minimizes heat exchange between the system and its surroundings). Let some solid $NaNO_3$ be added to the system. The $NaNO_3$ dissolves in the liquid water and lowers the freezing point of the water to some temperature T_f that is below 0°C. The solution and the ice at 0°C are no longer in equilibrium, and the ice starts to melt. Since the system is adiabatically enclosed, the system's internal energy U must remain constant and the energy to melt the ice must come at the expense of the kinetic energy of the molecules of the system. Statistical mechanics shows that the absolute temperature is proportional to the average molecular kinetic energy. The decrease in molecular kinetic energy therefore means that the system's temperature falls. The temperature will fall until it reaches the freezing point T_f of the solution and equilibrium is attained. The lowest temperature attainable by addition of $NaNO_3$ to ice + water is the eutectic temperature −18°C (point E), where ice, solid $NaNO_3$, and a saturated aqueous $NaNO_3$ solution exist in equilibrium; such a system will remain at −18°C until heat leaking into the system melts all the ice and the system moves into the region between the eutectic line and DE. Eutectic mixtures are used to provide a cold constant-temperature bath. For NaCl–H_2O, the eutectic temperature is −21°C; for $CaCl_2 \cdot 6H_2O$–H_2O it is −50°C.

Line ED in Fig. 12.21 shows the solubility of $NaNO_3$ in water to continually increase with increasing temperature. This behavior is commonly observed for solids in water, but there are exceptions. For example, the solubility of Li_2SO_4 in water decreases with increasing temperature for temperatures up to 160°C. The phase diagram looks like Fig. 12.22. Note that at certain compositions it is possible to take a solution of Li_2SO_4 in water, heat it and have Li_2SO_4 precipitate out, and then redissolve the Li_2SO_4 by further heating.

Since freezing-point depression and solubility are the same phenomenon, we can apply the freezing-point-depression equations to solubility. Equation (12.12)

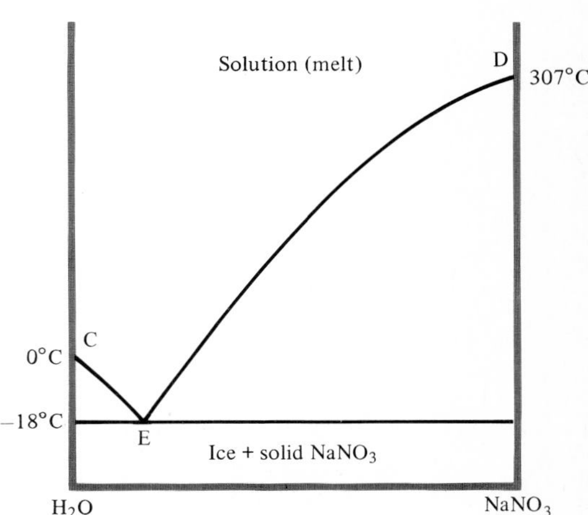

Figure 12.21
Solid–liquid phase diagram for $NaNO_3$–H_2O. (Not drawn to scale.)

relates the mole fraction x_A of A to the temperature T_f at which solid A is in equilibrium with the solution of A + B. Since A is usually used to refer to the solvent, we interchange A and B in (12.12) to get

$$R \ln \gamma_B x_B = \Delta\bar{H}_{fus,B}\left(\frac{1}{T^*_{f,B}} - \frac{1}{T}\right) - \Delta\bar{C}_{P,fus,B}\left(\frac{T - T^*_{f,B}}{T} - \ln\frac{T}{T^*_{f,B}}\right)$$

This equation gives the mole-fraction solubility x_B of solid B at temperature T in the solvent A and is exact except for the approximation of constant $\Delta C_{P,fus,B}$. Because of the unknown activity coefficient in this equation, the equation is not of great practical value. A very crude estimate for the solubility can be obtained by setting $\gamma_B = 1$. Also, Eqs. (12.14) and (12.15) show that the $\Delta\bar{C}_{P,fus}$ term is much smaller than the $\Delta\bar{H}_{fus}$ term (provided the solution is not greatly below the solute's freezing point), so we drop this term to give [cf. (12.58)]

$$R \ln x_B \approx \Delta\bar{H}_{fus,B}\left(\frac{1}{T^*_{f,B}} - \frac{1}{T}\right) \tag{12.60}$$

Equation (12.60) is a reasonable approximation for solutions of naphthalene in benzene but is worthless for most aqueous solutions.

12.11 THREE-COMPONENT SYSTEMS

For a three-component (or *ternary*) system, we have $f = 3 - p + 2 = 5 - p$. For $p = 1$, there are 4 degrees of freedom. To make a two-dimensional plot, we must hold two variables fixed (instead of one, as with binary systems). We shall hold both T and P fixed. For a one-phase system, the two variables will be taken as x_A and x_B, the mole fractions of components A and B. For multiphase systems, x_A and x_B will be taken as the *overall* mole fractions of A and B in the system. Once x_A and x_B are fixed, x_C is fixed. We could use a rectangular plot with x_A and x_B as the variables on the two axes (Prob. 12.33). However, Gibbs suggested the use of an equilateral-triangle plot, and this has become standard for ternary systems. The triangular plot allows all three values x_A, x_B, and x_C to be read off easily.

The triangular coordinate system is based on the following theorem. Let D be an arbitrary point inside an equilateral triangle. If perpendiculars are drawn from D to the sides of the triangle (Fig. 12.23*a*), the sum of the lengths of these three lines is a constant equal to the triangle's height h: $\overline{DE} + \overline{DF} + \overline{DG} = h$. Plane-geometry fans

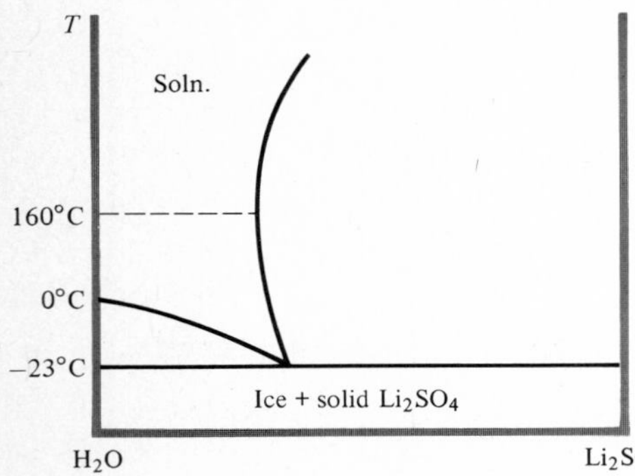

Figure 12.22
Solid–liquid phase diagram for Li_2SO_4–H_2O. (Not drawn to scale.)

will find suggestions for a proof of this in Prob. 12.34. We take the height h to be 100 units and take the lengths $\overline{DE}$, $\overline{DF}$, and $\overline{DG}$ equal to the mole percentages of components A, B, and C, respectively. (If more convenient, we can use weight percentages instead.)

Thus, the perpendicular distance from point D to the side of the triangle opposite the A vertex is the mole percent ($100x_A$) of component A at point D; similarly for components B and C. Any overall composition of the system can be represented by a point in (or on) the triangle, and we get Fig. 12.23*b*. In this figure, equally spaced lines have been drawn parallel to each side. On a line parallel to side BC (which is opposite to vertex A), the overall mole fraction of A is constant. The point marked with a dot represents 50 percent C, 25 percent A, 25 percent B. Along edge AC, the percentage of B present is zero; points on AC correspond to the binary system A + C. At vertex A, we have 100 percent A. At this point, the distance to the side opposite vertex A is a maximum. Note that once x_A and x_B are fixed, the location of the point in the triangle is fixed as the intersection of the two lines corresponding to the given values of x_A and x_B.

Instead of giving a comprehensive discussion of ternary phase diagrams, we shall consider only ternary liquid–liquid equilibrium. Consider the system acetone–water–diethyl ether ("ether") at 1 atm and 30°C. Under these conditions, water and acetone are completely miscible with each other, ether and acetone are completely miscible with each other, and water and ether are partly miscible. The ternary phase diagram is Fig. 12.24.

The region above curve CFKHD is a one-phase area. For a point in the region below this curve, the system consists of two liquid phases in equilibrium; the lines in this region are tie lines whose endpoints give the compositions of the two phases in equilibrium. (For a binary system, the tie lines on a T-x_A or P-x_A diagram are horizontal, since the two phases in equilibrium have the same T and P. On a ternary x_A-x_B-x_C triangular phase diagram, there is clearly no need for tie lines to be horizontal, and the phase diagram is incomplete unless tie lines are drawn in the two-phase regions.) The locations of the tie lines are determined by chemical analysis of pairs of phases in equilibrium. In Fig. 12.24, a system of overall composition G consists of a

A
F
G
D
B
E
C
(*a*)

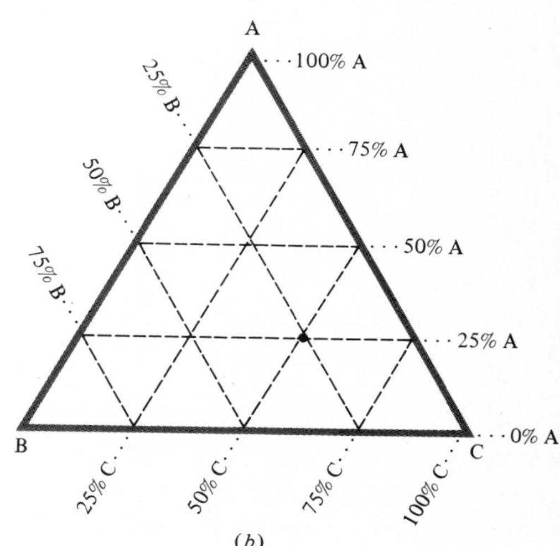

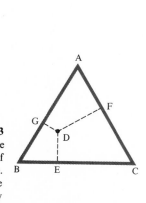

Figure 12.23
(*a*) $\overline{DE}$, $\overline{DF}$, and $\overline{DG}$ are perpendicular to the sides of the equilateral triangle. (*b*) Triangular coordinate system used in ternary phase diagrams.

water-rich, ether-poor phase α of composition F and an ether-rich, water-poor phase β of composition H. The slope of the tie line FGH shows that phase α has less acetone than phase β.

Point K, the limiting point approached by the tie lines as the two phases in equilibrium become more and more alike, is called the *plait point* or the *isothermal critical point*. Curve CFKHD is called the *binodal curve*.

What about the lever rule in ternary systems? The derivation of Eq. (12.52) is readily seen to be valid for *any* two-phase system; the number of components and the nature of the phases (solid, liquid, gas) are irrelevant. Rewriting (12.52) in a general form, we have

$$n_\alpha(x_A - x_{A,\alpha}) = n_\beta(x_{A,\beta} - x_A) \qquad \text{two-phase syst.} \qquad (12.61)$$

where n_α and n_β are the total number of moles (of all species) in phases α and β, and where x_A, $x_{A,\alpha}$, and $x_{A,\beta}$ are the overall mole fraction of A, the mole fraction of A in phase α, and the mole fraction of A in phase β. Figure 12.25 shows a typical tie line in

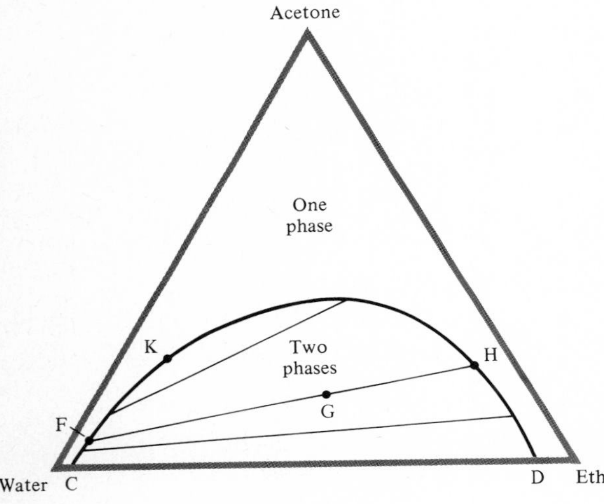

Figure 12.24
Liquid–liquid phase diagram for water–acetone–ether at 30°C and 1 atm. The coordinates are the mole fractions.

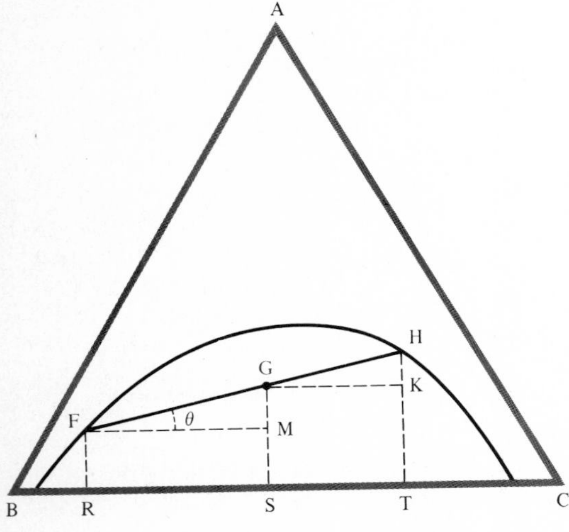

Figure 12.25
Derivation of the lever rule in a ternary system.

a two-phase region of a ternary phase diagram. The overall composition is point G, so the overall mole fraction x_A of A equals length $\overline{GS}$, the distance to the side opposite vertex A. Points F and H give the compositions of phases α and β, so $\overline{FR} = x_{A,\alpha}$ and $\overline{HT} = x_{A,\beta}$. Equation (12.61) becomes $n_\alpha(\overline{GS} - \overline{FR}) = n_\beta(\overline{HT} - \overline{GS})$, or

$$n_\alpha/n_\beta = \overline{HK}/\overline{GM} \tag{12.62}$$

where $\overline{GM}$ and $\overline{HK}$ are lengths marked in Fig. 12.25. From Fig. 12.25, we have $\sin\theta = \overline{HK}/\overline{GH} = \overline{GM}/\overline{FG}$. So $\overline{HK}/\overline{GM} = \overline{GH}/\overline{FG}$, and Eq. (12.62) becomes $n_\alpha/n_\beta = \overline{GH}/\overline{FG}$ or $\overline{FG}n_\alpha = \overline{GH}n_\beta$. Thus

$$l_\alpha n_\alpha = l_\beta n_\beta \tag{12.63}$$

where l_α and l_β are the tie-line lengths to the points that give the compositions of phases α and β. Equation (12.63) is identical to (12.54). The lever rule thus applies in two-phase regions of ternary (as well as binary) phase diagrams.

PROBLEMS

12.1 The vapor pressure of water at 110°C is 1489 torr. Find the vapor pressure at 110°C of a solution of 1.00 g of sucrose ($C_{12}H_{22}O_{11}$) in 100 g of water. State any approximations made.

12.2 (*a*) Beaker A contains 20 cm³ of pure H_2O; beaker B contains 20 cm³ of a 5 wt % NaCl solution. The beakers each have volumes of 400 cm³ and are in a sealed thermally conducting box. Describe the equilibrium state of this system. (*b*) Beaker A contains 0.0100 mol of sucrose dissolved in 100 g of water. Beaker B contains 0.0300 mol of sucrose in 100 g of water. Give the contents of each beaker at equilibrium if the beakers each have volumes of 400 cm³ and are in a sealed thermally conducting box.

12.3 Use data in Probs. 2.24 and 3.15 to verify Eq. (12.18).

12.4 The freezing point of a solution of 2.00 g of maltose in 98.0 g of water is −0.112°C. Estimate the molecular weight of maltose. (See Prob. 12.5 for a more accurate result.)

12.5 For aqueous solutions of maltose, the following freezing-point depressions (in °C) as a function of weight percent maltose in the solution are observed:

Wt % maltose	ΔT_f
3.00	−0.169
6.00	−0.352
9.00	−0.550
12.00	−0.765
15.00	−1.000

Plot the calculated molecular weights vs. weight percent maltose and extrapolate to zero concentration to find the molecular weight.

12.6 When 1.00 g of urea [$CO(NH_2)_2$] is dissolved in 200 g of solvent A, the A freezing point is lowered by 0.250°C. When 1.50 g of Y is dissolved in 125 g of the same solvent A, the A freezing point is lowered by 0.200°C. (*a*) Find the molecular weight of Y. (*b*) The freezing point of A is 12°C and its molecular weight is 200. Find $\Delta\bar{H}_{fus}$ of A.

12.7 The boiling point of $CHCl_3$ is 61.7°C. For a solution of 0.402 g of naphthalene ($C_{10}H_8$) in 26.6 g of chloroform, the boiling point is elevated by 0.455°C. Find $\Delta\bar{H}_{vap}$ of $CHCl_3$.

12.8 (*a*) Use (10.76) and (12.22) to verify that for a solution of an electrolyte, Eq. (12.16) with the quadratic term omitted becomes

$$\Delta T_f = -k_f \phi \nu m_i$$

where m_i is the electrolyte's stoichiometric molality. [Note the resemblance to (12.25), except for the presence of ϕ to correct for nonideality.] For historical reasons, the quantity $\phi\nu$ is called the *van't Hoff i factor*: $i \equiv \phi\nu$. (*b*) The freezing-point depression of an aqueous 4.00 wt % K_2SO_4 solution is −0.950°C. Calculate the practical osmotic coefficient ϕ in this solution at −1°C.

12.9 Phenol (C_6H_5OH) is partially dimerized in the solvent bromoform. When 2.58 g of phenol is dissolved in 100 g of bromoform, the bromoform freezing point is lowered by 2.37°C. Pure bromoform freezes at 8.3°C and has $k_f = 14.1$ °C kg mol⁻¹. Calculate the equilibrium constant K_m for the dimerization reaction of phenol in bromoform at 6°C, assuming an ideally dilute solution.

12.10 Suppose that 6.0 g of a mixture of naphthalene ($C_{10}H_8$) and anthracene ($C_{14}H_{10}$) is dissolved in 300 g of benzene. When the solution is cooled, it begins to freeze at a temperature 0.70°C below the freezing point (5.5°C) of pure benzene. Find the composition of the mixture, given that k_f for benzene is 5.1 °C kg mol^{-1}.

12.11 For an 8.00 wt % aqueous sucrose solution, calculate the freezing point using (*a*) Eq. (12.17) with $\gamma_A = 1$; (*b*) Eq. (12.20); (*c*) Eq. (12.21). (*d*) The observed freezing point of this solution is 0.485°C below that of water. Use Eqs. (12.17) and (12.18) to calculate γ_A. (Be careful about the sign of ΔT_f.) Then use (10.76) with $\nu = 1$ to calculate ϕ in this solution. Compare with the value of ϕ given by the equation in Prob. 10.13.

12.12 In a 0.300 mol/kg aqueous solution of sucrose ($C_{12}H_{22}O_{11}$) the sucrose concentration is 0.282 mol/dm^3 at 20°C. The density of water at 20°C is 0.998 g/cm^3. Calculate the osmotic pressure of this solution using (*a*) Eq. (12.31) with $\gamma_A = 1$; (*b*) Eq. (12.32); (*c*) Eq. (12.33); (*d*) Eq. (12.34). Compare with the observed value 7.61 atm.

12.13 Let the left chamber in Fig. 12.1 contain pure water and the right chamber 1.0 g of $C_{12}H_{22}O_{11}$ plus 100 g of water at 25°C. Estimate the height of the liquid in the right capillary tube at equilibrium. Assume that the volume of liquid in the capillary tube is negligible compared with that in the right chamber.

12.14 For a certain sample of a synthetic poly(amino acid) in water at 30°C (density 0.996 g/cm^3), osmotic-pressure determinations gave the following values for the difference in height Δh between the liquids in the capillary tubes in Fig. 12.1:

Δh/cm	2.18	3.58	6.13	9.22
$\rho_B/(g/dm^3)$	3.71	5.56	8.34	11.12

Convert the height readings to pressures and find the number average molecular weight of the polymer.

12.15 (*a*) The osmotic pressure of human blood is 7 atm at 37°C. Pretend that NaCl forms ideally dilute solutions in water and use (12.39) to estimate the molarity of a saline (NaCl) solution that is isotonic to blood at 37°C. Compare with the value 0.15 mol/dm^3 actually used for intravenous injections. (*b*) The principal solute molalities (in mol/kg) in seawater are NaCl, 0.460; $MgCl_2$, 0.034; $MgSO_4$, 0.019; $CaSO_4$, 0.009. Pretend that seawater is an ideally dilute solution, ignore ion pairing, and estimate the osmotic pressure of seawater at 20°C.

12.16 Use (12.31) and (10.76) to show that for an electrolyte solution the osmotic pressure is given by

$$\Pi = \phi RT\nu n_i/n_A\bar{V}_A^*$$

where ϕ is for the solution at $P + \Pi$. Comparison with Eq. (12.33) explains the name (practical) osmotic coefficient for ϕ.

12.17 (*a*) Solution 1 contains solvent A at mole fraction $x_{A,1}$ plus solute B. Solution 2 contains A at mole fraction $x_{A,2}$ (where $x_{A,2} < x_{A,1}$) plus B. A membrane permeable only to species A separates the two solutions. The solutions are assumed ideally dilute. Show that to achieve equilibrium, an osmotic pressure Π given by

$$\Pi = RT(x_{B,2} - x_{B,1})/\bar{V}_A^* \qquad \text{ideally dil. solns.}$$

must be applied to solution 2. Verify that this equation reduces to (12.32) when solution 1 is pure solvent. (*b*) Suppose a 0.100 mol/kg aqueous sucrose solution is separated from a 0.0200 mol/kg aqueous sucrose solution by a membrane permeable only to water. Calculate the value of Π needed to achieve equilibrium at 25°C.

12.18 Use Eq. (4.23) with $dw = -P^\alpha\, dV^\alpha - P^\beta\, dV^\beta$ and $dA = dA^\alpha + dA^\beta$, and substitute (4.87) at constant T for dA^α and for dA^β to show that the phase-equilibrium condition $\mu_i^\alpha = \mu_i^\beta$ holds even when $P^\alpha \neq P^\beta$.

12.19 What is the maximum number of phases that can coexist in a binary system?

12.20 From the data in Table 10.1, plot the acetone–chloroform liquid–vapor P-x_A phase diagram at 35°C.

12.21 For the system of Fig. 12.6, suppose that a liquid solution with A mole fraction 0.30 is distilled at the pressure of the diagram using a column with an efficiency of two theoretical plates. Give the composition of the first drop of distillate.

12.22 For the system of Fig. 12.6, suppose that a liquid solution with A mole fraction 0.30 is heated in a closed system held at the constant pressure of the diagram. (*a*) Give the composition of the first vapor formed. (*b*) Give the composition of the last drop of liquid vaporized. (*c*) Give the composition of each phase present when half the moles of liquid have been vaporized.

12.23 Use Fig. 12.10 to find the masses of water and

nicotine present in each phase if 10 g of nicotine and 10 g of water are mixed at 80°C and 1 atm.

12.24 Given the following melting points and heats of fusion: benzene, 5.5°C, 30.4 cal/g; cyclohexane (C_6H_{12}), 6.6°C, 7.47 cal/g, plot the T-x_A phase diagram for these two compounds and find the eutectic temperature and the eutectic composition. Assume that the liquid solutions are ideal and that no solid solutions are formed; neglect the temperature dependence of ΔH_{fus}. Compare your values with the experimental eutectic values $-42\frac{1}{2}$°C and $73\frac{1}{2}$ mole percent cyclohexane.

12.25 When melts of Zn + Mg are cooled, breaks and halts are observed at the following temperatures (all in °C):

Wt % Zn	Break	Halt
0	...	651°
10	623°	344°
20	566°	343°
30	530°	347°
40	443°	344°
50	356°	346°
60	437°	346°
70	517°	347°
80	577°	343°
84.3	...	595°
90	557°	368°
95	456°	367°
97	...	368°
97.5	379°	368°
100	...	419°

(Because of experimental errors, the temperature of the eutectic halt varies slightly from run to run. Ignore fluctuations of a couple of degrees in the eutectic temperature.) Plot the phase diagram of T vs. weight percent Zn and label all areas.

12.26 (*a*) Use the following facts to sketch the H_2O–NaCl T vs. weight percent NaCl solid–liquid phase diagram up to 100°C; label all areas as to the phases present. The components form the compound $NaCl \cdot 2H_2O$, which melts incongruently at the peritectic temperature of 0.1°C. The melting point of ice is (surprise!) 0.0°C. The eutectic temperature for liquid + $H_2O(s)$ + $NaCl \cdot 2H_2O(s)$ is -21°C, and the eutectic point occurs at 23 weight percent NaCl. The freezing point of a 13 wt % NaCl aqueous solution is -9°C. The solubility of NaCl in water is 26 g NaCl per 74 g H_2O at 0.1°C and increases to 28 g NaCl per 72 g H_2O at 100°C. (*b*) If an aqueous solution of NaCl is evaporated to dryness at 20°C, what solid(s) is (are) obtained? (*c*) Describe what happens when a system at 20°C with overall composition of 80 weight percent NaCl is slowly cooled to -10°C. Is any ice present at -10°C?

12.27 Sketch several cooling curves for the system Cu–Ag of Fig. 12.14 showing the different types of behavior observed.

12.28 Sketch several cooling curves for the system of Fig. 12.18*c* showing the different types of behavior observed.

12.29 Binary liquid-state partial miscibility corresponds to very large positive deviations from ideality, so the liquid–vapor phase-transition loop has a maximum on the P-x_A diagram and a minimum on the T-x_A diagram. The T-x_A liquid–vapor phase diagram when the liquids are partially miscible therefore shows the intersection of a miscibility gap with a phase-transition loop that has a minimum. This liquid–vapor phase diagram resembles Fig. 12.14, which also shows the intersection of a miscibility gap and a minimum-containing phase-transition loop. Sketch the appearance of a binary liquid–vapor T-x_A diagram for liquid-phase partial miscibility; label all areas and three-phase lines.

12.30 Bi and Te form the solid compound Bi_2Te_3, which melts congruently at about 600°C. Bi and Te melt at about 300 and 450°C, respectively. Solid Bi_2Te_3 is partially miscible at all temperatures with solid Bi and is partially miscible at all temperatures with solid Te. Sketch the appearance of the Bi–Te T-x_A solid–liquid phase diagram; label all regions.

12.31 The Fe–Au solid–liquid T-x_A phase diagram can be viewed as the intersection of a solid-phase miscibility gap with a solid–liquid phase-transition loop having a minimum at $x_{Au} = 0.8$. The miscibility gap intersects the phase-transition loop at $x_{Au} = 0.1$ and at $x_{Au} = 0.3$. Fe has a higher melting point than Au. Sketch the phase diagram and label all areas.

12.32 (*a*) The heat of fusion of naphthalene is 35.1 cal/g, and its melting point is 80°C. Estimate the mole-fraction solubility of naphthalene in benzene at 25°C and compare with the experimental value 0.296. (*b*) Estimate the mole-fraction solubility of naphthalene in toluene at 25°C and compare with the experimental value $x(C_{10}H_8) = 0.286$. (*c*) For anthracene ($C_{14}H_{10}$), $\Delta H_{fus} = 38.7$ cal/g, and the melting point is

216°C. Estimate the solubility of anthracene in benzene at 60°C.

12.33 For the ternary system of Fig. 12.24, set up rectangular coordinates with x(ether) on the y axis and $x(H_2O)$ on the x axis and sketch the general appearance of the phase diagram in these coordinates.

12.34 Prove that $\overline{DE} + \overline{DF} + \overline{DG} = h$ in Fig. 12.23*a*. *Hint:* Draw lines DA, DB, and DC, and recall that the area of a triangle equals half the product of the base and altitude.

12.35 (*a*) From Fig. 12.24, use a ruler to estimate the mole fractions in the phases present at point G. (*b*) Suppose the ternary system at point G has a total of 40 moles present. Find the number of moles of each component in each phase.

12.36 For the system water (1) plus ethyl acetate (2) plus acetone (3) at 30°C and 1 atm, the following mole-fraction compositions of pairs of liquid phases in equilibrium have been found:

Water-rich phase		Water-poor phase	
x_2	x_3	x_2	x_3
0.016	0.000	0.849	0.000
0.018	0.011	0.766	0.061
0.020	0.034	0.618	0.157
0.026	0.068	0.496	0.241
0.044	0.117	0.320	0.292
0.103	0.206	0.103	0.206

where the last set of data gives the isothermal critical solution point. (*a*) Plot the ternary phase diagram including tie lines; use commercially available triangular coordinate paper. (*b*) Suppose 0.10 mole of acetone, 0.20 mole of ethyl acetate, and 0.20 mole of water are mixed at 30°C and 1 atm. Find the mass of each component present in each phase at equilibrium.

12.37 A system consists of 1.00 mol of supercooled liquid water at −10°C in an adiabatic container; the pressure is kept fixed at 1 atm. An infinitesimally tiny crystal of ice is added to the system. (*a*) Give the equilibrium temperature and equilibrium amounts of ice and liquid water in the system. (*b*) Calculate ΔS.

12.38 Systems A and B are each at the same temperature T_1. The two systems are mixed in an adiabatically enclosed container. Is it possible for the final temperature to be less than T_1? If so, give one or more examples.

12.39 Use (12.14) multiplied by T_f^* and (8.41) with $x = 1 + \Delta T_f / T_f^*$ to derive (12.15).

12.40 Dry air at sea level has the following composition by volume: 78 percent N_2, 21 percent O_2, 1 percent Ar. Calculate the number average molecular weight of air.

12.41 The solid–liquid phase diagram of water–nitric acid shows formation of the congruently melting compounds (melting points in parentheses) $HNO_3 \cdot 3H_2O$ (−18°C) and $HNO_3 \cdot H_2O$ (−38°C). The melting point of HNO_3 is −41°C. As one goes across the phase diagram from H_2O to HNO_3, the eutectic temperatures are −43°C, −42°C, and −66°C. Draw the solid–liquid phase diagram.

13

SURFACE CHEMISTRY

13.1 THE INTERPHASE REGION

So far in this book, each phase of a thermodynamic system has been considered to be strictly homogeneous, with its intensive properties constant throughout the phase. However, when surface effects are considered, it is clear that a phase is not strictly homogeneous throughout. For example, consider a system composed of the phases α and β (Fig. 13.1*a*). Molecules at or near the region of contact of phases α and β have a different molecular environment than molecules in the interiors of α and β. Thus, if phase α is liquid cyclohexane and phase β is liquid water, molecules in the region of contact between α and β interact with both C_6H_{12} and H_2O molecules, whereas molecules within α interact only with C_6H_{12} molecules, and molecules within β interact only with H_2O molecules (since the two liquids are virtually immiscible).

The three-dimensional region of contact between phases α and β in which molecules interact with molecules of both phases is called the *interfacial layer*, *surface layer*, or *interphase region*. This region is a few molecules thick. (The term *interface* refers to the apparent two-dimensional geometrical boundary surface separating the two phases.) Figure 13.1*b* is a schematic drawing of a cross section of a two-phase system with a planar interface. All molecules between the planes VW and AB have the same environment and are part of the *bulk phase* α; all molecules between planes CD and RS have the same environment and are part of the bulk phase β. The interfacial layer consists of the molecules between plane AB and plane CD. The thickness of this layer, which has been greatly exaggerated, is actually of the order of magnitude 10 to 100 Å, depending on the size of the molecules and the nature of the intermolecular forces. (The thickness of the interfacial layer can be estimated by statistical-mechanical calculations and by examination of the polarization of light reflected from interfaces.) If the height of the system in Fig. 13.1*b* is 5 cm and the thickness of the interphase region is 50 Å $= 50 \times 10^{-8}$ cm, only 1 molecule in 10^7 is in the interphase region. Therefore the influence of surface effects on a system's properties is ordinarily small and can be neglected. In this chapter, we consider systems where surface effects are significant (e.g., colloidal systems, where the surface-to-volume ratio is high, and gas–solid systems, where substantial amounts of gas can be adsorbed at the solid's surface).

The interfacial layer is a transition region between the bulk phases α and β and is not homogeneous; instead its properties vary from those characteristic of the bulk phase α to those characteristic of the bulk phase β. For example, if β is a liquid solution and α is the vapor in equilibrium with the solution, approximate statistical-mechanical calculations and physical arguments indicate that the concentration c_i of component i will usually vary with z (the vertical coordinate in Fig. 13.1*b*) in one of the two ways shown in Fig. 13.2. The dashed lines mark the boundaries of the interfacial layer and correspond to planes AB and CD in Fig. 13.1*b*. For solid–solid, solid–liquid, and solid–gas interfaces, the transition between the bulk phases is usually more abrupt than for the liquid–vapor interface of Fig. 13.2.

Because of the change in intermolecular interactions, molecules in the inter-phase region have a different average molecular energy from those in either bulk phase. A change in the area of the interface between α and β will therefore change the system's internal energy U. For example, for a liquid in contact with its vapor (Fig. 13.3), molecules at the surface of the liquid experience fewer attractions from other liquid-phase molecules compared with molecules in the bulk liquid phase and so have a higher average energy than molecules in the bulk liquid phase. (The concentration of molecules in the vapor phase is so low that we can ignore interactions between vapor-phase molecules and molecules at the surface of the liquid.) It requires work to increase the area of the liquid–vapor interface in Fig. 13.3, since such an increase means fewer molecules in the bulk liquid phase and more in the surface layer. It is generally true that positive work is required to increase the area of an interface between two phases (see Sec. 13.3). For this reason, systems tend to assume a configuration of minimum surface area; thus an isolated drop of liquid assumes a spherical shape, since a sphere is the three-dimensional shape with a minimum ratio of surface area to volume.

Let $\mathscr{A}$ be the area of the interface between phases α and β. (A script letter is used to avoid confusion with the Helmholtz function A.) The number of molecules in

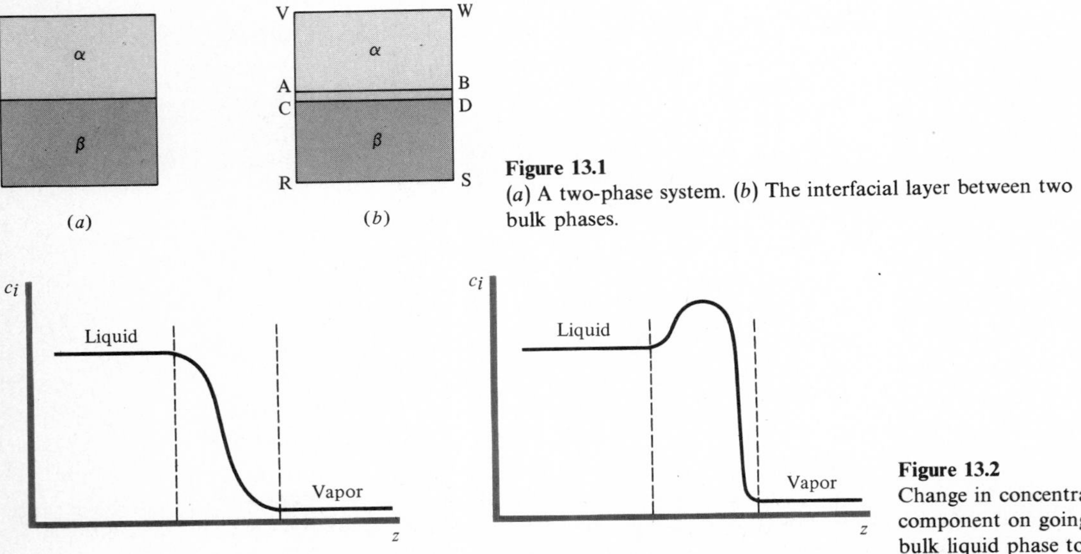

Figure 13.1
(*a*) A two-phase system. (*b*) The interfacial layer between two bulk phases.

Figure 13.2
Change in concentration of a component on going from the bulk liquid phase to the bulk vapor phase.

the interphase region is proportional to $\mathscr{A}$. Suppose we carry out a mechanically reversible process that increases the area of the interface by $d\mathscr{A}$. The increase in the number of molecules in the interphase region is proportional to $d\mathscr{A}$, and so the work needed to increase the interfacial area is proportional to $d\mathscr{A}$. Let the proportionality constant be symbolized by $\gamma^{\alpha\beta}$ (where the superscripts indicate that the value of this constant depends on the nature of the phases in contact). The reversible work needed to increase the interfacial area is then $\gamma^{\alpha\beta}\, d\mathscr{A}$. The quantity $\gamma^{\alpha\beta}$ is called the *interfacial tension* or the *surface tension*. (When one phase is a gas, the term surface tension is more commonly used.) Since it requires positive work to increase $\mathscr{A}$, the quantity $\gamma^{\alpha\beta}$ is positive.

In addition to the work $\gamma^{\alpha\beta}\, d\mathscr{A}$ to change the interfacial area, there is the work $-P\, dV$ associated with any volume change (where P is the pressure in each bulk phase and V is the system's total volume). Thus the work done on the closed system of phases α and β is

$$dw_{\text{rev}} = -P\, dV + \gamma^{\alpha\beta}\, d\mathscr{A} \qquad \text{plane interface} \qquad (13.1)^*$$

We shall take (13.1) as the definition of $\gamma^{\alpha\beta}$ for a closed two-phase system with a planar interface. (The reason for the restriction to a planar interface will become clear in the next section.) From (13.1), if the piston in Fig. 13.4 is slowly moved an infinitesimal distance, work $-P\, dV + \gamma^{\alpha\beta}\, d\mathscr{A}$ is done on the system.

The "surface tension of liquid α" means the interfacial tension $\gamma^{\alpha\beta}$ for the system of liquid α in equilibrium with its vapor β. Surface tensions of liquids are often measured against air; when phase β is an inert gas at low or moderate pressure, the value of $\gamma^{\alpha\beta}$ is usually nearly independent of the composition of β.

Since we shall be considering systems with only one interface, from here on, $\gamma^{\alpha\beta}$ will be symbolized simply by γ.

The surface tension γ has units of work (or energy) divided by area. The cgs unit of γ is thus ergs/cm^2, which equals dyn/cm (since 1 erg = 1 dyn cm). The SI unit of γ is J/m^2 = N/m. The reader can verify that

$$1 \text{ erg/cm}^2 = 1 \text{ dyn/cm} = 10^{-3} \text{ J/m}^2 = 10^{-3} \text{ N/m} = 1 \text{ mN/m} = 1 \text{ mJ/m}^2 \qquad (13.2)$$

For most organic and inorganic liquids, γ at room temperature ranges from 15 to 50 dyn/cm. For water, γ has the high value of 73 dyn/cm at 20°C, due to the strong intermolecular forces associated with hydrogen bonding. Liquid metals have very high surface tensions; that of Hg at 20°C is 490 dyn/cm; that of liquid Au at its melting point is 1100 dyn/cm. [Surface tensions of pure liquids are tabulated in J. J. Jasper, *J. Phys. Chem. Ref. Data*, **1**, 841 (1972).] For a liquid–liquid interface (with each liquid saturated with the other), γ is generally less than the surface tension of the pure liquid with the higher surface tension. Thus, γ for the H_2O–Hg interface is 400 dyn/cm at 20°C. (Measurement of γ is discussed in the next section.)

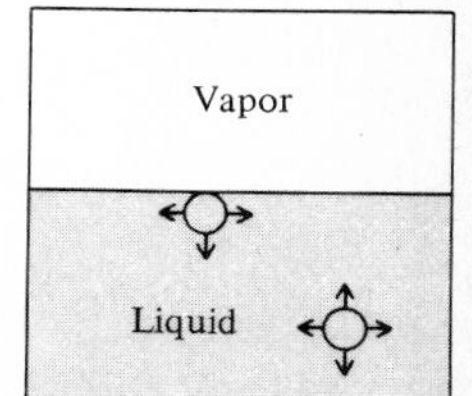

Figure 13.3
Attractive forces on molecules in a liquid.

Consider a liquid in equilibrium with its vapor. As the temperature is raised, the two phases become more and more alike until at the critical temperature T_c the liquid–vapor interface disappears and only one phase is present. At T_c, the value of γ must therefore become 0, and we expect that γ of a liquid will continually decrease as T is raised to the critical temperature. The following empirical equation (due to Katayama and Guggenheim) accurately reproduces the $\gamma(T)$ behavior of many liquids:

$$\gamma = \gamma_0(1 - T/T_c)^{11/9} \tag{13.3}$$

where γ_0 is an empirical parameter characteristic of a given liquid. (Similarly, if a two-phase liquid–liquid system whose overall composition corresponds to the upper critical solution temperature in Fig. 12.9 is heated, γ decreases to 0 at the temperature at which the liquids become completely miscible.)

Although the quantity P in (13.1) is the pressure in each of the bulk phases α and β of the system, because of the surface tension, P is not equal to the pressure exerted by the piston in Fig. 13.4 when the system and piston are in equilibrium. Let the system be contained in a rectangular box of dimensions l_x, l_y, and l_z, where the x, y, and z axes are shown in Fig. 13.4. Let the piston move a distance dl_y in the process of doing work dw_{rev} on the system, and let the piston exert a force F_{pist} on the system. From (2.25), the work done by the piston is $dw_{rev} = F_{pist}\, dl_y$. Use of (13.1) gives $F_{pist}\, dl_y = -P\, dV + \gamma\, d\mathscr{A}$. The system's volume is $V = l_x l_y l_z$, and $dV = l_x l_z\, dl_y$. The area of the interface between phases α and β is $\mathscr{A} = l_x l_y$, and $d\mathscr{A} = l_x\, dl_y$. Therefore $F_{pist}\, dl_y = -Pl_x l_z\, dl_y + \gamma l_x\, dl_y$ and

$$F_{pist} = -Pl_x l_z + \gamma l_x \tag{13.4}$$

The pressure P_{pist} exerted by the piston is $-F_{pist}/\mathscr{A}_{pist} = -F_{pist}/l_x l_z$, where $\mathscr{A}_{pist}$ is the piston's area. (F_{pist} is in the negative y direction and so is negative; pressure is a positive quantity, so the minus sign has been added.) Division of (13.4) by $\mathscr{A}_{pist} = l_x l_z$ gives

$$P_{pist} = P - \gamma/l_z \tag{13.5}$$

Of course, γ/l_z is ordinarily quite small compared with P (Prob. 13.6).

Since the piston and the system in Fig. 13.4 are in equilibrium, Eq. (13.4) shows that the system exerts a force $Pl_x l_z - \gamma l_x$ on the piston. The presence of the interface causes a force γl_x to be exerted by the system on the piston, and this force is in a direction opposite that associated with the system's pressure P. The quantity l_x is the length of the line of contact of the interface and the piston, so *γ is the force per unit length* exerted on the piston as a result of the existence of the interphase region. Mechanically, the system acts as if the two bulk phases were separated by a thin

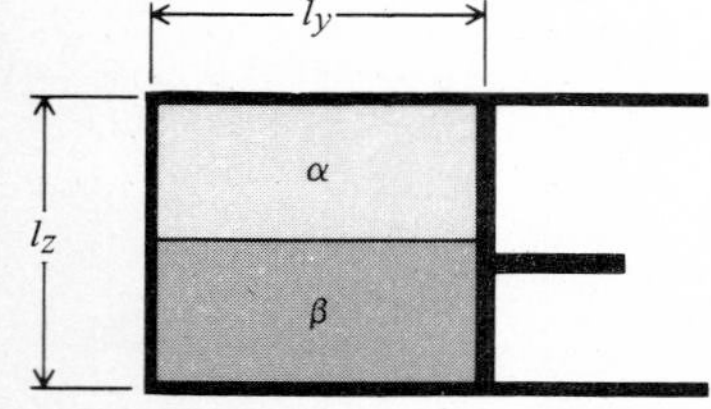

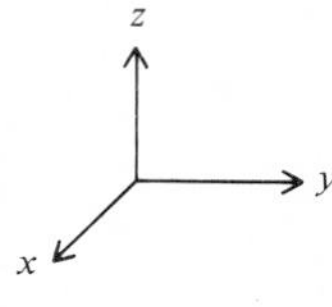

Figure 13.4
A two-phase system confined by a piston.

membrane under tension. (This is the origin of the name "surface tension" for γ.) Insects that skim over a water surface take advantage of surface tension.

Within the bulk phases α and β in Figs. 13.1 and 13.4, the pressure is uniform and equal to P in all directions. Within the interphase region, the pressure in the z direction equals P, but the pressure in the x and y directions is not equal to P. Instead the fact that the pressure (13.5) on the piston is less than the pressure P in the bulk phases tells us that P_y (the system's pressure in the y direction) in the interphase region is less than P. (By symmetry, $P_x = P_y$ in the interphase region.) The interphase region is not homogeneous, and the pressures P_x and P_y in this region are functions of the z coordinate. (Because the interphase region is extremely thin, it is an approximation to talk of a macroscopic property like pressure for this region.)

The relations to be developed in this chapter apply in principle to any kind of interface in equilibrium. However, when one of the phases is a solid, certain practical difficulties occur. When a fresh solid surface is formed, the immobility of the solid's molecules (or ions or atoms) prevents the attainment of equilibrium. Surfaces of solids commonly contain many imperfections and irregularities. Different faces of a crystal have different properties and different values of γ. Measurement of the surface tension of a solid is extremely difficult and has been done for only a few solids, usually at temperatures near their melting points, where the surface is fairly mobile. These measurements give solid surface tensions that are somewhat higher than for the corresponding liquid.

Surface effects are of tremendous biological and industrial importance. Many reactions occur most readily on the surfaces of catalysts, and knowledge of heterogeneous catalysis is important in the synthesis of industrial chemicals. Such subjects as lubrication, corrosion, adhesion, detergency, and reactions in electrochemical cells involve surface effects. Many industrial products are colloids (Sec. 13.6). The problem of how cell membranes function belongs to surface science (Sec. 13.7).

13.2 CURVED INTERFACES

When the interface between phases α and β is curved, the surface tension causes the equilibrium pressures in the bulk phases α and β to differ. This can be seen from Fig. 13.5*a*. If the lower piston is reversibly pushed in to force more of phase α into the conical region (while some of phase β is pushed out of the conical region through the top channel), the curved interface moves upward, thereby increasing the area $\mathscr{A}$ of the interface between α and β. Since it requires work to increase $\mathscr{A}$, it requires a greater force to push in the lower piston than to push in the upper piston (which

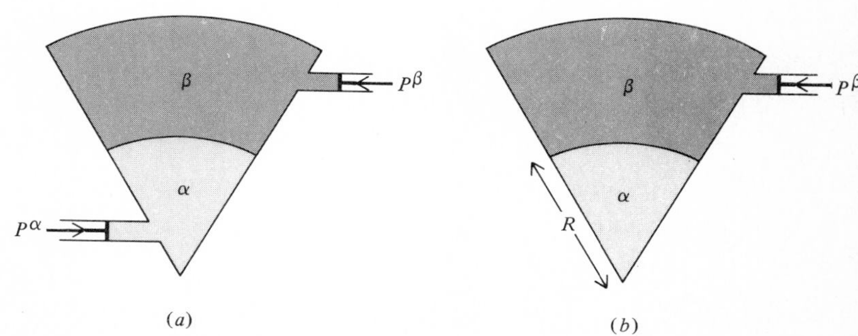

Figure 13.5
Two-phase systems having a curved interface.

would cause $\mathscr{A}$ to decrease). We have shown that $P^\alpha > P^\beta$, where α is the phase on the concave side of the curved interface. (Alternatively, if we imagine phases α and β to be separated by a thin membrane under tension, this hypothetical membrane would exert a net downward force on phase α, making P^α exceed P^β.)

To allow for this pressure difference, we rewrite the definition (13.1) of γ as

$$dw_{\text{rev}} = -P^\alpha\, dV^\alpha - P^\beta\, dV^\beta + \gamma\, d\mathscr{A} \qquad (13.6)^*$$

$$V = V^\alpha + V^\beta$$

In this equation, $-P^\alpha\, dV^\alpha$ is the P-V work done on the bulk phase α; V^α and V^β are the volumes of phases α and β, and V is the total volume of the system. Since the volume of the interphase region is negligible compared with that of a bulk phase, we have taken $V^\alpha + V^\beta = V$. [Because the interfacial layer does have a nonzero thickness, there is some choice in how we divide the system of volume V into regions of volumes V^α and V^β, and the value of γ in (13.6) depends on this choice of dividing surface. This dependence is of no consequence provided the thickness τ of the interfacial layer is much less than the distance R in Fig. 13.5*b*, which is usually the case. When the condition $\tau \ll R$ does not hold, then, according to Guggenheim, γ for a curved interface "can neither be uniquely defined nor accurately measured"; see *Guggenheim*, p. 52.]

To derive the relation between P^α and P^β, consider the slightly modified setup of Fig. 13.5*b*. We shall assume the interface to be a segment of a sphere. Let the piston be reversibly pushed in slightly, changing the system's total volume by dV. From the definition of work as the product of force and displacement, which equals

$$(\text{Force/area})(\text{displacement} \times \text{area}) = \text{pressure} \times \text{volume change}$$

the work done on the system by the piston is $-P_{\text{pist}}\, dV$. Since the system and piston are in equilibrium, P_{pist} equals P^β, and

$$dw_{\text{rev}} = -P^\beta\, dV = -P^\beta\, d(V^\alpha + V^\beta) = -P^\beta\, dV^\alpha - P^\beta\, dV^\beta \qquad (13.7)$$

Equating (13.7) and (13.6), we get

$$-P^\beta\, dV^\alpha - P^\beta\, dV^\beta = -P^\alpha\, dV^\alpha - P^\beta\, dV^\beta + \gamma\, d\mathscr{A}$$

$$P^\alpha - P^\beta = \gamma(d\mathscr{A}/dV^\alpha) \qquad (13.8)$$

Let R be the distance from the apex of the cone to the interface between α and β in Fig. 13.5*b*, and let the solid angle at the cone's apex be Ω. The total solid angle around a point in space is 4π steradians. Hence, V^α equals $\Omega/4\pi$ times the volume $\frac{4}{3}\pi R^3$ of a sphere of radius R, and $\mathscr{A}$ equals $\Omega/4\pi$ times the area $4\pi R^2$ of a sphere. (In Fig. 13.5*b*, all of phase α is within the cone.) We have

$$V^\alpha = \Omega R^3/3, \qquad \mathscr{A} = \Omega R^2$$

$$dV^\alpha = \Omega R^2\, dR, \qquad d\mathscr{A} = 2\Omega R\, dR$$

Substitution into (13.8) gives the desired result:

$$P^\alpha - P^\beta = \frac{2\gamma}{R} \qquad \text{spherical interface} \qquad (13.9)$$

Equation (13.9) was derived independently by Young and by Laplace around 1805. (Some of Thomas Young's other claims to fame are the discovery of the interference of light and work on deciphering the Rosetta stone.) As $R \to \infty$ in (13.9), the pressure difference goes to zero, as it should for a planar interface. The pressure-difference equation for a nonspherical curved interface is more complicated than (13.9) and is omitted.

One consequence of (13.9) is that the pressure inside a bubble of gas in a liquid is greater than the pressure of the liquid. For another consequence, see Prob. 13.21.

The pressure difference (13.9) becomes substantial only when R is small. For example, for a water–air interface at 20°C, $P^\alpha - P^\beta$ is 0.1 torr for $R = 1$ cm and is 10 torr for $R = 0.01$ cm.

Equation (13.9) is the basis for the *capillary-rise* method of measuring the surface tension of liquid–vapor and liquid–liquid interfaces. Here, a capillary tube is inserted in the liquid, and measurement of the height to which the liquid rises in the tube allows calculation of γ. You have probably observed that the water–air interface of an aqueous solution in a glass tube is curved rather than flat. The shape of the interface depends on the relative magnitudes of the adhesive forces between the liquid and the glass and the internal cohesive forces in the liquid. Let the liquid make a *contact angle* θ with the glass (Fig. 13.6). When the adhesive forces exceed the cohesive forces, θ lies in the range $0° \le \theta < 90°$ (Fig. 13.6*a*); when the cohesive forces exceed the adhesive forces, then $90° < \theta \le 180°$.

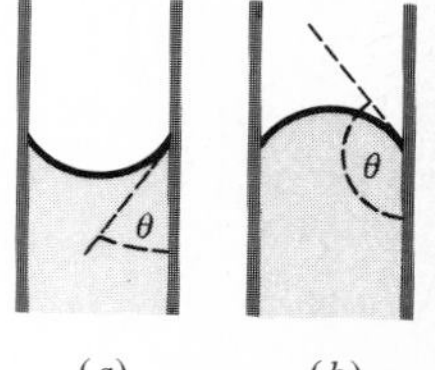

Figure 13.6
Contact angles between a liquid and a glass capillary tube.

Suppose that $0° \le \theta < 90°$. Figure 13.7*a* shows the situation immediately after a capillary tube has been inserted into a wide dish of liquid β. Points 1 and 6 are at the same height in phase α (which is commonly either air or vapor of liquid β), so $P_1 = P_6$. Points 2 and 5 are located an equal distance below points 1 and 6 in phase α, so $P_2 = P_5$. Points 2 and 3 are just above and just below the planar interface outside the capillary tube, so $P_2 = P_3$. Hence, $P_5 = P_3$. Because the interface in the capillary tube is curved, we know from (13.9) that $P_4 < P_5 = P_3$. Since $P_4 < P_3$, phase β is not in equilibrium, and fluid will flow from the high-pressure region around point 3 into the low-pressure region around point 4, causing fluid β to rise into the capillary tube.

The equilibrium condition is shown in Fig. 13.7*b*. Here, $P_1 = P_6$, and since points 8 and 5 are an equal distance below points 1 and 6, respectively, $P_8 = P_5$. Also, $P_3 = P_4$, since phase β is now in equilibrium. Subtraction gives $P_8 - P_3 = P_5 - P_4$. The pressures P_2 and P_3 are equal, so

$$P_8 - P_2 = P_5 - P_4 = (P_5 - P_7) + (P_7 - P_4) \tag{13.10}$$

where P_7 was added and subtracted. Equation (1.8) gives $P_2 - P_8 = \rho_\alpha gh$ and $P_4 - P_7 = \rho_\beta gh$, where ρ_α and ρ_β are the densities of phases α and β and h is the

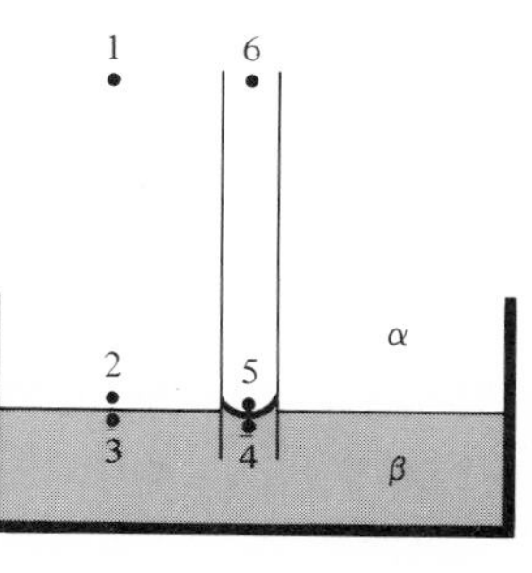

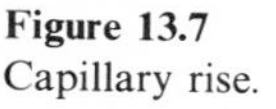

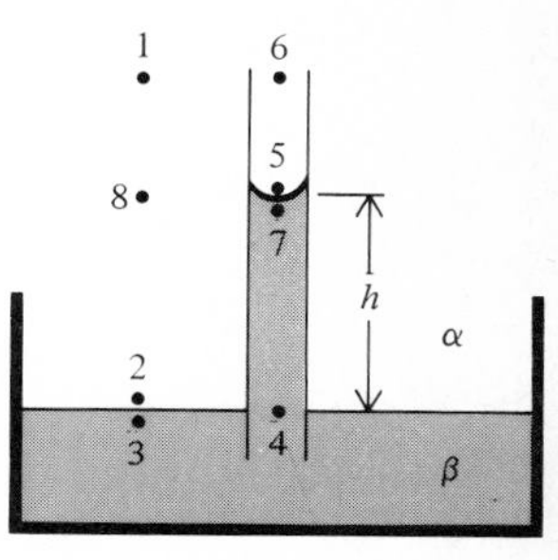

Figure 13.7
Capillary rise.

(*a*) (*b*)

capillary rise. Provided the capillary tube is narrow, the interface can be considered to be a segment of a sphere, and (13.9) gives $P_5 - P_7 = 2\gamma/R$, where R is the sphere's radius. Substitution in (13.10) gives $-\rho_\alpha gh = 2\gamma/R - \rho_\beta gh$ and

$$\gamma = \tfrac{1}{2}(\rho_\beta - \rho_\alpha)ghR \tag{13.11}$$

When phases β and α are a liquid and a gas, the contact angle on clean glass is usually 0 (liquid Hg is an exception). (For $\theta = 0$, the liquid is said to *completely wet* the glass.) With a zero contact angle and with a spherically shaped interface, the interface is a hemisphere, and the radius R becomes equal to the radius r of the capillary tube (Fig. 13.8*b*). Here,

$$\gamma = \tfrac{1}{2}(\rho_\beta - \rho_\alpha)ghr \qquad \text{for } \theta = 0 \tag{13.12}$$

For $\theta \neq 0$, we see from Fig. 13.8*a* that $r = R\cos\theta$, so $\gamma = \frac{1}{2}(\rho_\beta - \rho_\alpha)ghr/\cos\theta$. Since contact angles are difficult to measure accurately, the capillary-rise method is only accurate when $\theta = 0$.

For liquid mercury on glass, the liquid–vapor interface looks like Fig. 13.6*b* with $\theta \approx 140°$. Here, we get a capillary depression instead of a capillary rise.

In the determination of osmotic pressure in Fig. 12.1, a correction for capillary rise must be made.

For a water–air interface at 20°C in a capillary tube with inside diameter 0.20 mm, the capillary rise calculated from (13.12) is 15 cm. The large value of h is due to the very small diameter of the capillary tube.

The phenomenon of capillary action is familiar to all of us from such things as the spreading of a liquid dropped onto cloth or a paper napkin. The spaces between the fibers of the cloth or paper act as capillary tubes into which the liquid is drawn. When fabrics are made water-repellent, a chemical (e.g., a silicone polymer) is applied that makes the contact angle θ exceed 90°, so that water is not drawn into the fabric. Contact angles are of interest to ducks and swans; in water contaminated by oil, the contact angle is reduced below 90°, water is drawn into the feathers of these birds, and they sink.

13.3 THERMODYNAMICS OF CAPILLARY SYSTEMS

Any system in which surface effects are significant is called a *capillary system*. (This term is perhaps unsatisfactory since it suggests that the system is contained in a capillary tube, which is not necessarily the case.) There are two main approaches to

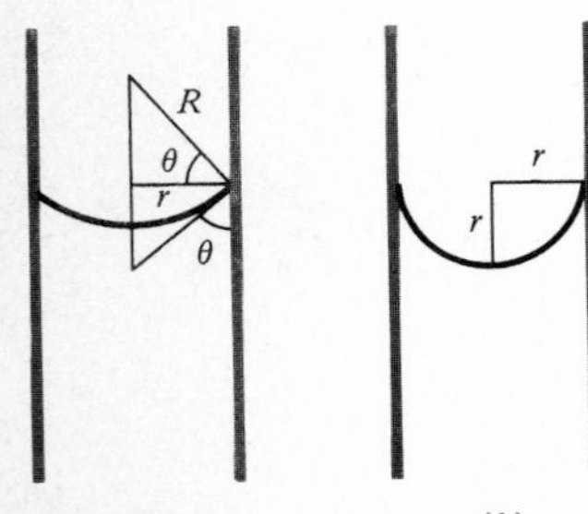

Figure 13.8
Contact angles: (*a*) $\theta \neq 0$; (*b*) $\theta = 0$.

the thermodynamics of capillary systems. Guggenheim in 1940 treated the interfacial layer as a three-dimensional thermodynamic phase having a certain volume, internal energy, entropy, etc. Gibbs in 1878 replaced the actual system by a hypothetical one in which the presence of the interphase region is allowed for by a two-dimensional surface phase that has zero volume but nonzero values of other thermodynamic properties. Compared with the Gibbs model, the Guggenheim method is easier to visualize and more closely corresponds to the actual physical situation. However, the Gibbs method is more widely used and is the one we shall adopt. [Commenting on the Gibbs treatment of capillary systems, Melrose noted that "no other important contribution of Gibbs to the understanding of the equilibrium states of heterogeneous substances has given rise to so many reservations and attempts to develop alternative treatments"; J. C. Melrose, *Pure Appl. Chem.*, **22**, 273 (1970).]

In the Gibbs approach, the actual system of Fig. 13.9*a* (which consists of the bulk phases α and β plus the interphase region) is replaced by the hypothetical model system of Fig. 13.9*b*. In the model system, phases α and β are separated by a surface of zero thickness, the *Gibbs dividing surface*; phases α and β on either side of this dividing surface are defined to have the same intensive properties as the bulk phases α and β, respectively, in the actual system. The location of the dividing surface in the model system is somewhat arbitrary but generally corresponds to a location within (or very close to) the interphase region of the actual system. Experimentally measurable quantities must be independent of the choice of location of the dividing surface, which is just a mental construct. To avoid certain subtleties associated with a curved interface, we restrict the treatment to a planar interface.

The Gibbs model ascribes to the dividing surface whatever values of thermodynamic properties are necessary to make the model system have the same total volume, internal energy, entropy, and amounts of components as the actual system has. We shall use a superscript σ to denote a thermodynamic property of the dividing surface. The dividing surface has zero thickness and zero volume: $V^\sigma = 0$. If V is the volume of the actual system, and V^α and V^β are the volumes of phases α and β in the model system, we require that $V = V^\alpha + V^\beta + V^\sigma$, so

$$V = V^\alpha + V^\beta \tag{13.13}$$

Let U^α_{bulk} and V^α_{bulk} be the energy and volume of the bulk phase α in the actual system. The intensive quantity $U^\alpha_{\text{bulk}}/V^\alpha_{\text{bulk}}$ is the energy per unit volume (the *energy density*) in the bulk phase α. By definition, the energy density in phase α of the model system equals the energy density $U^\alpha_{\text{bulk}}/V^\alpha_{\text{bulk}}$ in the bulk phase α of the actual system.

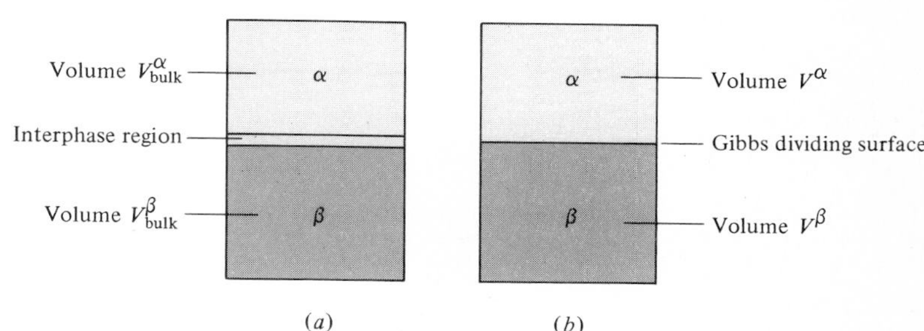

Figure 13.9
(*a*) A two-phase system. (*b*) The corresponding Gibbs model system.

Since phase α of the model system has volume V^α, the energy U^α of the model phase α is

$$U^\alpha = \left(\frac{U^\alpha_{\text{bulk}}}{V^\alpha_{\text{bulk}}}\right) V^\alpha \tag{13.14}$$

with a similar equation for the energy U^β of the model phase β. The total internal energy of the model system is $U^\alpha + U^\beta + U^\sigma$, where U^σ (called the *surface excess internal energy*) is the internal energy to be ascribed to the dividing surface. By definition, this total energy must equal the total internal energy U of the actual system:

$$U = U^\alpha + U^\beta + U^\sigma \qquad \text{or} \qquad U^\sigma = U - U^\alpha - U^\beta \tag{13.15}$$

Exactly the same arguments hold for entropy, so that

$$S^\alpha = (S^\alpha_{\text{bulk}}/V^\alpha_{\text{bulk}})V^\alpha, \qquad S^\beta = (S^\beta_{\text{bulk}}/V^\beta_{\text{bulk}})V^\beta \tag{13.16}$$

$$S^\sigma = S - S^\alpha - S^\beta \tag{13.17}$$

where S is the total entropy of the actual system and S^α, S^β, and S^σ are the entropies of the model phases α and β and the dividing surface.

The same arguments hold for the amount of component i, so that

$$n_i^\alpha = c_i^\alpha V^\alpha, \qquad n_i^\beta = c_i^\beta V^\beta \tag{13.18}$$

$$n_i = n_i^\alpha + n_i^\beta + n_i^\sigma \qquad \text{or} \qquad n_i^\sigma = n_i - n_i^\alpha - n_i^\beta \tag{13.19}$$

where c_i^α is the concentration of component i in the bulk phase α of the actual system (and, by definition, in phase α of the model system), n_i^α and n_i^β are the numbers of moles of component i in phases α and β of the model system, n_i^σ is the number of moles of component i in the dividing surface, and n_i is the total number of moles of i in the actual system (and in the model system). The quantity n_i^σ, called the *surface excess amount* of component i, can be positive, negative, or zero (see below). The definition

$$n_i^\sigma \equiv n_i - (n_i^\alpha + n_i^\beta) = n_i - (c_i^\alpha V^\alpha + c_i^\beta V^\beta) \tag{13.20}$$

states that the surface excess amount n_i^σ is the difference between the amount of i in the actual system and the amount of i that would be in the system if the homogeneity of the bulk phases α and β persisted right up to the dividing surface.

The value of n_i^σ depends on the location of the dividing surface, as we now show. Let the concentration c_i of component i in the actual system vary with the z coordinate as shown by the $c_i(z)$ curve of Fig. 13.10. The interphase region is between z_1 and z_2, and the dividing surface has been placed at z_0.

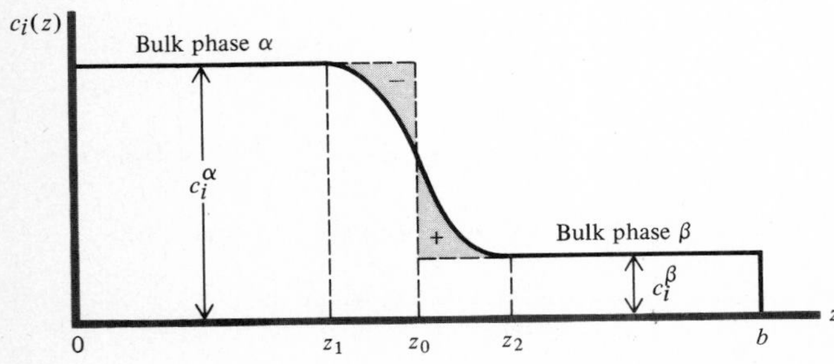

Figure 13.10
Variation of the concentration of component i with the z coordinate.

Imagine the system (which starts at $z = 0$ and extends to $z = b$) to be cut into infinitesimally thin slices taken parallel to the planar interface. Let a given slice contain dn_i moles of component i and have thickness dz, cross-sectional area $\mathscr{A}$, and volume $dV = \mathscr{A}\, dz$. Then $c_i = dn_i/dV = dn_i/(\mathscr{A}\, dz)$, and $dn_i = c_i \mathscr{A}\, dz$. The total number of moles n_i of i in the system is obtained by summing up the infinitesimal amounts dn_i for the infinite number of slices into which the system is cut. This sum is by definition the definite integral from 0 to b of $dn_i = c_i \mathscr{A}\, dz$:

$$n_i = \int_0^b c_i(z)\mathscr{A}\, dz = \int_0^{z_0} c_i(z)\mathscr{A}\, dz + \int_{z_0}^b c_i(z)\mathscr{A}\, dz \qquad (13.21)$$

In the model system of Fig. 13.9*b*, we have

$$n_i^\alpha = c_i^\alpha V^\alpha = c_i^\alpha \mathscr{A} z_0 = c_i^\alpha \mathscr{A} \int_0^{z_0} dz = \int_0^{z_0} c_i^\alpha \mathscr{A}\, dz \qquad (13.22)$$

$$n_i^\beta = c_i^\beta V^\beta = c_i^\beta \mathscr{A}(b - z_0) = c_i^\beta \mathscr{A} \int_{z_0}^b dz = \int_{z_0}^b c_i^\beta \mathscr{A}\, dz$$

since c_i^α and c_i^β are constants. Substitution into $n_i^\sigma = n_i - n_i^\alpha - n_i^\beta$ gives, after division by $\mathscr{A}$,

$$\frac{n_i^\sigma}{\mathscr{A}} = \int_0^{z_0} [c_i(z) - c_i^\alpha]\, dz + \int_{z_0}^b [c_i(z) - c_i^\beta]\, dz \qquad (13.23)$$

The first integral in (13.23) equals the negative of the shaded area to the left of z_0 in Fig. 13.10, and the second integral is equal to the shaded area to the right of z_0. A change in the location z_0 of the dividing surface therefore changes the value of n_i^σ. In Fig. 13.10, the positive and negative areas are roughly equal, so n_i^σ is approximately zero for this choice of dividing surface. If the dividing surface in Fig. 13.10 is moved to the right, then the negative area will exceed the positive area and n_i^σ becomes negative; if the dividing surface is moved to the left, then n_i^σ becomes positive.

Similar arguments show that U^σ and S^σ also depend on the location of the dividing surface. Since n_i^σ, U^σ, and S^σ are dependent on the location of the dividing surface, these quantities are not in general physically measurable. It should also be clear that the dividing surface is in general a hypothetical entity and is not intended to represent the actual interphase region.

The first law of thermodynamics is $dU = dq + dw$ for a closed system. For a reversible process $dq = T\, dS$. In a two-phase capillary system, Eq. (13.1) gives $dw_{\text{rev}} = -P\, dV + \gamma\, d\mathscr{A}$. Therefore

$$dU = T\, dS - P\, dV + \gamma\, d\mathscr{A} \qquad \text{rev. proc., closed syst., planar interface} \qquad (13.24)$$

For an open system, the arguments of Sec. 4.6 require that the terms

$$\sum_i \mu_i^\alpha\, dn_i^\alpha + \sum_i \mu_i^\beta\, dn_i^\beta + \sum_i \mu_i^\sigma\, dn_i^\sigma \qquad (13.25)$$

be added to (13.24), where μ_i^α, μ_i^β, and μ_i^σ are the chemical potentials of i in phase α, phase β, and the surface phase of the model system. At equilibrium, $\mu_i^\alpha = \mu_i^\beta = \mu_i^\sigma$; let μ_i denote the chemical potential of i anywhere in the system. The expression (13.25) becomes at equilibrium

$$\sum_i \mu_i\, dn_i^\alpha + \sum_i \mu_i\, dn_i^\beta + \sum_i \mu_i\, dn_i^\sigma = \sum_i \mu_i\, d(n_i^\alpha + n_i^\beta + n_i^\sigma) = \sum_i \mu_i\, dn_i$$

where (13.19) was used. Hence for an open two-phase system in equilibrium

$$dU = T\,dS - P\,dV + \gamma\,d\mathscr{A} + \sum_i \mu_i\,dn_i \qquad \text{rev. proc., planar interface} \tag{13.26}$$

The presence of the interface leads to the additional term $\gamma\,d\mathscr{A}$ in dU. [Actually, we have glossed over certain subtleties connected with the chemical potentials. A full and correct treatment is given in *Defay, Prigogine, Bellemans, and Everett*, pp. 52–53, 56–57, 59–60, and 71–72; their treatment shows that (13.26) is valid for a system in equilibrium.]

For phases α and β in the Gibbs model, Eq. (4.85) gives

$$dU^\alpha = T\,dS^\alpha - P\,dV^\alpha + \sum_i \mu_i\,dn_i^\alpha, \qquad dU^\beta = T\,dS^\beta - P\,dV^\beta + \sum_i \mu_i\,dn_i^\beta \tag{13.27}$$

Equation (13.15) gives $dU^\sigma = dU - dU^\alpha - dU^\beta$. Use of (13.26) and (13.27) together with $dS^\sigma = dS - dS^\alpha - dS^\beta$, $dV = dV^\alpha + dV^\beta$, and $dn_i^\sigma = dn_i - dn_i^\alpha - dn_i^\beta$ gives

$$dU^\sigma = T\,dS^\sigma + \gamma\,d\mathscr{A} + \sum_i \mu_i\,dn_i^\sigma \qquad \text{rev. proc.} \tag{13.28}$$

Equation (13.28) is now integrated for a process in which the size of the model system is increased at constant T, P, and concentrations in the phases, starting from state 1 and ending at state 2. Under these conditions, the intensive variables T, γ, and the μ_i's are constant and can be taken outside the integral sign. Therefore

$$\int_1^2 dU^\sigma = T\int_1^2 dS^\sigma + \gamma\int_1^2 d\mathscr{A} + \sum_i \mu_i \int_1^2 dn_i^\sigma \qquad \text{const. } T,\ P,\ \text{conc.}$$

$$U_2^\sigma - U_1^\sigma = T(S_2^\sigma - S_1^\sigma) + \gamma(\mathscr{A}_2 - \mathscr{A}_1) + \sum_i \mu_i(n_{i,2}^\sigma - n_{i,1}^\sigma)$$

Now let state 1 be the limiting state obtained as the size of the model system goes to zero. All extensive properties are zero in this state, so the terms with the subscript 1 drop out. State 2 is a general state, and omitting the superfluous subscript 2, we have

$$U^\sigma = TS^\sigma + \gamma\mathscr{A} + \sum_i \mu_i n_i^\sigma \tag{13.29}$$

Note the resemblance to (9.26), except that the surface phase has zero volume and an energy dependent on its area. Equation (13.29) is valid for any state of the system.

The total differential of (13.29) is

$$dU^\sigma = T\,dS^\sigma + S^\sigma\,dT + \gamma\,d\mathscr{A} + \mathscr{A}\,d\gamma + \sum_i \mu_i\,dn_i^\sigma + \sum_i n_i^\sigma\,d\mu_i \tag{13.30}$$

Equating the right sides of (13.28) and (13.30), we get

$$S^\sigma\,dT + \mathscr{A}\,d\gamma + \sum_i n_i^\sigma\,d\mu_i = 0 \tag{13.31}$$

The superalert reader may recall that a similar procedure was used in Chap. 9 and may realize that (13.31) is the analog of the Gibbs–Duhem equation (9.42) for the hypothetical surface phase in the Gibbs model system.

At constant temperature, (13.31) becomes $\mathscr{A}\,d\gamma = -\sum_i n_i^\sigma\,d\mu_i$, which is called the *Gibbs adsorption isotherm*. The *surface (excess) concentration* Γ_i^σ of component i is defined as

$$\Gamma_i^\sigma \equiv n_i^\sigma/\mathscr{A} \tag{13.32}$$

The Gibbs adsorption isotherm becomes

$$d\gamma = -\sum_i \Gamma_i^\sigma \, d\mu_i \qquad \text{const. } T \tag{13.33}$$

As noted earlier, the n_i^σ's (and hence the Γ_i^σ's) depend on the choice of dividing surface and are not experimentally observable quantities. To obtain physically meaningful quantities, we choose one particular dividing surface and refer the Γ_i^σ's to that surface. The dividing surface chosen is the one that makes n_1^σ (and hence Γ_1^σ) zero, where component 1 is a particular component of the system (usually the solvent). Let z_0 denote the position of this dividing surface, and let b denote the system's length in the direction perpendicular to the interface (as in Fig. 13.10). Use of (13.20) with i replaced by 1 gives

$$n_1^\sigma = n_1 - c_1^\alpha \mathscr{A} z_0 - c_1^\beta (b - z_0)\mathscr{A} = n_1 + z_0 \mathscr{A}(c_1^\beta - c_1^\alpha) - c_1^\beta V \tag{13.34}$$

since $\mathscr{A}b = V$. Setting $n_1^\sigma = 0$ and solving for z_0, we get

$$z_0 = \frac{1}{\mathscr{A}} \frac{c_1^\beta V - n_1}{c_1^\beta - c_1^\alpha} \tag{13.35}$$

Let $\Gamma_{i(1)}$ (called the *relative adsorption* of component i with respect to component 1) denote the value of $\Gamma_i^\sigma = n_i^\sigma/\mathscr{A}$ for the dividing surface that makes $n_1^\sigma = 0$. For component i, the equation analogous to (13.34) is

$$n_i^\sigma = n_i + \mathscr{A} z_0 (c_i^\beta - c_i^\alpha) - c_i^\beta V \tag{13.36}$$

Substitution of (13.35) for z_0 into (13.36) gives after division by $\mathscr{A}$

$$\Gamma_{i(1)} = \frac{1}{\mathscr{A}} \left[(n_i - c_i^\beta V) - (n_1 - c_1^\beta V) \frac{c_i^\beta - c_i^\alpha}{c_1^\beta - c_1^\alpha} \right] \tag{13.37}$$

Of course, by definition [and also from (13.37)] $\Gamma_{1(1)} = 0$. All quantities on the right side of (13.37) are experimentally measurable properties of the actual system and are independent of the location of the hypothetical dividing surface. Hence $\Gamma_{i(1)}$ is experimentally measurable.

For the dividing surface that makes n_1^σ and Γ_1^σ zero, the Gibbs adsorption isotherm (13.33) becomes

$$d\gamma = -\sum_{i \neq 1} \Gamma_{i(1)} \, d\mu_i \qquad \text{const. } T \tag{13.38}$$

All quantities in this equation are experimentally measurable. Applications of (13.38) are discussed later in this section.

The concept of the relative adsorption $\Gamma_{i(1)}$ was reached through the Gibbs model system, which does not correspond to the state of affairs in the actual system. Even though $\Gamma_{i(1)}$ in (13.37) is experimentally measurable, you might well wonder about its physical significance. We now show that frequently $\Gamma_{i(1)}$ has a simple physical interpretation.

Consider the very common case of a two-phase system in which the concentrations of components 1 and i in phase β are negligible compared with those in phase α:

$$c_1{}^\beta \ll c_1{}^\alpha \qquad \text{and} \qquad c_i{}^\beta \ll c_i{}^\alpha \tag{13.39}$$

Examples include:

1. A liquid–vapor system in which the vapor pressure is low or moderate; here, because of the low density of the vapor (typically one-thousandth that of the liquid), (13.39) holds.
2. A liquid–liquid system in which solvent 1 and solute i of phase α are virtually insoluble in phase β.
3. A solid–liquid system in which solvent 1 and solute i of the liquid are insoluble in the solid; this case is important in electrochemistry.

When (13.39) holds, one finds (Prob. 13.19) that Eq. (13.37) can be simplified to

$$\Gamma_{i(1)} = \frac{1}{\mathscr{A}}\left(n_i{}^s - n_1{}^s \frac{n^\alpha_{i,\text{bulk}}}{n^\alpha_{1,\text{bulk}}}\right) \qquad \text{when } c_i{}^\beta \ll c_i{}^\alpha,\ c_1{}^\beta \ll c_1{}^\alpha \tag{13.40}$$

where $n_i{}^s$ and $n_1{}^s$ are the numbers of moles of substances i and 1 in the interphase region of the actual system (not the model system) and $n^\alpha_{i,\text{bulk}}$ and $n^\alpha_{1,\text{bulk}}$ are the numbers of moles of i and 1 in the bulk phase α of the actual system.

For simplicity, let the interface have unit area. The ratio $n^\alpha_{i,\text{bulk}}/n^\alpha_{1,\text{bulk}}$ is independent of the size of the bulk phase α. Suppose we take a sample of bulk phase α of such size that the number of moles of solvent in it ($n^\alpha_{1,\text{bulk}}$) equals $n_1{}^s$, the number of moles of solvent in the interphase region. Equation (13.40) then reads $\Gamma_{i(1)} = \mathscr{A}^{-1}(n_i{}^s - n^\alpha_{i,\text{bulk}})$. Thus, when (13.39) holds, the relative adsorption $\Gamma_{i(1)}$ is numerically equal to the difference between the number of moles $n_i{}^s$ of solute i in an interphase-region portion that has unit area and the number of moles $n^\alpha_{i,\text{bulk}}$ of i in a bulk-phase-α sample that has the same number of moles of solvent as the interphase-region portion (Fig. 13.11). This is the desired physical interpretation of $\Gamma_{i(1)}$.

Equation (13.40) can be written as

$$\Gamma_{i(1)} = \frac{n_1{}^s}{\mathscr{A}}\left(\frac{c_i{}^s}{c_1{}^s} - \frac{c_i{}^\alpha}{c_1{}^\alpha}\right) \qquad \text{when } c_i{}^\beta \ll c_i{}^\alpha,\ c_1{}^\beta \ll c_1{}^\alpha \tag{13.41}$$

When the relative adsorption $\Gamma_{i(1)}$ is positive, the ratio of average concentrations $c_i{}^s/c_1{}^s$ in the interphase region is greater than the corresponding ratio $c_i{}^\alpha/c_1{}^\alpha$ in bulk phase α and component i is said to be (positively) adsorbed at the interface. When $\Gamma_{i(1)}$ is negative, i is negatively adsorbed at the interface. (*Adsorption* is the enrichment of a component in the interphase region compared with a bulk region.)

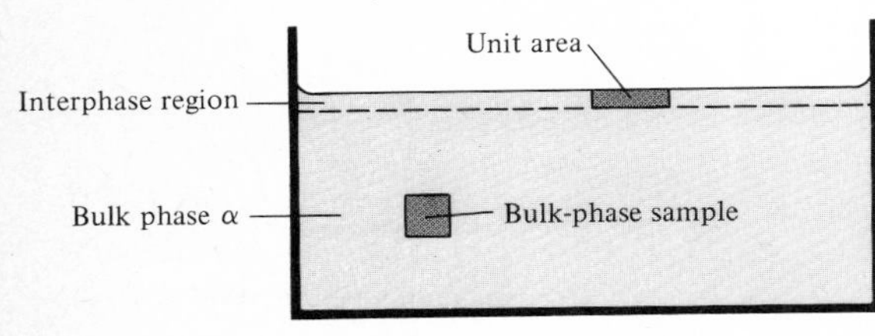

Figure 13.11
Samples used in the physical interpretation of the relative adsorption.

Having clarified the meaning of $\Gamma_{i(1)}$, we now return to the Gibbs adsorption isotherm. For a two-component system, (13.38) reads

$$d\gamma = -\Gamma_{2(1)}\, d\mu_2 \qquad \text{const. } T\text{, binary syst.} \tag{13.42}$$

At least one of the two bulk phases must be a solid or liquid; let us call this phase α. For a condensed phase, Eq. (10.4) gives $\mu_2 = \mu_2^{\circ,\alpha}(T, P) + RT \ln a_2^{\alpha}$. The pressure dependence of $\mu_2^{\circ,\alpha}$ is slight for a condensed phase; moreover, the surface tension is often measured in the presence of air at the constant pressure of 1 atm. Hence, at constant T, we can take $d\mu_2 = RT\, d \ln a_2^{\alpha}$, and (13.42) becomes

$$\Gamma_{2(1)} = -\frac{1}{RT}\left(\frac{\partial \gamma}{\partial \ln a_2^{\alpha}}\right)_T \qquad \text{binary syst.} \tag{13.43}$$

If phase α is dilute enough to be considered ideally dilute, and if the concentration scale is used for the solute 2, then Eq. (10.43) gives $a_2^{\alpha} = c_2^{\alpha}/c^{\circ}$ (where $c^{\circ} = 1$ mol/dm^3), and (13.43) becomes

$$\Gamma_{2(1)} = -\frac{1}{RT}\left(\frac{\partial \gamma}{\partial \ln c_2^{\alpha}/c^{\circ}}\right)_T \qquad \text{binary syst., ideally dil. soln.} \tag{13.44}$$

The slope of a plot of the solution's surface tension γ vs. $\ln (c_2^{\alpha}/c^{\circ})$ at a given temperature equals $-RT\Gamma_{2(1)}$ and allows calculation of $\Gamma_{2(1)}$. If the solution is not ideally dilute, activity-coefficient data are needed to find $\Gamma_{2(1)}$.

Equation (13.44) states that $\Gamma_{2(1)}$ is positive if the surface tension decreases with increasing solute concentration and is negative if γ increases with increasing c_2^{α}. The observed behavior of solutes in dilute aqueous solutions can be classified into three types (Fig. 13.12). Type I solutes produce a small rate of increase of γ with increasing concentration; examples include most inorganic salts and sucrose. The increase of γ for salt solutions can be rationalized by saying that the increased opportunity for attractions between oppositely charged ions in the bulk phase compared with the surface layer (Fig. 13.3) causes a depletion of ions in the surface layer and the negative adsorption increases γ.

Type II solutes give a substantial and steady rate of decrease of γ with increasing concentration; examples include the majority of organic compounds that have some water solubility. Water-soluble organic compounds usually contain a polar part (e.g., an OH or COOH group) and a nonpolar hydrocarbon part. Such molecules tend to accumulate in the surface layer, where they are oriented with their

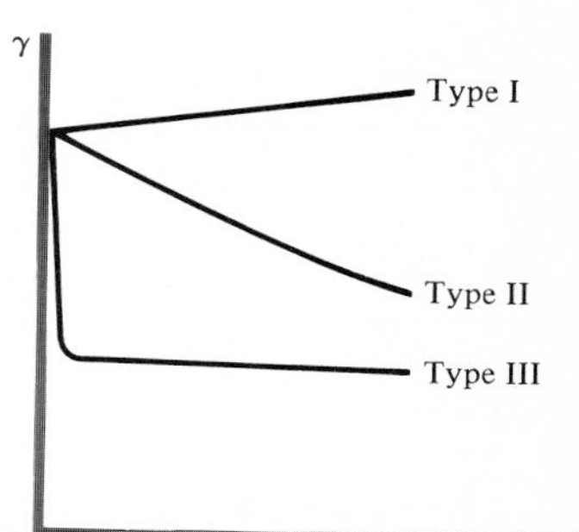

Figure 13.12
Surface tension vs. concentration for aqueous solutions.

polar parts pointing toward, and interacting with, the polar water molecules of the bulk solution and their nonpolar parts pointing away from the bulk solution (see Sec. 13.4 and Fig. 13.15). The resulting positive adsorption decreases γ.

For type III solutes, γ shows a very rapid drop followed by a sudden leveling off as the concentration is increased. Examples include salts of medium-chain-length organic acids (soaps, $RCOO^-Na^+$), alkyl sulfate salts ($ROSO_2O^-Na^+$), quaternary amine salts [$(CH_3)_3RN^+Cl^-$], alkyl sulfonate salts ($RSO_2O^-Na^+$), and polyoxyethylene compounds [$R(OCH_2CH_2)_nOH$, where n is 5 to 15]. Type III solutes are strongly adsorbed at the interface. (The leveling off of γ occurs at the critical micelle concentration; see Sec. 13.6.)

A solute that significantly lowers the surface tension is said to be a *surface-active agent* or *surfactant*. Type III solutes act as detergents and are surfactants par excellence; for example, γ is lowered from 72 to 39 dyn/cm in a 0.008 mol/dm^3 aqueous solution of $CH_3(CH_2)_{11}OSO_2O^-Na^+$ at 25°C. The lowering of γ helps remove oily dirt particles from solid surfaces.

We conclude this section by using (13.24) to show that γ must be positive. Consider an isolated two-phase system that has reached equilibrium. Experimental observations show that the equilibrium condition is one of minimum interfacial area $\mathscr{A}$. (If the equilibrium condition were one of maximum $\mathscr{A}$, each phase would break up into smaller and smaller drops which would intermingle with the drops of the other phase, and the phases would ultimately become miscible with each other, which contradicts the condition that we have a two-phase system.) For an isolated system, U and V are constant, and (13.24) becomes $T\,dS = -\gamma\,d\mathscr{A}$. Since $\mathscr{A}$ is minimized and S is maximized at equilibrium, a hypothetical increase in $\mathscr{A}$ at equilibrium must correspond to a decrease in S; with $d\mathscr{A}$ positive and dS negative, γ must be positive.

13.4 SURFACE FILMS ON LIQUIDS

In 1774 Benjamin Franklin read a paper to the Royal Society describing the result of pouring olive oil on the surface of a London pond: "the oil, though not more than a teaspoonful, produced an instant calm over a space several yards square, which spread amazingly, and extended itself gradually till it reached the leeside, making all that quarter of the pond, perhaps half an acre, as smooth as a looking glass." A calculation from Franklin's data gives the thickness of the oil film as 24 Å, which is the order of magnitude of the length of an olive-oil molecule and indicates a one-molecule-thick surface film.

Many water-insoluble organic compounds that contain a medium-length hydrocarbon chain with a polar group at one end are found to spread spontaneously over a water surface to give a surface film. Typical examples are $CH_3(CH_2)_{16}COOH$ (stearic acid), $CH_3(CH_2)_{11}OH$ (lauryl alcohol), and $CH_3(CH_2)_{14}COOC_2H_5$ (ethyl palmitate). Evidence (cited below) shows these films to be generally one molecule thick, and such surface films are called *spread monolayers*. Spread monolayers are formed by both solid and liquid compounds; the rate of spreading from solids is less than from liquids.

The substantial length of the hydrocarbon chain makes the solubility of these compounds in water extremely low. These compounds are solids or high-boiling liquids at room temperature and have very low room-temperature vapor pressures. Therefore the amounts of solute i present in the bulk phases (water and air) are negligible compared with the amount of i present in the interphase region as a monolayer. With $n_i^\alpha = 0 = n_i^\beta$, Eqs. (13.19) and (13.32) become $n_i^\sigma = n_i$ and

$\Gamma_i^\sigma = n_i/\mathscr{A}$, where n_i is (to a good approximation) the total number of moles of i in the system. For such systems, Γ_i^σ is independent of the location of the Gibbs dividing surface and has a direct physical significance. It follows that the relative adsorption $\Gamma_{i(1)}$ equals Γ_i^σ and is positive. Hence (Sec. 13.3) the surface tension is lowered by the presence of the film.

Surface films are studied with a *surface balance* (Fig. 13.13); a floating barrier (*float*) separates a clean water surface from a water surface containing the monolayer. The force on the float is measured by a torsion wire attached to it.

Let γ^* and γ be the surface tensions of pure water and of water covered with the monolayer, respectively. The discussion after Eq. (13.5) shows that the surface tension of the pure water produces a force γ^* per unit length that pulls the float toward the right in Fig. 13.13, and the surface tension of the monolayer-covered water produces a force γ per unit length that pulls the float toward the left. Since γ is less than γ^*, there is a net force per unit length on the float of $\gamma^* - \gamma$ toward the right; this force is called the *surface pressure* π:

$$\pi \equiv \gamma^* - \gamma \tag{13.52}$$

π has units of force per length, whereas an ordinary pressure has units of force per area.

Suppose the adjustable barrier in Fig. 13.13 is moved to the right. This decreases the area $\mathscr{A}$ available to the monolayer molecules and increases their adsorption $\Gamma_i^\sigma = n_i/\mathscr{A}$. The increased adsorption of these surface-active molecules further lowers γ and increases the surface pressure π. A typical π vs. $\mathscr{A}$ curve at a fixed temperature is shown in Fig. 13.14.

In Fig. 13.14, π increases very gradually with decreasing area until a point (labeled C) is reached, after which further decrease in $\mathscr{A}$ causes a sharp increase in π. It seems reasonable that at point C the film has been compressed sufficiently to bring the molecules of the film almost into contact with one another, so that the film now

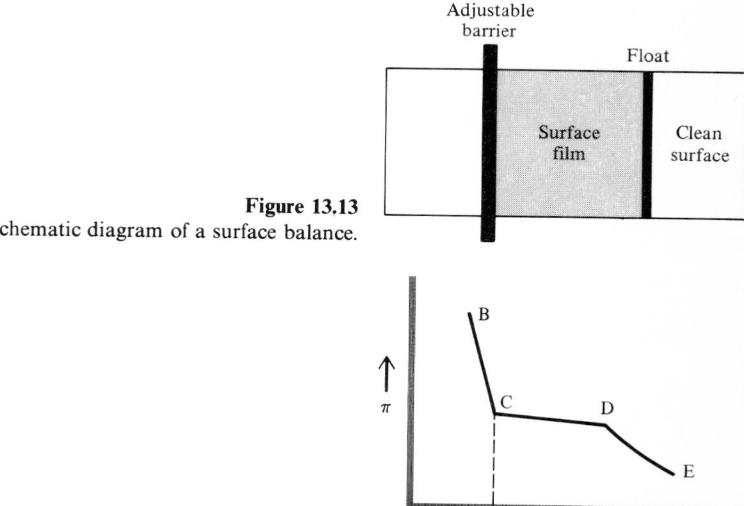

Figure 13.13
Schematic diagram of a surface balance.

Figure 13.14
Typical curve of surface pressure vs. area for a surface film.

strongly resists further compression. If the area $\mathscr{A}_0$ at point C (called the *Pockels point*) is divided by the number of molecules N_i of i present in the monolayer, we get an estimate of the cross-sectional area per molecule. Langmuir found that for each of the acids $CH_3(CH_2)_{14}COOH$, $CH_3(CH_2)_{16}COOH$, and $CH_3(CH_2)_{24}COOH$, $\mathscr{A}_0/N_i$ is 20 Å^2. This value agrees with that obtained by x-ray diffraction of crystals for the cross-sectional area of a straight-chain acid and confirms that the molecules are present as a monolayer. The independence of $\mathscr{A}_0/N_i$ of the chain length shows that at the Pockels point the monolayer molecules are oriented vertically; the polar COOH end points toward the water phase (and thus can interact with the polar water molecules), and the nonpolar hydrocarbon part points toward the vapor phase (Fig. 13.15). The polar part of the molecule is said to be *hydrophilic* ("water-loving") and the hydrocarbon part *hydrophobic* ("water-hating"). Molecules with both hydrophilic and hydrophobic groups are called *amphipathic* or *amphiphilic* (from the Greek *amphi*, "dual," *pathos*, "feeling," *philos*, "loving"); the polar part likes a polar solvent, and the nonpolar part likes a nonpolar solvent. The surfactant molecules mentioned in Sec. 13.3 are amphipathic.

Figure 13.15 Amphipathic molecules in a surface film at the Pockels point.

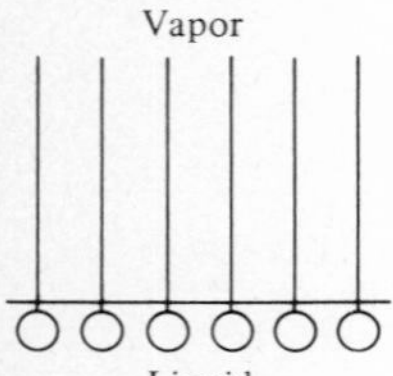

For the portion DE of the isotherm (Fig. 13.14) at low surface pressures, the acid molecules in the monolayer are reasonably far apart from one another, and there is little interaction between them. The state of the monolayer acid molecules along DE is analogous to that of a two-dimensional gas. At high surface pressures, along CB, the acid molecules are quite close together, and their state is analogous to that of a two-dimensional liquid. The approximately horizontal portion CD of the isotherm can be interpreted as states where some of the acid molecules are in a two-dimensional liquid state and the remainder in a two-dimensional gaseous state. Confirmation is provided by electron-microscope studies, which show the monolayer surface to be heterogeneous for states along CD. Note the resemblance of the curve BCDE to the isotherms of a three-dimensional fluid below its critical point (Fig. 8.2). For states along CD, it is likely that the hydrocarbon chains are not oriented vertically but flop over, so that part of each chain lies approximately parallel to the interface. Along CD, substantial numbers of water molecules separate acid molecules from one another, and the monolayer can be viewed as a two-dimensional solution of water and acid.

A practical application of monolayers is to reduce the rate of evaporation of water from reservoirs. Cetyl alcohol [$CH_3(CH_2)_{15}OH$] is the surfactant commonly used.

13.5 ADSORPTION OF GASES ON SOLIDS

In this section, the interphase region of an essentially involatile solid (A) in contact with a gas (B) is considered. The industrially important catalytic activity of such solids as finely divided Pt, Pd, and Ni results from adsorption of gases. Commonly used gases in adsorption studies include He, H_2, N_2, O_2, CO, CO_2, CH_4, C_2H_6, C_2H_4, NH_3, and SO_2; commonly used solids include metals, metal oxides, silica gel (SiO_2), and carbon in the form of charcoal. The solid on whose surface adsorption occurs is called the *adsorbent*. The adsorbed gas is the *adsorbate*. Adsorption occurs at the solid–gas interface, and is to be distinguished from absorption, in which the gas penetrates throughout the bulk solid phase. An example of absorption is the reaction of water vapor with anhydrous $CaCl_2$ to form a hydrate compound.

A complication in gas–solid studies is that the surfaces of solids are rough, and it is difficult to determine the surface area of a solid reliably.

Adsorption on solids is classified into *physical adsorption* (or *physisorption*) and *chemical adsorption* (or *chemisorption*); the dividing line between the two is not always sharp. In physical adsorption, the gas molecules are held to the solid's surface by relatively weak intermolecular van der Waals forces. In chemisorption, a chemical reaction occurs at the solid's surface, and the gas is held to the surface by relatively strong chemical bonds.

The enthalpy changes for chemisorption are usually substantially greater than for physical adsorption. For chemisorption, once a monolayer of adsorbed gas covers the solid's surface, no further chemical reaction between the gas (species B) and the solid (species A) can occur. For physical adsorption, once a monolayer has formed, intermolecular interactions between adsorbed B molecules in the monolayer and gas-phase B molecules can lead to formation of a second layer of adsorbed gas. The enthalpy change for formation of the first layer of physically adsorbed molecules is determined by solid–gas (A-B) intermolecular forces, whereas the enthalpy change for formation of the second, third, ... physically adsorbed layers is determined by B-B intermolecular forces and is about the same as the ΔH of condensation of gas B to a liquid. Although only one layer can be chemically adsorbed, physical adsorption of further layers on top of a chemisorbed monolayer sometimes occurs.

The chemical reactions that occur with chemisorption have been determined for several systems. When H_2 is chemisorbed on metals, H atoms are formed on the surface and bond to metal atoms, as evidenced by the fact that metals that chemisorb H_2 will catalyze the exchange reaction $H_2 + D_2 \rightarrow 2HD$. Chemisorption of C_2H_6 on metals occurs mainly by breakage of a C—H bond, and to a lesser extent by breakage of the C—C bond; the evidence for this is from a comparison of the rates of the metal-catalyzed exchange and cracking reactions $C_2H_6 + D_2 \rightarrow C_2H_5D + HD$ and $C_2H_6 + H_2 \rightarrow 2CH_4$; the chemisorbed structures are

$$\begin{array}{cc} & CH_3 \\ & | \\ H & CH_2 \\ | & | \\ -M- & M- \\ | & | \end{array} \qquad \text{and} \qquad \begin{array}{cc} H_3C & CH_3 \\ | & | \\ -M- & M- \\ | & | \end{array}$$

where M is a metal atom. Chemisorption of CO_2 on metal oxides probably occurs with formation of carbonate ions: $CO_2 + O^{2-} \rightarrow CO_3^{2-}$. Comparison of the infrared spectra of CO chemisorbed on metals with those of metal carbonyl compounds suggests that one or both of the following two kinds of bonding occur, depending on which metal is used:

$$\begin{array}{c} :\!O\!: \\ \| \\ C \\ / \quad \backslash \\ -M-M- \\ | \qquad | \end{array} \qquad \text{and} \qquad \begin{array}{cc} \ddot{O} & \ddot{O} \\ \equiv & \equiv \\ C & C \\ | & | \\ -M- & M- \\ | & | \end{array}$$

In adsorption studies, the amount of gas adsorbed at a given temperature is measured as a function of the gas pressure P. The amount of gas adsorbed is usually expressed as the volume v (corrected to 0°C and 1 atm) adsorbed per gram of adsorbent. A plot of v vs. P at constant T gives an *adsorption isotherm.* Figure 13.16 shows two typical isotherms.

For O_2 on charcoal at 90 K, the amount adsorbed increases with P until a limiting value is reached. This isotherm is said to be of type I, and its interpretation is that at the adsorption limit the solid's surface is covered by a monolayer of O_2 molecules, after which no further O_2 is adsorbed. The isotherm in Fig. 13.16*b* is of type II, and its interpretation is that after the formation of a monolayer of adsorbed gas is substantially complete, further increases in gas pressure cause formation of a second layer of adsorbed molecules, then a third layer, etc., so that we have multilayer adsorption. The majority of isotherms are of type II.

In 1918 Langmuir used a simple model of the solid surface to derive an equation for an isotherm. He assumed that the solid has a uniform surface, that there is no interaction between one adsorbed molecule and another, that the adsorbed molecules are localized at specific sites, and that only a monolayer can be adsorbed. At equilibrium, the rates of adsorption and desorption of molecules from the surface are equal. Let N be the number of adsorption sites on the bare solid surface. (For example, for chemisorption of CO_2 on a metal oxide to produce CO_3^{2-}, N is the number of oxide ions at the surface.) Let θ be the fraction of adsorption sites occupied by adsorbate at equilibrium. The rate of desorption is proportional to the number θN of adsorbed molecules and equals $k_d \theta N$, where k_d is a constant at fixed temperature. The adsorption rate is proportional to the number of collisions of gas-phase molecules with unoccupied adsorption sites (since only a monolayer can be formed); the number of collisions of gas molecules with the surface is proportional to the gas pressure P, and the number of unoccupied sites is $(1 - \theta)N$; the adsorption rate is therefore $k_a P(1 - \theta)N$, where k_a is a certain constant. Equating the desorption and adsorption rates, we get $k_d \theta N = k_a P(1 - \theta)N$, and

$$\theta = \frac{k_a P}{k_d + k_a P} = \frac{(k_a/k_d)P}{1 + (k_a/k_d)P} = \frac{bP}{1 + bP} \tag{13.53}$$

where $b \equiv k_a/k_d$. (b depends on temperature.) The fraction θ of sites occupied at pressure P equals v/v_m, where v is the volume adsorbed at P (as defined above) and v_m

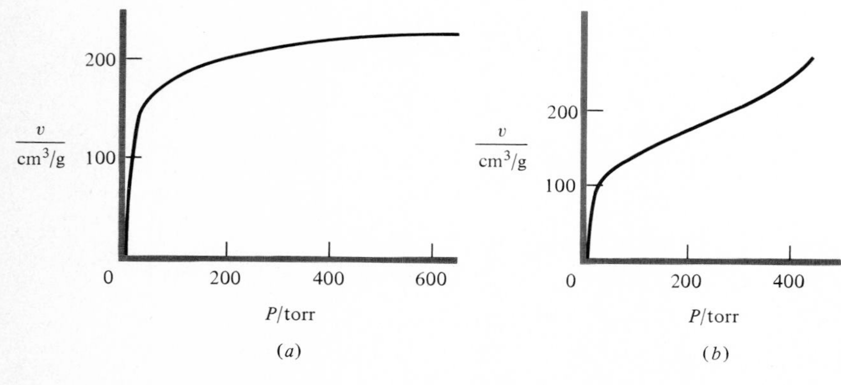

Figure 13.16
(a) Adsorption of O_2 on charcoal at 90 K. (b) Adsorption of N_2 on silica gel at 77 K.

is the volume adsorbed in the high-pressure limit when a monolayer covers the entire surface. Equation (13.53) becomes

$$v = \frac{cP}{1 + bP} \qquad \text{where } c \equiv v_m b \tag{13.54}$$

The shape of the *Langmuir isotherm* (13.54) resembles Fig. 13.16*a*.

To test whether (13.54) fits a given set of data, we take the reciprocal of each side to give $1/v = 1/v_m bP + 1/v_m$. A plot of $1/v$ vs. $1/P$ ought to give a straight line if the Langmuir isotherm is obeyed. The Langmuir isotherm is found to work reasonably well for many (but far from all) cases of chemisorption.

Most of the assumptions Langmuir used in his derivation are false. The surfaces of most solids are not uniform, and the desorption rate depends on the location of the adsorbed molecule. The force between adjacent adsorbed molecules is frequently substantial, as evidenced by changes in the heat of adsorption with increasing θ. There is substantial experimental and theoretical evidence that adsorbed molecules can move about on the surface; this mobility is much greater for physically adsorbed molecules than for chemisorbed ones and increases as T increases. Multilayer adsorption is quite common in physical adsorption. Thus Langmuir's derivation of (13.54) cannot be taken too seriously. Statistical-mechanical derivations of the Langmuir isotherm have been given, and these require fewer assumptions than Langmuir's derivation.

The *Freundlich isotherm*

$$v = kP^a \tag{13.55}$$

where k and a are constants (with $0 < a < 1$), was suggested on empirical grounds in the nineteenth century. Equation (13.55) gives $\log v = \log k + a \log P$; the constants k and a are evaluated from the intercept and slope of a plot of $\log v$ vs. $\log P$. Halsey and others showed that the Freundlich isotherm can be derived by modifying the Langmuir assumptions to allow for several kinds of adsorption sites on the solid, each kind having a different heat of adsorption. The Freundlich isotherm is not valid at very high pressures (note that it incorrectly predicts $v \to \infty$ as $P \to \infty$) but is frequently more accurate than the Langmuir isotherm in the intermediate-pressure range.

The Freundlich equation is often applied to adsorption of solutes from liquid solutions onto solids. Here, the solute's concentration c replaces P, and the mass adsorbed per unit mass of adsorbent replaces v.

The Langmuir and Freundlich equations are applicable to isotherms of type I only. In 1938 Brunauer, Emmett, and Teller modified Langmuir's assumptions to allow for multilayer adsorption (type II); their result is

$$\frac{P}{v(P^* - P)} = \frac{1}{v_m c} + \frac{c - 1}{v_m c}\frac{P}{P^*} \tag{13.56}$$

where v and v_m are defined above, c is a constant, and P^* is the vapor pressure of the adsorbate at the temperature of the experiment. (For $P \geq P^*$, the gas condenses to a liquid.) The constants c and v_m can be obtained from the slope and intercept of a plot of $P/v(P^* - P)$ vs. P/P^*. The Brunauer–Emmett–Teller (BET) equation fits many observed type II isotherms rather well, especially for the intermediate-pressure range. Once v_m has been obtained from the BET isotherm, the number of molecules needed to form a monolayer is known and the surface area of the solid adsorbent can be estimated by using an estimated value for the surface area occupied by one adsorbed molecule.

A solid–gas adsorption study requires an initially clean solid surface. To produce such a surface, the solid may be heated strongly in vacuum (a procedure called *outgassing*). However, such heating often does not desorb all the surface contamination. A better procedure is to vaporize the solid in vacuum and condense it as a thin, porous film on the surface of the glass container.

As Langmuir's kinetic derivation of (13.54) emphasizes, adsorption is a dynamical equilibrium between adsorbed molecules and gas-phase molecules. A statistical-mechanical treatment gives the following approximate equation for the average residence time τ of an adsorbed molecule: $\tau = \tau_0 e^{-Q/RT}$, where τ_0 is typically about 10^{-13} s and Q is the molar enthalpy change for adsorption. For physical adsorption, Q ranges from -1 to -10 kcal/mole; for chemisorption, Q ranges from -10 to -200 kcal/mole. At room temperature, the following τ-vs.-Q values are obtained:

Q/(kcal/mol)	-3	-10	-20	-40
τ/s	10^{-11}	10^{-6}	10^{2}	10^{16}

Since 10^{16} s is 10^9 years, τ is essentially infinite for $|Q|$ values above, say, 30 kcal/mol. Therefore (in contrast to physical adsorption), chemisorption is usually irreversible at room temperature, meaning that lowering the pressure does not cause desorption; to remove the chemisorbed gas, the solid must be heated.

Adsorption and absorption play key roles in chromatography. *Chromatography* is a separation method in which a multicomponent fluid phase moves past a stationary condensed phase. In *adsorption chromatography*, the stationary phase is a solid; in *partition chromatography*, it is a liquid. In *gas chromatography*, the mobile phase is a gas mixture; in *liquid chromatography*, it is a liquid solution.

In adsorption chromatography, the finely powdered solid adsorbent, for example, Al_2O_3, silica gel, or carbon, is packed in a column. In liquid chromatography, a concentrated solution of the sample to be separated is added to the top of the column, and when the sample has been adsorbed at the top, a large amount of a pure solvent is added, which flows through the column by gravity and carries the sample along. In gas chromatography, the sample is injected into a stream of flowing inert gas (the carrier gas). Molecules of the moving sample undergo many successive adsorptions and desorptions. Each component has a different value of the ratio of time spent adsorbed on the solid to time spent in solution; hence the components travel at different rates through the column and are separated.

In partition chromatography, a liquid solution or gas mixture flows past a stationary liquid phase. Separation results from the differing solubilities or partition coefficients of the components of the mobile phase in the stationary phase. (The *partition coefficient* for substance i between the immiscible phases α and β is the mole-fraction ratio x_i^α/x_i^β at equilibrium between the phases.) Here, the process is absorption (i.e., solution) in the stationary liquid phase (rather than adsorption). The stationary liquid phase (which is commonly water in liquid chromatography and a nonvolatile organic liquid in gas chromatography) is mixed with a finely powdered porous solid, and the mixture (which has a dry appearance) is packed into a column. The solid "supports" the liquid and keeps it stationary. Part of the stationary liquid is physically adsorbed on the solid as a multilayer, but most is present in tiny holes, crevices, and cavities in and between the solid particles. The general term for such tiny spaces is *pores*. Most of the liquid present in these pores behaves like bulk liquid. Common support solids include silica gel, cellulose powder, and various forms of powdered diatomaceous earth. [Diatomaceous earth (or diatomite) consists of skeletons of certain single-celled algae (diatoms). Its composition is mainly SiO_2 with some Al_2O_3 and other metal oxides.]

In *paper chromatography*, the sample mixture is placed as a spot on a strip of filter paper, and one end of the paper is dipped in an organic solvent. Capillary action causes the solvent to flow through the paper. The cellulose in filter paper contains a substantial amount of water present as adsorbed multilayers. Paper chromatography involves adsorption of compounds on cellulose molecules, adsorption on bound water molecules, and partition between loosely bound water and the flowing organic solvent.

In *thin-layer chromatography*, a finely powdered solid adsorbent is mixed with water and spread in a thin film on a glass plate, which is then dried. The sample is spotted on the plate, and then an organic solvent flows through the adsorbent by capillary action.

Chromatography was discovered in 1903 by the Russian botanist Tswett, who separated plant pigments using a flowing organic solvent and a column of powdered $CaCO_3$. The name "chromatography" arose from the colored bands Tswett obtained in the column (Greek *chromat-*, "color"). (Of course, the method is not limited to colored substances.) By coincidence, "Tswett" means "color" in Russian.

13.6 COLLOIDS

When an aqueous solution containing Cl^- ion is added to one containing Ag^+ ion, under certain conditions the solid AgCl precipitate may form as extremely tiny crystals that remain suspended in the liquid instead of settling out as a filterable precipitate. This is an example of a colloidal system.

Colloidal systems. A *colloidal system* consists of particles that have in at least one direction a dimension lying in the approximate range 30 to 10,000 Å and a medium in which the particles are dispersed. The particles are called *colloidal particles* or the *dispersed phase*; the medium is called the *dispersion medium* or the *continuous phase*. The colloidal particles may be in the solid, liquid, or gaseous state, or they may be individual molecules. The dispersion medium may be a solid, liquid, or gas. The term *colloid* can mean either the colloidal system of particles plus dispersion medium or just the colloidal particles.

A *sol* is a colloidal system whose dispersion medium is a liquid or gas. When the dispersion medium is a gas, the sol is called an *aerosol*. Fog is an aerosol with liquid particles. Smoke is an aerosol with liquid or solid particles. (Tobacco smoke has liquid particles.) A sol that consists of a liquid dispersed in a liquid is an *emulsion*. A sol that consists of solid particles suspended in a liquid is a *colloidal suspension*. (An example is the aqueous AgCl system mentioned above.)

A *foam* is a colloidal system in which gas bubbles are dispersed in a liquid or solid. Although the diameters of the bubbles usually exceed 10^4 Å, the distance between bubbles is usually less than 10^4 Å, so foams are classified as colloidal systems; in foams, the dispersion medium is in the colloidal state. Foams are familiar to anyone who uses soap, drinks beer, or goes to the beach. Pumice stone is a foam with air bubbles dispersed in rock of volcanic origin; its chemical composition varies.

Examples of colloidal systems with a solid dispersion medium are ruby glass, which consists of colloidal particles of CdS or Au dispersed in glass; opal, which consists of tiny drops of water in SiO_2; and pumice stone.

Colloidal systems can be classified into those in which the dispersed particles are single molecules (monomolecular particles) and those in which the particles are aggregates of many molecules (polymolecular particles). Colloidal dispersions of AgCl, As_2S_3, and Au in water contain polymolecular particles, and the system has two phases: water and the dispersed particles. The tiny size of the particles results in a very large interfacial area, and surface effects (e.g., adsorption on the colloidal particles) are of paramount importance in determining the system's properties. On the other hand, in a polymer solution (e.g., a solution of a protein in water) the colloidal particle is a single molecule, and the system has one phase; here, there are no interfaces, but solvation of the polymer molecules is significant. The large size of the solute molecules causes a polymer solution to resemble a colloidal dispersion of polymolecular particles in many of its properties (e.g., scattering of light, sedimentation in a centrifuge), so polymer solutions are classified as colloidal systems.

If the colloidal particles all have essentially the same size, the system is said to be *monodisperse*; otherwise it is *heterodisperse* (or *polydisperse*).

Lyophilic colloids. When a protein crystal is dropped into water, the polymer molecules spontaneously dissolve to produce a colloidal dispersion. Colloidal dispersions that can be formed by spontaneous dispersion of the dry bulk material of the colloidal particles in the dispersion medium are called *lyophilic* ("solvent-loving"). A lyophilic sol is thermodynamically more stable than the two-phase system of dispersion medium and bulk colloid material.

Certain compounds in solution yield lyophilic colloidal systems as a result of spontaneous association of their molecules to form particles of colloidal size. If one plots the osmotic pressure of an aqueous solution of a soap (a compound with the formula $RCOO^-M^+$, where R is a straight chain containing 10 to 20 carbons, and M is Na or K) as a function of the solute's stoichiometric concentration, one finds that at a certain concentration (called the *critical micelle concentration*, cmc) the solution shows a sharp drop in the slope of the Π-vs.-c curve. Starting at the cmc, the solution's light-scattering ability (turbidity) rises sharply. These facts indicate that above the cmc a substantial portion of the solute ions are aggregated to form units of colloidal size; such aggregates are called *micelles*. Dilution of the solution below the cmc eliminates the micelles, so micelle formation is reversible. Light-scattering data show that a micelle contains from 20 to a few hundred monomer units, depending on the compound and on the concentrations of other ions in the solution. Light-scattering and x-ray-diffraction data indicate an approximately spherical structure for most micelles. Figure 13.17 shows the structure of a soap micelle in aqueous solution. The hydrocarbon part of each monomer anion is directed toward the center, and the polar COO^- group is on the outside. A substantial fraction of the micelle's COO^- groups have solvated Na^+ ions bound to them (ion pairing, Sec. 10.7).

Although a micelle-containing system is sometimes treated as having two phases, it is best considered as a one-phase solution in which the reversible equilibrium $n\text{L} \rightleftharpoons \text{L}_n$ exists, where L is the monomer and L_n the micelle. That micelle

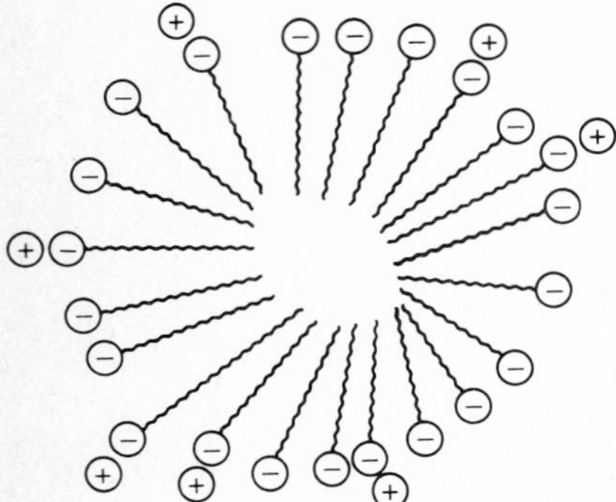

Figure 13.17
A soap micelle in aqueous solution.

formation does not correspond to separation of a second phase is shown by the fact that the cmc does not have a precisely defined value but corresponds to a narrow range of concentrations. Figure 13.18 shows the variation of monomer and micelle concentrations with the solute stoichiometric concentration. The rather sudden rise in micelle concentration at the cmc results from the large value of n; see Prob. 13.32. (The limit $n \to \infty$ would correspond to a phase change occurring at a precisely defined concentration to give a two-phase system.)

Some compounds give micelles with the same value of n; others give micelles with several values of n in the same solution.

Besides salts of fatty acids, the other type III surfactants listed in Sec. 13.3 also form micelles. The cmc is typically 10^{-2} to 10^{-3} mol/dm^3 for ionic micelle-forming surfactants and 10^{-4} mol/dm^3 for the (nonionic) polyoxyethylenes. Micelle formation may contribute to the detergent action of surfactants, since molecules of organic compounds can dissolve in the micelles.

The term *association colloid* is used when the colloidal particles are formed reversibly in solution as aggregates of small molecules, as in micelle formation.

Lyophobic colloids. When solid AgCl is brought in contact with water, it does not spontaneously disperse to form a colloidal system. Sols that cannot be formed by spontaneous dispersion are called lyophobic ("solvent-hating"). Lyophobic sols are thermodynamically unstable with respect to separation into two unmixed bulk phases (recall that the stable state of a system is one of minimum interfacial area), but the rate of separation may be extremely small. (Gold sols prepared by Faraday are on exhibit in the British Museum.)

The long life of lyophobic sols is commonly due to adsorbed ions on the colloidal particles; repulsion between like charges keeps the particles from aggregating. For example, when a sol of AgBr in water is prepared by the gradual addition of an $AgNO_3$ solution to an NaBr solution, more Br^- than Ag^+ ions are present; some of the excess Br^- ions are chemisorbed at Ag^+ ions on the surface of each colloidal AgBr crystal to produce a negatively charged particle. If more Ag^+ ions than Br^- ions are present, then Ag^+ ions are adsorbed to give positively charged colloidal particles. When a sol of colloidal sulfur in water is formed by the acidification of an aqueous $Na_2S_2O_3$ solution, the sulfur particles have adsorbed $S_2O_3^{2-}$ ions. When a gold sol is prepared by the reduction of $AuCl_3$ in aqueous solution with CH_2O, the gold particles contain chemisorbed Cl^- ions. The presence of adsorbed ions can be shown by the migration of the colloidal particles in an applied electric field (a phenomenon called *electrophoresis*).

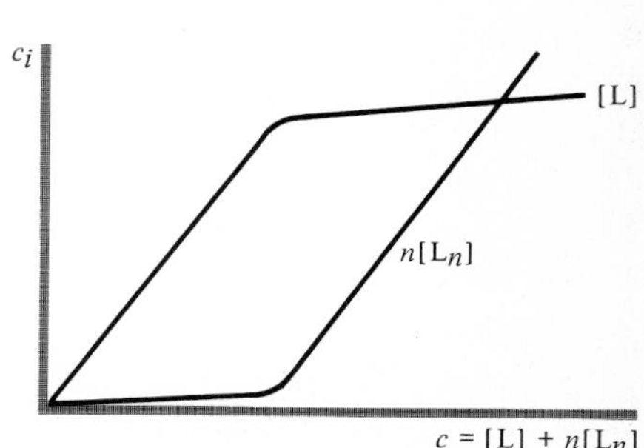

Figure 13.18
Monomer (L) and micelle (L_n) concentrations vs. stoichiometric concentration c.

Another way a lyophobic sol can be stabilized is by the presence of a polymer (e.g., the protein gelatin) in the solution. The large polymer molecules become adsorbed on and surround each colloidal particle, thereby preventing coagulation.

Many lyophobic colloids can be prepared by precipitation reactions. Precipitation in either very dilute or very concentrated solutions tends to produce colloids. Lyophobic sols can also be produced by mechanically breaking down a bulk substance into tiny particles and dispersing them in a medium. For example, emulsions can be prepared by vigorous shaking of two essentially immiscible liquids in the presence of an emulsifying agent (defined below).

Sedimentation. The particles in a noncolloidal suspension of a solid in a liquid will eventually settle out under the influence of gravity, a process called *sedimentation*. For colloidal particles whose size is well below 10^3 Å, accidental thermal convection currents and the random collisions between the colloidal particles and molecules of the dispersion medium prevent any significant amount of sedimentation. A sol with larger colloidal particles will show sedimentation with time.

Coagulation. The aggregation of colloidal particles of a lyophobic colloid to form larger particles is called *coagulation* or *flocculation*; when large enough particles are formed, they separate out due to sedimentation. Although a small amount of electrolyte is necessary to stabilize a lyophobic colloid, when sufficient electrolyte is added to the solution, coagulation occurs. If the sol particles are positively charged, they adsorb negative ions of the added electrolyte; random collisions between the neutral particles then produce coagulation. Even without addition of electrolyte, some degree of coagulation may occur with time.

Emulsions. The liquids in most emulsions are water and an oil, where "oil" denotes an organic liquid essentially immiscible with water. Such emulsions are classified as either oil-in-water (O/W) emulsions, in which water is the continuous phase and the oil is present as tiny droplets, or water-in-oil (W/O) emulsions, in which the oil is the continuous phase. Emulsions are lyophobic colloids. They are stabilized by the presence of an *emulsifying agent*, which is commonly an amphipathic species that forms a surface film at the interface between each colloidal droplet and the dispersion medium, thereby lowering the interfacial tension and preventing coagulation. The cleansing action of soaps and other detergents results in part from their acting as emulsifying agents to keep tiny droplets of grease suspended in water.

Milk is an O/W emulsion of butterfat in water; the emulsifying agent is the protein casein. The diameter of the colloidal droplets is about 10^5 Å in unhomogenized milk and 10^3 Å in homogenized milk. Many pharmaceutical preparations (salves, ointments) and cosmetics (cold cream, hair glop) are emulsions.

It has been suggested that the efficiency of automobile engines might be increased and pollution reduced by using an emulsion of gasoline and water as fuel.

Gels. A *gel* is a semirigid colloidal system of at least two components in which both components extend continuously throughout the system. An inorganic gel typically consists of water trapped within a three-dimensional network of tiny crystals of an inorganic solid. The crystals are held together by van der Waals forces, and the water is both adsorbed on the crystals and mechanically enclosed by them. [Recall the white, gelatinous precipitate of $Al(OH)_3$ obtained in the qualitative-analysis scheme.] In contrast to a gel, the solid particles in a colloidal suspension are well separated from one another and move about freely in the liquid.

When an aqueous solution of the protein gelatin is cooled, a polymer gel is formed. Here, water is trapped within a network formed by the long-chain polymer molecules. In this network, polymer chains are entangled with one another and are held together by van der Waals forces, by hydrogen bonds, and perhaps by some covalent bonds. (Include lots of sugar and some artificial flavor and color with the gelatin, and you've got Jell-O.) The polysaccharide agar forms a polymer gel with water that is used as a culture medium for bacteria. Blood clotting involves formation of a gel whose framework is the protein fibrinogen. Cellophane membranes used in osmotic-pressure determination are gels of cellulose molecules and water.

Some gels can be dried out to yield a highly porous solid called a *xerogel* (from the Greek *xeros*, "dry"). For example, when an aqueous sodium silicate solution is acidified, a silicic acid gel is formed that contains a three-dimensional network of Si—O chemical bonds; when this gel is dried, one obtains a porous xerogel of SiO_2, which is called *silica gel* but would more correctly be called silica xerogel.

13.7 BIOLOGICAL MEMBRANES

Each cell of an organism is surrounded by a thin ($\approx$100 Å thick) membrane (the *cell membrane* or *plasma membrane*), which actively regulates the flow of chemicals between the cell and its surroundings. Moreover, membranes surround many structures within the cell (e.g., the cell nucleus). The production of energy by cells involves a multistep process that couples the oxidation of glucose ($C_6H_{12}O_6$) to CO_2 and H_2O with the transformation of adenosine diphosphate (ADP) to adenosine triphosphate (ATP). The energy "stored" in ATP is then used in such processes as muscle contraction, protein synthesis, and active transport. Many steps of the oxidation of glucose are carried out by enzymes attached to the inner membranes of intracellular structures called mitochondria. In a nerve cell, the cell membrane conducts the nerve impulse. Thus, membranes play a key role in cell metabolism and transmission of information in organisms. (Cancer cells have very different surface properties than normal cells.)

Biological membranes are composed of phospholipids and proteins. A simple lipid is an ester of the trihydroxy alcohol glycerol (Fig. 13.19) with medium-chain-

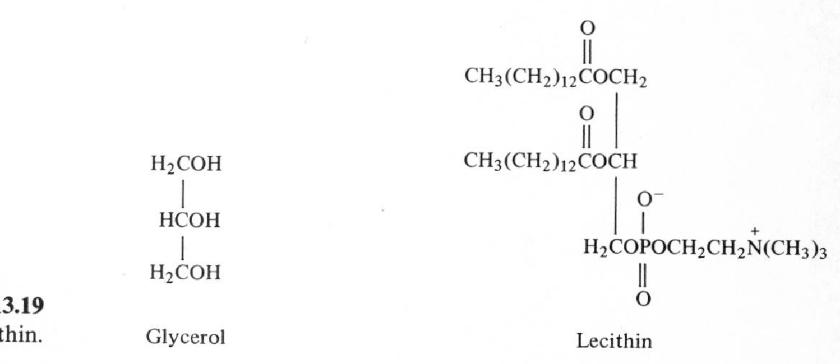

Figure 13.19
Glycerol and lecithin.

length acids. [Hydrolysis of a simple lipid with NaOH gives glycerol and salts of the acids (soaps).] Animal fats and vegetable oils are simple lipids. In a phospholipid, one of the fatty-acid groups is replaced by a phosphate-containing group. Figure 13.19 shows the structure of lecithin, one of the phospholipids found in membranes. Phospholipids are amphipathic; the two rather long hydrocarbon chains are hydrophobic, and the phosphate-containing part is hydrophilic. Proteins are polymers of amino acids. Active transport in membranes (Sec. 12.4) is probably due to the catalytic activity of membrane proteins.

In 1925 Gorter and Grendel extracted the membrane phospholipids from red blood cells, spread the lipids on water in a surface balance (Sec. 13.4), and measured the surface area for a close-packed lipid monolayer. They found an approximately 2 : 1 ratio of the monolayer area to the surface area of the blood cells and proposed that the cell membrane is a two-molecule-thick layer of phospholipids (called a bimolecular leaflet). In 1935 Davson and Danielli modified the Gorter–Grendel model to include a layer of protein molecules on each side of the lipid layer. The Davson–Danielli *paucimolecular model* (Latin *paucus*, "few") has been a very influential membrane model.

There is substantial experimental evidence both for and against the Davson–Danielli model. Because of the conflicting evidence, about 20 membrane models have been proposed since 1935. Stein and Danielli in 1956 modified the Davson–Danielli model to include aqueous pores in the membrane. Figure 13.20 show the Davson–Danielli–Stein model. The lipids' hydrophobic chains are in the middle of the membrane, and their hydrophilic parts point outward; protein molecules cover the outside and inside and line the pores.

Many of the other proposed models differ significantly from that of Davson and Danielli. The *fluid mosaic* membrane model [S. J. Singer and G. L. Nicolson, *Science*, **175**, 720 (1972)] uses a lipid bilayer but has the protein present as randomly distributed globular-shaped molecules (rather than spread out as monolayers); these globular protein molecules are partly embedded in, and partly protrude from, the membrane; some of the protein molecules span the entire thickness of the membrane.

Some workers argue that membrane structure differs from one cell type to another.

Since the structure of biological membranes has not been firmly established, their mode of action is still largely unknown.

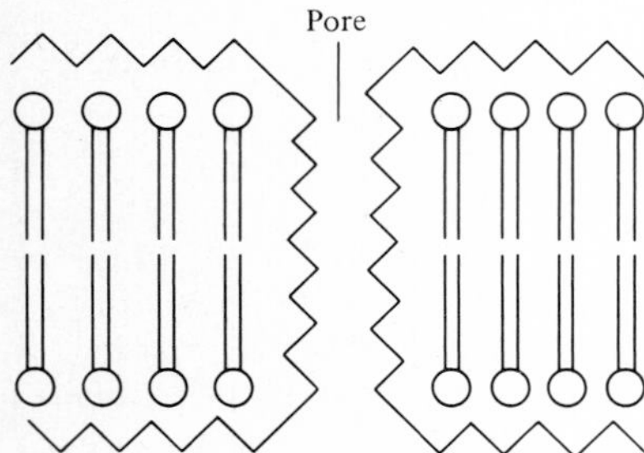

Figure 13.20
Davson–Danielli–Stein membrane model.

13.1 (*a*) Calculate the surface area of a 1.0-cm^3 sphere of gold. (*b*) Calculate the surface area of a colloidal dispersion of 1.0 cm^3 of gold in which each gold particle is a sphere of radius 300 Å.

13.2 Calculate the minimum work needed to increase the area of the surface of water from 2.0 to 5.0 cm^2 at 20°C. The surface tension of water is 73 dyn/cm at 20°C.

13.3 The surface tension of ethyl acetate at 0°C is 26.5 mN/m, and its critical temperature is 523.2 K. Estimate its surface tension at 50°C. (The experimental value is 20.2 mN/m.)

13.4 J. R. Brock and R. B. Bird [*Am. Inst. Chem. Eng. J.*, **1**, 174 (1955)] found that for liquids that are not highly polar or hydrogen-bonded, the constant γ_0 in (13.3) is usually well approximated by

$$\gamma_0 = (P_c/\text{atm})^{2/3}(T_c/\text{K})^{1/3}(0.432/Z_c - 0.951) \text{ dyn/cm}$$

where P_c, T_c, and Z_c are the critical pressure, temperature, and compressibility factor. For ethyl acetate, $P_c = 37.8$ atm, $T_c = 523.2$ K, and $Z_c = 0.252$. Calculate the percent error of the Brock–Bird–Katayama–Guggenheim predicted value of γ for ethyl acetate at 0°C. The experimental value is 26.5 dyn/cm.

13.5 Calculate the pressure inside a bubble of gas in water at 20°C if the pressure of the water is 760 torr and the bubble radius is 0.040 cm. (See Prob. 13.2.)

13.6 Calculate γ/l_z in Eq. (13.5) for the typical values $l_z = 10$ cm and $\gamma = 50$ dyn/cm; express your answer in atmospheres.

13.7 At 20°C the capillary rise at sea level for methanol in contact with air in a tube with inside diameter 0.350 mm is 3.33 cm. The contact angle is zero. The densities of methanol and air at 20°C are 0.7914 and 0.0012 g/cm^3. Find γ for CH_3OH at 20°C.

13.8 For the Hg–air interface on glass, $\theta = 140°$. Find the capillary depression of Hg in contact with air at 20°C in a glass tube with inside diameter 0.350 mm. For Hg at 20°C, $\rho = 13.59$ g/cm^3 and $\gamma = 490$ ergs/cm^2. (See Prob. 13.7.)

13.9 At 20°C the interfacial tension for the liquids *n*-hexane and water is 52.2 ergs/cm^2. The densities of *n*-hexane and water at 20°C are 0.6599 and 0.9982 g/cm^3. Assuming a zero contact angle, calculate the capillary rise at 20°C in a 0.350-mm inside diameter tube inserted into a two-phase *n*-hexane–water system.

13.10 (*a*) In (13.12), h is the height of the bottom of the meniscus. Hence, (13.12) neglects the pressure due to the small amount of liquid β above the bottom of the meniscus. Show that if this liquid is taken into account, then $\gamma = \frac{1}{2}(\rho_\beta - \rho_\alpha)gr(h + \frac{1}{3}r)$ for $\theta = 0$. (*b*) Rework Prob. 13.7 using this more accurate equation.

13.11 Two capillary tubes with inside radii 0.600 and 0.400 mm are inserted into a liquid with density 0.901 g/cm^3 in contact with air of density 0.001 g/cm^3. The difference between the capillary rises in the tubes is 1.00 cm. Find γ. (Assume a zero contact angle.)

13.12 For dilute solutions, the surface tension frequently varies linearly with solute concentration c: $\gamma = \gamma^* - bc$, where b is a constant. Show that in this case $\Gamma_{2(1)} = (\gamma^* - \gamma)/RT$.

13.13 At 21°C, surface tensions of aqueous solutions of $C_6H_5CH_2CH_2COOH$ vs. solute molality are:

10^3m/(mol/kg)	γ/(dyn/cm)
3.35	69.0
6.40	66.5
9.99	63.6
11.66	61.3
15.66	59.2
19.99	56.1
27.40	52.5
40.8	47.2

Find $\Gamma_{2(1)}$ for a solution with 20×10^{-3} mol of solute per kilogram of water. [Use the equation that corresponds to (13.44) for the molality scale.]

13.14 For aqueous solutions of saturated aliphatic acids at 18°C, the surface tensions are fitted by

$$\gamma = [73.0 - 29.9 \log (ac + 1)] \text{ dyn/cm}$$

where a is a constant and c is the acid's concentration. (*a*) Find $\Gamma_{2(1)}$ as a function of concentration. (*b*) Which gas–solid adsorption isotherm does the expression in (*a*) resemble? (*c*) For butanoic acid, $a = 19.64$ dm^3/mol. Plot $\Gamma_{2(1)}$ vs. c for aqueous solutions of this acid for the concentration range 0 to 1 mol/dm^3.

13.15 Show that for a binary solution of an electrolyte, the Gibbs adsorption isotherm is

$$\Gamma_{2(1)} = -\frac{1}{\nu RT}\left[\frac{\partial \gamma}{\partial \ln (\gamma_2 m_2/m^\circ)}\right]_T$$

where γ_2 and m_2 are the electrolyte's stoichiometric mean ionic activity coefficient and molality and ν and m° are defined by (10.61) and (10.55). The differences between this equation and (13.43) are the presence of ν

and the fact that activity coefficients cannot be neglected in dilute electrolyte solutions.

13.16 For stearic acid [$CH_3(CH_2)_{16}COOH$] the area per molecule at the Pockels point is 20 Å^2, and the density is 0.94 g/cm^3 at 20°C. Estimate the length of a stearic acid molecule.

13.17 One acre = 4057 m^2. One teaspoonful ≈ 4.8 cm^3. (*a*) From Franklin's data in Sec. 13.4. calculate the thickness of the olive-oil film. (Assume he used exactly 1 teaspoonful of oil.) (*b*) Olive oil is mainly glycerol trioleate [$(C_{17}H_{33}COO)_3C_3H_5$] with a density at room temperature of 0.90 g/cm^3. Calculate the area occupied by each olive-oil molecule in Franklin's film. (*c*) Calculate $\Gamma_{2(1)}$ for Franklin's monolayer.

13.18 A certain solution of solute i in water has mole fraction $x_i = 0.10$. The interphase region contains 2.0×10^{-8} mol of i and 45×10^{-8} mol of H_2O. The solution's surface area is 100 cm^2. Find $\Gamma_{i(1)}$.

13.19 (*a*) Show that when (13.39) holds, then $n_i - c_i^\beta V = n_{i,\text{bulk}}^\alpha + n_i^s$ and $n_1 - c_1^\beta V = n_{1,\text{bulk}}^\alpha + n_1^s$, where the quantities on the right are defined following Eq. (13.40). (Start by writing n_i as the sum of the moles of i in the two bulk phases and the interphase region.) (*b*) Substitute the result of (*a*) into (13.37) to derive (13.40).

13.20 Show that $\gamma = (\partial A/\partial \mathscr{A})_{T,V,n_i}$, where the Helmholtz function of a capillary system is $A \equiv U - TS$. [*Hint*: Use (13.26).]

13.21 From Eq. (13.9), a drop of liquid of radius r is at a higher pressure than the vapor it is in equilibrium with; this increased pressure affects the chemical potential of the liquid and raises its vapor pressure slightly. (*a*) Use the integrated form of Eq. (7.27) in Prob. 7.17 to show that the vapor pressure P_r of such a drop is

$$P_r = P \exp\,(2\gamma \bar{V}_l/rRT)$$

where $\bar{V}_l$ is the molar volume of the liquid and P its bulk vapor pressure. This is the *Kelvin equation.* (*b*) The vapor pressure and surface tension of water at 20°C are 17.535 torr and 73 dyn/cm. Calculate the 20°C vapor pressure of a drop of water of radius 1.00×10^{-5} cm.

13.22 For N_2 adsorbed on a certain sample of charcoal at −77°C, adsorbed volumes (recalculated to 0°C and 1 atm) per gram of charcoal vs. N_2 pressure are:

P/atm	3.5	10.0	16.7	25.7	33.5	39.2
v/(cm^3/g)	101	136	153	162	165	166

(*a*) Fit the data with the Langmuir isotherm and give the values of v_m and b. (*b*) Fit the data with the Freundlich isotherm and give the values of k and a. (*c*) Calculate v at 7.0 atm using both the Langmuir isotherm and the Freundlich isotherm.

13.23 The *Temkin isotherm* for gas adsorption on solids is $v = r \ln sP$, where r and s are constants. (*a*) What should be plotted against what to give a straight line if the Temkin isotherm is obeyed? (*b*) Fit the data of Prob. 13.22 to the Temkin isotherm and evaluate r and s.

13.24 Besides plotting $1/v$ vs. $1/P$, there is another way to plot the Langmuir isotherm to yield a straight line. What is this way?

13.25 For N_2 adsorbed on a certain sample of ZnO powder at 77 K, adsorbed volumes (recalculated to 0°C and 1 atm) per gram of ZnO vs. N_2 pressure are:

P/torr	v/(cm^3/g)	P/torr	v/(cm^3/g)	P/torr	v/(cm^3/g)
56	0.798	183	1.06	442	1.71
95	0.871	223	1.16	533	2.08
145	0.978	287	1.33	609	2.48

The normal boiling point of N_2 is 77 K. (*a*) Plot v vs. P and decide whether the Langmuir or the BET equation is more appropriate. (*b*) Use the equation you decided on in (*a*) to find the volume v_m needed to form a monolayer; also find the other constant in the isotherm equation. (*c*) Assume that an adsorbed N_2 molecule occupies an area of 16 Å^2 and calculate the surface area of 1.00 g of the ZnO powder.

13.26 Show that for $\theta \ll 1$, the Langmuir isotherm reduces to the Freundlich isotherm with $a = 1$.

13.27 A mixture of gases A and B is in equilibrium with a solid surface. Show that if the Langmuir assumptions are valid, then

$$\theta_A = \frac{b_A P_A}{1 + b_A P_A + b_B P_B} \qquad \frac{v}{v_m} = \frac{b_A P_A + b_B P_B}{1 + b_A P_A + b_B P_B}$$

where θ_A is the fraction of adsorption sites occupied by A molecules and b_A and b_B are constants.

13.28 (*a*) Write the BET isotherm in the form $v/v_m = f(P)$. (*b*) Show that if $P \ll P^*$ the BET isotherm reduces to the Langmuir isotherm.

13.29 When CO chemisorbed on W is heated gradually, a substantial amount of gas is evolved in the temperature range 400 to 600 K and a substantial amount is evolved in the range 1400 to 1800 K, with not much evolved at other temperatures. What does this suggest about CO chemisorbed on tungsten?

13.30 Lyophobic and lyophilic colloids are often called *irreversible* and *reversible* colloids, respectively. Explain the relevance of this terminology.

13.31 Around 1900, the American geochemist David Day passed samples of crude oil through a layer of powdered fuller's earth and obtained a partial separation into low-boiling and high-boiling fractions. What important feature of a chromatographic separation is missing from Day's experiment?

13.32 Let K_c° be the concentration-scale standard equilibrium constant for the equilibrium $n\text{L} = \text{L}_n$ between monomers and micelles in solution, where L is an uncharged species (e.g., a polyoxyethylene). (*a*) Let c be the stoichiometric concentration of the solute (i.e., the number of moles of monomer that are used to prepare a liter of solution) and let x be the concentration of micelles at equilibrium: $x = [\text{L}_n]$, where the brackets denote concentration. Show that $c = nx + (x/K_c^\circ)^{1/n}$. Assume all activity coefficients are 1. (*b*) Let f be the fraction of L present as monomer. Show that $f = 1 - nx/c$. (*c*) For $n = 50$ and $K_c^\circ = 10^{200}$, calculate and graph [L], $n[\text{L}_n]$, and f as functions of c. *Hint:* Calculate c for various assumed values of x, rather than the reverse. (*d*) If the cmc is taken as the value of c for which $f = 0.5$, give the value of the cmc.

14

ELECTROCHEMICAL SYSTEMS

14.1 ELECTROSTATICS

Before developing the thermodynamics of electrochemical systems, we review *electrostatics*, which is the physics of electric charges at rest. The laws of electricity contain different proportionality constants in different systems of units. Throughout this chapter, SI units are used for electrical quantities, and all equations are written in a form valid for SI units.

The SI unit of electric charge Q is the *coulomb* (C), defined in Sec. 16.5. Experiments by Charles Coulomb in 1785 showed that the magnitude of the force F that one electric charge Q_1 exerts on a second charge Q_2 is proportional to the magnitudes of the charges and inversely proportional to the square of the distance r between them: $F = KQ_1Q_2/r^2$, where K is a proportionality constant. The direction of **F** is along the line joining the charges. There are two kinds of charges, positive and negative; like kinds of charges repel each other, and unlike attract. In the SI system, the proportionality constant K is written as $1/4\pi\varepsilon_0$:

$$F = \frac{1}{4\pi\varepsilon_0}\frac{Q_1Q_2}{r^2} \tag{14.1*}$$

The constant ε_0 (the *permittivity of vacuum*) has been experimentally determined to be

$$\begin{aligned} \varepsilon_0 &= 8.854 \times 10^{-12}\ \text{C}^2\ \text{N}^{-1}\ \text{m}^{-2} = 8.854 \times 10^{-12}\ \text{C}^2\ \text{kg}^{-1}\ \text{m}^{-3}\ \text{s}^2 \\ 1/4\pi\varepsilon_0 &= 8.988 \times 10^9\ \text{N m}^2\ \text{C}^{-2} \end{aligned} \tag{14.2}$$

where (2.21) was used.

To avoid the notion of action at a distance, the concept of electric field is introduced. An electric charge Q_1 is said to produce an electric field in the space around itself, and this field exerts a force on any charge Q_2 that is present in the space around Q_1. The *electric field strength* **E** at a point P in space is defined as the electrical force per unit charge experienced by a test charge Q_t at rest at point P:

$$\mathbf{E} \equiv \mathbf{F}/Q_t \qquad \text{where } Q_t \text{ is part of syst.} \tag{14.3*}$$

Equation (14.3) associates a vector **E** with each point in space. (This definition of **E** is satisfactory and usable from a physicist's viewpoint but is perhaps inadequate from a philosopher's viewpoint, since it doesn't say what an electric field really "is.")

Equation (14.3) defines the electric field that exists at P when the charge Q_t is present in the system. However, the presence of Q_t may influence the surrounding charges, thereby making **E** depend on the nature of the test charge. For example, if Q_t is placed in or near a material body, it may alter the distribution of charges in the body. (See the discussion of polarization in Sec. 14.2.) Therefore, if the charge Q_t in (14.3) is not part of the system under discussion and we want to know what **E** is at a given point in the system (in the absence of Q_t), we rewrite (14.3) as

$$\mathbf{E} \equiv \lim_{Q_t \to 0} \mathbf{F}/Q_t \qquad \text{where } Q_t \text{ is not part of syst.} \tag{14.4}$$

An infinitesimally small test charge does not disturb the charge distribution in the system, so (14.4) gives the value of **E** in the absence of the test charge. (A practical difficulty is that it is experimentally impossible to let Q_t go to zero, since charges smaller than the proton charge do not exist. However, we can let Q_t go to zero in calculations.)

If Q_t is part of the system, then (14.3) gives the field at point P, including contributions to the field that arise from the effects of Q_t on the other charges in the system.

Let us find **E** in the space around a point charge Q if no other charges are present in the system. Let a tiny test charge dQ_t be put at a distance r from Q. Then the magnitude of **E** is given by (14.4) as $E = dF/dQ_t$, where the magnitude of the force on dQ_t is given by (14.1) as $dF = Q\,dQ_t/4\pi\varepsilon_0 r^2$. Hence the magnitude of **E** is

$$E = \frac{1}{4\pi\varepsilon_0}\frac{Q}{r^2} \tag{14.5}$$

The direction of **E** equals the direction of **F** on a positive test charge, so **E** at point P is in the direction of the line from the charge Q to point P. The vector **E** points outward if Q is positive; inward if Q is negative. Figure 14.1 shows **E** at several points around a positive charge. The arrows farther away from Q are shorter, since **E** falls off as $1/r^2$.

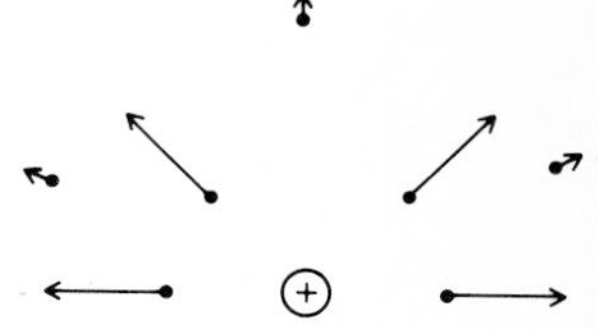

Figure. 14.1
The electric field vector at several points in the space around a positive charge.

Instead of describing things in terms of the electric field, it is often more convenient to use the electric potential ϕ. The *electric potential difference* $\phi_b - \phi_a$ between points b and a in an electric field is defined by

$$\phi_b - \phi_a \equiv \lim_{Q_t \to 0} w_{a\to b}/Q_t \equiv dw_{a\to b}/dQ_t \tag{14.6}$$

where $dw_{a\to b}$ is the reversible electrical work done by an external agent that moves an infinitesimal test charge dQ_t from a to b. (The word "reversible" indicates that the force exerted by the agent differs only infinitesimally from the force exerted by the system's electric field on dQ_t.) By assigning a value to the electric potential ϕ_a at point a we have then defined the *electric potential* ϕ_b at any point b. The most common convention is to choose point a at infinity (where the test charge interacts with no other charges) and to define ϕ at infinity as zero. Equation (14.6) then becomes

$$\phi_b \equiv \lim w_{\infty\to b}/Q_t \tag{14.7}$$

The SI unit of electric potential is the *volt* (V), defined as one joule per coulomb:

$$1 \text{ V} \equiv 1 \text{ J/C} = 1 \text{ N m C}^{-1} = 1 \text{ kg m}^2 \text{ s}^{-2} \text{ C}^{-1} \tag{14.8}$$

where (2.30) and (2.21) were used.

The SI unit of E in (14.4) is the newton per coulomb. Use of (14.8) gives $1 \text{ N/C} = 1 \text{ N V J}^{-1} = 1 \text{ N V N}^{-1}\text{m}^{-1} = 1 \text{ V/m}$, and E is usually expressed in volts per meter or volts per centimeter:

$$1 \text{ N/C} = 1 \text{ V/m} = 10^{-2} \text{ V/cm} \tag{14.9}$$

When we do reversible work $w_{\infty\to b}$ in moving the charge from infinity to b, we change its potential energy V by $w_{\infty\to b}$; thus, $\Delta V = V_b - V_\infty = V_b = w_{\infty\to b}$, where V_∞ has been taken as zero. Use of (14.7) gives

$$V_b = \phi_b Q_t \tag{14.10}$$

The electric field **E** is the force per unit charge. The electric potential ϕ is the potential energy per unit charge [Eqs. (14.3) and (14.10)].

Equation (2.39) gives $F_x = -\partial V/\partial x$. Division by Q_t gives $F_x/Q_t = -\partial(V/Q_t)/\partial x$. Use of (14.10) and the x component of (14.3) transforms this equation into $E_x = -\partial\phi/\partial x$. The same arguments hold for the y and z coordinates, so

$$E_x = -\partial\phi/\partial x, \qquad E_y = -\partial\phi/\partial y, \qquad E_z = -\partial\phi/\partial z \tag{14.11}$$

The electric field **E** at a point in space can be found if the electric potential ϕ is known as a function of x, y, and z. Conversely, $\phi(x, y, z)$ can be found from **E** by integration of (14.11); the integration constant is determined by setting $\phi = 0$ at some convenient location (usually infinity).

Let us find ϕ at a point P in the space around a point charge Q. Let the coordinate origin be at Q, and let the x axis run along the line from Q to P. The electric field is then in the x direction: $E = E_x = Q/4\pi\varepsilon_0 x^2$, $E_y = 0$, $E_z = 0$, where (14.5) was used. Substitution in the first equation of (14.11) and integration gives $\phi = -\int E_x\, dx = -\int (Q/4\pi\varepsilon_0 x^2)\, dx = Q/4\pi\varepsilon_0 x + c = Q/4\pi\varepsilon_0 r + c$, where r is the distance between charge Q and point P and c is an integration constant. In general, c could be a function of y and z, but the fact that $E_y = 0 = E_z$ together with

Eq. (14.11) requires that c be independent of y and z. Defining ϕ as zero at $r = \infty$, we get $c = 0$. Hence the potential due to a point charge Q is

$$\phi = \frac{1}{4\pi\varepsilon_0}\frac{Q}{r} \tag{14.12}$$

It is possible to show that (14.12) and (14.5) also hold for the field and electric potential outside a spherically symmetric charge distribution whose total charge is Q; here, r is the distance to the center of the charge distribution. (See *Halliday and Resnick*, sec. 28-6 for the proof.)

Experiment shows that the electric field of a system of charges equals the vector sum of the electric fields due to the individual charges; the electric potential equals the sum of the electric potentials due to the individual charges.

In discussing electric fields and electric potentials at a "point" in matter, we generally mean the average field and the average potential in a volume containing far, far fewer than 10^{23} molecules but far, far more than 1 molecule. (The electric field within a single molecule shows very sharp variations.)

Consider a single phase that is an electrical conductor (e.g., a metal, an electrolyte solution) and is in thermodynamic equilibrium. Since the phase is in equilibrium, no currents are flowing. (Flow of a noninfinitesimal current is an irreversible process, due to the heat generated by the current.) It follows that the electric field at all points in the interior of the phase must be zero. Otherwise, the charges of the phase would experience electrical forces, and a net current would flow. Since $\mathbf{E}$ is zero, Eq. (14.11) shows that ϕ is constant in the bulk phase of a conductor where no currents are flowing. If this phase has a net electrical charge, at equilibrium this charge will be distributed over the surface of the phase; this is so because the repulsion of like charges will cause them to move to the surface, where they are as far apart as possible.

14.2 DIPOLE MOMENTS AND POLARIZATION

An arrangement of charge occurring frequently is two charges Q and $-Q$ equal in magnitude and opposite in sign and separated by a distance d that is small compared with the distances from the charges to an observer. Such a combination is called an *electric dipole* (Fig. 14.2*a*). The *electric dipole moment* $\mathbf{p}$ of this combination is defined as a vector of magnitude

$$p \equiv Qd \tag{14.13}^*$$

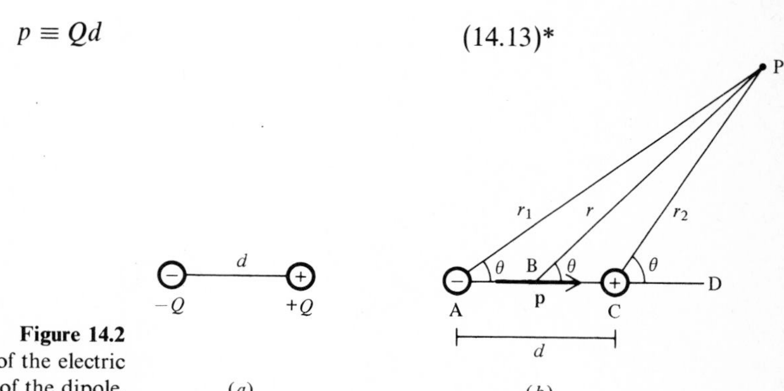

Figure 14.2
(*a*) An electric dipole. (*b*) Calculation of the electric potential of the dipole.

and direction from the negative charge to the positive charge. (Chemists often draw the direction incorrectly.)

Let us calculate the electric potential produced by a dipole. From Eq. (14.12) and Fig. 14.2*b*, we have

$$4\pi\varepsilon_0\,\phi = \frac{Q}{r_2} - \frac{Q}{r_1} = Q\frac{r_1 - r_2}{r_1 r_2} = Q\frac{r_1 - r_2}{r_1 r_2}\frac{r_1 + r_2}{r_1 + r_2} = Q\frac{(r_1{}^2 - r_2{}^2)}{r_1 r_2(r_1 + r_2)}$$

In accord with the above definition of an electric dipole, we assume that $r \gg d$, where r is the distance from the dipole to the point at which ϕ is being calculated. For $r \gg d$, it is clear from the figure that $r_1 \approx r_2 \approx r$. Hence, $r_1 + r_2 \approx 2r$ and $r_1 r_2(r_1 + r_2) \approx 2r^3$. Also, the angles PAB, PBC, and PCD are nearly equal to one another and have all been labeled θ. The law of cosines (see any high school geometry text) for the triangle PAC gives $r_2{}^2 = r_1{}^2 + d^2 - 2r_1 d \cos\theta$. Hence, $r_1{}^2 - r_2{}^2 = d(2r_1\cos\theta - d) \approx 2r_1 d\cos\theta \approx 2rd\cos\theta$, since $r_1 \gg d$. With these approximations, the above expression for ϕ becomes $4\pi\varepsilon_0\,\phi = Q(2rd\cos\theta/2r^3)$; use of (14.13) gives

$$\phi = \frac{1}{4\pi\varepsilon_0}\frac{p\cos\theta}{r^2} \qquad \text{for } r \gg d \tag{14.14}$$

Note the angular dependence of ϕ. Note also that ϕ for a dipole falls off as $1/r^2$, in contrast to ϕ for a single charge (a monopole), which falls off as $1/r$ [Eq. (14.12)]. Since the electric field is found by differentiation of ϕ [Eq. (14.11)], the electric field of a dipole falls off as $1/r^3$, compared with $1/r^2$ for a single charge. The force field of a dipole is of relatively short range compared with the long-range force field of a single charge.

Consider a distribution of several electrical charges Q_i, with the total charge being zero: $\sum_i Q_i = 0$. If one calculates the distribution's electric potential at any point whose distance from the distribution is much larger than the distance between any two charges of the distribution, one finds that the potential is given by Eq. (14.14) provided that the *electric dipole moment* of the charge distribution is defined as

$$\mathbf{p} \equiv \sum_i Q_i \mathbf{r}_i \tag{14.15}$$

where $\mathbf{r}_i$ is the vector from the origin (chosen arbitrarily) to charge Q_i. (Proof of this statement is given in E. R. Peck, *Electricity and Magnetism*, McGraw-Hill, 1953, p. 29.) Thus, the electric potential produced by a neutral molecule at a point well outside the molecule is given by (14.14) and (14.15); of course, the charges of the molecule are moving, so $\mathbf{r}_i$ must be interpreted as an average location for charge Q_i.

For a distribution that consists of a charge $-Q$ with (x, y, z) coordinates $(-\frac{1}{2}d, 0, 0)$ and a charge Q with coordinates $(\frac{1}{2}d, 0, 0)$, Eq. (14.15) gives $p_x = \sum_i Q_i x_i = -Q(-\frac{1}{2}d) + Q(\frac{1}{2}d) = Qd$ and $p_y = 0 = p_z$, in agreement with (14.13).

The sum in (14.15) can be written as a sum over the negative charges plus a sum over the positive charges. A molecule has a nonzero dipole moment if the effective centers of negative charge and positive charge do not coincide. Some molecules with $p \neq 0$ are HCl, H_2O (which is nonlinear), and CH_3Cl. Some molecules with $p = 0$ are H_2, CO_2 (which is linear), CH_4, and C_6H_6. A molecule is said to be *polar* or *nonpolar* according to whether $p \neq 0$ or $p = 0$, respectively.

When a molecule with a zero dipole moment is placed in an external electric field, the field shifts the centers of positive and negative charge, *polarizing* the molecule and giving it an *induced dipole moment* $\mathbf{p}_{ind}$. For example, if one or more positive charges are placed above the plane of a benzene molecule, the average positions of the electrons will shift upward, giving the molecule a dipole moment whose direction is perpendicular to the molecular plane. If the molecule has a nonzero dipole moment $\mathbf{p}$ (in the absence of any external field), the induced moment $\mathbf{p}_{ind}$ produced by the field will add to the permanent dipole moment $\mathbf{p}$. The induced dipole moment is proportional to the electric field $\mathbf{E}$ experienced by the molecule: $\mathbf{p}_{ind} = \alpha\mathbf{E}$, where the proportionality constant α is called the molecule's (*electric*) *polarizability*. (Actually, α is a function of direction in the molecule; e.g., the polarizability of HCl along the bond differs from its polarizability in a direction perpendicular to the bond axis. For liquids and gases, where the molecules are rotating rapidly, we use an α averaged over direction.)

The molecular polarizability increases with increasing number of electrons. From (14.15) and (14.3), the SI units of α are $(\text{C m})/(\text{N C}^{-1}) = \text{C}^2\ \text{m N}^{-1}$. From (14.2), the units of $\alpha/4\pi\varepsilon_0$ are m^3, which is a unit of volume. Polarizabilities are often tabulated as $\alpha/4\pi\varepsilon_0$ values. Some values are:

Molecule	He	H_2	H_2O	Ar	HCl	HI	C_6H_6
$\alpha(4\pi\varepsilon_0)^{-1}/\text{Å}^3$	0.20	0.80	1.4	1.6	2.6	5.1	10

A small volume in a piece of matter is said to be electrically *unpolarized* or *polarized* according to whether the net dipole moment of the volume is zero or nonzero, respectively. A small volume within the bulk phase of pure liquid water is unpolarized; the molecular dipoles are oriented randomly, so their vector sum is zero. In an electrolyte solution, the water in the immediate vicinity of each ion is polarized, due to the orientation of the H_2O dipoles and to the induced dipole moments.

Consider two flat, parallel metal plates with opposite charges that are equal in magnitude. This is a *capacitor* or *condenser* (Fig. 14.3). The electric field E in the region between the plates is constant. (See *Halliday and Resnick*, sec. 30-2, for the proof.) Let x be the direction perpendicular to the plates. Integration of $E_x = -\partial\phi/\partial x$ with $E_x = E = \text{const.}$ gives for the potential difference $\Delta\phi$ between the plates

$$|\Delta\phi| = Ed \tag{14.16}$$

where d is the distance between the plates.

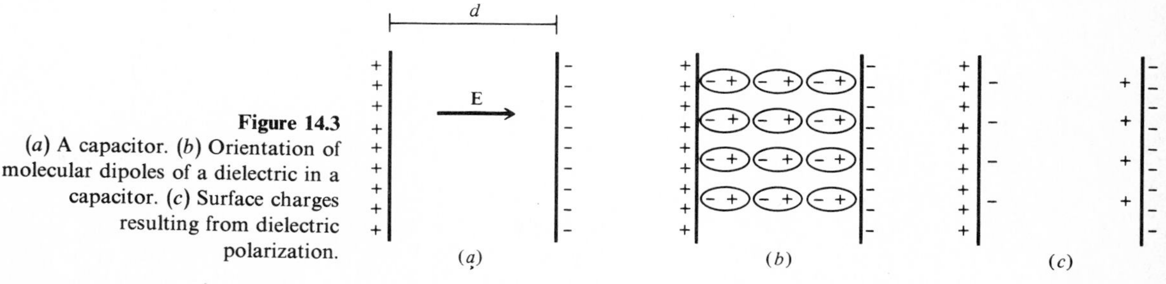

Figure 14.3
(*a*) A capacitor. (*b*) Orientation of molecular dipoles of a dielectric in a capacitor. (*c*) Surface charges resulting from dielectric polarization.

When a nonconducting substance (a *dielectric*) is placed between the plates, the nonconductor becomes polarized, due to the following two effects: (1) The electric field of the plates tends to orient the permanent dipoles of the dielectric so that the negative ends of the moments lie toward the positive plate. The degree of orientation shown in Fig. 14.3*b* is greatly exaggerated; the orientation is far from complete, since it is opposed by the random thermal motion of the molecules. (2) The electric field of the plates produces induced dipole moments $\mathbf{p}_{\text{ind}}$; these induced moments are all oriented with their negative ends toward the positive plate. For a dielectric whose molecules have zero permanent dipole moment, the *orientation polarization* (effect 1) is absent. The *induced* (or *distortion*) *polarization* (effect 2) is always present. For any small volume in the bulk phase of the polarized dielectric, the net charge is zero. However, due to the polarization, there is a negative charge at the dielectric surface in contact with the positive plate and a positive charge on the opposite surface of the dielectric (Fig. 14.3*c*). These surface charges partially nullify the effect of the charges on the metal plates, thereby reducing the electric field in the region between the plates and reducing the potential difference between the plates.

The *dielectric constant* (or *relative permittivity*) ε_r of a dielectric is defined by $\varepsilon_r \equiv E_0/E$, where E_0 and E are the electric fields in the space between the plates of a condenser when the plates are separated by a vacuum and by the dielectric, respectively. Equation (14.16) gives $E_0/E = \Delta\phi_0/\Delta\phi$, where $\Delta\phi_0$ and $\Delta\phi$ are the potential differences between the plates in the absence and in the presence of the dielectric. Thus

$$\varepsilon_r \equiv E_0/E = \Delta\phi_0/\Delta\phi \tag{14.17}$$

Let Q be the absolute value of the charge on one of the metal plates, and let $\mathscr{A}$ be its area. In the absence of a dielectric, the electric field between the plates is $E_0 = Q/\varepsilon_0 \mathscr{A}$ [eq. (30-5) in *Halliday and Resnick*]. With a dielectric between the plates, let Q_P be the absolute value of the charge on one surface of the polarized dielectric. Q_P neutralizes part of the charge on each plate, so the field is now $E = (Q - Q_P)/\varepsilon_0 \mathscr{A}$. Hence, $E_0/E = Q/(Q - Q_P) = \varepsilon_r$, where (14.17) was used. Thus

$$Q - Q_P = Q/\varepsilon_r \tag{14.18}$$

The deviation of ε_r from 1 is due to two effects: the induced polarization and the orientation of the permanent dipole moments. Therefore, ε_r is related to the molecular polarizability α and the molecular electric dipole moment p. Using the Boltzmann distribution law (Chap. 22) to describe the orientations of the dipoles in the applied electric field, one can show that for pure gases (polar or nonpolar) at low or moderate pressure and for nonpolar liquids or solids

$$\frac{\varepsilon_r - 1}{\varepsilon_r + 2}\frac{M}{\rho} = \frac{N_0}{3\varepsilon_0}\left(\alpha + \frac{p^2}{3kT}\right) \tag{14.19}$$

(For the derivation, see *McQuarrie*, chap. 13.) In (14.19), M is the molar mass (not the molecular weight), ρ is the density, k is Boltzmann's constant (3.57), N_0 is the Avogadro constant, and T is the temperature. Equation (14.19) is the *Debye–Langevin equation*; for the special case $p = 0$, it reduces to the Clausius–Mossotti equation.

The dielectric constant ε_r can be measured using Eq. (14.17). From (14.19), a plot of $M\rho^{-1}(\varepsilon_r - 1)/(\varepsilon_r + 2)$ vs. $1/T$ is a straight line with slope $N_0 p^2/9\varepsilon_0 k$ and intercept $N_0 \alpha/3\varepsilon_0$. Such a plot allows α and p of the gas to be determined.

The Debye equation is sometimes used to find the dipole moment of a polar species dissolved in a nonpolar solvent. Dipole moments obtained by this procedure are rather inaccurate.

Some dielectric constants for liquids at 25°C and 1 atm are:

CCl_4	C_6H_6	CH_3COOH	C_2H_5OH	CH_3OH	H_2O	HCN
2.2	2.3	6.2	24.3	32.6	78.4	107

The more polar the liquid, the larger its dielectric constant. The dielectric constant of a single crystal depends on its orientation in the capacitor. Values of ε_r for solids are usually given for a mixture of small crystals with random orientations. Some ε_r values for solids at 25°C and 1 atm are:

PbS	sucrose	NaCl	glasses	porcelain	wood
18	3	6	4–10	6–9	3–8

For gases, ε_r is quite close to 1, and $\varepsilon_r + 2$ can be taken as 3 in (14.19). Since ρ is proportional to P, Eq. (14.19) shows that $\varepsilon_r - 1$ increases essentially linearly with P at constant T. Some ε_r values at 20°C and 1 atm are 1.00054 for air, 1.00092 for CO_2, 1.0031 for HBr, and 1.0025 for *n*-pentane.

For polar liquids, ε_r decreases as T increases. At higher temperatures the random thermal motion decreases the orientation polarization. (For water, ε_r decreases from 80 to 64 as T increases from 20 to 70°C.) For nonpolar liquids, the orientation polarization is absent, so ε_r varies only slightly with T; this variation is due to the change of ρ with T in Eq. (14.19).

Consider two electric charges Q_1 and Q_2 immersed in a dielectric fluid with dielectric constant ε_r, and let the charges be separated by at least several molecules of dielectric. The charge Q_1 polarizes the dielectric in its immediate neighborhood. Let Q_1 be positive. The negative charges of the oriented dipoles immediately adjacent to Q_1 partially "neutralize" Q_1, giving it an effective charge $Q_{\text{eff}} = Q_1 - Q_P$, where $-Q_P$ is the charge on the spherical "surface" of the dielectric immediately surrounding Q_1. It turns out that Eq. (14.18) gives the correct result for the effective charge: $Q_{\text{eff}} = Q_1 - Q_P = Q_1/\varepsilon_r$. (See F. W. Sears, *Electricity and Magnetism*, Addison-Wesley, 1951, p. 190.) At points not too close to Q_1, the electric field E due to Q_1 and the induced charges surrounding it equals $Q_{\text{eff}}/4\pi\varepsilon_0 r^2$. Hence

$$E = \frac{1}{4\pi\varepsilon_0 \varepsilon_r} \frac{Q_1}{r^2} = \frac{1}{4\pi\varepsilon} \frac{Q_1}{r^2} \tag{14.20}$$

where the *permittivity* ε of the medium is defined as $\varepsilon \equiv \varepsilon_r \varepsilon_0$.

Now consider the force on Q_2. This force is due to (*a*) the charge Q_1, (*b*) the induced charge $-Q_P$ around Q_1, and (*c*) the induced charge around Q_2. The induced charge around Q_2 is spherically distributed around Q_2 and produces no net force on

Q_2. Hence, the force F on Q_2 is found from the field (14.20), which results from charges (a) and (b). Equations (14.3) and (14.20) give

$$F = \frac{1}{4\pi\varepsilon_0 \varepsilon_r} \frac{Q_1 Q_2}{r^2} \tag{14.21}$$

The force on Q_2 is reduced by the factor $1/\varepsilon_r$ compared with the force in vacuum. Note that the force between Q_1 and Q_2 is given by Coulomb's law (14.1) whether these charges are in vacuum or are in a dielectric medium. It is the induced charge $-Q_P$ in the neighborhood of Q_1 that reduces the force on Q_2.

Since intermolecular forces are electrical, the dielectric constant ε_r of a solvent influences equilibrium constants and reaction rate constants. Recall the much higher degree of ion pairing of electrolyte solutes in solvents with low ε_r values.

14.3 ELECTROCHEMICAL SYSTEMS

Each phase of the thermodynamic systems considered in previous chapters has been electrically neutral, and there have been no differences in electric potential between phases. However, when a system contains charged species and at least one charged species cannot penetrate all the phases of the system, some of the phases can become electrically charged.

For example, suppose a membrane permeable to K^+ ions but not to Cl^- ions separates an aqueous KCl solution from pure water. Diffusion of K^+ ions through the membrane will produce net charges on each phase and a potential difference between the phases (see Sec. 14.14). Another example is a piece of Zn dipping into an aqueous $ZnCl_2$ solution. Both the solution and the metal contain Zn^{2+} ions, and these can be transferred between the phases. The metal also contains electrons, and these are not free to enter the solution. Suppose the $ZnCl_2$ solution is extremely dilute. Then the initial rate at which Zn^{2+} ions leave the metal and enter the solution is greater than the rate at which Zn^{2+} ions enter the metal from the solution. This net loss of Zn^{2+} from the metal produces a negative charge (excess of electrons) on the Zn. The negative charge retards the rate of the process $Zn^{2+}(\text{metal}) \rightarrow Zn^{2+}(\text{soln.})$ and increases the rate of $Zn^{2+}(\text{soln.}) \rightarrow Zn^{2+}(\text{metal})$. Eventually an equilibrium is reached in which the rates of these opposing processes are equal. At equilibrium, the Zn has a net negative charge, and a potential difference exists between the Zn and the solution. Therefore, at any metal–solution interface at equilibrium, a potential difference $\Delta\phi$ exists. The magnitude and sign of $\Delta\phi$ depend on T, P, the nature of the metal, the nature of the solvent, and the concentrations of metal ions in solution.

Another example of an interphase potential difference is a piece of Cu in contact with a piece of Zn. Diffusion in solids is extremely slow at room temperature, so the Cu^{2+} and Zn^{2+} ions do not move between the phases to any significant extent. However, electrons are free to move from one metal to the other, and they do so, resulting at equilibrium in a net negative charge on the Cu and a net positive charge on the Zn. This charge can be detected by separating the metals and touching each of them to the terminal of an electroscope. (In an electroscope, two pieces of metal foil attached to the same terminal repel each other when they become charged.) The development of charge by two dissimilar metals in contact was first shown in experiments by Galvani and Volta in the 1790s. In one experiment, Galvani discharged this charge through the nerve of a dead frog's leg muscles, causing the muscles to contract.

The transfer of charge between two phases α and β produces a difference in electric potential between the phases at equilibrium: $\phi^\alpha \neq \phi^\beta$, where ϕ^α and ϕ^β are the potentials in the bulk phases α and β. (The electric potential in the bulk of a phase is sometimes called the *inner potential* or the *Galvani potential.*) We define an *electrochemical system* as a (heterogeneous) system in which there is a difference of electric potential between two or more phases.

Besides interphase charge transfer, there are other effects that contribute to $\phi^\alpha - \phi^\beta$. Consider again a piece of Zn dipping into a very dilute $ZnCl_2$ solution. The Zn loses Zn^{2+} ions to the solution, becoming negatively charged on its surface. Water is a polar molecule, and water molecules in the immediate vicinity of the Zn will tend to be oriented with their hydrogen atoms toward the Zn. The polarization of the water adjacent to the Zn affects $\phi^\alpha - \phi^\beta$. Also, the Zn^{2+} ions of the solution will be more strongly adsorbed in the interphase region (Sec. 13.1) than the Cl^- ions (since the Zn metal is negatively charged). This also affects $\phi^\alpha - \phi^\beta$.

It is possible for a difference in potential between phases to occur without charge transfer between the phases. Consider an aqueous solution of NaCl in equilibrium with liquid benzene. For simplicity, let us assume zero solubility of H_2O, Na^+, and Cl^- in benzene. The forces between benzene molecules and Na^+ ions near the benzene–aqueous solution interface differ from the forces between benzene and Cl^-. Hence, the extents of adsorption of Na^+ and of Cl^- in the interphase region will differ. Furthermore, the more strongly adsorbed ions will polarize benzene molecules near the interface. These effects produce a nonzero interphase potential difference. Another example is a two-phase system of liquid benzene plus liquid water. There will be a preferred orientation of the water molecules at the interface, due to the differing interactions between C_6H_6 molecules and the negative and positive ends of the H_2O dipoles. This makes $\phi^\alpha - \phi^\beta$ nonzero.

Thus, differences in potential between phases are the rule, rather than the exception, and most heterogeneous systems are electrochemical systems.

Our main interest will lie in electrochemical systems whose phases are electrical conductors. Such phases include metals, semiconductors, molten salts, and liquid solutions containing ions.

A significant point is that the potential difference $\Delta\phi$ between two phases in contact cannot be readily measured. Suppose, for example, we wanted to measure $\Delta\phi$ between a piece of Zn and an aqueous $ZnCl_2$ solution. If we make electrical contact with these phases with two wires of a voltmeter or potentiometer (Sec. 14.5), we create at least one new interface in the system, that between the voltmeter wire and the $ZnCl_2$ solution; hence, the potential difference measured by the meter includes the potential difference between the meter wire and the solution, and we have not measured what we set out to measure. The kind of potential difference that is readily measured is the potential difference between two phases having the same chemical composition. Attachment of voltmeter wires to these phases creates potential differences between the wires and the phases, but these potential differences are equal in magnitude and cancel each other (assuming the two wires are made of the same metal). [Oppenheim proposed an experiment to measure $\Delta\phi$ between a metal and a solution in equilibrium, but his method is not a practical one; see I. Oppenheim, *J. Phys. Chem.*, **68**, 2959 (1964).]

Although $\Delta\phi$ between phases in contact cannot readily be measured, it can be calculated from a statistical-mechanical model of the system. $\Delta\phi$ can be calculated if the distribution of charges and dipoles in the interphase region is known.

14.4 THERMODYNAMICS OF ELECTROCHEMICAL SYSTEMS

We now develop the thermodynamics of electrochemical systems composed of phases that are electrical conductors. The treatment applies only to systems in which there is at most an infinitesimal flow of current (since classical thermodynamics is inapplicable to irreversible processes).

In an electrochemical system, the electric potential ϕ differs between phases: $\phi^\alpha \neq \phi^\beta$ for phases α and β. As we go from phase α to phase β, the electric potential does not suddenly jump from ϕ^α to ϕ^β. Instead, in the interphase region (Sec. 13.1), there is a gradual transition from ϕ^α to ϕ^β (see Fig. 14.13). If z is the direction perpendicular to the α-β interface, the derivative $\partial\phi/\partial z$ is nonzero in the interphase region, so the electric field is nonzero in the interphase region [Eq. (14.11)]. An ion or electron with charge Q that moves from the bulk phase α to the bulk phase β experiences an electric force in the interphase region and has its electrical energy changed by $(\phi^\beta - \phi^\alpha)Q$ [Eq. (14.10)]. The energy of a charged particle therefore depends on the electrical state of the phase it is in, where the *electrical state* is defined by the value of ϕ in the phase.

The Gibbs equation (4.85) [and (13.27)] for a bulk phase α in thermal and mechanical equilibrium reads $dU^\alpha = T\,dS^\alpha - P\,dV^\alpha + \sum_i \mu_i^\alpha\,dn_i^\alpha$, where $\mu_i^\alpha \equiv (\partial U^\alpha/\partial n_i^\alpha)_{S^\alpha,V^\alpha,n^\alpha_{j\neq i}}$. This equation is applicable to a phase of an electrochemical system. However, the energies of the species that are electrically charged will depend on the electrical state of the phase; hence, the internal energy of the phase will depend on the electrical state as well as the chemical state of the phase. To emphasize this dependence of U^α on ϕ^α, we add a tilde to U^α, writing it as $\tilde{U}^\alpha$. The derivative $(\partial U^\alpha/\partial n_i^\alpha)$ will also depend on the electrical state of the phase if species i is charged, so we add a tilde to the symbol for this derivative, writing it as $\tilde{\mu}_i^\alpha$. (For uncharged species, $\partial U^\alpha/\partial n_i^\alpha$ is independent of the electrical state of α, but for notational convenience we add the tilde for all species, charged and uncharged.) Therefore

$$d\tilde{U}^\alpha = T\,dS^\alpha - P\,dV^\alpha + \sum_i \tilde{\mu}_i^\alpha\,dn_i^\alpha \qquad (14.22)^*$$

$$\tilde{\mu}_i^\alpha \equiv (\partial\tilde{U}^\alpha/\partial n_i^\alpha)_{S^\alpha,V^\alpha,n^\alpha_{j\neq i}} \qquad (14.23)$$

Rewriting (4.88) for phase α with tildes over G and μ_i, we have

$$d\tilde{G}^\alpha = -S^\alpha\,dT + V^\alpha\,dP + \sum_i \tilde{\mu}_i^\alpha\,dn_i^\alpha \qquad (14.24)^*$$

We call $\tilde{\mu}_i^\alpha$ the *electrochemical potential* of species i in phase α.

It is convenient to decompose the total internal energy $\tilde{U}^\alpha$ into an electrical part U^α_{el} and a nonelectrical, or chemical, part U^α_{chem}. We write $\tilde{U}^\alpha = U^\alpha_{\text{el}} + U^\alpha_{\text{chem}}$ and $d\tilde{U}^\alpha = dU^\alpha_{\text{el}} + dU^\alpha_{\text{chem}}$. To see how U^α_{el} and U^α_{chem} are to be defined, consider an open electrochemical system and let the amount of charge in phase α change by dQ^α, due to transport of electrons or ions into the phase. Equation (14.10) shows that the electrical potential energy of phase α will change by $\phi^\alpha\, dQ^\alpha$, where ϕ^α is the electric potential of phase α. Thus

$$d\tilde{U}^\alpha = dU^\alpha_{\text{chem}} + dU^\alpha_{\text{el}} = dU^\alpha_{\text{chem}} + \phi^\alpha\, dQ^\alpha \tag{14.25}$$

Combining (14.22) and (14.25), we get

$$d\tilde{U}^\alpha = dU^\alpha_{\text{chem}} + \phi^\alpha\, dQ^\alpha = T\, dS^\alpha - P\, dV^\alpha + \sum_i \tilde{\mu}_i^\alpha\, dn_i^\alpha \tag{14.26}$$

Let q_i be the charge on one particle of species i. For example, $q(H_2O) = 0$, $q(Na^+) = 1.6 \times 10^{-19}$ C, $q(e^-) = -1.6 \times 10^{-19}$ C, $q(SO_4^{2-}) = -3.2 \times 10^{-19}$ C, where e^- is an electron. The *charge number* z_i of species i is defined as $z_i \equiv q_i/e$, where $e = 1.6 \times 10^{-19}$ C is the proton charge. For example, $z(Na^+) = 1$, $z(PO_4^{3-}) = -3$, $z(e^-) = -1$. Let there be n_i^α moles of species i in phase α. The total charge on all particles of i in phase α is $N_0 n_i^\alpha q_i = N_0 n_i^\alpha z_i e = n_i^\alpha z_i \mathscr{F}$, where the *Faraday constant* $\mathscr{F}$ is defined by $\mathscr{F} \equiv N_0 e$. The Faraday constant is the charge per mole of protons. Use of $N_0 = 6.0220 \times 10^{23}$ mole^{-1} and $e = 1.60219 \times 10^{-19}$ C gives

$$\mathscr{F} \equiv N_0 e = 96{,}485 \text{ C mole}^{-1} \tag{14.27*}$$

The total charge of phase α is the sum of the charges of each species: $Q^\alpha = \sum_i n_i^\alpha z_i \mathscr{F}$, where $z_i\mathscr{F}$ is the molar charge on species i. We have $dQ^\alpha = \sum_i z_i \mathscr{F}\, dn_i^\alpha$. Equation (14.26) becomes

$$dU^\alpha_{\text{chem}} = T\, dS^\alpha - P\, dV^\alpha + \sum_i (\tilde{\mu}_i^\alpha - z_i \mathscr{F} \phi^\alpha)\, dn_i^\alpha \tag{14.28}$$

We define μ_i^α by

$$\mu_i^\alpha \equiv (\partial U^\alpha_{\text{chem}}/\partial n_i^\alpha)_{S^\alpha, V^\alpha, n^\alpha_{j\neq i}} = \tilde{\mu}_i^\alpha - z_i \mathscr{F} \phi^\alpha$$

where (14.28) was used. We have

$$\tilde{\mu}_i^\alpha = \mu_i^\alpha + z_i \mathscr{F} \phi^\alpha \tag{14.29*}$$

The quantity μ_i^α is the *chemical part of the electrochemical potential*, a name which is commonly shortened to the *chemical potential* of i in phase α. Note from (14.29) that μ_i^α is the value of $\tilde{\mu}_i^\alpha$ when the electric potential ϕ^α of phase α is zero. For uncharged species ($z_i = 0$), we have $\tilde{\mu}_i = \mu_i$.

Consider the conditions for phase equilibrium and reaction equilibrium in an electrochemical system. Equation (14.24) is really the same equation as (4.88), except that to emphasize that the Gibbs energy and its partial derivative with respect to n_i^α depend on the phase's electrical state, we added tildes to G^α and μ_i^α. Therefore, the phase- and reaction-equilibrium conditions (4.100) and (4.111) apply to an electrochemical system provided we put tildes over the μ's.

For phase equilibrium between phases α and β that each contain component i and that are in physical contact, Eq. (4.100) gives

$$\tilde{\mu}_i^\alpha = \tilde{\mu}_i^\beta \tag{14.30*}$$

Use of (14.29) gives $\mu_i^\alpha + z_i \mathscr{F} \phi^\alpha = \mu_i^\beta + z_i \mathscr{F} \phi^\beta$, or

$$\mu_i^\alpha - \mu_i^\beta = z_i \mathscr{F} (\phi^\beta - \phi^\alpha) \tag{14.31}$$

If substance i is absent from phase γ but present in phase α, then $\tilde{\mu}_i^\gamma$ need not equal $\tilde{\mu}_i^\alpha$ [Eq. (4.103)]. If i is present in phases α and δ but these phases are separated by a phase in which i is absent, then $\tilde{\mu}_i^\alpha$ need not equal $\tilde{\mu}_i^\delta$. An example is two pieces of metal dipping in the same solution but not in direct contact; $\tilde{\mu}$ of the electrons in one metal need not equal $\tilde{\mu}$ of the electrons in the second metal.

The reaction-equilibrium condition for the heterogeneous chemical reaction $\sum_i \nu_i A_i = 0$ [Eq. (4.107)] in an electrochemical system is given by (4.111) as

$$\sum_i \nu_i \tilde{\mu}_i = 0 \tag{14.32}$$

where the ν_i's are the stoichiometric coefficients.

Each $\tilde{\mu}_i$ in (14.32) contains the electric potential of the phase in which species A_i occurs, and the electric potentials of different phases differ. Consider the special case where all the charged species that participate in the reaction occur in the same phase, phase α. Then (14.32) and (14.29) give

$$0 = \sum_i \nu_i(\mu_i + z_i \mathscr{F}\phi^\alpha) = \sum_i \nu_i \mu_i + \mathscr{F}\phi^\alpha \sum_i \nu_i z_i \tag{14.33}$$

The total charge must be conserved in a chemical reaction. Hence, $\sum_i \nu_i z_i = 0$. [For example, for $2Fe^{3+}(aq) + Zn(s) = Zn^{2+}(aq) + 2Fe^{2+}(aq)$, we have $\sum_i \nu_i z_i = -2(3) - 1(0) + 1(2) + 2(2) = 0$.] Therefore (14.33) becomes

$$\sum_i \nu_i \mu_i = 0 \qquad \text{all charged species in same phase} \tag{14.34}$$

The value of ϕ^α is thus irrelevant when all charged species occur in the same phase. This makes sense, since the reference level of electric potential is arbitrary and we can take $\phi^\alpha = 0$ if we like.

In Chaps. 10 and 11, we considered chemical potentials and reaction equilibrium for ions in electrolyte solutions. All charged species were present in the same phase, so there was no need to consider the electrochemical potentials.

In an electrochemical system, the phases generally have nonzero net charges. Consider an isolated spherical phase of radius 5 cm, and let the phase contain a net charge due to an excess of 10^{-10} mole of Cu^{2+} ions. (The phase might be a piece of copper or an aqueous $CuSO_4$ solution.) The electric potential outside the phase is given by (14.12) as $\phi = Q/4\pi\varepsilon_0 r$. We have $Q = (10^{-10}\text{ mole})(2\mathscr{F}) = 19 \times 10^{-6}$ C. For $r = 5$ cm, the potential is [Eqs. (14.2) and (14.8)]

$$\phi = (19 \times 10^{-6}\text{ C})(9 \times 10^9\text{ N m}^2\text{ C}^{-2})/0.05\text{ m} = 3 \times 10^6\text{ V}$$

An excess of only 10^{-10} mole (6×10^{-9} g) of Cu^{2+} produces a potential of three million volts. The potential differences between phases of an electrochemical system are typically of the order of a few volts or less. We conclude that the net charges on phases of electrochemical systems are due to transfer of amounts of matter far too small to be detected chemically.

The systems considered in previous chapters had $\phi = 0$ in each phase, and their internal energy was entirely chemical: $U^\alpha = U^\alpha_{\text{chem}}$ and $U^\alpha_{\text{el}} = 0$. When $\phi^\alpha \neq 0$, the internal energy also contains an electrical term: $\tilde{U}^\alpha = U^\alpha_{\text{chem}} + U^\alpha_{\text{el}}$. The changes in chemical composition that accompany the development of interphase potential differences are extremely slight. We therefore expect that U^α_{chem} will change by an entirely negligible amount when we go from a system with $\phi = 0$ in every phase to the corresponding system with potential differences between the phases. (For example, U^{α}_{chem} of 85 g of a 0.50 mol/kg aqueous $CuSO_4$ solution at 25°C and 1 atm can be taken equal to U^{α}_{chem} of 85 g of a 0.50 mol/kg aqueous $CuSO_4$ solution in electrochemical equilibrium with a piece of Cu at 25°C and 1 atm.) The chemical potential μ_i^{α} in an electrochemical system was defined by the equation following (14.28) as $(\partial U^{\alpha}_{\text{chem}}/\partial n_i^{\alpha})_{S^{\alpha},\ldots}$. But U^{α}_{chem} in the electrochemical system equals U^{α} for the corresponding chemical system (with all ϕ's $= 0$), and μ_i^{α} for a chemical system was defined as $(\partial U^{\alpha}/\partial n_i^{\alpha})_{S^{\alpha},\ldots}$. Hence, the chemical potential μ_i^{α} of species i in phase α of an electrochemical system is equal to μ_i^{α} in the corresponding chemical system, and the expressions we developed for μ_i^{α} in earlier chapters also apply to electrochemical systems.

Suppose that the terminals α and δ of a battery are made of copper and that the potential difference between the terminals (the "voltage") is 2 V. Strictly speaking, the chemical compositions of the terminals differ, since the charges on the terminals differ. However, the difference in chemical composition is so slight that we can ignore it and take the chemical part of the electrochemical potential of a species as being the same in each terminal: $\mu_i^{\alpha} = \mu_i^{\delta}$.

14.5 GALVANIC CELLS

Galvanic cells. If we attach a piece of wire to a device that produces an electric current in the wire, we can use the current to do useful work. For example, we might put the current-carrying wire in a magnetic field; this produces a force on the wire, giving us a motor. For a wire of resistance R carrying a current I, there is an electric potential difference $\Delta\phi$ between its ends, where $\Delta\phi$ is given by "Ohm's law" [Eq. (16.59)] as $|\Delta\phi| = IR$. This difference of potential corresponds to an electric field in the wire, which causes electrons to flow. To generate a current in the wire, we require a device that will maintain an electric potential difference between its output terminals. Any such device is called a *seat* (or *source*) *of electromotive force* (emf). Attaching each end of a wire to a terminal of a seat of emf produces a current I in the wire (Fig. 14.4).

The *electromotive force* (emf) $\mathscr{E}$ of a seat of emf is defined as the potential difference between its terminals when the resistance R of the load attached to the terminals goes to infinity (and hence the current goes to zero). The emf is thus the open-circuit potential difference between the terminals. (The difference of potential $\Delta\phi$ between the terminals in Fig. 14.4 will depend on the value of the current I that flows through the circuit, because the seat of emf has an internal resistance R_{int} and the potential drop IR_{int} reduces $\Delta\phi$ between the terminals below the open-circuit $\Delta\phi$.)

One kind of seat of emf is an electric generator. Here, a mechanical force moves a metal wire through a magnetic field. This field exerts a force on the electrons in the metal, producing an electric current and a potential difference between the ends of the wire. An electric generator converts mechanical energy into electrical energy.

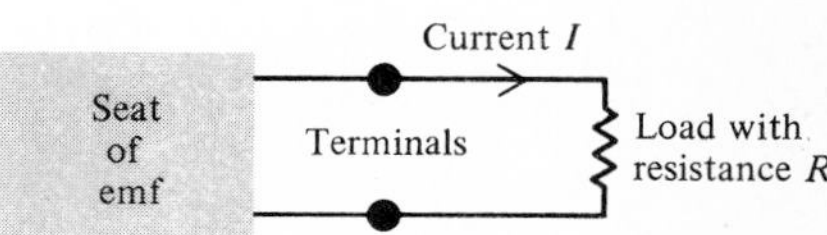

Figure 14.4
A seat of emf attached to a load.

Another kind of seat of emf is a *galvanic* (or *voltaic*) *cell.* This is a multiphase electrochemical system in which the interphase potential differences result in a net potential difference between the terminals. The potential differences between phases result from the transfer of chemical species between phases, and a galvanic cell converts chemical energy into electrical energy.

The phases of a galvanic cell must be electrical conductors; otherwise, a continuous current could not flow in Fig. 14.4.

Since only potential differences between chemically identical pieces of matter are readily measurable (Sec. 14.3), we specify that the two terminals of a galvanic cell are to be made of the same metal. Otherwise, we could not measure the cell emf, which is the open-circuit potential difference between the terminals.

The metal terminals of a galvanic cell are *electronic conductors*, meaning that the current is carried by electrons. Suppose all phases of the galvanic cell were electronic conductors. For example, the cell might be Cu′|Zn|Ag|Cu″, which is shorthand for a copper terminal Cu′ attached to a piece of Zn attached to a piece of Ag attached to a second copper terminal Cu″. Since electrons are free to move between all phases, the phase-equilibrium condition (14.30) shows that $\tilde{\mu}(e^-)$ (the electrochemical potential of electrons) is the same in all phases of the open-circuit cell. In particular, $\tilde{\mu}(e^- \text{ in Cu}') = \tilde{\mu}(e^- \text{ in Cu}'')$. Use of (14.29) gives $\mu(e^- \text{ in Cu}') - \mathscr{F}\phi(\text{Cu}') = \mu(e^- \text{ in Cu}'') - \mathscr{F}\phi(\text{Cu}'')$. Since the phases Cu′ and Cu″ have the same chemical composition, it follows (recall the last paragraph of Sec. 14.4) that $\mu(e^- \text{ in Cu}') = \mu(e^- \text{ in Cu}'')$. Therefore, $\phi(\text{Cu}') = \phi(\text{Cu}'')$. The terminals have the same open-circuit electric potential, and the cell emf is zero. We conclude that a galvanic cell must have at least one phase that is impermeable to electrons. This allows $\tilde{\mu}(e^-)$ to differ in the two terminals.

The current in the phase that is impermeable to electrons must be carried by ions. Most commonly, the ionic conductor in a galvanic cell is an electrolyte solution. Other possibilities include a molten salt and a solid salt at a temperature sufficiently high to allow ions to move through the solid at a useful rate.

Summarizing, a galvanic cell has terminals made of the same metal, has all phases being electrical conductors, has at least one phase that is an ionic conductor (but not an electronic conductor), and allows electric charge to be readily transferred between phases. We can symbolize the galvanic cell by T-E-I-E′-T′, where T and T′ are the terminals, I is the ionic conductor, and E and E′ are two pieces of metal (called *electrodes*) that make contact with the ionic conductor. The current is carried by electrons in T, T′, E, and E′ and by ions in I.

The Daniell cell. An example of a galvanic cell is the Daniell cell (Fig. 14.5), which was used in the early days of telegraphy. In this cell, a porous ceramic barrier separates a compartment containing a Zn rod in a $ZnSO_4$ solution from a compartment containing a Cu rod in a $CuSO_4$ solution. The Cu and Zn electrodes are attached to the wires Cu″ and Cu′, which are the terminals. The porous barrier prevents wholesale mixing of the solutions by convection currents but allows ions to pass from one solution to the other.

Consider first the open-circuit state, with the terminals not connected to a load (Fig. 14.5*a*). At the Zn electrode, an equilibrium is set up between Zn^{2+} ions in the solution and Zn^{2+} ions in the metal (as discussed in Sec. 14.3): $Zn^{2+}(Zn) = Zn^{2+}(aq)$. Adding this equation to the equation for the equilibrium between zinc ions and zinc atoms in the zinc metal, $Zn = Zn^{2+}(Zn) + 2e^-(Zn)$, we can write the equilibrium at the Zn–$ZnSO_4(aq)$ interface as $Zn = Zn^{2+}(aq) + 2e^-(Zn)$. Since the potential difference between the Zn electrode and the $ZnSO_4$ solution is not readily measurable, we don't know whether the equilibrium position for a given $ZnSO_4$ concentration leaves the Zn at a higher or lower potential than the solution. Let us assume that there is a net loss of Zn^{2+} to the solution, leaving a negative charge on the Zn and leaving the Zn at a lower potential than the solution: $\phi(\text{Zn}) < \phi(\text{aq. ZnSO}_4)$. Although the potential difference $\phi(\text{aq. ZnSO}_4) - \phi(\text{Zn})$ is not known, the emfs in Table 14.1 in Sec. 14.8 indicate that this potential difference is typically of the order of a volt or two. As noted in Sec. 14.4, the amount of Zn^{2+} transferred between the metal and the solution is far too small to be detected by chemical analysis.

A similar equilibrium occurs at the Cu–$CuSO_4(aq)$ interface. However, Cu is a less active metal than Zn and has much less tendency to go into solution. (If a Zn rod is dipped into a $CuSO_4$ solution, metallic Cu immediately plates out on the Zn and Zn goes into solution: $Cu^{2+} + Zn \rightarrow Cu + Zn^{2+}$. If a Cu rod is dipped into a $ZnSO_4$ solution, no detectable amount of Zn plates out on the Cu. The equilibrium constant for the reaction $Cu^{2+} + Zn = Cu + Zn^{2+}$ is extremely large.) We therefore expect that for comparable concentrations of $CuSO_4$ and $ZnSO_4$, the Cu electrode at equilibrium will have a smaller negative charge than the Zn electrode, and might even have a positive charge (corresponding to a net gain of Cu^{2+} ions from the solution). Let us therefore assume that the equilibrium electric potential of Cu is greater than that of the aqueous $CuSO_4$ solution: $\phi(\text{Cu}) > \phi(\text{aq. CuSO}_4)$.

At the junction between the Cu′ terminal and the Zn electrode, there is an equilibrium exchange of electrons, producing a potential difference between these phases. Since the potential difference between two phases of different composition is not readily measurable, the value of this potential difference is not known, but it is likely that $\phi(\text{Cu}') < \phi(\text{Zn})$.

There is no potential difference between the Cu electrode and the Cu″ terminal, since they are made of the same metal. More formally, $\mu(\text{e}^- \text{ in Cu}) = \mu(\text{e}^- \text{ in Cu}'')$; Eq. (14.31) then gives $\phi(\text{Cu}) = \phi(\text{Cu}'')$.

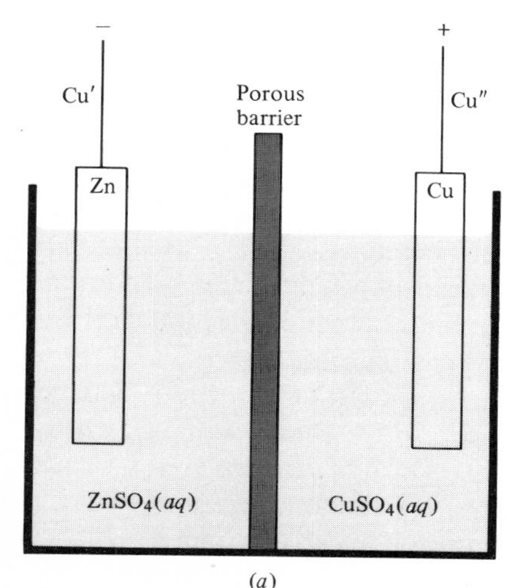

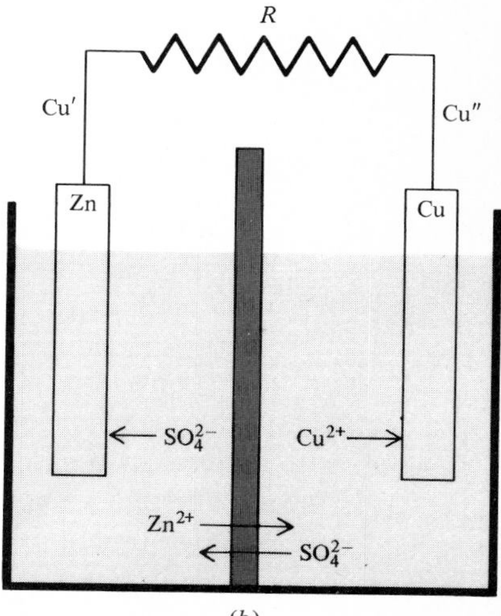

Figure 14.5
The Daniell cell.
(*a*) Open-circuit state.
(*b*) Closed-circuit state.

There is a potential difference at the junction of the $ZnSO_4$ and $CuSO_4$ solutions. However, this potential difference is small compared with the other interphase potential differences in the cell, and we shall neglect it for now, taking $\phi(\text{aq. } ZnSO_4) = \phi(\text{aq. } CuSO_4)$. (See Sec. 14.10 for discussion of this liquid-junction potential difference.)

The cell emf is defined as the open-circuit potential difference between the terminals: $\mathscr{E} \equiv \phi(\text{Cu}'') - \phi(\text{Cu}') = \phi(\text{Cu}) - \phi(\text{Cu}')$. Adding and subtracting $\phi(\text{aq. } CuSO_4)$, $\phi(\text{aq. } ZnSO_4)$, and $\phi(\text{Zn})$ on the right side of this equation, we get

$$\mathscr{E} = [\phi(\text{Cu}) - \phi(\text{aq. CuSO}_4)] + [\phi(\text{aq. CuSO}_4) - \phi(\text{aq. ZnSO}_4)] \\ + [\phi(\text{aq. ZnSO}_4) - \phi(\text{Zn})] + [\phi(\text{Zn}) - \phi(\text{Cu}')] \qquad (14.35)$$

The cell emf is the sum of the potential differences at the following interfaces between phases: Cu–$CuSO_4(aq)$, $CuSO_4(aq)$–$ZnSO_4(aq)$, $ZnSO_4(aq)$–Zn, Zn–Cu′. From the preceding discussion, the first term in brackets on the right of (14.35) is positive, the second term is negligible, the third term is positive, and the fourth term is positive. Therefore $\mathscr{E} \equiv \phi(\text{Cu}'') - \phi(\text{Cu}')$ is positive, and the terminal attached to the Cu electrode is at a higher potential than the terminal attached to Zn. This is indicated by the + and − signs in Fig. 14.5.

We can write down an expression for the open-circuit potential difference at each interface. Equation (14.31) applied to Cu^{2+} ions at the Cu–$CuSO_4(aq)$ interface gives

$$\phi(\text{Cu}) - \phi(\text{aq. CuSO}_4) = [\mu^{\text{aq}}(\text{Cu}^{2+}) - \mu^{\text{Cu}}(\text{Cu}^{2+})]/2\mathscr{F} \qquad (14.36)$$

where the superscripts aq and Cu indicate the aqueous $CuSO_4$ and the Cu phases. Note that $\Delta\phi$ at the Cu–$CuSO_4(aq)$ phase boundary is determined by the chemical potentials of Cu^{2+} in Cu and in aqueous $CuSO_4$ and these chemical potentials are independent of the electrical state of the phases. Thus, $\Delta\phi$ for these two phases is independent of the presence or absence of contacts with other phases.

Alternatively, we can consider the process at the Cu–$CuSO_4(aq)$ interface to be the electrochemical equilibrium $\text{Cu} = \text{Cu}^{2+}(aq) + 2\text{e}^-(\text{Cu})$. Equation (14.32) gives $\tilde{\mu}(\text{Cu}) = \tilde{\mu}^{\text{aq}}(\text{Cu}^{2+}) + 2\tilde{\mu}^{\text{Cu}}(\text{e}^-)$. Using (14.29), we get

$$\mu(\text{Cu}) = \mu^{\text{aq}}(\text{Cu}^{2+}) + 2\mathscr{F}\phi(\text{aq. CuSO}_4) + 2[\mu^{\text{Cu}}(\text{e}^-) - \mathscr{F}\phi(\text{Cu})]$$

$$\phi(\text{Cu}) - \phi(\text{aq. CuSO}_4) = [\mu^{\text{aq}}(\text{Cu}^{2+}) - \mu(\text{Cu}) + 2\mu^{\text{Cu}}(\text{e}^-)]/2\mathscr{F} \qquad (14.37)$$

Equations (14.36) and (14.37) are equivalent. This can be seen by use of (14.34) for the equilibrium $\text{Cu}^{2+}(\text{in Cu}) + 2\text{e}^-(\text{in Cu}) = \text{Cu}$.

An expression for the cell emf $\mathscr{E}$ can be found by substitution of (14.37) and similar equations for the other interphase potential differences into (14.35). The result would be the Nernst equations (14.51) and (14.55) below.

Now consider what happens when the circuit for the Daniell cell is completed by attaching a metal resistor R between the terminals (Fig. 14.5b). The Cu′ terminal (attached to Zn) is at a lower potential than the Cu″ terminal (attached to Cu), so electrons are forced to flow through R from Cu′ to Cu″. [From Eq. (14.6), positive work is required to move a positive charge from a point of low electric potential to one of high potential. Hence, positive charges spontaneously move from a region of high potential to a region of low potential. (For example, a positive charge spontaneously moves away from a second positive charge.) Electrons are negatively charged and therefore spontaneously move from low to high potential.] When electrons leave the Cu′ terminal, the equilibrium at the Cu′–Zn interface is disturbed, causing electrons to flow out of Zn into Cu′. This disturbs the equilibrium $Zn = Zn^{2+} + 2e^-(Zn)$ at the Zn–$ZnSO_4(aq)$ interface and causes more Zn to go into solution, leaving electrons behind on the Zn to make up for the electrons that are leaving the Zn. The flow of electrons into the Cu electrode from the external circuit causes Cu^{2+} ions from the $CuSO_4$ solution to combine with electrons in the Cu metal and deposit as Cu atoms on the Cu electrode: $Cu^{2+}(aq) + 2e^-(Cu) = Cu$.

In the region around the Cu electrode, the $CuSO_4$ solution is being depleted of positive ions (Cu^{2+}), while the region around the Zn electrode is being enriched in positive ions (Zn^{2+}). This produces a potential difference between these two regions of solution and causes a flow of positive ions through the solutions from the Zn electrode to the Cu electrode; simultaneously, negative ions move toward the Zn electrode (Fig. 14.5*b*). The current is carried through the solution by the Zn^{2+}, Cu^{2+}, and SO_4^{2-} ions.

During the operation of the cell, the electrochemical reactions $Zn \rightarrow Zn^{2+}(aq) + 2e^-(Zn)$ and $Cu^{2+}(aq) + 2e^-(Cu) \rightarrow Cu$ occur. We call these the *half-reactions* of the cell. There is also the electron-flow process $2e^-(Zn) \rightarrow 2e^-(Cu)$. Addition of this flow process and the two half-reactions gives the overall galvanic-cell reaction: $Zn + Cu^{2+}(aq) \rightarrow Zn^{2+}(aq) + Cu$. The Zn electrode plus its associated $ZnSO_4$ solution form a *half-cell*; likewise, Cu + aq. $CuSO_4$ form a second half-cell. So far, we have used the word "electrode" to mean the piece of metal that dips into a solution in a half-cell. Often, however, the term *electrode* is used to refer to a half-cell consisting of metal plus solution.

Oxidation is a loss of electrons. *Reduction* is a gain of electrons. The half-reaction $Zn \rightarrow Zn^{2+}(aq) + 2e^-(Zn)$ is an oxidation. The half-reaction $Cu^{2+}(aq) + 2e^-(Cu) \rightarrow Cu$ is a reduction. If we were to bring the species Cu, Zn, $Cu^{2+}(aq)$, and $Zn^{2+}(aq)$ in contact with one another, the oxidation–reduction (redox) reaction $Zn + Cu^{2+}(aq) \rightarrow Cu + Zn^{2+}(aq)$ would occur. In the Daniell cell, the oxidation and reduction parts of this reaction occur at different locations that are connected by a wire through which electrons are forced to flow. Separation of the oxidation and reduction half-reactions allows the chemical energy of the reaction to be converted into electrical energy.

We define the *anode* as the electrode at which oxidation occurs and the *cathode* as the electrode at which reduction occurs. In the Daniell cell, Zn is the anode.

The open-circuit condition (Fig. 14.5*a*) of the Daniell cell is not a stable situation. The slow diffusion of Cu^{2+} into the $ZnSO_4$ solution will eventually allow the Cu^{2+} ions to come in contact with the Zn electrode, causing the spontaneous redox reaction $Cu^{2+} + Zn \rightarrow Cu + Zn^{2+}$ to occur directly, without flow of electrons through a wire. This would destroy the cell. For this reason, the Daniell cell cannot be left sitting around on open circuit. Instead, a resistor is kept connected between the terminals. Note from Fig. 14.5*b* that as the cell operates, the electric field in the solution forces Cu^{2+} ions away from the $ZnSO_4$ solution, preventing them from getting at the Zn electrode. Many modern galvanic cells (batteries) have half-reactions that involve insoluble salts (see Sec. 14.12); this allows the cell to be kept on the shelf on open circuit.

Cell diagrams. A galvanic cell is represented by a diagram in which the following conventions are used. A vertical line indicates a phase boundary. The phase boundary between two miscible liquids is often indicated by a dashed or dotted vertical line. Two species present in the same phase are separated by a comma.

The diagram of the Daniell cell (Fig. 14.5) is

$$\mathrm{Cu'} \,|\, \mathrm{Zn} \,|\, \mathrm{ZnSO_4}(aq) \,\vdots\, \mathrm{CuSO_4}(aq) \,|\, \mathrm{Cu} \tag{14.38}$$

(The Cu″ terminal and the Cu electrode form a single phase.) The Cu′ terminal is often omitted from the cell diagram. For completeness, the $ZnSO_4$ and $CuSO_4$ molalities can be indicated in the diagram.

The following IUPAC conventions are used to define the cell emf and cell reaction for a given cell diagram: (A) The cell emf $\mathscr{E}$ is defined as

$$\mathscr{E} \equiv \phi_R - \phi_L \tag{14.39*}$$

where ϕ_R and ϕ_L are the open-circuit electric potentials of the terminals on the right side and left side of the cell diagram. ("Right" and "left" have nothing to do with the physical arrangement of the cell on the laboratory bench.) (B) The cell reaction is defined to involve oxidation at the electrode on the left side of the cell diagram and reduction at the electrode on the right.

For the cell diagram (14.38), Convention A gives $\mathscr{E} = \phi(\mathrm{Cu}) - \phi(\mathrm{Cu'})$. We saw earlier that $\phi(\mathrm{Cu})$ is greater than $\phi(\mathrm{Cu'})$, so $\mathscr{E}$ for (14.38) is positive. For $CuSO_4$ and $ZnSO_4$ molalities close to 1 mol/kg, experiment gives $\mathscr{E}_{(14.38)} = 1.1$ V. For (14.38), Convention B gives the half-reaction at the left electrode as $\mathrm{Zn} = \mathrm{Zn^{2+}} + 2e^-$ and that at the right electrode as $\mathrm{Cu^{2+}} + 2e^- = \mathrm{Cu}$. The overall reaction for (14.38) is $\mathrm{Zn} + \mathrm{Cu^{2+}} = \mathrm{Zn^{2+}} + \mathrm{Cu}$.

Suppose we had written the cell diagram as

$$\mathrm{Cu} \,|\, \mathrm{CuSO_4}(aq) \,\vdots\, \mathrm{ZnSO_4}(aq) \,|\, \mathrm{Zn} \,|\, \mathrm{Cu'} \tag{14.40}$$

Then Convention A gives $\mathscr{E}_{(14.40)} = \phi(\mathrm{Cu'}) - \phi(\mathrm{Cu})$. Since $\phi(\mathrm{Cu}) > \phi(\mathrm{Cu'})$, the emf for this diagram is negative: $\mathscr{E}_{(14.40)} = -1.1$ V. Convention B gives the half-reactions for (14.40) as $\mathrm{Cu} = \mathrm{Cu^{2+}} + 2e^-$ and $\mathrm{Zn^{2+}} + 2e^- = \mathrm{Zn}$. The overall reaction for (14.40) is $\mathrm{Zn^{2+}} + \mathrm{Cu} = \mathrm{Zn} + \mathrm{Cu^{2+}}$.

Note that a positive emf corresponds to the cell reaction's occurring spontaneously when the cell is connected to a load.

Measurement of cell emfs. The emf of a galvanic cell can be accurately measured using a *potentiometer* (Fig. 14.6). Here, the emf $\mathscr{E}_X$ of cell X is balanced by an

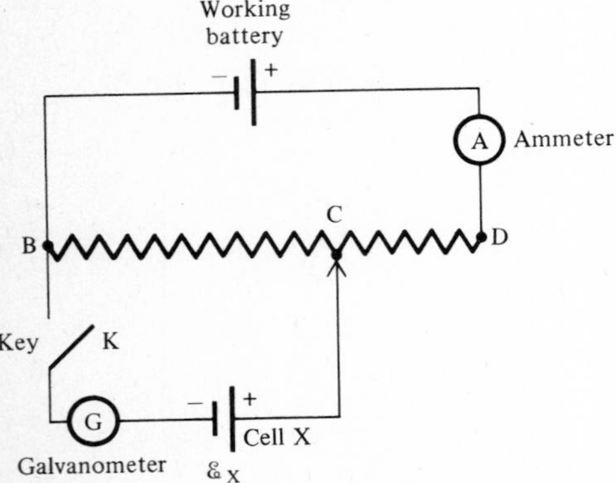

Figure 14.6
A potentiometer.

opposing potential difference $\Delta\phi_{opp}$, so as to make the current through the cell zero. Measurement of $\Delta\phi_{opp}$ gives $\mathscr{E}_X$.

The resistor between B and D is a uniform slide-wire of total resistance R. The contact point C_X is adjusted until the galvanometer G shows no deflection when the tap key K is closed, indicating zero current passing through the cell X. The positive terminal of the cell is at the same potential as C_X since it is connected to C_X by a wire of negligible resistance. The fact that no current flows through the cell when the key is closed indicates that the negative terminal of the cell is at the same potential as point B. Hence, when balance is achieved, the potential drop across the resistor C_XB equals the zero-current potential drop across the cell's terminals (which is the cell emf $\mathscr{E}_X$). Ohm's law (16.59) gives $\mathscr{E}_X = |\Delta\phi_{opp}| = IR_X$, where I is the current in the upper part of the circuit and R_X is the resistance of the wire between B and C_X: $R_X = (\overline{BC_X}/\overline{BD})R$. Measurement of I and R_X allows $\mathscr{E}_X$ to be found.

In practice, one balances the circuit twice, once with cell X and once with a standard cell S of accurately known emf $\mathscr{E}_S$ in place of X. Let R_S and R_X be the resistances needed to balance $\mathscr{E}_S$ and $\mathscr{E}_X$. Then $\mathscr{E}_S = IR_S$ and $\mathscr{E}_X = IR_X$. (Since no current flows through S or X, the current I is unchanged when the cell is changed.) We have $\mathscr{E}_X/\mathscr{E}_S = R_X/R_S$, which allows $\mathscr{E}_X$ to be found.

When the potentiometer in Fig. 14.6 is only infinitesimally out of balance, an infinitesimal current flows through the cell X. Equilibrium is maintained at each phase boundary in the cell, and the cell reaction occurs reversibly. The rate of the reversible cell reaction is infinitesimal, and it takes an infinite time to carry out a noninfinitesimal amount of reaction. When a noninfinitesimal current is drawn from the cell, as in Fig. 14.5*b*, the cell reaction occurs irreversibly.

Electrolytic cells. In a galvanic cell, a chemical reaction produces a flow of electric current; chemical energy is converted into electrical energy. In an *electrolytic cell*, a flow of current produces a chemical reaction; electrical energy from an external source is converted into chemical energy.

Figure 14.7 shows an electrolytic cell. Two Pt electrodes are attached to the terminals of a source of emf (e.g., a galvanic cell or a dc generator). The Pt electrodes dip into an aqueous NaOH solution. Electrons flow into the negative Pt electrode from the source of emf, and H_2 is liberated at this electrode: $2H_2O + 2e^- \rightarrow H_2 + 2OH^-$. At the positive electrode, O_2 is liberated: $4OH^- \rightarrow 2H_2O + O_2 + 4e^-$. Doubling the first half-reaction and adding it to the second, we get the overall electrolysis reaction: $2H_2O \rightarrow 2H_2 + O_2$.

The same definition of anode and cathode is used for electrolytic cells as for galvanic cells. Hence, the cathode in Fig. 14.7 is the negative electrode. (In a galvanic cell, the cathode is the positive electrode.)

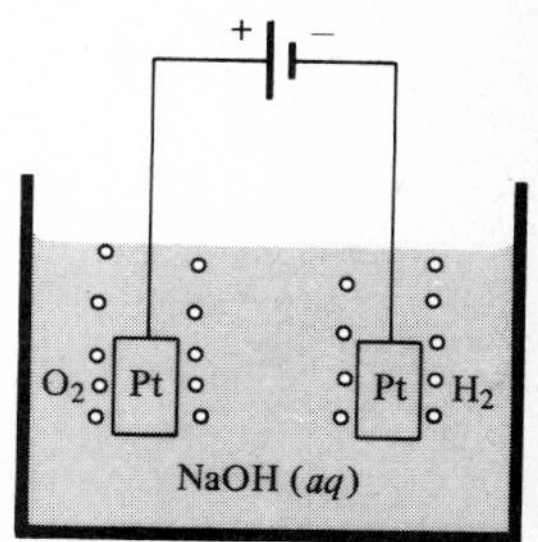

Figure 14.7
An electrolytic cell.

The elements Al, Na, and F_2 are commercially prepared by the electrolysis of molten Al_2O_3, molten NaCl, and liquid HF. Electrolysis is also used to plate one metal on another.

The term *electrochemical cell* indicates either a galvanic cell or an electrolytic cell.

14.6 TYPES OF REVERSIBLE ELECTRODES

Classical thermodynamics applies only to reversible processes. To apply thermodynamics to galvanic cells (Sec. 14.7), we require that the cell be reversible. Consider a cell with its emf exactly balanced in a potentiometer (Fig. 14.6). If the cell is reversible, the processes that occur in the cell when the contact point C_X is moved very slightly to the right are the exact reverse of the processes that occur when C_X is moved very slightly to the left.

For the Daniell cell, when C_X is moved slightly to the left, the potential drop across BC_X becomes slightly less than the cell's emf and the cell functions as a galvanic cell with Zn going into solution as Zn^{2+} at the zinc electrode and Cu^{2+} plating out as Cu at the copper electrode. When C_X is moved slightly to the right, the externally applied emf is slightly greater than the emf of the Daniell cell, so the direction of current flow through the cell is reversed; the cell then functions as an electrolytic cell, with Zn being plated out at the zinc electrode and Cu going into solution at the copper electrode; thus, the electrode reactions are reversed. Despite this, the Daniell cell is not reversible. The irreversibility arises at the liquid junction. (A *liquid junction* is the interface between two miscible electrolyte solutions.) When the Daniell cell functions as a galvanic cell, Zn^{2+} ions move into the $CuSO_4$ solution (Fig. 14.5*b*). However, when the cell's emf is overridden by an external emf that reverses the current direction, the reversal of current in the solution means that Cu^{2+} ions will move into the $ZnSO_4$ solution. Since these processes at the liquid junction are not the reverse of each other, the cell is irreversible.

To have reversibility at an electrode, significant amounts of all reactants and products of the electrode half-reaction must be present at the electrode. For example, if we had a cell one of whose electrodes was Zn dipping into an aqueous solution of NaCl, then when electrons are moving out of this electrode, the half-reaction is $Zn \rightarrow Zn^{2+}(aq) + 2e^-$, whereas when the potentiometer slide-wire is moved in the opposite direction and electrons are moving into the Zn electrode, the half-reaction is $2H_2O + 2e^- \rightarrow H_2 + 2OH^-$ (since there is no Zn^{2+} to plate out of the solution). Reversibility requires a significant concentration of Zn^{2+} in the solution around the Zn electrode.

The main types of reversible electrodes (half-cells) are:

1. Metal–metal-ion electrodes. Here, a metal M is in electrochemical equilibrium with a solution containing M^{z+} ions. The half-reaction is $M^{z+} + z_+e^- = M$. Examples include $Cu^{2+}|Cu$, $Hg_2^{2+}|Hg$, $Ag^+|Ag$, $Pb^{2+}|Pb$, and $Zn^{2+}|Zn$. Metals that react with the solvent can't be used. Group IA and IIA metals (Na, Ca, ...) react with water; zinc reacts with aqueous acidic solutions. For certain metals it is necessary to deaerate the cell with N_2 or He to prevent oxidation of the metal by dissolved O_2. Certain electrodes, for example, $Ni^{2+}|Ni$, exhibit irreversibility due to formation of a blocking surface film on the metal produced by hydrolysis.

2. Amalgam electrodes. An *amalgam* is a solution of a metal in liquid Hg. In an amalgam electrode, an amalgam of metal M is in equilibrium with a solution containing M^{z+} ions. The mercury does not participate in the electrode reaction, which is $M^{z+} + z_+e^- = M(Hg)$, where M(Hg) indicates M dissolved in Hg. Active metals like Na or Ca can be used in an amalgam electrode. The emf of a metal–metal-ion electrode is sensitive to such things as mechanical strains in the metal; a liquid-surface amalgam electrode gives a more reproducible emf.

3. Redox electrodes. Every electrode involves an oxidation-reduction half-reaction. However, custom dictates that the term "redox electrode" refer only to an electrode whose redox half-reaction is between two species present in the same solution; the metal that dips into this solution serves only to supply or accept electrons. For example, a Pt wire dipping into a solution containing Fe^{2+} and Fe^{3+} is a redox electrode whose half-reaction is $Fe^{3+} + e^- = Fe^{2+}$. The half-cell diagram is $Pt|Fe^{3+}, Fe^{2+}$. Another example is $Pt|MnO_4^-, Mn^{2+}$.

4. Metal–insoluble-salt electrodes. Here, a metal M is in contact with one of its very slightly soluble salts $M_{\nu_+}X_{\nu_-}$ and with a solution that is saturated with $M_{\nu_+}X_{\nu_-}$ and that contains a soluble salt (or acid) with the anion X^{z-}.

For example, the *silver–silver chloride* electrode (Fig. 14.8*a*) consists of Ag metal, solid AgCl, and a solution that contains Cl^- ions (from, say, KCl or HCl) and is saturated with AgCl. There are three phases present, and the electrode is usually symbolized by $Ag|AgCl(s)|Cl^-(aq)$. One way to prepare this electrode is by electrodeposition of a layer of Ag on a piece of Pt, followed by electrolytic conversion of part of the Ag to AgCl. The Ag is in electrochemical equilibrium with the Ag^+ in the solution: $Ag = Ag^+(aq) + e^-$. Since the solution is saturated with AgCl, any Ag^+ added to the solution reacts as follows: $Ag^+(aq) + Cl^-(aq) = AgCl(s)$. The net electrode half-reaction is the sum of these two reactions:

$$Ag(s) + Cl^-(aq) = AgCl(s) + e^- \qquad (14.41)$$

The *calomel electrode* (Fig. 14.8*b*) is $Hg|Hg_2Cl_2(s)|KCl(aq)$. The half-reaction is $2Hg + 2Cl^- = Hg_2Cl_2(s) + 2e^-$, which is the sum of $2Hg = Hg_2^{2+}(aq) + 2e^-$ and $Hg_2^{2+}(aq) + 2Cl^-(aq) = Hg_2Cl_2(s)$. (Calomel is Hg_2Cl_2.) When the solution is saturated with KCl, we have the *saturated calomel electrode*.

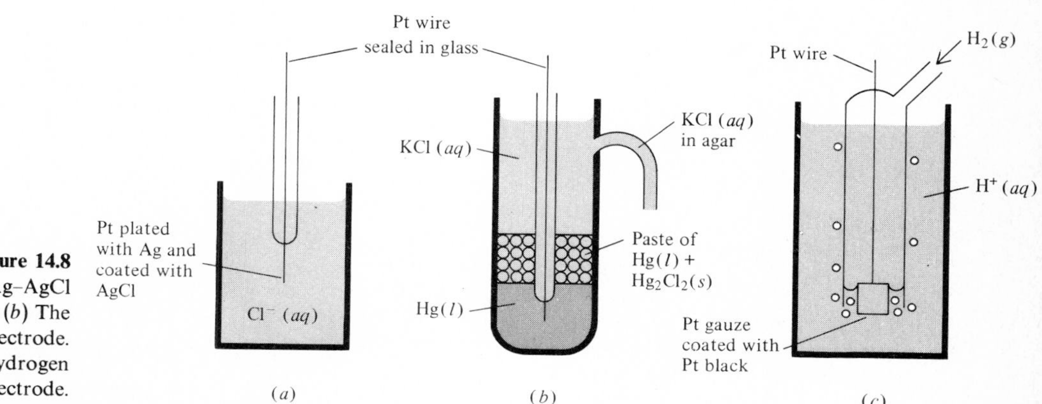

Figure 14.8 (*a*) The Ag–AgCl electrode. (*b*) The calomel electrode. (*c*) The hydrogen electrode.

The *lead–lead sulfate electrode* is $Pb|PbSO_4(s)|SO_4^{2-}(aq)$ with half-reaction $Pb + SO_4^{2-} = PbSO_4(s) + 2e^-$. This electrode is used in the lead storage battery.

The commonly used diagrams given above for metal–insoluble-salt half-cells are misleading. Thus, the diagram $Hg|Hg_2Cl_2(s)|KCl(aq)$ might seem to suggest that the Hg is not in contact with the aqueous solution, when in fact all three phases are in contact with one another. (In some methods of preparation of the Ag–AgCl electrode, the Ag is coated by a nonporous layer of AgCl, which blocks direct contact between the Ag and the solution. In this case, the current is conducted through the thin layer of solid AgCl by mobile Ag^+ ions.)

5. Gas electrodes. Here, a gas is in equilibrium with ions in solution. For example, the *hydrogen electrode* is $Pt|H_2(g)|H^+(aq)$, and its half-reaction is

$$H_2(g) = 2H^+(aq) + 2e^- \tag{14.42}$$

The hydrogen gas is bubbled over the Pt, which dips into an acidic solution (Fig. 14.8*c*). The Pt contains a coat of electrolytically deposited colloidal Pt particles (*platinum black*), which catalyze the forward and reverse reactions in (14.42), allowing the equilibrium to be rapidly established. The H_2 gas is chemisorbed as H atoms on the Pt: $H_2(g) = 2H(Pt) = 2H^+(aq) + 2e^-(Pt)$. "The hydrogen electrode is ... the primary standard with which all other electrodes are compared. ... It ... is the best electrode of all, capable of the highest degree of reproducibility and, contrary to statements frequently made, it is comparatively easy to prepare and use." (*Ives and Janz*, p. 71.)

The *chlorine electrode* is $Pt|Cl_2(g)|Cl^-(aq)$ with half-reaction $Cl_2 + 2e^- = 2Cl^-(aq)$.

A reversible oxygen electrode is extremely difficult to prepare, due to formation of an oxide layer on the metal and other problems.

6. Nonmetal nongas electrodes. The most important examples are the bromine and iodine electrodes: $Pt|Br_2(l)|Br^-(aq)$ and $Pt|I_2(s)|I^-(aq)$. In these electrodes, the solution is saturated with dissolved Br_2 or I_2.

7. Membrane electrodes. See Sec. 14.13.

A galvanic cell formed from half-cells that have different electrolyte solutions contains a liquid junction where these solutions meet and is therefore irreversible. An example is the Daniell cell (14.38). If we attempted to get around this irreversibility by having the Cu and Zn rods dip into a common solution that has both $CuSO_4$ and $ZnSO_4$, the Cu^{2+} ions would react with the Zn rod and the attempt would fail.

A reversible galvanic cell requires two half-cells that use the same electrolyte solution. An example is the cell

$$Pt|H_2(g)|HCl(aq)|AgCl(s)|Ag|Pt' \tag{14.43}$$

composed of a hydrogen electrode and an Ag–AgCl electrode, each dipping in the same HCl solution. (An advantage of metal–insoluble-salt electrodes is that they lend themselves to construction of cells without liquid junctions.)

There are certain subtleties connected with the cell (14.43). When the Ag–AgCl electrode is inserted into the HCl solution, the region of solution around the electrode becomes saturated with AgCl. Likewise, the region of solution around the $Pt|H_2$ electrode is saturated with dissolved H_2. Since diffusion in solution is relatively slow, there is virtually no dissolved AgCl in the region around the H_2 electrode and virtually no dissolved H_2 around the Ag–AgCl electrode. The composition of the HCl solution is not uniform, and the junctions between regions of different composition introduce some irreversibility in the cell. However, the solubilities of AgCl and H_2 in water are so small that this slight nonuniformity of composition has a completely negligible effect and can be ignored. For all practical purposes, the cell is reversible. Likewise, there is no need to worry that Ag^+ ions will diffuse to the H_2 electrode and react directly with H_2, according to $2Ag^+ + H_2 = 2Ag + 2H^+$. The extremely low concentration of Ag^+ in the solution makes the rate of this direct reaction negligible.

14.7 THERMODYNAMICS OF GALVANIC CELLS

In this section, we use thermodynamics to relate the emf of a reversible galvanic cell to the chemical potentials of the species in the cell reaction. Consider such a cell with its terminals on open circuit. For example, the cell might be

$$\mathrm{Pt}_L|\mathrm{H_2}(g)|\mathrm{HCl}(aq)|\mathrm{AgCl}(s)|\mathrm{Ag}|\mathrm{Pt}_R \tag{14.44}$$

where the subscripts L and R indicate the left and right terminals. The IUPAC conventions give the half-reactions and overall reaction as

$$\mathrm{H_2}(g) = 2\mathrm{H^+} + 2\mathrm{e^-}(\mathrm{Pt}_L)$$

$$[\mathrm{AgCl}(s) + \mathrm{e^-}(\mathrm{Pt}_R) = \mathrm{Ag} + \mathrm{Cl^-}] \times 2$$

$$2\mathrm{AgCl}(s) + \mathrm{H_2}(g) + 2\mathrm{e^-}(\mathrm{Pt}_R) = 2\mathrm{Ag} + 2\mathrm{H^+} + 2\mathrm{Cl^-} + 2\mathrm{e^-}(\mathrm{Pt}_L) \tag{14.45}$$

Since the terminals are on open circuit, flow of electrons from Pt_L to Pt_R cannot occur; hence, the electrons have been included in the overall reaction. We call (14.45) the cell's *electrochemical reaction*, to distinguish it from the cell's *chemical reaction*, which is

$$2\mathrm{AgCl}(s) + \mathrm{H_2}(g) = 2\mathrm{Ag} + 2\mathrm{H^+} + 2\mathrm{Cl^-} \tag{14.46}$$

When the open-circuit cell is assembled from its component phases, extremely small amounts of charge are transferred between phases, until electrochemical equilibrium is reached. For the open-circuit cell in equilibrium, let us imagine a process in which the cell's electrochemical reaction proceeds to the infinitesimal extent $d\xi$ at constant T and P. [For the cell (14.44) with electrochemical reaction (14.45), $2d\xi$ moles of AgCl, $d\xi$ moles of H_2, and $2d\xi$ moles of $e^-(Pt_R)$ would disappear, while $2d\xi$ moles of Ag, $2d\xi$ moles of H^+, $2d\xi$ moles of Cl^-, and $2d\xi$ moles of $e^-(Pt_L)$ would be formed.] For any constant-T-and-P infinitesimal process in a closed system at equilibrium, Eq. (4.21) (suitably modified) says that the infinitesimal change in the Gibbs energy (including the electrical part of the Gibbs energy) is zero: $d\tilde{G} = 0$. Equation (4.21) led to (4.111), which written with tildes becomes Eq. (14.32): $\sum_i \nu_i \tilde{\mu}_i = 0$, where the ν_i's are the stoichiometric coefficients in the cell's electrochemical reaction.

We break this sum into a sum over electrons plus a sum over all other species:

$$0 = \sum_i \nu_i \tilde{\mu}_i = \sum_{\mathrm{e^-}} \nu(\mathrm{e^-})\tilde{\mu}(\mathrm{e^-}) + \sum_i{}' \nu_i \tilde{\mu}_i \tag{14.47}$$

where the prime on the second sum indicates that it doesn't include electrons. [For example, for the cell reaction (14.45)

$$\sum_{\mathrm{e^-}} \nu(\mathrm{e^-})\tilde{\mu}(\mathrm{e^-}) = -2\tilde{\mu}(\mathrm{e^-}\text{ in }\mathrm{Pt}_R) + 2\tilde{\mu}(\mathrm{e^-}\text{ in }\mathrm{Pt}_L) \tag{14.48}$$

$$\sum_i{}' \nu_i \tilde{\mu}_i = -2\tilde{\mu}(\mathrm{AgCl}) - \tilde{\mu}(\mathrm{H_2}) + 2\tilde{\mu}(\mathrm{Ag}) + 2\tilde{\mu}(\mathrm{H^+}) + 2\tilde{\mu}(\mathrm{Cl^-})$$

where $\tilde{\mu}(e^- \text{ in } Pt_R)$ is the electrochemical potential of electrons in the right-hand terminal.]

Let T_R and T_L denote the right- and left-hand terminals of the cell. Let n, the *charge number of the cell reaction*, be defined as the number of electrons transferred for the cell electrochemical reaction as written. [For example, $n = 2$ for the cell reaction (14.45).] The charge number n is a positive, dimensionless number. The sum over electrons in (14.47) can be written [cf. (14.48)] as

$$\sum_{e^-} \nu(e^-)\tilde{\mu}(e^-) = -n\tilde{\mu}(e^- \text{ in } T_R) + n\tilde{\mu}(e^- \text{ in } T_L) \tag{14.49}$$

Use of (14.29) with $i = e^-$ and $z_i = -1$ gives

$$\sum_{e^-} \nu(e^-)\tilde{\mu}(e^-) = n[\mu(e^- \text{ in } T_L) - \mu(e^- \text{ in } T_R)] + n\mathscr{F}(\phi_R - \phi_L)$$

where ϕ_R and ϕ_L are the potentials of the right and left terminals. The terminals have the same chemical composition, so $\mu(e^- \text{ in } T_L) = \mu(e^- \text{ in } T_R)$. Hence

$$\sum_{e^-} \nu(e^-)\tilde{\mu}(e^-) = n\mathscr{F}(\phi_R - \phi_L) = n\mathscr{F}\mathscr{E} \tag{14.50}$$

where the definition (14.39) of the cell emf $\mathscr{E}$ was used.

Since the cell is reversible, there are no liquid junctions and all the ions in the cell reaction occur in the same phase, namely, the phase that is an ionic conductor (an electrolyte solution, a fused salt, etc.). Hence, the arguments that led to Eq. (14.34) show that the electrochemical potentials $\tilde{\mu}_i$ in the second sum on the right of (14.47) can be replaced by the chemical potentials: $\sum_i' \nu_i \tilde{\mu}_i = \sum_i' \nu_i \mu_i$. Substitution of this relation and of (14.50) into (14.47) gives

$$\sum_i{}' \nu_i \mu_i = -n\mathscr{F}\mathscr{E} \qquad \text{rev. cell} \tag{14.51}$$

Equation (14.51) relates the cell emf to the chemical potentials of the species in the cell's chemical reaction. For example, for the cell (14.44) with chemical reaction (14.46), we have

$$-2\mu(AgCl) - \mu(H_2) + 2\mu(Ag) + 2\mu(H^+) + 2\mu(Cl^-) = -2\mathscr{F}\mathscr{E}$$

The quantity $\sum_i' \nu_i \mu_i$ is called ΔG in many texts, and (14.51) written as $\Delta G = -n\mathscr{F}\mathscr{E}$. However, as noted in Sec. 11.9, the symbol ΔG has several meanings. A better designation for this sum is $(\partial G/\partial \xi)_{T,P}$, where G is the Gibbs energy of the species in the cell's chemical reaction.

We specified in Sec. 14.5 that the cell terminals must be made of the same metal. One might wonder whether the cell emf depends on the identity of this metal. For example, if Pd instead of Pt is used for the two terminals in (14.44), what happens to $\mathscr{E}$? Equation (14.51) shows that the cell emf depends only on the chemical potentials of the species involved in the cell chemical reaction. Hence $\mathscr{E}$ must be independent of what metal is used for the terminals T_L and T_R.

The definition of the activity a_i of species i gives [Eq. (11.2)] $\mu_i = \mu_i^\circ + RT \ln a_i$, where μ_i° is the chemical potential of i in its chosen standard state. Carrying out the same manipulations used to derive (6.17) and (11.3), we have

$$\sum_i{}' \nu_i \mu_i = \sum_i{}' \nu_i \mu_i^\circ + RT \sum_i{}' \nu_i \ln a_i = \Delta G^\circ + RT \ln \left[\prod_i{}' (a_i)^{\nu_i} \right] \tag{14.52}$$

where $\Delta G^\circ \equiv \sum_i' \nu_i \mu_i^\circ$ is the standard molar Gibbs energy change [Eq. (11.4)] for the cell's chemical reaction. Substitution in (14.51) gives

$$\mathscr{E} = -\frac{\Delta G^\circ}{n\mathscr{F}} - \frac{RT}{n\mathscr{F}} \ln \left[\prod_i{}' (a_i)^{\nu_i} \right] \tag{14.53}$$

The *standard potential* $\mathscr{E}^\circ$ of the cell's chemical reaction is defined as $\mathscr{E}^\circ \equiv -\Delta G^\circ / n\mathscr{F}$, so

$$\Delta G^\circ = -n\mathscr{F}\mathscr{E}^\circ \tag{14.54*}$$

($\mathscr{E}^\circ$ is also called the standard emf of the cell.) Hence

$$\mathscr{E} = \mathscr{E}^\circ - \frac{RT}{n\mathscr{F}} \ln \left[\prod_i{}' (a_i)^{\nu_i} \right] \tag{14.55*}$$

which is the famous (if not infamous) *Nernst equation.*

The preceding equations are ambiguous in that the scale for solute activities has not been specified. Activities a_i and standard-state chemical potentials μ_i° differ on the molality and concentration scales (Sec. 10.3). In work on electrochemical cells, the molality scale is generally used for all solutes in aqueous solutions. All solute activities in this chapter are on the molality scale.

If all chemical species in the cell are in their standard states, then all the a_i's are 1 and the log term in (14.55) vanishes, making $\mathscr{E}$ equal to $\mathscr{E}^\circ$. However, the molality-scale standard states of solutes are fictitious states, unattainable in reality. Hence, it is generally impossible to prepare a cell with all species in their standard states. Even though the species are not in their standard states, it still might happen that the activities of all species in the cell are 1, in which case $\mathscr{E} = \mathscr{E}^\circ$. This, however, is not a practical way to find $\mathscr{E}^\circ$, since activity coefficients are not generally known to high accuracy; one can't be sure that $\gamma_i m_i = 1$ mol/kg for each species in solution.

$\mathscr{E}^\circ$ can be calculated from (14.54) if the standard-state partial molar Gibbs energies of the chemical species in the cell reaction are known.

$\mathscr{E}^\circ$ can be determined from emf measurements on the cell by an extrapolation procedure. As an example, consider the cell (14.44) with overall chemical reaction [Eq. (14.46)] $2AgCl(s) + H_2(g) = 2Ag(s) + 2H^+(aq) + 2Cl^-(aq)$. The Nernst equation (14.55) gives

$$\mathscr{E} = \mathscr{E}^\circ - \frac{RT}{2\mathscr{F}} \ln \frac{[a(H^+)]^2[a(Cl^-)]^2[a(Ag)]^2}{[a(AgCl)]^2 a(H_2)} \tag{14.57}$$

At or near 1 atm, the activities of the pure solids Ag and AgCl are 1 (Sec. 11.4). Equation (10.127) gives $a(H_2) = f(H_2)/P^\circ$, where f is the H_2 fugacity and $P^\circ \equiv 1$ atm. For pressures close to 1 atm, we can replace $f(H_2)$ with $P(H_2)$ with negligible error (Prob. 14.15). Equations (10.43), (10.59), and the equation following (10.71) give for the molality-scale activities of the ions

$$a(H^+)a(Cl^-) = \gamma(H^+)m(H^+)\gamma(Cl^-)m(Cl^-)/(m^\circ)^2 = \gamma_+\gamma_-(\alpha m)^2/(m^\circ)^2 \tag{14.58}$$

$$a(H^+)a(Cl^-) = (\gamma_\pm)^2\alpha^2(m/m^\circ)^2 = \gamma_i{}^2(m/m^\circ)^2 \tag{14.59}$$

where $m^\circ \equiv 1$ mol/kg, m is the HCl stoichiometric molality, α is the fraction of H^+ ions that are not associated to ion pairs, and γ_i is the HCl stoichiometric activity coefficient. (For a 1 : 1 electrolyte like HCl in aqueous solution, α is essentially 1.)

Substituting for the activities of the species in (14.57), we get

$$\mathscr{E} = \mathscr{E}^\circ - \frac{RT}{2\mathscr{F}} \ln \frac{(\gamma_i m/m^\circ)^4}{P(\mathrm{H}_2)/P^\circ} \tag{14.60}$$

$$\mathscr{E} + \frac{2RT}{\mathscr{F}} \ln (m/m^\circ) - \frac{RT}{2\mathscr{F}} \ln [P(\mathrm{H}_2)/P^\circ] = \mathscr{E}^\circ - \frac{2RT}{\mathscr{F}} \ln \gamma_i \tag{14.61}$$

All quantities on the left side of (14.61) are known. In the limit $m \to 0$, the activity coefficient γ_i goes to 1 and ln γ_i goes to 0. Hence, extrapolation of the left side to $m = 0$ gives $\mathscr{E}^\circ$. At low molalities, ion pairing becomes negligible; with $\alpha = 1$, we see from (14.59) that γ_i becomes equal to $\gamma_\pm$. The Debye–Hückel equation (10.91) shows that ln $\gamma_\pm \propto m^{1/2}$ in very dilute solutions. Hence, for very low molalities a plot of the left side of (14.61) vs. $m^{1/2}$ gives a straight line whose intercept is $\mathscr{E}^\circ$.

The product $\prod_i' (a_i)^{\nu_i}$ in the Nernst equation (14.55) is not in general equal to the ordinary chemical equilibrium constant K_a that appears in (11.5). Although it is true that a reversible cell sitting on open circuit is in equilibrium, this equilibrium is an *electrochemical* equilibrium, and the cell's electrochemical reaction [e.g., (14.45)] involves transfer of electrons between phases that differ in electric potential. The activities in (14.55) are equal to whatever values are used when the cell is set up (since the attainment of electrochemical equilibrium between phases involves negligible changes in concentrations).

Let us rewrite the Nernst equation in terms of the chemical equilibrium constant K_a. Equations (14.54) and (11.3) give $\mathscr{E}^\circ = -\Delta G^\circ/n\mathscr{F} = (RT \ln K_a)/n\mathscr{F}$. We define the *activity quotient* Q_a as the product in brackets in (14.55). Equation (14.55) becomes $\mathscr{E} = (RT \ln K_a)/n\mathscr{F} - (RT \ln Q_a)/n\mathscr{F}$, or

$$\mathscr{E} = \frac{RT}{n\mathscr{F}} \ln \frac{K_a}{Q_a} \qquad \text{where } Q_a \equiv \prod_i{}' (a_i)^{\nu_i} \tag{14.62}$$

When Q_a equals K_a, the cell emf is zero. The greater the departure of Q_a from K_a, the greater the emf.

Let the cell diagram be written with the terminal that is at the higher potential put on the right. Hence, $\mathscr{E} = \phi_R - \phi_L$ is positive. From (14.62), Q_a is less than K_a. Suppose the cell is connected to a load (as in Fig. 14.5*b*). With $\phi_R > \phi_L$, electrons flow spontaneously through the load from the left terminal to the right terminal. Since electrons are flowing out of the left electrode, the reaction that occurs spontaneously at this electrode must be an oxidation (loss of electrons). But the convention for writing the half-cell reactions and overall reaction is to have oxidation occur at the left electrode. Hence, the direction of the spontaneous cell reaction is the same as the reaction written down on paper, and the effect of the actual cell reaction is to increase the concentrations of the products in the cell reaction as written and thus to increase Q_a. As Q_a increases toward K_a, the cell emf decreases, reaching zero when Q_a becomes equal to K_a.

The derivation of the Nernst equation assumed thermodynamic equilibrium, which means that the cell must be reversible. The emf given by the Nernst equation (14.55) is the sum of the potential differences at the phase boundaries of a cell without a liquid junction. When the cell has a liquid junction, the observed cell emf includes the additional potential difference between the two electrolyte solutions. [For example, see Eq. (14.35).] We call this additional potential difference the *liquid-junction potential* $\mathscr{E}_J$. Thus,

$$\mathscr{E}_J \equiv \phi_{\text{soln},R} - \phi_{\text{soln},L} \tag{14.63}$$

where $\phi_{\text{soln},R}$ is the potential of the electrolyte solution of the half-cell on the right of the cell diagram. For example, for the Daniell cell (14.38), $\mathscr{E}_J = \phi(\text{aq. } CuSO_4) - \phi(\text{aq. } ZnSO_4)$. The Nernst equation gives the sums of the potential differences at all interfaces except at the liquid junction. Hence, the emf $\mathscr{E}$ of a cell with a junction equals $\mathscr{E}_J + \mathscr{E}_{\text{Nernst}}$, where $\mathscr{E}_{\text{Nernst}}$ is given by (14.55). Therefore,

$$\mathscr{E} = \mathscr{E}_J + \mathscr{E}^\circ - \frac{RT}{n\mathscr{F}} \ln \left[\prod_i{}' (a_i)^{\nu_i} \right] \qquad \text{cell with liq. junct.} \tag{14.64}$$

Junction potentials are generally small, but they certainly cannot be neglected in accurate work. By connecting the two electrolyte solutions with a salt bridge, the junction potential can be minimized (but not completely eliminated). A *salt bridge* consists of a gel (Sec. 13.6) made by adding agar to a concentrated aqueous KCl solution. (The gel permits diffusion of ions but eliminates convection currents.) A cell with a salt bridge has two liquid junctions, the sum of whose potentials turns out to be quite small (see Sec. 14.10). A salt bridge is symbolized by two vertical lines (solid, dotted, or dashed, according to the whim of the writer). Thus, the diagram

$$\text{Au}_L | \text{Cu} | \text{CuSO}_4(aq) \,\vdots\vdots\, \text{ZnSO}_4(aq) | \text{Zn} | \text{Au}_R \tag{14.65}$$

symbolizes a Daniell cell with gold terminals and with a salt bridge separating the two solutions (Fig. 14.9).

The Nernst equation (14.55) and its modification (14.64) give the open-circuit potential difference (emf) of the cell terminals and also give the potential difference of the terminals when the cell is balanced in a potentiometer. However, these equations do not give the potential difference between the terminals in the highly irreversible situation where the cell is sending a noninfinitesimal current through a load. The potential difference between the terminals when current flows belongs to the subject of electrode kinetics. Unfortunately, the Nernst equation has frequently been used in irreversible situations, where it does not apply.

14.8 STANDARD ELECTRODE POTENTIALS

In this section, the term "electrode" is used as synonymous with "half-cell."

If we have 100 different electrodes, they can be combined to give 100(99)/2 = 4950 different galvanic cells. However, to determine $\mathscr{E}^\circ$ for these cells requires far

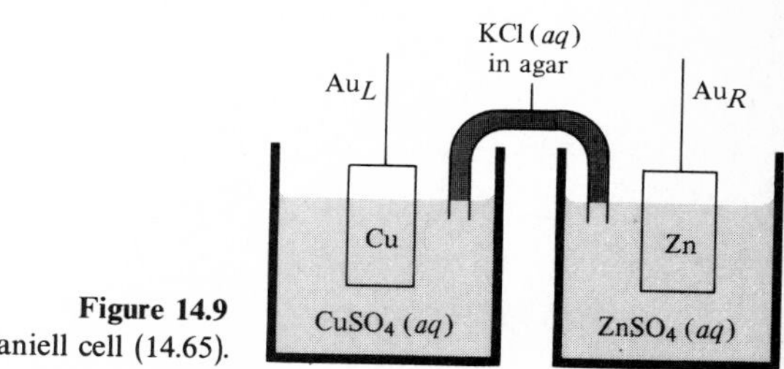

Figure 14.9
The Daniell cell (14.65).

fewer than 4950 measurements. All we have to do is pick one electrode as a reference and measure $\mathscr{E}^\circ$ for the 99 cells composed of the reference electrode and each of the remaining electrodes. These 99 $\mathscr{E}^\circ$ values allow calculation of all 4950 $\mathscr{E}^\circ$ values; this will be proved below.

Any reversible electrode can be used as the reference electrode. The reference electrode chosen for work in aqueous solutions is the hydrogen electrode $Pt|H_2(g)|H^+(aq)$. The *standard potential of an electrode reaction* (abbreviated to *standard electrode potential*) at temperature T (and pressure P) is defined to be the standard potential $\mathscr{E}^\circ$ for the cell at T and P that has the hydrogen electrode on the left of its diagram and the electrode in question on the right. For example, the standard electrode potential for the $Cu^{2+}|Cu$ electrode equals $\mathscr{E}^\circ$ for the cell

$$Cu'|Pt|H_2(g)|H^+(aq) \vdots Cu^{2+}(aq)|Cu \qquad (14.66)$$

which from (14.54) equals $-\Delta G^\circ/2\mathscr{F}$ for the chemical reaction $H_2(g) + Cu^{2+}(aq) = 2H^+(aq) + Cu$. Experiment gives $\mathscr{E}^\circ = 0.34$ V for this cell at 25°C and 1 atm. (Recall that the standard states of species in solution involve a variable pressure. Unless otherwise specified, a pressure of 1 atm will be understood.)

Suppose we have measured the standard electrode potentials of all electrodes of interest. We now ask for $\mathscr{E}^\circ$ for a cell composed of any two electrodes. For example, we might pick the electrodes $Cu^{2+}|Cu$ and $Fe^{3+}|Fe$ and ask for $\mathscr{E}^\circ$ of the cell

$$Cu|Cu^{2+} \vdots Fe^{3+}|Fe|Cu' \qquad (14.67)$$

To simplify the derivation, we shall write all cell reactions and half-reactions so as to make the cell-reaction charge number n equal to 1. (This is OK, since the potential difference between the two terminals is independent of the choice of stoichiometric coefficients in the cell reaction.) For the cell (14.67), the half-reactions are $\frac{1}{2}Cu = \frac{1}{2}Cu^{2+} + e^-$ and $\frac{1}{3}Fe^{3+} + e^- = \frac{1}{3}Fe$, and the cell reaction is

$$\tfrac{1}{2}Cu + \tfrac{1}{3}Fe^{3+} = \tfrac{1}{2}Cu^{2+} + \tfrac{1}{3}Fe \qquad (14.68)$$

Let $\mathscr{E}_R^\circ$ and $\mathscr{E}_L^\circ$ denote the standard electrode potentials of the electrodes on the right and left of (14.67); that is, $\mathscr{E}_R^\circ$ is $\mathscr{E}^\circ$ for the cell

$$Fe|Pt|H_2(g)|H^+ \vdots Fe^{3+}|Fe' \qquad (14.69)$$

and $\mathscr{E}_L^\circ$ is $\mathscr{E}^\circ$ for (14.66). The reaction of the cell (14.67) is the difference between the reactions of the cells (14.69) and (14.66):

$$\begin{array}{ll} \text{Cell (14.69):} & \tfrac{1}{3}Fe^{3+} + \tfrac{1}{2}H_2 = \tfrac{1}{3}Fe + H^+ \\ -\text{Cell (14.66):} & -(\tfrac{1}{2}Cu^{2+} + \tfrac{1}{2}H_2 = \tfrac{1}{2}Cu + H^+) \\ \hline \text{Cell (14.67):} & \tfrac{1}{2}Cu + \tfrac{1}{3}Fe^{3+} = \tfrac{1}{2}Cu^{2+} + \tfrac{1}{3}Fe \end{array}$$

Hence, ΔG° for the reaction of cell (14.67) is the difference between the ΔG°'s of the cells (14.69) and (14.66): $\Delta G^\circ_{(14.67)} = \Delta G^\circ_{(14.69)} - \Delta G^\circ_{(14.66)}$. Use of Eq. (14.54) with $n = 1$ gives $-\mathscr{F}\mathscr{E}^\circ_{(14.67)} = -\mathscr{F}\mathscr{E}^\circ_{(14.69)} + \mathscr{F}\mathscr{E}^\circ_{(14.66)}$. Division by $-\mathscr{F}$ gives $\mathscr{E}^\circ_{(14.67)} = \mathscr{E}^\circ_{(14.69)} - \mathscr{E}^\circ_{(14.66)}$, or

$$\mathscr{E}^\circ = \mathscr{E}_R^\circ - \mathscr{E}_L^\circ \qquad (14.70)^*$$

where $\mathscr{E}_R^\circ$ and $\mathscr{E}_L^\circ$ are the standard electrode potentials of the right and left half-cells of a cell whose standard emf is $\mathscr{E}^\circ$.

Although we considered a particular cell to arrive at (14.70), the same reasoning shows it to be valid for any cell. Equation (14.70) applies at any fixed temperature and allows $\mathscr{E}^\circ$ for any desired cell to be found from a tabulation of standard electrode potentials at the temperature of interest.

Table 14.1 lists some standard electrode potentials in aqueous solutions at 25°C and 1 atm. [For further values, see A. J. de Bethune and N. S. Loud, Electrode Potentials Table in C. A. Hampel (ed.), *Encyclopedia of Electrochemistry*, Reinhold, 1964. (This table does not use the IUPAC conventions.)]

Of course, the standard electrode potential for the hydrogen electrode is zero, since $\mathscr{E}^\circ$ is zero for the cell $Pt|H_2|H^+|H_2|Pt$. With all species at unit activity, the $Pt|H_2$ and $H_2|Pt$ interphase potential differences cancel each other, as do the $H_2|H^+$ and $H^+|H_2$ interphase potential differences. Alternatively, the chemical reaction for this cell is $0 = 0$, for which $\Delta G^\circ = 0$ and $\mathscr{E}^\circ = 0$.

Since the standard electrode potential of electrode i is defined for a cell with i on the right of the cell diagram, the potentials in Table 14.1 all correspond to a chemical reaction in which reduction occurs at electrode i. (Some older American texts list electrode potentials as oxidation potentials.)

Standard electrode potentials are sometimes called "single electrode potentials," but this name is highly misleading. Every number in Table 14.1 is an $\mathscr{E}^\circ$ value for a complete cell. For example, the value 0.34 V listed for the half-reaction $Cu^{2+} + 2e^- = Cu$ is $\mathscr{E}^\circ$ for the cell (14.66). Even if the potential difference across the interface (H_2 adsorbed on Pt)–$H^+(aq)$ with H_2 and H^+ at unit activity happened to be zero at 25°C, the listed value 0.34 V would not give the potential difference across the $Cu^{2+}(aq)$–Cu interface, since the cell (14.66) also contains a potential difference across the Cu′–Pt interface.

Example Vapor-pressure measurements give the stoichiometric activity coefficient of $CdCl_2$ in a 0.100 mol/kg aqueous $CdCl_2$ solution at 25°C and 1 atm as $\gamma_i = 0.228$. Find $\mathscr{E}^\circ$ and $\mathscr{E}$ at 25°C and 1 atm for the cell

$$Cu_L|Cd(s)|CdCl_2(aq,\ 0.100\ \text{mol/kg})|AgCl(s)|Ag(s)|Cu_R$$

Table 14.1. Standard electrode potentials in H_2O at 25°C and 1 atm

Half-cell reaction	$\mathscr{E}^\circ$/V	Half-cell reaction	$\mathscr{E}^\circ$/V
$K^+ + e^- = K$	−2.92	$2D^+ + 2e^- = D_2$	−0.003
$Ca^{2+} + 2e^- = Ca$	−2.87	$2H^+ + 2e^- = H_2$	0
$Na^+ + e^- = Na$	−2.71	$AgBr(c) + e^- = Ag + Br^-$	0.071
$Mg^{2+} + 2e^- = Mg$	−2.36	$AgCl(c) + e^- = Ag + Cl^-$	0.222
$Al^{3+} + 3e^- = Al$	−1.66	$Hg_2Cl_2(c) + 2e^- = 2Hg + 2Cl^-$	0.268
$Zn^{2+} + 2e^- = Zn$	−0.763	$Cu^{2+} + 2e^- = Cu$	0.34
$Ag_2S(c) + 2e^- = 2Ag + S^{2-}$	−0.66	$I_2(c) + 2e^- = 2I^-$	0.536
$Fe^{2+} + 2e^- = Fe$	−0.440	$Hg_2SO_4(c) + 2e^- = 2Hg + SO_4^{2-}$	0.615
$Cd^{2+} + 2e^- = Cd$	−0.403	$Fe^{3+} + e^- = Fe^{2+}$	0.77
$PbSO_4(c) + 2e^- = Pb + SO_4^{2-}$	−0.359	$Ag^+ + e^- = Ag$	0.799
$Sn^{2+} + 2e^- = Sn(\text{white})$	−0.14	$Br_2(l) + 2e^- = 2Br^-$	1.065
$Pb^{2+} + 2e^- = Pb$	−0.13	$Cl_2 + 2e^- = 2Cl^-$	1.359
$Fe^{3+} + 3e^- = Fe$	−0.04		

By convention, the left-hand electrode involves oxidation, so the half-reactions and overall chemical reaction are

$$Cd = Cd^{2+} + 2e^-$$

$$(AgCl + e^- = Ag + Cl^-) \times 2$$

$$Cd + 2AgCl = 2Ag + Cd^{2+} + 2Cl^-$$

Equation (14.70) and Table 14.1 give at 25°C:

$$\mathscr{E}^\circ = \mathscr{E}^\circ_R - \mathscr{E}^\circ_L = 0.222 \text{ V} - (-0.403 \text{ V}) = 0.625 \text{ V} \tag{14.71}$$

The Nernst equation (14.55) gives

$$\mathscr{E} = \mathscr{E}^\circ - \frac{RT}{2\mathscr{F}} \ln \frac{[a(\text{Ag})]^2 a(\text{Cd}^{2+})[a(\text{Cl}^-)]^2}{a(\text{Cd})[a(\text{AgCl})]^2}$$

$$\mathscr{E} = \mathscr{E}^\circ - (RT/2\mathscr{F}) \ln a(\text{Cd}^{2+})[a(\text{Cl}^-)]^2 \tag{14.72}$$

since the activities of the pure solids Ag, AgCl, and Cd are 1 at 1 atm. Equation (10.43) gives for the electrolyte $M_{\nu_+}X_{\nu_-}$

$$(a_+)^{\nu_+}(a_-)^{\nu_-} = (\gamma_+ m_+/m^\circ)^{\nu_+}(\gamma_- m_-/m^\circ)^{\nu_-} = (\gamma_\pm)^\nu (m_+/m^\circ)^{\nu_+}(m_-/m^\circ)^{\nu_-}$$

where (10.59) and (10.61) were used. This equals the quantity in brackets in (10.62), and using the equation following (10.69) and Eq. (10.66), we get

$$(a_+)^{\nu_+}(a_-)^{\nu_-} = (\nu_+)^{\nu_+}(\nu_-)^{\nu_-}(\gamma_i m_i/m^\circ)^{\nu_+ + \nu_-} \tag{14.73}$$

$$a(\text{Cd}^{2+})[a(\text{Cl}^-)]^2 = 1^1 \cdot 2^2 \cdot [(0.228)(0.100)]^3 = 4.74 \times 10^{-5}$$

Substitution in (14.72) gives

$$\mathscr{E} = 0.625 \text{ V} - \frac{(8.314 \text{ J mol}^{-1} \text{ K}^{-1})(298.15 \text{ K})}{2(96{,}485 \text{ C mol}^{-1})} \ln (4.74 \times 10^{-5})$$

$$\mathscr{E} = 0.625 \text{ V} - (-0.128 \text{ V}) = 0.753 \text{ V}$$

since 1 J/C = 1 V [Eq. (14.8)].

Suppose the cell reaction had been written $\frac{1}{2}Cd + AgCl = Ag + \frac{1}{2}Cd^{2+} + Cl^-$. Since $n = 1$ for this way of writing the reaction, the log term in the Nernst equation would be

$$-\frac{RT}{\mathscr{F}} \ln [a(\text{Cd}^{2+})]^{1/2} a(\text{Cl}^-) = -\frac{RT}{2\mathscr{F}} \ln a(\text{Cd}^{2+})[a(\text{Cl}^-)]^2$$

which is the same as in (14.72), as it must be (since the emf is independent of the choice of stoichiometric coefficients).

For the cell $Cu_L|Ag|AgCl(s)|CdCl_2(0.100 \text{ mol/kg})|Cd|Cu_R$ (which interchanges the electrodes compared with the diagram in the example), $\mathscr{E}^\circ$ would be -0.625 V and $\mathscr{E}$ would be -0.753 V.

Suppose we wanted to calculate the emf of the Daniell cell (14.65) with the assumption that the salt bridge makes the liquid-junction potentials negligible. The Nernst equation would contain the log of

$$a(\text{Cu}^{2+})/a(\text{Zn}^{2+}) = \gamma(\text{Cu}^{2+})m(\text{Cu}^{2+})/\gamma(\text{Zn}^{2+})m(\text{Zn}^{2+})$$

If both solutions were dilute, we could use the Debye–Hückel equation (or one of its extensions) to calculate the ionic activity coefficients. Also, we would have to know the equilibrium constants for ion-pair formation in $CuSO_4$ and $ZnSO_4$ solutions, so as to calculate the ionic molalities from the stoichiometric molalities of the salts. If the solutions are not dilute, we can't find the single-ion activity coefficients and hence can't calculate $\mathscr{E}$. If the $CuSO_4$ solution is dilute enough to allow $\gamma(Cu^{2+})$ to be calculated (and if the ion-pair-formation equilibrium constants are known), measurement of the cell emf allows the single-ion activity coefficient of Zn^{2+} to be found. However, this value will be somewhat in error, due to neglect of the junction potentials between the salt bridge and the two solutions.

The Nernst equation contains the term $-(RT/\mathscr{F})2.3026 \log Q_a$. At 25°C, one finds $2.3026RT/\mathscr{F} = 0.05916$ V.

14.9 CLASSIFICATION OF GALVANIC CELLS

To form a galvanic cell, we bring two half-cells together. If the electrochemical reactions in the half-cells differ, the overall cell reaction is a chemical reaction and the cell is a *chemical cell.* Examples are the cells (14.65) and (14.44). If the electrochemical reactions in the two half-cells are the same but one species B is at a different concentration in each half-cell, the cell will have a nonzero emf and its overall reaction will be a physical reaction that amounts to the transfer of B from one concentration to the other. This kind of cell is a *concentration cell.* An example is a cell composed of two chlorine electrodes with different pressures of Cl_2:

$$Pt_L|Cl_2(P_L)|HCl(aq)|Cl_2(P_R)|Pt_R \tag{14.74}$$

The half-reactions and overall reaction are

$$\begin{array}{r} 2Cl^- = Cl_2(P_L) + 2e^- \\ Cl_2(P_R) + 2e^- = 2Cl^- \\ \hline Cl_2(P_R) = Cl_2(P_L) \end{array}$$

where P_L is the Cl_2 pressure at the left electrode. Equation (14.70) gives $\mathscr{E}^\circ = \mathscr{E}_R^\circ - \mathscr{E}_L^\circ = 1.36\text{ V} - 1.36\text{ V} = 0$. (For any concentration cell, $\mathscr{E}^\circ$ is zero, since $\mathscr{E}_R^\circ$ equals $\mathscr{E}_L^\circ$.) The Nernst equation (14.55) with fugacities approximated by pressures gives for the cell (14.74)

$$\mathscr{E} = -(RT/2\mathscr{F}) \ln (P_L/P_R) \tag{14.75}$$

When $P_L = P_R$, the emf is zero.

A cell with a liquid junction involves transport of ions across the liquid junction and is therefore said to be a cell with *transference.* The Daniell cell (14.40) is a chemical cell with transference. The cell (14.43) is a chemical cell without transference. The cell (14.74) is a concentration cell without transference. An example of a concentration cell with transference is

$$Cu_L|CuSO_4(m_L)\,\vdots\,CuSO_4(m_R)|Cu_R \tag{14.76}$$

14.10 LIQUID-JUNCTION POTENTIALS

To see how a liquid-junction potential arises, consider the Daniell cell (14.40) with its emf balanced in a potentiometer, so no current flows. For simplicity, let the $CuSO_4$ and $ZnSO_4$ molalities be the same, giving equal SO_4^{2-} concentrations in the two solutions. At the junction between the solutions, ions from each solution diffuse into the other solution. It happens that Cu^{2+} ions in water are slightly more mobile than Zn^{2+} ions, so the Cu^{2+} ions diffuse into the $ZnSO_4$ solution faster than the Zn^{2+} ions diffuse into the $CuSO_4$ solution. This produces a small excess of positive charge on the $ZnSO_4$ side of the boundary and a small excess of negative charge on the $CuSO_4$ side. The negative charge on the $CuSO_4$ side speeds up the diffusion of the Zn^{2+} ions. The negative charge builds up until a steady state is reached with the Zn^{2+} and Cu^{2+} ions migrating at equal rates across the boundary. The steady-state charges on each side of the boundary produce a potential difference $\phi(\text{aq. } ZnSO_4) - \phi(\text{aq. } CuSO_4) \equiv \mathscr{E}_J$, which contributes to the measured cell emf.

In the cell (14.76), the Cu^{2+} and SO_4^{2-} ions tend to diffuse from the more concentrated to the more dilute solution. It happens that SO_4^{2-} ions diffuse faster than Cu^{2+} ions, so a slight negative charge is built up on the more dilute side of the junction.

In some cases, one can estimate liquid-junction potentials from emf measurements. For example, consider the cell

$$\text{Ag}\,|\,\text{AgCl}(s)\,|\,\text{LiCl}(m)\,\vdots\,\text{NaCl}(m)\,|\,\text{AgCl}(s)\,|\,\text{Ag} \qquad (14.77)$$

where $m(\text{LiCl}) = m(\text{NaCl})$. The half-reactions and overall reaction are

$$\text{Ag} + \text{Cl}^-(\text{in aq. LiCl}) = \text{AgCl} + e^-$$

$$\text{AgCl} + e^- = \text{Ag} + \text{Cl}^-(\text{in aq. NaCl})$$

$$\text{Cl}^-(\text{in aq. LiCl}) = \text{Cl}^-(\text{in aq. NaCl})$$

For this cell, $\mathscr{E}^\circ$ is zero, and Eq. (14.64) gives

$$\mathscr{E} = \mathscr{E}_J - \frac{RT}{\mathscr{F}} \ln \frac{\gamma(\text{Cl}^- \text{ in aq. NaCl})}{\gamma(\text{Cl}^- \text{ in aq. LiCl})}$$

At low molalities, $\gamma(\text{Cl}^-)$ will be very nearly the same in NaCl and LiCl solutions of equal molality (cf. the Debye–Hückel equation). Hence, to a good approximation, $\mathscr{E} = \mathscr{E}_J$, and the measured emf is that due to the liquid junction.

Some observed (approximate) liquid-junction potentials at 25°C for cells like (14.77) with various electrolyte pairs at $m = 0.01$ mol/kg are:

Electrolytes	HCl–NH_4Cl	HCl–LiCl	LiCl–NaCl	LiCl–CsCl
$\mathscr{E}_{J,\text{obs}}$/mV	27.0	33.8	−2.6	−7.8
$\mathscr{E}_{J,\text{calc}}$/mV	27.5	34.6	−2.5	−7.7

The $\mathscr{E}_{J,\text{calc}}$ values are theoretical values calculated from Eq. (16.103). (The larger values for junctions involving H^+ are due to the fact that the mobility of H^+ in water is much greater than that of all other cations; see Sec. 16.6.) We see that liquid-junction potentials are of the order of magnitude 10 or 20 mV (0.01 or 0.02 V). This is small but far from negligible, since cell emfs are routinely measured to 0.1 mV = 0.0001 V or better.

To see how effective a salt bridge is in reducing $\mathscr{E}_J$, consider the cell

$$\mathrm{Hg}|\mathrm{Hg_2Cl_2}(s)|\mathrm{HCl(0.1\ mol/kg)}\,\vdots\,\mathrm{KCl}(m)\,\vdots\,\mathrm{KCl(0.1\ mol/kg)}|\mathrm{Hg_2Cl_2}(s)|\mathrm{Hg}$$

where the KCl(m) solution is a salt bridge with molality m. When $m = 0.1$ mol/kg, the cell resembles the cell (14.77) and its emf (which is observed to be 27 mV) is a good approximation to $\mathscr{E}_J$ between 0.1 mol/kg HCl and 0.1 mol/kg KCl. Salt bridges use concentrated KCl solutions. When the KCl molality m is increased to 3.5 mol/kg, the cell emf drops to 1 mV, which is a good approximation to the sum of the junction potentials at the interfaces HCl(0.1 mol/kg)–KCl(3.5 mol/kg) and KCl(3.5 mol/kg)–KCl(0.1 mol/kg). We can expect that a cell with a concentrated KCl salt bridge will typically have a net junction potential of 1 or 2 mV.

The liquid-junction potential between a concentrated aqueous KCl solution and any dilute aqueous solution is quite small, for the following reasons. Because the KCl solution is concentrated, the junction potential is determined mainly by the ions of this solution, with the ions of the dilute solution making only a negligible contribution to $\mathscr{E}_J$. It happens that the mobilities of the K^+ and Cl^- ions in water are nearly equal, so these ions diffuse out of the salt bridge into the dilute solution at nearly equal rates and the junction potential is therefore quite small.

Most cells with salt bridges contain two junctions between concentrated KCl and dilute solutions, and here $\mathscr{E}_J$ is further reduced by a near cancellation of oppositely directed junction potentials. But even for a cell with a single con. KCl–dil. soln. junction, $\mathscr{E}_J$ will be quite small. Thus, $\mathscr{E}_J$ is quite small for the junction between a calomel electrode using concentrated KCl and any half-cell with low concentrations of ions.

14.11 APPLICATIONS OF EMF MEASUREMENTS

Determination of thermodynamic quantities. ΔG° for the cell reaction can be found from the experimentally determined standard emf: $\Delta G^\circ = -n\mathscr{F}\mathscr{E}^\circ$ [Eq. (14.54)]. From (9.34), we have $(\partial\mu_i^\circ/\partial T)_P = -\bar{S}_i^\circ$. Hence $[\partial(\Delta G^\circ)/\partial T]_P = -\Delta S^\circ$. Substitution of $-n\mathscr{F}\mathscr{E}^\circ$ for ΔG° gives

$$\Delta S^\circ = n\mathscr{F}\left(\frac{\partial\mathscr{E}^\circ}{\partial T}\right)_P \tag{14.78}$$

Evaluation of the temperature derivative of $\mathscr{E}^\circ$ allows the standard-state molar entropy change ΔS° of the cell's reaction to be found.

ΔH° can then be found from $\Delta G^\circ = \Delta H^\circ - T\,\Delta S^\circ$.

Since $\bar{C}_{P,i}^\circ = T(\partial\bar{S}_i^\circ/\partial T)_P$, we have $\Delta C_P^\circ = T[\partial(\Delta S^\circ)/\partial T]_P$ and (14.78) gives

$$\Delta C_P^\circ = n\mathscr{F}T(\partial^2\mathscr{E}^\circ/\partial T^2)_P \tag{14.79}$$

The temperature derivatives in (14.78) and (14.79) are found by measuring $\mathscr{E}^\circ$ at several temperatures and fitting the observed values to a truncated Taylor series of the form

$$\mathscr{E}^\circ = a + b(T - T_0) + c(T - T_0)^2 + d(T - T_0)^3 \tag{14.80}$$

where a, b, c, and d are constants and T_0 is some fixed temperature in the range of the measurements. Differentiation of (14.80) then allows ΔS°, ΔH°, and ΔC_P° to be calculated. Since each differentiation decreases the accuracy of the data, highly accurate $\mathscr{E}^\circ$ values are required for an accurate ΔC_P° to be found.

For example, the reaction $Cu + Zn^{2+} = Cu^{2+} + Zn$ occurs in the cell (14.65). Table 14.1 and Eq. (14.70) give $\mathscr{E}^\circ = -0.76\text{ V} - 0.34\text{ V} = -1.10\text{ V}$. Hence, for this redox reaction,

$$\Delta G^\circ = -n\mathscr{F}\mathscr{E}^\circ = -2(96{,}485\text{ C/mol})(-1.10\text{ V}) = 212\text{ kJ/mol} = 50.7\text{ kcal/mol}$$

since $1\text{ V} = 1\text{ J/C}$ [Eq. (14.8)] and $1\text{ cal} = 4.184\text{ J}$.

The chemical reaction $H_2 + 2AgCl = 2Ag + 2HCl(aq)$ [Eq. (14.46)] occurs in the cell (14.43). Measured $\mathscr{E}^\circ$ values for this cell in the temperature range 0 to 90°C at 1 atm are well fitted by Eq. (14.80) with

$$T_0 = 273.15\text{ K}, \qquad a = 0.23659\text{ V}, \qquad 10^4 b = -4.8564\text{ V/K}$$
$$10^6 c = -3.4205\text{ V/K}^2, \qquad 10^9 d = 5.869\text{ V/K}^3 \tag{14.81}$$

Equations (14.78) and (14.80) give

$$\Delta S^\circ = n\mathscr{F}[b + 2c(T - T_0) + 3d(T - T_0)^2] \tag{14.82}$$

Substitution of numerical values gives at 0°C

$$\Delta S^\circ_{273} = n\mathscr{F}b = 2(96{,}485\text{ C/mol})(-4.8564 \times 10^{-4}\text{ V/K}) = -93.71\text{ J mol}^{-1}\text{ K}^{-1}$$

(Recall the role of galvanic-cell measurements in establishing the third law of thermodynamics.)

Determination of equilibrium constants. Once $\mathscr{E}^\circ$ for a cell has been found, ΔG° and the equilibrium constant K_a of the cell's chemical reaction can be found from $\Delta G^\circ = -n\mathscr{F}\mathscr{E}^\circ$ and $\Delta G^\circ = -RT \ln K_a$ [Eq. (11.3)]. Combining these two equations, we have

$$\ln K_a = n\mathscr{F}\mathscr{E}^\circ/RT \tag{14.83}$$

Although the anode reaction in a cell is an oxidation and the cathode reaction is a reduction, the overall cell reaction is not necessarily an oxidation–reduction reaction (as can be seen from the AgCl example below), so (14.83) is not limited to redox reactions. The kinds of equilibrium constants that have been determined from cell emf measurements include redox K_a's, solubility products, dissociation constants of complex ions, the ionization constant of water, ionization constants of weak acids, and ion-pair-formation equilibrium constants.

For example, for the redox reaction $Zn + Cu^{2+} = Zn^{2+} + Cu$, data in Table 14.1 give $\mathscr{E}^\circ = 0.34\text{ V} - (-0.76\text{ V}) = 1.10\text{ V}$. Substitution in (14.83) gives at 25°C

$$\ln K_a = \frac{2(96{,}485\text{ C mol}^{-1})(1.10\text{ J C}^{-1})}{(8.314\text{ J mol}^{-1}\text{ K}^{-1})(298.15\text{ K})} = 85.6, \qquad K_a = 1.5 \times 10^{37}$$

(which could also have been found from the value $\Delta G^\circ = 212{,}000$ J/mol found above for the reverse reaction). At equilibrium, virtually no Cu^{2+} remains in solution.

Equation (14.83) gives $K_a = \exp(n\mathscr{F}\mathscr{E}^\circ/RT)$. For $n = 1$, we find that each difference of 0.1 V between the half-reaction standard potentials contributes a factor

49 to K_a at 25°C. The more positive $\mathscr{E}^\circ$ is, the larger K_a is. A large negative $\mathscr{E}^\circ$ indicates a very small K_a.

The more negative the reduction potential $\mathscr{E}^\circ$ for the half-reaction $M^{z+} + z_+e^- = M$, the greater the tendency for metal M to be oxidized. Thus, a metal will tend to replace in solution those metals which lie below it in Table 14.1. For example, Zn replaces Cu^{2+} from aqueous solutions ($Zn + Cu^{2+} = Zn^{2+} + Cu$). Metals lying above the hydrogen electrode in Table 14.1 replace H^+ from solution and dissolve readily in aqueous acids, generating H_2. Metals near the top of the table, for example, Na, K, Mg, Ca, replace H^+ from water.

To determine the solubility product K_{sp}° for AgCl, we need a cell with overall reaction $AgCl(s) = Ag^+(aq) + Cl^-(aq)$. Such a cell is

$$Ag|Ag^+ \vdots Cl^- |AgCl(s)|Ag \tag{14.84}$$

The half-reactions are $Ag = Ag^+ + e^-$ and $AgCl(s) + e^- = Ag + Cl^-$, and the overall reaction is

$$AgCl(s) = Ag^+ + Cl^- \tag{14.85}$$

At the anode, Ag is oxidized, and at the cathode Ag (in AgCl) is reduced, so the overall cell reaction is not a redox reaction. Table 14.1 and Eq. (14.83) give $\mathscr{E}^\circ = 0.222\text{ V} - 0.799\text{ V} = -0.577\text{ V}$, and $K_{sp}^\circ = 1.76 \times 10^{-10}$ at 25°C and 1 atm. Note that there is no need to set up and measure $\mathscr{E}^\circ$ for the cell (14.84), since its $\mathscr{E}^\circ$ can be found by combining measured standard electrode potentials of the $Ag^+|Ag$ and Ag–AgCl electrodes.

Use of emf data to determine other kinds of equilibrium constants is examined in Probs. 14.36 and 14.37.

Determination of activity coefficients. Since the emf of a cell depends on the activities of the ions in solution, it is easy to use measured emf values to calculate activity coefficients. For example, for the cell (14.44) with HCl as its electrolyte, the cell reaction is (14.46) and the emf is given by Eq. (14.60). Once $\mathscr{E}^\circ$ has been determined by extrapolation to $m = 0$, the stoichiometric activity coefficient γ_i of HCl(aq) at any molality m can be calculated from the measured emf $\mathscr{E}$ at that molality by using (14.60). Some results at 25°C and 1 atm are:

$m/(\text{mol kg}^{-1})$	0.001	0.01	0.10	1	2
$\gamma_i[\text{HCl}(aq)]$	0.966	0.905	0.796	0.809	1.009

Determination of pH. The symbol pb means $-\log b$, where b is some physical quantity: $\text{p}b \equiv -\log b$. For example,

$$\begin{aligned} \text{pc}(H^+) &\equiv -\log [c(H^+)/c^\circ], \qquad & \text{p}m(H^+) &\equiv -\log [m(H^+)/m^\circ] \\ \text{p}a_c(H^+) &\equiv -\log a_c(H^+), \qquad & \text{p}a_m(H^+) &\equiv -\log a_m(H^+) \end{aligned} \tag{14.86}$$

where the quantities $c^\circ \equiv 1\text{ mol/dm}^3$ and $m^\circ \equiv 1$ mol/kg have been inserted to make the arguments of the logarithms dimensionless (as they must be). In (14.86), $a_c(H^+)$ and $a_m(H^+)$ are the concentration-scale and molality-scale activities of H^+ [Eq. (10.43)].

Over the years, each of the quantities in (14.86) has been called "the pH" of a solution. The present definition of pH is none of these. Instead pH is defined operationally to yield a quantity that is easily and reproducibly measured and as closely equal to $-\log a_m(\mathrm{H}^+)$ as present theory allows.

To understand the current definition of pH, consider the cell

$$\mathrm{Pt}\,|\,\mathrm{H_2}(g)\,|\,\mathrm{soln.\ X}\,\vdots\,\mathrm{KCl(sat.)}\,|\,\mathrm{Hg_2Cl_2}(s)\,|\,\mathrm{Hg}\,|\,\mathrm{Pt}' \tag{14.87}$$

which consists of a saturated calomel electrode and a hydrogen electrode dipping into an aqueous solution X whose molality-scale activity of H^+ is $a_\mathrm{X}(\mathrm{H}^+)$. The cell reaction and emf $\mathscr{E}_\mathrm{X}$ [Eq. (14.64)] are

$$\tfrac{1}{2}\mathrm{H_2}(g) + \tfrac{1}{2}\mathrm{Hg_2Cl_2}(s) = \mathrm{Hg}(l) + \mathrm{H}^+(aq,\ \mathrm{X}) + \mathrm{Cl}^-(aq)$$

$$\mathscr{E}_\mathrm{X} = \mathscr{E}_{J,\mathrm{X}} + \mathscr{E}^\circ - RT\mathscr{F}^{-1}[\ln a_\mathrm{X}(\mathrm{H}^+) + \ln a(\mathrm{Cl}^-) - \tfrac{1}{2}\ln P(\mathrm{H_2})/P^\circ]$$

where $\mathscr{E}_{J,\mathrm{X}}$ is the junction potential between solution X and the saturated KCl solution. Provided the ionic strength of solution X is reasonably low, $\mathscr{E}_{J,\mathrm{X}}$ should be small, due to the concentrated KCl solution (see Sec. 14.10).

If a second cell is set up identical to (14.87) except that solution X is replaced by solution S, then the emf $\mathscr{E}_\mathrm{S}$ of this cell will be

$$\mathscr{E}_\mathrm{S} = \mathscr{E}_{J,\mathrm{S}} + \mathscr{E}^\circ - RT\mathscr{F}^{-1}[\ln a_\mathrm{S}(\mathrm{H}^+) + \ln a(\mathrm{Cl}^-) - \tfrac{1}{2}\ln P(\mathrm{H_2})/P^\circ]$$

where $a_\mathrm{S}(\mathrm{H}^+)$ is the activity of H^+ in solution S. Subtraction gives

$$\mathscr{E}_\mathrm{X} - \mathscr{E}_\mathrm{S} = \mathscr{E}_{J,\mathrm{X}} - \mathscr{E}_{J,\mathrm{S}} - RT\mathscr{F}^{-1}[\ln a_\mathrm{X}(\mathrm{H}^+) - \ln a_\mathrm{S}(\mathrm{H}^+)]$$

$$RT\mathscr{F}^{-1}\ln 10[-\log a_\mathrm{X}(\mathrm{H}^+) + \log a_\mathrm{S}(\mathrm{H}^+)] = \mathscr{E}_\mathrm{X} - \mathscr{E}_\mathrm{S} + \mathscr{E}_{J,\mathrm{S}} - \mathscr{E}_{J,\mathrm{X}}$$

$$\mathrm{p}a_\mathrm{X}(\mathrm{H}^+) = \mathrm{p}a_\mathrm{S}(\mathrm{H}^+) + \frac{\mathscr{E}_\mathrm{X} - \mathscr{E}_\mathrm{S}}{RT\mathscr{F}^{-1}\ln 10} + \frac{\mathscr{E}_{J,\mathrm{S}} - \mathscr{E}_{J,\mathrm{X}}}{RT\mathscr{F}^{-1}\ln 10} \tag{14.88}$$

where we used $\mathrm{p}a_\mathrm{X}(\mathrm{H}^+) \equiv -\log a_\mathrm{X}(\mathrm{H}^+)$ and Eq. (2.125).

If solutions X and S are reasonably similar, the junction potentials $\mathscr{E}_{J,\mathrm{X}}$ and $\mathscr{E}_{J,\mathrm{S}}$ will be approximately equal and the last term in (14.88) can be neglected. By analogy to (14.88) with the last term omitted, the pH of solution X is defined as

$$\mathrm{pH(X)} \equiv \mathrm{pH(S)} + \frac{\mathscr{E}_\mathrm{X} - \mathscr{E}_\mathrm{S}}{RT\mathscr{F}^{-1}\ln 10} \tag{14.89}$$

In this equation, pH(S) is the assigned pH value for a standard solution. The pH(S) values are numbers chosen to be as closely equal to $-\log a_\mathrm{S}(\mathrm{H}^+)$ as present knowledge allows. We have $a_\mathrm{S}(\mathrm{H}^+) = \gamma_\mathrm{S}(\mathrm{H}^+)m(\mathrm{H}^+)/m^\circ$. For a solution with low ionic strength, a reasonably accurate value of $\gamma_\mathrm{S}(\mathrm{H}^+)$ can be calculated from an extended form of the Debye–Hückel equation, allowing $-\log a_\mathrm{S}(\mathrm{H}^+)$ to be accurately estimated for the solution S of known composition. A list of assigned pH(S) values is given in *Bates*, p. 73. As an example, the assigned pH(S) value for an aqueous buffer solution with $m(\mathrm{Na_2CO_3}) = 0.025$ mol/kg and $m(\mathrm{NaHCO_3}) = 0.025$ mol/kg is 10.012 at 25°C and 1 atm.

The defined pH in Eq. (14.89) differs from $\mathrm{p}a_{m,\mathrm{X}}(\mathrm{H}^+)$ because the difference $\mathscr{E}_{J,\mathrm{S}} - \mathscr{E}_{J,\mathrm{X}}$ of junction potentials is not precisely zero and because the assigned standard pH(S) values are not precisely equal to $\mathrm{p}a_\mathrm{S}(\mathrm{H}^+)$. The current definition of pH is designed to give a quantity that is easily measured rather than a quantity that has a precise thermodynamic significance. For solutions of low ionic strength and pH in the range 2 to 12, it is believed that the operationally defined pH in (14.89) is within 0.02 of $-\log a_m(H^+)$.

In practical measurements of pH, it is convenient to replace the hydrogen electrode in (14.87) by a glass electrode (Sec. 14.13); this is done in commercially available pH meters.

Potentiometric titrations. In an acid–base titration, the pH changes rapidly as the neutralization point is reached; the slope of a plot of pH vs. volume of added reagent is a maximum at the endpoint. Monitoring the pH with a pH meter allows the endpoint to be determined; the solution being titrated is solution X in the cell (14.87).

Redox titrations can be done potentiometrically. For example, suppose a solution of Fe^{2+} is being titrated with $Ce(SO_4)_2$, according to $Fe^{2+} + Ce^{4+} = Fe^{3+} + Ce^{3+}$. We insert a Pt wire and the salt bridge of a calomel electrode into the solution to form the cell

$$Pt|Fe^{2+}, Fe^{3+} \vdots KCl(sat)|Hg_2Cl_2(s)|Hg|Pt' \qquad (14.90)$$

The emf of the cell contains a term proportional to $\log [a(Fe^{3+})/a(Fe^{2+})]$, so monitoring the emf allows the titration endpoint to be found. At the endpoint, the slope of a plot of cell emf vs. volume of added reagent is a maximum.

14.12 BATTERIES

Galvanic cells provide electric power for starting internal-combustion engines in automobiles, for running systems on space vehicles, and for such devices as flashlights, toys, heart pacers, electric automobiles, electronic calculators, and portable radios, TVs, and tape recorders.

The term *battery* means either a single galvanic cell or several galvanic cells connected in series (in which case the emfs are additive).

The lead storage battery used in cars consists of three or six galvanic cells in series and has an emf of 6 or 12 V. Each cell is

$$Pb|PbSO_4(s)|H_2SO_4(aq)|PbSO_4(s)|PbO_2(s)|Pb' \qquad (14.91)$$

The reactions are

$$Pb + HSO_4^- = PbSO_4(s) + H^+ + 2e^-$$

$$PbO_2 + 3H^+ + HSO_4^- + 2e^- = PbSO_4(s) + 2H_2O$$

$$\overline{Pb + PbO_2(s) + 2H^+ + 2HSO_4^- = 2PbSO_4(s) + 2H_2O}$$

The cell is reversible and is readily recharged.

The United States space program and the desire for electrically powered automobiles have spurred the development of many new batteries.

A *fuel cell* is a galvanic cell in which the reactants are continuously fed to each electrode from outside the cell. Figure 14.10 shows a hydrogen–oxygen fuel cell whose diagram is

$$C|H_2(g)|NaOH(aq)|O_2(g)|C' \qquad (14.92)$$

The electrodes are made of porous graphite (which is a good conductor). The H_2 and O_2 gases are continually fed in and diffuse into the electrode pores. The electrolyte solution also diffuses part way into the pores. Each electrode is impregnated with a catalyst to speed the oxidation or reduction half-reaction. Catalysts used include Pt, Ag, and metal oxides. In the anode pores, H_2 is oxidized to H^+, according to $H_2 \rightarrow 2H^+ + 2e^-$. The H^+ is neutralized by the OH^- of the electrolyte ($2H^+ + 2OH^- \rightarrow 2H_2O$), so the net anode reaction is $H_2 + 2OH^- \rightarrow 2H_2O + 2e^-$. At the cathode, oxygen is reduced: $O_2 + 2H_2O + 4e^- = 4OH^-$. The net reaction is $2H_2 + O_2 \rightarrow 2H_2O$. The half-reactions and overall reaction are the reverse of those in the electrolysis of water. The emf is 1 V. Oxygen is usually supplied as air.

Hydrogen–oxygen fuel cells are used in United States spacecraft to supply power for heat, light, and radio communication. The water produced by the cell reaction vaporizes out of the cell and is condensed and used for drinking water by the astronauts.

Fuel cells that use hydrocarbons (CH_4, C_2H_6, etc.) as fuel (instead of H_2) exist. A typical half-reaction is $C_3H_8 + 6H_2O \rightarrow 3CO_2 + 20H^+ + 20e^-$, and the overall reaction is $C_3H_8 + 5O_2 \rightarrow 3CO_2 + 4H_2O$. Unfortunately, the oxidation half-reaction for a hydrocarbon requires a Pt catalyst, which costs too much to make such cells economically practical.

When electric power is produced in a steam power plant that runs on heat supplied by burning a hydrocarbon fuel, the efficiency cannot exceed that of a reversible Carnot engine operating between the same temperatures as the heat engine; see Eqs. (3.7) and (3.17). A fuel cell is not a heat engine and is not subject to the Carnot limitation on efficiency. Thus, fuel-cell efficiencies are typically 60 to 70 percent, compared with 20 to 40 percent for internal-combustion engines and steam power plants. Fuel cells are also less polluting.

14.13 ION-SELECTIVE MEMBRANE ELECTRODES

An *ion-selective membrane electrode* contains a glass, crystalline, or liquid membrane whose nature is such that the potential difference between the membrane and an electrolyte solution it is in contact with is determined by the activity of one particular ion.

The oldest and most widely used membrane electrode is the *glass electrode* (Fig. 14.11*a*), whose essential component is a thin glass membrane of special composition. Glass contains a three-dimensional network of covalently bound Si and O atoms with a net negative charge, plus positive ions, for example, Na^+, Li^+, Ca^{2+}, in the spaces in the Si–O network. The positive ions of the alkali metals can move through the glass, giving it a very weak electrical conductivity. The thinness (0.005 cm) of the membrane reduces its resistance. Even so, the resistances of glass electrodes run 10^7 to 10^9 ohms. The high resistance makes emf measurements with a potentiometer (Fig. 14.6) inaccurate, due to the inability of the galvanometer to detect the extremely small currents involved, so an electronic voltmeter must be used.

An Ag–AgCl electrode plus an internal filling solution of aqueous HCl are sealed in as part of the glass electrode. (Sometimes a calomel electrode replaces the Ag–AgCl electrode. The glass electrode also works if one uses liquid Hg with a Pt terminal instead of the filling solution and Ag–AgCl electrode.)

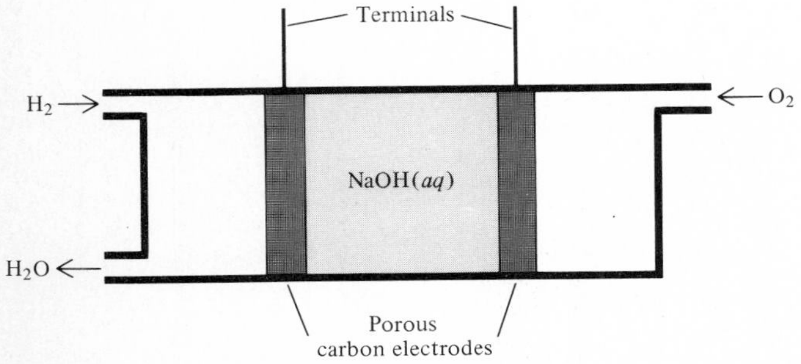

Figure 14.10
A fuel cell.

The main application of glass electrodes is to pH measurement. To measure the pH of solution X, we set up the following cell (Fig. 14.11*a*):

$$\mathrm{Pt}\,|\,\mathrm{Ag}\,|\,\mathrm{AgCl}(s)\,|\,\mathrm{HCl}(aq)\,|\,\mathrm{glass}\,|\,\mathrm{soln.\ X}\,\vdots\,\mathrm{KCl(sat.)}\,|\,\mathrm{Hg_2Cl_2}(s)\,|\,\mathrm{Hg}\,|\,\mathrm{Pt'}$$

Let this be cell X with emf $\mathscr{E}_X$.

Before a freshly made glass electrode is used, it is immersed in water for a few hours; monovalent cations, for example, Na^+, at and near the surface of the glass are replaced by H^+ ions from the water. When the glass electrode is immersed in solution X, an equilibrium between H^+ ions in solution and H^+ ions in the glass surface is set up. This charge transfer between glass and solution produces a potential difference between the glass and the solution. Equation (14.31) gives

$$\phi(\mathrm{X}) - \phi(\mathrm{glass}) = [\mu^{\mathrm{gl}}(\mathrm{H}^+) - \mu^{\mathrm{X}}(\mathrm{H}^+)]/\mathscr{F} \tag{14.93}$$

Use of $\mu_i = \mu_i^\circ + RT \ln a_i$ for H^+ in the glass and in the solution gives

$$\phi(\mathrm{X}) - \phi(\mathrm{glass}) = \frac{\mu^{\circ,\mathrm{gl}}(\mathrm{H}^+) - \mu^{\circ,\mathrm{aq}}(\mathrm{H}^+)}{\mathscr{F}} + \frac{RT}{\mathscr{F}}[\ln a^{\mathrm{gl}}(\mathrm{H}^+) - \ln a^{\mathrm{X}}(\mathrm{H}^+)] \tag{14.94}$$

where $\mu^{\circ,\mathrm{gl}}(\mathrm{H}^+)$ and $\mu^{\circ,\mathrm{aq}}(\mathrm{H}^+)$ are the standard-state chemical potentials of H^+ in glass and in water and $a^{\mathrm{gl}}(\mathrm{H}^+)$ and $a^{\mathrm{X}}(\mathrm{H}^+)$ are the activities of H^+ in the glass and in solution X. [Note the resemblance of (14.93) to (14.36).] The emf $\mathscr{E}_X$ of cell X equals (14.94) plus the $\Delta\phi$'s at all the other interfaces.

Let solution X be replaced by a standard solution S, to give cell S with emf $\mathscr{E}_S$. We have

$$\phi(\mathrm{S}) - \phi(\mathrm{glass}) = \frac{\mu^{\circ,\mathrm{gl}}(\mathrm{H}^+) - \mu^{\circ,\mathrm{aq}}(\mathrm{H}^+)}{\mathscr{F}} + \frac{RT}{\mathscr{F}}[\ln a^{\mathrm{gl}}(\mathrm{H}^+) - \ln a^{\mathrm{S}}(\mathrm{H}^+)] \tag{14.95}$$

If we assume that the liquid-junction potential $\mathscr{E}_{J,\mathrm{X}}$ between solution X and the calomel electrode equals the junction potential $\mathscr{E}_{J,\mathrm{S}}$ between solution S and the calomel electrode, then the $\Delta\phi$'s for the cells X and S are the same at all interfaces except at the glass–solution X or S interface. Hence $\mathscr{E}_X - \mathscr{E}_S$ equals (14.94) minus (14.95), and

$$\mathscr{E}_\mathrm{X} - \mathscr{E}_\mathrm{S} = RT\mathscr{F}^{-1}(\ln 10)[\log a^{\mathrm{S}}(\mathrm{H}^+) - \log a^{\mathrm{X}}(\mathrm{H}^+)]$$

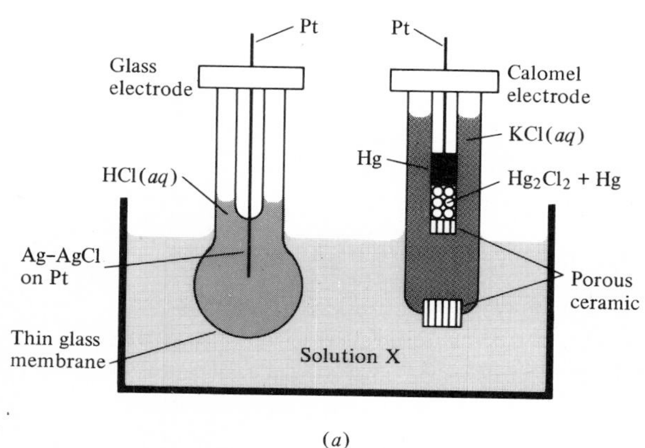

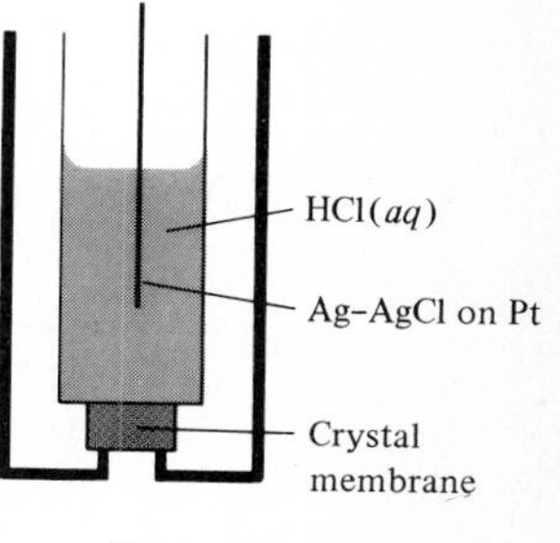

Figure 14.11 (*a*) Measurement of pH using a glass electrode. (*b*) A crystal-membrane electrode.

We now replace $\log a^{S}(H^+)$ by $-pH(S)$, where the defined pH(S) of the standard solution is designed to closely approximate $-\log a^{S}(H^+)$. Since we have used the same approximations as in Sec. 14.11 [namely, assuming equal liquid-junction potentials and using a defined pH(S) value], we replace $\log a^{X}(H^+)$ by $-pH(X)$, to get

$$pH(X) = pH(S) + (\mathscr{E}_X - \mathscr{E}_S)\mathscr{F}(RT \ln 10)^{-1} \tag{14.96}$$

which is the same as Eq. (14.89). Thus, a glass electrode can replace the hydrogen electrode in pH measurements.

It used to be thought that the glass electrode functioned as a membrane permeable to H^+, but radioactive-tracer studies proved that H^+ ions are not transported through the bulk of the membrane. The glass membrane works by ion exchange at its surface.

Glass is made by cooling a molten mixture of SiO_2 and metal oxides. By varying the composition of the glass, one can produce a glass electrode that is sensitive to an ion other than H^+. One particular glass electrode that is sensitive to H^+ is composed of 72 mole percent SiO_2, $21\frac{1}{2}$ percent Na_2O, and $6\frac{1}{2}$ percent CaO. By using glass composed of SiO_2, Al_2O_3, and Li_2O, we get a membrane sensitive to Na^+; this glass has sites on its surface that specifically adsorb Na^+ ions, and an expression like (14.94) holds with Na^+ replacing H^+. With various compositions, one can prepare a glass membrane electrode that is sensitive to any one of the following univalent cations: H^+, Na^+, K^+, Li^+, Cs^+, Rb^+, Ag^+, $NH_4{}^+$, Cu^+, and Tl^+.

The glass membrane can be replaced by a crystal of a salt that is "insoluble" in water and that has significant ionic conductivity at room temperature. Figure 14.11*b* shows a crystal-membrane electrode. As an example, a crystal of LaF_3 gives a membrane that is sensitive to F^-; there is an equilibrium between F^- adsorbed at the crystal surface and F^- in the solution the electrode dips into, and an equation like (14.94) holds with F^- replacing H^+ and $-\mathscr{F}$ replacing $\mathscr{F}$. A membrane of $Ag_2S(s)$ can be used to determine S^{2-} in solution. Membranes of AgX (where X = Cl, Br, or I) mixed with Ag_2S are used to determine halide ions.

Liquid-membrane electrodes use a solution of an organic salt held in an inert support of porous glass or plastic. For example, a Ca^{2+}-sensitive liquid-membrane electrode uses $Ca[(C_{10}H_{21}O)_2PO_2]_2$ (a calcium salt of a diester of phosphoric acid) in a suitable solvent.

Defined similar to pH are the quantities pCa, pCl, ..., which are designed to be very close approximations to $-\log a(Ca^{2+})$, $-\log a(Cl^-)$, ... Standard solutions with assigned values of pCa, pNa, pCl, and pF are listed in R. G. Bates and R. A. Robinson, *Pure Appl. Chem.*, **37**, 575 (1974).

Ion-selective membrane electrodes allow measurement of the activities of certain ions that are difficult to determine by traditional analytical methods. Examples are Na^+, K^+, Ca^{2+}, $NH_4{}^+$, Mg^{2+}, F^-, $NO_3{}^-$, and $ClO_4{}^-$. Although these electrodes respond to activities (and not concentrations or molalities), one can measure concentrations by using these electrodes in potentiometric titrations (Sec. 14.11).

Glass microelectrodes are used to measure activities of H^+, Na^+, and K^+ in biological tissues.

14.14 MEMBRANE EQUILIBRIUM

Consider two KCl solutions (α and β) separated by a membrane permeable to K^+ but impermeable to Cl^- and to the solvent(s). Let solution α be more concentrated than β. The K^+ ions will tend to diffuse through the membrane from α to β. This produces a net positive charge on the β side of the membrane and a net negative charge on the α side. The negative charge on solution α retards the diffusion of K^+ from α to β and speeds up diffusion of K^+ from β to α. Eventually an equilibrium is reached in which the rates of K^+ diffusion are equal. At equilibrium, solution β is at a higher electric potential than α, due to transfer of a chemically undetectable amount of K^+.

To derive an expression for the potential difference across the membrane, consider two electrolyte solutions α and β that are separated by a membrane permeable to ion k and possibly to some (but not all) of the other ions present; the membrane is impermeable to the solvent(s). At equilibrium, Eq. (14.30) gives $\tilde{\mu}_k^{\alpha} = \tilde{\mu}_k^{\beta}$. Equations (14.29) and (11.2) give $\tilde{\mu}_k^{\alpha} = \mu_k^{\alpha} + z_k \mathscr{F} \phi^{\alpha} = \mu_k^{\circ,\alpha} + RT \ln a_k^{\alpha} + z_k \mathscr{F} \phi^{\alpha}$. Hence

$$\mu_k^{\circ,\alpha} + RT \ln a_k^{\alpha} + z_k \mathscr{F} \phi^{\alpha} = \mu_k^{\circ,\beta} + RT \ln a_k^{\beta} + z_k \mathscr{F} \phi^{\beta}$$

$$\Delta\phi \equiv \phi^{\beta} - \phi^{\alpha} = -\frac{\mu_k^{\circ,\beta} - \mu_k^{\circ,\alpha}}{z_k \mathscr{F}} - \frac{RT}{z_k \mathscr{F}} \ln \frac{a_k^{\beta}}{a_k^{\alpha}} \tag{14.97}$$

$\Delta\phi$ is called the *membrane* (or *transmembrane*) *potential.* Note the resemblance of (14.97) to the Nernst equation (14.53). [In fact, we could apply (14.97) to the contact equilibrium between a metal electrode and a solution. For example, for a piece of Cu dipping into aqueous $CuSO_4$, the metal–solution interface can be viewed as a "membrane" that is permeable to Cu^{2+} but impermeable to SO_4^{2-}, e^-, and H_2O.]

If the solvents in solutions α and β are the same, then $\mu_k^{\circ,\alpha} = \mu_k^{\circ,\beta}$ and (14.97) becomes

$$\phi^{\beta} - \phi^{\alpha} = -\frac{RT}{z_k \mathscr{F}} \ln \frac{a_k^{\beta}}{a_k^{\alpha}} = \frac{RT}{z_k \mathscr{F}} \ln \frac{\gamma_k^{\alpha} m_k^{\alpha}}{\gamma_k^{\beta} m_k^{\beta}} \tag{14.98}$$

If the membrane is permeable to several ions, the equilibrium activities and the equilibrium value of $\Delta\phi$ must be such that (14.98) is satisfied for each ion that can pass through the membrane.

The above situation where the membrane is impermeable to the solvent is called *nonosmotic membrane equilibrium.* More commonly, the membrane is permeable to the solvent, as well as to one or more of the ions. The requirements of equal electrochemical potentials in the two phases for the solvent and for the permeating ions lead to a pressure difference between the two solutions at equilibrium. The equation for $\Delta\phi$ in *osmotic membrane equilibrium* (also called *Donnan membrane equilibrium*) is more complicated than (14.98). (See *Guggenheim*, sec. 8.08 for the treatment.) However, for dilute solutions, Eq. (14.98) turns out to be a good approximation for osmotic membrane equilibrium.

The potential difference between solutions α and β can be reasonably accurately measured by setting up the cell

$$Ag_L \,|\, AgCl(s) \,|\, KCl(aq) \,\vdots\, \alpha \,|\, \text{membrane} \,|\, \beta \,\vdots\, KCl(aq) \,|\, AgCl(s) \,|\, Ag_R \tag{14.99}$$

in which concentrated-KCl salt bridges connect each solution to an Ag–AgCl electrode. Provided α and β are reasonably dilute, the sum of the liquid-junction potentials will be small (1 or 2 mV); hence, to a good approximation, the cell emf is $\mathscr{E} \equiv \phi(\text{Ag}_R) - \phi(\text{Ag}_L) = \phi(\text{soln. }\beta) - \phi(\text{soln. }\alpha)$, since the potential differences at the interfaces to the left of α in the diagram cancel with those to the right of β.

Equation (14.98) is sometimes applied to cell-membrane potentials in living organisms. The K^+ molality in the fluid inside a resting nerve cell is 30 or 40 times that outside the cell. Due to the presence of Na^+, Cl^-, and other ions, the ionic strengths inside and outside the nerve cell are roughly equal, so we shall take $\gamma(K^+)$ to be the same on each side of the membrane. Equation (14.98) then gives at 25°C

$$\phi^{\text{int}} - \phi^{\text{ext}} = \frac{(8.31 \text{ J mol}^{-1} \text{ K}^{-1})(298 \text{ K})}{1(96{,}500 \text{ C mol}^{-1})} \ln \frac{1}{35} = -91 \text{ mV} \qquad (14.100)$$

where ϕ^{int} and ϕ^{ext} are the potentials inside and outside the cell. The observed transmembrane potential for a resting nerve cell is about -70 mV, which differs from (14.100). This is because the solutions in living organisms are not in equilibrium, and cell-membrane potentials cannot be understood solely in terms of the Nernst equation (14.98), which is an equilibrium equation. (At equilibrium, nothing is happening. If you're in equilibrium, you're dead.) For more on nerve-cell membrane potentials, see Sec. 14.16.

14.15 THE ELECTRICAL DOUBLE LAYER

We saw in Sec. 14.3 that the interphase region between two bulk phases usually contains a complex distribution of electric charge resulting from (*a*) charge transfer between phases, (*b*) unequal adsorption of positive and negative ions, (*c*) orientation of molecules with permanent dipole moments, and (*d*) induced polarization of molecules. For historical reasons, the charge distribution in the interphase region is called the *electrical double layer*.

We shall consider mainly the double layer at the interface between a metal electrode and an aqueous electrolyte solution, e.g., between Cu and $CuSO_4(aq)$. Suppose the electrode is positively charged due to a net gain of Cu^{2+} ions from the solution. In 1879 Helmholtz proposed that the charge distribution in the interphase region consists of a series of charges in the metal's surface and a series of opposite charges held fixed in the solution at a certain distance δ from the electrode (Fig. 14.12*a*). Gouy and Chapman rejected the Helmholtz model, arguing that the thermal motion of the ions in solution would cause them to be spread out diffusely rather than held motionless at a fixed distance from the electrode.

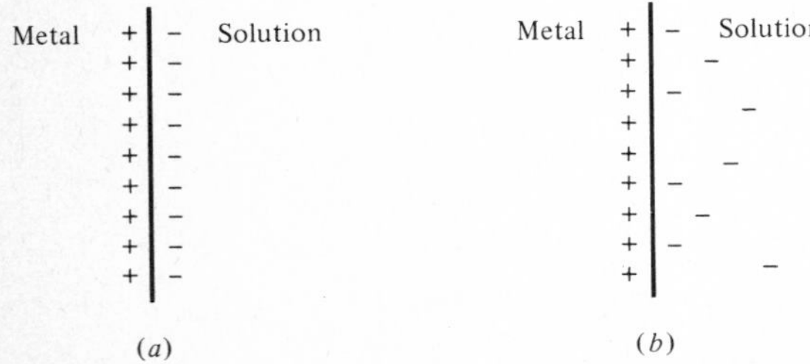

Figure 14.12
Electrical double layer. (*a*) Helmholtz model. (*b*) Stern model.

In 1924, Stern combined the Helmholtz and Gouy–Chapman models, proposing that some of the excess negative ions in the solution are adsorbed on the electrode and held at a fixed distance determined by the ionic radius while the remainder of the excess negative ions are distributed diffusely in the interphase region (Fig. 14.12*b*). Figure 14.13 shows the variation of the electric potential ϕ with distance from the electrode, as calculated from the Stern model.

It is believed that the Stern model is essentially correct. However, Stern did not explicitly consider the orientation of water dipoles at the electrode. Most of the electrode surface is covered with a layer of adsorbed water molecules. If the electrode is positively charged, most of the water molecules in the adsorbed layer will have their negative (oxygen) sides in contact with the electrode. This orientation of the water dipole moments affects ϕ in the interphase region.

The electric field in the electrode–solution interphase region is extremely high. Table 14.1 indicates that electrode–solution potential differences are typically about 1 V. The interphase region is of the order of 50 Å thick. Letting z be the direction perpendicular to the interface, we have $|E_z| = d\phi/dz \approx \Delta\phi/\Delta z \approx (1\text{ V})/(50 \times 10^{-8}\text{ cm}) = 2 \times 10^6\text{ V/cm}$.

Recall from Sec. 13.6 that the colloidal particles in a sol are charged. If the particles are positively charged, some negative charges in the dispersion medium will be adsorbed at each particle's surface and other negative charges will be distributed diffusely in the region near the particle. There is a double layer at each particle–solution interface.

14.16 BIOELECTROCHEMISTRY

Cytoplasm (the fluid in cells of living organisms) and the extracellular fluid that surrounds cells contain significant amounts of dissolved electrolytes. The total electrolyte molarity is typically 0.3 mol/dm^3 for mammalian fluids.

Figure 14.14 shows a piece of animal tissue pinned to the bottom of a chamber filled with a solution having the same composition as the extracellular fluid of the organism. By penetrating the membrane of a single cell with a micro salt bridge, one sets up the electrochemical cell (14.99), where phase β is the interior of the biological cell, the membrane is the cell's plasma membrane (Sec. 13.7), and phase α is the bathing solution. The potential difference measured is that between the interior and exterior of the biological cell and is the transmembrane potential. The membrane potential is displayed on an oscilloscope or a chart recorder. The micro salt bridge (often inaccurately called a microelectrode) consists of glass drawn to a fine tip

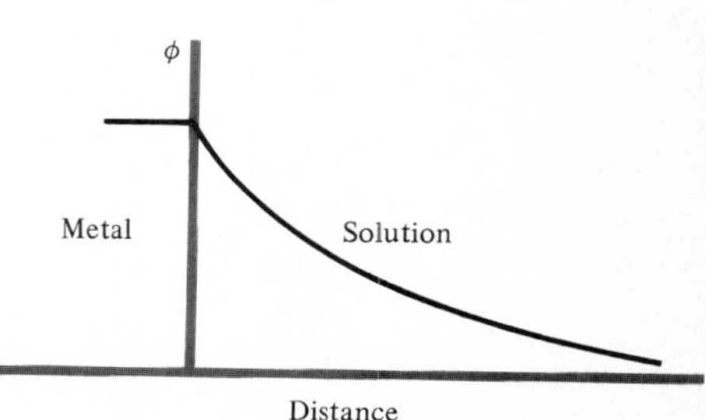

Figure 14.13
Electric potential vs. distance in the Stern model.

(about 5×10^{-5} cm in diameter) and filled with concentrated aqueous KCl. When the tip penetrates the cell membrane, the membrane seals around the glass.

It is found that cells show a potential difference $\phi^{int} - \phi^{ext}$ of -30 to -100 mV across their membranes, the cell's interior being at a lower potential than the exterior. Typical values are -90 mV for resting muscle cells, -70 mV for resting nerve cells, and -40 mV for liver cells. Since interphase potential differences exist in living organisms, living organisms meet the definition of electrochemical systems (Sec. 14.3).

Experiment shows that when an impulse propagates along a nerve cell or when a muscle cell contracts, the transmembrane potential $\phi^{int} - \phi^{ext}$ changes, becoming momentarily positive. Nerve impulses are transmitted by changes in nerve-cell membrane potentials. Muscles are caused to contract by changes in muscle-cell membrane potentials. Our perception of the external world through the senses of sight, hearing, touch, etc., our thought processes, and our voluntary and involuntary muscular contractions are all intimately connected to interphase potential differences. An understanding of life requires an understanding of how these potential differences are maintained and how they are changed. Our knowledge in this area is currently very fragmentary.

The existence of transmembrane potential differences means that there is an electrical double layer at the membrane of each cell. The double layer is approximately equivalent to a distribution of electric dipoles (Sec. 14.2) at the cell's surface. Consider the heart muscles. As these muscles undergo their various cycles of contraction and relaxation, the potential differences across their cell membranes are continually changing, and hence the total dipole moment of the heart is changing, as are the electric field and electric potential produced by the heart. An electrocardiogram (ECG) measures the difference in electric potential between several pairs of points on the surface of the body as a function of time. Changes in these potential differences arise from the changes in the heart dipole moment. [See R. K. Hobbie, *Am. J. Phys.*, **41**, 824 (1973) for details.]

An electromyogram (EMG) is similar to an ECG except that it is the electrical activity in skeletal muscles that is monitored.

An electroencephalogram (EEG) records the time-varying potential difference between two points on the scalp and reflects the electrical activity of nerve cells in the brain. Its interpretation is not simple, since the signal results from the activity of a large number of nerve cells.

Consider the transmembrane potential of a nerve cell. We noted in Sec. 14.14 that the equilibrium Nernst equation (14.98) is insufficient to describe the nonequilibrium situation in living organisms. In 1943, Goldman used a nonequilibrium

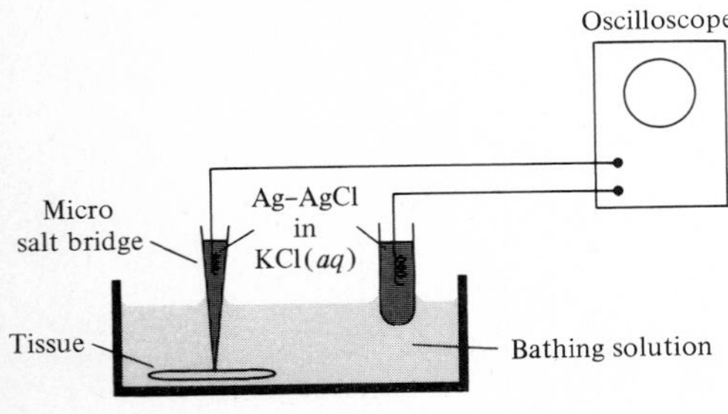

Figure 14.14
Measurement of the transmembrane potential.

approach together with the assumption of a linear variation of ϕ within the membrane to derive an expression for the transmembrane potential. This equation (in modified form) was subsequently used by Hodgkin and Katz in studies of nerve impulses. The Goldman–Hodgkin–Katz equation is

$$\phi^{\text{int}} - \phi^{\text{ext}} = \frac{RT}{\mathscr{F}} \ln \frac{P(\text{K}^+)[\text{K}^+]^{\text{ext}} + P(\text{Na}^+)[\text{Na}^+]^{\text{ext}} + P(\text{Cl}^-)[\text{Cl}^-]^{\text{int}}}{P(\text{K}^+)[\text{K}^+]^{\text{int}} + P(\text{Na}^+)[\text{Na}^+]^{\text{int}} + P(\text{Cl}^-)[\text{Cl}^-]^{\text{ext}}} \qquad (14.101)$$

where $[\text{K}^+]^{\text{ext}}$ and $[\text{K}^+]^{\text{int}}$ are the K^+ concentrations outside and inside the cell and $P(\text{K}^+)$ is the permeability of the membrane to K^+ ions. [$P(\text{K}^+)$ is defined as $D(\text{K}^+)/\tau$, where $D(\text{K}^+)$ is the diffusion coefficient of K^+ through the membrane of thickness τ. Diffusion coefficients are defined in Sec. 16.4.] For the derivation of (14.101), see *Eyring, Henderson, and Jost*, vol. IXB, chap. 11.

The six concentrations in (14.101) should actually be replaced by activities, but since the ionic strengths inside and outside the cell are approximately equal, all six activity coefficients are approximately equal and have been omitted. Note that if $P(\text{K}^+)$ were much greater than both $P(\text{Na}^+)$ and $P(\text{Cl}^-)$, Eq. (14.101) would reduce to the concentration-scale version of the Nernst equation (14.98) for K^+; similarly for the other two ions.

Tracer experiments show that a nerve-cell membrane is permeable to all three ions K^+, Na^+, and Cl^-. For a resting nerve cell of a squid, one finds $P(\text{K}^+)/P(\text{Cl}^-) \approx 2$ and $P(\text{K}^+)/P(\text{Na}^+) \approx 25$. For squid nerve cells, the observed concentrations in mmol/dm^3 are

$$\begin{array}{lll} [\text{K}^+]^{\text{int}} = 410 & [\text{Na}^+]^{\text{int}} = 49 & [\text{Cl}^-]^{\text{int}} = 40 \\ [\text{K}^+]^{\text{ext}} = 10 & [\text{Na}^+]^{\text{ext}} = 460 & [\text{Cl}^-]^{\text{ext}} = 540 \end{array} \qquad (14.102)$$

These three ions are the main inorganic ions. Inside the cell, there is also a substantial concentration of organic anions (charged proteins, organic phosphates, and anions of organic acids); these have quite low permeabilities.

We can use the Nernst equation (14.98) with activity coefficients omitted to see which ions are in electrochemical equilibrium across the membrane. The concentrations in (14.102) give the following equilibrium $\Delta\phi$ values at 25°C:

$$\Delta\phi_{\text{eq}}(\text{K}^+) = -95 \text{ mV}, \qquad \Delta\phi_{\text{eq}}(\text{Na}^+) = +57 \text{ mV}, \qquad \Delta\phi_{\text{eq}}(\text{Cl}^-) = -67 \text{ mV}$$

The observed transmembrane potential for a resting squid nerve cell is about -70 mV at 25°C. Hence, Cl^- is in electrochemical equilibrium, but K^+ and especially Na^+ are not.

For the -70 mV membrane potential, Eq. (14.98) gives the following equilibrium concentration ratios at 25°C: $c^{\text{ext}}/c^{\text{int}} = 1:15$ for $z = +1$ ions; $c^{\text{ext}}/c^{\text{int}} = 15:1$ for $z = -1$ ions. The actual concentration ratios are $1:41$ for K^+, $9:1$ for Na^+, and $14:1$ for Cl^-. Hence, Na^+ continuously flows spontaneously into the cell and K^+ flows spontaneously out. [We have $\tilde{\mu}^{\text{ext}}(\text{Na}^+) > \tilde{\mu}^{\text{int}}(\text{Na}^+)$.] To maintain the observed steady-state concentrations of Na^+ and K^+, there must be some sort of *active-transport* process that uses some of the cell's metabolic energy to continually "pump" Na^+ out of the cell and K^+ into it. The mechanism of this pump is unknown.

A nerve impulse is a brief (1 ms) change in the transmembrane potential. This change travels along the nerve fiber at a constant speed of 10^3 to 10^4 cm/s, depending on the species and the kind of nerve. The change in $\Delta\phi$ is initiated by a local increase in the membrane's permeability to Na^+, with $P(Na^+)/P(K^+)$ reaching perhaps 20. With $P(Na^+)$ much greater than both $P(K^+)$ and $P(Cl^-)$ in (14.101), the membrane potential moves toward the $\Delta\phi_{eq}(Na^+)$ value of +60 mV. The observed peak value of $\Delta\phi$ is +40 or 50 mV during the passage of a nerve impulse. After this peak is reached, $P(Na^+)$ decreases and $P(K^+)$ temporarily increases; hence, the potential returns toward its resting value of −70 mV. Because of the temporary increase in $P(K^+)$, the potential somewhat overshoots the resting value, ultimately returning to −70 mV as the ion permeabilities return to their resting values. The mechanism of the permeability changes is unknown.

A cell derives its energy from a complex series of metabolic reactions whose net result is to convert glucose ($C_6H_{12}O_6$) into CO_2 and H_2O. These reactions are redox reactions and occur at structures within the cell called mitochondria. A mitochondrion is surrounded by a double membrane. Whether a mitochondrial membrane potential is involved in metabolism is unknown.

14.17 ELECTROCHEMISTRY

In the broadest sense, all of chemistry is electrochemistry, since the forces within and between molecules are electrical. However, the term *electrochemistry* is generally understood to mean the study of ionic conductors and the transfer of charge between ionic and electronic conductors.

The science of ionic conductors (e.g., electrolyte solutions, molten salts) is called *ionics*. Sections 10.4 to 10.8 and 11.3 deal with the equilibrium aspects of ionics. Section 16.6 (on charge transport through electrolyte solutions) deals with the irreversible aspects of ionics.

The science of charge transfer between electronic conductors (metals) and ionic conductors is called *electrodics*. This chapter (14) deals mainly with the reversible (equilibrium) aspects of electrodics.

In recent years, electrochemistry has undergone a renaissance, as evidenced by advances in the theoretical and experimental study of electrode kinetics, the preparation of new kinds of ion-selective membrane electrodes, the development of new types of batteries and fuel cells, and progress in understanding electrochemical phenomena in living organisms.

PROBLEMS

14.1 Calculate the force that a He nucleus exerts on an electron 1.0 Å away.

14.2 Calculate the magnitude of the electric field of a proton at a distance of (*a*) 2.0 Å; (*b*) 4.0 Å.

14.3 Calculate the potential difference between two points that are 4.0 and 2.0 Å away from a proton.

14.4 The electric dipole moment of HCl is 3.57×10^{-30} C m, and its bond length is 1.30 Å. If we pretend that the molecule consists of charges $+\delta$ and $-\delta$ separated by 1.30 Å, find δ. Also, calculate δ/e, where e is the proton charge.

14.5 (*a*) For an electric dipole whose center is at the coordinate origin and whose charges lie on the z axis, show that the potential at point (x, y, z) is $\phi = (p/4\pi\varepsilon_0)(z/r^3)$, where $r = (x^2 + y^2 + z^2)^{1/2}$. (*b*) Find the expressions for the electric-field components E_x, E_y, and E_z of the dipole.

14.6 Calculate the work needed to increase the distance between a K^+ ion and a Cl^- ion from 10 to 100 Å in (*a*) a vacuum; (*b*) water at 25°C.

14.7 Theoretical calculations combined with experimental data indicate that for Li and Rb in contact at

25°C, the potential difference is $\phi(\text{Li}) - \phi(\text{Rb}) \approx 0.1$ V. Estimate the difference in chemical potential between electrons in Li and electrons in Rb.

14.8 Calculate the charge on (*a*) 3.00 moles of Hg_2^{2+} ions; (*b*) 0.600 mole of electrons.

14.9 Verify that Eqs. (14.36) and (14.37) are equivalent.

14.10 Give the charge number *n* for each of these reactions: (*a*) $H_2 + Br_2 = 2HBr$; (*b*) $\frac{1}{2}H_2 + \frac{1}{2}Br_2 = HBr$; (*c*) $2HBr = H_2 + Br_2$; (*d*) $3Zn + 2Al^{3+} = 3Zn^{2+} + 2Al$; (*e*) $Hg_2Cl_2 + H_2 = 2Hg + 2Cl^- + 2H^+$.

14.11 Use data in the Appendix to find $\mathscr{E}^\circ_{298}$ for $N_2O_4(g) + Cu^{2+}(aq) + 2H_2O(l) = Cu + 4H^+(aq) + 2NO_3^-(aq)$.

14.12 (*a*) Use data in the Appendix to find $\mathscr{E}^\circ_{298}$ for the reaction $3Cu^{2+} + 2Fe = 2Fe^{3+} + 3Cu$. (*b*) Use data in Table 14.1 to answer the question in (*a*).

14.13 Consider the cell

$$\text{M}_L \mid \text{Zn} \mid \text{Zn}^{2+}(aq) \,\vdots\, \text{Cu}^{2+}(aq) \mid \text{Cu} \mid \text{M}_R,$$

where M is any metal. Use (14.30) for electrons to show that

$$[\phi(\text{M}_R) - \phi(\text{Cu})] + [\phi(\text{Zn}) - \phi(\text{M}_L)] = [\mu(\text{e}^- \text{ in Zn}) - \mu(\text{e}^- \text{ in Cu})]/\mathscr{F}$$

The left side of this equation contains the terms that involve M in the expression for the cell's emf. But the quantities on the right are independent of the nature of M. Hence, the cell's emf is independent of what metal is used for the terminals.

14.14 For the cell (14.44), observed emfs at 60°C and 1 atm H_2 pressure as a function of HCl molality *m* are:

$m/(\text{mol kg}^{-1})$	0.001	0.002	0.005	0.1
$\mathscr{E}/\text{V}$	0.5953	0.5563	0.5052	0.3428

(*a*) Use a graphical method to find $\mathscr{E}^\circ$ at 60°C. (*b*) Calculate the 60°C HCl(*aq*) stoichiometric activity coefficients at $m = 0.005$ and 0.1 mol/kg.

14.15 For pressures up to 100 atm, $H_2(g)$ obeys the equation of state $P(\bar{V} - B) = RT$, where B is a function of T. At 25°C, $B = 13.99$ cm^3/mole. (*a*) Show that the fugacity of H_2 is $f = Pe^{BP/RT}$ for $P < 100$ atm. (*b*) Calculate $f(H_2)$ at 25°C and 1 atm. (*c*) Calculate the error in $\mathscr{E}_{298}$ of a cell that uses the hydrogen electrode at 25°C and 1 atm if f is replaced by P in the Nernst equation.

14.16 For the cell at 25°C and 1 atm

$$\text{Pt} \mid \text{Ag} \mid \text{AgCl}(s) \mid \text{HCl}(aq) \mid \text{Hg}_2\text{Cl}_2(s) \mid \text{Hg} \mid \text{Pt}'$$

(*a*) write the cell reaction; (*b*) use Table 14.1 to find the emf if the HCl molality is 0.100 mol/kg; (*c*) find the emf if the HCl molality is 1.00 mol/kg. (*d*) For this cell, $(\partial\mathscr{E}/\partial T)_P = 0.338$ mV/K at 25°C and 1 atm. Find ΔG°, ΔH°, and ΔS° for the cell reaction at 25°C.

14.17 (*a*) The standard emf $\mathscr{E}^\circ$ of the cell calomel | nonesuch composed of a calomel electrode and a nonesuch electrode is -1978 mV at 25°C. Find the standard electrode potential of the nonesuch electrode at 25°C. (*b*) At 43°C, the cell calomel | nonpareil has $\mathscr{E}^\circ = -0.80$ V and the cell nonesuch | calomel has $\mathscr{E}^\circ = 1.70$ V. Find $\mathscr{E}^\circ$ for the cell nonpareil | nonesuch at 43°C.

14.18 Suppose we add a pinch of salt (NaCl) to the $CuSO_4$ solution of the cell (14.65) thermostated at 25°C. (*a*) Is $\mathscr{E}$ changed? Explain. (*b*) Is $\mathscr{E}^\circ$ changed? Explain.

14.19 What values of the activity quotient Q_a are required for the cell (14.44) to have the following emfs at 25°C: (*a*) -1.00 V; (*b*) 1.00 V?

14.20 If the cell (14.44) has $a(\text{HCl}) = 1.00$, what value of $P(H_2)$ is needed to make the cell emf at 25°C equal to (*a*) -0.500 V; (*b*) 0.500 V?

14.21 For the cell

$$\text{Pt}_L \mid \text{Fe}^{2+}(a = 2.00), \text{Fe}^{3+}(a = 1.20) \,\vdots\vdots\, \text{I}^-(a = 0.100) \mid \text{I}_2(s) \mid \text{Pt}_R$$

(*a*) write the cell reaction; (*b*) calculate $\mathscr{E}_{298}$ assuming the net liquid-junction potential is negligible. (*c*) Which terminal is at the higher potential? (*d*) When the cell is connected to a load, into which terminal do electrons flow from the load?

14.22 For the cell

$$\text{Cu} \mid \text{CuSO}_4(1.00 \text{ mol/kg}) \mid \text{Hg}_2\text{SO}_4(s) \mid \text{Hg} \mid \text{Cu}'$$

(*a*) write the cell reaction; (*b*) calculate $\mathscr{E}$ at 25°C and 1 atm given that the stoichiometric activity coefficient of $CuSO_4$ is 0.043 for these conditions; (*c*) calculate the erroneous value of $\mathscr{E}$ that would be obtained if the $CuSO_4$ activity coefficient were taken as 1.

14.23 For the cell

$$\text{Ag}_L \mid \text{AgNO}_3(0.0100 \text{ mol/kg}) \,\vdots\vdots\, \text{AgNO}_3(0.0500 \text{ mol/kg}) \mid \text{Ag}_R$$

(*a*) use the Davies equation to find $\mathscr{E}_{298}$; neglect ion pairing and assume the salt bridge makes the net liquid-junction potential negligible. (*b*) Which terminal is at the higher potential? (*c*) When the cell is connected to a load, into which terminal do electrons flow from the load?

14.24 Find K_a at 25°C for $2H^+(aq) + D_2 = H_2 + 2D^+(aq)$ using data in Table 14.1.

14.25 Use data in Table 14.1 and the convention (10.104) to determine $\Delta G^\circ_{f,298}$ for (*a*) Na^+; (*b*) Cl^-; (*c*) Cu^{2+}.

14.26 Use data in Table 14.1 to calculate K°_{sp} of Ag_2S at 25°C.

14.27 For the cell $Pt|Fe|Fe^{2+} \vdots Fe^{2+}, Fe^{3+}|Pt'$, one finds $(\partial \mathscr{E}^\circ/\partial T)_P = 1.14$ mV/K at 25°C. (*a*) Write the cell reaction using the smallest possible whole numbers as the stoichiometric coefficients. (*b*) With the aid of data in Table 14.1, calculate ΔS°, ΔG°, and ΔH° for the cell reaction at 25°C.

14.28 Use Table 14.1 to calculate ΔG° and K_a for (*a*) $Cl_2 + 2Br^-(aq) = 2Cl^-(aq) + Br_2$; (*b*)

$$\tfrac{1}{2}Cl_2 + Br^-(aq) = Cl^-(aq) + \tfrac{1}{2}Br_2;$$

(*c*) $2Ag + Cl_2 = 2AgCl$; (*d*) $2AgCl = 2Ag + Cl_2$; (*e*) $3Fe^{2+}(aq) = Fe + 2Fe^{3+}(aq)$. Take $T = 298$ K.

14.29 Use data in Eq. (14.81) to find ΔG°, ΔH°, ΔS°, and ΔC°_P at 10°C for the reaction $H_2 + 2AgCl = 2Ag + 2HCl(aq)$.

14.30 The solubility product for AgI in water at 25°C is 8.2×10^{-17}. Use data in Table 14.1 to find $\mathscr{E}^\circ$ for the Ag–AgI electrode at 25°C.

14.31 The cell

$$Pt|H_2(1\text{ atm})|HBr(aq)|AgBr(s)|Ag|Pt'$$

at 25°C with HBr molality 0.100 mol/kg has $\mathscr{E} = 0.200$ V. Find the stoichiometric activity coefficient of HBr(*aq*) at this molality.

14.32 Use data in Table 14.1 to calculate $\Delta G^\circ_{f,298}$ of HCl(*aq*) and of $Cl^-(aq)$.

14.33 Using half-cells listed in Table 14.1, write the diagrams of three different chemical cells without transference that have HCl(*aq*) as the electrolyte.

14.34 Using half-cells listed in Table 14.1, write the diagram of a chemical cell without transference whose electrolyte is (*a*) KCl(*aq*); (*b*) $H_2SO_4(aq)$.

14.35 Calculate $\mathscr{E}_{298}$ for the cell

$$Cu_L|Zn|ZnCl_2(0.0100\text{ mol/kg})|AgCl(s)|Ag|Pt|Cu_R$$

given that the stoichiometric activity coefficient of $ZnCl_2$ is 0.708 at this molality and temperature.

14.36 Consider the cell at 1 atm H_2 pressure

$$Pt|H_2(g)|NaOH(m_1),\ NaCl(m_2)|AgCl(s)|Ag|Pt'$$

(*a*) Show that $\mathscr{E} = \mathscr{E}^\circ - RT\mathscr{F}^{-1}\ln a(H^+)a(Cl^-)$ and that

$$\mathscr{E} = \mathscr{E}^\circ - \frac{RT}{\mathscr{F}}\ln\frac{K^\circ_w a(H_2O)\gamma(Cl^-)m(Cl^-)}{\gamma(OH^-)m(OH^-)}$$

where K°_w is the ionization constant of water. (*b*) For this cell at 25°C, it is found that $\mathscr{E} - \mathscr{E}^\circ + RT\mathscr{F}^{-1}\ln[m(Cl^-)/m(OH^-)]$ approaches the limit 0.8279 V as the ionic strength goes to zero. Calculate K°_w at 25°C.

14.37 Consider the cell at 1 atm H_2 pressure

$$Pt|H_2(g)|HX(m_1),\ NaX(m_2),\ NaCl(m_3)|AgCl(s)|Ag|Pt'$$

where the anion X^- is acetate, $C_2H_3O_2^-$. (*a*) Show that

$$\mathscr{E} = \mathscr{E}^\circ - \frac{RT}{\mathscr{F}}\ln\frac{\gamma(Cl^-)m(Cl^-)\gamma(HX)m(HX)K^\circ_a}{\gamma(X^-)m(X^-)m^\circ}$$

where K°_a is the ionization constant of the weak acid HX and $m^\circ \equiv 1$ mol/kg. (*b*) The zero-ionic-strength limit of

$$\mathscr{E} - \mathscr{E}^\circ + RT\mathscr{F}^{-1}\ln[m(HX)m(Cl^-)/m(X^-)m^\circ]$$

at 25°C is 0.2814 V. Calculate K°_a for acetic acid at 25°C.

14.38 An excess of Sn powder is added to a 0.100 mol/kg aqueous $Pb(NO_3)_2$ solution at 25°C. Neglecting ion pairing and omitting activity coefficients, estimate the equilibrium molalities of Pb^{2+} and Sn^{2+}. Explain why omission of the activity coefficients is a reasonably good approximation in this case.

14.39 Calculate the emf of the cell (14.74) at 85°C if $P_L = 2521$ torr, $P_R = 666$ torr, and $m(HCl) = 0.200$ mol/kg.

14.40 Calculate $\mathscr{E}_{298}$ for the cell

$$Pt_L|Cl_2(580\text{ torr})|HCl(0.0100\text{ mol/kg})|H_2(860\text{ torr})|Pt_R$$

using an activity coefficient listed in Sec. 14.11. (*b*) Which terminal is at the higher potential?

14.41 Consider the cell

$$Ag_L|AgCl(s)|HCl(0.0100\text{ mol/kg}) \vdots HCl(0.100\text{ mol/kg})|AgCl(s)|Ag_R$$

The theoretical equation (16.103) gives the following estimate of the liquid-junction potential: $\mathscr{E}_J = -38$ mV. Use the single-ion form of the Davies equation to calculate the emf of this cell at 25°C.

14.42 Use the Davies equation to estimate $-\log a(H^+)$ in a 0.100 mol/kg aqueous HCl solution at 25°C. Compare with the assigned pH of 1.09 for this solution.

14.43 For the cell (14.87), the observed emf at 25°C was 612 mV. When solution X was replaced by a standard phosphate buffer solution whose assigned pH is 6.86, the emf was 741 mV. Find the pH of solution X.

14.44 A solution containing 0.100 mol/kg NaCl and 0.200 mol/kg KBr is separated by a membrane permeable only to Na^+ from a solution that is 0.150 mol/kg in $NaNO_3$ and 0.150 mol/kg in KNO_3. Calculate the transmembrane potential at 25°C; state and justify any approximations you make.

14.45 (*a*) Use Eq. (14.101) and data in and preceding Eq. (14.102) to calculate the membrane potential of a resting squid nerve cell at 25°C. Compare with the experimental value -70 mV. (*b*) Now repeat the calculation in (*a*) but omit the terms in (14.101) that involve Cl^-.

14.46 For $CCl_4(l)$ at 20°C and 1 atm, $\varepsilon_r = 2.24$ and $\rho = 1.59$ g/cm^3. Calculate α and $\alpha/4\pi\varepsilon_0$ for CCl_4.

14.47 (*a*) For $CH_4(g)$ at 0°C and 1.000 atm, $\varepsilon_r = 1.00094$. Calculate α and $\alpha/4\pi\varepsilon_0$ for CH_4. (*b*) Calculate ε_r for CH_4 at 100°C and 10.0 atm.

14.48 Values of $10^5(\varepsilon_r - 1)$ for $H_2O(g)$ at 1.000 atm as a function of T are:

T/K	384.3	420.1	444.7	484.1	522.0
$10^5(\varepsilon_r - 1)$	546	466	412	353	302

Use a graphical method to find the dipole moment and polarizability of H_2O.

15

KINETIC–MOLECULAR THEORY OF GASES

15.1 KINETIC–MOLECULAR THEORY OF GASES

Chapters 1 to 12 use mainly a macroscopic approach. Chapters 13 and 14 use both macroscopic and molecular approaches. The remaining chapters use mainly a molecular approach to physical chemistry.

This chapter and several sections of the next discuss the *kinetic–molecular theory of gases* (kinetic theory, for short). The kinetic theory of gases pictures a gas as composed of a very large number of molecules whose size is small compared with the average distance between molecules. The molecules move freely and rapidly through space. Although this picture seems obvious nowadays, it wasn't until about 1850 that the kinetic theory began to win acceptance.

Kinetic theory began with Daniel Bernoulli's 1738 derivation of Boyle's law using Newton's laws of motion applied to molecules. Bernoulli's work was ignored for over 100 years. In 1845 John Waterston submitted a paper to the Royal Society of England that correctly developed many of the concepts of kinetic theory. Waterston's paper was rejected as "nonsense." Joule's experiments demonstrating that heat is a form of energy made the ideas of kinetic theory seem plausible, and in the period from 1848 to 1898, Joule, Clausius, Maxwell, and Boltzmann developed the kinetic theory of gases.

From 1870 to 1910 a controversy raged between the school of Energetics and the school of Atomism. The Atomists (led by Boltzmann) held that atoms and molecules were real entities, while the Energetists (Mach, Ostwald, Duhem, et al.) denied the existence of atoms and molecules and argued that the kinetic theory of gases was at best a mechanical model that imitated the properties of gases but did not correspond to the true structure of matter. [The Freudian-oriented sociologist Lewis Feuer speculates that Mach's opposition to atomism was at an unconscious level an expression of revolt against his father; the shape of atoms unconsciously reminded Mach of testicles, and "a reality deatomized was a projection, we might infer, in which the father himself was unmanned." (L. Feuer, *Einstein and the Generations of Science*, Basic Books, 1974, p. 39.)]

The attacks on the kinetic theory of gases led Boltzmann to write in his 1898 book on kinetic theory: "I am conscious of being only an individual struggling weakly against the stream of time. But it still remains in my power to contribute in such a way that when the theory of gases is again revived, not too much will have to be rediscovered." (*Lectures on Gas Theory*, trans. S. G. Brush, University of California Press, 1964.) Some people have attributed Boltzmann's suicide in 1906 to depression resulting from attacks on the kinetic theory.

In 1905 Einstein applied kinetic theory to the Brownian motion of a tiny particle suspended in a fluid (Sec. 3.7). Einstein's theoretical equations were confirmed by Perrin's experiments in 1908, thereby convincing the Energetists of the reality of atoms and molecules.

The kinetic theory of gases uses a molecular picture to derive macroscopic properties of matter and is thus a branch of statistical mechanics.

The treatment in this chapter will be restricted to gases at low pressures (ideal gases). Since the molecules are far apart at low pressures, we ignore intermolecular forces (except at the moment of collision between two molecules; see Sec. 15.7). The kinetic theory of gases assumes the molecules to obey Newton's laws of motion. Actually, molecules obey quantum mechanics (Chap. 18). The use of classical mechanics leads to incorrect results for the heat capacities of gases (Sec. 15.10) but is an excellent approximation when dealing with properties like pressure and diffusion.

15.2 PRESSURE OF AN IDEAL GAS

The pressure exerted by a gas on the walls of its container is due to bombardment of the walls by the gas molecules. The number of molecules in a gas is huge (2×10^{19} in 1 cm^3 at 1 atm and 25°C), and the number of molecules that hit a container wall in a tiny time interval is huge [3×10^{17} impacts with a 1-cm^2 wall in 1 microsecond for O_2 at 1 atm and 25°C; see Eq. (15.57) below], so the individual impacts of molecules result in an apparently steady pressure on the wall.

Let the container be a rectangular box with sides of length l_x, l_y, and l_z. Let **v** be the *velocity* [Eq. (2.17)] of a given molecule. The components of **v** in the x, y, and z directions are v_x, v_y, and v_z. The particle's *speed* v is the magnitude (length) of the vector **v**. Application of the Pythagorean theorem twice in Fig. 15.1 gives $v^2 = \text{OC}^2 = \text{OB}^2 + v_z^2 = v_x^2 + v_y^2 + v_z^2$; thus

$$v^2 = v_x^2 + v_y^2 + v_z^2 \qquad (15.1)^*$$

The velocity **v** is a vector. The speed v and the velocity components v_x, v_y, v_z are scalars. A velocity component like v_x can be positive, negative, or zero (corresponding to motion in the positive x direction, motion in the negative x direction, or no motion in the x direction), but v must (by definition) be positive or zero.

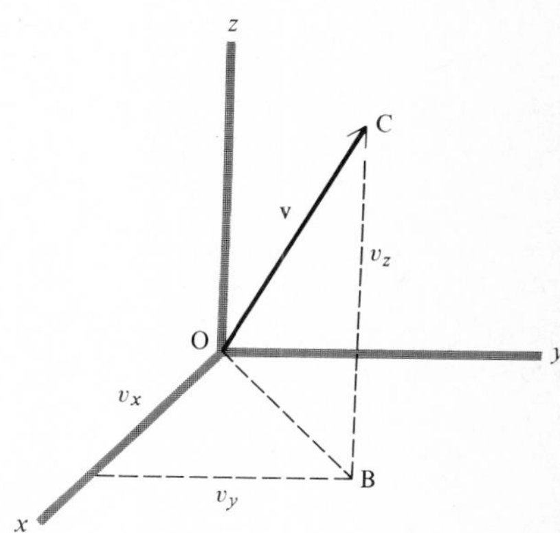

Figure 15.1
Velocity components.

The kinetic energy ε_{tr} associated with motion of a molecule through space is

$$\varepsilon_{tr} \equiv \tfrac{1}{2}mv^2 = \tfrac{1}{2}mv_x{}^2 + \tfrac{1}{2}mv_y{}^2 + \tfrac{1}{2}mv_z{}^2 \tag{15.2}$$

where m is the molecule's mass. We call ε_{tr} the *translational energy* of the molecule.

Let the gas be in thermodynamic equilibrium. Because the gas and its surroundings are in thermal equilibrium, there is no net transfer of energy between them. We therefore assume that in a collision with the wall, a gas molecule does not change its translational energy.

In reality, a molecule colliding with the wall may undergo a change in ε_{tr}. However, for each gas molecule that loses translational energy to the wall molecules in a wall collision, another gas molecule will gain translational energy in a wall collision. Besides translational energy, the gas molecules also have rotational and vibrational energy. In a wall collision, some of the molecule's translational energy may be transformed into rotational energy and vibrational energy or vice versa; on the average, such transformations balance out, and wall collisions cause no net transfer of energy between translation and vibration-rotation. Since pressure is a property averaged over many wall collisions, we assume that in any one wall collision, there is no change in the molecule's translational kinetic energy. Although this assumption is false, it is "true" averaged over all the molecules, and hence gives the correct result for the pressure.

Let $\langle F \rangle$ denote the average value of some property F. Before deriving the expression for the gas pressure, we prove the preliminary result that the average value of some function of time over the time interval from t_1 to t_2 is

$$\langle F(t) \rangle = \frac{1}{t_2 - t_1} \int_{t_1}^{t_2} F(t)\, dt \tag{15.3}$$

The average value of a quantity is the sum of its observed values divided by the total number of observed values:

$$\langle F \rangle = \frac{1}{n} \sum_{i=1}^{n} F_i \tag{15.4}$$

where F_i are the observed values. For the function $F(t)$, there are an infinite number of values, since there are an infinite number of times in the interval from t_1 to t_2. We therefore divide this interval into a very large number n of subintervals, each of duration Δt, and take the limit as $n \to \infty$ and $\Delta t \to 0$. Multiplying and dividing each term in (15.4) by Δt, we have

$$\langle F(t) \rangle = \lim_{n \to \infty} \frac{1}{n\, \Delta t} [F(t_1)\, \Delta t + F(t_1 + \Delta t)\, \Delta t + F(t_1 + 2\Delta t)\, \Delta t + \cdots + F(t_2)\, \Delta t]$$

The quantity in brackets is by Eq. (2.9) the definite integral of F from t_1 to t_2. Also, $n\, \Delta t = t_2 - t_1$. Hence (15.3) is valid.

Figure 15.2 shows a molecule i colliding with wall W, where W is parallel to the xz plane. Let i have the velocity components $v_{x,i}$, $v_{y,i}$, $v_{z,i}$ before the collision. For simplicity, we assume the molecule is reflected off the wall at the same angle it hit the wall. (Since the walls are not actually smooth but are made of molecules, this assumption does not reflect reality.) The collision thus changes $v_{y,i}$ to $-v_{y,i}$ and leaves $v_{x,i}$ and $v_{z,i}$ unchanged. This leaves the molecule's speed $v_i{}^2 = v_{x,i}^2 + v_{y,i}^2 + v_{z,i}^2$ unchanged and its translational energy $\tfrac{1}{2}mv_i{}^2$ unchanged.

To get the pressure on wall W, we need the average force exerted by the molecules on this wall. Consider the motion of molecule i. It collides with W and then moves to the right, eventually colliding with wall W′, then moves to the left to collide again with W, etc. (Collisions with the top, bottom, and side walls may occur between collisions with W and W′, but these collisions do not alter $v_{y,i}$.) For our purposes, one "cycle" of motion of molecule i will extend from a time t_1 that just precedes a collision with W to a time t_2 that just precedes the next collision with W. During the very short time that i is in process of colliding with W, Newton's second law (2.19) gives the y component of the force on i as

$$F_{y,i} = ma_{y,i} = m\frac{dv_{y,i}}{dt} = \frac{d}{dt}(mv_{y,i}) = \frac{dp_{y,i}}{dt} \qquad (15.5)$$

where the y component of the (linear) momentum is defined by $p_y \equiv mv_y$. [The *(linear) momentum* $\mathbf{p}$ is a vector defined by $\mathbf{p} \equiv m\mathbf{v}$.] Let the collision of i with W extend from time t' to t''. Equation (15.5) gives $dp_{y,i} = F_{y,i}\,dt$. Integrating from t' to t'', we get $p_{y,i}(t'') - p_{y,i}(t') = \int_{t'}^{t''} F_{y,i}\,dt$. (The integral in this equation is called the *impulse* of the force $F_{y,i}$.) The y momentum of i before the wall collision is $p_{y,i}(t') = mv_{y,i}$, and the y momentum after the collision is $p_{y,i}(t'') = -mv_{y,i}$. Hence, $-2mv_{y,i} = \int_{t'}^{t''} F_{y,i}\,dt$.

Let $F_{W,i}$ be the force on wall W due to collision with molecule i. Newton's third law (action = reaction) gives $F_{W,i} = -F_{y,i}$, so $2mv_{y,i} = \int_{t'}^{t''} F_{W,i}\,dt$. For times between t_1 and t_2 but outside the collision interval from t' to t'', the force $F_{W,i}$ is zero, since molecule i is not in the process of colliding with W during such times. Therefore the integration can be extended over the whole time interval t_1 to t_2 to get $2mv_{y,i} = \int_{t_1}^{t_2} F_{W,i}\,dt$. Use of (15.3) gives

$$2mv_{y,i} = \langle F_{W,i}\rangle(t_2 - t_1) \qquad (15.6)$$

where $\langle F_{W,i}\rangle$ is the average force exerted on wall W by molecule i.

The time $t_2 - t_1$ is the time needed for i to traverse a distance $2l_y$ in the y direction (so as to bring it back to W). Since $y = v_y t$, we have $t_2 - t_1 = 2l_y/v_{y,i}$. Equation (15.6) becomes $\langle F_{W,i}\rangle = mv_{y,i}^2/l_y$.

The time average of the total force on wall W is found by summing the average forces of the individual molecules. If there are N gas molecules present, then

$$\langle F_W\rangle = \sum_{i=1}^{N}\langle F_{W,i}\rangle = \sum_{i=1}^{N}\frac{mv_{y,i}^2}{l_y} = \frac{m}{l_y}\sum_{i=1}^{N} v_{y,i}^2$$

We shall see in Sec. 15.4 that the molecules do not all move at the same speed. The average value of $v_y{}^2$ for all the molecules is by definition [Eq. (15.4)] given by $\langle v_y{}^2\rangle = N^{-1}\sum_i v_{y,i}^2$. Hence, $\langle F_W\rangle = mN\langle v_y{}^2\rangle/l_y$.

The pressure P on W equals the average force $\langle F_W\rangle$ divided by the area $l_x l_z$ of W. Therefore

$$P = mN\langle v_y{}^2\rangle/V \qquad \text{ideal gas} \qquad (15.7)$$

where $V\ (= l_x l_y l_z)$ is the container volume.

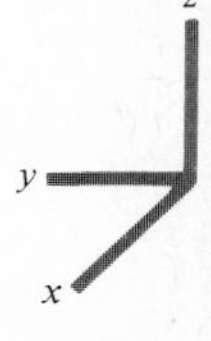

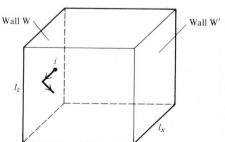

Figure 15.2
Molecule i colliding with container wall W.

There is nothing special about the y direction, so the properties of the gas must be the same in any direction. Therefore

$$\langle v_x^2 \rangle = \langle v_y^2 \rangle = \langle v_z^2 \rangle \tag{15.8}$$

Furthermore, $\langle v^2 \rangle$, the average of the square of the molecular speed, is [see Eqs. (15.1) and (15.4)]

$$\langle v^2 \rangle = \langle v_x^2 + v_y^2 + v_z^2 \rangle \equiv \frac{1}{N} \sum_{i=1}^{N} (v_{x,i}^2 + v_{y,i}^2 + v_{z,i}^2)$$

$$= \frac{1}{N} \sum_{i=1}^{N} v_{x,i}^2 + \frac{1}{N} \sum_{i=1}^{N} v_{y,i}^2 + \frac{1}{N} \sum_{i=1}^{N} v_{z,i}^2 \tag{15.9}$$

$$\langle v^2 \rangle = \langle v_x^2 \rangle + \langle v_y^2 \rangle + \langle v_z^2 \rangle = 3\langle v_y^2 \rangle \tag{15.10}$$

where (15.8) was used. Therefore (15.7) becomes

$$P = \frac{mN\langle v^2 \rangle}{3V} \qquad \text{ideal gas} \tag{15.11}$$

Since $mN/V = \rho$ (the gas density), we also have $P = \frac{1}{3}\rho\langle v^2 \rangle$. Equation (15.11) expresses the macroscopic property of pressure in terms of the molecular properties m, N, and $\langle v^2 \rangle$.

The translational kinetic energy ε_{tr} of molecule i is $\frac{1}{2}mv_i^2$. The average translational energy per molecule is

$$\langle \varepsilon_{tr} \rangle = \tfrac{1}{2}m\langle v^2 \rangle \tag{15.12}$$

This equation gives $\langle v^2 \rangle = 2\langle \varepsilon_{tr} \rangle / m$, so Eq. (15.11) can be written as $PV = \frac{2}{3}N\langle \varepsilon_{tr} \rangle$. The quantity $N\langle \varepsilon_{tr} \rangle$ is the total translational kinetic energy E_{tr} of the gas molecules. Hence

$$PV = \tfrac{2}{3}E_{tr} \qquad \text{ideal gas} \tag{15.13}$$

The treatment just given assumed a pure gas with all molecules having the same mass m. If instead we have a mixture of gases A, B, and C, then at low pressure the gas molecules act independently of one another and the pressure P is the sum of pressures due to each kind of molecule: $P = P_A + P_B + P_C$ (Dalton's law). From (15.11), $P_A = \frac{1}{3}N_A m_A \langle v_A^2 \rangle / V$, with similar equations for P_B and P_C.

15.3 TEMPERATURE

Consider two thermodynamic systems 1 and 2 in contact with each other. If the molecules of system 1 have an average translational kinetic energy $\langle \varepsilon_{tr} \rangle_1$ that is greater than the average translational energy $\langle \varepsilon_{tr} \rangle_2$ of system 2 molecules, the more energetic molecules of system 1 will tend to lose translational energy to the molecules of 2 in collisions with them. This transfer of energy at the molecular level will correspond to a flow of heat from 1 to 2 at the macroscopic level. Only if $\langle \varepsilon_{tr} \rangle_1$ equals $\langle \varepsilon_{tr} \rangle_2$ will there be no tendency for a net transfer of energy to occur in 1-2 collisions. But if there is no heat flow between 1 and 2, these systems are in thermal equilibrium and by the thermodynamic definition of temperature (Sec. 1.3) systems 1 and 2 have equal temperatures. Thus, when $\langle\varepsilon_{tr}\rangle_1 = \langle\varepsilon_{tr}\rangle_2$, we have $T_1 = T_2$; when $\langle\varepsilon_{tr}\rangle_1 > \langle\varepsilon_{tr}\rangle_2$, we have $T_1 > T_2$. We have shown that there must be a correspondence between $\langle\varepsilon_{tr}\rangle$ (the average translational energy per molecule) and the macroscopic property T (temperature). The system's temperature is therefore some function of the average translational energy per molecule:

$$T = T(\langle\varepsilon_{tr}\rangle) \tag{15.14}$$

[The ideal-gas kinetic–molecular equation (15.13) reads $PV = \frac{2}{3}E_{tr} = \frac{2}{3}N\langle\varepsilon_{tr}\rangle$. Since T is some function of $\langle\varepsilon_{tr}\rangle$, at constant temperature $\langle\varepsilon_{tr}\rangle$ is constant. Hence (15.13) says that PV of an ideal gas is constant at constant temperature. Thus Boyle's law has been derived from the kinetic–molecular theory.]

The equation relating T and $\langle\varepsilon_{tr}\rangle$ cannot be determined solely from the kinetic–molecular theory because the temperature scale is arbitrary and can be chosen in many different ways (Sec. 1.3). The choice of the temperature scale will determine the relation between $\langle\varepsilon_{tr}\rangle$ and T. We *defined* the absolute temperature T in Sec. 1.5 in terms of properties of ideal gases. The ideal-gas equation (1.18), $PV = nRT$, incorporates the definition of T. Comparison of $PV = nRT$ with $PV = \frac{2}{3}E_{tr}$ gives

$$E_{tr} = \tfrac{3}{2}nRT \tag{15.15}$$

Had some other definition of temperature been chosen, a different relation between E_{tr} and temperature would have been obtained.

We have $E_{tr} = N\langle\varepsilon_{tr}\rangle$. Also, the number of moles is $n = N/N_0$, where N_0 is the Avogadro constant. Equation (15.15) becomes $N\langle\varepsilon_{tr}\rangle = \frac{3}{2}NRT/N_0$, and $\langle\varepsilon_{tr}\rangle = \frac{3}{2}RT/N_0 = \frac{3}{2}kT$, where $k \equiv R/N_0$ is Boltzmann's constant [Eq. (3.57)]. Thus

$$\langle\varepsilon_{tr}\rangle = \tfrac{3}{2}kT \tag{15.16*}$$

Equation (15.16) is the explicit relation between absolute temperature and average molecular translational energy. Although we derived (15.16) by considering an ideal gas, the discussion at the beginning of this section indicates it to be valid for any system. [If system 1 is an ideal gas and system 2 is a general system, the relation $\langle\varepsilon_{tr}\rangle_1 = \langle\varepsilon_{tr}\rangle_2$ when $T_1 = T_2$ shows that (15.16) holds for the general system 2.] The absolute temperature (as defined by the ideal-gas scale and the thermodynamic scale) turns out to be directly proportional to the average translational kinetic energy per molecule: $T = \frac{2}{3}k^{-1}\langle\varepsilon_{tr}\rangle$. (See also Sec. 22.11.)

Besides translational energy, a molecule also has rotational, vibrational, and electronic energies. (These will be discussed in Chaps. 20 and 21.) A monatomic gas, for example, He or Ne, has no rotational or vibrational energy. We can therefore set the thermodynamic internal energy U of a monatomic gas equal to the total molecular translational energy E_{tr} plus the total molecular electronic energy E_{el}:

$$U = E_{tr} + E_{el} = \tfrac{3}{2}nRT + E_{el} \qquad \text{ideal monatomic gas} \tag{15.17}$$

The heat capacity at constant volume is given by (2.83) as $C_V = (\partial U/\partial T)_V$. Provided the temperature is not extremely high, the molecular electrons won't be excited to higher energy levels and the electronic energy will remain constant as T varies. Therefore $C_V = \partial U/\partial T = \partial E_{tr}/\partial T = \frac{3}{2}nR$, and the molar C_V is

$$\bar{C}_V = \tfrac{3}{2}R \qquad \text{ideal monatomic gas, } T \text{ not extremely high} \tag{15.18}$$

Use of (2.103) gives

$$\bar{C}_P = \tfrac{5}{2}R \qquad \text{ideal monatomic gas, } T \text{ not extremely high} \tag{15.19}$$

These equations are well obeyed by monatomic gases at low pressures. For example, for Ar at 1 atm, $\bar{C}_P$ values in cal mole^{-1} K^{-1} are 4.982 at -4°C and 4.970 at 540°C, compared with 4.968 calculated from (15.19). The small deviations are due to nonideality and disappear in the limit of zero pressure.

Equation (15.16) allows us to estimate how fast molecules move. We have $\frac{3}{2}kT = \langle\varepsilon_{tr}\rangle = \frac{1}{2}m\langle v^2\rangle$, so

$$\langle v^2\rangle = 3kT/m \tag{15.20}$$

The square root of $\langle v^2\rangle$ is called the *root-mean-square speed* v_{rms}:

$$v_{rms} \equiv \langle v^2\rangle^{1/2} \tag{15.21}$$

(We shall see in Sec. 15.5 that v_{rms} differs slightly from the average speed $\langle v\rangle$.) The quantity k/m in (15.20) equals $k/m = R/N_0 m = R/M$, since the molar mass M (Sec. 1.4) equals the mass of one molecule times the number of molecules per mole. (Once again let it be said that M is not the molecular weight; the molecular weight is dimensionless, whereas M has units of mass per mole.) Taking the square root of (15.20), we have

$$v_{rms} = \left(\frac{3RT}{M}\right)^{1/2} \tag{15.22}$$

Equation (15.22) need not be memorized since it can quickly be derived from (15.16). Equations (15.11) and (15.13) are also easily derived from (15.16).

It will be helpful to keep in mind the following notation to be used throughout this chapter:

m = mass of one gas molecule, M = molar mass of the gas

N = number of gas molecules, N_0 = the Avogadro constant

15.4 DISTRIBUTION OF MOLECULAR SPEEDS IN AN IDEAL GAS

We saw that v_{rms} equals $(3RT/M)^{1/2}$ for ideal-gas molecules. There is no reason to assume that all the molecules move at the same speed, and we now derive the distribution law for molecular speeds in an ideal gas in equilibrium at temperature T.

What is meant by the *distribution* of molecular speeds? One might answer that we want to know how many molecules have any given speed v. But this approach makes no sense. Thus, suppose we ask how many molecules have speed 585 m/s. The answer is zero, since the chance that any molecule has a speed of exactly 585.0000... m/s is vanishingly small. The only sensible approach is to ask how many molecules have a speed lying in some tiny range of speeds, for example, from 585.000 to 585.001 m/s.

We take an infinitesimal interval dv of speed, and we ask: How many molecules have a speed in the range from v to $v + dv$? Let this number be dN_v. (The number dN_v is infinitesimal compared with 10^{23} but large compared with 1.) The fraction of molecules having speeds in the range v to $v + dv$ is dN_v/N, where N is the total number of gas molecules. This fraction will obviously be proportional to the width of the infinitesimal interval of speeds: $dN_v/N \propto dv$. It will also depend on the location of the interval, i.e., on the value of v. (For example, the number of molecules with speeds in the range 627.400 to 627.401 m/s differs from the number with speeds in the range 585.000 to 585.001 m/s.) We can therefore write

$$dN_v/N = G(v)\, dv \tag{15.23}$$

where $G(v)$ is some to-be-determined function of v.

The function $G(v)$ is the *distribution function* for molecular speeds. $G(v)\, dv$ gives the fraction of molecules with speed in the range v to $v + dv$. The fraction dN_v/N is the *probability* that a molecule will have its speed between v and $v + dv$. Hence $G(v)\, dv$ is a probability. The distribution function $G(v)$ is also called a *probability density*, since it is a probability per unit interval of speed. (Probability densities also occur in quantum mechanics; see Chap. 18.)

Let $\Pr(v_1 \le v \le v_2)$ be the probability that a molecule's speed lies between v_1 and v_2. To find this probability (which equals the fraction of molecules with speeds in the range v_1 to v_2), we divide the interval from v_1 to v_2 into infinitesimal intervals each of width dv and sum the probabilities of being in each infinitesimal interval:

$$\Pr(v_1 \le v \le v_2) = G(v_1)\, dv + G(v_1 + dv)\, dv + G(v_1 + 2dv)\, dv + \cdots + G(v_2)\, dv$$

But the infinite sum of infinitesimals is the definite integral of $G(v)$ from v_1 to v_2 [see Eq. (2.9)], so

$$\Pr(v_1 \le v \le v_2) = \int_{v_1}^{v_2} G(v)\, dv \tag{15.24}$$

A molecule must have its speed in the interval $0 \le v \le \infty$, so the probability (15.24) becomes 1 when $v_1 = 0$ and $v_2 = \infty$. Hence, $G(v)$ must satisfy

$$\int_0^\infty G(v)\, dv = 1 \tag{15.25}$$

We now derive $G(v)$. This was first done by Maxwell in 1860. Amazingly enough, the only assumptions needed are that (1) the velocity distribution is independent of direction and (2) the value of v_y or v_z that a molecule has doesn't influence its value of v_x. Assumption 1 must be true because all directions of space are equivalent. (We assume no external electric or gravitational fields are present.) Assumption 2 seems plausible.

As an aid in finding $G(v)$, we first derive the distribution function for v_x, the x component of the velocity. Let g denote this function. By assumption 2, g depends only on v_x, so

$$dN_{v_x}/N = g(v_x)\, dv_x \tag{15.26}$$

where dN_{v_x} is the number of molecules whose x component of velocity lies between v_x and $v_x + dv_x$. (The functions g and G are different functions, so we use different symbols for them.) The range of v_x is $-\infty$ to $+\infty$, and, similar to (15.25), g must satisfy

$$\int_{-\infty}^{\infty} g(v_x)\, dv_x = 1 \tag{15.27}$$

There are also distribution functions for v_y and v_z. However, because of assumption 1, the functional form of the v_y and v_z distribution functions is the same as that for the v_x distribution. Therefore

$$dN_{v_y}/N = g(v_y)\,dv_y \qquad \text{and} \qquad dN_{v_z}/N = g(v_z)\,dv_z \tag{15.28}$$

where g is the same function in all three equations of (15.26) and (15.28).

We now ask: What is the probability that a molecule will simultaneously have its x component of velocity in the range v_x to $v_x + dv_x$, its y component of velocity in the range v_y to $v_y + dv_y$, and its z component of velocity in the range v_z to $v_z + dv_z$? By assumption 2, the probability for a given value of v_x is independent of v_y and v_z. Hence we are dealing with independent probabilities. The probability that all three of three independent events happen is the product of the probabilities of the three events. Hence the desired probability equals $g(v_x)\,dv_x \times g(v_y)\,dv_y \times g(v_z)\,dv_z$. Let $dN_{v_xv_yv_z}$ denote the number of molecules whose x, y, and z components of velocity all lie in the above ranges. Then

$$dN_{v_xv_yv_z}/N = g(v_x)g(v_y)g(v_z)\,dv_x\,dv_y\,dv_z \tag{15.29}$$

The function $G(v)$ in (15.23) is the distribution function for *speeds*. The function $g(v_x)g(v_y)g(v_z)$ in (15.29) is the distribution function for *velocities*; a vector like $\mathbf{v}$ is specified by giving its three components v_x, v_y, v_z, and the distribution function in (15.29) specifies these three components.

Let us set up a coordinate system whose axes give the values of v_x, v_y, and v_z (Fig. 15.3). (The "space" defined by this coordinate system is called *velocity space* and is an abstract mathematical space rather than a physical space.)

The probability $dN_{v_xv_yv_z}/N$ in (15.29) is the probability that a molecule has the tip of its velocity vector lying in a rectangular box located at (v_x, v_y, v_z) and having edges dv_x, dv_y, and dv_z (Fig. 15.3). By assumption 1, the velocity distribution is independent of direction. Hence the probability $dN_{v_xv_yv_z}/N$ cannot depend on the direction of the velocity vector but only on its magnitude (which is the speed v). (In other words, the probability that the tip of $\mathbf{v}$ lies in a tiny box with edges dv_x, dv_y, dv_z is the same for all boxes that lie at the same distance from the origin in Fig. 15.3. This makes sense because all directions in space are equivalent.) Hence the probability

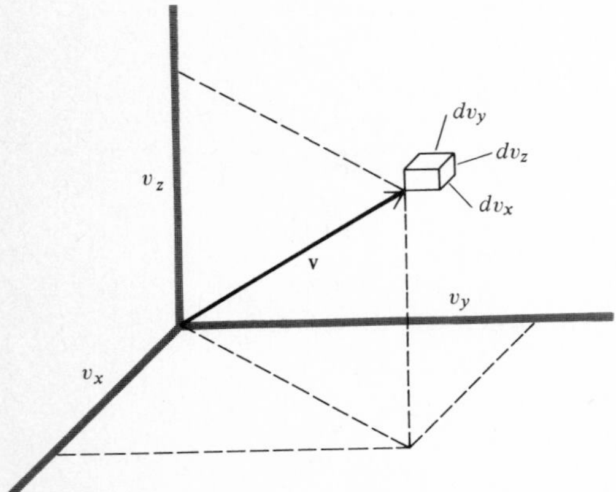

Figure 15.3
An infinitesimal box in velocity space.

density $g(v_x)g(v_y)g(v_z)$ in (15.29) must be a function of v only. Calling this function $\phi(v)$, we have

$$g(v_x)g(v_y)g(v_z) = \phi(v) \tag{15.30}$$

[$\phi(v)$ is not the same function as $G(v)$ in (15.23); see the following discussion for the relation between ϕ and G.] We also have $v^2 = v_x^2 + v_y^2 + v_z^2$, Eq. (15.1). Equations (15.30) and (15.1) are sufficient to determine g. [Before reading on, you might try and think of a function g that would have the property (15.30).]

To find g, we take $(\partial/\partial v_x)_{v_y,v_z}$ of (15.30), obtaining

$$g'(v_x)g(v_y)g(v_z) = \frac{d\phi(v)}{dv}\frac{\partial v}{\partial v_x}$$

where the chain rule was used to find $\partial\phi/\partial v_x$. From (15.1), we get $2v\,dv = 2v_x\,dv_x + 2v_y\,dv_y + 2v_z\,dv_z$, so $\partial v/\partial v_x = v_x/v$. [This also follows by direct differentiation of $v = (v_x^2 + v_y^2 + v_z^2)^{1/2}$.] We have

$$g'(v_x)g(v_y)g(v_z) = \phi'(v) \cdot v_x/v$$

Dividing this equation by $v_x g(v_x)g(v_y)g(v_z) = v_x\phi(v)$, we get

$$\frac{g'(v_x)}{v_x g(v_x)} = \frac{1}{v}\frac{\phi'(v)}{\phi(v)} \tag{15.31}$$

Since v_x, v_y, and v_z occur symmetrically in (15.30) and (15.1), taking $\partial/\partial v_y$ and $\partial/\partial v_z$ of (15.30) will give equations similar to (15.31):

$$\frac{g'(v_y)}{v_y g(v_y)} = \frac{1}{v}\frac{\phi'(v)}{\phi(v)} \quad \text{and} \quad \frac{g'(v_z)}{v_z g(v_z)} = \frac{1}{v}\frac{\phi'(v)}{\phi(v)} \tag{15.32}$$

Equations (15.31) and (15.32) give

$$\frac{g'(v_x)}{v_x g(v_x)} = \frac{g'(v_y)}{v_y g(v_y)} \equiv b \tag{15.33}$$

where we defined the quantity b. Since b equals $g'(v_y)/v_y g(v_y)$, b must be independent of v_x and v_z. But since b equals $g'(v_x)/v_x g(v_x)$, b must be independent of v_y and v_z. Hence b is independent of v_x, v_y, and v_z and is a constant.

Equation (15.33) reads $bv_x = (dg/dv_x)/g$, or $dg/g = bv_x\,dv_x$. Integration gives $\ln g = \frac{1}{2}bv_x^2 + c$, where c is an integration constant. Hence $g = e^{bv_x^2/2}e^c$, or

$$g = Ae^{bv_x^2/2} \tag{15.34}$$

where $A \equiv e^c$ is a constant. We have found the distribution function g for v_x, and as a check we note that it satisfies (15.30) since

$$g(v_x)g(v_y)g(v_z) = A^3 e^{b(v_x^2+v_y^2+v_z^2)/2} = A^3 e^{bv^2/2}$$

We still must evaluate the constants A and b in (15.34). To evaluate A, we substitute (15.34) into (15.27) to get

$$A\int_{-\infty}^{\infty} e^{bv_x^2/2}\,dv_x = 1 \tag{15.35}$$

(b must be negative; otherwise the integral doesn't exist.)

Table 15.1 lists some definite integrals useful in the kinetic theory of gases. (The derivation of these integrals is outlined in Probs. 15.14 to 15.16.) Recall that

$$n! \equiv n(n-1)(n-2)\cdots 2\cdot 1 \qquad \text{and} \qquad 0! \equiv 1 \tag{15.36}$$

where n is a positive integer. Integrals 2 and 5 in the table are special cases ($n = 0$) of integrals 3 and 6, respectively.

Using integral 1 with $n = 0$ and $a = -b/2$ and integral 2, we get from (15.35) that $2A\pi^{1/2}\frac{1}{2}(-b/2)^{-1/2} = 1$. Hence $A = (-b/2\pi)^{1/2}$, and (15.34) becomes

$$g(v_x) = (-b/2\pi)^{1/2}e^{bv_x{}^2/2} \tag{15.37}$$

To evaluate b, we use the fact that $\langle\varepsilon_{tr}\rangle = \frac{3}{2}kT$, Eq. (15.16). We have $\langle\varepsilon_{tr}\rangle = \frac{1}{2}m\langle v^2\rangle = \frac{3}{2}m\langle v_x{}^2\rangle$, where we used (15.10) with v_y replaced by v_x. Hence $\frac{3}{2}m\langle v_x{}^2\rangle = \frac{3}{2}kT$, and

$$\langle v_x{}^2\rangle = kT/m \tag{15.38}$$

We now calculate $\langle v_x{}^2\rangle$ from the distribution function (15.37) and compare the result with (15.38) to find b.

To evaluate $\langle v_x{}^2\rangle$, the following theorem is used. Let $g(w)$ be the distribution function for the continuous variable w; that is, the probability that this variable lies between w and $w + dw$ is $g(w)\,dw$. Then the average value of any function $f(w)$ is

$$\langle f(w)\rangle = \int_{w_{\min}}^{w_{\max}} f(w)g(w)\,dw \tag{15.39*}$$

where $w_{\min}$ and $w_{\max}$ are the minimum and maximum values of w. The proof of (15.39) follows.

We first consider a variable that takes on only discrete values (rather than a continuous range of values as is true for w). Suppose a class of 7 students takes a 5-question quiz, and the scores are 20, 40, 40, 80, 80, 80, 100. The average score $\langle s\rangle$ is

$$\begin{aligned}\langle s\rangle &= (20 + 40 + 40 + 80 + 80 + 80 + 100)/7\\ &= [0(0) + 1(20) + 2(40) + 0(60) + 3(80) + 1(100)]/7\\ &= \frac{1}{N}\sum_s n_s s = \sum_s \frac{n_s}{N} s\end{aligned}$$

Table 15.1. Integrals occurring in the kinetic theory of gases

Even powers of x	Odd powers of x
1. $\int_{-\infty}^{\infty} x^{2n}e^{-ax^2}\,dx = 2\int_0^{\infty} x^{2n}e^{-ax^2}\,dx$	**4.** $\int_{-\infty}^{\infty} x^{2n+1}e^{-ax^2}\,dx = 0$
2. $\int_0^{\infty} e^{-ax^2}\,dx = \frac{\pi^{1/2}}{2a^{1/2}}$	**5.** $\int_0^{\infty} xe^{-ax^2}\,dx = \frac{1}{2a}$
3. $\int_0^{\infty} x^{2n}e^{-ax^2}\,dx = \frac{(2n)!\,\pi^{1/2}}{2^{2n+1}n!\,a^{n+1/2}}$	**6.** $\int_0^{\infty} x^{2n+1}e^{-ax^2}\,dx = \frac{n!}{2a^{n+1}}$

where $a > 0$ and $n = 0, 1, 2, \ldots$

where the s's are the possible scores (0, 20, 40, 60, 80, 100), n_s is the number of people who got score s, N is the total number of people, and the sum goes over all the possible scores. If N is very large (as it is for molecules), then n_s/N is the probability $p(s)$ of getting score s. Therefore $\langle s \rangle = \sum_s p(s)s$.

Now suppose the average of the squares of the scores was wanted. Then

$$\langle s^2 \rangle = (20^2 + 40^2 + 40^2 + 80^2 + 80^2 + 80^2 + 100^2)/7$$

$$= [0(0)^2 + 1(20)^2 + 2(40)^2 + 0(60)^2 + 3(80)^2 + 1(100)^2]/7$$

$$= \frac{1}{N}\sum_s n_s s^2 = \sum_s \frac{n_s}{N} s^2 = \sum_s p(s)s^2$$

The same argument holds for the average value of any function of s. For example, $\langle 2s^3 \rangle = \sum_s p(s)2s^3$. If $f(s)$ is any function of s, then

$$\langle f(s) \rangle = \sum_s p(s)f(s) \qquad (15.40)^*$$

For a variable w that takes on a continuous range of values, Eq. (15.40) must be modified. The probability $p(s)$ of having the value s is replaced by the probability that w lies in the infinitesimal range from w to $w + dw$. This probability is $g(w)\,dw$, where $g(w)$ is the distribution function (probability density) for w. For a continuous variable, (15.40) becomes $\langle f(w) \rangle = \sum_w f(w)g(w)\,dw$. But the infinite sum over infinitesimal quantities is the definite integral over the full range of w. Hence we have shown the validity of (15.39).

It readily follows from (15.39) that the average of a sum equals the sum of the averages. If f_1 and f_2 are any two functions of w, then (Prob. 15.17)

$$\langle f_1(w) + f_2(w) \rangle = \langle f_1(w) \rangle + \langle f_2(w) \rangle \qquad (15.41)^*$$

However, the average of a product is not necessarily equal to the product of the averages (see Probs. 15.12, 15.33, and 15.34).

Returning to the evaluation of $\langle v_x{}^2 \rangle$ in (15.38), we use (15.39) with $w = v_x$, $f(w) = v_x{}^2$, and $g(v_x)$ given by Eq. (15.37) to get

$$\langle v_x{}^2 \rangle = \int_{-\infty}^{\infty} v_x{}^2 g(v_x)\,dv_x = \int_{-\infty}^{\infty} v_x{}^2 \left(\frac{-b}{2\pi}\right)^{1/2} e^{bv_x{}^2/2}\,dv_x$$

Integrals 1 and 3 with $n = 1$ and $a = -b/2$ in Table 15.1 give

$$\langle v_x{}^2 \rangle = 2\left(\frac{-b}{2\pi}\right)^{1/2} \frac{2!\,\pi^{1/2}}{2^3 1!\,(-b/2)^{3/2}} = \frac{1}{-b}$$

Comparison with (15.38) gives $-1/b = kT/m$, and $b = -m/kT$.

The distribution function (15.37) for v_x is therefore [see also (15.26)]

$$\frac{1}{N}\frac{dN_{v_x}}{dv_x} = g(v_x) = \left(\frac{m}{2\pi kT}\right)^{1/2} e^{-mv_x{}^2/2kT} \qquad (15.42)^*$$

Equation (15.42) looks complicated at first glance but is actually reasonably easy to remember since it has the form $g = \text{const.} \times e^{-\varepsilon_{tr,x}/kT}$, where $\varepsilon_{tr,x} = \frac{1}{2}mv_x{}^2$ is the kinetic energy of motion in the x direction. (The constant that multiplies the exponential is determined by the requirement that $\int_{-\infty}^{\infty} g\,dv_x = 1$.) Note the presence of kT as a characteristic energy in (15.42) and (15.16); this is a general occurrence in statistical mechanics.

The equations for $g(v_y)$ and $g(v_z)$ are obtained from (15.42) by replacing v_x by v_y and v_z.

Now that $g(v_x)$ has been found, it is a simple matter to find the distribution function $G(v)$ for the speed. $G(v)\,dv$ is the probability that a molecule's speed lies between v and $v + dv$; that is, $G(v)\,dv$ is the probability that in the v_x, v_y, v_z coordinate system (Fig. 15.3) the tip of the molecule's velocity vector **v** lies within a thin spherical shell of inner radius v and outer radius $v + dv$. Consider a tiny rectangular box lying within this shell and having edges dv_x, dv_y, and dv_z (Fig. 15.4). The probability of **v**'s being in this tiny box is given by (15.29) as

$$g(v_x)g(v_y)g(v_z)\,dv_x\,dv_y\,dv_z = (m/2\pi kT)^{3/2}e^{-m(v_x^2+v_y^2+v_z^2)/2kT}\,dv_x\,dv_y\,dv_z$$

$$g(v_x)g(v_y)g(v_z)\,dv_x\,dv_y\,dv_z = \left(\frac{m}{2\pi kT}\right)^{3/2} e^{-mv^2/2kT}\,dv_x\,dv_y\,dv_z \qquad (15.43)$$

where (15.42) and the analogous equations for $g(v_y)$ and $g(v_z)$ were used.

The probability that **v** lies in the thin spherical shell is the sum of the probabilities (15.43) for all the tiny rectangular boxes that make up the thin shell:

$$G(v)\,dv = \sum_{\text{shell}} \left(\frac{m}{2\pi kT}\right)^{3/2} e^{-mv^2/2kT}\,dv_x\,dv_y\,dv_z$$

$$= \left(\frac{m}{2\pi kT}\right)^{3/2} e^{-mv^2/2kT} \sum_{\text{shell}} dv_x\,dv_y\,dv_z$$

since the function $e^{-mv^2/2kT}$ is constant within the shell. (v varies only infinitesimally in the shell.) The quantity $dv_x\,dv_y\,dv_z$ is the volume of one of the tiny rectangular boxes, and the sum of these volumes over the shell is the volume of the shell. The shell has outer and inner radii $v + dv$ and v, so the shell volume is

$$\tfrac{4}{3}\pi(v+dv)^3 - \tfrac{4}{3}\pi v^3 = \tfrac{4}{3}\pi[v^3 + 3v^2\,dv + 3v(dv)^2 + (dv)^3] - \tfrac{4}{3}\pi v^3 = 4\pi v^2\,dv$$

since $(dv)^2$ and $(dv)^3$ are negligible compared with dv. (Of course, $4\pi v^2\,dv$ is the differential with respect to v of the spherical volume $\frac{4}{3}\pi v^3$.) Use of $4\pi v^2\,dv$ for the shell volume gives as the final result for the distribution function $G(v)$

$$\frac{dN_v}{N} = G(v)\,dv = \left(\frac{m}{2\pi kT}\right)^{3/2} e^{-mv^2/2kT} 4\pi v^2\,dv \qquad (15.44)$$

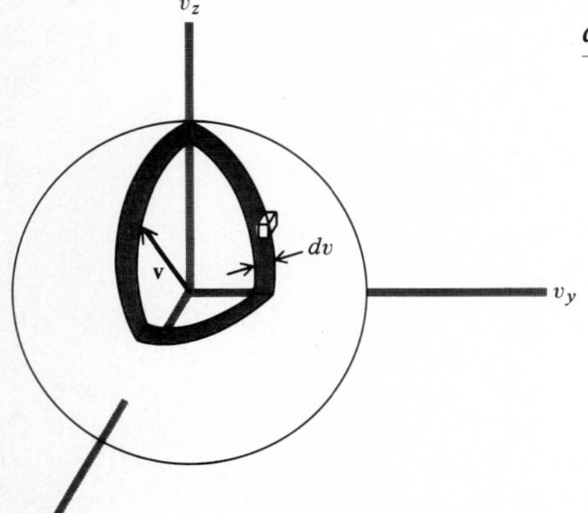

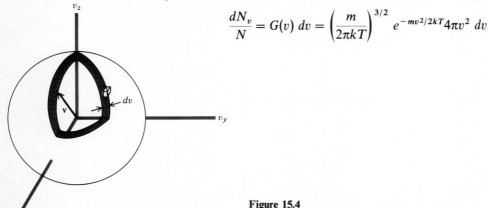

Figure 15.4
A thin spherical shell in velocity space.

Summarizing our results, we have shown that the fraction of ideal-gas molecules with x component of velocity in the range v_x to $v_x + dv_x$ is $dN_{v_x}/N = g(v_x)\,dv_x$, where $g(v_x)$ is given by (15.42); the fraction of molecules with speed in the range v to $v + dv$ is $dN_v/N = G(v)\,dv$, where $G(v)\,dv$ is given by (15.44). The relation between g and G is

$$G(v) = g(v_x)g(v_y)g(v_z) \cdot 4\pi v^2 \qquad (15.45)^*$$

where the factor $4\pi v^2$ comes from the volume $4\pi v^2\,dv$ of the thin spherical shell.

Equations (15.42) and (15.44) are the *Maxwell distribution laws* for v_x and v in an ideal gas. (Note that $m/k = N_0 m/N_0 k = M/R$, where M is the molar mass.)

The function $g(v_x)$ has the form $Ae^{-cv_x^2}$ and has its maximum at $v_x = 0$. Figure 15.5 plots $g(v_x)$ at two temperatures. This distribution law, called the *normal* or *gaussian* distribution, occurs in many contexts besides kinetic theory (e.g., the distribution of heights in a population, the distribution of random errors of measurement.)

The distribution function $G(v)$ for speeds is plotted in Fig. 15.6 at two temperatures. For very small v, the exponential $e^{-mv^2/2kT}$ is close to 1, and $G(v)$ increases as v^2. For very large v, the exponential factor dominates the v^2 factor, and $G(v)$ decreases rapidly with increasing v. As T increases, the distribution curve shifts to higher speeds.

The most probable value of v_x is zero (Fig. 15.5). Likewise, the most probable values of v_y and v_z are zero. Hence if we make a three-dimensional plot of the probabilities of the various values of v_x, v_y, v_z by putting dots at points in velocity space so that the density of dots in each region is proportional to the probability that $\mathbf{v}$ lies in that region, the maximum density of dots will occur at the origin ($v_x = v_y = v_z = 0$). Figure 15.7 shows a two-dimensional cross section of such a plot.

Although the most probable value of each component of the velocity is zero, Fig. 15.6 shows that the most probable value of the speed is not zero. This apparent contradiction is reconciled by realizing that although the probability density $g(v_x)g(v_y)g(v_z)$ is a maximum at the origin ($v_x = v_y = v_z = 0$) and decreases with

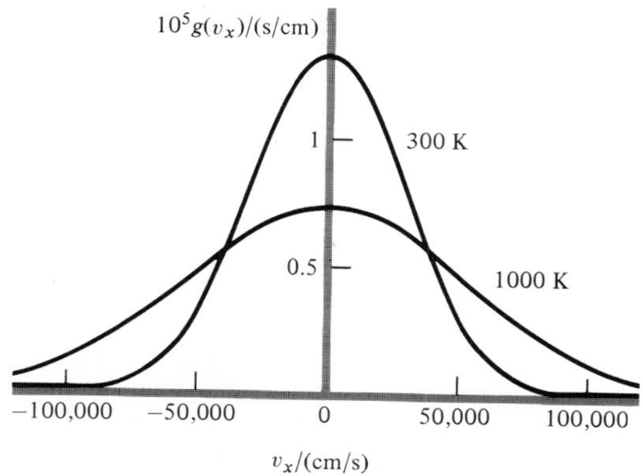

Figure 15.5
Distribution functions for v_x in N_2 at 300 and 1000 K. (N_2 has molecular weight 28. For H_2 with molecular weight 2, these curves apply at 21 and 71 K.)

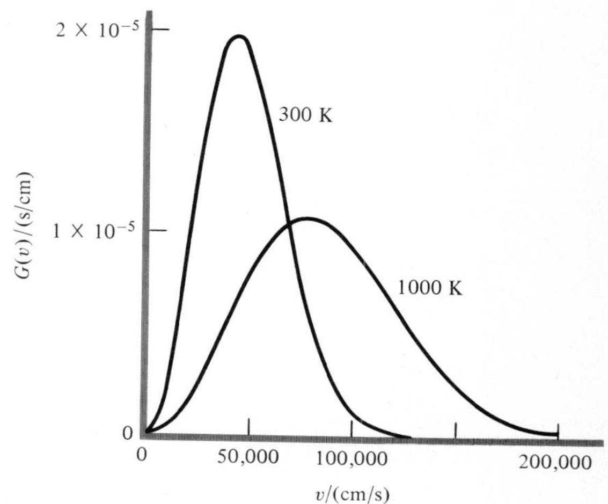

Figure 15.6
Distribution function for speeds in N_2 at 300 and 1000 K.

increasing v, the volume $4\pi v^2\ dv$ of the thin spherical shell increases with increasing v. These two opposing factors then give a nonzero most probable value for v. (These same ideas occur in the quantum mechanics of the hydrogen atom.)

Maxwell derived Eq. (15.44) in 1860. It took until 1955 for the first truly accurate direct experimental verification of this distribution law to be made. [R. C. Miller and P. Kusch, *Phys. Rev.*, **99**, 1314 (1955).] Miller and Kusch measured the distribution of speeds in a beam of gas molecules emerging from a small hole in an oven (Fig. 15.8). The rotating cylinder with spiral grooves acts as a speed selector, since only molecules with a certain value of v will pass through the grooves without striking the walls of the grooves. Changing the rate of cylinder rotation changes the speed selected. These workers found excellent agreement with the predictions of the Maxwell law. (Note that the velocity distribution in the beam is not Maxwellian: there are no negative values of v_x; moreover, there is a higher fraction of fast molecules in the beam than in the oven, since fast molecules hit the walls more often than slow ones and so are more likely to escape the oven. Measurement of the distribution of speeds in the beam allows calculation of the distribution in the oven.)

The derivation of the Maxwell distribution given above has the defect of not bringing out the physical process by which the distribution of speeds is attained. Thus, suppose two gas samples at different temperatures are mixed. Eventually equilibrium is reached at some intermediate temperature T', and a Maxwellian distribution characteristic of T' is established. The physical mechanism that brings about the Maxwellian distribution is the collisions between molecules. In 1872, Boltzmann derived the Maxwell distribution by a method based on the dynamics of molecular collisions.

15.5 APPLICATIONS OF THE MAXWELL DISTRIBUTION

The Maxwell distribution function $G(v)$ can be used to calculate the average value of any function of v, since $\langle f(v)\rangle = \int_0^\infty f(v)G(v)\ dv$ [Eq. (15.39)]. For example, the *average speed* $\langle v\rangle$ is

$$\langle v\rangle = \int_0^\infty vG(v)\ dv = 4\pi\left(\frac{m}{2\pi kT}\right)^{3/2}\int_0^\infty e^{-mv^2/2kT}v^3\ dv$$

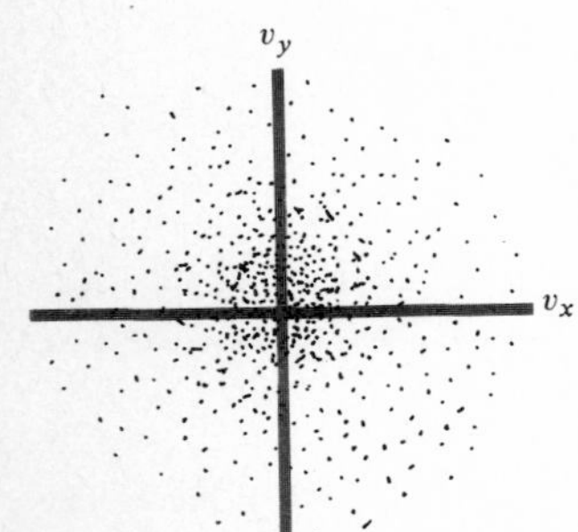

Figure 15.7
Probability-density plot for velocities.

where (15.44) was used. Use of integral 6 in Table 15.1 with $n = 1$ and $a = m/2kT$ gives

$$\langle v \rangle = 4\pi\left(\frac{m}{2\pi kT}\right)^{3/2}\frac{1}{2(m/2kT)^2} = \left(\frac{8kT}{\pi m}\right)^{1/2} = \left(\frac{8N_0kT}{\pi N_0 m}\right)^{1/2}$$

$$\langle v \rangle = \left(\frac{8RT}{\pi M}\right)^{1/2} \qquad (15.46)^*$$

We already know [Eqs. (15.22) and (15.21)] that $v_{rms} \equiv \langle v^2 \rangle^{1/2} = (3RT/M)^{1/2}$, but we could check this by calculating $\int_0^\infty v^2 G(v)\, dv$ (Prob. 15.10).

The *most probable speed* v_{mp} is the speed for which $G(v)$ is a maximum (see Fig. 15.6). Setting $dG(v)/dv = 0$, one finds (Prob. 15.9) $v_{mp} = (2RT/M)^{1/2}$.

The speeds v_{mp}, $\langle v \rangle$, v_{rms} stand in the ratio $2^{1/2} : (8/\pi)^{1/2} : 3^{1/2} =$ 1.414 : 1.596 : 1.732. Thus

$$v_{mp} : \langle v \rangle : v_{rms} = 1 : 1.128 : 1.225 \qquad (15.47)$$

Example Find $\langle v \rangle$ for (*a*) O_2 at 25°C and 1 atm; (*b*) H_2 at 25°C and 1 atm. The pressure is low enough to assume ideality. Substitution in (15.46) gives for O_2

$$\langle v \rangle = \left[\frac{8(8.314 \text{ J mole}^{-1}\text{ K}^{-1})(298 \text{ K})}{\pi(0.0320 \text{ kg mole}^{-1})}\right]^{1/2} = \left[\frac{8(8.314 \text{ kg m}^2\text{ s}^{-2})298}{3.14(0.0320 \text{ kg})}\right]^{1/2}$$

$$\langle v \rangle = 444 \text{ m/s} \qquad \text{for } O_2 \text{ at } 25°\text{C} \qquad (15.48)$$

where Eqs. (2.30) and (2.21) were used to write 1 J = 1 kg m^2 s^{-2}. Because the SI unit of the joule was used in R, the molar mass M was expressed in kg mole^{-1}. For the reader who prefers cgs units, the calculation would be

$$\langle v \rangle = \left[\frac{8(8.314 \times 10^7 \text{ ergs mole}^{-1}\text{ K}^{-1})(298 \text{ K})}{3.14(32.0 \text{ g mole}^{-1})}\right]^{1/2} = 44{,}400 \text{ cm/s} \quad (15.49)$$

since 1 J = 10^7 ergs and 1 erg = 1 g cm^2 s^{-2}. The speed 444 m/s is 993 mi/hr, so at room temperature the gas molecules are not sluggards.

For H_2, one obtains $\langle v \rangle = 1770$ m/s at 25°C. At the same temperature, the H_2 and O_2 molecules have the same average kinetic energy $\frac{3}{2}kT$. Hence the H_2 molecules must move faster on the average to compensate for their lighter mass. The H_2 molecule is one-sixteenth as heavy as an O_2 molecule and moves 4 times as fast, on the average.

At 25°C and 1 atm, the speed of sound in O_2 is 330 m/s and in H_2 is 1330 m/s,

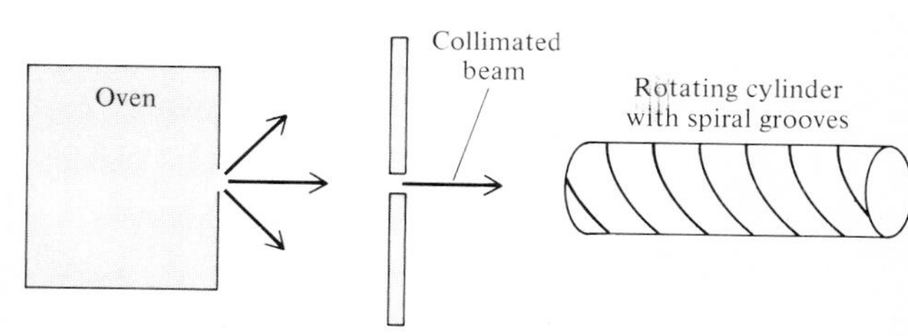

Figure 15.8
Apparatus for testing the Maxwell distribution law.

so $\langle v \rangle$ is the same order of magnitude as the speed of sound in the gas. This is reasonable since the propagation of sound is closely connected with the motions of the gas molecules. [It can be shown that the speed of sound in an ideal gas equals $(\gamma RT/M)^{1/2}$, where $\gamma \equiv C_P/C_V$; see *Zemansky*, sec. 5-9.]

The fraction of molecules whose x component of velocity lies between 0 and v_x' is given by (15.24) and (15.42) as

$$\frac{N(0 \le v_x \le v_x')}{N} = \left(\frac{m}{2\pi kT}\right)^{1/2} \int_0^{v_x'} e^{-mv_x^2/2kT}\, dv_x$$

where $N(0 \le v_x \le v_x')$ is the number of molecules with v_x in the range 0 to v_x'. Let $s \equiv (m/kT)^{1/2} v_x$. Then $ds = (m/kT)^{1/2}\, dv_x$, and

$$\frac{N(0 \le v_x \le v_x')}{N} = \frac{1}{(2\pi)^{1/2}} \int_0^u e^{-s^2/2}\, ds \qquad \text{where } u \equiv \left(\frac{m}{kT}\right)^{1/2} v_x' \tag{15.50}$$

The indefinite integral $\int e^{-s^2/2}\, ds$ cannot be expressed in terms of elementary functions. However, by expanding the integrand $e^{-s^2/2}$ in a Taylor series and integrating term by term, one can express the integral as an infinite series, thereby allowing the definite integral in (15.50) to be evaluated for any desired value of u (Prob. 15.18). The *Gauss error integral* $I(u)$ is defined as

$$I(u) \equiv \frac{1}{(2\pi)^{1/2}} \int_0^u e^{-s^2/2}\, ds \tag{15.51}$$

Equation (15.50) then reads $N(0 \le v_x \le v_x')/N = I[(m/kT)^{1/2} v_x']$. Some values of $I(u)$ are:

u	0	0.30	0.50	0.70	1.00	1.50	2.00	2.50	3.00
$I(u)$	0	0.12	0.19	0.26	0.34	0.43	0.477	0.494	0.4987

u	4.00	5.00	6.00	∞
$I(u)$	0.499968	0.4999997	0.4999999990	0.5

More extensive tables are given in statistics handbooks.

The fraction of molecules with speed v lying in the range 0 to v' can be expressed in terms of the function I (see Prob. 15.19). Because of the exponential falloff of $G(v)$ with increasing v, only a small fraction of molecules have speeds that greatly exceed v_{mp}. For example, the fraction of molecules with speeds exceeding $3v_{\text{mp}}$ is about 0.0004, only about 1 molecule in 10^{15} has a speed exceeding $6v_{\text{mp}}$, and it is virtually certain that in a mole of gas not a single molecule has a speed exceeding $9v_{\text{mp}}$.

The Maxwell distribution can be rewritten in terms of the translational kinetic energy $\varepsilon_{\text{tr}} = \frac{1}{2}mv^2$. We have $v = (2\varepsilon_{\text{tr}}/m)^{1/2}$ and $dv = (1/2m)^{1/2}\varepsilon_{\text{tr}}^{-1/2}\, d\varepsilon_{\text{tr}}$. If a molecule has a speed between v and $v + dv$, its translational energy is between $\frac{1}{2}mv^2$ and $\frac{1}{2}m(v + dv)^2 = \frac{1}{2}m[v^2 + 2v\, dv + (dv)^2] = \frac{1}{2}mv^2 + mv\, dv$; that is, its translational energy is between ε_{tr} and $\varepsilon_{\text{tr}} + d\varepsilon_{\text{tr}}$, where $\varepsilon_{\text{tr}} = \frac{1}{2}mv^2$ and $d\varepsilon_{\text{tr}} = mv\, dv$. Replacing v and dv in (15.44) by their equivalents gives as the distribution function for translational energy

$$\frac{dN_{\varepsilon_{\text{tr}}}}{N} = 2\pi \left(\frac{1}{\pi kT}\right)^{3/2} \varepsilon_{\text{tr}}^{1/2} e^{-\varepsilon_{\text{tr}}/kT}\, d\varepsilon_{\text{tr}} \tag{15.52}$$

where $dN_{\varepsilon_{tr}}$ is the number of molecules with translational energy between ε_{tr} and $\varepsilon_{tr} + d\varepsilon_{tr}$. Figure 15.9 plots the distribution function (15.52). Note the difference in shape between this figure and Fig. 15.6.

15.6 COLLISIONS WITH A WALL AND EFFUSION

A quantity of interest is the rate of collisions of gas molecules with a container wall. Let dN_W be the number of molecules that hit wall W in Fig. 15.2 in the infinitesimal time interval dt. Consider a molecule with y component of speed in the range v_y to $v_y + dv_y$. For this molecule to hit wall W in the interval dt, it (*a*) must be moving to the left (i.e., it must have $v_y > 0$) and (*b*) must be close enough to W to reach it in time dt or less—the molecule travels a distance $v_y\,dt$ in time dt and must be within this distance from W.

The number of molecules with y component of velocity in the range v_y to $v_y + dv_y$ is [Eq. (15.42)] $dN_{v_y} = Ng(v_y)\,dv_y$. The molecules are evenly distributed throughout the container, so the fraction of molecules within distance $v_y\,dt$ of W is $(v_y\,dt/l_y)$. The product $dN_{v_y}(v_y\,dt/l_y)$ gives the number of molecules that have y component of velocity in the range v_y to $v_y + dv_y$ and are within distance $v_y\,dt$ of W. This number is $[Ng(v_y)v_y/l_y]\,dv_y\,dt$. Dividing by the area $l_x l_z$ of the wall, we get as the number of collisions per unit area of W in time dt due to molecules with y velocity between v_y and $v_y + dv_y$

$$(N/V)g(v_y)v_y\,dv_y\,dt \tag{15.53}$$

where $V = l_x l_y l_z$ is the container volume. To get the total number of collisions with W in time dt, we sum (15.53) over all positive values of v_y and multiply by the area $\mathscr{A}$ of W. [A molecule with $v_y < 0$ does not satisfy requirement (*a*) above and cannot hit the wall in time dt; hence only positive v_y's are summed over.] The infinite sum of infinitesimal quantities is the definite integral over v_y from 0 to ∞, and the number of molecules that hit W in time dt is

$$dN_W = \mathscr{A}\frac{N}{V}\left[\int_0^\infty g(v_y)v_y\,dv_y\right]dt \tag{15.54}$$

Use of (15.42) and integral 5 in Table 15.1 gives

$$\int_0^\infty g(v_y)v_y\,dv_y = \left(\frac{m}{2\pi kT}\right)^{1/2}\int_0^\infty e^{-mv_y^2/2kT}v_y\,dv_y = \left(\frac{RT}{2\pi M}\right)^{1/2} = \tfrac{1}{4}\langle v\rangle$$

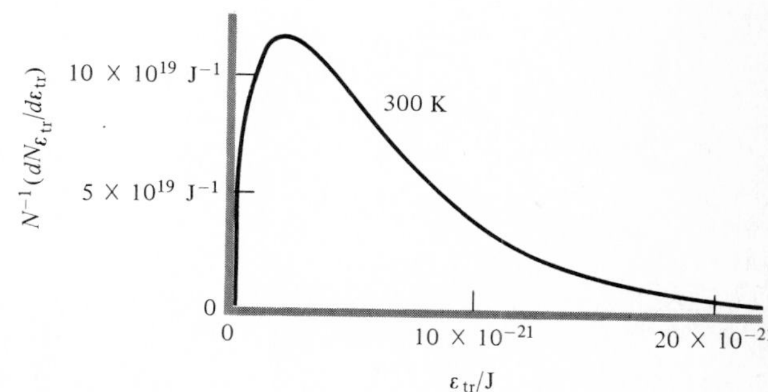

Figure 15.9
Distribution function for kinetic energy at 300 K. This curve applies to any gas.

where (15.46) for $\langle v \rangle$ was used. Also, since $PV = nRT = (N/N_0)RT$, the number of molecules per unit volume is

$$N/V = PN_0/RT \tag{15.55}$$

Equation (15.54) becomes

$$\frac{1}{\mathscr{A}}\frac{dN_W}{dt} = \frac{1}{4}\frac{N}{V}\langle v \rangle = \frac{1}{4}\frac{PN_0}{RT}\left(\frac{8RT}{\pi M}\right)^{1/2} \tag{15.56}$$

Example Calculate the number of wall collisions per second per square centimeter in O_2 at 25°C and 1.00 atm.

Using (15.48) for $\langle v \rangle$, we get

$$\frac{1}{\mathscr{A}}\frac{dN_W}{dt} = \frac{1}{4}\frac{(1.00 \text{ atm})(6.02 \times 10^{23} \text{ mole}^{-1})}{(82.06 \text{ cm}^3 \text{ atm mole}^{-1} \text{ K}^{-1})(298 \text{ K})} 4.44 \times 10^4 \text{ cm s}^{-1}$$

$$= 2.7 \times 10^{23} \text{ cm}^{-2} \text{ s}^{-1} \tag{15.57}$$

Suppose there is a *tiny* hole of area A in the wall and that outside the container is a vacuum. [If the hole is not tiny, the gas will escape rapidly, thereby destroying the Maxwellian distribution of velocities used to derive (15.56).] Molecules hitting the hole escape, and the rate of escape is given by (15.56) as

$$\frac{dN_A}{dt} = \frac{PN_0 A}{(2\pi MRT)^{1/2}} \tag{15.58}$$

Escape of a gas through a tiny hole is called *effusion*. The rate of effusion is proportional to $M^{-1/2}$ (*Graham's law of effusion*). Differences in effusion rates can be used to separate isotopic species, for example, $^{235}UF_6$ and $^{238}UF_6$.

Let λ be the average distance a gas molecule travels between collisions with other gas molecules. For (15.58) to apply, the diameter of the hole d_{hole} must be substantially less than λ. Otherwise, molecules will collide with one another in the vicinity of the hole, thereby developing a collective flow through the hole. This bulk flow arises because the escape of molecules through the hole depletes the region near the hole of gas molecules, lowering the pressure near the hole. Hence, molecules in the region of the hole experience fewer collisions on the side near the hole than on the side away from the hole and experience a net force toward the hole. (In effusion, there is no net force toward the hole.) Flow of gas (or liquid) due to a pressure difference is called *viscous flow*, *convective flow*, or *bulk flow* (see Sec. 16.3). Effusion is an example of *free-molecule* (or *Knudsen*) flow; here, λ is so large that intermolecular collisions can be ignored.

Another requirement for the applicability of (15.58) is that the wall be thin. Otherwise, escaping gas molecules colliding with the sides of the hole can be reflected back into the container.

In the Knudsen method for determining the vapor pressure of a solid, the weight loss due to effusion of vapor in equilibrium with the solid in a container with a tiny hole is measured. Since P is constant, we have $dN_A/dt = \Delta N_A/\Delta t$. Measurement of $\Delta N_A/\Delta t$ allows the vapor pressure P to be calculated from (15.58); see Prob. 15.23.

For a liquid in equilibrium with its vapor, the rate of evaporation of liquid molecules equals the rate of condensation of gas molecules. A reasonable assumption is that virtually every vapor molecule that strikes the liquid's surface condenses to liquid. Equation (15.58) with P equal to the vapor pressure allows calculation of the rate at which vapor molecules hit the surface and hence allows calculation of the rate at which liquid molecules evaporate (see Prob. 15.25). Measurements of the rates of evaporation of liquids into vacuum generally give good agreement with (15.58), thereby validating the assumption that virtually every vapor molecule hitting the liquid condenses.

15.7 MOLECULAR COLLISIONS AND MEAN FREE PATH

Kinetic theory allows the rate of intermolecular collisions to be calculated. We adopt the (rather crude) model of a molecule as a hard sphere of diameter d. We assume that no intermolecular forces exist except at the moment of collision, when the molecules bounce off each other like two colliding billiard balls. (At high gas pressures, intermolecular forces are substantial, and the equations derived in this section do not apply.) Intermolecular collisions are important in reaction kinetics, since molecules A and B must collide with each other in order to react. Intermolecular collisions also serve to maintain the Maxwell distribution of speeds.

The rigorous derivation of the collision rate is complicated, so we shall give only a nonrigorous treatment. (The full derivation is given in *Present*, sec. 5-2; and *Kauzmann*, chap. 5.)

Besides considering collisions in a pure gas, we shall also consider collisions in a mixture of gases A and B. For either pure gas A or a mixture of A and B, let z_{AA} be the number of collisions per unit time that one particular A molecule makes with other A molecules, and let Z_{AA} be the total number of A-A collisions per unit time and per unit volume of gas. For a mixture of A and B, let z_{AB} be the number of collisions per unit time that one particular A molecule makes with B molecules, and let Z_{AB} be the total number of A-B collisions per unit time per unit volume. Let N_A and N_B be the numbers of A and B molecules present.

To calculate z_{AB}, we pretend that all molecules in the gas mixture are at rest except one particular A molecule, which moves at the constant speed $\langle v_{AB} \rangle$, where $\langle v_{AB} \rangle$ is the average speed of A molecules relative to B molecules in the actual gas (with all molecules in motion). Let d_A and d_B be the diameters of A and B and r_A and r_B their radii. The moving A molecule will collide with a B molecule whenever the distance between the centers of the pair is within $\frac{1}{2}(d_A + d_B) = r_A + r_B$ (Fig. 15.10*a*). Imagine a cylinder of radius $r_A + r_B$ centered about the moving A molecule. In time dt, the moving molecule will travel a distance $\langle v_{AB} \rangle\, dt$ and will sweep out a cylinder of volume $\pi(r_A + r_B)^2 \cdot \langle v_{AB} \rangle\, dt \equiv V_{cyl}$. The moving molecule will collide with all B molecules whose centers lie within this cylinder. Since the stationary B molecules are uniformly distributed throughout the container volume V, the number of B

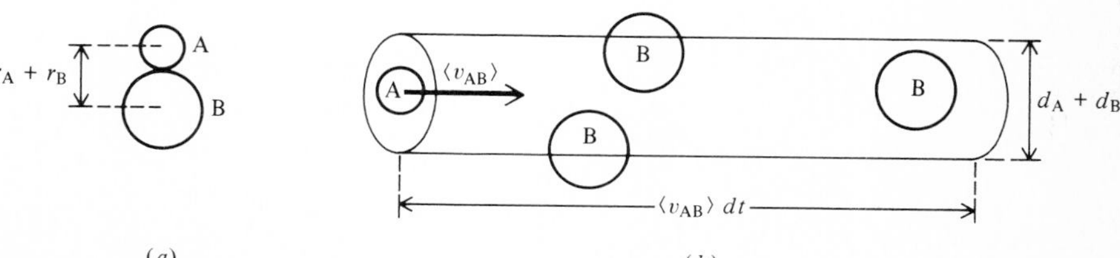

Figure 15.10
Colliding molecules.

molecules in the cylinder equals $(V_{cyl}/V)N_B$, and this is the number of collisions between a particular A molecule and B molecules in time dt. The number of such collisions per unit time is then $z_{AB} = (V_{cyl}/V)N_B/dt$, and

$$z_{AB} = (N_B/V)\pi(r_A + r_B)^2\langle v_{AB}\rangle \tag{15.59}$$

[Each time the moving A molecule collides with a stationary molecule, it bounces off and changes its direction of motion. Therefore the cylinder in Fig. 15.10*b* is actually not straight, but is bent at each point of collision. These bends have a negligible effect on the cylinder volume if (as is true in a low- or moderate-pressure gas) the average distance between collisions is large compared with $r_A + r_B$.]

To complete the derivation, we need $\langle v_{AB}\rangle$, the average speed of A molecules relative to B molecules. Figure 15.11*a* shows the displacement vectors $\mathbf{r}_A$ and $\mathbf{r}_B$ of molecules A and B from the coordinate origin. The position of A relative to B is specified by the vector $\mathbf{r}_{AB}$. These three vectors form a triangle, and, by the usual rule of vector addition, we have $\mathbf{r}_B + \mathbf{r}_{AB} = \mathbf{r}_A$, or $\mathbf{r}_{AB} = \mathbf{r}_A - \mathbf{r}_B$. Differentiating this equation with respect to time and using $\mathbf{v} \equiv d\mathbf{r}/dt$ [Eq. (2.17)], we get $\mathbf{v}_{AB} = \mathbf{v}_A - \mathbf{v}_B$. This equation shows that the vectors $\mathbf{v}_A$, $\mathbf{v}_B$, and $\mathbf{v}_{AB}$ form a triangle (Fig. 15.11*b*).

In an A-B collision, the molecules can approach each other at any angle from 0 to 180° (Fig. 15.12*a*). The average approach angle is 90°, so to calculate $\langle v_{AB}\rangle$, we imagine a 90° collision between an A molecule with speed $\langle v_A\rangle$ and a B molecule with speed $\langle v_B\rangle$, where these average speeds are given by (15.46). The triangle formed by the vectors in Fig. 15.11*b* is a right triangle for a 90° collision (Fig. 15.12*b*), and the Pythagorean theorem gives $\langle v_{AB}\rangle^2 = \langle v_A\rangle^2 + \langle v_B\rangle^2 = 8RT/\pi M_A + 8RT/\pi M_B$, where M_A and M_B are the molar masses of gases A and B. Substitution in (15.59) gives

$$z_{AB} = \pi(r_A + r_B)^2[\langle v_A\rangle^2 + \langle v_B\rangle^2]^{1/2}\frac{N_B}{V}$$

$$= \pi(r_A + r_B)^2\left[\frac{8RT}{\pi}\left(\frac{1}{M_A} + \frac{1}{M_B}\right)\right]^{1/2}\frac{N_B}{V} \tag{15.60}$$

To find z_{AA}, we put B = A in (15.60) to get

$$z_{AA} = 2^{1/2}\pi d_A{}^2\langle v_A\rangle\frac{N_A}{V} = 2^{1/2}\pi d_A{}^2\left(\frac{8RT}{\pi M_A}\right)^{1/2}\frac{P_A N_0}{RT} \tag{15.61}$$

This expression is valid whether A is a pure gas or a component of an ideal gas mixture.

The total A-B collision rate equals the collision rate of a particular A molecule

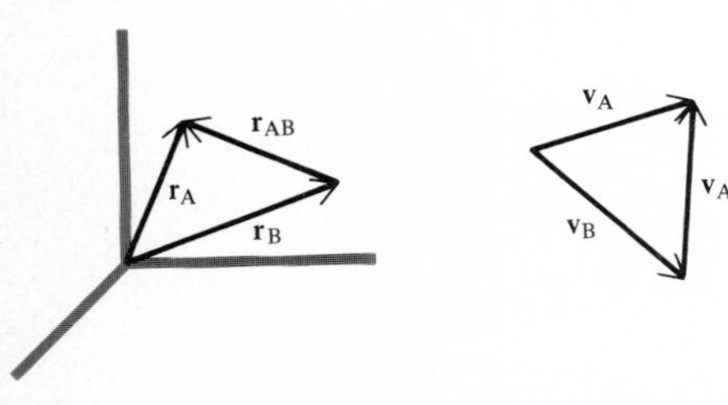

Figure 15.11
Molecular displacement vectors and velocity vectors.

with B molecules (z_{AB}) multiplied by the number of A molecules. Hence the total A-B collision rate per unit volume is $Z_{AB} = N_A z_{AB}/V$, and

$$Z_{AB} = \pi(r_A + r_B)^2 \left[\frac{8RT}{\pi}\left(\frac{1}{M_A} + \frac{1}{M_B}\right)\right]^{1/2} \left(\frac{N_A}{V}\right)\left(\frac{N_B}{V}\right) \tag{15.62}$$

If we were to calculate the total A-A collision rate by multiplying the A-A collision rate of one A molecule (z_{AA}) by the number of A molecules, we would be counting each A-A collision twice; e.g., the collision between molecules A_1 and A_2 would be counted once as one of the collisions made by A_1 and once as one of the collisions made by A_2. Hence, a factor of $\frac{1}{2}$ must be included, and the total A-A collision rate per unit volume is $Z_{AA} = \frac{1}{2}N_A z_{AA}/V$, or

$$Z_{AA} = \frac{1}{2^{1/2}}\pi d_A{}^2\langle v_A\rangle\left(\frac{N_A}{V}\right)^2 = \frac{1}{2^{1/2}}\pi d_A{}^2\left(\frac{8RT}{\pi M_A}\right)^{1/2}\left(\frac{P_A N_0}{RT}\right)^2 \tag{15.63}$$

Example For O_2 at 25°C and 1.00 atm, calculate z_{AA} and Z_{AA}. The bond distance in O_2 is 1.2 Å.

The O_2 molecule is neither hard nor spherical, but a reasonable estimate of the diameter d in the hard-sphere model might be twice the bond length: $d \approx 2.4$ Å. Equations (15.61) and (15.48) give the collision rate of one particular O_2 molecule as

$$z_{AA} \approx 2^{1/2}\pi(2.4 \times 10^{-8}\text{ cm})^2(4.44 \times 10^4\text{ cm s}^{-1}) \times \frac{(1.00\text{ atm})(6.02 \times 10^{23}\text{ mole}^{-1})}{(82.06\text{ cm}^3\text{ atm mole}^{-1}\text{ K}^{-1})(298\text{ K})}$$

$$z_{AA} \approx 2.8 \times 10^9\text{ collns./s} \qquad \text{for } O_2 \text{ at 25°C and 1 atm} \tag{15.64}$$

Even though the gas molecules are far apart from each other (compared with the molecular diameter), the very high average molecular speed causes very many collisions per second. Substitution in (15.63) gives the total number of collisions per second per cubic centimeter of gas as

$$Z_{AA} \approx 3.4 \times 10^{28}\text{ cm}^{-3}\text{ s}^{-1} \qquad \text{for } O_2 \text{ at 25°C and 1 atm} \tag{15.65}$$

The *mean free path* λ is the average distance a molecule travels between collisions. In a mixture of gases A and B, λ_A differs from λ_B. The gas molecules have a distribution of speeds. (More precisely, each species has its own distribution.) The speed of a given A molecule changes many times each second due to intermolecular collisions. Over a long time t, the average speed of a given A molecule is $\langle v_A\rangle$, the distance it travels is $\langle v_A\rangle t$, and the number of collisions it makes is $(z_{AA} + z_{AB})t$.

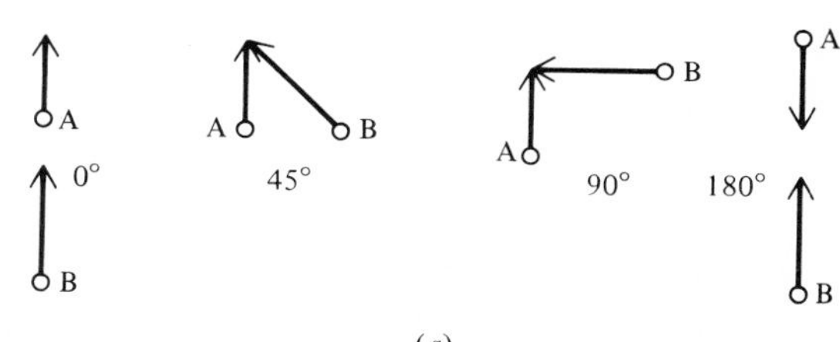

Figure 15.12
Colliding molecules.

Hence the average distance traveled by A between collisions is $\lambda_A = \langle v_A \rangle t/(z_{AA} + z_{AB})t$, and

$$\lambda_A = \langle v_A \rangle/(z_{AA} + z_{AB}) \tag{15.66}$$

where $\langle v_A \rangle$, z_{AA}, and z_{AB} are given by (15.46), (15.61), and (15.60).

In pure gas A, there are no A-B collisions, and $z_{AB} = 0$. Hence

$$\lambda = \frac{\langle v_A \rangle}{z_{AA}} = \frac{1}{2^{1/2}\pi d^2(N/V)} = \frac{1}{2^{1/2}\pi d^2}\frac{RT}{PN_0} \qquad \text{pure gas} \tag{15.67}$$

For O_2 at 25°C and 1 atm, use of (15.67), (15.64), and (15.48) gives

$$\lambda = \frac{4.44 \times 10^4 \text{ cm s}^{-1}}{2.8 \times 10^9 \text{ s}^{-1}} = 1.6 \times 10^{-5} \text{ cm} = 1600 \text{ Å} \qquad \text{for } O_2 \text{ at 25°C and 1 atm}$$

The average time between collisions is $\lambda/\langle v \rangle = 1/z_{AA}$, which equals 4×10^{-10} s for O_2 at 25°C and 1 atm.

Note that at 1 atm: (*a*) λ is small compared with macroscopic dimensions (1 cm), so the molecules collide with each other far more often than with the container walls; (*b*) λ is large compared with molecular dimensions (10^{-8} cm), so a molecule moves a distance of many molecular diameters before colliding with another molecule.

A good vacuum is 10^{-6} torr $\approx 10^{-9}$ atm. Since λ is inversely proportional to P, the mean free path in O_2 at 25°C and 10^{-9} atm is 1.6×10^{-5} cm $\times$ 10^9 = 16,000 cm = 0.1 mi, which is large compared with the usual container dimensions. In a good vacuum, the gas molecules collide far more often with the container walls than with one another. (When this condition holds, the gas is called a *Knudsen gas*.) At 10^{-9} atm and 25°C, a given O_2 molecule makes only an average of 2.8 collisions per second with other gas molecules.

The very high rate of intermolecular collisions in a gas at 1 atm raises several points. Chemical reactions between gases often proceed quite slowly. Equation (15.64) shows that if every A-B collision in a gas mixture caused a chemical reaction, gas-phase reactions at 1 atm would be over in a fraction of a second, which is contrary to experience. Usually, only a very small fraction of collisions result in reaction since the colliding molecules must have a certain minimum energy of relative motion in order to react; because of the exponential falloff of the distribution function $G(v)$ at high v, only a small fraction of molecules may have enough energy to react.

The direction of motion of a molecule changes at each collision, and the short mean free path ($\approx 10^{-5}$ cm) at 1 atm makes the molecular path resemble Fig. 3.12 for Brownian motion.

In deriving (15.7) for the gas pressure, we ignored intermolecular collisions and assumed that a given molecule changes its v_y value only when it collides with wall W or wall W′ in Fig. 15.2. Actually (unless P is extremely low), a gas molecule makes many, many collisions with other gas molecules between two successive wall collisions. These intermolecular collisions produce random changes in v_y of molecule i in Fig. 15.2. However, the derivation is based on an average over time, and its validity is not affected by these random changes in v_y.

15.8 THE BAROMETRIC FORMULA

For an ideal gas in the earth's gravitational field, the gas pressure decreases with increasing altitude. Consider a thin layer of gas at altitude z above the earth's surface (Fig. 15.13). Let the layer have thickness dz, mass dm, and cross-sectional area $\mathscr{A}$. The upward force F_{up} on this layer results from the pressure P of the gas just below the layer, and $F_{\text{up}} = P\mathscr{A}$. The downward force on the layer results from the gravitational force $dm\, g$ [Eq. (2.24)] and the pressure $P + dP$ of the gas just above the layer. Thus, $F_{\text{down}} = g\, dm + (P + dP)\mathscr{A}$. Since the gas layer is in mechanical equilibrium, these forces balance: $P\mathscr{A} = (P + dP)\mathscr{A} + g\, dm$, and

$$dP = -(g/\mathscr{A})\, dm \qquad (15.68)$$

The ideal-gas law gives $PV = P(\mathscr{A}\, dz) = (dm/M)RT$, so $dm = (PM\mathscr{A}/RT)\, dz$, and (15.68) becomes

$$dP/P = -(Mg/RT)\, dz \qquad (15.69)$$

Let us assume that T does not vary with altitude, and let us neglect the variation of the gravitational acceleration g with altitude. With these approximations, integration of (15.69) gives $\int_{P_0}^{P'} dP/P = -(Mg/RT)\int_0^{z'} dz$, where P_0 is the pressure at zero altitude (the earth's surface) and P' is the pressure at altitude z'. We get $\ln P'/P_0 = -Mgz'/RT$, or

$$P = P_0 e^{-Mgz/RT} \qquad \text{const. } T, g \qquad (15.70)$$

where we dropped the superfluous primes. The pressure decreases exponentially with increasing altitude. Since $P = \rho RT/M$, the gas density ρ decreases with increasing altitude.

In a mixture of ideal gases, each gas i will have its own molar mass M_i and its own distribution with altitude. The larger M_i is, the faster P_i and ρ_i decrease with increasing z. Thus the mole fractions of the light gases hydrogen and helium are far higher in the earth's upper atmosphere than at sea level.

Because the earth's atmosphere is far from isothermal, Eq. (15.70) is only a rough approximation to $P(z)$ of the atmosphere. Some values of the average atmospheric temperature T as a function of altitude are:

z/km	0	16	50	100	130	160	400
T/K	300	220	360	200	500	1000	1500

The temperature at 400 km varies from 700 K at night to 2000 K during the day. The high temperature of the upper atmosphere is due to absorption of sunlight.

Example A container of O_2 at 1.00 atm and 25°C is 100 cm high and is located in Chicago. Calculate the difference in pressure between the gas at the bottom of the container and the gas at the top.

Equation (15.70) gives $P_{\text{top}} = P_{\text{bot}} e^{-Mgz/RT}$, so $P_{\text{bot}} - P_{\text{top}} = P_{\text{bot}}(1 - e^{-Mgz/RT})$.

dz — $P + dP,\ z + dz$ / $P,\ z$

Figure 15.13
A thin layer of gas in a gravitational field.

Substitution of numerical values gives $Mgz/RT = 1.3 \times 10^{-4}$. Since Mgz/RT is very small, we can use the Taylor series $e^x = 1 + x + x^2/2! + \cdots \approx 1 + x$ to get

$$P_{\text{bot}} - P_{\text{top}} = P_{\text{bot}}[1 - (1 - Mgz/RT)] = P_{\text{bot}}Mgz/RT \tag{15.71}$$

$$\Delta P = (1.00 \text{ atm})(1.3 \times 10^{-4}) = 0.00013 \text{ atm} = 0.1 \text{ torr}$$

15.9 THE BOLTZMANN DISTRIBUTION LAW

Since $P = NRT/N_0 V$, the ratio P/P_0 in the barometric equation (15.70) equals $N(z)/N(0)$, where $N(z)$ and $N(0)$ are the numbers of molecules in thin layers at heights z and 0, respectively. Also, $Mgz/RT = N_0 mgz/N_0 kT = mgz/kT$, where m is the mass of a molecule. The quantity mgz equals $\varepsilon_p(z) - \varepsilon_p(0) \equiv \Delta\varepsilon_p$, where $\varepsilon_p(z)$ and $\varepsilon_p(0)$ are the potential energies of molecules at heights z and 0, respectively [see Eq. (2.47)]. Hence (15.70) can be written as

$$N(z)/N(0) = e^{-[\varepsilon_p(z) - \varepsilon_p(0)]/kT} = e^{-\Delta\varepsilon_p/kT} \tag{15.72}$$

The v_x distribution law (15.42) reads $dN_{v_x}/N = Ae^{-\varepsilon_{\text{tr}}/kT}\, dv_x$, where $\varepsilon_{\text{tr}} = \frac{1}{2}mv_x{}^2$. Let dN_1 and dN_2 be the numbers of molecules whose v_x values lie in the ranges $v_{x,1}$ to $v_{x,1} + dv_x$ and $v_{x,2}$ to $v_{x,2} + dv_x$, respectively. Then

$$dN_2/dN_1 = e^{-(\varepsilon_{\text{tr},2} - \varepsilon_{\text{tr},1})/kT} = e^{-\Delta\varepsilon_{\text{tr}}/kT} \tag{15.73}$$

Equations (15.72) and (15.73) are each special cases of the more general *Boltzmann distribution law*:

$$\frac{N_2}{N_1} = e^{-\Delta\varepsilon/kT} \qquad \text{where } \Delta\varepsilon \equiv \varepsilon_2 - \varepsilon_1 \tag{15.74*}$$

In this equation, N_1 is the number of molecules in state 1 and ε_1 is the energy of state 1; N_2 is the number of molecules in state 2. [The *state* of a particle is defined in classical mechanics by specifying its position and velocity (to within infinitesimal ranges).] Equation (15.74) is a result of statistical mechanics and will be derived in Chap. 22. If ε_2 is greater than ε_1, then $\Delta\varepsilon$ is positive and (15.74) says that N_2 is less than N_1. The number of molecules in a state decreases with increasing energy of the state.

The factor v^2 in the Maxwell distribution of speeds, Eq. (15.44), might seem to contradict the Boltzmann distribution (15.74), but this is not so. The quantity N_1 in (15.74) is the number of molecules in a given state, and many different states may have the same energy. Thus, molecules whose speeds are the same but whose velocities differ are in different states but have the same translational energy; the velocity vectors for such molecules point in different directions but have the same lengths, and their tips all lie in the thin spherical shell in Fig. 15.4. The factor $4\pi v^2$ in (15.44) arises from molecules having the same speed v but different velocities $\mathbf{v}$.

15.10 HEAT CAPACITIES OF POLYATOMIC MOLECULES

In Sec. 15.3 we used the kinetic theory to show that $\bar{C}_V = \frac{3}{2}R$ for an ideal monatomic gas. What about ideal polyatomic gases?

In addition to their translational energy, polyatomic molecules also have energy associated with rotational and vibrational motions. Thermal energy added to the gas can appear as rotational and vibrational (as well as translational) energy of the gas molecules, so the molar heat capacity $\bar{C}_V = (\partial \bar{U}/\partial T)_V$ of a polyatomic molecule exceeds that for a monatomic molecule. (There is also molecular electronic energy, but this is usually not excited except at very high temperatures, typically 10^4 K.) This section outlines the classical-mechanical kinetic-theory treatment of the contributions of molecular vibration and rotation to $\bar{U}$ and $\bar{C}_V$.

Let the molecular energy ε have the form

$$\varepsilon = a_1 p_1{}^2 + a_2 p_2{}^2 + \cdots + b_1 q_1{}^2 + b_2 q_2{}^2 + \cdots + \varepsilon_{el} \tag{15.75}$$

where the a's and b's are constants, the p's are momenta, the q's are coordinates, and ε_{el} is the electronic energy. (For example, a monatomic molecule has only translational and electronic energy, and its energy is

$$\varepsilon = \tfrac{1}{2}mv_x{}^2 + \tfrac{1}{2}mv_y{}^2 + \tfrac{1}{2}mv_z{}^2 + \varepsilon_{el} = p_x{}^2/2m + p_y{}^2/2m + p_z{}^2/2m + \varepsilon_{el} \tag{15.76}$$

where the momentum components are $p_x \equiv mv_x$, $p_y \equiv mv_y$, $p_z \equiv mv_z$. For a monatomic molecule, $a_1 = a_2 = a_3 = 1/2m$, $p_1 = p_x$, $p_2 = p_y$, $p_3 = p_z$, and $b_1 = b_2 = \cdots = 0$.) From (15.75), the molar internal energy $\bar{U}$ of a polyatomic ideal gas is

$$\bar{U} = N_0\langle\varepsilon\rangle = N_0(\langle a_1 p_1{}^2\rangle + \langle a_2 p_2{}^2\rangle + \cdots + \langle b_1 q_1{}^2\rangle + \langle b_2 q_2{}^2\rangle + \cdots + \langle\varepsilon_{el}\rangle) \tag{15.77}$$

since the average of a sum equals the sum of the averages [Eq. (15.41)].

Let w denote any one of the variables $p_1, p_2, \ldots, q_1, q_2, \ldots$ in (15.77), let c denote the coefficient of w^2, and let $\varepsilon_w \equiv cw^2$. To evaluate $\langle\varepsilon_w\rangle = \langle cw^2\rangle$, we need the distribution function $g(w)$ for w. Let dN_1 molecules have a value of w in the range w_1 to $w_1 + dw$, with a similar definition for dN_2. The definition of g gives $dN_1/N = g(w_1)\,dw$ and $dN_2/N = g(w_2)\,dw$, where N is the total number of gas molecules. Taking the ratio of these equations, we have $dN_2/dN_1 = g(w_2)/g(w_1)$. The Boltzmann distribution law (15.74) gives

$$\frac{dN_2}{dN_1} = \frac{g(w_2)}{g(w_1)} = e^{-[\varepsilon_w(w_2) - \varepsilon_w(w_1)]/kT} = \frac{e^{-\varepsilon_w(w_2)/kT}}{e^{-\varepsilon_w(w_1)/kT}} \tag{15.78}$$

The only way (15.78) can hold for all w_1 and w_2 values is for $g(w)$ to have the form

$$g(w) = Ae^{-\varepsilon_w/kT} \tag{15.79}$$

where the constant A is determined by the requirement that the total probability is 1:

$$\int_{w_{min}}^{w_{max}} g(w)\,dw = 1 \tag{15.80)*}$$

Equation (15.80) is called the *normalization* requirement, and A is the *normalization constant.*

The minimum and maximum values of w are $-\infty$ and ∞ whether w is a cartesian coordinate or a momentum, and Eqs. (15.39) and (15.79) give the average value of ε_w as

$$\langle\varepsilon_w\rangle = \langle cw^2\rangle = \int_{-\infty}^{\infty} cw^2 Ae^{-cw^2/kT}\,dw = cA\frac{\pi^{1/2}}{2(c/kT)^{3/2}} \tag{15.81}$$

where the Table 15.1 integrals 1 and 3 with $n = 1$ were used. To find A, we use (15.80):

$$1 = \int_{-\infty}^{\infty} g(w)\, dw = A \int_{-\infty}^{\infty} e^{-cw^2/kT}\, dw = \frac{\pi^{1/2}}{(c/kT)^{1/2}} A$$

where integrals 1 and 2 were used. Hence $A = (c/kT)^{1/2}\pi^{-1/2}$, and (15.81) becomes

$$\langle \varepsilon_w \rangle = \tfrac{1}{2}kT \tag{15.82}$$

Thus, each quadratic term in the molecular energy ε contributes $\frac{1}{2}kT$ to the average energy per molecule. The contribution of $\langle \varepsilon_w \rangle = \langle cw^2 \rangle$ to $\bar{U}$ is $N_0 \langle \varepsilon_w \rangle = \frac{1}{2}N_0 kT = \frac{1}{2}RT$. Since $\bar{C}_V = (\partial \bar{U}/\partial T)_V$, this term contributes $\frac{1}{2}R$ to $\bar{C}_V$.

We have proved the principle of *equipartition of energy*: Each term in the expression for the molecular energy ε that is quadratic in a coordinate or a momentum contributes $\frac{1}{2}RT$ to $\bar{U}$ and $\frac{1}{2}R$ to $\bar{C}_V$.

The molecular energy is the sum of translational, rotational, vibrational, and electronic energies:

$$\varepsilon = \varepsilon_{\text{tr}} + \varepsilon_{\text{rot}} + \varepsilon_{\text{vib}} + \varepsilon_{\text{el}} \tag{15.83}$$

The translational energy is given by (15.76) as $\varepsilon_{\text{tr}} = p_x{}^2/2m + p_y{}^2/2m + p_z{}^2/2m$. Since ε_{tr} has three terms, each quadratic in a momentum, the equipartition principle says that ε_{tr} contributes $3 \times \frac{1}{2}R = \frac{3}{2}R$ to $\bar{C}_V$. This result [which we found earlier; see Eq. (15.18)] is in agreement with data on monatomic gases.

The nature of rotational and vibrational motions will be considered in Chap. 21. At this point, we simply quote some results from that chapter, without justifying them. For a linear molecule, for example, H_2, CO_2, C_2H_2, the classical-mechanical expression for ε_{rot} has 2 quadratic terms and that for ε_{vib} has $2(3\mathcal{N} - 5)$ quadratic terms, where $\mathcal{N}$ is the number of atoms in the molecule. For a nonlinear molecule, for example, H_2O, CH_4, ε_{rot} has 3 quadratic terms and ε_{vib} has $2(3\mathcal{N} - 6)$ terms. A linear molecule has $3 + 2 + 2(3\mathcal{N} - 5) = 6\mathcal{N} - 5$ quadratic terms in ε, and $\bar{C}_V$ is predicted to be $(3\mathcal{N} - 2.5)R$ for an ideal gas of linear molecules (assuming T is not high enough to excite electronic energy). A nonlinear molecule has $3 + 3 + 2(3\mathcal{N} - 6) = 6\mathcal{N} - 6$ quadratic terms in ε, and $\bar{C}_V$ is predicted to be $(3\mathcal{N} - 3)R$ for an ideal gas of nonlinear molecules (except at extremely high T). These $\bar{C}_V$ values should apply at all temperatures (below, say, 10^4 K).

For CO_2 as an ideal gas, the equipartition theorem predicts $\bar{C}_V = 6.5R = 12.9$ cal mole^{-1} K^{-1}, independent of T. Experimental $\bar{C}_V$ values in cal mole^{-1} K^{-1} for CO_2 at low pressure are 6.9 at 300 K, 9.3 at 600 K, 11.0 at 1000 K, 12.4 at 2000 K, 12.7 at 2500 K.

Similar disagreements between experimental $\bar{C}_V$ values and those predicted by the equipartition theorem are found for other polyatomic molecules. In general, $\bar{C}_V$ increases with T and only attains the equipartition value at high temperatures. The experimental data show that the equipartition principle is *false*. The equipartition principle fails because it uses the classical-mechanical expression for the molecular energy, whereas molecular vibrations and rotations obey quantum mechanics, not classical mechanics. The quantum theory of heat capacities is discussed in Chaps. 22 and 24.

PROBLEMS

15.1 Calculate the total molecular translational energy at 25°C and 1 atm for 1.00 mole of (*a*) O_2; (*b*) CO_2.

15.2 Calculate the average translational energy of one molecule at 298°C and 1 atm for (*a*) O_2; (*b*) CO_2.

15.3 Calculate $\varepsilon_{tr}(100°C)/\varepsilon_{tr}(0°C)$ for an ideal gas.

15.4 Calculate $\gamma \equiv C_P/C_V$ for an ideal monatomic gas.

15.5 For 1.0 mol of O_2 at 25°C and 1 atm, calculate the number of molecules whose speed lies in the range 500.000 to 500.001 m/s.

15.6 For O_2 at 25°C and 1 atm, calculate the ratio of the probability that a molecule has its speed in an infinitesimal interval dv located at 500 m/s to the probability that its speed lies in an infinitesimal interval dv at 1500 m/s.

15.7 Calculate $v_{rms}(\text{Ne})/v_{rms}(\text{He})$ at 0°C.

15.8 For CO_2 at 500 K, calculate (*a*) v_{rms}; (*b*) $\langle v \rangle$; (*c*) v_{mp}.

15.9 Show that $v_{mp} = (2RT/M)^{1/2}$.

15.10 Use the Maxwell distribution to verify that $\langle v^2 \rangle = 3RT/M$.

15.11 Use the distribution function for v to find $\langle v^3 \rangle$ for ideal-gas molecules. Does $\langle v^3 \rangle$ equal $\langle v \rangle \langle v^2 \rangle$?

15.12 (*a*) Find $\langle v_x \rangle$ for ideal-gas molecules. Give a physical explanation of the result. (*b*) $\langle v_x{}^2 \rangle$ is given by (15.38). Explain why $\langle v_x \rangle^2 \neq \langle v_x{}^2 \rangle$.

15.13 Find $\langle v_x{}^4 \rangle$ for ideal-gas molecules.

15.14 (*a*) Use Fig. 15.5 to explain why $\int_{-\infty}^{\infty} e^{-ax^2}\,dx = 2\int_0^{\infty} e^{-ax^2}\,dx$. This integral is integral 1 in Table 15.1 with $n = 0$, and a similar argument shows the validity of integral 1. (*b*) Sketch the function xe^{-ax^2} and then explain why $\int_{-\infty}^{\infty} xe^{-ax^2}\,dx = 0$. A similar argument shows the validity of integral 4.

15.15 (*a*) Derive integral 5 in Table 15.1 by changing to a new variable. (*b*) Differentiate integral 5 with respect to a to show that $\int_0^{\infty} x^3 e^{-ax^2}\,dx = 1/2a^2$ [see Eq. (2.15)]. (*c*) Find $\int_0^{\infty} x^5 e^{-ax^2}\,dx$. Repeated differentiation gives integral 6.

15.16 (*a*) Derive integral 2 in Table 15.1 as follows. Let $J \equiv \int_0^{\infty} e^{-ax^2}\,dx$. Explain why $J = \int_0^{\infty} e^{-ay^2}\,dy$. We have $J^2 = \int_0^{\infty}\int_0^{\infty} e^{-a(x^2+y^2)}\,dx\,dy$. Switch to plane polar coordinates r, ϕ to evaluate J^2. (The area element in plane polar coordinates is $r\,dr\,d\phi$.) Then take the square root to get integral 2. (*b*) Differentiate integral 2 with respect to a to show that $\int_0^{\infty} x^2 e^{-ax^2}\,dx = \pi^{1/2}/4a^{3/2}$. Repeated differentiation gives integral 3.

15.17 Verify Eq. (15.41).

15.18 (*a*) Show that $(2\pi)^{1/2} I(u) = u - u^3/6 + u^5/40 - u^7/336 + \cdots$, where $I(u)$ is defined by (15.51). *Hint:* Use (8.42) with $x = -s^2/2$. (*b*) Use the series to verify that $I(0.30) = 0.12$.

15.19 (*a*) Use integration by parts to show that the fraction of molecules whose speed is in the range 0 to v' is

$$2I(2^{1/2}v'/v_{mp}) - 2(v'/v_{mp})\pi^{-1/2}e^{-(v'/v_{mp})^2}$$

where the function I is defined by (15.51). (*b*) Find the fraction of molecules whose speed exceeds $4.243v_{mp}$.

15.20 In the mythical world of Flatland, everything is two-dimensional. Find the expression for the probability that a molecule in a two-dimensional ideal gas has its speed in the range v to $v + dv$.

15.21 Find the most probable molecular translational energy ε_{mp} for an ideal gas.

15.22 Find the molecular formula of a hydrocarbon gas that effuses 0.872 times as fast as O_2 through a small hole (the temperatures and pressures being equal).

15.23 A container holding solid scandium in equilibrium with its vapor at 1690 K shows a weight loss of 10.5 mg in 49.5 min through a circular hole of diameter 0.1763 cm. (*a*) Find the vapor pressure of Sc at 1690 K in torr. (*b*) Is $\lambda \gg d_{hole}$?

15.24 Dry air contains 0.033 percent CO_2 by volume. Calculate the total mass of CO_2 that strikes 1 cm^2 of a green leaf in 1 s in dry air at 25°C and 1.00 atm.

15.25 The vapor pressure of Octoil

$$C_6H_4(COOC_8H_{17})_2$$

is 0.010 torr at 393 K. Calculate the number of Octoil molecules that evaporate into vacuum from a 1.0 cm^2 surface of the liquid at 393 K in 1.0 s; also calculate the mass that evaporates.

15.26 For 1.00 mole of a gas with molecular weight 50 and collision diameter 3.0 Å, calculate at 25°C and 1.00 atm (*a*) the number of collisions per second made by one molecule; (*b*) the number of collisions per second per cubic centimeter in the gas.

15.27 For a gas mixture at 25°C and 1.00 atm that contains 0.20 mole of gas A with molecular weight 40 and collision diameter 2.4 Å and 0.80 mole of gas B with molecular weight 80 and collision diameter 3.0 Å, calculate (*a*) the collision rate for one particular A molecule; (*b*) the collision rate for one particular B molecule; (*c*) the total number of A-B collisions in 1.00 s; (*d*) the total number of collisions in 1.00 s.

15.28 For a gas with collision diameter 3.0 Å, calculate the mean free path at 25°C and (*a*) 2.00 atm; (*b*) 1.00 torr.

15.29 The top of Pike's Peak is 14,100 ft above sea level. Neglecting the variation of T with altitude and using an average molecular weight of 29 for air, calculate the atmospheric pressure at the top of this mountain. Compare with the observed average value 446 torr.

15.30 Calculate the difference in barometer readings between the first and fourth floors of a building at sea level if each floor is 10 ft high. (See Prob. 15.29.)

15.31 For CH_4 at 400 K, what value of $\bar{C}_P$ is predicted by the equipartition principle?

15.32 Calculate v_{rms} at 25°C for a dust particle of mass 1.0×10^{-10} g suspended in air, assuming that the particle can be treated as a giant molecule.

15.33 Is $\langle v^2 \rangle$ equal to $\langle v \rangle^2$ for ideal-gas molecules?

15.34 Let s be the number obtained when a single cubic die is thrown. Assuming the die is not loaded, use (15.40) to calculate $\langle s \rangle$ and $\langle s^2 \rangle$. Is $\langle s \rangle^2$ equal to $\langle s^2 \rangle$?

15.35 The *standard deviation* σ_x of a variable x can be defined by $\sigma_x{}^2 \equiv \langle x^2 \rangle - \langle x \rangle^2$. (*a*) Show that $\sigma_{v_x} = (kT/m)^{1/2}$ for the Maxwell distribution of v_x. (*b*) What fraction of ideal-gas molecules have v_x within ± 1 standard deviation from the mean (average) value $\langle v_x \rangle$?

15.36 Apply the method used to derive (15.34) from (15.30) to show that (3.51) is the only function that satisfies (3.50).

16

TRANSPORT PROCESSES

16.1 KINETICS

So far, we have discussed only equilibrium properties of systems. Processes in systems in equilibrium are reversible and are comparatively easy to treat. This chapter and the next deal with nonequilibrium processes. Such processes are irreversible and are difficult to treat. The rate of a reversible process is infinitesimal. Irreversible processes occur at nonzero rates.

The branch of science that deals with rate processes is called *kinetics* (or *dynamics*). A system may be out of equilibrium because matter or energy or both are being transported between the system and its surroundings or between one part of the system and another. Such processes are *transport processes*, and the branch of kinetics that studies the rates and mechanisms of transport processes is *physical kinetics*. Even though neither matter nor energy is being transported through space, a system may be out of equilibrium because certain chemical species in the system are undergoing a chemical reaction to produce other species. The branch of kinetics that studies the rates and mechanisms of chemical reactions is *chemical kinetics* (or *reaction kinetics*). Physical kinetics is discussed in Chap. 16 and chemical kinetics in Chap. 17.

Consider some examples of transport processes. If unbalanced forces exist in the system, it is not in mechanical equilibrium and parts of the system move. The flow of fluids is the subject of *fluid dynamics* (or *fluid mechanics*). Some aspects of fluid dynamics are treated in Sec. 16.3 on viscosity. When an electric field is applied to a system, electrically charged particles (electrons and ions) experience a force and may move through the system, producing an electric current. Electrical conduction is studied in Secs. 16.5 and 16.6. If temperature differences exist between the system and surroundings or within the system, it is not in thermal equilibrium and heat energy flows. Thermal conduction is studied in Sec. 16.2. If differences in concentrations of substances exist between different regions of a solution, the system is not in material equilibrium and there is a flow of matter until the concentrations (and hence the chemical potentials) have been equalized. This flow differs from the bulk flow that arises from pressure differences and is called diffusion (Sec. 16.4).

We shall see that the laws describing fluid flow, electrical conduction, thermal conduction, and diffusion all have the same mathematical form, namely, that the rate of transport is proportional to the spatial derivative (gradient) of some property (see Sec. 16.7).

16.2 THERMAL CONDUCTIVITY

Figure 16.1 shows a substance in contact with two heat reservoirs at different temperatures. A steady state will eventually be reached in which there will be a uniform temperature gradient dT/dx in the substance. (The *gradient* of a quantity is its rate of change with respect to a spatial coordinate.) The rate of heat flow dq/dt across any plane perpendicular to the x axis and lying between the reservoirs will also be uniform. Experimental measurements show that the following law is obeyed:

$$\frac{dq}{dt} = -K\mathscr{A}\frac{dT}{dx} \tag{16.1}$$

where $\mathscr{A}$ is the substance's cross-sectional area in a plane perpendicular to the x axis and the proportionality constant K is its *thermal conductivity*. The minus sign occurs because dT/dx is positive but dq/dt is negative (the heat flows to the left in the figure). Equation (16.1) is *Fourier's law* of heat conduction. (This law also holds when the temperature gradient in the substance is nonuniform; in this case, dT/dx has different values at different places on the x axis, and dq/dt varies from place to place.)

Representative values of K in units of J K^{-1} cm^{-1} s^{-1} for substances at 25°C and 1 atm are:

Cu	Fe	NaCl	glass	pine wood	H_2O	CCl_4	N_2	CO_2
4	0.8	0.09	0.01	0.001	0.006	0.001	0.00026	0.00016

Metals are good conductors of heat; nonmetals are poor conductors; gases are very poor conductors due to the low density of molecules. The high thermal conductivity of metals is due to the electrical-conduction electrons, which move relatively freely through the metal.

Although the system in Fig. 16.1 is not in thermodynamic equilibrium, we assume that any tiny portion of the system can be assigned values of thermodynamic variables, such as T, U, S, and P, and that all the usual thermodynamic relations

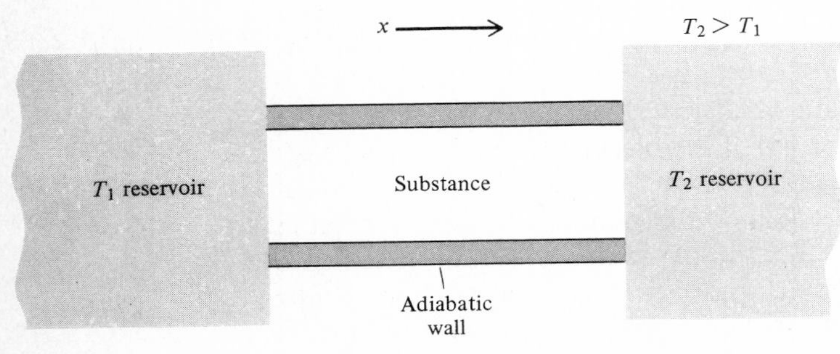

Figure 16.1
Conduction of heat through an insulated substance.

between such variables hold in each tiny subsystem. (Recall the discussion in Sec. 4.2.) This assumption, called the *principle of local state* (or the *hypothesis of local equilibrium*), holds well in most systems of interest.

The thermal conductivity K is a function of the local thermodynamic state of the system and therefore depends on T and P for a pure substance. For solids and liquids, K may either decrease or increase with increasing T. For gases, K increases with increasing T. The pressure dependence of K for gases is discussed later in this section.

The mechanism of the transport of heat by conduction is as follows. Molecules in a higher-temperature region have a higher average energy than molecules in an adjacent lower-temperature region. In intermolecular collisions, it is very probable for molecules with higher energy to lose energy to lower-energy molecules. This results in a flow of molecular energy from high-T to low-T regions. In gases, the molecules move relatively freely, and the flow of molecular energy in thermal conduction occurs by an actual transfer of molecules from one region of space to an adjacent region, where they undergo collisions. In liquids and solids, the molecules do not move freely, and the molecular energy is transferred bucket-brigade fashion by successive collisions between molecules in adjacent layers, without substantial transfer of molecules from one region to another.

Besides conduction, heat can be transferred by convection and by radiation. In *convection*, heat is transferred by a current of fluid moving between regions that differ in temperature. This bulk convective flow arises from differences in pressure or in density in the fluid and should be distinguished from the random molecular motion involved in thermal conduction in gases. In radiative transfer of heat, a warm body emits electromagnetic waves (Sec. 21.1), some of which are absorbed by a cooler body (e.g., the sun and the earth). Equation (16.1) assumes the absence of convection and radiation. In measuring K of fluids, great care must be taken to avoid convection currents.

Kinetic theory of thermal conductivity of gases. The kinetic theory of gases yields theoretical expressions for the thermal conductivity and other transport properties of gases, and the results are in reasonably good agreement with experiment. The rigorous kinetic-theory treatment of transport properties in gases is extremely complicated mathematically and physically. The rigorous equations underlying transport phenomena in gases were worked out in the 1860s and 1870s by Maxwell and by Boltzmann, but it wasn't until 1917 that Sydney Chapman and David Enskog, working independently, solved these equations. (The Chapman–Enskog theory is so severely mathematical that Chapman remarked that reading an exposition of the theory is "like chewing glass.") Instead of presenting rigorous analyses, this chapter gives very crude treatments based on the assumption of hard-sphere molecules with a mean free path given by (15.67). The mean-free-path method (given by Maxwell in 1860) leads to results that are qualitatively correct but quantitatively wrong.

We shall assume that the gas pressure is neither very high nor very low. Our treatment is based on collisions between two molecules and assumes no intermolecular forces except at the moment of collision. At high pressures, interactions between more than two molecules at a time and intermolecular forces in the intervals between collisions become important, and the mean-free-path formula (15.67) is inapplicable. At very low pressures, the mean free path λ becomes comparable to, or larger than, the dimensions of the container, and wall collisions become important. Thus our treatment applies only for pressures such that $d \ll \lambda \ll L$, where d is the molecular diameter and L is the smallest dimension of the container. In Sec. 15.7 we found λ to be about 10^{-5} cm at 1 atm and room temperature. Since λ is inversely proportional to pressure, λ is 10^{-7} cm at 10^2 atm and is 10^{-2} cm at 10^{-3} atm. Thus our treatment applies to the pressure range from 10^{-2} or 10^{-3} atm to 10^1 or 10^2 atm. We make the following assumptions: (1) The molecules are rigid, nonattracting spheres of diameter d. (2) Every molecule moves at speed $\langle v \rangle$ and travels a distance λ between successive collisions. (3) There is complete randomness of direction of molecular motion after a collision. (4) Complete adjustment of the molecular energy ε occurs at each collision; by this we mean that a gas molecule moving in the x direction and colliding with a molecule in a plane located at $x = x'$ will take on the average energy $\varepsilon(x')$ characteristic of molecules in the plane at x'.

Assumptions 1 and 2 are, of course, false. Assumption 3 is also inaccurate, in that after a collision, a molecule is somewhat more likely to be moving in or close to its original direction of motion than in other directions; this phenomenon is called the *persistence of velocities.* Assumption 4 is not bad for translational energy but is very inaccurate for rotational and vibrational energies.

Let a steady state be established in Fig. 16.1 and consider a plane perpendicular to the x axis and located at $x = x_0$ (Fig. 16.2). To calculate K, we must calculate the rate of flow of heat energy through this plane. Equation (15.56) gives the number of molecules moving into the x_0 plane from the left in time dt as

$$\tfrac{1}{4}(N/V)\langle v \rangle \mathscr{A}\, dt \tag{16.2}$$

These molecules traveled an average distance λ since their last collision. The molecules move into the x_0 plane at many different angles. By averaging over the angles, kinetic theory can be used to show that the average distance from the x_0 plane to the point of last collision is $\frac{2}{3}\lambda$ (see *Kennard*, pp. 139–140 for the proof). Figure 16.2 shows an "average" molecule moving into the x_0 plane from the left. Molecules moving into the x_0 plane from the left will (by assumption 4) have an average energy that is characteristic of molecules in the plane at $x_0 - \frac{2}{3}\lambda$. The energy flow through the x_0 plane due to molecules coming from the left is therefore

$$\varepsilon(x_0 - \tfrac{2}{3}\lambda) \times \tfrac{1}{4}\langle v \rangle \mathscr{A}(N/V)\, dt \tag{16.3}$$

where $\varepsilon(x_0 - \frac{2}{3}\lambda)$ is the average molecular energy in the plane at $x_0 - \frac{2}{3}\lambda$. Similarly, the energy flowing through the x_0 plane due to molecules coming from the right is

$$\varepsilon(x_0 + \tfrac{2}{3}\lambda) \times \tfrac{1}{4}\langle v \rangle \mathscr{A}(N/V)\, dt \tag{16.4}$$

[Actually, because of the temperature gradient, the mean speed $\langle v \rangle$ at $x_0 + \frac{2}{3}\lambda$ differs from that at $x_0 - \frac{2}{3}\lambda$. Likewise, N/V differs in these planes. However, since we are assuming no convection, there is no net flow of gas, and the net number of molecules crossing the plane is zero. Hence $\langle v \rangle(N/V)$ is the same in all planes.]

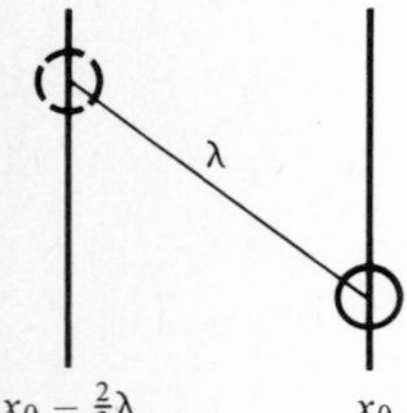

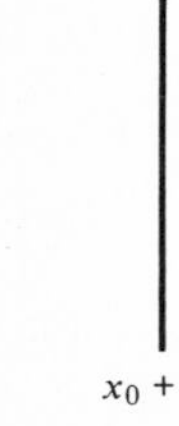

Figure 16.2
Three planes separated by $\frac{2}{3}\lambda$.

The net energy flow dU to the right in time dt is the difference between (16.3) and (16.4):

$$dU = dq = \tfrac{1}{4}\langle v\rangle \mathscr{A}(N/V)[\varepsilon(x_0 - \tfrac{2}{3}\lambda) - \varepsilon(x_0 + \tfrac{2}{3}\lambda)]\, dt \tag{16.5}$$

Since we assumed no intermolecular forces (except at the instant of collision), the total energy is the sum of the energies of the individual gas molecules and the local molar thermodynamic internal energy is $\bar{U} = N_0\,\varepsilon$, where N_0 is the Avogadro constant. Hence, $d\bar{U} = N_0\, d\varepsilon$. Equation (2.99) gives for an ideal gas $d\bar{U} = \bar{C}_V\, dT$. Equating these two expressions for $d\bar{U}$, we get $d\varepsilon = (\bar{C}_V/N_0)\, dT$. Taking $d\varepsilon$ to be the quantity in brackets in (16.5), we get

$$dq = \tfrac{1}{4}\langle v\rangle \mathscr{A}(N/V)(\bar{C}_V/N_0)[T(x_0 - \tfrac{2}{3}\lambda) - T(x_0 + \tfrac{2}{3}\lambda)]\, dt \tag{16.6}$$

We have $dT = (dT/dx)\, dx$. The quantity dT is the expression in brackets in (16.6). Hence $dx = (x_0 - \frac{2}{3}\lambda) - (x_0 + \frac{2}{3}\lambda) = -\frac{4}{3}\lambda$, and $dT = -(dT/dx)\frac{4}{3}\lambda$. Equation (16.6) becomes

$$dq = -\tfrac{1}{3}\langle v\rangle \mathscr{A}(N/N_0 V)\bar{C}_V \lambda (dT/dx)\, dt$$

We have $N/N_0 V = n/V = m/MV = \rho/M$, where n, m, ρ, and M are the number of moles of gas, the mass of gas, the gas density, and the gas molar mass. Therefore

$$\frac{dq}{dt} = -\frac{\langle v\rangle \rho \bar{C}_V \lambda}{3M}\mathscr{A}\frac{dT}{dx} \tag{16.7}$$

Comparison with Fourier's law (16.1) gives

$$K \approx \tfrac{1}{3}\bar{C}_V \lambda \langle v\rangle \rho/M \qquad \text{hard spheres} \tag{16.8}$$

Because of the crudity of assumptions 2 to 4, the numerical coefficient in this equation is wrong. A rigorous theoretical treatment for hard-sphere monatomic molecules gives

$$K = \frac{25\pi}{64}\frac{\bar{C}_V \lambda \langle v\rangle \rho}{M} \qquad \text{hard spheres, monatomic} \tag{16.9}$$

The rigorous extension of (16.9) to polyatomic gases is a very difficult problem that has not yet been fully solved. Experiments on intermolecular energy transfer show that rotational and vibrational energy is not as easily transferred in collisions as translational energy. The heat capacity $\bar{C}_V$ is the sum of a translational part and a vibrational and rotational part [see Sec. 15.10 and Eq. (15.18)]:

$$\bar{C}_V = \bar{C}_{V,\text{tr}} + \bar{C}_{V,\text{vib+rot}} = \tfrac{3}{2}R + \bar{C}_{V,\text{vib+rot}} \tag{16.10}$$

Because vibrational and rotational energy is less easily transferred than translational, it contributes less to K. Hence in the expression for K, the coefficient of $\bar{C}_{V,\text{vib+rot}}$ should be less than the value $25\pi/64$, which is correct for $\bar{C}_{V,\text{tr}}$ [Eq. (16.9)]. Eucken gave certain nonrigorous arguments for taking the coefficient of $\bar{C}_{V,\text{vib+rot}}$ as two-fifths that of $\bar{C}_{V,\text{tr}}$, and doing so leads to fair agreement with experiment. Therefore, for polyatomic molecules, $25\pi\bar{C}_V/64$ in (16.9) is replaced by

$$\frac{25\pi}{64}\bar{C}_{V,\text{tr}} + \frac{2}{5}\frac{25\pi}{64}\bar{C}_{V,\text{vib+rot}} = \frac{25\pi}{64}\frac{3R}{2} + \frac{2}{5}\frac{25\pi}{64}\left(\bar{C}_V - \frac{3R}{2}\right) = \frac{5\pi}{32}\left(\bar{C}_V + \frac{9}{4}R\right)$$

The thermal conductivity of a gas of polyatomic hard-sphere molecules is then predicted to be

$$K = \frac{5\pi}{32}\left(\bar{C}_V + \frac{9}{4}R\right)\frac{\lambda\langle v\rangle\rho}{M} \qquad \text{hard spheres} \tag{16.11}$$

[For a more rigorous and more accurate expression, see E. A. Mason and L. Monchick, *J. Chem. Phys.*, **36**, 1622 (1962); L. Monchick, A. N. G. Pereira, and E. A. Mason, ibid., **42**, 3241 (1965).]

Use of (16.11) to calculate K requires knowledge of the molecular diameter d, since λ depends on d. Even a truly spherical molecule like He does not have a well-defined size, so it is difficult to say what value of d should be used in (16.11). In the next section, we shall use experimental values of gas viscosities to get d values appropriate to the hard-sphere model.

To see how K in (16.11) depends on T and P, we use (15.67) and (15.46) for λ and $\langle v\rangle$ and the ideal-gas law $PM = \rho RT$ to write $\lambda \propto T/P$, $\langle v\rangle \propto T^{1/2}$, $\rho \propto P/T$. The heat capacity $\bar{C}_V$ varies slowly with T and extremely slowly with P. Hence (16.11) predicts $K \propto T^{1/2}P^0$. Surprisingly, K is predicted to be independent of pressure. As P increases, the number of heat carriers (i.e., molecules) per unit volume increases, thereby tending to increase K. However, this increase is nullified by the decrease of λ with increasing P. As λ decreases, each molecule goes a shorter average distance between collisions and is therefore less effective in transporting heat.

Experimental data show that K for gases does increase with increasing T but at a faster rate than the $T^{1/2}$ behavior predicted by the hard-sphere model. The reason for this lies in the crudity of the nonattracting-hard-sphere picture. Molecules are actually "soft" rather than hard; moreover, they attract one another over significant distances. Use of improved expressions for intermolecular forces gives better agreement with the observed temperature dependence of K. (See *Kauzmann*, pp. 218–231.)

The prediction that K is independent of P holds well, provided P is not too high or too low. (Recall the restriction $d \ll \lambda \ll L$.) Values of K as a function of P for Ne at 25°C are:

P/bars	1	10	20	50	100	200	500
$10^6 K/(\text{J cm}^{-1}\ \text{K}^{-1}\ \text{s}^{-1})$	489	490	494	503	517	545	634

(1 bar ≈ 1 atm.) Data for other gases indicate that K is essentially constant for pressures up to about 50 atm.

At very low pressures, the gas molecules in Fig. 16.1 travel back and forth between the reservoirs, making very few collisions with one another, and the expression (16.11) is inapplicable. For pressures low enough to make λ substantially larger than the separation between the heat reservoirs, the transfer of heat occurs by molecules moving directly from one reservoir to the other, and the rate of heat flow is proportional to the number of molecules colliding with each reservoir wall in unit time. Since the rate of wall collisions is directly proportional to the pressure, dq/dt becomes directly proportional to P at very low pressures and goes to zero as P goes to zero. Between the pressure range where dq/dt is independent of P and the range where it is proportional to P, there is a transition range in which K falls off from its moderate-pressure value. The falloff of K begins at 10 to 50 torr, depending on the gas.

Suppose the substance in Fig. 16.1 is a gas of monatomic molecules at a pressure low enough to permit collisions in the gas to be neglected. Let us make the rough approximation that at each collision with a reservoir wall a gas molecule takes on the average translational kinetic energy $\frac{3}{2}kT_2$ or $\frac{3}{2}kT_1$ appropriate to the temperature of the reservoir. Then in any tiny subvolume of the gas, molecules with negative v_x will have made their last collision with the T_2 reservoir and will have translational energy $\frac{3}{2}kT_2$, and molecules with positive v_x will have translational energy $\frac{3}{2}kT_1$. The principle of local state fails, since each tiny portion of gas is a mixture of molecules having two different temperatures. Also, the Maxwell distribution of velocities is inapplicable. The theoretical expression for dq/dt in this situation (*Kauzmann*, p. 206) does not have the form of Eq. (16.1), and Fourier's law does not apply.

Kinetic theory of thermal conductivity of liquids. Bridgman modified the kinetic-theory equation (16.8) to arrive at a theoretical expression for K of a liquid. Because the molecules in a liquid are close together, they do not move freely but vibrate in a "cage" formed by surrounding molecules. Bridgman assumed the thermal energy to be transferred at a speed equal to the speed of sound in the liquid. Bridgman's equation (see *Bird, Stewart, and Lightfoot*, p. 260 for the derivation) is

$$K = \frac{3R}{N_0^{1/3}\bar{V}^{2/3}}\left(\frac{\bar{C}_P}{\bar{C}_V \rho\kappa}\right)^{1/2} \tag{16.12}$$

where R is the gas constant and ρ, κ, and $\bar{V}$ are the density, isothermal compressibility, and molar volume of the liquid. This equation works surprisingly well, especially if the factor 3 is replaced by 2.8. (This is the well-known scientific principle of the fudge factor.)

The theory of thermal conduction in solids is complicated and is omitted.

16.3 VISCOSITY

This section deals with the bulk flow of fluids (liquids and gases) under a pressure gradient. Some fluids flow more easily than others. The property that characterizes a fluid's resistance to flow is its *viscosity* η (eta). We shall see that the speed of flow through a tube is directly proportional to the pressure gradient and inversely proportional to the viscosity [see Eq. (16.18) below].

To get a precise definition of η, consider a fluid flowing steadily between two large plane parallel plates (Fig. 16.3). Experiment shows that the speed v_y of the fluid flow at any point depends on the distance of that point from one of the plates. The speed is a maximum midway between the plates and decreases to zero at each plate. The arrows in the figure indicate the magnitude of v_y as a function of the vertical coordinate x. (The condition of zero flow speed at the boundary between a solid and a fluid, called the *no-slip condition*, is an experimental fact; an exception occurs for very high flow speeds in a very-low-density gas.) Adjacent horizontal layers of fluid

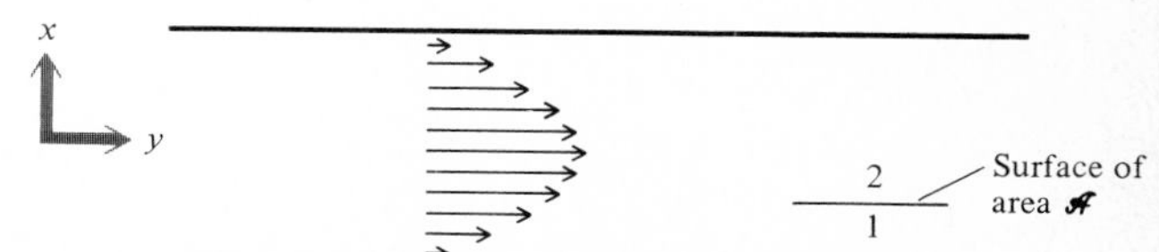

Figure 16.3
A fluid flowing between two planar plates.

flow at different speeds and "slide over" one another. As two adjacent layers slip past each other, each exerts a frictional resistive force on the other, and it is this internal friction in the fluid that gives rise to viscosity.

Consider an imaginary surface of area $\mathscr{A}$ drawn between and parallel to the plates (Fig. 16.3). Whether the fluid is at rest or in motion, the fluid on one side of this surface exerts a force in the x direction on the fluid on the other side; this force has the magnitude $P\mathscr{A}$, where P is the local pressure in the fluid. Moreover, because of the change of flow speed as x changes, one finds that the fluid on one side of the surface exerts a frictional force in the y direction on the fluid on the other side. Let F_y be the frictional force exerted by the slower-moving fluid on one side of the surface (side 1 in the figure) on the faster-moving fluid (side 2). Experiments on fluid flow show that F_y is proportional to the surface area of contact and to the gradient dv_y/dx of flow speed. The proportionality constant is the fluid's *viscosity* η:

$$F_y = -\eta\mathscr{A}\frac{dv_y}{dx} \tag{16.13}$$

The minus sign shows that the viscous force on the faster-moving fluid is in the direction opposite its motion. By Newton's third law of motion (action = reaction), the faster-moving fluid exerts a force $\eta\mathscr{A}(dv_y/dx)$ in the positive y direction on the slower-moving fluid. The viscous force tends to slow down the faster-moving fluid and speed up the slower-moving fluid, thereby tending to eliminate their relative motion.

Equation (16.13) is *Newton's law of viscosity*. Experiments show it to be well obeyed by gases and by most liquids (provided the flow rate is not too high). When Eq. (16.13) applies, we have *laminar* (or *streamline*) flow. At high rates of flow, (16.13) does not hold, and the flow is called *turbulent*. Both laminar flow and turbulent flow are types of bulk (or viscous) flow. In contrast, for flow of a gas at very low pressures, the mean free path is long, and the molecules flow independently of one another; this is *molecular flow*, and it is not a type of bulk flow.

Certain liquids do not obey (16.13) at any flow rate; they are called *non-Newtonian fluids*. For a non-Newtonian fluid, the viscosity, defined as $\eta \equiv -F_y\mathscr{A}^{-1}(dv_y/dx)^{-1}$, changes as dv_y/dx changes. (For Newtonian fluids, η is independent of dv_y/dx.) Examples of non-Newtonian fluids are certain colloidal suspensions and certain polymer solutions.

From (16.13), the SI units of η are $\text{N s m}^{-2} = \text{kg s}^{-1}\text{ m}^{-1}$ (since $1\text{ N} = 1\text{ kg m s}^{-2}$). The cgs units of η are $\text{dyn s cm}^{-2} = \text{g s}^{-1}\text{ cm}^{-1}$, and 1 dyn s cm^{-2} is called one *poise* (P):

$$1\text{ P} \equiv 1\text{ dyn s cm}^{-2} = 0.1\text{ N s m}^{-2} \tag{16.14}$$

since $1\text{ dyn} = 10^{-5}\text{ N}$.

Some values of η (in centipoises) at 25°C and 1 atm are:

Substance	C_6H_6	H_2O	H_2SO_4	olive oil	glycerol	O_2	CH_4
η/cP	0.6	0.9	19	80	954	0.02	0.01

Gases are much less viscous than liquids. The viscosity of liquids generally decreases rapidly with increasing temperature. (Molasses flows more readily at higher temperatures.) Values of η/cP for water at 1 atm are 1.8 at 0°C, 1.0 at 20°C, 0.5 at 55°C, and

0.3 at 100°C. For many liquids, the temperature dependence of η can be described by $\eta = Ae^{B/T}$, where A and B are positive constants characteristic of the liquid. The viscosity of liquids increases with increasing pressure.

Newton's viscosity law (16.13) allows the rate of flow of a fluid through a tube to be determined. Figure 16.4 shows a fluid flowing in a cylindrical tube. The pressure P_1 at the left end of the tube is greater than the pressure P_2 at the right end, and the pressure drops continually along the tube. The flow speed v_y is zero at the walls (the no-slip condition) and increases toward the center of the pipe. By the symmetry of the tube, v_y can depend only on the distance s from the tube's center: $v_y = v_y(s)$. The liquid flows in infinitesimally thin cylindrical layers, a layer with radius s flowing with speed $v_y(s)$.

Imagine a solid-cylindrical portion C of fluid of length dy, radius s, and axis coinciding with the pipe's axis (Fig. 16.4). Let P and $P + dP$ be the pressures at the left and right ends of C, respectively. Each thin cylindrical layer of fluid flows at constant speed and therefore is not being accelerated. With zero acceleration, the total force on any portion of fluid must be zero. The pressure at the left end of C exerts a force $P\pi s^2$, and the pressure at the right end exerts a force $-(P + dP)\pi s^2$. The negative sign indicates that the force is in the negative y direction. (Since the pressure decreases as y increases, dP is negative.) Also, the slower-moving fluid just outside C exerts a viscous force on C in the negative y direction; this force is given by (16.13) as $\eta\mathscr{A}(dv_y/ds) = \eta(2\pi s\, dy)(dv_y/ds)$. (The area of the curved surface of the cylinder C equals its circumference $2\pi s$ times its length dy.) Since v_y is a maximum at the center, dv_y/ds is negative. Setting the total force on C equal to zero, we get

$$-(P + dP)\pi s^2 + P\pi s^2 + \eta(2\pi s\, dy)(dv_y/ds) = 0 \tag{16.15}$$

$$\frac{dv_y}{ds} = \frac{s}{2\eta}\frac{dP}{dy} \tag{16.16}$$

which integrates to $v_y = (s^2/4\eta)(dP/dy) + c$. The integration constant c is evaluated from the no-slip condition: $v_y = 0$ at $s = r$, where r is the pipe's radius. Thus, $0 = (r^2/4\eta)(dP/dy) + c$, and $c = -(r^2/4\eta)(dP/dy)$. Hence, the function $v_y(s)$ is

$$v_y = \frac{1}{4\eta}(r^2 - s^2)\left(-\frac{dP}{dy}\right) \tag{16.17}$$

(Since dP/dy is negative, two minus signs were introduced to make all factors positive.)

Equation (16.17) shows that $v_y(s)$ is a parabolic function for laminar flow in a cylindrical pipe. The profile of v_y for laminar flow is shown in Fig. 16.5*a*. (For turbulent flow, the profile of v_y turns out to be something like Fig. 16.5*b*.)

To find the flow rate through a cross section of the pipe, consider the fluid in a thin shell between cylinders of radii s and $s + ds$. All fluid in this shell moves at the

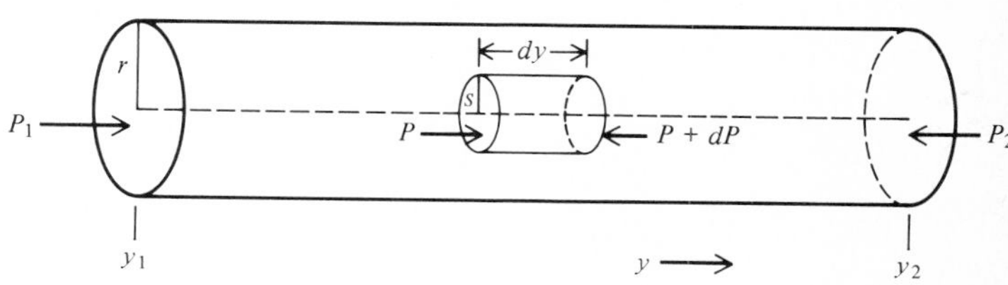

Figure 16.4
A fluid flowing in a cylindrical tube.

speed $v_y(s)$, and in time dt this fluid travels a distance $v_y(s)\,dt$. The volume of fluid in this shell that passes a given location in time dt is equal to the volume of a cylindrical shell with length $v_y(s)\,dt$ and inner and outer radii s and $s + ds$, and this volume is

$$[\pi(s + ds)^2 \cdot v_y(s)\,dt] - [\pi s^2 \cdot v_y(s)\,dt] = 2\pi s v_y(s)\,ds\,dt$$

since $(ds)^2$ is negligible compared with ds. To get the total volume dV that flows through a given cross section in time dt, we integrate over all shells from $s = 0$ to r; we have $dV = [\int_0^r 2\pi s v_y(s)\,ds]\,dt$. Use of (16.17) for v_y and of $dm = \rho\,dV$ (where ρ is the fluid density) gives as the rate of flow of mass through a fixed cross section:

$$\frac{dm}{dt} = \frac{\pi\rho}{2\eta}\left(-\frac{dP}{dy}\right)\int_0^r (r^2 s - s^3)\,ds = \frac{\pi r^4 \rho}{8\eta}\left(-\frac{dP}{dy}\right) \qquad (16.18)$$

Conservation of mass assures us that the mass passing any cross section in the tube in time dt is the same; thus, dm/dt is constant along the tube. For a liquid, ρ varies only slightly with pressure, and we can take ρ as constant along the tube. Hence for a liquid, (16.18) tells us that $-dP/dy$ is constant along the tube: $-dP/dy = a$, which integrates to $P = -ay + b$, where a and b are constants. Since P equals P_2 at y_2 and equals P_1 at y_1 (Fig. 16.4), we have $P_2 = -ay_2 + b$ and $P_1 = -ay_1 + b$. Subtraction gives $a = (P_1 - P_2)/(y_2 - y_1)$. Since $-dP/dy = a$, we have $-dP/dy = (P_1 - P_2)/(y_2 - y_1)$. Equation (16.18) becomes

$$\frac{dm}{dt} = \frac{\pi r^4 \rho}{8\eta}\frac{P_1 - P_2}{y_2 - y_1} \qquad \text{laminar flow of liquid} \qquad (16.19)$$

Since the flow rate is constant with time, (16.19) can also be written as

$$V = \frac{\pi r^4}{8\eta}\frac{P_1 - P_2}{y_2 - y_1}t \qquad \text{laminar flow of liquid} \qquad (16.20)$$

where V is the volume leaving the tube in time t.

For a gas, ρ is a strong function of pressure. The ideal-gas law gives $\rho = PM/RT$ (where M is the molar mass), and (16.18) becomes

$$\frac{dm}{dt} = \frac{\pi r^4 M}{8\eta RT}\left(-P\frac{dP}{dy}\right) \qquad (16.21)$$

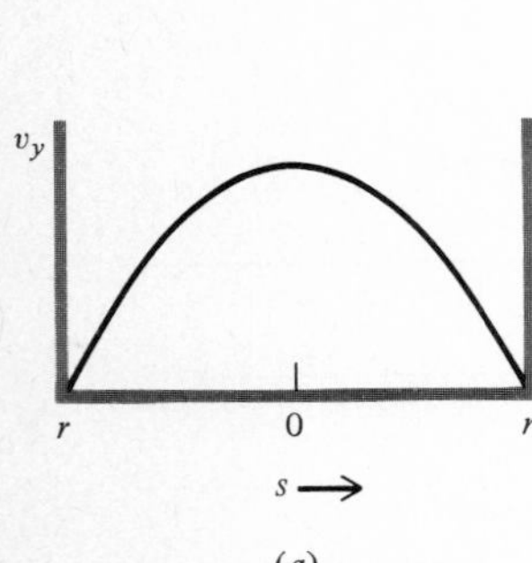

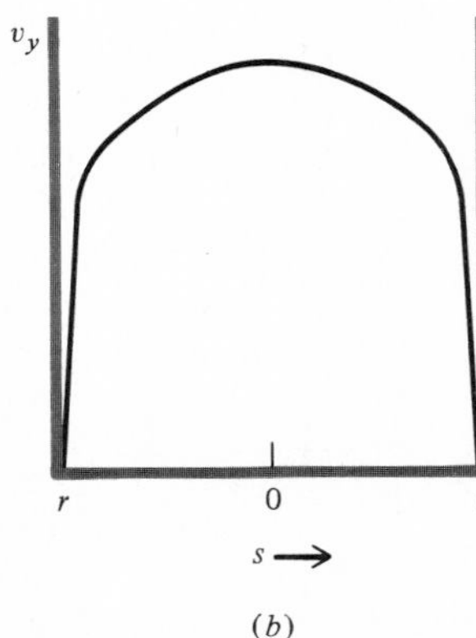

Figure 16.5
Velocity profiles for fluid flow in a cylindrical pipe: (*a*) laminar flow; (*b*) turbulent flow.

Since dm/dt is constant along the tube, $-P(dP/dy)$ is constant along the tube: $-P(dP/dy) = A$, which integrates to $P^2 = -2Ay + B$, where A and B are constants. We have $P_2^{\,2} = -2Ay_2 + B$ and $P_1^{\,2} = -2Ay_1 + B$. Subtraction gives $A = -(P_2^{\,2} - P_1^{\,2})/2(y_2 - y_1) = -P(dP/dy)$. Hence (16.21) becomes

$$\frac{dn}{dt} = \frac{\pi r^4}{16\eta RT}\frac{P_1^{\,2} - P_2^{\,2}}{y_2 - y_1} \qquad \text{laminar flow of ideal gas} \qquad (16.22)$$

where $n = m/M$ is the number of moles of gas.

Equations (16.20) and (16.22) are due to Poiseuille. Measurement of the flow rate through a capillary tube of known radius allows η of a liquid or gas to be found from these equations.

A second way to find η of a liquid is to measure the rate of fall of a spherical solid through the liquid. The layer of fluid in contact with the ball moves along with it (no-slip condition), and a gradient of speed develops in the fluid surrounding the sphere; this gradient generates a viscous force F resisting the sphere's motion. Stokes proved that the frictional force F on a solid sphere of radius r moving at speed v through a Newtonian fluid of viscosity η is

$$F = 6\pi\eta rv \qquad (16.23)$$

provided v is not too high. This equation applies to motion through a gas, provided r is much greater than the mean free path λ and there is no slip. [For a derivation of *Stokes' law* (16.23), see *Bird, Stewart, and Lightfoot*, pp. 132–133.]

A spherical body falling through a fluid experiences a downward gravitational force mg, an upward frictional force given by (16.23), and an upward buoyant force F_{buoy} that results from the greater fluid pressure below the body than above it [Eq. (1.8)]. To find F_{buoy}, imagine that the immersed object with volume V is replaced by fluid of equal volume. The buoyant force doesn't depend on the object being buoyed up, so the buoyant force on the fluid of volume V equals that on the original immersed object. However, the fluid is at rest, so the upward buoyant force on it equals the downward gravitational force, which is its weight. Therefore, an object of volume V immersed in a fluid is buoyed up by a force equal to the weight of fluid of volume V. This is *Archimedes' principle* (allegedly discovered while he was bathing).

Let m_{fl} be the mass of fluid of volume V. Equating the upward and downward forces on the falling sphere, we have $mg = 6\pi\eta rv + m_{\text{fl}}g$ and

$$6\pi\eta rv = (m - m_{\text{fl}})g = (\rho - \rho_{\text{fl}})gV = (\rho - \rho_{\text{fl}})g\tfrac{4}{3}\pi r^3$$

$$v = 2(\rho - \rho_{\text{fl}})gr^2/9\eta \qquad (16.24)$$

where ρ and ρ_{fl} are the densities of the sphere and the fluid, respectively. Measurement of the terminal speed of fall allows η to be found.

Another way besides flow rates of measuring the viscosity of a gas is to measure the rate of damping of oscillations of vanes suspended in the gas by a torsion wire.

Since $F_y = ma_y = m(dv_y/dt) = d(mv_y)/dt = dp_y/dt$, Newton's law of viscosity (16.13) can be written as

$$\frac{dp_y}{dt} = -\eta\mathscr{A}\frac{dv_y}{dx} \qquad (16.25)$$

where dp_y/dt is the time rate of change of the y component of momentum of a layer on one side of a surface in the fluid due to its interaction with fluid on the other side.

The molecular explanation of viscosity is that it is due to a transport of momentum across planes perpendicular to the x axis in Fig. 16.3. Molecules in adjacent layers of the fluid have different average values of p_y, since adjacent layers are moving at different speeds. In gases, the random molecular motion brings some molecules from the faster-moving layer into the slower-moving layer, where they collide with slower-moving molecules and impart extra momentum to them, thereby tending to speed up the slower layer. Similarly, slow-moving molecules moving into the faster layer tend to slow down this layer. In liquids, the momentum transfer between layers occurs mainly by collisions between molecules in adjacent layers, without actual transfer of molecules between layers.

Kinetic theory of gas viscosity. The kinetic-theory derivation of η for gases is very similar to the derivation of the thermal conductivity, except that momentum rather than energy is transported. Replacing dU by dp_y and ε by mv_y in (16.5), we get

$$dp_y = \tfrac{1}{4}\langle v\rangle \mathscr{A}(N/V)[mv_y(x_0 - \tfrac{2}{3}\lambda) - mv_y(x_0 + \tfrac{2}{3}\lambda)]\,dt \qquad (16.26)$$

In this equation, mv_y is the y momentum of one molecule, and dp_y is the total momentum flow across a surface of area $\mathscr{A}$ in time dt. We have $dv_y = (dv_y/dx)\,dx = -(dv_y/dx)\cdot\frac{4}{3}\lambda$, where $m\,dv_y$ is the quantity in brackets in (16.26). Also, $Nm/V = \rho$. Hence

$$dp_y/dt = -\tfrac{1}{3}\rho\langle v\rangle\lambda\mathscr{A}(dv_y/dx) \qquad (16.27)$$

Comparison with (16.25) gives

$$\eta \approx \tfrac{1}{3}\rho\langle v\rangle\lambda \qquad \text{hard spheres} \qquad (16.28)$$

Because of the crudity of assumptions 2 to 4, the coefficient in (16.28) is wrong. The rigorous result for hard-sphere molecules turns out to be

$$\eta = \frac{5\pi}{32}\rho\langle v\rangle\lambda = \frac{5}{16\pi^{1/2}}\frac{(MRT)^{1/2}}{N_0 d^2} \qquad \text{hard spheres} \qquad (16.29)$$

where (15.46) and (15.67) for $\langle v\rangle$ and λ, and $PM = \rho RT$ were used.

We can use (16.29) to calculate hard-sphere molecular diameters from experimental η values. Some results obtained using η at 0°C and 1 atm are:

Molecule	He	H_2	D_2	N_2	O_2	CH_4	C_2H_4	$n\text{-}C_4H_{10}$
d/Å	2.2	2.7	2.7	3.7	3.6	4.1	4.9	7.0

(16.30)

These values are certainly reasonable.

Equation (16.29) predicts the viscosity of a gas to increase with increasing temperature and to be independent of pressure. Both these predictions are quite surprising, in that (by analogy with liquids) one might expect the gas to flow more easily at higher T and less easily at higher P.

When Maxwell derived (16.28) in 1860, there were virtually no data on the temperature and pressure dependence of gas viscosities, so Maxwell and his wife Katherine (née Dewar) measured η as a function of T and P for gases. (In a postcard to a scientific colleague, Maxwell wrote: "My better $\frac{1}{2}$, who did all the real work of the kinetic theory is at present engaged in other researches. When she is done, I will let you know her answer to your enquiry [about experimental data].") The experimental results were that indeed η of a gas did increase with increasing T and was essentially independent of P. This provided strong early confirmation of the kinetic theory.

As with the thermal conductivity, η increases with T faster than the $T^{1/2}$ prediction of (16.29), due to the crudity of the hard-sphere model.

Data for η (in micropoises) as a function of P for N_2 at 25°C are:

P/atm	1	15	58	104	320	542
η/μP	190	191	198	209	274	351

As with K, the viscosity is essentially independent of P up to 50 or 100 atm. At very low pressures, where the mean free path is comparable to, or larger than, the dimensions of the container, Newton's viscosity law (16.13) does not hold. (See *Kauzmann*, p. 207.)

Combining (16.11) and the first equation in (16.29), we get $KM/\eta\bar{C}_V = 1 + 9R/4\bar{C}_V$. A good test of the kinetic theory of transport properties is to compare values of $KM/\eta\bar{C}_V$ with values of $1 + 9R/4\bar{C}_V$. Some results for gases at 0°C and 1 atm are 1.98 and 1.91 for O_2, 1.66 and 1.66 for CO_2, and 1.87 and 1.75 for CH_4, where the first value of each pair is $KM/\eta\bar{C}_V$.

The theory of viscosity of liquids is not as well developed as that of gases, and discussion is omitted. (See *Bird, Stewart, and Lightfoot*, pp. 26 to 29.)

Viscosity of polymer solutions. A molecule of a long-chain synthetic polymer usually exists in solution as a *random coil*. There is nearly free rotation about the single bonds of the chain, so we can crudely picture the polymer as composed of a large number of links with random orientations between adjacent links. This picture is essentially the same as the random motion of a particle undergoing Brownian motion, each "step" of Brownian motion corresponding to a chain link. A polymer random coil therefore resembles the path of a particle undergoing Brownian motion (Fig. 3.12). The degree of compactness of the coil depends on the relative strengths of the intermolecular forces between the polymer and solvent molecules as compared to the forces between two parts of the polymer chain. The degree of compactness therefore varies from solvent to solvent for a given polymer.

We can expect the viscosity of a polymer solution to depend on the size and shape (and hence on the molecular weight and the degree of compactness) of the polymer molecules in the solution. If we restrict ourselves to a given kind of synthetic polymer in a given solvent, then the degree of compactness remains the same, and the polymer molecular weight can be determined by viscosity measurements. Solutions of polyethylene $(CH_2CH_2)_n$ will show different viscosity properties in a given solvent depending on the degree of polymerization n.

The *relative viscosity* (or *viscosity ratio*) η_r of a polymer solution is defined as $\eta_r \equiv \eta/\eta_A$, where η and η_A are the viscosities of the solution and the pure solvent A. Note that η_r is a dimensionless number. Of course, η_r depends on concentration, approaching 1 in the limit of infinite dilution.

The *intrinsic viscosity* (or *limiting viscosity number*) $[\eta]$ of a polymer solution is defined by

$$[\eta] \equiv \lim_{\rho_B \to 0} \frac{\eta_r - 1}{\rho_B} \tag{16.31}$$

where $\rho_B \equiv m_B/V$ is the mass concentration of the polymer, m_B and V being the mass of polymer in the solution and the solution volume. One finds that $[\eta]$ depends on the solvent as well as the polymer. In 1942, Huggins showed that $(\eta_r - 1)/\rho_B$ is a linear function of ρ_B in dilute solutions, so a plot of $(\eta_r - 1)/\rho_B$ versus ρ_B allows one to obtain $[\eta]$ by extrapolation to $\rho_B = 0$.

Experimental data show that for a given kind of synthetic polymer in a given solvent, the following relation is well obeyed:

$$[\eta] = K(M_B/M^\circ)^a \tag{16.32}$$

where M_B is the molar mass of the polymer, K and a are empirical constants, and $M^\circ \equiv 1$ g/mole. (For example, for polyisobutylene in benzene at 24°C, one finds $a = 0.50$ and $K = 0.083$ cm^3/g.) Typically, a lies between 0.5 and 1.1, but can range as low as 0.2 and as high as 1.7. (Data on synthetic polymers are tabulated in J. Brandrup and E. H. Immergut, *Polymer Handbook*, 2d ed., Wiley-Interscience, 1975.) To apply (16.32), one must first determine K and a for the polymer and the solvent using a polymer sample whose molecular weight has been determined by some other method (e.g., from osmotic-pressure measurements). Once K and a are known, the molar mass of a given sample of that polymer can be quickly determined by viscosity measurements.

A particular protein has a definite molecular weight. In contrast, preparation of a synthetic polymer produces molecules with a distribution of molecular weights, since chain termination can occur with any length of chain. (One can use fractional crystallization from a solvent to separate polymer fractions having various ranges of molecular weights.)

Let n_i and x_i be the number of moles and the mole fraction of polymer species i with molar mass M_i present in a polymer sample. The *number average molar mass* M_n of the sample is defined by Eqs. (12.40) and (12.41) as $M_n \equiv m/n = \sum_i x_i M_i$, where the sum goes over all the polymer species and m and n are the total mass and the total number of moles of polymeric material.

In M_n, the molar mass of each species has a weighting factor given by x_i, its mole fraction; x_i is proportional to the relative number of i molecules present. In the *weight average molar mass* M_w, the molar mass of each species has a weighting factor given by w_i, its mass (or weight) fraction in the polymer mixture, where $w_i \equiv m_i/m$ (m_i is the mass of species i present in the mixture). Thus

$$M_w \equiv \sum_i w_i M_i = \frac{\sum_i m_i M_i}{\sum_i m_i} = \frac{\sum_i n_i M_i^2}{\sum_i n_i M_i} = \frac{\sum_i x_i M_i^2}{\sum_i x_i M_i}$$

For a polymer with a distribution of molecular weights, Eq. (16.32) yields a *viscosity average molar mass* M_v, where $M_v = [\sum_i w_i M_i^a]^{1/a}$. If a happens to be 1, then M_v equals M_w.

16.4 DIFFUSION AND SEDIMENTATION

Diffusion. Figure 16.6 shows two fluid phases 1 and 2 separated by a removable impermeable partition. The system is held at constant T and P. Each phase is a binary solution of substances A and B but with different initial concentrations ($c_{A,1} \neq c_{A,2}$ and $c_{B,1} \neq c_{B,2}$, where $c_{A,1}$ is the concentration of A in phase 1). We do not exclude the possibility that one or both phases might be pure A or pure B. When the partition is removed, the two phases are in contact, and the random molecular motion of A and B molecules will reduce and ultimately eliminate the concentration differences between the two solutions. This spontaneous decrease of concentration differences is *diffusion*.

Diffusion is a macroscopic motion of components of a system that arises from concentration differences. If $c_{A,1} < c_{A,2}$, there is a net flow of A from phase 2 to phase 1 (and a net flow of B from phase 1 to 2). This flow continues until the concentrations of A and B are constant throughout the cell. Diffusion should be distinguished from the macroscopic bulk flow that arises from pressure differences (Sec. 16.3). In bulk flow in the y direction (Fig. 16.4), the flowing molecules have an additional component of velocity v_y that is superimposed on the random distribution of velocities. In diffusion, all the molecules have only random velocities; however, because the concentration c_A on the right of a plane perpendicular to the diffusion direction is greater than the concentration to the left of this plane, more A molecules cross this plane from the right than from the left, giving a net flow of A from right to left. Figure 16.7 shows how the A concentration profile along the diffusion cell varies with time during a diffusion experiment.

Experiment shows that the following equations are obeyed in diffusion:

$$\frac{dn_A}{dt} = -D_{AB}\mathscr{A}\frac{dc_A}{dx} \quad \text{and} \quad \frac{dn_B}{dt} = -D_{BA}\mathscr{A}\frac{dc_B}{dx} \tag{16.33}$$

In (16.33) (which is *Fick's first law of diffusion*) dn_A/dt is the rate of flow of A (in moles per unit time) across a plane of area $\mathscr{A}$ perpendicular to the x axis, dc_A/dx is the rate of change of the (molar) concentration of A with respect to the x coordinate evaluated at that plane, and D_{AB} is a proportionality constant called the *diffusion coefficient*. The rate of diffusion is proportional to $\mathscr{A}$ and to the concentration gradient. As time goes on, dc_A/dx at a given plane changes, eventually becoming zero; diffusion then stops.

The diffusion coefficient D_{AB} is a function of the local state of the system and therefore depends on T, P, and the local composition of the solution. In a diffusion experiment, one measures the concentrations as a function of distance x at various times t. If the two solutions differ substantially in initial concentrations, then, since the diffusion coefficients are functions of concentration, D_{AB} varies substantially with

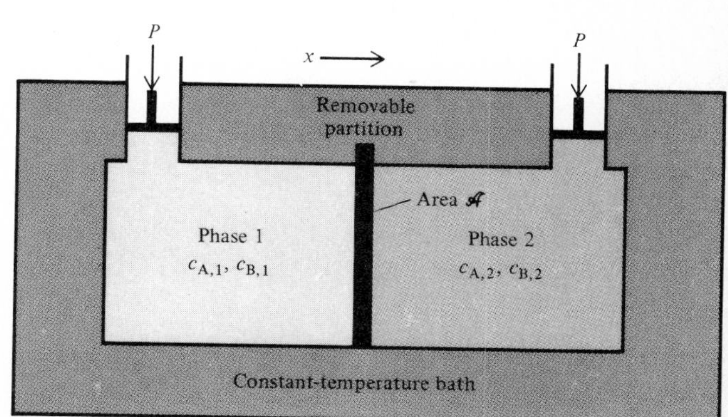

Figure 16.6
When the partition is removed, diffusion occurs.

distance x along the diffusion cell and with time as the concentrations change, so the experiment yields some sort of complicated average D_{AB} for the concentrations involved. [This average is called an *integral* diffusion coefficient, in contrast to D_{AB} in (16.33), which is sometimes called a *differential* diffusion coefficient.] If the initial concentrations in phase 1 are made close to those in phase 2, the variation of D_{AB} with concentration can be neglected and one obtains a D_{AB} value corresponding to the average composition of 1 and 2.

Instead of having phases 1 and 2 in direct contact, one sometimes separates them by a porous ceramic disk, through which diffusion occurs. The solutions on each side of the disk are continuously stirred, so that concentration gradients occur only within the disk. The disk has an effective area and an effective thickness that depend on the nature of its pores, and the apparatus must be calibrated with a system of known diffusion coefficient.

We can show that if solutions 1 and 2 mix with no volume change, then D_{AB} and D_{BA} in (16.33) are equal. For simplicity, we consider two stirred constant-T-and-P solutions with different compositions, the solutions being separated by a thin porous disk through which diffusion occurs; let dc_A/dx and dc_B/dx be constant within the disk. Equation (9.14) gives for the volume change dV of solution 1 that occurs in time dt

$$dV = \bar{V}_A\, dn_A + \bar{V}_B\, dn_B \qquad \text{const. } T, P \tag{16.34}$$

where $\bar{V}_A$ and $\bar{V}_B$ are the partial molar volumes of A and B in solution 1 and dn_A and dn_B are the number of moles of A and B that enter solution 1 in time dt. (Since A and B are diffusing in opposite directions, dn_A and dn_B have opposite signs.) We are assuming no volume change on mixing, so $dV = 0$. Solving the equations of (16.33) for dn_A and dn_B and substituting in (16.34), we get

$$dV = 0 = \bar{V}_A[-\mathscr{A}D_{AB}(dc_A/dx)\, dt] + \bar{V}_B[-\mathscr{A}D_{BA}(dc_B/dx)\, dt]$$

$$D_{AB}\bar{V}_A(dc_A/dx) + D_{BA}\bar{V}_B(dc_B/dx) = 0 \tag{16.35}$$

Dividing the equation $V = n_A\bar{V}_A + n_B\bar{V}_B$ by V, we get $1 = c_A\bar{V}_A + c_B\bar{V}_B$. Since there is no volume change on mixing, $\bar{V}_A$ and $\bar{V}_B$ are independent of the concentrations and hence are independent of the x coordinate. Differentiation with respect to x gives $0 = \bar{V}_A(dc_A/dx) + \bar{V}_B(dc_B/dx)$. Combining this relation with (16.35), we get

$$D_{AB} = D_{BA} \tag{16.36}$$

The derivation of (16.36) assumed no volume change. For gases, volume changes for constant-T-and-P mixing are negligible. For liquids, volume changes on mixing are not always negligible, but by having solutions 1 and 2 differ only slightly in composition we can satisfy the condition of negligible volume change.

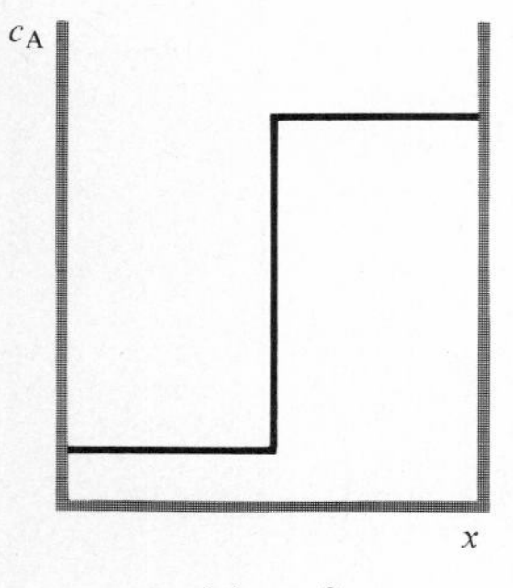

(*a*) $t = 0$

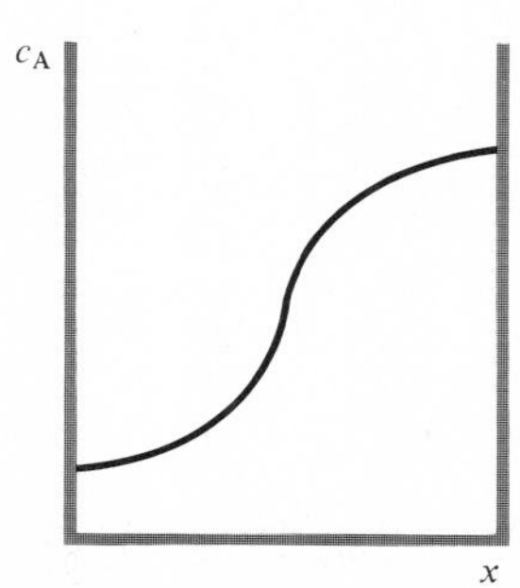

(*b*) Intermediate time

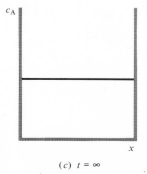

(*c*) $t = \infty$

Figure 16.7 Concentration profiles during a diffusion experiment.

For a given pair of gases, one finds that D_{AB} varies only slightly with composition, increases as T increases, and decreases as P increases. Values of D_{AB} for several gas pairs at 0°C and 1 atm are:

Gas pair	H_2–O_2	He–Ar	O_2–N_2	O_2–CO_2	CO_2–CH_4	CO–C_2H_4
$D_{AB}/(cm^2\ s^{-1})$	0.70	0.64	0.18	0.14	0.15	0.12

For liquids, D_{AB} varies strongly with composition and increases as T increases. Values of $10^5 D_{AB}$ for H_2O–C_2H_5OH at 25°C and 1 atm as a function of ethanol mole fraction are:

$x(C_2H_5OH)$	0	0.05	0.10	0.28	0.50	0.70	0.90	1
$10^5 D/(cm^2\ s^{-1})$	(1.24)	1.13	0.90	0.41	0.90	1.40	2.0	(2.4)

The values at $x = 0$ and 1 are extrapolations.

Let D_{BA}^{∞} denote the value of D_{BA} for a very dilute solution of solute B in solvent A. For example, data in the preceding table give $D_{H_2O,C_2H_5OH}^{\infty} = 2.4 \times 10^{-5}\ cm^2\ s^{-1}$ at 25°C and 1 atm. Some D^{∞} values at 25°C and 1 atm for the solvent H_2O are:

B	N_2	LiBr	NaCl	n-C_4H_9OH	sucrose	hemoglobin
$10^5 D_{B,H_2O}^{\infty}/(cm^2\ s^{-1})$	1.6	1.4	2.2	0.56	0.52	0.07

Diffusion coefficients in solids can be measured using chemical analysis or isotopic tracers. As with liquids, D_{AB} in solids depends on concentration and increases as T increases. The temperature dependence of D_{AB} in solids can usually be represented by an equation of the form $D_{AB} = ae^{-b/T}$, where a and b are positive constants. Some solid-phase diffusion coefficients at 1 atm are:

B-A	Bi–Pb	Sb–Ag	Al–Cu	Ni–Cu	Ni–Cu	Cu–Ni
Temperature	20°C	20°C	20°C	550°C	1025°C	1025°C
$D_{BA}^{\infty}/(cm^2\ s^{-1})$	10^{-16}	10^{-21}	10^{-30}	10^{-12}	10^{-9}	10^{-10}

Suppose the molecules of species A and B have the same size and the same mass, and that A-B intermolecular forces are identical to A-A and B-B intermolecular forces. (A close approximation to this case is if one atom of B is an isotope of the corresponding A atom, for example, $^{12}CH_4$ and $^{13}CH_4$.) The diffusion coefficient D_{AB} then becomes in effect a measure of how fast A molecules diffuse among themselves. This diffusion coefficient is the *self-diffusion coefficient* D_{AA}. Values of D_{AA} are determined using isotopic tracers. (The small difference in mass can usually be ignored, since experimental D values are not highly accurate.) Some D_{AA} values at 1 atm are:

Gas (0°C)	H_2	O_2	N_2	HCl	CO_2	C_2H_6	Xe
$D_{AA}/(cm^2\ s^{-1})$	1.5	0.19	0.15	0.12	0.10	0.09	0.05

Liquid (25°C)	H_2O	C_6H_6	Hg	CH_3OH	C_2H_5OH	n-C_3H_7OH
$10^5 D_{AA}/(cm^2\ s^{-1})$	2.4	2.2	1.7	2.3	1.0	0.6

Diffusion coefficients at 1 atm and 25°C are typically 10^{-1} cm^2 s^{-1} for gases and 10^{-5} cm^2 s^{-1} for liquids; they are extremely small for solids.

Distance traveled by diffusing molecules. One of the early objections to the kinetic theory of gases was that if gases really consisted of molecules moving about freely at supersonic speeds, mixing of gases should take place almost instantaneously. This does not occur. (If a chemistry lecturer generates Cl_2, it may take a couple of minutes for those in the back of the room to smell the gas.) The reason mixing of gases is slow relative to the speeds of gas molecules is that at ordinary pressures a gas molecule goes only a very short distance (about 10^{-5} cm at 1 atm and 25°C; see Sec. 15.7) before colliding with another molecule; at each collision, the direction of motion changes, so that each molecule has a zigzag path (Fig. 3.12); the net motion in any given direction is quite small because of these continual changes in direction.

The average distance traveled in any given direction in a given time by a molecule undergoing random diffusional motion can be found as follows. For a diffusing molecule, let Δx be the net displacement in the x direction that occurs in time t. Since the motion is random, the average value $\langle \Delta x \rangle$ is zero. We therefore consider $\langle (\Delta x)^2 \rangle$, the average of the square of the x displacement. Consider three planes (labeled L, M, and R, for left, middle, and right, in Fig. 16.8) perpendicular to the x axis and separated by a distance $\langle (\Delta x)^2 \rangle^{1/2}$. Let c_L be the average concentration of the diffusing species in the region between planes L and M. For simplicity, we take the concentration gradient dc/dx as constant. Therefore, the average concentration c_L equals the concentration midway between planes L and M. Similarly, the average concentration c_R between planes M and R equals the concentration midway between M and R. All molecules between L and M are within a distance $\langle (\Delta x)^2 \rangle^{1/2}$ of M and are therefore capable of crossing M in time t. However, half of them have positive values of Δx and half have negative values, so only half the molecules between L and M cross M in time t. The number of moles of the diffusing species between L and M equals $c_L V_L = c_L \mathscr{A} \langle (\Delta x)^2 \rangle^{1/2}$, where V_L is the volume between L and M and $\mathscr{A}$ is the area of the planes. The number of moles moving left to right through plane M in time t is therefore $\frac{1}{2} c_L \mathscr{A} \langle (\Delta x)^2 \rangle^{1/2}$. Similarly, $\frac{1}{2} c_R \mathscr{A} \langle (\Delta x)^2 \rangle^{1/2}$ moles move right to left through M in time t. The net rate of flow of the diffusing species through M is

$$dn/dt = \Delta n/\Delta t = \tfrac{1}{2}(c_L - c_R)\mathscr{A}\langle (\Delta x)^2 \rangle^{1/2}/t \tag{16.37}$$

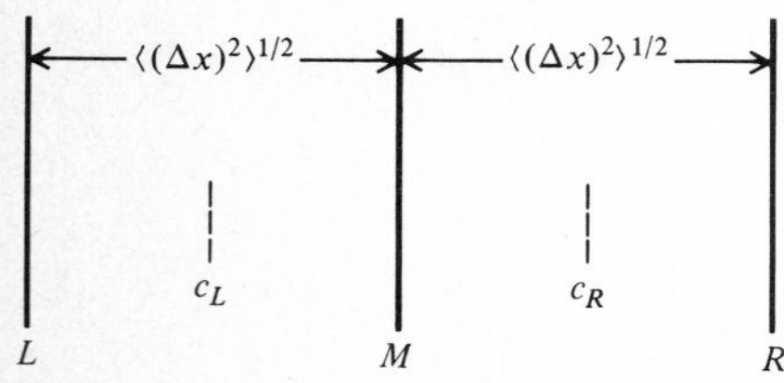

Figure 16.8
Planes in a system where diffusion is occurring.

The concentration gradient at M is (Fig. 16.8) $dc/dx = (c_R - c_L)/\langle(\Delta x)^2\rangle^{1/2}$. Substitution of these last two expressions into Fick's law (16.33) gives

$$(c_L - c_R)\mathscr{A}\langle(\Delta x)^2\rangle^{1/2}/2t = -D\mathscr{A}(c_R - c_L)/\langle(\Delta x)^2\rangle^{1/2}$$

$$\langle(\Delta x)^2\rangle = 2Dt \tag{16.38}$$

Equation (16.38) is the *Einstein–Smoluchowski equation*, derived by Einstein in 1905. (A more rigorous derivation than given above can be found in *Kennard*, pp. 286–287.)

The quantity

$$(\Delta x)_{\text{rms}} \equiv \langle(\Delta x)^2\rangle^{1/2} = (2Dt)^{1/2} \tag{16.39}$$

is the root-mean-square distance a diffusing molecule travels in the x direction in time t. Taking t to be 60 s and D to be 10^{-1}, 10^{-5}, and 10^{-20} cm^2 s^{-1}, we find the typical rms distances traveled in 1 min by molecules at room temperature and 1 atm to be only 3 cm in gases, 0.03 cm in liquids, and less than 1 Å in solids. In 1 min, a typical gas molecule of molecular weight 30 travels a total distance of 3×10^6 cm at room temperature and pressure [Eq. (15.48)], but its rms net displacement in any given direction is only 3 cm (due to collisions). Of course, there is a distribution of Δx values, and many molecules go shorter or longer distances than $(\Delta x)_{\text{rms}}$. This distribution turns out to be gaussian (Fig. 15.5), so a substantial fraction of molecules go 2 or 3 times $(\Delta x)_{\text{rms}}$, but a negligible fraction go 7 or 8 times $(\Delta x)_{\text{rms}}$.

If $(\Delta x)_{\text{rms}}$ is only 3 cm in 1 min in a gas at room T and P, why does a student in the back of the room smell the Cl_2 generated at the front of the room in only a couple of minutes? The answer is that under uncontrolled conditions, convection currents due to pressure and density differences are much more effective in mixing gases than diffusion.

Brownian motion. Diffusion results from the random thermal motion of molecules. This random motion can be observed indirectly by its effect on colloidal particles suspended in a fluid. These particles undergo a random Brownian motion (Sec. 3.7) due to microscopic fluctuations in pressure in the fluid. Brownian motion is the eternal dance of the molecules made visible. The colloidal particle can be considered to be a giant "molecule," and its Brownian motion is really a diffusion process.

Consider a colloidal particle of mass m in a fluid of viscosity η. The particle experiences a time-varying force $\mathbf{F}(t)$ due to random collisions with molecules of the fluid; let $F_x(t)$ be the x component of this random force. In addition, it experiences a frictional force $\mathbf{F}_{\text{fr}}$ that results from the liquid's viscosity and opposes the motion of the particle. $\mathbf{F}_{\text{fr}}$ is proportional to the particle's velocity $\mathbf{v}$. Let f (called the *friction coefficient*) be the negative of the proportionality constant. The x component of $\mathbf{F}_{\text{fr}}$ is then $F_{\text{fr},x} = -fv_x = -f(dx/dt)$. Newton's second law $F_x = ma_x = m(d^2x/dt^2)$ when multiplied by x gives

$$xF_x(t) - fx(dx/dt) = mx(d^2x/dt^2) \tag{16.40}$$

Einstein averaged (16.40) over a large number of colloidal particles; assuming that the colloidal particles have an average kinetic energy equal to the average translational energy $\frac{3}{2}kT$ of the molecules of the surrounding fluid [Eq. (15.16)], he found the particles' average square displacement in the x direction to increase linearly with time according to

$$\langle x^2 \rangle = 2kTf^{-1}t \tag{16.41}$$

The derivation of (16.41) from (16.40) is outlined in Prob. 16.17.

If the colloidal particles are spheres each with radius r, then Stokes' law (16.23) gives $F_{\text{fr},x} = 6\pi\eta r v_x$ and the friction coefficient is $f = 6\pi\eta r$. Equation (16.41) becomes

$$\langle x^2 \rangle = \frac{kT}{3\pi\eta r}t \qquad \text{spherical particles} \tag{16.42}$$

Equation (16.42) was derived by Einstein in 1905 and subsequently verified experimentally by Perrin. Measurement of $\langle x^2 \rangle$ for colloidal particles of known size allows $k = R/N_0$ to be calculated and hence allows Avogadro's number to be found.

Kinetic theory of diffusion in gases. The mean-free-path kinetic theory of diffusion in gases is similar to that of K and η, except that it is matter, rather than energy or momentum, that is transported. Consider first a mixture of A with an isotopic species $\text{A}^{\#}$ that has the same diameter and essentially the same mass as A. Let there be a concentration gradient $dc^{\#}/dx$ of $\text{A}^{\#}$. Molecules of $\text{A}^{\#}$ cross a plane at x_0 coming from the left and from the right. We take the concentration of $\text{A}^{\#}$ molecules crossing from either side as the concentration in the plane where (on the average) they made their last collision. These planes are at a distance $\frac{2}{3}\lambda$ from x_0 (Sec. 16.2). Since $N/V = N_0 n/V = N_0 c$, Eq. (16.2) gives the net number of molecules of $\text{A}^{\#}$ crossing the x_0 plane in time dt as

$$dN^{\#} = \tfrac{1}{4}\langle v \rangle \mathscr{A} N_0[c^{\#}(x_0 - \tfrac{2}{3}\lambda) - c^{\#}(x_0 + \tfrac{2}{3}\lambda)]\,dt \tag{16.43}$$

where $c^{\#}(x_0 - \frac{2}{3}\lambda)$ is the concentration of $\text{A}^{\#}$ at the $x_0 - \frac{2}{3}\lambda$ plane. But $dc^{\#} = (dc^{\#}/dx)\,dx = -(dc^{\#}/dx)\frac{4}{3}\lambda$, where $dc^{\#}$ is the term in brackets in (16.43). Therefore (16.43) becomes

$$\frac{dn^{\#}}{dt} = -\tfrac{1}{3}\lambda\langle v \rangle \mathscr{A}\frac{dc^{\#}}{dx} \tag{16.44}$$

Comparison with Fick's law (16.33) gives for the self-diffusion coefficient

$$D_{\text{AA}} \approx \tfrac{1}{3}\lambda\langle v \rangle \qquad \text{hard spheres} \tag{16.45}$$

As usual, the numerical coefficient is wrong, and a rigorous treatment gives for hard spheres

$$D_{\text{AA}} = \frac{3\pi}{16}\lambda\langle v \rangle = \frac{3}{8\pi^{1/2}}\left(\frac{RT}{M}\right)^{1/2}\frac{1}{d^2(N/V)} \qquad \text{hard spheres} \tag{16.46}$$

where (15.67) and (15.46) were used.

The simple mean-free-path treatment can be applied to a mixture of gases A and B to calculate the mutual diffusion coefficient D_{AB}. However, the simple treatment fails miserably in this case; it predicts D_{AB} to be a strong function of the A mole fraction, whereas experiment shows D_{AB} for gases to be almost independent of x_{A}. (The reasons for this failure are discussed in *Present*, pp. 50–51.) A rigorous treatment for hard spheres gives

$$D_{\text{AB}} = \frac{3}{8\pi^{1/2}}\left(\frac{RT}{2}\right)^{1/2}\left(\frac{1}{M_{\text{A}}} + \frac{1}{M_{\text{B}}}\right)^{1/2}\frac{1}{(r_{\text{A}} + r_{\text{B}})^2(N_{\text{A}} + N_{\text{B}})/V} \tag{16.47}$$

where r_A, N_A, and M_A are the radius of an A molecule, the number of A molecules, and the molar mass of A. Note the dependence on $r_A + r_B$, which is the diameter for an A-B collision. Equation (16.47) predicts D_{AB} to be independent of the relative proportions of A and B present.

Actually, Eqs. (16.9), (16.29), and (16.47) are white lies. The rigorous Chapman–Enskog theory of hard-sphere molecules gives small corrections to these equations. These corrections increase the coefficients in (16.9) and (16.29) by $1\frac{1}{2}$ and 1 percent, respectively, and add a term to (16.47) that makes D_{AB} vary by a few percent as x_A goes from 0 to 1. This small variation of D_{AB} with composition has been confirmed experimentally.

Using $PV = [(N_A + N_B)/N_0]RT$, we see from (16.47) that $D_{AB} \propto T^{3/2}/P$. Hence, D_{AB} increases with T and decreases with P. The inverse dependence on pressure is due to the increasing number of collisions at higher pressure. The inverse dependence on pressure is well obeyed. Data for N_2–CH_4 at 60°C are:

P/atm	6.8	34.0	68.1	136.2	170.2
$D_{AB}/(\text{cm}^2\ \text{s}^{-1})$	0.036	0.0068	0.0033	0.00155	0.00122

Theory of diffusion in liquids. Consider a very dilute solution of solute B in solvent A. The Einstein–Smoluchowski equation (16.38) gives the mean square x displacement of a B molecule in time t as $\langle(\Delta x)^2\rangle = 2D_{BA}^{\infty} t$, where D_{BA}^{∞} is the diffusion coefficient for a very dilute solution of B in A. Equation (16.41) gives $\langle(\Delta x)^2\rangle = (2kT/f)t$. Hence $(2kT/f)t = 2D_{BA}^{\infty} t$, or

$$D_{BA}^{\infty} = kT/f \tag{16.48}$$

where f is the friction coefficient for motion of B molecules in the solvent A. (Recall that $f \equiv F_{fr}/v$, where F_{fr} is the magnitude of the viscous force that resists the motion of a B molecule moving with speed v.) Equation (16.48) is the *Nernst–Einstein equation.*

Viscosity is a macroscopic property. Application of the concept of a viscous resisting force to the motion of a particle of colloidal size through a fluid is valid, but its application to the motion of individual molecules through a fluid is certainly stretching things (unless the solute molecules are much larger than the solvent molecules, e.g., a solution of a polymer in water). Therefore, Eq. (16.48) should be regarded as nonrigorous.

If we assume that the B molecules are spherical with radius r_B and further assume that Stokes' law (16.23) can be applied to the motion of B molecules through the fluid A, then $f = 6\pi\eta_A r_B$ and (16.48) becomes

$$D_{BA}^{\infty} \approx \frac{kT}{6\pi\eta_A r_B} \qquad \text{for B spherical, } r_B > r_A \tag{16.49}$$

Equation (16.49) is the *Stokes–Einstein equation.*

We can expect (16.49) to work best when r_B is substantially larger than r_A. Use of Stokes' law assumes that there is no slip at the surface of the diffusing particle. Fluid dynamics shows that when there is no tendency for the fluid to stick at the surface of the diffusing particle, Stokes' law is replaced by $F_{fr} = 4\pi\eta_A r_B v_B$. Experimental data on diffusion coefficients in solution indicate that for solute molecules of size comparable to that of the solvent molecules, the 6 in Eq. (16.49) should be replaced by a 4:

$$D_{BA}^{\infty} \approx \frac{kT}{4\pi\eta_A r_B} \qquad \text{for B spherical, } r_B \approx r_A \tag{16.50}$$

For $r_B < r_A$, data indicate that the 4 should be replaced by a smaller number.

A study of diffusion coefficients in water [J. T. Edward, *J. Chem. Educ.*, **47**, 261 (1970)] showed that Eqs. (16.49) and (16.50) work surprisingly well. The molecular radii r_B were calculated from the van der Waals radii of the atoms (Sec. 24.5).

For the self-diffusion coefficient D_{AA} in a liquid, we use (16.50). Let us suppose that the liquid volume can be divided into cubical cells, each cell having an edge length $2r_A$ and containing one spherical A molecule of radius r_A. Each cell has a volume $8r_A{}^3$, so the molar volume is $\bar{V}_A = 8r_A{}^3 N_0$. Hence, $r_A = \frac{1}{2}(\bar{V}_A/N_0)^{1/3}$, and (16.50) becomes

$$D_{AA} \approx \frac{kT}{2\pi\eta_A}\left(\frac{N_0}{\bar{V}_A}\right)^{1/3} \qquad \text{for A spherical} \tag{16.51}$$

an equation due to Li and Chang. Considering the crudity of the theory, Eq. (16.51) works well, predicting D_{AA} values with errors of about 10 percent. Some calculated and observed values (in parentheses) in 10^{-5} cm^2 s^{-1} for liquids at 25°C are: benzene, 2.1 (2.2); ethanol, 1.3 (1.0); acetone, 4.3 (4.8).

Sedimentation of polymer molecules in solution. Recall from Sec. 15.8 that the molecules of a gas in the earth's gravitational field show an equilibrium distribution in accord with the Boltzmann distribution law, the concentration of molecules decreasing with increasing altitude. A similar distribution holds for solute molecules in a solution in the earth's gravitational field. For a solution in which the distribution of solute molecules is initially uniform, there will be a net downward drift of solute molecules, until the equilibrium distribution is attained.

Consider a polymer molecule of mass M_B/N_0 (where M_B is the molar mass and N_0 the Avogadro constant) in a solvent of density less than that of the polymer. The polymer molecule will tend to drift downward (sediment). The downward force is the molecule's weight $M_B N_0^{-1} g$, where g is the gravitational acceleration. There is an upward viscous force fv_{sed}, where f is the friction coefficient and v_{sed} is the downward drift speed. Since there is a pressure gradient in the fluid, there is an upward buoyant force on the molecule that equals the weight of the displaced fluid (Sec. 16.3). The effective volume of the polymer molecule in solution depends on the solvent (Sec. 16.3) and we may take $\bar{V}_B/N_0$ as the effective volume of a molecule, where $\bar{V}_B$ is the partial molar volume of B in the solution. The buoyant force is therefore $(\rho\bar{V}_B/N_0)g$, where ρ is the density of the solvent. The molecular weight may not be known, so the partial molar volume may not be known. We therefore define the *partial specific volume* $\bar{v}_B$ as $\bar{v}_B \equiv (\partial V/\partial m_B)_{T,P,m_A}$, where V is the solution's volume, m_B is the mass of polymer in solution, and A is the solvent. Since $m_B = M_B n_B$, we have $\partial V/\partial m_B = (\partial V/\partial n_B)(\partial n_B/\partial m_B) = (\partial V/\partial n_B)/M_B$, or $\bar{v}_B = \bar{V}_B/M_B$. The buoyant force is then $\rho\bar{v}_B M_B N_0^{-1} g$.

The molecule will reach a terminal sedimentation speed v_{sed} at which the downward and upward forces balance:

$$M_B N_0^{-1} g = fv_{sed} + \rho\bar{v}_B M_B N_0^{-1} g$$

Although sedimentation of relatively large colloidal particles in the earth's gravitational field is readily observed (Sec. 13.6), the gravitational field is actually too weak to produce observable sedimentation of polymer molecules in solution. Instead, one uses an ultracentrifuge, a device that spins the polymer solution at very high speed (10^3 revolutions per second) and that is designed to minimize vibrations and convection currents.

A particle revolving at constant speed v in a circle of radius r is undergoing an acceleration v^2/r directed toward the center of the circle, a centripetal acceleration (*Halliday and Resnick*, Eq. 11-10). The speed is given by $v = r\omega$, where the angular speed ω is defined as $d\theta/dt$, where θ is the rotational angle in radians. The centripetal acceleration is therefore $r\omega^2$, where ω is 2π times the number of revolutions per unit time. The centripetal force is, by Newton's second law, $mr\omega^2$, where m is the particle's mass.

Just as a marble on a merry-go-round tends to move outward, so the protein molecules tend to sediment outward in the revolving tube in the ultracentrifuge. If we use a coordinate system that revolves along with the solution, then in this coordinate system, the centripetal acceleration $r\omega^2$ disappears, and in its place one must introduce a fictitious centrifugal force $mr\omega^2$ acting outward on the particle (*Halliday and Resnick*, Section 6-4 and Supplementary Topic I). The centrifugal force is a purely fictitious force that arises from the fact that in the revolving coordinate system, Newton's second law is not obeyed unless this fictitious force is introduced. ($F = ma$ holds only in a nonaccelerating coordinate system.)

Comparison of the fictitious centrifugal force $mr\omega^2$ in a centrifuge with the gravitational force mg in a gravitational field shows that $r\omega^2$ corresponds to g. Hence, replacing g in the last displayed equation by $r\omega^2$, we get

$$M_B N_0^{-1} r\omega^2 = fv_{sed} + \rho \bar{v}_B M_B N_0^{-1} r\omega^2$$

(The buoyant force arises, as in a gravitational field, from a pressure gradient in the fluid.)

The friction coefficient f can be found from diffusion data. The Nernst–Einstein equation (16.48) gives for a very dilute solution: $f = kT/D_{BA}^{\infty}$, where D_{BA}^{∞} is the infinite-dilution diffusion coefficient of the polymer in the solvent. Using the Nernst–Einstein equation and $R = N_0 k$, we find

$$M_B = \frac{RTv_{sed}^{\infty}}{D_{BA}^{\infty} r\omega^2(1 - \rho\bar{v}_B)}$$

Measurement of v_{sed} extrapolated to infinite dilution and of D_{BA}^{∞} allow the molar mass to be found. Special optical techniques are used to measure v_{sed} in the revolving solution.

The quantity $v_{sed}/r\omega^2$ is the *sedimentation coefficient* s; we have $s \equiv v_{sed}/r\omega^2$. For proteins, s at room temperature ranges from 10^{-13} to 10^{-11} s, increasing as M_B increases. (The quantity 10^{-13} second is called a *svedberg*, in honor of the developer of the ultracentrifuge.)

For a polymer containing a distribution of molecular weights, the molar mass given by a sedimentation-velocity measurement is close to the weight average molar mass M_w, but not identical to it.

16.5 ELECTRICAL CONDUCTIVITY

Electrical conduction is a transport phenomenon in which electrical charge (in the form of electrons or ions) moves through the system. The *electric current* I is defined as the rate of flow of charge through the conducting material:

$$I \equiv dQ/dt \tag{16.52*}$$

where dQ is the charge that passes through a cross section of the conductor in time dt. The *electric current density* j is defined as the electric current per unit cross-sectional area:

$$j \equiv I/\mathscr{A} \tag{16.53*}$$

where $\mathscr{A}$ is the conductor's cross-sectional area. The SI unit of current is the *ampere* (A) and equals one coulomb per second:

$$1 \text{ A} = 1 \text{ C/s} \tag{16.54*}$$

Although the charge Q is a more fundamental physical quantity than the current I, it is easier to measure current than charge. Hence the SI system takes the ampere as one of its fundamental units. The ampere is defined as that current which when flowing through two long, straight parallel wires exactly one meter apart will produce a force per unit length between the wires of exactly 2×10^{-7} N/m. (One current produces a magnetic field which exerts a force on the moving charges in the other wire.) The force between two current-carrying wires can be measured accurately using a current balance (see *Halliday and Resnick*, pp. 854–855).

The coulomb is then defined as the charge transported in one second by a current of one ampere: $1 \text{ C} \equiv 1 \text{ A s}$. Equation (16.54) then follows from this definition of the coulomb.

To avoid confusion, we shall use only SI units for electrical quantities in this chapter, and all electrical equations will be written in a form valid for SI units.

Charge flows because it experiences an electric force, so there must be an electric field $\mathbf{E}$ in a current-carrying conductor. The *conductivity* (or *specific conductance*) κ of a substance is defined by

$$\kappa \equiv j/E \qquad \text{or} \qquad j = \kappa E \tag{16.55*}$$

where E is the magnitude of the electric field. The reciprocal of the conductivity is the *resistivity* ρ:

$$\rho \equiv 1/\kappa \tag{16.56*}$$

The conductivity is a measure of the substance's response to an applied electric field.

Let the x direction be the direction of the electric field in the conductor. Equation (14.11) gives $E = -d\phi/dx$, where ϕ is the electric potential at a point in the conductor. Hence (16.55) can be written as $I/\mathscr{A} = -\kappa(d\phi/dx)$. Use of (16.52) gives

$$\frac{dQ}{dt} = -\kappa\mathscr{A}\frac{d\phi}{dx} \tag{16.57}$$

Current flows in a conductor only when there is a gradient of electric potential in the conductor. Such a gradient can be produced by attaching each end of the conductor to one of the two poles of a battery. Note the resemblance of (16.57) to the transport equations (16.1), (16.25), and (16.33) for thermal conduction, viscous flow, and diffusion.

Consider a current-carrying conductor that has a homogeneous composition and a constant cross-sectional area $\mathscr{A}$. Then the current density j will be constant at every point in the conductor. From (16.55), the field strength E is constant at every point, and the equation $E = -d\phi/dx$ integrates to $\phi(x_2) - \phi(x_1) = -E(x_2 - x_1)$. Hence $E = -d\phi/dx = -\Delta\phi/\Delta x$. Equation (16.55) becomes $I/\mathscr{A} = \kappa(-\Delta\phi)/\Delta x$. Let $\Delta x = l$, where l is the length of the conductor. Then $|\Delta\phi|$ is the magnitude of the potential difference between the ends of the conductor, and we have $|\Delta\phi| = Il/\kappa\mathscr{A}$ or

$$|\Delta\phi| = (\rho l/\mathscr{A})I \qquad (16.58)$$

The quantity $|\Delta\phi|$ is often called the "voltage," but the recommended name is the *electric tension.* The *resistance* R of the conductor is defined by

$$R \equiv |\Delta\phi|/I \qquad \text{or} \qquad |\Delta\phi| = IR \qquad (16.59)^*$$

Equation (16.58) gives

$$R = \rho l/\mathscr{A} \qquad (16.60)$$

The reciprocal of the resistance is the *conductance* G:

$$G \equiv 1/R \qquad (16.61)$$

From (16.59), R has units of volts per ampere. The SI unit of resistance is the *ohm* (symbol Ω, omega):

$$1\ \Omega \equiv 1\ \text{V/A} = 1\ \text{kg m}^2\ \text{s}^{-1}\ \text{C}^{-2} \qquad (16.62)$$

where (14.8) and (16.54) were used. From (16.60), the resistivity ρ has units of ohms times a length and is usually given in Ω cm or Ω m. The conductivity $\kappa = 1/\rho$ has units $\Omega^{-1}\ \text{cm}^{-1}$ or $\Omega^{-1}\ \text{m}^{-1}$. The unit Ω^{-1} is sometimes written as mho, which is ohm spelled backwards; however, the correct name for the reciprocal ohm is the *siemens* (S): $1\ \text{S} \equiv 1\ \Omega^{-1}$.

The conductivity κ (and its reciprocal, ρ) depend on the composition of the conductor but not on its dimensions. From (16.60), the resistance R (and its reciprocal, G) depend on the dimensions of the conductor as well as the material that composes it; R increases with increasing length of the conductor and decreases with increasing cross-sectional area.

For many substances, κ in (16.55) is independent of the magnitude of the applied electric field E (and hence is independent of the magnitude of the current density). Such substances are said to obey Ohm's law. *Ohm's law* is the statement that κ remains constant as E changes. For a substance that obeys Ohm's law, a plot of j vs. E is a straight line with slope κ. Metals obey Ohm's law. Solutions of electrolytes obey Ohm's law, provided E is not extremely high and provided steady-state conditions are maintained (see the next section). Many books state that Ohm's law is Eq. (16.59). This is inaccurate. Equation (16.59) is simply the *definition* of R, and this definition applies to all substances. Ohm's law is the statement that R is independent of $|\Delta\phi|$ (and of I) and does not apply to all substances.

Some resistivity and conductivity values for substances at 20°C and 1 atm are:

Substance	Cu	KCl(*aq*, 1 mol/dm^3)	CuO	glass
$\rho/(\Omega\ \text{cm})$	2×10^{-6}	9	10^5	10^{14}
$\kappa/(\Omega^{-1}\ \text{cm}^{-1})$	6×10^5	0.1	10^{-5}	10^{-14}

Metals have very low ρ values and very high κ values. Concentrated aqueous solutions of strong electrolytes have rather low ρ values. An *electrical insulator* (e.g., glass) is a substance with a very low κ. A *semiconductor*, for example, CuO, is a substance with κ intermediate between κ of metals and insulators. Semiconductors and insulators generally do not obey Ohm's law; their conductivity increases with increasing applied potential difference $|\Delta\phi|$.

16.6 ELECTRICAL CONDUCTIVITY OF ELECTROLYTE SOLUTIONS

Electrolysis. Figure 16.9 shows two metal electrodes at each end of a cell filled with a solution of electrolyte. A potential difference is applied to the electrodes by connecting them to a battery. Electrons carry the current through the metal wires and the metal electrodes. Ions carry the current through the solution. At each electrode–solution interface, an electrochemical reaction occurs that transfers electrons either to or from the electrode, thereby allowing charge to flow completely around the circuit. For example, if both electrodes are Cu and the electrolyte solute is $CuSO_4$, the electrode reactions are $Cu^{2+}(aq) + 2e^- \rightarrow Cu$ and $Cu \rightarrow Cu^{2+}(aq) + 2e^-$.

For 1 mole of Cu to be deposited from solution, 2 moles of electrons must flow through the circuit. (A mole of electrons is Avogadro's number of electrons.) If the current I is kept constant, the charge that flows is $Q = It$ [Eq. (16.52)]. Experiment shows that to deposit 1 mole of Cu requires the flow of 192,970 C, so the absolute value of the total charge on 1 mole of electrons is 96,485 C. The absolute value of the charge per mole of electrons is the *Faraday constant* $\mathscr{F} = 96{,}485$ C/mol. We have [Eq. (14.27)] $\mathscr{F} = N_0 e$, where e is the proton charge and N_0 is the Avogadro constant. To deposit 1 mole of the metal M from a solution containing the ion M^{z+} requires the flow of z_+ moles of electrons. Hence the mass m of metal M deposited by a flow of charge Q is

$$m = \frac{Q}{z_+ \mathscr{F}} M \qquad (16.63)$$

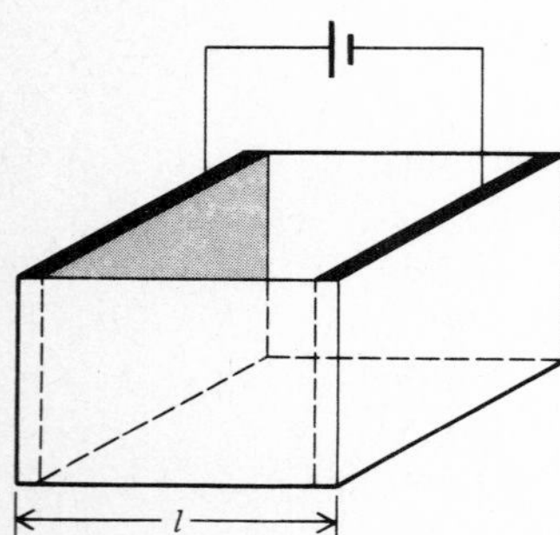

Figure 16.9
An electrolysis cell.

where M is the molar mass of the metal M. This equation embodies Faraday's laws of electrolysis.

The total charge that flows through a circuit during time t' is given by integration of (16.52) as $Q = \int_0^{t'} I\,dt$, which equals It' if I is constant. It isn't easy to keep I constant, and a good way to measure Q is to put an electrolysis cell in the circuit, weigh the metal deposited, and calculate Q from (16.63). Such a device is called a *coulometer*. Silver is the metal most often used.

Measurement of conductivity. The resistance R of an electrolyte solution cannot be reliably measured using direct current because changes in concentration of the electrolyte and buildup of electrolysis products at the electrodes change the resistance of the solution. To eliminate these effects, one uses an alternating current. Platinum electrodes coated with colloidal platinum black are used. The colloidal Pt adsorbs any gases produced during each half cycle of the alternating current. The conductivity cell (surrounded by a constant-T bath) is placed in one arm of a Wheatstone bridge (Fig. 16.10). The resistance R_3 is adjusted until no current flows through the detector between points C and D. These points are then at equal potential. From "Ohm's law" (16.59), we have $|\Delta\phi|_{AD} = I_1 R_1$, $|\Delta\phi|_{AC} = I_3 R_3$, $|\Delta\phi|_{DB} = I_1 R_2$, and $|\Delta\phi|_{CB} = I_3 R$. Since $\phi_D = \phi_C$, we have $|\Delta\phi|_{AC} = |\Delta\phi|_{AD}$ and $|\Delta\phi|_{CB} = |\Delta\phi|_{DB}$. Therefore $I_3 R_3 = I_1 R_1$, and $I_3 R = I_1 R_2$. Dividing the second equation by the first, we get $R/R_3 = R_2/R_1$, from which R can be found. [This discussion is oversimplified, since it ignores the capacitance of the conductivity cell; see J. Braunstein and G. D. Robbins, *J. Chem. Educ.*, **48**, 52 (1971).] R is found to be independent of the magnitude of the applied ac potential difference, so Ohm's law is obeyed.

Once R is known, the conductivity can be calculated from (16.60) and (16.56) as $\kappa = 1/\rho = l/\mathscr{A}R$, where $\mathscr{A}$ and l are the area of and the separation between the electrodes. The *cell constant* K_{cell} is defined as $l/\mathscr{A}$, and $\kappa = K_{\text{cell}}/R$. Instead of measuring l and $\mathscr{A}$, it is more accurate to determine K_{cell} for the apparatus by measuring R for a KCl solution of known conductivity. Accurate κ values for KCl at various concentrations have been determined by measurements in cells of accurately known dimensions.

It is necessary to use extremely pure solvent in conductivity work since traces of

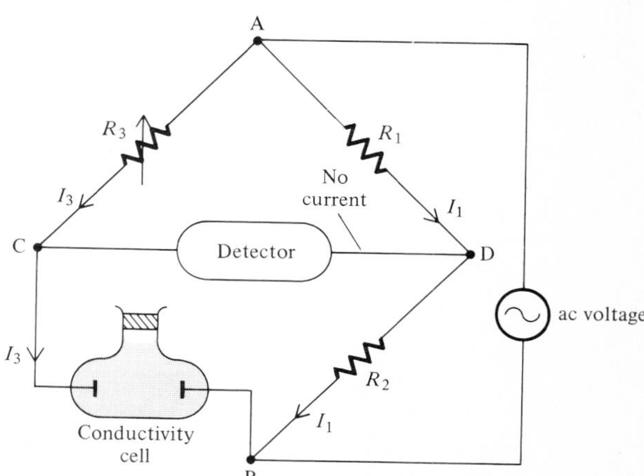

Figure 16.10
Measurement of conductivity of an electrolyte solution.

impurities can significantly affect κ. The conductivity of the pure solvent must be subtracted from that of the solution to get κ of the electrolyte.

Molar conductivity. Since the number of charge carriers per unit volume usually increases with increasing electrolyte concentration, the conductivity κ usually increases as the electrolyte's concentration increases. To get a measure of the current-carrying ability of a given amount of electrolyte, the *molar conductivity* Λ_m of an electrolyte is defined as

$$\Lambda_m \equiv \kappa/c \qquad (16.64)^*$$

where c is the electrolyte's stoichiometric concentration. For example, from the table in the last section, $\kappa = 0.1\ \Omega^{-1}\ \text{cm}^{-1}$ for a 1 mol/dm^3 aqueous KCl solution at 20°C and 1 atm, so for this solution

$$\Lambda_m(\text{KCl}) = \frac{0.1\ \Omega^{-1}\ \text{cm}^{-1}}{1\ \text{mol dm}^{-3}} = \frac{0.1\ \Omega^{-1}\ (10^{-2}\ \text{m})^{-1}}{1\ \text{mol}\ (10^{-1}\ \text{m})^{-3}} = 0.01\ \Omega^{-1}\ \text{m}^2\ \text{mol}^{-1}$$

which also equals $1 \times 10^2\ \Omega^{-1}\ \text{cm}^2\ \text{mol}^{-1}$.

For a strong electrolyte with no ion pairing, the concentration of ions is directly proportional to the stoichiometric concentration of the electrolyte, so it might be thought that dividing κ by c would give a quantity that is independent of concentration. However, Λ_m for NaCl, KBr, etc., in aqueous solution does vary somewhat with concentration. This is because interactions between ions affect the conductivity κ, and these interactions change as c changes. A detailed discussion is given later in this section.

Λ_m depends on the solvent as well as on the electrolyte. We shall consider mainly aqueous solutions.

Values of κ and Λ_m for KCl in aqueous solution at various concentrations at 25°C and 1 atm are:

$c/(\text{mol dm}^{-3})$	0	0.001	0.01	0.1	1
$\kappa/(\Omega^{-1}\ \text{cm}^{-1})$	0	0.000147	0.00141	0.0129	0.112
$\Lambda_m/(\Omega^{-1}\ \text{cm}^2\ \text{mol}^{-1})$	(150)	147	141	129	112

(16.65)

The Λ_m value at zero concentration is obtained by extrapolation. We shall use Λ_m^∞ to denote the infinite-dilution value: $\Lambda_m^\infty = \lim_{c\to 0} \Lambda_m$. Figure 16.11 plots κ and Λ_m vs. c for some electrolytes in aqueous solution.

For the electrolyte $M_{\nu_+}X_{\nu_-}$ yielding the ions M^{z_+} and X^{z_-} in solution, the quantity $\kappa/\nu_+ z_+ c$ is called the *equivalent conductivity* Λ_{eq}. (A solution containing 1 mole of completely dissociated electrolyte would contain $\nu_+ z_+$ moles of positive charge.) Thus, $\Lambda_{eq} = \Lambda_m/\nu_+ z_+$. For example, for $Cu_3(PO_4)_2$, we have $\nu_+ = 3$, $z_+ = 2$, and $\Lambda_{eq} = \Lambda_m/6$. Most tables in the literature list Λ_{eq}. The IUPAC has recommended discontinuing the use of equivalent conductivity, and we shall not use it in this book. (The concept of equivalents serves no purpose except to confuse chemistry students.)

[The subscripts m and eq can be dispensed with if the species to which Λ refers is given in parentheses. Thus, for $CuSO_4$ in water at 25°C and 1 atm, experiment gives $\Lambda_m^\infty = 267.2\ \Omega^{-1}\ \text{cm}^2\ \text{mol}^{-1}$. Since $\nu_+ z_+ = 2$, we have $\Lambda_{eq}^\infty = 133.6\ \Omega^{-1}\ \text{cm}^2$

equiv^{-1}. We therefore write $\Lambda^{\infty}(\mathrm{CuSO_4}) = 267.2\ \Omega^{-1}\ \mathrm{cm^2\ mol^{-1}}$ and $\Lambda^{\infty}(\frac{1}{2}\mathrm{CuSO_4}) = 133.6\ \Omega^{-1}\ \mathrm{cm^2\ mol^{-1}}$.]

Contributions of individual ions to the current. The current in an electrolyte solution must be the sum of the currents carried by the individual ions. Consider a solution with only two kinds of ions, positive ions with charge z_+e, and negative ions with charge z_-e, where e is the proton charge. When a potential difference is applied to the electrodes, the cations feel an electric field E, which accelerates them. The viscous frictional force exerted by the solvent on the ions is proportional to the speed of the ions and opposes their motion; this force increases as the ions are accelerated. When the viscous force balances the electric-field force, the cations are no longer accelerated and travel at a constant terminal speed v_+. We shall later see that the terminal speed is reached in about 10^{-13} s, which is virtually instantaneous.

Let there be N_+ cations in the solution. In time dt, the cations move a distance $v_+\,dt$ and all cations within this distance from the negative electrode will reach that electrode. The number of cations within this distance of the electrode is $(v_+\,dt/l)N_+$ (where l is shown in Fig. 16.9), and the positive charge dQ_+ crossing a plane parallel to the electrodes in time dt is therefore $dQ_+ = (z_+ev_+N_+/l)\,dt$. The current density j_+ due to the cations is $j_+ \equiv I_+/\mathscr{A} = \mathscr{A}^{-1}\,dQ_+/dt$, so $j_+ = z_+ev_+(N_+/V)$, where $V = \mathscr{A}l$ is the solution's volume. Similarly, the anions contribute a current density $j_- = |z_-|ev_-(N_-/V)$, where we adopt the convention that both v_+ and v_- are considered positive. We have $eN_+/V = en_+N_0/V = ec_+N_0 = c_+\mathscr{F}$, where n_+ is the number of moles of the cation M^{z+} in the solution and $c_+ = n_+/V$ is the concentration of M^{z+}. Hence $j_+ = z_+\mathscr{F}v_+c_+$. Similarly, $j_- = |z_-|\mathscr{F}v_-c_-$. The observed current density j is

$$j = j_+ + j_- = z_+\mathscr{F}v_+c_+ + |z_-|\mathscr{F}v_-c_- \tag{16.66}$$

If several kinds of ions are present in the solution, the current density j_B due to ion B and the total current density j are

$$j_B = |z_B|\mathscr{F}v_Bc_B \qquad \text{and} \qquad j = \sum_B j_B = \sum_B |z_B|\mathscr{F}v_Bc_B \tag{16.67}$$

j_B is proportional to the molar charge $z_B\mathscr{F}$, the drift speed v_B, and the concentration c_B.

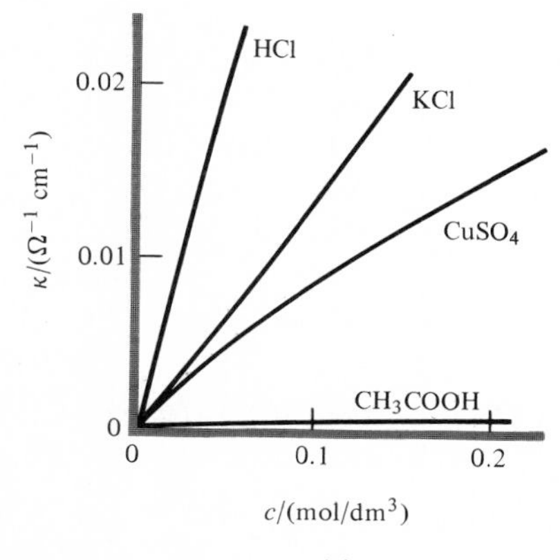

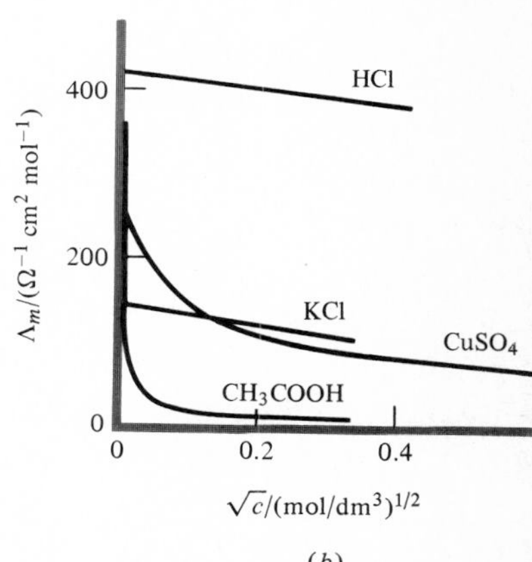

Figure 16.11
(a) Conductivity vs. concentration c for some electrolyte solutions at 25°C and 1 atm. (b) Molar conductivity vs. $c^{1/2}$ for these solutions.

The terminal speed v_B of an ion depends on the electric-field strength, the ion, the solvent, the temperature, the pressure, and the concentrations of all the ions in the solution.

Electric mobilities of ions. Since $j = \kappa E$, the conductivity of an electrolyte solution is [Eq. (16.67)] $\kappa = \sum_B |z_B|\mathscr{F}(v_B/E)c_B$. For a given solution with fixed values of the concentrations c_B, experiment shows Ohm's law to be obeyed, meaning that κ is independent of E. This implies that for fixed concentrations in the solution, each ratio v_B/E is equal to a constant that is characteristic of the ion B but independent of the electric-field strength E. We call this constant the *electric mobility* u_B of ion B:

$$u_B \equiv v_B/E \qquad \text{or} \qquad v_B = u_B E \tag{16.68*}$$

The terminal speed v_B of an ion is proportional to the applied field E, and the proportionality constant is the ion's mobility u_B.

The expression for κ becomes

$$\kappa = \sum_B |z_B|\mathscr{F} u_B c_B \tag{16.69}$$

For a solution with only two kinds of ions,

$$\kappa = z_+ \mathscr{F} u_+ c_+ + |z_-|\mathscr{F} u_- c_- \tag{16.70}$$

Each small portion of the conducting solution must remain electrically neutral, since even a tiny departure from electrical neutrality would generate a huge electric field (Sec. 14.4). Electroneutrality requires that $z_+ ec_+ + z_- ec_- = 0$, or $z_+ c_+ = |z_-|c_-$. Hence (16.70) and (16.66) become for a solution with two kinds of ions

$$\kappa = z_+ \mathscr{F} c_+(u_+ + u_-) \qquad \text{and} \qquad j = z_+ \mathscr{F} c_+(v_+ + v_-) \tag{16.71}$$

Ionic mobilities can be measured by the *moving-boundary method*. Figure 16.12 shows a solution of KCl placed over a solution of $CdCl_2$ in an electrolysis tube of uniform cross-sectional area $\mathscr{A}$. The solutions used must have an ion in common. When the current flows, the K^+ ions migrate upward to the negative electrode, as do the Cd^{2+} ions. For the experiment to work, the cations of the lower solution must have a lower mobility than the cations of the upper solution: $u(Cd^{2+}) < u(K^+)$.

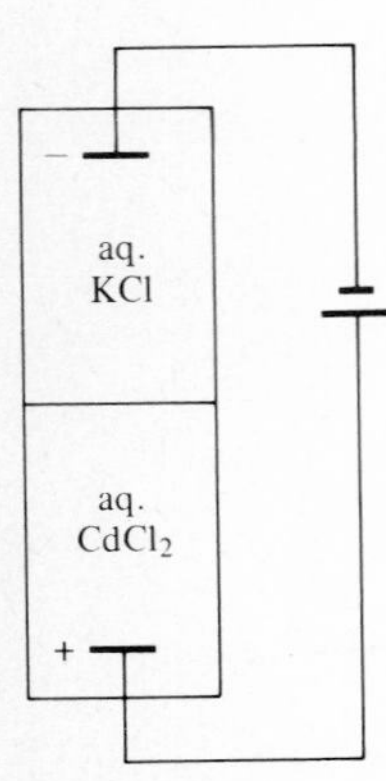

Figure 16.12
Moving-boundary apparatus.

The speed $v(K^+)$ of migration of the K^+ ions is found by measuring the distance x that the boundary moves in time t. (The boundary between the solutions is visible because of a difference in refractive index of the two solutions.) We have $v(K^+) = x/t$. The electric mobility $u(K^+)$ is given by (16.68) as $u(K^+) = v(K^+)/E$. The magnitude of the electric field in the KCl solution [see the discussion preceding (16.58)] is $E = |\Delta\phi|/l$, where $|\Delta\phi|$ is the potential difference across a length l of the KCl solution. Equation (16.58) gives $|\Delta\phi| = \kappa^{-1}lI/\mathscr{A}$, so

$$E = I/\kappa\mathscr{A} \tag{16.72}$$

Therefore

$$u(K^+) = x\kappa\mathscr{A}/It \tag{16.73}$$

where κ is the conductivity of the KCl solution (assumed known). The product It equals the charge Q that flows, and is measured by a coulometer.

Why does the boundary remain sharp, and why does the experiment measure $u(K^+)$ but not $u(Cd^{2+})$? Equation (16.68) gives

$$v(Cd^{2+}) = u(Cd^{2+})E(CdCl_2) \qquad \text{and} \qquad v(K^+) = u(K^+)E(KCl) \tag{16.74}$$

where $E(CdCl_2)$ and $E(KCl)$ are the electric-field strengths in the $CdCl_2$ and KCl solutions. Equation (16.72) gives $E \propto 1/\kappa$; therefore, $v \propto 1/\kappa$, where κ of each solution depends on the concentration of KCl or $CdCl_2$. By hypothesis, $u(Cd^{2+}) < u(K^+)$. Suppose that κ of the $CdCl_2$ solution is sufficiently low to make $v(Cd^{2+}) > v(K^+)$. Equation (16.74) then gives $E(CdCl_2) > E(KCl)$. Because of their higher speed, the Cd^{2+} ions start to move into the KCl solution. But as soon as they enter this solution, they are in a region of lower electric field, and their speed becomes $v(Cd^{2+}) = u(Cd^{2+})E(KCl)$. Since $u(Cd^{2+}) < u(K^+)$, their speed drops below that of K^+, and they are left behind, becoming part of the $CdCl_2$ solution, thereby increasing the $CdCl_2$ concentration just below the boundary; this increases κ and decreases E in this region. Hence $v(Cd^{2+})$ is reduced in the region just below the boundary. The increase in Cd^{2+} concentration just below the boundary must continue until $v(Cd^{2+})$ in this region has been reduced to the point where it equals $v(K^+)$ above the boundary. Thus, the K^+ ions above the boundary and the Cd^{2+} ions in the region immediately below the boundary travel at the same speed, and the boundary is preserved. The concentration of Cd^{2+} just below the boundary automatically adjusts itself (by a sort of feedback mechanism) to the value required to make $v(Cd^{2+})$ equal $v(K^+)$. A concentration gradient therefore develops in the $CdCl_2$ solution. The concentration of $CdCl_2$ and hence κ in the region just below the boundary is unknown, so only $u(K^+)$ can be found from the experiment.

If, on the other hand, the initial concentrations are such as to make $v(Cd^{2+}) < v(K^+)$, the Cd^{2+} ions will lag behind the K^+ ions, thereby decreasing the Cd^{2+} concentration just below the boundary; this decreases κ, increases E, and increases $v(Cd^{2+})$ in this region. The decrease in $c(Cd^{2+})$ continues until $v(Cd^{2+})$ has become equal to $v(K^+)$.

Readers should convince themselves that if the electric mobility of Cd^{2+} were greater than that of K^+, the automatic adjustment of $v(Cd^{2+})$ would not occur and the boundary would not be preserved.

To measure $u(Cl^-)$, we could use solutions of KCl and KNO_3.

To prevent convection currents, the lower solution should have a higher density than the upper solution.

Some observed mobilities as a function of electrolyte concentration for K^+ and Cl^- ions in aqueous KCl solutions at 25°C and 1 atm are:

$c/(\text{mol dm}^{-3})$	0	0.01	0.10	0.20	1.0
$10^5 u(K^+)/(\text{cm}^2\ \text{V}^{-1}\ \text{s}^{-1})$	(76.2)	71.8	65.4	62.9	56.6
$10^5 u(Cl^-)/(\text{cm}^2\ \text{V}^{-1}\ \text{s}^{-1})$	(79.1)	74.6	68.2	65.6	59.3

The values at $c = 0$ are extrapolations. Note the decrease in u of each ion as c increases. Each ion is surrounded by an "atmosphere" in which ions of opposite charge predominate. This atmosphere reduces the mobility of an ion, and the density of the atmosphere increases as c increases. (See also the discussion below of the Onsager equation.)

For a 0.20 mol/dm³ aqueous NaCl solution at 25°C and 1 atm, one finds $u(Cl^-) = 65.1 \times 10^{-5}\ \text{cm}^2\ \text{V}^{-1}\ \text{s}^{-1}$. This value differs slightly from the above-listed

$u(Cl^-)$ value in a 0.20 mol/dm^3 KCl solution, due to slight differences in Na^+–Cl^- interactions compared with K^+–Cl^- interactions.

Experimental electric mobilities extrapolated to infinite dilution for ions in water at 25°C and 1 atm are:

Ion	H_3O^+	Li^+	Na^+	Mg^{2+}	OH^-	Cl^-	Br^-	NO_3^-
$10^5u^\infty/(\text{cm}^2\ \text{V}^{-1}\ \text{s}^{-1})$	363	40.1	51.9	55.0	205	79.1	81.3	74.0

Since interionic forces vanish at infinite dilution, $u^\infty(Na^+)$ is the same for solutions of NaCl, Na_2SO_4, etc.

For small inorganic ions, u^∞ in aqueous solutions at 25°C and 1 atm usually lies in the range 40 to 80 × 10^{-5} cm^2 V^{-1} s^{-1}. However, H_3O^+ and OH^- show abnormally high mobilities in aqueous solution. These high mobilities are attributed to a special *jumping* mechanism that operates in addition to the usual motion through the solvent. A proton from an H_3O^+ ion can jump to a neighboring H_2O molecule, a process which has the same effect as the motion of H_3O^+ through the solution:

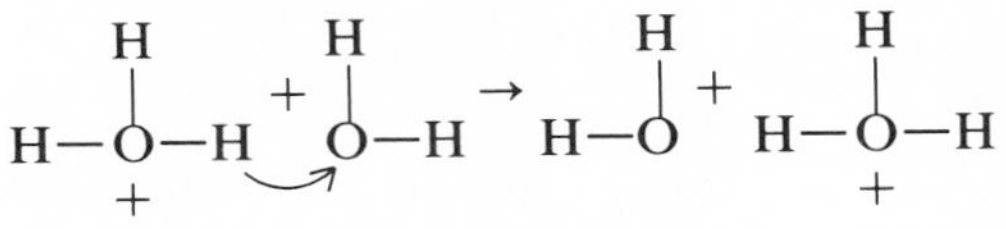

$$\mathrm{H{-}\overset{\displaystyle H}{\underset{+}{O}}{-}H \;\; \overset{\displaystyle H}{O}{-}H \;\rightarrow\; H{-}\overset{\displaystyle H}{O} \;\; H{-}\overset{\displaystyle H}{\underset{+}{O}}{-}H}$$

The high mobility of OH^- is due to a transfer of a proton from an H_2O molecule to an OH^- ion, a process which is equivalent to the motion of OH^- in the opposite direction:

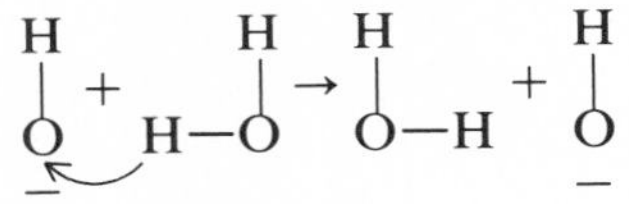

$$\mathrm{\overset{\displaystyle H}{\underset{-}{O}} \;\; H{-}\overset{\displaystyle H}{O} \;\rightarrow\; \overset{\displaystyle H}{O}{-}H \;\; \overset{\displaystyle H}{\underset{-}{O}}}$$

Example A typical electric-field strength for an electrolysis experiment is 10 V/cm. Calculate v for Mg^{2+} ions in this field in water at 25°C and 1 atm.

Equation (16.68) and the above table give

$$v = uE = (55 \times 10^{-5}\ \text{cm}^2\ \text{V}^{-1}\ \text{s}^{-1})(10\ \text{V/cm}) = 0.0055\ \text{cm/s}$$

In the absence of the field, the Mg^{2+} ions have an average translational kinetic energy of $\frac{3}{2}kT = \frac{1}{2}m\langle v^2\rangle$. Hence the rms speed of random thermal motion of the ions is $v_{\text{rms}} = (3RT/M)^{1/2}$, which equals 5×10^4 cm/s for Mg^{2+} at 25°C. The speed of migration toward the electrode is far, far smaller than the speed of random motion.

Ionic mobilities at infinite dilution can be estimated theoretically from the following approximate treatment. At extremely high dilution, interionic forces are negligible, so the only electric force an ion experiences is due to the applied electric field E. From (14.3), the electric force on an ion with charge $z_B e$ has the magnitude $|z_B|eE$. This force is opposed by the frictional force, whose magnitude is fv_B^∞, where f is the friction coefficient (Sec. 16.4). When the terminal speed has been reached, the electric and frictional forces balance: $|z_B|eE = fv_B^\infty$, and the terminal speed is $v_B^\infty = |z_B|eE/f$. The infinite-dilution mobility $u_B^\infty = v_B^\infty/E$ is then

$$u_B^\infty = |z_B|e/f \tag{16.75}$$

A rough estimate of the friction coefficient f can be obtained by assuming that the solvated ions are spherical and that Stokes' law (16.23) applies to their motion through the solvent. (Because the ions are solvated, they are substantially larger than the solvent molecules.) Stokes' law gives $f = 6\pi\eta r_B$, and

$$u_B^\infty \approx \frac{|z_B|e}{6\pi\eta r_B} \tag{16.76}$$

Equation (16.76) attributes the differences in infinite-dilution mobilities of ions entirely to the differences in their radii. (Of course, this equation can't be used for H_3O^+ or OH^-.)

The smaller values of u^∞ for cations than anions (H_3O^+ and OH^- excepted) indicate that cations are more hydrated than anions. The smaller size of cations produces a more intense electric field surrounding them, and they therefore hold on to more H_2O molecules than anions. The number of water molecules that move with an ion in solution is called the *primary hydration number* n_h of the ion. Some values of n_h estimated using electric mobilities and other methods are (*Bockris and Reddy*, p. 131):

Ion	Li^+	Na^+	K^+	F^-	Cl^-	Br^-	I^-
n_h	5	4	3	4	2	2	1

These values are estimated to be correct to ± 1.

Aside from the approximation of using Stokes' law, it is difficult to use (16.76) to predict u values because the radius r_B of the solvated ion is not accurately known. What is often done is to use (16.76) to calculate r_B from u_B^∞. For example, use of the above-listed mobility of Li^+ and the 25°C $\eta(H_2O)$ value in (16.101) gives

$$r(Li^+) \approx \frac{1(1.6 \times 10^{-19}\text{ C})}{6\pi(0.89 \times 10^{-3}\text{ kg s}^{-1}\text{ m}^{-1})[40 \times 10^{-5}(10^{-2}\text{ m})^2\text{ V}^{-1}\text{ s}^{-1}]}$$

$$\approx 2.4 \times 10^{-10}\text{ m} = 2.4\text{ Å}$$

A similar calculation gives $r(Na^+) \approx 1.8$ Å. The larger size of Li^+ can be attributed to its larger n_h value.

Stokes' law can be used to estimate how long it takes an ion to reach its terminal speed after the electric field is applied. From (16.76), the terminal speed equals $|z|eE/6\pi\eta r$. The force due to the electric field is $|z|eE$. If we neglect the frictional resistance, Newton's second law $F = ma = m\,dv/dt$ gives $|z|eE \approx m\,dv/dt$, which integrates to $v \approx |z|eEt/m$. Setting v equal to the terminal speed, we get $|z|eEt/m \approx |z|eE/6\pi\eta r$. The time needed to attain the terminal speed is then $t \approx m/6\pi\eta r$. For $m = 10^{-22}$ g, $\eta = 10^{-2}$ g s^{-1} cm^{-1}, and $r = 10^{-8}$ cm, we get $t \approx 10^{-13}$ s. (Since we neglected F_{fr}, the actual time required is somewhat longer.)

Transport numbers. The *transport number* (or *transference number*) t_B of ion B is defined as the fraction of the current that it carries:

$$t_B \equiv j_B/j \tag{16.77}^*$$

where the current density j_B of ion B and the total current density j are given by (16.67). Use of (16.67), (16.55), and (16.68) gives $t_B = |z_B|\mathscr{F}v_B c_B/\kappa E = |z_B|\mathscr{F}u_B c_B/\kappa$:

$$t_B = |z_B|\mathscr{F}c_B u_B/\kappa \qquad (16.78)$$

Thus, the transport number of an ion can be calculated from its mobility. The sum of the transport numbers of all the ionic species in solution must be 1.

For a solution containing only two kinds of ions, (16.77), (16.67), the electroneutrality condition $z_+c_+ = |z_-|c_-$, and (16.68) give

$$t_+ = \frac{j_+}{j_+ + j_-} = \frac{v_+}{v_+ + v_-} = \frac{u_+}{u_+ + u_-}, \qquad t_- = \frac{j_-}{j_+ + j_-} = \text{etc.} \qquad (16.79)$$

Transport numbers can be measured by the *Hittorf method* (Fig. 16.13). After electrolysis has proceeded for a while, one drains the solutions in each of the three compartments and analyzes them. The results allow t_+ and t_- to be found.

Figure 16.14 shows what happens in the electrolysis of $Cu(NO_3)_2$ with a Cu anode and an inert cathode. Let a total charge Q flow during the experiment. Then $Q/\mathscr{F}$ moles of electrons flow. The anode reaction is $Cu \rightarrow Cu^{2+}(aq) + 2e^-$, so $Q/2\mathscr{F}$ moles of Cu^{2+} enter the right compartment R from the anode. The total number of moles of charge on the ions that pass plane B during the experiment is $Q/\mathscr{F}$. The Cu^{2+} ions carry a fraction t_+ of the current, and the charge on the Cu^{2+} ions moving from R to M during the experiment is t_+Q. Hence, $t_+Q/2\mathscr{F}$ moles of Cu^{2+} pass out of R during the experiment. The overall change in the number of moles of Cu^{2+} in compartment R is $Q/2\mathscr{F} - t_+Q/2\mathscr{F}$:

$$\Delta n_R(Cu^{2+}) = (1 - t_+)Q/2\mathscr{F} = t_-Q/2\mathscr{F} \qquad (16.80)$$

Since NO_3^- carries a fraction t_- of the current, the charge on the nitrate ions moving from M to R during the experiment is t_-Q, and $t_-Q/\mathscr{F}$ moles of NO_3^- move into R:

$$\Delta n_R(NO_3^-) = t_-Q/\mathscr{F} \qquad (16.81)$$

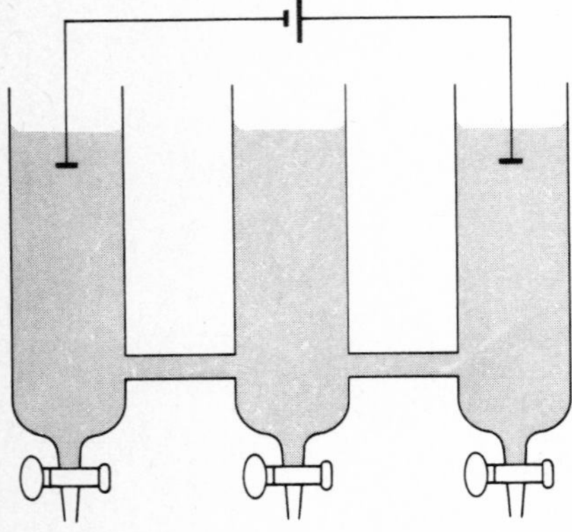

Figure 16.13
Hittorf apparatus.

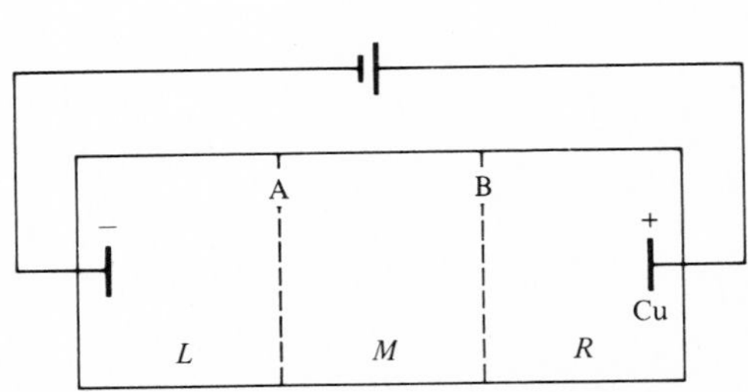

Figure 16.14
Electrolysis of $Cu(NO_3)_2$ with a Cu anode and an inert cathode.

Note that (16.80) and (16.81) are consistent with the requirement that R remain electrically neutral.

The charge on the Cu^{2+} passing plane A must equal that on the Cu^{2+} passing plane B. Hence $t_+ Q/2\mathscr{F}$ moles of Cu^{2+} move from M to L during the experiment, and there is no change in the number of moles of Cu^{2+} in M. Similarly for NO_3^-:

$$\Delta n_M(Cu^{2+}) = 0, \qquad \Delta n_M(NO_3^-) = 0 \tag{16.82}$$

The number of moles of NO_3^- passing through plane B equals the number passing through plane A, so $t_- Q/\mathscr{F}$ moles of NO_3^- move out of L into M:

$$\Delta n_L(NO_3^-) = -t_- Q/\mathscr{F} \tag{16.83}$$

We have $t_+ Q/2\mathscr{F}$ moles of Cu^{2+} moving from M into L. Also, the number of moles of Cu that plate out on the cathode must equal the number that were oxidized at the anode, namely $Q/2\mathscr{F}$. Hence

$$\Delta n_L(Cu^{2+}) = (t_+ - 1)Q/2\mathscr{F} = -t_- Q/2\mathscr{F} \tag{16.84}$$

The charge Q is measured with a coulometer, and chemical analysis gives the Δn's. Hence t_+ and t_- can be found from the above equations.

Note from Eqs. (16.80) to (16.84) that the concentration of $Cu(NO_3)_2$ increases in the region around the anode (compartment R), remains the same in the middle compartment, and decreases around the cathode (compartment L). The resulting concentration gradients cause diffusion. If significant diffusion occurs between compartments, this causes error in the results; however, diffusion in solution is slow.

Another source of error is mixing due to convection currents caused by heat produced by the current and by density changes accompanying concentration changes. Still another source of error is the concentration changes resulting from the water of hydration carried by the ions between compartments. With care, the method can give accurate results.

Since the mobilities u_+ and u_- depend on concentration and do not necessarily change at the same rate as c changes, the transport numbers t_+ and t_- depend on concentration. Some values for aqueous LiCl solutions at 25°C and 1 atm are:

$c/(\text{mol dm}^{-3})$	0	0.01	0.02	0.1	0.2	1
$t(Li^+)$	0.337	0.329	0.326	0.317	0.311	0.287
$t(Cl^-)$	0.663	0.671	0.674	0.683	0.689	0.713

The infinite-dilution values $t^\infty(Li^+)$ and $t^\infty(Cl^-)$ are found by extrapolation.

Observed t^∞ values lie between 0.3 and 0.7 for most ions. H_3O^+ and OH^- have unusually high t^∞ values in aqueous solution, due to their high mobilities. Some values for aqueous solutions at 25°C and 1 atm are

$$\begin{aligned} &\text{HCl:} && t^\infty(H^+) = 0.82, && t^\infty(Cl^-) = 0.18 \\ &\text{KCl:} && t^\infty(K^+) = 0.49, && t^\infty(Cl^-) = 0.51 \\ &\text{CaCl}_2\text{:} && t^\infty(Ca^{2+}) = 0.44, && t^\infty(Cl^-) = 0.56 \end{aligned} \tag{16.85}$$

In the preceding discussion, we considered only the ions Cu^{2+} and NO_3^-. However, in a $Cu(NO_3)_2$ solution that is not extremely dilute, there is a significant concentration of $Cu(NO_3)^+$ ion pairs, and these must carry part of the current. When a compartment is chemically analyzed for Cu^{2+}, it is the total amount of Cu present in solution that is determined, and individual amounts present as Cu^{2+} and $Cu(NO_3)^+$ are not found. Because of this complication, the values t_+ and t_- obtained in the Hittorf method are not, strictly speaking, the transport numbers of the actual ions. Instead they are what are called *ion-constituent transport numbers*. The *ion-constituent* Cu(II) exists in a $Cu(NO_3)_2$ solution as Cu^{2+} ions and as $Cu(NO_3)^+$ ions. Similarly, in the moving-boundary method, one obtains mobilities and transport numbers of ion-constituents. At infinite dilution, there is no ionic association, so t^∞ and u^∞ values do apply to the ions Cu^{2+} and $NO_3{}^-$. [See M. Spiro, *J. Chem. Educ.*, **33**, 464 (1956); M. Spiro in A. Weissberger and B. W. Rossiter, (eds.), *Physical Methods of Chemistry*, pt. IIA, chap. IV, Wiley-Interscience, 1971.]

Molar conductivities of ions. The molar conductivity of an electrolyte in solution is defined by (16.64) as $\Lambda_m \equiv \kappa/c$. By analogy, we define the *molar conductivity* $\lambda_{m,\mathrm{B}}$ of ion B as

$$\lambda_{m,\mathrm{B}} \equiv \kappa_\mathrm{B}/c_\mathrm{B} \tag{16.86*}$$

where κ_B is the contribution of ion B to the solution's conductivity and c_B is its concentration. (Note, however, that c_B is the actual concentration of ion B in the solution, whereas c is the stoichiometric concentration of the electrolyte.) Dividing the first equation in (16.67) by E and using (16.55), (16.68), and (16.86), we have

$$\kappa_\mathrm{B} = |z_\mathrm{B}|\mathscr{F}u_\mathrm{B}c_\mathrm{B} \tag{16.87}$$

$$\lambda_{m,\mathrm{B}} = |z_\mathrm{B}|\mathscr{F}u_\mathrm{B} \tag{16.88*}$$

The molar conductivity of an ion can therefore be found from its mobility. (The *equivalent conductivity* of ion B is defined as $\lambda_{\mathrm{eq,B}} = \lambda_{m,\mathrm{B}}/|z_\mathrm{B}| = \mathscr{F}u_\mathrm{B}$.)

Substitution of (16.64) and (16.86) in $\kappa = \sum_\mathrm{B}\kappa_\mathrm{B}$ gives

$$\Lambda_m = \frac{1}{c}\sum_\mathrm{B} c_\mathrm{B}\lambda_{m,\mathrm{B}} \tag{16.89}$$

which relates Λ_m of the electrolyte to the λ_m's of the ions. For a strong electrolyte $M_{\nu_+}X_{\nu_-}$ that is completely dissociated, (16.89) becomes

$$\Lambda_m = c^{-1}(c_+\lambda_{m,+} + c_-\lambda_{m,-}) = c^{-1}(\nu_+ c\lambda_{m,+} + \nu_- c\lambda_{m,-})$$

$$\Lambda_m = \nu_+\lambda_{m,+} + \nu_-\lambda_{m,-} \qquad \text{strong electrolyte, no ion pairs} \tag{16.90}$$

For example, $\Lambda_m(MgCl_2) = \lambda_m(Mg^{2+}) + 2\lambda_m(Cl^-)$, provided there are no ion pairs. For a weak acid HX whose degree of dissociation is α, Eq. (16.89) gives

$$\Lambda_m = c^{-1}(c_+\lambda_{m,+} + c_-\lambda_{m,-}) = c^{-1}(\alpha c\lambda_{m,+} + \alpha c\lambda_{m,-})$$

$$\Lambda_m = \alpha(\lambda_{m,+} + \lambda_{m,-}) \qquad \text{for 1 : 1 weak electrolyte} \tag{16.91}$$

For a weak acid in water, $\alpha^\infty \neq 1$ (Sec. 11.3); hence, $\Lambda_m^\infty \neq \lambda_{m,+}^\infty + \lambda_{m,-}^\infty$ for the weak acid HX in water.

Since the mobility $u(Cl^-)$ in an NaCl solution differs slightly from $u(Cl^-)$ in a KCl solution at nonzero concentrations, $\lambda_m(Cl^-)$ in NaCl and KCl solutions differ. However, in the limit of infinite dilution, interionic forces go to zero and the ions move independently. Hence $\lambda_m^\infty(\mathrm{Cl}^-)$ is the same for all chloride salts. Some data for $\lambda_m(\mathrm{Cl}^-)$ in units of Ω^{-1} cm^2 mol^{-1} for two salts in aqueous solutions at 25°C and 1 atm are:

c/(mol dm^{-3})	0	0.02	0.05	0.10	0.20
KCl(*aq*)	76.3	70.5	68.0	65.8	63.3
NaCl(*aq*)	76.3	70.5	67.9	65.6	62.8

Some λ^∞ values in water at 25°C and 1 atm are:

Cation	H_3O^+	NH_4^+	K^+	Na^+	Ag^+	Ca^{2+}	Mg^{2+}
$\lambda_m^\infty/(\Omega^{-1}$ cm^2 mol$^{-1})$	350	73.6	73.5	50.1	61.9	118.9	106.1

Anion	OH^-	Br^-	Cl^-	NO_3^-	CH_3COO^-	SO_4^{2-}
$\lambda_m^\infty/(\Omega^{-1}$ cm^2 mol$^{-1})$	198	78.4	76.3	71.4	40.8	160

The $|z_B|$ factor in (16.88) tends to make λ_m for $+2$ and -2 ions larger than for $+1$ and -1 ions.

From tabulated λ_m^∞ values, we can calculate Λ_m^∞ for a strong electrolyte as [Eq. (16.90)]:

$$\Lambda_m^\infty = \nu_+ \lambda_{m,+}^\infty + \nu_- \lambda_{m,-}^\infty \qquad \text{strong electrolyte} \qquad (16.92)$$

Mobilities u^∞ can be calculated from λ_m^∞ values using (16.88). From the u^∞ values, t^∞ can be calculated using (16.78).

Concentration dependence of molar conductivities. Some Λ_m data for NaCl and $HC_2H_3O_2$ in water at 25°C and 1 atm are:

c/(mol dm^{-3})	0	10^{-4}	10^{-3}	10^{-2}	10^{-1}
Λ_m(NaCl)/(Ω^{-1} cm^2 mol^{-1})	126.4	125.5	123.7	118.4	106.7
Λ_m(CH_3COOH)/(Ω^{-1} cm^2 mol^{-1})		134.6	49.2	16.2	5.2

Equation (16.89) shows that the molar conductivity $\Lambda_m = \kappa/c$ of an electrolyte changes with electrolyte concentration for two reasons: (*a*) The ionic concentrations c_B may not be proportional to the electrolyte stoichiometric concentration c, and (*b*) the ionic molar conductivities $\lambda_{m,B}$ change with concentration.

The sharp increase in Λ_m of a weak acid like acetic acid as c goes to zero (Fig. 16.11) is due mainly to the rapid increase in the degree of dissociation as c goes to zero; see Eq. (16.91). (This rapid increase in Λ_m makes extrapolation to $c = 0$ very difficult for weak electrolytes.) For strong electrolytes other than 1 : 1 electrolytes, part of the decrease of Λ_m with increasing c is due to formation of ion pairs, which reduces the ionic concentrations. However, even 1 : 1 electrolytes, which do not show significant ion pairing in water, show a decrease in Λ_m as c increases. This decrease arises from interionic forces. (One finds that for a strong electrolyte, a plot of Λ_m vs. $c^{1/2}$ is linear in the region of very high dilution, and this allows reliable extrapolation to $c = 0$.)

From (16.88), the ionic molar conductivity $\lambda_{m,\mathrm{B}}$ equals $|z_\mathrm{B}|\mathscr{F}u_\mathrm{B}$. If the mobility $u_\mathrm{B} = v_\mathrm{B}/E$ were independent of concentration, λ_m would be independent of c. However, the ion drift speed v_B depends on c for two reasons:

1. Each ion is surrounded in solution by an "atmosphere" in which ions of opposite charge predominate. Consider a positive ion. When the electric field is turned on, the positive ion moves one way and the negative ions in its atmosphere move the opposite way. The positive ion thus tends to move out of its atmosphere; the atmosphere is continually re-forming from new negative ions, but it takes a finite time to re-form the atmosphere, and this time lag between the motion of an ion and re-formation of its atmosphere produces an asymmetric distribution of the ions in the atmosphere (Fig. 16.15). The electrical attraction between the asymmetrical atmosphere and the ion reduces the speed of the ion. As c increases, the density of ions in the atmosphere increases and the ionic mobility is further reduced by this *asymmetry* (or *relaxation*) *effect*.

2. The viscous force opposing an ion's motion equals $fv_{\mathrm{B,rel}}$, where f is the friction coefficient and $v_{\mathrm{B,rel}}$ is the ion's speed relative to the solvent. The ions in the atmosphere are moving in the direction opposite the central ion and are carrying with them their molecules of solvation. Therefore, the solvent in the immediate vicinity of the central ion has a net motion in the direction opposite the ion's motion, and the ion's speed relative to the solvent is greater than its speed relative to the electrodes. This so-called *electrophoretic effect* retards the motion of ions.

The asymmetry effect decreases the electrical force on an ion, and the electrophoretic effect increases the viscous force. Both these effects decrease the drift speed v_B and the mobility u_B. Hence, $\lambda_{m,\mathrm{B}}$ and Λ_m decrease as c increases.

Debye and Hückel applied their theory of ionic interactions to calculate the electric mobility of ions in solution. Their treatment was improved by Onsager in 1927. The resulting *(Debye–Hückel–) Onsager limiting law* allows for the asymmetry and electrophoretic effects. For the special case of an electrolyte yielding two kinds of ions with $z_+ = |z_-|$, the Onsager equation for $\lambda_{m,+}$ is

$$\lambda_{m,+} = \lambda_{m,+}^{\infty} - \left[\frac{(2.92 \times 10^{-4})z_+^{\ 3}}{\eta(\varepsilon_r T/\mathrm{K})^{1/2}} \frac{\mathrm{C}^2}{\mathrm{mol\ m}} + \frac{5.80 z_+^{\ 3}\lambda_{m,+}^{\infty}}{10^{-5}(\varepsilon_r T/\mathrm{K})^{3/2}}\right]\left(\frac{2c_+}{c^\circ}\right)^{1/2}$$

where η and ε_r are the solvent's viscosity and dielectric constant, c_+ is the concentration of M^{z+}, and $c^\circ \equiv 1\ \mathrm{mol/dm^3}$ (for the derivation and the general form when $z_+ \neq |z_-|$, see *Eyring, Henderson, and Jost*, vol. IXA, chap. 1). The expression for $\lambda_{m,-}$ is obtained by replacing z_+ with $|z_-|$ and $\lambda_{m,+}^{\infty}$ with $\lambda_{m,-}^{\infty}$.

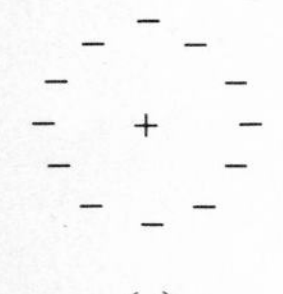
(*a*)

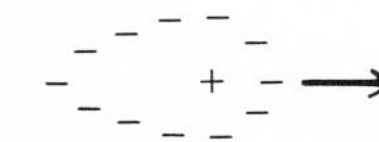
(*b*)

Figure 16.15
Ion atmosphere (*a*) without and (*b*) with electric field. The arrow shows the direction of motion of the cation.

Substitution of $\eta = 0.000890$ kg s^{-1} m^{-1}, $\varepsilon_r = 78.4$, and $T/\text{K} = 298.15$ for water at 25°C and 1 atm gives

$$\lambda_{m,+} = \lambda_{m,+}^{\infty} - (30.3z_+^3\ \Omega^{-1}\ \text{cm}^2\ \text{mol}^{-1} + 0.230z_+^3\lambda_{m,+}^{\infty})(c_+/c^\circ)^{1/2}$$
$$z_+ = |z_-| \qquad \text{for H}_2\text{O at 25°C} \tag{16.93}$$

For $z_+ = |z_-|$, we have $c_+ = c_-$, and substitution of (16.93) and the similar equation for $\lambda_{m,-}$ in (16.89) gives for H_2O at 25°C

$$\Lambda_m = (c_+/c)\{(\lambda_{m,+}^{\infty} + \lambda_{m,-}^{\infty}) - [az_+^3 + bz_+^3(\lambda_{m,+}^{\infty} + \lambda_{m,-}^{\infty})](c_+/c^\circ)^{1/2}\}$$
$$\text{where } a \equiv 60.6\ \Omega^{-1}\ \text{cm}^2\ \text{mol}^{-1},\ b \equiv 0.230,\ z_+ = |z_-| \tag{16.94}$$

For a strong electrolyte with no ion pairing, we have $c_+ = c$, and (16.94) becomes

$$\Lambda_m = \Lambda_m^{\infty} - (az_+^3 + bz_+^3\Lambda_m^{\infty})(c/c^\circ)^{1/2} \qquad \text{where } z_+ = |z_-| \tag{16.95}$$

where (16.92) was used. Note the $c^{1/2}$ dependence, in agreement with experimental data for strong electrolytes.

The Onsager equation is found to be well obeyed by solutions of 1 : 1 electrolytes with c_+ less than 0.002 mol/dm^3. The equation is also well obeyed by very dilute solutions of higher-valency electrolytes if ion pairing is taken into account in calculating c_+ and c_-.

Applications of conductivity. Substitution of (16.86) into $\kappa = \sum_B \kappa_B$ gives

$$\kappa = \sum_B \lambda_{m,B} c_B \tag{16.96}$$

Measurement of κ allows the endpoint of a titration to be found, since the plot of κ vs. volume of added reagent changes slope at the endpoint. For example, if an aqueous HCl solution is titrated with NaOH, κ decreases before the endpoint because H_3O^+ ions are being replaced by Na^+ ions, and κ increases after the endpoint due to the increase in Na^+ and OH^- concentrations.

Conductivity measurements can be used to follow the concentration changes during a chemical reaction between ions in solution, allowing the reaction rate to be followed.

Conductivity measurements can be used to determine certain ionic equilibrium constants. Consider a very dilute solution of the electrolyte MX in which there is an ionic equilibrium. Using the measured value of $\Lambda_m = \kappa/c$, we can solve the Onsager equation (16.94) for the ionic concentration c_+. From c_+, the electrolyte stoichiometric concentration c, and activity coefficients calculated using the Debye–Hückel equation (10.92), the ionic equilibrium constant K_c can be found (see Chap. 11). This method has been used to determine dissociation constants of weak acids, solubility-product constants of slightly soluble salts, the ionization constant of water, and association constants for ion-pair formation.

It is convenient to use (16.64) to rewrite (16.94) in the form

$$\kappa = c_+[\lambda_{m,+}^{\infty} + \lambda_{m,-}^{\infty} - S(c_+/c^\circ)^{1/2}] \qquad \text{where } z_+ = |z_-| \tag{16.97}$$

$$S \equiv az_+^3 + bz_+^3(\lambda_{m,+}^{\infty} + \lambda_{m,-}^{\infty}) \tag{16.98}$$

where S incorporates the Onsager corrections to the conductivity. Equation (16.97) is a cubic equation in $c_+^{1/2}$. For hand calculations, it is fastest to solve it by successive approximations. We rewrite the equation as

$$c_+ = \frac{\kappa}{\lambda_{m,+}^\infty + \lambda_{m,-}^\infty - S(c_+/c^\circ)^{1/2}} \qquad \text{where } z_+ = |z_-| \tag{16.99}$$

At the high dilutions to which the Onsager equation applies, the interionic-forces correction term $S(c_+/c^\circ)^{1/2}$ is much less than $\lambda_{m,+}^\infty + \lambda_{m,-}^\infty$, so as an initial approximation we can set $c_+ = 0$ in the denominator on the right of (16.99). Equation (16.99) is then used to calculate an improved value of c_+, which is then substituted in the right side of (16.99) to find a further improved c_+ value. The calculation is repeated until the answer converges.

Problems 16.42 to 16.45 outline applications of (16.99) to ionic equilibria.

To use (16.99), $\lambda_{m,+}^\infty$ and $\lambda_{m,-}^\infty$ must be known. They are found by extrapolation of mobility or transport-number measurements on strong electrolytes, as discussed earlier. See also Probs. 16.35 and 16.36.

Temperature dependence of molar conductivities. Molar conductivities of electrolytes and ions generally increase with temperature.

The temperature dependence of λ_m^∞ can be estimated by substituting the approximate equation (16.76) in (16.88) to give $\lambda_{m,\text{B}}^\infty \approx z_\text{B}{}^2\mathscr{F}e/6\pi\eta r_\text{B}$, or $\ln \lambda_{m,\text{B}}^\infty \approx \ln (z_\text{B}{}^2\mathscr{F}e/6\pi r_\text{B}) - \ln \eta$. Differentiation with respect to T gives

$$\frac{d \ln \lambda_m^\infty}{dT} \approx -\frac{d \ln \eta}{dT} = -\frac{1}{\eta}\frac{d\eta}{dT} \tag{16.100}$$

Viscosity values for water at 1 atm are:

Temp.	24°C	25°C	26°C
η/cP	0.9111	0.8904	0.8705

(16.101)

Over a short temperature interval, we have $d\eta/dT \approx \Delta\eta/\Delta T$. Hence

$$\frac{1}{\eta}\frac{d\eta}{dT} \approx \frac{1}{0.8904}\frac{0.8705 - 0.9111}{2 \text{ K}} = -0.023 \text{ K}^{-1} \qquad \text{for H}_2\text{O at 25°C} \tag{16.102}$$

Therefore (16.100) predicts $d \ln \lambda_m^\infty/dT \approx 0.023$ K^{-1} in water at 25°C. This compares with experimental values of 0.018 to 0.022 K^{-1} for most ions in water at 25°C, so the agreement is reasonably good. The ions H_3O^+ and OH^- are exceptional, having values of 0.014 and 0.016 K^{-1}, respectively. As noted earlier, these ions have a special mechanism of transport in water.

Liquid-junction potentials. We saw in Sec. 14.10 that $\mathscr{E}_J$, the potential difference between two electrolyte solutions 1 and 2 in contact, results from different rates of diffusion of the ions. Equations (16.48) and (16.75) show that the diffusion coefficient of ion B is proportional to its electric mobility u_B; also, u_B is proportional to the ionic molar conductivity $\lambda_{m,\text{B}}$ [Eq. (16.88)]. Hence, we expect that $\mathscr{E}_J$ will be a function of the u_B's or the $\lambda_{m,\text{B}}$'s. The theoretical treatment of $\mathscr{E}_J$ is fairly complicated, and we shall omit any derivations (see *Ives and Janz*, pp. 48–56). The accurate theoretical equation for $\mathscr{E}_J$ turns out to involve the activity coefficients of single ions, which are not readily measurable. By omitting the activity coefficients and assuming that the molality of each ion varies linearly with distance in the transition layer between the two bulk solutions 1 and 2, one obtains the *Henderson equation*

$$\mathscr{E}_J \equiv \phi_2 - \phi_1 \approx \frac{RT}{\mathscr{F}} \frac{\sum_B \lambda_{m,B} z_B^{-1}(m_{B,2} - m_{B,1})}{\sum_B \lambda_{m,B}(m_{B,2} - m_{B,1})} \ln \frac{\sum_B \lambda_{m,B} m_{B,1}}{\sum_B \lambda_{m,B} m_{B,2}} \tag{16.103}$$

where $m_{B,1}$ and $m_{B,2}$ are the molalities of ion B in the bulk solutions 1 and 2 and z_B is the charge number of ion B. The sums go over all ions in the solutions. The molar conductivities $\lambda_{m,B}$ in (16.103) are some sort of average values for the molality range $m_{B,1}$ to $m_{B,2}$, but since (16.103) is approximate, we might as well use the infinite-dilution values $\lambda_{m,B}^{\infty}$.

16.7 FLUXES AND GENERALIZED FORCES

A unified view of transport processes can be obtained by defining the *flux* J_W of the physical quantity W being transported as the rate of transport of W through unit area perpendicular to the direction of flow:

$$J_W \equiv \frac{1}{\mathscr{A}} \frac{dW}{dt}$$

For example, the heat flux J_q in thermal conduction, the matter flux $J_{n\text{A}}$ in diffusion, the momentum flux J_{p_y} in viscous fluid flow, and the charge flux J_Q in electrical conduction are

$$J_q = \frac{1}{\mathscr{A}} \frac{dq}{dt}, \qquad J_{n\text{A}} = \frac{1}{\mathscr{A}} \frac{dn_\text{A}}{dt}, \qquad J_{p_y} = \frac{1}{\mathscr{A}} \frac{dp_y}{dt}, \qquad J_Q = \frac{1}{\mathscr{A}} \frac{dQ}{dt}$$

Each of these fluxes is produced by a gradient of a physical quantity. The negative of each of these gradients is called a *generalized force* X. (Of course, X is not a force in the usual sense of the word.) The generalized forces for the above processes are

$$X_q \equiv -\frac{dT}{dx}, \qquad X_{n\text{A}} \equiv -\frac{dc_\text{A}}{dx}, \qquad X_{p_y} \equiv -\frac{dv_y}{dx}, \qquad X_Q \equiv -\frac{d\phi}{dx}$$

Fourier's law (16.1), Fick's law (16.33), Newton's viscosity law (16.25), and Ohm's law (16.57) can then be written as

$$J_q = L_q X_q, \qquad J_{n\text{A}} = L_{n\text{A}} X_{n\text{A}}, \qquad J_{p_y} = L_{p_y} X_{p_y}, \qquad J_Q = L_Q X_Q$$

where $\qquad L_q \equiv K, \qquad L_{n\text{A}} \equiv D_{\text{AB}}, \qquad L_{p_y} \equiv \eta, \qquad L_Q \equiv \kappa$

The proportionality constants (the L's) are called *phenomenological coefficients*.

The advantage of this formalism becomes especially clear for systems with two or more gradients. (For example, a system might have a temperature gradient and a concentration gradient.) Such systems are studied in *irreversible thermodynamics*, a subject outside the scope of this book (and the scope of my knowledge). (See *Zemansky*, secs. 9-13, 13-7, and 13-8; D. D. Fitts, *Nonequilibrium Thermodynamics*, McGraw-Hill, 1962.)

PROBLEMS

16.1 If the distance between the reservoirs in Fig. 16.1 is 200 cm, the reservoir temperatures are 325 and 275 K, the substance is an iron rod with cross-sectional area 24 cm^2, $K = 0.80$ J K^{-1} cm^{-1} s^{-1}, and a steady state is present, calculate (*a*) the heat that flows in 60 s; (*b*) ΔS_{univ} in 60 s.

16.2 (*a*) Show that Eq. (16.11) gives

$$K = \frac{5}{16}\left(\bar{C}_V + \frac{9}{4}R\right)\left(\frac{RT}{\pi M}\right)^{1/2}\frac{1}{N_0 d^2} \qquad (16.104)$$

(*b*) Use the d value in (16.30) to calculate the thermal conductivity of He at 1 atm and 0°C and at 10 atm and 100°C. The experimental value at 0°C and 1 atm is 1.4×10^{-3} J cm^{-1} K^{-1} s^{-1}.

16.3 Use (16.30), (16.104), and data in the Appendix to calculate the thermal conductivity of CH_4 at 25°C and 1 atm. (The experimental value is 34×10^{-5} J cm^{-1} K^{-1} s^{-1}.)

16.4 Use data in and preceding Eq. (4.75), and Eq. (16.12) (with 3 changed to 2.8) to estimate the thermal conductivity of water at 30°C and 1 atm. (The experimental value is 6.13 mJ cm^{-1} K^{-1} s^{-1}.)

16.5 (*a*) For laminar flow of a liquid in a pipe, sketch P vs. distance y along the length of the pipe. (*b*) Repeat (*a*) for an ideal gas.

16.6 The *Reynold's number Re* is defined by $Re \equiv \rho\langle v_y\rangle d/\eta$, where ρ and η are the density and viscosity of a fluid flowing with average speed $\langle v_y\rangle$ in a tube of diameter d. (The two-letter symbol Re stands for a single physical quantity.) Experience indicates that when $Re < 2000$, the flow is laminar. For water flowing in a pipe of diameter 1.00 cm, calculate the maximum value of $\langle v_y\rangle$ for laminar flow at 25°C. [See (16.101).]

16.7 For laminar flow of a fluid between two large, flat, parallel plates, show that the flow speed is $v_y = (b^2 - x^2)(2\eta)^{-1}(-dP/dy)$, where $2b$ is the separation between the plates and x is the distance from the plane midway between the plates.

16.8 (*a*) For a certain liquid flowing through a pipe of inside diameter 0.200 cm and length 24.0 cm, a volume of 148 cm^3 is discharged in 120 s when the pressure drop between the pipe ends is 32.0 torr. The liquid's density is 1.35 g/cm^3. Find the liquid's viscosity. (*b*) Calculate the Reynold's number (Prob. 16.6) and check that the flow is laminar.

16.9 The viscosity of O_2 at 0°C and pressures within an order of magnitude of 1 atm is 1.92×10^{-4} P. Calculate the flow rate (in g/s) of O_2 at 0°C through a tube of inside diameter 0.360 cm and length 220 cm when the inlet and outlet pressures are 2.00 and 1.00 atm.

16.10 Calculate the terminal speed of fall in water at 25°C of a steel ball of diameter 1.000 cm and density 7.82 g/cm^3. [See (16.101).]

16.11 The viscosity of CO_2 at 0°C and 1 atm is 1.34×10^{-4} P. Calculate the hard-sphere diameter of CO_2.

16.12 The viscosity of H_2 at 0°C and 1 atm is 8.53×10^{-5} P. Find the viscosity of D_2 at 0°C and 1 atm.

16.13 (*a*) For Sb diffusing into Ag at 20°C, how many years will it take for $(\Delta x)_{\text{rms}}$ to reach 1 cm? (See Sec. 16.4 for D.) (*b*) Repeat (*a*) for Al diffusing into Cu.

16.14 Calculate $(\Delta x)_{\text{rms}}$ for a sucrose molecule in a dilute aqueous solution at 25°C for times of (*a*) 1 min; (*b*) 1 hr; (*c*) 1 day (see Sec. 16.4 for D).

16.15 If r is the distance a diffusing molecule travels in time t, show that $r_{\text{rms}} = (6Dt)^{1/2}$.

16.16 Observations by Perrin on spherical particles of gamboge (a gum resin obtained from trees native to Cambodia) with average radius 2.1×10^{-5} cm suspended in water at 17°C (for which $\eta = 0.011$ P) gave $(\Delta x)_{\text{rms}}$ as 7.1, 10.6, and 11.3×10^{-4} cm for time intervals of 30, 60, and 90 s, respectively. Calculate values of Avogadro's number from these data.

16.17 (*a*) Verify that Eq. (16.40) can be written as $xF_x(t) - \frac{1}{2}f\,d(x^2)/dt = \frac{1}{2}m\,d^2(x^2)/dt^2 - m(dx/dt)^2$. (*b*) Take the average of the equation in (*a*) over many colloidal particles, noting that $\langle xF_x\rangle = 0$, because F_x and x vary independently of each other and each is as likely to be positive as negative. Show that $\langle m(dx/dt)^2\rangle = 2\langle\varepsilon_x\rangle = kT$, where $\langle\varepsilon_x\rangle$ is the particles' average kinetic energy in the x direction, and where it is assumed that $\langle\varepsilon\rangle = \frac{3}{2}kT$. (*c*) Show that $m\,ds/dt + fs = 2kT$, where $s \equiv d\langle x^2\rangle/dt$. (*d*) Show that the equation in (*c*) integrates to $2kT - fs = e^{-fc/m}e^{-ft/m}$, where c is an integration constant. The exponential $e^{-ft/m}$ is essentially zero for a finite time (see Prob. 16.25), so $s \equiv d\langle x^2\rangle/dt = 2kT/f$. Show that this equation integrates to (16.41) if we let $x = 0$ at $t = 0$.

16.18 Suppose the following Δx values (in μm) are found in observations over equal time intervals on particles undergoing Brownian motion: -5.3, $+3.4$, -1.9, -0.4, $+0.5$, $+3.1$, -0.2, -3.5, $+1.4$, $+0.3$, -1.0, $+2.6$. Calculate $\langle\Delta x\rangle$ and $(\Delta x)_{\text{rms}}$.

16.19 Use (16.30) to calculate D_{AA} for O_2 at 0°C and (*a*) 1.00 atm; (*b*) 10.0 atm. (The experimental value at 0°C and 1 atm is 0.19 cm^2/s.)

16.20 Calculate D_{AB} at 0°C and 1.00 atm for H_2–O_2 mixtures using data from (16.30). (The experimental value is 0.70 cm^2/s.)

16.21 Use (16.51) to estimate D_{AA} for H_2O at 25°C and 1 atm. [See (16.101).] The experimental value is 2.4×10^{-5} cm^2/s.

16.22 Calculate D^{∞}_{BA} for N_2 in water at 25°C and 1 atm; use (16.30) and (16.101) for data. (The experimental value is 1.6×10^{-5} cm^2/s.)

16.23 (*a*) Verify that the rigorous theory predicts for hard-sphere gas molecules $D_{AA} = 6\eta/5\rho$. (*b*) For Ne at 0°C and 1 atm, $\eta = 2.97 \times 10^{-4}$ P. Predict D_{AA} at 0°C and 1.00 atm. (The experimental value is 0.44 cm^2/s.)

16.24 Calculate D_{BA} for hemoglobin in water at 25°C, given that $\bar{V} = 48{,}000$ cm^3/mole for hemoglobin. (Assume the molecules are spherical and estimate the volume of a molecule as $\bar{V}/N_0$.) The experimental value is 7×10^{-7} cm^2/s. [See (16.101).]

16.25 In deriving Eq. (16.41) in Prob. 16.17, we set $e^{-ft/m}$ equal to 0. For a spherical colloidal particle of radius 10^{-5} cm and density 3 g/cm^3 undergoing Brownian motion in water at room temperature ($\eta = 0.01$ P), calculate $e^{-ft/m}$ for $t = 1$ s.

16.26 For a current of 1.0 A in a metal wire, how many electrons pass through a cross section in 1.0 s?

16.27 Calculate the resistance and the conductance at 20°C of a copper wire with length 250 cm and cross-sectional area 0.0400 cm^2, given that the resistivity of Cu at 20°C is 1.67×10^{-6} Ω cm.

16.28 Calculate the current in a 100-Ω resistor when the potential difference between its ends is 25 V.

16.29 Calculate the mass of Cu deposited in 30.0 min from a $CuSO_4$ solution by a 2.00-A current.

16.30 The following resistances are observed at 25°C in a conductivity cell filled with various solutions: 411.82 Ω for a 0.741913 wt % KCl solution; 10,875 Ω for a 0.001000 mol/dm^3 solution of MX_2; 368,000 Ω for the deionized water used to prepare the solutions. The conductivity of a 0.741913 percent KCl solution is known to be 0.012856 (U.S. int. Ω)$^{-1}$ cm^{-1} at 25°C. The U.S. international ohm is an obsolete unit equal to 1.000495 Ω. Calculate (*a*) the cell constant; (*b*) κ of MX_2 in a 25°C aqueous 10^{-3} mol/dm^3 solution; (*c*) Λ_m of MX_2 in this solution; (*d*) Λ_{eq} of MX_2 in this solution.

16.31 For a 5.000 mmol/dm^3 aqueous solution of $SrCl_2$ at 25°C, the conductivity is 1.242×10^{-3} Ω^{-1} cm^{-1}. For $SrCl_2$ in this solution, calculate (*a*) Λ_m; (*b*) Λ_{eq}.

16.32 The moving-boundary method was applied to a 0.02000 mol/dm^3 aqueous NaCl solution at 25°C using $CdCl_2$ as the following solution. For a current held constant at 1.600 mA, Longsworth found that the boundary moved 10.00 cm in 3453 s in a tube of average cross-sectional area 0.1115 cm^2. The conductivity of this NaCl solution at 25°C is 2.313×10^{-3} Ω^{-1} cm^{-1}. Calculate $u(Na^+)$ and $t(Na^+)$ in this solution.

16.33 (*a*) Show that the transport number t_B can be calculated from moving-boundary data using

$$t_B = |z_B| \mathscr{F} c_B \mathscr{A} x / Q$$

where Q is the charge that flows when the boundary moves a distance x. (*b*) The moving-boundary method was applied to a 33.27 mmol/dm^3 aqueous solution of $GdCl_3$ at 25°C using LiCl as the following solution. For a constant current of 5.594 mA, it took 4406 s for the boundary to travel between two marks on the tube; these marks were known to contain a volume of 1.111 cm^3 between them. Find the cation and anion transport numbers in this $GdCl_3$ solution.

16.34 A 0.14941 wt % aqueous KCl solution at 25°C was electrolyzed in a Hittorf apparatus using two Ag–AgCl electrodes. The cathode reaction was $AgCl(s) + e^- \rightarrow Ag(s) + Cl^-(aq)$; the anode reaction was the reverse of this. After the experiment it was found that 160.24 mg of Ag had been deposited in a coulometer connected in series with the apparatus and that the cathode compartment contained 120.99 g of solution that was 0.19404 percent KCl by weight. Calculate t_+ and t_- in the KCl solution used in the experiment. (Neglect the transport of water by ions.) (*Hint*: Use the fact that the mass of water in the cathode compartment remains constant.)

16.35 (*a*) The following Λ^{∞}_m values (in Ω^{-1} cm^2 mol^{-1}) are found for 25°C solutions in the solvent methanol: KNO_3, 114.5; KCl, 105.0; LiCl, 90.9. Using only these data, calculate Λ^{∞}_m for $LiNO_3$ in CH_3OH at 25°C. (*b*) The following Λ^{∞}_m values (in Ω^{-1} cm^2 mol^{-1}) are found for 25°C aqueous solutions: HCl, 426; NaCl, 126; $NaC_2H_3O_2$, 91. Using only these data, calculate $\lambda^{\infty}_{m,+} + \lambda^{\infty}_{m,-}$ for $HC_2H_3O_2$ in water at 25°C.

16.36 To find λ^{∞}_m values in a given solvent, we need only one accurate transport number in that solvent. For the solvent CH_3OH at 25°C, the cation transport number t^{∞}_+ in NaCl has been found to be 0.463. Observed Λ^{∞}_m values in CH_3OH at 25°C are:

Electrolyte	$\Lambda_m^\infty/(\Omega^{-1}\ cm^2\ mol^{-1})$
NaCl	96.9
$NaNO_3$	106.4
$LiNO_3$	100.2
NaCNS	107.0
HCl	192
$Ca(CNS)_2$	244

Find λ_m^∞ in CH_3OH at 25°C for Na^+, Cl^-, NO_3^-, Li^+, CNS^-, H_3O^+, and Ca^{2+}. (See also Prob. 16.39*a*.)

16.37 For ClO_4^- in water at 25°C, $\lambda_m^\infty = 67.4\ \Omega^{-1}\ cm^2\ mol^{-1}$. (*a*) Calculate $u^\infty(ClO_4^-)$ in water at 25°C. (*b*) Calculate the drift speed $v^\infty(ClO_4^-)$ in water at 25°C in a field of 24 V/cm. (*c*) Estimate the radius of the hydrated perchlorate ion.

16.38 From λ_m^∞ data tabulated in Sec. 16.6, calculate Λ_m^∞ for the following electrolytes in water at 25°C: (*a*) NH_4NO_3; (*b*) $(NH_4)_2SO_4$; (*c*) $MgSO_4$; (*d*) $Ca(OH)_2$.

16.39 (*a*) Show that for an electrolyte yielding only two kinds of ions,

$$t_+ = \frac{\lambda_{m,+}/z_+}{(\lambda_{m,+}/z_+) + (\lambda_{m,-}/|z_-|)} = \frac{\lambda_{eq,+}}{\lambda_{eq,+} + \lambda_{eq,-}}$$

(*b*) From λ_m^∞ data in Sec. 16.6, calculate $t^\infty(Mg^{2+})$ and $t^\infty(NO_3^-)$ for $Mg(NO_3)_2$ in water at 25°C.

16.40 Which term in (16.93) arises from the electrophoretic effect?

16.41 (*a*) Use the Onsager equation to calculate Λ_m and κ for a 0.00200 mol/dm^3 solution of KNO_3 in water at 25°C and 1 atm. (*b*) Find the resistance of this solution in a conductivity cell with electrodes of area 1.00 cm^2 and separation 10.0 cm. (*c*) Calculate t_+ in this solution. (See Prob. 16.39*a*.)

16.42 The conductivity of pure water at 25°C and 1 atm is $5.4_7 \times 10^{-8}\ \Omega^{-1}\ cm^{-1}$. [H. C. Duecker and W. Haller, *J. Phys. Chem.*, **66**, 225 (1962).] Use (16.99) to find K_c for the ionization of water at 25°C.

16.43 The conductivity of a saturated aqueous $CaSO_4$ solution at 25°C is $2.21 \times 10^{-3}\ \Omega^{-1}\ cm^{-1}$. (*a*) Use (16.99) to find the concentration-scale K_{sp} for $CaSO_4$ in water at 25°C. (*b*) Does the existence of $CaSO_4$ ion pairs in the solution cause error in the result of (*a*)?

16.44 The conductivity of a 0.001028 mol/dm^3 aqueous solution of $HC_2H_3O_2$ at 25°C is $4.95 \times 10^{-5}\ \Omega^{-1}\ cm^{-1}$. Use (16.99) to find K_c for the ionization of acetic acid in water at 25°C.

16.45 The conductivity of a $3.267 \times 10^{-4}\ mol/dm^3$ aqueous solution of $MgSO_4$ at 25°C is $4.06 \times 10^{-5}\ \Omega^{-1}\ cm^{-1}$. Use (16.99) to calculate K_c for the ion-pair-formation reaction $Mg^{2+} + SO_4^{2-} = MgSO_4(aq)$ at 25°C.

16.46 Verify that if the correction term $S(c_+/c^\circ)^{1/2}$ is omitted from (16.99), the degree of dissociation of the weak acid HX is given by $\alpha \approx \Lambda_m/(\lambda_{m,+}^\infty + \lambda_{m,-}^\infty)$, an equation due to Arrhenius.

16.47 Find Λ_m^∞ for $SrCl_2$ in water at 25°C from the following data on aqueous 25°C $SrCl_2$ solutions:

$c/(mol\ dm^{-3})$	$\Lambda_m/(\Omega^{-1}\ cm^2\ mol^{-1})$
0.00025	263.8
0.00050	260.7
0.0025	248.5
0.0050	240.5

16.48 Estimate λ_m^∞ for NO_3^- in water at 35°C and 1 atm from the tabulated 25°C value.

16.49 Use (16.103) to estimate the liquid-junction potential at 25°C between: (*a*) 0.100 mol/kg HCl and 0.0100 mol/kg HCl; (*b*) 0.0100 mol/kg HCl and 0.100 mol/kg HCl; (*c*) 0.100 mol/kg HBr and 0.200 mol/kg KCl; (*d*) 0.100 mol/kg HBr and 4.00 mol/kg KCl. (The second-listed solution is on the right of the cell diagram, and all solutions are in water.)

16.50 (*a*) Find M_n and M_w for a polymer sample that is an equimolar mixture of species with molecular weights 200,000 and 600,000. (*b*) Find M_n and M_w for a polymer sample that is a mixture of equal weights of species with molecular weights 200,000 and 600,000.

16.51 For solutions of polystyrene in benzene at 25°C, the following relative viscosities were measured as a function of polystyrene mass concentration ρ_B:

$\rho_B/(g/dm^3)$	1.000	3.000	4.500	6.00
η_r	1.157	1.536	1.873	2.26

For polystyrene in benzene at 25°C, the constants in (16.32) are $K = 0.034\ cm^3/g$ and $a = 0.65$. Find the viscosity average molecular weight of the polystyrene sample.

16.52 For human hemoglobin in water at 20°C, one finds $\bar{v}^\infty = 0.749\ cm^3/g$, $D^\infty = 6.9 \times 10^{-7}\ cm^2/s$, and $s^\infty = 4.47 \times 10^{-13}$ s. The density of water at 20°C is 0.998 g/cm^3. Calculate the molecular weight of human hemoglobin.

17

REACTION KINETICS

17.1 REACTION KINETICS

Reaction kinetics (also called *chemical kinetics* or *chemical dynamics*) studies the rates and mechanisms of chemical reactions. A reacting system is not in equilibrium, so reaction kinetics is not part of classical thermodynamics but is a branch of kinetics (Sec. 16.1). This chapter deals mainly with experimental aspects of reaction kinetics. The theoretical calculation of reaction rates is discussed in Chap. 23.

Applications of reaction kinetics abound. In the industrial synthesis of compounds, reaction rates are as important as equilibrium constants. The thermodynamic equilibrium constant tells us the maximum possible yield of NH_3 obtainable at any given T from N_2 and H_2, but if the reaction rate between N_2 and H_2 is too low, the reaction will not be economical to carry out. Frequently, in organic preparative reactions, several possible competing reactions can occur, and the relative rates of these reactions usually influence the yield of each product. Reaction rates are fundamental to the functioning of biological organisms. What happens to pollutants released to the atmosphere can only be understood by a kinetic analysis of atmospheric reactions. An automobile works because the rate of oxidation of hydrocarbons is negligible at room temperature but rapid at the elevated temperature of the engine. Many of the metals and plastics of modern technology are thermodynamically unstable with respect to oxidation, but the rate of this oxidation is slow at room temperature.

We begin with some definitions. A *homogeneous reaction* is one that occurs entirely in one phase. A *heterogeneous reaction* involves species present in two or more phases. Sections 17.1 to 17.15 deal with homogeneous reactions. Section 17.16 deals with heterogeneous reactions. Homogeneous reactions are divided into gas-phase reactions and reactions in (liquid) solutions. Sections 17.1 to 17.12 apply to both gas-phase and solution kinetics. Section 17.13 deals with aspects of kinetics unique to reactions in solution.

Consider the homogeneous reaction

$$aA + bB + \cdots \rightarrow eE + fF + \cdots \tag{17.1}$$

where $a, b, \ldots, e, f, \ldots$ are the coefficients in the balanced chemical equation and A, B, ..., E, F, ... are the chemical species. The rate at which any reactant disappears is proportional to its stoichiometric coefficient; thus

$$\frac{dn_A/dt}{dn_B/dt} = \frac{a}{b} \qquad \text{and} \qquad \frac{1}{a}\frac{dn_A}{dt} = \frac{1}{b}\frac{dn_B}{dt}$$

where t is the time and n_A is the number of moles of A. The *rate of reaction* J for (17.1) is defined as

$$J \equiv -\frac{1}{a}\frac{dn_A}{dt} = -\frac{1}{b}\frac{dn_B}{dt} = \cdots = \frac{1}{e}\frac{dn_E}{dt} = \frac{1}{f}\frac{dn_F}{dt} = \cdots \tag{17.2}$$

Since A is disappearing, dn_A/dt is negative and J is positive. At equilibrium, $J = 0$.

Actually, the relation $-a^{-1}\, dn_A/dt = e^{-1}\, dn_E/dt$ need not hold if the reaction has more than one step. For a multistep reaction, the reactant A may first be converted to some reaction intermediate rather than directly to a product. Thus the instantaneous relation between dn_A/dt and dn_E/dt may be complicated. If, as is frequently true, the concentrations of all reaction intermediates are very small throughout the reaction, their effect on the stoichiometry can be neglected.

Equation (4.110) gives $dn_A/(-a) = dn_A/\nu_A = d\xi$, where ξ is the extent of reaction and ν_A is the stoichiometric coefficient of A. Hence, (17.2) can be rewritten as $J \equiv d\xi/dt$.

The reaction rate J is an extensive quantity and depends on the system's size. The *rate of reaction per unit volume*, $r \equiv J/V$, is

$$r \equiv \frac{J}{V} = \frac{1}{V}\frac{1}{\nu_A}\frac{dn_A}{dt} \tag{17.3}$$

r is an intensive quantity and depends on T, P, and the concentrations in the homogeneous system. In most (but not all) systems studied, the volume is either constant or changes by a negligible amount; under these conditions

$$r = \frac{1}{\nu_A}\frac{d}{dt}\left(\frac{n_A}{V}\right) = \frac{1}{\nu_A}\frac{dc_A}{dt} = \frac{1}{\nu_A}\frac{d[\mathrm{A}]}{dt} \qquad \text{const. } V \tag{17.4}^*$$

where $c_A \equiv [\mathrm{A}]$ is the concentration of A. We shall always assume constant volume in this chapter. The quantities r and $d[\mathrm{A}]/dt$ have often been called "the rate of reaction" by chemists, but this nomenclature is not recommended by the IUPAC. Common units for r are mole dm^{-3} s^{-1} and mole cm^{-3} s^{-1}. As an example, for a constant-volume system where the reaction $N_2 + 3H_2 = 2NH_3$ is occurring, we have $r = -d[N_2]/dt = -\frac{1}{3}d[H_2]/dt = \frac{1}{2}d[NH_3]/dt$, provided any reaction intermediates have negligible concentrations.

For many reactions, r is found to have the form

$$r = k[\mathrm{A}]^{\alpha}[\mathrm{B}]^{\beta} \cdots [\mathrm{L}]^{\lambda} \tag{17.5}^*$$

where the exponents $\alpha, \beta, \ldots, \lambda$ are integers or half-integers ($\frac{1}{2}, \frac{3}{2}, \ldots$). The proportionality constant k, called the *rate constant*, is a function of temperature and pressure. The pressure dependence of k is small and is usually ignored. The reaction is said to have *order* α with respect to A, order β with respect to B, etc. The sum $\alpha + \beta + \cdots + \lambda \equiv n$ is the *overall order* (or simply the *order*) of the reaction. Since r

has units of concentration over time, k in (17.5) has units concentration^{1-n} time^{-1}. Most commonly, k is given in $(\text{dm}^3/\text{mole})^{n-1}\ \text{s}^{-1}$. A first-order ($n = 1$) rate constant has units s^{-1} and is independent of the units used for concentration.

The expression for r as a function of concentrations at a fixed temperature is called the *rate law*. Some experimentally observed rate laws for homogeneous reactions are

$$
\begin{array}{lll}
(1) & H_2 + Br_2 \rightarrow 2HBr & r = \dfrac{k[H_2][Br_2]^{1/2}}{1 + j[HBr]/[Br_2]} \\
(2) & 2N_2O_5 \rightarrow 4NO_2 + O_2 & r = k[N_2O_5] \\
(3) & H_2 + I_2 \rightarrow 2HI & r = k[H_2][I_2] \\
(4) & 2NO + O_2 \rightarrow 2NO_2 & r = k[NO]^2[O_2] \\
(5) & CH_3CHO \rightarrow CH_4 + CO & r = k[CH_3CHO]^{3/2} \\
(6) & 2H_2O_2 \xrightarrow{I^-} 2H_2O + O_2 & r = k[H_2O_2][I^-] \\
(7) & Hg_2^{2+} + Tl^{3+} \rightarrow 2Hg^{2+} + Tl^+ & r = k\dfrac{[Hg_2^{2+}][Tl^{3+}]}{[Hg^{2+}]}
\end{array}
\qquad (17.6)
$$

where the values of k depend on temperature and differ from one reaction to another; in reaction (1), j is a constant. Reactions (1) to (5) are gas phase; reactions (6) and (7) are in aqueous solution. For reaction (1), the concept of order is inapplicable. Reaction (5) has order $\frac{3}{2}$. In reaction (6), the species I^- speeds up the reaction but does not appear in the overall chemical equation and is therefore a *catalyst*. In reaction (7), the order with respect to Hg^{2+} is -1. Note from reactions (1) and (2) and (5) to (7) that the exponents in the rate law can differ from the coefficients in the balanced chemical equation. Rate laws must be determined from measurements of reaction rates and cannot be deduced from the reaction stoichiometry.

Actually, the use of concentrations in the rate law is strictly correct only for ideal systems. For nonideal systems, see Sec. 17.9.

Suppose all the concentrations in (17.5) have the order of magnitude 1 mole/dm^3. With the order-of-magnitude approximation $d[A]/dt \approx \Delta[A]/\Delta t$, Eqs. (17.4) and (17.5) give for 1 mole/dm^3 concentrations $\Delta[A]/\Delta t \approx k(\text{mole/dm}^3)^n$, where ν_A has been omitted since it doesn't affect the order of magnitude of things. For an appreciable amount of reaction, $\Delta[A]$ will have the same order of magnitude as [A], namely, 1 mole/dm^3. Hence, $1/k$ [multiplied by $(\text{dm}^3/\text{mole})^{n-1}$ to make things dimensionally correct] gives the order of magnitude of the time for an appreciable amount of reaction to occur when the concentrations have order of magnitude 1 mole/dm^3.

Equation (17.1) gives the overall stoichiometry of the reaction but does not tell us the process or *mechanism* by which the reaction actually occurs. For example, the I^--catalyzed decomposition of $H_2O_2(aq)$ [reaction (6) above] is believed to occur by the following two-step process:

$$
\begin{array}{c}
H_2O_2 + I^- \rightarrow H_2O + IO^- \\
IO^- + H_2O_2 \rightarrow H_2O + O_2 + I^-
\end{array}
\qquad (17.7)
$$

The overall stoichiometry of the two steps is $2H_2O_2 \rightarrow 2H_2O + O_2$ and does not contain the intermediate species IO^- or the catalyst I^- (which is consumed in the first step and regenerated in the second).

There is good evidence that the gas-phase decomposition of N_2O_5 with overall reaction $2N_2O_5 \rightarrow 4NO_2 + O_2$ occurs by the following multistep mechanism:

$$\begin{aligned} &\text{Step } (a)\text{:} \qquad && N_2O_5 \rightleftharpoons NO_2 + NO_3 \\ &\text{Step } (b)\text{:} \qquad && NO_2 + NO_3 \rightarrow NO + O_2 + NO_2 \\ &\text{Step } (c)\text{:} \qquad && NO + NO_3 \rightarrow 2NO_2 \end{aligned} \tag{17.8}$$

Here, there are two intermediates that do not appear in the final products, NO_3 and NO. Any proposed mechanism must add up to give the observed overall reaction stoichiometry. Step (c) consumes the NO molecule produced in step (b), so step (c) must occur once for each occurrence of (b). Steps (b) and (c) together consume two NO_3's; since step (a) produces only one NO_3, the forward reaction of step (a) must occur twice for each occurrence of steps (b) and (c). Taking 2 times step (a) plus 1 times step (b) plus 1 times step (c), we get $2N_2O_5 \rightarrow 4NO_2 + O_2$, as we should.

The number of times a given step in the mechanism occurs for each occurrence of the overall reaction as written is the *stoichiometric number* s of that step. For the overall reaction $2N_2O_5 \rightarrow 4NO_2 + O_2$, the stoichiometric numbers of steps (a), (b), and (c) in the mechanism (17.8) are 2, 1, and 1, respectively. (Don't confuse the stoichiometric number s of a step with the stoichiometric coefficient ν of a chemical species.)

Each step in the mechanism of a reaction is called an *elementary reaction.* A *simple reaction* consists of a single elementary step. A *complex reaction* consists of two or more elementary steps. The N_2O_5 decomposition reaction is complex. The Diels–Alder addition of ethylene to butadiene to give cyclohexene is believed to be simple, occurring as the single step $CH_2{=}CH_2 + CH_2{=}CHCH{=}CH_2 \rightarrow C_6H_{10}$.

The form of the rate law is a consequence of the mechanism of the reaction; see Sec. 17.6. For some reactions, the form of the rate law changes with temperature, indicating a change in mechanism. Sometimes one finds that data for a given homogeneous reaction are fitted by an expression like $r = k[A]^{1.38}$. This indicates that the reaction probably proceeds by two different simultaneously occurring mechanisms that produce different orders. Thus, it is likely that one could get as good or better a fit with an expression like $r = k'[A] + k''[A]^2$, where k' and k'' are constants.

An important concept is that of *pseudo order.* For the hydrolysis of sucrose,

$$C_{12}H_{22}O_{11} + H_2O \rightarrow \underset{\text{Glucose}}{C_6H_{12}O_6} + \underset{\text{Fructose}}{C_6H_{12}O_6} \tag{17.9}$$

one finds the rate per unit volume to be given by $r = k[C_{12}H_{22}O_{11}]$. However, since the solvent H_2O participates in the reaction, one would expect the rate law to have the form $r = k'[C_{12}H_{22}O_{11}]^s[H_2O]^v$. Because H_2O is always present in great excess, its concentration remains nearly constant during a given run and from one run to another. Hence $[H_2O]^v$ is essentially constant, and the rate law is apparently $r = k[C_{12}H_{22}O_{11}]$, where $k = k'[H_2O]^v$. This reaction is said to be *pseudo* first order. It is difficult to determine v, but kinetic data indicate $v \approx 6$. (This can be explained by a reaction mechanism that involves a hexahydrate of sucrose.)

Pseudo order is involved in catalyzed reactions. A catalyst affects the rate without being consumed during the reaction. The hydrolysis of sucrose is acid-catalyzed. During a given run, the H_3O^+ concentration remains fixed. However, when $[H_3O^+]$ is varied from one run to another, it is found that the rate is in fact first order with respect to H_3O^+. Thus, the correct rate law for the hydrolysis of sucrose is $r = k''[C_{12}H_{22}O_{11}][H_2O]^6[H_3O^+]$, and the reaction has order 8. During a given run, however, its apparent (or pseudo) order is 1.

17.2 MEASUREMENT OF REACTION RATES

To measure r, the reaction rate per unit volume [Eq. (17.4)], one must be able to follow the concentration of a reactant or product as a function of time. In the *chemical method*, one places several reaction vessels with identical initial compositions in a constant-temperature bath. At convenient intervals, one withdraws samples from the bath, slows down or stops the reaction, and rapidly analyzes the mixture chemically. Methods for slowing the reaction include cooling the sample to a low temperature, removing a catalyst, greatly diluting the reaction mixture, and adding a species that quickly combines with one of the reactants. Gas samples are frequently analyzed with a mass spectrometer or a gas chromatograph.

Physical methods are usually more accurate and less tedious than chemical methods. Here, one measures a physical property of the reacting system as a function of time. This allows the reaction to be followed continuously as it proceeds. For a gas-phase reaction with a change in total number of moles, the gas pressure P can be followed. (The danger in this procedure is that if side reactions occur, the total pressure will not correctly indicate the progress of the reaction being studied.) For a liquid-phase reaction that occurs with a measurable volume change, V can be followed by running the reaction in a dilatometer (a vessel capped with a graduated capillary tube). If one of the species has a characteristic spectroscopic absorption band, the intensity of this band can be followed. If one of the species is optically active, the optical rotation can be followed. [The first quantitative kinetics experiment was Wilhelmy's 1850 work on the hydrolysis of sucrose. All three sugars in (17.9) are optically active but have different specific rotations. Wilhelmy followed the reaction in a polarimeter.] Ionic reactions in solution can be followed by measuring the electrical conductivity. The refractive index of a liquid solution can be measured as a function of time.

Most commonly, the reactants are mixed and kept in a closed vessel; this is the *static method*. In the *flow method*, the reactants continuously flow into the reaction vessel (which is maintained at constant temperature), and products continuously flow out. After the reaction has run for a while, a steady state (Sec. 1.2) is reached in the reaction vessel and the concentrations at the outlet remain constant with time. The rate law and rate constant can be found by measuring the outlet concentrations for several different inlet concentrations and flow rates. (Flow systems are widely used in industrial chemical production.)

The above-mentioned "classical" methods of kinetics are limited to reactions with half-lives of at least a few seconds. (The half-life is the time it takes for the concentration of a reactant to be cut in half.) Many important chemical reactions have half-lives (for typical reactant concentrations) in the range 10^0 to 10^{-11} s, and may be called *fast reactions*. Examples include gas-phase reactions where one reactant is a free radical and most aqueous-solution reactions that involve ions. (A *free radical* is a species with one or more unpaired electrons; examples are CH_3 and Br.) Moreover, many reactions in biological systems are fast; e.g., the rate constants for formation of complexes between an enzyme and a small molecule (*substrate*) are typically 10^6 to 10^9 dm^3 mole^{-1} s^{-1}.

One way to study fast reactions is with *rapid-flow* methods. A schematic diagram of a liquid-phase *continuous-flow* system is given in Fig. 17.1. Reactants A and B are rapidly driven into the specially designed mixing chamber M by pushing in the plungers of the syringes. Typically, mixing occurs in $\frac{1}{2}$ to 1 millisecond. The reaction mixture then flows through the narrow observation tube. At point P along the tube one uses a physical method (usually measurement of light absorption at a wavelength at which one species absorbs) to determine the concentration of one species. (For gas-phase reactions, the syringes are replaced by bulbs of gases A and B, and flow is caused by pumping at the exit of the observation tube.)

Let the speed at which the mixture flows through the observation tube be v and the distance between the mixing chamber M and the observation point P be x. Then the familiar maxim "distance equals rate times time" gives the time t after the reaction started as $t = x/v$. For a typical flow speed of 1000 cm/s and a typical x value of 10 cm, observation at P gives the concentration of a species 10 ms after the reaction began. Because the mixture at point P is continuously replenished with newly mixed reactants, the concentrations of species remain constant at P. By varying the observation distance x and the flow speed v, one obtains reactant concentrations at various times.

A modification of the continuous-flow method is the *stopped-flow* method (Fig. 17.2). Here, the reactants are mixed at M and rapidly flow through the observation tube into the receiving syringe, driving its plunger against a barrier, thereby stopping the flow; this plunger also hits a switch which stops the motor-driven plungers and triggers the oscilloscope sweep. The light absorption at P is observed as a function of time using a photoelectric cell which converts the light signal to an electrical signal which is displayed on the oscilloscope screen. Because of the rapid mixing and flow and the short distance between M and P, the reaction is observed essentially from its start. The stopped-flow method is actually a static method with rapid mixing rather than a flow method.

The continuous-flow and stopped-flow methods are applicable to reactions with half-lives in the range 10^1 to 10^{-3} s.

The main limitation on the rapid-flow techniques is set by the time required to mix the reactants. The mixing problem is eliminated in *relaxation* methods (developed mainly by Eigen in the 1950s). Here, one takes a system that is in reaction equilibrium and suddenly changes one of the variables that determine the equilibrium position. By following the approach of the system to its new equilibrium

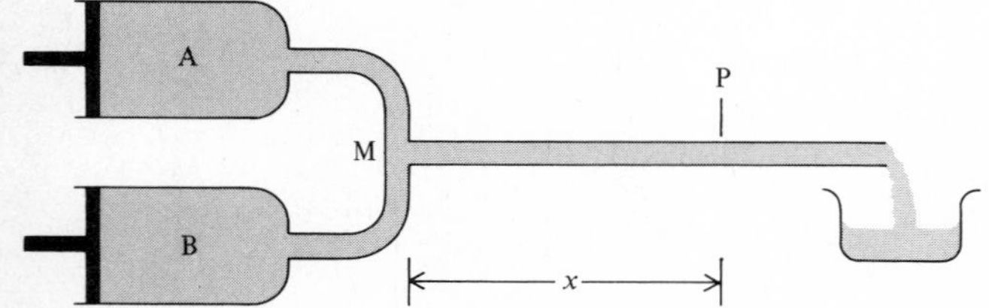

Figure 17.1
A continuous-flow system with rapid mixing of reactants.

position, one can determine rate constants. (Details of the calculation are given in Sec. 17.3.) Relaxation methods are useful mainly for liquid-phase reactions. The scientific meaning of "relaxation" is the approach of a system to a new equilibrium position after it has been perturbed.

The most common relaxation method is the *temperature-jump* (*T-jump*) *method*. Here, one abruptly discharges a high-voltage capacitor through the solution, raising its temperature from T_1 to T_2 in about 1 μs (1 microsecond = 10^{-6} s). Typically, $T_2 - T_1$ is 3 to 10 °C. Discharge of the capacitor triggers the oscilloscope sweep, which displays the light absorption of the solution as a function of time (as in the stopped-flow method); the oscilloscope trace is photographed. Another way to heat the solution in 1 μs is by a pulse of microwave radiation (recall microwave ovens); this method has the advantage of being applicable to nonconducting solutions and the disadvantage of producing only a small temperature rise ($\approx$ 1°C). Provided ΔH° is not zero, Eq. (11.44) shows that the equilibrium constant $K(T_2)$ differs from $K(T_1)$.

In the *pressure-jump method*, a sudden change in P shifts the equilibrium [see Eq. (11.45)]. In the *electric-field-jump method*, a suddenly applied electric field shifts the equilibrium of a reaction that involves a change in total dipole moment.

In the *ultrasonic* relaxation method, the small pressure and temperature oscillations produced by a high-frequency (10^4 to 10^9 cycles/s) sound wave perturb the equilibrium position of a reaction. If the reaction rate is much less than the period of the sound vibration, the reaction cannot respond to the rapid oscillations in the equilibrium constant and the solution behaves like a nonreacting mixture. If the reaction rate is much greater than the period of the sound vibration, the equilibrium oscillates in phase with the sound wave and this results in energy being absorbed from the sound wave by the system. The transition between these two extremes can be observed by measuring the sound absorption as a function of the sound-wave frequency; this allows the rate constants for the forward and reverse rèactions to be found.

A limitation on these relaxation methods is that the reaction must be reversible, with detectable amounts of all species present at equilibrium.

Rapid-flow and relaxation techniques have been used to measure the rates of proton-transfer (acid–base) reactions, electron-transfer (redox) reactions, complex-ion-formation reactions, ion-pair-formation reactions, and enzyme-substrate-complex formation reactions.

Eigen's relaxation methods apply rather small perturbations to a system and do not generate new chemical species. The flash-photolysis, pulse-radiolysis, and shock-tube methods apply a large perturbation to a system, thereby generating one or more reactive species whose reactions are then followed.

In *flash-photolysis* (Fig. 17.3), one exposes the system to a very-high-intensity, short-duration flash of visible and ultraviolet light. The method was developed by Norrish and Porter (who shared the 1967 Nobel prize in chemistry with Eigen) and is

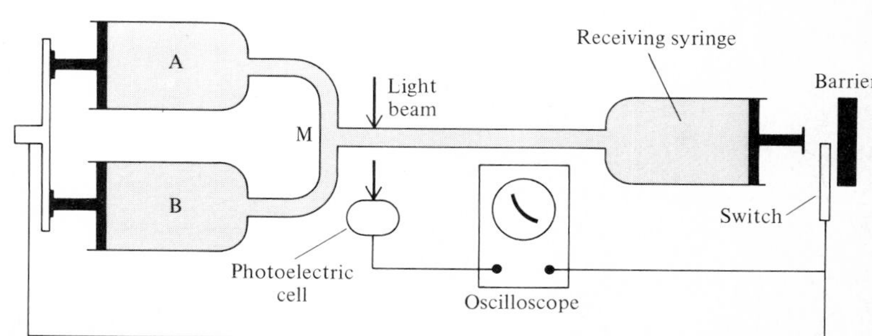

Figure 17.2
A stopped-flow system.

applicable to both gas and solution reactions. Molecules that absorb the light flash either dissociate to radicals or are excited to high-energy states. The reactions of these species are followed by measurement of light absorption; the oscilloscope sweep in Fig. 17.3 is triggered by a photoelectric cell that detects light from the flash. Species studied by flash photolysis include CH, CH_2, CH_3, CN, NH_2, and C_6H_5. With a flashlamp, the light flash lasts about 10 μs. Use of a laser instead of a flashlamp typically gives a pulse of 10 ns duration (1 nanosecond = 10^{-9} s). With special techniques, one can generate a laser light pulse with the incredibly short duration of 1 ps (1 picosecond = 10^{-12} s), thereby allowing rate processes in the picosecond range to be studied. Laser pulses also have the advantage of being monochromatic (i.e., having a single frequency). Picosecond spectroscopy is being used to study the mechanisms of photosynthesis and vision.

Pulse radiolysis is similar to flash photolysis, except that here a very short pulse of high-energy electrons (or x-rays) is passed through the sample, thereby ionizing some of the molecules. Light absorption is used to follow the reactions of the ionized species. Pulse radiolysis of water produces (among other things) hydrated electrons $e^-(aq)$, whose absorption spectrum can be observed. Several hundred reactions of the hydrated electron have been investigated.

In a *shock tube*, the reactant gas mixture at low pressure is separated by a thin diaphragm from an inert gas at high pressure. The diaphragm is punctured, causing a shock wave to travel down the tube; the sudden great increase in pressure and temperature produces excited states and free radicals. The reactions of these species are then followed by observation of their absorption spectra.

There are many pitfalls in kinetics work. A review of solution kinetics stated that "It is a sad fact that there are many kinetic data in the literature that are worthless and many more that are wrong in some important respect. These faulty data can be found in papers old and new, authored by chemists of small reputation and by some of the best known kineticists of all time...." (J. F. Bunnett in *Lewis*, p. 187.)

One must be sure the reactants and products are known. One must check for side reactions. The reagents and solvent must be carefully purified. Certain reactions are sensitive to trace impurities that are difficult to remove; e.g., dissolved O_2 from the air may strongly affect the rate and products of a free-radical reaction. Traces of metal ions catalyze certain reactions. Traces of water strongly affect certain reactions in nonaqueous solvents. To be sure that the rate law has been correctly determined, it is best to make a series of runs varying all concentrations and following the reaction as far to completion as possible. (Data that appear to lie on a straight line for the first

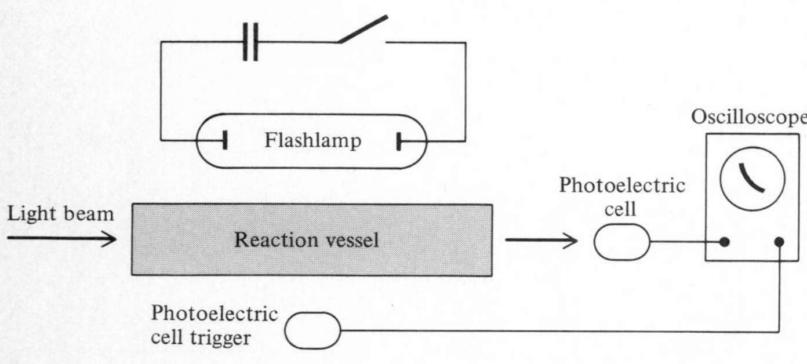

Figure 17.3
A flash-photolysis experiment.

50 percent of a reaction may deviate greatly from this line when the reaction is followed to 70 or 80 percent of completion.)

Because of the many possible sources of error, reported rate constants are frequently of low accuracy.

A list of available tabulations of rate constants is given in L. H. Gevantman and D. Garvin, *Int. J. Chem. Kinet.*, **5**, 213 (1973).

Benson has developed methods to estimate the order of magnitude of gas-phase rate constants. See S. W. Benson and D. M. Golden in *Eyring, Henderson, and Jost*, vol. VII, pp. 57–124; S. W. Benson, *Thermochemical Kinetics*, 2d ed., Wiley-Interscience, 1976.

17.3 INTEGRATION OF RATE LAWS

The rate law is a differential equation, which must be integrated to obtain concentrations of species as a function of time. In the following discussion, we assume that: (*a*) The reaction is carried out at constant temperature. With T constant, the rate constant k is constant. (*b*) The volume is constant. With V constant, the reaction rate per unit volume is given by (17.4). (*c*) The reaction is "irreversible," meaning that no significant amount of reverse reaction occurs. This will be true if the equilibrium constant is very large.

First-order reactions. Suppose the reaction $a\text{A} \rightarrow$ products is first order with $r = k[\text{A}]$. From (17.4) and (17.5), the rate law is

$$\frac{1}{\nu_a}\frac{d[\text{A}]}{dt} \equiv -\frac{1}{a}\frac{d[\text{A}]}{dt} = k[\text{A}] \tag{17.10}$$

Defining k_A as $k_\text{A} \equiv ak$, we have

$$d[\text{A}]/dt = -k_\text{A}[\text{A}] \qquad \text{where } k_\text{A} \equiv ak \tag{17.11}$$

The subscript in k_A reminds us that this rate constant refers to the rate of change of the concentration of A. Chemists are inconsistent in their definitions of rate constants, so in using measured k values one must be sure of the definition.

The variables in (17.11) are [A] and t. Rearranging this equation to separate [A] and t on opposite sides, we have $d[\text{A}]/[\text{A}] = -k_\text{A}\,dt$. Integration gives $\int_1^2 d[\text{A}]/[\text{A}] = -\int_1^2 k_\text{A}\,dt$, and

$$\ln\,([\text{A}]_2/[\text{A}]_1) = -k_\text{A}(t_2 - t_1) \tag{17.12}$$

Equation (17.12) is valid for any two times during the reaction. If state 1 is taken to be the state at the start of the reaction when $[\text{A}] = [\text{A}]_0$ and $t = 0$, then (17.12) becomes

$$\ln\frac{[\text{A}]}{[\text{A}]_0} = -k_\text{A}t \tag{17.13}$$

where [A] is the concentration at time t. Use of (2.120) gives $[\text{A}]/[\text{A}]_0 = e^{-k_\text{A}t}$, and

$$[\text{A}] = [\text{A}]_0 e^{-k_\text{A}t} \tag{17.14*}$$

For a first-order reaction, the concentration of A decreases exponentially with time (Fig. 17.4*a*).

If a reaction is first-order, (17.13) shows that a plot of ln $[A]_0/[A]$ vs. t gives a straight line of slope k_A. A plot of log [A] vs. t gives a straight line of slope $-k_A/2.303$.

The time required for [A] to drop to half its value is called the reaction's *half-life* $t_{1/2}$. Setting $[A] = \frac{1}{2}[A]_0$ and $t = t_{1/2}$ in (17.13) [or $[A]_2/[A]_1 = \frac{1}{2}$ and $t_2 - t_1 = t_{1/2}$ in (17.12)], we get $-k_A t_{1/2} = \ln \frac{1}{2} = \ln 1 - \ln 2 = -\ln 2 = -2.303 \log 2 = -0.693$. For a first-order reaction,

$$k_A t_{1/2} = 0.693 \qquad \text{first-order reaction} \tag{17.15}*$$

Second-order reactions. The two most common forms of second-order rate laws are $r = k[A]^2$ and $r = k[A][B]$, where A and B are two different reactants.

Suppose the reaction is $aA \rightarrow$ products and $r = k[A]^2$. Then $-a^{-1}\, d[A]/dt = k[A]^2$. Defining $k_A \equiv ak$ as in (17.11), we have

$$d[A]/dt = -k_A[A]^2 \qquad \text{and} \qquad \int_1^2 \frac{1}{[A]^2}\, d[A] = -k_A \int_1^2 dt$$

$$\frac{1}{[A]_1} - \frac{1}{[A]_2} = -k_A(t_2 - t_1) \qquad \text{or} \qquad \frac{1}{[A]} - \frac{1}{[A]_0} = k_A t \tag{17.16}$$

$$[A] = \frac{[A]_0}{1 + k_A t[A]_0}$$

From (17.16), a plot of $1/[A]$ vs. t gives a straight line of slope k_A.

The half-life is found by setting $[A] = \frac{1}{2}[A]_0$ and $t = t_{1/2}$ in (17.16), to give

$$t_{1/2} = 1/[A]_0 k_A \qquad \text{second-order reaction} \tag{17.17}$$

For a second-order reaction, $t_{1/2}$ depends on the initial A concentration, which is in contrast to a first-order reaction; $t_{1/2}$ doubles when the A concentration is cut in half. Thus it takes twice as long for the reaction to go from 50 to 75 percent completion as from 0 to 50 percent completion (Fig. 17.4*b*).

Now suppose the reaction is $aA + bB \rightarrow$ products with rate law $r = k[A][B]$. Then (17.4) with $\nu_A = -a$ and (17.5) give

$$\frac{1}{a}\frac{d[A]}{dt} = -k[A][B] \tag{17.18}$$

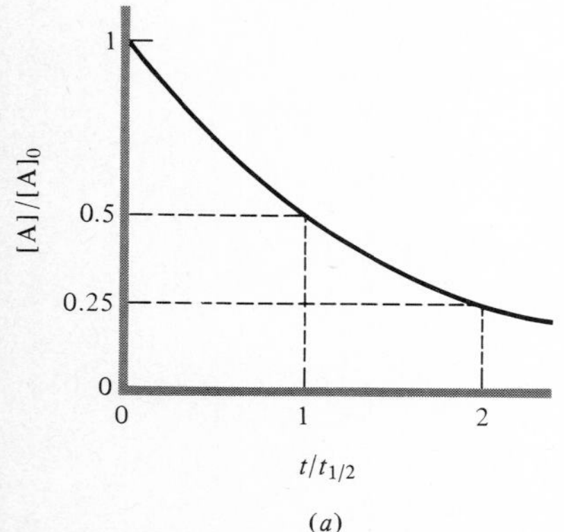

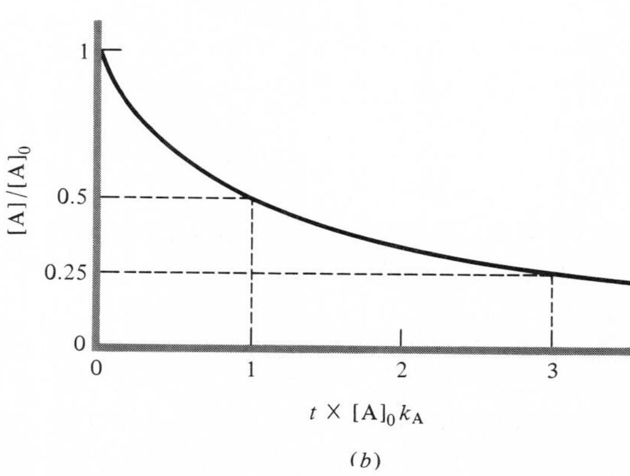

Figure 17.4 Reactant concentration vs. time in (*a*) a first-order reaction; (*b*) a second-order reaction.

To integrate this equation, we must relate [B] to [A]. Let $[A]_0$ and $[B]_0$ be the initial concentrations of A and B. Let x be the number of moles of A per unit volume that have reacted at time t. Then the concentration of A at t is $[A] = [A]_0 - x$. From the reaction's stoichiometry, $(b/a)x$ moles per unit volume of B will have reacted at t, so that $[B] = [B]_0 - (b/a)x$ or

$$[B] = [B]_0 - ba^{-1}[A]_0 + ba^{-1}[A] \tag{17.19}$$

Substituting (17.19) into (17.18) and integrating, we get

$$\int_1^2 \frac{1}{[A]([B]_0 - ba^{-1}[A]_0 + ba^{-1}[A])}\, d[A] = -\int_1^2 ak\, dt \tag{17.20}$$

A table of integrals gives

$$\int \frac{1}{x(p + sx)}\, dx = -\frac{1}{p} \ln \frac{p + sx}{x} \qquad \text{for } p \neq 0 \tag{17.21}$$

To verify this relation, differentiate the right side of (17.21). [The integral in (17.21) is derived by the method of partial fractions, explained in most calculus texts.] Using (17.21) with $p = [B]_0 - ba^{-1}[A]_0$, $s = ba^{-1}$, and $x = [A]$, we get for (17.20)

$$\frac{1}{[B]_0 - ba^{-1}[A]_0} \ln \frac{[B]_0 - ba^{-1}[A]_0 + ba^{-1}[A]}{[A]}\Bigg|_1^2 = ak(t_2 - t_1)$$

$$\frac{1}{a[B]_0 - b[A]_0} \ln \frac{[B]}{[A]}\Bigg|_1^2 = k(t_2 - t_1)$$

$$\frac{1}{a[B]_0 - b[A]_0} \ln \frac{[B]/[B]_0}{[A]/[A]_0} = kt \tag{17.22}$$

where (17.19) was used. In Eq. (17.22), [A] and [B] are the concentrations at time t, and $[A]_0$ and $[B]_0$ are the concentrations at time 0. A plot of the left side of (17.22) vs. t gives a straight line of slope k.

A special case of (17.18) is where A and B are initially present in stoichiometric proportions, so that $[B]_0/[A]_0 = b/a$. Equation (17.22) is clearly not applicable here, since $a[B]_0 - b[A]_0$ in (17.22) becomes zero. To deal with this special case, we recognize that B and A will remain in stoichiometric proportions throughout the reaction: $[B]/[A] = b/a$ at any time. [This follows from (17.19) with $[B]_0 = (b/a)[A]_0$.] Equation (17.18) becomes $(1/b[A]^2)\, d[A] = -k\, dt$. Integration gives a result similar to (17.16), namely,

$$\frac{1}{[A]} - \frac{1}{[A]_0} = bkt \tag{17.23}$$

Third-order reactions. The three most common third-order rate laws are $r = k[A]^3$, $r = k[A]^2[B]$, and $r = k[A][B][C]$. The gory details of the integrations are left as exercises for the reader (Prob. 17.7).

The rate law $d[A]/dt = -k_A[A]^3$ integrates to

$$\frac{1}{[A]^2} - \frac{1}{[A]_0^{\,2}} = 2k_A t \qquad \text{or} \qquad [A] = \frac{[A]_0}{(1 + 2k_A t[A]_0^{\,2})^{1/2}} \tag{17.24}$$

For the reaction $aA + bB \rightarrow$ products, the rate law $a^{-1}\, d[A]/dt = -k[A]^2[B]$ integrates to

$$\frac{1}{a[B]_0 - b[A]_0}\left(\frac{1}{[A]_0} - \frac{1}{[A]} + \frac{b}{a[B]_0 - b[A]_0}\ln\frac{[B]/[B]_0}{[A]/[A]_0}\right) = -kt \qquad (17.25)$$

where [B] is given by (17.19).

For $aA + bB + cC \rightarrow$ products, the rate law $a^{-1}\, d[A]/dt = -k[A][B][C]$ integrates to

$$\frac{a}{\delta_{ab}\delta_{ac}}\ln\frac{[A]}{[A]_0} + \frac{b}{(-\delta_{ab})\delta_{bc}}\ln\frac{[B]}{[B]_0} + \frac{c}{\delta_{ac}\delta_{bc}}\ln\frac{[C]}{[C]_0} = -kt \qquad (17.26)$$

$$\delta_{ab} \equiv a[B]_0 - b[A]_0, \qquad \delta_{ac} \equiv a[C]_0 - c[A]_0, \qquad \delta_{bc} \equiv b[C]_0 - c[B]_0$$

***n*th-order reaction.** Of the many possible nth-order rate laws, we consider only

$$d[A]/dt = -k_A[A]^n$$

Integration gives

$$\int_1^2 [A]^{-n}\, d[A] = -k_A \int_1^2 dt \qquad (17.27)$$

$$\frac{[A]^{-n+1} - [A]_0^{-n+1}}{-n+1} = -k_A t \qquad \text{for } n \neq 1$$

Multiplication of both sides by $(1-n)[A]_0^{n-1}$ gives

$$\left(\frac{[A]}{[A]_0}\right)^{1-n} = 1 + [A]_0^{n-1}(n-1)k_A t \qquad \text{for } n \neq 1 \qquad (17.28)$$

Setting $[A] = \frac{1}{2}[A]_0$ and $t = t_{1/2}$, we get as the half-life

$$t_{1/2} = \frac{2^{n-1} - 1}{(n-1)[A]_0^{n-1}k_A} \qquad \text{for } n \neq 1 \qquad (17.29)$$

For $n = 1$, integration of (17.27) gives a logarithm. Equations (17.14) and (17.15) give for this case

$$[A] = [A]_0 e^{-k_A t}, \qquad t_{1/2} = 0.693/k_A \qquad \text{for } n = 1 \qquad (17.30)$$

Note that (17.28) and (17.29) apply to all values of n except 1. In particular, these equations hold for $n = 0$, $n = \frac{1}{2}$, and $n = \frac{3}{2}$.

Reversible first-order reactions. So far, we have neglected the *reverse* (or *back*) reaction, an assumption which is strictly valid only if the equilibrium constant is infinite but which holds well during the early stages of a reaction. We now allow for the reverse reaction.

Let the balanced chemical reaction be $aA = cC$, and further let both the forward (f) and back (b) reactions be first order: $r_f = k_f[A]$ and $r_b = k_b[C]$. Then (assuming negligible concentrations of any intermediates) the rate law is

$$a^{-1}\, d[A]/dt = -k_f[A] + k_b[C] \qquad (17.31)$$

The same reasoning that gave (17.19) for the reaction $aA + bB \rightarrow$ products gives in this case (Prob. 17.9)

$$[C] = [C]_0 + ca^{-1}[A]_0 - ca^{-1}[A] \qquad (17.32)$$

Equation (17.31) becomes

$$d[A]/dt = k_b a[C]_0 + k_b c[A]_0 - (ak_f + ck_b)[A] \qquad (17.33)$$

Before integrating this equation, we simplify its appearance. In the limit as $t \rightarrow \infty$, the system reaches equilibrium, the rates of the forward and reverse reactions having become equal. At equilibrium the concentration of each species is constant, and $d[A]/dt$ is 0. Let $[A]_{eq}$ be the equilibrium concentration of A. Setting $d[A]/dt = 0$ and $[A] = [A]_{eq}$ in (17.33), we get

$$k_b a[C]_0 + k_b c[A]_0 = (ak_f + ck_b)[A]_{eq} \qquad (17.34)$$

Use of (17.34) in (17.33) gives $d[A]/dt = (ak_f + ck_b)([A]_{eq} - [A])$. Using the identity $\int (x + s)^{-1}\, dx = \ln (x + s)$ to integrate this equation, we get

$$\ln \frac{[A] - [A]_{eq}}{[A]_0 - [A]_{eq}} = -(ak_f + ck_b)t$$

$$[A] - [A]_{eq} = ([A]_0 - [A]_{eq})e^{-jt} \qquad \text{where } j \equiv ak_f + ck_b \qquad (17.35)$$

where $[A]_{eq}$ can be found from (17.34). Note the close resemblance to the first-order rate law (17.14). Equation (17.14) is a special case of (17.35) with $[A]_{eq} = 0$ and $k_b = 0$. A plot of [A] vs. t resembles Fig. 17.4*a* except that [A] approaches $[A]_{eq}$ rather than 0 as $t \rightarrow \infty$.

Discussion of opposing reactions with orders greater than 1 is omitted.

Consecutive first-order reactions. Frequently a product of one reaction becomes a reactant in a subsequent reaction. This is true in multistep reaction mechanisms. We shall consider only the simplest case, that of two consecutive irreversible first-order reactions: $A \rightarrow B$ with rate constant k_1 and $B \rightarrow C$ with rate constant k_2:

$$A \xrightarrow{k_1} B \xrightarrow{k_2} C$$

where for simplicity we have assumed stoichiometric coefficients of unity. The rates of change of the A, B, and C concentrations are

$$d[A]/dt = -k_1[A], \qquad d[B]/dt = k_1[A] - k_2[B], \qquad d[C]/dt = k_2[B] \quad (17.36)$$

Since B is formed by the first reaction and destroyed by the second reaction, the expression for $d[B]/dt$ has two terms. Let only A be present in the system at $t = 0$:

$$[A]_0 \neq 0, \qquad [B]_0 = 0, \qquad [C]_0 = 0 \qquad (17.37)$$

We have three coupled differential equations. The first equation in (17.36) is the same as (17.11), and use of (17.14) gives

$$[A] = [A]_0 e^{-k_1 t} \qquad (17.38)$$

Substitution of (17.38) into the second equation of (17.36) gives

$$d[B]/dt = k_1[A]_0 e^{-k_1 t} - k_2[B] \qquad (17.39)$$

The integration of (17.39) is a bit complicated. One procedure is outlined in Prob. 17.10. The result is

$$[\mathrm{B}] = \frac{k_1[\mathrm{A}]_0}{k_2 - k_1}(e^{-k_1 t} - e^{-k_2 t}) \tag{17.40}$$

To find [C], it is simplest to use conservation of matter. The total number of moles present is constant with time, so $[\mathrm{A}] + [\mathrm{B}] + [\mathrm{C}] = [\mathrm{A}]_0$. Use of (17.38) and (17.40) gives

$$[\mathrm{C}] = [\mathrm{A}]_0\left(1 - \frac{k_2}{k_2 - k_1}e^{-k_1 t} + \frac{k_1}{k_2 - k_1}e^{-k_2 t}\right) \tag{17.41}$$

Figure 17.5 plots [A], [B], and [C] for two values of k_2/k_1. Note the maximum in the intermediate species [B].

Competing first-order reactions. Frequently a species can react in different ways to give a variety of products. For example, toluene can be nitrated at the ortho, meta, or para positions. We shall consider the simplest case, that of two competing irreversible first-order reactions A → C and A → D:

$$\mathrm{A} \xrightarrow{k_1} \mathrm{C} \qquad \text{and} \qquad \mathrm{A} \xrightarrow{k_2} \mathrm{D} \tag{17.42}$$

where the stoichiometric coefficients are taken as 1 for simplicity. The rate law is

$$d[\mathrm{A}]/dt = -k_1[\mathrm{A}] - k_2[\mathrm{A}] = -(k_1 + k_2)[\mathrm{A}] \tag{17.43}$$

This equation is the same as (17.11) with k_A replaced by $k_1 + k_2$. Hence (17.14) gives $[\mathrm{A}] = [\mathrm{A}]_0 e^{-(k_1+k_2)t}$.

For the product C, we have

$$d[\mathrm{C}]/dt = k_1[\mathrm{A}] = k_1[\mathrm{A}]_0 e^{-(k_1+k_2)t}$$

$$[\mathrm{C}] = \frac{k_1[\mathrm{A}]_0}{k_1 + k_2}(1 - e^{-(k_1+k_2)t}) \tag{17.44}$$

since $[\mathrm{C}]_0 = 0$.

Similarly, integration of $d[\mathrm{D}]/dt = k_2[\mathrm{A}]$ (or use of the conservation equation $[\mathrm{A}] + [\mathrm{C}] + [\mathrm{D}] = [\mathrm{A}]_0$) gives

$$[\mathrm{D}] = \frac{k_2[\mathrm{A}]_0}{k_1 + k_2}(1 - e^{-(k_1+k_2)t}) \tag{17.45}$$

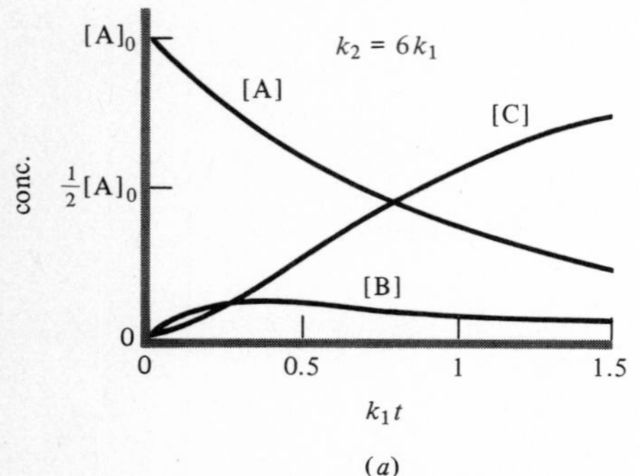

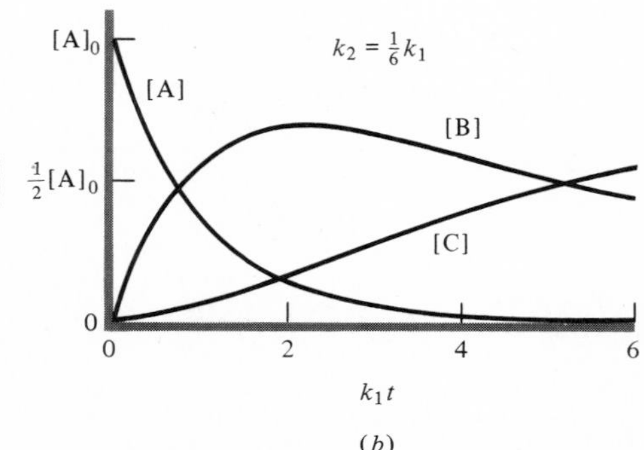

Figure 17.5 Reactant concentrations vs. time for the consecutive first-order reactions A → B → C.

Note the occurrence of the sum of the rate constants $k_1 + k_2$ in the exponentials for both [C] and [D].

Division of (17.44) by (17.45) gives at any time during the reaction

$$[\mathrm{C}]/[\mathrm{D}] = k_1/k_2 \tag{17.46}$$

The amounts of C and D obtained depend on the relative rates of the two competing reactions. Measurement of [C]/[D] allows k_1/k_2 to be determined.

In this example we assumed the competing reactions to be irreversible. In general, this will not be true, and we must also consider the reverse reactions

$$\mathrm{C} \xrightarrow{k_{-1}} \mathrm{A} \qquad \text{and} \qquad \mathrm{D} \xrightarrow{k_{-2}} \mathrm{A} \tag{17.47}$$

Moreover, the product C may well react to give the product D and vice versa:

$$\mathrm{C} \underset{k_{-3}}{\overset{k_3}{\rightleftharpoons}} \mathrm{D} \tag{17.48}$$

If we wait an infinite amount of time, the system will reach equilibrium and the ratio [C]/[D] will be determined by the ratio K_1/K_2 of concentration-scale equilibrium constants for the reactions in (17.42): $[\mathrm{C}]/[\mathrm{D}] = K_1/K_2$ at $t = \infty$ (where we assumed an ideal system). This situation is called *thermodynamic control* of products; here, the product with the most negative G° is favored. On the other hand, during the early stages of the reaction when any reverse reaction or interconversion of C and D can be neglected, Eq. (17.46) will apply and we have *kinetic control* of products. If the rate constants k_{-1} and k_{-2} for the reverse reactions (17.47) and the rate constants k_3 and k_{-3} for interconversion of products are all much, much less than k_1 and k_2 for the forward reactions (17.42), the products will be kinetically controlled even when A has been nearly all consumed. It frequently happens that $k_1/k_2 \gg 1$ and $K_1/K_2 \ll 1$, so C is favored kinetically and D is favored thermodynamically; the relative yield of products then depends on whether there is kinetic or thermodynamic control. "The history of organic chemistry is replete with examples of product analyses that were claimed to show which product was more stable or which rate of product formation was more rapid when the data were in fact irrelevant to the conclusions that were inferred." (R. M. Noyes in *Lewis*, p. 522.)

Relaxation. In relaxation methods (Sec. 17.2), a system in equilibrium suffers a small perturbation which changes the equilibrium constant. The system then relaxes to its new equilibrium position. We now integrate the rate law for a typical case.

Consider the reversible *elementary* reaction

$$\mathrm{A} + \mathrm{B} \underset{k_b}{\overset{k_f}{\rightleftharpoons}} \mathrm{C}$$

where the forward and reverse rate laws are $r_f = k_f[\mathrm{A}][\mathrm{B}]$ and $r_b = k_b[\mathrm{C}]$. (We shall see in Sec. 17.5 that for elementary reactions, the stoichiometric coefficients determine the reaction orders.) Suppose the temperature is suddenly changed to a new value. For all times after the T jump we have

$$d[\mathrm{A}]/dt = -k_f[\mathrm{A}][\mathrm{B}] + k_b[\mathrm{C}] \tag{17.49}$$

Let $[\mathrm{A}]_{\mathrm{eq}}$, $[\mathrm{B}]_{\mathrm{eq}}$, and $[\mathrm{C}]_{\mathrm{eq}}$ be the equilibrium concentrations at the new temperature T_2, and let $x \equiv [\mathrm{A}]_{\mathrm{eq}} - [\mathrm{A}]$. For every mole of A that reacts, 1 mole of B reacts and 1 mole of C is formed. Hence $[B]_{eq} - [B] = x$ and $[C]_{eq} - [C] = -x$. Also, $d[A]/dt = -dx/dt$. Equation (17.49) becomes

$$-dx/dt = -k_f([A]_{eq} - x)([B]_{eq} - x) + k_b([C]_{eq} + x)$$

$$dx/dt = k_f[A]_{eq}[B]_{eq} - k_b[C]_{eq} - xk_f([A]_{eq} + [B]_{eq} + k_b k_f^{-1} - x) \quad (17.50)$$

When equilibrium is reached, $d[A]/dt = 0$, and (17.49) gives

$$k_f[A]_{eq}[B]_{eq} - k_b[C]_{eq} = 0 \quad (17.51)$$

Furthermore, the perturbation is small, so the deviation x of [A] from its equilibrium value is small and $x \ll [A]_{eq} + [B]_{eq}$. Using (17.51) and neglecting the x in parentheses in (17.50), we get

$$dx/dt = -\tau^{-1}x \qquad \text{where } \tau \equiv \{k_f([A]_{eq} + [B]_{eq}) + k_b\}^{-1} \quad (17.52)$$

This integrates to $x = x_0 e^{-t/\tau}$, where x_0 is the value of x the instant after the T jump is applied at $t = 0$. Since $x = [A]_{eq} - [A] = [C] - [C]_{eq}$, we have

$$[A] - [A]_{eq} = ([A]_0 - [A]_{eq})e^{-t/\tau}$$

with similar equations holding for [B] and [C]. Thus the approach of each species to its new equilibrium value is first order with rate constant $1/\tau$ [cf. Eq. (17.14)]. This result holds when any elementary reaction is subjected to a *small* perturbation from equilibrium. (Of course, the definition of τ depends on the stoichiometry of the elementary reaction. See Prob. 17.12 for another example.) The constant τ is called the *relaxation time*; τ is the time it takes the deviation $[A] - [A]_{eq}$ to drop to $1/e$ of its initial value.

In a system with several reversible elementary reactions, there is a relaxation time for each reaction, and the mathematics is more complex. We omit discussion.

Example For $H^+ + OH^- = H_2O$, the relaxation time has been measured as 36 μs at 25°C. Find k_f.

The reaction has the form A + B = C, so the preceding treatment applies. Equations (17.52) and (17.51) give

$$\tau^{-1} = k_f([H^+]_{eq} + [OH^-]_{eq}) + k_b \qquad \text{and} \qquad k_b[H_2O]_{eq} = k_f[H^+]_{eq}[OH^-]_{eq}$$

Eliminating k_b and using $[OH^-]_{eq} = [H^+]_{eq}$, we get

$$\tau^{-1} = k_f(2[H^+]_{eq} + [H^+]_{eq}^2/[H_2O]_{eq})$$

Using $[H^+]_{eq} = 1.0 \times 10^{-7}$ mole/dm^3 and $[H_2O]_{eq} = 55.5$ mole/dm^3, we find $k_f = 1.4 \times 10^{11}$ dm^3 mole^{-1} s^{-1}.

17.4 DETERMINATION OF THE RATE LAW

This section discusses how the rate law is found from experimental data. We restrict the discussion to cases where the rate law has the form

$$r = k[A]^\alpha[B]^\beta \cdots [L]^\lambda \quad (17.53)$$

It is usually best to find the orders $\alpha, \beta, \ldots, \lambda$ first and then find the rate constant k. Four methods for finding the orders follow.

1. Half-life method. This method applies when the rate law has the form $r = k[\mathrm{A}]^n$. Then Eqs. (17.29) and (17.30) apply. If $n = 1$, then $t_{1/2}$ is independent of $[\mathrm{A}]_0$. If $n \neq 1$, then (17.29) gives

$$\log t_{1/2} = \log \frac{2^{n-1} - 1}{(n-1)k_\mathrm{A}} - (n-1) \log [\mathrm{A}]_0$$

A plot of $\log t_{1/2}$ vs. $\log [\mathrm{A}]_0$ gives a straight line of slope $1 - n$. (This statement is also valid for $n = 1$.) To use the method, one plots [A] vs. t for a run. One picks any [A] value, say [A]′, and finds the point where [A] has fallen to $\frac{1}{2}$[A]′; the time interval between these two points is $t_{1/2}$ for the initial concentration [A]′. One then picks another point [A]″ and determines $t_{1/2}$ for this A concentration. After repeating this process several times, one plots $\log t_{1/2}$ vs. the log of the corresponding initial A concentration and measures the slope.

The half-life method has the disadvantage that if data from a single run are used, the reaction must be followed to a high percentage of completion. An improvement is the use of the *fractional life* t_α, defined as the time required for $[\mathrm{A}]_0$ to fall to $\alpha[\mathrm{A}]_0$. (For the half-life, $\alpha = \frac{1}{2}$.) If $r = k[\mathrm{A}]^n$, the result of Prob. 17.14 shows that a plot of $\log t_\alpha$ vs. $\log [\mathrm{A}]_0$ is a straight line with slope $1 - n$. A convenient value of α is 0.75.

2. Powell-plot method. This method applies when the rate law has the form $r = k[\mathrm{A}]^n$. Let the dimensionless parameters α and ϕ be defined as

$$\alpha \equiv [\mathrm{A}]/[\mathrm{A}]_0, \qquad \phi \equiv k_\mathrm{A}[\mathrm{A}]_0^{n-1}t \tag{17.54}$$

(α is the fraction of A unreacted). Equations (17.28) and (17.13) become

$$\alpha^{1-n} - 1 = (n-1)\phi \qquad \text{for } n \neq 1$$

$$\ln \alpha = -\phi \qquad \text{for } n = 1$$

These equations are used to plot α vs. $\log \phi$ for commonly occurring values of n, to give a series of master curves (Fig. 17.6). From the data for a run in a kinetics

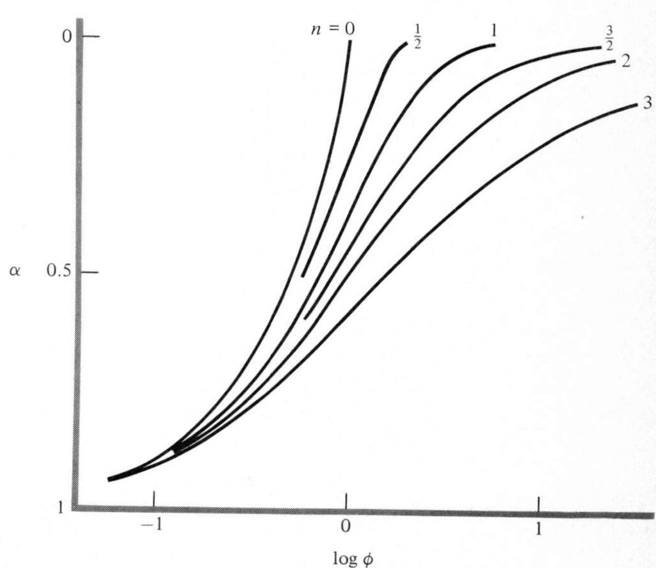

Figure 17.6
Powell-plot master curves for determination of reaction order. ($\alpha = 0$ is at the top.)

experiment, one plots α vs. log t on translucent graph paper having the same scale as the master plots. Since log ϕ differs from log t by log $k_A[A]_0^{n-1}$, which is a constant for a given run, the experimental curve will be shifted along the horizontal axis by a constant from the applicable master curve. One slides the experimental curve back and forth until it coincides with one of the master curves. This gives n.

It might seem that methods 1 and 2 are of little value since they apply only when $r = k[A]^n$. However, by taking the initial concentrations of reactants in stoichiometric proportions, the rate law (17.53) is reduced to this special form (provided the products do not appear in the rate law). Thus, if the reaction is $aA + bB \rightarrow$ products, and if $[A]_0 = as$, $[B]_0 = bs$, where s is a constant, then Eq. (4.108) gives $[A] = [A]_0 - a\xi = a(s - \xi)$ and $[B] = b(s - \xi)$; the rate law becomes $r = ka^\alpha b^\beta (s - \xi)^{\alpha+\beta} = \text{const.} \times [A]^{\alpha+\beta}$. Methods 1 and 2 will give the overall order when the reactants are mixed in stoichiometric proportions.

The Powell-plot method requires the initial investment of time needed to make the master plots, but once these are prepared, the method is quick, easy, and fun to use. Time can be saved by making the plots on semilogarithmic graph paper (4 cycles).

3. Initial-rate method. Here, one measures the initial rate r_0 for several runs, varying the initial concentration of one reactant at a time. Suppose we measure r_0 for the two different initial A concentrations $[A]'_0$ and $[A]''_0$ while keeping $[B]_0$, $[C]_0$, ... fixed. Then

$$r'_0 = k([A]'_0)^\alpha([B]'_0)^\beta \cdots \quad \text{and} \quad r''_0 = k([A]''_0)^\alpha([B]''_0)^\beta \cdots$$

$$\frac{r'_0}{r''_0} = \left(\frac{[A]'_0}{[A]''_0}\right)^\alpha \quad \text{and} \quad \alpha = \frac{\log (r'_0/r''_0)}{\log ([A]'_0/[A]''_0)}$$

The orders β, γ, ... are found similarly.

The initial rate r_0 can be found by plotting [A] vs. t and drawing the tangent line at $t = 0$.

4. Isolation method. Here, one makes the initial concentration of reactant A much less than the concentrations of all other species: $[B]_0 \gg [A]_0$, $[C]_0 \gg [A]_0$, etc. Thus, we can make $[A]_0 = 10^{-3}$ mole/dm^3 and all other concentrations at least 0.1 mole/dm^3. Then the concentrations of all reactants except A will be essentially constant with time. The rate law (17.53) becomes

$$r = k[A]^\alpha[B]_0{}^\beta \cdots [L]_0{}^\lambda = j[A]^\alpha \tag{17.55}$$

$$j \equiv k[B]_0{}^\beta \cdots [L]_0{}^\lambda \tag{17.56}$$

where j is essentially constant. The reaction has the pseudo order α under these conditions. One then analyzes the data from the run using method 1 or 2 to find α. To find β, we can make $[B]_0 \ll [A]_0$, $[B]_0 \ll [C]_0$, ... and proceed as we did in finding α. Alternatively, we can keep $[A]_0$ fixed at a value much less than all other concentrations but change $[B]_0$ to a new value $[B]'_0$; with only $[B]_0$ changed, we evaluate the apparent rate constant j' that corresponds to $[B]'_0$; Eq. (17.56) gives $j/j' = ([B]_0/[B]'_0)^\beta$, so β can be found from j and j'.

Many books suggest that the overall reaction order be obtained by trial and error, as follows. If the rate law is $r = k[A]^n$, one plots log [A] vs. t, $[A]^{-1}$ vs. t, and $[A]^{-2}$ vs. t and obtains a straight line for one of these plots according to whether

$n = 1$, 2, or 3, respectively. This method is dangerous to use, since it is often difficult to decide which of these plots is most nearly linear. (See Prob. 17.18.) Thus, the wrong order can be obtained, especially if the reaction has not been followed very far; see Fig. 17.6. Moreover, if the order is $\frac{3}{2}$, it is not hard to conclude erroneously that $n = 1$ or $n = 2$.

Once the order with respect to each species has been obtained by one of the above methods, the rate constant k is found from the slope of an appropriate graph. For example, if the rate law has been found to be $r = k[A][B]$, Eq. (17.22) shows that a plot of ln $[B]/[A]$ vs. t gives a straight line of slope k times $a[B]_0 - b[A]_0$. [If the isolation method has been used to find the orders, one can first find j in (17.55) from the slope of an appropriate plot and then calculate k from (17.56).] (The best straight-line fit can be found by a least-squares treatment; see *Shoemaker, Garland, and Steinfeld*, pp. 44–49 and 691–700.)

17.5 RATE LAWS AND EQUILIBRIUM CONSTANTS FOR ELEMENTARY REACTIONS

The examples in Eq. (17.6) show that the orders in the rate law of an overall reaction frequently differ from the stoichiometric coefficients. An overall reaction occurs as a series of elementary steps, these steps constituting the *mechanism* of the reaction. We now consider the rate law for an elementary reaction. (Sec. 17.6 shows how the rate law for an overall reaction follows from its mechanism.)

The number of molecules that react in an elementary step is the *molecularity* of the elementary reaction. Molecularity is defined only for elementary reactions and should not be used to describe overall reactions that consist of more than one elementary step. The elementary reaction A → products is *unimolecular*. The elementary reactions A + B → products and 2A → products are *bimolecular*. The elementary reactions A + B + C → products, 2A + B → products, and 3A → products are *trimolecular* (or *termolecular*). No elementary reactions involving more than three molecules are known, due to the very low probability of near-simultaneous collision of more than three molecules.

Consider a bimolecular elementary reaction A + B → products, where A and B may be the same or different molecules. Although not every collision between A and B will produce products, the rate of reaction per unit volume will be proportional to Z_{AB}, the number of A-B collisions per second per unit volume. Equations (15.62) and (15.63) show that in an ideal gas, Z_{AB} is proportional to $(N_A/V)(N_B/V)$, where N_A/V is the number of A molecules per unit volume. Therefore r, the reaction rate per unit volume for an elementary bimolecular ideal-gas reaction, will be proportional to $(N_A/V)(N_B/V) = N_0{}^2(n_A/V)(n_B/V) = N_0{}^2[A][B]$, where N_0 is the Avogadro constant. Calling the proportionality constant k, we have $r = k[A][B]$. Similarly, for a trimolecular elementary reaction in an ideal gas, the rate per unit volume will be proportional to the number of three-body collisions per unit volume and therefore will be proportional to $[A][B][C]$; see Sec. 23.1.

For the unimolecular ideal-gas reaction A → products, there is a fixed probability that any one A molecule will decompose or isomerize to products in unit time. Hence the number of molecules reacting in unit time is proportional to the number N_A present, and the rate of reaction per unit volume is proportional to N_A/V and hence to [A]; we have $r = k[A]$. (A fuller treatment of unimolecular reactions is given in Sec. 17.10.)

Similar considerations apply to reactions in an ideal or ideally dilute solution.

In summary, in an ideal system, the rate law for the elementary reaction $aA + bB \rightarrow$ products is $r = k[A]^a[B]^b$, where $a + b$ is 1, 2, or 3.

The rate law for elementary reactions in nonideal systems is discussed in Sec. 17.9. Kinetics data are usually not accurate enough to cause worry about deviations from ideality, except for ionic reactions. Thus, for an *elementary* reaction, the orders in the rate law are determined by the reaction's stoichiometry.

Most elementary reactions are unimolecular or bimolecular, trimolecular reactions being uncommon due to the low probability of three-body collisions.

We now examine the relation between the equilibrium constant for a reversible elementary reaction and the rate constants for the forward and reverse reactions. Consider the reversible *elementary* reaction

$$aA + bB \underset{k_b}{\overset{k_f}{\rightleftharpoons}} cC + dD$$

in an ideal system. The rate laws for the forward (f) and back (b) reactions are $r_f = k_f[A]^a[B]^b$ and $r_b = k_b[C]^c[D]^d$. At equilibrium these opposing rates must be equal: $r_{f,\text{eq}} = r_{b,\text{eq}}$, or

$$k_f([A]_{\text{eq}})^a([B]_{\text{eq}})^b = k_b([C]_{\text{eq}})^c([D]_{\text{eq}})^d \qquad \text{and} \qquad \frac{k_f}{k_b} = \frac{([C]_{\text{eq}})^c([D]_{\text{eq}})^d}{([A]_{\text{eq}})^a([B]_{\text{eq}})^b}$$

But the quantity on the right side of this last equation is the concentration-scale equilibrium constant K_c for the reaction. Hence

$$K_c = k_f/k_b \qquad \text{elementary reaction} \tag{17.57*}$$

If $k_f \gg k_b$, then $K_c \gg 1$ and the equilibrium position favors products. (The relation between k_f, k_b, and K_c for an overall reaction is discussed in Sec. 17.8.)

For a reaction in solution where the solvent is a reactant or product, the equilibrium constant K_c in (17.57) differs from the K_c used in thermodynamics. In thermodynamics, the mole-fraction scale (rather than the concentration-scale) is used for the solvent; for reactions in an ideally dilute solution, the solvent is usually omitted from the thermodynamic K_c (since its mole fraction is approximately 1). Therefore, if ν_A is the stoichiometric coefficient of the solvent A in the elementary reaction, then $K_{c,\text{kin}} = K_{c,\text{td}}[A]^{\nu_A}$, where $K_{c,\text{td}}$ is the thermodynamic K_c and $K_{c,\text{kin}}$ is the kinetic K_c [as in Eq. (17.57)].

17.6 REACTION MECHANISMS

The observed rate law provides information on the mechanism of a reaction, in that any proposed mechanism must yield the observed rate law. Usually an exact deduction of the rate law from the differential rate equations of a multistep mechanism is not possible, so one of two approximation methods is generally invoked, the rate-determining-step approximation or the steady-state approximation.

In the *rate-determining-step approximation* (also called the *rate-limiting-step approximation* or the *equilibrium approximation*), the reaction mechanism is assumed to consist of one or more reversible reactions that stay close to equilibrium during most of the reaction, followed by a relatively slow rate-determining step, which in turn is followed by one or more rapid reactions. (In special cases, there may be no equilibrium steps before the rate-determining step or no rapid reactions after the rate-determining step.)

As a simple example, consider the following mechanism composed of unimolecular (elementary) reactions

$$A \underset{k_{-1}}{\overset{k_1}{\rightleftharpoons}} B \underset{k_{-2}}{\overset{k_2}{\rightleftharpoons}} C \underset{k_{-3}}{\overset{k_3}{\rightleftharpoons}} D \qquad (17.58)$$

where step 2 ($B \rightleftharpoons C$) is assumed to be the rate-determining step. For this assumption to be valid, we must have $k_{-1} \gg k_2$. The slow rate of $B \rightarrow C$ compared with $B \rightarrow A$ ensures that most B molecules go back to A rather than going to C, thereby ensuring that step 1 ($A \rightleftharpoons B$) remains close to equilibrium. Furthermore, we must have $k_3 \gg k_2$ and $k_3 \gg k_{-2}$ to ensure that step 2 acts as a "bottleneck" and that product D is rapidly formed from C. The overall rate is then controlled by the rate-determining step $B \rightarrow C$. (Note that since $k_3 \gg k_{-2}$, the rate-limiting step is not in equilibrium.) Since we are examining the rate of the forward reaction $A \rightarrow D$, we further assume that $k_2[B] \gg k_{-2}[C]$. During the early stages of the reaction, the concentration of C will be low compared with B, and this condition will hold. Thus we neglect the reverse reaction for step 2. Since the rate-controlling step is taken to be essentially irreversible, it is irrelevant whether the rapid steps after the rate-limiting step are reversible or not. The observed rate law will depend only on the nature of the equilibria that precede the rate-determining step and on this step itself. (Examples are given below.) The relative magnitudes of k_1 and k_2 are irrelevant to the validity of the rate-limiting-step approximation. Hence, the rate constant k_2 of the rate-determining step might be larger than k_1.

An example is the Br^--catalyzed reaction

$$H^+ + HNO_2 + \phi NH_2 \xrightarrow{Br^-} \phi N_2{}^+ + 2H_2O$$

in water, where ϕ stands for the C_6H_5 group. The observed rate law is

$$r = k[H^+][HNO_2][Br^-] \qquad (17.59)$$

A proposed mechanism is

$$\begin{aligned} &H^+ + HNO_2 \underset{k_{-1}}{\overset{k_1}{\rightleftharpoons}} H_2NO_2{}^+ && \text{rapid equilib.} \\ &H_2NO_2{}^+ + Br^- \xrightarrow{k_2} ONBr + H_2O && \text{slow} \qquad (17.60) \\ &ONBr + \phi NH_2 \xrightarrow{k_3} \phi N_2{}^+ + H_2O + Br^- && \text{fast} \end{aligned}$$

The second step is rate-limiting. Since $k_3 \gg k_2$, we can take $d[\phi N_2{}^+]/dt$ as equal to the rate of formation of ONBr in step 2. Hence

$$r = k_2[H_2NO_2{}^+][Br^-] \qquad (17.61)$$

The species $H_2NO_2{}^+$ is a reaction intermediate, and we want to express r in terms of reactants and products. Since step 1 is in near equilibrium, Eq. (17.57) gives

$$K_{c,1} = \frac{k_1}{k_{-1}} = \frac{[H_2NO_2{}^+]}{[H^+][HNO_2]} \quad \text{and} \quad [H_2NO_2{}^+] = (k_1/k_{-1})[H^+][HNO_2]$$

Substitution in (17.61) gives

$$r = (k_1 k_2/k_{-1})[H^+][HNO_2][Br^-] \tag{17.62}$$

in agreement with (17.59). Since $k = k_1 k_2/k_{-1} = K_{c,1} k_2$, the observed rate constant contains the equilibrium constant for step 1 and the rate constant for the rate-determining step 2.

Note from Eq. (17.61) that the rate of the overall reaction equals the rate of the rate-determining step.

For the reverse overall reaction, the rate-determining step is the reverse of that for the forward reaction. For example, for the reverse of (17.58), the rate-determining step is $C \rightarrow B$. This follows from the above inequalities $k_{-2} \ll k_3$ (which ensures that step $D \rightleftharpoons C$ is in equilibrium) and $k_{-1} \gg k_2$ (which ensures that $B \rightarrow A$ is rapid).

We now discuss the steady-state approximation. Multistep reaction mechanisms usually involve one or more intermediate species that do not appear in the overall equation. For example, the postulated species $H_2NO_2{}^+$ in the above mechanism is such a *reaction intermediate*. Frequently, these intermediates are very reactive and therefore do not accumulate to any significant extent during the reaction; that is, $[I] \ll [R]$ and $[I] \ll [P]$ during most of the reaction, where I is an intermediate and R and P are reactants and products. Oscillations in the concentration of a species during a reaction are very rare, so we can assume that [I] will start at 0, rise to a maximum, $[I]_{max}$, and then fall back to 0. If [I] remains small during the reaction, $[I]_{max}$ will be small compared with $[R]_{max}$ and $[P]_{max}$ and the curves of [R], [I], and [P] vs. t will resemble those of Fig. 17.5*a*, where the reactant R is A, the intermediate I is B, and the product P is C. Note that [except for the initial period of time (called the *induction period*) when B is rising rapidly] the slope of the B curve is much less than the slopes of the A and C curves. In the R, I, P notation, we have $d[I]/dt \ll d[R]/dt$ and $d[I]/dt \ll d[P]/dt$. It is therefore frequently a good approximation to take $d[I]/dt = 0$ for each reactive intermediate. This is the *steady-state* (or *stationary-state*) *approximation.*

Let us apply the steady-state approximation to the mechanism (17.60), assuming that $H_2NO_2{}^+$ is a reactive intermediate but dropping the assumption that step 1 is in near equilibrium. The rate is given by (17.61). To find $[H_2NO_2{}^+]$, we set its time derivative equal to zero:

$$d[H_2NO_2{}^+]/dt = 0 = k_1[H^+][HNO_2] - k_{-1}[H_2NO_2{}^+] - k_2[H_2NO_2{}^+][Br^-]$$

$$[H_2NO_2{}^+] = \frac{k_1[H^+][HNO_2]}{k_{-1} + k_2[Br^-]}$$

Substitution in (17.61) gives

$$r = \frac{k_1 k_2[H^+][HNO_2][Br^-]}{k_{-1} + k_2[Br^-]} \tag{17.63}$$

which is the rate law predicted by the steady-state approximation. To obtain agreement with the observed rate law (17.59), it is necessary to further assume that $k_{-1} \gg k_2[Br^-]$, in which case, (17.63) reduces to (17.62). The assumption $k_{-1} \gg k_2[Br^-]$ means that the rate of reversion of $H_2NO_2{}^+$ back to H^+ and HNO_2 is much greater than its rate of reaction with Br^-, which is the condition for step 1 of (17.60) to be in near equilibrium, as in the rate-limiting-step approximation.

The steady-state approximation usually gives more complicated rate laws than the rate-limiting-step approximation. In a given reaction, one or the other or both or neither of these approximations may be valid. Noyes has analyzed the conditions of validity of each approximation. (R. M. Noyes, in *Lewis*, pp. 489–538.)

Computer programs have been developed to numerically integrate the differential equations of a multistep mechanism without any approximations. These programs have shown that the widely used steady-state approximation can sometimes lead to substantial error. [See L. A. Farrow and D. Edelson, *Int. J. Chem. Kinet.*, **6**, 787 (1974).]

The following rules help in finding mechanisms that fit an observed rate law. [See also J. O. Edwards, E. F. Greene, and J. Ross, *J. Chem. Educ.*, **45**, 381 (1968); H. Taube, ibid., **36**, 451 (1959); J. P. Birk, ibid., **47**, 805 (1970); J. F. Bunnett, in *Lewis*, pp. 399–402.] Rules 1 to 4 apply only when the rate-limiting-step approximation is valid.

1a. If the rate law is $r = k[A]^\alpha[B]^\beta \cdots [L]^\lambda$, where $\alpha, \beta, \ldots, \lambda$ are positive integers, the total composition of the reactants in the rate-limiting step is $\alpha A + \beta B + \cdots + \lambda L$. (Specification of the "total composition" of the reactants means specification of the total number of reactant atoms of each type and of the total charge on the reactants; however, the actual species that react in the rate-limiting step cannot be deduced from the rate law.)

The following example will clarify rule 1*a*. The gas-phase reaction $2NO + O_2 \rightarrow 2NO_2$ has the observed rate law $r = k[NO]^2[O_2]$. If we assume a mechanism with a rate-limiting step, the total reactant composition for the rate-limiting step is given by rule 1*a* as $2NO + O_2 = N_2O_4$. One plausible mechanism is

$$\begin{array}{ll} 2NO \rightleftharpoons N_2O_2 & \text{equilib.} \\ N_2O_2 + O_2 \rightarrow 2NO_2 & \text{slow} \end{array} \qquad (17.64)$$

A second mechanism is

$$\begin{array}{ll} NO + O_2 \rightleftharpoons NO_3 & \text{equilib.} \\ NO_3 + NO \rightarrow 2NO_2 & \text{slow} \end{array} \qquad (17.65)$$

A third mechanism is the one-step trimolecular reaction

$$2NO + O_2 \rightarrow 2NO_2 \qquad (17.66)$$

Each of these mechanisms adds up to the overall stoichiometry, and each has N_2O_4 as the total reactant composition in the rate-limiting step. The reader can verify that each mechanism leads to the rate law $r = k[NO]^2[O_2]$. Which of these mechanisms is the correct one is not known, each mechanism having its supporters.

1b. If the rate law is $r = k[A]^\alpha[B]^\beta \cdots [L]^\lambda/[M]^\mu[N]^\nu \cdots [R]^\rho$, where $\alpha, \beta, \ldots, \lambda, \mu, \nu, \ldots, \rho$, are positive integers, the total composition of the reactants in the rate-limiting step is $\alpha A + \beta B + \cdots + \lambda L - \mu M - \nu N - \cdots - \rho R$; moreover, the species $\mu M, \nu N, \ldots, \rho R$ appear as products in equilibria that precede the rate-limiting step, and these species do not enter into the rate-limiting step. (Rule 1*a* is a special case of rule 1*b* with $\mu = \nu = \cdots = \rho = 0$.)

For example, the reaction $Hg_2^{2+} + Tl^{3+} \rightarrow 2Hg^{2+} + Tl^+$ in aqueous solution has the rate law

$$r = k\frac{[Hg_2^{2+}][Tl^{3+}]}{[Hg^{2+}]} \tag{17.67}$$

According to rule 1*b*, the rate-limiting-step reactants have the total composition $Hg_2^{2+} + Tl^{3+} - Hg^{2+} = HgTl^{3+}$; the species Hg^{2+} is not a reactant in the rate-limiting step but is a product in an equilibrium that precedes the rate-limiting step. One possible mechanism (see also Prob. 17.29) is

$$\begin{aligned} Hg_2^{2+} &\underset{k_{-1}}{\overset{k_1}{\rightleftharpoons}} Hg^{2+} + Hg \qquad \text{equilib.} \\ Hg + Tl^{3+} &\xrightarrow{k_2} Hg^{2+} + Tl^+ \qquad \text{slow} \end{aligned} \tag{17.68}$$

The second step is rate-limiting, so $r = k_2[Hg][Tl^{3+}]$. To eliminate the reaction intermediate Hg, we use the equilibrium condition for the first step:

$$K_{c,1} = \frac{k_1}{k_{-1}} = \frac{[Hg^{2+}][Hg]}{[Hg_2^{2+}]} \qquad \text{and} \qquad [Hg] = \frac{k_1[Hg_2^{2+}]}{k_{-1}[Hg^{2+}]} \tag{17.69}$$

Therefore $r = k_1 k_2[Tl^{3+}][Hg_2^{2+}]/k_{-1}[Hg^{2+}]$, in agreement with (17.67).

An apparently puzzling thing about the rate law (17.67) is that it seems to predict $r = \infty$ at the start of the reaction when the concentration of the product Hg^{2+} is zero. Actually, Eq. (17.67) is not valid at the start of the reaction. In deriving (17.67) from the mechanism, we used the equilibrium expression (17.69). Hence (17.67) is valid only for times after the equilibrium $Hg_2^{2+} = Hg^{2+} + Hg$ has been established. Since this equilibrium is rapidly established compared with the rate-limiting second step, any deviation of the rate from (17.67) during the first few instants of the reaction will have no significant influence on the observed kinetics.

2. If (as is usually true) the order with respect to the solvent (S) is unknown, the total reactant composition of the rate-limiting step is $\alpha A + \beta B + \cdots + \lambda L - \mu M - \nu N - \cdots - \rho R + xS$, where the rate law is as in rule 1*b*, and x can be 0, ± 1, ± 2, ...

For example, the reaction $H_3AsO_4 + 3I^- + 2H^+ \rightarrow H_3AsO_3 + I_3^- + H_2O$ in aqueous solution has the rate law $r = k[H_3AsO_4][I^-][H^+]$, where the order with respect to H_2O is unknown. The total composition of the rate-limiting-step reactants is then $H_3AsO_4 + I^- + H^+ + xH_2O = AsIH_{4+2x}O_{4+x}$. Any value of x less than -2 would give a negative number of H atoms, so $x \geq -2$ and the rate-limiting-step reactants have one As atom, one I atom, and at least two O atoms.

3. If the rate law has the factor $[A]^{1/2}$, the mechanism involves splitting an A molecule into two species before the rate-limiting step.

An example is a reaction that is catalyzed by H^+, where the source of the H^+ is the ionization of a weak acid: $CH_3COOH = H^+ + CH_3COO^-$. Suppose the rate-limiting step is $H^+ + A \rightarrow B + C$, where the catalyst H^+ is regenerated in a subsequent rapid step. Then $r = k[H^+][A]$. Since H^+ is regenerated, its concentration remains essentially constant during the reaction and we have $[H^+] = [CH_3COO^-]$. Letting K_c be the acetic acid ionization constant, we have $K_c = [H^+]^2/[CH_3COOH]$, and $[H^+] = K_c^{1/2}[CH_3COOH]^{1/2}$. Therefore $r = kK_c^{1/2}[A][CH_3COOH]^{1/2}$.

Half-integral orders frequently occur in chain reactions (Sec. 17.12). The rate-limiting-step approximation is usually not applicable to chain reactions, but the half-integral orders still result from splitting of a molecule, usually as the first step in the chain reaction.

4. If the absolute value of the stoichiometric coefficient of a reactant exceeds its order in the rate law, there must be at least one step after the rate-limiting step.

An example is the reaction preceding (17.59), where the order with respect to ϕNH_2 is 0 in (17.59). Obviously, a step with ϕNH_2 as a reactant is needed after the rate-limiting step.

5. A rate law with a sum of terms in the denominator indicates a mechanism with one or more reactive intermediates to which the steady-state approximation is applicable (rather than a rate-limiting-step mechanism). An example would be the rate law (17.63).

There are usually a few plausible mechanisms that are compatible with a given observed rate law. Confirming evidence for a proposed mechanism can be obtained by detecting the postulated reaction intermediate(s). If the intermediates are relatively stable, the reaction can be slowed drastically (by cooling or dilution) and the reaction mixture analyzed chemically for the suspected intermediates. For example, rapid chilling of a bunsen-burner flame followed by chemical analysis shows the presence of CH_2O and peroxides, indicating that these species are intermediates in hydrocarbon combustion.

Usually, the intermediates are too unstable to be isolated. Reactive intermediates can frequently be detected spectroscopically. Thus, certain bands in the emission spectrum of an H_2–O_2 flame have been shown to be due to OH radicals. The absorption spectrum of the blue intermediate species NO_3 in the N_2O_5 decomposition [Eq. (17.8)] has been observed in shock-tube studies. Absorption spectra of many reactive species generated in flash photolysis have been observed.

Many gas-phase intermediates (for example, OH, CH_2, CH_3, H, O, C_6H_5) have been detected by allowing some of the reaction mixture to leak into a mass spectrometer. Species with unpaired electrons (free radicals) can be detected by electron-spin-resonance spectroscopy (Sec. 21.12). Frequently, radicals are "trapped" by condensing a mixture of the radical and an inert gas on a cold surface.

Evidence for a postulated reaction intermediate can often be obtained by seeing how the addition of a suitably chosen species affects the reaction rate and the products. For example, the hydrolysis of certain alkyl halides in the mixed solvent acetone–water according to $RCl + H_2O \rightarrow ROH + H^+ + Cl^-$ is found to have $r = k[RCl]$. The proposed mechanism is the slow, rate-determining step $RCl \rightarrow R^+ + Cl^-$, followed by the rapid step $R^+ + H_2O \rightarrow ROH + H^+$. Evidence for the existence of the carbonium ion R^+ is the fact that in the presence of N_3^-, the rate constant and rate law are unchanged, but substantial amounts of RN_3 are formed. The N_3^- combines with R^+ after the rate-determining dissociation of RCl and does not affect the rate.

Isotopically substituted species can help elucidate a mechanism. For example, isotopic tracers show that in a reaction between a primary or secondary alcohol and an organic acid to give an ester and water, the oxygen in the water usually comes from the acid: $R^{18}OH + R'C(O)^{16}OH \rightarrow R'C(O)^{18}OR + H^{16}OH$. The mechanism must involve breaking the C—OH bond of the acid.

The stereochemistry of reactants and products is an important clue to the reaction mechanism. For example, addition of Br_2 to a cycloalkene gives a product in which the two Br atoms are trans to each other; this indicates that Br_2 does not add to the double bond in a single elementary reaction. The relation between mechanism and stereochemistry is discussed in any modern organic chemistry text.

If the mechanism of a reaction has been determined, the mechanism of the reverse reaction is known since the reverse reaction must proceed by a series of steps that is the exact reverse of the forward mechanism (provided the conditions of temperature, solvent, etc., are unchanged). [This must be so, because the laws of particle motion are symmetric with respect to time reversal (Sec. 3.8).] Thus, if the mechanism of the reaction $H_2 + I_2 \rightarrow 2HI$ happened to be $I_2 \rightleftharpoons 2I$, followed by $2I + H_2 \rightarrow 2HI$, the decomposition of HI to H_2 and I_2 at the same temperature would proceed by the mechanism $2HI \rightarrow 2I + H_2$, followed by $2I \rightleftharpoons I_2$.

Once the mechanism of a reaction has been determined, it may be possible to use kinetic data to deduce rate constants for some of the elementary steps in the mechanism. In a reaction with a rate-limiting step, the observed rate constant k is usually the product of the rate constant of the rate-limiting step and the equilibrium constants of the steps preceding this step. Equilibrium constants for gas-phase reactions can frequently be found from thermodynamic data or statistical mechanics (Sec. 22.8), thereby allowing the rate constant of the rate-limiting step to be calculated. Knowledge of rate constants for elementary reactions allows the various theories of reaction rates to be tested.

It is wise to retain some skepticism towards proposed reaction mechanisms. The reaction $H_2 + I_2 \rightarrow 2HI$ was found to have the rate law $r = k[H_2][I_2]$ by Bodenstein in the 1890s. Until 1967, most kineticists believed the mechanism to be the single bimolecular step $H_2 + I_2 \rightarrow 2HI$. In 1967 Sullivan presented strong experimental evidence that the mechanism involves I atoms and might well consist of the rapid equilibrium $I_2 = 2I$ followed by the rate-limiting trimolecular reaction $2I + H_2 \rightarrow 2HI$. (The reader can verify that this mechanism leads to the observed rate law.) Sullivan's work convinced most people that the one-step bimolecular mechanism was wrong. However, further theoretical discussion of Sullivan's data has led some workers to argue that these data are not inconsistent with the bimolecular mechanism. Thus, the mechanism of the H_2–I_2 reaction is now an open question. [See G. G. Hammes and B. Widom, *J. Am. Chem. Soc.*, **96**, 7621 (1974); R. M. Noyes, ibid., **96**, 7623.]

The largest compilation of proposed reaction mechanisms is *Bamford and Tipper*, vols. 4–13.

17.7 TEMPERATURE DEPENDENCE OF RATE CONSTANTS

Rate constants depend strongly on temperature, typically increasing rapidly with increasing T (Fig. 17.7*a*). A rough rule, valid for many reactions in solution, is that near room temperature k doubles or triples for each 10 °C increase in temperature.

In 1889, Arrhenius pointed out that the $k(T)$ data for many reactions fit the expression

$$k = Ae^{-E_a/RT} \tag{17.70*}$$

where A and E_a are constants characteristic of the reaction. E_a is the *Arrhenius activation energy* and A is the *preexponential factor* (or the *Arrhenius A factor*). The units of A are the same as those of k. The units of E_a are the same as those of RT, namely energy per mole; E_a is usually expressed in kcal/mole or kJ/mole. [Arrhenius arrived at (17.70) by arguing that the temperature dependence of rate constants would probably resemble the temperature dependence of equilibrium constants. By analogy to (6.45) and the equation in Prob. 6.5, Arrhenius wrote $d \ln k/dT = E_a/RT^2$, which integrates to (17.70) if E_a is assumed independent of T.]

Taking logs of (17.70), we get

$$\ln k = \ln A - \frac{E_a}{RT} \qquad \text{or} \qquad \log k = \log A - \frac{E_a}{2.303RT} \tag{17.71}$$

If the Arrhenius equation is obeyed, a plot of $\log k$ vs. $1/T$ is a straight line with slope $-E_a/2.303R$ and intercept $\log A$. This allows E_a and A to be found. A typical experimental error in E_a is 1 kcal/mole and in A is a factor of 3 or 4.

Equation (17.70) is found to hold well for nearly all elementary homogeneous reactions and for most complex reactions. A simple interpretation of (17.70) is that two colliding molecules require a certain minimum kinetic energy of relative motion to break the appropriate bonds and allow new compounds to be formed. (For a unimolecular reaction, a certain minimum energy is needed to isomerize or decompose the molecule; the source of this energy is collisions; see Sec. 17.10.) The Boltzmann distribution law (15.52) contains a factor $e^{-\varepsilon/kT}$, and one finds (Sec. 23.1) that the fraction of collisions in which the relative kinetic energy of the molecules along the line of the collision exceeds the value ε_a is proportional to $e^{-\varepsilon_a/kT} = e^{-E_a/RT}$, where $E_a = N_0 \varepsilon_a$ is the molecular kinetic energy expressed on a per mole basis.

Note from (17.70) that a low activation energy means a fast reaction and a high activation energy means a slow reaction. The rapid increase in k as T increases is due mainly to the increase in the number of collisions whose energy exceeds the activation energy.

In the Arrhenius expression (17.70), both A and E_a are constants. Sophisticated

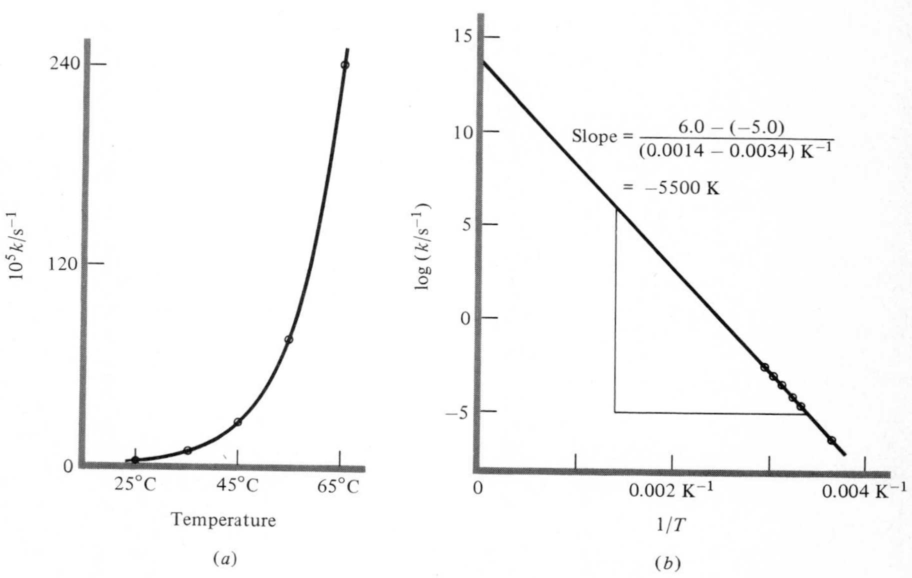

Figure 17.7 (*a*) Rate constant vs. temperature for the first-order decomposition reaction $2N_2O_5 \rightarrow 4NO_2 + O_2$. (*b*) Arrhenius plot for this reaction. From Eq. (17.71), the intercept gives $13.5 = \log (A/s^{-1})$ and $A = 3 \times 10^{13}\ s^{-1}$; the slope gives $-5500\ K = -E_a/2.30R$ and $E_a = 25$ kcal/mole. Note the long extrapolation needed to find A.

theories of reaction rates (Chap. 23) yield an equation similar to (17.70), except that A and E_a both depend on temperature. When $E_a \gg RT$ (which is true for most chemical reactions), the temperature dependences of E_a and A are usually too small to be detected by the rather inaccurate kinetic data available, unless a wide temperature range is studied.

The general definition of the *activation energy* E_a of any rate process, applicable whether or not E_a varies with T, is

$$E_a \equiv RT^2 \frac{d \ln k}{dT} \tag{17.72}$$

(which resembles the equation in Prob. 6.5). If E_a is independent of T, integration of (17.72) yields (17.70), where A is also independent of T. Whether or not E_a depends on T, the *preexponential factor* A for any rate process is defined [in analogy with (17.70)] as

$$A \equiv ke^{E_a/RT} \tag{17.73}$$

From (17.73), we get $k = Ae^{-E_a/RT}$, a generalized version of (17.70), in which both A and E_a may depend on T.

Observed activation energies lie in the range 0 to 80 kcal/mole for most elementary chemical reactions and tend to be lower for bimolecular than unimolecular reactions. (Unimolecular decompositions of compounds with strong bonds have very high E_a values; for example, E_a is 100 kcal/mole for the gas-phase decomposition $CO_2 \rightarrow CO + O$.) An upper limit on observed E_a values is set by the fact that reactions with extremely high activation energies will be too slow to observe.

For unimolecular reactions, A is typically 10^{12} to 10^{15} s^{-1}. For bimolecular reactions, A is typically 10^8 to 10^{12} dm^3 $mole^{-1}$ s^{-1}.

Recombination of two radicals to form a stable polyatomic molecule requires no bonds to be broken, and most such reactions have zero activation energies. Examples are $2CH_3 \rightarrow C_2H_6$ and $CH_3 + Cl \rightarrow CH_3Cl$. (For the recombination of atoms, see Sec. 17.11.) With zero activation energy, the rate constant is essentially independent of T.

Example Calculate E_a for a reaction whose rate at room temperature is doubled by a 10 °C increase in T. Then repeat the calculation for a reaction whose rate is tripled.

Equation (17.70) gives

$$\frac{k(T_2)}{k(T_1)} = \frac{Ae^{-E_a/RT_2}}{Ae^{-E_a/RT_1}} = \exp\left(\frac{E_a}{R}\frac{T_2 - T_1}{T_1 T_2}\right)$$

Taking logs, we get

$$\begin{aligned} E_a &= RT_1 T_2(\Delta T)^{-1} \ln [k(T_2)/k(T_1)] \\ &= (1.987 \text{ cal mole}^{-1} \text{ K}^{-1})(298 \text{ K})(308 \text{ K})(10 \text{ K})^{-1} \ln (2 \text{ or } 3) \end{aligned}$$

$$E_a = \begin{cases} 13 \text{ kcal/mole} & \text{for doubling} \\ 20 \text{ kcal/mole} & \text{for tripling} \end{cases}$$

Example Calculate the room-temperature ratio of the rate constants for two reactions that have the same A value but have E_a values that differ by (a) 1 kcal/mole; (b) 10 kcal/mole.

Equation (17.70) gives

$$\frac{k_1}{k_2} = \frac{Ae^{-E_{a,1}/RT}}{Ae^{-E_{a,2}/RT}} = \exp \frac{E_{a,2} - E_{a,1}}{RT}$$

$$= \exp \frac{1 \text{ kcal mole}^{-1} \quad \text{or} \quad 10 \text{ kcal mole}^{-1}}{(1.987 \times 10^{-3} \text{ kcal mole}^{-1} \text{ K}^{-1})(298 \text{ K})}$$

$$= \begin{cases} 5.4 & \text{for part } (a) \\ 2 \times 10^7 & \text{for part } (b) \end{cases}$$

Each 1 kcal/mole decrease in E_a multiplies the room-temperature rate by 5.4.

The rates of many physiological processes vary with T according to the Arrhenius equation. Examples include the rate of chirping of tree crickets ($E_a = 12$ kcal/mole), the rate of firefly light flashes ($E_a = 12$ kcal/mole), and the frequency of human alpha brain waves ($E_a = 7$ kcal/mole) as measured as a function of body temperature. [See K. J. Laidler, *J. Chem. Educ.*, **49**, 343 (1972).]

Let k_f and k_b be the forward and reverse rate constants of an *elementary* reaction, and let $E_{a,f}$ and $E_{a,b}$ be the corresponding activation energies. Equation (17.57) gives $k_f/k_b = K_c$, where K_c is the concentration-scale equilibrium constant of the reaction. Hence, $\ln k_f - \ln k_b = \ln K_c$. Differentiation with respect to T gives

$$d \ln k_f/dT - d \ln k_b/dT = d \ln K_c/dT \tag{17.74}$$

Equation (17.72) gives $d \ln k_f/dT = E_{a,f}/RT^2$ and $d \ln k_b/dT = E_{a,b}/RT^2$. The result of Prob. 6.5 gives $d \ln K_c/dT = \Delta U^\circ/RT^2$ for an ideal-gas reaction, where ΔU° is the change in standard-state molar internal energy for the reaction and is related to ΔH° by (5.15). Hence, for an ideal-gas elementary reaction, Eq. (17.74) becomes

$$E_{a,f} - E_{a,b} = \Delta U^\circ \qquad \text{elementary reaction} \tag{17.75}$$

Figure 17.8 illustrates Eq. (17.75) for positive and negative ΔU° values.

For reactions in solution, things are a bit more complicated. The K_c in (17.74) is $K_{c,\text{kin}}$ and includes the concentration of the solvent A raised to its stoichiometric coefficient ν_A (recall the discussion at the end of Sec. 17.5). From the relation $K_{c,\text{kin}} = K_{c,\text{td}}[\text{A}]^{\nu_A}$ and the equation in Prob. 11.20c, we get for a reaction in solution $\partial \ln K_{c,\text{kin}}/\partial T = \Delta H^\infty/RT^2 - \alpha_A \sum_i \nu_i$, where α_A is the thermal expansivity of the solvent and the sum goes over all species. The term $\alpha_A \sum_i \nu_i$ is generally quite small compared with $\Delta H^\infty/RT^2$. Moreover, ΔH^∞ (which is the change in standard-state enthalpy values using the molality scale for solutes)

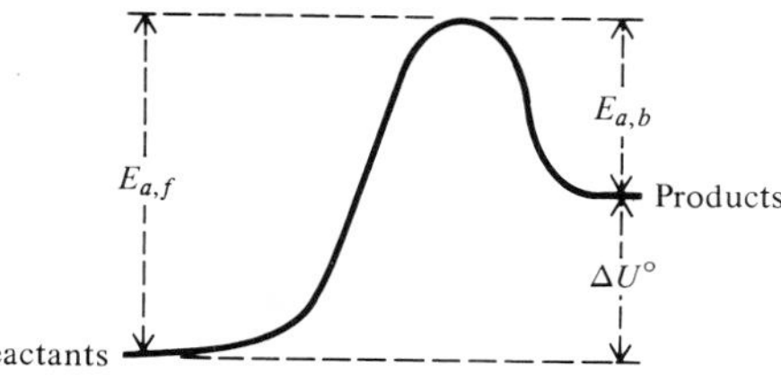

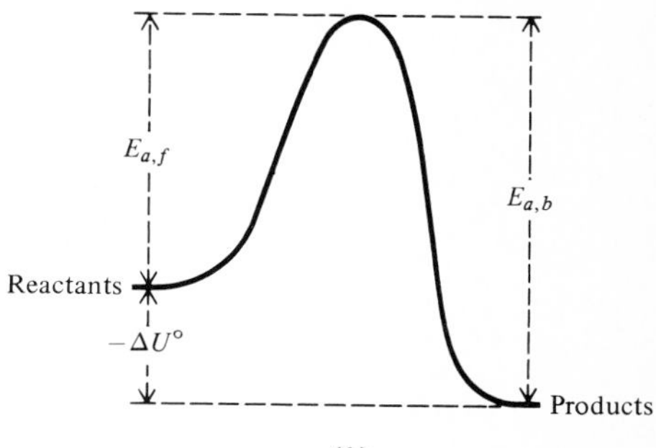

Figure 17.8 Relation between forward and back activation energies and ΔU°.

differs negligibly from ΔU° for a condensed-phase reaction. Hence we can take $d \ln K_c/dT = \Delta U^\circ/RT^2$ for reactions in solution, with no significant error. (Of course, in the spirit of neglecting the pressure dependence of rate constants, we take K_c to depend on T only.) Thus Eq. (17.75) holds well for reactions in solution. To calculate ΔU° for a reaction in solution, we use the standard-state enthalpy values for the species in the relevant solvent.

Now consider the temperature dependence of the rate constant k of an overall reaction composed of several elementary steps. If the rate-limiting-step approximation is valid, k will typically have the form $k_1 k_2/k_{-1}$, where k_1 and k_{-1} are the forward and reverse rate constants of the equilibrium step preceding the rate-determining step 2. [See Eq. (17.62).] Using the Arrhenius equation, we have

$$k = \frac{k_1 k_2}{k_{-1}} = \frac{A_1 e^{-E_{a,1}/RT} A_2 e^{-E_{a,2}/RT}}{A_{-1} e^{-E_{a,-1}/RT}} = \frac{A_1 A_2}{A_{-1}} e^{-(E_{a,1} - E_{a,-1} + E_{a,2})/RT}$$

Therefore, writing k in the form $k = Ae^{-E_a/RT}$, we have for the overall activation energy $E_a = E_{a,1} - E_{a,-1} + E_{a,2}$.

If a reaction proceeds by two competing mechanisms, the overall rate constant will not obey the Arrhenius equation. For example, suppose the reactions in (17.42) are elementary. Then Eq. (17.43) gives the overall rate constant k as $k = k_1 + k_2$, and $k = A_1 e^{-E_{a,1}/RT} + A_2 e^{-E_{a,2}/RT}$, which does not have the form (17.70) unless $E_{a,1}$ and $E_{a,2}$ happen to be equal. (A competing mechanism for a homogeneous gas-phase reaction may be a heterogeneous reaction catalyzed by the walls of the container.) With two competing mechanisms, Eq. (17.72) shows that the one with the higher E_a will increase in rate more rapidly as T increases. At low T, the mechanism with the lower E_a will predominate; at intermediate T, both mechanisms will contribute; at high T, the mechanism with the higher E_a will predominate (provided it has a substantially higher A factor). A plot of log k vs. $1/T$ will be linear at low T and at high T, with a nonlinear transition region at intermediate T. From the high-T and low-T slopes, the activation energy of each mechanism can be determined.

17.8 RELATION BETWEEN RATE CONSTANTS AND EQUILIBRIUM CONSTANTS FOR COMPLEX REACTIONS

For an elementary reaction, K_c equals k_f/k_b [Eq. (17.57)]. For a complex reaction consisting of several elementary steps, this simple relation need not hold. A thorough discussion of the relation between k_f, k_b, and K_c is complicated [see R. K. Boyd, *Chem. Rev.*, **77**, 93 (1977)], so the treatment of this section is restricted to reactions where the rate-limiting-step approximation is valid (and where the system is ideal).

Let the overall reaction be $aA + bB = cC + dD$. When the rate-limiting-step approximation holds, the forward and reverse rates per unit volume will have the forms

$$r_f = k_f[A]^{\alpha_f}[B]^{\beta_f}[C]^{\gamma_f}[D]^{\delta_f} \quad \text{and} \quad r_b = k_b[A]^{\alpha_b}[B]^{\beta_b}[C]^{\gamma_b}[D]^{\delta_b}$$

[Both reactants and products can occur in the forward rate law, as shown by the example (17.68).] r_f is the observed rate per unit volume when the concentrations of products are much less than those of reactants; r_b is the rate per unit volume when the concentrations of reactants are much less than those of products.

Before presenting the general relation between k_f, k_b, and K_c, we look at a specific example. The reaction

$$2Fe^{2+} + 2Hg^{2+} = 2Fe^{3+} + Hg_2^{2+} \tag{17.76}$$

in an aqueous perchloric acid solution is found to have the forward rate law $r_f = k_f[Fe^{2+}][Hg^{2+}]$. A plausible mechanism compatible with this rate law is

$$\begin{aligned} Fe^{2+} + Hg^{2+} &\underset{k_{-1}}{\overset{k_1}{\rightleftharpoons}} Fe^{3+} + Hg^+ \qquad \text{slow} \\ 2Hg^+ &\underset{k_{-2}}{\overset{k_2}{\rightleftharpoons}} Hg_2^{2+} \qquad \text{rapid} \end{aligned} \tag{17.77}$$

where step 1 is rate-limiting. Then

$$r_f = k_1[Fe^{2+}][Hg^{2+}] = k_f[Fe^{2+}][Hg^{2+}] \tag{17.78}$$

(Since we are examining the forward rate with product Fe^{3+} present in only small amounts, we don't include the reverse of the rate-limiting step 1; see the discussion in Sec. 17.6. Likewise in considering the reverse reaction below, we only consider the reverse of step 1 with rate constant k_{-1}.)

For the reverse of reaction (17.76), the mechanism is the reverse of (17.77), namely, the rapid equilibrium $Hg_2^{2+} = 2Hg^+$ followed by the rate-limiting step $Fe^{3+} + Hg^+ \rightarrow Fe^{2+} + Hg^{2+}$ with rate constant k_{-1}. The reverse rate is $r_b = k_{-1}[Fe^{3+}][Hg^+]$. From the equilibrium step, we get the relation $[Hg^+] = (k_{-2}/k_2)^{1/2}[Hg_2^{2+}]^{1/2}$. Hence

$$r_b = k_{-1}(k_{-2}/k_2)^{1/2}[Fe^{3+}][Hg_2^{2+}]^{1/2} = k_b[Fe^{3+}][Hg_2^{2+}]^{1/2} \tag{17.79}$$

At equilibrium, $r_f = r_b$, and (17.78) and (17.79) give

$$\frac{k_f}{k_b} = \frac{[Fe^{3+}]_{eq}([Hg_2^{2+}]_{eq})^{1/2}}{[Fe^{2+}]_{eq}[Hg^{2+}]_{eq}} = K_c^{1/2} \tag{17.80}$$

where K_c is the equilibrium constant for (17.76).

In 1957 Horiuti proved that for a reaction with a rate-limiting step,

$$k_f/k_b = K_c^{1/s} \tag{17.81}$$

where s is the stoichiometric number (Sec. 17.1) of the rate-limiting step. For the reaction (17.76), the rate-limiting step is step 1 of (17.77); step 1 must occur twice for each occurrence of (17.76) since step 1 consumes one Fe^{2+} ion and the overall reaction (17.76) consumes two Fe^{2+} ions. Thus $s = 2$ for the mechanism (17.77) and the overall reaction (17.76); Eq. (17.81) becomes $k_f/k_b = K_c^{1/2}$, as found above. Only when $s = 1$ is $K_c = k_f/k_b$. For a proof of (17.81), see J. Horiuti and T. Nakamura, *Adv. Catal.*, **17**, 1 (1967).

The relation between the equilibrium constant for the overall reaction and the rate constants of the elementary steps of the mechanism (valid whether or not the rate-limiting-step approximation applies) is given in Prob. 17.47.

17.9 THE RATE LAW IN NONIDEAL SYSTEMS

For an elementary reaction in an *ideal* system, the rates of the forward and reverse reactions contain the concentrations of the reacting species, as does the equilibrium constant K_c. In a *nonideal* system, the equilibrium constant for the elementary reaction $aA + bB = cC + dD$ is [Eq. (11.5)]

$$K_a = (a_{C,eq})^c(a_{D,eq})^d/(a_{A,eq})^a(a_{B,eq})^b$$

where $a_{A,eq}$ is the activity of A at equilibrium. It would seem reasonable therefore that the rate laws for an elementary reaction in a nonideal system would be

$$r_f \stackrel{?}{=} k_f a_A{}^a a_B{}^b \qquad \text{and} \qquad r_b \stackrel{?}{=} k_b a_C{}^c a_D{}^d \tag{17.82}$$

Setting $r_f = r_b$ at equilibrium and using (17.82), we get $K_a = k_f/k_b$.

In the 1920s it was generally believed that (17.82) was correct. However, suppose that instead of (17.82) we write

$$r_f = k_f Y a_A{}^a a_B{}^b \qquad \text{and} \qquad r_b = k_b Y a_C{}^c a_D{}^d \qquad \text{elem. react.} \tag{17.83}$$

where Y is some unspecified function of T, P, and the concentrations. Then at equilibrium, $r_f = r_b$, and (17.83) also leads to $K_a = k_f/k_b$, since Y cancels in the numerator and denominator. In fact, kinetic data on ionic reactions in aqueous solution clearly show that (17.82) is wrong and that the correct form of the rate law is (17.83).

Thus, in a nonideal solution, the rate law for the elementary reaction $aA + bB \rightarrow$ products is

$$r = k^\infty Y(\gamma_A[A])^a(\gamma_B[B])^b \equiv k_{app}[A]^a[B]^b \tag{17.84}$$

where the γ's are the concentration-scale activity coefficients, Y is a parameter that depends on T, P, and the concentrations, and the apparent rate constant is defined by $k_{app} \equiv k^\infty Y(\gamma_A)^a(\gamma_B)^b$. (The reason for the ∞ superscript on k will become clear shortly.) In the limit of an infinitely dilute solution, ideal behavior is reached, and r must equal $k^\infty[A]^a[B]^b$; since the γ's become 1 at infinite dilution, Y in (17.84) must go to 1 in the infinite-dilution limit. Therefore, the true rate constant k^∞ can be determined by measuring k_{app} as a function of concentration and extrapolating to infinite dilution. Once k^∞ is known, Y can be calculated for any solution composition from $Y = k_{app}/k^\infty(\gamma_A)^a(\gamma_B)^b$.

In Sec. 23.8, we shall give a physical interpretation of Y. The accuracy of rate data is generally too low to detect deviations from ideality except for ionic reactions.

17.10 UNIMOLECULAR REACTIONS

Most elementary reactions are either bimolecular ($A + B \rightarrow$ products) or unimolecular ($A \rightarrow$ products). Unimolecular reactions are either *isomerizations*, for example, *cis*-CHCl=CHCl $\rightarrow$ *trans*-CHCl=CHCl, or *decompositions*, for example, $CH_3CH_2I \rightarrow CH_2{=}CH_2 + HI$. It is easy to understand how a bimolecular elementary reaction occurs: the molecules A and B collide, and if their relative kinetic energy exceeds the activation energy, the collision can lead to the breaking of bonds and the formation of new bonds. But what about a unimolecular reaction? Why should a molecule spontaneously break apart or isomerize? It would seem reasonable that an A molecule would acquire the necessary activation energy by collision with another molecule; however, a collisional activation would seem to imply second-order kinetics, in contrast to the observed first-order kinetics of unimolecular reactions. The answer to this problem was given by the physicist F. A. Lindemann in 1922.

Lindemann became Lord Cherwell in later life, and was chief scientific advisor to Churchill during World War II. Lindemann's role in the strategic bombing of German cities is discussed in C. P. Snow, *Science and Government* and *Appendix to Science and Government*, Harvard University Press, 1961 and 1962. In his youth, Lindemann won the tennis championship of Sweden.

Lindemann proposed the following detailed mechanism to explain the unimolecular reaction A → B (+ C).

$$\begin{aligned} A + M &\underset{k_{-1}}{\overset{k_1}{\rightleftharpoons}} A^* + M \\ A^* &\xrightarrow{k_2} B(+C) \end{aligned} \tag{17.85}$$

In this scheme, A* is an A molecule that has sufficient vibrational energy to decompose or isomerize (i.e., its vibrational energy exceeds the activation energy for the reaction A → products). A* is called an *energized* molecule. [The species A* is not an activated complex (this term will be defined in Chap. 23) but is simply an A molecule in a high vibrational energy level.] The energized species A* is produced by collision of A with an M molecule (step 1); in this collision, kinetic energy of M is transferred into vibrational energy of A. Any molecule M can excite A to a higher vibrational level; thus M might be another A molecule, or a product molecule, or a molecule of a species present in the gas or solution but not appearing in the overall unimolecular reaction A → products. Once A* has been produced, it can either (*a*) be deenergized back to A by a collision in which the vibrational energy of A* is transferred to kinetic energy of an M molecule (step −1), or (*b*) be transformed to products B + C by having the extra vibrational energy break the appropriate chemical bond(s) to cause decomposition or isomerization (step 2).

The reaction rate per unit volume is $r = d[B]/dt = k_2[A^*]$. Applying the steady-state approximation to the reactive species A*, we have

$$d[A^*]/dt = 0 = k_1[A][M] - k_{-1}[A^*][M] - k_2[A^*]$$

$$[A^*] = \frac{k_1[A][M]}{k_{-1}[M] + k_2}$$

Substituting in $r = k_2[A^*]$, we get

$$r = \frac{k_1 k_2[A][M]}{k_{-1}[M] + k_2} \tag{17.86}$$

The rate law (17.86) has no definite order.

There are two limiting cases for (17.86), depending on the relative sizes of the terms in the denominator. If $k_{-1}[M] \gg k_2$, the k_2 term in the denominator can be dropped to give

$$r = (k_1 k_2/k_{-1})[A] \qquad \text{for } k_{-1}[M] \gg k_2 \tag{17.87}$$

If $k_2 \gg k_{-1}[M]$, the $k_{-1}[M]$ term is omitted and

$$r = k_1[A][M] \qquad \text{for } k_2 \gg k_{-1}[M] \tag{17.88}$$

In gas-phase reactions, Eq. (17.87) may be called the high-pressure limit, since at high pressures, the concentration [M] is large and $k_{-1}[M]$ is much larger than k_2. Equation (17.88) is the low-pressure limit.

The high-pressure rate law (17.87) is first order. The low-pressure rate law (17.88) is second order but is more subtle than it looks. The concentration [M] is the total concentration of all species present. If the overall reaction is an isomerization, $A \rightarrow B$, the total concentration [M] remains constant as the reaction progresses and one observes pseudo-first-order kinetics. If the overall reaction is a decomposition, $A \rightarrow B + C$, then [M] increases as the reaction progresses; however, the decomposition products B and C are generally less effficient (i.e., have smaller k_1 values) than A in energizing A, and this approximately compensates for the increase in [M]; thus, $k_1[M]$ remains approximately constant during the reaction, and again we have pseudo-first-order kinetics.

In the high-pressure limit $k_{-1}[M] \gg k_2$, the rate of the deenergization reaction $A^* + M \rightarrow A + M$ is much greater than the rate of $A^* \rightarrow B + C$, and steps 1 and -1 are essentially in equilibrium; the unimolecular step 2 is then rate-controlling, and we get first-order kinetics [Eq. (17.87)]. In the low-pressure limit $k_{-1}[M] \ll k_2$, the reaction $A^* \rightarrow B + C$ is much faster than the deenergization reaction, the rate-limiting step is the bimolecular energization reaction $A + M \rightarrow A^* + M$ (which is relatively slow due to the low values of [M] and [A]), and we get second-order kinetics [Eq. (17.88)].

A key idea in the Lindemann mechanism is the time lag that exists between the energization of A to A* and the decomposition of A* to products; this time lag allows A* to be deenergized back to A, and the near equilibrium of steps 1 and -1 produces first-order kinetics. In the limit of zero lifetime of A*, the reaction would become $A + M \rightarrow B\,(+\,C)$ and would be second order. Note also that in the limit $k_2 \rightarrow \infty$, Eq. (17.86) gives second-order kinetics. The vibrationally excited species A* has a finite (i.e., nonzero) lifetime because the molecule has several bonds, and it takes time for the vibrational energy to concentrate in the particular chemical bond that breaks in the reaction A → products. It follows that a molecule with only one bond (for example, I_2) cannot decompose by a unimolecular reaction (see also Sec. 17.11).

The experimental unimolecular rate constant k_{uni} is defined by $r = k_{\text{uni}}[A]$, where r is the observed rate per unit volume. Equation (17.86) gives

$$k_{\text{uni}} = \frac{k_1 k_2[M]}{k_{-1}[M] + k_2} = \frac{k_1 k_2}{k_{-1} + k_2/[M]} \tag{17.89}$$

The high-pressure limit of k_{uni} is $k_{\text{uni},P=\infty} = k_1 k_2/k_{-1}$. As the initial pressure P_0 for a run is decreased, k_{uni} decreases (since [M] decreases). At very low initial pressures, $k_{\text{uni}} = k_1[M]$, and k_{uni} decreases linearly with decreasing P_0. This predicted falloff of k_{uni} with decreasing P_0 has been experimentally confirmed for most unimolecular gas-phase reactions by measuring the initial rate r_0 as a function of the initial gas pressure P_0. Figure 17.9 shows a typical result. The pressure at which significant falloff of k_{uni} from its high-pressure value begins is usually in the range 10 to 200 torr.

The Lindemann mechanism also applies to reactions in liquid solutions. However, in solutions, it is not possible to observe a falloff of k_{uni} since the presence of the solvent keeps [M] high. Thus the rate law is (17.87) in solution.

Steps 1 and -1 of the Lindemann mechanism (17.85) are not elementary *chemical* reactions (since no new compounds are formed) but are elementary *physical* reactions in which energy is transferred. Of course, such energy-transfer processes occur continually in any system. The reasons for considering steps 1 and -1 in addition to the unimolecular elementary chemical reaction of step 2 are (*a*) to explain how collisional activation can produce first-order kinetics and (*b*) to deal with the low-P falloff of gas-phase unimolecular rate constants. Unless one is dealing with a gas-phase system in the low-P range, it is not necessary to consider steps 1 and -1 explicitly, and a unimolecular reaction can be written simply as $A \rightarrow$ products.

17.11 TRIMOLECULAR REACTIONS

Trimolecular (elementary) reactions are quite uncommon. The best examples of gas-phase trimolecular reactions are the recombinations of two atoms to form a diatomic molecule. The energy released on formation of the chemical bond becomes vibrational energy of the diatomic molecule, and unless a third body is present to carry away this energy, the molecule will dissociate back to atoms during its first vibration. Thus, the recombination of two I atoms to form an I_2 molecule occurs as the single elementary step

$$I + I + M \rightarrow I_2 + M \tag{17.90}$$

where M can be any atom or molecule. The observed rate law is $r = k[I]^2[M]$. (A reaction like $CH_3 + CH_3 \rightarrow C_2H_6$ does not require a third body, since the extra vibrational energy attained on formation of the C_2H_6 molecule can be distributed among vibrations of several bonds and no bond vibration need be energetic enough to break that bond.) The species M in (17.90) is sometimes called a "chaperon," but this name is inappropriate since the commonly understood function of a chaperon is to prevent union rather than promote it.

The rate constant for (17.90) has been measured as a function of T in flash-photolysis experiments. Since no bonds are broken in (17.90), one would expect it to have zero activation energy. Actually, the rate constant *decreases* with increasing T, indicating a negative E_a [see Eq. (17.72)]. Increasing the temperature increases the trimolecular collision rate; however, as the energy of a trimolecular collision increases, the probability decreases that a given $I + I + M$ collision will result in transfer of energy to M and concomitant formation of I_2. The activation energy for

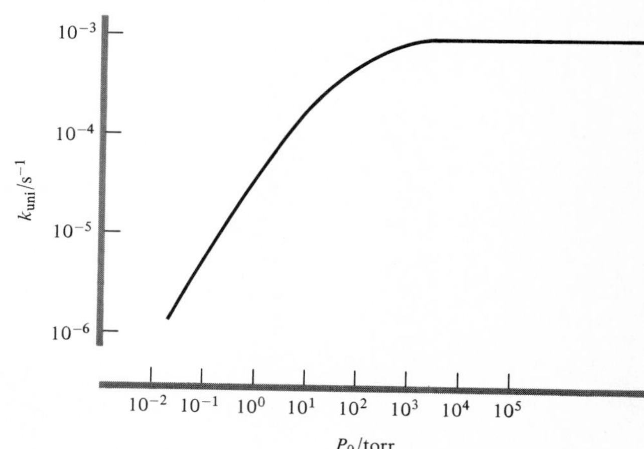

Figure 17.9
Observed rate constant for the unimolecular reaction $CH_3NC \rightarrow CH_3CN$ at 230°C as a function of initial pressure P_0. The scales are logarithmic.

the recombination of atoms $(A + B + M \rightarrow AB + M)$ is typically -1 to -4 kcal/mole, depending on the nature of A and M.

The decomposition of I_2 (or any diatomic molecule) must occur by the reverse of (17.90) (see Sec. 17.6). Thus, the diatomic molecule AB decomposes by the bimolecular reaction $AB + M \rightarrow A + B + M$, where the atoms A and B may be the same (as in I_2) or different (as in HCl). Since E_a for (17.90) is slightly negative, Eq. (17.75) shows that E_a for the decomposition of a diatomic molecule is slightly less than ΔU° for the decomposition. For decomposition of a polyatomic molecule to two radicals, E_a equals ΔU° since E_a for the recombination is zero (as noted in Sec. 17.7).

A recombination reaction to give a triatomic molecule often requires a third body M to carry away energy. A triatomic molecule has only two bonds, and the extra vibrational energy produced by recombination might rapidly concentrate in one bond and dissociate the molecule unless a third body is present. For example, the recombination of O_2 and O is $O + O_2 + M \rightarrow O_3 + M$. Reversal of this elementary reaction shows that O_3 decomposes by a bimolecular step (see Prob. 17.31). Molecules with several bonds can decompose by a unimolecular reaction and do not require a third body when they are formed in a recombination reaction.

The gas-phase reactions of NO with Cl_2, Br_2, and O_2 are kinetically third order. Some people believe the mechanism to be a single elementary trimolecular step (for example, $2NO + Cl_2 \rightarrow 2NOCl$), but others consider the mechanism to be two bimolecular steps, as in (17.64) or (17.65).

In solutions, trimolecular (elementary) reactions are also uncommon.

17.12 CHAIN REACTIONS

A *chain reaction* contains a series of steps in which a reactive intermediate is consumed, reactants are converted to products, and the intermediate is regenerated. Regeneration of the intermediate allows this cycle to be repeated over and over again. Thus a small amount of intermediate leads to production of a large amount of product.

One of the best-understood chain reactions is that between H_2 and Br_2. The overall stoichiometry is $H_2 + Br_2 = 2HBr$. The observed rate law for this gas-phase reaction in the temperature range 500 to 1500 K is

$$r = \frac{1}{2}\frac{d[\mathrm{HBr}]}{dt} = \frac{k[\mathrm{H_2}][\mathrm{Br_2}]^{1/2}}{1 + j[\mathrm{HBr}]/[\mathrm{Br_2}]} \tag{17.91}$$

where k and j are constants. Since $\nu = 2$ for HBr, we include a factor $\frac{1}{2}$ in (17.91) [see Eq. (17.4)]. The constant j has only a very slight temperature dependence and equals 0.12. During the early stages of the reaction, we have $[\mathrm{HBr}]/[\mathrm{Br_2}] \ll 1$, and the second term in the denominator can be neglected; hence $r \approx k[\mathrm{H_2}][\mathrm{Br_2}]^{1/2}$ for $[\mathrm{HBr}] \ll [\mathrm{Br_2}]$. Since an increase in [HBr] decreases r in (17.91), the product HBr is said to *inhibit* the reaction.

The $\frac{1}{2}$-power dependence of r on $[\mathrm{Br_2}]$ suggests that the mechanism involves splitting a Br_2 molecule (recall rule 3 in Sec. 17.6). Br_2 can only split into two Br atoms. A Br atom can then react with H_2 to give HBr and H. Each H atom produced can then react with Br_2 to give HBr and Br, thereby regenerating the reactive intermediate Br. The mechanism of the reaction is thus believed to be

$$\begin{aligned} &Br_2 + M \underset{k_{-1}}{\overset{k_1}{\rightleftharpoons}} 2Br + M \\ &Br + H_2 \underset{k_{-2}}{\overset{k_2}{\rightleftharpoons}} HBr + H \\ &H + Br_2 \xrightarrow{k_3} HBr + Br \end{aligned} \qquad (17.92)$$

In step 1, a Br_2 molecule collides with any species M, thereby gaining the energy to dissociate into two Br atoms. Step -1 is the reverse process, in which two Br atoms recombine to form Br_2, the third body M being needed to carry away part of the bond energy released (see Sec. 17.11). Step 1 is the *chain-initiation step*, since it generates the chain-carrying reactive species Br. Step -1 is the *chain-termination* (or *chain-breaking*) *step*, since it removes Br.

Steps 2 and 3 form a *chain* that consumes Br, converts H_2 and Br_2 into HBr, and regenerates Br (Fig. 17.10). Steps 2 and 3 are *chain-propagating steps*. Step -2 ($HBr + H \rightarrow Br + H_2$) is an *inhibiting step*, since it destroys the product HBr and therefore decreases r. Note from steps -2 and 3 that HBr and Br_2 compete for H atoms; this competition leads to the $j[HBr]/[Br_2]$ term in the denominator of r. For each Br atom produced by step 1, we get many repetitions of steps 2 and 3 (we shall see below that $k_1 \ll k_2$ and $k_1 \ll k_3$). The reactive intermediates H and Br that occur in the chain-propagating steps are called *chain carriers*. Adding steps 2 and 3, we get $Br + H_2 + Br_2 \rightarrow 2HBr + Br$, which agrees with the overall stoichiometry $H_2 + Br_2 \rightarrow 2HBr$.

The mechanism (17.92) gives the rate of product formation as

$$d[HBr]/dt = k_2[Br][H_2] - k_{-2}[HBr][H] + k_3[H][Br_2] \qquad (17.93)$$

This expression contains the free-radical intermediates H and Br. Applying the steady-state approximation to these two reactive intermediates, we get

$$d[H]/dt = 0 = k_2[Br][H_2] - k_{-2}[HBr][H] - k_3[H][Br_2] \qquad (17.94)$$

$$\begin{aligned} d[Br]/dt = 0 = {} & 2k_1[Br_2][M] - 2k_{-1}[Br]^2[M] - k_2[Br][H_2] \\ & + k_{-2}[HBr][H] + k_3[H][Br_2] \end{aligned} \qquad (17.95)$$

The factors of 2 in (17.95) arise because Eq. (17.4) gives, for step 1, $\frac{1}{2}d[Br]/dt = k_1[Br_2][M]$, and similarly for step -1. Adding (17.94) and (17.95), we get

$$0 = 2k_1[Br_2][M] - 2k_{-1}[Br]^2[M]$$

$$[Br] = (k_1/k_{-1})^{1/2}[Br_2]^{1/2} \qquad (17.96)$$

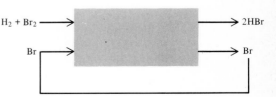

Figure 17.10
Scheme for the $H_2 + Br_2 \rightarrow 2HBr$ chain reaction.

Substituting (17.96) into (17.94) and solving for [H], we get

$$[\mathrm{H}] = \frac{k_2(k_1/k_{-1})^{1/2}[\mathrm{Br_2}]^{1/2}[\mathrm{H_2}]}{k_3[\mathrm{Br_2}] + k_{-2}[\mathrm{HBr}]} = \frac{k_2(k_1/k_{-1})^{1/2}[\mathrm{H_2}][\mathrm{Br_2}]^{-1/2}}{k_3 + k_{-2}[\mathrm{HBr}]/[\mathrm{Br_2}]} \tag{17.97}$$

By substituting (17.97) and (17.96) into (17.93), we can find $d[\mathrm{HBr}]/dt$ as a function of $[\mathrm{H_2}]$, $[\mathrm{Br_2}]$, and [HBr]. To avoid the tedious algebra involved, we note from (17.94) that $k_2[\mathrm{Br}][\mathrm{H_2}] = k_{-2}[\mathrm{HBr}][\mathrm{H}] + k_3[\mathrm{H}][\mathrm{Br_2}]$. Use of this expression for $k_2[\mathrm{Br}][\mathrm{H_2}]$ in (17.93) gives $d[\mathrm{HBr}]/dt = 2k_3[\mathrm{H}][\mathrm{Br_2}]$; use of (17.97) for [H] then gives the desired result:

$$r = \frac{1}{2}\frac{d[\mathrm{HBr}]}{dt} = \frac{k_2(k_1/k_{-1})^{1/2}[\mathrm{H_2}][\mathrm{Br_2}]^{1/2}}{1 + (k_{-2}/k_3)[\mathrm{HBr}]/[\mathrm{Br_2}]} \tag{17.98}$$

which happily agrees with the observed form of the rate law, Eq. (17.91). We have

$$k = k_2(k_1/k_{-1})^{1/2} \qquad \text{and} \qquad j = k_{-2}/k_3 \tag{17.99}$$

The activation energies for all five steps in this mechanism can be estimated. The trimolecular recombination reaction, step -1, must have an E_a that is essentially zero or even slightly negative (see Sec. 17.11); we thus take $E_{a,-1} \approx 0$. For step 1, thermodynamic data in the Appendix give $\Delta U° = 45$ kcal/mole, and Eq. (17.75) therefore gives $E_{a,1} \approx 45$ kcal/mole. The ratio k_1/k_{-1} in (17.99) is the equilibrium constant $K_{c,1}$ for the dissociation reaction $\mathrm{Br_2} = 2\mathrm{Br}$, and $K_{c,1}(T)$ can be found from thermodynamic data. The rate constant k in (17.91) is known as a function of T from measurement of r at different temperatures. The first equation in (17.99) then allows us to find the elementary rate constant k_2 as a function of T. Use of (17.72) then gives the activation energy for step 2. The result is $E_{a,2} = 18$ kcal/mole. Thermodynamic data give $\Delta U°$ for reaction 2 as 17 kcal/mole; Eq. (17.75) then gives $E_{a,-2} = 1$ kcal/mole. For the constant j, we have from (17.99) and (17.70):

$$j = k_{-2}/k_3 = (A_{-2}/A_3)e^{(E_{a,3} - E_{a,-2})/RT}$$

Since j is observed to be essentially independent of T, we must have $E_{a,3} = E_{a,-2} =$ 1 kcal/mole. Note that $E_{a,1}$ (45 kcal/mole) is much greater than $E_{a,2}$ (18 kcal/mole) and $E_{a,3}$ (1 kcal/mole).

In the mechanism (17.92), the chain reaction is *thermally* initiated by heating the reaction mixture to a temperature at which some $\mathrm{Br_2}$–M collisions have enough relative kinetic energy to dissociate $\mathrm{Br_2}$ to the chain carrier Br. The $\mathrm{H_2}$–$\mathrm{Br_2}$ chain reaction can also be initiated *photochemically* (at lower temperatures than required for the thermal reaction) by absorption of light, which dissociates $\mathrm{Br_2}$ to 2Br. Still another way to initiate a chain reaction is by addition of a substance (called an *initiator*) that reacts to produce chain carriers. For example, Na vapor added to an $\mathrm{H_2}$–$\mathrm{Br_2}$ mixture will react with $\mathrm{Br_2}$ to give the chain carrier Br: $\mathrm{Na} + \mathrm{Br_2} \rightarrow \mathrm{NaBr} + \mathrm{Br}$.

Since each atom or molecule of chain carrier produces many product molecules, a small amount of any substance that destroys chain carriers will slow down (or *inhibit*) a chain reaction to a marked extent. For example, NO can combine with the chain-carrying radical $\mathrm{CH_3}$ to give $\mathrm{CH_3NO}$. $\mathrm{O_2}$ is an inhibitor in the $\mathrm{H_2}$–$\mathrm{Cl_2}$ chain reaction since it combines with Cl atoms to give $\mathrm{ClO_2}$.

Each of the steps 2 and 3 in the chain of the $\mathrm{H_2}$–$\mathrm{Br_2}$ reaction consumes one chain carrier and produces one chain carrier. In certain chain reactions, the chain produces more chain carriers than it consumes; this is a *branching chain reaction.* For a branching chain reaction, the reaction rate may increase rapidly as the reaction proceeds, and this increase may lead to an explosion; obviously, the steady-state approximation doesn't apply in such a situation. One of the most studied branching chain reactions is the combustion of hydrogen: $2H_2 + O_2 \rightarrow 2H_2O$. The chain-branching steps include $H + O_2 \rightarrow OH + O$ and $O + H_2 \rightarrow OH + H$. Each of these reactions produces two chain carriers and consumes only one. At a given temperature, the O_2–H_2 chain reaction may or may not produce an explosion, depending on the pressure. At low pressures, many of the chain-carrying radicals are destroyed by collisions with the container walls (where they are adsorbed) before they collide and react with O_2 or H_2; this prevents explosion. At higher pressure, the reaction mixture explodes. At still higher pressures, the trimolecular reaction $H + O_2 + M \rightarrow HO_2 + M$ occurs to a significant degree and reduces the concentration of the chain carrier H, thereby preventing explosion. At yet higher pressures, the mixture explodes.

Branching chain reactions occur in the oxidation of gaseous hydrocarbons. The $Pb(C_2H_5)_4$ that is added to gasoline reacts with radicals, thereby breaking the chain and regulating the oxidation to avoid engine "knocking." Formation of atmospheric smog involves chain reactions that oxidize hydrocarbons; initiation occurs photochemically when the pollutant gas NO_2 absorbs light and dissociates to give chain-carrying O atoms. The O then attacks pollutant gaseous hydrocarbons (originating from automobiles and industrial processes), yielding such noxious chemicals as aldehydes and peroxynitrates [formula $RC(O)OONO_2$]; oxygen atoms also combine with O_2 to give O_3.

Chain reactions also occur in solutions, e.g., in certain addition polymerizations.

In chain reactions, the carriers are usually free radicals (including atoms).

17.13 REACTIONS IN LIQUID SOLUTIONS

Most of the ideas of the previous sections of this chapter apply to both gas-phase and liquid-phase kinetics. We now examine aspects of reaction kinetics unique to reactions in solutions.

Solvent effects on rate constants. The difference between a gas-phase and a liquid-phase reaction is the presence of the solvent. The reaction rate can depend strongly on the solvent used. For example, rate constants at 25°C for the second-order substitution reaction $CH_3I + Cl^- \rightarrow CH_3Cl + I^-$ in three different amide solvents are:

Solvent	$HC(O)NH_2$	$HC(O)N(H)CH_3$	$HC(O)N(CH_3)_2$
$k/(\text{dm}^3 \text{ mole}^{-1} \text{ s}^{-1})$	5×10^{-5}	1.4×10^{-4}	4×10^{-1}

Thus the rate constant k for a given reaction is a function of the solvent as well as the temperature.

Solvent effects on reaction rates have many sources. The reacting species are usually *solvated* (i.e., are bound to one or more solvent molecules), and the degree of solvation changes with change in solvent, thus affecting k. Certain solvents may catalyze the reaction. Most reactions in solution involve ions or polar molecules, and here the electrostatic forces between the reactants depend on the solvent dielectric constant. The rate of very fast reactions in solution may be limited by the rate at which two reactant molecules can diffuse through the solvent to encounter each other, and here the solvent's viscosity influences k. Hydrogen bonding between solvent and a reactant can affect k.

For a reaction that can occur by two competing mechanisms, the rates of these mechanisms may be affected differently by a change in solvent, so the mechanism can differ from one solvent to another.

For certain unimolecular reactions and certain bimolecular reactions between nonpolar species, the rate constant is essentially unchanged on going from one solvent to another. For example, rate constants at 50°C for the bimolecular Diels–Alder dimerization of cyclopentadiene ($2C_5H_6 \rightarrow C_{10}H_{12}$) in the gas phase and in three solvents are:

Medium	gas	CS_2	C_6H_6	C_2H_5OH
$k/(\text{dm}^3\ \text{mole}^{-1}\ \text{s}^{-1})$	6×10^{-6}	6×10^{-6}	10×10^{-6}	20×10^{-6}

When the solvent is a reactant, it is usually not possible to determine the order with respect to solvent.

Ionic reactions. In gas-phase kinetics, reactions involving ions are rare. In solution, ionic reactions are abundant. The difference is due to solvation of ions in solution, which sharply reduces ΔH° (and hence ΔG°) for ionization.

Ionic gas-phase reactions occur when the energy needed to ionize molecules is supplied by outside sources. In a mass spectrometer, bombardment by an electron beam knocks electrons out of gas-phase molecules, and the kinetics of ionic reactions in gases can be studied in a mass spectrometer. In the earth's upper atmosphere, absorption of light produces O_2^+, N_2^+, O^+, and He^+ ions, which then undergo various reactions.

For ionic reactions in solution, activity coefficients must be used in analyzing kinetic data (Sec. 17.9). Since the activity coefficients are frequently unknown, one often adds a substantial amount of an inert salt to keep the ionic strength (and hence the activity coefficients) essentially constant during the reaction. The apparent rate constant obtained then depends on the ionic strength.

Many ionic reactions in solution are extremely fast; see the discussion below of diffusion-controlled reactions.

Encounters, collisions, and the cage effect. In a gas at low or moderate pressure, the molecules are far apart and move freely between collisions. In a liquid, there is little empty space between molecules, and the molecules cannot move freely. Instead, a given molecule can be viewed as being surrounded by a "cage" formed by other molecules. A given molecule vibrates against the "walls" of this cage many times before it "squeezes" through the closely packed surrounding molecules and diffuses out of the cage. A liquid's structure thus resembles somewhat the structure of a solid.

This reduced mobility in liquids hinders two reacting solute molecules A and B from getting to each other in solution. However, once A and B do meet, they will be surrounded by a cage of solvent molecules that keeps them close together for a relatively long time, during which they collide repeatedly with each other (and with the cage walls of solvent molecules). A process in which A and B diffuse together to become neighbors is called an *encounter*. Each encounter in solution involves many *collisions* between A and B while they remain trapped in the solvent cage (Fig. 17.11). (In a gas, there is no distinction between a collision and an encounter.)

Various theoretical estimates indicate that two molecules in a solvent cage will collide on the order of 20 to 200 times before they diffuse out of the cage. (In liquids a collision is not as well defined a process as in a gas.) The number of collisions per encounter will be greater the greater the viscosity of the solvent. Although the number of encounters per unit volume per second between pairs of solute molecules in a liquid solution is much less than the corresponding number of collisions in a gas, the compensating effect of a large number of collisions per encounter in solution makes the collision number roughly the same in solution as in a gas at comparable concentrations of A and B. Direct evidence for this is the near constancy of rate constants for nonpolar reactions on going from the gas phase to a solution (see the above data on the C_5H_6 dimerization). Although the collision number is about the same in a gas and in solution, the pattern of collisions is quite different, with collisions in solution grouped into sets.

What experimental evidence exists for the cage effect? In 1936 Rabinowitch and Wood placed balls on a tray that was shaken mechanically. They found that when the number of spheres was increased to the point where there was not much empty space (as would be true in a liquid), the collisions between a given pair of marked spheres (representing two solute molecules) occurred in sets, with short time intervals between successive collisions of any one set and long intervals between successive sets of collisions (Fig. 17.12).

Playing around with marbles is all well and good, but one would like more direct evidence. In 1961, Lyon and Levy photochemically decomposed mixtures of the isotopic species CH_3NNCH_3 and CD_3NNCD_3. Absorption of light dissociates the molecules to N_2 and $2CH_3$ (or $2CD_3$). The methyl radicals then combine to give ethane. When the reaction was carried out in the gas phase, the ethane formed consisted of CH_3CH_3, CH_3CD_3, and CD_3CD_3 in proportions indicating random mixing of CH_3 and CD_3 before recombination. When the reaction was carried out in the inert solvent isooctane, only CH_3CH_3 and CD_3CD_3 were obtained; the absence of CH_3CD_3 showed that the solvent cage kept the two methyl radicals formed from a given parent molecule together until they recombined.

Diffusion-controlled reactions. Suppose the activation energy for the bimolecular elementary reaction $A + B \rightarrow$ products in solution is very low, so that there is a substantial probability for reaction to occur at each collision. Since each encounter in solution consists of about 100 collisions, it may well be that A and B will react

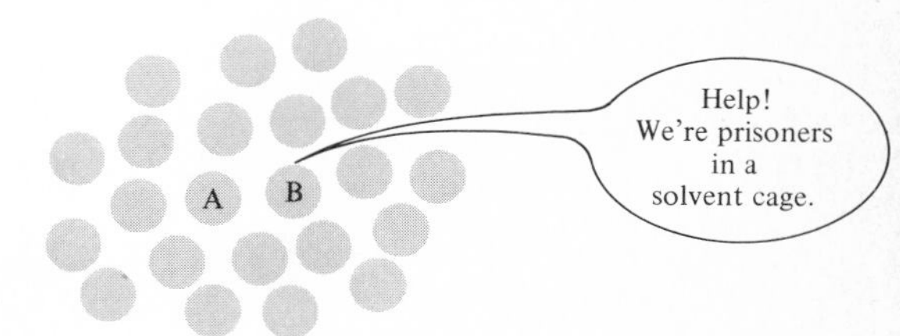

A
B
Help!
We're prisoners
in a
solvent cage.

Figure 17.11
Molecules A and B in a solvent cage.

every time they encounter each other. The reaction rate will then be given by the number of A-B encounters per second, and the rate will be determined solely by how fast A and B can diffuse toward each other through the solvent. A reaction that occurs whenever A and B encounter each other is called a *diffusion-controlled reaction.*

In 1917 Smoluchowski derived the following theoretical expression for the rate constant k_D of a diffusion-controlled reaction:

$$k_D = 4\pi N_0(r_A + r_B)(D_A + D_B) \qquad \text{where A} \neq \text{B, nonionic} \qquad (17.100)$$

In this equation, N_0 is the Avogadro constant, r_A and r_B are the radii of A and B (for simplicity, the molecules A and B are taken as spheres), and D_A and D_B are the diffusion coefficients (Sec. 16.4) of A and B in the solvent. A derivation of (17.100) is given in Sec. 23.8.

The rate of the diffusion-controlled elementary reaction A + B → products is $r = -d[A]/dt = k_D[A][B]$ when A and B are different species. However, when A and B are the same species, the definition (17.4) gives the rate of 2A → products as $r = -\frac{1}{2}d[A]/dt = k_D[A]^2$; because of the factor $\frac{1}{2}$ in the definition of r, a factor of $\frac{1}{2}$ must be added to (17.100) when A and B are the same species:

$$k_D = 2\pi N_0(r_A + r_B)(D_A + D_B) \qquad \text{where A} = \text{B, nonionic} \qquad (17.101)$$

where $r_A = r_B$ and $D_A = D_B$.

Equations (17.100) and (17.101) apply when A and B are uncharged. However, if A and B are ions, the strong Coulomb attraction or repulsion will clearly affect the encounter rate. Debye in 1942 showed that for ionic diffusion-controlled reactions in very dilute solutions

$$k_D = 4\pi N_0(D_A + D_B)(r_A + r_B)\frac{W}{e^W - 1} \qquad \text{where A} \neq \text{B, ionic} \qquad (17.102)$$

$$W \equiv \frac{z_A z_B e^2}{4\pi\varepsilon_0 \varepsilon_r kT(r_A + r_B)} \qquad (17.103)$$

In the definition (17.103) of W, SI units are used, ε_r is the solvent dielectric constant, z_A and z_B are the charge numbers of A and B, k is Boltzmann's constant, e is the proton charge, and ε_0 is the permittivity of vacuum. The quantity $r_A + r_B$ is the same as a in the Debye–Hückel equation (10.89). Since a typically ranges from 3 to 8 Å, a reasonable value for $r_A + r_B$ is 5 Å $= 5 \times 10^{-10}$ m. Using the SI values of k, e, and ε_0 and the value $\varepsilon_r = 78.4$ for water at 25°C, one finds for H_2O at 25°C and $r_A + r_B = 5$ Å:

z_A, z_B	1, 1	2, 1	2, 2	1, −1	2, −1	2, −2	3, −1
$W/(e^W - 1)$	0.45	0.17	0.019	1.9	3.0	5.7	4.3

Time ⟶

Figure 17.12
Collision pattern in a liquid.

Equations (17.100) to (17.102) set upper limits on the rate constants for reactions in solution. To test them, we need reactions with essentially zero activation energy. Recombination of two radicals to form a stable molecule has $E_a \approx 0$. Using flash photolysis, one can produce such species in solution and measure their recombination rate. A precise test of the equations for k_D is usually not possible since the diffusion coefficients of such radicals in solution are not known; however, D_A and D_B can be estimated by analogy with stable species having similar structures. Recombination reactions studied include $I + I \rightarrow I_2$ in CCl_4, $OH + OH \rightarrow H_2O_2$ in H_2O, and $2CCl_3 \rightarrow C_2Cl_6$ in cyclohexene. One finds that recombination rates in solution are generally in good agreement with the theoretically calculated values. (See Prob. 17.55 for an example.)

To decide whether a reaction is diffusion-controlled, one compares the observed k with k_D calculated from one of the above equations. Rate constants for many very fast reactions in solution have been measured using relaxation techniques. All reactions of H_3O^+ with strong bases (for example, OH^-, $C_2H_3O_2^-$) are found to be diffusion-controlled. The majority of reactions of $e^-(aq)$ (produced by pulse radiolysis) are diffusion-controlled.

An especially interesting reaction is $H_3O^+ + OH^- \rightarrow 2H_2O$, for which $k = 1.4 \times 10^{11}$ dm^3 mole^{-1} s^{-1} at 25°C. Using the known diffusion coefficients of the ions, one finds that (17.102) gives the correct rate constant if $r_A + r_B$ is taken as 8 Å. Now 8 Å is far larger than the sum of the H_3O^+ and OH^- radii but is about right if the reacting species are $H_3O^+(H_2O)_3 = H_9O_4^+$ and $OH^-(H_2O)_3 = H_7O_4^-$. The H_3O^+ ion solvates three water molecules since each hydrogen can hydrogen bond to the oxygen of a water molecule. In OH^-, the oxygen has three unshared electron pairs and therefore forms hydrogen bonds with three water molecules. The species $H_9O_4^+$ and $H_7O_4^-$ have been observed in a mass spectrometer. Transfer of the proton from $H_9O_4^+$ to $H_7O_4^-$ probably involves shifts in hydrogen bonds (see fig. 3.1 in *Hague*). The reaction $H_3O^+ + OH^- \rightarrow 2H_2O$ is the fastest known bimolecular reaction in aqueous solution. This reaction proceeds even faster in ice, due to the better-developed hydrogen-bonding structure in ice. [The self-ionization reaction (11.13) also occurs in ice, although to a considerably smaller extent than in water.]

Equations (17.100) to (17.102) for k_D can be simplified by assuming the applicability of the Stokes–Einstein equation (16.49) relating the diffusion coefficient of a sphere to the viscosity of the medium it moves through: $D_A \approx kT/6\pi\eta r_A$, and $D_B \approx kT/6\pi\eta r_B$, where η is the solvent's viscosity and k is Boltzmann's constant. (It is stretching things rather far to apply an η value measured using macroscopic spheres to molecular species; hence the $\approx$ sign.) Equation (17.100) becomes

$$k_D \approx \frac{2RT}{3\eta}\frac{(r_A + r_B)^2}{r_A r_B} = \frac{2RT}{3\eta}\left(2 + \frac{r_A}{r_B} + \frac{r_B}{r_A}\right) \qquad \text{where A} \neq \text{B, nonionic}$$

The value of $2 + r_A/r_B + r_B/r_A$ is rather insensitive to the ratio r_A/r_B; since the treatment is approximate, we might as well set $r_A = r_B$. Hence

$$k_D \approx \begin{cases} 8RT/3\eta & \text{where A} \neq \text{B, nonionic} \qquad (17.104) \\ 4RT/3\eta & \text{where A} = \text{B, nonionic} \qquad (17.105) \end{cases}$$

For water at 25°C, $\eta = 8.90 \times 10^{-4}$ kg m^{-1} s^{-1}, and substitution in (17.104) gives $k_D \approx 0.7 \times 10^{10}$ dm^3 mole^{-1} s^{-1} for a nonionic diffusion-controlled reaction with A $\neq$ B. The $W/(e^W - 1)$ factor multiplies k_D by 2 to 10 for oppositely charged ions and by 0.5 to 0.01 for like charged ions. Thus, k_D is 10^8 to 10^{11} dm^3 mole^{-1} s^{-1} in water at 25°C, depending on the charges and sizes of the reacting species.

The majority of reactions in liquid solution are not diffusion-controlled; instead only a small fraction of encounters lead to reaction. Such reactions are called *chemically controlled* since their rate depends on the probability that an encounter leads to chemical reaction.

Activation energies. Gas-phase reactions are commonly studied at temperatures up to 1500 K, whereas reactions in solution are studied up to 400 or 500 K. Hence, reactions with high activation energies will proceed at negligible rates in solution. Therefore, most reactions observed in solution have activation energies in the range 2 to 35 kcal/mole (compared with -3 to 80 or 100 kcal/mole for gas-phase reactions). The 10 °C doubling or tripling rule (Sec. 17.7) indicates that many reactions in solution have activation energies in or near the range 13 to 20 kcal/mole.

For a nonionic diffusion-controlled reaction, Eqs. (17.104), (17.105), and (17.72) show that E_a depends on $\eta^{-1}\, d\eta/dT$. For water at 25°C, one finds (Prob. 17.60) a theoretical prediction of $E_a \approx 4\frac{1}{2}$ kcal/mole.

17.14 CATALYSIS

A *catalyst* is a substance that increases the rate of a reaction and can be recovered chemically unchanged at the end of the reaction. The rate of a reaction depends on the rate constants in the elementary steps of the reaction mechanism. A catalyst provides an alternate mechanism that is faster than the mechanism in the absence of the catalyst. Moreover, although the catalyst participates in the mechanism, it must be regenerated. A simple scheme for a catalyzed reaction is

$$\begin{aligned} R_1 + C &\rightarrow I + P_1 \\ I + R_2 &\rightarrow P_2 + C \end{aligned} \tag{17.106}$$

where C is the catalyst, R_1 and R_2 are reactants, P_1 and P_2 are products, and I is an intermediate. The catalyst is consumed to form an intermediate which then reacts to regenerate the catalyst and give products. The mechanism (17.106) has an overall activation energy that is less than that for the mechanism in the absence of C. An example of (17.106) is the mechanism (17.7). In (17.7), R_1 and R_2 are H_2O_2 and H_2O_2; the catalyst C is I^-; the intermediate I is IO^-; the product P_1 is H_2O and P_2 is $H_2O + O_2$. Another example is (17.110), below, in which the catalyst is Cl and the intermediate is ClO. In many cases, the catalyzed mechanism consists of several steps with more than one intermediate.

The equilibrium constant for the overall reaction $R_1 + R_2 = P_1 + P_2$ is determined by $\Delta G°$ (according to $\Delta G° = -RT \ln K$) and is therefore independent of the reaction mechanism. Hence a catalyst cannot alter the equilibrium position of a reaction. This being so, a catalyst for a forward reaction must be a catalyst for the reverse reaction also. [Note that reversing the mechanism (17.106) gives a mechanism whereby catalyst is consumed to produce an intermediate which then reacts to regenerate the catalyst.] Since the hydrolysis of esters is catalyzed by H_3O^+, the esterification of alcohols must also be catalyzed by H_3O^+. An enzyme that catalyzes the hydrolysis of proteins to amino acids must also catalyze the polymerization of amino acids to proteins.

The subject of catalysis is divided into *homogeneous catalysis*, in which the catalyzed reaction occurs in one phase, and *heterogeneous* (or *contact*) *catalysis*, in which the reaction occurs at the interface between two phases. Heterogeneous catalysis is discussed in Sec. 17.16.

The rate law for a catalyzed reaction frequently has the form

$$r = k_0[\mathrm{A}]^{\alpha}\cdots[\mathrm{L}]^{\lambda} + k_{\mathrm{cat}}[\mathrm{A}]^{\alpha'}\cdots[\mathrm{L}]^{\lambda'}[\mathrm{cat.}]^{\sigma} \tag{17.107}$$

where k_0 is the rate constant in the absence of catalyst ([cat.] = 0) and k_{cat} is the rate constant for the catalyzed mechanism. The order with respect to catalyst is commonly 1. If the lowering of E_a is substantial, the first term in (17.107) is negligible compared with the second (unless [cat.] is extremely small). The activation energies for the uncatalyzed and catalyzed reactions can be found from the temperature dependences of k_0 and k_{cat}. The reaction $2H_2O_2(aq) \rightarrow 2H_2O + O_2$ has the following activation energies: 18 kcal/mole when uncatalyzed; 14 kcal/mole when catalyzed by I^-; 12 kcal/mole when catalyzed by colloidal Pt particles; 6 kcal/mole when catalyzed by the enzyme liver catalase. From the example in Sec. 17.7, a decrease from 18 to 6 kcal/mole in E_a speeds up the room-temperature rate by $(5.4)^{12} = 6 \times 10^8$, assuming the A factor is not significantly changed.

Many reactions in solution are catalyzed by acids or bases or both. A reaction that is catalyzed by H_3O^+ but not by other Brønsted acids (proton donors) is said to be *specifically hydrogen-ion-catalyzed*. A reaction that is catalyzed by any Brønsted acid is subject to *general acid catalysis*. Similar definitions hold for base catalysis. Since water can act as either a Brønsted acid or a Brønsted base, it catalyzes reactions that are subject to general acid or general base catalysis. The hydrolysis of esters is catalyzed by H_3O^+ and by OH^- but not by other acids or bases, and this reaction exhibits specific acid–base catalysis. The rate law for ester hydrolysis generally has the form

$$r = k_0[\mathrm{RCOOR}] + k_{\mathrm{H}^+}[\mathrm{H_3O^+}][\mathrm{RCOOR}] + k_{\mathrm{OH}^-}[\mathrm{OH^-}][\mathrm{RCOOR}] \tag{17.108}$$

where the rate constants include the concentration of water raised to unknown powers.

An *autocatalytic reaction* is one where a product speeds up the reaction. An example is the H_3O^+-catalyzed hydrolysis of esters, $\mathrm{RCOOR'} + H_2O \rightarrow \mathrm{RCOOH} + \mathrm{R'OH}$; H_3O^+ from the ionization of the product RCOOH increases the H_3O^+ concentration as the reaction proceeds, and this tends to speed up the reaction. Another kind of autocatalysis occurs in the elementary reaction $\mathrm{A} + \mathrm{B} \rightarrow \mathrm{C} + 2\mathrm{A}$. The rate law is $r = k[\mathrm{A}][\mathrm{B}]$. During the reaction, the A concentration increases, and this increase acts to offset the decrease in r produced by the decrease in [B]. A spectacular example is an atomic bomb; here A is a neutron (see Sec. 17.17).

A *negative catalyst* (or *inhibitor*) is a substance that slows down the rate of a reaction when added in small quantities. Negative catalysts act by destroying a catalyst present in the system.

In 1971 the United States Senate voted to end support for the Boeing supersonic transport (SST) airplane. A major factor in the decision was concern that large amounts of nitrogen oxides ($NO_x \equiv NO$ and NO_2) injected into the stratosphere by a fleet of SSTs might substantially reduce the amount of stratospheric ozone. (The NO_x is formed by combustion of atmospheric N_2.) The stratosphere is the layer of the atmosphere that extends from 10 to 50 km. (The Anglo–French Concorde cruises at 17 km.) There is a small (a few parts per million) but significant amount of O_3 in the stratosphere, formed when O_2 absorbs ultraviolet radiation and dissociates to O atoms, which combine with O_2 to give O_3. The O_3 can break down to O_2 by absorption of ultraviolet radiation and by reaction with O. The result of these reactions is an approximately steady-state concentration of O_3 in the stratosphere. NO catalyzes the decomposition of stratospheric O_3, as follows:

$$\begin{aligned} NO + O_3 &\rightarrow NO_2 + O_2 \\ NO_2 + O &\rightarrow NO + O_2 \end{aligned} \tag{17.109}$$

The net reaction is $O_3 + O \rightarrow 2O_2$. Depletion of ozone is undesirable since it would increase the amount of ultraviolet radiation reaching us, thereby increasing the incidence of skin cancer and perhaps altering the climate.

Chlorine atoms also catalyze the decomposition of stratospheric O_3:

$$\begin{aligned} Cl + O_3 &\rightarrow ClO + O_2 \\ ClO + O &\rightarrow Cl + O_2 \end{aligned} \tag{17.110}$$

The chlorofluorocarbons $CFCl_3$ and CF_2Cl_2 have been used as propellants for aerosol sprays in deodorants, hair sprays, etc. When released into the atmosphere, they diffuse to the stratosphere, where they produce Cl atoms by absorbing ultraviolet radiation. In 1977, the U.S. government announced the phasing out of chlorofluorocarbon aerosol propellants.

A good reference on stratospheric pollution is H. S. Johnston, *Ann. Rev. Phys. Chem.*, **26**, 315 (1975).

17.15 ENZYME CATALYSIS

Most of the reactions that occur in living organisms are catalyzed by molecules called *enzymes*. Enzymes are proteins with molecular weights ranging from 10^4 to 10^6. An enzyme is specific in its action; many enzymes catalyze only the conversion of a particular reactant to a particular product (and the reverse reaction); other enzymes catalyze only a certain class of reactions (e.g., ester hydrolysis). Enzymes lower activation energies very substantially, and in their absence most biochemical reactions occur at negligible rates. The molecule an enzyme acts on is called the *substrate*. The substrate binds to a specific *active site* on the enzyme to form an *enzyme–substrate complex*; while bound to the enzyme, the substrate is converted to product, which is then released from the enzyme. Some physiological poisons act by binding to the active site of an enzyme, thereby blocking (or *inhibiting*) the action of the enzyme. The structure of an inhibitor may resemble the structure of the enzyme's substrate. (Cyanide acts by blocking the enzyme cytochrome oxidase.)

The single-celled *E. coli* (a bacterium which flourishes in human colons) contains about 3000 different enzymes and a total of 10^6 enzyme molecules.

There are many possible schemes for enzyme catalysis, but we shall consider only the simplest mechanism, which is

$$\mathrm{E + S \underset{k_{-1}}{\overset{k_1}{\rightleftharpoons}} ES \underset{k_{-2}}{\overset{k_2}{\rightleftharpoons}} E + P} \tag{17.111}$$

where E is the (free) enzyme, S is the substrate, ES is the enzyme–substrate complex, and P is the product. The overall reaction is S → P. The enzyme is consumed in step 1 and regenerated in step 2.

In most experimental studies of enzyme kinetics, the enzyme concentration is much less than the substrate concentration: $[\mathrm{E}] \ll [\mathrm{S}]$. Hence the concentration of the intermediate ES is much less than that of S, and the steady-state approximation can be used for ES:

$$d[\mathrm{ES}]/dt = 0 = k_1[\mathrm{E}][\mathrm{S}] - k_{-1}[\mathrm{ES}] - k_2[\mathrm{ES}] + k_{-2}[\mathrm{E}][\mathrm{P}]$$

If $[\mathrm{E}]_0$ is the initial enzyme concentration, then $[\mathrm{E}]_0 = [\mathrm{E}] + [\mathrm{ES}]$. Since the enzyme concentration [E] during the reaction is generally not known while $[\mathrm{E}]_0$ is known, we replace [E] by $[\mathrm{E}]_0 - [\mathrm{ES}]$:

$$0 = ([\mathrm{E}]_0 - [\mathrm{ES}])(k_1[\mathrm{S}] + k_{-2}[\mathrm{P}]) - (k_{-1} + k_2)[\mathrm{ES}] \tag{17.112}$$

$$[\mathrm{ES}] = \frac{k_1[\mathrm{S}] + k_{-2}[\mathrm{P}]}{k_{-1} + k_2 + k_1[\mathrm{S}] + k_{-2}[\mathrm{P}]}[\mathrm{E}]_0 \tag{17.113}$$

The reaction rate per unit volume is $r = -d[\mathrm{S}]/dt$, and (17.111) gives

$$r = k_1[\mathrm{E}][\mathrm{S}] - k_{-1}[\mathrm{ES}] = k_1([\mathrm{E}]_0 - [\mathrm{ES}])[\mathrm{S}] - k_{-1}[\mathrm{ES}]$$

$$r = k_1[\mathrm{E}]_0[\mathrm{S}] - (k_1[\mathrm{S}] + k_{-1})[\mathrm{ES}] \tag{17.114}$$

(Note that since the concentration of the intermediate ES is very small, we have $-d[\mathrm{S}]/dt = d[\mathrm{P}]/dt$.) Substitution of (17.113) into (17.114) gives

$$r = \frac{k_1 k_2[\mathrm{S}] - k_{-1}k_{-2}[\mathrm{P}]}{k_1[\mathrm{S}] + k_{-2}[\mathrm{P}] + k_{-1} + k_2}[\mathrm{E}]_0$$

Usually, the reaction is followed only to a few percent completion and the initial rate determined. Setting the product concentration [P] equal to 0 and [S] equal to $[\mathrm{S}]_0$, we get as the initial rate r_0

$$r_0 = \frac{k_1 k_2[\mathrm{S}]_0[\mathrm{E}]_0}{k_1[\mathrm{S}]_0 + k_{-1} + k_2} = \frac{k_2[\mathrm{E}]_0[\mathrm{S}]_0}{K_M + [\mathrm{S}]_0} \tag{17.115}$$

where the *Michaelis constant* K_M is defined by $K_M \equiv (k_{-1} + k_2)/k_1$. ($K_M$ is not an equilibrium constant.) The reciprocal of (17.115) is

$$\frac{1}{r_0} = \frac{K_M}{k_2[\mathrm{E}]_0}\frac{1}{[\mathrm{S}]_0} + \frac{1}{k_2[\mathrm{E}]_0} \tag{17.116}$$

Equation (17.115) is the *Michaelis–Menten equation*, and (17.116) is the *Lineweaver–Burk equation.* One measures r_0 for several $[\mathrm{S}]_0$ values with $[\mathrm{E}]_0$ held fixed. The constants k_2 and K_M are found from the slope and intercept of a plot of $1/r_0$ vs. $1/[\mathrm{S}]_0$, since $[\mathrm{E}]_0$ is known. (Strictly speaking, r_0 is not the rate at $t = 0$, since there is a short induction period before steady-state conditions are established. However, the induction period is generally too short to detect.)

Figure (17.13) plots r_0 in (17.115) against $[S]_0$ for fixed $[E]_0$. In the limit of high concentration of substrate, virtually all the enzyme is in the form of the ES complex, and the rate becomes a maximum that is independent of substrate concentration; Eq. (17.115) gives $r_{max} = k_2[E]_0$ for $[S]_0 \gg K_M$. At low substrate concentrations, Eq. (17.115) gives $r = (k_2/K_M)[E]_0[S]_0$, and the reaction is second order. Equation (17.115) predicts that r is always proportional to $[E]_0$ provided that $[E]_0 \ll [S]_0$ so steady-state conditions hold.

The quantity $r_{max}/[E]_0$ is the *turnover number* of the enzyme. The turnover number is the maximum number of moles of product produced in unit time by 1 mole of enzyme and is also the maximum number of molecules of product produced in unit time by one enzyme molecule. From the preceding paragraph, $r_{max} = k_2[E]_0$, so the turnover number for the simple model (17.111) is k_2. Turnover numbers for enzymes range from 10^{-2} to 10^6 molecules per second, with 10^3 s^{-1} being typical. One molecule of the enzyme carbonic anhydrase will dehydrate 600,000 H_2CO_3 molecules per second. (The equilibrium $H_2CO_3 = H_2O + CO_2$ is important in the excretion of CO_2 from the capillaries of the lungs.) For comparison, a typical turnover rate in heterogeneous catalysis (Sec. 17.16) is 1 s^{-1}; one of the highest turnover rates in heterogeneous catalysis is the 10^4 to 10^5 s^{-1} rate in the Pt-catalyzed conversion of ortho H_2 to para H_2.

Although many experimental studies of enzyme kinetics give a rate law in agreement with the Michaelis–Menten equation, the mechanism (17.111) is grossly oversimplified. For one thing, there is much evidence that while the substrate is bound to the enzyme, it generally undergoes a chemical change before being released as product. Hence a better model is

$$E + S \rightleftharpoons ES \rightleftharpoons EP \rightleftharpoons E + P \qquad (17.117)$$

The model (17.117) gives a rate law that has the same form as the Michaelis–Menten equation, but the constants k_2 and K_M are replaced by constants with different significances. (See *Lewis*, pp. 647–648.) Another defect of (17.111) is that it takes the catalytic reaction as $S \rightleftharpoons P$, whereas most enzyme-catalyzed reactions involve two substrates and two products: $A + B \rightleftharpoons P + Q$. The enzyme then has two active sites, one for each substrate. With two substrates, there are many possible mechanisms. For details, see *Lewis*, pp. 656–659.

Enzyme reactions are quite fast but can be studied using "classical" methods

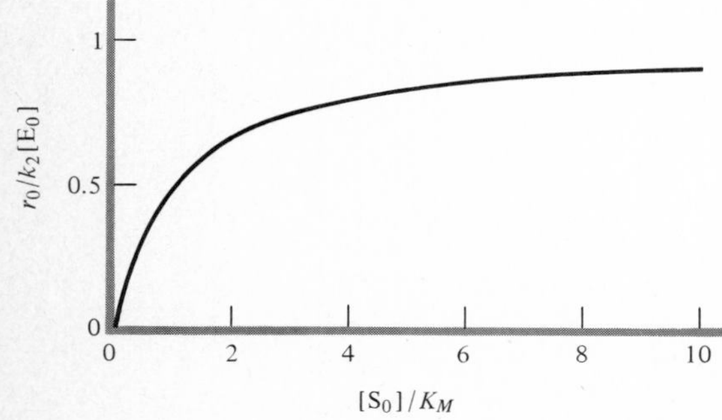

Figure 17.13
Initial rate vs. initial substrate concentration for the Michaelis–Menten mechanism.

by keeping [E] and [S] very low. Typical values are [E] = 10^{-9} mole dm^{-3} and [S] = 10^{-5} mole dm^{-3}. The ratio [S]/[E] must be large to assure steady-state conditions. Modern methods of studying fast reactions (e.g., rapid flow, relaxation) provide more information than the classical methods, in that rate constants for individual steps in a multistep mechanism can be determined.

17.16 HETEROGENEOUS CATALYSIS

The majority of industrial chemical reactions are run in the presence of solid catalysts. Examples are the Fe-catalyzed synthesis of NH_3 from N_2 and H_2; the SiO_2/Al_2O_3-catalyzed cracking of relatively high-molecular-weight hydrocarbons to gasoline; the Pt-catalyzed (or V_2O_5-catalyzed) oxidation of SO_2 to SO_3 [which is then reacted with water to produce H_2SO_4, the leading industrial chemical (1976 United States production, 66×10^9 lb)]. The liquid catalyst H_3PO_4 (distributed on diatomaceous earth) is used in the polymerization of alkenes.

Solid-state catalysis can lower activation energies substantially. For $2HI \rightarrow H_2 + I_2$, the activation energy is 44 kcal/mole for the uncatalyzed homogeneous reaction, is 25 kcal/mole when catalyzed by Au, and is 14 kcal/mole when catalyzed by Pt. The A factor is also changed.

For a solid catalyst to be effective, one or more of the reactants must be chemisorbed on the solid. Physical adsorption is believed to be of little or no importance in heterogeneous catalysis.

The mechanisms of only a few heterogeneously catalyzed reactions are known. In writing such mechanisms, the adsorption site is commonly indicated by a star. Deuterium tracer studies indicate that the mechanism of hydrogenation of ethylene on a metallic catalyst is probably

$$H_2C{=}CH_2(g) + 2* \rightleftharpoons \underset{*}{\underset{|}{H_2C}}{-}\underset{*}{\underset{|}{CH_2}} \qquad \text{and} \qquad H_2(g) + 2* \rightleftharpoons 2\underset{*}{\underset{|}{H}}$$

$$\underset{*}{\underset{|}{H_2C}}{-}\underset{*}{\underset{|}{CH_2}} + \underset{*}{\underset{|}{H}} \rightleftharpoons \underset{*}{\underset{|}{H_2C}}{-}CH_3 + 2*$$

$$\underset{*}{\underset{|}{H}} + \underset{*}{\underset{|}{H_2C}}{-}CH_3 \rightarrow H_3C{-}CH_3(g) + 2*$$

where each star is a metal atom on the solid's surface.

Most heterogeneous catalysts are metals, metal oxides, or acids. Common metal catalysts include Fe, Co, Ni, Pd, Pt, Cr, Mn, W, Ag, and Cu. Many metallic catalysts are Group VIII transition metals with partly vacant d orbitals that can be used in bonding to the chemisorbed species. Common metal oxide catalysts are Al_2O_3, Cr_2O_3, V_2O_5, ZnO, NiO, and Fe_2O_3. Common acid catalysts are H_3PO_4 and H_2SO_4.

To increase the exposed surface area, the catalyst is often distributed on the surface of a porous *support* (or *carrier*). Common supports are silica gel (SiO_2), alumina (Al_2O_3), carbon (in the form of charcoal), and diatomaceous earth. The support may be inert or may contribute catalytic activity, depending on the reaction.

The activity of a catalyst may be increased and its lifetime extended by addition of small amounts (5 or 10 percent) of substances called *promoters*. The iron catalyst used in NH_3 synthesis contains small amounts of the oxides of K, Ca, Al, Si, Mg, Ti, Zr, and Va; the Al_2O_3 acts as a barrier that prevents the tiny crystals of Fe from joining together (sintering); formation of larger crystals decreases the surface area and the catalytic activity.

Small amounts of certain substances that bond strongly to the catalyst can inactivate (or *poison*) it. These poisons may be present as impurities in the reactants or be formed as reaction by-products. Catalytic poisons include compounds of S, N, and P having lone pairs of electrons (for example, H_2S, CS_2, HCN, PH_3, CO) and certain metals (for example, Hg, Pb, As). Because lead is a catalytic poison, lead-free gasoline must be used in cars equipped with catalytic afterburners used to remove pollutants from the exhaust.

The amount of poison needed to eliminate the activity of a catalyst is usually much less than needed to cover the catalyst's surface completely. This indicates that the catalyst's activity is largely confined to a small fraction of surface sites, called *active sites* (or *active centers*). The surface of a solid is not smooth and uniform but is rough on an atomic scale. The surface of a metal catalyst contains steplike jumps that join relatively smooth planes; there is evidence that hydrocarbon bonds break mainly at these steps and not on the smooth planes (*Chem. Eng. News*, Dec. 8, 1975, p. 23.)

The following five steps are involved in fluid-phase reactions catalyzed by solids: (1) diffusion of reactant molecules to the solid's surface; (2) chemisorption of reactants on the surface; (3) chemical reaction of adsorbed molecules on the surface; (4) desorption of product molecules from the surface; (5) diffusion of product molecules into the bulk fluid.

A general treatment involves the rates for all five steps and is complicated. In many cases, one of these steps is much slower than all the others, and only the rate of the slow step need be considered. Diffusion (steps 1 and 5) is generally fast in gases. We restrict the treatment to solid-catalyzed reactions of gases where step 3 is much slower than all other steps.

Step 3 may consist of more than one elementary chemical reaction. In the $C_2H_4 + H_2$ example above, step 3 involves two elementary reactions. Since the detailed mechanism of the surface reaction is usually unknown, we adopt the simplifying assumption of taking step 3 to consist of a single (unimolecular or bimolecular) elementary reaction [or a slow (rate-determining) elementary reaction followed by one or more rapid steps]. This assumption may be compared with the assumption of the grossly oversimplified mechanism (17.111) for enzyme catalysis.

Since we are assuming the adsorption and desorption rates to be much greater than the chemical-reaction rate for each species, adsorption–desorption equilibrium is maintained for each species during the reaction; we can therefore use the Langmuir isotherm, which is derived by equating the adsorption and desorption rates for a given species (see Sec. 13.5). Of course, the Langmuir isotherm assumes a uniform surface, which is far from true in heterogeneous catalysis, so use of the Langmuir isotherm is one more gross oversimplification in the treatment.

The reaction rate J of a heterogeneously catalyzed reaction is defined by (17.2) as $\nu_B^{-1}\, dn_B/dt$, where ν_B is the stoichiometric coefficient of any species B in the overall reaction. Since the chemical reaction occurs on the catalyst's surface, J will clearly be proportional to the catalyst's surface area $\mathscr{A}$. Let r_s be the *reaction rate per unit surface area* of the catalyst. Then

$$r_s \equiv \frac{J}{\mathscr{A}} \equiv \frac{1}{\mathscr{A}} \frac{1}{\nu_B} \frac{dn_B}{dt} \tag{17.118}$$

(If $\mathscr{A}$ is unknown, one uses the reaction rate per unit mass of catalyst.)

Suppose the elementary reaction on the surface is the unimolecular step $A \rightarrow C + D$. Then r_s, the reaction rate per unit surface area, will be proportional to the number of adsorbed A molecules per unit surface area ($n_A/\mathscr{A}$), and this in turn will be proportional to θ_A, the fraction of adsorption sites occupied by A molecules. Therefore, $r_s = k\theta_A$, where k is a rate constant with units moles cm^{-2} s^{-1}. Since the products C and D might compete with A for adsorption sites, we use the form of the Langmuir isotherm that applies when more than one species is adsorbed. The result of Prob. 13.27 generalized to several adsorbed species is

$$\theta_A = \frac{b_A P_A}{1 + \sum_i b_i P_i}$$

where the sum goes over all species. The rate law $r_s = k\theta_A$ becomes

$$r_s = k \frac{b_A P_A}{1 + b_A P_A + b_C P_C + b_D P_D} \tag{17.119}$$

If the products are very weakly adsorbed ($b_C P_C$ and $b_D P_D \ll 1 + b_A P_A$), then

$$r_s = k \frac{b_A P_A}{1 + b_A P_A} \tag{17.120}$$

The low-pressure and high-pressure limits of (17.120) are

$$r_s = \begin{cases} k b_A P_A & \text{at low } P \\ k & \text{at high } P \end{cases}$$

At low P, the reaction is first order; at high P, zero order. At high pressure, the surface is fully covered with A, so an increase in P_A has no effect on the rate. [Note the resemblance to the low-substrate and high-substrate limits of the Michaelis–Menten equation (17.115). In fact, Eqs. (17.115) and (17.120) have essentially the same form. Compare also Figs. 17.13 and 13.16*a*. In both enzyme catalysis and heterogeneous catalysis, there is binding to a limited number of active sites.]

The W-catalyzed decomposition of PH_3 at 700°C follows the rate law (17.120), becoming first order at 10^{-2} torr and zero order at 1 torr. The decomposition of NH_3 on Pt obeys the rate law $r_s = kP(NH_3)/P(H_2)$; this rate law follows from (17.119) if it is assumed that $b(H_2)P(H_2)$ is much larger than all the other terms in the denominator. The product H_2 is strongly adsorbed and competes with NH_3 for the active sites, thereby inhibiting the reaction.

If the elementary surface reaction is bimolecular, $A + B \rightarrow C + D$, then r_s will be proportional to the product $(n_A/\mathscr{A})(n_B/\mathscr{A})$. Hence, $r_s = k\theta_A\theta_B$. Use of the Langmuir isotherm for θ_A and θ_B gives

$$r_s = k \frac{b_A b_B P_A P_B}{(1 + b_A P_A + b_B P_B + b_C P_C + b_D P_D)^2} \tag{17.121}$$

A particularly interesting case of (17.121) is when the reactant B is adsorbed much more strongly than all other species: $b_B P_B \gg 1 + b_A P_A + b_C P_C + b_D P_D$. Then, $r_s = kb_A P_A / b_B P_B$, and the *reactant* B inhibits the reaction. This seemingly paradoxical situation can be understood by realizing that when reactant B is much more strongly adsorbed than reactant A, the fraction of surface occupied by A goes to zero; hence $r_s = k\theta_A \theta_B$ goes to zero. The maximum rate occurs when the two reactants are equally adsorbed. An example of inhibition by a reactant is the Pt-catalyzed reaction $2CO + O_2 \rightarrow 2CO_2$, whose rate is inversely proportional to the CO pressure. (CO also binds more strongly than O_2 to the Fe atom in hemoglobin and so is a physiological poison.)

Equations (17.119) and (17.121) and their various limiting cases fit the data of many (but far from all) heterogeneously catalyzed reactions. However, the model used to obtain these equations is likely to be wrong in most cases.

17.17 NUCLEAR REACTIONS

The decay of a radioactive isotope follows first-order kinetics. Each nucleus of a particular radioactive isotope has a certain probability of breaking down in unit time. The number $-dN$ of nuclei that decay in a small time dt is therefore proportional to the number N of radioactive nuclei currently present in the system and also to the length of the small time interval: $-dN \propto N\,dt$, or

$$dN/dt = -\lambda N \tag{17.122}$$

where the proportionality constant λ is called the *decay constant*. (The number of radioactive nuclei is decreasing with time, so dN is negative.) For radioactive decay, there is no point in dividing by the sample's volume, as is customary for chemical reactions. Also, it is more convenient to express the decay rate in terms of the number of elementary particles (atoms or nuclei) than in terms of moles. Equation (17.122) has the same form as (17.11), and by analogy to (17.14), Eq. (17.122) integrates to

$$N = N_0 e^{-\lambda t} \tag{17.123}$$

where N_0 (not to be confused with the Avogadro constant) is the number of radioactive nuclei present at $t = 0$.

Since the kinetics is first order, the isotope's half-life is given by Eq. (17.15) as

$$t_{1/2} = 0.693/\lambda \tag{17.124}$$

Some half-lives and decay modes are:

	^1_0n	^{12}N	^{14}C	^{40}K	^{238}U
Half-life	15 min	0.01 s	5700 yr	1.3×10^9 yr	4.5×10^9 yr
Decay mode	β^-	β^+	β^-	β^-(90%), EC(10%), $\beta^+(10^{-4}\%)$	α

An α particle consists of two protons and two neutrons and is a ^4He nucleus. The decay of ^{238}U is $^{238}_{92}U \rightarrow {}^4_2He + {}^{234}_{90}Th$. The ^{234}Th nucleus is itself unstable and decays by beta emission to ^{234}Pa, which is unstable; the ultimate decay product of ^{238}U is ^{206}Pb. (The *mass number* A of a nucleus equals the number of protons plus neutrons and is written as a left superscript. The *atomic number Z* equals the number of protons and is written as a left subscript.)

A β^- particle is an electron. Note that the free neutron is unstable, decaying into a proton, an electron, and an antineutrino (symbol $\bar{\nu}$): ${}^1_0\text{n} \to {}^1_1\text{p} + {}^{\ 0}_{-1}\text{e} + {}^0_0\bar{\nu}$. The antineutrino, first detected in 1956, has zero charge and zero rest mass and travels at the speed of light. The emission of an electron from a nucleus can be described as the breakdown of a neutron in the nucleus into a proton (which remains in the nucleus) and an electron and antineutrino (which are ejected from the nucleus). The decay equation for ${}^{14}\text{C}$ is ${}^{14}_{\ 6}\text{C} \to {}^{14}_{\ 7}\text{N} + {}^{\ 0}_{-1}\text{e} + {}^0_0\bar{\nu}$. Note that the symbol ${}^{14}\text{N}$ in this equation stands for a *nucleus* of ${}^{14}\text{N}$; emission of a beta particle by ${}^{14}\text{C}$ produces a ${}^{14}\text{N}^+$ ion, which eventually picks up an electron from the surroundings to become a neutral ${}^{14}\text{N}$ atom.

A β^+ particle is a positron. A positron has the same mass as an electron and a charge equal in magnitude but opposite in sign to the electron charge. The positron is the *antiparticle* of the electron. When a positron collides with an electron, they annihilate each other, producing two high-energy photons. (Photons are discussed in Sec. 18.2.) Positrons are not constituents of ordinary matter. Several man-made nuclei decay by positron emission; an example is ${}^{12}\text{N} \to {}^{12}_{\ 6}\text{C} + {}^0_1\text{e} + {}^0_0\nu$, where ν is a neutrino. (The particles ν and $\bar{\nu}$ are antiparticles of each other.)

In *electron capture* (EC), a nucleus captures one of the orbital electrons of the atom, thereby converting a proton into a neutron. (We shall see in Sec. 19.3 that *s* electrons have some probability of being at the nucleus.) The equation for electron capture in ${}^{40}\text{K}$ is ${}^{40}_{19}\text{K} + {}^{\ 0}_{-1}\text{e} \to {}^{40}_{18}\text{Ar} + {}^0_0\nu$. Note from the above table that ${}^{40}\text{K}$ decays by three competing reactions.

Just as the electrons in atoms can exist in excited states (Chap. 19), the nucleus possesses excited states. An excited nucleus can lose energy by emission of a high-energy photon (a gamma ray) or by transferring energy to one of the orbital electrons, causing this electron to be emitted from the atom. This process is called *internal conversion.* In gamma emission and in internal conversion, the charge and mass number of the nucleus remain unchanged; the nucleus simply goes from a high energy level to a lower one.

The *activity A* of a radioactive sample is defined as the number of disintegrations per second: $A \equiv -dN/dt$. Equation (17.122) gives

$$A = \lambda N \tag{17.125}$$

where N is the number of radioactive nuclei present. Use of this equation in (17.123) gives

$$A = A_0 e^{-\lambda t} \tag{17.126}$$

Example

A 1.00-g sample of ${}^{226}\text{Ra}$ emits 3.7×10^{10} alpha particles per second. Find λ and $t_{1/2}$. Find A after 999 years.

We have

$$N = 1.00\ \text{g}\ \frac{1\ \text{mole Ra}}{226\ \text{g}}\ \frac{6.02 \times 10^{23}\ \text{atoms}}{1\ \text{mole}} = 2.66 \times 10^{21}\ \text{atoms}$$

$$\lambda = A/N = (3.7 \times 10^{10}\ \text{s}^{-1})/(2.66 \times 10^{21}) = 1.39 \times 10^{-11}\ \text{s}^{-1}$$

$$t_{1/2} = 0.693/\lambda = 5.0 \times 10^{10}\ \text{s} = 1600\ \text{yr}$$

$$A = A_0 e^{-\lambda t} = (3.7 \times 10^{10}\ \text{s}^{-1}) \exp[(-1.39 \times 10^{-11}\ \text{s}^{-1})(3.15 \times 10^{10}\ \text{s})]$$
$$= 2.4 \times 10^{10}\ \text{dis/s}$$

Radioactive decay is a probabilistic process, and the discussion of Sec. 3.7 shows that fluctuations of the order of $A^{1/2}$ are to be expected in the observed activity. The activity of a 10-mg sample of ^{238}U is only 100 dis/s, and easily observed fluctuations of 10 dis/s, or 10 percent, occur in the decay rate. Anyone who has ever heard a Geiger counter clicking will have noted the fluctuations in counting rate. In a chemical reaction, the number of particles reacting per unit time is extremely large, and fluctuations in rate are unobservable.

The age of uranium-bearing rocks can be determined by measuring the $^{238}U/^{206}Pb$ ratio and the isotopic composition of the lead. Decay of ^{40}K to ^{40}Ar can also be used to date rocks. (The isotope ^{40}K makes up 0.12 percent of naturally occurring K.) The oldest rocks found on earth are 4.0×10^9 years old. The oldest moon rocks are 4.6×10^9 years old, and this is also the age of most meteorites. The best estimate of the age of the earth (and the rest of the solar system) is 4.6×10^9 years. The earth is believed to have melted early in its history, due to heat released by radioactive decay; hence the oldest earth rocks are younger than the earth's age.

Carbon 14 is continuously formed in the earth's atmosphere by collisions of cosmic-ray neutrons with nitrogen nuclei: $^{14}_{7}N + ^{1}_{0}n \rightarrow ^{14}_{6}C + ^{1}_{1}p$. The ^{14}C is radioactive, with half-life 5700 years. The CO_2 that plants ingest contains about 1 part per 10^{12} $^{14}CO_2$, and this ^{14}C is incorporated into the bodies of animals when they eat plants. When an organism dies, it no longer ingests ^{14}C, and the amount of ^{14}C present decreases by radioactive decay. Comparison of the ^{14}C activity of the remains of an organism with the ^{14}C activity of living organisms allows the organism's time of death to be determined. The time limits on the method are 500 to 50,000 years. ^{14}C dating has been of considerable help to archeologists in determining the chronology of ancient civilizations (Egypt, Mesopotamia, etc.). Linen wrappings from the Dead Sea Scrolls were dated at 1900 ± 200 years. In 1969 the Vatican submitted for ^{14}C dating a piece of wood from a throne that some believed to be the throne of St. Peter. The throne turned out to be only 1100 years old and therefore was not St. Peter's.

Nuclear fission and fusion. When a ^{235}U nucleus captures a neutron, the ^{236}U nucleus formed undergoes *fission* within 10^{-11} s into fragments that consist of two nuclei of intermediate atomic number, two or three neutrons, and gamma rays. Which nuclei are formed varies from one fission event to another. The isothermal change $\Delta\bar{U}$ in internal energy is -5×10^9 kcal/mole for ^{235}U fission, which is far greater than the $\pm 10^2$ kcal/mole values typical of chemical reactions. (The isothermal ΔU in nuclear reactions is the ΔU for reactant and product nuclei at rest.) The $\Delta\bar{U}$ values in nuclear reactions are large enough to correspond to a measurable change in rest mass; $\Delta\bar{U}$ can be calculated from Einstein's equation (5.2) as $c^2\ \Delta m_0$, where Δm_0 is the change in rest mass.

Because neutron-induced ^{235}U fission produces more than one neutron, the possibility of a branching-chain fission reaction exists. The ^{235}U isotope makes up 0.7 percent of uranium occurring on earth. When ^{235}U fission occurs in ordinary uranium, the neutrons produced by fission are more likely to be captured by ^{238}U nuclei (99.3 percent abundance) than to be captured by ^{235}U to propagate the chain reaction. If the ^{235}U isotope is separated from the ^{238}U, fission in a relatively pure ^{235}U sample may produce a branching chain reaction leading to explosion. Such an explosion will not occur in a small piece of ^{235}U, since too many fission-produced neutrons will escape out the piece before being captured by ^{235}U. The minimum mass of ^{235}U needed to sustain a chain reaction is the *critical mass*. (This is 52 kg for 94 percent pure ^{235}U.) In the atomic bomb dropped on Hiroshima, pieces of ^{235}U, each with mass less than the critical mass, were driven together by a surrounding charge of chemical explosive to form a piece with mass greater than the critical mass.

In a *nuclear reactor*, a ^{235}U fission reaction is sustained in naturally occurring uranium (or in uranium that is only partly enriched in ^{235}U) by embedding pieces of the uranium in a *moderator*. Collisions of fission-produced neutrons with nuclei of the moderator slow the neutrons. Slow neutrons have a much smaller probability of being captured by ^{238}U than fast neutrons, so the slow neutrons can propagate the ^{235}U fission reaction. Commonly used moderators are graphite and heavy water (D_2O), each of which has a low probability for neutron capture. A nuclear reactor also contains control rods of a neutron-capturing material (B or Cd); the control rods are inserted far enough into the reactor for the fission reaction to occur at a steady rate, without exploding.

The energy released by the fission reaction appears mainly as kinetic energy of the fragment nuclei. These nuclei collide with surrounding matter, heating up the reactor core. The thermal energy produced is removed by circulating a coolant fluid through the reactor core; this energy is used to boil water to steam; the steam runs a turbine generator.

In *nuclear fusion*, light nuclei collide and fuse to heavier ones. Because of the Coulomb repulsion between nuclei, fusion has a very high activation energy and occurs only at temperatures exceeding 10^6 K. The energy of most stars comes from fusion of hydrogen to helium in their cores. The dominant mechanism of H fusion in the sun is believed to be

$$\begin{aligned}
&{}^1_1\text{H} + {}^1_1\text{H} \rightarrow {}^2_1\text{H} + {}^0_1\text{e} + \nu \\
&{}^2_1\text{H} + {}^1_1\text{H} \rightarrow {}^3_2\text{He} + \gamma \\
&{}^3_2\text{He} + {}^3_2\text{He} \rightarrow {}^4_2\text{He} + 2\ {}^1_1\text{H} \\
&{}^0_1\text{e} + {}_{-1}^{\ 0}\text{e} \rightarrow 2\gamma
\end{aligned} \tag{17.127}$$

where the last reaction is electron–positron annihilation. The overall reaction is (Prob. 17.68)

$$4\ {}^1_1\text{H} + 2\ {}_{-1}^{\ 0}\text{e} \rightarrow {}^4_2\text{He} + 2\nu + 6\gamma \tag{17.128}$$

The isothermal $\Delta\bar{U}$ for (17.128) is 6×10^8 kcal/mole. In stars hotter than the sun, the dominant H-to-He fusion mechanism is believed to involve ^{12}C as a catalyst. Fusion of He to heavier elements occurs in red-giant stars. Most cosmologists believe that only H and He existed immediately after the Big Bang that started the universe (Sec. 3.8), so all elements heavier than He originated in stars; the nuclei of carbon, oxygen, nitrogen, iron, etc., present in our bodies were produced in stars older than the sun.

Nuclear fusion occurs in hydrogen bombs. H-bomb ingredients include D, T, and ^6Li [where D (deuterium) is ^2H and T (tritium) is ^3H]. Some of the reactions that occur are $D + D \rightarrow {}^3He + n$, $D + D \rightarrow T + H$, $T + D \rightarrow {}^4He + n$, and ${}^6Li + n \rightarrow {}^4He + T$. The high temperature needed for fusion is attained in an H-bomb by a nuclear-fission explosion.

Much effort is being expended to find ways of using deuterium, tritium, and lithium fusion reactions as a source of power.

PROBLEMS

17.1 For the gas-phase reaction

$$2N_2O_5 \rightarrow 4NO_2 + O_2,$$

the rate constant k is 1.73×10^{-5} s^{-1} at 25°C. The observed rate law is given in Eq. (17.6). (*a*) Calculate r and J for this reaction in a 12.0-dm^3 container with $P(N_2O_5) = 0.10$ atm at 25°C. (*b*) Calculate $d[N_2O_5]/dt$ for the conditions of part (*a*). (*c*) Calculate the number of N_2O_5 molecules that decompose in 1 s for the conditions of (*a*). (*d*) What are k, r, and J for the conditions of (*a*) if the reaction is written $N_2O_5 \rightarrow 2NO_2 + \frac{1}{2}O_2$?

17.2 In gas-phase kinetics, pressures instead of concentrations are sometimes used in rate laws. Suppose that for $aA \rightarrow$ products, one finds that $-a^{-1}dP_A/dt = k_P P_A{}^n$, where k_P is a constant and P_A is the partial pressure of A. (*a*) Show that $k_P = k(RT)^{1-n}$. (*b*) Is this relation valid for any nth-order reaction? (*c*) Calculate k_P for a gas-phase reaction having $k = 2.00 \times 10^{-4}$ dm^3 mole^{-1} s^{-1} at 400 K.

17.3 Reactions 1 and 2 are each first order, and $k_1 > k_2$ at a certain temperature T. Must r_1 be greater than r_2 at T?

17.4 For the mechanism

$$A + B \rightarrow C + D$$

$$2C \rightarrow F$$

$$F + B \rightarrow 2A + G$$

(*a*) give the stoichiometric number of each step and give the overall reaction; (*b*) classify each species as reactant, product, intermediate, or catalyst.

17.5 The first-order reaction $2A \rightarrow 2B + C$ is 35 percent complete after 325 s. (*a*) Find k and k_A, where k_A is defined in (17.11). (*b*) How long will it take for the reaction to be 70 percent complete? 90 percent complete?

17.6 (*a*) Use information in Prob. 17.1 to calculate the half-life for the N_2O_5 decomposition at 25°C. (*b*) Calculate $[N_2O_5]$ after 24.0 hr if $[N_2O_5]_0 = 0.010$ mole dm^{-3} and the system is at 25°C.

17.7 (*a*) Verify Eq. (17.24). (*b*) Verify Eq. (17.25); use a table of integrals to help you.

17.8 For the gas-phase reaction $2NO_2 + F_2 \rightarrow 2NO_2F$, the rate constant k is 38 dm^3 mole^{-1} s^{-1} at 27°C. The reaction is first order in NO_2 and first order in F_2. Calculate the number of moles of NO_2, F_2, and NO_2F present after 10.0 s if 2.00 moles of NO_2 is mixed with 3.00 moles of F_2 in a 400-dm^3 vessel at 27°C.

17.9 Verify Eq. (17.32).

17.10 (*a*) The differential equation $dy/dx = f(x) + g(x)y$, where f and g are functions of x, has as its solution

$$y = e^{w(x)}\left[\int e^{-w(x)}f(x)\,dx + c\right]$$

$$\text{where } w(x) \equiv \int g(x)\,dx$$

and where c is an arbitrary constant. Prove this result by substituting the proposed solution into the differential equation. (*b*) Use the result of (*a*) to solve the differential equation (17.39); use (17.37) to evaluate c.

17.11 Suppose the rate law for $A + B \rightarrow 2B + C$ is $r = k[A][B]$. (*a*) Show that the integrated rate law for this autocatalytic reaction is

$$\frac{1}{[A]_0 + [B]_0} \ln \frac{[B]/[B]_0}{[A]/[A]_0} = kt$$

(Use a table of integrals if necessary.) (*b*) Sketch [B] vs. t for $[A]_0 = 0.1$ mole dm^{-3}, $[B]_0 = 0.001$ mole dm^{-3}, and $k = 0.1$ dm^3 mole^{-1} s^{-1}. Where else have you seen this curve?

17.12 For the elementary reaction $A \rightleftharpoons 2C$, show that if a system in equilibrium is subjected to a small perturbation, then $[A] - [A]_{eq}$ is given by the equation following (17.52) if τ is defined as $\tau^{-1} \equiv k_f + 4k_b[C]_{eq}$.

17.13 Does the term "reversible" have the same meaning in kinetics as in thermodynamics?

17.14 Show that if $r = k[A]^n$, then

$$\log t_\alpha = \log \frac{\alpha^{1-n} - 1}{(n-1)k_A} - (n-1)\log [A]_0 \qquad \text{for } n \neq 1$$

$$t_\alpha = -(\ln \alpha)/k_A \qquad \text{for } n = 1$$

where t_α is the fractional life.

17.15 The rate constant for the elementary gas-phase reaction $N_2O_4 \rightarrow 2NO_2$ is 4.8×10^4 s^{-1} at 25°C. (*a*) Use data in the Appendix to calculate the rate constant at 25°C for $2NO_2 \rightarrow N_2O_4$. (*b*) Use the result of Prob. 17.12 to calculate the relaxation time in an equilibrium mixture of NO_2 and N_2O_4 at 1.00 atm and 25°C.

17.16 For the decomposition of $(CH_3)_2O$ (species A) at 777 K, the time required for $[A]_0$ to fall to $0.69[A]_0$ as a function of $[A]_0$ is:

$10^3[A]_0$/(mole/dm^3)	8.13	6.44	3.10	1.88
$t_{0.69}$/s	590	665	900	1140

(a) Find the order of the reaction. (b) Find k_A in $d[A]/dt = -k_A[A]^n$.

17.17 For a sample of 81 males diagnosed as having congestive heart failure, the number N alive at time t after the diagnosis was reported in a 1971 study as:

t/yr	0	1	3	5	7	9
N	81	64	44	31	24	15

(a) Find the order and rate constant for the reaction: man with congestive heart failure → dead man. (b) Estimate the number alive after 12 years.

17.18 It was noted in Sec. 17.4 that the trial-and-error method of determining reaction orders is poor. Data for the decomposition of $(CH_3)_3COOC(CH_3)_3(g)$, species A, at 155°C are:

t/min	10^3[A]/(mole dm^{-3})
0	6.35
3.0	5.97
6.0	5.64
9.0	5.31
12.0	5.02
15.0	4.74
18.0	4.46
21.0	4.22

(a) Plot log 10^3[A] vs. t and $(10^3[A])^{-1}$ vs. t and see if you can decide which plot is more nearly linear. (b) Make a Powell plot and see if this allows the order to be determined.

17.19 The reaction

$$n\text{-}C_3H_7Br + S_2O_3^{2-} \rightarrow C_3H_7S_2O_3^{-} + Br^{-}$$

in aqueous solution is first order in C_3H_7Br and first order in $S_2O_3^{2-}$. At 37.5°C, the following data were obtained:

[$S_2O_3^{2-}$]/(mmol dm^{-3})	t/s
96.6	0
90.4	1,110
86.3	2,010
76.6	5,052
66.8	11,232

The initial C_3H_7Br concentration was 39.5 mmol dm^{-3}. Find the rate constant using a graphical method.

17.20 At $t = 0$, butadiene was introduced into an empty vessel at 326°C and the dimerization reaction

$2C_4H_6 \rightarrow C_8H_{12}$ followed by monitoring the pressure P. The following data were obtained:

t/s	P/torr	t/s	P/torr	t/s	P/torr
0	632.0	1,751	535.4	5,403	453.3
367	606.6	2,550	509.3	7,140	432.8
731	584.2	3,652	482.8	10,600	405.3
1,038	567.3				

(a) Find the reaction order using a Powell plot or the fractional-life method. (b) Evaluate the rate constant.

17.21 For the reaction A + B → C + D, a run with $[A]_0 = 400$ mmole dm^{-3} and $[B]_0 = 0.400$ mmole dm^{-3} gave the following data:

t/s	[C]/(mmol dm^{-3})
0	0
120	0.200
240	0.300
360	0.350
∞	0.400

and a run with $[A]_0 = 0.400$ mmole dm^{-3} and $[B]_0 = 1000$ mmole dm^{-3} gave the following data:

$10^{-3}t$/s	[C]/(mmol dm^{-3})
0	0
69	0.200
208	0.300
485	0.350
∞	0.400

Find the rate law and the rate constant. (Note that the numbers have been chosen to make determination of the orders simple.)

17.22 For the reaction A → products, data for a run with $[A]_0 = 0.600$ mole dm^{-3} are:

t/s	$[A]/[A]_0$	t/s	$[A]/[A]_0$
0	1	400	0.511
100	0.829	600	0.385
200	0.688	1000	0.248
300	0.597		

(a) Find the order of the reaction. (b) Find the rate constant.

17.23 For the reaction 2A + B → C + D + 2E, data

for a run with $[A]_0 = 800$ mmole dm^{-3} and $[B]_0 =$ 2.00 mmole dm^{-3} are:

t/s	$[B]/[B]_0$	t/s	$[B]/[B]_0$
0	1	30,000	0.582
8,000	0.836	50,000	0.452
14,000	0.745	90,000	0.318
20,000	0.680		

and data for a run with $[A]_0 = 600$ mmole dm^{-3} and $[B]_0 = 2.00$ mmole dm^{-3} are:

t/s	$[B]/[B]_0$	t/s	$[B]/[B]_0$
0	1	50,000	0.593
8,000	0.901	90,000	0.453
20,000	0.787		

Find the rate law and the rate constant.

17.24 For the reaction $OCl^- + I^- \rightarrow OI^- + Cl^-$ in aqueous solution at 25°C, initial rates r_0 as a function of initial concentrations (where $c^\circ \equiv 1$ mol/dm^3) are:

$10^3[ClO^-]/c^\circ$	4.00	2.00	2.00	2.00
$10^3[I^-]/c^\circ$	2.00	4.00	2.00	2.00
$10^3[OH^-]/c^\circ$	1000	1000	1000	250
$10^3 r_0/(c^\circ\, s^{-1})$	0.48	0.50	0.24	0.94

(*a*) Find the rate law and the rate constant. (*b*) Devise a mechanism consistent with the observed rate law.

17.25 The gas-phase reaction $2NO_2Cl \rightarrow 2NO_2 + Cl_2$ has $r = k[NO_2Cl]$. Devise a mechanism consistent with this rate law.

17.26 The gas-phase reaction $2NO_2 + F_2 \rightarrow 2NO_2F$ has $r = k[NO_2][F_2]$. Devise a mechanism consistent with this rate law.

17.27 The gas-phase reaction

$$XeF_4 + NO \rightarrow XeF_3 + NOF$$

has $r = k[XeF_4][NO]$. Devise a mechanism consistent with this rate law.

17.28 The gas-phase reaction $2Cl_2O + 2N_2O_5 \rightarrow 2NO_3Cl + 2NO_2Cl + O_2$ has the rate law $r = k[N_2O_5]$. Devise a mechanism consistent with this rate law.

17.29 For the reaction $Hg_2^{2+} + Tl^{3+} \rightarrow 2Hg^{2+} + Tl^+$, devise another mechanism besides (17.68) that gives the observed rate law (17.67).

17.30 Explain why it is virtually certain that the homogeneous gas-phase reaction $2NH_3 \rightarrow N_2 + 3H_2$ does not occur by a one-step mechanism.

17.31 The gas-phase decomposition of ozone, $2O_3 \rightarrow 3O_2$, is believed to have the following mechanism:

$$O_3 + M \underset{k_{-1}}{\overset{k_1}{\rightleftharpoons}} O_2 + O + M$$

$$O + O_3 \xrightarrow{k_2} 2O_2$$

where M is any molecule. (*a*) Verify that $d[O_2]/dt = 2k_2[O][O_3] + k_1[O_3][M] - k_{-1}[O_2][O][M]$. Write down a similar expression for $d[O_3]/dt$. (*b*) Use the steady-state approximation for [O] to simplify the expressions in (*a*) to $d[O_2]/dt = 3k_2[O_3][O]$ and $d[O_3]/dt = -2k_2[O_3][O]$. (*c*) Show that when the steady-state approximation for [O] is substituted into either $d[O_2]/dt$ or $d[O_3]/dt$, one obtains

$$r = \frac{k_1 k_2[O_3]^2}{k_{-1}[O_2] + k_2[O_3]/[M]}$$

(*d*) Assume step 1 is in near equilibrium, so that step 2 is rate-determining, and derive an expression for r starting from one of the expressions in (*a*). (*e*) Under what condition does the steady-state approximation reduce to the equilibrium approximation?

17.32 (*a*) Apply the steady-state approximation to the N_2O_5 decomposition mechanism (17.8) and show that $r = k[N_2O_5]$, where $k = k_a k_b/(k_{-a} + 2k_b)$. *Hint:* Use the steady-state approximation for both intermediates. (*b*) Apply the rate-determining-step approximation to the N_2O_5 mechanism, assuming that step b is slow compared with steps $-a$ and c. (*c*) Under what condition does the rate law in (*a*) reduce to that in (*b*)? (*d*) The rate constant for the reaction in Prob. 17.28 is numerically equal to the rate constant for the N_2O_5 decomposition. Devise a mechanism for the reaction in Prob. 17.28 that will explain this fact.

17.33 The reaction $2DI \rightarrow D_2 + I_2$ has $k = 1.2 \times 10^{-3}$ dm^3 mole^{-1} s^{-1} at 660 K and $E_a = 42.4$ kcal/mole. Calculate k at 720 K for this reaction.

17.34 Rate constants for the gas-phase reaction $H_2 + I_2 \rightarrow 2HI$ at various temperatures are:

T/K	$10^3 k/(\text{dm}^3\ \text{mole}^{-1}\ \text{s}^{-1})$
599	0.54
629	2.5
647	5.2
666	14
683	25
700	64

Find E_a and A from a graph.

17.35 For the reaction $2HI \rightarrow H_2 + I_2$, values of k are 1.2×10^{-3} and 3.0×10^{-5} dm^3 $mole^{-1}$ s^{-1} at 700 and 629 K, respectively. Estimate E_a and A.

17.36 What value of k is predicted by the Arrhenius equation for $T \rightarrow \infty$? Is this result physically reasonable?

17.37 The number of chirps per minute of a snowy tree cricket (*Oecanthus fultoni*) at several temperatures is 178 at 25.0°C, 126 at 20.3°C, and 100 at 17.3°C. (*a*) Find the activation energy for the chirping process. (*b*) Find the number of chirps to be expected at 14.0°C. Compare the result with the well-known rule that the Fahrenheit temperature equals 40 plus the number of cricket chirps in 15 seconds.

17.38 From the results of Probs. 17.34 and 17.35, find $\Delta U°$ and $\Delta H°$ for $H_2 + I_2 = 2HI$ in the neighborhood of 650 K.

17.39 For the gas-phase reaction

$$2N_2O_5 \rightarrow 4NO_2 + O_2,$$

$$k = 2.05 \times 10^{13} \exp(-24{,}650 \text{ cal mole}^{-1}/RT) \text{ s}^{-1}$$

(*a*) Give the values of A and E_a. (*b*) Find $k(0°C)$. (*c*) Find $t_{1/2}$ at $-50°C$, 0°C, and 50°C.

17.40 For a system with the two competing elementary reactions in (17.42), show that the observed activation energy is

$$E_a = \frac{k_1(T)E_{a,1} + k_2(T)E_{a,2}}{k_1(T) + k_2(T)}$$

17.41 For the elementary gas-phase reaction $CO + NO_2 \rightarrow CO_2 + NO$, one finds that $E_a = 27.8$ kcal/mole. Use data in the Appendix to find E_a for the reverse reaction.

17.42 For the mechanism (1) $A + B \rightleftharpoons C + D$; (2) $2C \rightarrow G + H$, step 2 is rate-determining. Given the activation energies $E_{a,1} = 30$ kcal/mole, $E_{a,-1} = 24$ kcal/mole, and $E_{a,2} = 49$ kcal/mole, find E_a for the overall reaction.

17.43 $\Delta G°_{f,700}$ of HI is -2.81 kcal/mole. For $H_2 + I_2 \rightarrow 2HI$, use data in Probs. 17.34 and 17.35 to find (*a*) the stoichiometric number of the rate-determining step; (*b*) K_c at 629 K.

17.44 For the reaction

$$BrO_3^- + 3SO_3^{2-} \rightarrow Br^- + 3SO_4^{2-}$$

in aqueous solution, one finds $r = k[BrO_3^-][SO_3^{2-}] \times [H^+]$. Give the expression for the rate law of the reverse reaction if the rate-determining step has stoichiometric number (*a*) 1; (*b*) 2.

17.45 For a gas-phase reaction whose rate-limiting step has stoichiometric number s, show that $E_{a,f} - E_{a,b} = \Delta U°/s$.

17.46 Show that (17.81) is valid when the designations of forward and reverse reactions are interchanged.

17.47 When an overall reaction is at equilibrium, the forward rate of a given elementary step must equal the reverse rate of that step. Also, the elementary steps multiplied by their stoichiometric numbers add to the overall reaction. Use these facts to show that the equilibrium constant K_c for an overall reaction whose mechanism has m elementary steps is related to the elementary rate constants by $K_c = \prod_{i=1}^{m} (k_i/k_{-i})^{s_i}$, where k_i, k_{-i}, and s_i are the forward and reverse rate constants and the stoichiometric number for the ith elementary step.

17.48 For the N_2O_5 mechanism (17.8), what is the rate law for the reverse reaction $4NO_2 + O_2 \rightarrow 2N_2O_5$ if step (*b*) is the rate-determining step?

17.49 For the unimolecular isomerization of cyclopropane to propylene, values of k_{uni} vs. initial pressure P_0 at 470°C are:

P_0/torr	110	211	388	760
$10^5 k_{uni}/s^{-1}$	9.58	10.4	10.8	11.1

Take the reciprocal of Eq. (17.89) and plot these data in a way that gives a straight line. From the slope and intercept, evaluate $k_{uni,P=\infty}$ and the Lindemann parameters k_1 and k_{-1}/k_2.

17.50 An oversimplified version of the CH_3CHO decomposition mechanism is

$$CH_3CHO \xrightarrow{(1)} CH_3 + CHO$$

$$CH_3 + CH_3CHO \xrightarrow{(2)} CH_4 + CH_3CO$$

$$CH_3CO \xrightarrow{(3)} CO + CH_3$$

$$2CH_3 \xrightarrow{(4)} C_2H_6$$

(The CHO reacts to form minor amounts of various species.) (*a*) Identify the initiation, propagation, and termination steps. (*b*) What is the overall reaction (neglecting minor products formed in initiation and termination steps)? (*c*) Show that $r = k[CH_3CHO]^{3/2}$, where $k = k_2(k_1/2k_4)^{1/2}$.

17.51 In the treatment of the $H_2 + Br_2$ chain reaction, the following elementary reactions were not considered: (I) $H_2 + M \rightarrow 2H + M$; (II) $Br + HBr \rightarrow H + Br_2$; (III) $H + Br + M \rightarrow HBr + M$. Use qualitative reasoning involving activation energies and concentrations to explain why the rate of each of these reactions is negligible compared with the rates of those in (17.92).

17.52 For the reversible reaction $CO + Cl_2 = COCl_2$, the mechanism is believed to be

Step 1: $Cl_2 + M \rightleftharpoons 2Cl + M$

Step 2: $Cl + CO + M \rightleftharpoons COCl + M$

Step 3: $COCl + Cl_2 \rightleftharpoons COCl_2 + Cl$

(*a*) Identify the initiation, propagation, and termination steps. (*b*) Assume steps 1 and 2 each to be in equilibrium, and find the rate law for the *forward* reaction. (*c*) What is the rate law for the reverse reaction?

17.53 Let E_a be the activation energy for the rate constant k in (17.99). (*a*) Relate E_a to $E_{a,1}$, $E_{a,-1}$, and $E_{a,2}$. (*b*) Measurement of $k(T)$ gives $E_a = 40.6$ kcal/mole and $A = 1.6 \times 10^{11}$ dm$^{3/2}$ mole$^{-1/2}$ s^{-1}. Use data in the Appendix to evaluate $E_{a,2}$ and find an expression for the elementary rate constant k_2 as a function of T.

17.54 For the photolysis of $CH_3NNC_2H_5$, what products will be obtained if the reaction is carried out (*a*) in the gas phase; (*b*) in solution in an inert solvent?

17.55 For I in CCl_4 at 25°C, the diffusion coefficient is estimated to be 4.2×10^{-5} cm^2 s^{-1}, and the radius of I is about 2 Å. Calculate k_D for $I + I \rightarrow I_2$ in CCl_4 at 25°C and compare with the observed value 0.8×10^{10} dm^3 mole^{-1} s^{-1}.

17.56 For the reaction α-glucose $\rightarrow$ β-glucose in aqueous solution, the rate law has the form (17.108) with RCOOR replaced by α-glucose. At 25°C, $k_O = 4 \times 10^{-4}$ s^{-1}, $k_{H^+} = 13 \times 10^{-3}$ dm^3 mole^{-1} s^{-1}, and $k_{OH^-} = 4 \times 10^2$ dm^3 mole^{-1} s^{-1}. For an α-glucose concentration of 0.1 mole dm^{-3}, plot log r vs. p[H^+], where $p[H^+] \equiv -\log([H^+]/c^\circ)$.

17.57 The reaction $CO_2(aq) + H_2O \rightarrow H^+ + HCO_3^-$ catalyzed by the enzyme bovine carbonic anhydrase was studied in a stopped-flow apparatus at pH 7.1 and temperature 0.5°C. For an initial enzyme concentration of 2.8×10^{-9} mole dm^{-3}, initial rates as a function of $[CO_2]_0$ are:

$[CO_2]_0$/(mmole dm^{-3})	r_0/(mmole dm^{-3} s^{-1})
1.25	0.028
2.50	0.048
5.00	0.080
20.0	0.15_5

Find k_2 and K_M from a Lineweaver–Burk plot.

17.58 Observed half-lives for the W-catalyzed decomposition of NH_3 at 1100°C as a function of initial NH_3 pressure P_0 for a fixed mass of catalyst and a fixed container volume are 7.6, 3.7, and 1.7 min for P_0 values of 265, 130, and 58 torr, respectively. Find the reaction order.

17.59 It is believed that N_2 and H_2 are chemisorbed on Fe as N and H atoms, which then react stepwise to give NH_3. What would be the stoichiometric number of the rate-determining step in the Fe-catalyzed synthesis of NH_3 if the rate-determining step were: (*a*) $N_2 + 2* \rightarrow 2N*$; (*b*) $H_2 + 2* \rightarrow 2H*$; (*c*) $N* + H* \rightarrow *NH + *$; (*d*) $*NH + H* \rightarrow *NH_2 + *$; (*e*) $*NH_2 + H* \rightarrow *NH_3 + *$; (*f*) $*NH_3 \rightarrow NH_3 + *$? Rate measurements using isotopic tracers indicate that the stoichiometric number of the rate-determining step is probably 1 for the NH_3 synthesis on iron. What does this indicate about the rate-determining step? (Write the overall reaction with the smallest possible integers.)

17.60 (*a*) Show that for a nonionic diffusion-controlled reaction, $E_a \approx RT - RT^2\eta^{-1}\, d\eta/dT$. (*b*) Use (16.102) to calculate E_a for such a reaction in water at 25°C.

17.61 Show that for a half-reaction at an electrode of a galvanic or electrolytic cell, the reaction rate per unit surface area is $r_s = j/n\mathcal{F}$, where n is the number of electrons in the half-reaction and $j \equiv I/\mathcal{A}$ is the current density.

17.62 The molar rest masses of ^2H, ^3He, and n are 2.014102, 3.016030, and 1.008665 g, respectively. Calculate the isothermal $\Delta\bar{U}$ for the fusion reaction $D + D \rightarrow {}^3He + n$.

17.63 A sample of 0.420 mg of $^{233}UF_6$ shows an activity of 9.88×10^4 counts per second. Find $t_{1/2}$ of ^{233}U.

17.64 The half-life of ^3H is 12.4 years. Calculate the activity of 20.0 g of HNO_3 containing 0.200 mole percent $^3HNO_3$.

17.65 The half-life of ^{14}C is 5700 years. The activity of carbon in living beings is 12.5 counts per minute per gram of carbon. (*a*) Calculate the percent of carbon in living beings that is ^{14}C. (*b*) Calculate the activity of carbon from the remains of an organism that died 50,000 years ago. (*c*) Find the age of wood from an

Egyptian tomb that shows an activity of 7.0 counts per minute per gram of carbon.

17.66 Explain why when a sample is dated using ^{14}C, the activity is counted for a period of 2 or 3 days, rather than for, say, 1 hour.

17.67 The decay of ^{40}K is 90 percent by β^- emission and 10 percent by electron capture. The ^{40}K half-life is 1.26×10^9 years. Calculate λ for β^- decay and λ for EC in ^{40}K.

17.68 (*a*) Verify that (17.128) is the overall reaction for the fusion mechanism (17.127). Give the stoichiometric number of each step. (*b*) The sun radiates 3.9×10^{26} J/s. How many moles of ^4He are produced each second in the sun? (*c*) The earth is on the average 1.5×10^8 km from the sun. Find the number of neutrinos that hit a square centimeter of the earth in 1 s. Consider the square centimeter to be perpendicular to the earth–sun line. (Experiments indicate the actual neutrino flux from the sun to be far less than the theoretically calculated value, so a revision in current views of solar structure or neutrino properties may be required.)

17.69 The uranium present on earth today is 99.28 percent ^{238}U and 0.72 percent ^{235}U. The half-lives are 4.51×10^9 years for ^{238}U and 7.0×10^8 years for ^{235}U. How long ago was this uranium 50 percent ^{238}U and 50 percent ^{235}U? (Isotopic abundances are given on an atom percent basis.)

17.70 The radius of an atomic nucleus with mass number A is $r = 1.4 \times 10^{-13} A^{1/3}$ cm. Calculate the density of an atomic nucleus.

18

QUANTUM MECHANICS

18.1 BLACKBODY RADIATION AND ENERGY QUANTIZATION

Classical physics is the physics developed before 1900. It consists of classical (Newtonian) mechanics (Sec. 2.2), Maxwell's theory of electricity, magnetism, and electromagnetic radiation (Sec. 21.1), thermodynamics, and the kinetic theory of gases (Chaps. 15 and 16). In the late nineteenth century, some physicists believed that the theoretical structure of physics was complete, but in the last quarter of the nineteenth century, various experimental results were obtained that could not be explained by classical physics. These results led to the development of quantum theory and the theory of relativity. An understanding of atomic structure, chemical bonding, and molecular spectroscopy must be based on quantum theory, which is the subject of this chapter. (We shall not need to consider relativity theory explicitly, since the valence electrons in atoms and molecules move at average speeds much less than the speed of light.)

One failure of classical physics was the incorrect $\bar{C}_V$ values of polyatomic molecules predicted by the kinetic theory of gases (Sec. 15.10). A second failure was the inability of classical physics to explain the observed frequency distribution of radiant energy emitted by a hot solid.

When a solid is heated, it emits light. The classical picture of light is that it is a wave consisting of oscillating electric and magnetic fields, an *electromagnetic wave.* (See Sec. 21.1 for a fuller discussion.) The frequency ν and wavelength λ of an electromagnetic wave traveling through vacuum are related by $\lambda\nu = c$, where $c = 3.0 \times 10^{10}$ cm/s is the speed of light in vacuum. The human eye is sensitive to electromagnetic waves whose wavelengths lie in the range 4000 Å (violet light) to 7500 Å (red light), which corresponds to frequencies in the range 7×10^{14} to 4×10^{14} cycles/s. However, electromagnetic radiation can have any frequency (see Fig. 21.2). We shall use the term "light" as synonymous with electromagnetic radiation, not restricting it to visible light.

Different solids emit radiation at different rates at the same temperature. To simplify things, one deals with the radiation emitted by a blackbody. A *blackbody* is a body that absorbs all the electromagnetic radiation that falls on it. A good approximation to a blackbody is a cavity with a tiny hole. Radiation that enters the hole is repeatedly reflected within the cavity (Fig. 18.1); at each reflection, a certain fraction of the radiation is absorbed by the cavity walls, and the large number of reflections leads to virtually all the incident radiation's being absorbed. When the cavity is heated, its walls emit light, a tiny portion of which escapes through the hole.

It isn't hard to show that the rate of radiation emitted per unit surface area of a blackbody is a function of only its temperature and is independent of the material of which the blackbody is made. (See *Zemansky*, sec. 4-14 for a proof.) (It can also be shown that the rate of radiation emission by unit area of any body at temperature T cannot exceed that from a blackbody at T; see *Zemansky*, sec. 4-15.)

By using a prism to separate the various frequencies emitted by the cavity, one can measure the amount of blackbody radiant energy emitted in a given narrow frequency range. Let the frequency distribution of the emitted blackbody radiation be described by the function $R(\nu)$, where $R(\nu)\,d\nu$ is the energy with frequency in the range ν to $\nu + d\nu$ that is radiated per unit time and per unit surface area. (Recall the discussion of distribution functions in Sec. 15.4.) Figure 18.2 shows experimentally observed $R(\nu)$ curves at various temperatures. As T increases, the maximum in $R(\nu)$ shifts to higher frequencies. When a metal rod is heated, it first glows red, then orange-yellow, then white, then blue-white. (White light is a mixture of all colors.) Our bodies are not hot enough to emit visible light, but we do emit infrared radiation. This emission is used in the infrared thermography method of detecting breast-cancer tumors; the tumor is at a slightly higher temperature than normal tissue and emits more infrared radiation.

In June 1900, Lord Rayleigh attempted to derive the theoretical expression for the function $R(\nu)$. Using the equipartition-of-energy theorem (Sec. 15.10), he found that classical physics predicted $R(\nu) = (2\pi kT/c^2)\nu^2$, where k and c are Boltzmann's constant and the speed of light. But this result is physically absurd, since it predicts that the amount of energy radiated would increase without limit as ν increases. In actuality, $R(\nu)$ reaches a maximum and then falls off to zero as ν increases (Fig. 18.2). Thus, classical physics fails to predict the spectrum of blackbody radiation.

On Oct. 19, 1900, the German physicist Max Planck announced to the German Physical Society his discovery of an empirical formula that gave a highly accurate fit to the observed curves of blackbody radiation. Planck's formula was $R(\nu) = c_1\nu^3/(e^{c_2\nu/T} - 1)$, where c_1 and c_2 are constants with certain numerical values. Planck

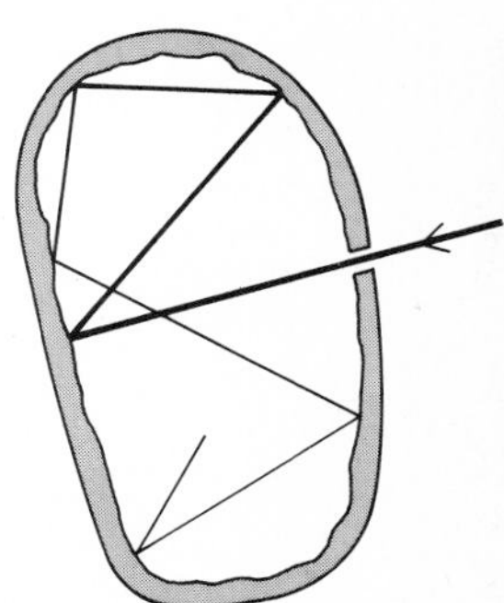

Figure 18.1
A cavity acting as a blackbody.

obtained this formula by trial and error and at this time had no theory to explain it. On Dec. 14, 1900, Planck again appeared before the German Physical Society, and presented a theory that yielded the blackbody-radiation formula he had found empirically a few weeks earlier. Planck's theory gave the constants c_1 and c_2 as $c_1 = 2\pi h/c^2$ and $c_2 = h/k$, where h was a new constant in physics. Planck's theoretical expression for the frequency distribution of blackbody radiation is then

$$R(\nu) = \frac{2\pi h\nu^3}{c^2} \frac{1}{e^{h\nu/kT} - 1} \tag{18.1}$$

To obtain (18.1), Planck introduced the hypothesis that the radiating atoms or molecules of the blackbody could emit and absorb electromagnetic energy of frequency ν only in amounts of $h\nu$, where h is a constant (later called *Planck's constant*) having the dimensions of energy × time. If ΔE is the energy change in a blackbody atom due to emission of electromagnetic radiation of frequency ν, then $\Delta E = h\nu$. This assumption then leads to Eq. (18.1). (For Planck's derivation, see M. Jammer, *The Conceptual Development of Quantum Mechanics*, McGraw-Hill, 1966, sec. 1.2.) Planck obtained a numerical value of h by fitting the formula (18.1) to the observed blackbody curves. The modern value is

$$h = 6.626 \times 10^{-27} \text{ erg} \cdot \text{s} = 6.626 \times 10^{-34} \text{ J} \cdot \text{s} \tag{18.2)*$$

In classical physics, energy takes on a continuous range of values, and a system can lose or gain any amount of energy. Planck proposed that a blackbody atom radiating light of frequency ν was restricted to emitting an amount of energy given by $h\nu$. Planck called this definite amount of energy a *quantum* of energy (the Latin word *quantum* means "how much"). In classical physics, energy is a continuous variable. In quantum physics, the energy of a system is *quantized*, meaning that the energy can take on only certain values. Planck introduced the idea of energy quantization in one particular case, that of the emission of blackbody radiation. In the years 1900–1926, the concept of energy quantization was gradually extended to all microscopic systems.

It is likely that Planck's assumption of energy quantization was originally intended only as a calculational device, and Planck planned to take the limit $h \to 0$ at the end of the derivation. He found, however, that taking this limit gave the wrong result but with $h \neq 0$ he obtained the correct formula (18.1). [Expanding the exponential in (18.1) in a Taylor series, we get $-1 + e^{h\nu/kT} = -1 + 1 + h\nu/kT +$

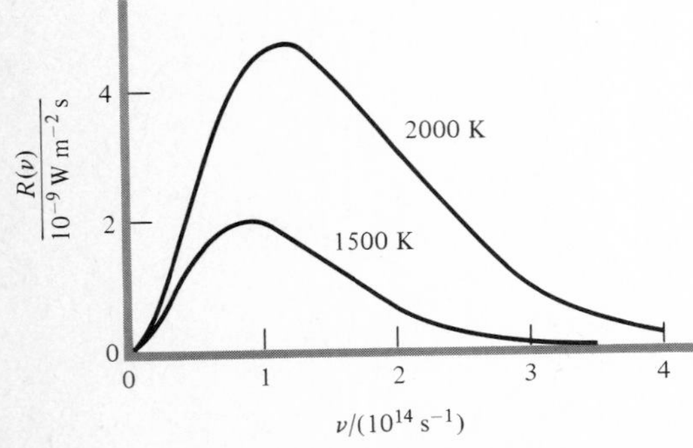

Figure 18.2
Frequency distribution of blackbody radiation at two temperatures.

$h^2\nu^2/2k^2T^2 + \cdots$, which goes to $h\nu/kT$ as $h \to 0$. Hence, (18.1) approaches $2\pi\nu^2kT/c^2$ as $h \to 0$, which is the (erroneous) classical result of Rayleigh.]

The concept of quantization of energy was a revolutionary departure from classical physics, and most physicists were very reluctant to accept this idea. One of the most reluctant was Planck himself, whose conservative temperament was offended by energy quantization. In the years following 1900, Planck tried repeatedly to derive (18.1) without using energy quantization, but he failed.

18.2 THE PHOTOELECTRIC EFFECT AND PHOTONS

The person who recognized the value of Planck's idea was Einstein, who applied the concept of energy quantization to electromagnetic radiation and showed that this allowed the experimental observations in the photoelectric effect to be explained.

In the *photoelectric effect*, a beam of electromagnetic radiation (light) shining on a metal surface causes the metal to emit electrons; electrons absorb energy from the light beam, thereby acquiring enough energy to escape from the metal. [A practical application is the photoelectric cell, used to measure light intensities, to prevent elevator doors from crushing people, in smoke detectors (light scattered by smoke particles causes electron emission, which sets off an alarm), in solar-energy cells, etc.]

Experimental work by Lenard and others around 1900 had shown that: (*a*) Electrons are emitted only when the frequency of the light exceeds a certain minimum frequency ν_0 (the *threshold frequency*); the value of ν_0 differs for different metals and lies in the ultraviolet for most metals. (*b*) Increasing the intensity of the light increases the number of electrons emitted but does not affect the kinetic energy of the emitted electrons. (*c*) Increasing the frequency of the radiation increases the kinetic energy of the emitted electrons.

In 1905, Lenard received the Nobel physics prize for his work on cathode rays. In 1933, Lenard wrote: "The most important example of the dangerous influence of Jewish circles in the study of nature has been provided by Einstein with his mathematically botched-up theories." In his book *German Physics*, published in the mid-1930s, Lenard wrote: "German physics? one asks. I might also have said Aryan physics or the physics of the Nordic species of man; the physics of those who have explored the depths of reality, seekers of truth, the physics of the very founders of science."

Lenard's observations on the photoelectric effect cannot be understood using the classical picture of light as a wave. The energy in a wave is proportional to its intensity but is independent of its frequency, so one would expect the kinetic energy of the emitted electrons to increase with an increase in light intensity and to be independent of the light's frequency. Moreover, the wave picture of light would predict the photoelectric effect to occur at any frequency, provided the light is sufficiently intense.

In 1905, Einstein explained the photoelectric effect by extending Planck's concept of energy quantization to electromagnetic radiation. (Planck had applied energy quantization to the emission process but had considered the electromagnetic radiation to be a wave.) Einstein proposed that in addition to having wavelike properties, light could also be considered to consist of particlelike entities (quanta), each quantum of light having an energy $h\nu$, where h is Planck's constant and ν is the frequency of the light. These entities were later named *photons*, and we have for the energy of a photon

$$E_{\text{photon}} = h\nu \qquad (18.3)^*$$

The energy in a light beam is the sum of the energies of the individual photons and is therefore quantized.

Let electromagnetic radiation of frequency ν fall on a metal. The photoelectric effect occurs when an electron in the metal is hit by a photon; the photon disappears and its energy $h\nu$ is transferred to the electron. Part of the energy absorbed by the electron is used to overcome the forces holding the electron in the metal, and the remainder appears as kinetic energy of the emitted electron. Conservation of energy therefore gives

$$h\nu = W + \tfrac{1}{2}mv^2 \tag{18.4}$$

where the *work function* W is the minimum energy needed by an electron to escape the metal and $\frac{1}{2}mv^2$ is the kinetic energy of the free electron. [The valence electrons in metals have a distribution of energies (Sec. 24.9), so some electrons need more energy than others to leave the metal; the emitted electrons therefore show a distribution of kinetic energies, and $\frac{1}{2}mv^2$ in (18.4) is the maximum kinetic energy of emitted electrons.]

Einstein's equation (18.4) explains all the observations in the photoelectric effect. If the light frequency is such that $h\nu < W$, a photon does not have sufficient energy to allow an electron to escape the metal and no photoelectric effect occurs. The minimum frequency ν_0 at which the effect occurs is given by $h\nu_0 = W$. (The work function W differs for different metals, being lowest for the alkali metals.) Equation (18.4) shows the kinetic energy of the emitted electrons to increase with ν and to be independent of the light intensity. An increase in intensity with no change in frequency increases the energy of the light beam and hence increases the number of photons per unit volume in the light beam, thereby increasing the rate of emission of electrons.

Einstein's theory of the photoelectric effect agreed with the qualitative observations of Lenard, but it wasn't until 1916 that R. A. Millikan made an accurate quantitative test of Eq. (18.4). The difficulty in testing (18.4) is the necessity of maintaining a very clean surface of the metal. Millikan found accurate agreement between (18.4) and experiment.

At first physicists were very reluctant to accept Einstein's hypothesis of photons. Light shows the phenomena of diffraction and interference (*Halliday and Resnick*, chaps. 43 and 44), and these effects are exhibited only by waves, not by particles. As late as 1913, Planck and Nernst wrote: "That [Einstein] may sometimes have missed the target in his speculations, as, for example, in his theory of light quanta, cannot really be held against him. For every innovation in science entails risk."

Eventually, physicists became convinced that the photoelectric effect could be understood only by viewing light as being composed of photons. However, diffraction and interference can be understood only by viewing light as a wave and not as a collection of particles. Thus, light seems to exhibit a dual nature, behaving like waves in some situations and like particles in other situations. This apparent duality is logically contradictory, since the wave and particle models are mutually exclusive. Particles are localized in space, but waves are not. The photon picture gives a quantization of the light energy, but the wave picture does not. In Einstein's equation $E_{\text{photon}} = h\nu$, the quantity E_{photon} is a particle concept, but the frequency ν is a wave concept, so this equation is, in a sense, self-contradictory. An explanation of these apparent contradictions is given in Sec. 18.4.

In 1907, Einstein applied the concept of energy quantization to the vibrations of the atoms in a solid, thereby showing that the heat capacity of a solid goes to zero as T goes to zero, a result in agreement with experiment but in disagreement with the classical equipartition theorem. (See Sec. 24.10 for details.)

18.3 THE BOHR THEORY OF THE HYDROGEN ATOM

The next major application of energy quantization was Niels Bohr's 1913 theory of the hydrogen atom. A heated gas of hydrogen atoms emits electromagnetic radiation containing only certain distinct frequencies (Fig. 21.19). During 1885 to 1910, Balmer, Rydberg, and others found that the following empirical formula correctly reproduces the observed H-atom spectral frequencies:

$$\frac{\nu}{c} = \frac{1}{\lambda} = R\left(\frac{1}{n_b^2} - \frac{1}{n_a^2}\right) \qquad n_b = 1, 2, 3, \ldots; \quad n_a = 2, 3, \ldots; \quad n_a > n_b \quad (18.5)$$

where the *Rydberg constant* R equals 109,677.6 cm^{-1}. There was no explanation for this formula until Bohr's work.

If one accepts Einstein's equation $E_{\text{photon}} = h\nu$, the fact that only certain frequencies of light are emitted by H atoms indicates that, contrary to classical ideas, a hydrogen atom can exist only in certain energy states. Bohr therefore postulated that the energy of a hydrogen atom is quantized: (1) An atom can take on only certain distinct energies $E_1, E_2, E_3, \ldots$. Bohr called these allowed states of constant energy the *stationary states* of the atom. (This term is not meant to imply that the electron is at rest in a stationary state.) Bohr further assumed that: (2) An atom in a stationary state does not emit electromagnetic radiation. To explain the line spectrum of hydrogen, Bohr assumed that: (3) When an atom makes a transition from a stationary state with energy E_a to a lower-energy stationary state with energy E_b, it emits a photon of light. Since $E_{\text{photon}} = h\nu$, conservation of energy gives

$$E_a - E_b = h\nu \qquad (18.6)^*$$

where $E_a - E_b$ is the energy difference between the atomic states involved in the transition and ν is the frequency of the light emitted. Similarly, an atom can make a transition from a lower-energy to a higher-energy state by absorbing a photon of frequency given by (18.6). The Bohr theory provided no description of the transition process between two stationary states; somehow, the electron made a very rapid "jump" between states *a* and *b*. (Of course, transitions between stationary states can occur by means other than absorption or emission of electromagnetic radiation; e.g., an atom can gain or lose electronic energy in a collision with another atom.)

Equations (18.5) and (18.6) give $E_a - E_b = Rhc(1/n_b^2 - 1/n_a^2)$, which strongly indicates that the energies of the H-atom stationary states are given by $E = -Rhc/n^2$, with $n = 1, 2, 3, \ldots$. Bohr then introduced further postulates to derive a theoretical expression for the Rydberg constant. He assumed that (4) the electron in an H-atom stationary state moves in a circle around the nucleus and obeys the laws of classical mechanics. The energy of the electron is the sum of its kinetic energy and the potential energy of the electron–nucleus electrostatic attraction. Classical mechanics shows that the energy depends on the radius of the orbit. Since the energy is quantized, only certain orbits are allowed. Bohr used one final postulate to select the allowed orbits. Most books give this postulate as (5) the allowed orbits are those for which the electron's angular momentum mvr equals $nh/2\pi$, where m and v are the electron's mass and speed, r is the radius of the orbit, and $n = 1, 2, 3, \ldots$. Actually, Bohr used a different postulate which is less arbitrary than 5 but perhaps less simple to state. The postulate Bohr used is equivalent to 5, and is omitted here. (If you're curious, see *Karplus and Porter*, sec. 1.4.)

With his postulates, Bohr derived the following expression for the H-atom energy levels: $E = -2\pi^2 me'^4/h^2n^2$, where e' is the proton charge in statcoulombs (Sec. 19.1). Hence, Bohr predicted that $Rhc = 2\pi^2 me'^4/h^2$ and $R = 2\pi^2 me'^4/h^3c$. Substitution of the values of m, e', h, and c gave a result in good agreement with the experimental value of the Rydberg constant, indicating that the Bohr model gave the correct energy levels of H.

Although the Bohr theory is of great historical importance for the development of quantum theory, postulates 4 and 5 are in fact totally false, and the Bohr theory was superseded in 1926 by the Schrödinger equation, which provides a correct picture of electronic behavior in atoms and molecules. Since the Bohr-theory derivation of the H-atom energy levels is based partly on false premises, we omit giving it. (Although postulates 4 and 5 are false, postulates 1, 2, and 3 are consistent with quantum mechanics.)

18.4 THE DE BROGLIE HYPOTHESIS

In the years 1913 to 1925, attempts were made to apply the Bohr theory to atoms with more than one electron and to molecules. However, all attempts to derive the spectra of such systems using extensions of the Bohr theory failed. It gradually became clear that there was a fundamental error in the Bohr theory. (The fact that the Bohr theory works for H is something of an accident.)

A key idea toward resolving these difficulties was advanced by the French physicist Louis de Broglie (1892–   ) in 1923. The fact that a heated gas of atoms or molecules emits radiation of only certain frequencies shows that the energies of atoms and molecules are quantized, only certain energy values being allowed. Quantization of energy does not occur in classical mechanics; a particle can have any energy in classical mechanics. (The introduction of energy quantization in the Bohr theory was quite arbitrary; Bohr gave no reason why only certain orbits and energies were allowed.) Quantization does occur in wave motion. For example, a string held fixed at each end has quantized modes of vibration (Fig. 18.3). The string can vibrate at its fundamental frequency ν, at its first overtone frequency 2ν, at its second overtone frequency 3ν, etc.; frequencies lying between these integral multiples of ν are not allowed. De Broglie therefore proposed that just as light shows both wave and particle aspects, matter also has a "dual" nature. As well as showing particlelike behavior, an electron would also show wavelike behavior, the wavelike behavior manifesting itself in the quantized energy levels of electrons in atoms and molecules. Holding the ends of a string fixed quantizes its vibrational frequencies; similarly, confining an electron in an atom quantizes its energies.

Figure 18.3
Fundamental and overtone vibrations of a string.

De Broglie obtained an equation for the wavelength λ to be associated with a material particle by reasoning in analogy with photons. We have $E_{\text{photon}} = h\nu$. Einstein's special theory of relativity gives the photon energy as $E_{\text{photon}} = mc^2$, where c is the speed of light. (In this equation, m is the relativistic mass of the photon. A photon has zero rest mass, but photons always move at speed c in vacuum and are never at rest. At speed c, the photon has a nonzero mass m.) Equating the two expressions for E_{photon}, we get $h\nu = mc^2$. But $\nu = c/\lambda$, where λ is the wavelength of the light. Hence, $hc/\lambda = mc^2$ and $\lambda = h/mc$ for a photon. By analogy, de Broglie proposed that a material particle with mass m and speed v would have a wavelength λ given by

$$\lambda = h/mv \tag{18.7}$$

Note that $mv = p$, where p is the particle's momentum.

The de Broglie wavelength of an electron moving at 1.0×10^8 cm/s is

$$\lambda = \frac{6.6 \times 10^{-27} \text{ erg s}}{(9.1 \times 10^{-28} \text{ g})(1.0 \times 10^8 \text{ cm/s})} = 7 \times 10^{-8} \text{ cm} = 7 \text{ Å}$$

This wavelength is of the order of magnitude of molecular dimensions and indicates that wave effects are important in electronic motions in atoms and molecules. For a macroscopic particle of mass 1.0 g moving at 1.0 cm/s, a similar calculation gives $\lambda = 7 \times 10^{-27}$ cm. The extremely small size of λ (which results from the smallness of Planck's constant h in comparison with mv) indicates that quantum effects are unobservable for the motion of macroscopic objects.

De Broglie's bold hypothesis was experimentally confirmed in 1927 by Davisson and Germer, who observed diffraction effects when an electron beam was reflected from a crystal of Ni; G. P. Thomson observed diffraction effects when electrons were passed through a thin sheet of metal. See Fig. 18.4. Similar diffraction effects have been observed with neutrons, protons, helium atoms, and hydrogen molecules, indicating that the de Broglie hypothesis applies to all material particles, not just electrons. Some applications of the wavelike behavior of microscopic particles are the electron microscope and the use of electron diffraction and neutron diffraction to obtain molecular structures (Sec. 24.8).

Electrons show particlelike behavior in some experiments (e.g., the cathode-ray experiments of J. J. Thomson, Sec. 19.2) and wavelike behavior in other experiments. As noted in Sec. 18.2, the wave and particle models are incompatible with each other. An entity cannot be both a wave and a particle. How can we explain the apparently contradictory behavior of electrons? The source of the difficulty is the attempt to

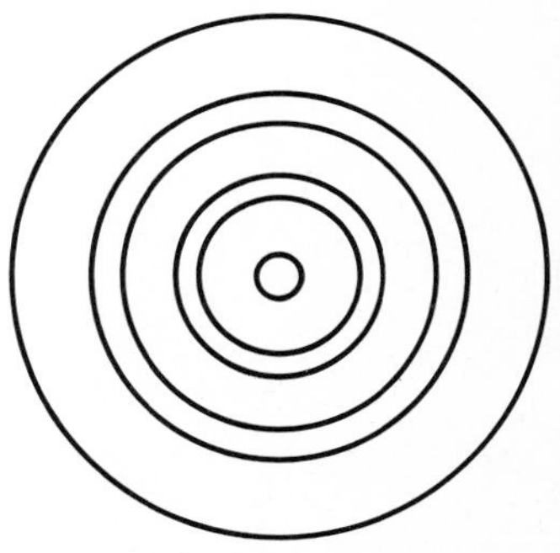

Figure 18.4
Diffraction rings observed when electrons are passed through a thin polycrystalline metal sheet.

describe microscopic entities like electrons by using concepts developed from our experience in the macroscopic world. The particle and wave concepts were developed from observations on large-scale objects, and there is no guarantee that they will be fully applicable on the microscopic scale. Under certain experimental conditions, an electron behaves like a particle; under other conditions, it behaves like a wave. However, an electron is neither a particle nor a wave: it is something that cannot be adequately described in terms of a model we can visualize.

A similar situation holds for light, which shows wave properties in some situations and particle properties in others. Light originates in the microscopic world of atoms and molecules and cannot be fully understood in terms of models visualizable by the human mind.

Although both electrons and light exhibit an apparent "wave–particle duality," there are significant differences between these entities. Light travels at speed c in vacuum, and photons have zero rest mass. Electrons always travel at speeds less than c and have a nonzero rest mass.

18.5 THE UNCERTAINTY PRINCIPLE

The apparent wave–particle duality of matter and of radiation imposes certain limitations on the information we can obtain about a microscopic system. Consider a microscopic particle traveling in the y direction. Suppose we measure the x coordinate of the particle by having it pass through a narrow slit of width w and fall on a fluorescent screen (Fig. 18.5). If we see a spot on the screen, we can be sure the particle went through the slit; hence, we have measured the x coordinate at the time of passing the slit to an accuracy w. Before the measurement, the particle had zero velocity v_x and zero momentum $p_x = mv_x$ in the x direction. Because the microscopic particle has wavelike properties, it will be diffracted at the slit. [Photographs of electron-diffraction patterns at a single slit and at multiple slits are given in C. Jönsson, *Am. J. Phys.*, **42**, 4 (1974).]

Diffraction is the bending of a wave around an obstacle. A classical particle would go straight through the slit, and a beam of such particles would show a spread

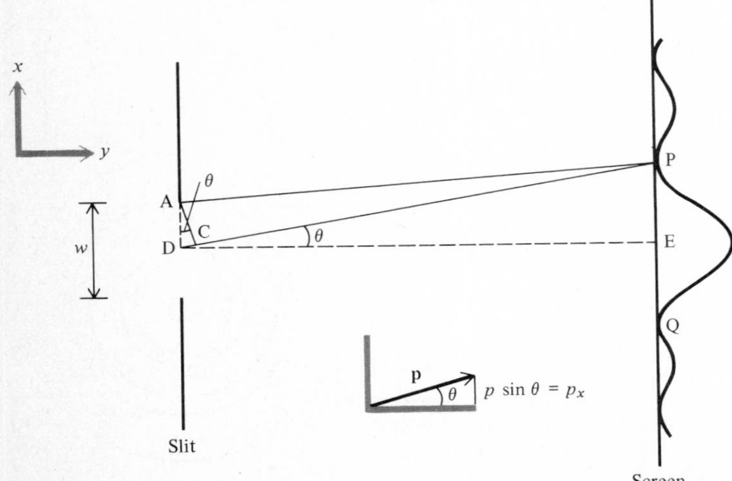

Figure 18.5
Diffraction at a slit.

of length w in where they hit the screen. A wave passing through the slit will spread out to give a diffraction pattern; the curve in Fig. 18.5 shows the intensity of the wave at various points on the screen; the maxima and minima on this curve result from constructive and destructive interference between waves originating from various parts of the slit. *Interference* results from the superposition of two waves traveling through the same region of space; when the waves are in phase (crests occurring together), constructive interference occurs, with the amplitudes adding to give a stronger wave; when the waves are out of phase (crests of one wave coinciding with troughs of the second wave), destructive interference occurs and the intensity is diminished.

The first minima (points P and Q) in the single-slit diffraction pattern occur at places on the screen where waves originating from the top of the slit travel one-half wavelength less or more than waves originating from the middle of the slit. These waves are then exactly out of phase and cancel each other. Similarly, waves originating from a distance d below the top of the slit cancel with waves originating a distance d below the center of the slit. The condition for the first diffraction minimum is then $\overline{DP} - \overline{AP} = \frac{1}{2}\lambda = \overline{CD}$ in Fig. 18.5, where C is located so that $\overline{CP} = \overline{AP}$. Because the distance from the slit to the screen is much greater than the slit width, angle APC is nearly zero and angles PAC and ACP are each nearly 90°. Hence, angle ACD is essentially 90°. Angles PDE and DAC each equal 90° minus angle ADC; these two angles are therefore equal, and have been marked θ. We have $\sin\theta = \overline{DC}/\overline{AD} = \frac{1}{2}\lambda/\frac{1}{2}w = \lambda/w$. The angle θ at which the first diffraction minimum occurs is given by $\sin\theta = \lambda/w$.

Coming back to the microscopic particle passing through the slit, diffraction at the slit will cause the particle to change its direction of motion. A particle diffracted by angle θ and hitting the screen at P or Q will have an x component of momentum $p_x = p \sin\theta$ at the slit (Fig. 18.5), where p is the particle's momentum. The intensity curve in Fig. 18.5 shows that the particle is most likely to be diffracted by an angle lying in the range $-\theta$ to $+\theta$, where θ is the angle to the first diffraction minimum. Hence, the position measurement results in an uncertainty in the p_x value given by $p\sin\theta - (-p\sin\theta) = 2p\sin\theta$. We write $\Delta p_x = 2p\sin\theta$, where Δp_x gives the uncertainty in our knowledge of p_x at the slit. We saw above that $\sin\theta = \lambda/w$, so $\Delta p_x = 2p\lambda/w$. The de Broglie relation (18.7) gives $\lambda = h/p$, so $\Delta p_x = 2h/w$. The uncertainty in our knowledge of the x coordinate is given by the slit width, so $\Delta x = w$. Therefore, $\Delta x\, \Delta p_x = 2h$.

Before the measurement, we had no knowledge of the particle's x coordinate, but we knew that it was traveling in the y direction and hence had $p_x = 0$. (Thus, before the measurement, $\Delta x = \infty$ and $\Delta p_x = 0$.) The slit of width w gave the x coordinate to an uncertainty w ($\Delta x = w$) but introduced an uncertainty $\Delta p_x = 2h/w$ in p_x. By reducing the slit width w, we can measure the x coordinate as accurately as we please, but as $\Delta x = w$ becomes smaller, $\Delta p_x = 2h/w$ becomes larger. The more we know about x, the less we know about p_x. The measurement introduces an uncontrollable and unpredictable disturbance in the system, changing p_x by an unknown amount.

Although we have analyzed only one experiment, analysis of many other experiments leads to the same conclusion, namely, that the product of the uncertainties in x and p_x of a particle is of the order of magnitude of Planck's constant or greater:

$$\Delta x\, \Delta p_x \gtrsim h \tag{18.8*}$$

This is the *uncertainty principle*, discovered by Heisenberg in 1927. [A general proof of (18.8) that starts from the postulates of quantum mechanics was given by Robertson in 1929; see *Gatz*, sec. 6*e*.] Of course, we have $\Delta y\ \Delta p_y \gtrsim h$ and $\Delta z\ \Delta p_z \gtrsim h$.

The small size of h makes the uncertainty principle of no consequence for macroscopic particles.

18.6 QUANTUM MECHANICS

The fact that electrons and other microscopic "particles" show wavelike as well as particlelike behavior indicates that electrons do not obey classical mechanics. Classical mechanics was formulated from the observed behavior of macroscopic objects and does not apply to microscopic particles. The form of mechanics obeyed by microscopic systems is called *quantum mechanics*, since a key feature of this mechanics is the quantization of energy. The laws of quantum mechanics were discovered by Heisenberg, Born, and Jordan in 1925 and by Schrödinger in 1926. Before discussing these laws, we consider some aspects of classical mechanics.

The motion of a one-particle, one-dimensional classical-mechanical system is governed by Newton's second law $F = ma = m\ d^2x/dt^2$. To obtain the particle's position x as a function of time, this differential equation must be integrated twice with respect to time. The first integration gives dx/dt, and the second integration gives x. Each integration introduces an arbitrary integration constant. Therefore, integration of $F = ma$ gives an equation for x that contains two unknown constants c_1 and c_2; we have $x = f(t, c_1, c_2)$, where f is some function. To evaluate c_1 and c_2, we need two pieces of information about the system. If we know that at a certain time t_0, the particle was at the position x_0 and had speed v_0, then c_1 and c_2 can be evaluated from the equations $x_0 = f(t_0, c_1, c_2)$ and $v_0 = f'(t_0, c_1, c_2)$, where f' is the derivative of f with respect to t. Thus, provided we know the force F and the particle's initial position and velocity (or momentum), we can use Newton's second law to predict the position of the particle at any future time. A similar conclusion holds for a three-dimensional, many-particle classical system.

The *state* of a system in classical mechanics is defined by specifying all the forces acting and all the positions and velocities (or momenta) of the particles. We saw in the preceding paragraph that knowledge of the present state of a classical-mechanical system allows its future state to be predicted with certainty.

The Heisenberg uncertainty principle, Eq. (18.8), shows that simultaneous specification of position and momentum is impossible for a microscopic particle. Hence, the very knowledge needed to specify the classical-mechanical state of a system is unobtainable in quantum theory. The state of a quantum-mechanical system must therefore involve less knowledge about the system than in classical mechanics.

In quantum mechanics, the *state* of a system is defined by a mathematical function Ψ (capital psi) called the *state function* or the *time-dependent wave function*. Ψ is a function of the coordinates of the particles of the system and (since the state may change with time) is also a function of time. For example, for a two-particle system, $\Psi = \Psi(x_1, y_1, z_1, x_2, y_2, z_2, t)$, where x_1, y_1, z_1 are the coordinates of particle 1, etc. The state function is in general a complex quantity; that is, $\Psi = f + ig$, where f and g are real functions of the coordinates and time and $i \equiv \sqrt{-1}$. The state function is an abstract entity, but we shall later see how Ψ is related to physically measurable quantities.

The state function changes with time. For an n-particle system, quantum mechanics postulates that the equation governing how Ψ changes with t is

$$-\frac{\hbar}{i}\frac{\partial\Psi}{\partial t} = -\frac{\hbar^2}{2m_1}\left(\frac{\partial^2\Psi}{\partial x_1^{\,2}} + \frac{\partial^2\Psi}{\partial y_1^{\,2}} + \frac{\partial^2\Psi}{\partial z_1^{\,2}}\right) - \cdots - \frac{\hbar^2}{2m_n}\left(\frac{\partial^2\Psi}{\partial x_n^{\,2}} + \frac{\partial^2\Psi}{\partial y_n^{\,2}} + \frac{\partial^2\Psi}{\partial z_n^{\,2}}\right) + V\Psi \tag{18.9}$$

In this equation, $\hbar$ (h bar) is Planck's constant divided by 2π,

$$\hbar \equiv h/2\pi \tag{18.10*}$$

i is $\sqrt{-1}$, $m_1, \ldots, m_n$ are the masses of particles $1, \ldots, n$, x_1, y_1, z_1 are the spatial coordinates of particle 1, and V is the potential energy of the system expressed as a function of the particles' coordinates and the time. V is derived from the forces acting in the system; see Eqs. (2.39), (2.45), and (2.46).

Equation (18.9) is a complicated partial differential equation. For most of the problems dealt with in this book, it will not be necessary to use (18.9), so don't panic.

The concept of the state function Ψ and Eq. (18.9) were introduced by the Austrian physicist Erwin Schrödinger (1887–1961) in 1926. Equation (18.9) is the *time-dependent Schrödinger equation*. Schrödinger was inspired by the de Broglie hypothesis to search for a mathematical equation that would resemble the differential equations that govern wave motion and would have solutions giving the allowed energy levels of a quantum system. Using $\lambda = h/p$ and certain plausibility arguments, Schrödinger proposed Eq. (18.9) and the related time-independent equation (18.19) below. Since chemistry students are usually not familiar with the partial differential equations of wave motion, such plausibility arguments have been omitted. It should be emphasized that these arguments can at best make the Schrödinger equation seem plausible. They can in no sense be used to derive or prove the Schrödinger equation. The Schrödinger equation is a fundamental postulate of quantum mechanics and cannot be derived. The reason we believe it to be true is that its predictions give excellent agreement with experimental results.

In 1925, several months before Schrödinger's work, Werner Heisenberg (1901–1976), Max Born (1882–1970), and Pascual Jordan developed a form of quantum mechanics that is based on mathematical entities called matrices. (A matrix is a rectangular array of numbers; matrices are added and multiplied according to certain rules.) The *matrix mechanics* of these workers turns out to be fully equivalent to the Schrödinger form of quantum mechanics (which is often called *wave mechanics*). We shall not discuss matrix mechanics.

Schrödinger also contributed to statistical mechanics, relativity, and the theory of color vision. In later life, he became interested in Eastern philosophy. In an epilogue to his 1944 book *What Is Life?*, Schrödinger wrote: "So let us see whether we cannot draw the correct, non-contradictory conclusion from the following two premises: (i) My body functions as a pure mechanism according to the Laws of Nature. (ii) Yet I know, by incontrovertible direct experience, that I am directing its motions.... The only possible inference from these two facts is, I think, that I—I in the widest meaning of the word, that is to say, every conscious mind that has ever said or felt 'I'—am the person, if any, who controls the 'motion of the atoms' according to the Laws of Nature."

The time-dependent Schrödinger equation (18.9) contains the first derivative of Ψ with respect to t, and a single integration with respect to time gives us Ψ. Integration of (18.9) therefore introduces only one integration constant, which can be evaluated if Ψ is known at some initial time t_0. Therefore, knowing the initial quantum-mechanical state $\Psi(x_1, \ldots, z_n, t_0)$, we can use (18.9) to predict the future quantum-mechanical state. The time-dependent Schrödinger equation is the quantum-mechanical analog of Newton's second law, which allows the future state of a classical-mechanical system to be predicted from its present state. We shall soon see, however, that knowledge of the state in quantum mechanics usually involves a knowledge of only probabilities, rather than certainties, as in classical mechanics.

What is the relation between quantum mechanics and classical mechanics? Experiment shows that macroscopic bodies obey classical mechanics (provided their speed is much less than the speed of light). We therefore expect that in the classical-mechanical limit of taking $h \to 0$, the time-dependent Schrödinger equation ought to reduce to Newton's second law. This was shown by Ehrenfest in 1927; for an outline of the proof, see *Levine*, *Quantum Chemistry*, prob. 7.19; this reference will hereinafter be referred to as *Q. C.*

(Paul Ehrenfest made several contributions to quantum theory and, in collaboration with his wife Tatiana, to statistical mechanics. In 1933, Ehrenfest shot and blinded his young son and then shot and killed himself.)

What is the meaning of the state function Ψ? Schrödinger originally conceived of Ψ as the amplitude of some sort of wave that was associated with the system. It soon became clear that this interpretation was wrong. (For example, for a two-particle system, Ψ is a function of the six spatial coordinates $x_1, y_1, z_1, x_2, y_2, z_2$, whereas a wave moving through space is a function of only three spatial coordinates.) The correct physical interpretation of Ψ was given by Max Born in 1926. Born postulated that $|\Psi|^2$ gives the probability density for finding the particles at given locations in space. (Probability densities for molecular speeds were discussed in Sec. 15.4.) To be more precise, suppose a one-particle system has the state function $\Psi(x, y, z, t')$ at time t'. Then $|\Psi(x, y, z, t')|^2\, dx\, dy\, dz$ is the probability at time t' that the particle has its three spatial coordinates in the ranges x to $x + dx$, y to $y + dy$, and z to $z + dz$; this is the probability that a measurement of position at time t' will show the particle to be in a tiny rectangular box located at point (x, y, z) in space and having edges dx, dy, and dz.

The state function Ψ is a complex quantity, and $|\Psi|$ is the absolute value of Ψ. Let $\Psi = f + ig$, where f and g are real and $i \equiv \sqrt{-1}$. The *absolute value* of Ψ is defined by $|\Psi| \equiv (f^2 + g^2)^{1/2}$. For a real quantity, g is zero, and the absolute value becomes $(f^2)^{1/2}$, which is the usual meaning of absolute value for a real quantity. The *complex conjugate* Ψ^* of Ψ is defined by $\Psi^* \equiv f - ig$; to get Ψ^*, we just replace i by $-i$ wherever it occurs. Note that $\Psi^*\Psi = (f - ig)(f + ig) = f^2 - i^2g^2 = f^2 + g^2 = |\Psi|^2$. Hence, instead of $|\Psi|^2$, we can write $\Psi^*\Psi$. The quantity $|\Psi|^2 = \Psi^*\Psi = f^2 + g^2$ is real and nonnegative, as a probability density must be.

For a two-particle system, $|\Psi(x_1, y_1, z_1, x_2, y_2, z_2, t')|^2\, dx_1\, dy_1\, dz_1\, dx_2\, dy_2\, dz_2$ is the probability that at time t', particle 1 is in a tiny rectangular box located at point (x_1, y_1, z_1) and having dimensions dx_1, dy_1, dz_1 and particle 2 is simultaneously in a box at (x_2, y_2, z_2) with dimensions dx_2, dy_2, dz_2. Born's postulated interpretation of Ψ gives results fully in agreement with experiment.

For a one-particle, one-dimensional system, $|\Psi(x, t)|^2\, dx$ is the probability that the particle is between x and $x + dx$ at time t. The probability that it is in the region between a and b is found by summing the infinitesimal probabilities over the interval from a to b to give the definite integral $\int_a^b |\Psi|^2\, dx$. The probability for finding the particle somewhere on the x axis must be 1. Hence, $\int_{-\infty}^{\infty} |\Psi|^2\, dx = 1$. When Ψ satisfies this equation, it is said to be *normalized*. The normalization condition for a one-particle, three-dimensional system is

$$\int_{-\infty}^{\infty}\int_{-\infty}^{\infty}\int_{-\infty}^{\infty} |\Psi(x, y, z, t)|^2\, dx\, dy\, dz = 1 \qquad (18.11)$$

For an n-particle, three-dimensional system, the integral of $|\Psi|^2$ over all $3n$ coordinates $x_1, \ldots, z_n$, each integrated from $-\infty$ to ∞, equals 1.

The normalization requirement is often written

$$\int |\Psi|^2 \, d\tau = 1 \qquad (18.12)^*$$

where $\int d\tau$ is a shorthand notation that stands for the definite integral over the full ranges of all the spatial coordinates of the system. For a one-particle, three-dimensional system, $\int d\tau$ implies a triple integral over x, y, and z from $-\infty$ to ∞ for each coordinate [Eq. (18.11)].

By substitution, it is easy to see that if Ψ is a solution of (18.9), then so is $c\Psi$, where c is an arbitrary constant. Thus, there is always an arbitrary multiplicative constant in each solution to (18.9). The value of this constant is chosen so as to satisfy the normalization requirement (18.12).

From the state function Ψ, we can calculate the probabilities of the various possible outcomes when a measurement of position is made on the system. In fact, Born's work is more general than this. It turns out that Ψ gives information on the outcome of a measurement of *any* property of the system, not just position. For example, if Ψ is known, it is possible to calculate the probability of each possible outcome when a measurement of p_x, the x component of momentum, is made. The same is true for a measurement of energy, or angular momentum, etc. The procedure for calculating these probabilities from Ψ is discussed in *Q. C.*, sec. 7.6.

The state function Ψ is not to be thought of as any sort of physical wave. Instead Ψ is an abstract mathematical entity that gives information about the state of the system. Everything that can be known about the system in a given state is contained in the state function Ψ. Instead of saying "the state described by the function Ψ," we can just as well say "the state Ψ." The information given by Ψ is the probabilities for the possible outcomes of measurements of physical properties of the system.

The state function Ψ describes a physical system. In Chaps. 18 to 21, the system will usually be a particle, atom, or molecule. One can also consider the state function of a system that contains a large number of molecules, e.g., a mole of some compound; this will be done in Chap. 22 on statistical mechanics.

Classical mechanics is a deterministic theory in that it allows us to predict the exact paths taken by the particles of the system and tells us where they will be at any future time. In contrast, quantum mechanics gives only the probabilities for finding the particles at various locations in space. The concept of a path for a particle becomes rather fuzzy in a time-dependent quantum-mechanical system and disappears entirely in a time-independent quantum-mechanical system.

Some philosophers have used the Heisenberg uncertainty principle and the nondeterministic nature of quantum mechanics as arguments in favor of human free will.

The probabilistic nature of quantum mechanics disturbed many physicists, including Einstein, Schrödinger, and de Broglie. (Einstein wrote in 1926: "Quantum mechanics ... says a lot, but does not really bring us any closer to the secret of the Old One. I, at any rate, am convinced that He does not throw dice.") These scientists believed that quantum mechanics does not furnish a complete description of physical reality. However, all attempts to replace quantum mechanics by an underlying causal, deterministic theory have failed. There appears to be a fundamental randomness in nature at the microscopic level.

18.7 THE TIME-INDEPENDENT SCHRÖDINGER EQUATION

For an isolated atom or molecule, the forces acting depend only on the coordinates of the charged particles of the system and are independent of time; hence, the potential energy is independent of t. For systems where V is independent of time, the time-dependent Schrödinger equation has solutions of the form $\Psi(x_1, \ldots, z_n, t) = f(t)\psi(x_1, \ldots, z_n)$, where ψ (lowercase psi) is a function of the $3n$ coordinates of the n particles and f is a certain function of time. We shall demonstrate this for a one-particle, one-dimensional system.

For such a system, Eq. (18.9) becomes

$$-\frac{\hbar^2}{2m}\frac{\partial^2\Psi}{\partial x^2} + V(x)\Psi = -\frac{\hbar}{i}\frac{\partial\Psi}{\partial t} \tag{18.13}$$

Let us look for solutions of (18.13) that have the form

$$\Psi(x, t) = f(t)\psi(x) \tag{18.14}$$

We have $\partial^2\Psi/\partial x^2 = f(t)\, d^2\psi/dx^2$ and $\partial\Psi/\partial t = \psi(x)\, df/dt$. Substitution into (18.13) followed by division by $f\psi = \Psi$ gives

$$-\frac{\hbar^2}{2m}\frac{1}{\psi(x)}\frac{d^2\psi}{dx^2} + V(x) = -\frac{\hbar}{i}\frac{1}{f(t)}\frac{df(t)}{dt} \equiv E \tag{18.15}$$

where we defined the parameter E by $E \equiv -(\hbar/i)f'(t)/f(t)$.

From the definition of E, it is equal to a function of t only and hence is independent of x. However, (18.15) shows that $E = -(\hbar^2/2m)\psi''(x)/\psi(x) + V(x)$, which is a function of x only and is independent of t. Hence, E is independent of t as well as independent of x and must therefore be a constant. Since the constant E has the same dimensions as V, it has the dimensions of energy. Quantum mechanics postulates that E is in fact the energy of the system.

Equation (18.15) gives $df/f = -(iE/\hbar)\, dt$, which integrates to $\ln f = -iEt/\hbar + C$. Hence, $f = e^C e^{-iEt/\hbar} = Ae^{-iEt/\hbar}$, where $A \equiv e^C$ is an arbitrary constant. The constant A can be included as part of the $\psi(x)$ factor in (18.14), so we omit it from f. Thus

$$f(t) = e^{-iEt/\hbar} \tag{18.16}$$

Equation (18.15) also gives

$$-\frac{\hbar^2}{2m}\frac{d^2\psi(x)}{dx^2} + V(x)\psi(x) = E\psi(x) \tag{18.17}$$

which is the (*time-independent*) *Schrödinger equation* for a one-particle, one-dimensional system. Equation (18.17) can be solved for ψ when the potential-energy function $V(x)$ has been specified.

For an n-particle, three-dimensional system, the same procedure that led to Eqs. (18.14), (18.16), and (18.17) gives

$$\Psi = e^{-iEt/\hbar}\psi(x_1, y_1, z_1, \ldots, x_n, y_n, z_n) \tag{18.18}$$

where the function ψ is found by solving

$$-\frac{\hbar^2}{2m_1}\left(\frac{\partial^2\psi}{\partial x_1^2} + \frac{\partial^2\psi}{\partial y_1^2} + \frac{\partial^2\psi}{\partial z_1^2}\right) - \cdots - \frac{\hbar^2}{2m_n}\left(\frac{\partial^2\psi}{\partial x_n^2} + \frac{\partial^2\psi}{\partial y_n^2} + \frac{\partial^2\psi}{\partial z_n^2}\right) + V\psi = E\psi \tag{18.19*}$$

The solutions ψ to the time-independent Schrödinger equation (18.19) are the (*time-independent*) *wave functions*. States for which Ψ is given by (18.18) are called *stationary states*. We shall see that for a given system there are many different solutions to (18.19), different solutions corresponding to different values of the energy E. In general, quantum mechanics gives only probabilities and not certainties for the outcome of a measurement; however, when a system is in a stationary state, a measurement of its energy is certain to give the particular energy value that corresponds to the wave function ψ of the system. Different systems have different forms for the potential-energy function $V(x_1, \ldots, z_n)$, and this leads to different sets of allowed wave functions and energies when (18.19) is solved for different systems. (All this will be made clearer by the examples in the next few sections.)

For a stationary state, the probability density $|\Psi|^2$ becomes

$$|\Psi|^2 = |f\psi|^2 = (f\psi)^* f\psi = f^*\psi^* f\psi = e^{iEt/\hbar}\psi^* e^{-iEt/\hbar}\psi = \psi^*\psi = |\psi|^2$$

where we used the identity $(f\psi)^* = f^*\psi^*$ (Prob. 18.33). Hence, for a stationary state, $|\Psi|^2 = |\psi|^2$, which is independent of time. For a stationary state, the probability density and the energy are constant with time. (There is no implication, however, that the particles of the system are at rest in a stationary state.)

It turns out that the probabilities for the outcomes of measurements of any physical property involve $|\Psi|$, and since $|\Psi| = |\psi|$, these probabilities are independent of time for a stationary state. Thus, the $e^{-iEt/\hbar}$ factor in (18.18) is of no consequence, and the essential part of the state function for a stationary state is the time-independent wave function $\psi(x_1, \ldots, z_n)$. For a stationary state, the normalization condition (18.12) becomes $\int |\psi|^2 d\tau = 1$.

The time-independent Schrödinger equation (18.19) contains two sets of unknown quantities, the wave functions ψ and the allowed energies E. It is not possible to solve the single equation (18.19) for the two kinds of unknowns ψ and E. To obtain the allowed energy levels, quantum mechanics postulates that not all functions ψ that satisfy (18.19) are allowed as wave functions for the system. In addition to being a solution of (18.19), a wave function must meet the following three conditions: (*a*) The wave function must be single-valued. (*b*) The wave function must be continuous. (*c*) The wave function must be quadratically integrable. Condition (*a*) means that ψ has one and only one value at each point in space. The function of Fig. 18.6*a*, which is multiple-valued at some points, is not a possible wave function for a one-particle, one-dimensional system. Condition (*b*) means that ψ makes no sudden jumps in value. A function like that in Fig. 18.6*b* is ruled out. Condition (*c*) means that the integral over all space $\int |\psi|^2 \, d\tau$ is a finite number. The function x^2 (Fig. 18.6*c*) is not quadratically integrable, since $\int_{-\infty}^{\infty} x^4 \, dx = (x^5/5)|_{-\infty}^{\infty} = \infty - (-\infty) = \infty$. Condition (*c*) allows the wave function to be multiplied by a constant that normalizes it,

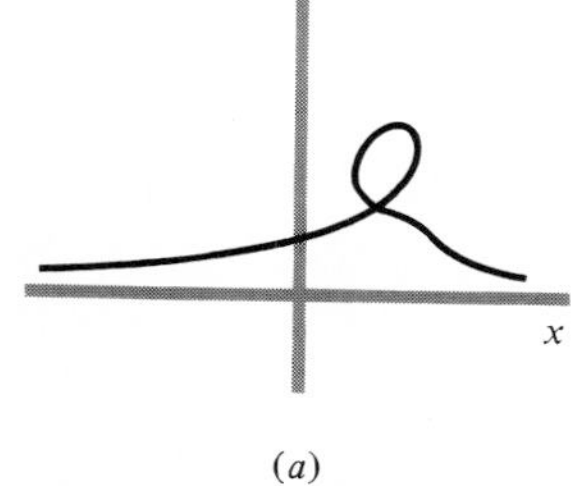

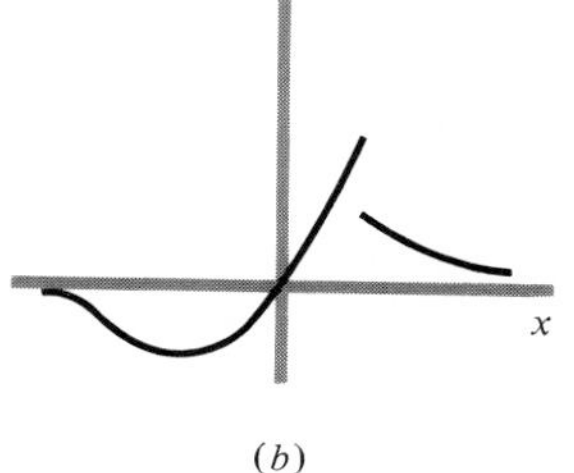

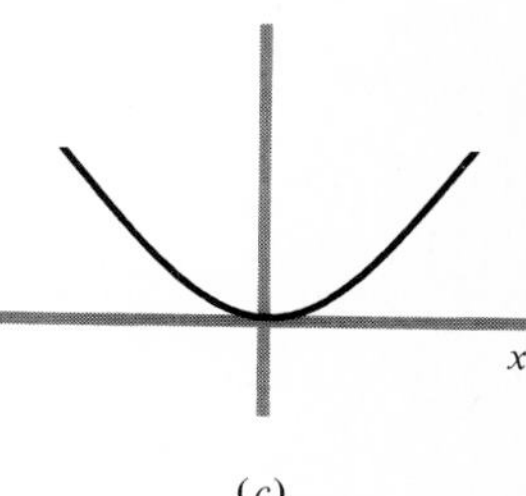

Figure 18.6
(*a*) A multivalued function. (*b*) A discontinuous function. (*c*) A function that is not quadratically integrable.

i.e., that makes $\int |\psi|^2\, d\tau = 1$. A function obeying conditions (*a*), (*b*), and (*c*) is said to be *well behaved*.

Since E occurs as an undetermined parameter in the Schrödinger equation (18.19), the solutions ψ that are found by solving (18.19) will depend on E as a parameter: $\psi = \psi(x_1, \ldots, z_n; E)$. It turns out that ψ is well behaved only for certain particular values of E, and it is these values that are the allowed energy levels. An example is given in the next section.

We shall mainly be interested in the stationary states of atoms and molecules, since these give the allowed energy levels. For a collision between two molecules or for a molecule exposed to the time-varying electric and magnetic fields of electromagnetic radiation, the potential energy V depends on time, and one must deal with the time-dependent Schrödinger equation and with nonstationary states.

18.8 THE PARTICLE IN A ONE-DIMENSIONAL BOX

To illustrate some of the ideas of quantum mechanics, we shall apply the Schrödinger equation to an extremely simple system, a particle in a one-dimensional box. By this is meant a single particle of mass m moving in one dimension x and subject to the potential-energy function of Fig. 18.7. The potential energy is zero for x between 0 and a (region II) and is infinite elsewhere (regions I and III):

$$V = \begin{cases} 0 & \text{for } 0 \le x \le a \\ \infty & \text{for } x < 0 \text{ and for } x > a \end{cases}$$

This potential energy confines the particle to move in the region between 0 and a on the x axis. (No real system has a V as simple as that of Fig. 18.7.)

We restrict ourselves to considering the states of constant energy, the stationary states. For these states, the (time-independent) wave functions ψ are found by solving the Schrödinger equation (18.19), which reads for a one-particle, one-dimensional system

$$-\frac{\hbar^2}{2m}\frac{d^2\psi}{dx^2} + V\psi = E\psi \tag{18.20}$$

Since a particle can't have infinite energy, there must be zero probability for finding the particle in regions I and III, where V is infinite. Therefore, $|\psi|^2$ and hence ψ must be zero in these regions: $\psi_{\text{I}} = 0$ and $\psi_{\text{III}} = 0$, or

$$\psi = 0 \qquad \text{for } x < 0 \text{ and for } x > a \tag{18.21}$$

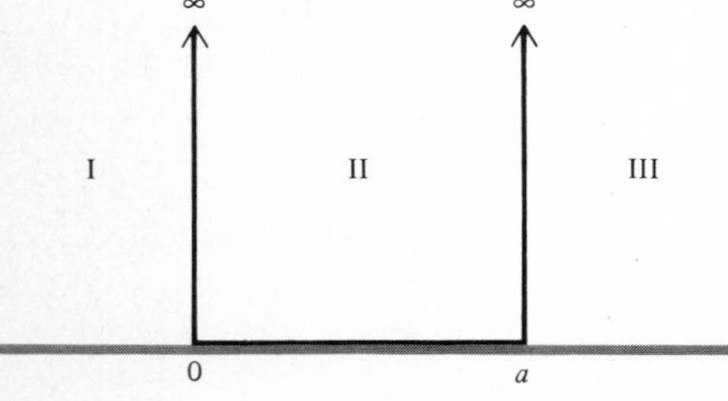

Figure 18.7
Potential energy for a particle in a one-dimensional box.

Inside the box (region II), V is zero and (18.20) becomes

$$\frac{d^2\psi}{dx^2} = -\frac{2mE}{\hbar^2}\psi \qquad \text{for } 0 \le x \le a \tag{18.22}$$

To solve this equation, we need a function whose second derivative gives us the same function back again, but multiplied by a constant. Two functions that behave this way are the sine function and the cosine function, so let us try as a solution $\psi = A \sin rx + B \cos sx$, where A, B, r, and s are constants. Differentiation gives $d^2\psi/dx^2 = -Ar^2 \sin rx - Bs^2 \cos sx$. Substitution of the trial solution in (18.22) gives

$$-Ar^2 \sin rx - Bs^2 \cos sx = -2mE\hbar^{-2}A \sin rx - 2mE\hbar^{-2}B \cos sx \tag{18.23}$$

If we take $r = s = (2mE)^{1/2}\hbar^{-1}$, Eq. (18.23) is satisfied. The solution of (18.22) is therefore

$$\psi = A \sin\,[(2mE)^{1/2}\hbar^{-1}x] + B \cos\,[(2mE)^{1/2}\hbar^{-1}x] \qquad \text{for } 0 \le x \le a \tag{18.24}$$

[A more formal derivation than we have given shows that (18.24) is indeed the general solution of the differential equation (18.22).]

As noted in Sec. 18.7, not all solutions of the Schrödinger equation are acceptable wave functions. Only well-behaved functions are allowed. The solution of the particle-in-a-box Schrödinger equation is the function defined by (18.21) and (18.24), where A and B are arbitrary constants of integration. For this function to be continuous, the wave function inside the box must go to zero at the two ends of the box, since ψ equals zero outside the box. We must require that ψ in (18.24) go to zero as $x \to 0$ and as $x \to a$. Setting $x = 0$ and $\psi = 0$ in (18.24), we get $0 = A \sin 0 + B \cos 0$, so $B = 0$. Hence,

$$\psi = A \sin\,(2mE)^{1/2}\hbar^{-1}x \tag{18.25}$$

Setting $x = a$ and $\psi = 0$ in (18.25), we get $0 = \sin\,(2mE)^{1/2}\hbar^{-1}a$. The sine function vanishes at $0, \pm\pi, \pm 2\pi, \ldots, \pm n\pi$, so we must have

$$(2mE)^{1/2}\hbar^{-1}a = \pm n\pi \tag{18.26}$$

Substitution of (18.26) in (18.25) gives $\psi = A \sin\,(\pm n\pi x/a) = \pm A \sin\,(n\pi x/a)$, since $\sin\,(-z) = -\sin z$. Use of $-n$ multiplies ψ by -1. Since A is arbitrary, this doesn't give a different solution than the $+n$ solution, so there is no need to consider the $-n$ values. Also, the value $n = 0$ must be ruled out, since it would make $\psi = 0$ everywhere, meaning there is no probability of finding the particle in the box. The allowed wave functions are therefore

$$\psi = A \sin\,(n\pi x/a) \qquad \text{where } n = 1, 2, 3, \ldots$$

The allowed energies are found by solving (18.26) for E to get

$$E = \frac{n^2h^2}{8ma^2}, \qquad n = 1, 2, 3, \ldots \tag{18.27*}$$

where $\hbar \equiv h/2\pi$ was used. It is only these values of E that make ψ a well-behaved (continuous) function. Confining the particle to be between 0 and a requires that ψ be zero at $x = 0$ and $x = a$, and this quantizes the energy. An analogy is the quantization of the vibrational modes of a string that occurs when the string is held fixed at both ends.

The magnitude of A in ψ is found from the normalization condition $\int |\psi|^2 \, d\tau = 1$. Since $\psi = 0$ outside the box, we have

$$1 = \int_{-\infty}^{\infty} |\psi|^2 \, dx = \int_0^a |\psi|^2 \, dx = |A|^2 \int_0^a \sin^2\left(\frac{n\pi x}{a}\right) dx$$

A table of integrals gives $\int \sin^2 cx \, dx = x/2 - (1/4c) \sin 2cx$, and we find $|A| = (2/a)^{1/2}$. The *normalization constant* A can be taken as any number having absolute value $(2/a)^{1/2}$. We could take $A = (2/a)^{1/2}$, or $A = -(2/a)^{1/2}$, or $A = i(2/a)^{1/2}$ [where $i = \sqrt{-1}$], etc. Choosing $A = (2/a)^{1/2}$, we get

$$\psi = \left(\frac{2}{a}\right)^{1/2} \sin \frac{n\pi x}{a} \qquad \text{where } n = 1, 2, 3, \ldots \tag{18.28}$$

The state functions for the stationary states of the particle in a box are given by (18.14), (18.16), and (18.28) as $\Psi = e^{-iEt/\hbar}(2/a)^{1/2} \sin(n\pi x/a)$, where $E = n^2h^2/8ma^2$, and $n = 1, 2, 3, \ldots$.

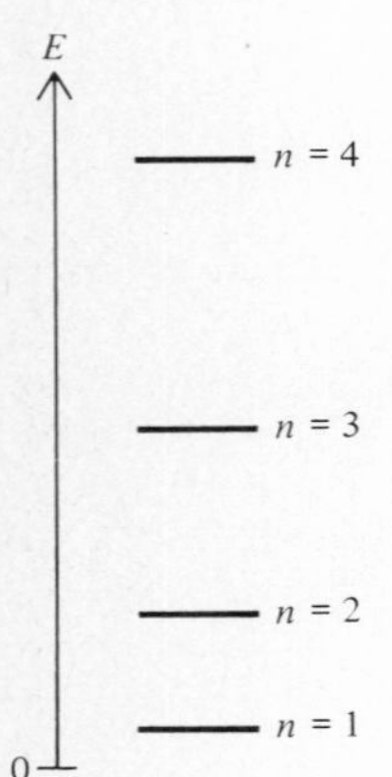

Figure 18.8
Lowest four energy levels of a particle in a one-dimensional box.

Let us contrast the quantum-mechanical and classical pictures. Classically, the particle can rattle around in the box with any nonnegative energy; $E_{\text{classical}}$ can be any number from zero on up. (The potential energy is zero in the box, so the particle's energy is entirely kinetic. Its speed v can have any nonnegative value, so $\frac{1}{2}mv^2$ can have any nonnegative value.) Quantum-mechanically, the energy can take on only the values (18.27); see Fig. 18.8. The energy is quantized in quantum mechanics, whereas it is continuous in classical mechanics.

Classically, the minimum energy is zero. Quantum-mechanically, the particle in a box has a minimum energy that is greater than zero. This energy, $h^2/8ma^2$, is the *zero-point energy*. Its existence is a consequence of the uncertainty principle. Suppose the particle could have zero energy. Since its energy is entirely kinetic, its speed v_x and momentum $mv_x = p_x$ would then be zero. With p_x known to be precisely zero, the uncertainty Δp_x is zero, and the uncertainty principle (18.8) gives $\Delta x = \infty$. However, we know the particle to be somewhere between $x = 0$ and $x = a$, so Δx cannot exceed a. Hence, a zero energy is impossible for a particle in a box.

The stationary states of a particle in a box are specified by giving the value of the integer n in (18.28). n is called a *quantum number*. The lowest-energy state ($n = 1$) is the *ground state*.

Let us check the uncertainty principle for the ground state. We take $\Delta x = a$. The ground-state energy is entirely kinetic, so $h^2/8ma^2 = mv_x^2/2 = m^2v_x^2/2m = p_x^2/2m$. Solving for p_x, we get $p_x = \pm h/2a$, corresponding to the particle moving to the right or the left. (Actually, other values of p_x are possible, but this subtlety will be ignored; see *Q. C.*, prob. 7.30.) The momentum is therefore uncertain by $h/2a - (-h/2a) = h/a$, so $\Delta p_x = h/a$. Therefore, $\Delta x \, \Delta p_x = h$, in agreement with (18.8).

Figure 18.9 plots the wave functions ψ and the probability densities $|\psi|^2$ for the first three particle-in-a-box stationary states.

Classically, all locations for the particle in the box are equally likely. Quantum-mechanically, the probability density is not uniform but shows oscillations. In the limit of very high quantum number n, the oscillations in $|\psi|^2$ come closer and closer together and ultimately become undetectable; this corresponds to the classical result of uniform probability density. The relation $8ma^2E/h^2 = n^2$ shows that for a macroscopic system (E, m, and a having macroscopic magnitudes), n is very large, so the limit of large n is the classical limit.

A point at which $\psi = 0$ is called a *node*. The number of nodes increases by 1 for each increase in n. The existence of nodes is surprising from a classical viewpoint. For example, for the $n = 2$ state, it is hard to understand how the particle can be found in the left half of the box or in the right half of the box but never at the center. The behavior of microscopic particles cannot be rationalized in terms of a visualizable model.

The wave functions ψ and probability densities $|\psi|^2$ are spread out over the length of the box, much like a wave (compare Figs. 18.9 and 18.3). However, quantum mechanics does not assert that the particle itself is spread out like a wave; a measurement of position will give a definite location for the particle. It is the probability function ψ that is spread out in space and obeys a wave equation.

Example Find the wavelength of the light emitted when a 1×10^{-27} g particle in a 3-Å one-dimensional box jumps from the $n = 2$ to the $n = 1$ level.

Equations (18.6) and (18.27) give $h\nu = E_a - E_b = 4h^2/8ma^2 - h^2/8ma^2$, so $\nu = 3h/8ma^2$. Use of $\lambda = c/\nu$ gives

$$\lambda = \frac{8ma^2c}{3h} = \frac{8(1 \times 10^{-27}\text{ g})(3 \times 10^{-8}\text{ cm})^2(3 \times 10^{10}\text{ cm/s})}{3(6.6 \times 10^{-27}\text{ erg s})} = 1 \times 10^{-5}\text{ cm}$$

Note that the mass m is that of an electron and the wavelength (1000 Å) lies in the ultraviolet.

If ψ_i and ψ_j are particle-in-a-box wave functions with quantum numbers n_i and n_j, one finds (Prob. 18.35) that

$$\int_0^a \psi_i^* \psi_j \, d\tau = 0 \qquad \text{for } n_i \neq n_j \tag{18.29}$$

where $\psi_i = (2/a)^{1/2} \sin (n_i \pi x/a)$ and $\psi_j = (2/a)^{1/2} \sin (n_j \pi x/a)$. The functions f and g are said to be *orthogonal* when $\int f^* g \, d\tau = 0$, where the integral goes over the full

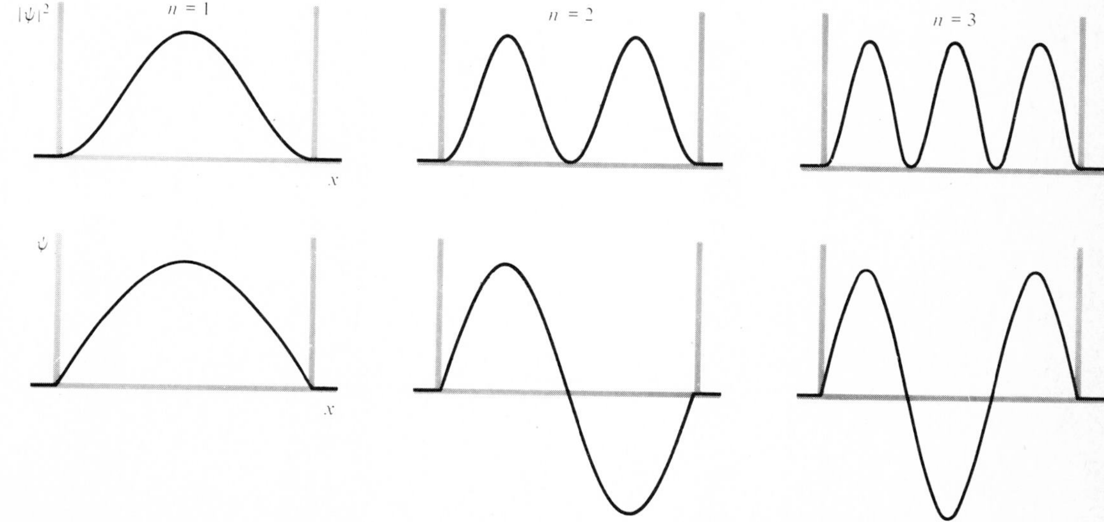

Figure 18.9
Wave functions and probability densities for the lowest three particle-in-a-box stationary states.

range of the spatial coordinates. One can show that two wave functions that correspond to different energy levels of a quantum-mechanical system are orthogonal (*Q. C.*, sec. 7.2).

18.9 THE PARTICLE IN A THREE-DIMENSIONAL BOX

The particle in a three-dimensional box is a single particle of mass m confined to remain within the volume of a box by an infinite potential energy outside the box. The simplest shape of box to deal with is a rectangular parallelepiped. The potential energy is therefore $V = 0$ for points such that $0 \le x \le a$, $0 \le y \le b$, and $0 \le z \le c$ and $V = \infty$ elsewhere. The dimensions of the box are a, b, and c. In later chapters, this system will be used to give the energy levels for translational motion of ideal-gas molecules.

Let us solve the time-independent Schrödinger equation for the stationary-state wave functions and energies. Since $V = \infty$ outside the box, ψ is zero outside the box, just as for the corresponding one-dimensional problem. Inside the box, $V = 0$, and the Schrödinger equation (18.19) becomes

$$-\frac{\hbar^2}{2m}\left(\frac{\partial^2\psi}{\partial x^2} + \frac{\partial^2\psi}{\partial y^2} + \frac{\partial^2\psi}{\partial z^2}\right) = E\psi \tag{18.30}$$

Let us assume that solutions of (18.30) exist that have the form $X(x)Y(y)Z(z)$, where $X(x)$ is a function of x only, and Y and Z are functions of y and z. For an arbitrary partial differential equation, it is not in general possible to find solutions in which the variables are present in separate factors. However, it can be proved mathematically that if we succeed in finding well-behaved solutions to (18.30) that have the form $X(x)Y(y)Z(z)$, then there are no other well-behaved solutions, so we shall have found the general solution of (18.30). Our assumption is then

$$\psi = X(x)Y(y)Z(z) \tag{18.31}$$

Partial differentiation of (18.31) gives $\partial^2\psi/\partial x^2 = X''(x)Y(y)Z(z)$, $\partial^2\psi/\partial y^2 = X(x)Y''(y)Z(z)$, and $\partial^2\psi/\partial z^2 = X(x)Y(y)Z''(z)$. Substitution in (18.30) followed by division by $X(x)Y(y)Z(z) = \psi$ gives

$$-\frac{\hbar^2}{2m}\frac{X''(x)}{X(x)} - \frac{\hbar^2}{2m}\frac{Y''(y)}{Y(y)} - \frac{\hbar^2}{2m}\frac{Z''(z)}{Z(z)} = E \tag{18.32}$$

Let $E_x \equiv -(\hbar^2/2m)X''(x)/X(x)$. Then (18.32) gives

$$E_x \equiv -\frac{\hbar^2}{2m}\frac{X''(x)}{X(x)} = E + \frac{\hbar^2}{2m}\frac{Y''(y)}{Y(y)} + \frac{\hbar^2}{2m}\frac{Z''(z)}{Z(z)} \tag{18.33}$$

From its definition, E_x is a function of x only. However, the relation $E_x = E + \hbar^2 Y''/2mY + \hbar^2 Z''/2mZ$ shows E_x to be independent of x. Hence E_x is a constant, and we have from (18.33)

$$-(\hbar^2/2m)X''(x) = E_x X(x) \qquad \text{for } 0 \le x \le a \tag{18.34}$$

Equation (18.34) is the same as the Schrödinger equation (18.22) for a particle in a one-dimensional box if X and E_x in (18.34) are identified with ψ and E in (18.22). Moreover, the condition that $X(x)$ be continuous requires that $X(x) = 0$ at $x = 0$

and at $x = a$, since the three-dimensional wave function is zero outside the box. These are the same requirements that ψ in (18.22) must satisfy. Hence, the well-behaved solutions of (18.34) and (18.22) are the same. Replacing ψ and E in (18.27) and (18.28) by X and E_x, we get

$$X(x) = \left(\frac{2}{a}\right)^{1/2} \sin\frac{n_x \pi x}{a}, \qquad E_x = \frac{n_x^2 h^2}{8ma^2}, \qquad n_x = 1, 2, 3, \ldots \tag{18.35}$$

where the quantum number is called n_x.

Equation (18.32) is symmetric with respect to x, y, and z, so the same reasoning that gave (18.35) gives

$$Y(y) = \left(\frac{2}{b}\right)^{1/2} \sin\frac{n_y \pi y}{b}, \qquad E_y = \frac{n_y^2 h^2}{8mb^2}, \qquad n_y = 1, 2, 3, \ldots \tag{18.36}$$

$$Z(z) = \left(\frac{2}{c}\right)^{1/2} \sin\frac{n_z \pi z}{c}, \qquad E_z = \frac{n_z^2 h^2}{8mc^2}, \qquad n_z = 1, 2, 3, \ldots \tag{18.37}$$

where, by analogy to (18.33),

$$E_y \equiv -\frac{\hbar^2}{2m}\frac{Y''(y)}{Y(y)}, \qquad E_z \equiv -\frac{\hbar^2}{2m}\frac{Z''(z)}{Z(z)} \tag{18.38}$$

Equations (18.31) and (18.35) to (18.37) give the wave functions as

$$\psi = \left(\frac{8}{abc}\right)^{1/2} \sin\frac{n_x \pi x}{a} \sin\frac{n_y \pi y}{b} \sin\frac{n_z \pi z}{c} \tag{18.39}$$

Equations (18.32), (18.33), and (18.38) give $E = E_x + E_y + E_z$, and use of (18.35) to (18.37) gives the allowed energy levels as

$$E = \frac{h^2}{8m}\left(\frac{n_x^2}{a^2} + \frac{n_y^2}{b^2} + \frac{n_z^2}{c^2}\right) \tag{18.40}$$

The quantities E_x, E_y, and E_z are the kinetic energies associated with motion in the x, y, and z directions.

The procedure used to solve (18.30) is called *separation of variables*. The conditions under which it works are discussed in Sec. 18.11.

There are three quantum numbers because this problem is a three-dimensional one. The quantum numbers n_x, n_y, and n_z vary independently of one another. The state of the particle in the box is specified by giving the values of n_x, n_y, and n_z. The ground state is $n_x = 1$, $n_y = 1$, $n_z = 1$.

18.10 DEGENERACY

Suppose the sides of the box of the last section have equal lengths: $a = b = c$. Then (18.39) and (18.40) become

$$\psi = (2/a)^{3/2} \sin(n_x \pi x/a) \sin(n_y \pi y/b) \sin(n_z \pi z/c) \tag{18.41}$$

$$E = (n_x^2 + n_y^2 + n_z^2)h^2/8ma^2 \tag{18.42}$$

Let us use numerical subscripts on ψ to specify the n_x, n_y, and n_z values. The lowest-energy state is ψ_{111} with $E = 3h^2/8ma^2$. The states ψ_{211}, ψ_{121}, and ψ_{112} each have the energy $6h^2/8ma^2$. Even though they have the same energy, these are different states. With $n_x = 2$, $n_y = 1$, $n_z = 1$ in (18.41), we get a different wave function than with $n_x = 1$, $n_y = 2$, $n_z = 1$. The ψ_{211} state has zero probability density of finding the particle at $x = a/2$ (see Fig. 18.9), but the ψ_{121} state has a maximum probability density at $x = a/2$.

The terms *state* and *energy level* have different meanings in quantum mechanics. A stationary state is specified by giving the wave function ψ. Each different ψ is a different state. An energy level is specified by giving the value of the energy. Each different value of E is a different energy level. The three different particle-in-a-box states ψ_{211}, ψ_{121}, and ψ_{112} belong to the same energy level, $6h^2/8ma^2$. Figure 18.10 shows the lowest few states and energy levels of a particle in a cubic box.

An energy level that corresponds to more than one state is said to be *degenerate*. The number of different states belonging to the level is the *degree of degeneracy* of the level. The level $6h^2/8ma^2$ is threefold degenerate. The particle-in-a-box degeneracy arises when the dimensions of the box are made equal. Degeneracy usually arises from the symmetry of the system.

18.11 OPERATORS

Operators. Quantum mechanics is most conveniently formulated in terms of operators. An *operator* is a rule for transforming a given function into another function. For example, the operator d/dx transforms a function into its first derivative: $(d/dx)f(x) = f'(x)$. Let $\hat{A}$ symbolize an arbitrary operator. (We shall use a circumflex to denote an operator.) If $\hat{A}$ transforms the function $f(x)$ into the function $g(x)$, we write $\hat{A}f(x) = g(x)$. If $\hat{A}$ is the operator d/dx, then $g(x) = f'(x)$. If $\hat{A}$ is the operator "multiplication by $3x^2$," then $g(x) = 3x^2 f(x)$. If $\hat{A} = \log$, then $g(x) = \log f(x)$.

The *sum* of two operators $\hat{A}$ and $\hat{B}$ is defined by $(\hat{A} + \hat{B})f(x) \equiv \hat{A}f(x) + \hat{B}f(x)$. For example, $(\log + d/dx)f(x) = \log f(x) + f'(x)$. The *square* of an operator is defined by $\hat{A}^2 f(x) \equiv \hat{A}[\hat{A}f(x)]$. For example,

$$(d/dx)^2 f(x) = (d/dx)[(d/dx)f(x)] = (d/dx)[f'(x)] = f''(x) = (d^2/dx^2)f(x).$$

Therefore, $(d/dx)^2 = d^2/dx^2$. The *product* of two operators is defined by $(\hat{A}\hat{B})f(x) \equiv \hat{A}[\hat{B}f(x)]$. For example, if $\hat{A} = 3x^2 \times$, and $\hat{B} = d/dx$, then $\hat{A}\hat{B}f(x) = 3x^2[(d/dx)f(x)] = 3x^2 f'(x)$. (See also Prob. 18.15.)

In quantum mechanics, each physical property of a system has a corresponding

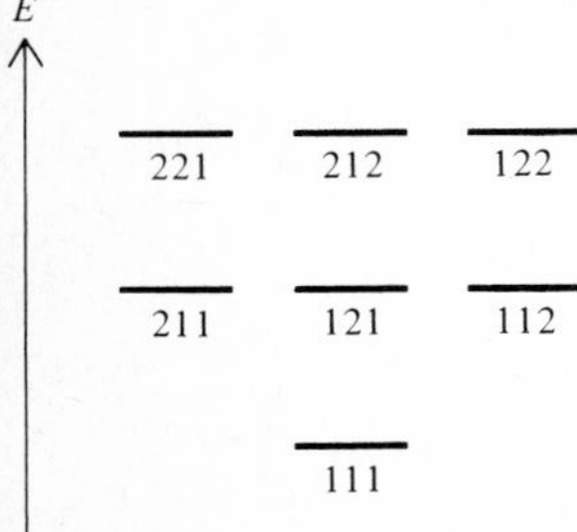

Figure 18.10
Lowest few states for a particle in a cubic box.

operator. The operator that corresponds to p_x, the x component of momentum of a particle, is postulated to be $(\hbar/i)(\partial/\partial x)$, with similar operators for p_y and p_z. Thus,

$$\hat{p}_x = \frac{\hbar}{i}\frac{\partial}{\partial x}, \qquad \hat{p}_y = \frac{\hbar}{i}\frac{\partial}{\partial y}, \qquad \hat{p}_z = \frac{\hbar}{i}\frac{\partial}{\partial z} \tag{18.43}*$$

where $\hat{p}_x$ is the quantum-mechanical operator for the property p_x and $i \equiv \sqrt{-1}$. The operator that corresponds to the x coordinate of a particle is multiplication by x, and the operator that corresponds to $f(x, y, z)$, where f is any function, is multiplication by that function. Thus,

$$\hat{x} = x\,\times, \qquad \hat{y} = y\,\times, \qquad \hat{z} = z\,\times, \qquad \hat{f}(x, y, z) = f(x, y, z)\,\times \tag{18.44}*$$

To find the operator that corresponds to any other physical property, we write down the classical-mechanical expression for that property as a function of cartesian coordinates and corresponding momenta and then replace the coordinates and momenta by their corresponding operators (18.43) and (18.44). For example, the energy of a one-particle system is the sum of its kinetic and potential energies: $E = T + V = \frac{1}{2}m(v_x^2 + v_y^2 + v_z^2) + V(x, y, z, t)$. To express E as a function of the momenta and coordinates, we note that $p_x = mv_x$, $p_y = mv_y$, $p_z = mv_z$. Therefore,

$$E = \frac{1}{2m}(p_x^2 + p_y^2 + p_z^2) + V(x, y, z, t) \equiv H \tag{18.45}*$$

The expression for the energy as a function of coordinates and momenta is called the system's *Hamiltonian H* [after William Rowan Hamilton (1805–1865), who reformulated Newton's second law in terms of H]. Equation (18.43) gives $\hat{p}_x^2 = [(\hbar/i)(\partial/\partial x)]^2 = (\hbar^2/i^2)\,\partial^2/\partial x^2 = -\hbar^2\,\partial^2/\partial x^2$. Hence, $\hat{p}_x^2/2m = -(\hbar^2/2m)\,\partial^2/\partial x^2$. From (18.44), the potential-energy operator is simply multiplication by $V(x, y, z, t)$. (Time is a parameter in quantum mechanics, and there is no time operator.) The energy operator, or *Hamiltonian operator*, for a one-particle system is therefore

$$\hat{E} = \hat{H} = -\frac{\hbar^2}{2m}\left(\frac{\partial^2}{\partial x^2} + \frac{\partial^2}{\partial y^2} + \frac{\partial^2}{\partial z^2}\right) + V(x, y, z, t)\,\times \tag{18.46}$$

To save time in writing, the *Laplacian operator* ∇^2 is defined by $\nabla^2 \equiv \partial^2/\partial x^2 + \partial^2/\partial y^2 + \partial^2/\partial z^2$, and the one-particle Hamiltonian operator is written

$$\hat{H} = -(\hbar^2/2m)\nabla^2 + V \tag{18.47}$$

where the multiplication sign after V is understood.

For a many-particle system, we have $\hat{p}_{x,1} = (\hbar/i)\,\partial/\partial x_1$ for particle 1, and the Hamiltonian operator is readily found to be

$$\hat{H} = -\frac{\hbar^2}{2m_1}\nabla_1^2 - \frac{\hbar^2}{2m_2}\nabla_2^2 - \cdots - \frac{\hbar^2}{2m_n}\nabla_n^2 + V(x_1, \ldots, z_n, t) \tag{18.48}*$$

$$\nabla_1^2 \equiv \frac{\partial^2}{\partial x_1^2} + \frac{\partial^2}{\partial y_1^2} + \frac{\partial^2}{\partial z_1^2} \tag{18.49}*$$

with similar definitions for $\nabla_2^2, \ldots, \nabla_n^2$. The terms on the right of (18.48) are the operators for the kinetic energy of particle 1, the kinetic energy of particle 2, ..., the kinetic energy of particle n, and the potential energy of the system.

From (18.48), we see that the time-dependent Schrödinger equation (18.9) can be written as

$$-\frac{\hbar}{i}\frac{\partial \Psi}{\partial t} = \hat{H}\Psi \tag{18.50}$$

and the time-independent Schrödinger equation (18.19) can be written as

$$\hat{H}\psi = E\psi \qquad (18.51)^*$$

where V in (18.51) is independent of time. Since there is a whole set of allowed stationary-state wave functions and energies, it might be better to write (18.51) as $\hat{H}\psi_j = E_j\psi_j$, where the subscript j labels the various wave functions and energies.

When an operator $\hat{B}$ applied to the function f gives the function back again but multiplied by the constant c, that is, when $\hat{B}f = cf$, one says that f is an *eigenfunction* of $\hat{B}$ with *eigenvalue* c. The wave functions ψ in (18.51) are eigenfunctions of the Hamiltonian operator, the eigenvalues being the allowed energies E.

Average values. From (15.39), the average value of x for a one-particle, one-dimensional quantum-mechanical system equals $\int_{-\infty}^{\infty} xg(x)\,dx$, where $g(x)$ is the probability density for finding the particle between x and $x + dx$. But the Born postulate (Sec. 18.6) gives $g(x) = |\Psi(x)|^2$. Hence, $\langle x\rangle = \int_{-\infty}^{\infty} x|\Psi(x)|^2\,dx$. Since $|\Psi|^2 = \Psi^*\Psi$, we have $\langle x\rangle = \int_{-\infty}^{\infty} \Psi^* x\Psi\,dx = \int_{-\infty}^{\infty} \Psi^*\hat{x}\Psi\,dx$, where (18.44) was used. Quantum mechanics postulates that for any physical property M, the average value for a system whose state function is Ψ is given by

$$\langle M\rangle = \int \Psi^*\hat{M}\Psi\,d\tau \tag{18.52}$$

where $\hat{M}$ is the operator for the property M and the integral goes over all space. For example, Eqs. (18.52) and (18.43) give $\langle p_x\rangle = (\hbar/i)\int \Psi^*(\partial\Psi/\partial x)\,d\tau$.

The average value of M is the average of the results of a very large number of measurements of M made on identical systems, each of which is in the same state Ψ just before the measurement.

For a stationary state, Ψ equals $e^{-iEt/\hbar}\psi$ [Eq. (18.18)]. Since $\hat{M}$ doesn't affect the $e^{-iEt/\hbar}$ factor, we have

$$\Psi^*\hat{M}\Psi = e^{iEt/\hbar}\psi^*\hat{M}e^{-iEt/\hbar}\psi = e^{iEt/\hbar}e^{-iEt/\hbar}\psi^*\hat{M}\psi = \psi^*\hat{M}\psi$$

Hence, for a stationary state,

$$\langle M\rangle = \int \psi^*\hat{M}\psi\,d\tau \tag{18.53}$$

Separation of variables. Let $q_1, q_2, \ldots, q_r$ be the coordinates of a system. (For example, for a two-particle system, $q_1 = x_1, q_2 = y_1, \ldots, q_6 = z_2$.) Suppose the Hamiltonian operator has the form

$$\hat{H} = \hat{H}_1 + \hat{H}_2 + \cdots + \hat{H}_r \tag{18.54}$$

where the operator $\hat{H}_1$ involves only the coordinate q_1, the operator $\hat{H}_2$ involves only q_2, etc. An example is the particle in a three-dimensional box, where $\hat{H} = \hat{H}_x + \hat{H}_y + \hat{H}_z$, with $\hat{H}_x \equiv -(\hbar^2/2m)\,\partial^2/\partial x^2$, etc. We saw in Sec. 18.9 that for this case, $\psi = X(x)Y(y)Z(z)$ and $E = E_x + E_y + E_z$, where $\hat{H}_x X(x) = E_x X(x)$, $\hat{H}_y Y(y) = E_y Y(y)$, $\hat{H}_z Z(z) = E_z Z(z)$ [Eqs. (18.34) and (18.38)].

An extension of the arguments of Sec. 18.9 shows that when $\hat{H}$ has the form (18.54), the wave functions and energies are given by (see *Q. C.*, sec. 6.2 for the proof)

$$\psi = f_1(q_1)f_2(q_2)\cdots f_r(q_r) \tag{18.55}$$

$$E = E_1 + E_2 + \cdots + E_r \tag{18.56}$$

where $E_1, E_2, \ldots$ and the functions $f_1, f_2, \ldots$ are found by solving

$$\hat{H}_1 f_1 = E_1 f_1,\ \hat{H}_2 f_2 = E_2 f_2,\ \ldots,\ \hat{H}_r f_r = E_r f_r \tag{18.57}$$

The equations in (18.57) are, in effect, separate Schrödinger equations, one for each coordinate.

Noninteracting particles. An important case where separation of variables applies is a system of n noninteracting particles. For such a system, the energy is the sum of the energies of the individual particles, so the classical Hamiltonian H and the quantum-mechanical Hamiltonian operator $\hat{H}$ have the forms $H = H_1 + H_2 + \cdots + H_n$ and $\hat{H} = \hat{H}_1 + \hat{H}_2 + \cdots + \hat{H}_n$, where $\hat{H}_1$ involves only the coordinates of particle 1, $\hat{H}_2$ involves only the coordinates of particle 2, etc. Here, by analogy to (18.55) to (18.57), we have

$$\psi = f_1(x_1, y_1, z_1)f_2(x_2, y_2, z_2)\cdots f_n(x_n, y_n, z_n) \tag{18.58}$$

$$E = E_1 + E_2 + \cdots + E_n \tag{18.59}$$

$$\hat{H}_1 f_1 = E_1 f_1,\ \hat{H}_2 f_2 = E_2 f_2,\ \ldots,\ \hat{H}_n f_n = E_n f_n \tag{18.60}$$

There is a separate Schrödinger equation for each particle, the wave function is the product of wave functions of the individual particles, and the energy is the sum of the energies of the individual particles.

Degeneracy. By a *linear combination* of the functions $g_1, g_2, \ldots, g_k$, one means a function of the form $c_1 g_1 + c_2 g_2 + \cdots + c_k g_k$, where the c's are constants. It can be proved (see Prob. 18.29) that any linear combination of two or more stationary-state wave functions that belong to the same degenerate energy level is an eigenfunction of the Hamiltonian with the same energy value as that of the degenerate level. In other words, if $\hat{H}\psi_1 = E_1\psi_1$ and $\hat{H}\psi_2 = E_1\psi_2$, then $\hat{H}(c_1\psi_1 + c_2\psi_2) = E_1(c_1\psi_1 + c_2\psi_2)$. The linear combination $c_1\psi_1 + c_2\psi_2$ (when multiplied by a normalization constant) is therefore also a valid wave function.

Note that this theorem does not apply to wave functions belonging to two different energy levels. If $\hat{H}\psi_5 = E_5\psi_5$ and $\hat{H}\psi_6 = E_6\psi_6$ with $E_5 \neq E_6$, then $c_1\psi_5 + c_2\psi_6$ is not an eigenfunction of $\hat{H}$.

In earlier chapters of this book, most of the results were proved. Because the mathematical derivations of many of the results of quantum mechanics are lengthy and complicated, many results in Chaps. 18 to 21 will be stated without proof; references to proofs will be given for the interested reader.

18.12 THE ONE-DIMENSIONAL HARMONIC OSCILLATOR

The one-dimensional harmonic oscillator is a useful model for treating the vibrations of a diatomic molecule (Sec. 21.3). Before examining the quantum mechanics of a harmonic oscillator, we review the classical treatment (*Halliday and Resnick*, chap. 15). Consider a particle of mass m that moves in one dimension and is attracted to the coordinate origin by a force proportional to its displacement from the origin: $F = -kx$, where k is called the *force constant*. When x is positive, the force is in the $-x$ direction, and when x is negative, F is in the $+x$ direction. (A physical example is a mass attached to a frictionless spring, x being the displacement from the equilibrium position.) From (2.39), $F = -dV/dx$, where V is the potential energy. Hence $-dV/dx = -kx$, and $V = \frac{1}{2}kx^2 + c$. The choice of zero of potential energy is arbitrary. Choosing the integration constant c as zero, we have (Fig. 18.11)

$$V = \tfrac{1}{2}kx^2 \tag{18.61}$$

Newton's second law $F = ma$ gives $m\, d^2x/dt^2 = -kx$. The solution to this differential equation is

$$x = A \sin\,[(k/m)^{1/2}t + b] \tag{18.62}$$

as can be verified by substitution in the differential equation (Prob. 18.22). In (18.62), A and b are integration constants. The maximum and minimum values of the sine function are $+1$ and -1, so the particle's x coordinate oscillates back and forth between $+A$ and $-A$. (A is the *amplitude* of the motion.)

The *period* τ of the oscillator is the time required for one complete cycle of oscillation. For one cycle of oscillation, the argument of the sine function in (18.62) must increase by 2π, since 2π is the period of a sine function. Hence the period satisfies $(k/m)^{1/2}\tau = 2\pi$, and $\tau = 2\pi(m/k)^{1/2}$. The *frequency* ν is the reciprocal of the period and equals the number of vibrations per second ($\nu = 1/\tau$) and

$$\nu = \frac{1}{2\pi}\left(\frac{k}{m}\right)^{1/2} \tag{18.63*}$$

The energy of the harmonic oscillator is $E = T + V = \frac{1}{2}mv_x{}^2 + \frac{1}{2}kx^2$. Use of (18.62) for x and of $v_x = dx/dt = (k/m)^{1/2}A \cos\,[(k/m)^{1/2}t + b]$ leads to (Prob. 18.34)

$$E = \tfrac{1}{2}kA^2 \tag{18.64}$$

Equation (18.64) shows that the classical energy can have any nonnegative value. As the particle oscillates, its kinetic energy and potential energy continually change, but the total energy remains constant at $\frac{1}{2}kA^2$.

Classically, the particle is limited to the region $-A \le x \le A$. When the particle reaches $x = A$ or $x = -A$, its speed is zero (since it reverses its direction of motion at

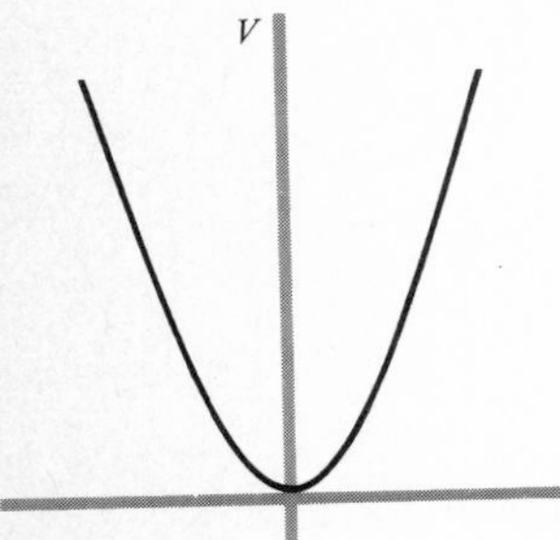

Figure 18.11
Potential energy for a one-dimensional harmonic oscillator.

$+A$ and $-A$) and its potential energy is a maximum, being equal to $\frac{1}{2}kA^2$. If the particle were to move beyond $x = \pm A$, its potential energy would increase above $\frac{1}{2}kA^2$. This is impossible for a classical particle: the total energy is $\frac{1}{2}kA^2$ and the kinetic energy is nonnegative, so the potential energy ($V = E - T$) cannot exceed the total energy.

Now for the quantum-mechanical treatment. Substitution of $V = \frac{1}{2}kx^2$ in (18.20) gives the time-independent Schrödinger equation as

$$-\frac{\hbar^2}{2m}\frac{d^2\psi}{dx^2} + \tfrac{1}{2}kx^2\psi = E\psi \qquad (18.65)$$

Fortunately, the mathematics involved in the solution of this equation is too complicated for an undergraduate physical-chemistry course, so we can omit the solution process (see *Q. C.*, chap. 4) and simply quote the results. One finds that well-behaved solutions ψ to (18.65) exist only for the following values of E:

$$E = (v + \tfrac{1}{2})h\nu \qquad \text{where } v = 0, 1, 2, \ldots \qquad (18.66)^*$$

where the vibrational frequency ν is given by (18.63) and the quantum number v takes on nonnegative integral values. [Don't confuse the typographically similar symbols ν (nu) and v (vee).] The energy is quantized. The allowed energy levels are equally spaced. The zero-point energy is $\frac{1}{2}h\nu$. (For a collection of harmonic oscillators in thermal equilibrium, all the oscillators will fall to the ground state as the temperature goes to absolute zero; hence the name zero-point energy.)

The well-behaved solutions to (18.65) turn out to have the form $e^{-\alpha x^2/2}$ times a polynomial of degree v in x, where

$$\alpha \equiv 2\pi\nu m/\hbar \qquad (18.67)$$

The explicit forms of the lowest few wave functions are

$$\psi_0 = (\alpha/\pi)^{1/4}e^{-\alpha x^2/2} \qquad (18.68)$$

$$\psi_1 = (4\alpha^3/\pi)^{1/4}xe^{-\alpha x^2/2} \qquad (18.69)$$

$$\psi_2 = (\alpha/4\pi)^{1/4}(2\alpha x^2 - 1)e^{-\alpha x^2/2}$$

where the subscript on ψ gives the value of v. Figure 18.12 plots ψ for $v = 0, 1, 2,$ and 3. As with the particle in a one-dimensional box, the number of nodes increases by 1 for each increase in the quantum number. Note the qualitative resemblances of the wave functions in Figs. 18.12 and 18.9.

The harmonic-oscillator wave functions fall off exponentially to zero as $x \to \pm\infty$. Note, however, that even for very large values of x, the wave function ψ and the probability density $|\psi|^2$ are not zero; there is some probability of finding the particle at an indefinitely large value of x. For a classical-mechanical harmonic oscillator with energy $(v + \frac{1}{2})h\nu$, Eq. (18.64) gives $(v + \frac{1}{2})h\nu = \frac{1}{2}kA^2$, and $A = [(2v + 1)h\nu/k]^{1/2}$. A classical oscillator is confined to the region $-A \le x \le A$. However, a quantum-mechanical oscillator has some probability of being found in the *classically forbidden* regions $x > A$ and $x < -A$, where the potential energy is greater than the particle's total energy. This penetration into classically forbidden regions is called *tunneling*.

18.13 TWO-PARTICLE PROBLEMS

Consider a two-particle system where the coordinates of the particles are x_1, y_1, z_1 and x_2, y_2, z_2. The *relative* (or *internal*) *coordinates* x, y, z are defined by

$$x \equiv x_2 - x_1, \qquad y \equiv y_2 - y_1, \qquad z \equiv z_2 - z_1 \tag{18.70}$$

These are the coordinates of particle 2 in a coordinate system whose origin is attached to particle 1 and moves with it.

In many cases, the potential energy V of the system depends only on the relative coordinates x, y, z. [For example, if the particles are electrically charged, the Coulomb's law potential-energy of interaction between the particles depends only on the distance r between them, and $r = (x^2 + y^2 + z^2)^{1/2}$.] Let us assume that $V = V(x, y, z)$. Let X, Y, Z be the coordinates of the center of mass of the system; X is given by $(m_1 x_1 + m_2 x_2)/(m_1 + m_2)$, where m_1 and m_2 are the masses of the particles (*Halliday and Resnick*, sec. 9-1). If one expresses the classical energy (i.e., the classical Hamiltonian) of the system in terms of the coordinates x, y, z, X, Y, Z, instead of $x_1, y_1, z_1, x_2, y_2, z_2$, it turns out (see Prob. 18.26) that

$$H = \left[\frac{1}{2\mu}(p_x^2 + p_y^2 + p_z^2) + V(x, y, z)\right] + \left[\frac{1}{2M}(p_X^2 + p_Y^2 + p_Z^2)\right] \tag{18.71}$$

where M is the total mass of the system ($M = m_1 + m_2$), the *reduced mass* μ is defined by

$$\mu \equiv \frac{m_1 m_2}{m_1 + m_2} \tag{18.72}^*$$

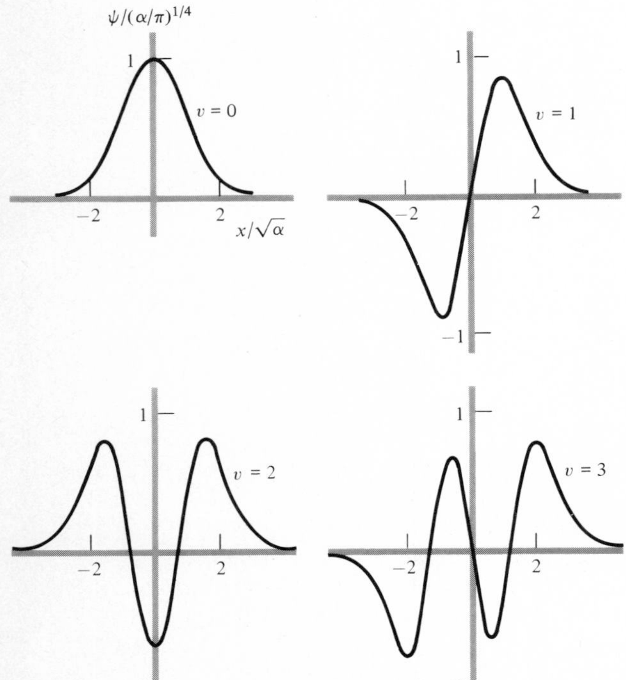

Figure 18.12
Wave functions for the lowest four harmonic-oscillator states.

and the momenta in (18.71) are defined by

$$p_x \equiv \mu v_x, \qquad p_y \equiv \mu v_y, \qquad p_z \equiv \mu v_z$$
$$p_X \equiv M v_X, \qquad p_Y \equiv M v_Y, \qquad p_Z \equiv M v_Z \tag{18.73}$$

where $v_x = dx/dt$, etc., and $v_X = dX/dt$, etc.

Equation (18.45) shows that the Hamiltonian (18.71) is the sum of a Hamiltonian for a fictitious particle of mass μ and coordinates x, y, z that has the potential energy $V(x, y, z)$ and a Hamiltonian for a second fictitious particle of mass $M = m_1 + m_2$ and coordinates X, Y, Z that has $V = 0$. Moreover, there is no term for any interaction between these two fictitious particles. Hence, Eqs. (18.59) and (18.60) show that the quantum-mechanical energy E of the two-particle system is given by $E = E_\mu + E_M$, where E_μ and E_M are found by solving $\hat{H}_\mu \psi_\mu(x, y, z) = E_\mu \psi_\mu(x, y, z)$ and $\hat{H}_M \psi_M(X, Y, Z) = E_M \psi_M(X, Y, Z)$. The Hamiltonian operator $\hat{H}_\mu$ is formed from the terms in the first pair of brackets in (18.71), and $\hat{H}_M$ is formed from the terms in the second pair of brackets.

Introduction of the relative coordinates x, y, z and the center-of-mass coordinates X, Y, Z reduces the two-particle problem to two separate one-particle problems. We solve a Schrödinger equation for a fictitious particle of mass μ moving subject to the potential energy $V(x, y, z)$, and we solve a separate Schrödinger equation for a fictitious particle of mass M. The Hamiltonian $\hat{H}_M$ involves only kinetic energy; if the two particles are confined to a box, we can use the particle-in-a-box energies (18.40) for E_M. The energy E_M is translational energy of the two-particle system as a whole. The Hamiltonian $\hat{H}_\mu$ involves the kinetic energy and potential energy of motion of the particles relative to each other, so E_μ is the energy associated with this relative or "internal" motion.

18.14 THE TWO-PARTICLE RIGID ROTOR

The two-particle rigid rotor consists of particles of masses m_1 and m_2 constrained to remain a fixed distance d from each other. (This is a useful model for treating the rotation of a diatomic molecule; see Sec. 21.3.) The system's energy is wholly kinetic, and $V = 0$. Since $V = 0$ is a special case of V being a function of only the relative coordinates of the particles, the results of the last section apply. The quantum-mechanical energy is the sum of the translational energy of the system as a whole and the energy of internal motion. The interparticle distance is constant, so the internal motion consists entirely of changes in the spatial orientation of the interparticle axis; i.e., the internal motion is a rotation of the two-particle system.

Solution of the Schrödinger equation for internal motion is complicated, so we shall just quote the results without proof. (For a derivation, see *Q. C.*, sec. 6.3.) The allowed rotational energies turn out to be

$$E_{\text{rot}} = J(J+1)\frac{\hbar^2}{2I} \qquad \text{where } J = 0, 1, 2, \ldots \tag{18.74}^*$$

where the *moment of inertia* I is given by

$$I = \mu d^2 \tag{18.75}$$

with $\mu = m_1 m_2/(m_1 + m_2)$. The spacing between adjacent rotational energy levels increases with increasing quantum number J. There is no zero-point energy.

The rotational wave functions are most conveniently expressed in terms of the angles θ and ϕ that give the spatial orientation of the rotor (Fig. 18.13). One finds $\psi_{\text{rot}} = \Theta_{JM_J}(\theta)\Phi_{M_J}(\phi)$, where Θ_{JM_J} is a function of θ whose form depends on the two quantum numbers J and M_J and Φ_{M_J} is a function of ϕ whose form depends on M_J. (These functions won't be given here but will be discussed in Sec. 19.3.)

Ordinarily, the wave function for internal motion of a two-particle system is a function of three coordinates. However, since the interparticle distance is held fixed in this problem, ψ_{rot} is a function of only two coordinates, θ and ϕ. Since there are two coordinates, there are two quantum numbers, J and M_J. The possible values of M_J turn out to range from $-J$ to J in steps of 1:

$$M_J = -J, \ -J+1, \ldots, J-1, J \qquad (18.76)^*$$

For example, if $J = 2$, then $M_J = -2, -1, 0, 1, 2$. For a given J, there are $2J + 1$ values of M_J. The quantum numbers J and M_J determine the rotational wave function, but E_{rot} depends only on J. Hence, each rotational level is $(2J + 1)$-fold degenerate.

18.15 APPROXIMATION METHODS

For a many-electron atom or molecule, the interelectronic repulsion terms in the potential energy V make it impossible to solve the Schrödinger equation (18.19) exactly. One must resort to approximation methods.

The most widely used approximation method is the *variation method*. From the postulates of quantum mechanics, one can deduce the following theorem (for the proof, see *Q. C.*, sec. 8.1). Let $\hat{H}$ be the time-independent Hamiltonian operator of a quantum-mechanical system. If ϕ is any normalized, well-behaved function, then

$$\int \phi^* \hat{H} \phi \, d\tau \geq E_{gs} \qquad \text{for } \phi \text{ normalized} \qquad (18.77)$$

where E_{gs} is the true ground-state energy of the system and the integral goes over all space.

To apply the variation method, one takes many different normalized, well-behaved functions $\phi_1, \phi_2, \ldots$, and for each of them one computes the *variational integral* $\int \phi^* \hat{H} \phi \, d\tau$. The variation theorem (18.77) shows that the function giving the

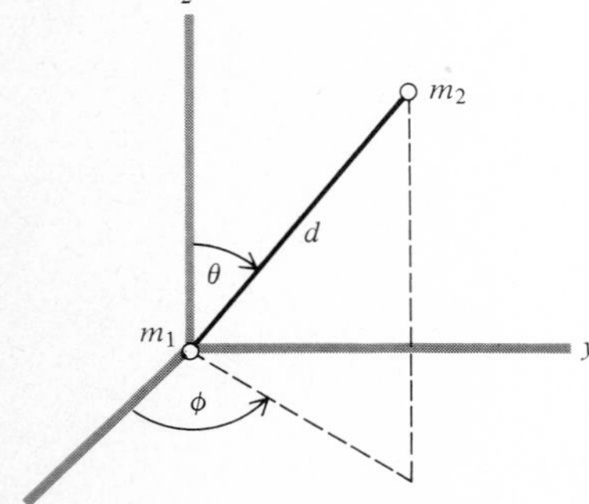

Figure 18.13
A two-particle rigid rotor.

lowest value of $\int \phi^* \hat{H} \phi \, d\tau$ provides the closest approximation to the ground-state energy. This function can serve as an approximation to the true ground-state wave function and can be used to compute approximations to ground-state molecular properties in addition to the energy (e.g., dipole moment).

Suppose we were lucky enough to guess the true ground-state wave function ψ_{gs}. Substitution of $\phi = \psi_{gs}$ in (18.77) and use of (18.51) and (18.12) gives the variational integral as

$$\int \psi_{gs}^* \hat{H} \psi_{gs} \, d\tau = \int \psi_{gs}^* E_{gs} \psi_{gs} \, d\tau = E_{gs} \int \psi_{gs}^* \psi_{gs} \, d\tau = E_{gs}$$

We would then get the true ground-state energy.

If the variation function ϕ is not normalized, it must be multiplied by a normalization constant N before being used in (18.77). The normalization condition is $1 = \int |N\phi|^2 \, d\tau = |N|^2 \int |\phi|^2 \, d\tau$. Hence,

$$|N|^2 = \frac{1}{\int |\phi|^2 \, d\tau} \tag{18.78}$$

Use of the normalized function $N\phi$ in place of ϕ in (18.77) gives $\int N^* \phi^* \hat{H}(N\phi) \, d\tau = |N|^2 \int \phi^* \hat{H} \phi \, d\tau \geq E_{gs}$. [$\hat{H}$ contains things like $\partial^2/\partial x^2$ and $V \times$, and these operators don't affect the constant N, so $H(N\phi) = NH\phi$.] Use of (18.78) gives

$$\frac{\int \phi^* \hat{H} \phi \, d\tau}{\int \phi^* \phi \, d\tau} \geq E_{gs} \tag{18.79}$$

where ϕ need not be normalized.

Example Devise a trial variation function for the particle in a one-dimensional box and use it to estimate E_{gs}.

The particle in a box is exactly soluble, and there is no need to resort to an approximate method. For instructional purposes, let's pretend we don't know how to solve the particle-in-a-box Schrödinger equation. We know that the true ground-state wave function is zero outside the box, so we take the variation function ϕ to be zero outside the box. Equations (18.77) and (18.79) are valid only if ϕ is a well-behaved function, and this requires that ϕ be continuous. For ϕ to be continuous at the ends of the box, it must be zero at $x = 0$ and at $x = a$, where a is the box length. Perhaps the simplest way to get a function that vanishes at 0 and a is to take $\phi = x(a - x)$ for the region inside the box; as noted above, $\phi = 0$ outside the box. Since we made no effort to normalize ϕ, Eq. (18.79) must be used. For the particle in a box, $V = 0$ and $\hat{H} = -(\hbar^2/2m) \, d^2/dx^2$ inside the box. Taking the second derivative of ϕ, we get $\hat{H}\phi = -(\hbar^2/2m)(-2) = \hbar^2/m$. Therefore, $\int \phi^* \hat{H} \phi \, d\tau = \int_0^a x(a - x) \times (\hbar^2/m) \, dx = \hbar^2 a^3/6m$. Also, $\int \phi^* \phi \, d\tau = \int_0^a x^2(a - x)^2 \, dx = a^5/30$. Hence, the variation theorem (18.79) becomes $(\hbar^2 a^3/6m) \div (a^5/30) \geq E_{gs}$, or $E_{gs} \leq 5h^2/4\pi^2 ma^2 = 0.12665h^2/ma^2$. From Eq. (18.27), the true ground-state energy is $E_{gs} = h^2/8ma^2 = 0.125h^2/ma^2$. The variation function $x(a - x)$ gives a 1.3 percent error in E_{gs}.

One usually chooses a variation function that has one or more undetermined parameters and varies these parameters to minimize the variational integral.

A common form for variational functions in atomic and molecular quantum mechanics is the *linear variation function* $\phi = c_1 f_1 + c_2 f_2 + \cdots + c_n f_n$, where $f_1, \ldots, f_n$ are functions and $c_1, \ldots, c_n$ are variational parameters whose values are determined by minimizing the variational integral. Let W be the left side of (18.79). Then the conditions for a minimum in W are $\partial W/\partial c_1 = 0$, $\partial W/\partial c_2 = 0, \ldots,$ $\partial W/\partial c_n = 0$. These conditions lead to a set of equations that allows the c's to be found. (See *Q. C.*, sec. 8.5 for details.) It turns out that there are n different sets of coefficients $c_1, \ldots, c_n$ that satisfy $\partial W/\partial c_1 = \cdots = \partial W/\partial c_n = 0$, so we end up with n different variational functions $\phi_1, \ldots, \phi_n$ and n different values for the variational integral $W_1, \ldots, W_n$, where $W_1 = \int \phi_1^* \hat{H}\phi_1 \, d\tau / \int \phi_1^* \phi_1 \, d\tau$, etc. If these W's are numbered in order of increasing energy, it can be shown that $W_1 \geq E_{gs}$, $W_2 \geq E_{gs+1}$, etc., where E_{gs}, $E_{gs+1}, \ldots$ are the true energies of the ground state, the first excited state, etc. Thus, use of the linear variation function $c_1 f_1 + \cdots + c_n f_n$ gives us approximations to the energies and wave functions of the lowest n states of the system. (In using this method, one deals separately with wave functions of different symmetry.)

Another important approximation method is *perturbation theory*. Suppose the Hamiltonian $\hat{H}$ for the system we are interested in has the form $\hat{H} = \hat{H}^0 + \hat{H}'$, where $\hat{H}^0$ is the Hamiltonian of a system whose Schrödinger equation $\hat{H}^0\psi_n^{(0)} = E_n^{(0)}\psi_n^{(0)}$ is exactly soluble, and where $\hat{H}'$ (called the *perturbation*) is small compared with $\hat{H}^0$. One can then derive expressions that relate the energy E_n of a state of the system with Hamiltonian $\hat{H}$ to the energy of the corresponding state of the $\hat{H}^0$ system. To the lowest order of approximation, it turns out (see *Q. C.*, sec. 9.2) that

$$E_n \approx E_n^{(0)} + \int \psi_n^{(0)*} \hat{H}' \psi_n^{(0)} \, d\tau \tag{18.80}$$

For example, suppose a system has $\hat{H} = -(\hbar^2/2m)\, d^2/dx^2 + \frac{1}{2}kx^2 + cx^4$, where $c \ll k$. Let $\hat{H}^0 = -(\hbar^2/2m)\, d^2/dx^2 + \frac{1}{2}kx^2$ and $\hat{H}' = cx^4$. Then $\hat{H}^0$ is a harmonic-oscillator Hamiltonian. The ground-state energy of the system with Hamiltonian $\hat{H}$ is given by (18.80) as $E_{gs} \approx \frac{1}{2}h\nu + c\int_{-\infty}^{\infty} \psi_0^* x^4 \psi_0 \, dx$, where ψ_0 is given by (18.68). Similar equations hold for excited-state energies.

PROBLEMS

18.1 (*a*) Let ν_{max} be the frequency at which the blackbody-radiation function (18.1) is a maximum. Show that $\nu_{max} = kTx/h$, where x is the nonzero solution of $x + 3e^{-x} = 3$. Since x is a constant, ν_{max} increases linearly with T. (*b*) Use a calculator with an e^x key to solve the equation in (*a*) by trial and error. To save time, use interpolation after you have found the successive integers that x lies between. (*c*) Calculate ν_{max} for a blackbody at 300 K and at 3000 K. Refer to Fig. 21.2 to state in which portions of the electromagnetic spectrum these frequencies lie. (*d*) For a certain star, ν_{max} is observed to be $3.4 \times 10^{14}\ s^{-1}$. Assume the star is a blackbody and estimate its surface temperature.

18.2 (*a*) Use the fact that $\int_0^\infty [z^3/(e^z - 1)]\, dz = \pi^4/15$ to show that the total radiant energy emitted per second by unit area of a blackbody is $2\pi^5k^4T^4/15c^2h^3$. Note that this quantity is proportional to T^4 (*Stefan's law*). (*b*) The sun's diameter is 1.4×10^9 m, and its surface temperature is 5800 K. Assume the sun is a blackbody and estimate the rate of energy loss by radiation from the sun. (*c*) Use $E = mc^2$ to calculate the relativistic mass of the photons lost by radiation from the sun in 1 year.

18.3 The work function of K is 2.2 eV and that of Ni is 5.0 eV, where 1 eV = 1.60×10^{-12} erg. (*a*) Calculate the threshold frequencies and wavelengths for these two metals. (*b*) Will violet light of wavelength 4000 Å cause the photoelectric effect in K? In Ni?

18.4 Calculate the energy of a photon of red light of wavelength 7000 Å.

18.5 A 100-W sodium-vapor lamp emits yellow light of wavelength 5900 Å. Calculate the number of photons emitted per second.

18.6 Millikan found the following data for the photoelectric effect in Na:

$10^{12}T_{max}$/ergs	3.41	2.56	1.95	0.75
λ/Å	3125	3650	4047	5461

where T_{max} is the maximum kinetic energy of emitted electrons and λ is the wavelength of the incident radiation. Plot T_{max} vs. ν; from the slope and intercept, calculate h and the work function for Na.

18.7 Calculate the de Broglie wavelength of (*a*) a neutron moving at 6.0×10^6 cm/s; (*b*) a 50-g particle moving at 120 cm/s.

18.8 Calculate the relativistic mass of a photon of wavelength (*a*) 10,000 Å; (*b*) 1000 Å.

18.9 A beam of electrons traveling at 6.0×10^8 cm/s falls on a slit of width 2400 Å. The diffraction pattern is observed on a screen 40 cm from the slit. The x and y axes are defined as in Fig. 18.5. Find (*a*) the angle θ to the first diffraction minimum; (*b*) the width of the central maximum of the diffraction pattern on the screen; (*c*) the uncertainty Δp_x at the slit.

18.10 Calculate the wavelength of the photon emitted when a 1.0×10^{-27} g particle in a box of length 6.0 Å jumps from the $n = 5$ to $n = 4$ level.

18.11 (*a*) For a particle in the stationary state n of a one-dimensional box of length a, find the probability that the particle is in the region $0 \le x \le a/4$. (*b*) Calculate this probability for $n = 1$, 2, and 3.

18.12 For a 1.0×10^{-26} g particle in a box of length 2.0 Å whose ends are at $x = 0$ and $x = 2.000$ Å, calculate the probability that the particle's x coordinate is between 1.6000 and 1.6001 Å if (*a*) $n = 1$; (*b*) $n = 2$.

18.13 Sketch ψ and $|\psi|^2$ for the $n = 4$ and $n = 5$ states of a particle in a one-dimensional box.

18.14 For a particle in a cubic box of edge a: (*a*) How many states have energies in the range 0 to $16h^2/8ma^2$? (*b*) How many energy levels lie in this range?

18.15 If $\hat{A} = 3x^2 \times$ and $\hat{B} = d/dx$, then $\hat{A}\hat{B}f(x) = 3x^2 f'(x)$. Find $\hat{B}\hat{A}f(x)$. Is $\hat{B}\hat{A}f = \hat{A}\hat{B}f$?

18.16 Find the quantum-mechanical operator for (*a*) $p_x{}^3$; (*b*) $p_z{}^4$.

18.17 (*a*) Which of the functions sin $3x$, 6 cos $4x$, $5x^3$, $1/x$, $3e^{-5x}$, ln $2x$ are eigenfunctions of d^2/dx^2? (*b*) For each eigenfunction, state the eigenvalue.

18.18 For a particle in a one-dimensional-box stationary state, show that (*a*) $\langle p_x \rangle = 0$; (*b*) $\langle x \rangle = a/2$; (*c*) $\langle x^2 \rangle = a^2(1/3 - 1/2n^2\pi^2)$.

18.19 Calculate the frequency of radiation emitted when a harmonic-oscillator of frequency 6.0×10^{13} s^{-1} jumps from the $v = 8$ to the $v = 7$ level.

18.20 Draw rough sketches of ψ^2 for the $v = 0$, 1, 2, and 3 harmonic oscillator states.

18.21 Find the most probable value(s) of x for a harmonic oscillator in the state (*a*) $v = 0$; (*b*) $v = 1$.

18.22 Verify by substitution that (18.62) satisfies the differential equation $m\, d^2x/dt^2 = -kx$.

18.23 A mass of 45 g on a spring oscillates at the frequency of 2.4 vibrations per second with an amplitude 4.0 cm. (*a*) Calculate the force constant of the spring. (*b*) What would be the quantum number v if the system were treated quantum-mechanically?

18.24 A three-dimensional harmonic oscillator has $V = \frac{1}{2}k_x x^2 + \frac{1}{2}k_y y^2 + \frac{1}{2}k_z z^2$, where the three force constants k_x, k_y, k_z are not necessarily equal. (*a*) Write down the expression for the energy levels of this system. Define all symbols. (*b*) What is the zero-point energy?

18.25 Calculate $\langle x^2 \rangle$ using the particle-in-a-box trial variation function in the example of Sec. 18.15. Compare with the true ground state $\langle x^2 \rangle$ (Prob. 18.18).

18.26 Substitute (18.72), (18.73), and $M = m_1 + m_2$ into (18.71) and verify that H reduces to $p_1{}^2/2m_1 + p_2{}^2/2m_2 + V$, where p_1 is the momentum of particle 1.

18.27 (*a*) Apply the variation function $x^2(a - x)^2$ to the particle in a box and estimate the ground-state energy. Calculate the percent error in E_{gs}. (*b*) Explain why the function x^2 (for x between 0 and a) cannot be used as a variation function for the particle in a box.

18.28 Use perturbation theory to find an approximate expression for the energy of the lowest state of the one-particle, one-dimensional system with $V = \frac{1}{2}kx^2 + cx^4$. (See Table 15.1.)

18.29 $\hat{B}$ is a *linear operator* if it meets the following two requirements: $\hat{B}(cf) = c\hat{B}f$ and $\hat{B}(f + g) = \hat{B}f + \hat{B}g$, where c is an arbitrary constant and f and g are arbitrary functions. (*a*) Verify that $\hat{H}$ in (18.48) is linear. (*b*) Use the linearity of $\hat{H}$ to prove that a linear combination of two stationary-state wave functions belonging to the same degenerate level is an eigenfunction of $\hat{H}$.

18.30 Verify that ψ_0 in (18.68) is a solution of the Schrödinger equation (18.65).

18.31 Verify that ψ_1 in (18.69) is normalized. (See Table 15.1.)

18.32 For the ground state of a harmonic oscillator, calculate (*a*) $\langle x \rangle$; (*b*) $\langle x^2 \rangle$; (*c*) $\langle p_x \rangle$. (See Table 15.1.)

18.33 Prove that $(fg)^* = f^*g^*$, where f and g are complex quantities.

18.34 Verify Eq. (18.64).

18.35 Verify Eq. (18.29).

19

ATOMIC STRUCTURE

19.1 UNITS

The forces in atoms and molecules are electrical. Chapter 14 used SI units for electrical quantities. Although SI electrical units could be used in discussing atoms and molecules, it is traditional to use a system called *gaussian units*. The gaussian system uses cgs units (cm, g, s) for mechanical quantities; in gaussian units, the Coulomb's law force between two charges separated by a distance r in vacuum is written

$$F = Q_1' Q_2' / r^2 \tag{19.1}$$

In this equation, F is in dynes, r is in centimeters, and Q_1' and Q_2' are in *statcoulombs* (sometimes called esu, electrostatic units of charge). The primes on the charges indicate the use of gaussian units. In the gaussian system, electrical charge is not taken as a fundamental quantity; instead Eq. (19.1) defines the statcoulomb (statC) in terms of other units. We have 1 dyn = (1 statC)^2/cm^2, and use of (2.22) gives

$$1 \text{ statC} = 1 \text{ g}^{1/2} \text{ cm}^{3/2} \text{ s}^{-1} \tag{19.2}$$

To find the relation between coulombs (C) and statcoulombs, consider two 1-C charges separated by 1 m in vacuum. From (14.1), the force between them is $F = 1 \text{ C}^2/(4\pi\varepsilon_0 \text{ m}^2)$. Use of (14.2) and (2.23) gives $F = 8.9876 \times 10^9 \text{ N} = 8.9876 \times 10^{14}$ dyn. Using this F in (19.1), we have $8.9876 \times 10^{14} \text{ dyn} = Q'^2/(100 \text{ cm})^2$, and $Q' = 2.9979 \times 10^9$ statC. Hence,

$$1 \text{ C} = 2.9979 \times 10^9 \text{ statC} \tag{19.3}$$

The numerical coefficient in (19.3) is the same as that in the speed of light. [Strictly speaking, the equals sign in (19.3) is incorrect. The coulomb and statcoulomb have different units from each other; Eq. (19.3) should be interpreted to mean that a charge of one coulomb in SI units corresponds to a charge of 3×10^9 statC in gaussian units.]

Comparison of (19.1) with the SI equation $F = Q_1 Q_2/4\pi\varepsilon_0 r^2$ shows that to convert an equation in gaussian units to the corresponding SI equation, the gaussian charge Q' must be replaced by $Q/(4\pi\varepsilon_0)^{1/2}$; similar replacements must be made for other electric and magnetic quantities. A table on the inside back cover lists these replacements. From (14.10) and (14.12), the potential energy of the interaction between two charges separated by r in vacuum is in SI units: $V = Q_1 Q_2/4\pi\varepsilon_0 r$. In gaussian units, this becomes

$$V = Q'_1 Q'_2/r \qquad (19.4)^*$$

In Chaps. 19 and 20 we shall write Coulomb's law in the form (19.1). This equation can be interpreted as being in gaussian units by thinking of Q'_1 and Q'_2 as being in statcoulombs. Alternatively, Eqs. (19.1) and (19.4) can be regarded as SI equations in which Q'_1 and Q'_2 are abbreviations for $Q_1/(4\pi\varepsilon_0)^{1/2}$ and $Q_2/(4\pi\varepsilon_0)^{1/2}$, where Q_1 and Q_2 are in coulombs:

$$Q' \equiv Q/(4\pi\varepsilon_0)^{1/2} \qquad (19.5)$$

Atomic and molecular energies are very small. A convenient unit to express these energies is the *electron volt* (eV), defined as the energy acquired by an electron accelerated through a potential difference of one volt. From (14.10), the magnitude of this energy is $e(1\ \text{V})$, where e is the magnitude of the electron's charge. Substitution of (19.7) for e gives $1\ \text{eV} = (1.6022 \times 10^{-19}\ \text{C})\ (1\ \text{V})$; use of (14.8) and (2.32) gives

$$1\ \text{eV} = 1.6022 \times 10^{-19}\ \text{J} = 1.6022 \times 10^{-12}\ \text{erg} \qquad (19.6)$$

19.2 HISTORICAL BACKGROUND

In a low-pressure gas-discharge tube, bombardment of the negative electrode (the cathode) by positive ions causes the cathode to emit what nineteenth-century physicists called *cathode rays*. In 1897, J. J. Thomson measured the deflection of cathode rays in simultaneously applied electric and magnetic fields of known strengths. His experiment allowed calculation of the charge-to-mass ratio Q/m of the cathode-ray particles. (See *Halliday and Resnick*, sec. 33-8 for the details.) Thomson found that Q/m was independent of the metal used for the cathode, and his experiments are generally regarded as marking the discovery of the electron. [G. P. Thomson, who was one of the first people to observe diffraction effects with electrons (Sec. 18.4), was J. J.'s son. It has been said that J. J. Thomson got the Nobel prize for proving the electron to be a particle and G. P. Thomson got the Nobel prize for proving the electron to be a wave.]

We shall use the symbol e to stand for the magnitude of the charge on the proton or electron. The electron charge is then $-e$; the proton charge is e. Thomson found $e/m = 1.7 \times 10^8$ C/g. The modern value is 1.7588×10^8 C/g.

The first accurate measurement of the electron's charge was made by R. A. Millikan in the period 1909–1913. Millikan observed the motion of charged oil drops in oppositely directed electric and gravitational fields. When the forces balance each other, the drop becomes motionless and Eqs. (14.3) and (2.24) give $|QE| = Mg$, where Q and M are the charge and mass of the oil drop, E is the electric field strength, and g is the gravitational acceleration. Millikan found that all observed values of Q satisfied $|Q| = ne$, where n was a small integer and was clearly the number of extra electrons or the number of missing electrons on the charged oil drop. The smallest observed value of $|Q|$ could then be taken as the magnitude of the charge of the electron. Millikan obtained $e' = 4.77 \times 10^{-10}$ statC. The currently accepted value of the proton charge is

$$e' = 4.8032 \times 10^{-10} \text{ statC}, \qquad e = 1.6022 \times 10^{-19} \text{ C} \tag{19.7}$$

where the prime indicates gaussian units.

From the values of e and the Faraday constant, an accurate value of the Avogadro constant can be obtained. Equation (14.27) gives $N_0 = \mathscr{F}/e = (96{,}485 \text{ C mole}^{-1})/(1.6022 \times 10^{-19} \text{ C}) = 6.022 \times 10^{23} \text{ mole}^{-1}$.

From the values of e and e/m, the electron (rest) mass m can be found. The modern value is $m = 9.1095 \times 10^{-28}$ g. The mass of an ^{1}H atom is $1.0078 \text{ g}/6.022 \times 10^{23} = 1.6735 \times 10^{-24}$ g. This is 1837 times the electron's mass, and so a proton is 1836 times as heavy as an electron. Nearly all the mass of an atom is in its nucleus.

The existence of the atomic nucleus was demonstrated by the 1909–1911 experiments of Rutherford, Geiger, and Marsden, who allowed a beam of alpha particles (He^{2+} nuclei) to fall on a very thin gold foil. Although most of the alpha particles passed nearly straight through the foil, a few were deflected through large angles. Since the very light electrons of the gold atoms cannot significantly deflect the alpha particles (in a collision between a truck and a bicycle, it is the bicycle that gets deflected), one need only consider the force between the alpha particle and the positive charge of a gold atom. The Coulomb's law force between this positive charge and the alpha particle is given by (19.1). To get a force sufficiently large to produce the observed large deflections, Rutherford found that r in (19.1) had to be in the range 10^{-12} to 10^{-13} cm. This is so much less than the known radius of an atom (10^{-8} cm) that Rutherford concluded in 1911 that the positive charge of an atom was not distributed throughout the atom but was concentrated in a tiny central region, the nucleus. Rutherford pictured the electrons as moving about the nucleus, much as the planets circle the sun.

In 1913, Bohr proposed his theory of the hydrogen atom (Sec. 18.3). By the early 1920s, physicists realized that the Bohr theory was not correct.

In January 1926, Erwin Schrödinger formulated the Schrödinger equation. He solved the time-independent Schrödinger equation for the hydrogen atom in his first paper on quantum mechanics, obtaining energy levels in agreement with the observed spectrum. In 1929, Hylleraas used the quantum-mechanical variational method (Sec. 18.15) to obtain a ground-state energy for helium in accurate agreement with experiment.

19.3 THE HYDROGEN ATOM

The hydrogen atom is a two-particle system in which a nucleus and an electron interact according to Coulomb's law. Instead of dealing only with the H atom, we shall consider the slightly more general problem of the *hydrogenlike atom*; this is an atom with one electron and Z protons in the nucleus. The values $Z = 1, 2, 3, \ldots$ give the species H, He^+, Li^{2+}, ... With the nuclear charge Q_1' set equal to Ze' and the electron charge Q_2' set equal to $-e'$, Eq. (19.4) gives the potential energy as $V = -Ze'^2/r$.

The potential-energy function depends only on the relative coordinates of the two particles, and so the conclusions of Sec. 18.13 apply. The total energy E_{tot} of the atom is the sum of the translational energy of the atom and the energy of internal motion. The translational energy levels can be taken as the particle-in-a-box levels (18.40); the box is the container holding the gas of H atoms. We now focus on the energy E of internal motion. The Hamiltonian H for the internal motion is given by the terms in the first pair of brackets in (18.71), and the corresponding Hamiltonian operator $\hat{H}$ for the internal motion is

$$\hat{H} = -\frac{\hbar^2}{2\mu}\left(\frac{\partial^2}{\partial x^2} + \frac{\partial^2}{\partial y^2} + \frac{\partial^2}{\partial z^2}\right) - \frac{Ze'^2}{r} \tag{19.8}$$

where x, y, z are the coordinates of the electron relative to the nucleus and $r = (x^2 + y^2 + z^2)^{1/2}$. The reduced mass μ is given by (18.72), m_1 and m_2 being the nuclear and electron masses. For an H atom, $m_{\text{nucleus}} = 1836m$ (where m is the electron mass), and $\mu = 1836m^2/1837m = 0.99946m$. The reduced mass differs only slightly from the electron mass.

The H-atom Schrödinger equation $\hat{H}\psi = E\psi$ is difficult to solve in cartesian coordinates but is relatively easy to solve if one uses spherical polar coordinates. The spherical polar coordinates r, θ, ϕ of the electron relative to the nucleus are defined in Fig. 19.1. (Math books usually interchange θ and ϕ.) The projection of r on the z axis is $r \cos \theta$, and its projection on the xy plane is $r \sin \theta$. The relation between cartesian and spherical polar coordinates is therefore

$$x = r \sin \theta \cos \phi, \qquad y = r \sin \theta \sin \phi, \qquad z = r \cos \theta \tag{19.9}$$

Of course, $x^2 + y^2 + z^2 = r^2$. The ranges of the coordinates are

$$0 \le r \le \infty, \qquad 0 \le \theta \le \pi, \qquad 0 \le \phi \le 2\pi \qquad \text{(19.10)*}$$

To solve the H-atom Schrödinger equation, one transforms the partial derivatives in (19.8) to derivatives with respect to r, θ, and ϕ and then uses the separation-of-variables procedure (Sec. 18.11). The details are omitted. (See *Q. C.*, chap. 6.) One finds that the allowed wave functions have the form

$$\psi = R_{nl}(r)\Theta_{lm}(\theta)\Phi_m(\phi) \tag{19.11}$$

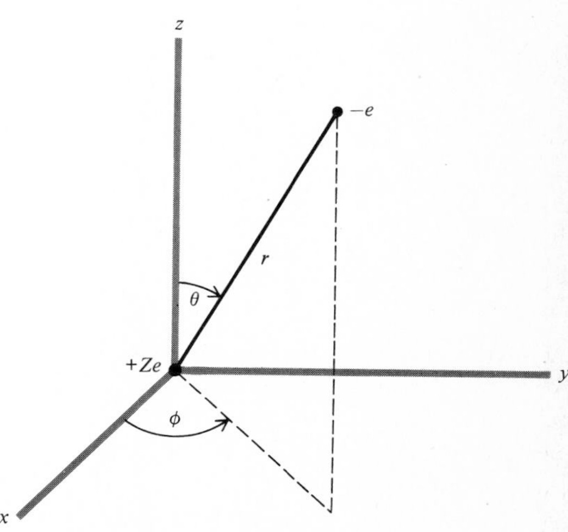

Figure 19.1
Coordinates of the electron relative to the nucleus in a hydrogenlike atom.

Since there are three variables, the solutions involve three quantum numbers: the *principal quantum number* n, the *angular-momentum quantum number* l, and the *magnetic quantum number* m. For ψ to be well behaved, the quantum numbers are restricted to the values

$$n = 1, 2, 3, \ldots \tag{19.12*}$$

$$l = 0, 1, 2, \ldots, n-1 \tag{19.13*}$$

$$m = -l, -l+1, \ldots, l-1, l \tag{19.14*}$$

For example, for $n = 2$, l can be 0 or 1. For $l = 0$, m is 0. For $l = 1$, m can be -1, 0, or 1.

The radial factor $R_{nl}(r)$ is a function of r whose nature depends on the quantum numbers n and l. The theta factor depends on l and m. The phi factor is

$$\Phi_m(\phi) = (2\pi)^{-1/2} e^{im\phi} \tag{19.15}$$

where $i = \sqrt{-1}$.

The following letter code is often used to specify the l value of an electron:

l value	0	1	2	3	4	5
Code letter	s	p	d	f	g	h

(19.16)*

The value of n is given as a prefix to the l code letter, and the m value is added as a subscript. Thus, $2s$ denotes the $n = 2$, $l = 0$ state; $2p_{-1}$ denotes the $n = 2$, $l = 1$, $m = -1$ state.

The allowed energy levels turn out to be

$$E = -\frac{Z^2}{n^2}\frac{e'^2}{2a} \qquad \text{where } a \equiv \frac{\hbar^2}{\mu e'^2} \tag{19.17}$$

where $n = 1, 2, 3, \ldots$. Also, all values $E \geq 0$ are allowed, corresponding to an ionized atom. The potential-energy function $-Ze'^2/r$ in (19.8) takes the zero level of energy at infinite separation of the electron and the nucleus. Hence, states in which the electron is bound to the nucleus have negative energies; the bound-state energies are (19.17). Figure 19.2 shows some of the allowed energy levels and the potential-energy function.

The hydrogenlike-atom energy levels (19.17) depend only on n, but the wave functions (19.11) depend on all three quantum numbers n, l, and m. Therefore, there is degeneracy. For example, the $n = 2$ H-atom level is fourfold degenerate (spin considerations omitted), the wave functions (states) $2s$, $2p_1$, $2p_0$, and $2p_{-1}$ all having the same energy.

The defined quantity a in (19.17) has the dimensions of length. For a hydrogen atom, substitution of numerical values gives (Prob. 19.11) $a = 0.5295$ Å.

If the reduced mass μ in the definition of a is replaced by the electron mass m, we get the *Bohr radius* a_0:

$$a_0 \equiv \hbar^2/me'^2 = 0.5292 \text{ Å} \tag{19.18}$$

(a_0 was the radius of the $n = 1$ circle in the Bohr theory.)

Setting $n = 1$ and $Z = 1$ in (19.17), we get the hydrogen-atom ground-state energy as $E_{gs} = -e'^2/2a$. The calculation of E_{gs} in gaussian units is

$$E_{gs} = -\frac{(4.803 \times 10^{-10}\ \text{statC})^2}{2(0.5295 \times 10^{-8}\ \text{cm})} = -2.179 \times 10^{-11}\ \text{erg}$$

where (19.7) and (19.2) were used. To do the calculation in SI units, we substitute $e' = e/(4\pi\varepsilon_0)^{1/2}$ [Eq. (19.5)] and use (19.7) and (14.2) to get

$$E_{gs} = -\frac{(1.6022 \times 10^{-19}\ \text{C})^2}{4\pi(8.854 \times 10^{-12}\ \text{C}^2\ \text{J}^{-1}\ \text{m}^{-1})2(0.5295 \times 10^{-10}\ \text{m})} = -2.179 \times 10^{-18}\ \text{J}$$

Use of (19.6) gives for an H atom

$$E_{gs} = -e'^2/2a = -13.60\ \text{eV} \tag{19.19}$$

$|E_{gs}|$ is the minimum energy needed to remove the electron from a hydrogen atom and is the *ionization energy* of H. The *ionization potential* of H is 13.60 V.

From (19.19), the energy levels (19.17) can be written as

$$E = -(Z^2/n^2)(13.60\ \text{eV}) \tag{19.20*}$$

Quantum chemists often use a system called *atomic units*, in which energies are reported in hartrees and distances in bohrs. These quantities are defined as 1 bohr $\equiv a_0 = 0.5292$ Å and 1 hartree $\equiv e'^2/a_0 = 27.212$ eV. The ground-state energy of H would be $-\frac{1}{2}$ hartree if a were approximated by a_0.

The first few $R_{nl}(r)$ and $\Theta_{lm}(\theta)$ factors in the wave functions (19.11) are

$$R_{1s} = 2(Z/a)^{3/2}e^{-Zr/a}$$
$$R_{2s} = 2^{-1/2}(Z/a)^{3/2}(1 - Zr/2a)e^{-Zr/2a} \tag{19.21}$$
$$R_{2p} = (24)^{-1/2}(Z/a)^{5/2}re^{-Zr/2a}$$
$$\Theta_{s0} = 1/\sqrt{2}, \qquad \Theta_{p0} = \tfrac{1}{2}\sqrt{6}\cos\theta, \qquad \Theta_{p1} = \Theta_{p-1} = \tfrac{1}{2}\sqrt{3}\sin\theta$$

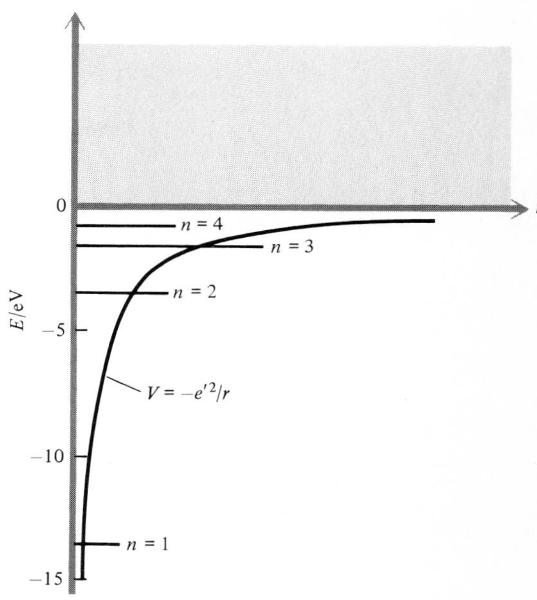

Figure 19.2
Energy levels and potential-energy function of the hydrogen atom. The shading indicates that all positive energies are allowed.

where the code (19.16) was used for l. The general form of R_{nl} is $r^l e^{-Zr/na}$ times a polynomial of degree $n - l - 1$ in r. As n increases, the exponential dies off more slowly, so that the average radius $\langle r \rangle$ of the atom increases as n increases. For the ground state, one finds (Prob. 19.13) $\langle r \rangle = 3a/2Z$, which is 0.79 Å for H. [In (19.21) and (19.15), e is the base of natural logarithms, and not the proton charge.]

Figure 19.3 shows some plots of $R_{nl}(r)$. The radial factor in ψ has $n - l - 1$ nodes (not counting the node at the origin for $l \neq 0$).

For s states ($l = 0$), Eqs. (19.15) and (19.21) give the angular factor in ψ as $1/\sqrt{4\pi}$, which is independent of θ and ϕ. For s states, ψ depends only on r and is therefore said to be *spherically symmetric*. For $l \neq 0$, the angular factor is not constant, and ψ is not spherically symmetric. Note from Fig. 19.3 that R_{nl} and hence ψ is nonzero at the nucleus ($r = 0$) for s states.

Angular momentum. The quantum numbers l and m are related to the angular momentum of the electron. The (linear) momentum $\mathbf{p}$ of a particle of mass m and velocity $\mathbf{v}$ is defined classically by $\mathbf{p} \equiv m\mathbf{v}$. (Don't confuse the mass m with the m quantum number.) Let $\mathbf{r}$ be the vector from the origin of a coordinate system to the particle. The particle's *angular momentum* $\mathbf{L}$ with respect to the coordinate origin is defined classically as a vector of length $rp \sin \beta$ (where β is the angle between $\mathbf{r}$ and $\mathbf{p}$) and direction perpendicular to both $\mathbf{r}$ and $\mathbf{p}$. (More succinctly, $\mathbf{L} \equiv \mathbf{r} \times \mathbf{p}$, where $\times$ indicates the vector cross product.)

To deal with angular momentum in quantum mechanics, one must use the quantum-mechanical operators for the components of the $\mathbf{L}$ vector. We shall omit the quantum-mechanical treatment (see *Q. C.*, chaps. 5 and 6) and simply state the results. It should be noted that there are two kinds of angular momentum in quantum mechanics. *Orbital* angular momentum is the quantum-mechanical analog of the classical quantity $\mathbf{L}$ and is due to the motion of a particle through space. In

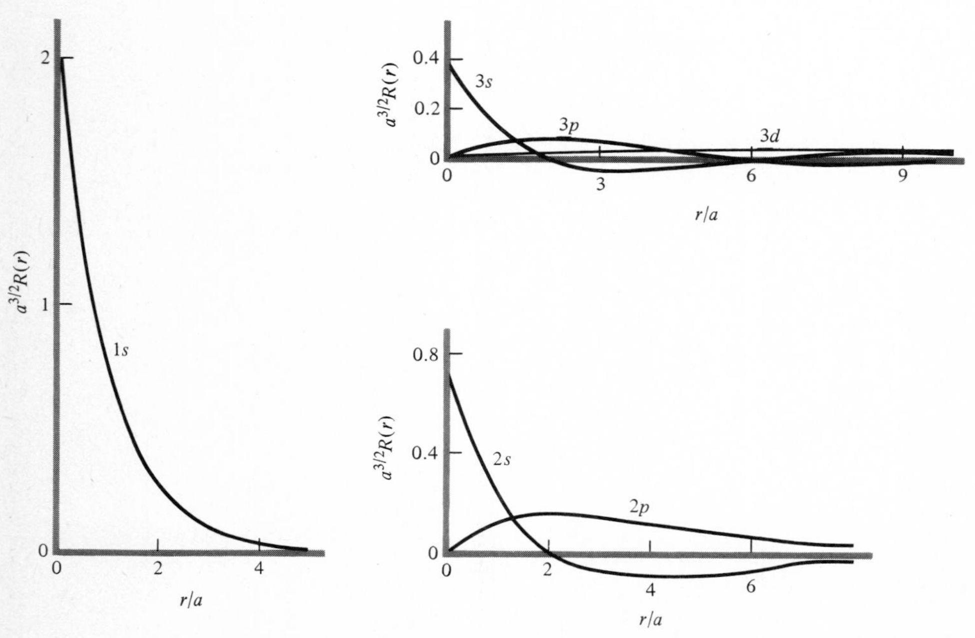

Figure 19.3 Radial factors in some hydrogen-atom wave functions.

addition to orbital angular momentum, many particles possess an intrinsic angular momentum, called *spin* angular momentum; this will be discussed in the next section.

The magnitude of the electron's orbital angular momentum **L** about the nucleus can be shown to equal $[l(l+1)]^{1/2}\hbar$. The component of **L** along the z axis can be shown to equal $m\hbar$. For s states ($l = 0$), the electronic orbital angular momentum is zero (a result quite difficult to understand classically). For p states ($l = 1$), the magnitude of **L** is $\sqrt{2}\,\hbar$, and L_z can be $\hbar$, 0, or $-\hbar$. The three possible orientations of **L** for $l = 1$ are shown in Fig. 19.4.

When an external magnetic field is applied to a hydrogen atom, the energies of the states depend on the m quantum number as well as on n.

The H-atom quantum numbers l and m are analogous to the two-particle-rigid-rotor quantum numbers J and M_J (Sec. 18.14). The functions Θ and Φ in the two-particle-rigid-rotor wave functions ψ_{rot} are the same functions as Θ and Φ in the H-atom wave functions (19.11).

Real wave functions. The ψ function (19.11) plus (19.15) is complex. Chemists often find it convenient to work with real wave functions instead. To get real functions, we use the theorem (Sec. 18.11) that any linear combination of two wave functions having the same energy is a valid wave function with that same energy.

From (19.11), (19.15), and (19.21), the complex $2p$ functions are

$$2p_{+1} = be^{-Zr/2a}r \sin\theta\, e^{i\phi}, \qquad 2p_{-1} = be^{-Zr/2a}r \sin\theta\, e^{-i\phi}$$

$$2p_0 = ce^{-Zr/2a}r \cos\theta$$

where $b \equiv (1/8\pi^{1/2})(Z/a)^{5/2}$ and $c \equiv \pi^{-1/2}(Z/2a)^{5/2}$. The $2p_0$ function is real as it stands. Equation (19.9) gives $r\cos\theta = z$, so the $2p_0$ function is also written as $2p_z \equiv 2p_0 = cze^{-Zr/2a}$. (Don't confuse the nuclear charge Z with the z spatial coordinate.) The $2p_{+1}$ and $2p_{-1}$ functions are each eigenfunctions of the Hamiltonian operator with the same energy eigenvalue, so we can take any linear combination of them and have a valid wave function. As a preliminary, we note that

$$e^{i\phi} = \cos\phi + i \sin\phi \qquad (19.22)^*$$

and $e^{-i\phi} = (e^{i\phi})^* = \cos\phi - i\sin\phi$. [For a proof of (19.22), see Prob. 19.18.] We define the linear combinations $2p_x$ and $2p_y$ as

$$2p_x \equiv (2p_1 + 2p_{-1})/\sqrt{2}, \qquad 2p_y \equiv (2p_1 - 2p_{-1})/i\sqrt{2} \qquad (19.23)$$

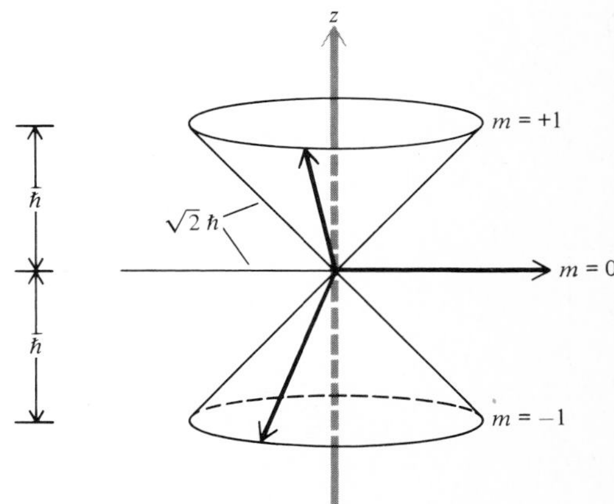

Figure 19.4
Allowed spatial orientations of the electronic orbital angular momentum for $l = 1$.

(The $1/\sqrt{2}$ factors normalize these functions.) Using (19.22) and its complex conjugate, we find (Prob. 19.19) that

$$2p_x = cxe^{-Zr/2a}, \qquad 2p_y = cye^{-Zr/2a} \tag{19.24}$$

where $c \equiv \pi^{-1/2}(Z/2a)^{5/2}$. The $2p_x$ and $2p_y$ functions have the same n and l values as the $2p_1$ and $2p_{-1}$ functions (namely, $n = 2$ and $l = 1$) but do not have a definite value of m. Similar linear combinations give real wave functions for higher H-atom states. Table 19.1 lists the $n = 1$ and $n = 2$ real hydrogenlike functions.

Orbitals. An *orbital* is a one-electron spatial wave function. Since a hydrogenlike atom has one electron, all the hydrogenlike wave functions are orbitals. The use of (one-electron) orbitals in many-electron atoms is considered later in this chapter.

The shape of an orbital is defined as a surface of constant probability density that encloses some large fraction (say 90 percent) of the probability for finding the electron. The probability density is $|\psi|^2$. When $|\psi|^2$ is constant, so is $|\psi|$. Hence, $|\psi|$ is constant on the surface of an orbital.

For an s orbital, ψ depends only on r, and $|\psi|$ is constant on the surface of a sphere with center at the nucleus. An s orbital has a spherical shape. The volume element in spherical polar coordinates (see any calculus text) is

$$d\tau = r^2 \sin\theta \, dr \, d\theta \, d\phi \tag{19.25*}$$

(This is the volume of an infinitesimal solid for which the spherical polar coordinates lie in the ranges r to $r + dr$, θ to $\theta + d\theta$, and ϕ to $\phi + d\phi$.) If we take an orbital surface to enclose 90 percent of the probability density, the radius r_{1s} of the 1s orbital satisfies

$$\int_0^{r_{1s}} |R_{1s}(r)|^2 r^2 \, dr \int_0^{\pi} |\Theta_{1s}(\theta)|^2 \sin\theta \, d\theta \int_0^{2\pi} |\Phi_{1s}(\phi)|^2 \, d\phi = 0.9$$

[see Eqs. (19.10), (19.11), and (19.25)]. One finds $r_{1s} = 1.4$ Å for H (Prob. 19.20).

Consider the shapes of the real 2p orbitals. The $2p_z$ orbital is $2p_z = cze^{-Zr/2a}$, where c is a constant. The $2p_z$ function is zero in the xy plane (where $z = 0$), is positive above this nodal plane, and negative below it. A detailed investigation (*Q. C.*, sec. 6.6) gives Fig. 19.5 as the cross section of a $2p_z$ orbital in the xz plane. The three-dimensional shape is obtained by rotating this cross section about the z axis; this gives two distorted ellipsoids, one above and one below the xy plane. The ellipsoids do not touch each other. (This is obvious from the fact that ψ has opposite signs on each ellipsoid.) The plus and minus signs in Fig. 19.5 give the signs of ψ (and have nothing to do with electrical charge). The absolute value $|\psi|$ is the same on each ellipsoid.

Table 19.1. Real hydrogenlike wave functions for $n = 1$ and $n = 2$

$$1s = \pi^{-1/2}(Z/a)^{3/2}e^{-Zr/a}$$
$$2s = \tfrac{1}{4}(2\pi)^{-1/2}(Z/a)^{3/2}(2 - Zr/a)e^{-Zr/2a}$$
$$2p_x = \tfrac{1}{4}(2\pi)^{-1/2}(Z/a)^{5/2}re^{-Zr/2a} \sin\theta \cos\phi$$
$$2p_y = \tfrac{1}{4}(2\pi)^{-1/2}(Z/a)^{5/2}re^{-Zr/2a} \sin\theta \sin\phi$$
$$2p_z = \tfrac{1}{4}(2\pi)^{-1/2}(Z/a)^{5/2}re^{-Zr/2a} \cos\theta$$

Figure 19.6 shows the shapes of some hydrogenlike orbitals. The $2p_x$, $2p_y$, and $2p_z$ orbitals have the same shape but different orientations in space; the two distorted ellipsoids are located on the x axis for the $2p_x$ orbital, on the y axis for the $2p_y$ orbital, and on the z axis for the $2p_z$ orbital. The $3p_z$ orbital has a spherical node (shown by the dashed line). The $3d_{z^2}$ orbital has two nodal cones (dashed lines). The other four $3d$ orbitals have the same shape as one another but different orientations; each of these orbitals has two nodal planes separating the four lobes.

Probability density. The electron probability density for the hydrogenlike-atom ground state (Table 19.1) is $|\psi_{1s}|^2 = (Z^3/\pi a^3)e^{-2Zr/a}$. The $1s$ probability density is a maximum at the nucleus ($r = 0$). Figure 19.7 is a schematic indication of this, the

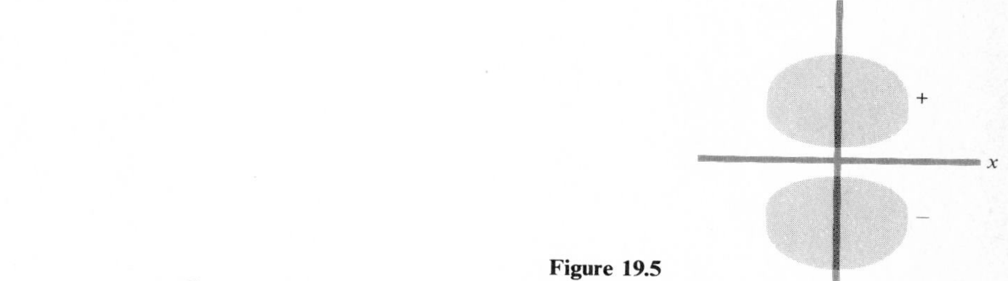

Figure 19.5
Cross section of the $2p_z$ orbital in the xz plane.

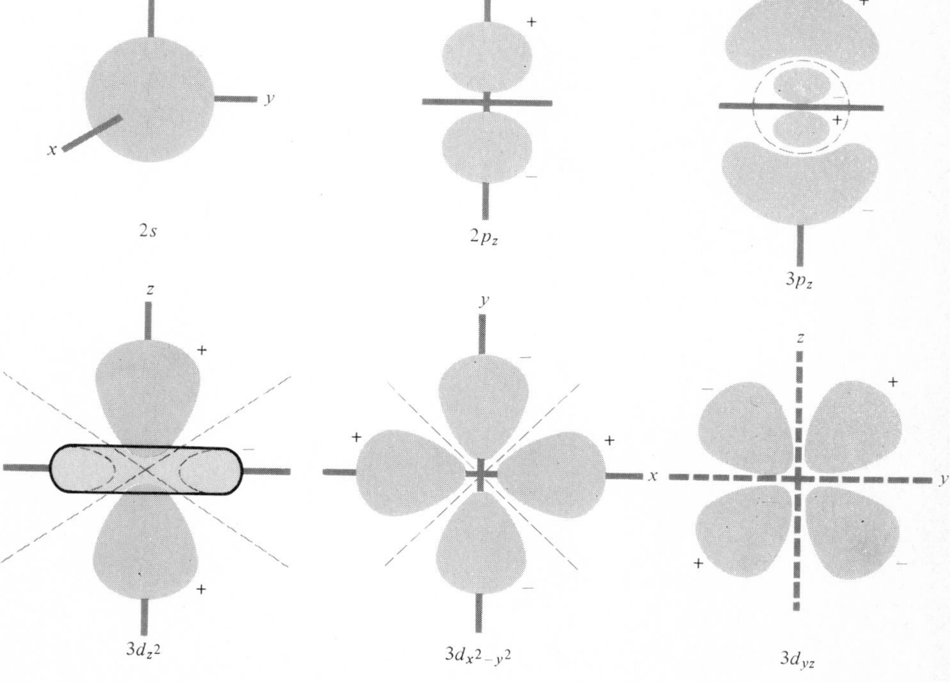

Figure 19.6
Shapes of some hydrogen-atom orbitals. (Not drawn to scale.) Note the different orientation of the axes in the $3d_{x^2-y^2}$ figure compared with the others. Not shown are the $3d_{xy}$ and $3d_{xz}$ orbitals; these have their lobes between the x and y axes and the x and z axes, respectively.

density of the dots indicating the relative probability densities in various regions. Since $|\psi_{1s}|^2$ is nonzero everywhere, the electron can be found at any location in the atom (in contrast to the Bohr theory, where it had to be at a fixed r). Figure 19.7 also indicates the variation in probability density for the $2s$ and $2p_z$ states. Note the nodal sphere in the $2s$ function.

Radial distribution function. Suppose we ask for the probability $p(r \to r + dr)$ that the electron–nucleus distance is between r and $r + dr$. This is the probability of finding the electron in a thin spherical shell whose center is at the nucleus and whose inner and outer radii are r and $r + dr$. For an s orbital, ψ is independent of θ and ϕ and so is essentially constant in the thin shell. Hence, the desired probability is found by multiplying $|\psi_s|^2$ (the probability per unit volume) by the volume of the thin shell. This volume is $\frac{4}{3}\pi(r + dr)^3 - \frac{4}{3}\pi r^3 = 4\pi r^2\,dr$, where the terms in $(dr)^2$ and $(dr)^3$ are negligible. Hence, for an s state $p(r \to r + dr) = 4\pi r^2 |\psi_s|^2\,dr$.

For a non-s state, ψ depends on the angles, so $|\psi|^2$ is not constant in the thin shell. Let us divide the shell into tiny volume elements such that the spherical polar coordinates range from r to $r + dr$, from θ to $\theta + d\theta$, and from ϕ to $\phi + d\phi$ in each tiny element. The volume $d\tau$ of each such element is given by (19.25), and the probability of the electron's being in an element is $|\psi|^2\,d\tau = |\psi|^2 r^2 \sin\theta\,dr\,d\theta\,d\phi$. To find $p(r \to r + dr)$ we must sum these infinitesimal probabilities over the thin shell. Since the shell goes over the full range of θ and ϕ, the desired sum is the definite integral over the angles. Hence, $p(r \to r + dr) = \int_0^{2\pi} \int_0^{\pi} |\psi|^2 r^2 \sin\theta\,dr\,d\theta\,d\phi$. Use of (19.11) gives

$$p(r \to r + dr) = |R|^2 r^2\,dr \int_0^{\pi} |\Theta|^2 \sin\theta\,d\theta \int_0^{2\pi} |\Phi|^2\,d\phi$$

$$p(r \to r + dr) = [R_{nl}(r)]^2 r^2\,dr \tag{19.26}$$

since the multiplicative constants in the Θ and Φ functions have been chosen to normalize Θ and Φ: $\int_0^{\pi} |\Theta|^2 \sin\theta\,d\theta = 1$ and $\int_0^{2\pi} |\Phi|^2\,d\phi = 1$ (Prob. 19.23). Equation (19.26) holds for both s and non-s states. The function $[R(r)]^2 r^2$ in (19.26) is the *radial distribution function* and is plotted in Fig. 19.8 for several states. For the ground state, the radial distribution function is a maximum at $r = a/Z$ (Prob. 19.21), which is 0.53 Å for H.

For the hydrogen-atom ground state, the probability density $|\psi|^2$ is a maximum at the origin (nucleus), but the radial distribution function $R^2 r^2$ is zero at the nucleus, due to the r^2 factor; the most probable value of r is 0.53 Å. A little thought shows that these facts are not contradictory. In finding $p(r \to r + dr)$ we find the probability that the electron is in a thin shell; this thin shell ranges over all values of θ and ϕ and so is composed of many volume elements. As r increases, the thin-shell volume $4\pi r^2\,dr$ increases. This increase, combined with the decrease in $|\psi|^2$ as r

1s

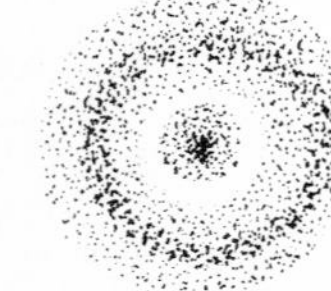

2s

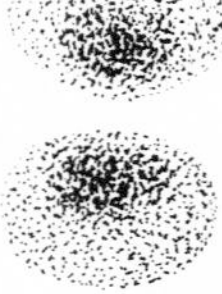

$2p_z$

Figure 19.7
Probability densities in three hydrogen-atom states. (Not drawn to scale.)

increases, gives a maximum in $p(r \to r + dr)$ for a value of r between 0 and ∞. The radial distribution function is zero at the nucleus because the thin-shell volume $4\pi r^2\, dr$ is zero here. (Note the resemblance to the discussion of the distribution function for speeds in a gas; Sec. 15.4.)

19.4 ELECTRON SPIN

The Schrödinger equation is a nonrelativistic equation. There are certain relativistic phenomena that the Schrödinger equation fails to take into account. In 1928, the British physicist P. A. M. Dirac discovered the correct relativistic quantum-mechanical equation for a one-electron system. Dirac's relativistic equation predicts the existence of electron spin. (Electron spin was first proposed by Uhlenbeck and Goudsmit in 1925 to explain certain observations in atomic spectra.) In the nonrelativistic Schrödinger version of quantum mechanics that we are using, the existence of electron spin must be tacked on to the theory as an additional postulate.

What is spin? Spin is an intrinsic (built-in) angular momentum possessed by elementary particles. This intrinsic angular momentum is in addition to the orbital angular momentum (Sec. 19.3) the particle has due to its motion through space. In a crude way, one can think of this intrinsic (or spin) angular momentum as being due to the particle's spinning about its own axis, but this picture should not be considered to represent reality; spin is a nonclassical effect.

Quantum mechanics shows that the magnitude of the orbital angular momentum $\mathbf{L}$ of any particle can take on only the values $[l(l+1)]^{1/2}\hbar$, where $l = 0, 1, 2, \ldots$; the z component L_z can take on only the values $m\hbar$, where $m = -l, \ldots, +l$. (We mentioned this in connection with the electron in the H atom, but it is a generally valid result.)

Let $\mathbf{S}$ be the spin-angular-momentum vector of an elementary particle. By analogy to orbital angular momentum, we postulate that the magnitude of $\mathbf{S}$ is $[s(s+1)]^{1/2}\hbar$ and that S_z, the component of the spin angular momentum along the z axis, can take on only the values

$$m_s\hbar \qquad \text{where } m_s = -s, -s+1, \ldots, s-1, s \tag{19.27}$$

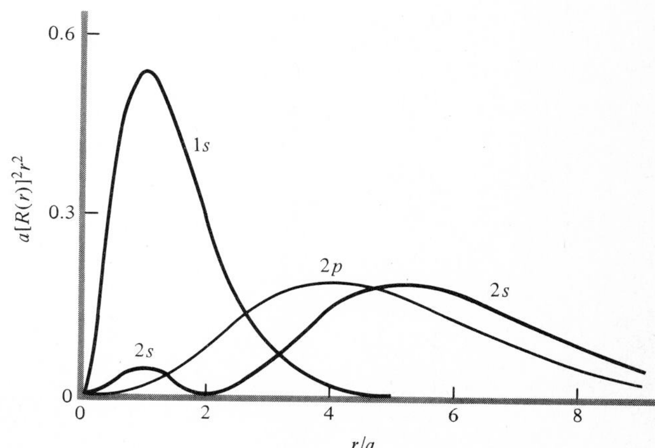

Figure 19.8
Radial distribution functions for some hydrogen-atom states.

The spin-angular-momentum quantum numbers s and m_s are analogous to the orbital-angular-momentum quantum numbers l and m, respectively. The analogy is not complete, in that it is an observed fact that a given species of elementary particle can have only one value for s and this value may be half-integral ($\frac{1}{2}, \frac{3}{2}, \ldots$) as well as integral ($0, 1, \ldots$). Experiment shows that electrons, protons, and neutrons (the elementary particles of greatest interest to chemists) all have $s = \frac{1}{2}$. Therefore, $m_s = -\frac{1}{2}$ or $+\frac{1}{2}$ for these particles. Figure 19.9 shows the orientations of the spin-angular-momentum vector for these two spin states. Chemists often use the symbols $\uparrow$ and $\downarrow$ to indicate the $m_s = +\frac{1}{2}$ and $m_s = -\frac{1}{2}$ states, respectively.

The wave function is supposed to describe the state of the system as fully as possible. An electron has two possible spin states, namely, $m_s = +\frac{1}{2}$ and $m_s = -\frac{1}{2}$, and the wave function should indicate which spin state the electron is in. We therefore postulate the existence of two spin functions α and β that indicate the electron's spin state: α means that m_s is $+\frac{1}{2}$; β means that m_s is $-\frac{1}{2}$. The spin functions α and β can be considered to be functions of some hypothetical internal coordinate ω of the electron: $\alpha = \alpha(\omega)$ and $\beta = \beta(\omega)$. Since little is known of the internal structure of an electron (or even whether it has an internal structure), ω is purely hypothetical.

For a one-electron system, the spatial wave function $\psi(x, y, z)$ is multiplied by either α or β to form the complete wave function including spin. To a very good approximation, the spin has no effect on the energy of a one-electron system. For the hydrogen atom, the electron spin simply doubles the degeneracy of each level. For the H-atom ground level, there are two possible wave functions, $1s\alpha$ and $1s\beta$, where $1s = \pi^{-1/2}(Z/a)^{3/2}e^{-Zr/a}$. A one-electron wave function like $1s\alpha$ or $1s\beta$ that includes both spatial and spin functions is called a *spin-orbital.*

With the inclusion of spin in the wave function, ψ of an n-electron system becomes a function of $4n$ variables: $3n$ spatial variables and n spin variables. The normalization condition (18.12) must be amended to include an integration (or summation) over the spin variables as well as over the spatial variables. (See *Q. C.*, sec. 10.1.)

19.5 THE HELIUM ATOM AND THE PAULI PRINCIPLE

The helium atom. The helium atom consists of two electrons and a nucleus (Fig. 19.10). Separation of the translational energy of the atom as a whole from the internal motion is more complicated than for a two-particle problem and won't be

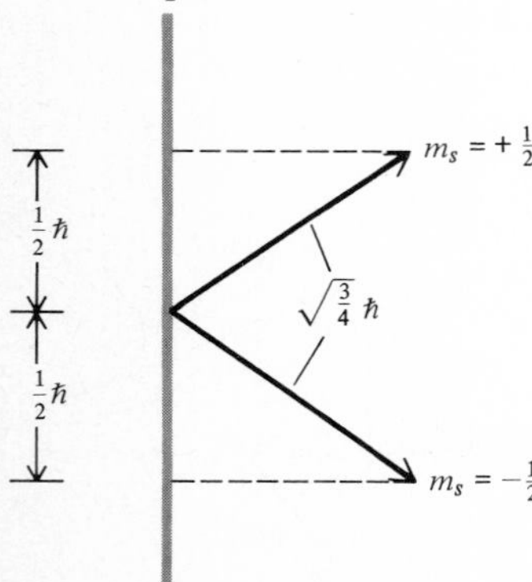

Figure 19.9
Orientations of the electron spin vector **S** with respect to the z axis. (For $m_s = +\frac{1}{2}$, the vector **S** must lie on the surface of a cone about the z axis; similarly for $m_s = -\frac{1}{2}$.)

gone into here. We shall just assume that it is possible to separate off the translational motion from the internal motions.

The Hamiltonian operator for the internal motions in a heliumlike atom is

$$\hat{H} = -\frac{\hbar^2}{2m}\nabla_1^{\ 2} - \frac{\hbar^2}{2m}\nabla_2^{\ 2} - \frac{Ze'^2}{r_1} - \frac{Ze'^2}{r_2} + \frac{e'^2}{r_{12}} \qquad (19.28)$$

The first term is the operator for the kinetic energy of electron 1; in this term, $\nabla_1^{\ 2} \equiv \partial^2/\partial x_1^{\ 2} + \partial^2/\partial y_1^{\ 2} + \partial^2/\partial z_1^{\ 2}$ (where x_1, y_1, z_1 are the coordinates of electron 1, the origin being taken at the nucleus) and m is the electron mass. (It would be more accurate to replace m by the reduced mass μ, but μ differs almost negligibly from m for He and heavier atoms.) The second term is the operator for the kinetic energy of electron 2, and $\nabla_2^{\ 2} \equiv \partial^2/\partial x_2^{\ 2} + \partial^2/\partial y_2^{\ 2} + \partial^2/\partial z_2^{\ 2}$. The third term is the potential energy of interaction between electron 1 and the nucleus and is obtained by putting $Q_1' = -e'$ and $Q_2' = Ze'$ in (19.4); for helium, the atomic number Z is 2; in this term, r_1 is the distance between electron 1 and the nucleus: $r_1^{\ 2} = x_1^{\ 2} + y_1^{\ 2} + z_1^{\ 2}$. The fourth term is the potential energy of interaction between electron 2 and the nucleus. The last term is the potential energy of interaction between electrons 1 and 2 separated by distance r_{12} and is found by putting $Q_1' = Q_2' = -e'$ in (19.4).

The Schrödinger equation is $\hat{H}\psi = E\psi$, where ψ is a function of the spatial coordinates of the electrons relative to the nucleus: $\psi = \psi(x_1, y_1, z_1, x_2, y_2, z_2)$, or $\psi = \psi(r_1, \theta_1, \phi_1, r_2, \theta_2, \phi_2)$ if spherical polar coordinates are used. Electron spin is being ignored for now and will be taken care of later.

Because of the interelectronic repulsion term e'^2/r_{12}, the helium-atom Schrödinger equation can't be solved exactly. As a crude approximation, we can ignore the e'^2/r_{12} term. The Hamiltonian then has the approximate form $\hat{H}_{\text{approx}} = \hat{H}_1 + \hat{H}_2$, where $\hat{H}_1 \equiv -(\hbar^2/2m)\nabla_1^{\ 2} - Ze'^2/r_1$ is a hydrogenlike Hamiltonian for electron 1 and $\hat{H}_2 \equiv -(\hbar^2/2m)\nabla_2^{\ 2} - Ze'^2/r_2$ is a hydrogenlike Hamiltonian for electron 2. Since $\hat{H}_{\text{approx}}$ is the sum of Hamiltonians for two noninteracting particles, Eqs. (18.58) to (18.60) apply. These equations give $E \approx E_1 + E_2$ and $\psi \approx \psi_1(r_1, \theta_1, \phi_1) \times \psi_2(r_2, \theta_2, \phi_2)$, where $\hat{H}_1\psi_1 = E_1\psi_1$ and $\hat{H}_2\psi_2 = E_2\psi_2$. Since $\hat{H}_1$ and $\hat{H}_2$ are hydrogenlike Hamiltonians, E_1 and E_2 are hydrogenlike energies and ψ_1 and ψ_2 are hydrogenlike wave functions (orbitals).

Let us check the accuracy of this approximation. Equations (19.17) to (19.20) give $E_1 = -(Z^2/n_1^{\ 2})(e'^2/2a_0) = -(Z^2/n_1^{\ 2})(13.6\text{ eV})$, where n_1 is the principal quantum number of electron 1 and a has been replaced by the Bohr radius a_0, since the reduced mass μ was replaced by the electron mass in (19.28). A similar equation holds for E_2. For the helium ground state, the principal quantum numbers of the electrons are $n_1 = 1$ and $n_2 = 1$; also, $Z = 2$. Hence, $E \approx E_1 + E_2 = -4(13.6\text{ eV}) - 4(13.6\text{ eV}) = -108.8$ eV. The experimental first and second ionization potentials of

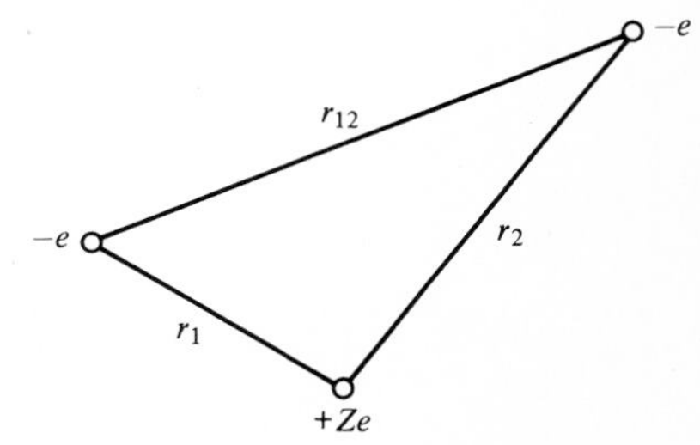

Figure 19.10
The heliumlike atom.

He are 24.6 and 54.4 V, so the ground-state energy is -79.0 eV. (The first and second ionization energies are the energy changes for the processes $He \rightarrow He^+ + e^-$ and $He^+ \rightarrow He^{2+} + e^-$, respectively.) The approximate result -108.8 eV is grossly in error, as might be expected from the fact that the e'^2/r_{12} term we ignored is not small.

The approximate ground-state wave function is given by Eq. (18.58) and Table 19.1 as

$$\psi \approx (Z/a_0)^{3/2}\pi^{-1/2}e^{-Zr_1/a_0} \cdot (Z/a_0)^{3/2}\pi^{-1/2}e^{-Zr_2/a_0} \tag{19.29}$$

with $Z = 2$. We shall abbreviate (19.29) to

$$\psi \approx 1s(1)1s(2) \tag{19.30}$$

where $1s(1)$ indicates that electron 1 is in a $1s$ hydrogenlike orbital (one-electron spatial wave function).

Two-electron spin functions. To be fully correct, electron spin must be included in the wave function. One's first impulse might be to write down the following four spin functions for two-electron systems:

$$\alpha(1)\alpha(2), \qquad \beta(1)\beta(2), \qquad \alpha(1)\beta(2), \qquad \beta(1)\alpha(2) \tag{19.31}$$

where the notation $\beta(1)\alpha(2)$ means electron 1 has its spin quantum number m_{s1} equal to $-\frac{1}{2}$ and electron 2 has $m_{s2} = +\frac{1}{2}$. However, the last two functions in (19.31) are invalid because they distinguish between the electrons. Electrons are identical to one another, and there is no way of experimentally determining which electron has $m_s = +\frac{1}{2}$ and which has $m_s = -\frac{1}{2}$. In classical mechanics, we can distinguish two identical particles from each other by following their paths. However, the Heisenberg uncertainty principle makes it impossible to follow the path of a particle in quantum mechanics. Hence, the wave function must not distinguish between the electrons. Thus, the fourth spin function in (19.31), which says that electron 1 has spin β and electron 2 has spin α, cannot be used. Instead of the third and fourth spin functions in (19.31), it turns out (see below for the justification) that one must use the functions $2^{-1/2}[\alpha(1)\beta(2) - \beta(1)\alpha(2)]$ and $2^{-1/2}[\alpha(1)\beta(2) + \beta(1)\alpha(2)]$. For each of these functions, electron 1 has both spin α and spin β, and so does electron 2. The $2^{-1/2}$ in these functions is a normalization constant. The proper two-electron spin functions are therefore

$$\alpha(1)\alpha(2), \qquad \beta(1)\beta(2), \qquad 2^{-1/2}[\alpha(1)\beta(2) + \beta(1)\alpha(2)] \tag{19.32}$$

$$2^{-1/2}[\alpha(1)\beta(2) - \beta(1)\alpha(2)] \tag{19.33}$$

The three spin functions in (19.32) are unchanged when electrons 1 and 2 are interchanged. For example, interchanging the electrons in the third function gives $2^{-1/2}[\alpha(2)\beta(1) + \beta(2)\alpha(1)]$, which equals the original function. These three spin functions are said to be *symmetric* with respect to electron interchange. The spin function (19.33) is multiplied by -1 when the electrons are interchanged, since interchange gives $2^{-1/2}[\alpha(2)\beta(1) - \beta(2)\alpha(1)] = -2^{-1/2}[\alpha(1)\beta(2) - \beta(1)\alpha(2)]$. The function (19.33) is *antisymmetric*.

The Pauli principle. Before including spin in the ground-state helium wave function, we shall introduce some definitions and one final quantum-mechanical postulate.

A particle whose spin quantum number s is half-integral ($\frac{1}{2}$ or $\frac{3}{2}$ or $\frac{5}{2}$ or ...) is called a *fermion* (after the Italian–American physicist Enrico Fermi). A particle whose s is integral (0 or 1 or 2 or ...) is called a *boson* (after the Indian physicist S. N. Bose). Electrons have $s = \frac{1}{2}$ and are fermions.

Since two identical particles cannot be distinguished from each other in quantum mechanics, interchange of two identical particles in the wave function must leave all physically observable properties unchanged. In particular, the probability density $|\psi|^2$ must be unchanged. We therefore expect that ψ itself would be multiplied by either $+1$ or -1 by such an interchange or relabeling. It turns out that only one of these possibilities occurs, depending on the nature of the particles. Experimental evidence shows the validity of the following statement:

The complete wave function (including both spatial and spin coordinates) of a system of identical fermions must be antisymmetric with respect to interchange of all the coordinates (spatial and spin) of any two particles. For a system of identical bosons, the complete wave function must be symmetric with respect to such interchange.

This fact, discovered by Dirac and by Heisenberg in 1926, is called the *Pauli principle*. In 1940, Pauli deduced the Pauli principle from relativistic quantum field theory. In the nonrelativistic version of quantum mechanics that we are using, the Pauli principle must be regarded as an additional postulate.

We are now ready to include spin in the ground-state He wave function. The approximate ground-state spatial function $1s(1)1s(2)$ of (19.30) is symmetric with respect to electron interchange, since $1s(2)1s(1) = 1s(1)1s(2)$. The Pauli principle demands that the complete wave function be antisymmetric. To get an antisymmetric ψ, we must multiply $1s(1)1s(2)$ by the antisymmetric function (19.33). [Use of the symmetric spin functions in (19.32) would give a symmetric wave function, which is forbidden for fermions.] Hence, with inclusion of spin, the approximate ground-state He wave function becomes

$$\psi \approx 1s(1)1s(2) \cdot 2^{-1/2}[\alpha(1)\beta(2) - \beta(1)\alpha(2)] \tag{19.34}$$

Interchange of the electrons multiplies ψ by -1. Note that the two electrons in the $1s$ orbital have opposite spins.

The wave function (19.34) can be written as the determinant

$$\psi \approx \frac{1}{\sqrt{2}} \begin{vmatrix} 1s(1)\alpha(1) & 1s(1)\beta(1) \\ 1s(2)\alpha(2) & 1s(2)\beta(2) \end{vmatrix} \tag{19.35}$$

A second-order determinant is defined by

$$\begin{vmatrix} a & b \\ c & d \end{vmatrix} \equiv ad - bc \tag{19.36}$$

Use of (19.36) in (19.35) gives (19.34).

The justification for replacing the third and fourth spin functions in (19.31) by the linear combinations in (19.32) and (19.33) is that the latter two functions are the only normalized linear combinations of $\alpha(1)\beta(2)$ and $\beta(1)\alpha(2)$ which are either symmetric or antisymmetric with respect to electron interchange and which therefore don't distinguish between the electrons.

Improved ground-state wave functions for helium. For one- and two-electron systems, the wave function is a product of a spatial factor and a spin factor. The atomic Hamiltonian (to a very good approximation) contains no terms involving spin. Because of these facts, the spin part of the wave function need not be explicitly included in calculating the energy of one- and two-electron systems and will be omitted in the calculations below. (See also *Q. C.*, sec. 10.5.)

We saw above that ignoring the e'^2/r_{12} term in $\hat{H}$ and taking E as the sum of two hydrogenlike energies gave a 38 percent error in the ground-state He energy. To improve on this dismal result, we can use the variational method. The most obvious choice of variational function is (19.29) and (19.30), which is a normalized product of hydrogenlike 1s orbitals. The variational integral in (18.77) is then $W = \int 1s(1)1s(2)\hat{H}1s(1)1s(2)\,d\tau$, where $\hat{H}$ is the true Hamiltonian (19.28) and $d\tau = d\tau_1\,d\tau_2$, with $d\tau_1 = r_1^2 \sin\theta_1\,dr_1\,d\theta_1\,d\phi_1$. Since e'^2/r_{12} is part of $\hat{H}$, the effect of the interelectronic repulsion will be included in an average way. Evaluation of the variational integral is complicated and is omitted here. (See *Q. C.*, sec. 9.4.) The result is $W = -74.8$ eV, which is reasonably close to the true ground-state energy -79.0 eV.

A further improvement is to use a variational function having the same form as (19.29) and (19.30) but with the nuclear charge Z replaced by a variational parameter ζ (zeta). We then vary ζ to minimize the variational integral $W = \int \phi^*\hat{H}\phi\,d\tau$, where the normalized variation function ϕ is $(\zeta/a_0)^3\pi^{-1}e^{-\zeta r_1/a_0}e^{-\zeta r_2/a_0}$. Substitution of (19.28) with $Z = 2$ and evaluation of the integrals leads to $W = (\zeta^2 - 27\zeta/8)e'^2/a_0$. (See *Q. C.*, sec. 9.4 for the details.) The minimization condition $\partial W/\partial\zeta = 0$ then gives the optimum value of ζ as $27/16 = 1.6875$. With this value of ζ, we get $W = -2.848e'^2/a_0 = -2.848(2 \times 13.6\text{ eV}) = -77.5$ eV, where (19.19) was used. This result is only 2 percent above the true ground-state energy -79.0 eV.

The parameter ζ is called an *orbital exponent*. The fact that ζ is less than the atomic number $Z = 2$ can be attributed to the "shielding" of one electron from the nucleus by the other electron. When electron 1 is between electron 2 and the nucleus, the repulsion between electrons 1 and 2 subtracts from the attraction between electron 2 and the nucleus. Thus, ζ can be viewed as the "effective" nuclear charge for the 1s electrons. Since both electrons are in the same orbital, the screening effect is not great and ζ is only 0.31 less than Z.

By using complicated variational functions, workers have obtained agreement to 1 part in 2 million between the theoretical and the experimental ionization energies of ground-state He. (See *Q. C.*, sec. 9.4.)

Excited-state wave functions for helium. We saw that the approximation of ignoring the interelectronic repulsion in the Hamiltonian gives the helium wave functions as products of two hydrogenlike functions. The hydrogenlike 2s and 2p orbitals have the same energy, and we might expect the approximate spatial wave functions for the lowest excited energy level of He to be $1s(1)2s(2)$, $1s(2)2s(1)$, $1s(1)2p_x(2)$, $1s(2)2p_x(1)$, $1s(1)2p_y(2)$, $1s(2)2p_y(1)$, $1s(1)2p_z(2)$, $1s(2)2p_z(1)$, where $1s(2)2p_z(1)$ is a function with electron 2 in the 1s orbital and electron 1 in the $2p_z$ orbital. Actually, these functions are incorrect, in that they distinguish between the electrons. As we did above with the spin functions, we must take linear combinations to give functions that don't distinguish between the electrons. Analogous to the linear combinations in (19.32) and (19.33), the correct normalized approximate spatial functions are

$$2^{-1/2}[1s(1)2s(2) + 1s(2)2s(1)] \tag{19.37}$$

$$2^{-1/2}[1s(1)2s(2) - 1s(2)2s(1)] \tag{19.38}$$

$$2^{-1/2}[1s(1)2p_x(2) + 1s(2)2p_x(1)] \quad \text{etc.} \tag{19.39}$$

$$2^{-1/2}[1s(1)2p_x(2) - 1s(2)2p_x(1)] \quad \text{etc.} \tag{19.40}$$

where each etc. indicates two similar functions with $2p_x$ replaced by $2p_y$ and by $2p_z$.

If ψ_k is the true wave function of state k of a system, then $\hat{H}\psi_k = E_k\psi_k$, where E_k is the energy of state k. We therefore have $\int \psi_k^* \hat{H}\psi_k \, d\tau = \int \psi_k^* E_k \psi_k \, d\tau = E_k \int \psi_k^* \psi_k \, d\tau = E_k$, since ψ_k is normalized. This result suggests that if we have an approximate wave function $\psi_{k,\text{approx}}$ for state k, an approximate energy can be obtained by replacing ψ_k with $\psi_{k,\text{approx}}$ in the integral:

$$E_k \approx \int \psi_{k,\text{approx}}^* \hat{H} \psi_{k,\text{approx}} \, d\tau \tag{19.41}$$

where $\hat{H}$ is the true Hamiltonian, including the interelectronic repulsion term(s).

Use of the eight approximate functions (19.37) to (19.40) in Eq. (19.41) then gives approximate energies for these states. Because of the difference in signs, the two states (19.37) and (19.38) that arise from the $1s2s$ configuration will clearly have different energies. The state (19.38) turns out to be lower in energy. Our approximate wave functions are real, and the contribution of the e'^2/r_{12} interelectronic repulsion term in $\hat{H}$ to the integral in Eq. (19.41) is $\int \psi_{k,\text{approx}}^2 e'^2/r_{12} \, d\tau$. One finds that the integral $\int (19.37)^2 e'^2/r_{12} \, d\tau$ differs in value from $\int (19.39)^2 e'^2/r_{12} \, d\tau$, where "(19.37)" and "(19.39)" stand for the functions in Eqs. (19.37) and (19.39). (For the details, see *Q. C.*, sec. 9.7.) Hence, the $1s2p$ state in (19.39) differs in energy from the corresponding $1s2s$ state in (19.37). Likewise, the states (19.38) and (19.40) differ in energy from each other. Since the $2p_x$, $2p_y$, and $2p_z$ orbitals have the same shape, changing $2p_x$ to $2p_y$ or $2p_z$ in (19.39) or (19.40) doesn't affect the energy.

Thus, the states of the $1s2s$ configuration give two different energies and the states of the $1s2p$ configuration give two different energies, for a total of four different energies (Fig. 19.11). The $1s2s$ states turn out to lie lower in energy than the $1s2p$ states. Although the $2s$ and $2p$ orbitals have the same energy in one-electron (hydrogenlike) atoms, the interelectronic repulsion(s) in atoms with two or more electrons remove the $2s$-$2p$ degeneracy. The reason the $2s$ orbital lies below the $2p$ orbital can be seen from Figs. 19.8 and 19.3. The $2s$ orbital has more probability density near the nucleus than the $2p$ orbital. Thus, a $2s$ electron is more likely than a $2p$ electron to penetrate within the probability density of the $1s$ electron. When it does so penetrate, it is no longer shielded from the nucleus and feels the full nuclear charge, thereby lowering its energy. Similar penetration effects remove the l degeneracy in higher orbitals. For example, $3s$ lies lower than $3p$, which lies lower than $3d$.

What about electron spin? The function (19.37) is symmetric with respect to electron interchange and so must be combined with the antisymmetric two-electron spin function (19.33) to give an overall ψ that is antisymmetric: $\psi \approx (19.37) \times (19.33)$. The function (19.38) is antisymmetric and so must be combined with one of the symmetric spin functions in (19.32); because there are three symmetric spin functions, inclusion of spin in (19.38) gives three different wave functions, each having the same spatial factor. Since the spin factor doesn't affect the energy, there is a threefold spin degeneracy associated with the function (19.38). Similar considerations hold for the $1s2p$ states.

Figure 19.11 shows the energies and some of the approximate wave functions for the states arising from the $1s2s$ and $1s2p$ configurations. The labels ^3S, ^1S, ^3P, ^1P are explained below. The atomic energies shown in Fig. 19.11 are called *terms*, rather than energy levels, for a reason to be explained later. The ^3S term of the $1s2s$ configuration is threefold degenerate, due to the three symmetric spin functions. The ^1S term is onefold degenerate (i.e., nondegenerate), since there is only one wave function for this term. The approximate wave functions for the ^3P term are obtainable from those of the ^3S term by replacing $2s$ by $2p_x$, by $2p_y$, and by $2p_z$. Each of these replacements gives three wave functions (due to the three symmetric spin functions), so the ^3P term is ninefold degenerate. The ^1P term is threefold degenerate, since three functions are obtained by replacement of $2s$ in the ^1S function with $2p_x$, with $2p_y$, and with $2p_z$.

The magnitude of the total electronic orbital angular momentum of an atomic state has the possible values $[L(L+1)]^{1/2}\hbar$, where the quantum number L can be 0, 1, 2, ... The value of L is indicated by a code similar to (19.16), except that capital letters are used. For the $1s2s$ configuration, both electrons have $l = 0$; hence, the total orbital angular momentum is zero, and the code letter S is used, as in Fig. 19.11. For the $1s2p$ configuration, one electron has $l = 0$ and one has $l = 1$; hence, the total-orbital-angular-momentum quantum number L equals 1, and the code letter P is used.

The total electronic spin angular momentum of an atom (or molecule) is the vector sum of the spin angular momenta of the individual electrons. The magnitude of the total electronic spin angular momentum has the possible values $[S(S+1)]^{1/2}\hbar$, where the total electronic spin quantum number S can be $0, \frac{1}{2}, 1, \frac{3}{2}, \ldots$. (Don't confuse the spin quantum number S with the orbital-angular-momentum code letter S.) The component of the total electronic spin angular momentum along the z axis has the possible values $M_S\hbar$, where $M_S = -S, -S+1, \ldots, S-1, S$.

For a two-electron system, each electron has spin quantum number $s = \frac{1}{2}$, and the total spin quantum number S can be 0 or 1, depending on whether the two electron spin vectors point in opposite directions or in approximately the same direction. [For $S = 1$, the total spin angular momentum is $(1 \cdot 2)^{1/2}\hbar = 1.414\hbar$. The spin angular momentum of each electron is $(\frac{1}{2} \cdot \frac{3}{2})^{1/2}\hbar = 0.866\hbar$. The algebraic sum of the spin angular momenta of the individual electrons is $0.866\hbar + 0.866\hbar = 1.732\hbar$, which is greater than the magnitude of the total spin angular momentum. Hence, the two spin-angular-momentum vectors of the electrons cannot be exactly parallel; see, for example, Fig. 19.12.]

For $S = 0$, M_S must be zero, and there is only one possible spin state. This spin state corresponds to the antisymmetric spin function (19.33).

For $S = 1$, M_S can be -1, 0, or $+1$. The $M_S = -1$ spin state arises when each electron has $m_s = -\frac{1}{2}$, and so corresponds to the symmetric spin function $\beta(1)\beta(2)$ in (19.32). The $M_S = +1$ spin state corresponds to the function $\alpha(1)\alpha(2)$. For the $M_S = 0$ state, one electron must have $m_s = +\frac{1}{2}$ and the other $m_s = -\frac{1}{2}$; this is the function $2^{-1/2}[\alpha(1)\beta(2) + \beta(1)\alpha(2)]$ in (19.32). Although the z components of the two

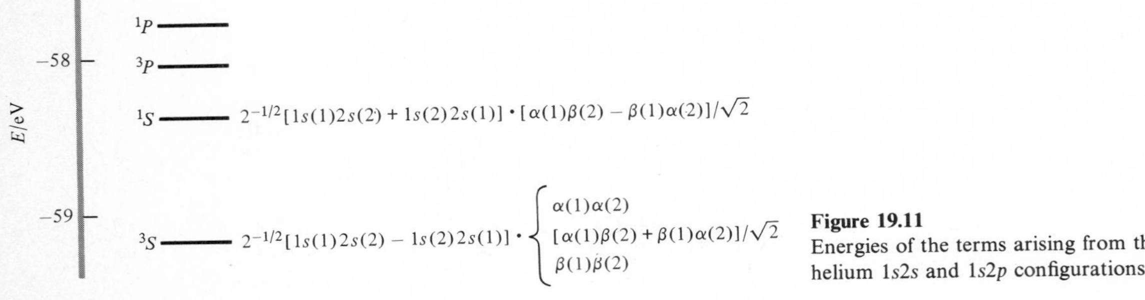

Figure 19.11
Energies of the terms arising from the helium $1s2s$ and $1s2p$ configurations.

electron spins are in opposite directions, the two spin vectors can still add to give a total electronic spin with $S = 1$, as shown in Fig. 19.12*a*. The three spin functions in (19.32) thus correspond to $S = 1$.

The quantity $2S + 1$ (where S is the total spin quantum number) is called the *multiplicity* of an atomic term and is written as a left superscript to the code letter for L. The lowest term in Fig. 19.11 has spin wave functions which correspond to total spin quantum number $S = 1$; hence, $2S + 1 = 3$ for this term, and the term is designated ^3S (read "triplet ess"). The second-lowest term in Fig. 19.11 has the $S = 0$ spin function and so has $2S + 1 = 1$; this is a ^1S ("singlet ess") term.

Note that the triplet term of the $1s2s$ configuration lies lower than the singlet term. The same is true for the terms of the $1s2p$ configuration. This illustrates *Hund's rule: For a set of terms arising from a given electron configuration, the lowest-lying term is generally the one with the maximum multiplicity*. There are a number of exceptions to Hund's rule. [The theoretical basis for Hund's rule is the subject of current research; see R. L. Snow and J. L. Bills, *J. Chem. Educ.*, **51**, 585 (1974) and references cited therein.]

The maximum multiplicity corresponds to the maximum number of electrons with parallel spins. Two electrons are said to have *parallel* spins when their spin-angular-momentum vectors point in approximately the same direction, as, for example, in Fig. 19.12*a*. Two electrons have *antiparallel* spins when their spin vectors point in opposite directions to give a net spin angular momentum of zero, as in Fig. 19.12*b*. The ^3S and ^1S terms of the $1s2s$ configuration can be represented by the diagrams

$$^3S\!: \ \underset{1s}{\uparrow} \quad \underset{2s}{\uparrow} \qquad\qquad {}^1S\!: \ \underset{1s}{\uparrow} \quad \underset{2s}{\downarrow}$$

where the spins are parallel in ^3S and antiparallel in ^1S.

It should be mentioned that the Hamiltonian (19.28) is not quite complete in that it omits a term called the *spin–orbit interaction* arising from the interaction between the spin and orbital motions of the electrons. (See *Q. C.*, sec. 11.6.) The spin–orbit interaction is quite small (except in heavy atoms), but it causes a partial removal of the degeneracy of a term, splitting an atomic term into a number of closely spaced energy levels. For example, the ^3P term in Fig. 19.11 is split into three closely spaced levels; the other three terms are each slightly shifted in energy by the spin–orbit interaction but are not split. Because of this spin–orbit splitting, the energies shown in Fig. 19.11 do not quite correspond to the actual pattern of atomic energy levels, and the energies in this figure are therefore called terms rather than energy levels.

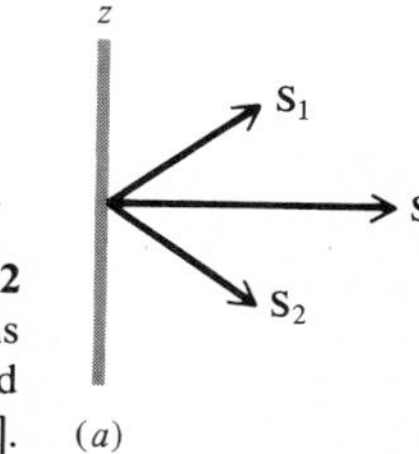

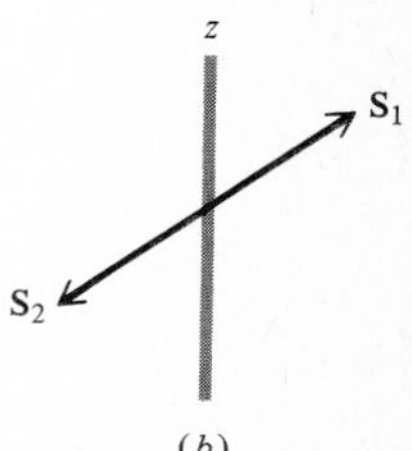

Figure 19.12
Spin orientations corresponding to the spin functions (*a*) $2^{-1/2}[\alpha(1)\beta(2) + \beta(1)\alpha(2)]$ and (*b*) $2^{-1/2}[\alpha(1)\beta(2) - \beta(1)\alpha(2)]$.

Finally, we emphasize that the helium wave functions (19.34) and (19.37) to (19.40) are only approximations, since at best they take account of the interelectronic repulsion in only an average way. Thus, to say that the helium ground state has the electron configuration $1s^2$ is only approximately true. The use of orbitals (one-electron wave functions) in many-electron atoms is only an approximation. (See also Sec. 19.7.)

19.6 MANY-ELECTRON ATOMS AND THE PERIODIC TABLE

Lithium and the Pauli exclusion principle. As we did with helium, we can omit the interelectronic repulsion terms $e'^2/r_{12} + e'^2/r_{13} + e'^2/r_{23}$ from the lithium-atom Hamiltonian to give an approximate Hamiltonian that is the sum of three hydrogen-like Hamiltonians. The approximate wave function is then the product of hydrogen-like (one-electron) wave functions. For the ground state we might expect the approximate wave function $1s(1)1s(2)1s(3)$. However, we have not taken account of electron spin or the Pauli principle. The symmetric spatial function $1s(1)1s(2)1s(3)$ must be multiplied by an antisymmetric three-electron spin function. One finds, however, that it is quite impossible to write an antisymmetric spin function for three electrons. With three or more electrons, the antisymmetry requirement of the Pauli principle cannot be satisfied by writing a wave function that is the product of separate spatial and spin factors.

The clue to constructing an antisymmetric wave function for three or more electrons lies in Eq. (19.35), which shows that the ground-state wave function of helium can be written as a determinant. The reason a determinant gives an antisymmetric wave function follows from the theorem that: Interchange of two rows of a determinant changes the sign of the determinant. (For a proof, see *Sokolnikoff and Redheffer*, app. A.) Interchange of rows 1 and 2 of the determinant in (19.35) amounts to interchanging the electrons. Thus a determinantal ψ is multiplied by -1 by such an interchange and therefore satisfies the Pauli antisymmetry requirement.

Let f, g, and h be three spin-orbitals. (Recall that a spin-orbital is the product of a spatial orbital and a spin factor; Sec. 19.4.) We can get an antisymmetric three-electron wave function by writing the following determinant (called a *Slater determinant*, after the American physicist who pointed out its use in constructing antisymmetric wave functions):

$$\frac{1}{\sqrt{6}}\begin{vmatrix} f(1) & g(1) & h(1) \\ f(2) & g(2) & h(2) \\ f(3) & g(3) & h(3) \end{vmatrix} \tag{19.42}$$

(The $1/\sqrt{6}$ is a normalization constant; there are six terms in the expansion of this determinant.)

A third-order determinant is defined by

$$\begin{vmatrix} a & b & c \\ d & e & f \\ g & h & i \end{vmatrix} \equiv a\begin{vmatrix} e & f \\ h & i \end{vmatrix} - b\begin{vmatrix} d & f \\ g & i \end{vmatrix} + c\begin{vmatrix} d & e \\ g & h \end{vmatrix}$$

$$= aei - ahf - bdi + bgf + cdh - cge \tag{19.43}$$

where (19.36) was used. (The second-order determinant that multiplies a in the expansion is found by striking out the row and the column that contains a in the third-order determinant; similarly for the multipliers of $-b$ and c.) The reader can verify that interchange of two rows multiplies the determinant's value by -1.

To get an antisymmetric approximate wave function for Li, we use (19.42). Let us try to put all three electrons in the 1s orbital by taking the spin-orbitals to be $f = 1s\alpha$, $g = 1s\beta$, and $h = 1s\alpha$. The determinant (19.42) becomes

$$\frac{1}{\sqrt{6}}\begin{vmatrix} 1s(1)\alpha(1) & 1s(1)\beta(1) & 1s(1)\alpha(1) \\ 1s(2)\alpha(2) & 1s(2)\beta(2) & 1s(2)\alpha(2) \\ 1s(3)\alpha(3) & 1s(3)\beta(3) & 1s(3)\alpha(3) \end{vmatrix} \tag{19.44}$$

Expansion of this determinant using (19.43) shows it to equal zero. This can be seen without multiplying out the determinant by using the following theorem (*Sokolnikoff and Redheffer*, app. A): If two columns of a determinant are identical, the determinant equals zero. The first and third columns of (19.44) are identical, and so (19.44) vanishes. If any two of the spin-orbitals f, g, h in (19.42) are the same, two columns of the determinant are the same and the determinant vanishes. Of course, zero is ruled out as a possible wave function (since there would then be no probability of finding the electrons).

The Pauli-principle requirement that the electronic wave function be antisymmetric thus leads to the conclusion that:

No more than one electron can occupy a given spin-orbital.

This is the *Pauli exclusion principle* (first stated by Pauli in 1925). An orbital (or one-electron spatial wave function) is defined by giving its three quantum numbers (n, l, m in an atom); a spin-orbital is defined by giving the three quantum numbers of the orbital and the m_s quantum number ($+\frac{1}{2}$ for spin function α, $-\frac{1}{2}$ for β). Thus, in an atom, the exclusion principle requires that no two electrons have the same values for all four quantum numbers n, l, m, and m_s.

The antisymmetry requirement holds for any system of identical fermions (Sec. 19.5), so in a system of identical fermions each one-particle state (spin-orbital) can hold no more than one fermion. In contrast, ψ is symmetric for bosons, so there is no limit to the number of bosons that can occupy a given one-particle state.

Coming back to the Li ground state, we can put two electrons with opposite spins in the 1s orbital ($f = 1s\alpha$, $g = 1s\beta$), but to avoid violating the exclusion principle, the third electron must go in the 2s orbital ($h = 2s\alpha$ or $2s\beta$). The approximate ground-state wave function is therefore

$$\psi \approx \frac{1}{\sqrt{6}}\begin{vmatrix} 1s(1)\alpha(1) & 1s(1)\beta(1) & 2s(1)\alpha(1) \\ 1s(2)\alpha(2) & 1s(2)\beta(2) & 2s(2)\alpha(2) \\ 1s(3)\alpha(3) & 1s(3)\beta(3) & 2s(3)\alpha(3) \end{vmatrix} \tag{19.45}$$

When (19.45) is multiplied out, it becomes a sum of six terms, each containing a spatial and a spin factor, so that ψ cannot be written as a single spatial factor times a single spin factor. Of course, because the 2s electron could have been given spin β, the ground state is doubly degenerate.

Since all the electrons are s electrons with $l = 0$, the total-orbital-angular-momentum quantum number L is 0. The 1s electrons have antiparallel spins, so the total electronic spin of the atom is due to the 2s electron and the total-electronic-spin quantum number S is $\frac{1}{2}$. The multiplicity $2S + 1$ is 2, and the ground term of Li is designated ^2S.

A variational treatment using (19.45) would replace Z in the $1s$ function in Table 19.1 by a parameter Z_1 and Z in the $2s$ function by a parameter Z_2. These parameters are "effective" atomic numbers that allow for electron screening. One finds the optimum values to be $Z_1 = 2.69$ and $Z_2 = 1.78$. As expected, the $2s$ electron is much better screened from the $Z = 3$ nucleus than the $1s$ electrons. The calculated variational energy turns out to be -201.2 eV, compared with the true ground-state energy -203.5 eV (*Q. C.*, sec. 10.9).

The periodic table. A qualitative and semiquantitative understanding of atomic structure can be obtained from the orbital approximation. As we did with He and Li, we write an approximate wave function that assigns the electrons to hydrogenlike spin-orbitals. In each orbital, the nuclear charge is replaced by a variational parameter that represents an effective nuclear charge Z_{eff} and allows for electron screening. To satisfy the Pauli principle, the wave function is written as a Slater determinant. (For some atomic states, the wave function must be written as a linear combination of a few Slater determinants, but we won't worry about this complication.)

Since an electron has two possible spin states (α or β), the exclusion principle requires that no more than two electrons occupy the same orbital in an atom (or molecule). Two electrons in the same orbital must have antiparallel spins, and such electrons are said to be *paired*. A set of orbitals with the same n value and the same l value constitutes a *subshell*. The lowest few subshells are $1s$, $2s$, $2p$, $3s$, ... An s subshell has $l = 0$ and $m = 0$ and hence can hold at most two electrons without violating the exclusion principle. A p subshell has $l = 1$ and the three possible m values $-1, 0, +1$; hence, a p subshell has a capacity of 6 electrons; d and f subshells hold a maximum of 10 and 14 electrons, respectively.

The hydrogenlike energy formula (19.20) can be modified to crudely approximate the energy ε of a given atomic orbital as

$$\varepsilon \approx -(Z_{\text{eff}}^2/n^2)(13.6 \text{ eV}) \tag{19.46}$$

where n is the principal quantum number and the effective nuclear charge Z_{eff} differs for different subshells in the same atom. We write $Z_{\text{eff}} = Z - s$, where Z is the atomic number and the *screening constant* s for a given subshell is the sum of contributions from other electrons in the atom.

R. Latter used an approximate method to calculate orbital energies for neutral atoms (Fig. 19.13). The scales on this figure are logarithmic. Note that the energy of an orbital depends strongly on the atomic number, decreasing with increasing Z, as would be expected from Eq. (19.46). As mentioned earlier, the l degeneracy that exists for $Z = 1$ is removed in many-electron atoms. For most values of Z, the $3s$ and $3p$ orbitals are much closer together than the $3p$ and $3d$ orbitals, and we get the familiar stable octet of outer electrons (ns^2np^6). For Z between 7 and 21, the $4s$ orbital lies below the $3d$ (s orbitals are more penetrating than d orbitals), but for $Z > 21$, the $3d$ lies lower.

We saw that the ground-state configuration of Li is $1s^22s$, where the superscripts give the numbers of electrons in each subshell and a superscript of 1 is understood on the $2s$ subshell. We expect Li to readily lose one electron (the $2s$ electron) to form the Li^+ ion, and this is the observed chemical behavior.

The ground-state configurations of Be and B are $1s^22s^2$ and $1s^22s^22p$, respectively. For C, the ground-state configuration is $1s^22s^22p^2$. We saw that a given electron configuration can give rise to more than one atomic term; e.g., the He $1s2s$ configuration produces the two terms ^3S and ^1S. Figuring out the terms arising from the $1s^22s^22p^2$ configuration is complicated and is omitted. Hund's rule tells us that the lowest-lying term will have the two $2p$ spins parallel:

$$\underset{1s}{\underline{\uparrow\downarrow}} \quad \underset{2s}{\underline{\uparrow\downarrow}} \quad \underbrace{\underline{\uparrow} \;\; \underline{\uparrow} \;\; \underline{}}_{2p}$$

(Note that putting the two $2p$ electrons in different orbitals minimizes the electrostatic repulsion between them.) The $2p$ subshell is filled at $_{10}$Ne, whose electron configuration is $1s^22s^22p^6$. Like helium, neon does not form chemical compounds.

Sodium has the ground-state configuration $1s^22s^22p^63s$, and its chemical and physical properties resemble those of Li (ground-state configuration $1s^22s$), its predecessor in Group IA of the periodic table. The periodic table is a consequence of

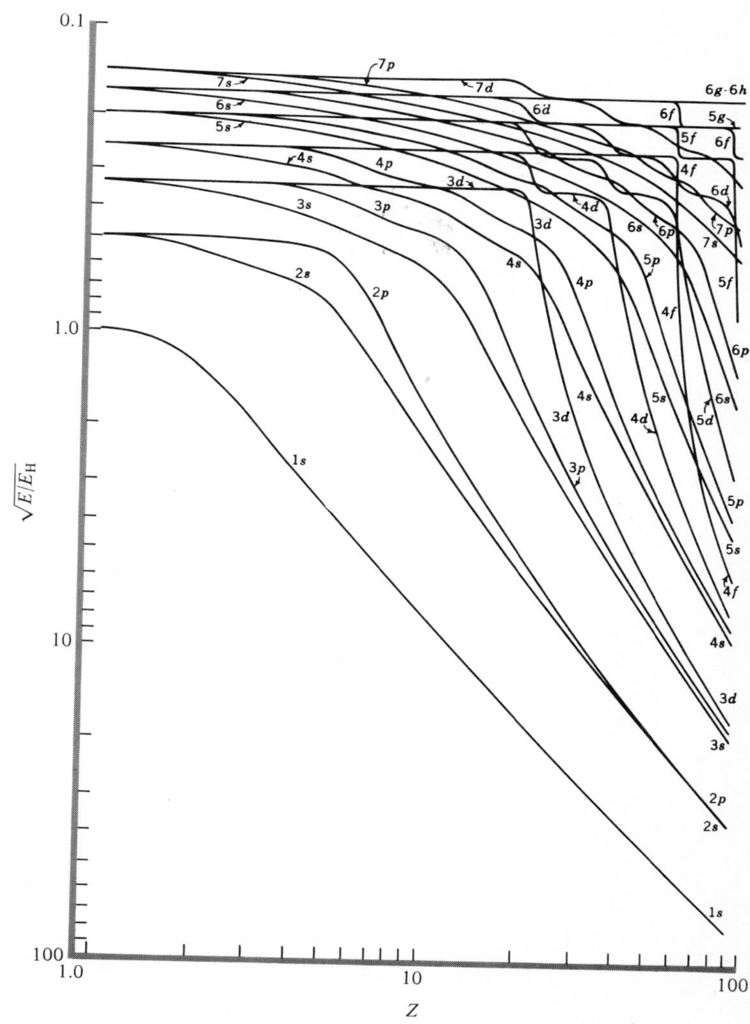

Figure 19.13
Approximate orbital energies vs. atomic number Z in neutral atoms. $E_H = -13.6$ eV, the energy of the ground-state hydrogen atom. [*Redrawn by M. Kasha from R. Latter, Phys. Rev.*, **99**, *510 (1955).*]

the hydrogenlike energy-level pattern, the allowed electronic quantum numbers, and the exclusion principle. The third period ends with Ar, whose ground-state configuration is $1s^2 2s^2 2p^6 3s^2 3p^6$.

For $Z = 19$ and 20, the 4*s* subshell lies below the 3*d* (Fig. 19.13) and K and Ca have the outer electron configurations 4*s* and $4s^2$, respectively. With $Z = 21$, the 3*d* subshell begins to fill, giving the first series of *transition elements*. The 3*d* subshell is filled at $_{30}$Zn, outer electron configuration $3d^{10}4s^2$ (3*d* now lies lower than 4*s*). Filling the 4*p* subshell then completes the fourth period.

The rare earths (lanthanides) and actinides in the sixth and seventh periods correspond to filling the 4*f* and 5*f* subshells.

The order of filling of subshells in the periodic table is given by the $n + l$ rule: Subshells fill in order of increasing $n + l$ values; for subshells with equal $n + l$ values, the one with the lower n fills first.

Niels Bohr was the person who rationalized the periodic table in terms of filling the atomic energy levels, and the familiar long form of the periodic table is due to him.

Atomic properties. The *first, second, third, ... ionization energies* of atom A are the energies required for the processes $A \rightarrow A^+ + e^-$, $A^+ \rightarrow A^{2+} + e^-$, $A^{2+} \rightarrow A^{3+} + e^-$, ..., where A, A^+, etc., are isolated atoms or ions in their ground states. Ionization energies are traditionally expressed in eV. The corresponding numbers in volts are called the ionization potentials. Some first, second, and third ionization energies in eV are [C. E. Moore, Ionization Potentials and Ionization Limits, *Nat. Bur. Stand. U.S. Publ. NSRDS-NBS 34*, 1970]:

H	He	Li	Be	B	C	N	O	F	Ne	Na
13.6	24.6	5.4	9.3	8.3	11.3	14.5	13.6	17.4	21.6	5.1
	54.4	75.6	18.2	25.2	24.4	29.6	35.1	35.0	41.0	47.3
		122.5	153.9	37.9	47.9	47.4	54.9	62.7	63.4	71.6

Note the low value for removal of the 2*s* electron from Li and the high value for removal of a 1*s* electron from Li^+. Ionization energies clearly show the "shell" structure of atoms.

The *electron affinity* of atom A is the energy released in the process $A + e^- \rightarrow A^-$. Some values in eV are [H. Hotop and W. C. Lineberger, *J. Phys. Chem. Ref. Data*, **4**, 539 (1975)]:

H	He	Li	Be	B	C	N	O	F	Ne	Na
0.8	< 0	0.6	< 0	0.3	1.3	-0.1	1.5	3.4	< 0	0.5

Metals have lower ionization potentials and lower electron affinities than nonmetals.

The motion of an electric charge produces a magnetic field. The orbital angular momentum of atomic electrons with $l \neq 0$ therefore produces a magnetic field. The "spinning" of an electron about its own axis is a motion of electric charge and also produces a magnetic field; because of the existence of electron spin, an electron acts like a tiny magnet. (Magnetic interactions between electrons are much smaller than the electrical forces and can be neglected in the Hamiltonian except in very precise calculations.) The magnetic fields of electrons with opposite spins cancel each other. It follows that an atom in a state with $L \neq 0$ and/or $S \neq 0$ produces a magnetic field and is said to be *paramagnetic*. In a magnetized piece of iron, the majority of electron spins are aligned in the same direction to produce the observed magnetic field.

The radius of an atom is not a well-defined quantity, as is obvious from Figs. 19.7 and 19.8. From observed bond lengths in molecules and interatomic distances in crystals, various kinds of atomic radii can be deduced. (See Secs. 20.1 and 24.5.) Atomic radii decrease going across a given period and increase going down a given group.

The energies of excited states of most atoms of the periodic table have been determined from atomic spectral data and are tabulated in C. E. Moore, Atomic Energy Levels, *Nat. Bur. Stand. U.S. Circ.* 467, vols. I, II, and III, 1949, 1952, and 1958.

19.7 HARTREE–FOCK AND CONFIGURATION-INTERACTION WAVE FUNCTIONS

Wave functions like (19.34) for He and (19.45) for Li are approximations. How can these approximate wave functions be improved? One way is by not restricting the one-electron spatial functions to hydrogenlike functions. Instead, for the He ground state, we take as a trial variational function $\phi(1)\phi(2)$ times the antisymmetric spin function (19.33), and we look for the function ϕ that minimizes the variational integral, where ϕ need not be a hydrogenlike 1*s* orbital but can have any form. For the Li ground state, we use a function like (19.45) but with the 1*s* and 2*s* functions replaced by unknown functions f and g, and we look for those functions f and g that minimize the variational integral. These variational functions are still antisymmetrized products of one-electron spin-orbitals, so the functions ϕ, f, and g are still atomic orbitals.

In the period 1927–1930, the English physicist Hartree and the Russian physicist Fock developed a systematic procedure for finding the best possible forms for the orbitals. A variational wave function that is an antisymmetrized product of the best possible orbitals is called a *Hartree–Fock wave function*. For each state of a given system, there is a single Hartree–Fock wave function. Hartree and Fock showed that the Hartree–Fock orbitals ϕ_i satisfy the equation

$$\hat{F}\phi_i = \varepsilon_i \phi_i \tag{19.47}$$

where the Hartree–Fock operator $\hat{F}$ is a complicated operator whose form we omit. Each of the spatial orbitals ϕ_i is a function of the three spatial coordinates; ε_i is the energy of orbital *i*.

Originally, Hartree–Fock orbitals were calculated numerically, and the results expressed as a table of values of ϕ_i at various points in space. In 1951, Roothaan showed that the most convenient way to express Hartree–Fock orbitals is as linear combinations of a set of functions called *basis functions*. A set of functions is said to be a *complete set* if *any* well-behaved function can be written as a linear combination of the members of the complete set. If the functions $g_1, g_2, g_3, \ldots$ form a complete set, then an arbitrary well-behaved function f can be expressed as $f = \sum_k c_k g_k$, where the coefficients c_k are constants that depend on what the function f is. It generally requires an infinite number of functions $g_1, g_2, \ldots$ to have a complete set (but not every infinite set of functions is complete). The basis functions g_k used to express the Hartree–Fock orbitals ϕ_i must be a complete set. We have $\phi_i = \sum_k b_k g_k$. An orbital ϕ_i is specified by stating what the basis set is and giving the coefficients b_k. Roothaan showed how to calculate the b_k's.

The complete set of basis functions used in atomic Hartree–Fock calculations is the set of *Slater-type orbitals* (STOs). An STO has the form $G_n(r)\Theta_{lm}(\theta)\Phi_m(\phi)$, where $\Theta_{lm}(\theta)$ and $\Phi_m(\phi)$ are the same functions as in the hydrogenlike orbitals (19.11); the radial factor has the form $G(r) = Nr^{n-1}e^{-\zeta r/a_0}$, where N is a normalization constant, n is the principal quantum number, and ζ is a variational parameter (the *orbital exponent*). The function $G(r)$ differs from a hydrogenlike radial factor in containing r^{n-1} in place of r^l times a polynomial in r. It has been shown that the set of STOs with n, l, and m given by (19.12) to (19.14) and with all possible positive values of ζ forms a complete set. Although, in principle, one needs an infinite number of basis functions to express a Hartree–Fock orbital, in practice, each atomic Hartree–Fock orbital can be very accurately approximated using only a few well-chosen STOs. For example, for the helium ground state, Clementi expressed the Hartree–Fock orbital of the electrons as a linear combination of only five 1*s* STOs, these five STOs differing in their values of ζ.

To solve (19.47) for the Hartree–Fock orbitals of an atom or molecule requires a tremendous amount of computation, and it wasn't until the advent of large, high-speed computers in the 1960s that such calculations became practicable. Hartree–Fock wave functions have been computed for the ground states and certain excited states of the first 54 atoms of the periodic table.

Although a Hartree–Fock wave function is an improvement on one that uses hydrogenlike orbitals, it is still only an approximation to the true wave function. The Hartree–Fock wave function assigns each electron pair to its own orbital; the forms of these orbitals are computed to take interelectronic repulsions into account in an average way. However, electrons are not smeared out into a static distribution of charge but interact with one another instantaneously. An orbital wave function cannot account for these instantaneous interactions, so the true wave function cannot be expressed as an antisymmetrized product of orbitals.

For helium, the use of a hydrogenlike 1*s* orbital with a variable orbital exponent gives a ground-state energy of -77.5 eV (Sec. 19.5) compared with the true value -79.0 eV. The Hartree–Fock wave function for the helium ground state gives an energy of -77.9 eV, which is still in error by 1.1 eV. The energy error of the Hartree–Fock wave function is called the *correlation energy*, since it results from the fact that the Hartree–Fock wave function neglects the instantaneous correlations in the motions of the electrons; electrons repel one another and correlate their motions to avoid being close together.

The most common method used to improve a Hartree–Fock wave function is *configuration interaction* (CI). A by-product of the calculation of the Hartree–Fock ground-state wave function of an atom or molecule is expressions for the unoccupied excited-state orbitals. It is possible to show that the set of functions obtained by making all possible assignments of electrons to the available orbitals is a complete set. Hence, the true wave function ψ of the ground state can be expressed as $\psi = \sum_j a_j \psi_{\mathrm{orb},j}$, where $\psi_{\mathrm{orb},j}$ are approximate orbital wave functions that differ in the assignment of electrons to orbitals; each $\psi_{\mathrm{orb},j}$ is a Slater determinant of spin-orbitals. The functions $\psi_{\text{orb},j}$ are called *configuration state functions* (or *configurations*). One uses a variational procedure to find the values of the coefficients a_j that minimize the variational integral; this type of calculation is called *configuration interaction* (CI). For the helium ground state, the term with the largest coefficient in the CI wave function will be a Slater determinant with both electrons in orbitals resembling 1*s* orbitals, but Slater determinants with electrons in 2*s*-like and higher orbitals will also contribute. CI calculations are extremely difficult, since it often requires a linear combination of thousands of configuration state functions to give an accurate representation of ψ. Most CI calculations use only a relatively few configuration state functions and therefore only give approximations to ψ.

PROBLEMS

19.1 (*a*) Calculate the electrostatic potential energy of two electrons separated by 3.0 Å in vacuum. Express your answer in joules, in ergs, and in electron volts. (*b*) Calculate the electrostatic potential energy in eV of a system of two electrons and a proton in vacuum if the electrons are separated by 3.0 Å and the electron–proton distances are 4.0 and 5.0 Å.

19.2 A particle has a charge-to-mass ratio of 6.0×10^8 C/g and a charge of 4.5×10^{-17} C. Find its mass.

19.3 Explain why the observed charge-to-mass ratio of electrons decreases when the electrons are accelerated to very high speeds.

19.4 Calculate the de Broglie wavelength of electrons accelerated through 100 V.

19.5 What fraction of the volume of an atom of radius 10^{-8} cm is occupied by its nucleus if the nuclear radius is 10^{-12} cm?

19.6 The density of gold is 19.3 g/cm^3. If gold atoms were cubes, what would the length of each side of a cubic atom be?

19.7 Give the allowed values of (*a*) l for $n = 5$ and (*b*) m if $l = 5$.

19.8 Omitting spin considerations, give the degeneracy of the hydrogenlike energy level with (*a*) $n = 1$; (*b*) $n = 2$; (*c*) $n = 3$.

19.9 Calculate the ionization potential in V of (*a*) He^+; (*b*) Li^{2+}.

19.10 Calculate the wavelength of the photon emitted when an electron jumps from the $n = 3$ to $n = 2$ level of a hydrogen atom.

19.11 Calculate a in Eq. (19.17).

19.12 Positronium is a species consisting of an electron bound to a positron (Sec. 17.17). Calculate its ionization potential.

19.13 Show that $\langle r \rangle = 3a/2Z$ for a ground-state hydrogenlike atom. (Use a table of integrals.)

19.14 Calculate the angles the three angular-momentum vectors make with the z axis in Fig. 19.4.

19.15 (*a*) From the definition of angular momentum in Sec. 19.3, show that for a particle of mass m moving on a circle of radius r, the magnitude of the angular momentum with respect to the circle's center is mvr. (*b*) What is the direction of the **L** vector for this system?

19.16 Calculate the magnitude of the ground-state orbital angular momentum of the electron in a hydrogen atom according to (*a*) quantum mechanics; (*b*) the Bohr theory.

19.17 Calculate the magnitude of the orbital angular momentum of a 3*p* electron in a hydrogenlike atom.

19.18 Use the Taylor-series expansions about $\phi = 0$ for $e^{i\phi}$, $\sin \phi$, and $\cos \phi$ to verify (19.22).

19.19 Verify Eq. (19.24).

19.20 Calculate the radius of the 1*s* orbital in H using the 90 percent probability definition. (You will have to use trial and error to solve the equation for r_{1s}.)

19.21 Show that the maximum in the radial distribution function of a hydrogenlike atom is at a/Z.

19.22 For a hydrogen atom in a 1*s* state, calculate the probability that the electron is between 0 and 2.00 Å from the nucleus.

19.23 Verify that $\int_0^{2\pi} |\Phi|^2 \, d\phi = 1$, where Φ is given by (19.15).

19.24 Verify that the 1*s* wave function in Table 19.1 is an eigenfunction of the hydrogenlike Hamiltonian operator. (Use the chain rule to find the partial derivatives.)

19.25 Calculate $\langle r \rangle$ for a hydrogen atom in a $2p_z$ state.

19.26 Find the expression for the average potential energy $\langle V \rangle$ for a ground-state hydrogenlike atom.

19.27 Calculate the angles between the spin vectors and the z axis in Fig. 19.9.

19.28 Write down the Hamiltonian for the internal motion in Li.

19.29 State whether each of these functions is symmetric, antisymmetric, or neither: (*a*) $f(1)g(2)$; (*b*) $g(1)g(2)$; (*c*) $f(1)g(2) - g(1)f(2)$; (*d*) $r_1^2 - 2r_1r_2 + r_2^2$; (*e*) $(r_1 - r_2)e^{-br_{12}}$, where r_{12} is the distance between particles 1 and 2.

19.30 A student does a variational calculation on the ground state of He and finds that the variational integral equals -86.7 eV. Explain why it is certain that the student made an error.

19.31 For a particle with $s = 3/2$: (*a*) sketch the possible orientations of the **S** vector with the z axis; (*b*) calculate the smallest possible angle between **S** and the z axis.

19.32 For a system of two electrons in a one-dimensional box, write down the approximate wave functions (interelectronic repulsion ignored) including spin for states that have one electron with $n = 1$ and one electron with $n = 2$. Which of these states has (have) the lowest energy?

19.33 Give the term symbol for the term arising from each of the following H-atom electron configurations: (*a*) 1*s*; (*b*) 3*p*; (*c*) 3*d*.

19.34 Give the values of L and S for a ^4F term.

19.35 Write down an approximate wave function for the Be ground state.

19.36 Which of the first 10 elements in the periodic table have paramagnetic ground states?

19.37 Calculate the eighteenth ionization potential of Ar.

19.38 Use the ionization-potential data in Sec. 19.6 to calculate Z_{eff} for the 2*s* electrons in (*a*) Li; (*b*) Be.

19.39 Suppose the electron had spin quantum number $s = \frac{3}{2}$. What would be the ground-state configurations of atoms with 3, 9, and 17 electrons?

19.40 For each pair, state which would have the higher first ionization potential (refer to a periodic table): (*a*) Na, K; (*b*) K, Ca; (*c*) Cl, Br; (*d*) Br, Kr.

19.41 Use Fig. 19.13 to calculate Z_{eff} for the 1*s*, 2*s*, and 2*p* electrons in Ne.

20

MOLECULAR ELECTRONIC STRUCTURE

20.1 CHEMICAL BONDS

A full and correct treatment of molecules must be based on quantum mechanics. Indeed, it is not possible to understand the stability of a covalent bond as in H_2 without quantum mechanics. Because of the mathematical difficulties involved in the application of quantum mechanics to molecules, chemists have developed a variety of empirical concepts to describe bonding. This section discusses some of these concepts. Section 20.2 describes how the molecular Schrödinger equation is separated into Schrödinger equations for electronic motion and for nuclear motion. The one-electron molecule H_2^+ is discussed in Sec. 20.3 to develop some ideas about electron orbitals in molecules. The major approximation method used in describing molecular electronic structure is the molecular-orbital method, developed in Secs. 20.4 to 20.6. Section 20.7 shows how molecular properties are calculated from electronic wave functions.

Bond radii. The length of a bond in a molecule is the distance between the nuclei of the two atoms forming the bond. Spectroscopic and diffraction methods (Chaps. 21 and 24) allow bond lengths to be measured accurately. The length of a given kind of bond is found to be approximately constant from molecule to molecule. (For example, the carbon–carbon single-bond length in most nonconjugated molecules lies in the range 1.53 to 1.54 Å.) Moreover, one finds that the bond length d_{AB} between atoms A and B is *approximately* equal to $\frac{1}{2}(d_{AA} + d_{BB})$, where d_{AA} and d_{BB} are the typical A—A and B—B bond lengths. For example, let A and B be Cl and C. The bond length in Cl_2 is 1.99 Å, and $\frac{1}{2}(d_{AA} + d_{BB}) = \frac{1}{2}(1.99 + 1.54)$ Å $= 1.76$ Å, in good agreement with the observed bond length 1.76_6 Å in CCl_4.

One can therefore take $\frac{1}{2}d_{AA}$ as the *bond radius* r_A for atom A and use a table of bond radii to estimate the bond length d_{AB} as $r_A + r_B$. Double and triple bonds are shorter than the corresponding single bonds, and so different bond radii are used for single, double, and triple bonds. Some bond radii (due mainly to Pauling) in angstroms (Å) are:

	H	C	N	O	F	P	S	Cl	Br	I
Single	0.30	0.77	0.70	0.66	0.64	1.10	1.04	0.99	1.14	1.33
Double		0.67	0.60	0.56	0.60	1.00	0.94	0.89	1.04	1.23
Triple		0.60	0.55							

(The observed bond length 0.74 Å in H_2 would indicate $r_A = 0.37$ Å for H, but the listed value 0.30 Å works much better in predicting bond lengths between H and other elements.)

When there is a substantial electronegativity difference between atoms A and B, the observed bond length is often shorter than $r_A + r_B$. Various correction terms have been suggested to allow for electronegativity differences, but none of them work very well.

The carbon–carbon bond length in benzene is 1.40 Å. This lies between the carbon–carbon single-bond length 1.54 Å and double-bond length 1.34 Å, which indicates that the benzene bonds are intermediate between single and double bonds. In the conjugated molecule butadiene ($CH_2CHCHCH_2$) the single-bond length is only 1.46 Å, whereas the double bonds show the normal length 1.34 Å.

Bond energies. Section 5.7 explained how experimental ΔH°_{298} values for gas-phase atomization processes can be used to give average bond energies. Some average bond energies are listed in Table 20.1. Double and triple bonds are stronger than single bonds. The N—N, O—O, and N—O single bonds are quite weak.

Bond energies can be used to *estimate* ΔH°_{298} for a reaction involving only gases. The two-step process

$$\text{Gaseous reactants} \rightarrow \text{gaseous atoms} \rightarrow \text{gaseous products}$$

gives $\Delta H^\circ_{298} = \Delta H^\circ_{\text{at},298,\text{re}} - \Delta H^\circ_{\text{at},298,\text{pr}}$, where $\Delta H^\circ_{\text{at,re}}$ and $\Delta H^\circ_{\text{at,pr}}$, the heats of atomization of the reactants and products, can be found by adding up the bond energies. Corrections for such things as strain energies in small-ring compounds, resonance energies in conjugated compounds, and steric energies in bulky compounds are often included.

Table 20.1. Average bond energies in kcal/mole[a]

C—H	C—C	C—O	C—N	C—S	C—F	C—Cl	C—Br	C—I
99.₂	82.₂	84	70	62	105	78	66	57
N—H	O—H	S—H	S—S	N—O	O—O	N—N	N—Cl	H—H
93	111	88	64	42	34	38	48	104
C=C	C=O	C=N	N=N	C≡C	C≡N	N≡N		
147	173	147	100	194	213	226		

[a] Data from L. Pauling, *General Chemistry*, 3d ed., Freeman, 1970, p. 913.

Tabulated bond energies are on a per mole basis. To convert to a per molecule basis, we divide by the Avogadro constant. One kcal/mole corresponds to 1.6606×10^{-21} cal per molecule. Using (2.65) and (19.6), we find that 1 kcal/mole corresponds to 0.04336 eV per molecule. Thus,

$$1 \text{ eV/molecule corresponds to } 23.06 \text{ kcal/mole} \tag{20.1}$$

Bond moments. The dipole moment **p** of a charge distribution is defined by Eq. (14.15). Molecular dipole moments can be determined by microwave spectroscopy (Sec. 21.6) or by dielectric-constant measurements (Sec. 14.2). From (14.15) the SI unit of p is the coulomb-meter (C m). The corresponding cgs gaussian electric dipole moment p' has units of statcoulomb-centimeters. Equation (19.3) shows that

$$1 \text{ C m corresponds to } 2.9979 \times 10^{11} \text{ statC cm} \tag{20.2}$$

For example, p of HCl is 3.57×10^{-30} C m, and so $p'(\text{HCl}) = 1.07 \times 10^{-18}$ statC cm. Most commonly, p' is expressed in debyes (D), where

$$1 \text{ D} \equiv 10^{-18} \text{ statC cm} \tag{20.3}$$

Thus, $p'(\text{HCl}) = 1.07$ D.

A rough estimate of the dipole moment of a molecule can be obtained by taking the *vector* sum of assigned *bond dipole moments* for the individual bonds. Some bond moments in debyes are:

H—O	H—N	H—C	C—Cl	C—Br	C—O	C=O	C—N	C≡N
1.5	1.3	(0.4)	1.5	1.4	0.8	2.5	0.5	3.5

where the first-listed atom is the positive end of the bond moment. The value for H—C is an assumed one, and the other moments involving C depend on the magnitude and sign of this assumed value. The polarity of the H—C bond moment is unsettled at present. Arguments for the assumed polarity $H^+—C^-$ are given in S. W. Benson and M. Luria, *J. Am. Chem. Soc.*, **97**, 704 (1975).

The H—O and H—N bond moments are calculated from the observed dipole moments of H_2O and NH_3 without explicitly considering the contributions of the lone pairs to the dipole moment. (Their contributions are absorbed into the values calculated for the OH and NH moments.) For example, the observed p' of H_2O is 1.85 D, and the bond angle is 104.5°; Fig. 20.1 gives $2p'_{\text{OH}} \cos 52.2° = 1.85$ D, and the O—H bond moment is $p'_{\text{OH}} = 1.5$ D. The other moments listed are calculated from

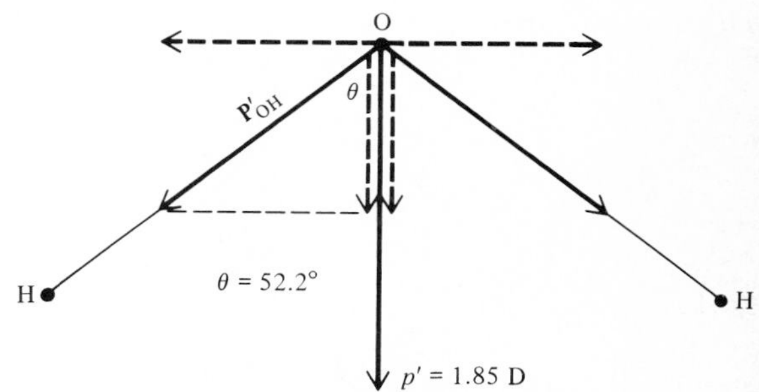

Figure 20.1
Calculation of the OH bond moment in H_2O. The dashed vectors are the bond-moment components along **p** and perpendicular to **p**.

the experimental dipole moments and geometries of CH_3Cl, CH_3Br, CH_3OH, $(CH_3)_2CO$, $(CH_3)_3N$, and CH_3CN using the assumed CH bond moment and the OH and NH moments.

A useful shortcut in doing bond-moment calculations is to note that the vector sum of the three CH bond moments of a tetrahedral CH_3 group equals the moment of one CH bond. This follows from the zero dipole moment of the tetrahedral molecule methane (HCH_3).

Electronegativity. The electronegativity of an element is a measure of the ability of an atom of that element to attract the electrons in a bond. The degree of polarity of an A—B bond is related to the difference in the electronegativities of the atoms forming the bond.

Several different electronegativity scales have been proposed. The best-known is the Pauling scale, based on bond energies [see Prob. 20.7 and A. L. Allred, *J. Inorg. Nucl. Chem.*, **17**, 215 (1961)]. The Allred–Rochow scale [A. L. Allred and E. G. Rochow, *J. Inorg. Nucl. Chem.*, **5**, 264, 269 (1958)] defines the electronegativity x_A of element A as

$$x_A \equiv 0.359 Z_{eff}/(r_A/\text{Å})^2 + 0.744 \tag{20.4}$$

where r_A is the bond radius of A and Z_{eff} is the effective nuclear charge [Eq. (19.46)] that would act on an electron added to the valence shell of a neutral A atom. The quantity $Z_{eff}e'^2/r_A{}^2$ (where e' is the proton charge in gaussian units) is the average force exerted by atom A on an added electron. The constants 0.359 and 0.744 were chosen to make the scale as consistent as possible with values on the Pauling electronegativity scale.

Simons et al. used quantum-mechanical calculations to define an electronegativity scale. [G. Simons, M. E. Zandler, and E. R. Talaty, *J. Am. Chem. Soc.*, **98**, 7869 (1976).] The floating-orbital method of calculating approximate molecular wave functions represents a chemical bond by an orbital located on the bond axis; the position of this orbital is varied to minimize the variational integral. If the orbital for the A—B bond ends up closer to B than to A, then B is more electronegative than A. Simons et al. took the electronegativity difference between A and B as

$$x_B - x_A \equiv 2.273(R_A - R_B)/(R_A + R_B) \tag{20.5}$$

where R_A and R_B are the distances from the center of the floating orbital to atoms A and B.

Some electronegativities on the Pauling, Allred–Rochow, and Simons scales are given in Table 20.2. Allred–Rochow electronegativities of metals run from 0.9 for Cs and Fr to 1.8 for Sb; most transition metals have values in the range 1.2 to 1.7. Fluorine is the most electronegative element, followed by oxygen and then nitrogen. Electronegativities tend to decrease going down a given group in the periodic table (due to the increasing distance of the valence electrons from the nucleus) and to increase going across a given period (due mainly to the increasing Z_{eff} resulting from the lesser screening by electrons added to the same shell).

Although the electronegativity concept is not a precise one, electronegativities on the Pauling, Allred–Rochow, and Simons scales generally agree quite well. The major exception is hydrogen, which is special in having no inner-shell electrons. The Simons and Allred–Rochow electronegativities of H are substantially greater than the Pauling value and are not considered reliable.

Table 20.2 Some Pauling (P), Allred–Rochow (A), and Simons (S) electronegativities

	Li	Be	B	C	N	O	F	H
P	1.0	1.6	2.0	2.5	3.0	3.4	4.0	2.2
A	1.0	1.5	2.0	2.5	3.1	3.5	4.1	
S	1.0	1.5	1.8	2.4	3.2	3.5	4.0	
	Na	Mg	Al	Si	P	S	Cl	
P	0.9	1.3	1.6	1.9	2.2	2.6	3.2	
A	1.0	1.2	1.5	1.7	2.1	2.4	2.8	
S	0.9	1.2	1.4	1.6	2.1	2.5	2.8	

20.2 THE BORN–OPPENHEIMER APPROXIMATION

All molecular properties are, in principle, calculable by solving the Schrödinger equation for the molecule. In 1929, Dirac wrote: "The underlying physical laws necessary for the mathematical theory of ... the whole of chemistry are thus completely known, and the difficulty is only that the exact application of these laws leads to equations much too complicated to be soluble." Because of the great mathematical difficulties involved in solving the molecular Schrödinger equation, one must make approximations. Until about 1960, the level of approximations was such that the calculations gave only qualitative and not quantitative information. Since that time, the use of computers has made molecular wave-function calculations accurate enough to give reliable quantitative information in many cases.

The Hamiltonian for a molecule is

$$\hat{H} = \hat{T}_N + \hat{T}_e + \hat{V}_{NN} + \hat{V}_{Ne} + \hat{V}_{ee} \tag{20.6}$$

where $\hat{T}_N$ stands for the kinetic-energy operators for the nuclei, $\hat{T}_e$ stands for the kinetic-energy operators for the electrons, $\hat{V}_{NN}$ is the potential energy of repulsion between the nuclei, $\hat{V}_{Ne}$ is the potential energy of attraction between the electrons and the nuclei, and $\hat{V}_{ee}$ is the potential energy of repulsion between the electrons.

The molecular Schrödinger equation $\hat{H}\psi = E\psi$ is extremely complicated, and it would be almost hopeless to attempt an exact solution, even for rather small molecules. Fortunately, the fact that nuclei are much heavier than electrons allows the use of a very accurate approximation that greatly simplifies things. In 1927, Max Born and J. Robert Oppenheimer showed that it is an excellent approximation to treat the electronic and nuclear motions separately. The mathematics of the Born–Oppenheimer approximation is complicated, and so we shall give only a qualitative physical discussion.

Because of their much greater masses, the nuclei move far more slowly than the electrons, and the electrons carry out many "cycles" of motion in the time it takes the nuclei to move a short distance. The electrons see the heavy, slow-moving nuclei as almost stationary point charges, whereas the nuclei see the fast-moving electrons as essentially a three-dimensional distribution of charge.

One therefore assumes a fixed configuration of the nuclei, and for this configuration one solves an electronic Schrödinger equation to find the molecular electronic energy and wave function. This process is repeated for many different fixed nuclear configurations to give the electronic energy as a function of the positions of the nuclei. The nuclear configuration that corresponds to the minimum value of the electronic energy is the equilibrium geometry of the molecule. Having found how the electronic energy varies as a function of the nuclear configuration, one then uses this electronic energy function as the potential-energy function in a Schrödinger equation for the nuclear motion, thereby obtaining the molecular vibrational and rotational energy levels for a given electronic state.

The electronic Schrödinger equation is formulated for a fixed set of locations for the nuclei. Hence, the nuclear kinetic-energy operator $\hat{T}_N$ is omitted from the Hamiltonian, and the electronic Hamiltonian $\hat{H}_e$ and electronic Schrödinger equation are

$$\hat{H}_e = \hat{T}_e + \hat{V}_{Ne} + \hat{V}_{ee} + \hat{V}_{NN} \tag{20.7}$$

$$\hat{H}_e \psi_e = E_e \psi_e \tag{20.8}$$

E_e is the electronic energy, including the energy V_{NN} of nuclear repulsion. Note that V_{NN} in (20.7) is a constant, since the nuclei are held fixed. The electronic wave function ψ_e is a function of the $3n$ spatial and n spin coordinates of the n electrons of the molecule.

Consider a diatomic (two-atom) molecule with nuclei A and B with atomic numbers Z_A and Z_B. The spatial configuration of the nuclei is specified by the distance R between the two nuclei. The potential-energy operator $\hat{V}_{Ne}$ depends on R as a parameter, as does the internuclear repulsion $\hat{V}_{NN}$, which equals $Z_A Z_B e'^2/R$ [Eq. (19.4)]. Hence, at each value of R, we get a different electronic wave function and energy; these quantities depend on R as a parameter and vary continuously as R varies. We therefore have $\psi_e = \psi_e(q_1, \ldots, q_n; R)$ and $E_e = E_e(R)$, where q_n stands for the spatial coordinates and spin coordinate of electron n. For a polyatomic molecule, ψ_e and E_e will depend parametrically on the locations of all the nuclei:

$$\psi_e = \psi_e(q_1, \ldots, q_n; Q_1, \ldots, Q_{\mathcal{N}}), \qquad E_e = E_e(Q_1, \ldots, Q_{\mathcal{N}})$$

where the Q's are the coordinates of the $\mathcal{N}$ nuclei.

Of course, a molecule has many different possible electronic states. For each such state, there is a different electronic wave function and energy, which vary as the nuclear configuration varies. Figure 20.2 shows $E_e(R)$ curves for the ground electronic state and some excited states of H_2. Since $E_e(R)$ is the potential-energy function for motion of the nuclei, a state with a minimum in the $E_e(R)$ curve is a bound state, with the atoms bonded to each other. For an electronic state with no minimum, $E_e(R)$ increases continually as R decreases; this means that the atoms repel each other as they come together, and this is not a bound state. The colliding atoms simply bounce off each other. The two lowest electronic states in Fig. 20.2 each dissociate to two ground-state ($1s$) hydrogen atoms. The ground electronic state of H_2 dissociates to (and arises from) $1s$ H atoms with opposite electronic spins, whereas the repulsive first excited electronic state arises from $1s$ H atoms with parallel electron spins.

The internuclear distance R_e at the minimum in the E_e curve of a bound electronic state is the equilibrium bond distance for that state. (Because of molecular zero-point vibrations, R_e is not quite the same as the observed bond length.) As R goes to zero, E_e goes to infinity, due to the internuclear repulsion V_{NN}. As R goes to

Table 20.3. Diatomic-molecule D_e and R_e values

	H_2^+	H_2	He_2^+	Li_2	C_2	N_2	O_2	F_2
D_e/eV	2.8	4.75	3	1.1	6.4	9.9	5.2	1.6
R_e/Å	1.06	0.74	1.1	2.7	1.24	1.10	1.21	1.42
	CH	CO	NaCl	OH	HCl	CaO	NaH	NaK
D_e/eV	3.6	11.2	4.3	4.6	4.6	4.0	2	0.6
R_e/Å	1.12	1.13	2.36	0.97	1.27	1.82	1.89	

infinity, E_e goes to the sum of the energies of the separated atoms into which the molecule decomposes. The difference $E_e(\infty) - E_e(R_e)$ is the *equilibrium dissociation energy* D_e of the molecule (Fig. 20.2). Some D_e and R_e values (determined by spectroscopic methods) for the ground electronic states of diatomic molecules are given in Table 20.3.

The physical reason for the stability of a covalent bond is not a fully settled question. A somewhat oversimplified statement is that the stability is due to the decrease in the average potential energy of the electrons forming the bond. This decrease results from the greater electron–nuclear attractions in the molecule compared with those in the separated atoms. The electrons in the bond can feel the simultaneous attractions of two nuclei. This decrease in electronic potential energy outweighs the increases in interelectronic repulsions and internuclear repulsions that occur as the atoms come together.

An ionic molecule like NaCl is held together by the Coulombic attraction

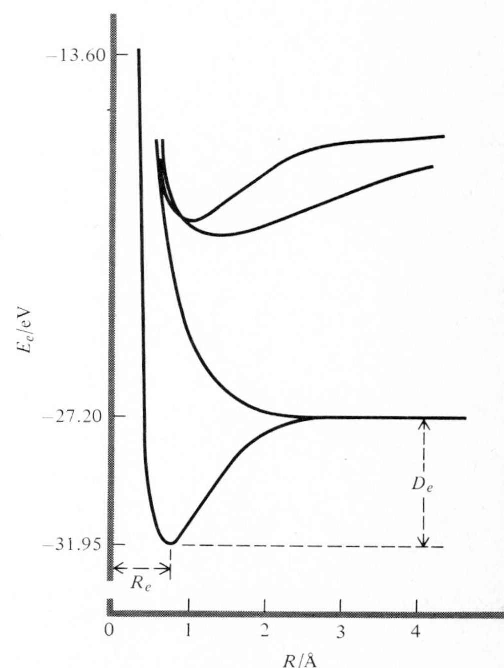

Figure 20.2
Potential-energy curves for the lowest few electronic states of H_2. R_e and D_e of the ground electronic state are indicated.

between the ions. Solid NaCl consists of an array of alternating Na^+ and Cl^- ions, and one can't pick out individual NaCl molecules. However, gas-phase NaCl consists of individual NaCl molecules. (In aqueous solution, hydration of the ions makes the separated hydrated ions more stable than Na^+Cl^- molecules.) Ionic molecules dissociate to neutral atoms in the gas phase. Consider, for example, NaCl. The ionization potential of Na is 5.1 eV, whereas the electron affinity of Cl is only 3.6 eV. Hence, isolated Na and Cl atoms are more stable than isolated Na^+ and Cl^- ions. Thus as the internuclear distance R increases, the bonding in NaCl shifts from ionic to covalent at very large R.

A simple model of spherical ions allows D_e of an ionic molecule like NaCl to be calculated reasonably accurately. From (19.4) the energy needed to separate two nonoverlapping spherical Na^+ and Cl^- ions from $R_e = 2.36$ Å to ions at infinity is e'^2/R_e, which equals [Eqs. (19.7), (19.2), and (19.6)]

$$(4.80 \times 10^{-10} \text{ statC})^2/(2.36 \times 10^{-8} \text{ cm}) = 9.76 \times 10^{-12} \text{ erg} = 6.09 \text{ eV}$$

Addition of an electron to Na^+ lowers the energy by the Na ionization potential 5.14 eV, and removal of an electron from Cl^- raises the energy by the Cl electron affinity 3.61 eV. Hence, the nonoverlapping-spherical-ion model gives the energy needed to dissociate NaCl into Na + Cl as 6.09 eV − 5.14 eV + 3.61 eV = 4.56 eV, which is only 6 percent away from the experimental value $D_e = 4.29$ eV. The error results from neglect of the repulsion between the slightly overlapping electron probability densities of the Na^+ and Cl^- ions, which makes the molecule less stable than calculated. (See Prob. 20.11.) This repulsion becomes especially important at internuclear separations less than R_e.

We can also estimate the dipole moment using this model. For spherical charges separated by 2.36 Å, Eq. (14.15) gives (it is simplest to take the origin at the center of one charge)

$$p' = e'R_e = (4.80 \times 10^{-10} \text{ statC})(2.36 \times 10^{-8} \text{ cm}) = 11.3 \text{ D}$$

where (20.3) was used. This value is not far from the experimental NaCl dipole moment 9.0 D. The error can be attributed to polarization of one ion by the other.

The ionic bonding in NaCl can be contrasted with the nonpolar covalent bonding in H_2 and other homonuclear diatomic molecules. Here, there is equal sharing of the bonding electrons. For a diatomic molecule formed from different nonmetals (for example, HCl, BrCl) or from different metals (for example, NaK), the bonding is polar covalent, the more electronegative atom having a greater share of the electrons and a partial negative charge. Bonds between metals with relatively high electronegativities and nonmetals are sometimes polar covalent, rather than ionic, as noted in Sec. 10.4.

We now resume consideration of the Born–Oppenheimer approximation. Having solved the electronic Schrödinger equation (20.8) to obtain the function $E_e(Q_1, \ldots, Q_{\mathcal{N}})$, we use this as the potential-energy function in the Schrödinger equation for nuclear motion:

$$(\hat{T}_N + E_e)\psi_N \equiv \hat{H}_N\psi_N = E\psi_N \tag{20.9}$$

The Hamiltonian $\hat{H}_N$ for nuclear motion equals the nuclear kinetic-energy operators $\hat{T}_N$ plus the electronic energy function E_e, so E in (20.9) includes both electronic and nuclear energies and is the total energy of the molecule. The nuclear wave function ψ_N is a function of the $3\mathcal{N}$ spatial and $\mathcal{N}$ spin coordinates of the $\mathcal{N}$ nuclei.

E_e is the potential energy for nuclear vibration. As the relatively sluggish nuclei vibrate, the rapidly moving electrons almost instantaneously adjust their wave function ψ_e and energy E_e to follow the nuclear motion. The electrons act somewhat like springs connecting the nuclei; as the internuclear distances change, the energy stored in the "springs" (i.e., in the electronic motions) changes.

The nuclear kinetic-energy operator $\hat{T}_N$ involves both vibrational and rotational kinetic energy. (Rotational motion does not change the electronic energy E_e.) We shall deal with nuclear vibrations and rotations in the next chapter. The remainder of this chapter deals with the electronic wave function and energy.

The Born–Oppenheimer treatment shows that the complete molecular wave function ψ is to a very good approximation equal to the product of electronic and nuclear wave functions: $\psi = \psi_e \psi_N$.

In addition to making the Born–Oppenheimer approximation, one usually neglects relativistic effects in treating molecules. These effects are generally slight.

20.3 THE HYDROGEN MOLECULE ION

The simplest molecule is H_2^+, which consists of two protons and one electron. The spectrum of H_2^+ can be observed in gas discharge tubes.

Adopting the Born–Oppenheimer approximation, we hold the nuclei at a fixed distance R and deal with the electronic Schrödinger equation (20.8). The electronic Hamiltonian (including nuclear repulsion) for H_2^+ is given by Eqs. (20.7), (19.4), and (18.48) as

$$\hat{H}_e = -\frac{\hbar^2}{2m}\nabla^2 - \frac{e'^2}{r_A} - \frac{e'^2}{r_B} + \frac{e'^2}{R} \tag{20.10}$$

where r_A and r_B are the distances from the electron to nuclei A and B and R is the internuclear distance (Fig. 20.3). The first term on the right of (20.10) is the operator for the kinetic energy of the electron; the second and third terms are the potential energies of attraction between the electron and the nuclei; the last term is the repulsion between the nuclei. There is only one electron, and so there is no interelectronic repulsion.

The electronic Schrödinger equation $\hat{H}_e \psi_e = E_e \psi_e$ can be solved exactly for H_2^+, but the solutions are complicated. For our modest purposes, an approximate treatment will suffice. The lowest electronic state of H_2^+ will dissociate to a ground-state (1*s*) H atom and a proton as R goes to infinity. Suppose the electron in H_2^+ is close to nucleus A and rather far from nucleus B. The H_2^+ electronic wave function

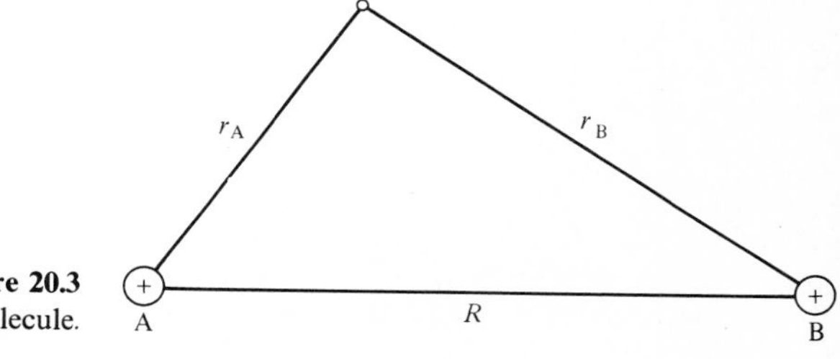

Figure 20.3
The H_2^+ molecule.

should then resemble a ground-state H-atom wave function for atom A; that is, ψ_e will be approximately given by the function (Table 19.1)

$$1s_A \equiv (1/a_0)^{3/2}\pi^{-1/2}e^{-r_A/a_0} \tag{20.11}$$

where the Bohr radius a_0 is used since the nuclei are fixed. Similarly, when the electron is close to nucleus B, ψ_e can be roughly approximated by

$$1s_B \equiv (1/a_0)^{3/2}\pi^{-1/2}e^{-r_B/a_0}$$

This then suggests as an approximate wave function for the H_2^+ ground electronic state:

$$\phi = c_A 1s_A + c_B 1s_B = a_0^{-3/2}\pi^{-1/2}(c_A e^{-r_A/a_0} + c_B e^{-r_B/a_0}) \tag{20.12}$$

which is a linear combination of the $1s_A$ and $1s_B$ atomic orbitals. When the electron is very close to nucleus A, then r_A is much less than r_B and e^{-r_A/a_0} is much greater than e^{-r_B/a_0}; hence, the $1s_A$ term in (20.12) dominates, and the wave function resembles that of an H atom at nucleus A, as it should. Similarly for the electron close to nucleus B.

The function (20.12) can be regarded as a variation function and the constants c_A and c_B chosen to minimize the variational integral $W = \int \phi^* \hat{H} \phi \, d\tau$. The function (20.12) is a linear combination of two functions, and, as noted in Sec. 18.15, the conditions $\partial W/\partial c_A = 0 = \partial W/\partial c_B$ will be satisfied by two sets of values of c_A and c_B. These sets will yield approximate wave functions and energies for the lowest two electronic states of H_2^+. We need not go through the details of evaluating W and setting $\partial W/\partial c_A = 0 = \partial W/\partial c_B$, since the fact that the nuclei are identical requires that the electron probability density be the same on each side of the molecule. Restricting ourselves to a real variation function, the electron probability density is $\phi^2 = c_A^2(1s_A)^2 + c_B^2(1s_B)^2 + 2c_A c_B 1s_A 1s_B$. To have ϕ^2 be the same at corresponding points on each side of the molecule, we must have either $c_B = c_A$ or $c_B = -c_A$. For $c_B = c_A$, we have

$$\phi = c_A(1s_A + 1s_B), \qquad \phi^2 = c_A^2(1s_A^2 + 1s_B^2 + 2 \cdot 1s_A 1s_B) \tag{20.13}$$

For $c_B = -c_A$,

$$\phi' = c_A'(1s_A - 1s_B), \qquad \phi'^2 = c_A'^2(1s_A^2 + 1s_B^2 - 2 \cdot 1s_A 1s_B) \tag{20.14}$$

The constants c_A and c_A' are evaluated by requiring that ϕ and ϕ' be normalized. The normalization constants differ for the functions in (20.13) and (20.14).

The normalization condition for the function in (20.13) is

$$1 = \int \phi^2 \, d\tau = c_A^2 \left(\int 1s_A^2 \, d\tau + \int 1s_B^2 \, d\tau + 2 \int 1s_A 1s_B \, d\tau \right)$$

The H-atom wave functions are normalized, so $\int 1s_A^2 \, d\tau = \int 1s_B^2 \, d\tau = 1$. Defining the *overlap integral* S as $S \equiv \int 1s_A 1s_B \, d\tau$, we get $1 = c_A^2(2 + 2S)$ and $c_A = (2 + 2S)^{-1/2}$. Similarly, one finds $c_A' = (2 - 2S)^{-1/2}$. Hence the normalized approximate wave functions for the lowest two H_2^+ electronic states are

$$\phi = (2 + 2S)^{-1/2}(1s_A + 1s_B), \qquad \phi' = (2 - 2S)^{-1/2}(1s_A - 1s_B) \tag{20.15}$$

For completeness, each spatial function should be multiplied by a one-electron spin function, either α or β.

The value of the overlap integral $\int 1s_A 1s_B \, d\tau$ depends on how much the electron probability densities of the functions $1s_A$ and $1s_B$ overlap each other. Only regions of space where both $1s_A$ and $1s_B$ are of significant magnitude will contribute substantially to S. The main contribution to S therefore comes from the region between the nuclei. The value of S clearly depends on the internuclear distance R. For $R = 0$, we have $1s_A = 1s_B$ and $S = 1$. For $R = \infty$, there is no overlap between the $1s_A$ and $1s_B$ atomic orbitals, and $S = 0$. For R between 0 and ∞, S is between 0 and 1. Evaluation of S (*Q.C.*, sec. 13.4) gives $S = e^{-R/a_0}(1 + R/a_0 + R^2/3a_0^2)$.

The probability density ϕ^2 in (20.13) can be written as $c_A^2(1s_A^2 + 1s_B^2)$ plus $2c_A^2 1s_A 1s_B$. The $1s_A^2 + 1s_B^2$ part of ϕ^2 is proportional to the probability density due to two separate noninteracting 1s H atoms. The term $2c_A^2 1s_A 1s_B$ is of significant magnitude only in regions where both $1s_A$ and $1s_B$ are reasonably large; this term therefore increases the electron probability density in the region between the nuclei. This buildup of probability density between the nuclei (at the expense of regions outside the internuclear region) causes the electron to feel the attractions of both nuclei at once, thereby lowering its average potential energy and providing a stable covalent bond. The bonding is due to the overlap of the atomic orbitals $1s_A$ and $1s_B$.

Figure 20.4*a* graphs ϕ and ϕ^2 of (20.13) for points along the line joining the nuclei. The probability-density buildup between the nuclei is evident.

For the function ϕ' of (20.14), the term $-2c_A'^2 1s_A 1s_B$ diminishes the electron probability density between the nuclei. At any point on a plane midway between the nuclei and perpendicular to the internuclear axis we have $r_A = r_B$ and $1s_A = 1s_B$. Hence $\phi' = 0 = \phi'^2$ on this plane, which is a *nodal plane* for the function ϕ'. Figure 20.4*b* shows ϕ' and ϕ'^2 for points along the internuclear axis.

The functions ϕ and ϕ' depend on the internuclear distance R, since r_A and r_B in $1s_A$ and $1s_B$ depend on R (see Fig. 20.3). The variational integral W is therefore a function of R. When W is evaluated for ϕ and ϕ', one finds that ϕ gives a curve with a minimum (like the lowest curve in Fig. 20.2, except that D_e is less than and R_e is greater than in H_2); in contrast, the ϕ'-vs.-R curve has no minimum and resembles the second lowest curve in Fig. 20.2. These facts are understandable in terms of the preceding electron probability-density discussion.

The true values of R_e and D_e for H_2^+ are 1.06 Å and 2.8 eV. The function (20.13) gives $R_e = 1.32$ Å and $D_e = 1.8$ eV, which is rather poor. Substantial improvement can be obtained if a variational parameter ζ is included in the exponentials, so that $1s_A$ and $1s_B$ become proportional to $e^{-\zeta r_A/a_0}$ and $e^{-\zeta r_B/a_0}$. One then finds $R_e = 1.07$ Å and $D_e = 2.35$ eV.

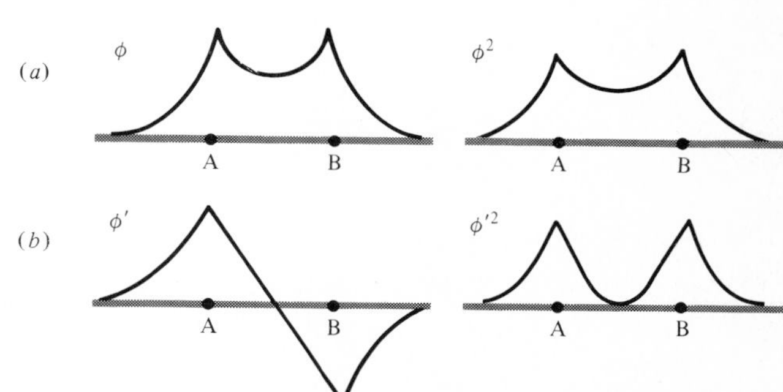

Figure 20.4
Graphs of (*a*) ground state and (*b*) first excited state H_2^+ approximate wave functions for points along the internuclear axis. (Not to scale.)

Recall that an orbital is a one-electron spatial wave function. H_2^+ has but one electron, and the approximate wave functions ϕ and ϕ' in (20.13) and (20.14) are approximations to the orbitals of the two lowest electronic states of H_2^+. An orbital for an atom is called an atomic orbital (AO). An orbital for a molecule is a *molecular orbital* (MO). Just as the wave function of a many-electron atom can be approximated by use of AOs, the wave function of a many-electron molecule can be approximated by use of MOs. Each MO can hold two electrons of opposite spin.

The situation is more complicated for molecules than for atoms, in that the number of nuclei vary from molecule to molecule. Whereas hydrogenlike orbitals with effective nuclear charges are useful for all many-electron atoms, the H_2^+-like orbitals with effective nuclear charges are directly applicable only to molecules with two identical nuclei, i.e., *homonuclear* diatomic molecules. We shall later see, however, that since molecules are composed of bonds, and since (with some exceptions) each bond is between two atoms, we can construct an approximate molecular wave function using bond orbitals (and lone-pair and inner-shell orbitals), where the bond orbitals resemble diatomic-molecule MOs.

Let us consider further excited electronic states of H_2^+. We expect such states to dissociate to a proton and a $2s$ or $2p$ or $3s$ or ... H atom. Therefore, analogous to the functions (20.13) and (20.14), we write as approximate wave functions (molecular orbitals) for excited H_2^+ states

$$N(2s_A + 2s_B), \quad N(2s_A - 2s_B), \quad N(2p_{xA} + 2p_{xB}), \quad N(2p_{xA} - 2p_{xB}), \quad \text{etc.} \tag{20.16}$$

where the normalization constant N differs for different states. (Actually, because of the degeneracy of the $2s$ and $2p$ states in the H atom, we should expect extensive mixing together of $2s$ and $2p$ AOs in the H_2^+ MOs. Since we are mainly interested in H_2^+ MOs for use in many-electron molecules, and since the $2s$ and $2p$ levels are not degenerate in many-electron atoms, we shall ignore such mixing, at least for now.) A wave function like $2s_A + 2s_B$ expresses the fact that there is a 50-50 probability which nucleus the electron will go with when the molecule dissociates ($R \to \infty$).

The MOs in (20.15) and (20.16) are linear combinations of atomic orbitals and hence are called LCAO MOs. (There is no necessity for MOs to be expressed as linear combinations of AOs, but this approximate form is a very convenient one.) Let us see what these MOs look like.

Addition of $1s_A$ to $1s_B$ gives a buildup of probability density between the nuclei, whereas the linear combination $1s_A - 1s_B$ has a nodal plane between the nuclei.

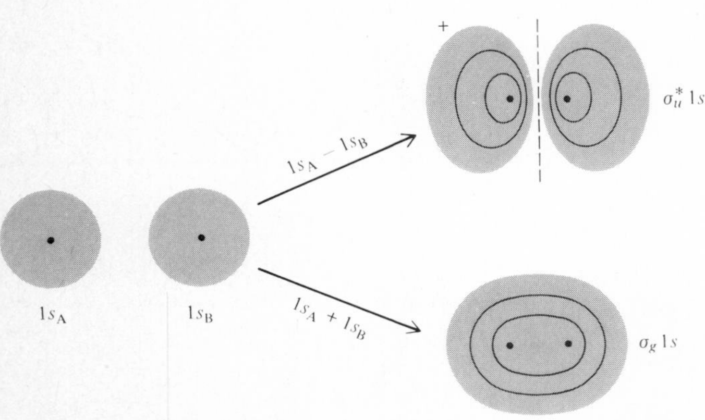

Figure 20.5
Formation of homonuclear diatomic MOs from $1s$ AOs. The dashed line indicates a nodal plane.

Figure 20.5 shows contours of constant probability density for the two MOs formed from 1*s* AOs. The three-dimensional shape of these orbitals is obtained by rotating the contours about the line joining the nuclei.

A word about terminology. The component of electronic orbital angular momentum along the internuclear axis of H_2^+ can be shown to have the possible values $m\hbar$, where $m = 0, \pm 1, \pm 2, \ldots$. (Unlike the H atom, there is no l quantum number in H_2^+, since the magnitude of the total electronic orbital angular momentum is not fixed in H_2^+. This is related to the fact that there is spherical symmetry in H but only axial symmetry in H_2^+.) The following code letters are used to indicate the $|m|$ value:

$\lvert m \rvert$	0	1	2	3	...
Letter	σ	π	δ	ϕ	...

(20.17)

(These are the Greek equivalents of *s, p, d, f*.) The AOs $1s_A$ and $1s_B$ have zero electronic orbital angular momentum along the molecular axis, and so the two MOs formed from these AOs have $m = 0$ and from (20.17) are σ (sigma) MOs. We call these the $\sigma_g 1s$ MO and the $\sigma_u^* 1s$ MO. The 1*s* indicates that they originate from separated-atom 1*s* AOs. The star indicates the *antibonding* character of the $1s_A - 1s_B$ MO, associated with the nodal plane and the charge depletion between the nuclei.

The *g* subscript (from the German *gerade*, meaning "even") indicates that the orbital has the same value at two points that are on diagonally opposite sides of the center of the molecule and are equidistant from the center. The *u* subscript (*ungerade*, "odd") indicates that the values of the orbital differ by a factor -1 at two such points.

The linear combinations $2s_A + 2s_B$ and $2s_A - 2s_B$ give the $\sigma_g 2s$ and $\sigma_u^* 2s$ MOs, whose shapes resemble those of the $\sigma_g 1s$ and $\sigma_u^* 1s$ MOs.

Let the molecular axis be the *z* axis. Because of the opposite signs of the right lobe of $2p_{zA}$ and the left lobe of $2p_{zB}$ (Fig. 20.6), the linear combination $2p_{zA} + 2p_{zB}$ has a nodal plane midway between the nuclei, as indicated by the dashed line; the charge depletion between the nuclei makes this an antibonding MO. The linear combination $2p_{zA} - 2p_{zB}$ gives charge buildup between the nuclei and is a bonding MO. The $2p_z$ AO has atomic quantum number $m = 0$ (Sec. 19.3) and hence has a *z* component of electronic orbital angular momentum equal to zero. The MOs formed from $2p_z$ AOs are therefore σ MOs, the $\sigma_g 2p$ and $\sigma_u^* 2p$ MOs. Their three-dimensional shapes are obtained by rotating the contours in Fig. 20.6 about the internuclear (*z*) axis.

Formation of homonuclear diatomic MOs from the $2p_x$ AOs is shown in Fig. 20.7. The $2p_x$ AO is a linear combination of $m = 1$ and $m = -1$ AOs [Eq. (19.23)] and has $|m| = 1$. Hence, the MOs made from the $2p_x$ AOs have $|m| = 1$ and are

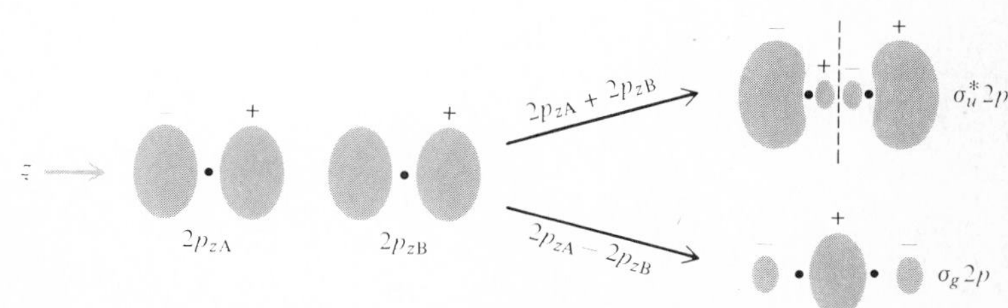

Figure 20.6
Formation of homonuclear diatomic MOs from $2p_z$ AOs.

π MOs [Eq. (20.17)]. The linear combination $N(2p_{xA} + 2p_{xB})$ has charge buildup in the internuclear regions above and below the z axis and is therefore bonding. This MO has opposite signs at the diagonally opposite points c and d in Fig. 20.7 and hence is a u MO, the $\pi_u 2p_x$ MO. The linear combination $N(2p_{xA} - 2p_{xB})$ gives the antibonding $\pi_g^* 2p_x$ MO.

The σ MOs in Figs. 20.5 and 20.6 are symmetric about the internuclear axis; the orbital shapes are figures of rotation about the z axis. In contrast, the $\pi_u 2p_x$ and $\pi_g^* 2p_x$ MOs consist of blobs of probability density above and below the yz plane (which is a nodal plane for these MOs).

The linear combinations $2p_{yA} + 2p_{yB}$ and $2p_{yA} - 2p_{yB}$ give the $\pi_u 2p_y$ and $\pi_g^* 2p_y$ MOs. These MOs have the same shapes as the $\pi_u 2p_x$ and $\pi_g^* 2p_x$ MOs but are rotated by 90° about the internuclear axis compared with the $\pi 2p_x$ MOs. Since they have the same shapes, the $\pi_u 2p_x$ and $\pi_u 2p_y$ MOs have the same energy; likewise, the $\pi_g^* 2p_x$ and $\pi_g^* 2p_y$ MOs have the same energy (see Fig. 20.8).

The σ MOs have no nodal planes containing the internuclear axis. (Of course, some σ MOs have a nodal plane or planes perpendicular to the internuclear axis.) Each π MO has one nodal plane containing the internuclear axis. (This is true provided one uses the real $2p$ AOs to form the MOs, as we have done.) It turns out that δ MOs have two nodal planes containing the internuclear axis (see Fig. 20.19c). We shall later use the number of nodal planes to classify bond orbitals in polyatomic molecules.

20.4 THE SIMPLE MO METHOD FOR DIATOMIC MOLECULES

MOs for homonuclear diatomic molecules. Just as we constructed approximate wave functions for many-electron atoms by feeding electrons two at a time into hydrogenlike AOs, we shall construct approximate wave functions for many-electron homonuclear diatomic molecules by feeding electrons two at a time into H_2^+-like MOs. Figure 20.8 shows the lowest-lying H_2^+-like MOs (Sec. 20.3). Similar to AO energies (Fig. 19.13), the energies of these MOs vary from molecule to molecule; they also vary with varying internuclear distance in the same molecule. The energy order shown in the figure is the order in which the MOs are filled in going through the periodic table. The AOs at the sides of each MO are those used to form the MO. Note that each pair of AOs leads to formation of two MOs, a bonding MO with energy lower than that of the AOs and an antibonding MO with energy higher than that of the AOs.

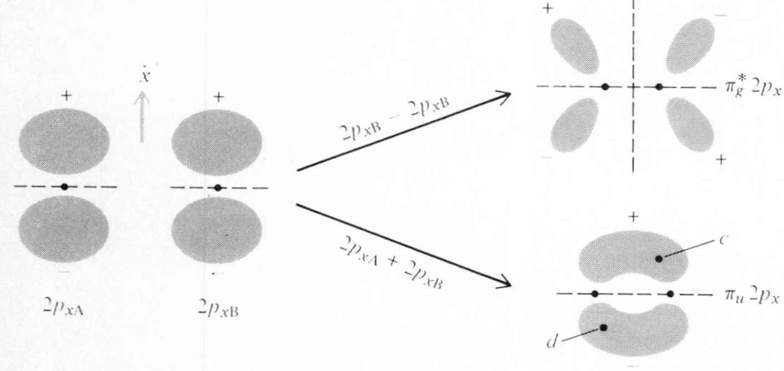

Figure 20.7
Formation of homonuclear diatomic MOs from $2p_x$ AOs.

The hydrogen molecule. H_2 consists of two protons (A and B) and two electrons (1 and 2). The electronic Hamiltonian (including nuclear repulsion) is given by Eqs. (20.7), (19.4), and (18.48) as

$$\hat{H}_e = -\frac{\hbar^2}{2m}\nabla_1^2 - \frac{\hbar^2}{2m}\nabla_2^2 - \frac{e'^2}{r_{1A}} - \frac{e'^2}{r_{1B}} - \frac{e'^2}{r_{2A}} - \frac{e'^2}{r_{2B}} + \frac{e'^2}{r_{12}} + \frac{e'^2}{R} \tag{20.18}$$

where r_{1A} is the distance between electron 1 and nucleus A, r_{12} is the distance between the electrons, and R is the distance between the nuclei. The first two terms are the kinetic-energy operators for electrons 1 and 2, the next four terms are the potential energy of attractions between the electrons and the nuclei, e'^2/r_{12} is the potential energy of repulsion between the electrons, and e'^2/R is the internuclear repulsion. R is held fixed.

Because of the interelectronic repulsion term e'^2/r_{12}, the electronic Schrödinger equation $\hat{H}_e\psi_e = E_e\psi_e$ cannot be solved exactly for H_2. If this term is ignored, we get an approximate electronic Hamiltonian that is the sum of two H_2^+-like electronic Hamiltonians, one for electron 1 and one for electron 2. [This isn't quite true, because the internuclear repulsion e'^2/R is the same in (20.10) and (20.18). However, e'^2/R is a constant and therefore only shifts the energy by e'^2/R but does not affect the wave functions; see Prob. 20.16.] The approximate electronic wave function for H_2 is then the product of two H_2^+-like electronic wave functions, one for each electron. (This is exactly analogous to approximating the He wave function by the product of two H-like wave functions.)

The first function in Eq. (20.15) is an approximate wave function for the H_2^+ ground electronic state, so that our MO approximation to the H_2 ground-electronic-state spatial wave function is

$$\sigma_g 1s(1) \cdot \sigma_g 1s(2) = N[1s_A(1) + 1s_B(1)] \cdot [1s_A(2) + 1s_B(2)] \tag{20.19}$$

where the normalization constant N is $(2 + 2S)^{-1}$. The numbers in parentheses refer to the electrons. Thus, $1s_A(2)$ is proportional to e^{-r_{2A}/a_0}. The MO wave function (20.19) is analogous to the He ground-state AO wave function $1s(1)1s(2)$ in Eq. (19.30); the MO $\sigma_g 1s$ replaces the AO $1s$.

To take care of spin and the Pauli principle, the symmetric two-electron spatial function (20.19) must be multiplied by the antisymmetric spin function (19.33). The

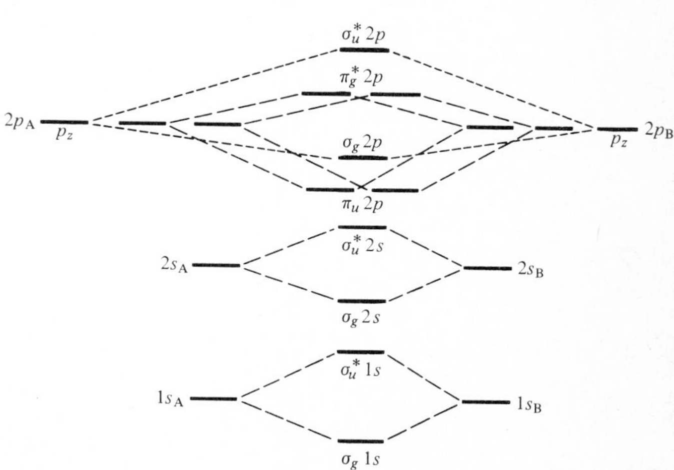

Figure 20.8
Lowest-lying homonuclear diatomic MOs.

approximate MO ground-state wave function for H_2 is then

$$\sigma_g 1s(1)\sigma_g 1s(2)2^{-1/2}[\alpha(1)\beta(2) - \beta(1)\alpha(2)] = \frac{1}{\sqrt{2}}\begin{vmatrix} \sigma_g 1s(1)\alpha(1) & \sigma_g 1s(1)\beta(1) \\ \sigma_g 1s(2)\alpha(2) & \sigma_g 1s(2)\beta(2) \end{vmatrix} \qquad (20.20)$$

where we introduced our old friend, the Slater determinant. Just as the ground-state electronic configuration of He is $1s^2$, the ground-state electronic configuration of H_2 is $(\sigma_g 1s)^2$; compare Eqs. (20.20) and (19.35).

We have put each electron in H_2 into an MO. Doing so allows for interelectronic repulsion at best only in an average way, so the treatment is an approximate one. We are using the crudest possible version of the MO approximation.

Using the approximate wave function (20.20), one evaluates the variational integral W to get W as a function of R. Since evaluation of molecular quantum-mechanical integrals is complicated, we shall just quote the results. With inclusion of a variable orbital exponent, the function (20.20) gives a $W(R)$ curve with a minimum at $R = 0.73$ Å, which is close to the observed value $R_e = 0.74$ Å in H_2. The calculated D_e is 3.49 eV, in not very good agreement with the experimental value 4.75 eV. (This is the main failing of the MO method; molecular dissociation energies are not accurately calculated.)

We approximated the $\sigma_g 1s$ MO in the H_2 ground-state approximate wave function (20.19) by the linear combination $N(1s_A + 1s_B)$. To improve the MO wave function, we can look for the best possible form for the $\sigma_g 1s$ MO, still writing the spatial wave function as the product of an orbital for each electron. The best possible MO wave function is the Hartree–Fock wave function (Secs. 19.7 and 20.5). Finding the Hartree–Fock wave function for H_2 is not too difficult. The H_2 Hartree–Fock wave function gives $R_e = 0.73$ Å and $D_e = 3.64$ eV; D_e is still substantially in error. As noted earlier, the Hartree–Fock wave function is not the true wave function, due to neglect of electron correlation.

In the 1960s, Kolos and Wolniewicz used very complicated variational functions that go beyond the Hartree–Fock approximation. With the inclusion of relativistic corrections and corrections for deviations from the Born–Oppenheimer approximation, they calculated $D_0/hc = 36{,}117.9\ \text{cm}^{-1}$ for H_2. (D_0 differs from D_e by the zero-point vibrational energy; see Chap. 21.) At the time the calculation was completed, the experimental D_0 was $36{,}114 \pm 1\ \text{cm}^{-1}$, and the 4 cm^{-1} discrepancy was a source of embarrassment to the theoreticians. Finally, a reinvestigation of the spectrum of H_2 showed that the experimental result was in error and gave the new experimental value $36{,}117.3 \pm 1\ \text{cm}^{-1}$, in excellent agreement with the value calculated from quantum mechanics.

What about excited electronic states for H_2? The lowest-lying excited H_2 MO is the $\sigma_u^* 1s$ MO. Just as the lowest excited electronic configuration of He is $1s2s$, the lowest excited electronic configuration of H_2 is $(\sigma_g 1s)(\sigma_u^* 1s)$, with one electron in each of the MOs $\sigma_g 1s$ and $\sigma_u^* 1s$. Like the He $1s2s$ configuration, the $(\sigma_g 1s)(\sigma_u^* 1s)$ H_2 configuration gives rise to two terms, a singlet with total spin quantum number $S = 0$ and a triplet with total spin quantum number $S = 1$. In accord with Hund's rule, the triplet lies lower and is therefore the lowest excited electronic level of H_2. In analogy with (19.38), the triplet has the MO wave functions

$$2^{-1/2}[\sigma_g 1s(1)\sigma_u^* 1s(2) - \sigma_g 1s(2)\sigma_u^* 1s(1)] \times \text{spin function} \qquad (20.21)$$

where the spin function is one of the three symmetric spin functions (19.32). With one electron in a bonding orbital and one in an antibonding orbital, we would expect no net bonding. This is borne out by experiment and by accurate theoretical calculations, which show the $E_e(R)$ curve to have no minimum (Fig. 20.2).

The H_2 levels (20.20) and (20.21) both dissociate to two H atoms in 1*s* states. The bonding level (20.20) has the electrons paired with opposite spins and a net spin of zero. The repulsive level (20.21) has the electrons unpaired with approximately parallel spins. Whether two approaching 1*s* H atoms attract or repel each other depends on whether their spins are antiparallel or parallel.

Other homonuclear diatomic molecules. The simple MO treatment of He_2 feeds the four electrons into the two lowest available MOs to give the ground-state configuration $(\sigma_g 1s)^2(\sigma_u^* 1s)^2$. The MO wave function is a Slater determinant with four rows and four columns. With two bonding and two antibonding electrons, we expect no net bonding and no stability for the ground electronic state. This is in agreement with experiment. When two ground-state He atoms approach each other, the electronic energy curve $E_e(R)$ resembles the second lowest curve of Fig. 20.2. Since $E_e(R)$ is the potential energy for nuclear motion, two $1s^2$ He atoms strongly repel each other. (Actually, in addition to the strong, relatively short-range repulsion, there is a very weak attraction at relatively large values of R that produces a very slight minimum in the He-He potential-energy curve. This attraction is responsible for the liquefaction of He at very low temperature. See Sec. 22.10 on intermolecular forces for further discussion.)

Similar to the repulsion between two $1s^2$ He atoms is the observed repulsion whenever two closed-shell atoms or molecules approach each other closely. This repulsion is important in chemical kinetics since it is the source of the activation energy of chemical reactions (see Sec. 23.2). Part of this repulsion is attributable to the Coulomb repulsion between electrons, but a major part of the repulsion is a consequence of the Pauli principle, as we now show. Let $\psi(q_1, q_2, q_3, \ldots)$ be the wave function for a system of electrons, where q_1 stands for the four coordinates (three spatial and one spin) of electron 1. The Pauli antisymmetry principle (Sec. 19.5) requires that interchange of the coordinates of electrons 1 and 2 multiply ψ by -1. Hence, $\psi(q_2, q_1, q_3, \ldots) = -\psi(q_1, q_2, q_3, \ldots)$. Now suppose that electrons 1 and 2 have the same spin coordinate (both α or both β) and the same spatial coordinates. Then $q_1 = q_2$, and $\psi(q_1, q_1, q_3, \ldots) = -\psi(q_1, q_1, q_3, \ldots)$. Hence, $2\psi(q_1, q_1, \ldots) = 0$, and $\psi(q_1, q_1, q_3, \ldots) = 0$.

The vanishing of $\psi(q_1, q_1, q_3, \ldots)$ shows that there is zero probability for two electrons to have the same spatial and spin coordinates. Two electrons having the same spin have zero probability of being at the same point in space. Moreover, because ψ is a continuous function, the probability that two electrons with the same spin will approach each other closely must be very small. Electrons with the same spin tend to avoid each other and act as if they repelled each other over and above the Coulomb repulsion. This apparent extra repulsion of electrons with like spins is called the *Pauli repulsion*. (It is sometimes mistakenly said that the Pauli repulsion is due to magnetic forces between spins. This is not so. Magnetic forces are weak and generally can be neglected in atoms and molecules.) The Pauli repulsion is not a real physical force; it is an apparent force that is a consequence of the antisymmetry requirement of the wave function.

When two $1s^2$ He atoms approach, the antisymmetry requirement causes an apparent Pauli repulsion between the spin-α electron on one atom and the spin-α electron on the other atom; likewise for the spin-β electrons. As the He atoms approach each other, there is a depletion of electron probability density in the region between the nuclei (and a corresponding buildup of probability density in regions outside the nuclei) and the atoms repel each other.

We can form the ground-state electron configurations of Li_2, Be_2, etc., by filling in the homonuclear diatomic MOs in Fig. 20.8 (see Prob. 20.13). For example, O_2 has 16 electrons, and the ground-state configuration is

$$(\sigma_g 1s)^2(\sigma_u^* 1s)^2(\sigma_g 2s)^2(\sigma_u^* 2s)^2(\pi_u 2p)^4(\sigma_g 2p)^2(\pi_g^* 2p)^2$$

Figure 20.9 Occupied valence MOs in the O_2 ground electronic state. (Not to scale.)

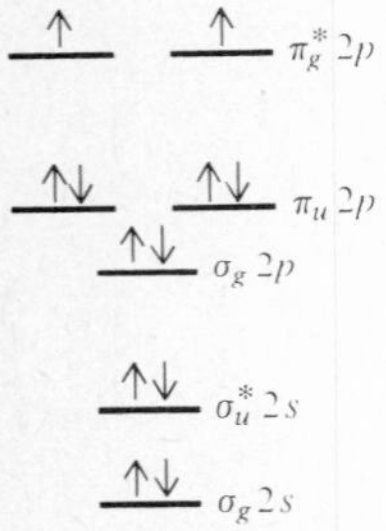

$\pi_g^* 2p$ ↑ ↑
$\pi_u 2p$ ↑↓ ↑↓
$\sigma_g 2p$ ↑↓
$\sigma_u^* 2s$ ↑↓
$\sigma_g 2s$ ↑↓

Figure 20.9 shows the distribution of the valence electrons in MOs. (Spectroscopic evidence indicates that in O_2 the $\sigma_g 2p$ MO is slightly lower than the $\pi_u 2p$ MOs.) Note that in accord with Hund's rule of maximum multiplicity for the ground state, the two antibonding π electrons are placed in separate orbitals to allow a triplet ground state. This is in accord with the observed paramagnetism of ground-state O_2. In O_2, there are four more bonding electrons than antibonding electrons, and so the MO theory predicts a double bond (composed of one σ bond and one π bond) for O_2. Note the higher D_e for O_2 compared with the single-bonded species F_2 and Li_2 (Table 20.3). The double bond makes R_e of O_2 less than R_e of Li_2. (R_e of O_2 is greater than R_e of H_2 because of the presence of the inner-shell $1s$ electrons on the O atoms.)

In O_2, the high nuclear charge draws the $1s$ orbitals on each atom in close to the nuclei, and there is virtually no overlap between these AOs. Therefore, the $\sigma_g 1s$ and $\sigma_u^* 1s$ MO energies in O_2 are each nearly the same as the $1s$ AO energy in an O atom. Inner-shell electrons play no real part in chemical bonding, other than to screen the valence electrons from the nuclei.

The F_2 MO configuration is $\ldots(\sigma_g 2s)^2(\sigma_u^* 2s)^2(\sigma_g 2p)^2(\pi_u 2p)^4(\pi_g^* 2p)^4$. The $(\pi_g^*)^4$ electrons deplete the probability density between the nuclei, thereby canceling the buildup of probability density due to the $(\pi_u)^4$ electrons; this gives the effect of four lone pairs of π electrons, two lone pairs on each atom. Likewise, the $(\sigma_g 2s)^2$ and $(\sigma_u^* 2s)^2$ electrons act like two lone pairs, one pair on each atom. This leaves the $(\sigma_g 2p)^2$ electrons to provide the single bond. (This is in accord with the Lewis dot formula $:\ddot{\underset{..}{F}}:\ddot{\underset{..}{F}}:$.)

Heteronuclear diatomic molecules. The MO method feeds the electrons of a heteronuclear diatomic molecule into molecular orbitals. In the crudest approximation, each bonding MO is taken as a linear combination of two AOs, one from each atom. In constructing MOs, one uses the principle that only AOs of reasonably similar energies contribute substantially to a given MO.

As an example, consider HF. Figure 19.13 shows that the energy of a $2p$ AO in $_9F$ is reasonably close to the $1s$ AO energy in H, but the $2s$ AO in F is substantially lower in energy than the $1s$ H AO. (The logarithmic scale makes the fluorine $2s$ level appear closer to the $2p$ level than it actually is.) The $2p$ AO in F lies somewhat lower than the $1s$ AO in H because the five $2p$ electrons in F screen one another rather poorly, giving a large Z_{eff} for the $2p$ electrons [Eq. (19.46)]; this large Z_{eff} makes F more electronegative than H [Eq. (20.4)].

Let the HF molecular axis be the z axis, and let F$2p$ and H$1s$ denote a $2p$ AO on F and a $1s$ AO on H. The F$2p_z$ AO has quantum number $m = 0$ and has no nodal plane containing the internuclear axis. The overlap of this AO with the H1s AO, which also has $m = 0$ and no nodal plane containing the z axis, therefore gives rise to a σ MO (Fig. 20.10). We therefore form the linear combination $c_1\text{H}1s + c_2\text{F}2p_z$. Minimization of the variational integral will lead to two sets of values for c_1 and c_2, one set giving a bonding MO and the other an antibonding MO:

$$\sigma = c_1\text{H}1s + c_2\text{F}2p_z \quad \text{and} \quad \sigma^* = c_1'\text{H}1s - c_2'\text{F}2p_z \tag{20.22}$$

The σ MO in (20.22) has c_1 and c_2 both positive and is bonding because of the charge buildup between the nuclei. The antibonding σ^* MO in (20.22) has opposite signs for the coefficients of the AOs and so has charge depletion between the nuclei; this MO is unoccupied in the HF ground state. (The g, u designation is not applicable to heteronuclear diatomics.)

In contrast to the F2p_z AO, the F2p_x and F2p_y AOs have $|m| = 1$ and have one nodal plane containing the internuclear (z) axis. These AOs will therefore be used to form π MOs in HF. Since H has no valence-shell AOs with $|m| = 1$, the π MOs in HF will consist entirely of F AOs, and we have $\pi_x = \text{F}2p_x$ and $\pi_y = \text{F}2p_y$ as the π MOs in HF.

The 1s and 2s AOs in F are too low in energy to take a substantial part in the bonding and therefore form nonbonding σ MOs in HF. (Don't confuse a nonbonding MO with an antibonding MO. A nonbonding MO shows neither charge depletion nor charge buildup between the nuclei.)

In the standard notation for heteronuclear diatomic molecules, the lowest σ MO is called the 1σ MO, the next lowest σ MO is called the 2σ MO, etc. The lowest π energy level is called the 1π level, etc. In our rather crude approximation, we have as the occupied MOs in hydrogen fluoride

$$\begin{gathered} 1\sigma = \text{F}1s, \qquad 2\sigma = \text{F}2s, \qquad 3\sigma = c_1\text{H}1s + c_2\text{F}2p_z \\ 1\pi_x = \text{F}2p_x, \qquad 1\pi_y = \text{F}2p_y \end{gathered} \tag{20.23}$$

where $1\pi_x$ and $1\pi_y$ have the same energy. Since F is more electronegative than H, we expect $|c_2| > |c_1|$ in the 3σ MO; the electrons of the bond are more likely to be found close to F than to H.

Figure 20.11 shows the energy-level scheme for HF in the simple approximation (20.23). The 1π MOs are lone-pair AOs on F and have the same energy as F2p AOs; the 2σ MO is also a lone-pair orbital.

An H atom is special since it has no p valence orbitals. Consider the general case of a polar-covalent heteronuclear diatomic molecule AB, where both A and B are from the second or a higher period and hence have s and p valence levels. Let B be more electronegative than A. We draw Fig. 20.12 similar to Figs. 20.8 and 20.11 to show the formation of valence MOs from the ns and np valence AOs of A and the $n's$ and $n'p$ valence AOs of B. (n and n' are the principal quantum numbers of the valence electrons and equal the periods of A and B in the periodic table.) The MO shapes are

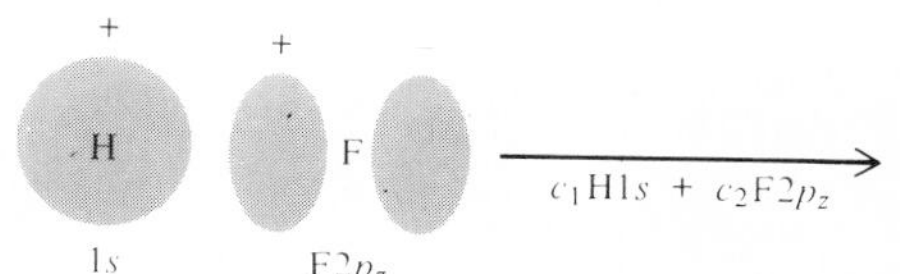

Figure 20.10
Formation of the bonding MO in HF.

similar to those in Figs. 20.5 to 20.7 for homonuclear diatomics, except that in each bonding MO the probability density is greater around the more electronegative element B than around A and each bonding MO contour is therefore larger around B than A. In each antibonding MO, the probability density is larger around A, since more of the atom-B AO has been "used up" in forming the corresponding bonding MO.

To get the valence MO configuration of molecules like CN, NO, CO, or FCl, we feed the valence electrons into the MOs of Fig. 20.12. For example, CO has 10 valence electrons and has the configuration $(\sigma_s)^2(\sigma_s^*)^2(\pi)^4(\sigma_p)^2$. With six more bonding than antibonding electrons, the molecule has a triple bond (composed of one σ and two π bonds), in accord with the dot structure :C≡O:. The lowest two MOs in CO are the 1σ and 2σ MOs, formed from linear combinations of C1s and O1s AOs, so that the complete MO configuration of CO in the standard notation is $1\sigma^2 2\sigma^2 3\sigma^2 4\sigma^2 1\pi^4 5\sigma^2$.

20.5 SCF, HARTREE–FOCK, AND CI WAVE FUNCTIONS

The best possible wave function with electrons assigned to orbitals is the Hartree–Fock wave function. Starting in the 1960s, the use of electronic computers allowed Hartree–Fock wave functions for many molecules to be calculated. The Hartree–Fock orbitals ϕ_i of a molecule must be found by solving the Hartree–Fock equations (19.47): $\hat{F}\phi_i = \varepsilon_i \phi_i$. As is done for atoms, each Hartree–Fock MO is expressed as a linear combination of a set of functions called basis functions. If sufficient basis functions are included, one can get MOs that differ negligibly from the true Hartree–Fock MOs. Any functions whatever can be used as basis functions, so long as they form a complete set (as defined in Sec. 19.7). However, molecules are made of atoms, and it is most convenient to use atomic orbitals as the basis functions. Each MO is then written as a linear combination of the basis-set AOs, and the coefficients of each AO are found by solving the Hartree–Fock equations.

To have an accurate representation of an MO requires that the MO be expressed as a linear combination of a complete set of functions. This means that all the AOs of a given atom, whether occupied or unoccupied in the free atom, contribute to

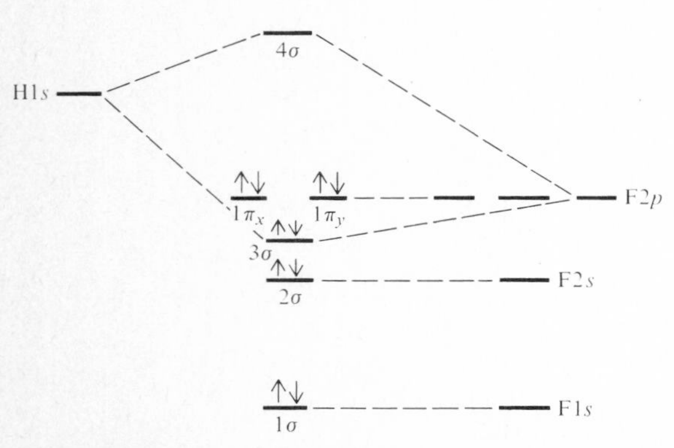

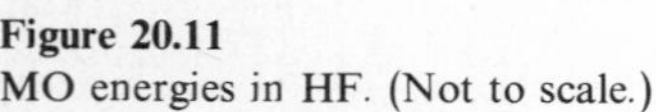

Figure 20.11
MO energies in HF. (Not to scale.)

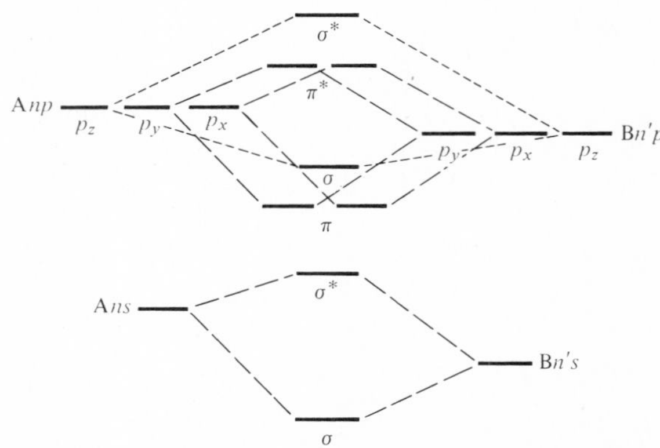

Figure 20.12
MOs formed from valence s and p AOs of atoms A and B.

the MOs. To simplify the calculation, one frequently solves the Hartree–Fock equations (19.47) using in the basis set only those AOs from each atom whose principal quantum number does not exceed the principal quantum number of the atom's valence electrons. Such a basis set limited to inner-shell and valence-shell AOs is called a *minimal basis set*. Use of a minimal basis set gives only an approximation to the Hartree–Fock MOs. Any wave function found by solving the Hartree–Fock equations is called a *self-consistent-field* (SCF) wave function; only if the basis set is very large is an SCF wave function accurately equal to the Hartree–Fock wave function.

Which AOs contribute to a given MO is determined by the symmetry properties of the MO. For example, we saw in the previous section that MOs of a diatomic molecule can be classified as σ, π, δ, ... according to whether they have 0, 1, 2, ... nodal planes containing the internuclear axis; only AOs that have 0 such nodal planes contribute to a σ MO; only AOs that have 1 such nodal plane contribute to a π MO; etc. In the previous section, we took each diatomic MO as a linear combination of only two AOs. This is the crudest approximation and does not give an accurate representation of MOs. In actuality, all σ AOs of the two atoms contribute to each σ MO; similarly for π MOs. (By a σ AO, we mean one with no nodal plane containing the internuclear axis.)

Consider, for example, a minimal-basis-set calculation of HF. The valence electron in H has $n = 1$, so we use only the H1s AO. The valence electrons in F have $n = 2$, so we use the 1s, 2s, 2p_x, 2p_y, and 2p_z AOs of F. This gives a total of six basis functions. The H1s, F1s, F2s, and F2p_z AOs each have 0 nodal planes containing the internuclear (z) axis, so each σ MO of the HF molecule is a linear combination of these four AOs. Solution of the Hartree–Fock equations using this minimal basis set gives the occupied σ MOs as [B. J. Ransil, *Rev. Mod. Phys.*, **32**, 245 (1960)]

$$
\begin{aligned}
1\sigma &= 1.000(\mathrm{F}1s) + 0.012(\mathrm{F}2s) + 0.002(\mathrm{F}2p_z) - 0.003(\mathrm{H}1s) \\
2\sigma &= -0.018(\mathrm{F}1s) + 0.914(\mathrm{F}2s) + 0.090(\mathrm{F}2p_z) + 0.154(\mathrm{H}1s) \\
3\sigma &= -0.023(\mathrm{F}1s) - 0.411(\mathrm{F}2s) + 0.711(\mathrm{F}2p_z) + 0.516(\mathrm{H}1s)
\end{aligned}
\qquad (20.24)
$$

The 1σ MO has a significant contribution only from the F1s AO. The 2σ MO has a significant contribution only from the F2s AO. This is in accord with the simple arguments that led to the very approximate MOs in (20.23). The 3σ MO has its largest contributions from H1s and F2p_z but [unlike the crude approximation of (20.23)] also has a significant contribution from the F2s AO. The mixing together of two or more AOs on the same atom is called *hybridization*.

The F2p_x and F2p_y AOs each have one nodal plane containing the internuclear axis, and these form the occupied π MOs of HF:

$$1\pi_x = \mathrm{F}2p_x, \qquad 1\pi_y = \mathrm{F}2p_y \qquad (20.25)$$

The 1π level is doubly degenerate, so any two linear combinations of the orbitals in (20.25) could be used. (Recall the theorem about degenerate levels in Sec. 18.11.)

For the homonuclear molecule F_2, a minimal-basis-set SCF calculation (Ransil, op. cit.) gives the MO we previously called the $\sigma_g 2p$ MO as $-0.005(1s_A + 1s_B) - 0.179(2s_A + 2s_B) + 0.648(2p_{zA} - 2p_{zB})$. This can be compared with the simple expression $N(2p_{zA} - 2p_{zB})$ used earlier. When a large enough basis set is used to give the Hartree–Fock MOs, this F_2 MO is found to have small contributions also from 3*s*, 3*dσ*, and 4*fσ* AOs, where 3*dσ* and 4*fσ* signify AOs with no nodal planes containing the molecular axis.

To reach the true wave function, one must go beyond the Hartree–Fock approximation and introduce configuration interaction (Sec. 19.7). The CI wave function is a linear combination of a large number of Slater determinants, each determinant corresponding to a different occupancy of the orbitals. Accurate CI wave functions are extremely difficult to calculate, even for rather small molecules.

20.6 THE MO TREATMENT OF POLYATOMIC MOLECULES

As with diatomic molecules, one expresses the MOs of a polyatomic molecule as linear combinations of basis functions. Most commonly, AOs of the atoms forming the molecule are used as the basis functions. To find the coefficients in the linear combinations, one solves the Hartree–Fock equations (19.47). Which AOs contribute to a given MO is determined by the symmetry of the molecule.

The BeH_2 molecule. We shall apply the MO method to BeH_2. Since the valence shell of Be has $n = 2$, a minimal-basis-set calculation uses the Be1*s*, Be2*s*, Be$2p_x$, Be$2p_y$, Be$2p_z$ AOs and the H_A1s and H_B1s AOs, where H_A and H_B are the two H atoms. The molecule has six electrons, and these will fill the lowest three MOs in the ground state.

Accurate theoretical calculations show that the equilibrium geometry is linear and symmetric (HBeH), and we shall assume this structure. Each MO of this linear molecule can be classified as σ, π, δ, ... according to whether it has 0, 1, 2, ... nodal planes containing the internuclear axis. Further, since the molecule has a center of symmetry at the Be nucleus, we can classify each MO as *g* or *u* (as we did with homonuclear diatomics), according to whether it has the same or opposite signs on diagonally opposite sides of the Be atom.

The Be1*s* AO has a much lower energy than all the other AOs in the basis set (Fig. 19.13), so the lowest MO will be nearly identical to the Be1*s* AO. The function Be1*s* has no nodal planes containing the internuclear axis and is a σ function; it also has *g* symmetry. We therefore write

$$1\sigma_g = \text{Be}1s \tag{20.26}$$

where the 1 indicates this is the lowest σ_g MO.

The 2*s* and 2*p* valence AOs of Be and the 1*s* valence AOs of H_A and H_B have comparable energies and will be combined to form the remaining occupied MOs. In forming these MOs one must take the symmetry of the molecule into account. An MO without either *g* or *u* symmetry could not be a solution of the BeH_2 Hartree–Fock equations. (The proof of this requires group theory and is omitted.) For a BeH_2 MO to have *g* or *u* symmetry (i.e., for the square of the MO to have the same value at

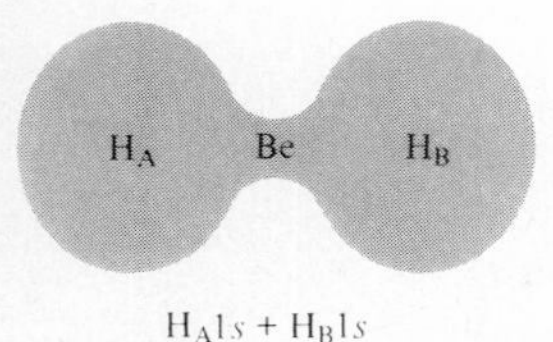

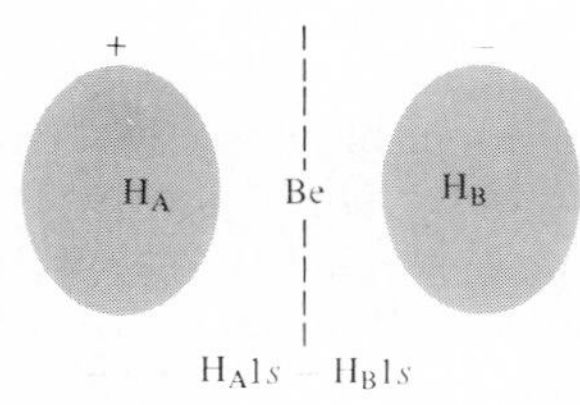

Figure 20.13
Linear combinations of H-atom 1*s* AOs in BeH_2 that have suitable symmetry.

corresponding diagonally opposite points on each side of the central Be atom) the squares of the coefficients of the H_A1s and H_B1s AOs must be equal in each BeH_2 MO. Just as the $1s_A$ and $1s_B$ AOs in a homonuclear diatomic molecule's MOs occur as the linear combinations $1s_A + 1s_B$ and $1s_A - 1s_B$, the H_A1s and H_B1s AOs in BeH_2 can only occur as the linear combinations $H_A1s + H_B1s$ and $H_A1s - H_B1s$ in the BeH_2 MOs that satisfy the Hartree–Fock equations.

Both these linear combinations have no nodal plane containing the internuclear axis; hence these linear combinations will contribute to σ MOs. The linear combination $H_A1s + H_B1s$ has equal values at points diagonally opposite the center of the molecule (Fig. 20.13) and hence has σ_g symmetry. The linear combination $H_A1s - H_B1s$ has opposite signs at points diagonally opposite the molecular center and thus has σ_u symmetry.

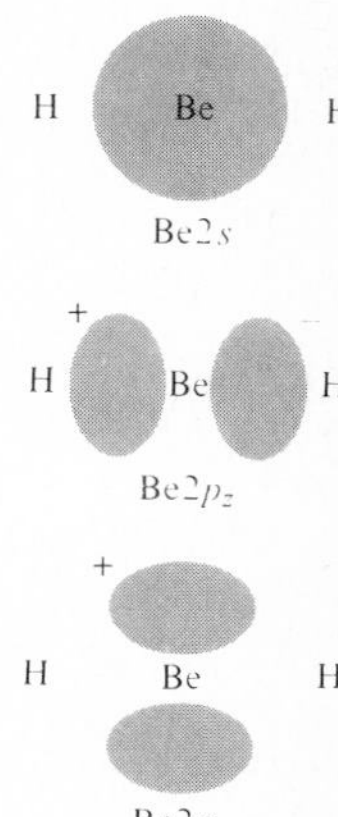

Figure 20.14
Be AOs in BeH_2.

What about the Be AOs? The Be2s AO has σ_g symmetry. Calling the internuclear axis the z axis, we see from Fig. 20.14 that the $Be2p_z$ AO has σ_u symmetry. The $Be2p_x$ and $Be2p_y$ AOs each have π_u symmetry.

Combining AOs that have σ_g symmetry and comparable energies, we form a σ_g MO as follows:

$$2\sigma_g = c_1\text{Be}2s + c_2(\text{H}_A1s + \text{H}_B1s) \tag{20.27}$$

The 2 in $2\sigma_g$ indicates this is the second lowest σ_g MO, the lowest being (20.26). (The very-low-energy Be1s AO will make a very slight contribution to $2\sigma_g$, which we shall neglect.) With c_1 and c_2 both positive, this MO has probability-density buildup between Be and H_A and between Be and H_B and is therefore bonding (Fig. 20.15).

Similarly, we form a bonding σ_u MO as (Fig. 20.15)

$$1\sigma_u = c_3\text{Be}2p_z + c_4(\text{H}_A1s - \text{H}_B1s) \tag{20.28}$$

with c_3 and c_4 both positive. This bonding MO has its energy below the energies of the $Be2p_z$ and H1s AOs from which it is formed.

The coefficients c_1, c_2, c_3, c_4 are found by solving the Hartree–Fock equations. The $Be2p_x$ and $Be2p_y$ AOs go to form two π_u MOs:

$$1\pi_{u,x} = \text{Be}2p_x, \qquad 1\pi_{u,y} = \text{Be}2p_y \tag{20.29}$$

These two MOs are degenerate and constitute the $1\pi_u$ energy level. These nonbonding π MOs will have nearly the same energy as the $Be2p_x$ and $Be2p_y$ AOs and hence

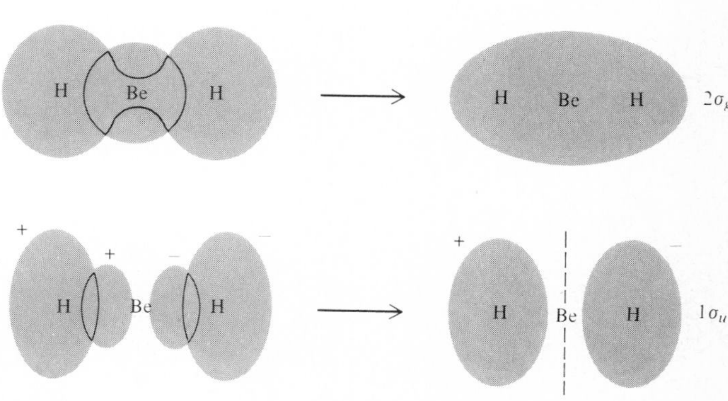

Figure 20.15
Formation of bonding MOs in BeH_2.

will lie above the bonding $2\sigma_g$ and $1\sigma_u$ MOs. The π MOs are therefore unoccupied in the ground state of this six-electron molecule.

A minimal-basis-set calculation on BeH_2 [R. G. A. R. Maclagan and G. W. Schnuelle, *J. Chem. Phys.*, **55**, 5431 (1971)] gave the following expressions for the occupied MOs:

$$1\sigma_g = 1.00(\text{Be}1s) + 0.016(\text{Be}2s) - 0.002(\text{H}_\text{A}1s + \text{H}_\text{B}1s)$$
$$2\sigma_g = -0.09(\text{Be}1s) + 0.40(\text{Be}2s) + 0.45(\text{H}_\text{A}1s + \text{H}_\text{B}1s) \qquad (20.30)$$
$$1\sigma_u = 0.44(\text{Be}2p_z) + 0.44(\text{H}_\text{A}1s - \text{H}_\text{B}1s)$$

The $1\sigma_g$ MO is essentially a Be1s AO, as anticipated. The $2\sigma_g$ and $1\sigma_u$ MOs have essentially the forms of (20.27) and (20.28).

There are also two antibonding MOs $3\sigma_g^*$ and $2\sigma_u^*$ corresponding to the two bonding MOs (20.27) and (20.28):

$$3\sigma_g^* = c_1'\text{Be}2s - c_2'(\text{H}_\text{A}1s + \text{H}_\text{B}1s), \qquad 2\sigma_u^* = c_3'\text{Be}2p_z - c_4'(\text{H}_\text{A}1s - \text{H}_\text{B}1s) \qquad (20.31)$$

Figure 20.16 sketches the AO and MO energies for BeH_2. (Of course, this molecule has many higher unoccupied MOs that are not shown in the figure. These MOs are formed from higher AOs of Be and the H's.) The BeH_2 ground-state configuration is $(1\sigma_g)^2(2\sigma_g)^2(1\sigma_u)^2$. There are four bonding electrons and hence two bonds. (The $1\sigma_g$ electrons are nonbonding inner-shell electrons.)

Note that a bonding MO has a lower energy than the AOs from which it is formed, an antibonding MO has a higher energy than the AOs from which it is formed, and a nonbonding MO has approximately the same energy as the AO or AOs from which it is formed.

Localized MOs. The BeH_2 bonding MOs $2\sigma_g$ and $1\sigma_u$ in Fig. 20.15 are each delocalized over the entire molecule. The two electrons in the $2\sigma_g$ MO move over the entire molecule, as do the two in the $1\sigma_u$ MO. This is rather puzzling to a chemist, who likes to think in terms of individual bonds: H—Be—H or H : Be : H. The existence of bond energies, bond moments, and bond vibrational frequencies (Sec. 21.8) that are roughly the same for a given kind of bond in different molecules shows

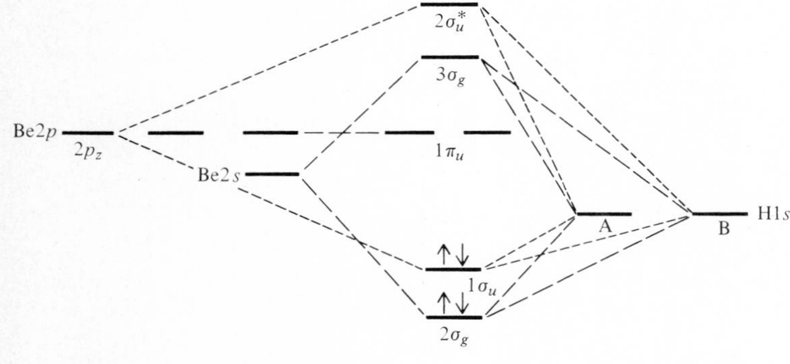

Figure 20.16
Energy-level scheme for BeH_2 MOs. (Not to scale.)

there is much validity in the picture of individual bonds. How can we reconcile the existence of individual bonds with the delocalized MOs found by solving the Hartree–Fock equations?

Actually, we *can* use the MO method to arrive at a picture in accord with chemical experience, as we now show. The MO wave function for BeH_2 is a 6×6 Slater determinant (Sec. 19.6). The first two rows of this Slater determinant are

$$\begin{matrix} 1\sigma_g(1)\alpha(1) & 1\sigma_g(1)\beta(1) & 2\sigma_g(1)\alpha(1) & 2\sigma_g(1)\beta(1) & 1\sigma_u(1)\alpha(1) & 1\sigma_u(1)\beta(1) \\ 1\sigma_g(2)\alpha(2) & 1\sigma_g(2)\beta(2) & 2\sigma_g(2)\alpha(2) & 2\sigma_g(2)\beta(2) & 1\sigma_u(2)\alpha(2) & 1\sigma_u(2)\beta(2) \end{matrix}$$

The third row involves electron 3, etc. Each column has the same spin-orbital. Now it is a well-known theorem (*Sokolnikoff and Redheffer*, app. A) that addition of a constant times one column of a determinant to another column leaves the determinant unchanged in value. For example, if we add thrice column 1 of the determinant in (19.36) to column 2, we get

$$\begin{vmatrix} a & b + 3a \\ c & d + 3c \end{vmatrix} = a(d + 3c) - c(b + 3a) = ad - bc = \begin{vmatrix} a & b \\ c & d \end{vmatrix}$$

Thus, if we like, we can add a multiple of one column of the Slater determinant MO wave function to another column without changing the wave function. This addition will "mix" together different MOs, since each column is a different spin-orbital. We can therefore take linear combinations of MOs to form new MOs without changing the overall wave function. Of course, the new MOs should each be normalized and for computational convenience should also be orthogonal to one another.

The BeH_2 MOs $1\sigma_g$, $2\sigma_g$, and $1\sigma_u$ satisfy the Hartree–Fock equations (19.47) and have the symmetry of the molecule. Because they have the molecular symmetry, they are delocalized over the whole molecule. (More accurately, the $2\sigma_g$ and $1\sigma_u$ MOs are delocalized, but the inner-shell $1\sigma_g$ is localized on the central Be atom.) These delocalized MOs satisfying the Hartree–Fock equations are called the *canonical* MOs. The canonical MOs are unique (except for the possibility of taking linear combinations of degenerate MOs).

As shown above, we can take linear combinations of the canonical MOs to form a new set of MOs that will give the same overall wave function. The new MOs will not individually be solutions of the Hartree–Fock equations $\hat{F}\phi_i = \varepsilon_i \phi_i$, but the overall wave function formed from these MOs will have the same energy and the same total probability density as the wave function formed from the canonical MOs.

Of the many possible sets of MOs that can be formed, we want to find a set that will have each MO classifiable as one of the following: a bonding (b) orbital localized between two atoms, an inner-shell (i) orbital, or a lone-pair (l) orbital. We call such a set of MOs *localized* MOs. Each localized MO will not have the symmetry of the molecule, but the localized MOs will correspond closely to a chemist's picture of bonding. (Since localized MOs are not eigenfunctions of the Hartree–Fock operator $\hat{F}$, in a certain sense each such MO does not correspond to a definite orbital energy; however, one can calculate an average energy of a localized MO by averaging over the orbital energies of the canonical MOs that form the localized MO.)

Consider BeH_2. The $1\sigma_g$ canonical MO is an inner-shell (i) AO on Be and can therefore be taken as one of the localized MOs: $i(\text{Be}) = 1\sigma_g = \text{Be}1s$. The $2\sigma_g$ and $1\sigma_u$ canonical MOs are delocalized. Figure 20.15 shows that the $1\sigma_u$ MO has opposite signs in the two halves of the molecule, whereas $2\sigma_g$ is essentially positive throughout the molecule. Hence, by taking linear combinations that are the sum and difference of these two canonical MOs, we shall get MOs that are each largely localized between only two atoms (Fig. 20.17). Thus, we take the localized bonding MOs b_1 and b_2 as

$$b_1 = 2^{-1/2}(2\sigma_g + 1\sigma_u), \qquad b_2 = 2^{-1/2}(2\sigma_g - 1\sigma_u) \tag{20.32}$$

where the $2^{-1/2}$ is a normalization constant. The b_1 localized MO corresponds to a bond between Be and H_A; the b_2 MO gives the Be—H_B bond.

Using these localized MOs, we write the BeH_2 MO wave function as a 6×6 Slater determinant whose first row is

$$i(1)\alpha(1) \quad i(1)\beta(1) \quad b_1(1)\alpha(1) \quad b_1(1)\beta(1) \quad b_2(1)\alpha(1) \quad b_2(1)\beta(1)$$

This localized-MO wave function is mathematically identical to the wave function that uses delocalized (canonical) MOs.

Equation (20.32) expresses the localized bonding MOs b_1 and b_2 in terms of the canonical MOs. Substitution of (20.27) and (20.28) into (20.32) gives

$$\begin{aligned} b_1 &= 2^{-1/2}[c_1\text{Be}2s + c_3\text{Be}2p_z + (c_2 + c_4)\text{H}_\text{A}1s + (c_2 - c_4)\text{H}_\text{B}1s] \\ b_2 &= 2^{-1/2}[c_1\text{Be}2s - c_3\text{Be}2p_z + (c_2 - c_4)\text{H}_\text{A}1s + (c_2 + c_4)\text{H}_\text{B}1s] \end{aligned} \tag{20.33}$$

As a rough approximation, we see from (20.27), (20.28), and (20.30) that $c_2 \approx c_4$ and $c_1 \approx c_3$. These approximations give

$$\begin{aligned} b_1 &\approx 2^{-1/2}[c_1(\text{Be}2s + \text{Be}2p_z) + 2c_2\text{H}_\text{A}1s] \\ b_2 &\approx 2^{-1/2}[c_1(\text{Be}2s - \text{Be}2p_z) + 2c_2\text{H}_\text{B}1s] \end{aligned} \tag{20.34}$$

The approximate MOs (20.34) are each fully localized between Be and one H atom, but the more accurate expressions (20.33) show that the Be—H_A bonding MO b_1 has a small contribution from the H_B1s AO and so is not fully localized between the two atoms forming the bond.

Note that [unlike the canonical MOs (20.27) and (20.28)] the b_1 and b_2 localized MOs each have the Be2s and Be$2p_z$ AOs mixed together, or *hybridized*. The precise degree of hybridization depends on the values of c_1 and c_3 in (20.33). In the approximation of (20.34), the MOs b_1 and b_2 would each contain equal amounts of the Be2s and Be$2p_z$ AOs. The two normalized linear combinations

$$2^{-1/2}(2s + 2p_z) \qquad \text{and} \qquad 2^{-1/2}(2s - 2p_z) \tag{20.35}$$

are called *sp* hybrid AOs. Comparison of (20.30) with (20.27) and (20.28) gives $c_1 = 0.40$ and $c_3 = 0.44$, so c_1 and c_3 are not precisely equal, but they are nearly

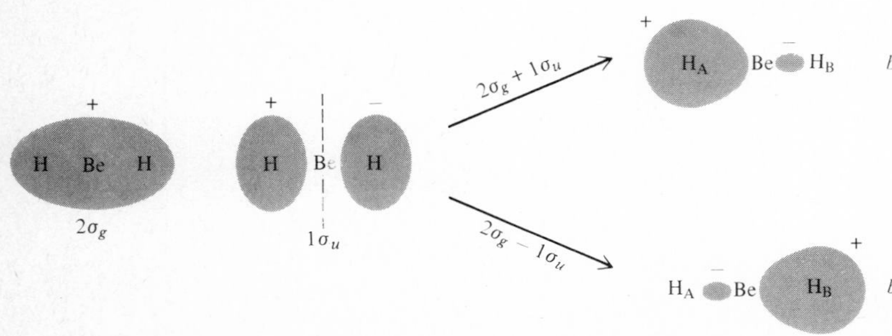

Figure 20.17
Formation of localized bonding MOs in BeH_2 from linear combinations of delocalized (canonical) MOs.

equal. Thus, the Be AOs in the BeH_2 bonding MOs are not precisely *sp* hybrids but are nearly *sp* hybrids.

Note from Eq. (20.33) and Fig. 20.17 that the localized bonding MOs b_1 and b_2 in BeH_2 are *equivalent* to each other. By this we mean that b_1 and b_2 have the same shapes and are interchanged by a rotation that interchanges the two equivalent chemical bonds in BeH_2. If we rotate b_1 and b_2 180° about an axis through Be and perpendicular to the molecular axis (thereby interchanging H_A1s and H_B1s and changing $Be2p_z$ to $-Be2p_z$), then b_1 is changed to b_2, and vice versa. Because of the symmetry of the molecule, b_1 and b_2 ought to be equivalent orbitals. It is possible to show that the linear combinations in (20.32) are the only linear combinations of the $2\sigma_g$ and $1\sigma_u$ canonical MOs that meet the requirements of being normalized, equivalent, and orthogonal.

We arrived at the approximately *sp* hybrid Be AOs in the localized BeH_2 bonding MOs by first finding the delocalized canonical MOs and then transforming to localized MOs. A somewhat simpler (and more approximate) approach is often preferred by chemists for qualitative discussions of bonding. In this procedure, one omits consideration of the canonical MOs; instead, one forms the required hybrid AOs on the free Be atom and then uses these hybrids to form localized bonding MOs with the H1*s* AOs. For BeH_2 with its 180° bond angle, we need two equivalent hybrid AOs on Be that point in opposite directions. The valence AOs of Be are 2*s* and 2*p*. Figure 20.18 shows that the linear combinations (20.35) give two equivalent hybridized AOs oriented 180° apart. It is possible to show that the *sp* hybrids (20.35) are the only linear combinations that give AOs at 180° that are equivalent, normalized, and orthogonal in the free atom. We then overlap each of these *sp* hybrid AOs with an H1*s* AO to form the two bonds; this gives the localized bonding MOs of Eq. (20.34).

Although the *sp* hybrids (20.35) are the only linear combinations of 2*s* and $2p_z$ that give equivalent orbitals in the free Be atom, we must expect that in the BeH_2 molecule the interaction between the Be hybrids and the H atoms will alter the nature of these hybrids somewhat. What is really wanted is equivalent, normalized, orthogonal MOs in the BeH_2 molecule and not equivalent, normalized, orthogonal AOs in the Be atom. The equivalent MOs in BeH_2 are (20.33), and, as noted above, these are not precisely *sp* hybrids but only approximately *sp* hybrids. Another approximation of using *sp* Be hybrids to form the MOs (20.34) is the neglect of the small contribution of the H_B1s AO to the bonding MO between Be and H_A; this is the term $(c_2 - c_4)H_B1s = 0.01(H_B1s)$ in (20.33).

Energy-localized MOs. For BeH_2, the symmetry of the molecule allows one to determine what linear combination of canonical MOs to use to get localized bonding

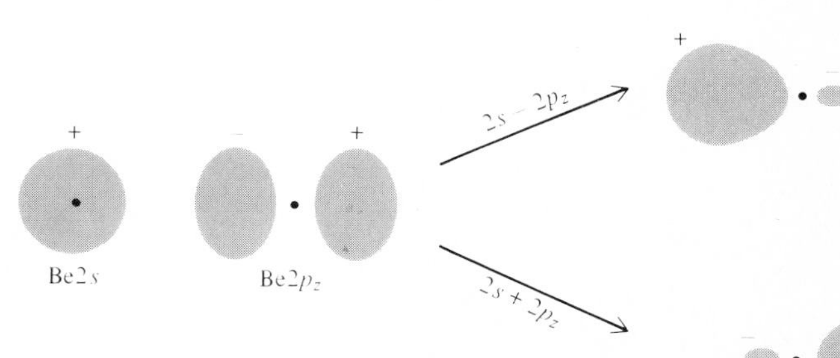

Figure 20.18
Formation of *sp* hybrid AOs in the Be atom.

MOs. For less symmetric molecules, one can't use symmetry, since the localized MOs need not be equivalent to one another. A variety of criteria have been suggested for finding localized MOs from the canonical MOs. The most widely accepted approach is that of Edmiston and Ruedenberg, who defined the *energy-localized* MOs as those orthogonal MOs that minimize the total of the Coulombic repulsions between the various pairs of localized MOs considered as charge distributions in space. This gives localized MOs that are separated as far as possible from one another.

In most cases, the energy-localized MOs agree with what one would expect from the Lewis dot formula. For example, for H_2O, the energy-localized MOs turn out to be one inner-shell MO, two bonding MOs, and two lone-pair MOs, in agreement with the dot formula H:$\ddot{\underset{..}{O}}$:H. The inner-shell MO is nearly identical to the O1*s* AO. One bonding energy-localized MO is largely localized in the O—H_A region, and the other is largely localized in the O—H_B region. The angle between the bonding localized MOs is 103°, which is nearly the same as the 104.5° experimental bond angle in water. The angle between the lone-pair localized orbitals is 114°. Each bonding localized MO is mainly a linear combination of 2*s* and 2*p* oxygen AOs and a 1*s* hydrogen AO. Each lone-pair MO is mainly a hybrid of 2*s* and 2*p* AOs on oxygen. [See *Q. C.*, sec. 15.5 and W. von Niessen, *Theor. Chim. Acta*, **29**, 29 (1973).]

Sigma, pi, and delta bonds. In most cases, each localized bonding MO of a molecule contains substantial contributions from AOs of only two atoms, the atoms forming the bond. In analogy with the classification used for diatomic molecules, each localized bonding MO of a polyatomic molecule is classified as $\sigma, \pi, \delta, \ldots$ according to whether the MO has, 0, 1, 2, ... nodal planes containing the axis between the two bonded atoms. The BeH_2 MOs b_1 and b_2 in Fig. 20.17 are clearly σ MOs. One finds that nearly always a single bond between two atoms corresponds to a σ localized MO. Nearly always, a double bond between two atoms is composed of one σ localized MO and one π localized MO. Nearly always, a triple bond is composed of one σ-bond orbital and two π-bond orbitals. A quadruple bond is composed of one σ bond, two π bonds, and one δ bond.

A σ bond is formed by overlap of two AOs that have no nodal planes containing the bond axis. Figure 20.19*a* shows some kinds of AO overlap that produce σ

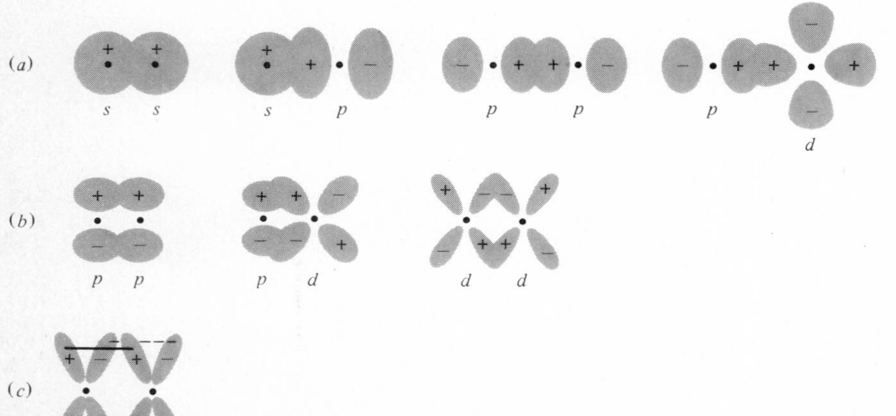

Figure 20.19
Overlap of AOs to form (*a*) σ bonds; (*b*) π bonds; (*c*) a δ bond. The lobes in (*c*) are in front of and behind the plane of the paper.

localized bond MOs. Figure 20.19*b* shows some overlaps that lead to π bonds. Figure 20.19*c* shows formation of a δ bond.

Chemists have been familiar with σ and π bonds since the 1930s. In 1964, Cotton pointed out that the $Re_2Cl_8^{2-}$ ion has a quadruple bond between the two Re atoms, as evidenced by an abnormally short Re—Re bond distance; this bond is composed of one σ bond, two π bonds, and one δ bond, the δ bond being formed by overlap of two $d_{x^2-y^2}$ AOs, one on each Re atom. Several other transition-metal species with quadruple bonds are known. [See F. A. Cotton, *Chem. Soc. Rev.*, **4**, 27 (1975).]

Methane, ethylene, and acetylene. For CH_4, the canonical occupied MOs are found to consist of an MO that is a nearly pure C1*s* AO and four delocalized bonding MOs that each extend over much of the molecule. When the canonical MOs are transformed to energy-localized MOs, the localized MOs are found to consist of an inner-shell MO that is essentially a pure C1*s* AO and four localized bonding MOs, each bonding MO pointing toward one of the H atoms of the tetrahedral molecule. The localized bonding MO between C and atom H_A is [R. M. Pitzer, *J. Chem. Phys.*, **46**, 4871 (1967)]

$$0.02(\mathrm{C}1s) + 0.292(\mathrm{C}2s) + 0.277(\mathrm{C}2p_x + \mathrm{C}2p_y + \mathrm{C}2p_z) + 0.57(\mathrm{H_A}1s) - 0.07(\mathrm{H_B}1s + \mathrm{H_C}1s + \mathrm{H_D}1s)$$

The carbon 2*s* and 2*p* AOs make nearly equal contributions, and the hybridization on carbon is approximately sp^3. It would be exactly sp^3 if the coefficient of C2*s* equaled that of the C2*p* AOs, but in fact there is a bit more than 25 percent *s* character in the localized bond MO. Atom H_A is in the positive octant of space, and the combination $\mathrm{C}2p_x + \mathrm{C}2p_y + \mathrm{C}2p_z$ (which is proportional to $x + y + z$) has its maximum probability density along the line running through C and H_A and on both sides of the C nucleus. Addition of C2*s* to $\mathrm{C}2p_x + \mathrm{C}2p_y + \mathrm{C}2p_z$ cancels most of the probability density on the side of C that is away from H_A and reinforces the probability density in the region between C and H_A. (This is the same thing that occurs in Fig. 20.18 for the BeH_2 *sp* hybrids.) Overlap of the C hybrid AO with the $\mathrm{H_A}1s$ AO then forms the bond. Each bonding localized MO has no nodal planes containing the axis between the bonded atoms and is a σ MO.

Consider ethylene ($H_2C{=}CH_2$). The molecule is planar, with the bond angles at each carbon close to 120°. A minimal basis set consists of four H1*s* AOs and two each of C1*s*, C2*s*, $\mathrm{C}2p_x$, $\mathrm{C}2p_y$, and $\mathrm{C}2p_z$. Let the molecular plane be the *yz* plane. One way to form localized MOs for C_2H_4 is to use linear combinations of the C2*s*, $\mathrm{C}2p_y$, and $\mathrm{C}2p_z$ AOs at each carbon to form three sp^2 hybrid AOs at each carbon; these hybrids make 120° angles with one another. Overlap of two of the three sp^2 hybrids at each carbon with H1*s* AOs forms the C—H single bonds, and these are σ bonds. Overlap of the third sp^2 hybrid of one carbon with the third sp^2 hybrid of the second carbon gives a σ bonding MO between the two carbons (Fig. 20.20*a*). Overlap of the $2p_x$ AOs of each carbon gives a localized π bonding MO between the carbons (Fig. 20.20*b*); this π MO has a nodal plane coinciding with the molecular plane and containing the C—C axis. Ethylene has 16 electrons. Four of them fill the two localized inner-shell MOs, each of which is a 1*s* AO on one of the carbons; eight electrons fill the four C—H bond MOs; two electrons fill the C—C σ-bond MO, and two fill the C—C π-bond MO. (Unlike that in diatomic molecules, the ethylene

π-bond MO is nondegenerate.) In this picture, the carbon–carbon double bond consists of one σ and one π bond.

The above description of localized MOs for C_2H_4 is the traditional one. However, actual calculation of the energy-localized MOs in ethylene shows that two bent, equivalent "banana" bonds (Fig. 20.21) between the two carbons are more localized than the traditional σ-π description. (See *Q. C.*, sec. 15.6.)

The traditional description of HC≡CH uses two *sp* hybrids on each carbon to overlap the H1*s* AOs and to form a σ bond between the carbons. The linear combinations $C_A 2p_x + C_B 2p_x$ and $C_A 2p_y + C_B 2p_y$ (where the z axis is the molecular axis) give two π bonds between C_A and C_B. In this picture, the triple bond consists of one σ bond and two π bonds. Again, actual calculation shows the energy-localized MOs to consist of three equivalent bent banana bonds.

Benzene. Benzene (C_6H_6) is a regular hexagon with 120° bond angles. We can use three sp^2 hybrid AOs at each carbon to form localized σ bonds with one hydrogen and with two adjacent carbons. This leaves a $2p_z$ AO at each carbon (where the z axis is perpendicular to the molecular plane). For benzene, two equivalent Lewis dot formulas can be written; the carbon–carbon single bonds and double bonds are interchanged in the two formulas. Moreover, ΔH of formation of benzene (Prob. 20.21) and the chemical behavior of benzene differ from what is expected of a species with localized double bonds. Hence, it would not be suitable to form three localized π MOs by pairwise interactions of the six $2p_z$ AOs. Instead, all six $2p_z$ AOs must be considered to interact with one another, and one uses delocalized (canonical) MOs to form the π bonds. The six $2p_z$ AOs give six delocalized π MOs, three bonding and three antibonding. These six MOs are linear combinations of the six $2p_z$ AOs; their forms are fully determined by the symmetry of benzene. (See Prob. 20.27 and *Q. C.*, sec. 15.8.)

Three of the four valence electrons of each carbon go into the three bonding MOs formed from the sp^2 hybrids, leaving one electron at each carbon to go into the π MOs. These remaining six electrons fill the three bonding π MOs. Each π MO has a nodal plane that coincides with the molecular plane (since each $2p_z$ AO has such a plane). This nodal plane is analogous to the nodal plane of a localized π bond joining two doubly bonded atoms (e.g., as in ethylene), and so these benzene MOs are called π MOs.

A similar situation holds for other planar conjugated organic compounds. (A *conjugated* molecule has a framework consisting of alternating carbon–carbon single and double bonds.) One can form in-plane localized σ-bond MOs using sp^2 hybrids on each carbon, but one uses delocalized (canonical) MOs for the π MOs.

Three-center bonds. The molecule B_2H_6 has 12 valence electrons, which is not enough to allow one to write a Lewis dot structure with two electrons shared be-

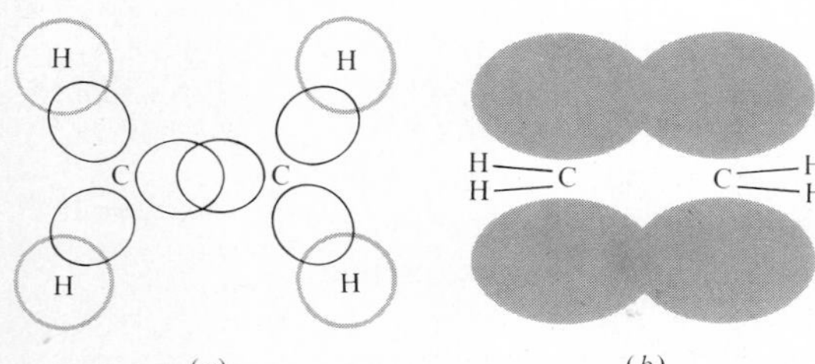

Figure 20.20
Bonding in ethylene: (*a*) σ bonds. (*b*) π bond.

tween each pair of bonded atoms. Calculation of the energy-localized MOs of B_2H_6 shows that two of the localized MOs each extend over two B atoms and one H atom to give two three-center bonds. [E. Switkes et al., *J. Chem. Phys.*, **51**, 2085 (1969).] The H atoms of the three-center bonds lie above and below the plane of the remaining atoms and midway between the borons. Three-center bonds also occur in higher boron hydrides.

Ligand-field theory. The application of MO theory to transition-metal complexes gives what is called *ligand-field theory*. Discussion is omitted. (See *Gray*, chap. 5.)

Canonical vs. localized MOs. For accurate quantitative calculations of molecular properties, one solves the Hartree–Fock equations (19.47) and obtains the canonical (delocalized) MOs. Since each canonical MO corresponds to a definite orbital energy, these MOs are also useful in discussing transitions to excited electronic states and ionization.

For qualitative discussion of bonding in the ground electronic state of a molecule, it is usually simplest to describe things in terms of localized bonding MOs constructed from suitably hybridized AOs of pairs of bonded atoms, lone-pair MOs, and inner-shell MOs. One can usually get a reasonably good idea of the localized MOs without going through the very difficult computations involved in first finding the canonical MOs and then using the Edmiston–Ruedenberg criterion to transform the canonical MOs to localized MOs. Localized MOs are approximately transferable from molecule to molecule; e.g., the C—H localized MOs in CH_4 and C_2H_6 are very similar to each other. Canonical MOs are not transferable.

20.7 CALCULATION OF MOLECULAR PROPERTIES

We now show how various molecular properties can be calculated from approximate molecular wave functions.

Molecular geometry. The equilibrium geometry of a molecule is the spatial configuration of the nuclei for which the electronic energy (including nuclear repulsion) E_e in (20.8) is a minimum. To determine the equilibrium geometry theoretically, one must calculate the molecular wave function at many different configurations of the nuclei, varying the bond distances and angles to find the minimum-energy configuration. For a molecule with a reasonably large number of atoms, such a calculation is extremely time-consuming even on the fastest computers, since the wave-function calculation must be repeated many times. (Theoreticians frequently cheat a bit by using the experimentally determined geometry and only calculating ψ at this geometry. They then use this ψ to estimate other molecular properties.)

One finds that even though the Hartree–Fock MO wave function differs significantly from the true wave function, it gives generally accurate bond distances and bond angles. Some examples of calculated Hartree–Fock (or approximate Hartree–Fock) geometries are:

H_2O: $r(\text{OH}) = 0.94\ (0.96)$, HOH angle $= 106.6°\ (104.5°)$

H_2CO: $r(\text{CH}) = 1.10\ (1.12)$, $r(\text{CO}) = 1.22\ (1.21)$, HCH angle $= 114.8°\ (116.5°)$

C_6H_6: $r(\text{CC}) = 1.39\ (1.40)$, $r(\text{CH}) = 1.08\ (1.08)$

where the experimental values are in parentheses and bond lengths are in angstroms.

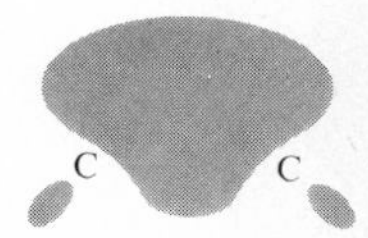

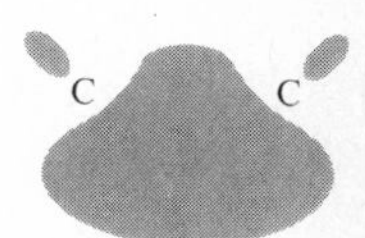

Figure 20.21 Equivalent "banana" bonds in ethylene. The view is the same as in Fig. 20.20*b*.

Dipole moments. Equation (14.15) is the classical expression for the dipole moment of a charge distribution. To calculate the dipole moment of a molecule from its equilibrium-geometry electronic wave function ψ, we use the right side of (14.15) as an operator in the average-value expression (18.53) to get

$$\mathbf{p}' = \int \psi^* \sum_i Q_i' \mathbf{r}_i \psi \, d\tau = \sum_i Q_i' \int |\psi|^2 \mathbf{r}_i \, d\tau \tag{20.36}$$

where the sum goes over all the electrons and nuclei, the integral is over the electronic coordinates, and Q_i' is the charge (in gaussian units) on particle i. Evaluation of (20.36) once ψ is known is easy. The hard thing is to get a reasonably accurate approximation to ψ.

One finds that Hartree–Fock MO wave functions give generally accurate molecular dipole moments. Some values of p'_{HF} and p'_{exp}, the Hartree–Fock and the experimental dipole moment, are:

	HCN	H_2O	LiH	NaCl	CO	NH_3
p'_{HF}/D	3.29	2.03	6.00	9.18	−0.11	1.66
p'_{exp}/D	2.98	1.85	5.88	9.00	+0.27	1.48

The experimental CO dipole moment has the carbon negative (Prob. 20.9); the calculated Hartree–Fock CO dipole moment is in the wrong direction. A calculation using a CI wave function gives the proper polarity for CO.

Dissociation energies. To calculate D_e theoretically, one subtracts the calculated Hartree–Fock molecular energy at the equilibrium geometry from the Hartree–Fock energies of the separated atoms that form the molecule. Hartree–Fock wave functions give poor D_e values. Some results for molecular binding energies are:

	H_2	BeO	N_2	CO	F_2	CO_2	H_2O	N_2O
$D_{e,HF}/\text{eV}$	3.64	2.0	5.3	7.9	−1.4	11.3	6.9	4.0
$D_{e,\text{exp}}/\text{eV}$	4.75	4.7	9.9	11.2	1.6	16.8	10.1	11.7

The Hartree–Fock wave functions predict the separated atoms F + F to be more stable than the F_2 molecule at R_e.

Ionization energies. The molecular ionization energy I is the energy needed to remove the most loosely held electron from the molecule. T. C. Koopmans (cowinner of the 1975 Nobel prize in economics) proved in 1933 that the energy needed to remove an electron from an orbital of a closed-shell atom or molecule is well approximated by minus the Hartree–Fock orbital energy ε_i in (19.47). Hence, the molecular ionization energy can be estimated by taking $-\varepsilon_i$ of the highest occupied MO. One finds pretty good agreement between Koopmans' theorem Hartree–Fock ionization potentials and experimental ionization potentials. Some results are:

	N_2	H_2O	C_6H_6
I_{HF}/eV	17.4	13.8	9.1
I_{exp}/eV	15.6	12.6	9.3

Rotational barriers. The equilibrium conformation of ethane, $H_3C{-}CH_3$, has the hydrogens of one CH_3 group staggered with respect to the hydrogens of the other CH_3 group. The barrier to internal rotation about the single bond in ethane is quite low, being 0.13 eV, which corresponds to 3 kcal/mole. (The barrier can be determined experimentally from thermodynamic data or the infrared spectrum.) To calculate the rotational barrier B theoretically, one calculates wave functions and energies for the staggered and eclipsed geometries and takes the difference. One finds that Hartree–Fock and approximate Hartree–Fock wave functions give pretty accurate rotational barriers. Some results (values in kcal/mole) are:

	C_2H_6	CH_3CHO	CH_3OH	CH_3NH_2
B_{exp}	2.9	1.2	1.1	2.0
B_{calc}	3.2	0.8	1.1	2.4

The reason the Hartree–Fock method does well on barrier calculations is that no bonds are broken or formed in going from the staggered to the eclipsed conformation and the correlation energy (which is the energy error in the Hartree–Fock method) is nearly the same for the two conformations. In contrast, when a molecule dissociates, one or more bonds are broken and the correlation energy changes substantially; the Hartree–Fock wave function is therefore inadequate to deal with dissociation. Configuration-interaction wave functions are needed to calculate dissociation energies.

Electron probability density. Let $\rho(x, y, z)\,dx\,dy\,dz$ be the probability of finding an electron of a many-electron molecule in the box in space located at x, y, z and having edges dx, dy, dz; by "an electron" we mean any electron, not a particular one. The electron probability density $\rho(x, y, z)$ can be calculated theoretically by integrating the electronic wave function ψ_e over the spin coordinates of all electrons and over the spatial coordinates of all but one electron and multiplying the result by the number of electrons in the molecule (see *Q. C.*, sec. 13.11 for details). One can determine ρ experimentally by analyzing x-ray diffraction data of crystals (Chap. 24).

20.8 THE VSEPR METHOD

An accurate theoretical determination of a molecule's geometry requires a vast amount of computation. At present, calculation of the Hartree–Fock wave function and energy as functions of the nuclear configuration is beyond the capabilities of computers when the molecule is reasonably large. The VSEPR (valence-shell electron-pair-repulsion) method provides a simple, reasonably accurate means of predicting molecular shapes. The method is based on ideas stated by Sidgwick and Powell in 1940 and has been further developed and widely applied by Gillespie.

According to the VSEPR method, the geometry is determined by the number of pairs of valence electrons around the central atom of the molecule. Consider H_2O, for example. The electron-dot formula is H : $\ddot{\underset{\cdot\cdot}{O}}$: H. The four pairs of valence electrons around the central oxygen atom can be considered to reside in four localized MOs (Sec. 20.6), two of which are lone-pair and two of which are bonding. Because of electrostatic repulsions and especially the Pauli repulsion between electrons of like spin, there are strong repulsions between the electron pair of one localized MO and the pairs in the other localized MOs. The energetically most favorable geometry will therefore have the four localized MOs separated spatially as much as possible from one another. The most separated configuration for four localized MOs around a central atom turns out to occur when the MOs point to the four corners of a tetrahedron (Fig. 20.22). We therefore expect the four valence-electron pairs around oxygen to be approximately tetrahedrally disposed. Since the tetrahedral bond angle is $109\frac{1}{2}°$, we expect H_2O to be bent, with a $109\frac{1}{2}°$ bond angle.

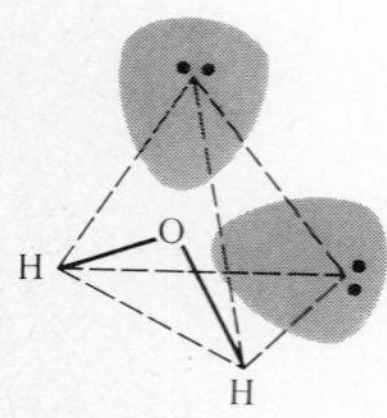

Figure 20.22 Valence electron pairs in H_2O. The O is at the center of the tetrahedron.

This prediction can be refined a bit by realizing that the lone-pair localized MOs in H_2O are not exactly equivalent to the bonding localized MOs. The electrons in each bonding MO are strongly attracted to two nuclei (the O and one H), whereas the electrons in a lone-pair MO are attracted strongly to only one nucleus (the O). Hence, the lone-pair MOs will be more spread out ("fatter") than the bonding MOs, and the electron pairs in the lone-pair MOs exert greater repulsions than the pairs in the bonding MOs. (Recall the 114° angle between the energy-localized lone pairs compared with the 103° angle between the energy-localized bonding pairs; Sec. 20.6.) The fat lone-pair MOs will therefore push the bonding MOs together a bit, thereby reducing the bond angle somewhat below $109\frac{1}{2}°$. The experimental angle of $104\frac{1}{2}°$ bears this out.

Figure 20.23

:—— A ——:

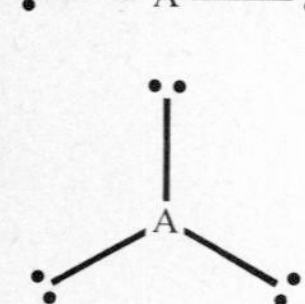

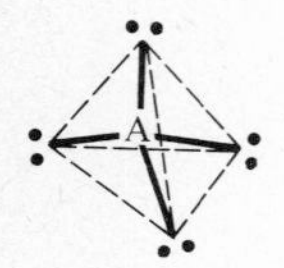

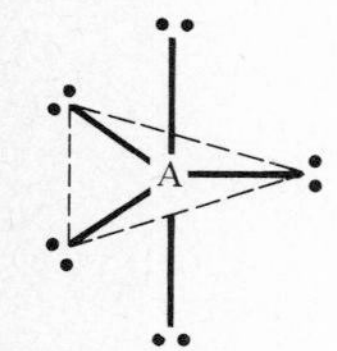

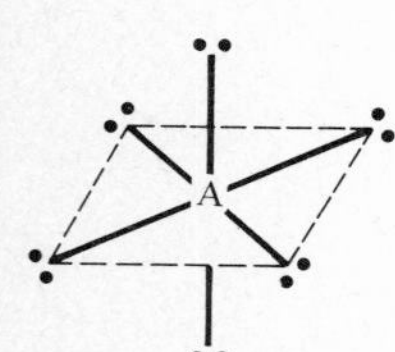

For :NH_3, there are again four valence-electron pairs around the central atom, but there is only one fat lone pair to push the bonding MOs together. Hence we expect an angle closer to $109\frac{1}{2}°$ than in H_2O. The observed angle is 107°.

For CH_4, there are four pairs of valence electrons around carbon. There are no lone pairs, and we expect a tetrahedral angle, as is observed.

Although the electron pairs around O and N in H_2O and NH_3 are approximately tetrahedrally disposed, the geometries of these molecules are described as bent and trigonal pyramidal, respectively. The geometry refers to the arrangement of atoms in space and does not include the lone pairs.

For two, three, four, five, and six valence-electron pairs around a central atom, the arrangements that give the maximum separation are linear, trigonal planar, tetrahedral, trigonal bipyramidal, and octahedral, respectively (Fig. 20.23).

To apply the VSEPR method to molecules with double or triple bonds, one pretends that each multiple bond consists of one pair of electrons. For example, for $H_2C{=}CH_2$, one considers each carbon to be surrounded by three electron pairs, and the VSEPR method predicts 120° bond angles at each carbon. To refine this prediction, we note that the carbon–carbon bond actually consists of two electron pairs, which will exert greater repulsions on the C—H pairs than the C—H pairs will exert on each other. Hence we expect each HCH angle to be a bit less than 120°. The observed HCH angle is 117°.

To determine the molecular geometry using the VSEPR method, we (*a*) write the Lewis dot structure and count the number of valence-electron pairs around the central atom, (*b*) arrange these pairs in space according to Fig. 20.23, and (*c*) take account of the extra repulsions due to lone pairs or multiple-bond pairs to refine the bond angles.

For two, three, four, and six pairs, all the positions around the central atom in Fig. 20.23 are equivalent, and it doesn't matter where a single lone pair is put. However, for five pairs, the two axial and three equatorial positions are not equivalent. To decide where lone pairs go, we note that the three equatorial pairs make 120° angles with one another, whereas an axial pair makes 90° angles with each of the three equatorial pairs. Since the pair–pair Pauli repulsion falls off rapidly with increasing angle, and since lone pairs exert the greatest repulsions, the lone pairs always go in the equatorial positions, thereby minimizing the number of 90° repulsions between lone pairs and other pairs.

Let SN (the *steric number*) be the number of valence-electron pairs around the central atom and LP be the number of lone pairs around this atom.

For SN = 2 and LP = 0, the molecule is linear. Examples are BeH_2 and CO_2. (Each double bond in CO_2 contributes 1 to the steric number.)

For SN = 3 and LP = 0, the molecule is trigonal planar (120° bond angles). An example is BF_3. For SN = 3 and LP = 1, the molecule is bent with a bond angle a bit less than 120°. An example is SO_2 (bond angle $119\frac{1}{2}°$), whose electron dot structure has one lone pair on S.

For SN = 4 and LP = 0, the molecule is tetrahedral with (essentially) $109\frac{1}{2}°$ bond angles. An example is $HCCl_3$, where the ClCCl angle is $110\frac{1}{2}°$. For SN = 4 and LP = 1, the molecule is trigonal pyramidal with bond angles slightly less than $109\frac{1}{2}°$ (for example, NH_3). For SN = 4 and LP = 2, the molecule is bent with angle somewhat less than $109\frac{1}{2}°$ (for example, H_2O).

For SN = 5 and LP = 0, the molecule is trigonal bipyramidal (for example, PCl_5). For SN = 5 and LP = 1, the lone pair goes in the equatorial position to give the shape shown in Fig. 20.24*a*. An example is SF_4. For SN = 5 and LP = 2, the two lone pairs go in equatorial positions to give the T shape shown in Fig. 20.24*b*. An example is ClF_3. For SN = 5 and LP = 3, the molecule is linear; for example, XeF_2; Fig. 20.24*c*. (The eight xenon valence electrons plus one valence electron from each F give five pairs around xenon, three lone pairs and two bonding pairs.)

For SN = 6 and LP = 0, the molecule is octahedral (for example, SF_6). For SN = 6 and LP = 1, the molecule is square pyramidal (for example, IF_5). For SN = 6 and LP = 2, the lone pairs go trans to each other to give a square-planar molecule (for example, XeF_4). See Fig. 20.25.

The VSEPR method does not apply to transition-metal compounds unless the metal has a filled, half-filled, or empty *d* subshell.

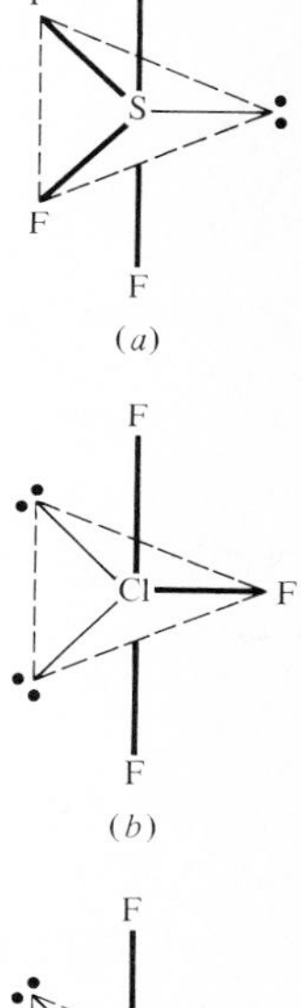

Figure 20.24

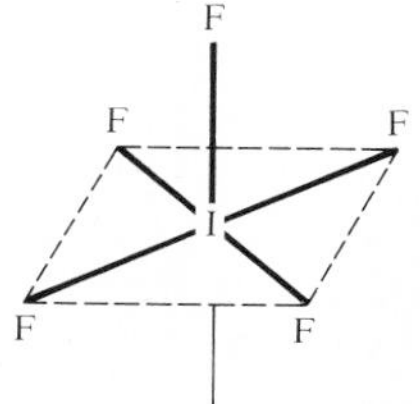

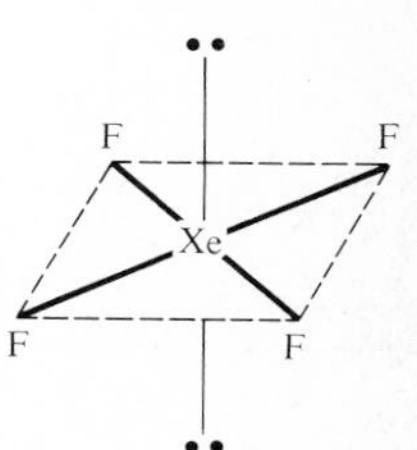

Figure 20.25

There are a very few compounds where the VSEPR method fails. Thus, BaF_2 is bent, rather than having the expected linear structure.

Some references on VSEPR are R. J. Gillespie, *J. Chem. Educ.*, **40**, 295 (1963); **47**, 18 (1970); **51**, 367 (1974); R. J. Gillespie, *Molecular Geometry*, Van Nostrand Reinhold, 1972.

20.9 SEMIEMPIRICAL METHODS

Quantum-mechanical methods of treating molecules are classified as *ab initio* or *semiempirical*. An *ab initio* calculation uses the true molecular Hamiltonian and does not use empirical data in the calculation. The Hartree–Fock method calculates the antisymmetrized product Φ of spin-orbitals that minimizes the variational integral $\int \Phi^* \hat{H} \Phi \, d\tau$, where $\hat{H}$ is the true molecular Hamiltonian; hence, the Hartree–Fock method is an *ab initio* one. Of course, because of the restricted form of Φ, the Hartree–Fock method does not give the true wave function. A CI calculation based on Hartree–Fock orbitals is also an *ab initio* calculation and can give the true wave function if sufficient configuration state functions are included.

A semiempirical method uses a simpler Hamiltonian than the true one, uses empirical data to assign values to some of the integrals that occur in the calculation, and neglects some of the integrals. The reason for resorting to semiempirical methods is that accurate *ab initio* calculations on reasonably large molecules cannot be done at present. Semiempirical methods were originally developed for conjugated organic molecules and later were extended to encompass all molecules.

For a planar (or near planar) conjugated organic compound, each MO can be classified as σ or π. Each σ MO is unchanged on reflection in the molecular plane (which is not a nodal plane for a σ MO), whereas each π MO changes sign on reflection in the molecular plane (which is a nodal plane for each π MO). (Recall the discussion of benzene in Sec. 20.6.) The σ MOs have electron probability density strongly concentrated in the region of the molecular plane. The π MOs have blobs of probability density above and below the molecular plane. The σ MOs can be taken as either delocalized or localized. However, the π MOs in a conjugated compound are generally best taken as delocalized. In conjugated molecules, the highest-energy occupied MOs are usually π MOs.

Because of the different symmetries of σ and π MOs, one can make the *approximation* of treating the π electrons separately from the σ electrons. One imagines the σ electrons to produce some sort of effective potential in which the π electrons move.

The simplest semiempirical treatment of conjugated molecules is the free-electron molecular-orbital (FE MO) method. The FE MO method deals only with the π electrons. It assumes that each π electron is free to move along the length of the molecule (potential energy $V = 0$) but cannot move beyond the ends of the molecule (potential energy $V = \infty$). This is the particle-in-a-box potential energy, and the FE MO method feeds electrons into particle-in-a-one-dimensional-box MOs.

Let us apply the FE MO method to the ions

$$(CH_3)_2\overset{+}{N}{=}CH({-}CH{=}CH)_k{-}\ddot{N}(CH_3)_2$$

where $k = 0, 1, 2, \ldots$. Each ion has an equivalent Lewis structure with the charge on the right-hand nitrogen and all carbon–carbon single and double bonds interchanged. Hence, all the carbon–carbon bond lengths are equal, and the π electrons, which form the second bond of each double bond, are reasonably free to move along the molecule. If we assume a carbon–carbon and a conjugated carbon–nitrogen bond distance of 1.40 Å (as in benzene) and add in an extra bond length at each end of the ion, the length a of the "box" is $a = (2k + 4)(1.40\ \text{Å})$. There are two π electrons in each double bond, plus the lone pair on nitrogen (which takes part in the π bonding). There are thus $2k + 4$ pi electrons. We feed these into the lowest $k + 2$ free-electron pi MOs (Figs. 18.8 and 18.9). These MOs have particle-in-a-box quantum numbers $n = 1, 2, 3, \ldots, k + 2$. The lowest-energy (longest-wavelength) electronic absorption band results from excitation of a pi electron from the $n = k + 2$ to the $n = k + 3$ level. Equations (18.6) and (18.27) give

$$\Delta E = h\nu = \frac{hc}{\lambda} = \frac{h^2}{8ma^2}[(k+3)^2 - (k+2)^2] = \frac{h^2}{8ma^2}(2k+5)$$

Substitution of $a = (2k + 4)(1.40\ \text{Å})$ and numerical values for h and the electron mass gives

$$\lambda = \frac{8mca^2}{h(2k+5)} = \frac{(2k+4)^2}{2k+5} \times 646\ \text{Å} \tag{20.37}$$

A comparison of experimental values with those calculated from (20.37) follows:

k	0	1	2	3	4	5	6
$\lambda_{\text{calc}}/\text{Å}$	2070	3320	4590	5870	7160	8440	9730
$\lambda_{\text{exp}}/\text{Å}$	2240	3125	4160	5190	6250	7345	8480

Considering the extreme crudity of the FE MO method, the results are not too bad.

For the conjugated polyenes $CH_2{=}(CH{-}CH{=})_kCH_2$, one finds that the FE MO method does not predict the absorption bands very well. Since the bonds in the polyenes are not all equivalent, we should expect the FE MO method to work poorly.

Better results for the polyenes can be obtained with the Hückel method (developed in the 1930s). The Hückel MO method deals only with the pi electrons. It takes each pi MO as a linear combination of the $2p_z$ AOs of the conjugated carbon atoms (where the z axis is perpendicular to the molecular plane). These linear combinations are used in the variational integral, which is expressed as a sum of integrals involving the various AOs. The Hückel method approximates many of these integrals as zero and leaves others as parameters whose values are picked to give the best fit to experimental data. Details may be found in most quantum-chemistry texts. The Hückel method has been a mainstay of theoretically inclined organic chemists for many years, but the development of improved semiempirical methods (discussed below) has made the Hückel method largely obsolete.

For certain purposes, all one is interested in is the relative signs of the AOs that contribute to the pi MOs. (An example is the Woodward–Hoffmann rules for deducing the steric course of certain organic reactions; see any modern organic-chemistry text.) Although the FE MO method works poorly for quantitative calculations on linear polyenes, we can use it to deduce the signs of the AOs in the pi MOs.

Consider butadiene ($CH_2{=}CH{-}CH{=}CH_2$), for example. We take the pi MOs as linear combinations of the four $2p_z$ carbon AOs. (This is a minimal-basis-set treatment; Sec. 20.5.) Let p_1, p_2, p_3, p_4 denote these AOs. Figure 18.9 shows that the lowest pi MO will have no nodes perpendicular to the molecular plane, the next lowest pi MO will have one such node (located at the midpoint of the molecule), etc. To form the lowest pi MO, we must therefore combine the four $2p_z$ AOs all with the same signs: $c_1 p_1 + c_2 p_2 + c_3 p_3 + c_4 p_4$, where the c's are all positive. For the purpose of determining the relative signs, we won't worry about the fact that c_1 and c_2 differ in value (since the end and interior carbons are not equivalent); we shall simply write $p_1 + p_2 + p_3 + p_4$ for the lowest pi MO. To have a single node in the center of the molecule, we must take $p_1 + p_2 - p_3 - p_4$ as the second lowest pi MO; this is the highest occupied pi MO in the ground state, since there are four pi electrons in butadiene. To get two symmetrically placed nodes, we take $p_1 - p_2 - p_3 + p_4$ as the third lowest pi MO. To get three nodes, we take $p_1 - p_2 + p_3 - p_4$ as the fourth lowest pi MO. See Fig. 20.26.

An improved version of the Hückel method, applicable to both conjugated and nonconjugated molecules, is the *extended Hückel* (EH) *method*, developed in the 1950s and 1960s by Wolfsberg and Helmholz and by Hoffmann. The EH method treats all the valence electrons of a molecule and neglects fewer integrals than the Hückel method. The calculations of the EH method are relatively easy to perform (thanks to the many simplifying approximations made), and the method has been applied with some success to calculating equilibrium conformations of organic compounds.

The Hückel and extended Hückel methods are quite crude, in that they use a very simplified Hamiltonian that contains no repulsion terms between electrons. Several improved semiempirical theories have been developed that include electron repulsions in the Hamiltonian. These include the PPP, CNDO, INDO, and MINDO theories, discussed below. These theories treat only the valence electrons. They use a less approximate Hamiltonian than the Hückel Hamiltonian, in that some of the electron repulsions are included. These theories solve equations resembling the Hartree–Fock equations (19.47) to find self-consistent-field MOs, but since an approximate Hamiltonian is used and rather drastic approximations are made for many of the integrals that occur, the MOs found are only approximations to Hartree–Fock MOs.

The Pariser–Parr–Pople (PPP) theory was developed in the 1950s to deal with conjugated molecules. It treats only the π electrons. The theory has had pretty good success in predicting the electronic spectra of conjugated compounds.

The *complete neglect of differential overlap* (CNDO) *method* and the *intermediate neglect of differential overlap* (INDO) *method* were developed by Pople and coworkers in the 1960s. (The names indicate the nature of the approximations made in the theories.) These are generalizations of the PPP method and apply to both

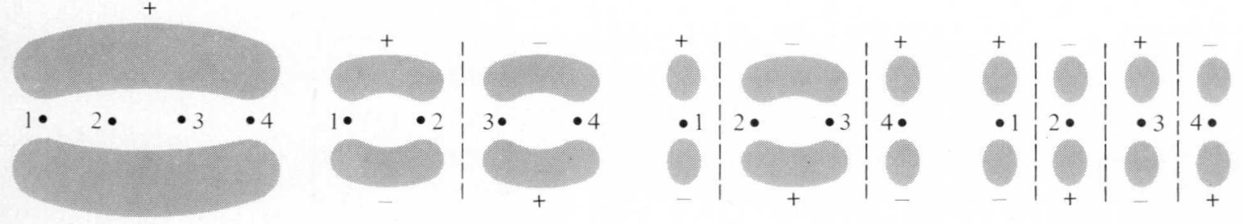

Figure 20.26 Rough sketches of the four lowest π MOs in butadiene.

conjugated and nonconjugated molecules. The CNDO and INDO methods give reasonably accurate molecular geometries and dipole moments but poor dissociation energies. This is to be expected, since these methods are approximations to the Hartree–Fock method.

The MINDO method (modified INDO) was developed by Dewar and coworkers in the period 1969–1975 and has evolved through the successive versions MINDO/1, MINDO/2, MINDO/2′, and MINDO/3. Dewar's aim was not to have a method giving approximations to Hartree–Fock results but one giving accurate molecular geometries and accurate dissociation energies. It might seem unreasonable to expect a theory that involves more approximations than the Hartree–Fock method to succeed in an area (calculation of dissociation energies) where the Hartree–Fock method fails. However, by choosing the values of the parameters that go into the MINDO method so as to reproduce the heats of atomization of a number of compounds, Dewar was able to build in compensation for the neglect of electron correlation that occurs in the Hartree–Fock theory. The MINDO/3 method gives generally accurate molecular geometries and gives molecular heats of atomization with a typical error of 6 kcal/mole ($\frac{1}{4}$ eV per molecule), which is a significant achievement considering the complete failure of the Hartree–Fock method in predicting dissociation energies.

One can usually estimate the heat of formation of a gas-phase compound to within 2 kcal/mole using purely empirical methods such as the Benson method of group contributions (Sec. 5.7). Hence, the real usefulness of MINDO does not lie in predicting heats of formation. Rather, the theory gives some promise of allowing one to calculate the potential-energy function for the motion of the nuclei during the course of a chemical reaction; see Sec. 23.2.

The MINDO method has been criticized by some advocates of *ab initio* methods. Some references on this controversy are R. C. Bingham, M. J. S. Dewar, and D. H. Lo, *J. Am. Chem. Soc.*, **97**, 1285 (1975); M. J. S. Dewar et al., ibid., 4540 (especially the appendix); J. A. Pople, ibid., 5306; W. J. Hehre, ibid., 5308; M. J. S. Dewar, ibid., 6591.

20.10 THE VALENCE-BOND METHOD

So far our discussion of molecular electronic structure has been based on the MO approximation. Historically, the first quantum-mechanical treatment of molecular bonding was the 1927 Heitler–London treatment of H_2. Their approach was extended by Slater and especially by Pauling to give the *valence-bond* (VB) *method* of treating molecules.

Heitler and London started with the idea that a ground-state H_2 molecule is formed from two 1*s* H atoms. If all interactions between the H atoms were ignored, the wave function for the system of two atoms would be the product of the separate wave functions of each atom. Hence, the first approximation to the H_2 spatial wave function is $1s_A(1)1s_B(2)$, where the function $1s_A(1)$ equals a constant times e^{-r_{1A}/a_0}. This product wave function is unsatisfactory since it distinguishes between the identical electrons, saying that electron 1 is on nucleus A and electron 2 is on nucleus B. To take care of electron indistinguishability, we must write the approximation to the ground-state H_2 spatial wave function as the linear combination $N'[1s_A(1)1s_B(2) + 1s_A(2)1s_B(1)]$. This function is symmetric with respect to electron interchange and hence requires the antisymmetric two-electron spin function (19.33). The ground-state H_2 Heitler–London VB wave function is therefore

$$N'[1s_A(1)1s_B(2) + 1s_A(2)1s_B(1)] \cdot 2^{-1/2}[\alpha(1)\beta(2) - \beta(1)\alpha(2)] \qquad (20.38)$$

Introducing a variable orbital exponent and using (20.38) in the variational integral, one finds a predicted D_e of 3.78 eV compared with the experimental value 4.75 eV (and the Hartree–Fock value 3.64 eV).

The Heitler–London function (20.38) is a linear combination of two determinants:

$$N\begin{vmatrix} 1s_A(1)\alpha(1) & 1s_B(1)\beta(1) \\ 1s_A(2)\alpha(2) & 1s_B(2)\beta(2) \end{vmatrix} - N\begin{vmatrix} 1s_A(1)\beta(1) & 1s_B(1)\alpha(1) \\ 1s_A(2)\beta(2) & 1s_B(2)\alpha(2) \end{vmatrix} \qquad (20.39)$$

The two determinants differ by giving different spins to the AOs $1s_A$ and $1s_B$ involved in the bonding.

When multiplied out, the MO spatial function (20.19) equals

$$N[1s_A(1)1s_A(2) + 1s_B(1)1s_B(2) + 1s_A(1)1s_B(2) + 1s_B(1)1s_A(2)]$$

Because of the terms $1s_A(1)1s_A(2)$ and $1s_B(1)1s_B(2)$, the MO function gives a 50 percent probability that an H_2 molecule will dissociate to $H^- + H^+$, and a 50 percent probability for dissociation to $H + H$. In actuality, a ground-state H_2 molecule always dissociates to two neutral hydrogen atoms. This incorrect dissociation prediction is related to the poor dissociation energies predicted by the MO method. In contrast, the VB function (20.38) correctly predicts dissociation to $H + H$.

If, instead of the symmetric spatial function in (20.38), one uses the antisymmetric spatial function $N[1s_A(1)1s_B(2) - 1s_A(2)1s_B(1)]$ multiplied by one of the three symmetric spin functions in (19.32), one gets the VB functions for the first excited electronic level (a triplet level) of H_2. The minus sign produces charge depletion between the nuclei, and the atoms repel each other as they come together.

To apply the VB method to polyatomic molecules, one writes down all possible ways of pairing up the unpaired electrons of the atoms forming the molecule. Each way of pairing gives one of the *resonance structures* of the molecule. For each resonance structure, one writes down a function (called a *bond eigenfunction*) resembling (20.39), and the molecular wave function is taken as a linear combination of the bond eigenfunctions. The coefficients in the linear combination are found by minimizing the variational integral. Besides covalent pairing structures, one also includes ionic structures. For example, for H_2, the only covalent pairing structure is H—H, but one also has the ionic resonance structures $H^+ \quad H^-$ and $H^- \quad H^+$. These ionic structures correspond to the bond eigenfunctions $1s_A(1)1s_A(2)$ and $1s_B(1)1s_B(2)$. By symmetry, the two ionic structures contribute equally, so with inclusion of ionic structures, the VB spatial wave function for H_2 becomes

$$c_1[1s_A(1)1s_B(2) + 1s_B(1)1s_A(2)] + c_2[1s_A(1)1s_A(2) + 1s_B(1)1s_B(2)]$$

One says there is *ionic–covalent resonance.* (Of course, one expects $c_2 \ll c_1$ for this nonpolar molecule.)

In many cases, one uses hybrid atomic orbitals to form the bond eigenfunctions. For example, for the tetrahedral molecule CH_4, one combines four sp^3 hybrid AOs on carbon with the $1s$ AOs of the hydrogens.

For polyatomic molecules, the VB wave function becomes cumbersome. For example, CH_4 has four bonds, and the bond eigenfunction corresponding to the single most important resonance structure (the one with each H1*s* AO paired with one of the carbon sp^3 hybrids) turns out to be a linear combination of $2^4 = 16$ determinants. Inclusion of other resonance structures further complicates the wave function.

The calculations of the VB method turn out to be far more difficult than those of the MO method. The various MO approaches have completely overshadowed the VB method when it comes to actual computation of molecular wave functions and properties. However, the language of VB theory provides organic chemists with a simple qualitative tool for rationalizing many observed trends.

20.11 FUTURE PROSPECTS

The use of electronic computers has brought truly remarkable advances in the ability of quantum chemists to deal with problems of real chemical interest. For example, *ab initio* calculations are now being used to study chemisorption on metal catalysts and hydration of ions in solution. [See H. F. Schaefer, *Ann. Rev. Phys. Chem.*, **27**, 261 (1976).] Quantum-mechanical calculations have "now become accepted by organic, inorganic, and physical chemists as a legitimate tool for the study of significant chemical problems." (ibid.) Whereas quantum-mechanical calculations used to be confined to small molecules and were published in journals read mainly by physical chemists and chemical physicists, such calculations now deal with medium-sized and even fairly large molecules and appear regularly in the *Journal of the American Chemical Society*, read by all kinds of chemists.

To what extent theoretical calculations can replace experiment is not clear at present. Two different viewpoints are summarized in the following statements: "Can quantum mechanical calculations replace experiment? My own answer to this is: at present, seldom." [E. B. Wilson, *Pure Appl. Chem.*, **47**, 41 (1976).] "We can calculate everything." [E. Clementi, quoted in R. G. Parr, *Proc. Natl. Acad. Sci. U.S.*, **72**, 763 (1975).]

PROBLEMS

20.1 Estimate the bond lengths in (*a*) CH_3OH; (*b*) HCN.

20.2 Explain why the observed boron–fluorine bond length in BF_3 is substantially less than the sum of the B and F single-bond radii.

20.3 Use average bond energies to estimate ΔH°_{298} for the following gas-phase reactions: (*a*) $C_2H_2 + 2H_2 \rightarrow C_2H_6$; (*b*) $N_2 + 3H_2 \rightarrow 2NH_3$. Compare with the true values found from data in the Appendix.

20.4 The dipole moments of CH_3F and CH_3I are 1.85 and 1.62 D, respectively. Use the H—C bond moment listed in Sec. 20.1 to estimate the C—F and C—I bond moments.

20.5 Use bond moments to estimate the dipole moments of (*a*) CH_3Cl; (*b*) CH_3CCl_3; (*c*) $CHCl_3$; (*d*) $Cl_2C{=}CH_2$. (Assume tetrahedral angles at singly bonded carbons and 120° angles at doubly bonded carbons.) Compare with the experimental values, which are (*a*) 1.87 D; (*b*) 1.78 D; (*c*) 1.01 D; (*d*) 1.34 D.

20.6 What would the bond moment of C≡N be if one assumed the H—C moment was 0.4 but had the polarity $H^-{-}C^+$?

20.7 Pauling observed that the A—B average bond energy generally exceeds the mean of the A—A and B—B average bond energies and this difference increases with increasing polarity of the A—B bond. The

Pauling electronegativity scale defines the electronegativity difference between elements A and B as $|x_A - x_B| \equiv 0.208(\Delta_{AB}/\text{kcal mol}^{-1})^{1/2}$, where

$$\Delta_{AB} \equiv E(\text{A—B}) - \tfrac{1}{2}[E(\text{A—A}) + E(\text{B—B})],$$

where the E's are average single-bond energies. The electronegativity of H is arbitrarily set at 2.2. Use Table 20.1 and data in the Appendix to compute $|x_A - x_B|$ for the following pairs of elements: (*a*) C, H; (*b*) C, O; (*c*) C, Cl. Compare with the electronegativity differences found from the Pauling values in Table 20.2. (The discrepancies are due to use of average bond energies different from those listed in Table 20.1.)

20.8 (*a*) Write a Lewis dot formula for H_2SO_4 that has eight electrons around S. (*b*) What formal charge does this dot formula give the S atom? (The formal charge is found by dividing the electrons of each bond equally between the two bonded atoms.) How reasonable is this formal charge? (*c*) Write the Lewis dot formula for SF_6. (*d*) Write a dot formula for H_2SO_4 that gives S a zero formal charge. (*e*) Explain why the observed sulfur–oxygen bond lengths in SO_4^{2-} in metal sulfates are 1.5 to 1.6 Å whereas the sum of the single-bond radii of S and O is 1.70 Å.

20.9 Draw the Lewis dot structure of CO. What is the formal charge (Prob. 20.8*b*) on carbon?

20.10 (*a*) The KF molecule has $R_e = 2.17$ Å. The ionization potential of K is 4.34 V, and the electron affinity of F is 3.40 eV. Use the model of nonoverlapping spherical ions to estimate D_e of KF. (The experimental value is 5.18 eV.) (*b*) Estimate the dipole moment of KF. (The experimental value is 8.60 D.) (*c*) Explain why KCl has a larger dipole moment than KF.

20.11 For an ionic molecule like NaCl, the electronic energy $E_e(R)$ equals the Coulomb's law potential energy $-e'^2/R$ plus a term that allows for the Pauli-principle repulsion due to the overlap of the ions' probability densities. This repulsion term can be very crudely estimated by the function B/R^{12}, where B is a positive constant. (See the discussion of the Lennard-Jones potential in Sec. 22.10.) Thus,

$$E_e \approx B/R^{12} - e'^2/R$$

for an ionic molecule. (*a*) Use the fact that E_e is a minimum at $R = R_e$ to show that $B = R_e^{11}e'^2/12$. (*b*) Use the above expression for E_e and the Na and Cl ionization potential and electron affinity to estimate D_e for NaCl ($R_e = 2.36$ Å). (*c*) The experimental D_e of NaCl is 4.29 eV. Does the function B/R^{12} overestimate or underestimate the Pauli repulsion? What value of m gives agreement with the observed D_e if the Pauli repulsion is taken as A/R^m, where A and m are constants?

20.12 Write down the MO wave function for the (repulsive) ground electronic state of He_2.

20.13 Give the ground-state MO electronic configurations for each of the following: (*a*) He_2^+; (*b*) Li_2; (*c*) Be_2; (*d*) C_2; (*e*) N_2. Which of these species are paramagnetic?

20.14 The bond order is one-half the difference between the number of bonding and antibonding electrons. Give the bond order of each molecule in Prob. 20.13.

20.15 Use the MO electron configurations to predict which of each of the following sets has the highest D_e: (*a*) N_2 or N_2^+; (*b*) O_2, O_2^+, or O_2^-.

20.16 Let ψ be an eigenfunction of $\hat{H}$; that is, let $\hat{H}\psi = E\psi$. Show that $(\hat{H} + c)\psi = (E + c)\psi$, where c is any constant. Hence, ψ is an eigenfunction of $\hat{H} + c$ with eigenvalue $E + c$.

20.17 For each of the species NF, NF^+, and NF^-, use the MO method to (*a*) write the valence-electron configuration; (*b*) find the bond order (Prob. 20.14); (*c*) decide whether the species is paramagnetic.

20.18 Sketch the two antibonding MOs (20.31) of BeH_2.

20.19 Let the line between atom A and atom B of a polyatomic molecule be the z axis. For each of the following atom A atomic orbitals, state whether it will contribute to a σ, π, or δ localized MO in the molecule: (*a*) s; (*b*) p_x; (*c*) p_y; (*d*) p_z; (*e*) d_{z^2}; (*f*) $d_{x^2-y^2}$; (*g*) d_{xy}; (*h*) d_{xz}; (*i*) d_{yz}.

20.20 (*a*) For H_2CO, list all the AOs that go into a minimal-basis-set MO calculation. (*b*) Use these AOs to form localized MOs for H_2CO. For each localized MO, state which AOs make the main contributions to it, state whether it is inner-shell, lone-pair, or bonding, and state whether it is σ or π. Take the z axis along the CO bond and the x axis perpendicular to the molecule. Use the σ-π description of the double bond.

20.21 (*a*) Use average-bond-energy data and data on C(g) and H(g) in the Appendix to estimate $\Delta H^\circ_{f,298}$ of $C_6H_6(g)$ on the assumption that benzene contains three carbon–carbon single bonds and three carbon–carbon double bonds. Compare the result with the experimental value. (*b*) Repeat (*a*) for cyclohexene(g) (one double bond).

20.22 Predict the geometry of (*a*) $TeBr_2$; (*b*) I_3^-; (*c*) $HgCl_2$; (*d*) $SnCl_2$; (*e*) XeF_2; (*f*) ClO_2^-.

20.23 Predict the geometry of (*a*) BrF_3; (*b*) GaI_3; (*c*) H_3O^+; (*d*) PCl_3.

20.24 Predict the geometry of (*a*) SnH_4; (*b*) SeF_4; (*c*) XeF_4; (*d*) BH_4^-; (*e*) BrF_4^-.

20.25 Predict the geometry of (*a*) $AsCl_5$; (*b*) BrF_5; (*c*) $SnCl_6^{2-}$.

20.26 Predict the geometry of (*a*) O_3; (*b*) NO_3^-; (*c*) SO_3; (*d*) SO_2; (*e*) SO_2Cl_2; (*f*) $SOCl_2$; (*g*) IO_3^-; (*h*) SOF_4; (*i*) XeO_3; (*j*) $XeOF_4$.

20.27 Let $p_1, \ldots, p_6$ be the $2p_z$ AOs of the carbons in benzene. The unnormalized forms of the occupied π MOs in benzene are $p_1 + p_2 + p_3 + p_4 + p_5 + p_6$, $p_2 + p_3 - p_5 - p_6$, and

$$2p_1 - p_2 - p_3 + 2p_4 - p_5 - p_6.$$

(*a*) Sketch these MOs. (*b*) Which of the three is lowest in energy?

21

SPECTROSCOPY AND PHOTOCHEMISTRY

Most of our experimental information on the energy levels of atoms and molecules comes from spectroscopy, the study of the absorption and the emission of electromagnetic radiation (light) by matter. Section 21.1 examines the nature of light. Section 21.2 is a general discussion of the interaction of radiation and matter. This is followed by Secs. 21.3 to 21.9 on the rotational and vibrational spectra of diatomic and polyatomic molecules. Electronic spectra are considered in Sec. 21.10, and magnetic resonance spectra in Secs. 21.11 and 21.12. Closely related to spectroscopy is photochemistry, the study of reactions caused or catalyzed by light; Sec. 21.13 on photochemistry concludes the chapter.

21.1 ELECTROMAGNETIC RADIATION

In 1801 Thomas Young observed interference of light when a light beam was diffracted at two pinholes, thereby showing the wave nature of light. A wave involves a vibration in space and in time, and so the question arises: What is the physical quantity that is doing the vibrating in a light wave? The answer was provided by James Clerk Maxwell.

In the 1860s Maxwell systematized the known laws of electricity and magnetism by showing that these laws can all be derived from a set of four relatively simple differential equations. These four equations (called *Maxwell's equations*) interrelate the electric and magnetic field vectors **E** and **B**, the electric charge, and the electric current. (**E** is defined in Sec. 14.1; **B** is defined in Sec. 21.11. The explicit forms of Maxwell's equations are given in *Halliday and Resnick*, suppl. top. V.) Maxwell's equations are the fundamental equations of electricity and magnetism, just as Newton's laws are the fundamental equations of classical mechanics.

In addition to containing all the laws of electricity and magnetism known in the 1860s, Maxwell's equations predicted something that was unknown at that time, namely, that an accelerated electric charge will emit energy in the form of electromagnetic waves traveling at a speed v_{em} in vacuum, where

$$v_{em} = (4\pi\varepsilon_0 \times 10^{-7} \text{ N s}^2 \text{ C}^{-2})^{-1/2} \tag{21.1}$$

[Recall that ε_0 (the permittivity of vacuum) occurs in Coulomb's law (14.1).] Substitution of the experimental value (14.2) of ε_0 gives

$$v_{em} = [4\pi(8.854 \times 10^{-19})\ \text{s}^2\ \text{m}^{-2}]^{-1/2} = 2.998 \times 10^8\ \text{m/s}$$

which equals the experimentally observed speed of light in vacuum. Maxwell therefore proposed that light consists of electromagnetic waves.

Maxwell's prediction of the existence of electromagnetic waves was confirmed by Hertz in 1887. Hertz produced electromagnetic waves by the oscillations of electrons in the metal wires of a tuned ac circuit; he detected these waves using a loop of wire (just as the antenna in your radio or TV set detects electromagnetic waves emitted by the transmitters of radio and TV stations). The oscillating electric field of the electromagnetic wave exerts a time-varying force on the electrons in the wires of the detector circuit, thereby producing an alternating current in these wires.

Maxwell's equations show electromagnetic waves to consist of oscillating electric and magnetic fields. The electric and magnetic field vectors **E** and **B** are perpendicular to each other and are perpendicular to the direction of travel of the wave. Figure 21.1 shows an electromagnetic wave traveling in the y direction. The vectors shown give the values of **E** and **B** at points on the y axis at one instant of time. As time passes, the crests (peaks) and troughs (valleys) move to the right. This figure is not a complete description of the wave, since it gives the values of **E** and **B** only at points lying on the y axis. To describe an electromagnetic wave fully we must give the values of six numbers (the three components of **E** and the three components of **B**) at every point in the region of space through which the wave is moving.

The wave shown in Fig. 21.1 is *plane-polarized*, meaning that the **E** vectors at various points in the wave all lie in the same plane. Such a plane-polarized wave would be produced by the back-and-forth oscillation of electrons in a straight wire. The light emitted by a collection of heated atoms or molecules (e.g., sunlight) is *unpolarized*, with the electric-field vectors pointing in different directions at different points in space. This is because the molecules act independently of one another and the radiation produced has random orientations of the **E** vector at various points in space. (Of course, for an unpolarized wave, **E** is still perpendicular to the direction of travel.)

The *wavelength* λ of a wave is the distance between successive crests. One *cycle* is that portion of the wave which lies between two successive crests (or between any two successive points having the same phase). The *frequency* ν of the wave is the number of cycles passing a given point per unit time. If ν cycles pass a given point in unit time, the time it takes one cycle to pass a given point is $1/\nu$ and a crest has

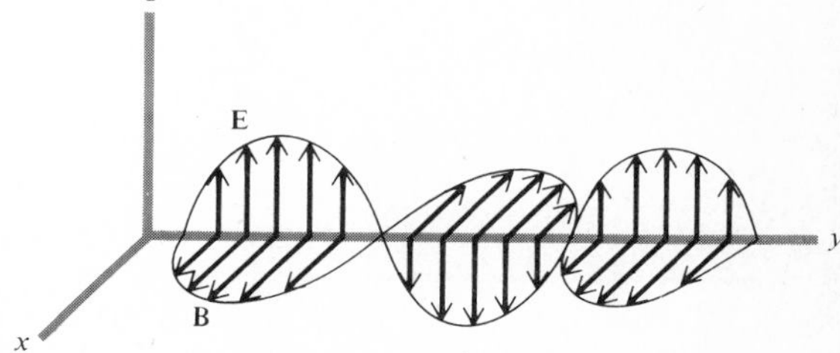

Figure 21.1
A portion of an electromagnetic wave ($1\frac{1}{2}$ cycles are shown).

therefore traveled a distance λ in time $1/\nu$. If c is the speed of the wave, then since distance = rate × time, we have $\lambda = c(1/\nu)$ or

$$\lambda\nu = c \tag{21.2}^*$$

The frequency is commonly given in the units s^{-1}. The SI system of units defines the frequency unit of one reciprocal second as one *hertz* (Hz): $1 \text{ Hz} \equiv 1 \text{ s}^{-1}$. Various multiples of the hertz are sometimes used, e.g., the kilohertz (kHz), megahertz (MHz), and gigahertz (GHz):

$$1 \text{ Hz} \equiv 1 \text{ s}^{-1}, \qquad 1 \text{ kHz} = 10^3 \text{ Hz}, \qquad 1 \text{ MHz} = 10^6 \text{ Hz}, \qquad 1 \text{ GHz} = 10^9 \text{ Hz}$$

Common units of λ include the *angstrom* (Å), the *micrometer* (μm), and the *nanometer* (nm):

$$1 \text{ Å} \equiv 10^{-8} \text{ cm} = 10^{-10} \text{ m}, \qquad 1 \ \mu\text{m} = 10^{-6} \text{ m}, \qquad 1 \text{ nm} = 10^{-9} \text{ m} = 10 \text{ Å}$$

(The micrometer was formerly called the micron.)

The *wave number* σ of a wave is the reciprocal of the wavelength:

$$\sigma \equiv 1/\lambda \tag{21.3}^*$$

Most commonly, σ is expressed in cm^{-1}.

The human eye is sensitive to electromagnetic radiation with λ in the range 4000 Å (violet light) to 7500 Å (red light), but there is no upper or lower limit to the values of λ and ν of an electromagnetic wave. Figure 21.2 shows the *electromagnetic spectrum*, the range of frequencies and wavelengths of electromagnetic waves. For convenience, the electromagnetic spectrum is divided into various regions, but there is no sharp boundary between adjacent regions.

All frequencies of electromagnetic radiation (light) travel at the same speed $c = 3 \times 10^{10}$ cm/s in vacuum. Let c_B be the speed of light in substance B. One finds that c_B depends on the nature of the substance B and on the frequency of the light. The ratio c/c_B for a given frequency of light is the *refractive index* n_B of substance B for that frequency:

$$n_B \equiv c/c_B \tag{21.4}$$

Some values of n_B at 25°C and 1 atm for yellow light of vacuum wavelength 5893 Å ("sodium D light") follow:

Air	H_2O	C_6H_6	C_2H_5OH	CS_2	CH_2I_2	NaCl	glass
1.0003	1.33	1.50	1.36	1.63	1.75	1.53	1.5–1.9

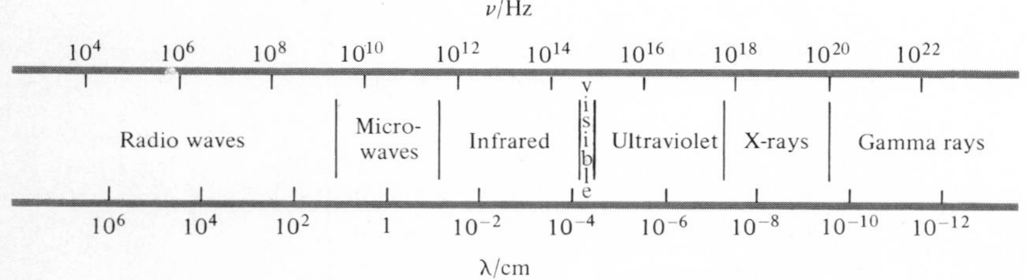

Figure 21.2
The electromagnetic spectrum. The scale is logarithmic.

For quartz at 18°C and 1 atm, n decreases from 1.57 to 1.45 as the vacuum wavelength of the light increases from 1850 to 8000 Å. Organic chemists use n as a conveniently measured property to help characterize a liquid.

When a light beam passes obliquely from one substance to another, it is bent or *refracted*, due to the difference in speeds in the two substances. (Since $c = \lambda\nu$, if c changes, either λ or ν or both must change. It turns out that λ changes but ν stays the same as the wave goes from one medium to another.) The amount of refraction depends on the ratio of the speeds of the light in the two substances and hence depends on the refractive indices of the substances. The fact that n of a substance depends on ν means that light of different frequencies is refracted by different amounts. This allows us to separate or *disperse* an electromagnetic wave containing many frequencies into its component frequencies. An example is the dispersion of white light into the colors red, orange, yellow, green, blue, and violet by a glass prism or by water droplets in the atmosphere to give a rainbow.

So far in this section, we have presented the "classical" picture of light. However, in 1905, Einstein proposed that the interaction between light and matter could best be understood by postulating a particlelike aspect to light, each quantum (photon) of light having an energy $h\nu$ (Sec. 18.2). The direction of increasing frequency in Fig. 21.2 is thus the direction of increasing energy of the photons.

Although the classical picture of electromagnetic radiation as being produced by an accelerated electric charge is appropriate for the production of radio waves by electrons moving more or less freely in a metal wire (such electrons have a continuous range of allowed energies), the emission and absorption of radiation by atoms and molecules can generally be understood only by using quantum physics. The quantum theory of radiation pictures a photon as being produced or absorbed when an atom or molecule jumps between two different allowed energy levels.

21.2 SPECTROSCOPY

In *spectroscopy* one studies the electromagnetic radiation (light) emitted or absorbed by a sample. The collection of frequencies absorbed by a sample is its *absorption spectrum*; the frequencies emitted constitute the *emission spectrum*. A *line spectrum* contains only discrete frequencies. A *continuous spectrum* contains a continuous range of frequencies. One finds that a heated solid commonly gives a continuous emission spectrum; an example is the blackbody radiation spectrum (Fig. 18.2). A heated gas that is not at very high pressure gives a line spectrum, corresponding to transitions between the quantum-mechanically allowed energy levels of the individual molecules of the gas.

When a sample of molecules is exposed to electromagnetic radiation, the electric field of the radiation exerts a time-varying force on the electrical charges (electrons and nuclei) of each molecule. To treat the interaction of radiation and matter, one uses quantum mechanics, in particular, the time-dependent Schrödinger equation (18.9). Since the mathematics is rather complicated, we shall just quote the results, omitting the derivations. (See *Levine, Molecular Spectroscopy*, chap. 3; this reference will hereinafter be referred to as *M. S.*)

The quantum-mechanical treatment shows that a molecule in the stationary state m that is exposed to electromagnetic radiation may absorb a photon of frequency ν and make a transition to a higher-energy state n if the radiation's frequency ν satisfies $E_n - E_m = h\nu$. This is in agreement with Eq. (18.6), given by Bohr. A molecule in stationary state n in the absence of radiation can spontaneously jump to a lower stationary state m, emitting a photon whose frequency satisfies $E_n - E_m = h\nu$. This is *spontaneous emission* of radiation.

Exposing a molecule in state n to electromagnetic radiation whose frequency satisfies $E_n - E_m = h\nu$ will increase the probability that the molecule will jump to the lower state m with emission of a photon of frequency ν. Emission due to exposure to electromagnetic radiation is called *stimulated emission*. The radiation of stimulated emission is emitted in phase with the radiation causing this emission. Lasers use stimulated emission to produce a very intense, highly directional beam of nearly monochromatic (i.e., single-frequency) light.

The quantum-mechanical treatment shows that the probability of absorption or emission between the stationary states m and n is proportional to the square of the magnitude of the integral

$$\mathbf{p}_{mn} \equiv \int \psi_m^* \hat{\mathbf{p}} \psi_n \, d\tau \qquad \text{where } \hat{\mathbf{p}} = \sum_i Q_i \mathbf{r}_i \tag{21.5}$$

where the integration is over the full range of electronic and nuclear coordinates, and where the sum goes over all the charged particles in the molecule. In the Born–Oppenheimer approximation, the stationary-state wave functions ψ_m and ψ_n are each the product of electronic and nuclear wave functions. $\hat{\mathbf{p}}$ is the electric dipole-moment operator and occurs in Eq. (20.36).

For pairs of states for which the integral $\mathbf{p}_{mn}$ equals zero, the probability of a radiative transition is zero, and the transition is said to be *forbidden*. For example, consider a particle with charge Q in a one-dimensional box of length a. There is only one particle, so that the sum in (21.5) contains only one term. For this one-dimensional problem, the displacement from the origin equals the x coordinate. The wave functions are given by (18.28). Hence

$$p_{mn} = \frac{2Q}{a} \int_0^a x \sin \frac{m\pi x}{a} \sin \frac{n\pi x}{a} \, dx \tag{21.6}$$

Using the identities $\sin r \sin s = \frac{1}{2} \cos (r - s) - \frac{1}{2} \cos (r + s)$ and $\int x \cos cx \, dx = c^{-2} \cos cx + (x/c) \sin cx$, one finds (Prob. 21.3)

$$p_{mn} = \frac{Qa}{\pi^2} \left\{ \frac{\cos [(m - n)\pi] - 1}{(m - n)^2} - \frac{\cos [(m + n)\pi] - 1}{(m + n)^2} \right\} \tag{21.7}$$

We have $\cos \theta = 1$ for $\theta = 0, \pm 2\pi, \pm 4\pi, \pm 6\pi, \ldots$ and $\cos \theta = -1$ for $\theta = \pm\pi, \pm 3\pi, \pm 5\pi, \ldots$. If m and n are both even numbers or both odd numbers, then $m - n$ and $m + n$ are even numbers and p_{mn} equals zero. If m is even and n is odd or vice versa, then $m - n$ and $m + n$ are odd and p_{mn} is nonzero. Hence, radiative transitions between particle-in-a-box states m and n are allowed only if the change $m - n$ in quantum numbers is an odd number. This is the *selection rule* for the particle in a box. The selection rule(s) for a system tell which transitions are allowed. A particle in a box in the $n = 1$ ground state can absorb radiation of appropriate frequency to jump to $n = 2$ or $n = 4$, etc., but cannot make a radiative transition to $n = 3$ or $n = 5$, etc. The selection rule is $\Delta n = \pm 1, \pm 3, \pm 5, \ldots$.

The selection rule for the one-dimensional harmonic oscillator (Sec. 18.12) turns out to be (*M. S.*, sec. 3.3) $\Delta v = \pm 1$. Thus, a harmonic oscillator in the $v = 2$ state can only go to $v = 3$ or $v = 1$ by absorption or emission of a photon.

The selection rule for the two-particle rigid rotor (Sec. 18.14) is found to be (*M. S.*, sec. 4.4) $\Delta J = \pm 1$. For the rigid rotor (and for the harmonic oscillator) only a small fraction of possible radiative transitions are allowed.

We shall see that electronic states of a molecule are more widely spaced than vibrational states, which, in turn, are more widely spaced than rotational states. Transitions between molecular electronic states correspond to absorption in the ultraviolet (UV) and visible regions. Vibrational transitions correspond to absorption in the infrared (IR) region. Rotational transitions correspond to absorption in the microwave region. (See Fig. 21.2.)

The experimental techniques for absorption spectroscopy in the UV, visible, and IR regions are similar. Here, one passes a beam of light containing a continuous range of frequencies through the sample, disperses the radiation using a prism or diffraction grating, and at each frequency compares the intensity of the transmitted light with the intensity of a reference beam that did not pass through the sample. A more detailed description of an IR spectrometer is given in Sec. 21.8.

The techniques of microwave spectroscopy are described in Sec. 21.6.

The absorption of UV, visible, and IR light by a liquid or gaseous sample is often expressed by the Beer–Lambert law, which we now derive.

Consider a beam of light passing through a sample of pure substance B or of B dissolved in a solvent that does not absorb radiation and does not interact strongly with B. The beam may contain a continuous range of wavelengths, but we shall focus attention on the radiation whose wavelength lies in the very narrow range from λ to $\lambda + d\lambda$. (The wavelength λ is the wavelength measured in vacuum or in air; not the wavelength in the absorbing sample.)

Let $I_{\lambda,0}$ be the intensity of the radiation with wavelength in the range λ to $\lambda + d\lambda$ that is incident on the sample, and let I_λ be the intensity of this radiation after it has gone through a length x of the sample. The *intensity* is defined as the energy per unit time that falls on unit area perpendicular to the beam. The intensity is proportional to the number of photons incident on unit area in unit time. Let N_λ photons of wavelength between λ and $\lambda + d\lambda$ fall on unit area of the sample in unit time, and let dN_λ be the number of such photons absorbed by a thickness dx of the sample (Fig. 21.3). The probability that a given photon will be absorbed in the thickness dx is dN_λ/N_λ; this probability is proportional to the number of B molecules in dx and hence to the number of moles dn_B of B in this layer. The number of moles dn_B is proportional to the molar concentration c_B of B and to the layer thickness dx. Hence, $dN_\lambda/N_\lambda \propto c_B\, dx$.

Figure 21.3
Radiation incident on and emergent from a thin slice of sample.

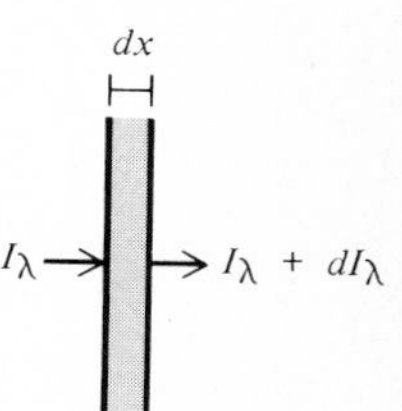

Let dI_λ be the change in light intensity at wavelength λ due to passage through the layer of thickness dx. Then $dI_\lambda \propto -dN_\lambda$. (The minus sign arises because dN_λ was defined as positive and dI_λ is negative.) Also, $I_\lambda \propto N_\lambda$. Hence, $dI_\lambda/I_\lambda \propto -dN_\lambda/N_\lambda$, and $dI_\lambda/I_\lambda \propto -c_B\, dx$. Letting α_λ be the proportionality constant and integrating along the length of the sample, we have

$$dI_\lambda/I_\lambda = -\alpha_\lambda c_B\, dx \qquad \text{and} \qquad \int_{I_{\lambda,0}}^{I_\lambda} \frac{dI_\lambda}{I_\lambda} = -\alpha_\lambda c_B \int_0^x dx \tag{21.8}$$

$$\ln \frac{I_\lambda}{I_{\lambda,0}} = 2.303 \log \frac{I_\lambda}{I_{\lambda,0}} = -\alpha_\lambda c_B x \tag{21.9}$$

Letting $\varepsilon_\lambda \equiv \alpha_\lambda/2.303$ and defining the *absorbance A* at wavelength λ as $\log (I_{\lambda,0}/I_\lambda)$, we have

$$A \equiv \log \frac{I_{\lambda,0}}{I_\lambda} = \varepsilon_\lambda c_B x \tag{21.10}$$

which is the *Beer–Lambert law.*

Equations (21.9) and (21.10) can be written as

$$I_\lambda = I_{\lambda,0} e^{-\alpha_\lambda c_B x} = I_{\lambda,0} 10^{-\varepsilon_\lambda c_B x} \tag{21.11}$$

The quantity $I_\lambda/I_{\lambda,0}$ is the *transmittance T* of the sample at wavelength λ.

The quantity ε_λ is the *molar absorption coefficient* or *molar absorptivity* (formerly called the molar extinction coefficient) of substance B at wavelength λ. Most commonly, the concentration c_B is expressed in mol/dm^3 and the path length x in centimeters, so that ε is commonly given in dm^3 mol^{-1} cm^{-1}. Figure 21.20 below shows ε as a function of λ for benzene over a range of UV frequencies. Since the vertical scale in this figure is logarithmic, ε varies over an enormous range.

When several different absorbing species B, C, ... are present, and when there are no strong interactions between the species, we have $dI_\lambda/I_\lambda = -(\alpha_B c_B + \alpha_C c_C + \cdots)\, dx$ and (21.10) becomes

$$A \equiv \log (I_{\lambda,0}/I_\lambda) = (\varepsilon_{\lambda,B} c_B + \varepsilon_{\lambda,C} c_C + \cdots)x \tag{21.12}$$

where $\varepsilon_{\lambda,B}$ is the molar absorption coefficient of B at wavelength λ. If the molar absorption coefficients are known for B, C, ... at several wavelengths, measurement of $I_{\lambda,0}/I_\lambda$ at several wavelengths allows a mixture of unknown composition to be analyzed.

The energy of the absorbed radiation is usually dissipated by intermolecular collisions to translational, rotational, and vibrational energies of the molecules, thereby increasing the temperature of the sample. Some of the absorbed energy may be radiated by the excited molecules (fluorescence and phosphorescence; Sec. 21.13). This occurs especially in low-pressure gases, where the average time between collisions is much greater than in liquids. The relative amount of emission depends on the average time between collisions compared with the average lifetimes of the various excited states. Sometimes the absorbed radiation leads to a chemical reaction (Sec. 21.13).

21.3 ROTATION AND VIBRATION OF DIATOMIC MOLECULES

We now consider nuclear motion in an isolated diatomic molecule. By "isolated" we mean that interactions with other molecules are slight enough to neglect. This condition is well met in a gas at low pressure.

Translation, rotation, and vibration. The Schrödinger equation (20.9) for nuclear motion in a molecule is $(\hat{T}_N + E_e)\psi_N \equiv \hat{H}_N\psi_N = E\psi_N$, where E is the total energy, $\hat{T}_N$ is the operator for the kinetic energies of the nuclei, and E_e is the electronic energy (including nuclear repulsion) as a function of the spatial configuration of the nuclei.

For a diatomic molecule composed of atoms A and B with masses m_A and m_B, we have $\hat{T}_N = -(\hbar^2/2m_A)\nabla_A{}^2 - (\hbar^2/2m_B)\nabla_B{}^2$ and $E_e = E_e(R)$, where R is the internuclear distance. The potential energy $E_e(R)$ for nuclear motion is a function of only the relative coordinates of the two nuclei; therefore the results of Sec. 18.13 apply, and

$$E = E_{tr} + E_{int} \tag{21.13}$$

$$[-(\hbar^2/2\mu)\nabla^2 + E_e(R)]\psi_{int} = E_{int}\psi_{int} \tag{21.14}$$

The total molecular energy is the sum of E_{tr}, the translational energy of the molecule as a whole, and E_{int}, the energy of the relative or internal motion of the two atoms. The operator ∇^2 equals $\partial^2/\partial x^2 + \partial^2/\partial y^2 + \partial^2/\partial z^2$, where x, y, z are the coordinates of one nucleus relative to the other; ψ_{int} is a function of x, y, and z. The reduced mass μ is given by Eq. (18.72) as

$$\mu = m_A m_B/(m_A + m_B)$$

The allowed translational energies E_{tr} can be taken as those of a particle in a three-dimensional box, Eq. (18.40). The box is the container in which the gas molecules are confined.

The kinetic energy of internal motion can be divided into kinetic energy of rotation and kinetic energy of vibration. Rotation involves a change in the orientation of the molecular axis in space while the internuclear distance R remains fixed. Vibration involves a change in the internuclear distance. Since R is fixed for rotation, the potential energy $E_e(R)$ is associated with the vibration of the molecule. To a good approximation, the rotational and vibrational motions can be treated separately. (See *M. S.*, sec. 4.1, for details.)

Even in the ground vibrational state, there is a vibrational motion (due to the zero-point vibrational energy), so that the diatomic molecule is not really rigid. However, the nuclei vibrate back and forth about the equilibrium internuclear distance R_e, so to a good approximation, we can treat the molecule as a two-particle rigid rotor with separation R_e between the nuclei. From (18.74) and (18.75) the rotational energy of a *diatomic* molecule is (approximately)

$$E_{rot} \approx \frac{J(J+1)\hbar^2}{2I_e}, \qquad I_e = \mu R_e^2, \qquad J = 0, 1, 2, \ldots \tag{21.15}$$*

$$E_{rot} \approx B_e hJ(J+1) \qquad \text{where } B_e \equiv h/8\pi^2 I_e \tag{21.16}$$

I_e is the *equilibrium moment of inertia*. B_e is the *equilibrium rotational constant*. Note that R_e differs for different electronic states of the same molecule (see, for example, Fig. 20.2).

The rotational wave function ψ_{rot} depends on the angles θ and ϕ defining the orientation of the molecule in space (Fig. 18.13). The vibrational wave function ψ_{vib} depends on the internuclear distance R. In the process of separating off the rotational motion and converting (21.14) from a three- to a one-dimensional Schrödinger equation one finds that an extra factor of $1/R$ is introduced (see *M. S.*, sec. 4.1), so that $\psi_{int} = \psi_{rot}(\theta, \phi)\psi_{vib}(R)/R$.

With rotational motion separated off, Eq. (21.14) becomes

$$\left[-\frac{\hbar^2}{2\mu}\frac{d^2}{dR^2} + E_e(R)\right]\psi_{vib}(R) = (E_{int} - E_{rot})\psi_{vib}(R) \tag{21.17}$$

The potential-energy function $E_e(R)$ differs for each different electronic state and is known accurately for only a few states of a relatively few molecules. The nuclei vibrate about the equilibrium distance R_e, and it is useful to expand $E_e(R)$ in a Taylor series about R_e. Equation (8.37) gives

$$E_e(R) = E_e(R_e) + E_e'(R_e)(R - R_e) + \tfrac{1}{2}E_e''(R_e)(R - R_e)^2 + \tfrac{1}{6}E_e'''(R_e)(R - R_e)^3 + \cdots$$

where $E_e'(R_e)$ equals $dE_e(R)/dR$ evaluated at $R = R_e$, with similar definitions for $E_e''(R_e)$, etc. Since we are considering a bound state with a minimum in E_e at $R = R_e$ (Fig. 20.2), $E_e'(R_e) = 0$. This eliminates the second term in the expansion. The vibrations will be confined to distances R reasonably close to R_e, so the terms involving $(R - R_e)^3$ and higher powers will be smaller than the term involving $(R - R_e)^2$; we shall neglect these terms. Therefore

$$E_e(R) \approx E_e(R_e) + \tfrac{1}{2}E_e''(R_e)(R - R_e)^2 \tag{21.18}$$

The quantity $E_e(R_e)$ is a constant for a given electronic state and will be called the *equilibrium electronic energy* E_{el} of the electronic state:

$$E_{\text{el}} \equiv E_e(R_e) \tag{21.19}$$

For the H_2 ground electronic state, the equilibrium dissociation energy is $D_e = 4.75$ eV, so $E_e(R_e) = E_{\text{el}}$ is 4.75 eV below the energy of the dissociated molecule. The energy of two ground-state H atoms is given by (19.19) as $2(-13.60 \text{ eV}) = -27.20$ eV, so $E_{\text{el}} = -31.95$ eV for the H_2 ground electronic state.

Let $x \equiv R - R_e$. Then $d^2/dx^2 = d^2/dR^2$. Substitution of (21.18) and (21.19) into (21.17) and rearrangement gives as the approximate vibrational Schrödinger equation

$$\left[-\frac{\hbar^2}{2\mu}\frac{d^2}{dx^2} + \tfrac{1}{2}E_e''(R_e)x^2\right]\psi_{\text{vib}} = (E_{\text{int}} - E_{\text{rot}} - E_{\text{el}})\psi_{\text{vib}} \tag{21.20}$$

Equation (21.20) has the same form as the Schrödinger equation (18.65) for a one-dimensional harmonic oscillator. Hence, the solutions to (21.20) are the harmonic-oscillator wave functions. The quantity $E_{\text{int}} - E_{\text{rot}} - E_{\text{el}}$ in (21.20) corresponds to the harmonic-oscillator energy E in (18.65); therefore $E_{\text{int}} - E_{\text{rot}} - E_{\text{el}} = (v + \tfrac{1}{2})h\nu_e$. Combining this equation with (21.13), we have

$$E = E_{\text{tr}} + E_{\text{rot}} + E_{\text{vib}} + E_{\text{el}} \tag{21.21*}$$

$$E_{\text{vib}} \approx (v + \tfrac{1}{2})h\nu_e, \qquad v = 0, 1, 2, \ldots \qquad \text{diatomic molecule} \tag{21.22*}$$

The force constant k in (18.65) corresponds to $E_e''(R_e)$ in (21.20). Also, m in (18.65) corresponds to μ in (21.20). Therefore the *equilibrium vibrational frequency* ν_e of the diatomic molecule is [Eq. (18.63)]

$$\nu_e = \frac{1}{2\pi}\left(\frac{k_e}{\mu}\right)^{1/2} = \frac{1}{2\pi}\left[\frac{E_e''(R_e)}{\mu}\right]^{1/2} \tag{21.23}$$

From (18.61), the *equilibrium force constant* k_e has units of force over length. The literature usually gives k_e in millidynes per angstrom (mdyn/Å); the SI unit is newtons per meter. From Prob. 21.10, 1 mdyn/Å = 100 N/m.

Anharmonicity. The approximation (21.18) for the potential energy of nuclear motion leads to harmonic-oscillator vibrational energy levels. Comparing the lowest curve in Fig. 20.2 with the harmonic-oscillator parabolic curve in Fig. 18.11, we see that for $R \gg R_e$, the parabolic approximation (21.18) is poor. The potential energy

$E_e(R)$ is not really a harmonic-oscillator potential energy, and this *anharmonicity* adds correction terms to the approximate vibrational-energy expression (21.22). A perturbation treatment (*M. S.*, sec. 4.2) shows that the main correction term to (21.22) is

$$-h\nu_e x_e(v + \tfrac{1}{2})^2 \tag{21.24}$$

where the *anharmonicity constant* $\nu_e x_e$ is nearly always positive and depends on $E_e'''(R_e)$ and $E^{\text{iv}}(R_e)$.

A harmonic oscillator has equally spaced energy levels. With inclusion of the correction term (21.24) the spacing between the adjacent levels v and $v + 1$ becomes

$$(v + \tfrac{3}{2})h\nu_e - h\nu_e x_e(v + \tfrac{3}{2})^2 - [(v + \tfrac{1}{2})h\nu_e - h\nu_e x_e(v + \tfrac{1}{2})^2] = h\nu_e - 2h\nu_e x_e(v + 1)$$

Hence, the spacing between adjacent vibrational levels decreases as v increases. A harmonic oscillator has an infinite number of vibrational levels. However, a diatomic-molecule bound electronic state has only a finite number of vibrational levels, since once the vibrational energy exceeds the equilibrium dissociation energy D_e, the molecule dissociates. Figure 21.4 shows the vibrational levels of the H_2 ground electronic state, which has 15 bound vibrational levels ($v = 0$ through 14). Note the decreasing spacing as v increases. The zero level of energy has been taken at the minimum of the potential-energy curve. The $v = 14$ level lies a mere 150 cm^{-1} below $E_e(\infty)/hc$.

Spectroscopic dissociation energy. The lowest value of molecular rotational energy is zero, since $E_{\text{rot}} = 0$ for $J = 0$ in (21.15). However, the lowest vibrational energy is nonzero. In the harmonic-oscillator approximation, the $v = 0$ ground vibrational state has the energy $\frac{1}{2}h\nu_e$. With inclusion of the anharmonicity correction (21.24), the vibrational ground-state energy is $\frac{1}{2}h\nu_e - \frac{1}{4}h\nu_e x_e$. The dissociation energy measured from the lowest vibrational state is called the *spectroscopic dissociation energy* D_0 (Fig. 21.4) and is somewhat less than the equilibrium dissociation energy D_e, due to the zero-point vibrational energy:

$$D_e = D_0 + \tfrac{1}{2}h\nu_e - \tfrac{1}{4}h\nu_e x_e \tag{21.25}$$

For the H_2 ground electronic state, D_0 is 4.48 eV, compared with 4.75 eV for D_e.

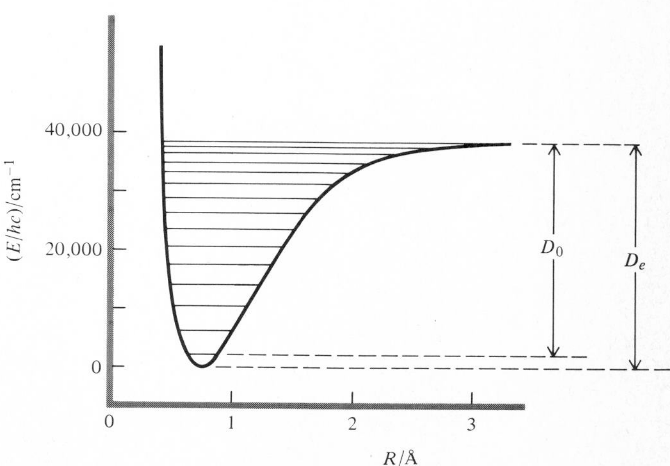

Figure 21.4
Ground-electronic-state vibrational levels of H_2. [Data from *S. Weissman et al., J. Chem. Phys.*, **39,** *2226 (1963)*.]

Consider the gas-phase reaction $H_2(g) \rightarrow 2H(g)$. The standard-state thermodynamic properties refer to ideal gases, and there are no intermolecular interactions in ideal gases. At absolute zero, the molecules will all be in the lowest available energy level. For H_2, this is the $v = 0$, $J = 0$ level of the ground electronic state. Hence, the thermodynamic quantity ΔU_0° for $H_2(g) \rightarrow 2H(g)$ will equal $N_0 D_0$, where N_0 is the Avogadro constant. Observation of D_0 for the H_2 ground electronic state from spectroscopy experiments gives $D_0 = 4.4780$ eV. Using (20.1), we get $\Delta U_0^\circ = 103.26$ kcal/mole.

Consider now ΔU_{298}° for $H_2(g) \rightarrow 2H(g)$. The molecules will be distributed among their various translational, rotational, vibrational, and electronic states in accord with the Boltzmann distribution law (15.74). Because of the close spacing between adjacent translational levels and between adjacent rotational levels (see below), it is a good approximation to use the classical equipartition results (Sec. 15.10) of $\frac{3}{2}kT$ for the average molecular translational energy and kT for the average rotational energy of a diatomic molecule. (See Chap. 22 for a fuller justification of this.) Moreover, the large spacing between vibrational energy levels and between electronic energy levels (see below) keeps nearly all the H_2 molecules in the $v = 0$ vibrational level of the ground electronic state at room temperature. The H atoms have only translational energy, and the translational energy of 2 moles of H(g) is $3RT = 1.78$ kcal/mol at 298.15 K. The translational plus rotational energy of 1 mole of $H_2(g)$ is $\frac{3}{2}RT + RT = 1.48$ kcal/mol at 298.15 K. We therefore expect $\Delta U_{298}^\circ/(\text{kcal mol}^{-1}) = 103.26 + 1.78 - 1.48 = 103.56$ for the dissociation of H_2. The Appendix gives $\Delta H_{f,298}^\circ$ of H(g) as 52.10 kcal/mol. Multiplication by 2 and use of (5.15) gives $\Delta U_{298}^\circ = 103.61$ kcal/mol for $H_2 \rightarrow 2H$, in pretty good agreement with the calculated value.

Vibration–rotation interaction. Equation (21.21) neglects the interaction between vibration and rotation. Because of the anharmonicity of the potential-energy curve $E_e(R)$, the average distance R_{av} between the nuclei increases as the vibrational quantum number increases (see Fig. 21.4). This increase in R_{av} increases the effective moment of inertia $I = \mu R_{av}^2$ and decreases the rotational energies (which are proportional to $1/I$). To allow for this effect, one adds the term $-h\alpha_e(v + \frac{1}{2})J(J + 1)$ to the energy, where the *vibration–rotation coupling constant* α_e is a positive number.

Centrifugal distortion. Since the molecule is not really a rigid rotor, there is a tendency for the internuclear distance to increase slightly as the rotational energy increases, a phenomenon called *centrifugal distortion*. Centrifugal distortion increases the moment of inertia and hence decreases the rotational energy below that of a rigid rotor. One finds that the term $-hDJ^2(J + 1)^2$ must be added to the rotational energy, where the *centrifugal-distortion constant* D is a very small positive constant. (Don't confuse D with the dissociation energy.)

Internal energy of a diatomic molecule. With inclusion of anharmonicity, vibration–rotation interaction, and centrifugal distortion, the expression for the internal energy of a diatomic molecule becomes

$$E_{int} = E_{el} + h\nu_e(v + \tfrac{1}{2}) - h\nu_e x_e(v + \tfrac{1}{2})^2 + hB_e J(J + 1)$$
$$- h\alpha_e(v + \tfrac{1}{2})J(J + 1) - hDJ^2(J + 1)^2 \qquad (21.26)$$

The total energy equals $E_{int} + E_{tr}$.

Level spacings and the Boltzmann distribution. The translational energy levels for a particle in a cubic box of volume V are [Eq. (18.42)] $(n_x^2 + n_y^2 + n_z^2)h^2/8mV^{2/3}$. The spacing between adjacent translational levels with quantum numbers n_x, n_y, n_z and $n_x + 1$, n_y, n_z is $(2n_x + 1)h^2/8mV^{2/3}$, since $(n_x + 1)^2 - n_x^2 = 2n_x + 1$. A typical molecule has translational energy of roughly $\frac{1}{2}kT$ associated with motion in each direction (Chap. 15). The equation $\frac{1}{2}kT = n_x^2h^2/8mV^{2/3}$ gives for a molecule of molecular weight 100 in a 1-cm^3 box at room temperature: $n_x = 8 \times 10^8$; this gives a spacing of 5×10^{-23} erg $= 3 \times 10^{-11}$ eV between adjacent translational levels. This energy gap is so tiny that for all practical purposes, we can consider the translational energy levels of the gas molecules to be continuous rather than discrete. The very close spacing results from the macroscopic size of V.

The spacing between the rotational levels with quantum numbers J and $J + 1$ is given by (21.15) as $(J + 1)\hbar^2/\mu R_e^2$, since $(J + 1)(J + 2) - J(J + 1) = 2(J + 1)$. For the CO ground electronic state, R_e is 1.1 Å, and

$$\mu = \frac{[(12 \text{ g mol}^{-1})/N_0][(16 \text{ g mol}^{-1})/N_0]}{(28 \text{ g mol}^{-1})/N_0} = 1.1 \times 10^{-23} \text{ g}$$

This gives $\hbar^2/\mu R_e^2 = 8 \times 10^{-16}$ erg $= 0.0005$ eV. The spacing between adjacent CO rotational levels is $J + 1$ times this number, where $J = 0, 1, 2, \ldots$.

The spacing between the harmonic-oscillator vibrational levels is $h\nu_e$. Experimental observations on vibrational spectra show that ν_e is typically 10^{13} to 10^{14} s^{-1}, so the vibrational spacing is typically 7×10^{-14} to 7×10^{-13} erg or 0.04 to 0.4 eV.

Observations on electronic absorption spectra show that the spacing $E_{\text{el},2} - E_{\text{el},1}$ between the ground and first excited electronic levels is typically 2 to 6 eV.

Thus, electronic energy differences are substantially greater than vibrational energy differences, which in turn are much greater than rotational energy differences.

The Boltzmann distribution law (15.74) gives the population ratio for states i and j as $N_i/N_j = e^{-(E_i - E_j)/kT}$. At room temperature, $kT = 4.1 \times 10^{-14}$ erg $=$ 0.026 eV. The typical molecular rotational-level spacing ΔE_{rot} is substantially less than kT, so $e^{-\Delta E_{\text{rot}}/kT}$ is close to 1 and many excited rotational levels are populated at room temperature.

The vibrational frequency ν_e equals $(1/2\pi)(k_e/\mu)^{1/2}$. For heavy diatomic molecules (for example, Br_2 and I_2), the large value of μ gives a ν_e of roughly 10^{13} s^{-1} and a spacing $h\nu_e$ of roughly 0.04 eV, which is comparable to kT at room temperature. Hence, for heavy diatomic molecules, there is significant occupation of one or more excited vibrational levels at room temperature. However, relatively light diatomic molecules (for example, H_2, HCl, CO, and O_2) have ν_e values of roughly 10^{14} s^{-1} and vibrational spacings of roughly 0.4 eV. This is substantially greater than kT at room temperature, so $e^{-\Delta E_{\text{vib}}/kT}$ is very small and nearly all the molecules are in the ground vibrational level at room temperature.

Since ΔE_{el} is substantially greater than kT at room temperature, there is generally no occupation of excited electronic states at room temperature.

Since ν_e [which depends on $E_e''(R_e)$] and R_e each differ for different electronic states of the same molecule, each electronic state of a molecule has its own set of vibrational and rotational levels. Figure 21.5 shows some of the vibration–rotation levels of one electronic state of a diatomic molecule. The dots indicate higher rotational levels of each vibrational level.

Figure 21.5
Diatomic-molecule vibration–rotation levels.

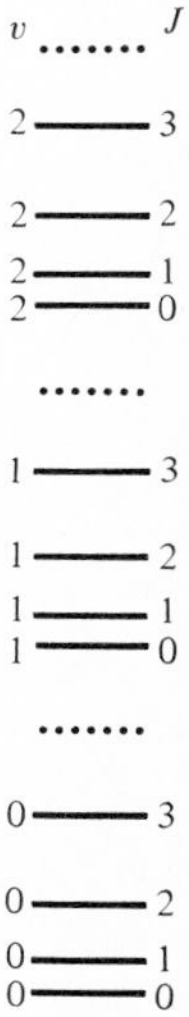

Selection rules. We now consider the selection rules for transitions between different vibration–rotation levels of the same electronic state of a diatomic molecule. (Transitions between different electronic states are considered in Sec. 21.10.) The Born–Oppenheimer approximation gives the molecular wave function as the product of electronic and nuclear wave functions: $\psi = \psi_e \psi_N$. For a transition between states ψ and ψ' with no change in electronic state ($\psi_e = \psi_e'$), the integral (21.5) determining the selection rules is

$$\iint \psi_e^* \psi_N^* \hat{\mathbf{p}} \psi_e \psi_N' \, d\tau_e \, d\tau_N = \int \psi_N^* \psi_N' \left(\int \psi_e^* \hat{\mathbf{p}} \psi_e \, d\tau_e \right) d\tau_N = \int \psi_N^* \psi_N' \mathbf{p} \, d\tau_N$$

where (20.36) was used to introduce the electric dipole moment $\mathbf{p}$ of the electronic state. The dipole moment $\mathbf{p}$ is directed along the molecular axis. The magnitude p of $\mathbf{p}$ depends on the internuclear distance. (The experimentally observed value is an average over the molecular zero-point vibration.) One expands p in a Taylor series about its value at R_e:

$$p(R) = p(R_e) + p'(R_e)(R - R_e) + \tfrac{1}{2}p''(R_e)(R - R_e)^2 + \cdots \qquad (21.27)$$

where $p(R_e)$ is virtually the same as the experimentally observed dipole moment. One then substitutes (21.27) and the expressions for ψ_N and ψ_N' into the above integral and evaluates this integral. (See *M. S.*, sec. 4.4 for details.)

The selection rules turn out to be

$$\Delta J = \pm 1 \qquad (21.28)$$

$$\Delta v = 0, \ \pm 1 \ (\pm 2, \ \pm 3, \ldots) \qquad (21.29)$$

$$\Delta v = 0 \text{ not allowed if } p(R_e) = 0 \qquad \Delta v = \pm 1 \text{ not allowed if } p'(R_e) = 0 \qquad (21.30)$$

The parentheses in (21.29) indicate that the $\Delta v = \pm 2, \pm 3, \ldots$ transitions are far less probable than the $\Delta v = 0$ and ± 1 transitions. [If the molecule were a harmonic oscillator, and if terms after the $p'(R_e)(R - R_e)$ term in (21.27) were negligible, only $\Delta v = 0, \pm 1$ transitions would occur.]

Rotational spectra. Transitions with no change in electronic state and with $\Delta v = 0$ give the *pure-rotation spectrum* of the molecule. These transitions correspond to photons with energies in the microwave and far-IR regions. (The far-IR region is the portion of the IR region bordering the microwave region in Fig. 21.2.) Equation (21.30) shows that a diatomic molecule has a pure-rotation spectrum only if it has a nonzero electric dipole moment. A homonuclear diatomic molecule (for example, H_2, N_2, Cl_2) has no pure-rotation spectrum.

The pure-rotation spectrum is usually observed as an absorption spectrum. The absorption transitions all have $\Delta J = +1$ [Eq. (21.28)]. Because the rotational levels are not equally spaced, and because many excited rotational levels are populated at room temperature, there will be several lines in the pure-rotation spectrum. From (18.6), the absorption frequencies are $\nu = (E_{J+1} - E_J)/h$, and use of (21.26) with centrifugal distortion neglected gives for the frequency of the transition between levels J and $J + 1$:

$$\nu = B_e(J + 1)(J + 2) - \alpha_e(v + \tfrac{1}{2})(J + 1)(J + 2) - [B_e J(J + 1) - \alpha_e(v + \tfrac{1}{2})J(J + 1)]$$

$$\nu = 2(J + 1)[B_e - \alpha_e(v + \tfrac{1}{2})] \equiv 2(J + 1)B_v \qquad (21.31)$$

where $J = 0, 1, 2, \ldots$, and where the mean rotational constant B_v for states with vibrational quantum number v is defined as

$$B_v \equiv B_e - \alpha_e(v + \tfrac{1}{2}) \tag{21.32}$$

Tables sometimes list $B_0 = B_e - \frac{1}{2}\alpha_e$, instead of B_e.

The wave numbers of the pure-rotational transitions are $\sigma = 1/\lambda = \nu/c = 2(J + 1)B_v/c$. We shall use a tilde to indicate division of a molecular constant by c. Thus, $\sigma = 2(J + 1)\tilde{B}_v$, where $\tilde{B}_v \equiv B_v/c$. In particular [Eq. (21.16)],

$$\tilde{B}_0 \equiv B_0/c = h/8\pi^2 I_0 c \tag{21.33}$$

where I_0 is the moment of inertia averaged over the zero-point vibration. (In the literature, the tilde is often omitted.)

In the majority of cases, only the $v = 0$ vibrational level is significantly populated at room temperature, and the pure-rotation frequencies (21.31) become $2(J + 1)B_0$; the pure-rotation spectrum is a series of equally spaced lines at $2B_0, 4B_0, 6B_0, \ldots$ (if centrifugal distortion is neglected). See Fig. 21.6.

If there is appreciable population of excited vibrational levels, each rotational transition shows one or more nearby satellite lines that are due to transitions between rotational levels of the $v = 1$ vibrational level, between rotational levels of the $v = 2$ vibrational level, etc. These satellites are much weaker than the main line because the population of vibrational levels falls off rapidly as v increases.

Since the moments of inertia of different isotopic species of the same molecule differ, each isotopic species has its own pure-rotation spectrum. For $^{12}C^{16}O$, some observed pure-rotational transitions (all for $v = 0$) are:

$J \to J + 1$	$0 \to 1$	$1 \to 2$	$2 \to 3$	$3 \to 4$	$4 \to 5$
ν/MHz	115,271	230,538	345,796	461,041	576,268

The slight decreases in the successive spacings are due to centrifugal distortion. For $^{13}C^{16}O$ and $^{12}C^{18}O$, the $J = 0 \to 1$ pure-rotation transitions occur at 110,201 and 109,782 MHz, respectively.

From the observed frequency of a pure-rotational transition of a heteronuclear diatomic molecule, one can use (21.31) to calculate B_0; from B_0 one gets I_0, and from $I_0 = \mu R_0^2$ one gets R_0, the internuclear distance averaged over the zero-point vibration. If vibrational satellites are observed, they can be used to find α_e and then (21.32) allows calculation of B_e and hence of the equilibrium internuclear distance R_e.

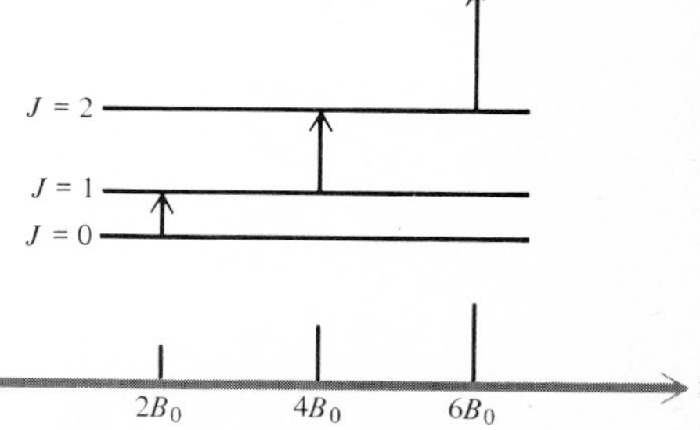

Figure 21.6
Diatomic-molecule pure-rotational absorption transitions.

Vibration–rotation spectra. Transitions with no change in electronic state and with $\Delta v \neq 0$ give the vibration–rotation spectrum of the molecule. These transitions involve infrared photons. Equation (21.30) shows that a diatomic molecule has a vibration–rotation spectrum only if the change in dipole moment dp/dR is nonzero at R_e. When a homonuclear diatomic molecule vibrates, its p remains zero; for a heteronuclear diatomic molecule, vibration changes p. Hence, only heteronuclear diatomics show an IR vibration–rotation spectrum.

The vibration–rotation spectrum is usually observed as an absorption spectrum. From (21.29) the absorption transitions have $\Delta v = +1$ $(+2, +3, \ldots)$, where the transitions in parentheses are much less probable. Since $\Delta J = 0$ is not allowed by the selection rule (21.28), there is no pure-vibration spectrum for a diatomic molecule; when v changes, J also changes.

Vibrational levels are much more widely spaced than rotational levels, so the IR vibration–rotation spectrum consists of a series of bands; each band corresponds to a transition between two particular vibrational levels v'' and v' and consists of a series of lines, each line corresponding to a different change in rotational state. (See Fig. 21.7.) We shall first ignore the rotational structure of the bands and shall calculate the frequency of a *hypothetical* transition where v changes, but J is 0 in both the initial and final states; this gives the position of the *band origin*.

IR spectroscopists commonly work with wave numbers rather than frequencies. The wave number σ_0 of the band origin for an absorption transition from vibrational level v'' to v' is $\sigma_0 = 1/\lambda = \nu/c = (E_{v'} - E_{v''})/hc$. Use of (21.26) with $J = 0$ gives

$$\sigma_0 = \tilde{\nu}_e(v' - v'') - \tilde{\nu}_e x_e[v'(v' + 1) - v''(v'' + 1)] \tag{21.34}$$

$$\tilde{\nu}_e \equiv \nu_e/c, \qquad \tilde{\nu}_e x_e \equiv \nu_e x_e/c \tag{21.35}$$

Throughout this chapter, a double prime will be used on quantum numbers of the lower state and a single prime on quantum numbers of the upper state.

Usually, most of the molecules are in the $v = 0$ vibrational level at room temperature, and the strongest band is the $v = 0 \rightarrow 1$ band, called the *fundamental band*. The $v = 0 \rightarrow 2$ band (the *first overtone*) is much weaker than the fundamental band. From (21.34), the fundamental, first overtone, second overtone, ... occur at $\tilde{\nu}_e - 2\tilde{\nu}_e x_e$, $2\tilde{\nu}_e - 6\tilde{\nu}_e x_e$, $3\tilde{\nu}_e - 12\tilde{\nu}_e x_e$, ... Because $\tilde{\nu}_e x_e \ll \tilde{\nu}_e$, the first overtone is at roughly twice the frequency of the fundamental band. The wave number of the fundamental band is called $\tilde{\nu}_0$:

$$\tilde{\nu}_0 = \tilde{\nu}_e - 2\tilde{\nu}_e x_e \qquad \text{and} \qquad \nu_0 = \nu_e - 2\nu_e x_e \tag{21.36}$$

The symbols ω_e, ω_0, and $\omega_e x_e$ are often used for $\tilde{\nu}_e$, $\tilde{\nu}_0$, and $\tilde{\nu}_e x_e$, respectively.

Some band origins of $^1H^{35}Cl$ IR bands are:

$v'' \rightarrow v'$	$0 \rightarrow 1$	$0 \rightarrow 2$	$0 \rightarrow 3$	$0 \rightarrow 4$	$0 \rightarrow 5$
σ_0/cm^{-1}	2886.0	5668.0	8346.8	10,922.8	13,396.2

If the vibrational frequency of the molecule is relatively low, or if the gas is heated, one observes absorption transitions originating from states with $v'' > 0$; these are called *hot bands*.

We now consider the rotational structure of IR bands of diatomic molecules. Vibration–rotation transitions with $\Delta J = +1$ give the *R branch* lines of the band; transitions with $\Delta J = -1$ give the *P branch*. Let J'' be the rotational quantum number of the lower vibration–rotation level and J' the rotational quantum number of the upper level. The wave numbers of the vibration–rotation lines are $\sigma = (E_{J'v'} - E_{J''v''})/hc$. For R branch lines, $J' = J'' + 1$, and use of (21.26) with centrifugal distortion neglected gives (Prob. 21.15)

$$\sigma_R = \sigma_0 + [2\tilde{B}_e - \tilde{\alpha}_e(v' + v'' + 1)](J'' + 1) - \tilde{\alpha}_e(v' - v'')(J'' + 1)^2 \qquad (21.37)$$

where $J'' = 0, 1, 2, \ldots$ and σ_0 is given by (21.34). For P branch lines, $J' = J'' - 1$, and we find

$$\sigma_P = \sigma_0 - [2\tilde{B}_e - \tilde{\alpha}_e(v' + v'' + 1)]J'' - \tilde{\alpha}_e(v' - v'')J''^2 \qquad (21.38)$$

where $J'' = 1, 2, 3, \ldots$. ($J'' = 0$ is excluded, since J' cannot be -1.)

If α_e were entirely negligible compared with B_e, then (21.37) and (21.38) would read

$$\sigma_R = \sigma_0 + 2\tilde{B}_e(J'' + 1) \qquad \text{where } J'' = 0, 1, 2, \ldots \qquad (21.39)$$

$$\sigma_P = \sigma_0 - 2\tilde{B}_e J'' \qquad \text{where } J'' = 1, 2, 3, \ldots \qquad (21.40)$$

This would give a series of equally spaced lines on either side of σ_0 (with no line at σ_0, since $J = 0 \to 0$ is forbidden). Since α_e is not entirely negligible, the actual R branch line spacings decrease as J increases, and the P branch line spacings increase as J increases.

Figure 21.7 shows the $v = 0 \to 1$ vibration–rotation band of $^1H^{35}Cl$.

Note that the line intensities in each branch in Fig. 21.7 first increase as J increases and then decrease at high J. The explanation for this is as follows. The Boltzmann distribution law $N_i/N_j = e^{-(E_i - E_j)/kT}$ gives the populations of the quantum-mechanical *states* i and j and not the populations of the quantum-mechanical energy levels. For each rotational energy level, there are $2J + 1$ rotational states, corresponding to the $2J + 1$ values of the M_J quantum number (Sec. 18.14), which does not affect the energy. Thus, the population of each rotational energy level J is $2J + 1$ times the population of the quantum-mechanical rotational state with quantum numbers J and M_J. For low J, this $2J + 1$ factor outweighs the decreasing exponential in the Boltzmann distribution law, so the populations of

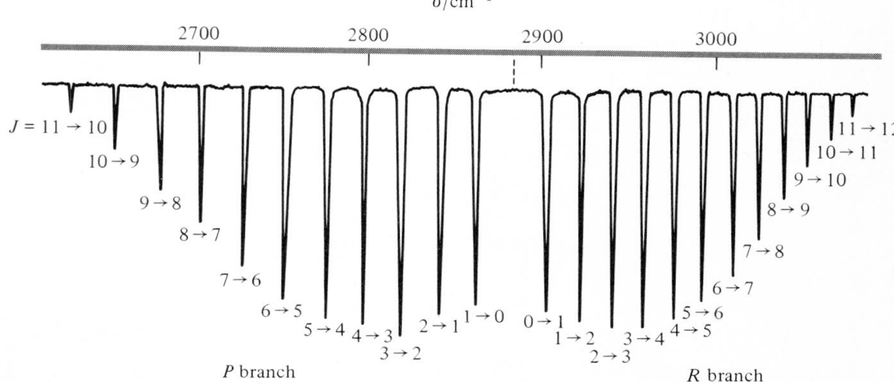

Figure 21.7 The $v = 0 \to 1$ vibration band of $^1H^{35}Cl$. The dashed line indicates the position of the band origin ($\sigma_0 = 2886$ cm^{-1}). (In a sample of naturally occurring HCl, each line is a doublet, due to the presence of $^1H^{37}Cl$.) It is conventional to display IR spectra with absorption intensity increasing downward.

rotational levels at first increase as J increases; for high J, the exponential dominates the $2J + 1$ factor, and the rotational populations decrease as J increases.

Measurement of the wave numbers of the lines of a vibration–rotation band makes it possible for σ_0, $\tilde{B}_e$, and $\tilde{\alpha}_e$ to be found from Eqs. (21.37) and (21.38). Knowing σ_0 for at least two bands, we can use (21.34) to find $\tilde{\nu}_e$ and $\tilde{\nu}_e x_e$. The rotational constant $\tilde{B}_e$ allows the moment of inertia and hence the internuclear distance to be found. The vibrational frequency ν_e gives the force constant of the bond.

Vibrational and rotational constants for some diatomic molecules are listed in Table 21.1. Homonuclear diatomics have no pure-rotational (microwave) or vibration–rotation (IR) spectra, and their constants are found from electronic spectra (Sec. 21.10) or Raman spectra (Sec. 21.9). The values listed are for the ground electronic states, except that the CO* values are for one of the lower excited electronic states of CO. Note that $D_e/hc > \tilde{\nu}_e \gg \tilde{B}_e$, in accord with the earlier conclusion that electronic energies are greater than vibrational energies, which are greater than rotational energies. Note also the large force constants for N_2 and CO, which have triple bonds. Other things being equal, the stronger the bond, the larger the force constant and the shorter the bond length.

The centrifugal-distortion constant D is of significant magnitude only for light molecules. Some values of $\tilde{D} \equiv D/c$ for ground electronic states are 0.05 cm^{-1} for $^1H_2$, 0.002 cm^{-1} for $^1H^{19}F$, 6×10^{-6} cm^{-1} for $^{14}N_2$, and 5×10^{-9} cm^{-1} for $^{127}I_2$.

The Appendix lists masses of some common isotopes.

21.4 MOLECULAR SYMMETRY

In Chap. 20, we used the symmetry of AOs to decide which AOs contribute to a given MO. The full application of symmetry to quantum chemistry requires the branch of mathematics called *group theory*. Group theory helps in finding the MOs of symmetric molecules, in determining selection rules, and in determining the vibrations of polyatomic molecules. We shall not develop group theory in this book but will give a short discussion of molecular symmetry elements as preparation for discussing the nuclear motion of polyatomic molecules.

Table 21.1. Constants of some diatomic molecules[a]

Molecule	D_e/hc / cm^{-1}	R_e / Å	k_e / N/m	$\tilde{\nu}_e$ / cm^{-1}	$\tilde{B}_e$ / cm^{-1}	$\tilde{\alpha}_e$ / cm^{-1}	$\tilde{\nu}_e x_e$ / cm^{-1}
$^1H_2$	38,297	0.741	576	4403.2	60.85	3.06	121.3
$^1H^{35}Cl$	37,220	1.274	516	2991.1	10.594	0.31	52.8
$^{14}N_2$	79,870	1.098	2294	2358.0	1.998	0.018	14.1
$^{12}C^{16}O$	90,542	1.128	1902	2169.8	1.931	0.018	13.3
$^{12}C^{16}O^*$	29,522	1.370	537	1152.6	1.310	0.017	7.3
$^{127}I_2$	12,560	2.667	172	214.5	0.0374	0.0001	0.6
$^{23}Na^{35}Cl$	34,630	2.361	110	366	0.2180	0.002	1.0
$^{12}C^1H$	29,400	1.120	448	2859.1	14.448	0.53	63.3

[a] Data mainly from S. Bourcier (ed.), *Spectroscopic Data Relative to Diatomic Molecules*, vol. 17 of *Tables of Constants and Numerical Data*, Pergamon, 1970.

A molecule has an *n-fold axis of symmetry*, symbolized by C_n, if rotation by $2\pi/n$ radians ($1/n$th of a complete rotation) about this axis results in a nuclear configuration that is indistinguishable from the original one. For example, the bisector of the bond angle in HOH is a C_2 axis, since rotation by $2\pi/2$ radians (180°) about this axis merely interchanges the two equivalent H atoms (Fig. 21.8); the hydrogens in the figure have been labeled H_a and H_b, but in reality they are physically indistinguishable from each other.

The NH_3 molecule has a C_3 axis passing through the N nucleus and the midpoint of the triangle formed by the three H nuclei (Fig. 21.8); rotation by $2\pi/3$ radians (120°) about this axis sends equivalent hydrogens into one another. The hexagonal molecule benzene (C_6H_6) has a C_6 axis perpendicular to the molecular plane and passing through the center of the molecule; rotation by $2\pi/6$ radians (60°) about this axis sends equivalent nuclei into one another. This C_6 axis is also a C_3 axis and a C_2 axis, since 120° and 180° rotations about it send equivalent atoms into one another. Benzene has six other C_2 axes, each lying in the molecular plane; three of these go through two diagonally opposite carbons, and three bisect opposite pairs of carbon–carbon bonds.

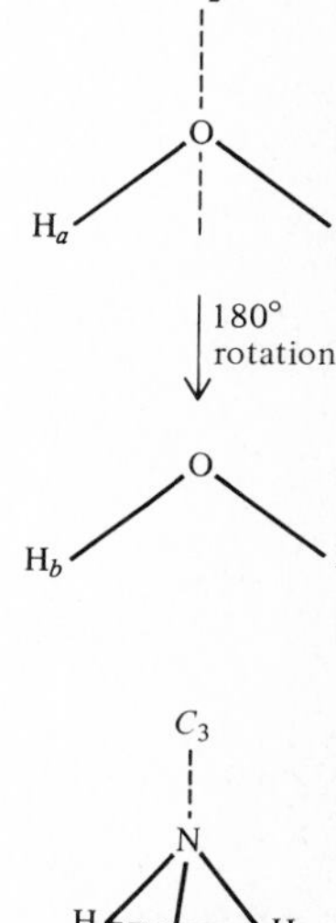

Figure 21.8

A molecule has a *plane of symmetry*, symbolized by σ, if reflection of all nuclei through this plane sends the molecule into a configuration physically indistinguishable from the original one. Any planar molecule (for example, H_2O) has a plane of symmetry coinciding with the molecular plane, since reflection in the molecular plane leaves all nuclei unchanged in position. H_2O also has a second plane of symmetry; this lies perpendicular to the molecular plane (Fig. 21.9); reflection through this second plane interchanges the equivalent hydrogens.

NH_3 has three planes of symmetry; each symmetry plane passes through the nitrogen and one hydrogen and bisects the angle formed by the nitrogen and the other two hydrogens. Reflection in one of these planes leaves N and one H unchanged in position and interchanges the other two H's.

A molecule has a *center of symmetry*, symbol i, if inversion of each nucleus through this center results in a configuration indistinguishable from the original one. By *inversion* through point P we mean moving a nucleus at x, y, z to the location $-x$, $-y$, $-z$ on the opposite side of P, where the origin is at P. Neither H_2O nor NH_3 has a center of symmetry; *p*-dichlorobenzene has a center of symmetry, but *m*-dichlorobenzene does not (Fig. 21.10). A molecule with a center of symmetry cannot have a dipole moment.

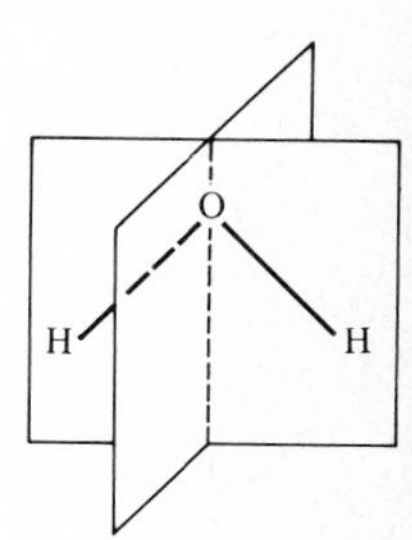

Figure 21.9
Symmetry planes in H_2O.

A molecule has an *n-fold improper axis* (or *rotation–reflection axis*), symbol S_n, if rotation by $2\pi/n$ radians about the axis followed by reflection in a plane perpendicular to this axis sends the molecule into a configuration indistinguishable from the original one. Figure 21.11 shows a 90° rotation about an S_4 axis in CH_4 followed by a reflection in a plane perpendicular to this axis. The final configuration has hydrogens at the same locations in space as the original configuration. (Note that the 90° rotation alone produces a configuration of the hydrogens that is distinguishable from

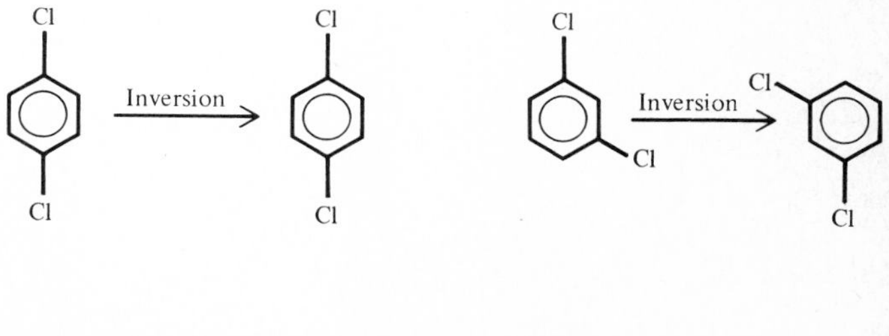

Cl Cl Inversion Cl Cl Cl Cl Inversion Cl Cl

Figure 21.10
Inversion in *p*-dichlorobenzene and in *m*-dichlorobenzene.

the original one, so the S_4 axis in methane is not a C_4 axis.) Methane has two other S_4 axes; each S_4 axis goes through the centers of opposite faces of the cube in which the molecule has been inscribed. The C_6 axis in benzene is also an S_6 axis.

Associated with each symmetry element of a molecule is a set of *symmetry operations*. For example, consider the C_4 axis of the square-planar molecule XeF_4. We can rotate the molecule by 90°, 180°, 270°, and 360° about this axis to give configurations indistinguishable from the original one (Fig. 21.12). The operation of rotating the molecule by 90° about the C_4 axis is symbolized by $\hat{C}_4$. (Note the circumflex.) Since a 180° rotation can be viewed as two successive 90° rotations, we symbolize a 180° rotation by $\hat{C}_4{}^2$, which equals $\hat{C}_4\hat{C}_4$. (Recall that multiplication of two operators means applying the operators in succession.) Similarly, 270° and 360° rotations are denoted by $\hat{C}_4{}^3$ and $\hat{C}_4{}^4$. Since a 360° rotation brings every nucleus back to its original location, we write $\hat{C}_4{}^4 = \hat{E}$, where $\hat{E}$ is the *identity operation*, defined as "do nothing." The operation $\hat{C}_4{}^5$, which is a 450° rotation about the C_4 axis, is the same as the 90° $\hat{C}_4$ rotation and is not counted as a new symmetry operation.

Since two successive reflections in the same plane bring the nuclei back to their original locations, we have $\hat{\sigma}^2 = \hat{E}$. Likewise, $\hat{\imath}^2 = \hat{E}$.

Since each symmetry operation must leave the location of the molecular center of mass unchanged, the symmetry elements of a molecule all intersect at the center of mass.

The set of symmetry operations of a molecule forms what mathematicians call a *group*.

21.5 ROTATION OF POLYATOMIC MOLECULES

As with diatomic molecules, it is usually a good approximation to take the energy of a polyatomic molecule as the sum of translational, rotational, vibrational, and electronic energies. We now consider the rotational energy levels.

Classical mechanics of rotation. The classical-mechanical treatment of the rotation of a three-dimensional body is complicated, and we shall just quote the results without proof.

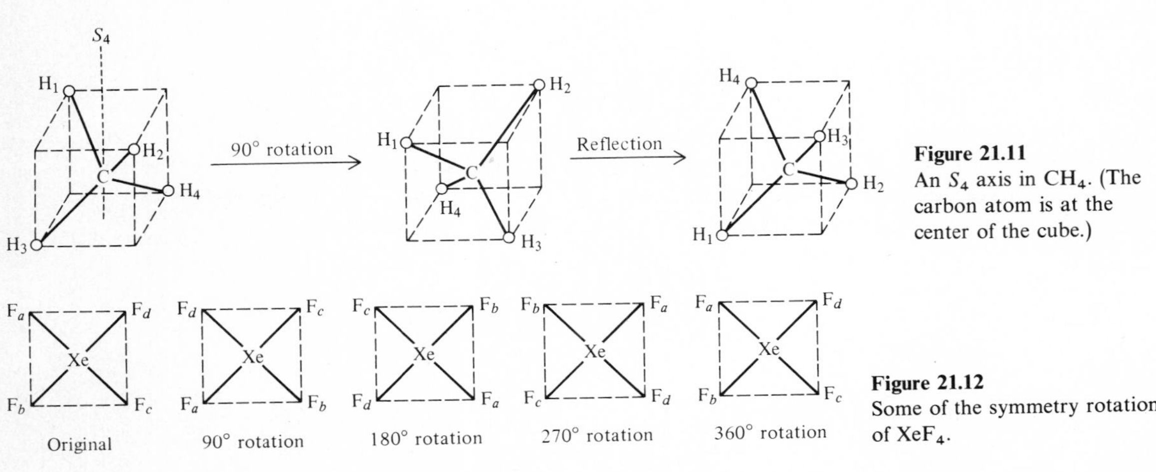

Figure 21.11 An S_4 axis in CH_4. (The carbon atom is at the center of the cube.)

Figure 21.12 Some of the symmetry rotations of XeF_4.

The *moment of inertia* I_x of a system of mass points $m_1, m_2, \ldots$ about an arbitrary axis x is defined by

$$I_x \equiv \sum_i m_i r_{x,i}^2 \tag{21.41}$$

where $r_{x,i}$ is the perpendicular distance from mass m_i to the x axis. Consider a set of three mutually perpendicular axes x, y, and z. The *products of inertia* I_{xy}, I_{xz}, and I_{yz} for the x, y, z system are defined by the sums $I_{xy} \equiv -\sum_i m_i x_i y_i$, etc. Any three-dimensional body possesses three mutually perpendicular axes a, b, and c passing through the center of mass and having the property that the products of inertia I_{ab}, I_{ac}, and I_{bc} are each zero for these axes; these three axes are called the *principal axes of inertia* of the body. The moments of inertia I_a, I_b, and I_c calculated with respect to the principal axes are called the *principal moments of inertia* of the body.

Symmetry aids in locating the principal axes. It can be shown that (*M. S.*, sec. 5.4): A molecular symmetry axis must coincide with one of the principal axes. A molecular symmetry plane must contain two of the principal axes and be perpendicular to the third.

Consider the square-planar molecule XeF_4 (Fig. 21.13), for example. The center of mass is at the Xe nucleus, and the three principal axes of inertia must intersect at this point. One of the principal axes coincides with the C_4 symmetry axis perpendicular to the molecular plane. The other two principal axes are perpendicular to the C_4 axis and lie in the molecular plane. They can be taken to coincide with the two C_2 axes that pass through the four F's, or they can be taken to coincide with the two C_2 axes that bisect the FXeF angles. (For highly symmetric molecules, the orientation of the principal axes may not be unique.)

Figure 21.13

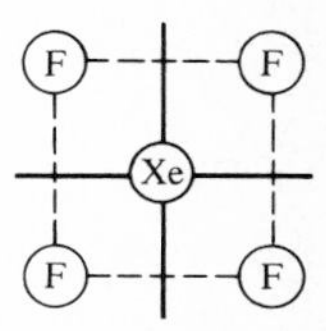

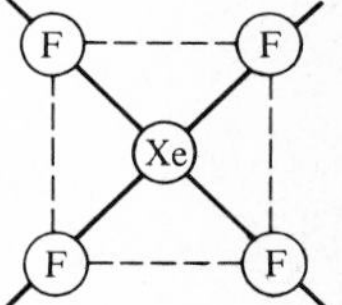

For molecules without symmetry, finding the principal axes is more complicated. See *M. S.*, sec. 5.2.

The classical-mechanical expression for the rotational energy of a body turns out to be

$$E_{\text{rot}} = P_a^2/2I_a + P_b^2/2I_b + P_c^2/2I_c \tag{21.42}$$

where P_a, P_b, P_c are the components of the rotational angular momentum about the a, b, c principal axes. Note the presence of three terms, each quadratic in a momentum, a fact mentioned in Sec. 15.10 on equipartition of energy.

If all the mass points lie on the same line (as in a linear molecule), this line is one of the principal axes (since it is a symmetry axis). The rotational angular momentum of the body about this axis is zero, since the distance between all the masses and the axis is zero. Hence, one of the three terms in (21.42) is zero, and a linear molecule has only two quadratic terms in its classical rotational energy.

The principal axes are labeled so that $I_a \le I_b \le I_c$.

A body is classified as a *spherical*, *symmetric*, or *asymmetric top*, according to whether three, two, or none of the principal moments I_a, I_b, I_c are equal:

Spherical top: $I_a = I_b = I_c$

Symmetric top: $I_a = I_b \neq I_c$ or $I_a \neq I_b = I_c$

Asymmetric top: $I_a \neq I_b \neq I_c$

It can be shown (*M. S.*, sec. 5.2) that a molecule with one C_n or S_n axis with $n \ge 3$ is a symmetric top. A molecule with two or more noncoincident C_n or S_n axes with $n \geq 3$ is a spherical top. NF_3, with one C_3 axis, is a symmetric top. XeF_4, with one C_4 axis, is a symmetric top. CCl_4, with four noncoincident C_3 axes (one through each C—Cl bond), is a spherical top. H_2O, with no C_3 or higher axis, is an asymmetric top.

Quantum-mechanical rotational energy levels. (For proofs of the following results, the reader is referred to *M. S.*, chap. 5.)

The *rotational constants* A, B, C of a polyatomic molecule are defined by [cf. (21.16)]

$$A \equiv h/8\pi^2 I_a, \qquad B \equiv h/8\pi^2 I_b, \qquad C \equiv h/8\pi^2 I_c \tag{21.43}$$

The rotational energy levels of a spherical top are

$$E_{\text{rot}} = J(J+1)\hbar^2/2I = BhJ(J+1), \qquad J = 0, 1, 2, \ldots \tag{21.44}$$

where $I \equiv I_a = I_b = I_c$. The spherical-top rotational wave functions involve the three quantum numbers J, K, and M, where K and M each range from $-J$ to J in integral steps. Each spherical-top rotational level is $(2J+1)^2$-fold degenerate, corresponding to the $(2J+1)^2$ different choices for K and M for a fixed J.

The rotational levels of a symmetric top with $I_b = I_c$ are

$$E_{\text{rot}} = \frac{J(J+1)\hbar^2}{2I_b} + K^2\hbar^2\left(\frac{1}{2I_a} - \frac{1}{2I_b}\right) = BhJ(J+1) + (A-B)hK^2$$
$$J = 0, 1, 2, \ldots; \qquad K = -J, -J+1, \ldots, J-1, J \tag{21.45}$$

There is also an M quantum number that does not affect E_{rot}. If $I_a = I_b$, then I_a and A in (21.45) are replaced by I_c and C. The quantity $K\hbar$ is the component of the rotational angular momentum along the molecular symmetry axis.

The axis of a linear molecule is a C_∞ axis, since rotations by $2\pi/n$ radians, where $n = 2, 3, \ldots, \infty$, about this axis are symmetry operations. Hence, a linear molecule is a symmetric top. (This is also obvious from the fact that the moments of inertia about all axes that pass through the center of mass and are perpendicular to the molecular axis are equal.) Because all the nuclei lie on the C_∞ symmetry axis, there can be no rotational angular momentum about this axis and K must be zero for a linear molecule. Hence, Eq. (21.45) gives $E_{\text{rot}} = J(J+1)\hbar^2/2I_b$ for a linear molecule, where I_b is the moment of inertia about an axis through the center of mass and perpendicular to the molecular axis. [A special case is a diatomic molecule, Eq. (21.15).]

The rotational levels for an asymmetric top are extremely complicated and follow no simple pattern.

Selection rules. The selection rules for the pure-rotational (microwave) spectra of polyatomic molecules are as follows.

Just as for diatomic molecules, a polyatomic molecule must have a nonzero dipole moment to undergo a pure-rotational transition with absorption or emission of radiation. Because of their high symmetry, all spherical tops (for example, CCl_4 and SF_6) and some symmetric tops (for example, C_6H_6 and XeF_4) have no dipole moment and exhibit no pure-rotation spectrum.

For a symmetric top with a dipole moment (for example, CH_3F), the pure-rotational selection rules are

$$\Delta J = \pm 1, \qquad \Delta K = 0 \tag{21.46}$$

Use of (21.46), (21.45), and (18.6) gives the frequency of the pure-rotational $J \rightarrow J + 1$ absorption transition as

$$\nu = 2B(J + 1), \qquad J = 0, 1, 2, \ldots \qquad \text{symmetric top} \tag{21.47}$$

The microwave spectrum of a symmetric top consists of a series of equally spaced lines at $2B$, $4B$, $6B$, ... (provided centrifugal distortion is negligible).

The most populated vibrational state has all vibrational quantum numbers equal to zero, and the rotational constant B determined from the microwave spectrum is an average over the zero-point vibrations and is designated B_0. (If excited vibrational levels are significantly populated at room temperature, vibrational satellites are observed, as discussed for diatomic molecules.)

For asymmetric tops, the selection rules are complicated, and will not be discussed.

21.6 MICROWAVE SPECTROSCOPY

Except for light diatomic molecules, pure-rotational spectra occur in the microwave portion of the electromagnetic spectrum. Figure 21.14 sketches a highly simplified version of a microwave spectrometer. A special electronic tube called a klystron generates virtually monochromatic microwave radiation, whose frequency can be varied with a tuning screw on the klystron. The radiation is transmitted through a hollow metal pipe called a *waveguide*. A portion of waveguide sealed at both ends with mica windows is the absorption cell. The absorption cell is filled with low-pressure (0.01 to 0.1 torr) vapor of the molecule to be investigated. (At medium and high pressures, intermolecular interactions broaden the rotational absorption lines, giving an essentially continuous rotational absorption spectrum.) The microwave radiation is detected with a metal-rod antenna mounted in the waveguide and connected to a semiconductor diode.

Any substance with a dipole moment and with sufficient vapor pressure can be studied. Use of a waveguide heated to 1000 K allows the rotational spectra of the alkali halides to be observed. Very large molecules are difficult to study; such molecules have many low-energy vibrational levels that are significantly populated at room temperature; this produces a microwave spectrum with so many lines that it is extremely hard to figure out which rotational transitions the various lines correspond to. Some relatively large molecules whose microwave spectra have been successfully investigated are *o*-xylene, azulene, β-fluoronaphthalene, and $C_6H_6Cr(CO)_3$.

The microwave spectrum of a symmetric top is very simple [Eq. (21.47)] and yields the rotational constant B_0. The microwave spectrum of an asymmetric top is quite complicated, but once one has assigned several lines to transitions between specific rotational levels, one can calculate the three rotational constants A_0, B_0, C_0 and the corresponding principal moments of inertia.

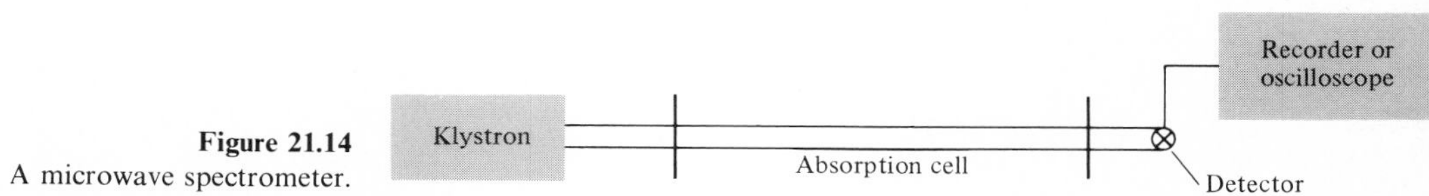

Figure 21.14
A microwave spectrometer.

The principal moments of inertia depend on the bond distances and angles. Knowledge of one moment of inertia for a symmetric top or three moments of inertia for an asymmetric top is generally not enough information to determine the structure fully. One therefore prepares isotopically substituted species of the molecule and observes their microwave spectra to find their moments of inertia. The molecular geometry is determined by the nuclear configuration that minimizes the energy E_e in the electronic Schrödinger equation $\hat{H}_e \psi_e = E_e \psi_e$. The terms in the electronic Hamiltonian $\hat{H}_e$ are independent of the nuclear masses. Hence, isotopic substitution does not affect the equilibrium geometry. (This isn't quite true, due to very slight deviations from the Born–Oppenheimer approximation.) When sufficient isotopically substituted species have been studied, the complete molecular structure can be determined. Some examples of molecular structures determined by microwave spectroscopy (bond distances in angstroms) are

CH_2F_2: $R(\text{CH}) = 1.09$, $R(\text{CF}) = 1.36$, $\angle\text{HCH} = 113.7°$, $\angle\text{FCF} = 108.3°$

CH_3OH: $R(\text{CH}) = 1.09$, $R(\text{CO}) = 1.43$, $R(\text{OH}) = 0.94$, $\angle\text{HCH} = 108.6°$, $\angle\text{COH} = 108.5°$

C_6H_5F: average $R(\text{CC}) = 1.39$, $R(\text{CH}) = 1.08$, $R(\text{CF}) = 1.35$

H_2S: $R(\text{SH}) = 1.34$, $\angle\text{HSH} = 92.1°$

SO_2: $R(\text{SO}) = 1.43$, $\angle\text{OSO} = 119.3°$

O_3: $R(\text{OO}) = 1.27$, $\angle\text{OOO} = 116.8°$

A comprehensive compilation of structures of gas-phase molecules is Landolt-Börnstein, New Series, Group II, vol. 7, J. H. Callomon et al., *Structure Data of Free Polyatomic Molecules*, Springer-Verlag, 1976. Structures, vibrational frequencies, and rotational constants of molecules with 3 to 12 atoms are tabulated in G. Herzberg, *Electronic Spectra and Electronic Structure of Polyatomic Molecules*, Van Nostrand, 1966, app. VI.

Besides molecular structures, one can also obtain molecular dipole moments from microwave spectra. An insulated metal plate is inserted lengthwise in the waveguide. Application of a voltage to this plate subjects the gas molecules to an electric field, thereby shifting their rotational energies. (A shift in molecular energy levels due to application of an external electric field is called a *Stark effect*.) The magnitudes of these shifts depend on the components of the molecular electric dipole moment. Microwave spectroscopy gives quite accurate dipole moments. Moreover, it gives the components of the dipole-moment vector along the principal axes of inertia, so the orientation of the dipole-moment vector is known. Some dipole moments (in debyes) determined by microwave spectroscopy are (R. D. Nelson et al., Selected Values of Electric Dipole Moments for Molecules in the Gas Phase, *Natl. Bur. Stand. U.S. Publ.* NSRDS-NBS 10, 1967; A. L. McClellan, *Tables of Experimental Dipole Moments*, vol. 1, Freeman, 1963, vol. 2, Rahara Enterprises, 1974):

C_3H_8	$HC(CH_3)_3$	H_2O_2	H_2O	H_2S	azulene	NaCl	KCl	HCl	FCl	CH_3D
0.08	0.13	2.2	1.85	0.97	0.80	9.0	10.3	1.12	0.88	0.006

The nonzero dipole moments of the saturated hydrocarbons $H_2C(CH_3)_2$ and $HC(CH_3)_3$ are due to deviations from the $109\frac{1}{2}°$ tetrahedral bond angle and to polarization of the electron probability densities of the CH_3, CH_2, and CH groups by the unsymmetrical electric fields in the molecules [S. W. Benson and M. Luria, *J. Am. Chem. Soc.*, **97**, 704 (1975)]. The dipole moment of CH_3D is due to deviations from the Born–Oppenheimer approximation.

About 1000 molecules have been studied by microwave spectroscopy. References to work done before 1973 may be found in *Landolt-Börnstein Tables*, New Series, Group II, vols. 4 and 6: B. Starck, *Molecular Constants from Microwave Spectroscopy*, Springer-Verlag, 1976, and J. Demaison et al., *Molecular Constants*, Springer-Verlag, 1974.

21.7 VIBRATION OF POLYATOMIC MOLECULES

The nuclear wave function ψ_N of a molecule containing $\mathcal{N}$ atoms is a function of the $3\mathcal{N}$ coordinates needed to specify the locations of the nuclei. The nuclear motions are translations, rotations, and vibrations. The nuclear translational wave function is a function of three coordinates, the x, y, and z coordinates of the center of mass. The rotational wave function of a linear molecule is a function of the two angles θ and ϕ (Fig. 18.13) needed to specify the spatial orientation of the molecular axis. To specify the orientation of a nonlinear molecule, one chooses some axis in the molecule, gives θ and ϕ for this axis, and gives the angle of rotation of the molecule itself about this axis. Thus, ψ_{rot} depends on three angles for a nonlinear molecule. The nuclear vibrational wave function therefore depends on $3\mathcal{N} - 3 - 2 = 3\mathcal{N} - 5$ coordinates for a linear molecule and $3\mathcal{N} - 3 - 3 = 3\mathcal{N} - 6$ coordinates for a nonlinear molecule.

We first discuss the classical-mechanical treatment of molecular vibration and then the quantum-mechanical treatment.

Classical mechanics of vibration. Consider a molecule with $\mathcal{N}$ nuclei vibrating about their equilibrium positions subject to the potential energy E_e, where E_e is a function of the nuclear coordinates and is found by solving the electronic Schrödinger equation. For small vibrations, we can expand E_e in a Taylor series and neglect terms higher than quadratic (as we did for a diatomic molecule). Substituting this expansion into Newton's second law, one finds (*M. S.*, sec. 6.2) that any classical molecular vibration can be expressed as a linear combination of what are called normal modes of vibration.

In a given *normal mode*, each nucleus vibrates back and forth about its equilibrium position at the same frequency as every other nucleus; the nuclei all vibrate in phase, meaning that each nucleus passes through its equilibrium position at the same time; however, the vibrational amplitudes of different nuclei may differ. A molecule has $3\mathcal{N} - 6$ or $3\mathcal{N} - 5$ normal modes, depending on whether it is nonlinear or linear, respectively.

For a diatomic molecule, $3\mathcal{N} - 5 = 1$. The single normal mode involves the back and forth vibration of the two atoms along the internuclear axis. For a heteronuclear diatomic molecule, the vibrational amplitude of the heavier atom is less than that of the lighter atom.

Each normal mode has its own vibrational frequency. The forms and frequencies of the normal modes depend on the molecular geometry, the nuclear masses, and the force constants (the second derivatives of E_e with respect to the nuclear coordinates). If these quantities are known, one can find the normal modes and their frequencies.

(For each normal mode, the vibrational Hamiltonian has a kinetic-energy term that is quadratic in a momentum and a potential-energy term that is quadratic in a coordinate. This fact was used in the Sec. 15.10 discussion of energy equipartition.)

Figure 21.15 shows the normal modes of the linear molecule CO_2, which has $3(3) - 5 = 4$ normal modes. In the symmetric stretching vibration labeled ν_1, the carbon nucleus is motionless. In the asymmetric stretch ν_3, all the nuclei vibrate. The bending mode ν_{2b} is the same as ν_{2a} rotated by 90° about the molecular axis; clearly, these two modes have the same frequency. Bond-bending frequencies are generally lower than bond-stretching frequencies. For CO_2, observation of IR and Raman spectra gives

$$\tilde{\nu}_1 = 1340 \text{ cm}^{-1}, \qquad \tilde{\nu}_2 = 667 \text{ cm}^{-1}, \qquad \tilde{\nu}_3 = 2349 \text{ cm}^{-1}$$

Figure 21.16 shows the three normal modes of the nonlinear molecule H_2O. Here, ν_1 and ν_3 are stretching vibrations, and ν_2 is a bending vibration.

Quantum mechanics of vibration. Neglecting terms higher than quadratic in the E_e expansion and expressing the vibrations in terms of normal modes, one finds (*M. S.*, sec. 6.4) that the Hamiltonian for vibration becomes the sum of harmonic-oscillator vibrational Hamiltonians, one for each normal mode. Hence, the quantum-mechanical vibrational energy of a polyatomic molecule is the sum of $3\mathcal{N} - 6$ or $3\mathcal{N} - 5$ harmonic-oscillator energies, one for each normal mode:

$$E_{\text{vib}} \approx \sum_{i=1}^{3\mathcal{N}-6} (\upsilon_i + \tfrac{1}{2})h\nu_i, \qquad \upsilon_i = 0, 1, 2, \ldots \tag{21.48*}$$

where ν_i is the frequency of the ith normal mode and υ_i is its quantum number. (For linear molecules, the upper limit is $3\mathcal{N} - 5$.) The $3\mathcal{N} - 6$ or $3\mathcal{N} - 5$ vibrational quantum numbers vary independently of one another. The ground vibrational level has all υ_i's equal to zero and has the zero-point energy $\frac{1}{2}\sum_i h\nu_i$ (anharmonicity neglected).

One finds that the most probable vibration–rotation absorption transitions are ones in which one vibrational quantum number υ_j changes by $+1$ with all others (υ_k, $k \neq j$) unchanged. However, for the transition $\upsilon_j \rightarrow \upsilon_j + 1$ to occur, the jth normal vibration must change the molecular dipole moment. Since the harmonic-oscillator

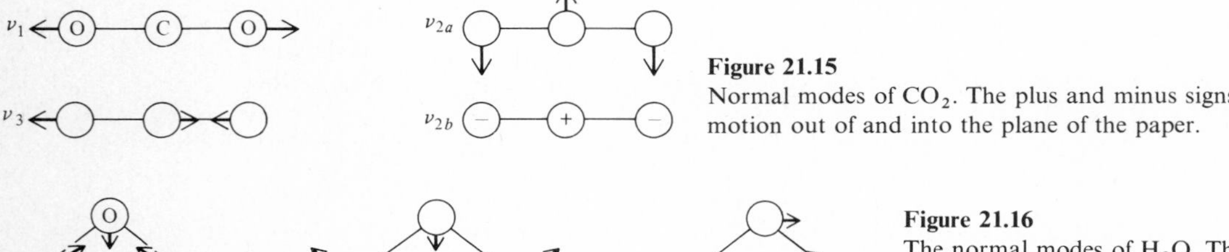

Figure 21.15
Normal modes of CO_2. The plus and minus signs indicate motion out of and into the plane of the paper.

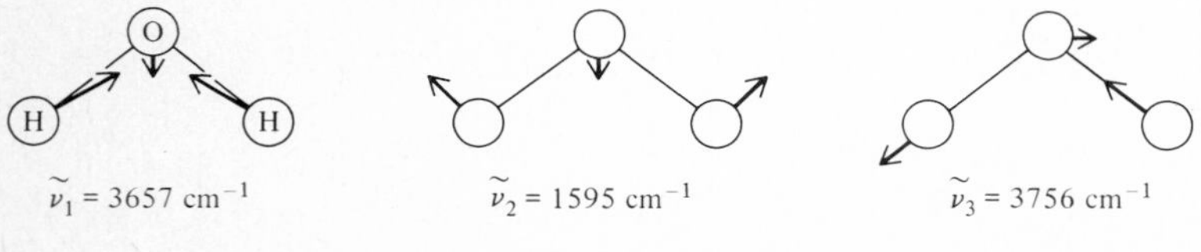

$\tilde{\nu}_1 = 3657 \text{ cm}^{-1}$ $\quad \tilde{\nu}_2 = 1595 \text{ cm}^{-1}$ $\quad \tilde{\nu}_3 = 3756 \text{ cm}^{-1}$

Figure 21.16
The normal modes of H_2O. The heavy oxygen atom has a much smaller vibrational amplitude than the light hydrogen atoms.

levels are spaced by $h\nu_j$, the transition $v_j \to v_j + 1$ produces a band in the IR spectrum at the frequency ν_j ($\nu_{\text{light}} = \Delta E/h = h\nu_j/h = \nu_j$).

For CO_2, the symmetric stretch ν_1 leaves the dipole moment unchanged, and no IR band is observed at ν_1. One says this vibration is *IR-inactive*. The vibrations ν_2 and ν_3 each change the dipole moment and are IR-active.

The most populated vibrational level has all the v_i's equal to 0. A transition from this level to a level with $v_j = 1$ and all other vibrational quantum numbers equal to zero gives an IR *fundamental band*. Transitions where one v_j changes by 2 or more with all others unchanged give *overtone bands*. Transitions where two vibrational quantum numbers change give *combination bands*. Overtone and combination bands are substantially weaker than fundamental bands.

The frequencies of the IR fundamental bands give (some of) the *fundamental vibration frequencies* of the molecule. Because of anharmonicity, the fundamental frequencies differ somewhat from the equilibrium vibrational frequencies [cf. Eq. (21.36)]. Usually, insufficient data are available to determine the anharmonicity corrections, and so one works with the fundamental frequencies.

For H_2O, all three normal modes are IR-active. Wave numbers of some IR band origins for H_2O vapor follow. Also listed are the band intensities (s = strong, m = medium, w = weak) and the quantum numbers $v_1' v_2' v_3'$ of the upper vibrational level. In all cases, the lower level is the 000 ground state.

σ_0/cm^{-1}	1595(s)	3152(m)	3657(s)	3756(s)	5331(m)	6872(w)
$v_1' v_2' v_3'$	010	020	100	001	011	021

The three strongest bands are the fundamentals and give the fundamental vibration wave numbers as $\tilde{\nu}_1 = 3657\ \text{cm}^{-1}$, $\tilde{\nu}_2 = 1595\ \text{cm}^{-1}$, and $\tilde{\nu}_3 = 3756\ \text{cm}^{-1}$. Because of anharmonicity, the overtone transition $000 \to 020$ occurs at a bit less than twice the frequency of the $000 \to 010$ fundamental band. The combination band at $6872\ \text{cm}^{-1}$ has its wave number approximately equal to $2\tilde{\nu}_2 + \tilde{\nu}_3 = 6946\ \text{cm}^{-1}$.

Molecules with more than, say, five atoms generally have one or more vibrational modes with frequencies low enough to have excited vibrational levels significantly populated at room temperature. This gives hot bands in the IR absorption spectrum.

As with diatomic molecules, gas-phase IR bands of polyatomic molecules show a rotational structure. For certain vibrational transitions, the $\Delta J = 0$ transition is allowed, giving a line (called the *Q branch*) at the band origin. (See *M. S.*, sec. 6.5 for details.)

21.8 INFRARED SPECTROSCOPY

Figure 21.17 outlines a double-beam IR spectrometer. The radiation source is an electrically heated rod that emits continuous radiation. The sample may be gaseous, liquid, or solid. If the sample is in solution, the reference cell is filled with pure solvent; otherwise, it is empty. The chopper causes the sample beam and the reference beam to fall alternately on the prism. The prism disperses the radiation into its component frequencies. The rotating mirror changes the frequency of the radiation falling on the detector. The motor that drives this mirror also drives the chart paper on which the spectrum is recorded as a function of frequency. If the sample does not absorb at a given frequency, the sample and reference beams falling on the detector are equally intense and the ac amplifier receives no ac signal. At a frequency for which the sample absorbs, the light intensity reaching the detector varies at a frequency equal to the chopper frequency; the detector then puts out an ac signal whose strength depends on the intensity of absorption. IR detectors used include thermocouples, temperature-dependent resistors, and photoconductive materials.

Gas-phase IR bands under high resolution consist of closely spaced lines due to the various changes in the rotational quantum numbers. In liquids and solids, rotational structure is not observed, and each band appears as a broad absorption. In solids, the molecules do not rotate. In liquids (and in high-pressure gases), the strong intermolecular interactions continually shift the rotational energy levels, thus broadening the rotational lines sufficiently to merge them into a single broad absorption.

The spacings between the rotational lines of an IR band depend on the moments of inertia, so analysis of the rotational structure of vibration–rotation bands allows molecular structures to be determined. Usually, isotopic species must also be studied to give information sufficient for a structural determination. Structures determined from IR spectra are not as accurate as those found from microwave spectra, but IR spectroscopy has the advantage of allowing structures of nonpolar molecules to be determined; nonpolar molecules (other than diatomics) have some IR-active vibrations.

Some nonpolar-molecule structures determined from IR spectra are

$$C_2H_6: \quad R(CH) = 1.10 \text{ Å}, \quad R(CC) = 1.54 \text{ Å}, \quad \angle HCC = 110°$$

$$C_2H_4: \quad R(CH) = 1.09 \text{ Å}, \quad R(CC) = 1.34 \text{ Å}, \quad \angle HCC = 121°$$

$$C_2H_2: \quad R(CH) = 1.06 \text{ Å}, \quad R(CC) = 1.20 \text{ Å}, \quad \angle HCC = 180°$$

Organic chemists use infrared spectroscopy to help identify an unknown compound. Although most or all of the atoms are vibrating in each normal mode, certain normal modes may involve mainly motion of only a small group of atoms bonded together, with the other atoms vibrating only slightly. For example, in aldehydes and ketones, there is one normal mode which is mainly a C=O stretching vibration, and its frequency is approximately the same in most aldehydes and ketones. Normal-mode analysis shows that two bonded atoms A and B in an organic compound will exhibit a characteristic vibrational frequency provided that either the force constant of the A—B bond differs greatly from the other force constants in the

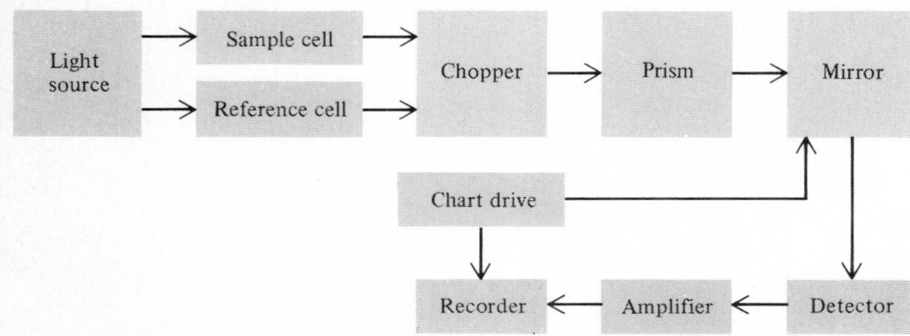

Figure 21.17
A double-beam IR spectrometer.

molecule or there is a large difference in mass between A and B. Double and triple bonds have much larger force constants than single bonds, so one observes characteristic vibrational frequencies for C=O, C=C, C≡C, and C≡N bonds. Since hydrogens are much lighter than carbons, one observes characteristic vibrational frequencies for OH, NH, and CH groups. In contrast, vibrations involving C—C, C—N, and C—O single bonds occur over a very wide range of frequencies.

Some characteristic IR wave numbers in cm^{-1} for stretching vibrations are:

OH	NH	CH	C≡C	C=C	C=O
3200–3600	3100–3500	2700–3300	2100–2250	1620–1680	1650–1850

The complete IR spectrum is a highly characteristic property of a compound and has been compared to a fingerprint.

21.9 RAMAN SPECTROSCOPY

Raman spectroscopy is quite different from absorption spectroscopy in that it studies light scattered by a sample rather than light absorbed or emitted. Consider a collision between a photon and a molecule in state *a*. If the energy of the photon corresponds to the energy difference between state *a* and a higher level, the photon may be absorbed, the molecule making a transition to the higher level. No matter what the energy of the photon is, the photon–molecule collision may *scatter* the photon, meaning that the photon's direction of motion is changed. Although most of the scattered photons undergo no change in frequency and energy (Rayleigh scattering), a small fraction of the scattered photons exchange energy with the molecule during the collision process. The resulting increase or decrease in energy of the scattered photons is the *Raman effect*, first observed by an Indian physicist in 1928.

Let ν_a and ν_b be the frequencies of the incident photon and the Raman-scattered photon, respectively, and let E_a and E_b be the energies of the molecule before and after it scatters the photon. Conservation of energy gives $h\nu_a + E_a = h\nu_b + E_b$, or

$$\Delta E \equiv E_b - E_a = h(\nu_a - \nu_b) \tag{21.49}$$

The energy difference ΔE is the difference between two stationary-state energies of the molecule, so measurement of the Raman shifts $\nu_b - \nu_a$ gives molecular energy-level spacings.

In Raman spectroscopy, one exposes the sample to monochromatic radiation of any convenient frequency ν_a. The Raman-effect lines are extremely weak (most of the scattered radiation is Rayleigh-scattered with no frequency change). Hence, one uses the very intense light of a laser beam as the exciting radiation. The scattered light is observed at right angles to the laser beam.

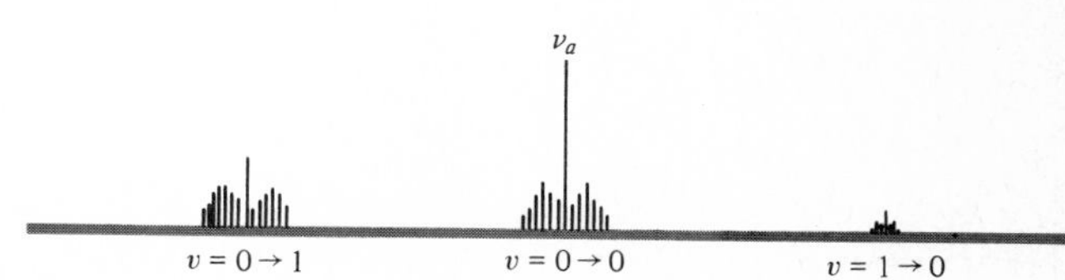

Figure 21.18
Raman spectrum of a diatomic molecule.

Figure 21.18 shows the pattern of Raman lines for a diatomic molecule. The strong central line at ν_a is due to light scattered with no frequency change. On either side of ν_a and close to ν_a are lines corresponding to pure-rotational transitions in the molecule; the lines with $\nu_b > \nu_a$ result from transitions where J decreases, and those with $\nu_b < \nu_a$ result from increases in J. On the low-frequency side of ν_a is a band of lines corresponding to $v = 0 \rightarrow 1$ vibration–rotation transitions in the molecule. If there is significant population of the $v = 1$ vibrational level, there is a weak band of lines on the high-frequency side, corresponding to $v = 1 \rightarrow 0$ vibration–rotation transitions.

Investigation of the Raman selection rules shows that spherical tops exhibit no pure-rotational Raman spectra, but all symmetric and asymmetric tops show pure-rotational Raman spectra. The pure-rotational Raman spectrum can yield the structure of a nonpolar molecule (for example, F_2, C_6H_6) that is not a spherical top. For benzene, the rotational Raman spectrum gives $R(\text{CC}) = 1.397$ Å and $R(\text{CH}) = 1.08$ Å.

For a polyatomic molecule, the Raman spectrum shows several $\Delta v_j = 1$ fundamental bands, each corresponding to a Raman-active vibration. A normal mode is Raman-active if it changes the molecular polarizability (Sec. 14.2). Since this requirement is different from that for IR activity, the Raman spectrum often allows the frequencies of IR-inactive bands to be found. An example is the ν_1 vibration of CO_2, which is IR-inactive but Raman-active.

21.10 ELECTRONIC SPECTROSCOPY

Electronic spectra involving transitions of valence electrons occur in the visible and UV regions and are studied both in absorption and emission. In emission spectroscopy, the molecules are excited in a flame or gas-discharge tube. A photoelectric cell or photographic plate is the detector.

Atomic spectra. For the hydrogen atom, the n selection rule is found to be $\Delta n =$ any value. Use of $E_a - E_b = h\nu$ and the energy-level formula (19.17) gives the spectral wave numbers as

$$\frac{1}{\lambda} = \frac{\nu}{c} = \frac{2\pi^2 \mu e'^4}{ch^3}\left(\frac{1}{n_b^{\,2}} - \frac{1}{n_a^{\,2}}\right) \equiv R\left(\frac{1}{n_b^{\,2}} - \frac{1}{n_a^{\,2}}\right) \tag{21.50}$$

where the reduced mass depends on the electron and proton masses: $\mu = m_e m_p/(m_e + m_p)$ [Eq. (18.72)]. Substitution of numerical values for the physical constants gives $R = 109{,}678\ \text{cm}^{-1}$, in agreement with the observed values (18.5).

The H-atom spectrum consists of several series of lines, each ending in a continuous band. Transitions between n values of 2 and 1, 3 and 1, 4 and 1, ... give the Lyman series in the UV (Fig. 21.19). As n_a in (21.50) goes to infinity, the Lyman

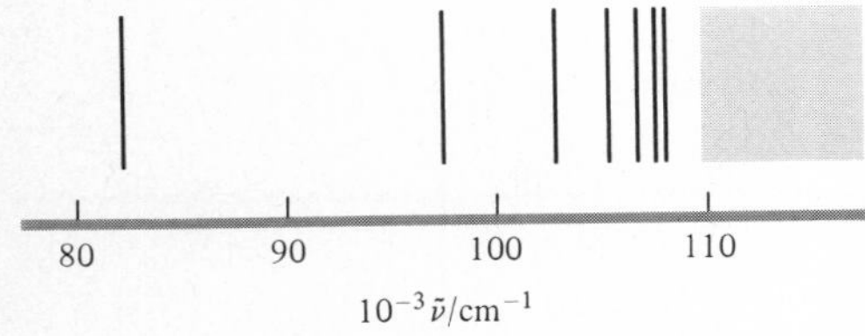

Figure 21.19
The Lyman series of H. Only the first seven lines are shown.

series lines converge to the limiting value $1/\lambda = R = 109{,}678\ \text{cm}^{-1}$, corresponding to $\lambda = 912\ \text{Å} = 91.2$ nm. Beyond this limit is a continuous absorption or emission, due to transitions between ionized H atoms and ground-state H atoms; the energy of an ionized atom takes on a continuous range of positive values. The position of the Lyman series limit allows the ionization potential of H to be found.

Transitions between the H levels 3 and 2, 4 and 2, 5 and 2, ... give the Balmer series, which lies in the visible region. Transitions between 4 and 3, 5 and 3, 6 and 3, ... give the Paschen series in the IR.

Spectra of many-electron atoms are quite complicated, due to the many terms and levels arising from a given electron configuration. Once the spectrum has been unraveled, the atomic energy levels can be found.

For inner-shell electrons, the effective nuclear charge Z_{eff} is close to the atomic number; Eq. (19.46) shows that the differences between these inner-shell energies increase rapidly with increasing atomic number. For atoms beyond the second period, these energy differences correspond to x-ray photons. X-rays are produced when a beam of high-energy electrons penetrates a metal target. The deceleration of the electrons as they penetrate the target produces a continuous x-ray emission spectrum. In addition, an electron in the beam that collides with an inner-shell electron of a target atom can knock this electron out of the atom; the spontaneous jump of a higher-level atomic electron into the vacancy thereby created will produce an x-ray photon of frequency corresponding to the energy jump. (Recall Moseley's work on x-ray spectra and atomic numbers.)

Since each element has lines at frequencies characteristic of that element, atomic absorption and emission spectra are used to analyze for most chemical elements. For example, Ca, Mg, Na, K, and Pb in blood samples can be determined by atomic absorption spectroscopy.

Molecular electronic spectra. If ψ'' and ψ' are the lower and upper states in a molecular electronic transition, the transition frequencies are given by

$$h\nu = (E'_{\text{el}} - E''_{\text{el}}) + (E'_{\text{vib}} - E''_{\text{vib}}) + (E'_{\text{rot}} - E''_{\text{rot}}) \tag{21.51}$$

Each term in parentheses is substantially smaller than the preceding term.

The electronic selection rules are rather complicated. Perhaps the most important one is that the integral (21.5) vanishes unless

$$\Delta S = 0 \tag{21.52}$$

where S is the total electronic spin quantum number. (Actually, this selection rule does not hold rigorously, and transitions with $\Delta S \neq 0$ are sometimes observed, but they are generally weak.) The ground electronic state of most molecules has all electron spins paired and so has $S = 0$ (a singlet state). Some exceptions are O_2 (a triplet ground level with $S = 1$) and NO_2 (the odd number of electrons gives $S = \frac{1}{2}$ and a doublet ground level).

An electronic transition consists of a series of *bands*, each band corresponding to a transition between a given pair of vibrational levels. Gas-phase spectra under high resolution may show each band to consist of closely spaced lines arising from transitions between different rotational levels; for relatively heavy molecules, the close spacings between the rotational levels usually makes it impossible to resolve the rotational lines. In pure liquids and liquid solutions, no rotational structure is observed. Also, intermolecular interactions in the liquid phase often broaden the vibrational bands sufficiently to merge them into a single broad absorption band for each electronic transition. Figure 21.20 shows the electronic absorption spectrum of gas-phase benzene.

Analysis of the rotational lines of an electronic absorption band allows the molecular vibrational and rotational constants to be obtained. Since all molecules show electronic absorption spectra, this allows one to obtain R_e and ν_e of homonuclear diatomics (which show no pure-rotational or vibration–rotation spectra). Excited electronic states often have quite different geometries from the ground electronic state. For example, HCN is nonlinear in some excited states. Molecular dissociation energies and spacings between electronic energy levels are also obtained from electronic spectra.

Substances that absorb visible light are colored. Conjugated organic compounds often show electronic absorption in the visible region, due to excitation of a pi electron to an antibonding pi orbital. (In the particle-in-a-box model, the energy spacings are proportional to $1/a^2$, where a is the box length, so that as the length of the conjugated chain increases, the lowest absorption frequency moves from the UV into the visible region.) Transition-metal ions (for example, Cu^{2+}, MnO_4^-) frequently show visible absorption due to electronic transitions.

Certain groups called *chromophores* give characteristic electronic absorption bands. For example, the C═O group in the majority of aldehydes and ketones produces a weak electronic absorption in the region 270 to 295 nm with $\varepsilon = 10$ to 30 $dm^3\ mol^{-1}\ cm^{-1}$ and a strong band around 180 to 195 nm with $\varepsilon = 10^3$ to 10^4 $dm^3\ mol^{-1}\ cm^{-1}$. The weak band is due to an $n \rightarrow \pi^*$ transition, and the strong band is due to a $\pi \rightarrow \pi^*$ transition, where n signifies an electron in a nonbonding (lone-pair) orbital on oxygen, π is an electron in the π MO of the double bond, and π^* is an electron in the corresponding antibonding π MO.

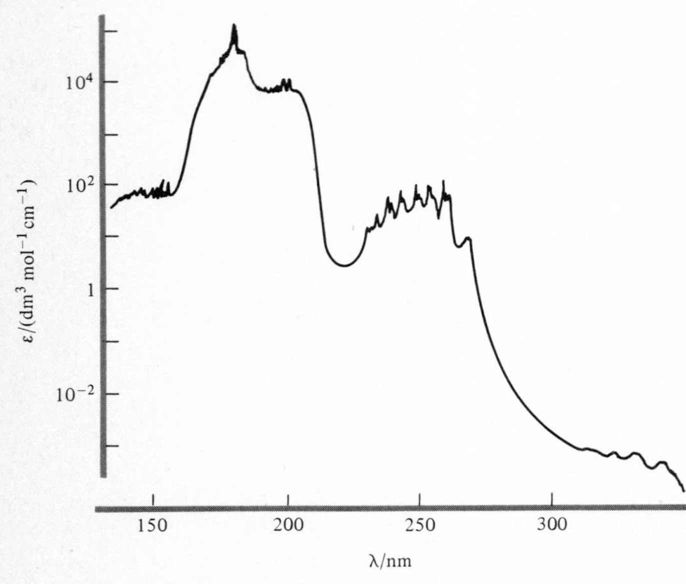

Figure 21.20
Sketch of the electronic absorption spectrum of gas-phase benzene. The vertical scale is logarithmic.

The magnetic field. A magnetic field is produced by the motion of electric charge. Examples include the motion of electrons in a wire, the motion of electrons in free space, and the "spinning" of an electron about its own axis. The fundamental magnetic field vector **B** is called the *magnetic induction* or the *magnetic flux density*. (There is a second magnetic field vector **H** named the *magnetic field strength*. It used to be thought that **H** was the fundamental magnetic vector, but it is now known that **B** is. Really, **B** ought to be called the magnetic field strength, but it's too late to correct this injustice by giving **B** its proper name.)

The definition of **B** is as follows. Imagine a positive test charge Q_t moving through point P in space with velocity **v**. If for arbitrary directions of **v** we find that a force $\mathbf{F}_\perp$ perpendicular to **v** acts on Q_t at P, we say that there is a magnetic field **B** at point P. One finds that there is one direction of **v** that makes $\mathbf{F}_\perp$ equal zero, and the direction of **B** is defined to coincide with this particular direction of **v**. The magnitude of **B** at point P is then defined by $B \equiv F_\perp/(Q_t v \sin\theta)$, where θ is the angle between **v** and **B**. (For $\theta = 0$, $F_\perp$ becomes 0.) Thus

$$F_\perp = Q_t vB \sin\theta \tag{21.53}$$

(In terms of vectors, $\mathbf{F}_\perp = Q_t\mathbf{v} \times \mathbf{B}$, where $\mathbf{v} \times \mathbf{B}$ is the vector cross product; the magnetic force is perpendicular to both **v** and **B**.) If an electric field is also present, it exerts a force $Q_t\mathbf{E}$ in addition to the magnetic force $\mathbf{F}_\perp$.

Equation (21.53) is written in SI units. The SI unit of B is the *tesla* (T), also called the Wb/m^2, where Wb stands for weber. From (21.53) and (2.21)

$$1\ \text{T} \equiv 1\ \text{N C}^{-1}\ \text{m}^{-1}\ \text{s} = 1\ \text{kg s}^{-1}\ \text{C}^{-1} \tag{21.54}$$

Magnetic fields are produced by electric currents. Experiment shows that the magnetic field at a distance r from a very long straight wire in vacuum carrying a current I is proportional to I and inversely proportional to r; that is, $B = kI/r$, where k is a constant. (This is Ampère's law.) The unit of electric current, the ampere (A), is defined (Sec. 16.5) so as to give k the value 2×10^{-7} T m A^{-1}. The constant k is also written as $\mu_0/2\pi$, where μ_0 is called the *permeability of vacuum*. Thus

$$\mu_0 \equiv 4\pi \times 10^{-7}\ \text{T m A}^{-1} = 4\pi \times 10^{-7}\ \text{N C}^{-2}\ \text{s}^2 \tag{21.55}$$

where (21.54) and 1 A = 1 C/s were used.

Equation (21.1) for the speed of light in vacuum can be written as

$$c = (\varepsilon_0 \mu_0)^{-1/2} \tag{21.56}$$

Consider a tiny loop of current I flowing in a circle of area A. At distances large compared with the radius of the loop, one finds (*Halliday and Resnick*, sec. 34-6) that the magnetic field produced by this current loop has the same mathematical form as the electric field of an electric dipole (Sec. 14.2) except that the electric dipole moment **p** is replaced by the *magnetic dipole moment* **μ**, where **μ** is a vector with magnitude IA and direction perpendicular to the plane of the current loop; thus

$$\mu \equiv IA \tag{21.57}$$

The tiny current loop is called a *magnetic dipole*.

A magnetic dipole acts like a tiny magnet with a north pole on one side of the current loop and a south pole on the other side. A bar magnet suspended in an external magnetic field has a preferred minimum-energy orientation in the field. Thus, a compass needle orients itself in the earth's magnetic field with one particular end of the needle pointing toward the earth's north magnetic pole. To turn the needle away from this orientation requires an input of energy. In general, a magnetic dipole **μ** has a minimum-energy orientation in an externally applied magnetic field **B**; the potential energy V of interaction between **μ** and the external field **B** can be shown to be (*Halliday and Resnick*, sec. 33-4)

$$V = -\mu B \cos\theta \tag{21.58}$$

where θ is the angle between **μ** and **B**. The minimum-energy orientation has **μ** and **B** in the same direction, so that $\theta = 0$ and $V = -\mu B$. The maximum-energy orientation has **μ** and **B** in opposite directions ($\theta = 180°$) and $V = \mu B$. The zero of potential energy has been arbitrarily chosen to make $V = 0$ at $\theta = 90°$.

Chemists commonly use gaussian units in dealing with magnetic phenomena. In gaussian units, the defining equation (21.53) is written as $F_\perp = (Q_t' vB' \sin\theta)/c$, where B' is the magnetic field in gaussian units and c is the speed of light. The gaussian unit of B' is the gauss (G). Thus

$$1 \text{ G} = 1 \text{ dyn statC}^{-1} = 1 \text{ g}^{1/2} \text{ s}^{-1} \text{ cm}^{-1/2} \tag{21.59}$$

where (19.2) was used.

Equating the SI and gaussian expressions for $F_\perp$ and using (19.5), we get $B' = Bc(4\pi\varepsilon_0)^{1/2}$. Use of (21.56) then gives

$$B' = Bc(4\pi\varepsilon_0)^{1/2} = B(4\pi/\mu_0)^{1/2} \tag{21.60}$$

where μ_0 is given by (21.55). One finds that when $B = 1$ T, then $B' = 10^4$ G (Prob. 21.39), so that

$$1 \text{ T corresponds to } 10{,}000 \text{ G} \tag{21.61}$$

The relation between the magnetic dipole moment μ' in gaussian units and μ in SI units turns out to be

$$\mu' = (\mu_0/4\pi)^{1/2}\mu \tag{21.62}$$

Nuclear spins and magnetic moments. Nuclei, like electrons, have a spin angular momentum **I**. The nuclear spin is the resultant of the spin and orbital angular momenta of the neutrons and protons that compose the nucleus. A nucleus has two spin quantum numbers, I and M_I. (These are analogous to s and m_s for an electron.) The magnitude of the nuclear spin angular momentum is

$$|\mathbf{I}| = [I(I+1)]^{1/2}\hbar \tag{21.63}$$

and the possible values of the z component of **I** are

$$I_z = M_I\hbar \qquad \text{where } M_I = -I, -I+1, \ldots, I-1, I \tag{21.64}$$

The quantum number I can be integral or half-integral. [See Table (21.67).]

A moving charge produces a magnetic field. We can crudely picture spin as due to a particle rotating about one of its own axes. Hence, we expect any charged particle with spin to act as a tiny magnet. The magnetic properties of a particle with spin can be described in terms of the particle's magnetic dipole moment **μ**.

Consider a particle of charge Q and mass m moving in a circle of radius r with speed v. The time for one complete revolution is $t = 2\pi r/v$, and the current flow is

$I = Q/t = Qv/2\pi r$. From (21.57), the magnetic moment is $\mu = \pi r^2 I = Qvr/2$. The particle's orbital angular momentum (Sec. 19.3) is $L = mvr$, and so the magnetic moment can be written as $\mu = QL/2m$. The angular-momentum vector $\mathbf{L}$ is perpendicular to the circle, as is the magnetic-moment vector $\boldsymbol{\mu}$. Therefore

$$\boldsymbol{\mu} = Q\mathbf{L}/2m \tag{21.65}$$

A nucleus has a spin angular momentum $\mathbf{I}$, and the nuclear magnetic (dipole) moment $\boldsymbol{\mu}_N$ is given by an equation resembling (21.65). However, instead of using the charge and mass of the nucleus, it is more convenient to use the proton charge and mass e and m_p; moreover, because of the composite structure of the nucleus, an extra numerical factor g_N must be included. Thus

$$\boldsymbol{\mu}_N = g_N \frac{e}{2m_p}\mathbf{I} \tag{21.66}$$

where g_N is the *nuclear g factor*. Present theories of nuclear structure cannot predict g_N values; they must be determined experimentally. Some values of g_N and I are:

Nucleus	^{1}H	^{2}H	^{12}C	^{13}C	^{16}O	^{19}F
I	$\frac{1}{2}$	1	0	$\frac{1}{2}$	0	$\frac{1}{2}$
g_N	5.5856	0.8574	—	1.4048	—	5.257

(21.67)

A larger table is given in the Appendix. (The fact that g_N for ^{1}H is not a simple number indicates that the proton has an internal structure.)

Equation (21.66) is in SI units (as indicated by the absence of primes). Using (21.62), (19.5), and (21.56), we find in gaussian units $\boldsymbol{\mu}'_N = (g_N e'/2m_p c)\mathbf{I}$.

From Eqs. (21.66) and (21.63) the magnitude of the magnetic moment of a nucleus is

$$\mu_N = |g_N|(e/2m_p)[I(I+1)]^{1/2}\hbar \equiv |g_N|\beta_N[I(I+1)]^{1/2}$$

where the *nuclear magneton* β_N is a physical constant defined as $\beta_N \equiv e\hbar/2m_p$. Substitution of numerical values gives

$$\beta_N = \frac{(1.6022 \times 10^{-19}\text{ C})(6.626 \times 10^{-34}\text{ J}\cdot\text{s})}{4\pi(1.6726 \times 10^{-27}\text{ kg})}$$

$$\beta_N \equiv e\hbar/2m_p = 5.051 \times 10^{-27}\text{ J/T} \tag{21.68}$$

where (21.54) was used. In gaussian units,

$$\beta'_N = e'\hbar/2m_p c = 5.051 \times 10^{-24}\text{ erg/G} \tag{21.69}$$

Nuclear-magnetic-resonance (NMR) spectroscopy. In NMR spectroscopy, one applies an external magnetic field $\mathbf{B}$ to a sample containing nuclei with nonzero spin. For simplicity, let us initially consider a single isolated nucleus with magnetic dipole moment $\boldsymbol{\mu}_N$. The energy of the nuclear magnetic dipole in the applied field $\mathbf{B}$ depends on the orientation of $\boldsymbol{\mu}_N$ with respect to $\mathbf{B}$ and is given by (21.58) and (21.66) as

$$E = -\mu_N B\cos\theta = -g_N(e/2m_p)|\mathbf{I}|B\cos\theta \tag{21.70}$$

where $|\mathbf{I}|$ is the magnitude (length) of the spin-angular-momentum vector **I** and θ is the angle between **B** and **I**. Let the direction of the applied field be called the z direction. Figure 21.21 shows that $|\mathbf{I}| \cos \theta$ equals I_z, the z component of **I**. Hence, $E = -g_N(e/2m_p)I_z B$. However, only certain orientations of **I** (and the associated magnetic moment $\boldsymbol{\mu}_N$) in the field are allowed by quantum mechanics; I_z is quantized with the possible values $M_I \hbar$ [Eq. (21.64)]. The nuclear magnetic moment in the applied magnetic field therefore has the following set of quantized energy levels:

$$E = -g_N(e\hbar/2m_p)BM_I = -g_N \beta_N BM_I, \qquad M_I = -I, \ldots, +I \tag{21.71}$$

where (21.68) was used.

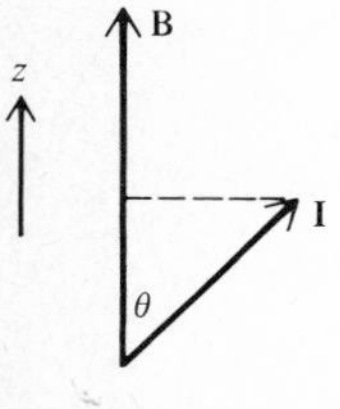

Figure 21.21

Figure 21.22 shows the allowed nuclear-spin energy levels of the nuclei ^{1}H (with $I = \frac{1}{2}$) and ^{2}H (with $I = 1$) as a function of the applied magnetic field. As B increases, the spacing between levels increases. In the absence of an external magnetic field, all orientations of the spin have the same energy.

By exposing the sample to electromagnetic radiation of appropriate frequency, one can observe transitions between these nuclear-spin energy levels. The selection rule is found to be (*M. S.*, sec. 8.3)

$$\Delta M_I = \pm 1 \tag{21.72}$$

The absorption frequency ν satisfies $h\nu = |\Delta E| = |g_N|\beta_N B|\Delta M_I| = |g_N|\beta_N B$, and

$$\nu = |g_N|\beta_N B/h \tag{21.73}$$

(g_N is negative for some nuclei.) Although there are $2I + 1$ different energy levels for the nuclear magnetic dipole in the field, the selection rule (21.72) allows only transitions between adjacent levels, which are equally spaced. A collection of identical noninteracting nuclei therefore gives a single NMR absorption frequency.

A typical magnetic field easily attained in the laboratory is 1 T (10,000 G). For the ^{1}H nucleus (a proton) in this field,

$$\nu = 5.586(5.051 \times 10^{-27} \text{ J/T})(1 \text{ T})/(6.626 \times 10^{-34} \text{ J s}) = 42.6 \text{ MHz}$$

This is in the radio-frequency (rf) portion of the electromagnetic spectrum. (FM radio stations broadcast from 88 to 108 MHz.) Some NMR frequencies at $B = 1$ T for other nuclei are 6.54 MHz for ^{2}H, 10.71 MHz for ^{13}C, and 40.1 MHz for ^{19}F.

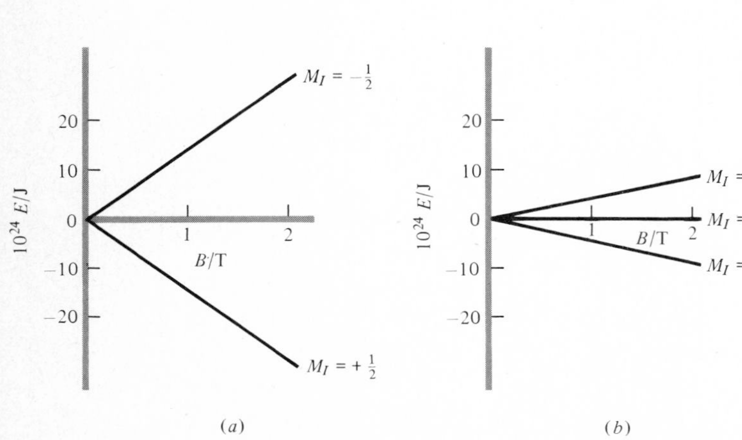

Figure 21.22
Nuclear-spin energy levels vs. applied magnetic field for (*a*) ^{1}H; (*b*) ^{2}H.

Nuclear magnetic resonance in bulk matter was first observed by Bloch and by Purcell in 1945. In his Nobel prize acceptance speech, Purcell said: "I remember, in the winter of our first experiments, just seven years ago, looking on snow with new eyes. There the snow lay around my doorstep—great heaps of protons quietly precessing in the earth's magnetic field. To see the world for a moment as something rich and strange is the private reward of many a discovery." [*Science*, **118**, 431 (1953).]

Nuclei with $I = 0$ (for example, ^{12}C, ^{16}O, ^{32}S) have no magnetic dipole moment and no NMR spectrum. Nuclei with $I \geq 1$ have something called an electric quadrupole moment, which broadens the NMR absorption lines, eliminating the chemically interesting details. Thus, one generally studies only nuclei with $I = \frac{1}{2}$. The most studied nucleus is the proton, ^1H. In the 1970s, improved instrumentation led to great interest in ^{13}C NMR spectra (see below).

In spectroscopy one usually varies the frequency ν of the incident electromagnetic radiation until absorption is observed. However, in NMR spectroscopy it is technically more convenient to keep ν fixed and to vary the spacing between the levels by varying the magnitude B of the applied field until (21.73) is satisfied and absorption occurs.

Figure 21.23 shows a simplified version of an NMR spectrometer. The sample is usually a liquid. An electromagnet or permanent magnet applies a uniform magnetic field B_0. The value of B_0 is varied over a narrow range by varying the current through coils around the magnet poles; this current also drives the chart paper on which the spectrum is recorded. The sample tube is spun rapidly to average out any inhomogeneities in $\mathbf{B}_0$. A coil connected to an rf transmitter (oscillator) and wound around the sample exposes the sample to electromagnetic radiation of fixed frequency. This coil is an inductor whose inductance depends in part on what is inside the coil. The coil is part of a carefully tuned rf circuit in the transmitter. When the sample absorbs energy, its characteristics change, thereby detuning the transmitter circuit and decreasing the transmitter output. This decrease is recorded as the NMR signal. (An alternative setup uses a separate detector coil at right angles to the transmitter coil. The flipping of the spins caused by absorption induces a current in the detector coil.)

As we shall see below, the NMR spectrum of a molecule contains lines at several frequencies, so that it takes a significant amount of time to scan through the

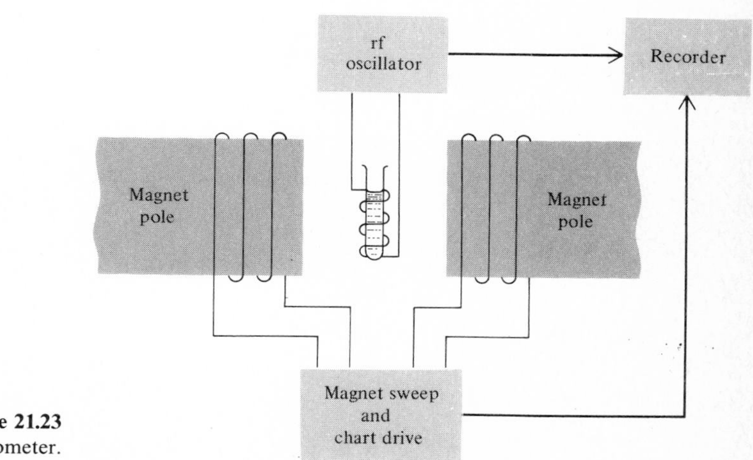

Figure 21.23
An NMR spectrometer.

spectrum. There is great interest in observing ^{13}C NMR spectra of organic compounds, since this provides information on the "backbone" of organic compounds. The isotope ^{13}C is present in only 1 percent natural abundance, which makes the ^{13}C NMR absorption signals very weak. One can scan the spectrum repeatedly and feed the results into a computer that adds the results of successive scans, thereby enhancing the signal. However, a sufficiently large number of scans takes a long time (several days) and is impractical. To overcome this difficulty, one uses Fourier-transform NMR spectroscopy. Here, instead of being exposed to rf radiation of a single frequency, the sample is irradiated with a very short pulse of rf radiation; the pulse contains a mixture of all frequencies, so that the entire spectrum is observed at once. A computer is used to calculate the NMR spectrum from the observed absorption from the pulse. The results of many successive pulses are then added together to give the ^{13}C NMR spectrum.

If equation (21.73) gave the NMR absorption frequencies for nuclei in molecules, NMR would be of no interest to chemists. However, the actual magnetic field experienced by a nucleus in a molecule differs very slightly from the applied field B_0, due to the magnetic field produced by the molecular electrons. Most ground-state molecules have all electrons paired, which makes the total electronic spin and orbital angular momentum equal to zero; with zero electronic angular momentum, the electrons produce no magnetic field. However, when an external magnetic field is applied to a molecule, this changes the electronic wave function slightly, thereby producing a slight contribution of the electronic motions to the magnetic field at each nucleus. (This effect is similar to the polarization of a molecule produced by an applied electric field.) The magnetic field of the electrons usually opposes the applied magnetic field B_0 and is proportional to B_0; the electronic contribution to the magnetic field at a given nucleus i is $-\sigma_i B_0$, where the proportionality constant σ_i is called the *shielding* (or *screening*) *constant* for nucleus i. The value of σ_i at a given nucleus depends on the electronic environment of the nucleus. For molecular protons, σ_i usually lies in the range 1×10^{-5} to 4×10^{-5}; for heavier nuclei (which have more electrons than H), σ_i may be 10^{-4} or 10^{-3}.

For each of the six protons in benzene (C_6H_6), σ_i is the same, since each proton has the same electronic environment. For chlorobenzene (C_6H_5Cl), there are three different values of σ_i for the protons, one value for the two ortho protons, one value for the two meta protons, and one value for the para proton. For CH_3CH_2Br, there is one value of σ_i for the CH_3 protons and a different value for the CH_2 protons; the low barrier to internal rotation about the carbon–carbon single bond makes the electronic environment of all three methyl protons the same.

Addition of the electronic contribution $-\sigma_i B_0$ to the applied field B_0 gives the magnetic field B_i experienced by nucleus i as $B_i = B_0(1 - \sigma_i)$. Substitution in (21.73) gives as the NMR frequencies of a molecule

$$\nu_i = |g_i|\beta_N h^{-1}(1 - \sigma_i)B_0 \tag{21.74}$$

where g_i is the g value of nucleus i. Since one usually holds the frequency fixed and varies B_0, the values of the applied field B_0 at which absorption occurs are

$$B_{0,i} = \frac{h\nu_{\text{spec}}}{|g_i|\beta_N(1 - \sigma_i)} \tag{21.75}$$

where ν_{spec} is the fixed spectrometer frequency.

Different kinds of nuclei (^{1}H, ^{13}C, ^{19}F, etc.) have quite different g values, so their NMR absorption lines occur at very different frequencies. In a given NMR experiment, one examines the NMR spectrum of only one kind of nucleus. From here on, we shall consider only proton NMR spectra.

Each chemically different kind of proton in a molecule has a different value of σ_i, and we get a different NMR absorption frequency. Thus, C_6H_6 shows one NMR peak, C_6H_5Cl shows three NMR peaks, and CH_3CH_2Cl shows two NMR peaks (spin–spin coupling neglected; see below). The relative intensities of the peaks are proportional to the number of protons producing the absorption. For CH_3CH_2OH, the three peaks have a 3 : 2 : 1 ratio (Fig. 21.24).

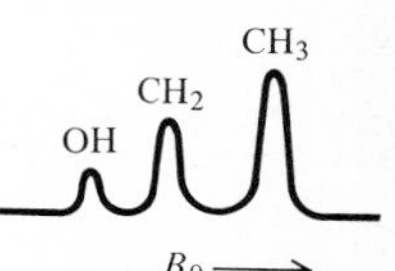

Figure 21.24 The low-resolution proton NMR spectrum of pure liquid CH_3CH_2OH.

The variation in ν_i in (21.74) or $B_{0,i}$ in (21.75) due to variation in the chemical (i.e., electronic) environment of the nucleus is called the chemical shift. The *chemical shift* δ_i of proton i is defined by

$$\delta_i \equiv (\sigma_{\text{ref}} - \sigma_i) \times 10^6 \tag{21.76}$$

where σ_{ref} is the shielding constant for the protons of the reference compound tetramethylsilane (TMS), $(CH_3)_4Si$; all the TMS protons are equivalent, and TMS shows a single proton NMR peak. The factor 10^6 is included to give δ a convenient magnitude. Note that σ and δ are dimensionless. Also, the proportionality constants σ_i and σ_{ref} in (21.76) are molecular properties that are independent of the applied field B_0 and the spectrometer frequency ν_{spec}. Hence, δ_i is independent of B_0 and ν_{spec}. The chemical shift δ can be expressed as (Prob. 21.38)

$$\delta_i = \frac{B_{0,\text{ref}} - B_{0,i}}{B_{0,\text{ref}}} \times 10^6 \tag{21.77}$$

where $B_{0,\text{ref}}$ and $B_{0,i}$ are the applied fields at which NMR absorption occurs for the reference nucleus and for nucleus i.

One finds that δ for the protons of a given kind of chemical group differs only slightly from compound to compound. Some typical proton δ values follow:

RCH_2R'	RCH_3	RNH_2	ROH	$RC(O)CH_2R'$	C≡CH	OCH_3	ArH	RC(O)H	RCOOH
1.1–1.5	0.8–1.2	1–4	2–6	2.5–2.9	2–3	3–4	6–9	9–11	10–13

where R and Ar are aliphatic and aromatic groups. Chemical shifts are affected by intermolecular interactions, so one usually observes the proton NMR spectrum of an organic compound in a dilute solution of an inert solvent (for example, CCl_4). The large δ range for alcohols is due to hydrogen bonding, the extent of which varies depending on the alcohol concentration.

NMR is an invaluable tool for structural determination, as you have no doubt seen in your organic-chemistry course.

Proton NMR spectra are more complex than we have so far indicated, due to the existence of nuclear *spin–spin coupling*. Each nucleus with spin $I \neq 0$ has a nuclear magnetic moment, and the magnetic field of this magnetic moment can affect the magnetic field experienced by a neighboring nucleus, thereby slightly changing the frequency at which the neighboring nucleus will undergo NMR absorption. Because of the rapid molecular rotation in liquids and gases, the direct nuclear spin–spin interaction averages to zero (*M. S.*, sec. 8.7). However, there is an additional, indirect interaction between the nuclear spins that is transmitted through the bonding electrons; this interaction is unaffected by molecular rotation and causes splitting of the NMR peaks. The magnitude of the indirect spin–spin interaction depends on the number of bonds between the nuclei involved. For protons separated by four or more bonds, this interaction is usually negligible. The magnitude of the spin–spin interaction between nuclei i and k is proportional to a quantity J_{ik}, called the *spin–spin coupling constant*; the interaction adds the term $hJ_{ik}(\hat{I}_{x,i}\hat{I}_{x,k} + \hat{I}_{y,i}\hat{I}_{y,k} + \hat{I}_{z,i}\hat{I}_{z,k})/\hbar^2$ to the NMR Hamiltonian operator, where $\hat{I}_{x,i}$ is the operator for the x component of the spin angular momentum of nucleus i. Since $I_{x,i}$ has the same dimensions as $\hbar$, the quantity hJ_{ik} must have units of energy; hence, J_{ik} has units of frequency.

Some typical proton–proton J values in hertz are:

HC—CH	C=CH$_2$	*cis*-HC=CH	*trans*-HC=CH	HCOH	HCC(O)H
5 to 9	−3 to +3	5 to 12	12 to 19	5 to 10	1 to 3

The nonequivalent protons in CH_3CH_2Br are separated by three bonds (H—C_a, C_a—C_b, and C_b—H), and J for these protons is 7.2 Hz. The nonequivalent protons in CH_3OCH_2Br are separated by four chemical bonds, and J is negligible for them.

The correct derivation of the NMR energy levels and frequencies allowing for spin–spin coupling requires a complicated quantum-mechanical treatment (often best done on a computer). Fortunately, for many compounds, a simple, approximate treatment allows the spectrum to be accurately calculated. This approximation is valid provided both the following conditions are satisfied: (1) The differences between NMR resonance frequencies of chemically different sets of protons are all much larger than the spin–spin coupling constants between nonequivalent protons. (2) There is only one coupling constant between any two sets of chemically equivalent protons.

The above table of δ values shows that in many cases the difference $\delta_i - \delta_j$ for chemically nonequivalent protons is equal to or greater than 1. Let us assume this difference equals 1.5. Then (21.76) gives $\sigma_i - \sigma_j = 1.5 \times 10^{-6}$. From (21.74), the difference between the NMR absorption frequencies of nuclei i and j is

$$\nu_i - \nu_j = h^{-1}\beta_N B_0 g_p(\sigma_j - \sigma_i) \tag{21.78}$$

Many commercially available proton NMR spectrometers have $\nu_{\text{spec}} = 60$ MHz; this corresponds to $B_0 = 1.41$ T (Prob. 21.34). Substitution of numerical values in (21.78) gives $\nu_i - \nu_j = 90$ Hz. This is substantially greater than J_{ij}, which is typically 10 Hz. Hence, condition 1 is met in many organic compounds. However, there are many cases where condition 1 is not met. For example, in CHR=CHR′, the protons have δ_i very close to δ_j, and the simple treatment cannot be used.

We shall illustrate the simple treatment by applying it to CH_3CH_2OH. Consider first the CH_3 protons. They are separated by four bonds from the OH proton, and so the spin–spin interaction between these two groups is negligible. Since the methyl protons are separated by three bonds from the CH_2 protons, the spin–spin interaction between CH_2 and CH_3 protons splits the CH_3 peak. One can prove from quantum mechanics that when the simple treatment applies, the spin–spin interactions between equivalent protons don't affect the spectrum; hence, we can ignore the spin–spin interactions between one methyl proton and another. Only the CH_2 protons affect the CH_3 peak.

Since $I = \frac{1}{2}$ for a proton, each CH_2 proton can have $M_I = +\frac{1}{2}$ or $-\frac{1}{2}$; we use up and down arrows to symbolize these proton spin states. The two CH_2 protons can have the following possible nuclear-spin alignments:

$$\underset{(a)}{\uparrow\uparrow} \qquad \underset{(b)}{\uparrow\downarrow} \qquad \underset{(c)}{\downarrow\uparrow} \qquad \underset{(d)}{\downarrow\downarrow} \tag{21.79}$$

Since the two CH_2 protons are indistinguishable, one actually takes symmetric and antisymmetric linear combinations of (*b*) and (*c*). States (*a*), (*d*), and the symmetric linear combination of (*b*) and (*c*) are analogous to the electron spin functions (19.32); the antisymmetric linear combination of (*b*) and (*c*) is analogous to (19.33). The two linear combinations of (*b*) and (*c*) each have a total M_I of 0. State (*a*) has a total M_I of 1. State (*d*) has a total M_I of -1. In a sample of ethanol, 25 percent of the molecules will have the CH_2 proton spins aligned as in (*a*), 50 percent as in (*b*) or (*c*), and 25 percent as in (*d*). Alignments (*b*) and (*c*) do not affect the magnetic field experienced by the CH_3 protons, whereas alignments (*a*) and (*d*) either increase or decrease this field. The CH_2 protons therefore split the CH_3 NMR absorption peak into a triplet (Fig. 21.25). The spacing between the lines of the triplet equals the coupling constant J between the CH_2 and CH_3 protons. [Although one usually varies the magnetic field and keeps the frequency fixed, observed splittings in teslas are converted to hertz by multiplying by $g_p\beta_N/h$; Eq. (21.73).] The preceding discussion shows that the intensity ratios of the members of the triplet is 1 : 2 : 1.

Now consider the CH_2 peak. The possible alignments of the CH_3 proton spins are

$$\underset{(a)}{\uparrow\uparrow\uparrow} \quad \underset{(b)}{\uparrow\uparrow\downarrow} \quad \underset{(c)}{\uparrow\downarrow\uparrow} \quad \underset{(d)}{\downarrow\uparrow\uparrow} \quad \underset{(e)}{\uparrow\downarrow\downarrow} \quad \underset{(f)}{\downarrow\uparrow\downarrow} \quad \underset{(g)}{\downarrow\downarrow\uparrow} \quad \underset{(h)}{\downarrow\downarrow\downarrow}$$

States (*b*), (*c*), and (*d*) have the same total M_I. States (*e*), (*f*), and (*g*) have the same total M_I. The CH_3 protons therefore act to split the CH_2 absorption into a quartet with 1 : 3 : 3 : 1 intensity ratios. The CH_2 protons are separated by three bonds from the OH proton, so we must also consider the effect of the OH proton. A trace of

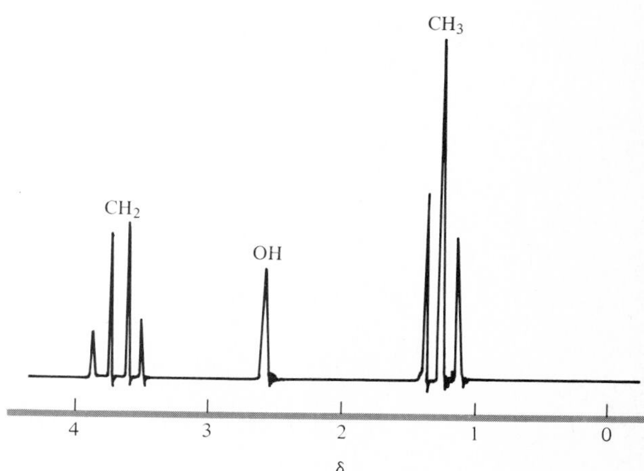

Figure 21.25
The high-resolution proton NMR spectrum of a dilute solution of CH_3CH_2OH in CCl_4 with a trace of acid. The different position of the OH resonance compared with that in Fig. 21.24 is explained by hydrogen bonding in pure liquid ethanol.

H_3O^+ or OH^- (including that coming from H_2O) will catalyze a rapid exchange of the OH protons between different ethanol molecules; this exchange eliminates the spin–spin interaction between the CH_2 protons and the OH proton, and the CH_2 peak remains a quartet. In pure ethanol, this exchange does not occur, and the OH proton acts to split each member of the CH_2 quartet into a doublet (corresponding to the OH proton spin states ↑ and ↓); the CH_2 absorption becomes an octet (eight lines) for pure ethanol. (These eight lines are so closely spaced that it may be difficult to resolve all of them.)

In ethanol containing a trace of acid or base, the OH proton NMR peak is a singlet. In pure ethanol, the OH absorption is split into a triplet by the CH_2 protons.

In general, n equivalent protons act to split the absorption peak of a set of adjacent protons into $n + 1$ lines, provided conditions 1 and 2 are met.

What about spin–spin splittings from other nuclei? Since ^{12}C, ^{16}O, and ^{32}S each have $I = 0$, these nuclei don't split proton NMR peaks. ^{14}N has $I = 1$; ^{35}Cl, ^{37}Cl, ^{79}Br, and ^{81}Br each have $I = \frac{3}{2}$; ^{127}I has $I = \frac{5}{2}$. It turns out that nuclei with $I > \frac{1}{2}$ generally don't split proton NMR peaks (see *M. S.*, sec. 8.10 for the reason why). ^{19}F has $I = \frac{1}{2}$ and does split proton NMR peaks.

For reasonably large organic compounds, the chances are good that two or more nonequivalent sets of protons will have similar δ values, making the simple treatment invalid. The spectrum then becomes very complicated and is difficult to interpret. To overcome this difficulty, one can use a spectrometer with higher values of B_0 and ν_{spec}. Note from (21.78) that $\nu_i - \nu_j$ increases as B_0 increases, so that at sufficiently high B_0, condition 1 is satisfied and the simple treatment applies. Values of ν_{spec} for commercially available proton NMR spectrometers include 60, 100, 220, and 300 MHz.

The use of NMR to detect cancer tumors is being investigated.

21.12 ELECTRON-SPIN-RESONANCE SPECTROSCOPY

In electron-spin-resonance (ESR) spectroscopy, one observes transitions between the quantum-mechanical energy levels of an unpaired electron spin magnetic moment in an external magnetic field. Most ground-state molecules have all electron spins paired, and such molecules show no ESR spectrum. One observes ESR spectra from free radicals such as H, CH_3, $(C_6H_5)_3C$, and $C_6H_6^-$, from transition-metal ions with unpaired electrons, and from excited triplet states of organic compounds.

An electron has spin-angular-momentum quantum numbers $s = \frac{1}{2}$ and $m_s = +\frac{1}{2}$ and $-\frac{1}{2}$. Relativistic quantum mechanics and experiment show that the electron g value is 2.0023, which we shall approximate as 2; thus, $g_e = 2$. Analogous to (21.66), the magnetic dipole moment of an electron is

$$\boldsymbol{\mu}_e = (-e/m)\mathbf{S} \tag{21.80}$$

where $-e$ is the electron charge, m is the electron mass, and $\mathbf{S}$ is the electron spin-angular-momentum vector; the magnitude of $\mathbf{S}$ is $[s(s + 1)]^{1/2}\hbar = \frac{1}{2}\sqrt{3}\,\hbar$. An electron spin magnetic moment has two energy levels in an applied field B, corresponding to the two orientations $m_s = +\frac{1}{2}$ and $-\frac{1}{2}$. The ESR transition frequency is given by the equation analogous to (21.73), namely,

$$\nu = 2\beta_e B h^{-1} \tag{21.81}$$

where the *Bohr magneton* β_e is defined analogously to (21.68) as

$$\beta_e \equiv e\hbar/2m = 9.274 \times 10^{-24}\ \text{J/T} \tag{21.82}$$

For a field of 1 T (10,000 G), Eq. (21.81) gives the ESR frequency as 28,000 MHz, which is in the microwave region. Since the electron mass is 1/2000th the proton mass, ESR frequencies are much higher than NMR frequencies.

Interaction between the electron spin magnetic moment and the nuclear spin magnetic moments splits the ESR absorption peak into several lines. (See *M. S.*, sec. 8.11 for details.)

ESR spectroscopy is a good way to detect free-radical intermediates in chemical reactions.

One can use ESR spectroscopy to obtain information on biological molecules by bonding an organic free radical to the macromolecule under study, a procedure called *spin labeling*. (See *Chang*, chap. 6.)

21.13 PHOTOCHEMISTRY

Photochemistry. *Photochemistry* is the study of chemical reactions produced by light. Absorption of a photon of light of sufficient energy may raise a molecule to an excited electronic state. In this high-energy state, it will be more likely to undergo a chemical reaction than in the ground electronic state. In *photochemical* reactions, the activation energy is supplied by absorption of light. In contrast, the reactions studied in Chap. 17 are *thermal* reactions, in which the activation energy is supplied by intermolecular collisions.

Equations (18.3) and (19.6) give the following energies E for a photon of wavelength λ (Prob. 21.41):

λ/nm	200 (UV)	400 (violet)	700 (red)	1000 (IR)
E/eV	6.2	3.1	1.8	1.2

(21.83)

Since it usually takes at least $1\frac{1}{2}$ or 2 eV to put a molecule into an excited electronic state, photochemical reactions are initiated by UV or visible light. The energy of 1 mole of photons (sometimes called an *einstein*) is $N_0 h\nu$. We find (Prob. 21.41):

λ/nm	200	400	700	1000
$N_0 h\nu$/(kcal mole^{-1})	143	71	41	29

Ordinarily, there is a one-to-one correspondence between the number of photons absorbed and the number of molecules making a transition to an excited electronic state; this is the *Stark–Einstein law* of photochemistry.

In exceptional circumstances, this law is violated. A high-power laser beam provides a very high density of photons; there is some probability that a molecule will be hit almost simultaneously by two laser-beam photons, so that one sometimes observes transitions in which a single molecule absorbs two photons at once. There are also unusual examples where a single photon can excite two molecules in contact with each other; see E. A. Ogryzlo, *J. Chem. Educ.*, **42**, 647 (1965).

Photochemical reactions are of tremendous biological importance. Most plant and animal life on earth depends on photosynthesis, a process in which green plants synthesize carbohydrates from CO_2 and water:

$$6CO_2 + 6H_2O \rightarrow C_6H_{12}O_6(\text{glucose}) + 6O_2 \tag{21.84}$$

The reverse of this reaction provides energy for plants, for animals that eat plants, and for animals that eat animals that eat plants. For (21.84), we find $\Delta G^\circ = +688$ kcal/mole, so the equilibrium lies far to the left in the absence of light. The presence of light and of the green pigment chlorophyll makes reaction (21.84) possible. Chlorophyll contains a conjugated ring system that allows it to absorb visible radiation; the main absorption peaks of chlorophyll are at 450 nm (blue) and 650 nm (red). Photosynthesis requires about eight photons per molecule of CO_2 consumed. Photosynthesis is a multistep process whose details are only partly understood.

The process of vision depends on photochemical reactions. One of the pigments in the retina is the compound rhodopsin, a combination of the protein opsin and the vitamin A–related compound retinene. Retinene contains five conjugated double bonds. Absorption of visible light by rhodopsin causes the retinene to isomerize and dissociate from the protein opsin. The mechanism of vision is only partly understood.

Other important photochemical reactions are the formation of ozone from O_2 in the earth's upper atmosphere, the formation of photochemical smog from automobile exhausts, the reactions in photography, and the formation of vitamin D and skin cancer by sunlight. There is some evidence that light striking the retina can affect the pineal gland (which is the legendary "third eye" of Eastern mythology).

Photochemical reactions are more selective than thermal reactions. By using monochromatic light, we can excite one particular species in a mixture to a higher electronic state. In contrast, heating a sample increases the translational energies of all species. Organic chemists use photochemical reactions as a tool in syntheses.

Certain chemical reactions produce products in excited electronic states; decay of these excited states may then produce emission of light, a process called *chemiluminescence*. Fireflies and many deep-sea fish show chemiluminescence. In a sense, chemiluminescence is the reverse of a photochemical reaction.

Consequences of light absorption. Let A* and A_0 denote an A molecule in an excited electronic state and in the ground electronic state, respectively. The initial absorption of radiation is $A_0 + h\nu \rightarrow A^*$. In most cases, the ground electronic state is a singlet with all electron spins paired. The selection rule (21.52) then shows that the excited electronic state A* is also a singlet.

Following light absorption, many things can happen.

The A* molecule is usually produced in an excited vibrational level [vibrational quantum number(s) > 0]. Intermolecular collisions (especially collisions with the solvent if the reaction is in solution) can transfer this extra vibrational energy to other molecules, causing A* to lose most of its vibrational energy and attain an equilibrium population of vibrational levels (a process called *vibrational relaxation*).

The A* molecule can lose its electronic energy by spontaneously emitting a photon, thereby falling to a lower singlet state, which may be the ground electronic state: $A^* \rightarrow A_0 + h\nu$. Emission of radiation by an electronic transition in which the total electronic spin doesn't change ($\Delta S = 0$) is called *fluorescence*. Fluorescence is favored in very low-pressure gases, where the time between collisions is relatively long. A typical lifetime of an excited singlet electronic state is 10^{-8} s in the absence of collisions.

The A* molecule can transfer its electronic excitation energy to another molecule during a collision, thereby returning to the ground electronic state, a process called *radiationless deactivation*: $A^* + B \rightarrow A_0 + B$, where A_0 and B on the right have extra translational, rotational, and vibrational energies.

The A* molecule (especially after undergoing vibrational relaxation) can make a radiationless transition to a different excited electronic state: $A^* \rightarrow A^{*\prime}$. Conservation of energy requires that A* and A*′ have the same energy. Generally, the molecule A*′ has a lower electronic energy and a higher vibrational energy than the A* molecule.

If A* and A*′ are both singlet states (or both triplet states), then the radiationless process $A^* \rightarrow A^{*\prime}$ is called *internal conversion*. If A* is a singlet electronic state and A*′ is a triplet (or vice versa), then $A^* \rightarrow A^{*\prime}$ is called *intersystem crossing*. (Recall that a triplet state has two unpaired electrons and total electronic spin quantum number $S = 1$.)

Suppose A*′ is a triplet electronic state. It can lose its electronic excitation energy and return to the ground electronic state during an intermolecular collision or by intersystem crossing to form A_0 in a high vibrational energy level. In addition, A*′ can emit a photon and fall to the singlet ground state A_0. Emission of radiation with $\Delta S \neq 0$ is called *phosphorescence*. Phosphorescence violates the selection rule (21.52) and has a very low probability of occurring. The lifetime of the lowest-lying excited triplet electronic state is typically 10^{-3} to 1 s in the absence of collisions.

Figure 21.26 summarizes the preceding processes.

Besides the above physical processes, absorption of light can cause several kinds of chemical processes.

Since A* is often formed in a high vibrational level, the A* molecule may have enough vibrational energy to dissociate: $A^* \rightarrow R + S$. The decomposition products R and S may react further, especially if they are free radicals. If A* is a diatomic molecule with vibrational energy exceeding the dissociation energy D_e of the excited electronic state, then dissociation occurs in the time it takes one molecular vibration to occur, 10^{-13} s. For a polyatomic molecule with sufficient vibrational energy to break a bond, dissociation may take a while to occur; there are many vibrational modes, and it requires time for vibrational energy to flow into the bond to be broken. Excitation of a diatomic molecule to a repulsive electronic state [one with no mini-

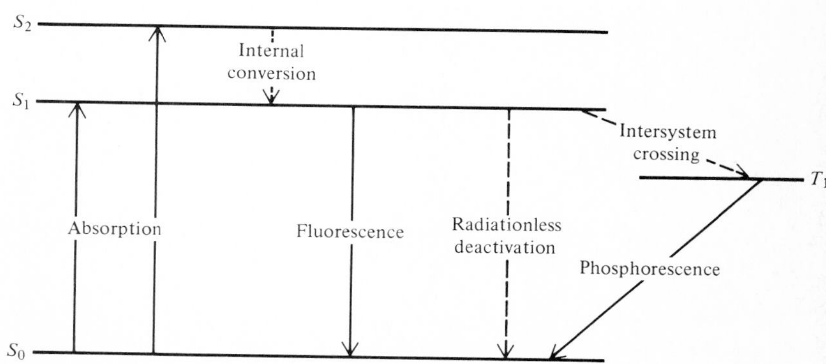

Figure 21.26
Photophysical processes. Dashed arrows indicate radiationless transitions. S_0 is the ground (singlet) electronic state; S_1 and S_2 are the lowest two excited singlet electronic states; T_1 is the lowest triplet electronic state. (For simplicity, vibration–rotation levels are omitted.)

mum in the $E_e(R)$ curve] always causes dissociation; excitation of a diatomic molecule to a bound excited electronic state with a minimum in the $E_e(R)$ curve causes dissociation if the vibrational energy of the excited molecule exceeds D_e. (See Fig. 21.27.) Sometimes a molecule undergoes internal conversion from a bound excited state to a repulsive excited electronic state, which then dissociates.

The vibrationally excited A* molecule may isomerize: A* → P. (Many cis–trans isomerizations can be carried out photochemically.)

The A* molecule may collide with a B molecule, the excitation energy of A* providing the activation energy for a bimolecular chemical reaction: A* + B → R + S.

The A* molecule may transfer its energy to another species C in a collision; species C then undergoing a chemical reaction. Thus, A* + C → A + C*, followed by C* + D → products; alternatively, A* + C → A + P + R. This process is *photosensitization.* The species A functions as a photochemical catalyst. An example is photosynthesis, where the photosensitizer is chlorophyll.

Of course, all these chemical processes can be preceded by internal conversion or intersystem crossing, A* → A*′, so that it is A*′ that undergoes the chemical process.

The many possible chemical and physical processes make it hard to deduce the precise sequence of events in a photochemical reaction.

Photochemical kinetics. The most common setup for kinetic study of a photochemical reaction exposes the sample to a continuous beam of nearly monochromatic radiation. (Of course, only radiation that is absorbed is effective in producing reaction. For example, exposing acetaldehyde to 400-nm radiation will have no effect, since radiation with wavelength less than 350 nm is required to excite acetaldehyde to a higher electronic level.) According to the Beer–Lambert law (21.11), the intensity I of radiation varies over the length of the reaction cell; convection currents (and perhaps stirring) are usually sufficient to maintain a near uniform concentration of reactants over the cell length, despite the variation of I.

As in any kinetics experiment, one follows the concentration of a reactant or product as a function of time. In addition, one measures the rate of absorption of light energy by comparing the energies reaching radiation detectors after the beam passes through two side-by-side cells, one filled with the reaction mixture and one empty (or filled with solvent only). One type of radiation detector is a photoelectric cell; the photocell is calibrated using a standard light source of known intensity.

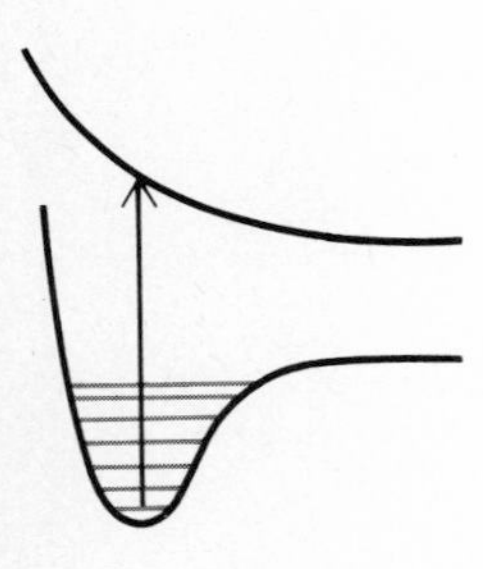

(*a*)

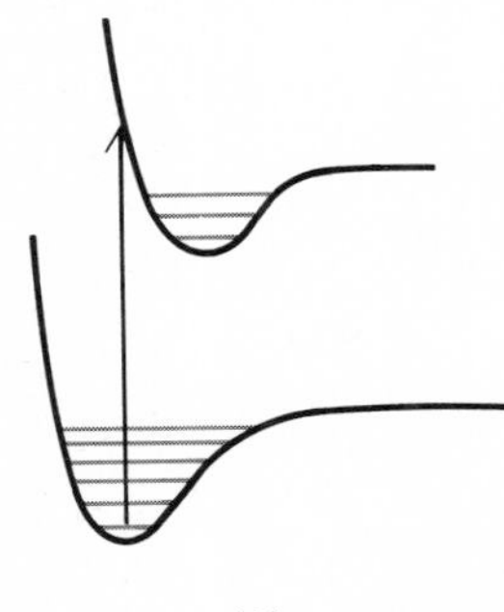

(*b*)

Figure 21.27
Electronic absorptions in a diatomic molecule that always lead to dissociation.

The initial step in a photochemical reaction is

$$(1) \qquad A + h\nu = A^* \tag{21.85}$$

For the elementary process (21.85), the rate per unit volume is $r_1 \equiv d[A^*]/dt$, where $[A^*]$ is the molar concentration of A^*. From the Stokes–Einstein law, r_1 equals $\mathscr{I}_a$, where $\mathscr{I}_a$ is defined as the number of moles of photons absorbed per second and per unit volume: $r_1 = \mathscr{I}_a$. (We assume that A is the only species absorbing radiation.) Let the reaction cell have length l, cross-sectional area $\mathscr{A}$, and volume $V = \mathscr{A}l$. Let I_0 and I_l be the intensities of the (monochromatic) beam as it enters the cell and as it leaves the cell, respectively. The intensity I is the energy that falls on unit cross-sectional area per unit time, so the radiation energy incident per second on the cell is $I_0 \mathscr{A}$ and the energy emerging per second is $I_l \mathscr{A}$. The energy absorbed per second in the cell is $I_0 \mathscr{A} - I_l \mathscr{A}$. Dividing by the energy $N_0 h\nu$ per mole of photons and by the cell volume, we get $\mathscr{I}_a$, the number of moles of photons absorbed per unit volume per second:

$$r_1 = \mathscr{I}_a = \frac{I_0 \mathscr{A} - I_l \mathscr{A}}{V N_0 h\nu} = \frac{I_0}{l N_0 h\nu}(1 - e^{-\alpha[A]l}) \tag{21.86}$$

where the Beer–Lambert law (21.11) was used. In (21.86), $\alpha = 2.303\varepsilon$, where ε is the molar absorption coefficient of A at the wavelength used in the experiment. $\mathscr{I}_a$ has units of moles (of photons) divided by (volume $\times$ time).

The *quantum yield* Φ_X of a photochemical reaction is the number of moles of product X formed divided by the number of moles of photons absorbed. Division of numerator and denominator by volume and time gives

$$\Phi_X = \frac{d[X]/dt}{\mathscr{I}_a} \tag{21.87}$$

Quantum yields vary from 0 to 10^6. Quantum yields less than 1 are due to deactivation of A^* molecules by the various physical processes discussed above and to recombination of fragments of dissociation. The quantum yield of the photochemical reaction $H_2 + Cl_2 \rightarrow 2HCl$ with 400-nm radiation is typically 10^5. Absorption of light by Cl_2 puts it into an excited electronic state that "immediately" dissociates to Cl atoms; the Cl atoms then start a chain reaction (Sec. 17.12), yielding many, many HCl molecules for each Cl atom formed.

As an example of photochemical kinetics, we consider the dimerization of anthracene ($C_{14}H_{10}$), which occurs when a solution of anthracene in benzene is irradiated with UV light. A simplified version of the accepted mechanism is

$$\begin{array}{lll} (1) & A + h\nu \rightarrow A^* & r_1 = \mathscr{I}_a \\ (2) & A^* + A \rightarrow A_2 & r_2 = k_2[A^*][A] \\ (3) & A^* \rightarrow A + h\nu' & r_3 = k_3[A^*] \\ (4) & A_2 \rightarrow 2A & r_4 = k_4[A_2] \end{array}$$

where A is anthracene. Step (1) is absorption of a photon by anthracene to raise it to an excited electronic state; Eq. (21.86) gives $r_1 = \mathscr{I}_a$. Step (2) is dimerization. Step (3) is fluorescence. Step (4) is a unimolecular decomposition of the dimer.

The rate r per unit volume for the overall reaction $2A \rightarrow A_2$ is

$$r = d[A_2]/dt = k_2[A^*][A] - k_4[A_2] \tag{21.88}$$

Application of the steady-state approximation to the reactive intermediate A* gives

$$d[\mathrm{A}^*]/dt = 0 = \mathscr{I}_a - k_2[\mathrm{A}][\mathrm{A}^*] - k_3[\mathrm{A}^*] \tag{21.89}$$

which gives $[\mathrm{A}^*] = \mathscr{I}_a/(k_2[\mathrm{A}] + k_3)$. Substitution in (21.88) gives

$$r = \frac{k_2[\mathrm{A}]\mathscr{I}_a}{k_2[\mathrm{A}] + k_3} - k_4[\mathrm{A}_2] \tag{21.90}$$

Note that $\mathscr{I}_a$ depends on [A] in a complicated way [Eq. (21.86)].

The quantum yield is given by (21.87) as

$$\Phi_{\mathrm{A}_2} = \frac{d[\mathrm{A}_2]/dt}{\mathscr{I}_a} = \frac{r}{\mathscr{I}_a} = \frac{k_2[\mathrm{A}]}{k_2[\mathrm{A}] + k_3} - \frac{k_4}{\mathscr{I}_a}[\mathrm{A}_2] \tag{21.91}$$

If $k_4 = 0$ (no reverse reaction) and $k_3 = 0$ (no fluorescence), then Φ becomes 1. The first fraction on the right can be written as $k_2/(k_2 + k_3/[\mathrm{A}])$; an increase in [A] increases Φ, since it increases r_2 (dimerization) compared with r_3 (fluorescence). This is the observed behavior. A typical Φ for this reaction is 0.2.

Instead of dealing with the individual rates of all the physical and chemical processes that follow absorption of radiation, one often adopts the simplifying approach of writing the initial step of the reaction as

$$\text{(I)} \qquad \mathrm{B} + h\nu \rightarrow \mathrm{R} + \mathrm{S} \qquad r_{\mathrm{I}} = \phi\mathscr{I}_a$$

Here, R and S are the first chemically different species formed following the absorption of radiation by B. Step I really summarizes several processes, namely, absorption of radiation by B to give B*, deactivation of B* by collisions and fluorescence, decomposition (or isomerization) of B* to R and S, and recombination of R and S immediately after their formation (recall the cage effect). The quantity ϕ, called the *primary quantum yield*, varies between 0 and 1. The greater the degree of collisional and fluorescent deactivation of B*, the smaller ϕ is. For absorption by gas-phase diatomic molecules that leads to dissociation, the dissociation is so rapid that deactivation is usually negligible and $\phi \approx 1$.

The photostationary state. Consider a chemical reaction in equilibrium in an isolated system. Suppose we now expose the system to a beam of radiation of fixed intensity and containing a frequency at which one of the reactants absorbs. This will increase the rate of the forward reaction, throwing the system out of equilibrium. Eventually a state will be reached in which the forward and reverse rates are again equal. This state will have a different composition from the original equilibrium state and is a *photostationary state.* It is a steady state (Sec. 1.2) rather than an equilibrium state because removal of the system from its surroundings (the radiation beam) will alter the system's properties. An important photostationary state is the ozone layer in the earth's atmosphere (Sec. 17.14).

As an example, let us consider the dimerization of anthracene, $2\mathrm{A} = \mathrm{A}_2$. In the absence of radiation, the rate of the forward reaction is negligible, but for educational purposes, let us assume a forward bimolecular reaction with rate constant k_5 in the absence of radiation:

$$\text{(5)} \qquad 2\mathrm{A} \rightarrow \mathrm{A}_2 \qquad r_5 = k_5[\mathrm{A}]^2$$

In the presence of radiation, reactions (1) to (5) all occur, where (1) to (4) are listed preceding (21.88). With inclusion of reaction (5), Eq. (21.88) becomes $r =$

$k_2[A^*][A] - k_4[A_2] + k_5[A]^2$. Reaction (5) does not involve the species A*, so (21.89) still holds and we get

$$r = \frac{k_2[A]\mathscr{I}_a}{k_2[A] + k_3} - k_4[A_2] + k_5[A]^2 \tag{21.92}$$

For the photostationary state in the presence of radiation, $r = 0$ and (21.92) gives

$$[A_2] = \frac{k_2[A]\mathscr{I}_a}{k_4 k_2[A] + k_3 k_4} + \frac{k_5[A]^2}{k_4} \tag{21.93}$$

In the absence of radiation, $\mathscr{I}_a = 0$ and (21.93) becomes $[A_2] = k_5[A]^2/k_4$, or $[A_2]/[A]^2 = k_5/k_4 = K$, where K is the equilibrium constant. The concentrations of A_2 at $r = 0$ differ in the presence and in the absence of radiation.

PROBLEMS

21.1 Find the frequency, wavelength, and wave number of light with photons of energy 1.00 eV per photon.

21.2 Find the speed and wavelength of sodium D light in water at 25°C. For data, see the material following Eq. (21.4).

21.3 Verify Eq. (21.7).

21.4 Write down an equation relating the transmittance T and the absorbance A.

21.5 Ethylene has a UV absorption peak at 162 nm with $\varepsilon = 1.0 \times 10^4$ dm^3 mol^{-1} cm^{-1}. Calculate the transmittance of 162 nm radiation through a sample of ethylene gas at 25°C and 10 torr for a cell length of (*a*) 1.0 cm; (*b*) 10 cm.

21.6 Methanol has a UV absorption peak at 184 nm with $\varepsilon = 150$ dm^3 mol^{-1} cm^{-1}. Calculate the transmittance of 184 nm radiation through a 0.001 mol dm^{-3} solution of methanol in a nonabsorbing solvent for a cell length of (*a*) 1.0 cm; (*b*) 10 cm.

21.7 For a 6.0×10^{-4} mol dm^{-3} solution of a certain compound, one finds that 21.0 percent of the incident radiation at 3600 Å is transmitted through a 1.00-cm-long absorption cell. Calculate the absorbance and the molar absorption coefficient at 3600 Å.

21.8 A solution of 2.00 g of a compound transmits 60.0 percent of the 4300-Å light incident on a 3.00-cm-long cell. What percent of 4300-Å light will be transmitted by a solution of 4.00 g of this compound in this same cell?

21.9 At 330 nm, the ion $Fe(CN)_6^{3-}(aq)$ has $\varepsilon = 800$ dm^3 mol^{-1} cm^{-1}, and $Fe(CN)_6^{4-}(aq)$ has $\varepsilon = 320$ dm^3 mol^{-1} cm^{-1}. The reduction of $Fe(CN)_6^{3-}$ to $Fe(CN)_6^{4-}$ is being followed spectrophotometrically in a 1.00-cm-long cell. The solution has an initial $Fe(CN)_6^{3-}$ concentration of 1.00×10^{-3} mol dm^{-3} and no $Fe(CN)_6^{4-}$. After 340 s, the absorbance is 0.701. Calculate the percent of $Fe(CN)_6^{3-}$ that has reacted.

21.10 Verify that 1 mdyn/Å = 100 N/m.

21.11 Use data in Table 21.1 to calculate D_0 for the ground electronic state of (*a*) $^{14}N_2$; (*b*) $^{12}C^{16}O$.

21.12 (*a*) Explain why D_e and k_e for $D^{35}Cl$ are essentially the same as D_e and k_e for $H^{35}Cl$ but D_0 for these two species differs. (*b*) Use data in Table 21.1 to calculate D_0 for each of these two species. Neglect the difference in $\tilde{\nu}_e x_e$ for the two species.

21.13 The $J = 0 \to 1$, $v = 0 \to 0$ transition for $^{12}C^{16}O$ and $^{13}C^{16}O$ occurs at 115,271 and 110,201 MHz, respectively. Calculate the bond distances R_0 in each of these molecules. (Neglect centrifugal distortion.)

21.14 The $J = 2 \to 3$ pure-rotational transition for the ground vibrational state of $^{39}K^{37}Cl$ occurs at 22,410 MHz. Neglecting centrifugal distortion, predict the frequency of the $J = 0 \to 1$ pure-rotational transition of (*a*) $^{39}K^{37}Cl$; (*b*) $^{39}K^{35}Cl$.

21.15 Verify Eqs. (21.37) and (21.38).

21.16 For $^{16}O_2$, $\tilde{\nu}_e = 1580$ cm^{-1}. Find k_e for $^{16}O_2$.

21.17 (*a*) From the IR data following Eq. (21.36), calculate $\tilde{\nu}_e$ and $\tilde{\nu}_e x_e$ for $^1H^{35}Cl$. (*b*) Use the results of (*a*) to predict σ_0 for the $v = 0 \to 6$ transition of this molecule.

21.18 Use data in Table 21.1 to calculate the relative populations at 300 K of the $J = 0$ through $J = 6$ rotational levels of the $v = 0$ vibrational level of $^1H^{35}Cl$.

21.19 List all the symmetry elements present in (*a*) H_2S; (*b*) CF_3Cl; (*c*) XeF_4; (*d*) PCl_5; (*e*) IF_5; (*f*) benzene; (*g*) *p*-dibromobenzene.

21.20 Without doing any calculations, describe as fully as you can the locations of the principal axes of inertia of (*a*) BF_3; (*b*) H_2O; (*c*) CO_2.

21.21 Classify each of these as a spherical, symmetric, or asymmetric top: (*a*) SF_6; (*b*) IF_5; (*c*) H_2S; (*d*) PF_3; (*e*) benzene; (*f*) CO_2; (*g*) $^{35}ClC^{37}Cl_3$.

21.22 For CF_3I, the rotational constants are $\tilde{A} = 0.1910\ cm^{-1}$ and $\tilde{B} = 0.05081\ cm^{-1}$. (*a*) Calculate E_{rot}/h for the $J = 0$ and $J = 1$ rotational levels. (*b*) Calculate the two lowest microwave absorption frequencies.

21.23 Give the number of normal vibrational modes of (*a*) SO_2; (*b*) C_2F_2; (*c*) CCl_4.

21.24 In addition to the values listed in Sec. 21.7, some other σ_0 values for IR absorption bands of H_2O vapor are 7252 and 8807 cm^{-1}. Give the upper vibrational level for each of these bands. (The lower level is 000 in both cases.)

21.25 Use data in Sec. 21.7 to calculate the zero-point vibrational energy of (*a*) CO_2; (*b*) H_2O.

21.26 From the CH and C=O stretching frequencies listed in Sec. 21.8 estimate the force constants for stretching vibrations of these bonds.

21.27 Calculate the wavelength of the series limit of the Balmer lines of the hydrogen-atom spectrum.

21.28 Calculate the wavelengths of the first three lines in the Paschen series of the hydrogen-atom spectrum.

21.29 Calculate the wavelength of the $n = 2 \rightarrow 1$ transition in Li^{2+}.

21.30 Calculate the force on an electron moving at 3.0×10^8 cm/s through a magnetic field of 1.5 T if the angle between the electron's velocity vector and the magnetic field is (*a*) 0°; (*b*) 45°; (*c*) 90°; (*d*) 180°.

21.31 Calculate the magnetic dipole moment of a particle with charge 2.0×10^{-16} C moving on a circle of radius 25 Å with speed 2.0×10^7 cm/s.

21.32 The nucleus ^{11}B has $I = \frac{3}{2}$ and $g_N = 1.792$. Calculate the energy levels of a ^{11}B nucleus in a magnetic field of (*a*) 1.50 T; (*b*) 15,000 G.

21.33 Use data in Prob. 21.32 to calculate the NMR absorption frequency of ^{11}B in a magnetic field of (*a*) 1.50 T; (*b*) 2.00 T.

21.34 Calculate the value of B in a proton magnetic resonance spectrometer that has ν_{spec} equal to (*a*) 60 MHz; (*b*) 300 MHz.

21.35 Calculate the ratio of the populations of the two nuclear-spin energy levels of a proton in a field of 1.41 T at 25°C.

21.36 For an applied field of 1.41 T (the field used in a 60-MHz spectrometer), calculate the difference in NMR absorption frequencies for two protons whose δ values differ by 1.0.

21.37 For each of the following state how many proton NMR peaks occur, the relative intensity of each peak, and whether each peak is a singlet, doublet, triplet, etc.: (*a*) benzene; (*b*) acetaldehyde; (*c*) C_2H_6; (*d*) $CH_3CH_2OCH_2CH_3$; (*e*) $(CH_3)_2CHBr$; (*f*) methyl acetate.

21.38 Verify Eq. (21.77). (Use the fact that $\sigma \ll 1$.)

21.39 Verify that 1 T corresponds to 10^4 G.

21.40 Calculate the electron-spin-resonance frequency in a field of 25,000 G.

21.41 Verify the calculations in (21.83) and in the table following it.

21.42 In a certain photochemical reaction using 464 nm radiation, the incident-light power was 0.00155 W, and the system absorbed 74.4 percent of the incident light; 6.80×10^{-6} mole of product was produced during an exposure of 110 s. Find the quantum yield.

21.43 The photochemical decomposition of HI proceeds by the following mechanism:

$$HI + h\nu \rightarrow H + I$$

$$H + HI \rightarrow H_2 + I$$

$$I + I + M \rightarrow I_2 + M$$

where the rate of the first step is $\phi \mathscr{I}_a$ with $\phi = 1$. (*a*) Show that $-d[HI]/dt = 2\mathscr{I}_a$. Hence the quantum yield with respect to HI is 2. (*b*) How many HI molecules will be decomposed when 1.00 kcal of 250 nm radiation is absorbed?

21.44 A polymer solution scatters light, due to random fluctuations in the local concentrations of polymer molecules. The intensity I of transmitted light at a given wavelength λ is given by an equation similar to the Beer–Lambert law (21.11): $I = I_0 e^{-\tau x}$, where I_0 is the intensity of incident light of wavelength λ and τ is the *turbidity* of the solution for wavelength λ.

Let $\rho_B \equiv m_B/V$ be the mass concentration of polymer in solution, where m_B is the mass of polymeric material and V is the solution volume. Debye showed that when the size of the polymer molecule is substantially less than the wavelength of the light being scattered, then [P. Debye, *J. Phys. Colloid Chem.*, **51**, 18 (1947)]:

$$\frac{H\rho_B}{\tau} = \frac{1}{M_{w,B}} + a\rho_B + b\rho_B^2 + \cdots,$$

$$H \equiv \frac{32\pi^3 n_A}{3\lambda^4 N_0}\left(\frac{dn}{d\rho_B}\right)^2$$

where n_A and n are the refractive indices of the pure solvent A and of the polymer solution, $M_{w,B}$ is the polymer weight average molar mass, and a and b are constants. A plot of $H\rho_B/\tau$ versus ρ_B in dilute solutions is a straight line with intercept $1/M_{w,B}$. (When the size of the polymer molecule is comparable to λ, a more complicated procedure must be used to find M_w.)

For an aqueous solution of the enzyme lysozyme at 25°C, observations using light with $\lambda = 436$ nm gave $dn/d\rho_B = 0.1955$ cm^3/g, $n_A = 1.337$, and the following τ vs. ρ_B values:

$10^4\tau/\text{cm}^{-1}$	4.80	9.92	15.54
$\rho_B/(\text{g/dm}^3)$	4.00	8.00	12.00

Calculate the molecular weight of lysozyme. (Lysozyme occurs in tears.)

22

STATISTICAL MECHANICS

22.1 STATISTICAL MECHANICS

The aim of *statistical mechanics* is to deduce the macroscopic properties of matter from the properties of the molecules composing the system. Typical macroscopic properties are entropy, internal energy, heat capacity, surface tension, dielectric constant, viscosity, electrical conductivity, and chemical reaction rate. Molecular properties include molecular masses, molecular geometries, intramolecular forces (which determine the molecular vibration frequencies), and intermolecular forces. Because of the huge number of molecules in a macroscopic system, one uses statistical methods instead of attempting to consider the motion of each molecule in the system.

This chapter is restricted to *equilibrium statistical mechanics* (also called *statistical thermodynamics*), which deals with systems in thermodynamic equilibrium. *Nonequilibrium statistical mechanics* (whose theory is not as well developed as that of equilibrium statistical mechanics) deals with transport properties and chemical reaction rates. Very crude treatments of transport properties were given in Chap. 16. A statistical-mechanical theory of reaction rates is given in Chap. 23. Application of equilibrium statistical mechanics to solids is considered in Chap. 24.

Statistical mechanics originated in the work of Maxwell and Boltzmann on the kinetic theory of gases (1860–1900). Major advances in the theory and the methods of practical calculation were made by Gibbs in his 1902 book *Elementary Principles in Statistical Mechanics* and by Einstein in a series of papers (1902–1904). Since quantum mechanics had not yet been discovered, these workers assumed that the system's molecules obeyed classical mechanics. This led to incorrect results in some cases; e.g., calculated heat capacities of polyatomic gases did not agree with experiment (Sec. 15.10). When quantum mechanics was discovered, the necessary modifications in statistical mechanics were easily made.

Statistical mechanics deals with both the microscopic (molecular) level and the macroscopic level, and it is important to define our terminology clearly. The word *system* in this chapter refers only to a macroscopic thermodynamic system. The fundamental microscopic entities that compose a (thermodynamic) system will be called *molecules* or *particles*. (In some cases, these entities are not actually molecules; e.g., one can apply statistical mechanics to the conduction electrons in a metal or to the photons in electromagnetic radiation.)

The term *state of a system* has two meanings in statistical mechanics. The *thermodynamic state* of a system is specified by giving the values of enough macroscopic parameters to characterize the system. (For example, we might have 24.0 g of benzene plus 2.87 g of toluene at 72°C and 3.65 atm; or we might have 18.0 g of H_2O at 54°C in a volume of 17.2 cm^3.) On the other hand, we can talk about the *quantum state* of the system. By this we mean the following. Consider as an example the 18.0 g of H_2O at 54°C and 17.2 cm^3. We set up and solve the Schrödinger equation for the system composed of the 6×10^{23} H_2O molecules to obtain a set of allowed wave functions and energy levels. (In practice, this is an impossible task, but we can imagine doing so in principle.) At any instant of time, the system will be in a definite quantum state j characterized by a certain wave function ψ_j (which is a function of a huge number of spatial and spin coordinates), an energy E_j, and a set of quantum numbers. (Actually, the system might well be in a nonstationary, time-dependent state, but to simplify the arguments, we shall not consider this possibility.)

The term *macrostate* means the thermodynamic state of a system. The term *microstate* means the quantum state of a system. Because of the very large number of particles in a system, there are a huge number of different microstates that are compatible with a given macrostate. (For example, suppose we have a fixed amount of an ideal monatomic gas whose internal energy and volume are fixed. The translational energy levels (18.40) lie extremely close together [see the discussion after (21.26)], and there are a huge number of ways we can populate these levels and still end up with the same total energy.) The macrostate is experimentally observable; the microstate is usually not observable.

If the molecules of the system do not interact with one another, we can also refer to the quantum states available to each molecule. We call these the *molecular states*. For example, consider a pure ideal monatomic gas. Its macrostate can be specified by giving the thermodynamic variables T, P, and n (or the variables T, V, and n, or some other set of three thermodynamic properties). Its microstate is specified by saying how many molecules are in each of the available molecular translational quantum states (18.39), where n_x, n_y, and n_z each go from 1 on up. A molecular state is specified by giving the values of three numbers: n_x, n_y, n_z. In contrast, it takes a huge set of numbers to specify the system's microstate.

We now proceed to relate macrostates to microstates, in order to calculate macroscopic properties from molecular properties.

22.2 THE CANONICAL ENSEMBLE

Suppose we measure a macroscopic property of an equilibrium system, e.g., the pressure. It takes time to make the measurement, and the observed pressure is a time average over the impacts of individual molecules on the walls. [Equation (15.57) shows that for a gas at 1 atm there are about 10^{17} impacts every microsecond on a 1-cm^2 wall area.] To calculate the value of the macroscopic property, we would have to take a time average over the changes in the microstate of the system. In some simple cases, the time-averaging calculation can be performed—recall the kinetic-theory calculation of ideal-gas pressure in Sec. 15.2. In general, however, it is not feasible to do a time-average calculation. Instead, one resorts to what is called an *ensemble*.

The canonical ensemble. An ensemble is a hypothetical collection of an infinite number of noninteracting systems, each of which is in the same macrostate (thermodynamic state) as the system of interest. Although the members of the ensemble are macroscopically identical, they show a wide variety of microstates (since many different microstates are compatible with a given macrostate). It is postulated that:

The measured time average of a macroscopic property in the system of interest is equal to the average value of that property in the ensemble.

Consider a single equilibrium thermodynamic system whose volume, temperature, and composition are held fixed. The system is enclosed in walls that are rigid (V constant), impermeable (composition constant), and thermally conducting, and the system is immersed in an extremely large constant-temperature bath. Due to thermal interactions with the bath, the system's microstate changes from moment to moment, and its pressure and quantum-mechanical energy fluctuate. Of course, these fluctuations are generally far too small to be detectable macroscopically, because of the large number of molecules in the system (Sec. 3.7).

We want to calculate such thermodynamic properties as pressure, internal energy, and entropy for this system. To do so, we imagine an ensemble, each of whose systems has the same temperature, volume, and composition as the system of interest; each system of the ensemble sits in a very large constant-temperature bath (Fig. 22.1). An ensemble of systems each having fixed T, V, and composition is called a *canonical ensemble*. (Canonical means standard, basic, authoritative. The name was chosen by Gibbs.)

To make things less abstract, consider an example. Suppose we are interested in calculating the thermodynamic properties of 1.00 millimole of H_2 at 0°C in a 2.00-dm^3 box:

$$1.00 \text{ mmole } H_2(273 \text{ K}, 2.00 \text{ dm}^3) \tag{22.1}$$

We would imagine an infinitely large number of macroscopic copies of this system, each copy containing 1.00 mmol of H_2 in a 2.00-dm^3 box that is immersed in a very large bath at 0°C.

We shall take an average at a fixed time over the microstates of the systems in the ensemble. The possible microstates are found by solving the Schrödinger equation $\hat{H}\psi_j = E_j\psi_j$ for the system. The possible wave functions ψ_j ($j = 1, 2, \ldots$) and

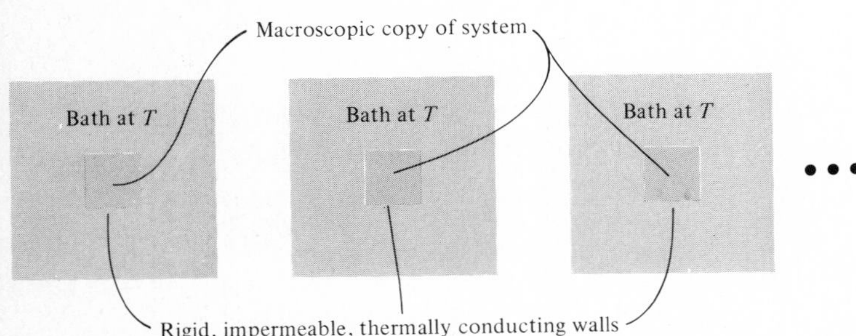

Figure 22.1
A canonical ensemble. (The dots indicate the infinite number of systems and baths.)

quantum energies E_j will depend on the composition of the system (since the number of molecules and the intermolecular and intramolecular forces depend on the composition) and will depend on the system's volume [recall from (18.42) that the translational energy of a particle in a cubic box depends on the volume of the box]. We have

$$E_j = E_j(V, N_A, N_B, \ldots) \qquad (22.2)$$

where the system's composition is specified by giving N_A, N_B, ..., the number of molecules of each chemical species A, B, ... in the system. (Although our procedure is applicable to a multiphase system, for simplicity we shall consider only one-phase systems.) Note that E_j does not depend on the system's temperature. Temperature is a macroscopic nonmechanical property and does not appear in the quantum-mechanical Hamiltonian of the system or in the condition that the wave functions be well behaved.

We postulated above that any macroscopic property of the system is to be calculated as an average over the ensemble at a fixed time. For example, the thermodynamic internal energy U will equal the average energy of the systems in the ensemble: $U = \langle E_j \rangle$. (Because of the above-mentioned fluctuations, E_j will not be precisely the same for different systems in the ensemble.) Similar to Eq. (15.40), we write

$$U = \langle E_j \rangle = \sum_j p_j E_j \qquad (22.3)$$

where p_j is a probability. There are two ways this equation can be interpreted. If the sum is taken over the different possible energy values E_j of the system, then p_j is the probability that a system in the ensemble has energy E_j. Alternatively, if the sum is taken over the different possible quantum states of the system, then p_j is the probability that a system is in the microstate j (whose energy is E_j). We shall use the second interpretation. Thus, the *symbol* $\sum_j$ *indicates a sum over the allowed quantum states of the system* (and not a sum over energy levels); p_j *is the probability that the system is in the quantum state* j (and not the probability that the system has energy E_j). A sum over states differs from a sum over energy levels because there may be several different quantum states that have the same energy, a condition called degeneracy (Sec. 18.10). For the system (22.1), each system energy level is highly degenerate: there are 6×10^{20} molecules, each of which has translational energies in the x, y, and z directions, rotational energy, and vibrational energy; there are a huge number of ways a fixed amount of energy can be distributed among the various molecular quantum states.

To find U in (22.3), we need the probabilities p_j (and the quantum-mechanical energies E_j). We postulate that:

For a thermodynamic system of fixed volume, composition, and temperature, all quantum states that have equal energy have equal probability of occurring.

(This postulate and the postulate of equality of ensemble averages and time averages are the two fundamental postulates of statistical mechanics.) If microstates i and k have the same energy ($E_i = E_k$), then p_i equals p_k. Hence, p_j in (22.3) can depend only on the energy of state j:

$$p_j = f(E_j) \qquad (22.4)$$

Evaluation of p_j. We now find how the probability p_j of being in microstate j depends on the energy E_j of the microstate. To find the function f, imagine that a second system of fixed volume, temperature, and composition is put into each bath of the ensemble (Fig. 22.2). Let I and II denote the two systems in each bath. All the systems labeled I are macroscopically identical to one another. All the systems labeled II are macroscopically identical. However, systems I and II are not necessarily identical to each other; they can differ in volume and composition. We have

$$p_{\mathrm{I},j} = f(E_{\mathrm{I},j}) \qquad \text{and} \qquad p_{\mathrm{II},k} = g(E_{\mathrm{II},k}) \tag{22.5}$$

where $p_{\mathrm{I},j}$ is the probability that system I is in the microstate j (whose energy is $E_{\mathrm{I},j}$) and $p_{\mathrm{II},k}$ is the probability that system II is in the microstate k (whose energy is $E_{\mathrm{II},k}$), and where f and g are (at this point) unknown functions. Since systems I and II may differ, f and g are not necessarily the same function.

If we like, we can consider I and II to form a single composite system (I + II) for which

$$p_{\mathrm{I+II},i} = h(E_{\mathrm{I+II},i}) \tag{22.6}$$

where $p_{\mathrm{I+II},i}$ is the probability that system I + II is in the microstate i (whose energy is $E_{\mathrm{I+II},i}$) and h is some function. The subsystems I and II are independent of each other, so $E_{\mathrm{I+II},i} = E_{\mathrm{I},j} + E_{\mathrm{II},k}$. Also, since the probability that two independent events both happen is the product of the probabilities of each event, we have $p_{\mathrm{I+II},i} = p_{\mathrm{I},j}\,p_{\mathrm{II},k}$. Substitution of (22.5) and (22.6) gives

$$h(E_{\mathrm{I},j} + E_{\mathrm{II},k}) = f(E_{\mathrm{I},j})g(E_{\mathrm{II},k}) \tag{22.7}$$

Equation (22.7) has the form

$$h(x + y) = f(x)g(y) \qquad \text{where } x \equiv E_{\mathrm{I},j},\ y \equiv E_{\mathrm{II},k} \tag{22.8}$$

$$h(z) = f(x)g(y) \qquad \text{where } z \equiv x + y \tag{22.9}$$

We solve (22.9) for f by the same procedure used to derive (15.34) from (15.30). Partial differentiation of (22.9) with respect to x and with respect to y gives

$$\frac{dh}{dz} = \frac{df(x)}{dx}g(y) \qquad \text{and} \qquad \frac{dh}{dz} = f(x)\frac{dg(y)}{dy}$$

[where we used $\partial h/\partial x = (dh/dz)(\partial z/\partial x)$, $\partial h/\partial y = (dh/dz)(\partial z/\partial y)$, and $\partial z/\partial x = 1 = \partial z/\partial y$]. Hence, $f'(x)g(y) = f(x)g'(y)$, and

$$\frac{g'(y)}{g(y)} = \frac{f'(x)}{f(x)} \equiv -\beta \tag{22.10}$$

where β was defined as $-f'(x)/f(x)$. Since the function $f'(x)/f(x)$ is independent of y, β is independent of y. The left side of (22.10) is independent of x, so β is independent

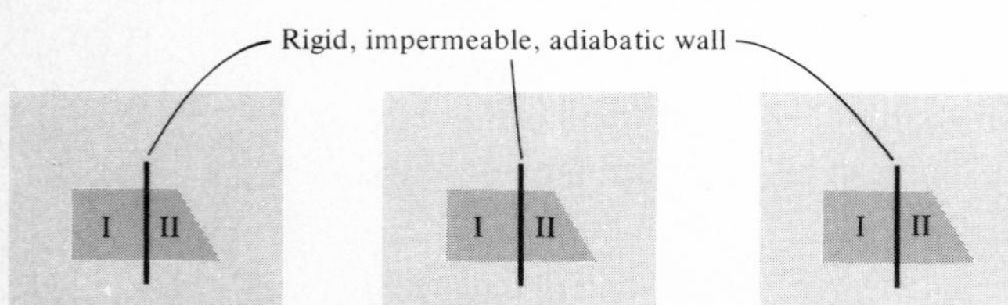

Figure 22.2
A canonical ensemble for a composite system composed of two noninteracting parts I and II.

of x. Hence, β is a constant. We have $df(x)/f(x) = -\beta\,dx$, which integrates to $\ln f = -\beta x + \text{const}$; thus $f = e^{-\beta x}e^{\text{const}} = ae^{-\beta x}$, where $a \equiv e^{\text{const}}$ is a constant. Similarly, $g'(y)/g(y) = -\beta$ integrates to $g(y) = ce^{-\beta y}$, where c is another constant. We have

$$f(x) = ae^{-\beta x} \quad \text{and} \quad g(y) = ce^{-\beta y} \tag{22.11}$$

Use of Eqs. (22.5) and (22.8) in (22.11) gives

$$p_{\text{I},j} = ae^{-\beta E_{\text{I},j}} \quad \text{and} \quad p_{\text{II},k} = ce^{-\beta E_{\text{II},k}} \tag{22.12}$$

To complete things, we must find a, β, and c. We know that these quantities are constants for a system of fixed temperature, volume, and composition. Hence, they might depend on one or more of T, V, and composition. However, they cannot depend on the microstate energies $E_{\text{I},j}$ (or $E_{\text{II},k}$), since β, a, and c are independent of x and y in the above derivation and x and y are $E_{\text{I},j}$ and $E_{\text{II},k}$.

Consider β. Equation (22.10) or (22.12) shows that any two systems I and II in the same constant-temperature bath have the same value of β. Two systems in the same bath have a common value of T but can differ in volume and in composition. We conclude that β must be a function of T only: $\beta = \phi(T)$, where ϕ is the same function for any two systems. In contrast, a can depend on all three of temperature, volume, and composition.

Use of two systems was a temporary device, and we now go back to the ensemble with a single system in each bath. We thus write (22.12) as

$$p_j = ae^{-\beta E_j} \tag{22.13}$$

where β is a function of T and a is a function of T, V, and composition.

To evaluate a, we use the fact that the total probability must be 1: $\sum_j p_j = 1 = a \sum_j e^{-\beta E_j}$, and

$$a = \frac{1}{\sum_j e^{-\beta E_j}} \tag{22.14}$$

Equation (22.13) becomes

$$p_j = \frac{e^{-\beta E_j}}{\sum_k e^{-\beta E_k}} \equiv \frac{e^{-\beta E_j}}{Z} \tag{22.15}$$

The summation index is a dummy variable (Sec. 2.1), so any letter can be used for it. To avoid later confusion, j in (22.14) was changed to k.

The sum in (22.15) plays a star role in statistical mechanics and is called the *canonical partition function* of the system (or the partition function at constant T, V, and composition):

$$Z \equiv \sum_j e^{-\beta E_j} \qquad (22.16)^*$$

where the sum goes over all possible quantum states of the system for a given composition and volume. E_j is the quantum-mechanical energy of the macroscopic system when it is in microstate j. From (22.15), the terms $e^{-\beta E_j}$ in the partition function govern how the systems of the ensemble are distributed or "partitioned" among the possible quantum states of the system.

Having evaluated a, we turn our attention to β. We know that $\beta = \phi(T)$. What is the function ϕ? One way to answer this is to recall (Sec. 1.3) that the choice of temperature scale is arbitrary. Hence we could, if we like, define a statistical-mechanical temperature T_{sm} by choosing some convenient functional form for ϕ. Instead of doing so, we shall stick with the ideal-gas thermodynamic scale of absolute T and determine ϕ as a function of the thermodynamic T.

Evaluation of U and P. To help find $\phi(T)$, we shall get expressions for U and P, the internal energy and the pressure. Equations (22.3) and (22.15) give

$$U = \sum_j p_j E_j = \frac{\sum_j E_j e^{-\beta E_j}}{\sum_k e^{-\beta E_k}} = \frac{\sum_j E_j e^{-\beta E_j}}{Z} \tag{22.17}$$

The canonical partition function Z is a function of β and the quantum-mechanical energy levels E_j, which depend on V and composition [Eq. (22.2)]. Hence

$$Z = Z(\beta, V, N_{\text{A}}, N_{\text{B}}, \ldots) = Z(T, V, N_{\text{A}}, N_{\text{B}}, \ldots) \tag{22.18}$$

since β is a function of T. Partial differentiation of (22.16) gives

$$\left(\frac{\partial Z}{\partial \beta}\right)_{V,N_{\text{B}}} = \sum_j \left(\frac{\partial e^{-\beta E_j}}{\partial \beta}\right)_{V,N_{\text{B}}} = -\sum_j E_j e^{-\beta E_j} \tag{22.19}$$

since E_j is independent of T and hence of β. In (22.19) the subscript N_{B} indicates constant values of each of the composition variables $N_{\text{A}}, N_{\text{B}}, N_{\text{C}}, \ldots$. Equation (22.17) can thus be written as

$$U = -\frac{1}{Z}\left(\frac{\partial Z}{\partial \beta}\right)_{V,N_{\text{B}}} = -\left(\frac{\partial \ln Z}{\partial \beta}\right)_{V,N_{\text{B}}} \tag{22.20}$$

which is the desired expression for U.

Now consider the system's pressure P. (Don't confuse P for pressure with p for probability.) Just as there are energy fluctuations among systems of the ensemble, there are also pressure fluctuations. Let P_j be the pressure in a system whose microstate is j. Our averaging postulate gives, similar to (22.3),

$$P = \sum_j p_j P_j \tag{22.21}$$

What is P_j? Consider a single adiabatically enclosed system in quantum state j with pressure P_j and quantum energy E_j. Its thermodynamic energy is $U = E_j$. Imagine a reversible change in volume by dV while the system remains in state j. The quantum-mechanical expression for the energy change is $(\partial E_j/\partial V)_{N_{\text{B}}}\,dV$. [For example, consider a system consisting of two noninteracting distinguishable particles 1 and 2 having only translational energy and confined to a cubic box of volume V. The system's quantum energy E_j is the sum of the energies of the noninteracting particles; the particle-in-a-box formula (18.42) gives

$$E_j = \frac{h^2}{8mV^{2/3}}(n_{x,1}^2 + n_{y,1}^2 + n_{z,1}^2 + n_{x,2}^2 + n_{y,2}^2 + n_{z,2}^2)$$

$$\left(\frac{\partial E_j}{\partial V}\right)_{N_{\text{B}}} dV = -\frac{2}{3}\,\frac{h^2}{8mV^{5/3}}(n_{x,1}^2 + \cdots + n_{z,2}^2)\,dV$$

where, since the system's quantum state doesn't change, the quantum numbers $n_{x,1}, \ldots, n_{z,2}$ of the two particles stay fixed. (Of course, actual thermodynamic systems have far more than two particles.)]

The thermodynamic expression for the adiabatic energy change is $dU = dw_{\text{rev}} = -P_j\, dV$. (We restrict ourselves to systems capable of P-V work only.) Equating the thermodynamic and quantum-mechanical expressions, we have $-P_j\, dV = (\partial E_j/\partial V)_{N_B}\, dV$, and

$$P_j = -(\partial E_j/\partial V)_{N_B} \tag{22.22}$$

Use of (22.22) and (22.15) in (22.21) gives

$$P = \sum_j p_j P_j = -\frac{1}{Z}\sum_j e^{-\beta E_j}\left(\frac{\partial E_j}{\partial V}\right)_{N_B} \tag{22.23}$$

Partial differentiation of $Z = \sum_j e^{-\beta E_j}$ gives

$$\frac{\partial Z}{\partial V} = \sum_j \frac{\partial e^{-\beta E_j}}{\partial V} = \sum_j \frac{\partial e^{-\beta E_j}}{\partial E_j}\frac{\partial E_j}{\partial V} = -\sum_j \beta e^{-\beta E_j}\frac{\partial E_j}{\partial V}$$

where constant T and composition are understood. (Recall that β is a function of T only.) Hence, Eq. (22.23) can be written as

$$P = \frac{1}{\beta Z}\left(\frac{\partial Z}{\partial V}\right)_{T,N_B} = \frac{1}{\beta}\left(\frac{\partial \ln Z}{\partial V}\right)_{T,N_B} \tag{22.24}$$

which is the desired expression for the pressure.

Evaluation of β. To find β, we evaluate $(\partial U/\partial V)_{T,N_B}$ using (22.20) for U and (22.24) for P:

$$\left(\frac{\partial U}{\partial V}\right)_T = -\left[\frac{\partial}{\partial V}\left(\frac{\partial \ln Z}{\partial \beta}\right)_V\right]_T = -\left[\frac{\partial}{\partial \beta}\left(\frac{\partial \ln Z}{\partial V}\right)_T\right]_V$$

$$= -\left[\frac{\partial}{\partial \beta}(\beta P)\right]_V = -P - \beta\left(\frac{\partial P}{\partial \beta}\right)_V \tag{22.25}$$

where we reversed the order of differentiation, and where constant composition is understood in (22.25) and the following few equations.

The thermodynamic identity (4.71) gives at constant composition

$$\left(\frac{\partial U}{\partial V}\right)_T = T\left(\frac{\partial P}{\partial T}\right)_V - P = -\frac{1}{T}\left[\frac{\partial P}{\partial(1/T)}\right]_V - P \tag{22.26}$$

where we used $\partial P/\partial T = [\partial P/\partial(1/T)][\partial(1/T)/\partial T] = -[\partial P/\partial(1/T)]/T^2$.

Equating (22.25) and (22.26), we get

$$-\beta\left(\frac{\partial P}{\partial \beta}\right)_V = -T^{-1}\left[\frac{\partial P}{\partial(1/T)}\right]_V$$

Let $Y \equiv 1/T$. Then $\beta(\partial P/\partial \beta)_V = Y(\partial P/\partial Y)_V$, and

$$\frac{\beta}{Y} = \left(\frac{\partial \beta}{\partial P}\right)_V\left(\frac{\partial P}{\partial Y}\right)_V = \left(\frac{\partial \beta}{\partial Y}\right)_V = \frac{d\beta}{dY}$$

since β and Y are each functions of T only. We have $dY/Y = d\beta/\beta$, which integrates to $\ln Y = \ln \beta + \text{const}$, so $Y = e^{\ln \beta}e^{\text{const}} = \beta k$, where $k \equiv e^{\text{const}}$ is a constant. Since

$Y = 1/T$, our final result is

$$\beta = \frac{1}{kT} \tag{22.27*}$$

We saw earlier that β is the same for any two systems in thermal equilibrium. Hence k must be a universal constant. In Sec. 22.6 we shall find that k is Boltzmann's constant, Eq. (3.57).

Equations (22.20) and (22.24) for U and P become

$$U = -\frac{\partial \ln Z}{\partial \beta} = -\frac{\partial \ln Z}{\partial T}\frac{dT}{d\beta} = -\frac{\partial \ln Z}{\partial T}\left(-\frac{1}{k\beta^2}\right) = kT^2\left(\frac{\partial \ln Z}{\partial T}\right)_{V,N_B} \tag{22.28}$$

$$P = kT\left(\frac{\partial \ln Z}{\partial V}\right)_{T,N_B} \tag{22.29}$$

Equation (22.29) allows the system's equation of state to be calculated from its canonical partition function.

The probability that a system of fixed volume, temperature, and composition is in quantum state j is given by (22.15) as $p_j = e^{-E_j/kT}/Z$, where Z is a constant at fixed T, V, and composition. Let $p(E_i)$ be the probability that this system has quantum energy E_i. As noted earlier, the quantum energy levels of a system with a large number of molecules are highly degenerate, with many different quantum states having the same energy. Let W_i be the number of quantum states that have energy E_i; that is, W_i is the degeneracy of the level E_i. The probability of being in any one of the states having energy E_i is $e^{-E_i/kT}/Z$. Hence, the probability $p(E_i)$ equals $W_i e^{-E_i/kT}/Z$:

$$p(E_i) = W_i e^{-E_i/kT}/Z \tag{22.30}$$

The degeneracy W_i is a sharply increasing function of the system's energy E_i, since as E_i increases, the number of ways of distributing the energy among the molecules increases rapidly. The exponential function $e^{-E_i/kT}$ is a sharply decreasing function of E_i, especially since Boltzmann's constant k is a very small number. (The system energy E_i is of order of magnitude RT, and $-RT/kT = -N_0$.) The product of a sharply increasing function and a sharply decreasing function in (22.30) gives a probability $p(E_i)$ that is peaked extremely narrowly about a single value, which is the observed thermodynamic internal energy U (Fig. 22.3).

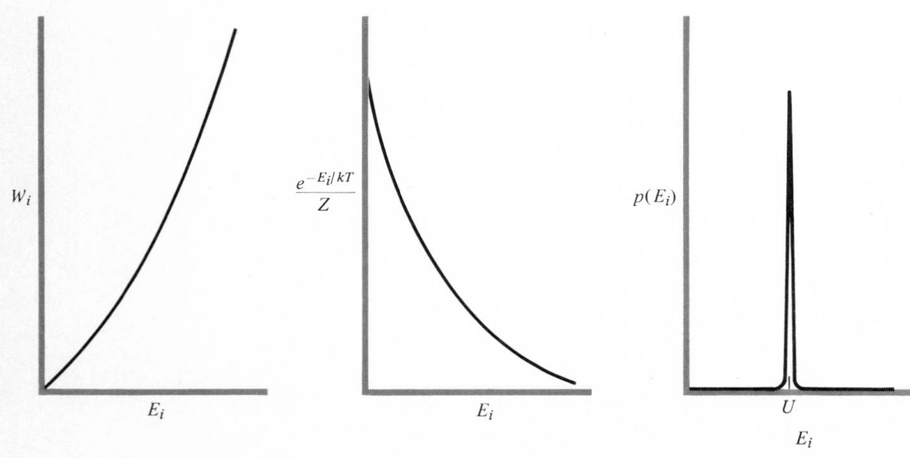

Figure 22.3
The probability $p(E_i) = W_i \times (e^{-E_i/kT}/Z)$ that the system have energy E_i is very sharply peaked about U.

Evaluation of S. There remains but one task—to find the statistical-mechanical expression for the entropy. For a reversible process in a system of fixed composition and capable of P-V work only, we have $dU = T\,dS - P\,dV$; solving for dS, we have (at constant composition)

$$dS = T^{-1}\,dU + PT^{-1}\,dV = d(T^{-1}U) + T^{-2}U\,dT + PT^{-1}\,dV$$

since $d(T^{-1}U) = -T^{-2}U\,dT + T^{-1}\,dU$. Substitution of (22.28) and (22.29) for U and P gives

$$dS = d(T^{-1}U) + k\left(\frac{\partial \ln Z}{\partial T}\right)_{V,N_B} dT + k\left(\frac{\partial \ln Z}{\partial V}\right)_{T,N_B} dV \qquad \text{const. } N_B$$

From (22.18), ln Z is a function of T, V, and composition. At constant composition, we have

$$d \ln Z = \left(\frac{\partial \ln Z}{\partial T}\right)_{V,N_B} dT + \left(\frac{\partial \ln Z}{\partial V}\right)_{T,N_B} dV \qquad \text{const. } N_B$$

Hence, $dS = d(T^{-1}U) + k\,d \ln Z = d(T^{-1}U + k \ln Z)$. Integration gives

$$S = T^{-1}U + k \ln Z + C \qquad \text{const. } N_B \tag{22.31}$$

Since the derivation of (22.31) assumed constant composition, the integration "constant" C might be a function of composition: $C = f(N_A, N_B, \ldots)$. However, by considering a more general kind of ensemble than the canonical ensemble (namely, an ensemble in which the composition of each system can vary), it is possible to show that C is independent of composition. Then we have $S = U/T + k \ln Z + C$. Since the constant C is the same for any system no matter what its composition, and since only entropy changes are measurable, a sensible thing to do is to take C as zero. (See also Sec. 22.9 for more discussion on this point.) Therefore

$$S = \frac{U}{T} + k \ln Z = kT\left(\frac{\partial \ln Z}{\partial T}\right)_{V,N_B} + k \ln Z \tag{22.32}$$

where (22.28) was used for U. Equation (22.32) is the desired expression for S.

An alternative expression for S is often convenient in theoretical discussions. Equation (22.15) gives $p_j = e^{-\beta E_j}/Z$, so $\ln p_j = -\beta E_j - \ln Z$. Consider the sum $-\sum_j p_j \ln p_j$. Substitution for $\ln p_j$ gives $-\sum_j p_j \ln p_j = \beta \sum_j p_j E_j + \ln Z \sum_j p_j = k^{-1}T^{-1}U + \ln Z$, where (22.3) and $\sum_j p_j = 1$ were used. Comparison with (22.32) gives

$$S = -k \sum_j p_j \ln p_j \tag{22.33}$$

The Helmholtz energy is $A = U - TS$. Use of (22.32) for S gives $A = U - T(U/T + k \ln Z) = -kT \ln Z$:

$$A = -kT \ln Z \tag{22.34}^*$$

The Gibbs energy is $G = A + PV$, and can be found from (22.34) and (22.29).

The chemical potential of species B is given by (4.87) as $\mu_B = (\partial A/\partial n_B)_{T,V,n_{C\neq B}}$, where $n_B = N_B/N_0$ is the number of moles of B and N_0 is the Avogadro constant. Use of (22.34) gives

$$\mu_B = -kT\left(\frac{\partial \ln Z}{\partial n_B}\right)_{T,V,n_{C\neq B}} = -RT\left(\frac{\partial \ln Z}{\partial N_B}\right)_{T,V,N_{C\neq B}} \tag{22.35}$$

since $dn_B = dN_B/N_0$ and we are anticipating the result (Sec. 22.6) that k is Boltzmann's constant: $N_0 k = R$.

If the formula $A = -kT \ln Z$ is remembered, the equations for P, U, S, G, and μ_B can be quickly derived, as follows. The Gibbs equation (4.87) reads $dA = -S\,dT - P\,dV + \sum_B \mu_B\,dn_B$. Hence, $S = -(\partial A/\partial T)_{V,N_B}$, $P = -(\partial A/\partial V)_{T,N_B}$, and $\mu_B = (\partial A/\partial n_B)_{T,V,n_{C\neq B}}$. Partial differentiation of (22.34) then gives (22.32) for S, (22.29) for P, and (22.35) for μ_B. Equation (22.28) for U is then derived from $U = A + TS$. The reason A is simply related to the canonical partition function is that the canonical ensemble consists of systems of constant T, V, and composition, and these are the "natural" variables for A.

Summary. The key statistical-mechanical formulas for calculating thermodynamic properties are

$$Z \equiv \sum_j e^{-E_j/kT} \tag{22.36*}$$

$$P = kT\left(\frac{\partial \ln Z}{\partial V}\right)_{T,N_B} \tag{22.37}$$

$$U = kT^2\left(\frac{\partial \ln Z}{\partial T}\right)_{V,N_B} \tag{22.38}$$

$$S = U/T + k \ln Z \tag{22.39}$$

$$A = -kT \ln Z \tag{22.40*}$$

$$\mu_B = -RT\left(\frac{\partial \ln Z}{\partial N_B}\right)_{T,V,N_{C\neq B}} \tag{22.41}$$

[If species B is electrically charged, μ_B in (22.41) is to be replaced by the electrochemical potential $\tilde{\mu}_B$ (Sec. 14.4); this is true of all equations containing μ_B in this chapter.]

Equations (22.36) to (22.41) are valid for gases, liquids, and solids and for solutions as well as pure substances.

22.3 CANONICAL PARTITION FUNCTION FOR A SYSTEM OF NONINTERACTING PARTICLES

Once a system's canonical partition function Z has been found by summation of $e^{-E_j/kT}$ over all possible quantum states of the system, all the system's thermodynamic properties (P, U, S, A, μ_B, μ_C, ...) are readily found. However, the existence of forces between molecules makes Z extremely difficult to evaluate, since it is extremely hard to solve the Schrödinger equation for N interacting molecules to find the E_j's. For a system with no intermolecular forces, we can readily calculate Z.

Let the Hamiltonian operator $\hat{H}$ for the system be the sum of separate terms for the individual molecules, with no interaction terms between the molecules: $\hat{H} = \hat{H}_1 + \hat{H}_2 + \cdots + \hat{H}_N$, where there are N molecules. Then the system's energy E_j is the sum of an energy for each molecule [Eq. (18.59)]:

$$E_j = \varepsilon_{1,r} + \varepsilon_{2,s} + \cdots + \varepsilon_{N,w} \tag{22.42}$$

where $\varepsilon_{1,r}$ is the energy of molecule 1 when the system is in state j, and r denotes the quantum state of molecule 1. (We use an epsilon to indicate the energy of a single molecule, to avoid confusion with E_j, which is the quantum-mechanical energy of a thermodynamic system containing N molecules.) From (18.60), the allowed energies for molecule 1 are found from the one-molecule Schrödinger equation $\hat{H}_1 \psi_{1,r} = \varepsilon_{1,r} \psi_{1,r}$. [When the molecules exert forces on one another, Eq. (22.42) does not hold.]

The canonical partition function (22.36) for a system of noninteracting molecules is

$$Z = \sum_j e^{-\beta E_j} = \sum_j e^{-\beta(\varepsilon_{1,r} + \varepsilon_{2,s} + \cdots + \varepsilon_{N,w})} \tag{22.43}$$

First consider the case where the molecules are distinguishable from one another by being confined to different locations in space. This would occur in a crystal. (Of course, the molecules in a crystal do interact with each other, but we'll worry about that later; Sec. 24.10.) For distinguishable molecules, the state of the system is defined by giving the quantum state of each molecule; molecule 1 is in state r, molecule 2 in state s, etc. Hence, to sum over all possible quantum states j of the system, we sum separately over all the possible states of each molecule, and (22.43) becomes

$$Z = \sum_r e^{-\beta\varepsilon_{1,r}} \sum_s e^{-\beta\varepsilon_{2,s}} \cdots \sum_w e^{-\beta\varepsilon_{N,w}} \tag{22.44}$$

We define the *molecular partition functions* $z_1, z_2, \ldots$ as

$$z_1 \equiv \sum_r e^{-\beta\varepsilon_{1,r}}, \quad z_2 \equiv \sum_s e^{-\beta\varepsilon_{2,s}}, \quad \ldots$$

where $\varepsilon_{1,r}$ is the energy of molecule 1 in the molecular quantum state r, and the first sum goes over the available quantum states of molecule 1. Equation (22.44) becomes

$$Z = z_1 z_2 \cdots z_N \qquad \text{localized noninteracting molecules} \tag{22.45}$$

If all the molecules happen to be the same kind, the set of molecular quantum states is the same for each molecule and $z_1 = z_2 = \cdots = z_N$. Thus

$$Z = z^N, \qquad \text{identical loc. nonint. molecules} \tag{22.46}$$

$$z \equiv \sum_s e^{-\beta\varepsilon_s} \tag{22.47}$$

(To avoid confusion between Z and z in handwritten equations, use a script lowercase zee for the molecular partition function.)

If the molecules are not all alike but there are N_A molecules of species A, N_B molecules of species B, etc., (22.45) becomes

$$Z = (z_A)^{N_A}(z_B)^{N_B} \cdots \qquad \text{loc. nonint. molecules} \tag{22.48}$$

$$z_A = \sum_r e^{-\beta\varepsilon_{A,r}}, \quad z_B = \sum_s e^{-\beta\varepsilon_{B,s}}, \quad \ldots$$

Now suppose that the molecules are not confined to fixed locations and so are not distinguishable by position. We first consider the case where all the nonlocalized molecules are of the same species. (This corresponds to a pure ideal gas.) With all molecules moving through the entire volume of the system and having the same structure, there is no way whatever of distinguishing between the molecules. Hence, a situation where molecule 1 is in state r and molecule 2 is in state s corresponds to the same quantum state as the situation where molecule 1 is in state s and molecule 2 is in state r (and the states of the other molecules are unchanged). Recall that for the helium atom [Eqs. (19.37) and (19.38)], we could not say that electron 1 was in the 1*s* orbital and electron 2 in the 2*s* orbital or that electron 1 was in the 2*s* orbital and electron 2 in the 1*s* orbital. Instead a state of He with the electron configuration 1*s*2*s* corresponds to a wave function containing the terms $1s(1)2s(2)$ and $2s(1)1s(2)$; each electron is in both orbitals. Similarly, the wave function for a given quantum state of the gas of N identical nonlocalized molecules contains terms in which all the molecules are permuted among all the occupied molecular states. The microstate of the system depends not on which molecules are in which molecular states but on how many molecules are in each of the available molecular states.

How do we get Z when the noninteracting molecules are indistinguishable? Let us assume that there are many, many more molecular states available than there are molecules in the gas. (This assumption will be justified below.) Then the probability that two or more gas molecules are in the same molecular state is extremely small, and we shall assume that no two molecules are in the same molecular state. If we were to use Eq. (22.44) or its equivalent (22.46) for Z, we would be counting each microstate of the system too many times. For example, suppose the system consists of three identical molecules, and let the system microstate j have one molecule in molecular state r, one molecule in t, and one in w. The system's wave function for state j would contain the term $\psi_r(1)\psi_t(2)\psi_w(3)$ and the five other terms that correspond to the $3! = 6$ permutations of molecules 1, 2, and 3 among the molecular states r, t, and w [cf. Eq. (19.45)].

Equation (22.44), which sums separately over all molecular states for each molecule, counts $\psi_r(1)\psi_t(2)\psi_w(3)$ as a separate state from $\psi_t(1)\psi_r(2)\psi_w(3)$ and from the other four permutations, since it contains the separate terms

$$e^{-\beta\varepsilon_{1,r}}e^{-\beta\varepsilon_{2,t}}e^{-\beta\varepsilon_{3,w}}, \quad e^{-\beta\varepsilon_{1,t}}e^{-\beta\varepsilon_{2,r}}e^{-\beta\varepsilon_{3,w}}, \quad \ldots$$

These six terms are numerically equal, since each contains the sum $\varepsilon_r + \varepsilon_t + \varepsilon_w$. Thus, Eq. (22.46) has $3! = 6$ numerically equal terms where there should be only one term, and this situation holds for each quantum state of the system [*provided* there is a vanishing probability that two molecules have the same molecular state; for $\psi_r(1)\psi_r(2)\psi_w(3)$, there are only two other possible permutations of the molecules among the occupied molecular states]. The correct value of Z for our hypothetical three-particle system can be obtained by dividing (22.46) by 3!. For a system of N indistinguishable noninteracting particles, the correct canonical partition function is obtained by dividing by $N!$:

$$Z = \frac{z^N}{N!} \qquad \text{for } \langle N_r \rangle \ll 1, \text{ nonloc. ident. nonint. molecules} \qquad (22.49)^*$$

$$z \equiv \sum_r e^{-\beta\varepsilon_r} \qquad (22.50)^*$$

where $\langle N_r \rangle$ is the average number of molecules in the molecular quantum state r and the inequality must hold for each available molecular state.

Now suppose that all the molecules are not the same kind but there are N_A molecules of species A, N_B of B, etc. Then we can distinguish A molecules from B molecules but cannot distinguish two A molecules from each other. We must correct (22.44) for permutations of A molecules with other A molecules by dividing by $N_A!$, for permutations of B molecules with other B molecules by dividing by $N_B!$, etc. Instead of (22.48), we get

$$Z = \frac{z_A^{N_A}}{N_A!}\frac{z_B^{N_B}}{N_B!}\cdots \quad \text{nonloc. nonint. molecules} \tag{22.51}$$

provided $\langle N_{A,r}\rangle \ll 1$, $\langle N_{B,s}\rangle \ll 1$, etc.

Let us show that the assumption $\langle N_r\rangle \ll 1$ is justified. Consider an ideal gas. The molecular energy is the sum of translational, rotational, vibrational, and electronic energies. We shall count only the translational states; inclusion of rotational, vibrational, and electronic states would only strengthen the result.

For a cubic box of volume V, Eq. (18.42) gives the translational energy of a molecule as $\varepsilon_{tr} = (h^2/8mV^{2/3})(n_x^2 + n_y^2 + n_z^2)$. We shall calculate the number of translational states whose energy is less than some maximum value ε_{max}. For $\varepsilon_{tr} \le \varepsilon_{max}$, the translational quantum numbers must satisfy the inequality

$$n_x^2 + n_y^2 + n_z^2 \le 8mV^{2/3}h^{-2}\varepsilon_{max} \tag{22.52}$$

We know that the average translational energy per molecule is $\frac{3}{2}kT$ [Eq. (15.16)] and that most molecules have ε_{tr} within 2 or 3 times this average value. We therefore take $\varepsilon_{max} = 3kT$ as the "maximum" ε_{tr} for a typical molecule.

Figure 22.4 shows axes whose coordinates are the quantum numbers n_x, n_y, and n_z. Imagine a dot placed at each point whose coordinates are integers. The distance of a dot from the origin is $r = (n_x^2 + n_y^2 + n_z^2)^{1/2}$. The number of quantum states with $\varepsilon_{tr} \le \varepsilon_{max}$ equals the number of dots in one-eighth of a sphere whose radius r_{max} is [Eq. (22.52)]

$$r_{max} = (8mV^{2/3}h^{-2}\varepsilon_{max})^{1/2} = (24mV^{2/3}h^{-2}kT)^{1/2}$$

(We take one-eighth of a sphere because n_x, n_y, and n_z must all be positive; only the first octant of space in Fig. 22.4 has dots.) Figure 22.4 shows eight dots that form a cube whose edge length is 1. We can imagine many such cubes to fill up the first octant out to r_{max}. The cube shown has eight dots, but each dot is shared by eight cubes—four at the same altitude and four more either immediately above or below. The density of translation states is thus 8/8 = 1 per unit volume of space. Therefore,

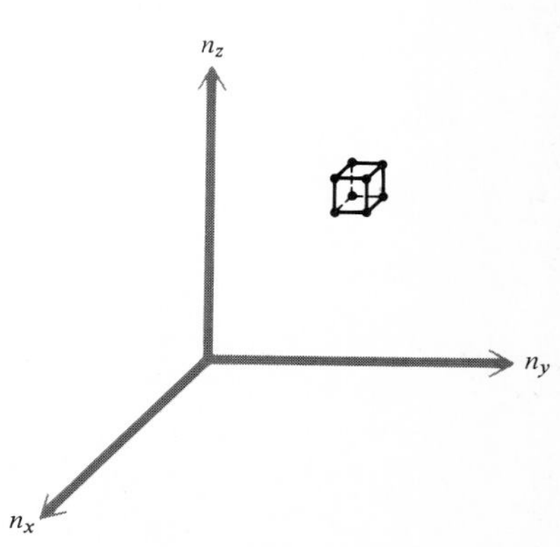

Figure 22.4
Coordinate system used to calculate the number of translational states with energy less than ε_{max}.

the number of molecular states with $\varepsilon_{tr} \le \varepsilon_{max}$ equals one-eighth the volume of a sphere of radius r_{max}. The number of available molecular translational states is then

$$\frac{1}{8}\frac{4\pi r_{max}^3}{3} = \frac{\pi}{6}(24mV^{2/3}h^{-2}kT)^{3/2} \approx 60\left(\frac{mkT}{h^2}\right)^{3/2} V$$

For $\langle N_r \rangle \ll 1$ to hold, we must have $60(mkT/h^2)^{3/2}V \gg N$, where N is the number of molecules. This inequality reads

$$\frac{1}{60}\left(\frac{h^2}{mkT}\right)^{3/2}\frac{N}{V} \ll 1 \tag{22.53}$$

A typical system is 1 mole of O_2 at 0°C and 1 atm, for which $m = [32/(6 \times 10^{23})]$ g, $N = 6 \times 10^{23}$, and $V = 22{,}400$ cm^3. One finds 4×10^{-8} for the left side of (22.53). Hence, $\langle N_r \rangle \ll 1$ is well satisfied. The translational levels are spaced extremely close together, so there are far more translational states available than there are molecules in the gas.

The inequality (22.53) does not hold for (*a*) extremely low temperatures, (*b*) extremely high densities, (*c*) extremely small particle masses. Physical examples of these conditions are (*a*) liquid helium at 2 or 3 K, (*b*) white dwarf stars and neutron stars (pulsars), (*c*) the conduction electrons in metals.

The last stage in the life of many stars is gravitational collapse to a white dwarf star or a neutron star (or a black hole, if such things exist). The atomic structure of matter is crushed and destroyed in these stars. White dwarfs (typical radius 10,000 miles, compared with 1,000,000 miles for the sun) do not contain atoms but are composed of atomic nuclei (mainly He^{2+}) and free electrons. Their density is typically 10^6 g/cm^3. A neutron star (typical radius 10 miles) is produced when gravitational collapse proceeds to the point where the electrons and protons are forced to unite to form neutrons. The density of a neutron star is about the same as the density of matter in an atomic nucleus, 10^{14} g/cm^3 (Prob. 17.70).

The conduction electrons in a metal are relatively free to move through the metal and can be approximately treated as a "gas" of electrons. Assuming one conduction electron per metal atom and a molar volume of 20 cm^3, we find that the left side of (22.53) equals 700 at 0°C. The assumption $\langle N_r \rangle \ll 1$ fails, due mainly to the small mass of electrons and, to a lesser extent, to the higher density in the electron "gas" than in an ordinary gas.

When the condition $\langle N_r \rangle \ll 1$ is not met, Eqs. (22.49) and (22.51) don't hold. The proper result for Z depends on whether the particles of the gas are bosons or fermions (Sec. 19.5). The boson and fermion expressions for Z in terms of molecular energy levels are given in Prob. 22.17.

22.4 CANONICAL PARTITION FUNCTION OF A PURE IDEAL GAS

In Secs. 22.4, 22.6, and 22.7, we restrict ourselves to a pure ideal gas (all molecules alike). (The thermodynamic functions of an ideal gas mixture are easily obtained as the sum of those for each gas in the mixture; Prob. 6.22.)

In an ideal gas (no intermolecular forces) the system energy is the sum of molecular energies, and the canonical partition function is given by (22.49) as $Z = z^N/N!$, where z is the molecular partition function. Since it is ln Z (rather than Z

itself) that occurs in Eqs. (22.37) to (22.41) for the thermodynamic properties, we take $\ln Z$:

$$\ln Z = \ln (z^N/N!) = N \ln z - \ln N! \qquad \text{pure ideal gas} \tag{22.54}$$

It's a bit tiresome to evaluate a number like $\ln (10^{23}!)$. Fortunately, an excellent approximation is available. *Stirling's formula*, valid for large N, is

$$N! = (2\pi)^{1/2} N^{N+1/2} e^{-N} \left(1 + \frac{1}{12N} + \frac{1}{288N^2} - \cdots\right) \qquad \text{for } N \text{ large} \tag{22.55}$$

(For a partial derivation, see *Sokolnikoff and Redheffer*, p. 46.) Taking the natural log, we have

$$\ln N! = \tfrac{1}{2} \ln 2\pi + (N + \tfrac{1}{2}) \ln N - N + \ln (1 + 1/12N + \cdots)$$

Since N is something like 10^{23}, we have $N + \frac{1}{2} \approx N$, $\frac{1}{2} \ln 2\pi - N \approx -N$, and $\ln (1 + 1/12N + \cdots) \approx \ln 1 = 0$. Hence

$$\ln N! \approx N \ln N - N \qquad \text{for } N \text{ large} \tag{22.56}$$

To show how well (22.56) works, the following table lists some values of $\ln N!$ and of $N \ln N - N$:

N	$\ln N!$	$N \ln N - N$	Error
10^2	363.7	360.5	−0.8%
10^3	5912.1	5907.8	−0.07%
10^4	82,108.9	82,103.4	−0.007%
10^5	1,051,299	1,051,293	−0.0006%

Each additional power of 10 in N decreases the percent error by roughly a factor of 10, so the percent error is entirely negligible for the usual numbers of molecules in a thermodynamic system.

Now consider the ideal-gas molecular partition function $z = \sum_r e^{-\varepsilon_r/kT}$. It is usually a good approximation to write the molecular energy as the sum of translational, rotational, vibrational, and electronic energies [Eq. (21.21)]

$$\varepsilon_r = \varepsilon_{\text{tr},s} + \varepsilon_{\text{rot},t} + \varepsilon_{\text{vib},v} + \varepsilon_{\text{el},u} \tag{22.57}$$

where the subscripts s, t, v, u indicate the translational, rotational, vibrational, and electronic states. The quantum numbers of the four kinds of energies vary independently of one another. Hence

$$z = \sum_r e^{-\beta\varepsilon_r} = \sum_s \sum_t \sum_v \sum_u e^{-\beta(\varepsilon_{\text{tr},s} + \varepsilon_{\text{rot},t} + \varepsilon_{\text{vib},v} + \varepsilon_{\text{el},u})}$$

$$= \sum_s e^{-\beta\varepsilon_{\text{tr},s}} \sum_t e^{-\beta\varepsilon_{\text{rot},t}} \sum_v e^{-\beta\varepsilon_{\text{vib},v}} \sum_u e^{-\beta\varepsilon_{\text{el},u}}$$

$$z = z_{\text{tr}} z_{\text{rot}} z_{\text{vib}} z_{\text{el}} \tag{22.58}$$

$$\ln z = \ln z_{\text{tr}} + \ln z_{\text{rot}} + \ln z_{\text{vib}} + \ln z_{\text{el}} \tag{22.59}$$

$$z_{\text{tr}} \equiv \sum_s e^{-\beta\varepsilon_{\text{tr},s}}, \qquad z_{\text{rot}} \equiv \sum_t e^{-\beta\varepsilon_{\text{rot},t}}, \qquad z_{\text{vib}} \equiv \text{etc.} \tag{22.60}$$

When the molecular energy is the sum of different kinds of energies, the molecular partition function factors into a product of partition functions, one for each kind of energy. [Actually, Eq. (22.58) is a lie, but this will be cleared up in Sec. 22.6.]

Substitution of (22.59) and (22.56) into (22.54) gives for an ideal gas

$$\ln Z = N \ln z_{\text{tr}} + N \ln z_{\text{rot}} + N \ln z_{\text{vib}} + N \ln z_{\text{el}} - N(\ln N - 1) \quad (22.61)$$

The thermodynamic internal energy is given by (22.38) as $U = kT^2(\partial \ln Z/\partial T)_{V,N}$. Partial differentiation of (22.61) gives for an ideal gas

$$U = NkT^2 \left[\left(\frac{\partial \ln z_{\text{tr}}}{\partial T} \right)_V + \frac{d \ln z_{\text{rot}}}{dT} + \frac{d \ln z_{\text{vib}}}{dT} + \frac{d \ln z_{\text{el}}}{dT} \right]$$

$$U = U_{\text{tr}} + U_{\text{rot}} + U_{\text{vib}} + U_{\text{el}} \quad (22.62)$$

$$U_{\text{tr}} \equiv NkT^2 \left(\frac{\partial \ln z_{\text{tr}}}{\partial T} \right)_V, \qquad U_{\text{rot}} \equiv NkT^2 \frac{d \ln z_{\text{rot}}}{dT}, \qquad U_{\text{vib}} \equiv \text{etc.} \quad (22.63)$$

(Only the translational energy depends on the volume.)

The entropy is given by (22.39) as $S = U/T + k \ln Z$. Use of (22.62) and (22.61) gives for a pure ideal gas

$$S = S_{\text{tr}} + S_{\text{rot}} + S_{\text{vib}} + S_{\text{el}} \quad (22.64)$$

$$S_{\text{tr}} \equiv U_{\text{tr}}/T + Nk \ln z_{\text{tr}} - Nk(\ln N - 1) \quad (22.65)$$

$$S_{\text{rot}} \equiv U_{\text{rot}}/T + Nk \ln z_{\text{rot}}, \qquad S_{\text{vib}} \equiv U_{\text{vib}}/T + Nk \ln z_{\text{vib}}, \qquad \text{etc.} \quad (22.66)$$

The translational contribution to the entropy differs in form from the rotational, vibrational, and electronic contributions because it incorporates the $-\ln N!$ term. The $N!$ in Z is due to the indistinguishability of the molecules, which results from their being nonlocalized. Hence it is appropriate to include this term in S_{tr}.

Detailed application of these ideal-gas formulas is given in Secs. 22.6 and 22.7.

22.5 THE BOLTZMANN DISTRIBUTION LAW FOR NONINTERACTING MOLECULES

Consider a system of noninteracting identical molecules, and let the possible quantum states available to each molecule be $r, s, t, u, \ldots$. Since there are no intermolecular interactions, the quantum energy of the system can be written as

$$E_j = N_{r,j}\varepsilon_r + N_{s,j}\varepsilon_s + \cdots$$

where $N_{r,j}, N_{s,j}, \ldots$ are the numbers of molecules in the molecular states $r, s, \ldots$ when the system is in microstate j, and $\varepsilon_r, \varepsilon_s, \ldots$ are the energies of the molecular states $r, s, \ldots$. (To avoid confusion, quantum states of the system are labeled $i, j, k, \ldots$, whereas quantum states of a single molecule are labeled $r, s, t, \ldots$.)

Let us calculate $\langle N_s \rangle$, the average number of molecules in the molecular state s when the system is in a given macrostate. As we did with U and P, we average over the microstates in the canonical ensemble, and we have

$$\langle N_s \rangle = \sum_j p_j N_{s,j}$$

similar to (22.3). Equation (22.15) gives the probability p_j as $p_j = e^{-\beta E_j}/Z$. Hence

$$\langle N_s \rangle = \frac{\sum_j N_{s,j} e^{-\beta(N_{r,j}\varepsilon_r + N_{s,j}\varepsilon_s + \cdots)}}{Z} \tag{22.67}$$

where the above expression for E_j was used.

Now consider the partial derivative $\partial Z/\partial \varepsilon_s$. We have

$$\frac{\partial Z}{\partial \varepsilon_s} = \frac{\partial}{\partial \varepsilon_s} \sum_j e^{-\beta(N_{r,j}\varepsilon_r + N_{s,j}\varepsilon_s + \cdots)} = -\beta \sum_j N_{s,j} e^{-\beta(N_{r,j}\varepsilon_r + N_{s,j}\varepsilon_s + \cdots)}$$

Hence, the numerator in (22.67) equals $-(1/\beta)(\partial Z/\partial \varepsilon_s)$, and

$$\langle N_s \rangle = -\frac{1}{\beta}\frac{1}{Z}\frac{\partial Z}{\partial \varepsilon_s} = -\frac{1}{\beta}\frac{\partial \ln Z}{\partial \varepsilon_s} \tag{22.68}$$

where the partial derivative is taken at constant β (that is, constant T) and constant $\varepsilon_{r \neq s}$.

For a system of noninteracting identical molecules, Eqs. (22.49) and (22.46) give (provided $\langle N_r \rangle \ll 1$ for all r): $Z = z^N/(N!)$ and $\ln Z = N \ln z - (\ln N!)$, where the $N!$ is present or absent according to whether the molecules are nonlocalized or localized. In either case,

$$\frac{\partial \ln Z}{\partial \varepsilon_s} = N\frac{\partial \ln z}{\partial \varepsilon_s} = \frac{N}{z}\frac{\partial z}{\partial \varepsilon_s} = \frac{N}{z}\frac{\partial(e^{-\beta\varepsilon_r} + e^{-\beta\varepsilon_s} + \cdots)}{\partial \varepsilon_s} = -\beta\frac{N}{z}e^{-\beta\varepsilon_s}$$

where (22.50) was used for z. Substitution in (22.68) gives

$$\frac{\langle N_s \rangle}{N} = \frac{e^{-\varepsilon_s/kT}}{z} = \frac{e^{-\varepsilon_s/kT}}{\sum_r e^{-\varepsilon_r/kT}} \qquad \text{for } \langle N_r \rangle \ll 1 \tag{22.69*}$$

Note that $\langle N_s \rangle/N$ is the probability that a molecule picked at random is in the molecular state r. This probability decreases exponentially as the energy ε_r of state r increases.

Equation (22.69) is the *Boltzmann* (or *Maxwell–Boltzmann*) *distribution law*. This equation is easily written down if it is remembered that $\langle N_s \rangle = \text{const} \times e^{-\varepsilon_s/kT}$ and $N = \sum_r \langle N_r \rangle$.

Equation (22.69) for state r is $\langle N_r \rangle/N = e^{-\varepsilon_r/kT}/z$, so

$$\frac{\langle N_r \rangle}{\langle N_s \rangle} = e^{-(\varepsilon_r - \varepsilon_s)/kT} \equiv e^{-\Delta\varepsilon/kT} \qquad \text{for } \langle N_r \rangle \ll 1 \tag{22.70}$$

Often we are interested in the average number of molecules having a given energy rather than the average number in a given molecular state. Several different molecular states may have the same energy (degeneracy, Sec. 18.10). If the molecular states s, t, and u all have the same energy ($\varepsilon_s = \varepsilon_t = \varepsilon_u$) and no other states have this energy, then the energy level ε_s is threefold degenerate and the number of molecules having energy ε_s equals $N_s + N_t + N_u = 3N_s$, where (22.69) was used. In general, if the energy level ε_s is g_s-fold degenerate, (22.69) gives

$$\frac{\langle N(\varepsilon_s) \rangle}{N} = \frac{g_s e^{-\varepsilon_s/kT}}{z} \qquad \text{for } \langle N_r \rangle \ll 1 \tag{22.71}$$

$$\frac{\langle N(\varepsilon_r) \rangle}{\langle N(\varepsilon_s) \rangle} = \frac{g_r}{g_s} e^{-(\varepsilon_r - \varepsilon_s)/kT} \qquad \text{for } \langle N_r \rangle \ll 1 \tag{22.72*}$$

where $\langle N(\varepsilon_s)\rangle$ is the average number of molecules having energy ε_s. The degeneracy g_s is sometimes called the *statistical weight* of the energy level ε_s.

The molecular partition function $z = \sum_r e^{-\varepsilon_r/kT}$ is a sum over all one-molecule states. The term $e^{-\varepsilon_r/kT}$ is equal for two or more states that have the same energy. Hence, if we write z as a sum over molecular energy values (rather than as a sum over molecular states), we have

$$z = \sum_{m(\text{levels})} g_m e^{-\varepsilon_m/kT} \tag{22.73}$$

where the sum is over the energy levels and g_m is the degeneracy of level m.

The above equations apply when the noninteracting molecules are all alike. When several species are present, each species has its own set of molecular energy levels and Z is given by (22.51) or (22.48). A derivation similar to the above gives

$$\frac{\langle N_{\text{B},r}\rangle}{N_\text{B}} = \frac{e^{-\varepsilon_{\text{B},r}/kT}}{z_\text{B}} \qquad \text{for } \langle N_{\text{B},r}\rangle \ll 1 \tag{22.74}$$

where $\langle N_{\text{B},r}\rangle$ is the average number of species B molecules in the B molecular state r (whose energy is $\varepsilon_{\text{B},r}$), and N_B is the total number of B molecules in the system.

Let us show that the Boltzmann distribution law can be applied to each kind of energy—translational, rotational, vibrational, and electronic. Equation (22.69) with s changed to r and Eqs. (22.57) and (22.58) give

$$\langle N_r\rangle = \frac{Ne^{-\varepsilon_r/kT}}{z} = \frac{Ne^{-\beta\varepsilon_{\text{tr},s}}e^{-\beta\varepsilon_{\text{rot},t}}e^{-\beta\varepsilon_{\text{vib},v}}e^{-\beta\varepsilon_{\text{el},u}}}{z_\text{tr}z_\text{rot}z_\text{vib}z_\text{el}}$$

Let $\langle N_{\text{vib},v}\rangle$ be the average number of molecules in the vibrational state v, without regard to what translational, rotational, and electronic states these molecules are in. To find $\langle N_{\text{vib},v}\rangle$ we must add up all the $\langle N_r\rangle$'s that correspond to different translational, rotational, and electronic states s, t, and u but the same vibrational state. Hence, we must sum $\langle N_r\rangle$ over s, t, and u keeping the vibrational state v fixed:

$$\langle N_{\text{vib},v}\rangle = \frac{Ne^{-\beta\varepsilon_{\text{vib},v}}}{z_\text{tr}z_\text{rot}z_\text{vib}z_\text{el}} \sum_s e^{-\beta\varepsilon_{\text{tr},s}} \sum_t e^{-\beta\varepsilon_{\text{rot},t}} \sum_u e^{-\beta\varepsilon_{\text{el},u}}$$

$$\frac{\langle N_{\text{vib},v}\rangle}{N} = \frac{e^{-\varepsilon_{\text{vib},v}/kT}}{z_\text{vib}} = \frac{e^{-\varepsilon_{\text{vib},v}/kT}}{\sum_v e^{-\varepsilon_{\text{vib},v}/kT}} \qquad \text{for } \langle N_r\rangle \ll 1 \tag{22.75}$$

where (22.60) was used.

Equation (22.75) has the same form as (22.69). Similar equations hold for translational, rotational, and electronic populations. The condition for the validity of (22.75) is $\langle N_r\rangle \ll 1$, but $\langle N_r\rangle \ll 1$ does not require that $\langle N_{\text{vib},v}\rangle$ be much less than 1. If two molecules are in the same vibrational state v but are in different translational states, the molecular state r differs for these molecules. The huge number of available translational states allows many molecules to have the same vibrational quantum number without violating $\langle N_r\rangle \ll 1$.

Some examples of the Boltzmann distribution are Eq. (15.52) for the Maxwell distribution of kinetic energy in gases and Eq. (15.72) for the distribution of potential energy of gas molecules in a gravitational field. The Boltzmann distribution plays a key role in the distribution of ions around a given ion in an electrolyte solution [note the kT in the Debye–Hückel equation (10.82)], in the degree of orientation polarization that occurs when a dielectric composed of polar molecules is placed in an electric field [note the kT in (14.19)], and in the distribution of double-layer ions and dipoles near a charged electrode (Sec. 14.15).

To get a feeling for the Boltzmann distribution, we shall apply it to the molecular vibrational energy levels of an ideal diatomic gas. In the harmonic-oscillator approximation, the vibrational energy of a diatomic molecule is given by (21.22) as $\varepsilon_{\text{vib}} = (v + \frac{1}{2})h\nu$, where $v = 0, 1, 2, \ldots$. The zero level of energy is arbitrary, and we choose to measure the energy starting from the ground state $v = 0$; thus, we subtract $\frac{1}{2}h\nu$ from each level, and write $\varepsilon_{\text{vib}} = vh\nu$. [If this seems disturbing, note that the numerator and each term in the denominator of (22.75) contain the factor $e^{-h\nu/2kT}$ and this factor cancels.] The levels are equally spaced, the spacing being $\Delta\varepsilon_v = h\nu = \varepsilon_{\text{vib}}/v$. Thus (22.75) can be written as

$$\frac{\langle N_v \rangle}{N} = \frac{e^{-v\,\Delta\varepsilon_v/kT}}{\sum_{v=0}^{\infty} e^{-v\,\Delta\varepsilon_v/kT}} = \frac{e^{-v\,\Delta\varepsilon_v/kT}}{z_{\text{vib}}} \tag{22.76}$$

where $\langle N_v \rangle$ is the average number of molecules in vibrational state v.

The populations of vibrational levels thus depend on the ratio $\Delta\varepsilon_v/kT$ and on the quantum number v. Suppose $h\nu = 4kT$, so that $\Delta\varepsilon_v/kT = 4$. The molecular vibrational partition function in (22.76) is then $z_{\text{vib}} = e^0 + e^{-4} + e^{-8} + e^{-12} + \cdots = 1.01866$. [The terms decrease rapidly, so it is easy enough to sum the series on a calculator. However, time can be saved by noting that z_{vib} is a geometric series whose sum is $1/(1 - e^{-4})$; see Eq. (22.89).] Equation (22.76) with $\Delta\varepsilon_v/kT = 4$ gives the fractional populations of the $v = 0, 1, 2, \ldots$ levels as $e^0/z_{\text{vib}}, e^{-4}/z_{\text{vib}}, e^{-8}/z_{\text{vib}}, \ldots$. The calculated percent populations are:

$\Delta\varepsilon_v/kT = 4$

v	0	1	2	3
%	98.2	1.8	0.03	0.0006

Similarly, for $\Delta\varepsilon_v/kT = 1$ and $\Delta\varepsilon_v/kT = 0.2$, one finds:

$\Delta\varepsilon_v/kT = 1$

v	0	1	2	3	4	5	6	7
%	63.2	23.3	8.6	3.1	1.2	0.4	0.2	0.06

$\Delta\varepsilon_v/kT = 0.2$

v	0	1	2	3	4	5	⋯	10	⋯	15
%	18.1	14.8	12.2	9.9	8.1	6.7	⋯	2.5	⋯	0.9

When the vibrational energy spacing is substantially greater than kT, the molecules pile up in the ground vibrational level; this is the low-temperature behavior. When $\Delta\varepsilon_v \ll kT$ (high T), the distribution tends toward uniformity (Fig. 22.5).

What is the physical reason behind the Boltzmann distribution? For a fixed system energy, any one given distribution of the total system energy among the molecules is as likely to occur as any other distribution. (This follows from our postulate in Sec. 22.2 that the probability of a microstate is a function only of the system energy.) The probability that a molecule has a given energy is thus proportional to the number of ways of distributing the rest of the energy among the other molecules. As the amount of energy given one particular molecule increases, the number of ways of distributing the remaining energy decreases, so the probability $\langle N_s \rangle / N$ in (22.69) decreases with increasing ε_s.

When the condition $\langle N_r \rangle \ll 1$ is not obeyed, the Boltzmann distribution doesn't hold. The average populations of the one-particle states then depend on whether the particles are bosons or fermions. There is no limit on the number of bosons that can have the same molecular state, but for fermions, each molecular state can hold no more than one fermion (Sec. 19.6). For a system containing only one species of particle, the correct result for the average population of state r turns out to be (for a derivation, see *Andrews*, chap. 9)

$$\langle N_r \rangle_{\mathrm{BE}}^{\mathrm{FD}} = \frac{1}{e^{-\mu/N_0 kT} e^{\varepsilon_r/kT} \pm 1} \qquad \text{nonloc. ident. nonint. particles} \qquad (22.77)$$

where μ is the chemical potential of the species and the upper sign is for fermions and the lower sign for bosons. (For electrons and other charged species, μ is replaced by the electrochemical potential $\tilde{\mu}$.) Note that with the plus sign, the denominator exceeds the numerator and $\langle N_r \rangle$ is less than 1, as it should be for fermions.

Bosons are said to obey *Bose–Einstein statistics.* Fermions obey *Fermi–Dirac statistics.* Particles that obey the Boltzmann distribution law (22.69) are said to obey *Boltzmann* (or *Maxwell–Boltzmann*) statistics and are sometimes called *boltzons.* Of course, all particles are in reality either bosons or fermions, and there are no boltzons. However, when $\langle N_r \rangle \ll 1$, Eq. (22.77) reduces to the Boltzmann distribution law (22.69), and so fermions and bosons behave like boltzons when there are many more states available than there are particles.

When $\langle N_r \rangle \ll 1$, the denominator of (22.77) must be very large compared with the numerator, which is 1; hence the ± 1 in the denominator can be neglected, to give $\langle N_r \rangle \approx e^{\mu/N_0 kT} e^{-\varepsilon_r/kT}$. In Prob. 22.6 it is shown that $\mu = -N_0 kT \ln(z/N)$, so that $e^{\mu/N_0 kT} = e^{-\ln(z/N)} = N/z$. Hence $\langle N_r \rangle \approx (N/z)e^{-\varepsilon_r/kT}$, which is Eq. (22.69).

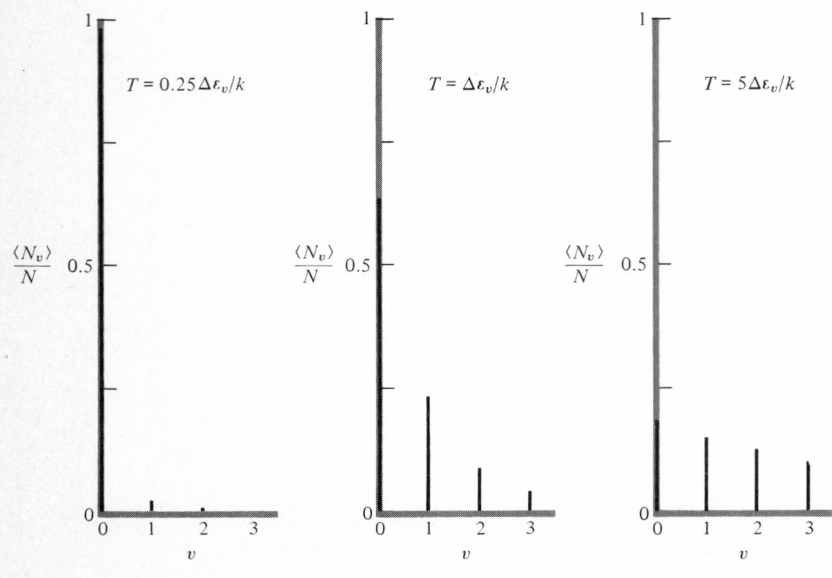

Figure 22.5
Fractional populations of the four lowest one-dimensional harmonic-oscillator vibrational levels for three different values of $kT/\Delta\varepsilon_v$.

22.6 STATISTICAL THERMODYNAMICS OF IDEAL DIATOMIC AND MONATOMIC GASES

We now explicitly evaluate Z for a pure ideal gas of diatomic or monatomic molecules. To find Z in (22.61), we must find the translational, rotational, vibrational, and electronic partition functions.

Translational partition function. Equation (22.60) gives $z_{tr} = \sum_s e^{-\beta\varepsilon_{tr,s}}$. For the translational energies, we use the energies (18.40) of a particle in a rectangular box with sides a, b, c; thus, $8m\varepsilon_{tr}/h^2 = n_x{}^2/a^2 + n_y{}^2/b^2 + n_z{}^2/c^2$, where each quantum number goes from 1 to ∞, independently of the others. To sum over all translational quantum states, we sum over all the quantum numbers:

$$z_{tr} = \sum_{n_x=1}^{\infty}\sum_{n_y=1}^{\infty}\sum_{n_z=1}^{\infty} e^{-(\beta h^2/8m)(n_x{}^2/a^2+n_y{}^2/b^2+n_z{}^2/c^2)}$$

$$z_{tr} = \sum_{n_x=1}^{\infty} e^{-(\beta h^2/8ma^2)n_x{}^2} \sum_{n_y=1}^{\infty} e^{-(\beta h^2/8mb^2)n_y{}^2} \sum_{n_z=1}^{\infty} e^{-(\beta h^2/8mc^2)n_z{}^2} \tag{22.78}$$

The sums in (22.78) can be evaluated exactly, but to do so involves advanced mathematics. Instead, we shall use an approximate treatment that is simple and highly accurate. The approximation we make is to replace each sum by an integral. This procedure is accurate provided the terms in the sum change very little from one term to the next. Consider $\sum_{n=0}^{\infty} f(n)$, where $f(n) \approx f(n+1)$ for all n. We have

$$\sum_{n=0}^{\infty} f(n) = f(0) + f(1) + \cdots = f(0)\int_0^1 dn + f(1)\int_1^2 dn + \cdots$$

$$= \int_0^1 f(0)\,dn + \int_1^2 f(1)\,dn + \cdots$$

Since the function $f(n)$ varies very slowly with n, we can set $f(0) \approx f(n)$ for n in the range 0 to 1 and $f(1) \approx f(n)$ for n in the range 1 to 2, etc. Therefore $\sum_{n=0}^{\infty} f(n) \approx \int_0^1 f(n)\,dn + \int_1^2 f(n)\,dn + \cdots = \int_0^{\infty} f(n)\,dn$.

Because $f(n)$ varies slowly, we have $f(0) \approx f(1) \approx f(2) \approx \cdots$. Hence, many terms contribute substantially to the sum, and the relative contribution of any one term is small and can be neglected. If we like, we can start the sum at $n = 1$ instead of $n = 0$. Likewise, we can start the integration at $n = 1$ instead of $n = 0$. Therefore

$$\sum_{n=0}^{\infty} f(n) \approx \sum_{n=1}^{\infty} f(n) \approx \int_0^{\infty} f(n)\,dn \approx \int_1^{\infty} f(n)\,dn \tag{22.79}$$

provided $|[f(n+1) - f(n)]/f(n)| \ll 1$. Whether the lower limit is taken as 0 or 1 is simply a matter of which integral is easier to evaluate.

Let us check that the approximation (22.79) is applicable to the sums in (22.78). For the n_x sum we have

$$\frac{f(n_x+1) - f(n_x)}{f(n_x)} = \frac{e^{-\beta\varepsilon(n_x+1)} - e^{-\beta\varepsilon(n_x)}}{e^{-\beta\varepsilon(n_x)}} = e^{-\Delta\varepsilon/kT} - 1 \tag{22.80}$$

where $\varepsilon(n_x) = (h^2/8ma^2)n_x{}^2$ and $\Delta\varepsilon$ is the spacing between adjacent levels of x translational energy. The translational levels are extremely close together, and for typical conditions one finds $\Delta\varepsilon/kT \approx 10^{-9}$ (Prob. 22.7). Hence, $e^{-\Delta\varepsilon/kT} - 1 = e^{-10^{-9}} - 1 = (1 - 10^{-9} + \cdots) - 1 \approx -10^{-9}$ [where the Taylor series (8.42) was used]. Thus f decreases by 1 part in 10^9 for each increase of 1 in n_x, and replacement of the sum by an integral is eminently justified. More generally, we have shown that the molecular-partition-function sum $\sum_s e^{-\varepsilon_s/kT}$ can be replaced by an integral whenever $\Delta\varepsilon_s \ll kT$, where $\Delta\varepsilon_s$ is the spacing between adjacent levels of the kind of energy being considered (translational, rotational, etc.).

Use of (22.79) and of integral 2 in Table 15.1 gives for the first sum in (22.78)

$$\sum_{n_x=1}^{\infty} e^{-(\beta h^2/8ma^2)n_x^2} = \int_0^{\infty} e^{-(\beta h^2/8ma^2)n_x^2}\, dn_x = \frac{1}{2}\left(\frac{8m\pi}{\beta h^2}\right)^{1/2} a$$

Similarly, the n_y and n_z sums equal $\frac{1}{2}(8m\pi/\beta h^2)^{1/2}b$ and $\frac{1}{2}(8m\pi/\beta h^2)^{1/2}c$. Thus $z_{\text{tr}} = \frac{1}{8}(8m\pi/\beta h^2)^{3/2}abc$. Since $\beta = 1/kT$ and $abc = V$ (the volume), we have

$$z_{\text{tr}} = (2\pi mkT/h^2)^{3/2}V \tag{22.81}$$

$$\ln z_{\text{tr}} = \tfrac{3}{2}\ln(2\pi mk/h^2) + \tfrac{3}{2}\ln T + \ln V \tag{22.82}$$

Rotational partition function. From Secs. 21.3 and 18.14, the rotational quantum state of a diatomic molecule is defined by the quantum numbers J and M_J, and the energy levels (in the rigid-rotor approximation) are $\varepsilon_{\text{rot}} = (\hbar^2/2I)J(J+1)$, where I is the moment of inertia [Eq. (21.15)] and J goes from 0 to ∞. For J fixed, M_J takes on the $2J+1$ integral values from $-J$ to $+J$. Since ε_{rot} is independent of M_J, each level is $(2J+1)$-fold degenerate. To evaluate z_{rot} by summing over all rotational states as in Eq. (22.50) we must sum over both J and M_J. It's a bit easier to use a sum over energy levels instead of states. The energy levels are defined by J, and the degeneracy is $2J+1$; so Eq. (22.73) gives

$$z_{\text{rot}} = \sum_{J=0}^{\infty}(2J+1)e^{-(\hbar^2/2IkT)J(J+1)} = \sum_{J=0}^{\infty}(2J+1)e^{-(\Theta_{\text{rot}}/T)J(J+1)}$$

where the *characteristic rotational temperature* Θ_{rot} was defined as

$$\Theta_{\text{rot}} \equiv \hbar^2/2Ik = \tilde{B}hc/k \tag{22.83}$$

where (21.33) was used for the rotational constant $\tilde{B}$. The parameter Θ_{rot} has the dimensions of temperature but is not a temperature in the physical sense of the word.

If Θ_{rot}/T is small, the rotational levels are closely spaced compared with kT and we can approximate the sum by an integral (as we did for z_{tr}). Thus

$$z_{\text{rot}} \approx \int_0^{\infty}(2J+1)e^{-(\Theta_{\text{rot}}/T)J(J+1)}\, dJ = \int_0^{\infty} e^{-(\Theta_{\text{rot}}/T)w}\, dw$$

where we made the change of variable $w \equiv J(J+1) = J^2 + J$, $dw = (2J+1)\,dJ$. Using $\int e^{-bw}\,dw = -b^{-1}e^{-bw}$, we get

$$z_{\text{rot}} \approx T/\Theta_{\text{rot}} \qquad \text{heteronuclear, } \Theta_{\text{rot}} \ll T \tag{22.84}$$

Equation (22.84) is valid only for heteronuclear diatomic molecules. For homonuclear diatomics, the Pauli principle (Sec. 19.5) requires that ψ_N (the wave function for nuclear motion) be symmetric with respect to interchange of the identical nuclei if the nuclei are bosons or antisymmetric if the nuclei are fermions. This restriction on ψ_N cuts the number of available quantum states in half compared with a heteronuclear diatomic, for which there is no symmetry restriction. (For example, if the identical nuclei are bosons, a symmetric nuclear-spin function requires a symmetric rotational wave function and an antisymmetric nuclear-spin function requires an antisymmetric rotational function, but the combination of a symmetric spin function with an antisymmetric rotational function or an antisymmetric spin function with a symmetric rotational function is excluded.) Chemists customarily ignore the nuclear-spin quantum states in calculating z (see Sec. 22.9 for an explanation of why this is OK), but for consistency between homonuclear and heteronuclear partition functions, one must allow for the reduction in quantum states by dividing z_{rot} by 2 for a homonuclear diatomic. (This explanation is a bit vague; a better and more explicit discussion is given in *McQuarrie*, pp. 104–105.) Therefore, $z_{rot} \approx T/2\Theta_{rot}$ for a homonuclear diatomic.

To have a single formula for both homonuclear and heteronuclear diatomics, we define the *symmetry number* σ as 2 for a homonuclear diatomic and 1 for a heteronuclear diatomic, and we include a factor $1/\sigma$ in z_{rot}:

$$z_{rot} \approx \frac{T}{\sigma\Theta_{rot}} = \frac{2IkT}{\sigma\hbar^2} \qquad \text{for } T \gg \Theta_{rot} \tag{22.85}$$

Some values of Θ_{rot} in kelvins for ground electronic states are:

Molecule	H_2	$H^{35}Cl$	CO	N_2	$^{35}Cl_2$	Na_2
Θ_{rot}/K	85.3	15.02	2.77	2.862	0.351	0.222

The light molecule H_2 has a small moment of inertia and hence a relatively high Θ_{rot}. Except for H_2 (and its isotopic species D_2 and HD), Θ_{rot} for diatomics is much less than T at temperatures that chemists commonly deal with, so we can use the approximation (22.85).

Actually, the derivation of (22.85) was a bit phony. It is true that the rotational spacing is generally small compared with kT; however, the $2J + 1$ factor in the z_{rot} sum causes substantial term-to-term variation for the low-J terms, and we have not really justified replacing the sum by an integral. A more rigorous derivation is given in Prob. 22.18, where it is shown that

$$z_{rot} = \frac{T}{\sigma\Theta_{rot}}\left[1 + \frac{1}{3}\frac{\Theta_{rot}}{T} + \frac{1}{15}\left(\frac{\Theta_{rot}}{T}\right)^2 + \frac{4}{315}\left(\frac{\Theta_{rot}}{T}\right)^3 + \cdots\right] \tag{22.86}$$

For $\Theta_{rot} \ll T$, this accurate formula reduces to (22.85).

Vibrational partition function. The harmonic-oscillator approximation gives the vibrational energies of a diatomic molecule as $(v + \frac{1}{2})h\nu$, where v goes from 0 to ∞ and there is no degeneracy (Sec. 21.3). The choice of zero level of energy is arbitrary, and it is customary in molecular statistical mechanics to take the energy zero at the lowest available energy level of the molecule. This level has zero rotational energy and $\frac{1}{2}h\nu$ vibrational energy. We therefore write $\varepsilon_{vib} = (v + \frac{1}{2})h\nu - \frac{1}{2}h\nu = vh\nu$, where ε_{vib} is measured from the $v = 0$ level. The expression for the thermodynamic internal energy that is obtained from the partition function will then be U relative to the lowest molecular energy level. (See also Prob. 22.3.) (To be fully consistent, we should have subtracted the zero-point translational energy from ε_{tr}, but this is so small as to be utterly negligible.)

We have

$$z_{\text{vib}} = \sum_{v} e^{-\beta\varepsilon_{\text{vib},v}} = \sum_{v=0}^{\infty} e^{-vh\nu/kT} = \sum_{v=0}^{\infty} e^{-v\Theta_{\text{vib}}/T}$$

$$\Theta_{\text{vib}} \equiv h\nu/k = \tilde{\nu}hc/k \tag{22.87}$$

where Θ_{vib} is the *characteristic vibrational temperature* and (21.35) was used.

Some values of Θ_{vib} for ground electronic states are:

Molecule	H_2	$H^{35}Cl$	N_2	CO	$^{35}Cl_2$	I_2
$\Theta_{\text{vib}}/\text{K}$	5990	4152	3352	3084	795	307

(22.88)

The generally high values of Θ_{vib} compared with ordinary temperatures show that the vibrational levels are not closely spaced compared with kT. (Recall from Sec. 21.3 that for most diatomic molecules there is very little occupancy of excited vibrational levels at room temperature.) Hence, the vibrational sum cannot be replaced by an integral. This is no cause for alarm, because the sum is easily evaluated exactly.

Recall from high-school algebra the following formula for the sum of a geometric series:

$$1 + x + x^2 + x^3 + \cdots = \sum_{n=0}^{\infty} x^n = \frac{1}{1-x} \qquad \text{for } |x| < 1 \tag{22.89}$$

[A derivation of (22.89) is given in Prob. 22.31.] The sum in z_{vib} corresponds to (22.89) with $x = e^{-h\nu/kT}$ and $n = v$. Hence

$$z_{\text{vib}} = \frac{1}{1 - e^{-h\nu/kT}} \tag{22.90}$$

The rigid-rotor and harmonic-oscillator expressions for ε_{rot} and ε_{vib} are only approximations, since they omit the effects of anharmonicity, centrifugal distortion, and vibration–rotation interaction (Sec. 21.3). Our expressions for z_{rot} and z_{vib} are therefore approximations. Fortunately, the corrections introduced by anharmonicity, etc., are generally quite small (except at high temperatures) and need only be considered in highly precise work. (For details of these corrections, see *McQuarrie*, prob. 6-24; *Davidson*, pp. 116–119.) We should point out that ν in (22.90) is to be taken as ν_0 [Eq. (21.36)], the fundamental frequency corresponding to the $v = 0 \rightarrow 1$ transition (rather than as the equilibrium vibrational frequency ν_e). Likewise, I in (22.85) is I_0, the moment of inertia averaged over the zero-point vibrations, and is calculated from the rotational constant B_0 [Eq. (21.32) with $v = 0$].

Electronic partition function. We shall calculate z_{el} as a sum over electronic energy levels (rather than states), so that we include the degeneracy g_{el} of each electronic level. Let the electronic energy levels be numbered 0, 1, 2, ... in order of increasing energy, and let their degeneracies be $g_{\text{el},0}, g_{\text{el},1}, g_{\text{el},2}, \ldots$. As noted above, we take the zero level of energy at the lowest available level (the $J = 0$, $v = 0$ level of the ground electronic level); hence the energy of the ground electronic level is taken as zero: $\varepsilon_{\text{el},0} = 0$. The energy of each excited electronic level is then measured relative to the ground electronic level. With $\varepsilon_{\text{el},0} = 0$, we have

$$z_{\text{el}} = g_{\text{el},0} + g_{\text{el},1}e^{-\beta\varepsilon_{\text{el},1}} + g_{\text{el},2}e^{-\beta\varepsilon_{\text{el},2}} + \cdots \tag{22.91}$$

Since there is no general formula for the ε_{el}'s, the series is added term by term using spectroscopically observed electronic energies.

For nearly all diatomic molecules, $\varepsilon_{\text{el},1}$ is very much greater than kT at room temperature (recall that $\varepsilon_{\text{el}} > \varepsilon_{\text{vib}} > \varepsilon_{\text{rot}}$), and all terms in (22.91) after the first term contribute negligibly for all temperatures up to 5000 or 10,000 K. Hence

$$z_{\text{el}} = g_{\text{el},0} \qquad T \text{ not extremely high} \tag{22.92}$$

The main exception to (22.92) is NO, which has a very-low-lying excited electronic state (see *Kestin and Dorfman*, p. 261). (O_2 has an excited electronic state that contributes significantly to z_{el} above 1500 K.)

For most diatomics, the ground electronic level is nondegenerate: $g_{\text{el},0} = 1$. An important exception is O_2, for which $g_{\text{el},0} = 3$, due to spin degeneracy; recall that O_2 has a triplet ground level. Another exception is NO, which has an odd number of electrons; here, $g_{\text{el},0} = 2$, due to the two possible orientations of the spin of the unpaired electron. (The general rule for g_{el} of a diatomic molecule is given in *Hirschfelder, Curtiss, and Bird*, p. 119.)

Monatomic molecules with filled subshells (Be, He, Ne, Ar, ...) have $g_{\text{el},0} = 1$. For H, Li, Na, K, ..., the spin degeneracy of the odd electron gives $g_{\text{el},0} = 2$. For F, Cl, Br, and I, $g_{\text{el},0} = 4$ and there is a low-lying excited state that contributes to z_{el}. (See *McQuarrie*, sec. 5-2 for more details.)

In deriving $z = z_{\text{tr}} z_{\text{rot}} z_{\text{vib}} z_{\text{el}}$ [Eq. (22.58)] we assumed the four kinds of energies to be independent of one another and summed separately over each kind of energy. Actually, this assumption is false. The bond distance and force constant of a diatomic molecule change from one electronic state to another, so each electronic state has a different vibrational frequency and a different moment of inertia. To allow for this, $z_{\text{rot}} z_{\text{vib}} z_{\text{el}}$ must be replaced by

$$g_{\text{el},0} z_{\text{vib},0} z_{\text{rot},0} + g_{\text{el},1} e^{-\beta\varepsilon_{\text{el},1}} z_{\text{vib},1} z_{\text{rot},1} + \cdots$$

where $z_{\text{vib},0}$ and $z_{\text{rot},0}$ are calculated using the vibrational frequency and moment of inertia of the ground electronic level, $z_{\text{vib},1}$ and $z_{\text{rot},1}$ use the parameters of the first excited electronic level, etc. Since the contributions of excited electronic levels are generally very small (except at very high T), it is usually an adequate approximation to ignore the change in z_{vib} and z_{rot} from one electronic state to another. The accurate expression then reduces to the form $z_{\text{vib}} z_{\text{rot}} z_{\text{el}}$ used earlier.

Equation of state. Having found the molecular partition function $z = z_{\text{tr}} z_{\text{rot}} z_{\text{vib}} z_{\text{el}}$, we are now ready, willing, and able to calculate the thermodynamic properties of an ideal gas in terms of molecular properties. We begin with the pressure. Equation (22.37) reads $P = kT(\partial \ln Z/\partial V)_{T,N}$. For an ideal gas, $\ln Z$ is given by (22.61). The molecular vibrational, rotational, and electronic energies all depend on properties of the gas molecules but are independent of V; in contrast, ε_{tr} does depend on V. Hence, only z_{tr} is a function of V, as can be verified from (22.91), (22.90), (22.85), and (22.82). Use of (22.61) gives

$$P = kT\left(\frac{\partial \ln Z}{\partial V}\right)_{T,N} = kT\left[\frac{\partial(N \ln z_{\text{tr}})}{\partial V}\right]_{T,N} = NkT\left(\frac{\partial \ln z_{\text{tr}}}{\partial V}\right)_T$$

Equation (22.82) gives $(\partial \ln z_{\text{tr}}/\partial V)_T = 1/V$. Hence, $P = NkT/V$, or $PV = NkT$. Since $N = N_0 n$ (where N_0 is the Avogadro constant and n the number of moles of gas), we have $PV = nN_0kT$. The absolute temperature scale was defined in Sec. 1.5 to make $PV = nRT$ hold. Hence, $N_0k = R$, and $k = R/N_0$. This agrees with Eq. (3.57) and shows that k in (22.27) is Boltzmann's constant: $\beta = 1/kT = N_0/RT$. Also, $Nk = NR/N_0 = nR$:

$$Nk = nR \tag{22.93}$$

Internal energy. We now calculate the internal energy of an ideal diatomic gas. Since we took the zero level of energy at the lowest available molecular energy level, we shall be finding $U - U_0$, where U_0 is the thermodynamic internal energy of a hypothetical ideal gas in which every molecule is in the lowest translational, rotational, vibrational, and electronic state. This would be the internal energy at absolute zero if the gas didn't condense and if the molecules were not fermions. Thus Eq. (22.62) reads $U - U_0 = U_{\text{tr}} + U_{\text{rot}} + U_{\text{vib}} + U_{\text{el}}$. From (22.63), (22.93), and (22.82), we have $U_{\text{tr}} = nRT^2(\partial \ln z_{\text{tr}}/\partial T)_V = nRT^2(3/2T) = 3nRT/2$. Similarly, U_{rot}, U_{vib}, and U_{el} are found from (22.63), (22.85), (22.90), and (22.92).

The results for a diatomic ideal gas at temperatures that are not extremely high or low are (Prob. 22.19)

$$U - U_0 = U_{\text{tr}} + U_{\text{rot}} + U_{\text{vib}} + U_{\text{el}} \tag{22.94}$$

$$U_{\text{tr}} = \tfrac{3}{2}nRT \tag{22.95}$$

$$U_{\text{rot}} = nRT \tag{22.96}$$

$$U_{\text{vib}} = nR\frac{h\nu}{k}\frac{1}{e^{h\nu/kT} - 1} = nR\Theta_{\text{vib}}\frac{1}{e^{\Theta_{\text{vib}}/T} - 1} \tag{22.97}$$

$$U_{\text{el}} = 0 \tag{22.98}$$

A monatomic gas has no rotational or vibrational degrees of freedom and has $\bar{U} - \bar{U}_0 = \bar{U}_{\text{tr}} = \tfrac{3}{2}RT$, in agreement with the classical-kinetic-theory result (15.17).

Heat capacity. Differentiation of (22.94) with respect to T at constant V and n gives

$$C_V = C_{V,\text{tr}} + C_{V,\text{rot}} + C_{V,\text{vib}} + C_{V,\text{el}} \tag{22.99}$$

where $C_{V,\text{tr}} \equiv (\partial U_{\text{tr}}/\partial T)_{V,n}$, etc. Differentiation of (22.95) to (22.98) gives for an ideal gas of diatomic molecules at moderate temperatures:

$$C_{V,\text{tr}} = \tfrac{3}{2}nR \tag{22.100}$$

$$C_{V,\text{rot}} = nR \tag{22.101}$$

$$C_{V,\text{vib}} = nR\left(\frac{\Theta_{\text{vib}}}{T}\right)^2 \frac{e^{\Theta_{\text{vib}}/T}}{(e^{\Theta_{\text{vib}}/T} - 1)^2} \tag{22.102}$$

$$C_{V,\text{el}} = 0 \tag{22.103}$$

Note that U and C_V of the ideal gas are functions of T only, in agreement with (2.98) and (2.100).

Equations (22.100) and (22.101) for $\bar{C}_{V,\text{tr}}$ and $\bar{C}_{V,\text{rot}}$ agree with the classical-statistical-mechanical equipartition theorem of $\tfrac{1}{2}R$ for each quadratic term in the energy (Sec. 15.10), but $\bar{C}_{V,\text{vib}}$ does not agree with the equipartition theorem. We obtained z_{tr} and z_{rot} by using the fact that the translational and rotational energy spacings were very small compared with kT, so that the sums over discrete energy levels could be replaced by integrals over a continuous range of energy. Since continuous values for energy correspond to classical mechanics, we obtained the classical results for $C_{V,\text{tr}}$ and $C_{V,\text{rot}}$. However, vibrational levels are not closely spaced compared with kT, and the classical equipartition theorem fails for $C_{V,\text{vib}}$.

Figure 22.6 plots $\bar{C}_{V,\text{vib}}$ vs. T. At high temperature, $\bar{C}_{V,\text{vib}}$ goes to the classical equipartition value R (see Prob. 22.20).

The result $C_{V,\text{rot}} = nR$ applies only at temperatures for which $T \gg \Theta_{\text{rot}}$. At low temperatures, we must use the series (22.86) for z_{rot} and calculate U_{rot} and $C_{V,\text{rot}}$ from (22.86). At very low temperatures, (22.86) fails, and z_{rot} must be calculated by direct term-by-term summation. We omit details but simply plot $\bar{C}_{V,\text{rot}}$ vs. T in Fig. 22.7. The classical value R is reached at $T \approx 1.5\Theta_{\text{rot}}$.

For a monatomic gas with no low-lying excited electronic states, there is only a translational contribution to C_V, and $\bar{C}_V = \frac{3}{2}R$, in agreement with (15.18).

Entropy. Equations (22.64) to (22.66) allow the entropy of an ideal gas to be calculated. Using these equations, (22.95) to (22.98), (22.82), (22.85), (22.90), (22.92), and $PV = NkT$, we readily find for an ideal diatomic gas (Prob. 22.21)

$$S_{\text{tr}} = \frac{5}{2}nR + nR \ln \left[\frac{(2\pi m)^{3/2}}{h^3} \frac{(kT)^{5/2}}{P} \right] \tag{22.104}$$

$$S_{\text{rot}} = nR + nR \ln \frac{T}{\sigma\Theta_{\text{rot}}} \tag{22.105}$$

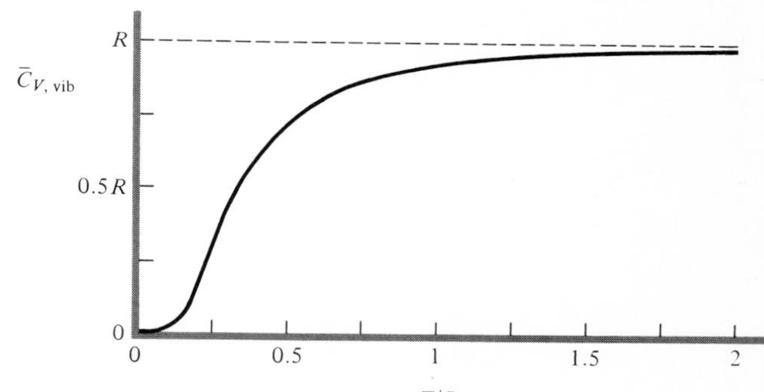

Figure 22.6
The vibrational contribution to $\bar{C}_V$ of a diatomic gas. (The dashed line is the classical equipartition result of R at all temperatures.)

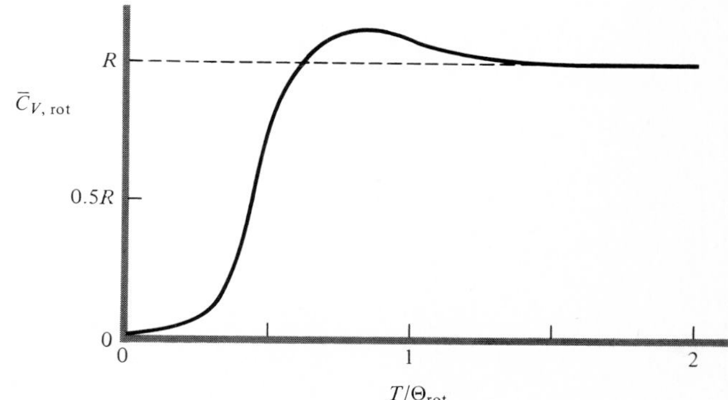

Figure 22.7
The rotational contribution to $\bar{C}_V$ of a diatomic gas. (The dashed line is the classical equipartition result.)

$$S_{\text{vib}} = nR\frac{\Theta_{\text{vib}}}{T}\frac{1}{e^{\Theta_{\text{vib}}/T}-1} - nR\ln\left(1-e^{-\Theta_{\text{vib}}/T}\right) \tag{22.106}$$

$$S_{\text{el}} = nR\ln g_{\text{el},0} \tag{22.107}$$

where Θ_{rot} and Θ_{vib} are given by (22.83) and (22.87). S is the sum of (22.104) to (22.107) (provided T is not extremely high or low).

Substitution of $P = (P/P^\circ)(1{,}013{,}250 \text{ dyn cm}^{-2})$, where $P^\circ \equiv 1$ atm (see Sec. 1.5), $m = (M_r/N_0)$ g/mol, where M_r is the (dimensionless) molecular weight, $T = (T/\text{K})$ K (where K is one kelvin), and the cgs values of h, k, and N_0 in (22.104) gives S_{tr} in a form convenient for calculation (Prob. 22.22):

$$\bar{S}_{\text{tr}} = R[1.5\ln M_r + 2.5\ln(T/\text{K}) - \ln(P/P^\circ) - 1.1649] \tag{22.108}$$

For a monatomic gas with a nondegenerate ground electronic state, $S_{\text{rot}} = S_{\text{vib}} = S_{\text{el}} = 0$, and $S = S_{\text{tr}}$.

Example Calculate $\bar{S}^\circ_{298}$ of N_2.

Even though N_2 is slightly nonideal at 25°C and 1 atm, the standard state is the hypothetical ideal gas at 1 atm, so the above ideal-gas equations apply. Tables earlier in this section give $\Theta_{\text{rot}} = 2.862$ K and $\Theta_{\text{vib}} = 3352$ K for N_2; also, $g_{\text{el},0} = 1$. Substitution in (22.105) to (22.108) gives at 298.15 K

$$\bar{S}^\circ_{\text{tr}} = (1.987 \text{ cal mol}^{-1}\text{ K}^{-1})(1.5\ln 28.013 + 2.5\ln 298.15 - \ln 1 - 1.1649)$$

$$= 35.92 \text{ cal mol}^{-1}\text{ K}^{-1}$$

$$\bar{S}^\circ_{\text{rot}} = (1.987 \text{ cal mol}^{-1}\text{ K}^{-1})\left[1 + \ln\frac{298.15\text{ K}}{2(2.862\text{ K})}\right] = 9.84 \text{ cal mol}^{-1}\text{ K}^{-1}$$

$$\bar{S}^\circ_{\text{vib}} = 0.0003 \text{ cal mol}^{-1}\text{ K}^{-1}$$

$$\bar{S}^\circ_{\text{el}} = 0$$

$$\bar{S}^\circ_{298} = 45.76 \text{ cal mol}^{-1}\text{ K}^{-1}$$

For comparison, the experimental value (found by the methods of Sec. 5.5) is 45.9 cal mol^{-1} K^{-1}. (Further comparisons between theoretically calculated and experimental entropies are given in Sec. 22.9.)

The calculations of this example reflect a truly impressive synthesis of quantum mechanics, statistical mechanics, and thermodynamics.

22.7 STATISTICAL THERMODYNAMICS OF IDEAL POLYATOMIC GASES

The separation of the molecular energy ε into translational, rotational, vibrational, and electronic contributions holds well for the ground electronic states of most polyatomic molecules, so we have $z = z_{\text{tr}} z_{\text{rot}} z_{\text{vib}} z_{\text{el}}$ for an ideal gas of polyatomic molecules.

Translational partition function. Since ε_{tr} has the same form for polyatomics as for diatomics, Eq. (22.81) gives z_{tr} for polyatomic molecules.

Rotational partition function. For a linear polyatomic molecule, ε_{rot} and the rotational quantum numbers are the same as for a diatomic molecule, so (22.85) gives z_{rot} for a linear polyatomic molecule. For a linear molecule, $\sigma = 2$ if there is a center of symmetry (for example, HCCH, OCO), and $\sigma = 1$ if there is no center of symmetry (for example, HCCF, OCS).

For a nonlinear molecule, the exact rotational energies can be found if the molecule is a spherical or symmetric top (Sec. 21.5), but there is no simple algebraic formula for the quantum-mechanical rotational energies of an asymmetric top. Hence, we have a problem in evaluating z_{rot}. We noted in Sec. 22.6 that replacing the sum in z by an integral amounts to treating the system classically. The classical-mechanical formula for the partition function will be given in Sec. 22.11. Using this formula and the known classical-mechanical expression for ε_{rot} of a polyatomic molecule, one finds for any nonlinear polyatomic molecule (see *McClelland*, sec. 11.6 for the derivation)

$$z_{\text{rot}} = \frac{\pi^{1/2}}{\sigma}\left(\frac{2kT}{\hbar^2}\right)^{3/2}(I_a I_b I_c)^{1/2} \qquad \text{nonlinear} \qquad (22.109)$$

where I_a, I_b, and I_c are the molecule's principal moments of inertia (Sec. 21.5). Equation (22.109) holds provided T is not extremely low.

The *symmetry number* σ in z_{rot} is one more than the number of ways we can start with an initial orientation of the molecule and rotate it into an indistinguishable orientation. For example, $\sigma = 8$ for the square-planar molecule XeF_4 (Fig. 22.8). The second, third, and fourth orientations in Fig. 22.8 are obtained by 90, 180, and 270° rotations about the C_4 axis; the fifth orientation is obtained from the first by a 180° rotation about the C_2 axis passing through atoms 1 and 3; the sixth, seventh, and eighth orientations are obtained by 90, 180, and 270° rotations applied to the fifth orientation. For CH_3Cl, $\sigma = 3$. The factor $1/\sigma$ arises for the same reason as with diatomic molecules: the Pauli principle restricts the number of possible quantum states, so that for a given nuclear-spin wave function, only $1/\sigma$ of the rotational wave functions are allowed.

Vibrational partition function. Equation (21.48) gives the harmonic-oscillator approximation to the vibrational energy of a polyatomic molecule as the sum of vibrational energies associated with the $3\mathcal{N} - 6$ (or $3\mathcal{N} - 5$) normal modes of vibration. The vibrational quantum numbers vary independently of one another. As usual, when the energy is the sum of independent energies, the partition function is the product of partition functions, one for each kind of energy:

$$z_{\text{vib}} = \prod_{s=1}^{3\mathcal{N}-6} z_{\text{vib},s} = \prod_{s=1}^{3\mathcal{N}-6} \frac{1}{1 - e^{-h\nu_s/kT}} \qquad (22.110)$$

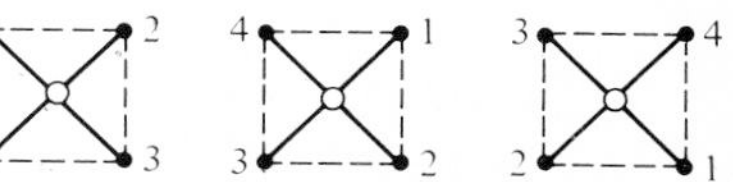

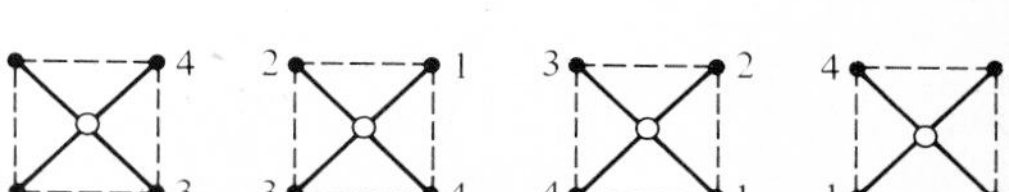

Figure 22.8
The eight indistinguishable orientations of XeF_4 obtainable from one another by rotations.

where ν_s is the frequency of the sth normal mode and (22.90) was used for the partition function of a single vibrational mode. (If the molecule is linear, change $3\mathcal{N} - 6$ to $3\mathcal{N} - 5$ here and below.)

Internal rotation. The molecular partition function must be modified for molecules with internal rotation (for example, CH_3CH_3, CH_3OH, CH_3CCCH_3). See *McClelland*, secs. 5.6, 10.5, and 10.6 for details.

Electronic partition function. For nearly all stable molecules, $g_{el,0}$ is 1, and there are no low-lying excited electronic states. Hence, z_{el} can usually be taken as 1. For species with an odd number of electrons (for example, NO_2 and the CH_3 radical), $g_{el,0}$ is 2 (due to spin degeneracy).

Equation of state. As with diatomic molecules, only z_{tr} is a function of volume, and the equation of state of an ideal polyatomic gas is $PV = NkT$.

Internal energy. Equation (22.63) gives (Prob. 22.22)

$$U_{tr} = \tfrac{3}{2}nRT, \qquad U_{el} = 0 \tag{22.111}$$

$$U_{rot} = \begin{cases} \tfrac{3}{2}nRT & \text{nonlinear} \\ nRT & \text{linear} \end{cases} \tag{22.112}$$

$$U_{vib} = nR \sum_{s=1}^{3\mathcal{N}-6} \frac{\Theta_{vib,s}}{e^{\Theta_{vib,s}/T} - 1} \qquad \text{where } \Theta_{vib,s} \equiv h\nu_s/k \tag{22.113}$$

The sum of these four energies is $U - U_0$.

Heat capacity. Differentiation of the U's gives

$$C_{V,tr} = \tfrac{3}{2}nR, \qquad C_{V,el} = 0 \tag{22.114}$$

$$C_{V,rot} = \begin{cases} \tfrac{3}{2}nR & \text{nonlinear} \\ nR & \text{linear} \end{cases} \tag{22.115}$$

$$C_{V,vib} = \sum_{s=1}^{3\mathcal{N}-6} C_{V,vib,s} \tag{22.116}$$

where $C_{V,vib,s}$ is given by (22.102) with Θ_{vib} replaced by $\Theta_{vib,s}$. Large polyatomic molecules often have low-frequency vibrations that contribute to C_V at room temperature.

Entropy. Application of Eqs. (22.64) to (22.66) shows that S_{tr} is given by (22.104) and (22.108), that S_{rot} for a linear molecule is given by (22.105), and that

$$S_{rot} = \tfrac{3}{2}nR + nR \ln z_{rot} \qquad \text{nonlinear} \tag{22.117}$$

$$S_{vib} = \sum_{s=1}^{3\mathcal{N}-6} S_{vib,s}, \qquad S_{el} = nR \ln g_{el,0} \tag{22.118}$$

where $S_{vib,s}$ is gotten by replacing Θ_{vib} by $\Theta_{vib,s}$ in (22.106) and z_{rot} is given by (22.109).

As a reminder, the equations of Secs. 22.6 and 22.7 apply only to ideal gases.

One final point. Molecular properties (molecular weights, vibrational frequencies, moments of inertia), partition functions, and thermodynamic properties differ slightly for different isotopic species. To calculate the thermodynamic properties of CH_3Cl using statistical mechanics, one does the calculations for $^{35}ClCH_3$ and for $^{37}ClCH_3$ and then takes a weighted average based on the percent abundances of ^{35}Cl and ^{37}Cl.

22.8 IDEAL-GAS THERMODYNAMIC PROPERTIES AND EQUILIBRIUM CONSTANTS

We have seen that $U - U_0$, S, and C_V of an ideal gas are readily calculated using statistical mechanics. The molecular properties needed for these calculations are: (*a*) the molecular weight (which occurs in z_{tr}), (*b*) the molecular geometry (which is needed to calculate the moments of inertia in z_{rot}), (*c*) the molecular vibration frequencies (which occur in z_{vib}), and (*d*) the degeneracy of the ground electronic level and the energies and degeneracies of any low-lying excited electronic levels (which occur in z_{el}). This information is obtained spectroscopically (Chap. 21). Hence, from observations on a molecule's spectrum, we can calculate its thermodynamic properties. For molecules of reasonably simple structure, the gas-phase thermodynamic properties calculated by statistical mechanics are usually more accurate than the values determined from calorimetric measurements (Chap. 5), and many of the values listed in the Appendix are theoretical statistical-mechanical values. For liquids and solids, the intermolecular forces make the statistical-mechanical values less accurate than the calorimetric values.

We now examine the relation between the tabulated gas-phase thermodynamic quantities ΔH_f°, S°, and ΔG_f° and the calculated statistical-mechanical quantities $U - U_0$ and S°. (The standard state of any gas is the hypothetical ideal gas at 1 atm, so the ideal-gas formulas of Secs. 22.6 and 22.7 are applicable.)

As a preliminary, consider the gas-phase reaction $\sum_B \nu_B B = 0$, where ν_B is the stoichiometric coefficient for species B. The reaction's standard change in internal energy at temperature T is $\Delta U_T^\circ \equiv \sum_B \nu_B \bar{U}_{B,T}^\circ$, where $\bar{U}_{B,T}^\circ$ is the standard molar internal energy of B at T. Let ΔU_0° denote ΔU° for the gas-phase reaction at $T = 0$ K. We have

$$\Delta U_T^\circ = \Delta(\bar{U}_T^\circ - \bar{U}_0^\circ) + \Delta U_0^\circ \tag{22.119}$$

$$\Delta(\bar{U}_T^\circ - \bar{U}_0^\circ) \equiv \sum_B \nu_B(\bar{U}_{B,T}^\circ - \bar{U}_{B,0}^\circ), \qquad \Delta U_0^\circ \equiv \sum_B \nu_B \bar{U}_{B,0}^\circ \tag{22.120}$$

Statistical mechanics gives $\bar{U}_T^\circ - \bar{U}_0^\circ$ for each gas, and hence allows $\Delta(\bar{U}_T^\circ - \bar{U}_0^\circ)$ to be calculated. To find ΔU_T° in (22.119) we therefore need ΔU_0°. This is found from the following hypothetical two-step ideal-gas process at $T = 0$:

$$\text{Reactants} \xrightarrow{(a)} \text{gaseous atoms} \xrightarrow{(b)} \text{products} \tag{22.121}$$

For example, for the reaction $CH_4 + 2F_2 \rightarrow CH_2F_2 + 2HF$, the process (22.121) is

$$CH_4(g) + 2F_2(g) \xrightarrow{(a)} C(g) + 4H(g) + 4F(g) \xrightarrow{(b)} CH_2F_2(g) + 2HF(g)$$

ΔU_0° equals $\Delta U_{0,a}^\circ + \Delta U_{0,b}^\circ$ in (22.121). At $T = 0$, all the species are in their ground electronic, vibrational, and rotational states. Hence the energy required for the $T = 0$ process $CH_4(g) \rightarrow C(g) + 4H(g)$ is N_0 times the dissociation energy D_0 of a CH_4

molecule. Therefore

$$\Delta U^\circ_{0,a}/N_0 = D_0(\mathrm{CH_4}) + 2D_0(\mathrm{F_2}), \qquad \Delta U^\circ_{0,b}/N_0 = -D_0(\mathrm{CH_2F_2}) - 2D_0(\mathrm{HF})$$

$$\Delta U^\circ_0 = N_0[D_0(\mathrm{CH_4}) + 2D_0(\mathrm{F_2}) - D_0(\mathrm{CH_2F_2}) - 2D_0(\mathrm{HF})]$$

Generalizing this result, we have for the gas-phase reaction $\sum_\mathrm{B} \nu_\mathrm{B}\mathrm{B} = 0$

$$\Delta U^\circ_0 = -N_0\,\Delta D_0 \qquad \text{where } \Delta D_0 \equiv \sum_\mathrm{B} \nu_\mathrm{B} D_{0,\mathrm{B}} \tag{22.122}$$

Standard enthalpies of formation. $\Delta H^\circ_{f,T}$ of a gaseous compound is readily calculated from $\Delta U^\circ_{f,T}$ using Eq. (5.15). If the elements forming the compound are all gaseous at T and 1 atm, then $\bar{U}^\circ_T - \bar{U}^\circ_0$ for each element and for the compound can be calculated from statistical mechanics; these values give $\Delta(\bar{U}^\circ_T - \bar{U}^\circ_0)$ for the formation reaction [Eq. (22.120)]; $\Delta U^\circ_{f,0}$ is calculated from (22.122), and then $\Delta U^\circ_{f,T}$ is found from (22.119).

If one or more of the elements are not gaseous, additional data are needed to find $\Delta U^\circ_{f,T}$. For example, suppose we want $\Delta U^\circ_{f,T}$ of $\mathrm{CH_4}$, according to $\mathrm{C}(gr) + 2\mathrm{H_2}(g) \to \mathrm{CH_4}(g)$. For the solid graphite, $U^\circ_T - U^\circ_0$ is more accurately found from experimental heat-capacity data than from statistical-mechanical calculations; since $(\partial H/\partial T)_P = C_P$, we have $\bar{H}^\circ_T - \bar{H}^\circ_0 = \int_0^T \bar{C}^\circ_P\,dT$. Also, since step (a) in (22.121) involves vaporization of solid graphite to $\mathrm{C}(g)$, instead of $N_0 D_0$ we use $\Delta\bar{U}^\circ_0$ of vaporization of graphite.

Standard entropies. In Sec. 22.9 we shall show that the convention of setting $C = 0$ in (22.31) agrees with the entropy convention used in thermodynamics (Sec. 5.5). Hence, the thermodynamic S° of any gas is found by setting $P = 1$ atm in S_tr and using (22.64).

Standard Gibbs energies of formation. We have $\Delta G^\circ_{f,T} = \Delta H^\circ_{f,T} - T\,\Delta S^\circ_{f,T}$. Calculation of $\Delta H^\circ_{f,T}$ is discussed above. $\Delta S^\circ_{f,T}$ is found from the entropies of the compound and of its elements. If these are all gases, their entropies can all be calculated by statistical mechanics. For solid and liquid elements, the experimental S must be used.

Standard heat capacities. The equations in Secs. 22.6 and 22.7 give C_V. The ideal-gas equation $\bar{C}_P = \bar{C}_V + R$ then gives $\bar{C}_P$.

Equilibrium constants. For the ideal-gas reaction $\sum_\mathrm{B} \nu_\mathrm{B}\mathrm{B} = 0$, we have $\Delta G^\circ = -RT \ln K^\circ_P$. To calculate K°_P using statistical mechanics, one calculates $\Delta G^\circ_{f,T}$ of each gas (as discussed above), calculates ΔG°_T as $\sum_\mathrm{B} \nu_\mathrm{B}\,\Delta G^\circ_{f,\mathrm{B},T}$, and then calculates K°_P from ΔG°_T.

For theoretical discussions, it is useful to express the equilibrium constant in terms of the partition functions of the reactants and products. We have $A \equiv U - TS$. Since it is $U - U_0$ that is calculated from the partition function, it follows that Eq. (22.40) must be modified to read $A - U_0 = -kT \ln Z$, to reflect our choice of zero energy at the lowest available molecular energy level. Since $\mu_\mathrm{B} = (\partial A/\partial n_\mathrm{B})_{T,V,n_{\mathrm{C}\neq\mathrm{B}}}$, and since U of an ideal gas mixture is independent of V and P, partial differentiation of A with respect to n_B gives $\mu_\mathrm{B} = \bar{U}_{\mathrm{B},0} - RT(\partial \ln Z/\partial N_\mathrm{B})_{T,V,N_{\mathrm{C}\neq\mathrm{B}}}$, where $\bar{U}_{\mathrm{B},0}$ is the partial molar internal energy of B at $T = 0$. Since the mixture is ideal, $\bar{U}_{\mathrm{B},0}$ equals $\bar{U}$ of pure ideal gas B at $T = 0$. Using

(22.51) for Z and (22.56) for $\ln N_B!$, we get

$$\mu_B = \bar{U}_{B,0} - RT(\partial/\partial N_B)[N_B \ln z_B - (N_B \ln N_B - N_B) + \cdots]$$

$$\mu_B = \bar{U}_{B,0} - RT \ln (z_B/N_B) \qquad \text{ideal gas mixture} \tag{22.123}$$

The equilibrium condition $\sum_B \nu_B \mu_B = 0$ [Eq. (4.111)] gives

$$0 = \sum_B \nu_B \bar{U}_{B,0} - RT \sum_B \nu_B \ln \frac{z_B}{N_B} = \Delta U_0^\circ - RT \ln \left[\prod_B \left(\frac{z_B}{N_B}\right)^{\nu_B}\right]$$

$$\frac{\Delta U_0^\circ}{RT} = \ln \left[\prod_B \left(\frac{z_B/N_0 V}{N_B/N_0 V}\right)^{\nu_B}\right] = \ln \left[\prod_B \left(\frac{z_B/N_0 V}{c_B}\right)^{\nu_B}\right]$$

$$e^{\Delta U_0^\circ/RT} = \prod_B \left(\frac{z_B/N_0 V}{c_B}\right)^{\nu_B} = \frac{\prod_B (z_B/N_0 V)^{\nu_B}}{\prod_B (c_B)^{\nu_B}}$$

$$K_c \equiv \prod_B (c_B)^{\nu_B} = e^{-\Delta U_0^\circ/RT} \prod_B \left(\frac{z_B}{N_0 V}\right)^{\nu_B} \qquad \text{ideal gases} \tag{22.124}$$

where $\Delta U_0^\circ \equiv \sum_B \nu_B \bar{U}_{B,0}$ is the change in molar internal energy at $T = 0$ [and can be calculated from (22.122) as $\Delta U_0^\circ/RT = -\Delta D_0/kT$], where $c_B \equiv n_B/V = N_B/N_0 V$ is the equilibrium concentration of B, and where K_c is the concentration-scale equilibrium constant [with units of $(\text{moles/dm}^3)^{\Delta|\nu|}$].

Equations (22.58) and (22.81) show that z_B is proportional to V. Hence, z_B/V is independent of V and is a function of T only (as is K_c). The exponential factor in (22.124) corrects for the use of different zero levels of energy in z of each species. The $N_0 V$ converts from numbers of molecules N_B to concentrations c_B.

Equation (22.124) is actually an example of the Boltzmann distribution law. To see this, consider the special case of the isomerization reaction $A \rightleftharpoons B$. The Boltzmann distribution law (22.74) gives the average number of molecules in quantum state r of species B in an ideal gas mixture at equilibrium as $\langle N_{B,r}\rangle = N_B e^{-\beta\varepsilon_{B,r}}/z_B$, where N_B is the total number of B molecules. Let r be the ground state (state 0) of B. By convention, in calculating z_B we set the zero level of energy at the ground state of B, so that $\varepsilon_{B,0} = 0$. Hence, $\langle N_{B,0}\rangle = N_B/z_B$, and $N_B = \langle N_{B,0}\rangle z_B$. Similarly, $N_A = \langle N_{A,0}\rangle z_A$, where z_A is calculated taking $\varepsilon_{A,0} = 0$. Thus $N_B/N_A = \langle N_{B,0}\rangle z_B/\langle N_{A,0}\rangle z_A$. But the Boltzmann distribution gives $\langle N_{B,0}\rangle/\langle N_{A,0}\rangle = e^{-\beta(\varepsilon_{B,0}-\varepsilon_{A,0})} = e^{-\Delta\varepsilon_0/kT} = e^{-\Delta U_0^\circ/RT}$, and so

$$\frac{N_B}{N_A} = e^{-\Delta U_0^\circ/RT}\frac{z_B}{z_A} \qquad \text{or} \qquad \frac{c_B}{c_A} = e^{-\Delta U_0^\circ/RT}\frac{z_B}{N_0 V}\frac{N_0 V}{z_A} \tag{22.125}$$

which is (22.124). An extension of these arguments shows that for the general reaction $0 = \sum_B \nu_B B$, the Boltzmann distribution law yields (22.124). (The derivation for the A + B = C case is given in *Knox*, sec. 11.2.)

Consider the isotope-exchange reaction $^{35}Cl_2 + {}^{37}Cl_2 = 2\,{}^{35}Cl^{37}Cl$. The molecular weights, vibrational frequencies, moments of inertia, and D_0 values for all three species are nearly the same, and we shall neglect the small differences in them. Hence, $\Delta D_0 \approx 0 \approx \Delta U_0^\circ$, and (22.124) gives

$$K_c \approx \frac{[z(^{35}Cl^{37}Cl)]^2}{z(^{35}Cl_2)z(^{37}Cl_2)}$$

One's first impulse might be to set all three z's equal to one another. However, the symmetry numbers differ: $\sigma = 2$ for $^{37}Cl_2$ and for $^{35}Cl_2$, but $\sigma = 1$ for $^{35}Cl^{37}Cl$. Since z_{rot} contains a factor $1/\sigma$, we have $z(^{35}Cl_2) \approx z(^{37}Cl_2) \approx \frac{1}{2}z(^{35}Cl^{37}Cl)$. Hence, $K_c \approx 4$. This result might seem surprising, but it can be readily understood. Imagine reaching into a bag containing equal numbers of ^{35}Cl and ^{37}Cl atoms, pulling out one atom with each hand, and joining them to form a Cl_2 molecule. There are four possible outcomes: ^{35}Cl, ^{35}Cl; ^{37}Cl, ^{37}Cl; ^{35}Cl, ^{37}Cl; ^{37}Cl, ^{35}Cl; where the first-listed atom is the one in your left hand. Since these outcomes are equally likely, the number of $^{35}Cl^{37}Cl$ molecules will be twice the number of $^{37}Cl_2$ molecules and twice the number of $^{35}Cl_2$ molecules. Hence, $K = 4$. (Note also that the reaction can be written as $^{35}Cl_2 + {}^{37}Cl_2 = {}^{35}Cl^{37}Cl + {}^{37}Cl^{35}Cl$.)

For the isotope-exchange reaction $H_2 + D_2 = 2HD$, the molecular weights, etc., differ substantially, and these differences cannot be ignored. Hence, K differs substantially from 4. At 25°C, the value calculated from (22.124) is 3.28, in good agreement with the experimental value 3.3.

22.9 ENTROPY AND THE THIRD LAW OF THERMODYNAMICS

So far, our discussion has been based on the canonical ensemble (Sec. 22.2), an ensemble of systems with fixed composition, volume, and temperature. The canonical ensemble was introduced into statistical mechanics by Gibbs (and independently by Einstein). Before Gibbs' work, Maxwell and Boltzmann had used a *microcanonical ensemble*, which consists of an infinitely large number of macroscopically identical isolated systems; being isolated, the systems of the microcanonical ensemble have fixed composition, volume, and internal energy. The formulas derived from the canonical ensemble are generally much easier to apply than those found from the microcanonical ensemble, but the microcanonical ensemble leads to an interesting formula for entropy, which we now derive.

All systems in a microcanonical ensemble have the same quantum energy E (but are not necessarily in the same quantum state because of degeneracy). The systems of a canonical ensemble show a distribution of energy values, this distribution being sharply peaked at the equilibrium thermodynamic internal energy U (Fig. 22.3). We can mentally construct a microcanonical ensemble by selecting at one instant of time just those systems of a canonical ensemble whose quantum energy E equals the thermodynamic internal energy U and enclosing each such system in an adiabatic, rigid, impermeable container. (The name "microcanonical ensemble" indicates that the members of this ensemble are a subset of the members of a canonical ensemble.)

The probability function (22.15) for being in quantum state j is $e^{-\beta E_j}/Z$. Since all systems of the microcanonical ensemble have the same E_j, each microstate that occurs in this ensemble has an equal probability. This also follows directly from the second postulate of statistical mechanics: the probability of a microstate depends only on the system energy. Let W be the number of quantum states with energy E, where E is the energy of each system in the microcanonical ensemble. W is the degeneracy of the level E. Let p_j be the microcanonical-ensemble probability for one of the quantum states (state j) whose energy is E. We know that p_j is the same for all the W states that occur in this ensemble and that the p_j's sum to 1. Hence, $p_j = 1/W$ for the microcanonical ensemble.

Equation (22.33) for the entropy becomes

$$S = -k \sum_{j=1}^{W} p_j \ln p_j = -k \sum_{j=1}^{W} \frac{1}{W} \ln \frac{1}{W}$$

There are W terms in the sum, and each term equals $(1/W) \ln (1/W)$. Hence the sum equals $\ln (1/W) = -\ln W$. We have

$$S = k \ln W \qquad \text{isolated syst.} \tag{22.126}$$

The entropy of an isolated system is proportional to the log of W, where W is the number of possible microstates for a system of fixed composition, V, and U. Equation (22.126) is *Boltzmann's principle* and is carved on his tombstone in Vienna. This equation is a more explicit formulation of Eq. (3.52). [We were rather vague about the meaning of the probability in (3.52).] Equation (22.126) is generally not as useful for practical calculations as the canonical-ensemble formula $S = U/T + k \ln Z$.

Let us estimate the magnitude of W. The entropy of a thermodynamic system containing 1 mole is of the order of R (where R is the gas constant). Equation (22.126) then gives the order of magnitude of W as

$$e^{S/k} \approx e^{R/k} = e^{N_0} \approx e^{10^{23}} \approx 10^{10^{22}} \approx 10^{10,000,000,000,000,000,000,000}$$

(Recall from Fig. 22.3 that the degeneracy W is a very sharply increasing function of the system's quantum energy.)

One further point about the microcanonical ensemble. Because the system energies in the canonical ensemble are sharply peaked at U, there is no significant macroscopic difference between a canonical and a microcanonical ensemble. The observed macroscopic properties of an isolated system are indistinguishable from its macroscopic properties when it is in a constant-temperature bath. The system's energy fluctuations in the bath are far too small to be detected.

We now examine the conventions used for entropy. The statistical-mechanical result (22.31) reads $S = U/T + k \ln Z + C$, and we set the integration constant C equal to zero for all systems. In Sec. 5.5 we adopted the thermodynamic convention that $\lim_{T\to 0} S = 0$ for each element. This convention plus the third law of thermodynamics led to the result $\lim_{T\to 0} S = 0$ for every pure substance in internal equilibrium. We must now see whether the statistical-mechanical convention ($C = 0$) is consistent with the thermodynamic convention ($S_0 = 0$ for elements).

As T goes to 0, all the systems in a canonical ensemble fall into the system's lowest available quantum energy level and we can consider the ensemble to be a microcanonical one. Let W_0 be the degeneracy of the lowest quantum energy level of the system. Then (22.126) gives

$$\lim_{T\to 0} S = k \ln W_0$$

For this statistical-mechanical result to be consistent with the thermodynamic result $\lim_{T\to 0} S = 0$ for a pure substance, the degeneracy W_0 of the ground level of a one-component system would have to be 1. Actually, W_0 is not 1 for a pure substance, for two reasons:

1. Each atomic nucleus has spin quantum numbers I and M_I (Sec. 21.11). I is a constant for a given nucleus, but M_I takes on the $2I + 1$ values from $-I$ to $+I$. In the absence of an external magnetic field, the $2I + 1$ states corresponding to different M_I values have the same energy [note that $E = 0$ for $B = 0$ in (21.71)]. Thus there is a nuclear-spin degeneracy of $2I + 1$ for each atom. The M_I quantum numbers of the atoms vary independently of one another, so that the total ground-level nuclear-spin degeneracy is the product $\prod_a (2I_a + 1)$, where I_a is the nuclear spin of atom a and the product goes over all atoms. For example, for a crystal containing N atoms each with $I = 1$, we get a nuclear-spin contribution to W_0 of 3^N and a contribution to S_0 of $k \ln 3^N = Nk \ln 3$.

2. When chemists talk of "pure" FCl, they mean a mixture of $75\frac{1}{2}$ percent $^{19}F^{35}Cl$ and $24\frac{1}{2}$ percent $^{19}F^{37}Cl$. These isotopic species are distinguishable from each other, and there is an entropy of mixing associated with mixing the pure isotopic species to get the naturally occurring isotopic mixture. Since isotopic species form ideal solutions, ΔS_{mxg} is given by (9.73). In a crystal of naturally occurring FCl, there is a degeneracy associated with the various permutations of the $F^{35}Cl$ and $F^{37}Cl$ molecules among different locations in the crystal.

To make the statistical-mechanical and thermodynamic entropies agree, chemists have adopted the convention of ignoring contributions to S made by nuclear spins and by isotopic mixing. Recall that we didn't sum over different nuclear-spin quantum states in calculating partition functions in Secs. 22.6 and 22.7 and we ignored the entropy of isotopic mixing. There is no harm in using these conventions because nuclear spins are not changed in chemical reactions and the amount of isotopic fractionation that occurs in chemical reactions is ordinarily negligible. With these conventions, we can expect W_0 to be 1 for a perfect crystal of a pure substance, so that $\lim_{T\to 0} S = k \ln 1 = 0$ for the statistical-mechanical entropy of a pure substance at absolute zero, in agreement with the thermodynamic result.

Note that statistical-mechanical entropies are not absolute entropies, since they ignore the contributions of nuclear spins and isotopic mixing. Even if these contributions were included, we still would not have absolute entropies, since the statistical-mechanical formulas for S are based on the arbitrary convention of setting $C = 0$ in (22.31).

A comparison of calorimetrically determined thermodynamic (td) entropies (Sec. 5.5) and spectroscopically determined statistical-mechanical (sm) entropies (Secs. 22.6 and 22.7) follows:

Molecule	N_2	Cl_2	H_2S	CO_2	CH_3Cl	CH_3NO_2	$C_6H_6(g)$
$\bar{S}^\circ_{298,td}$	45.9	53.3	49.1	51.1	55.9	65.7	64.5
$\bar{S}^\circ_{298,sm}$	45.8	53.3	49.2	51.1	56.0	65.7	64.4

where the units of $\bar{S}$ in this and the following table are cal mol^{-1} K^{-1}. The entropies agree within the limits of experimental error.

There are, however, several gases for which S_{td} and S_{sm} disagree. Examples are:

Molecule	CO	N_2O	NO	$H_2O(g)$	CH_3D	H_2
$\bar{S}^\circ_{298,td}$	46.2	51.4	49.7	44.3	36.7	29.7
$\bar{S}^\circ_{298,sm}$	47.2	52.5	50.3	45.1	39.5	31.2

To see the reason for these discrepancies, consider CO. There are two possible orientations for each CO molecule in the crystal (either CO or OC). The dipole moment of CO is very small (0.1 D), and so the difference in energy $\Delta\varepsilon$ for these two orientations is very small. When the crystal is formed at the CO normal melting point (66 K), $\Delta\varepsilon/kT$ is very small and the Boltzmann factor $e^{-\Delta\varepsilon/kT} \approx e^0 = 1$; hence the crystal is formed with approximately equal numbers of CO molecules with each orientation. As T is decreased toward zero, $\Delta\varepsilon/kT$ becomes large and $e^{-\Delta\varepsilon/kT} \rightarrow e^{-\infty} = 0$. Hence, if thermodynamic equilibrium were maintained, all the CO molecules would adopt the orientation with the lower energy. However, to have the incorrectly oriented CO molecules rotate 180° in the crystal requires a substantial activation energy, which is not available to the molecules at low T. The CO molecules remain locked into their nearly random orientations as T is lowered. The thermodynamic determination of S from observed C_P values is based on the third law, which applies only to systems in equilibrium. A CO crystal at 10 or 15 K is not in true thermodynamic equilibrium, and the calorimetrically measured entropy S_{td} is therefore in error. The correct entropy is that calculated by statistical mechanics; it is $\bar{S}^\circ_{sm}$ that is listed in tables of thermodynamic properties.

The error in $\bar{S}^\circ_{td}$ of CO can be estimated as minus the entropy of mixing of $\frac{1}{2}$ mole of carbon monoxide having the CO orientation with $\frac{1}{2}$ mole of carbon monoxide having the OC orientation. Equation (9.73) gives $-R(\frac{1}{2}\ln\frac{1}{2} + \frac{1}{2}\ln\frac{1}{2}) = R\ln 2 = 1.4$ cal mol^{-1} K^{-1}, in rough agreement with the observed value 1.0 cal mol^{-1} K^{-1}. The difference between these two numbers indicates that a partial degree of ordering does occur in the CO crystal.

The discrepancies for N_2O, NO, and CH_3D arise from the same reason as for CO. The discrepancy for H_2O is due to a randomness in the positions of H atoms in hydrogen bonds. H_2 is a special case, and will not be discussed here; see *McClelland*, sec. 8.4.

One final comment. Although the entropy of a thermodynamic system is calculable from molecular properties, entropy is not a molecular property. Entropy has meaning only for a large collection of molecules. Individual molecules don't have entropy.

22.10 INTERMOLECULAR FORCES

The general statistical-mechanical formulas of Sec. 22.2 were applied to ideal gases (which have no intermolecular forces) in Secs. 22.3 to 22.8. Before treating systems with intermolecular forces, we discuss the nature of these forces.

In solids, liquids, and nonideal gases, the Hamiltonian of the system contains the potential energy $\mathscr{V}$ of interaction between the molecules. $\mathscr{V}$ is a function of the distances between the molecules and for polar molecules is also a function of the molecular orientations in space.

In most statistical-mechanical treatments, the force between molecules 1 and 2 is assumed to be unaffected by the nearby presence of a third molecule. This is an approximation because molecule 3 will polarize molecules 1 and 2 (Sec. 14.2), thereby changing the 1-2 intermolecular force. In gases not at high densities, there is little probability of three gas molecules' being simultaneously close together. Ignoring three-body forces is a very good approximation for low- and medium-density gases but not so good for solids, liquids, and high-density gases.

In dealing with gases and liquids, one often makes the simplifying approximation of averaging over the orientation dependence of $\mathscr{V}$, converting $\mathscr{V}$ into a function solely of the distances between molecules. (This approximation is inapplicable to solids composed of polar molecules, since the molecules are held in fixed orientations in the solid.)

With the neglect of three-body forces, $\mathscr{V}$ is a sum of pairwise potential energies of interaction v_{ij} between molecules i and j. With averaging over the orientation dependence, v_{ij} depends only on the distance r_{ij} between i and j: $v_{ij} = v_{ij}(r_{ij})$. For example, for a system with three molecules, $\mathscr{V} = v_{12}(r_{12}) + v_{13}(r_{13}) + v_{23}(r_{23})$; if the molecules are identical, v_{12}, v_{13}, and v_{23} are the same function. For a system of N molecules

$$\mathscr{V} \approx \sum_{i=1}^{N} \sum_{j>i} v_{ij}(r_{ij}) \tag{22.127}$$

The two-molecule potential energy v_{ij} could be calculated in principle by solving the Schrödinger equation for a system consisting of molecules i and j and averaging the result over all orientations. Such a calculation is extremely difficult. Instead of quantum mechanics, one uses an approximate model of a molecule as an electric dipole having a certain electric polarizability (Sec. 14.2), and one uses classical electrostatics to calculate v_{ij}.

The force $F(r)$ and potential energy $v(r)$ between two bodies are related by Eq. (2.39) as $F(r) = -dv(r)/dr$. For two ions, $F(r)$ is given by Coulomb's law (14.1), and $v(r) = z_1 z_2 e^2/4\pi\varepsilon_0 r$ [Eq. (19.4)]. Forces between ions occur in electrolyte solutions, ionic solids, and molten salts. In this section, we are considering neutral molecules.

If molecules 1 and 2 have permanent electric dipole moments p_1 and p_2, the force one exerts on the other will depend on p_1, p_2, the separation r, and the relative orientation of the two dipoles. Since we are ignoring the orientation dependence of v, we must average v over all orientations. If the two molecular dipoles were oriented completely randomly with respect to each other, their average interaction energy would be zero because repulsive orientations would occur just as often as attractive orientations. However, the Boltzmann factor $e^{-v/kT}$ favors attractive orientations (which have lower energies) over repulsive ones. (Recall the phenomenon of orientation polarization; Sec. 14.2.) The orientation-averaged potential energy of two dipoles calculated with allowance for the Boltzmann distribution turns out to be

$$v_{\text{d-d}}(r) = -\frac{2}{3kT}\frac{p_1'^2 p_2'^2}{r^6} \tag{22.128}$$

where p_1' and p_2' are the dipole moments in cgs gaussian units (statC cm) and v is in ergs. (For conversion of the equations in this section to SI units, see Sec. 19.1 and inside the back cover.) As $T \to \infty$, the $e^{-\Delta\varepsilon/kT}$ Boltzmann factor goes to 1, all orientations become equally likely, and $v_{\text{d-d}} \to 0$. The term (22.128) is the *dipole–dipole* contribution to v. [For derivations of (22.128) to (22.130), see *Hirschfelder, Curtiss, and Bird*, secs. 13.3 and 13.5.].

Besides $v_{\text{d-d}}$ other interactions contribute to the intermolecular potential v. The permanent dipole moment of one molecule will induce a dipole moment in a second molecule (whether or not the second molecule has a permanent dipole moment), and the interaction between the permanent moment of one molecule and the induced moment of the second gives the *dipole–induced-dipole* contribution to v. This turns out to be

$$v_{\text{d-id}} = -\frac{p_1'^2\alpha_2' + p_2'^2\alpha_1'}{r^6} \tag{22.129}$$

where α_1' and α_2' are the polarizabilities of molecules 1 and 2 in cgs gaussian units (cm^3). Note the absence of kT in (22.129); the induced dipoles are born oriented, as noted in Sec. 14.2.

Even if neither molecule has a permanent dipole moment, there will still be an attractive force between the molecules. (This must be so; otherwise, gases like He or N_2 would not condense to liquids.) The electrons (and to a lesser extent the nuclei) are in continual motion within a molecule. The permanent dipole moment p is calculated using the average locations of the charges. If p is zero, the time-average charge distribution must be completely symmetric; however, the charge distribution at any instant in time need not be. (For example, at a given instant, both electrons in a He atom might be on the same side of the nucleus.) The instantaneous dipole of a molecule induces a dipole in a nearby molecule. The interaction between the instantaneous dipole moment and the induced dipole moment produces a net attraction, whose form was calculated by London in 1930 using quantum mechanics. This *London* (or *dispersion*) energy is (approximately)

$$v_{\text{disp}} \approx -\frac{3I_1I_2}{2(I_1+I_2)}\frac{\alpha_1'\alpha_2'}{r^6} \tag{22.130}$$

where I_1 and I_2 are the ionization energies of molecules 1 and 2. (The order of magnitude of I is 10 eV for most molecules.)

The net long-range attractive potential energy for two neutral molecules is the sum $v_{\text{d-d}} + v_{\text{d-id}} + v_{\text{disp}}$. Each term is proportional to $1/r^6$. The attractive force (called the *van der Waals force*) that results from these three effects is proportional to $1/r^7$ and falls off much faster with distance than a Coulomb's law force ($1/r^2$). Table 22.1 gives minus the coefficient of $1/r^6$ in (22.128) to (22.130) for several pure substances at 25°C.

Table 22.1 Contributions to the intermolecular potential energy of like molecules at 25°C

				$-10^{60}vr^6/(\text{erg cm}^6)$		
Molecule	p'/D	$\alpha'/\text{Å}^3$	I/eV	Dipole–dipole	Dipole–induced dipole	Dispersion
Ar	0	1.63	15.8	0	0	50
N_2	0	1.76	15.6	0	0	58
C_6H_6	0	9.89	9.2	0	0	1086
C_3H_8	0.08	6.29	11.1	0.0008	0.09	528
HCl	1.08	2.63	12.7	22	6	106
CH_2Cl_2	1.60	6.48	11.3	106	33	570
SO_2	1.63	3.72	12.3	114	20	205
H_2O	1.85	1.59	12.6	190	11	38
HCN	2.98	2.59	13.8	1277	46	111

Note that, except for small, highly polar molecules (for example, H_2O and HCN), the dominant term in the van der Waals attraction is the dispersion energy. Also, $v_{d\text{-}id}$ is always quite small. The polarizability (which determines v_{disp}) increases with increasing number of electrons. Values of α' for the halogens and some straight-chain hydrocarbons are:

Molecule	F_2	Cl_2	Br_2	I_2	CH_4	C_2H_6	C_3H_8
$\alpha'/Å^3$	1.3	4.6	6.7	10.2	2.6	4.5	6.3

The outer electrons in I_2 are farther from the nuclei than in Br_2 and are therefore more easily distorted. The increase in α increases v_{disp} in (22.130), so that F_2 and Cl_2 are gases at room temperature, Br_2 is a liquid, and I_2 a solid. The increase in boiling point as n increases in the series C_nH_{2n+2} is due to the increase in α and v_{disp}.

Just as molecular dipole moments can be estimated as the vector sum of bond moments, the molecular polarizability can be estimated as the sum of bond polarizabilities. Some bond-polarizability values are:

Bond	C—C	C=C	C—H	C—Cl	C=O
$\alpha'/Å^3$	0.64	1.66	0.65	2.61	1.16

Besides the terms (22.128) to (22.130) another important intermolecular attraction is the *hydrogen bond*. This is an attraction between an electronegative atom in one molecule and a hydrogen atom bound to an electronegative atom in a second molecule. The electronegative atoms involved are F, O, N, and, to a lesser extent, Cl and S. Species showing substantial hydrogen bonding include HF, H_2O, NH_3, CH_3OH, CH_3NH_2, and CH_3COOH. In water, each OH bond is highly polar, and the small, positive H of one H_2O molecule is strongly attracted to the negative O of a nearby H_2O molecule, as symbolized by $HOH\cdots OH_2$. The magnitude of the hydrogen-bond energy (2 to 10 kcal/mole at the distance of lowest potential energy) is significantly greater than that of the energies (22.128) to (22.130) (which run 0.1 to 2 kcal/mole for molecules that are not large) but substantially less than the energy of an intramolecular covalent bond (30 to 230 kcal/mole, Sec. 20.1). The relatively high melting and boiling points of H_2O and NH_3 are due to hydrogen bonding.

Because of its occurrence in such key species as water, proteins, and DNA, the hydrogen bond has been intensively studied experimentally and theoretically. A full quantum-mechanical understanding of the hydrogen bond has not yet been achieved. The hydrogen bond cannot be explained solely as an electrostatic attraction but has a significant amount of covalent character due to sharing of lone-pair electrons on N, O, or F with H. [For further discussion, see the references cited in L. C. Allen, *J. Am. Chem. Soc.*, **97**, 6921 (1975).] The hydrogen bond is a relatively strong, highly directional interaction, and we shall assume the absence of hydrogen bonding in the discussion below.

So far we have considered the attractive forces that occur at relatively large separations between molecules. At small intermolecular distances, the molecules exert strong repulsions on each other, due mainly to the Pauli-exclusion-principle repulsion between the overlapping electron probability densities (Sec. 20.4). The very steep short-range repulsion can be crudely approximated by an inverse power of r:

$$v_{\text{rep}} \approx A/r^n$$

where A is a positive constant and n is a large integer (8 to 18). The relative incompressibility of liquids and solids is due to v_{rep}. (To experience v_{rep}, bang your fist on a table.)

In addition, there are intermediate-range quantum-mechanical interactions whose calculation is extremely difficult.

If the intermolecular potential energy v is approximated as the sum of the short-range (v_{rep}) and long-range ($v_{\text{d-d}} + v_{\text{d-id}} + v_{\text{disp}}$) potentials, we get a function with the form $v = A/r^n - B/r^6$. Perhaps the most widely used intermolecular potential is the *Lennard-Jones 6-12 potential* (Prob. 22.47)

$$v = 4\varepsilon\left[\left(\frac{\sigma}{r}\right)^{12} - \left(\frac{\sigma}{r}\right)^{6}\right] \tag{22.131}$$

where ε is the depth of the minimum in the potential and σ is the intermolecular distance at which $v = 0$ (Fig. 22.9*a*). For $r < \sigma$, the $1/r^{12}$ term dominates, and v increases steeply as r decreases. For $r > \sigma$, the $-1/r^6$ term dominates, and v decreases as r decreases. The parameter σ is an approximation to the sum of the average radii of the two molecules. Note the resemblance to Fig. 21.4, which is an *intra*molecular potential. The Lennard-Jones potential is not too bad but is not an accurate representation of intermolecular interactions, even for rare-gas atoms. (The physicist John E. Jones changed his surname to Lennard-Jones when he married Kathleen Mary Lennard in 1925.)

The crudest approximation to intermolecular potentials is the *hard-sphere* potential function (Fig. 22.9*b*)

$$v = \begin{cases} 0 & \text{for } r \geq d \\ \infty & \text{for } r < d \end{cases} \tag{22.132}$$

where d is the sum of the radii of the colliding molecules, considered to be infinitely hard "billiard balls." The hard-sphere potential is often used in statistical mechanics, not because it accurately represents intermolecular forces but because it greatly simplifies calculations.

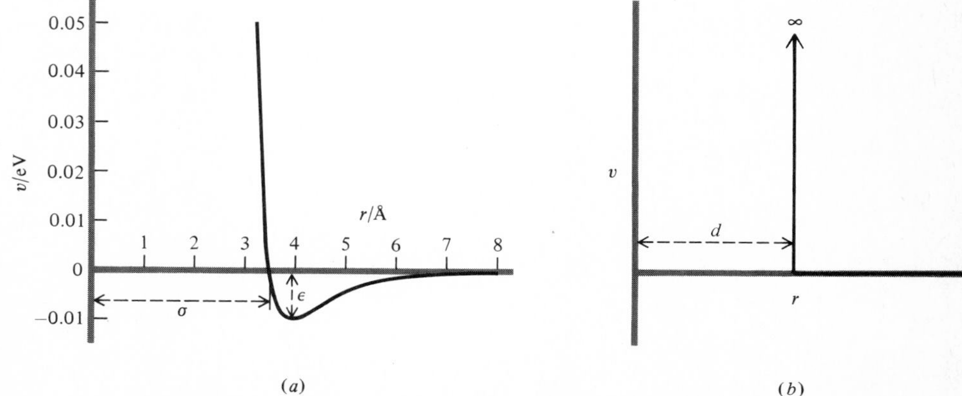

Figure 22.9 (*a*) The Lennard-Jones intermolecular potential for Ar. (*b*) The hard-sphere intermolecular potential.

We shall see in Sec. 22.11 that the second virial coefficient $B_2(T)$ in the virial equation of state (8.5) for gases can be expressed in terms of the intermolecular potential. If v is taken to be a Lennard-Jones function, the best fit to experimental $B_2(T)$ data for several gases is obtained with the following parameters:

Molecule	$d_{vis}/Å$	$\sigma/Å$	ε/eV	ε/kT_b
Ar	3.7	3.5	0.010	1.3
Xe	4.9	4.1	0.019	1.3
CH_4	4.1	3.8	0.013	1.3
CO_2	4.6	4.3	0.017	
C_6H_6	7.4	8.6	0.021	0.7

The d_{vis} values are hard-sphere diameters calculated from gas viscosities using the kinetic-theory equation (16.29) and are in rough agreement with the Lennard-Jones σ values. T_b is the normal boiling point. Interestingly, $\varepsilon \approx 1.3kT_b$ for the nonpolar, spherical molecules Ar, Xe, and CH_4, but this relation doesn't hold for C_6H_6, where v depends on the relative molecular orientation and the Lennard-Jones potential is a poor approximation. When $\varepsilon = \frac{3}{2}kT = 1.5kT$, the mean molecular translational energy equals the maximum intermolecular attraction energy, and so we should expect the Lennard-Jones ε to correlate with the boiling point.

Note that ε is far less than the dissociation energies D_0 of diatomic molecules, which run 1 to 10 eV. The van der Waals attraction between two molecules is far weaker than chemical-bonding forces. Also, σ (which runs 3 to 6 Å for molecules that are not very large) is substantially greater than chemical-bond distances (1 to 3 Å). Hydrogen bonds run 2 to 3 Å for the distance of minimum potential energy.

Equations (22.128) and (22.129) for $v_{\text{d-d}}$ and $v_{\text{d-id}}$ treat the molecule as an electric dipole, and their derivation is based on Eq. (14.14) for the electric potential of a dipole. Since (14.14) applies only at large distances from a system of charges, Eqs. (22.128) and (22.129) are not accurate for molecules that approach each other closely. These inaccuracies are partly compensated for by the use of empirically adjusted values of σ and ε in the Lennard-Jones potential. Note also that the orientation-averaged $v_{\text{d-d}}$ in (22.128) is temperature-dependent, but this is ignored in the Lennard-Jones potential.

For polar molecules, a substantial improvement on the Lennard-Jones potential is obtained by dropping the approximation that v depends only on the intermolecular distance r. A commonly used orientation-dependent potential is the *Stockmayer potential* $v = v_{LJ} - (p_1'^2 p_2'^2/r^3) f(\text{angles})$. Here v_{LJ} is the Lennard-Jones function (22.131) and includes the effects of v_{disp}, $v_{\text{d-id}}$, and v_{rep}; the second term is the electrostatic interaction between dipole moments p_1' and p_2' calculated without averaging over orientations [and is derived from (14.14)]. The function $f(\text{angles})$ is a function of the angles that define the relative orientation of the two molecular dipoles:

$$f(\text{angles}) = 2 \cos \theta_1 \cos \theta_2 - \sin \theta_1 \sin \theta_2 \cos (\phi_1 - \phi_2).$$

To define the angles, let the z axis be the line joining molecules 1 and 2. Then θ_1 and θ_2 are the angles made by the dipole-moment vectors $\mathbf{p}_1$ and $\mathbf{p}_2$ with the z axis, and

ϕ_1 and ϕ_2 are the angles of rotation of $\mathbf{p}_1$ and $\mathbf{p}_2$ about the z axis (as in spherical polar coordinates).

The van der Waals attractions in Ar_2 are sufficiently strong for a small concentration of Ar_2 molecules to exist in argon gas at temperatures below 100 K. The Ar_2 dissociation energy and internuclear distance are $D_e = 0.012$ eV and $R_e = 3.76$ Å, and Ar_2 has several bound vibrational levels existing in the potential of Fig. 22.9*a*. Van der Waals molecules detected by spectroscopic methods include Ne_2, Ar_2, Kr_2, Xe_2, Ar–N_2, $(O_2)_2$, $(N_2)_2$, and Ar–HCl. See B. L. Blaney and G. E. Ewing, *Ann. Rev. Phys. Chem.*, **27**, 553 (1976).

22.11 STATISTICAL MECHANICS OF FLUIDS

Ideal gases were treated in Secs. 22.3 to 22.8. We now consider nonideal gases and liquids. (Section 24.10 treats the statistical mechanics of solids.)

The potential energy $\mathscr{V}$ of intermolecular interactions makes it impossible to write the canonical partition function Z as a product of molecular partition functions. Moreover, it is a hopeless task to try and solve the Schrödinger equation for the entire system of 10^{23} molecules to obtain the system quantum energies E_j. One therefore resorts to the approximation of using classical statistical mechanics to evaluate part of Z.

In earlier sections we noted that when the spacing between the levels of a given kind of energy is much less than kT ($\Delta\varepsilon \ll kT$), that kind of energy can be treated classically, replacing the sum over states by an integral. Let H be the Hamiltonian (Sec. 18.11) for the entire system. Suppose that H can be written as

$$H = H_{\text{cl}} + H_{\text{qu}} \tag{22.133}$$

where H_{cl} contains the energies that can be treated classically and H_{qu} contains the energies that must be treated quantum mechanically. Further, suppose that the terms in H_{cl} are independent of those in H_{qu}, and vice versa. Because of the assumed independence of the terms in H_{cl} and H_{qu}, we have

$$Z = Z_{\text{scl}} Z_{\text{qu}} \tag{22.134}$$

where Z_{scl} is calculated classically using integration over states and Z_{qu} is calculated quantum mechanically using summation over states. The subscript scl stands for semiclassical and will be explained later.

Except at very low temperatures, the translational and rotational levels are closely spaced compared with kT, so H_{cl} contains the translational and rotational energies. The intermolecular potential $\mathscr{V}$ is a function of the coordinates of the centers of mass of the molecules (since the distances between molecules depend on these coordinates) and is a function of the angles defining the spatial orientations of the molecules (since the force between polar molecules depends on their orientation). These are the same coordinates that describe molecular translations and rotations, so $\mathscr{V}$ is part of H_{cl}. The vibrational and electronic levels are certainly not closely spaced compared with kT and are included in H_{qu} of the fluid. Thus

$$H_{\text{cl}} = H_{\text{tr}} + H_{\text{rot}} + \mathscr{V}, \qquad H_{\text{qu}} = H_{\text{vib}} + H_{\text{el}} \tag{22.135}$$

The restoring forces in chemical bonds within a molecule are much stronger than intermolecular forces (as noted in the last section), so the molecular vibrations are not substantially affected by intermolecular forces. Also, rotational and vibrational motions are approximately independent of each other. Although the intermolecular potential depends on the electronic states of the interacting molecules, at ordinary temperatures there is negligible occupation of excited electronic states. For these reasons, the separation (22.135) into noninteracting classical and quantum-mechanical motions is reasonably accurate.

We need the expression for Z_{scl} in (22.134). This is obtained by taking the limit of the quantum-mechanical canonical partition function for the relevant energies as $\Delta\varepsilon/kT \to 0$. The derivation is very complicated (see *Hill*, sec. 22-6 or *McQuarrie*, sec. 10-7), and we omit it, simply quoting the result. For a fluid containing N molecules, all of the same species, one obtains

$$Z_{\text{scl}} = \frac{1}{N!\,h^{fN}} \int \cdots \iint \cdots \int e^{-H_{\text{cl}}/kT}\, dq_1 \cdots dq_{fN}\, dp_1 \cdots dp_{fN} \qquad (22.136)$$

where f is the number of degrees of freedom of a molecule that are being treated classically and the p's and q's are explained below. We are treating the translational and rotational coordinates classically. Each molecule has 3 translational degrees of freedom (since three coordinates are needed to specify its center-of-mass coordinates) and 2 or 3 rotational degrees of freedom, depending on whether the molecule is linear or nonlinear (since two or three angles are needed to specify the molecule's spatial orientation). Thus, $f = 5$ for a linear molecule, and $f = 6$ for a nonlinear molecule.

In Eq. (22.136) q_1 to q_{fN} are the translational and rotational coordinates of the N molecules. These coordinates consist of the $3N$ translational cartesian coordinates $x_1, y_1, z_1, \ldots, x_N, y_N, z_N$ of the centers of mass of the N molecules, and the $2N$ or $3N$ angles of rotation for the N molecules. This gives a total of $5N$ or $6N$ coordinates: $fN = 5N$ or $6N$. The quantities $p_1, \ldots, p_{fN}$ are the momenta that correspond to these coordinates. For the translational coordinate x_1 of molecule 1, the corresponding momentum is the linear momentum $p_{x,1} \equiv mv_{x,1}$. For a rotational coordinate (i.e., angle), the corresponding momentum is the angular momentum of rotation involving that angle.

H_{cl} is a function of the fN coordinates and the fN momenta. The integration in (22.136) consists of $2fN$ definite integrals over the full ranges of $q_1, \ldots, q_{fN}, p_1, \ldots, p_{fN}$. Each cartesian coordinate ranges from 0 to a or b or c, where a, b, and c are the x, y, and z dimensions of the container. Each linear momentum and each angular momentum ranges from $-\infty$ to ∞. The angular coordinates range from 0 to π or 0 to 2π, depending on the angle. For simplicity, the integration limits have been omitted in (22.136).

Recall that the state of a system in classical mechanics is defined by the coordinates and momenta of all the particles (Sec. 18.6). Hence, the quantum-mechanical summation of $e^{-\beta E_j}$ over states is replaced in (22.136) by a classical-mechanical integration of $e^{-\beta H_{\text{cl}}}$ over coordinates and momenta. The $N!$ arises from the indistinguishability of the molecules. The h^{-fN} factor cannot be understood classically, since h occurs only in quantum mechanics. (A rough explanation is that the h^{-fN} factor arises from the uncertainty principle, i.e., exact simultaneous specification of coordinates and momenta is impossible, and the uncertainty product $\Delta p\, \Delta q \approx h$ for each of the fN coordinate and momentum pairs.) The classical canonical partition function used by Gibbs in his 1902 book does not contain the h^{-fN} factor. Because of the h^{-fN} in (22.136), we use the designation semiclassical rather than classical.

Equation (22.136) is for a pure fluid. For a fluid mixture, $N!\,h^{fN}$ is replaced by $\prod_{\text{B}} N_{\text{B}}!\,h^{f_{\text{B}}N_{\text{B}}}$, where N_{B} is the number of molecules of species B, where f_{B} is the number of classical degrees of freedom (5 or 6) of a B molecule, and where the product goes over all species present. In the rest of this section we shall assume a pure fluid, but the generalization to a mixture is readily made.

H_{qu} in (22.135) contains the vibrational and electronic energies. These have been taken to be independent of intermolecular interactions. Hence the system's vibrational and electronic energy levels are the sums of the vibrational and electronic energies of the isolated molecules, and the work of Secs. 22.6 and 22.7 applies. We have (for a pure substance)

$$Z_{qu} = z_{vib}^N z_{el}^N \tag{22.137}$$

where z_{vib} is given by (22.110) or (22.90) and z_{el} by (22.92).

H_{cl} in (22.135) is

$$H_{cl} = H_{tr,1} + \cdots + H_{tr,N} + H_{rot,1} + \cdots + H_{rot,N} + \mathscr{V} \tag{22.138}$$

$H_{tr,1}$ is the translational kinetic energy of molecule 1, and is given by (18.45) as

$$H_{tr,1} = (p_{x,1}^2 + p_{y,1}^2 + p_{z,1}^2)/2m \tag{22.139}$$

$H_{rot,1}$ is the classical Hamiltonian for rotation of molecule 1 and is a function of the two or three rotational angular momenta and the rotational angles. The intermolecular potential energy $\mathscr{V}$ is a function of the center-of-mass coordinates of each molecule and of the rotational angles specifying the molecular orientations.

Since the $3N$ linear momenta $p_{x,1}, \ldots, p_{z,N}$ occur only in H_{tr}, we can readily integrate $e^{-\beta H_{tr}}$ in (22.136) and (22.135) over these momenta; each integral over a linear momentum has the form $\int_{-\infty}^{\infty} e^{-\beta p_x^2/2m}\, dp_x$ and gives a factor of $(2\pi mkT)^{1/2} = (2\pi mkT/h^2)^{1/2}h$ (Prob. 22.55*a*). Since there are $3N$ such integrals, Eq. (22.136) becomes (Prob. 22.55*b*)

$$Z_{scl} = \frac{1}{N!\, h^{(\mathscr{f}-3)N}}\left(\frac{2\pi mkT}{h^2}\right)^{3N/2} Z_{rot+\mathscr{V}} = \frac{1}{N!\, h^{(\mathscr{f}-3)N}}\left(\frac{z_{tr}}{V}\right)^N Z_{rot+\mathscr{V}} \tag{22.140}$$

where we defined

$$Z_{rot+\mathscr{V}} \equiv \int \cdots \iiint e^{-\beta(H_{rot,1}+\cdots+H_{rot,N}+\mathscr{V})}\, dx_1 \cdots dz_N\, d(\text{angs})\, dp_{\text{angs}} \tag{22.141}$$

and where (22.81) was used for the molecular translational partition function z_{tr}. In (22.141), angs and p_{angs} stand for the $2N$ or $3N$ rotational angles and the corresponding rotational angular momenta.

The next step is to integrate $e^{-\beta H_{rot}}$ over the angular momenta p_{angs} in (22.141). We omit the details (*McClelland*, sec. 11.6) and just give the final result. One finds

$$Z_{scl} = \frac{1}{N!}\left(\frac{z_{tr}}{V}\right)^N z_{rot}^N Z_{con} \tag{22.142}$$

where the *configuration integral* Z_{con} is defined as

$$Z_{con} = \frac{1}{(4\pi \text{ or } 8\pi^2)^N}\int \cdots \iint e^{-\beta\mathscr{V}} \sin\theta_1 \cdots \sin\theta_N\, dx_1 \cdots dz_N\, d(\text{angs}) \tag{22.143}$$

In (22.142) z_{rot} is the molecular rotational partition function, given by (22.85) or (22.109). In (22.143) one uses 4π for linear molecules and $8\pi^2$ for nonlinear molecules.

For a linear molecule, $d(\text{angs}) = d\theta_1\, d\phi_1 \cdots d\theta_N\, d\phi_N$, where θ_1 and ϕ_1 are the polar coordinate angles that give the spatial orientation of the axis of molecule 1 (Fig. 18.13). For a nonlinear molecule, $d(\text{angs}) = d\theta_1\, d\phi_1\, d\chi_1 \cdots d\theta_N\, d\phi_N\, d\chi_N$, where θ_1 and ϕ_1 give the spatial orientation of some chosen axis in molecule 1 and χ_1 (which ranges from 0 to 2π) is the angle of rotation of the molecule about that axis.

Substituting (22.142) and (22.137) into (22.134), we get as the canonical partition function of a pure fluid

$$Z = \frac{1}{N!}\left(\frac{z_{\text{tr}}}{V}\right)^N z_{\text{rot}}^N z_{\text{vib}}^N z_{\text{el}}^N Z_{\text{con}} \qquad \text{pure fluid} \tag{22.144}$$

This differs from the ideal-gas Z of Eqs. (22.49) and (22.58) solely in the presence of the factor Z_{con}/V^N. We can write

$$\ln Z = \ln Z_{\text{id}} + \ln Z_{\text{con}} - N \ln V \tag{22.145}$$

where Z_{id} is for the corresponding ideal gas. Equation (22.145) is the key result of this section.

Once Z has been found, the thermodynamic properties of the fluid are readily calculated from Z by using Eqs. (22.37) to (22.41). For example, $A = -kT \ln Z$ and $P = kT(\partial \ln Z/\partial V)_{T,N}$. Since z_{rot}, z_{vib}, z_{el}, and z_{tr}/V [Eq. (22.81)] are independent of V, partial differentiation of (22.144) gives as the equation of state

$$P = kT\left(\frac{\partial \ln Z_{\text{con}}}{\partial V}\right)_{T,N} \tag{22.146}$$

The obstacle to finding Z for a fluid is, of course, the difficulty in evaluating Z_{con}, which involves the intermolecular potential $\mathscr{V}$.

For a nonideal gas or a liquid, one approximates $\mathscr{V}$ as the sum of pairwise interactions, $\mathscr{V} = \sum\sum v_{ij}$ [Eq. (22.127)], chooses some form for v_{ij} (e.g., the Lennard-Jones potential or the Stockmayer potential), and attempts to evaluate Z_{con}.

Since direct evaluation of Z_{con} is extremely difficult, various mathematical gymnastics are used to simplify the problem. One approach leads to the following expression for Z_{con} as an infinite series (for a partial derivation, see *Kestin and Dorfman*, chap. 7 or *Jackson*, sec. 4.5):

$$\ln Z_{\text{con}} = N \ln V - N\left[\frac{N}{VN_0}B_2(T) + \frac{1}{2}\left(\frac{N}{VN_0}\right)^2 B_3(T) + \frac{1}{3}\left(\frac{N}{VN_0}\right)^3 B_4(T) + \cdots\right] \tag{22.147}$$

where $VN_0/N = \bar{V}$ is the molar volume and $B_2, B_3, \ldots$ are functions of temperature that can be expressed as certain integrals involving $e^{-\beta v_{ij}}$. [The general expressions for B_2 and B_3 are given in E. A. Mason and T. H. Spurling, *The Virial Equation of State*, Pergamon, 1969, eqs. (2.5.25) and (2.5.26).]

If v_{12} depends only on the intermolecular distance r, it turns out that

$$B_2(T) = -2\pi N_0 \int_0^\infty (e^{-v(r)/kT} - 1)r^2\, dr \qquad \text{if } v = v(r) \tag{22.148}$$

where v is the intermolecular potential for two molecules separated by r.

Example Find B_2 for the hard-sphere potential (22.132).

Since the function (22.132) has no angular dependence, Eq. (22.148) applies. Breaking the integration range into two parts, we have $B_2 = -2\pi N_0 \int_0^d (e^{-\infty} - 1)r^2\,dr - 2\pi N_0 \int_d^\infty (e^0 - 1)r^2\,dr$. The first integral is $-\int_0^d r^2\,dr = -d^3/3$, and the second integral is $\int_d^\infty 0\,dr = 0$. Hence, $B_2 = 2\pi N_0 d^3/3 = 4N_0 V_{\text{molec}}$, where $V_{\text{molec}} = \frac{4}{3}\pi(d/2)^3$ is the volume of one molecule.

Substitution of (22.147) into (22.146) gives for the equation of state

$$P = \frac{RT}{\bar{V}}\left[1 + \frac{B_2(T)}{\bar{V}} + \frac{B_3(T)}{\bar{V}^2} + \cdots\right] \tag{22.149}$$

which is the virial equation (8.5).

Expressions for the thermodynamic functions $U - U_0$, S, $G - U_0$, etc. are readily calculated from (22.147) and (22.144); see Prob. 22.54 for the results.

If the intermolecular pair potential v is known (say from a quantum-mechanical calculation), the virial coefficients $B_2, B_3, \ldots$ can (in principle) be calculated and the thermodynamic properties are then found from (22.149) and the equations of Prob. 22.54. Conversely, the experimentally measured virial coefficients (determined from P-V-T data) can provide information on v.

Figure 22.10 shows the results obtained for the second virial coefficient $B_2(T)$ when $v(r)$ in (22.148) is taken as the Lennard-Jones potential (22.131). (For the mathematics, see *Hirschfelder, Curtiss, and Bird*, pp. 163, 1114, and 1119.) From this curve, B_2 can be found at any temperature if the Lennard-Jones parameters ε and σ are known. Conversely, one can use experimental $B_2(T)$ values to find the values of ε and σ that give the best fit to the curve. [Due to quantum effects neglected in the classical expression (22.148), Fig. 22.10 is inaccurate for the light gases H_2 and He at low temperatures.] For nonpolar gases that are approximately spherical (for example, Ar, N_2, CH_4), observed $B_2(T)$ values fit the Lennard-Jones curve in Fig. 22.10

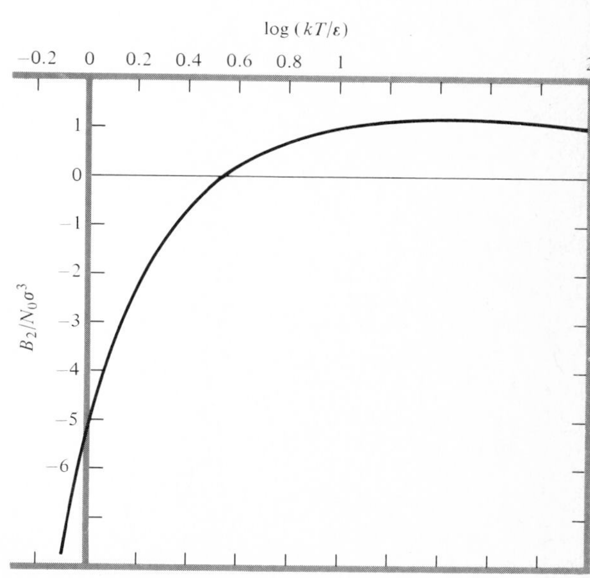

Figure 22.10
The second virial coefficient for the Lennard-Jones potential.

very well. Note that B_2 is negative at low temperatures and positive at high T. At low T ($kT \le \varepsilon$), the depth ε of the potential well is of the same magnitude as, or larger than, the average molecular kinetic energy, and the molecules spend a significant amount of time in the region where v is negative; this reduces the observed pressure, so B_2 in (22.149) is negative. At high temperatures, there is little time spent in the negative-v region, and the repulsive part of the potential dominates, making P greater than the ideal-gas value.

Using reasonable potentials to calculate the virial coefficients, one finds that for gases at low or moderate densities, the successive terms in the virial expansion (22.147) rapidly decrease in magnitude and only the first few terms need be included. However, for gases at very high densities (low $\bar{V}$) and for liquids, the successive terms do not decrease, and the series does not converge; the virial expansion fails, and one must use a different approach for liquids (see Sec. 24.12).

Let us evaluate the average translational energy for a molecule in the fluid. Using the log of the partition function (22.144) in Eq. (22.38), we get

$$U = NkT^2\left(\frac{\partial \ln z_{\text{tr}}}{\partial T}\right)_V + NkT^2\frac{d \ln z_{\text{rot}}}{dT} + NkT^2\frac{d \ln z_{\text{vib}}}{dT}$$

$$+ NkT^2\frac{d \ln z_{\text{el}}}{dT} + kT^2\left(\frac{\partial \ln Z_{\text{con}}}{\partial T}\right)_{V,N}$$

Clearly, the first term on the right is the total translational energy, the second term is the total rotational energy, etc. Thus, $U = U_{\text{tr}} + U_{\text{rot}} + U_{\text{vib}} + U_{\text{el}} + U_{\text{intermol}}$, where U_{intermol} is the contribution of the intermolecular potential energies. Using (22.81) for z_{tr}, we get

$$U_{\text{tr}} = NkT^2\left(\frac{\partial \ln z_{\text{tr}}}{\partial T}\right)_V = \tfrac{3}{2}NkT \tag{22.150}$$

Hence, the average translational energy per molecule is $\langle\varepsilon_{\text{tr}}\rangle = U_{\text{tr}}/N = \frac{3}{2}kT$. This holds for any fluid, not just an ideal gas [Eq. (15.16)].

Moreover, because the fluid's partition function (22.144) is the product of independent translational, rotational, vibrational, electronic, and intermolecular partition functions, the Boltzmann distribution law can be applied to each kind of energy. In particular, the distribution of translational energies is given by the Maxwell distribution (15.52). The Maxwell distribution of speeds, Eq. (15.44), holds in any fluid, not just in an ideal gas.

PROBLEMS

22.1 What are the units of Z?

22.2 Verify that $G = kTV^2[\partial(V^{-1} \ln Z)/\partial V]_{T,N_B}$.

22.3 The choice of zero level of energy is arbitrary. Show that if a constant b is added to each of the system's possible energies E_j in (22.36), then (*a*) P in (22.37) is unchanged; (*b*) U is increased by b; (*c*) S is unchanged; (*d*) A is increased by b.

22.4 For neon gas in a 10-cm^3 box at 300 K, calculate the number of available translational states with energy less than $3kT$.

22.5 Use Eq. (22.55) including the three terms in parentheses to estimate (*a*) 10!; (*b*) 100!; (*c*) 1000!. The accurate values are 3,628,800, $9.3326215444 \times 10^{157}$, and $4.0238726008 \times 10^{2567}$.

22.6 Show that for a system of noninteracting, nonlocalized molecules: $\mu_B = -RT \ln (z_B/N_B)$.

22.7 Most molecules in a gas have $\varepsilon_{\text{tr}} \le 3kT$ and $\varepsilon_{\text{tr},x} \le kT$, where $\varepsilon_{\text{tr},x}$ is the x component of translational energy. For N_2 in an 8-cm^3 cubic box at 0°C, calculate $\Delta\varepsilon_x/kT$, where $\Delta\varepsilon_x$ is the spacing between adjacent x translational energies when $\varepsilon_{\text{tr},x}$ equals kT.

22.8 The mean translational energy of molecules in a

fluid is $\frac{3}{2}kT$. The relative populations of energy levels are determined by kT. Calculate kT in eV, kT/hc in cm^{-1}, and RT in kcal/mole at room temperature.

22.9 Suppose a system consists of N noninteracting, nonlocalized, identical particles and that each particle has available to it only two quantum states, whose energies are $\varepsilon_1 = 0$ and $\varepsilon_2 = a$. (*a*) Find expressions for z, Z, U, C_V, and S. (*b*) If $a = 1.0 \times 10^{-13}$ erg and $N = 6.0 \times 10^{23}$, calculate z, S, and U at 400 K. (*c*) Find the limiting values of U and of C_V as $T \to \infty$. Give physical explanations for your results in terms of the populations of the levels. (*d*) Find the limiting value of S as $T \to \infty$. [*Note:* The two quantum states referred to are the quantum states for the internal motion (Sec. 21.3) in each particle. In addition, each nonlocalized particle has a huge number of available translational states that allow $\langle N_r \rangle \ll 1$ to be satisfied. The problem therefore calculates the contributions of only the internal motions to the thermodynamic properties. The $1/N!$ should be omitted from Z, since the $1/N!$ belongs as part of the translational factor in Z, which is not being considered in this problem.]

22.10 For N_2, the fundamental vibrational frequency is $\nu_0 = 6.98 \times 10^{13}\ s^{-1}$. Calculate the ratio of the $v = 1$ to $v = 0$ populations at (*a*) 25°C; (*b*) 800°C; (*c*) 3000°C.

22.11 For N_2, the rotational constant is $B_0 = 5.96 \times 10^{10}\ s^{-1}$. Calculate the ratio of the $J = 1$ to $J = 0$ populations at (*a*) 200 K; (*b*) 600 K. (*c*) What is the limiting value of this ratio as $T \to \infty$?

22.12 For a collection of harmonic oscillators, calculate the percent of oscillators in the $v = 0$ level when $h\nu/kT$ equals (*a*) 10; (*b*) 3; (*c*) 2; (*d*) 1; (*e*) 0.1.

22.13 For a system of noninteracting molecules, let the molecular energy levels (not states) be $\varepsilon_1, \varepsilon_2, \varepsilon_3, \ldots$. If $\varepsilon_2 > \varepsilon_1$, is it possible for a system in equilibrium to have more molecules in the level ε_2 than in ε_1? Explain.

22.14 In this problem, $\langle N_0 \rangle, \langle N_1 \rangle, \ldots$ are the average number of molecules in the $v = 0, v = 1, \ldots$ vibrational levels of a diatomic molecule. (*a*) Measurement of the intensities of spectroscopic absorption lines for a certain sample of $I_2(g)$ gives $\langle N_1 \rangle / \langle N_0 \rangle = 0.528$ and $\langle N_2 \rangle / \langle N_0 \rangle = 0.279$. Do these data indicate an equilibrium vibrational distribution? (*b*) For I_2, $\nu = 6.39 \times 10^{12}\ s^{-1}$. Calculate the temperature of the I_2 sample. (*c*) For a certain gas of diatomic molecules (not I_2), one finds $\langle N_1 \rangle / \langle N_0 \rangle = 0.340$. Find $\langle N_3 \rangle / \langle N_0 \rangle$. Explain why your result is only an approximation.

22.15 A certain system is composed of 1 mole of identical, noninteracting, nonlocalized molecules. Each molecule has available to it only three energy levels, whose energies and degeneracies are: $\varepsilon_1 = 0$, $g_1 = 1$; $\varepsilon_2/k = 100$ K, $g_2 = 3$; $\varepsilon_3/k = 300$ K, $g_3 = 5$. (k is Boltzmann's constant.) (*a*) Calculate z at 200 K. (*b*) Calculate the average number of molecules in each level at 200 K. (*c*) Calculate the average number of molecules in each level in the limit $T \to \infty$. (See the note at the end of Prob. 22.9.)

22.16 (*a*) Show that $\ln N! = \sum_{x=1}^{N} \ln x$. (*b*) Explain why this sum can be approximated by an integral if N is large. (*c*) Use (22.79) to show that $\ln N! \approx N \ln N - N$ for large N.

22.17 The correct expression for the canonical partition function of a system of identical nonlocalized noninteracting particles is [*Mandl*, eqs. (11.15) and (11.31)]

$$Z_{\mathrm{BE}}^{\mathrm{FD}} = e^{-\beta\mu N/N_0} \prod_r \left(1 \pm e^{\beta(\mu/N_0 - \varepsilon_r)}\right)^{\pm 1}$$

where $\beta = 1/kT$, μ is the chemical potential, the product goes over the molecular quantum states, the upper signs are for fermions, and the lower signs are for bosons. For $\langle N_r \rangle \ll 1$, we see from (22.77) that $e^{-\beta(\mu/N_0 - \varepsilon_r)} \gg 1$. Show that $Z_{\mathrm{BE}}^{\mathrm{FD}}$ reduces to (22.49) when $\langle N_r \rangle \ll 1$. *Hint:* Take $\ln Z_{\mathrm{BE}}^{\mathrm{FD}}$ and use (8.41); then use (22.77) (with the ± 1 neglected) twice.

22.18 (*a*) The *Euler–Maclaurin summation formula* is

$$\sum_{n=a}^{\infty} f(n) = \int_a^{\infty} f(n)\, dn + \frac{f(a)}{2} - \frac{f'(a)}{12} + \frac{f'''(a)}{720} - \frac{f^{(v)}(a)}{30{,}240} + \cdots$$

where $f^{(v)}$ is the fifth derivative of $f(n)$ evaluated at $n = a$. Use this formula to derive Eq. (22.86). (*b*) Calculate the percent error in z_{rot} of H_2 at 0°C when (22.85) is used. (*c*) Repeat (*b*) for N_2.

22.19 Verify Eqs. (22.96) to (22.98) for the contributions to U.

22.20 Show that $C_{V,\mathrm{vib}}$ in (22.102) goes to nR as $T \to \infty$.

22.21 Verify Eqs. (22.104) to (22.107) for the contributions to S.

22.22 (*a*) Verify (22.108). (*b*) Verify (22.111) to (22.113).

22.23 Calculate $\bar{S}^\circ_{298}$ for He, Ne, Ar, Kr, and Xe.

22.24 Calculate $\bar{S}^\circ_{298}$ for a gas of H atoms.

22.25 Calculate $\bar{U}^\circ_{298} - \bar{U}^\circ_0$ for He, Ne, and Ar.

22.26 Calculate $\bar{C}^\circ_{V,298}$ and $\bar{C}^\circ_{P,298}$ for He, Ne, and Ar.

22.27 For Ar at 25°C, Probs. 22.23 and 22.25 give $\bar{S}^\circ_{298} = 37.0$ cal mol^{-1} K^{-1} and $\bar{U}^\circ_{298} - \bar{U}^\circ_0 = 889$ cal mol^{-1}. Calculate Z and z for 1 mole of the ideal gas Ar at 25°C and 1 atm.

22.28 For HF, $\tilde{\nu}_0 = 3959\ \text{cm}^{-1}$, and $\tilde{B}_0 = 20.56\ \text{cm}^{-1}$. For HF, calculate (*a*) $\bar{S}^\circ_{298}$; (*b*) $\bar{C}^\circ_{V,298}$; (*c*) $\bar{C}^\circ_{P,298}$.

22.29 For I_2, $\nu_0 = 6.395 \times 10^{12}\ \text{s}^{-1}$, and the internuclear distance is 2.67 Å. For $I_2(g)$, calculate (*a*) $\bar{U}^\circ_{500} - \bar{U}^\circ_0$; (*b*) $\bar{H}^\circ_{500} - \bar{U}^\circ_0$; (*c*) $\bar{S}^\circ_{500}$; (*d*) $\bar{G}^\circ_{500} - \bar{U}^\circ_0$.

22.30 The far-IR spectrum of HCl shows a series of lines with the approximately constant spacing 20.9 cm^{-1}. The near-IR spectrum shows a strong absorption band at 2885 cm^{-1}. Calculate $\bar{S}^\circ_{298}$ of HCl.

22.31 (*a*) Let $s = 1 + x + x^2 + \cdots$. Show that $s - xs = 1$. Then solve for s to get the geometric-series formula (22.89). (*b*) By considering the function $1/(1 - x)$ devise another derivation for (22.89).

22.32 (*a*) Find the expression for z_{vib} of a diatomic molecule if the zero of energy is taken at the bottom of the potential-energy curve, so that $\varepsilon_{\text{vib}} = (v + \frac{1}{2})h\nu$. (*b*) Then find U_{vib} and compare with (22.97). Does your result agree with Prob. 22.3*b*?

22.33 Sketch $\bar{C}_V$ vs. T for a typical ideal diatomic gas.

22.34 Calculate the numerical values of z_{tr}, z_{rot}, z_{vib}, and z_{el} for 1.00 mole of CO at 25°C and 1 atm given that $\Theta_{\text{rot}} = 2.77$ K and $\Theta_{\text{vib}} = 3084$ K.

22.35 In the example in Sec. 22.6 we found that $\bar{S}^\circ_{\text{vib},298}$ is negligible for N_2. However, $\bar{S}^\circ_{\text{vib},298}$ is not completely negligible for F_2 or for FCl. Explain this in terms of the bonds in the molecules. Explain why $\bar{S}^\circ_{\text{vib},298}$ is much smaller for HF than for F_2.

22.36 Give the symmetry number for (*a*) BF_3; (*b*) H_2O; (*c*) HCN; (*d*) CH_4; (*e*) C_2H_2; (*f*) C_2H_4.

22.37 For H_2S the fundamental vibrational wave numbers are 2615, 1183, and 2628 cm^{-1}, and the rotational constants $\tilde{A}_0$, $\tilde{B}_0$, and $\tilde{C}_0$ are 10.37, 8.99, and 4.73 cm^{-1}. Calculate $\bar{S}^\circ_{298}$ of H_2S.

22.38 For CO_2 the fundamental vibrational wave numbers are 1340, 667, 667, and 2349 cm^{-1}, and the rotational constant is $\tilde{B}_0 = 0.390\ \text{cm}^{-1}$. Calculate $\bar{S}^\circ_{298}$ of CO_2.

22.39 Give the high-temperature values of $\bar{C}^\circ_P$ and $\bar{C}^\circ_V$ for (*a*) CO_2; (*b*) SO_2; (*c*) C_2H_4. (Interpret "high temperature" to mean a temperature not high enough to give a significant population of excited electronic levels.)

22.40 (*a*) Replace the sum in the equation preceding (22.87) by an integral and perform the integration to get the semiclassical expression for z_{vib} of a diatomic molecule. (*b*) Verify that in the limit $T \to \infty$, Eq. (22.90) reduces to the result of (*a*).

22.41 Equation (22.104) predicts that $S_{\text{tr}} \to -\infty$ as $T \to 0$. But from the third law of thermodynamics we know that $S \to 0$ as $T \to 0$ for a pure substance. Explain this apparent contradiction.

22.42 Without doing elaborate calculations, estimate K°_P for each of the following isotope-exchange reactions: (*a*) $B^{35}Cl_3 + {}^{37}Cl_2 = B^{35}Cl_2{}^{37}Cl + {}^{35}Cl^{37}Cl$, (*b*) $2NH_3 + ND_3 = 3NH_2D$, (*c*) $PH_2D + H_2O = PH_3 + HDO$.

22.43 Use data in Table 21.1 to calculate Θ_{rot} and Θ_{vib} for N_2.

22.44 Write Eqs. (22.128) to (22.130) in SI units.

22.45 Verify the values listed in Table 22.1 for $v_{\text{d-d}}$, $v_{\text{d-id}}$, and v_{disp} of CH_2Cl_2.

22.46 Use the Lennard-Jones parameters listed in Sec. 22.10 to estimate the force between two CH_4 molecules separated by (*a*) 8 Å; (*b*) 5 Å; (*c*) 3 Å.

22.47 With n in $v_{\text{rep}} \approx A/r^n$ set equal to 12, we get $v = A/r^{12} - B/r^6$. (*a*) Let σ be the value of r at which $v = 0$. Show that $v = B\sigma^6/r^{12} - B/r^6$. (*b*) Let r_{min} be the value of r at which v is a minimum. Show that $r_{\text{min}} = 2^{1/6}\sigma$. (*c*) Let $\varepsilon \equiv v(\infty) - v(r_{\text{min}})$ be the depth of the potential well. Show that $B = 4\sigma^6\varepsilon$. Hence v is given by the Lennard-Jones potential (22.131). (*d*) Verify that (22.131) can be written as $v/\varepsilon = (r_{\text{min}}/r)^{12} - 2(r_{\text{min}}/r)^6$.

22.48 Use the result of Prob. 22.47*b* and data in Sec. 22.10 to calculate r_{min} in the Lennard-Jones potential for (*a*) Ar; (*b*) C_6H_6.

22.49 (*a*) The normal boiling point of Ne is 27.1 K. Estimate the Lennard-Jones ε for Ne. (The experimental value is 4.92×10^{-15} erg.) (*b*) The critical temperature of C_2H_6 is 305 K. Estimate the Lennard-Jones ε for C_2H_6. (The experimental value is 3.18×10^{-14} erg.)

22.50 (*a*) Sketch $F(r)$ vs. r, where $F(r)$ is the force between two nonpolar molecules. (*b*) Let b be the value of r at which $F = 0$. Express b in terms of the Lennard-Jones σ.

22.51 An intermolecular potential intermediate in crudity between the hard-sphere and the Lennard-Jones potentials is the *square-well potential*, defined by $v = \infty$ for $r < \sigma$, $v = -\varepsilon$ for $\sigma \le r \le a$, and $v = 0$ for $r > a$. (*a*) Sketch this potential. (*b*) Calculate $B_2(T)$ in (22.148) for this potential.

22.52 (*a*) Show that $Z_{\text{con}} = V^N$ for an ideal gas in a rectangular box with edges a, b, c. (Consider separately the cases of linear and nonlinear molecules.) (b) Show that (22.146) gives $P = NkT/V$ for an ideal gas.

22.53 The Lennard-Jones parameters for N_2 are $\sigma = 3.74$ Å and $\varepsilon = 1.31 \times 10^{-14}$ erg. Find B_2 for N_2 at (a) 100 K, (b) 300 K, (c) 500 K. (The experimental values are -149, -4, and 17 cm^3/mole.) (d) Estimate the compressibility factor $Z \equiv P\bar{V}/RT$ for N_2 at 100 K and 3.0 atm.

22.54 Use Eqs. (22.147) and (22.145) to show that for a nonideal gas

$$U = U_{id} - nRT^2\left(\frac{1}{\bar{V}}\frac{dB_2}{dT} + \frac{1}{2\bar{V}^2}\frac{dB_3}{dT} + \cdots\right)$$

$$S = S_{id} - nR\left[\frac{1}{\bar{V}}\left(B_2 + T\frac{dB_2}{dT}\right) + \frac{1}{2\bar{V}^2}\left(B_3 + T\frac{dB_3}{dT}\right) + \cdots\right]$$

$$G = G_{id} + nRT\left(\frac{2}{1}\frac{B_2}{\bar{V}} + \frac{3}{2}\frac{B_3}{\bar{V}^2} + \cdots\right)$$

22.55 (a) Verify that $\int_{-\infty}^{\infty} e^{-\beta p_x^2/2m}\, dp_x = (2\pi mkT)^{1/2}$. (See Table 15.1.) ($b$) Derive (22.140) from (22.136).

THEORIES OF REACTION RATES

This chapter discusses the theoretical calculation of rate constants of elementary chemical reactions. If a reaction mechanism (Sec. 17.6) consists of several steps, the theories of this chapter can be applied to each elementary step to calculate its rate constant.

Sections 23.1 to 23.7 deal with gas-phase reactions. Reactions in solution are considered in Sec. 23.8.

23.1 HARD-SPHERE COLLISION THEORY OF GAS-PHASE REACTIONS

The hard-sphere collision theory (developed around 1920 by several people) uses the following assumptions to arrive at an expression for the rate constant of an elementary bimolecular gas-phase reaction. (*a*) The molecules are hard spheres. (*b*) For a reaction to occur between molecules A and B, the two molecules must collide. (*c*) Not all collisions produce reaction; instead reaction occurs if and only if the relative translational kinetic energy along the line of centers of the colliding molecules exceeds a *threshold energy* ε_{thr} (Fig. 23.1). (*d*) The Maxwell–Boltzmann equilibrium distribution of molecular speeds is maintained during the reaction.

Assumption (*a*) is a crude approximation; polyatomic molecules are not spherical but have a structure, and the hard-sphere potential function (Fig. 22.9*b*) is a grotesque caricature of intermolecular interactions. Assumption (*c*) seems reasonable, since a certain minimum energy is needed to initiate the breaking of the relevant

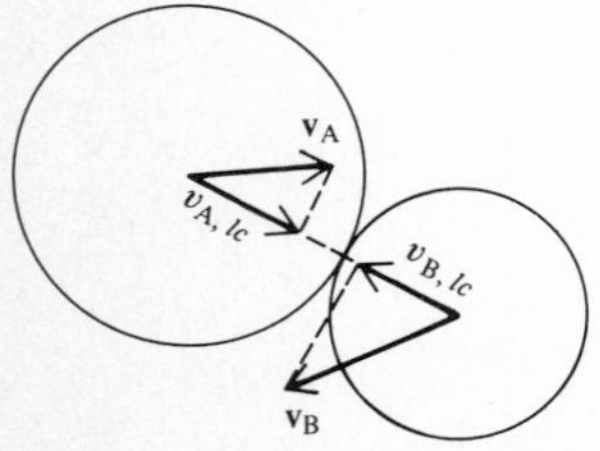

Figure 23.1
Line-of-centers components $v_{A,lc}$ and $v_{B,lc}$ of the velocities of two colliding molecules.

bond(s) to cause reaction. We shall see in Sec. 23.3 that assumption (*c*) is somewhat inaccurate.

What about assumption (*d*)? According to assumption (*c*), it is the fast-moving molecules that react. Hence the gas mixture is continually being depleted of high-energy reactant molecules. The equilibrium distribution of molecular speeds is maintained by molecular collisions. In most reactions the threshold energy (which we shall soon see is nearly the same as the experimental activation energy) is many times greater than the mean molecular translational energy $\frac{3}{2}kT$. (Typical gas-phase bimolecular activation energies are 3 to 30 kcal/mole, compared with 0.9 kcal/mole for $\frac{3}{2}RT$ at room temperature.) Consequently, only a tiny fraction of collisions produces reaction. Since the collision rate is usually far greater than the rate of depletion of high-energy reactant molecules, the redistribution of energy by collisions is able to maintain the Maxwell distribution of speeds during the reaction. For the opposite extreme, where $\varepsilon_{\text{thr}} \approx 0$, virtually all A-B collisions lead to reaction, so the mixture is being depleted of low-energy as well as high-energy molecules. A careful study [B. M. Morris and R. D. Present, *J. Chem. Phys.*, **51**, 4862 (1969)] showed that for all values of $\varepsilon_{\text{thr}}/kT$, the corrections to the rate due to departure from the Maxwell distribution are slight. The maximum error occurs when $\varepsilon_{\text{thr}}/kT$ equals 5, where there is a 6 percent error in the rate; bimolecular recombination reactions of radicals have $\varepsilon_{\text{thr}} \approx 0$, and the correction for such reactions is found to be negligible.

To calculate the gas-phase reaction rate, one simply calculates the rate of collisions where the line-of-centers relative translational energy ε_{lc} exceeds ε_{thr}. Equation (15.62) gives Z_{AB}, the total number of A-B collisions per unit time per unit volume when the gas is in equilibrium. Z_{AB} was found from the Maxwell distribution of speeds, so, by assumption (*d*) above, we use (15.62) in the reacting system. The reaction rate per unit volume equals Z_{AB} multiplied by the fraction ϕ of collisions for which $\varepsilon_{lc} \geq \varepsilon_{\text{thr}}$. Since the rigorous calculation of this fraction is complicated (see *Frost and Pearson*, chap. 4), we shall give only a semiplausible argument, rather than a derivation.

There are two components of velocity involved (namely, the component of velocity of each molecule along the line of centers), so we might suspect that ϕ equals the fraction of molecules in a hypothetical two-dimensional gas whose translational energy $\varepsilon = \varepsilon_x + \varepsilon_y = \frac{1}{2}mv_x^2 + \frac{1}{2}mv_y^2 = \frac{1}{2}mv^2$ exceeds ε_{thr}. A derivation similar to the one that gave (15.44) shows that the fraction of molecules with speed between v and $v + dv$ in a two-dimensional gas is (Prob. 15.20)

$$(m/2\pi kT)e^{-mv^2/2kT} \times 2\pi v\, dv = (m/kT)e^{-mv^2/2kT}v\, dv \tag{23.1}$$

Use of $\varepsilon = \frac{1}{2}mv^2$ and $d\varepsilon = mv\, dv$ gives the fraction of molecules with energy between ε and $\varepsilon + d\varepsilon$ as $(1/kT)e^{-\varepsilon/kT}\, d\varepsilon$. Integration of this expression from ε_{thr} to ∞ gives the fraction with translational energy exceeding ε_{thr} as $e^{-\varepsilon_{\text{thr}}/kT}$. The rigorous derivation shows that ϕ does equal $e^{-\varepsilon_{\text{thr}}/kT}$.

Hence, the number of A molecules reacting per unit volume per second in the elementary bimolecular reaction $\text{A} + \text{B} \rightarrow \text{products}$ is $Z_{\text{AB}}e^{-E_{\text{thr}}/RT}$, where $E_{\text{thr}} \equiv N_0\varepsilon_{\text{thr}}$ is the threshold energy on a per mole basis. The reaction rate per unit volume r is defined in terms of moles of A, so $r = Z_{\text{AB}}e^{-E_{\text{thr}}/RT}/N_0$. Since $r = k[\text{A}][\text{B}]$, the predicted rate constant is

$$k = \frac{Z_{\text{AB}}e^{-E_{\text{thr}}/RT}}{N_0[\text{A}][\text{B}]} \tag{23.2}$$

Use of (15.62) with $N_A/V = N_0 n_A/V = N_0[A]$ and $N_B/V = N_0[B]$ gives

$$k = N_0 \pi (r_A + r_B)^2 \left[\frac{8RT}{\pi}\left(\frac{1}{M_A} + \frac{1}{M_B}\right)\right]^{1/2} e^{-E_{thr}/RT} \qquad \text{for A} \neq \text{B} \qquad (23.3)$$

For the bimolecular reaction 2A → products, the rate per unit volume is given by (17.4) and (17.5) as $r \equiv -\frac{1}{2}d[A]/dt = k[A]^2$. The molar rate per unit volume of disappearance of A is $-d[A]/dt = 2Z_{AA}e^{-E_{thr}/RT}/N_0$. The factor 2 appears because two molecules of A disappear at each reactive collision. Hence $k = Z_{AA}e^{-E_{thr}/RT}/N_0[A]^2$. Substitution of (15.63) with $N_A/V = N_0[A]$ gives

$$k = \frac{1}{2^{1/2}} N_0 \pi d_A{}^2 \left(\frac{8RT}{\pi M_A}\right)^{1/2} e^{-E_{thr}/RT} \qquad (23.4)$$

Equations (23.3) and (23.4) have the form $\ln k = \text{const} + \frac{1}{2}\ln T - E_{thr}/RT$. The definition (17.72) gives the activation energy as $E_a \equiv RT^2\, d\ln k/dT = RT^2(1/2T + E_{thr}/RT^2)$:

$$E_a = E_{thr} + \tfrac{1}{2}RT \qquad (23.5)$$

Substitution of (23.5) and (23.3) in $A \equiv ke^{E_a/RT}$ [Eq. (17.73)] gives the preexponential factor as

$$A = N_0 \pi (r_A + r_B)^2 \left[\frac{8RT}{\pi}\left(\frac{1}{M_A} + \frac{1}{M_B}\right)\right]^{1/2} e^{1/2} \qquad \text{for A} \neq \text{B} \qquad (23.6)$$

Since $\frac{1}{2}RT$ is small, the hard-sphere threshold energy is nearly the same as the activation energy. The simple collision theory provides no means of calculating E_{thr} but gives only the Arrhenius factor A. Because of the crudities of the theory, the predicted $T^{1/2}$ dependence of A should not be taken seriously.

Example The bimolecular (elementary) reaction $CO + O_2 \rightarrow CO_2 + O$ has an observed activation energy of 51.0 kcal/mole and an Arrhenius factor 3.5×10^9 dm^3 s^{-1} mole^{-1} for the temperature range 2400 to 3000 K. Viscosity measurements and Eq. (16.29) give the hard-sphere diameters as $d(O_2) = 3.6$ Å and $d(CO) = 3.7$ Å. Calculate the hard-sphere collision-theory A factor and compare with the experimental value.

Equation (23.6) gives at the mean temperature 2700 K:

$$A = (6.0 \times 10^{23}\ \text{mole}^{-1})(3.14)(3.6_5 \times 10^{-8}\ \text{cm})^2$$

$$\times \left[\frac{8(8.3 \times 10^7\ \text{ergs/mole K})(2700\ \text{K})}{3.14}\left(\frac{\text{mole}}{28\ \text{g}} + \frac{\text{mole}}{32\ \text{g}}\right) 2.72\right]^{1/2}$$

$$A = 8.1 \times 10^{14}\ \text{cm}^3\ \text{s}^{-1}\ \text{mole}^{-1} = 8.1 \times 10^{11}\ \text{dm}^3\ \text{s}^{-1}\ \text{mole}^{-1}$$

(Note that use of 2400 or 3000 K instead of 2700 K would change A only slightly.) The calculated Arrhenius factor is 230 times the experimental A value. Since experimental A values are accurate to a factor of 3 or 4, the discrepancy between theory and experiment cannot be blamed on experimental error.

For most reactions, the calculated A values are much higher than the observed values. Hence, in the 1920s the hard-sphere collision theory was modified by adding a factor p to the right sides of Eqs. (23.2) to (23.4) and (23.6). The factor p is called the *steric* (or *probability*) *factor*. The argument is that the colliding molecules must be properly oriented for collision to produce a reaction; p (which lies between 0 and 1) supposedly represents the fraction of collisions in which the molecules have the right orientation. For example, for $CO + O_2 \rightarrow CO_2 + O$, we would expect the reaction to occur if the carbon end of the CO hits the O_2 but not if the oxygen end of CO hits O_2.

The hard-sphere collision theory provides no way to calculate p theoretically. Instead, p is found from the ratio of the experimental A to the collision-theory A. Thus, for $CO + O_2 \rightarrow CO_2 + O$, $p = 1/230 = 0.0043$. With this approach, p is simply a fudge factor that brings theory into agreement with experiment.

The idea of a proper orientation being needed for reaction is valid, but p also includes contributions that arise from using a hard-sphere potential and from ignoring molecular vibration and rotation. Typical p values range from 1 to 10^{-6}, tending to be smaller for reactions involving larger molecules, as would be expected if p were due (at least in part) to orientational requirements.

For trimolecular reactions, we need the three-body collision rate. Since the probability that three hard spheres collide at precisely the same instant is vanishingly small, a three-body collision is defined as one in which the three molecules are within a specified short distance from one another. With this definition, the trimolecular collision rate per unit volume is found to be proportional to [A][B][C]. We omit the expression for k.

23.2 POTENTIAL-ENERGY SURFACES

We have seen that the hard-sphere collision theory of chemical kinetics doesn't give accurate rate constants. A correct theory must use the true intermolecular forces between the reacting molecules and must take into account the internal structure of the molecules and their vibrations and rotations. In chemical reactions, bonds are being formed and broken, so we must consider the forces acting on the atoms in the molecules. During a molecular collision, the force on a given atom depends on both the intramolecular forces (which determine the vibrational motions in the molecule) and the intermolecular forces (Sec. 22.10). We cannot deal separately with each of the colliding molecules but must consider the two molecules to form a single quantum-mechanical entity, which we shall call a *supermolecule*. (The supermolecule is not to be thought of as having any permanence or stability; it exists only during the collision process.)

The force on a given atom in the supermolecule is determined by the potential energy V of the supermolecule; Eq. (2.39) gives $F_{x,a} = -\partial V/\partial x_a$, where $F_{x,a}$ is the x component of the force on atom a and V is the potential energy for atomic motion in the supermolecule. (The phrase "atomic motion" in the last sentence can be replaced by "nuclear motion," since the electrons follow the nuclear motion almost perfectly; Sec. 20.2.) The supermolecule's potential energy V is determined the same way the potential energy for nuclear vibrational motion in an ordinary molecule is calculated. Using the Born–Oppenheimer approximation, we solve the electronic Schrödinger equation $\hat{H}_e \psi_e = \varepsilon_e \psi_e$ [Eq. (20.8)] for a fixed nuclear configuration. The energy ε_e equals V at the chosen nuclear configuration. By varying the nuclear configuration, we get V as a function of the nuclear coordinates.

If the supermolecule has $\mathcal{N}$ atoms, there are $3\mathcal{N}$ nuclear coordinates. As with an ordinary molecule, there are 3 translational and 3 rotational degrees of freedom, which leave V unchanged (since they leave all internuclear distances unchanged), so V is a function of $3\mathcal{N} - 6$ variables. If V were a function of two variables x and y, we could plot V in three-dimensional space; this plot is a surface (the *potential-energy surface*) whose distance above the xy plane at the point $x = a$, $y = b$ equals $V(a, b)$. Since V is usually a function of far more than two variables, such a plot usually cannot be made. Nevertheless, the function V is still called the *potential-energy surface*, no matter how many variables it depends on.

Consider some examples. The simplest bimolecular collision is that between two atoms. Here V is a function of only one variable, the interatomic distance R. (The supermolecule has $\mathcal{N} = 2$ and is linear. There are only 2 rotational degrees of freedom, and $3\mathcal{N} - 3 - 2 = 1$.) The function $V(R)$ is the familiar potential-energy curve for the diatomic molecule formed by the two atoms. For example, for two colliding H atoms, the supermolecule is H_2. Its $V(R)$ is given by the second lowest curve in Fig. 20.2 if the electron spins are parallel and by the lowest curve if the spins are antiparallel. Since there are three two-electron spin functions that have total electron-spin quantum number $S = 1$ and only one spin function that has $S = 0$ [Eqs. (19.32) and (19.33)], two H atoms will repel each other in 75 percent of collisions. Even when the spins are antiparallel, a stable molecule will not be formed, since a third body is required to carry away some of the bond energy to prevent dissociation (Sec. 17.11).

Two colliding ground-state He atoms always repel each other (except for the very weak van der Waals attraction at relatively large R). A rough approximation to $V(R)$ is the Lennard-Jones potential.

Two colliding atoms are a very special case. Consider the general features of V when two stable closed-shell polyatomic molecules collide. At relatively large distances, the rather weak van der Waals attraction (Sec. 22.10) causes V to decrease slightly as the molecules approach each other. When they are close enough for their electron probability densities to overlap substantially, the Pauli-principle repulsive force sets in, causing V to increase substantially. If the colliding molecules are not oriented properly to react, or if they don't have sufficient relative kinetic energy to overcome the intermolecular repulsion, the short-range repulsion causes them to bounce off each other without reacting. If, however, the molecules A and B are properly oriented and have sufficient relative kinetic energy to approach closely enough, a new chemical bond can be formed between them, usually accompanied by the simultaneous breaking of one or more bonds in the original molecules, thereby yielding the products C and D. The Pauli-principle repulsion between C and D then causes them to move away from each other, the potential energy V decreasing as this happens. During the course of the elementary reaction, the supermolecule's potential energy first decreases slightly, then rises to a maximum, and then falls off.

The most studied potential-energy surface is that for the reaction $H + H_2 \rightarrow H_2 + H$. The reaction can be studied experimentally using isotopes ($D + H_2 \rightarrow DH + H$) or ortho and para H_2 (H + para $H_2 \rightarrow$ ortho $H_2 + H$). [In ortho H_2, the nuclear (i.e., proton) spins are parallel; in para H_2, they are antiparallel.] The supermolecule is H_3. The first quantum-mechanical calculation of the H_3 potential surface was made by Eyring and Polanyi in 1931, but reasonably accurate results were not obtained until the 1960s. [The Hungarian-born physician, physical chemist, economist, political scientist, and philosopher Michael Polanyi (1891–1976) made important contributions to the fields of adsorption on solids, x-ray crystallography, dislocations in solids, rate processes, and polymerization mechanisms. The American physical chemist Henry Eyring is closely identified with the activated-complex theory (Sec. 23.4).] The first highly accurate *ab initio* calculation of the H_3 potential surface (including nonlinear geometries) was by Liu and Siegbahn. [B. Liu, *J. Chem. Phys.*, **58**, 1925 (1973); P. Siegbahn and B. Liu, to be published.] The Liu–Siegbahn surface is estimated to be in error by less than 0.04 eV (1 kcal/mole) everywhere.

It may seem surprising that it took until the 1970s for a highly accurate H_3 calculation to be made. After all, the rotational barrier in C_2H_6 (a much larger species than H_3) was accurately calculated in the 1960s. There are two reasons why H_3 is a harder problem than C_2H_6. No bonds are broken or formed in going from staggered to eclipsed C_2H_6, and the Hartree–Fock approximate wave function is adequate to give an accurate barrier. In H_3, bonds are being formed and broken, and the Hartree–Fock wave function turns out to be poor in giving energy differences between points on the potential surface. One must do a configuration-interaction calculation, which is a far more difficult task than a Hartree–Fock calculation. Moreover, to find the C_2H_6 barrier, only two calculations are needed, one at the staggered and one at the eclipsed configuration. To find the H_3 potential surface, Liu and Siegbahn performed calculations at hundreds of geometries. (Once V has been calculated at a sufficient number of points, one can fit an algebraic equation to the calculated V values; this allows V and its partial derivatives to be found everywhere on the surface.)

For H_3, V is a function of $9 - 6 = 3$ variables. These can be taken as two interatomic distances R_{ab}, R_{bc} and an angle θ (Fig. 23.2). The reaction is

$$H_a + H_bH_c \rightarrow H_aH_b + H_c \tag{23.7}$$

The electron probability density of H_2 is greatest in the region between the nuclei, and the overlap between the electron probability densities of H_a and the H_2 molecule is a minimum when $\theta = 180°$; we therefore expect the Pauli repulsion to be minimized when the H_a atom approaches along the H_bH_c axis. (Recall the steric factor of simple collision theory.) Accurate quantum-mechanical calculations bear this out.

Since V is a function of three variables, the potential-energy "surface" must be plotted in four dimensions. If we (temporarily) restrict ourselves to a fixed value of θ, then V is a function of R_{ab} and R_{bc} only. We plot R_{ab} and R_{bc} on the two horizontal axes and $V(R_{ab}, R_{bc})$ for a fixed θ on the vertical axis. Figure 23.3 shows the $V(R_{ab}, R_{bc})$ surface for $\theta = 180°$. A contour map of this surface is shown in Fig. 23.4.

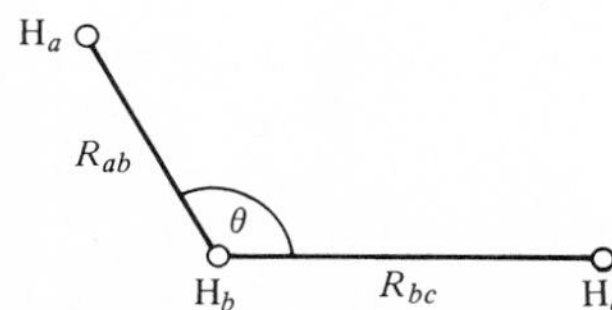

Figure 23.2
Variables for the H + H_2 reaction.

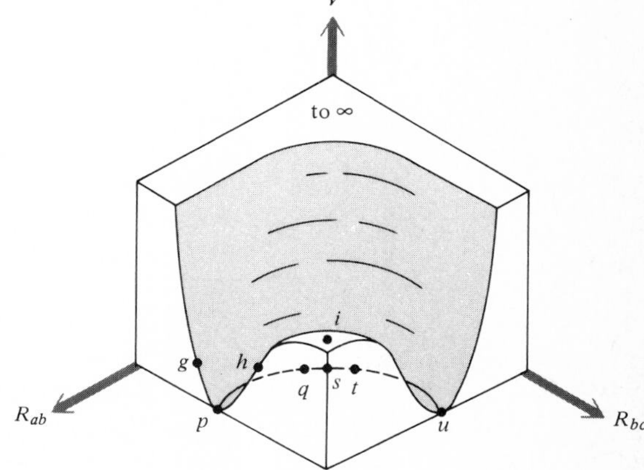

Figure 23.3
The potential-energy surface for the H + H_2 reaction for $\theta = 180°$.

Plots for other values of θ have the same general appearance. The solid lines in Fig. 23.4 are lines of constant V, each labeled with its V value.

At point p, the distance R_{bc} equals the equilibrium bond length (0.74 Å) in H_2, and R_{ab} is large, indicating that atom H_a is far from molecule H_bH_c. Point p corresponds to the reactants $H_a + H_bH_c$. Point u, where R_{bc} is large and $R_{ab} = 0.74$ Å, corresponds to the products $H_{ab} + H_c$. The energy of the reactants $H_a + H_bH_c$ has been taken as zero, so points p and u each have zero energy. (This is not true in general for the reaction A + BC → AB + C, since the energy of AB will usually differ from that of BC.)

Point i has both R_{ab} and R_{bc} very large and corresponds to three widely separated H atoms: $H_a + H_b + H_c$. The region around point i in Fig. 23.3 is a nearly flat plateau. The potential energy varies hardly at all here, since the atoms are so far apart that changing R_{ab} or R_{bc} doesn't affect V. The potential energy at i is 4.75 eV ($109\frac{1}{2}$ kcal/mole) above that at p, this being the equilibrium dissociation energy D_e of H_2. The region around i plays no part in the reaction (23.7). (This would be the reactant region for the trimolecular reaction $H + H + H \rightarrow H_2 + H$.)

Along a line joining points g, p, and h, the distance R_{ab} between H_a and molecule H_bH_c is fixed, but R_{bc} in H_bH_c is being varied. This generates the ground-state diatomic potential-energy curve of H_2 (Fig. 21.4), which can be seen in Fig. 23.3. V increases in going from p to g or from p to h, so point p lies at the bottom of a valley. Point u lies at the bottom of a second valley at right angles to the valley of p.

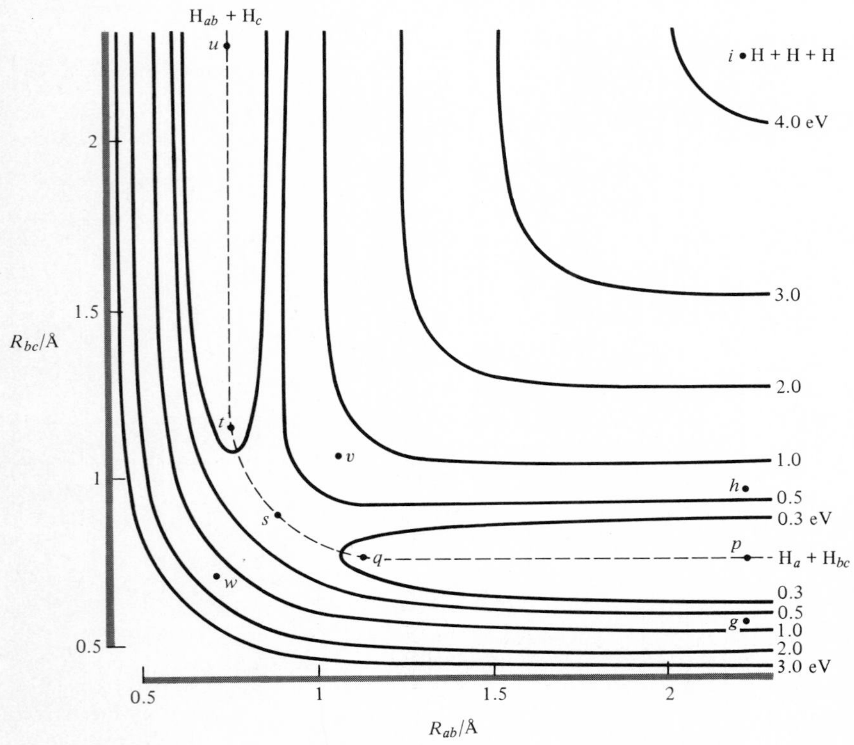

Figure 23.4
Contour map of the $H + H_2$ potential-energy surface for $\theta = 180°$. Note that this diagram starts at $R_{ab} = R_{bc} = 0.5$ Å rather than at zero. [Data from R. N. Porter and M. Karplus, *J. Chem. Phys.*, **40,** 1105 (1964).]

We now look for the path of minimum potential energy that connects reactants to products on the potential-energy surface. This is called the *minimum-energy path.* (It is often called the "reaction path" or the "reaction coordinate," but these names are open to objection. The term "reaction path" is misleading, in that the reacting molecules do not precisely follow the minimum-energy path; see Sec. 23.3. The term "reaction coordinate" has a meaning slightly different from the minimum-energy path; see Sec. 23.4.) The minimum-energy path is the dotted line *pqstu* in Figs. 23.3 and 23.4. Figure 23.5 shows the configurations of the H_3 supermolecule for some points on the minimum-energy path.

At point q, the atom H_a has approached fairly close to the molecule H_bH_c, and the bond distance R_{bc} has lengthened a bit, indicating a slight weakening of the bond. Point s has $R_{ab} = R_{bc}$ and corresponds to the H_a—H_b bond's being half formed and the H_b—H_c bond's being half broken. At point t, the atom H_c is retreating from the newly formed H_aH_b molecule. At point u, the reaction is over. Figures 23.3 and 23.4 show that the potential energy along the minimum-energy path increases from p to q to s, reaches a maximum at s, and then decreases from s to t to u (Fig. 23.6*a*). (Actually, there is an initial decrease of V between points p and q due to the van der Waals attraction, but this is too slight to be visible in the figures.)

Point s is the maximum point on the minimum-energy path and is what mathematicians call a *saddle point*, since the surface around s resembles a saddle (Fig. 23.6*b*). We have $V_s < V_v$ and $V_s < V_w$, but $V_s > V_q$ and $V_s > V_t$. A hiker starting at point p and facing toward q is in a deep valley with walls rising to infinity on the left

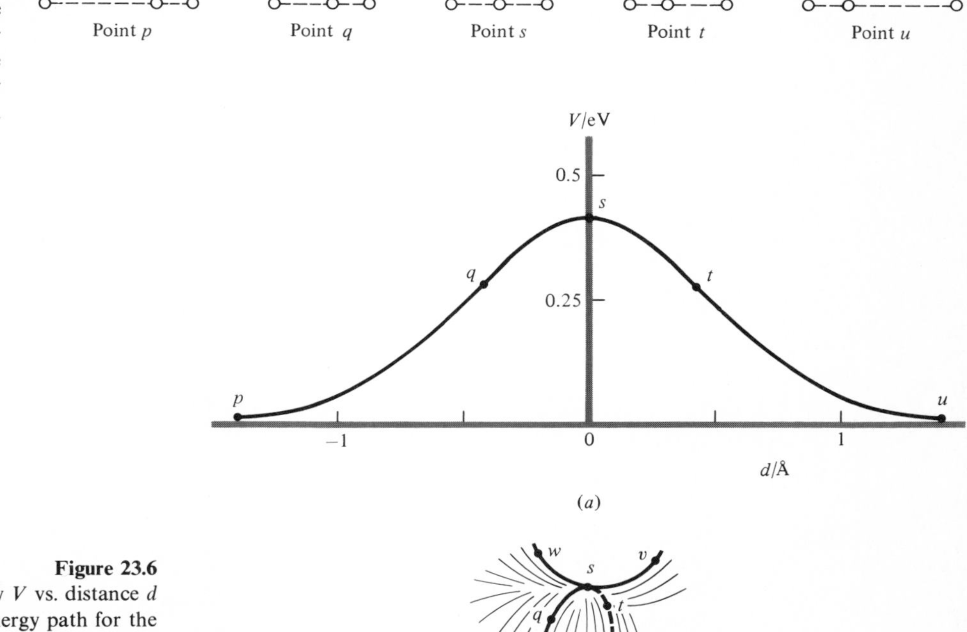

Figure 23.5
Configurations of the H_3 supermolecule for various points on the minimum-energy path.

Figure 23.6
(*a*) Potential energy V vs. distance d along the minimum-energy path for the $H + H_2$ reaction. (*b*) The potential-energy surface in the region of the saddle point s.

and to a high plateau on the right. As he walks from p to s, his elevation rises gradually from 0 to 0.4 eV (10 kcal/mole) [compared with 4.7 eV (110 kcal/mole) for the plateau height]. The region around s is a pass connecting the reactant valley to the product valley.

For the present, let us ignore the rotation and vibration of H_2 and use classical mechanics to consider a linear collision between an H atom and an H_2 molecule. The total energy of the colliding species in Fig. 23.5 remains fixed (conservation of energy). As H_a and H_bH_c approach each other, their potential energy increases (due to the Pauli-principle repulsion), so their kinetic energy decreases. If the relative translational kinetic energy of the colliding molecules is insufficient to allow the supermolecule to climb the potential hill to point s, the repulsion causes H_a to bounce off without reacting. If the relative kinetic energy is high enough, the supermolecule can go over point s and yield products (point u). An analogy is a ball rolled on the potential-energy surface from point p toward s. Whether or not it gets over the saddle point at s depends on its initial kinetic energy.

Since the colliding reactants need not have $\theta = 180°$, all values of θ from 0 to 180° must be considered. For each value of θ, one can do quantum-mechanical calculations of $V(R_{ab}, R_{bc})$. Contour plots of V for various values of θ resemble Fig. 23.4, and can be found in R. N. Porter and M. Karplus, *J. Chem. Phys.*, **40**, 1105 (1964). Drawing a graph like Fig. 23.6*a* for each value of θ, one finds the potential-energy maximum E_{max} to be surmounted in going from reactants to products at various θ values to be:

θ	180°	112°	60°	45°	0°
E_{max}/(kcal mole^{-1})	10	18	64	34	10

As noted earlier, reaction occurs most readily for a linear approach.

A plot of $V(R_{ab}, R_{bc}, \theta)$ requires four dimensions and is not readily made. Instead, we can set up coordinates with the R_{ab} and R_{bc} axes horizontal and the θ axis vertical and plot contours like Fig. 23.4 at various values of θ. Figure 23.7 indicates such a contour plot. (The meaning of the critical dividing surface is discussed in Sec. 23.4.) The angle θ may change during a given H_a-H_bH_c collision, and the full potential-energy function $V(R_{ab}, R_{bc}, \theta)$ is required to deal with an arbitrary collision process. Since a linear approach requires the least energy for reaction to occur, most collisions that lead to reaction will have θ reasonably close to 180°.

Since $\theta = 180°$ is the energetically most favored angle, point s in Fig. 23.4 (which is for $\theta = 180°$) lies at a minimum with respect to variations in θ, as well as with respect to variations along the line vsw. The configuration of the colliding molecules at the saddle point s is called the *transition state*. For $H + H_2$, the transition state is linear and symmetric, with each H—H distance equal to 0.93 Å (compared with $R_e = 0.74$ Å in H_2). The potential-energy difference between the transition state and the reactants (omitting zero-point vibrational energy) is the (*classical*) *barrier height* ε_b; that is, $\varepsilon_b \equiv V_s - V_p$. For $H_2 + H$, *ab initio* quantum-mechanical calculations give $\varepsilon_b \approx 0.42_5$ eV and $E_b \equiv N_0 \varepsilon_b \approx 9.8$ kcal/mole (with an estimated error of ± 0.3 kcal/mole). We shall see later that the barrier height is approximately (but not exactly) equal to the activation energy of the reaction.

The term "transition state" should not mislead one into thinking that the supermolecule at point s has any sort of stability. The transition state is just one particular point on the continuous path from reactants to products. (H_3 calculations made in the 1930s showed a slight dip in V in the region of point s to give a shallow "well" or "basin," but this dip results from errors in the crude quantum-mechanical approximations used. Accurate calculations show that there is no dip.)

The 10 kcal/mole barrier height for $H_a + H_bH_c \rightarrow H_aH_b + H_c$ is much less than the 110 kcal/mole needed to break the bond in H_2 ($H_2 \rightarrow 2H$). In general, observed bimolecular activation energies are only a fraction of the energy needed to break the relevant bond because simultaneous formation of a new chemical bond compensates for the breaking of the old bond. As the H_b—H_c bond is breaking, the H_a—H_b bond is forming. (Of course, when no new bonds are formed, E_a is quite high. Thus, the E_a values for unimolecular decompositions are high and are approximately equal to the energy of the bond broken if the products have no bonds not present in the reactants.)

Figure 23.8*a* shows potential-energy contours found from *ab initio* calculations for the $F + H_2 \rightarrow H + HF$ reaction (and for the reverse reaction $H + HF \rightarrow H_2 + F$) for $\theta = 180°$, the energetically most favored approach angle. The figure is less symmetric than Fig. 23.4 for H_3, since the products differ from the reactants. The calculated barrier height is 1.7 kcal/mole for the forward reaction. (The experimental activation energy is 1.7 kcal/mole.) Since the calculated energy change for the reaction is -34.4 kcal/mole, the calculated barrier height for the reverse reaction is 36.1 kcal/mole (Fig. 23.8*b*).

The calculated saddle-point (transition-state) geometry has $R(\text{HF}) = 1.54$ Å and $R(\text{HH}) = 0.77$ Å, compared with $R(\text{HH}) = 0.74$ Å in the isolated H_2 molecule and $R(\text{HF}) = 0.93$ Å in HF. The transition state occurs early in the reaction, with the H—H bond distance only slightly elongated and the F atom relatively far from the H_2 molecule. The transition state for $F + H_2 \rightarrow H + HF$ resembles the reactants much more than the products. For the reverse reaction, the transition state (which is

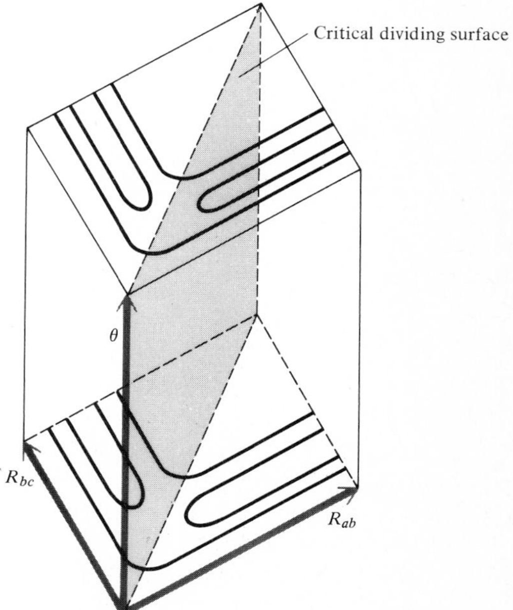

Figure 23.7
Potential-energy contour plots for the $H + H_2$ reaction at various values of θ. (There is a set of contours for each θ value, but to keep the figure simple, only two sets of contours are shown.)

the same as for the forward reaction) resembles the products. (In 1955, Hammond postulated that in an exothermic reaction the transition state will be likely to resemble the reactants, while in an endothermic reaction it will resemble the products. See *Lewis*, pp. 466–468.)

The rate constant and the detailed course of an elementary chemical reaction depend on the shape of the entire potential-energy surface (see Sec. 23.3). However, the main features of an elementary reaction can be determined if the barrier height and the structure of the transition state are known. The barrier height does not differ greatly from the activation energy, so a reaction with a low barrier height is rapid whereas one with a high barrier is slow.

The transition state occurs at the maximum point on the minimum-energy path between reactants and products. The energy of the transition state relative to that of the reactants determines the barrier height. The geometrical structure of the transition state determines the stereochemistry of the reaction products. For example, consider the elementary bimolecular reaction $I^- + CRR'R''Br \rightarrow CRR'R''I + Br^-$ (in the jargon of organic chemists, an S_N2 reaction, a bimolecular nucleophilic substitution). The I^- might attack the alkyl bromide on the same side of the molecule as the Br or on the opposite side (Fig. 23.9). The transition states are shown in braces. (These transition states are not reaction intermediates but are simply one point on the continuous path from reactants to products.) The product obtained in one case is the mirror image of that in the other case. For attack by I^- on the side opposite the Br, the transition state is expected to be of lower energy than that formed in attack on the same side as Br; this is because for "backside" attack, the carbon can undergo gradual rehybridization to sp^2 AOs, which bond the three R groups, leaving a carbon p orbital to partially bond both the I and the Br in the transition state. The barrier should be lower for backside attack. This is borne out by the observation that the

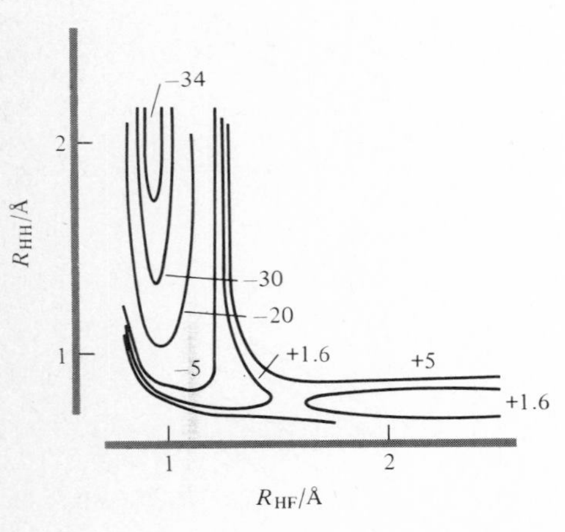

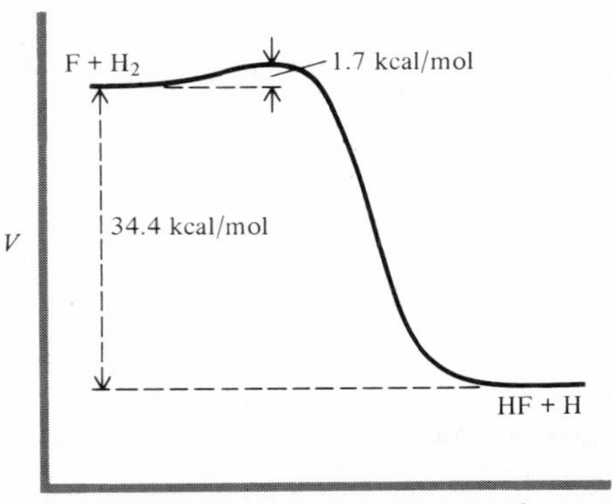

Figure 23.8
(*a*) Potential-energy contours for the $F + H_2 \rightarrow HF + H$ reaction for $\theta = 180°$. Energies are in kcal/mole. The zero of energy is taken at the separated reactants. (Adapted from C. F. Bender, S. V. O'Neil, P. K. Pearson, and H. F. Schaefer, *Science*, **176**, 1412–1414, 30 June 1972. Original figure copyright 1972 by the American Association for the Advancement of Science.) (*b*) Barrier height for this reaction.

product is stereochemically inverted with respect to the reactant. The transition-state structure also explains why this reaction is very slow when the R groups are bulky. Large R groups mean a large Pauli-principle repulsion between I^- and the alkyl halide, and hence a greater barrier height and a slower reaction. This example shows why organic chemists spend so much time talking about transition-state structures. (A model of the S_N2 transition state is on the roof of the Chemistry Conference Building at Stanford University.)

For a reaction between two five-atom molecules, the potential-energy surface is a function of 24 variables. Clearly, the calculation of an accurate *ab initio* potential-energy surface for such a reaction is a fantastically difficult task. Things can be simplified by the fact that only a small number of bond distances and angles change significantly during the reaction, so the number of variables can be substantially reduced. Even so, an accurate *ab initio* calculation of V is still extremely difficult and has been accomplished for only a few reactions.

Semiempirical calculations require much less computer time and storage capacity than *ab initio* calculations. The MINDO method (Sec. 20.9) gives generally reliable ΔH° values for chemical reactions and so might be useful for calculation of reaction potential-energy surfaces. Many potential-energy surfaces, barrier heights, and transition-state structures have been calculated using MINDO, but it is too early to assess the reliability of such surfaces.

Since reliable theoretical calculation of barrier heights is very difficult, several empirical and semiempirical schemes have been proposed to estimate barrier heights. The bond-energy–bond-order (BEBO) method of Johnston works well in many cases. (See *Johnston*, pp. 179–183, 339–347, chap. 11.) Benson has discussed methods of estimating E_a and A. (See S. W. Benson and D. M. Golden, chap. 2 in vol. VII of *Eyring, Henderson, and Jost*; S. W. Benson, *Thermochemical Kinetics*, 2d ed., Wiley-Interscience, 1976.)

The Woodward–Hoffmann rules use MO theory to decide which stereochemical path has the lower barrier height in various organic reactions. These rules are discussed in most current undergraduate organic-chemistry texts. Pearson has used MO concepts to decide whether the activation energies of various elementary inorganic reactions are high or low. (See R. G. Pearson, *Symmetry Rules for Chemical Reactions*, Wiley-Interscience, 1976.) Theorems that restrict the symmetry of the transition state in a given reaction have been developed; see R. E. Stanton and J. W. McIver, *J. Am. Chem. Soc.*, **97**, 3632 (1975); J. W. McIver, *Acc. Chem. Res.*, **7**, 72 (1974).

Potential-energy surfaces for unimolecular and trimolecular reactions are discussed in Secs. 23.6 and 23.7.

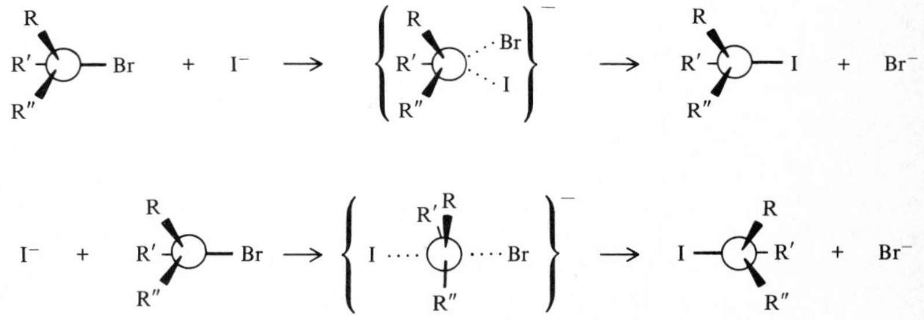

Figure 23.9
Two possible mechanisms for the reaction $I^- + RR'R''CBr \rightarrow RR'R''CI + Br^-$.

23.3 MOLECULAR REACTION DYNAMICS

Molecular reaction dynamics studies what happens at the molecular level during an elementary chemical reaction. The kinetics methods discussed in Chap. 17 can tell us that the rate constant for the elementary reaction $Cl + H_2 \rightarrow HCl + H$ is given by $k = Ae^{-E_a/RT}$, where $A = 1.2 \times 10^{10}$ dm^3 mole^{-1} s^{-1} and $E_a = 4.3$ kcal/mole in the temperature range 250 to 450 K, but this statement gives little information about the details of the elementary reaction. Molecular dynamics considers such questions as: How does the probability for reaction vary with the angle the incoming Cl atom makes with the H—H line? How does this probability vary with the relative translational energy of the reactants? How is the product HCl distributed among its various translational, rotational, and vibrational states? How can quantum mechanics and statistical mechanics be used to theoretically calculate the rate constant at a given temperature?

The initial development of molecular reaction dynamics began in the 1930s with work of Eyring, M. Polanyi, and Hirschfelder, but it wasn't until the 1960s that new experimental techniques and the availability of electronic computers for theoretical calculations allowed reliable information to be obtained. With these developments, chemists are beginning to understand what happens in an elementary chemical reaction.

Trajectory calculations. Suppose that the potential-energy surface for a given gas-phase elementary chemical reaction has been accurately calculated using quantum mechanics. How does one use this surface to calculate the rate constant at a given temperature?

The probability that two colliding gas molecules react depends on the initial translational, rotational, and vibrational states of the molecules. One picks a pair of initial states for the reactant molecules and solves the time-dependent Schrödinger equation (18.9) to calculate the probability that reaction occurs for these initial states. The potential-energy V occurs in the Hamiltonian operator $\hat{H}$ in (18.9), so that V is needed for this calculation. Since the gas molecules are distributed among many translational, rotational, and vibrational states, the process of solving the time-dependent Schrödinger equation must be repeated for sufficiently many different initial states to give a statistically representative sample of all possible initial states. Then one uses the Boltzmann distribution to calculate the relative number of collisions that occur for each set of initial states at temperature T, and a weighted average of the reaction probabilities is taken, thereby giving the rate constant at T. (This procedure can be viewed as the rigorously correct version of the simple collision theory of Sec. 23.1.)

Solving the time-dependent Schrödinger equation is an extremely difficult task, and only a very few such calculations of rate constants have been made. Fortunately, there is an approximate approach that is much easier. Instead of using quantum mechanics to deal with the collision process, one uses classical mechanics. One chooses a pair of initial states for the reactant molecules (rotational and vibrational energies, relative translational energy, approach angle). The forces are obtained from the potential-energy surface ($F_{x,a} = -\partial V/\partial x_a$, etc.). Newton's second law $F = ma$ is numerically integrated (on a computer) to give the locations of the atoms as functions of time. The path of a particle is called its trajectory in classical mechanics, and such calculations are called *trajectory calculations*. After trajectories have been calculated for a representative set of initial conditions, suitable averaging using the Boltzmann distribution at the temperature of interest gives the rate constant. Of course, the motions of atoms obey quantum mechanics, not classical mechanics. However, comparisons of the results of quantum-mechanical and classical calculations indicate that the classical-trajectory results are accurate except when very light species are involved. (As a particle's mass increases, its behavior approaches classical behavior.) Classical mechanics is completely deterministic, and the probability that a given initial state will lead to reaction is either 0 or 1. This is not so in quantum mechanics. In particular, there is some probability that molecules having relative kinetic energy less than the potential-energy barrier height will tunnel (Sec. 18.12) through the barrier and yield products. For reactions involving one of the species e^-, H^+, H, and H_2 (including reactions that transfer e^-, H^+, or H between two heavy molecules), tunneling is important and can easily multiply the rate constant by a factor of 3 or more. For heavier species, tunneling is unimportant. Methods for correcting classical-trajectory results to allow for tunneling are being developed. [See W. H. Miller, *Adv. Chem. Phys.*, **25**, 69 (1974).]

Consider the atom–diatomic-molecule reaction A + BC → AB + C. Figure 23.10 shows two typical classical trajectories for a linear collision plotted on the $\theta = 180°$ contour map. (The angle θ may well change during a collision, and the trajectory must be plotted on a figure like Fig. 23.7. Here we restrict ourselves to collisions in which θ stays constant at 180°.) In Fig. 23.10*a*, molecule BC is in the $v = 0$ vibrational level. Because of the zero-point vibration in BC, the trajectory of the supermolecule oscillates about the minimum-energy path (the dashed line), and the supermolecule is not likely to pass over the barrier precisely at the saddle point. Another reason the minimum-energy path is not precisely followed is that many reactive collisions will have $\theta \neq 180°$. The wide vibrations of the product AB in Fig. 23.10*a* indicate that in this collision AB has been produced in an excited vibrational level. Figure 23.10*b* shows a nonreactive collision trajectory.

For the reaction H_2 + H, Karplus and coworkers calculated thousands of classical trajectories on a reasonably accurate semiempirical potential surface and calculated rate constants at several temperatures [M. Karplus, R. N. Porter, and R.

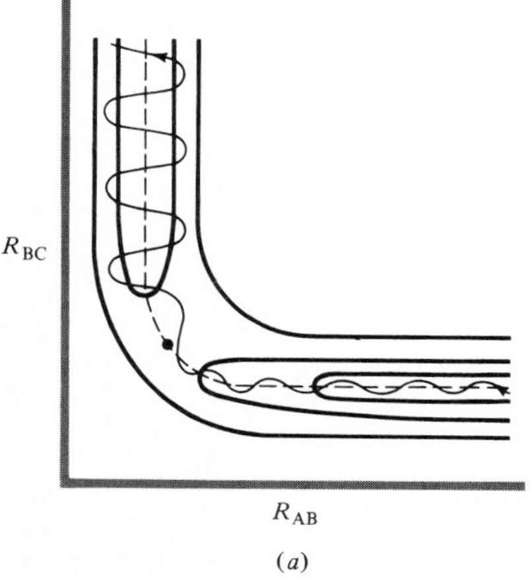

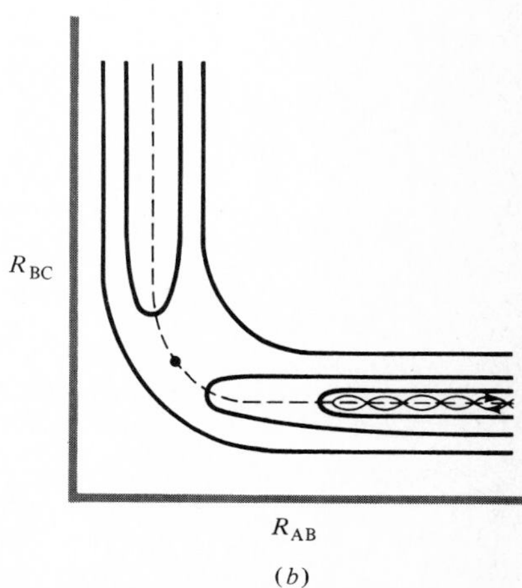

Figure 23.10 Classical trajectories for the reaction A + BC → AB + C. The saddle point is indicated by a dot. (*a*) A reactive collision. (*b*) A nonreactive collision.

D. Sharma, *J. Chem. Phys.*, **43**, 3259 (1965)]. Figure 23.11 shows the results of two such calculations.

In Fig. 23.11*a*, the distances R_{ab} and R_{ac} first decrease (indicating that atom H_a is approaching molecule H_bH_c) and then increase (indicating that H_a is moving away from H_bH_c); no reaction has occurred. The continual fluctuation in R_{bc} is due to zero-point vibration in molecule H_bH_c. In Fig. 23.11*b*, atom H_a again approaches H_bH_c, colliding with it at 3×10^{-14} s. Now, however, R_{bc} goes to infinity after 3×10^{-14} s, indicating that the $H_b—H_c$ bond has broken, and R_{ab} fluctuates about the equilibrium H_2 bond distance 0.74 Å, indicating that atom H_a has bonded to H_b. The reaction $H_a + H_bH_c \rightarrow H_aH_b + H_c$ has occurred. Note that at 4×10^{-14} s atom H_c is closer to H_b in H_aH_b, but at 5×10^{-14} s it is closer to H_a. This indicates that the product molecule H_aH_b is rotating. The initial state of the reactants had $J = 0$, and so part of the relative translational energy of the reactants has gone into rotational energy of a product. Trajectory calculations thus show how the energy of the products is distributed among rotational, vibrational, and relative translational energies.

Since most collisions are nonlinear (and since θ changes during rotation), one needs to know V as a function of all three variables R_{ab}, R_{bc}, and θ in order to calculate k from $H_2 + H$ trajectory calculations.

The $H + H_2$ calculations of Karplus, Porter, and Sharma showed that vibrational energy of the reactants can contribute to the energy needed to overcome the barrier but that rotational energy does not contribute. Although $\theta = 180°$ is the energetically most favored approach angle, the large number of nonlinear approaches leads to an average approach angle of 160° for reactive collisions.

Rate constants calculated from classical trajectories agree reasonably well with experimental rate constants. The main sources of error in the calculated k values are inaccuracies in the potential-energy surface and tunneling.

In the hard-sphere collision theory (Sec. 23.1), the probability p_r of a reactive collision varies with the line-of-centers component of the relative translational kinetic energy of the colliding molecules according to Fig. 23.12*a*. When the hard-sphere p_r is calculated as a function of ε_{rel}, the total relative translational kinetic energy of the molecules, one obtains Fig. 23.12*b*. [See E. F. Greene and A. Kupperman, *J. Chem. Educ.*, **45**, 361 (1968) for the derivation.] The results of trajectory calculations on potential-energy surfaces and of molecular-beam experiments (see below) show

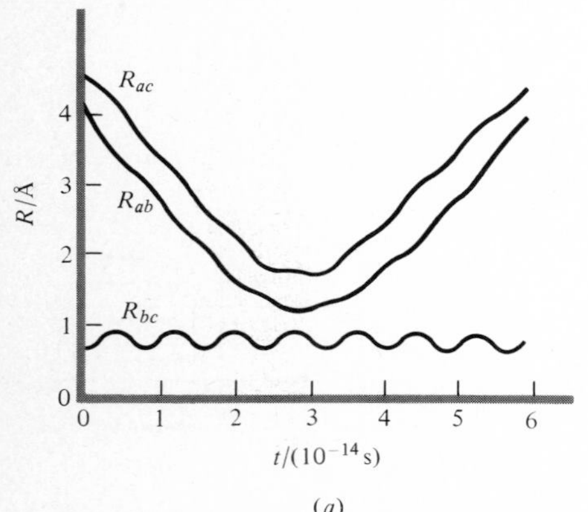

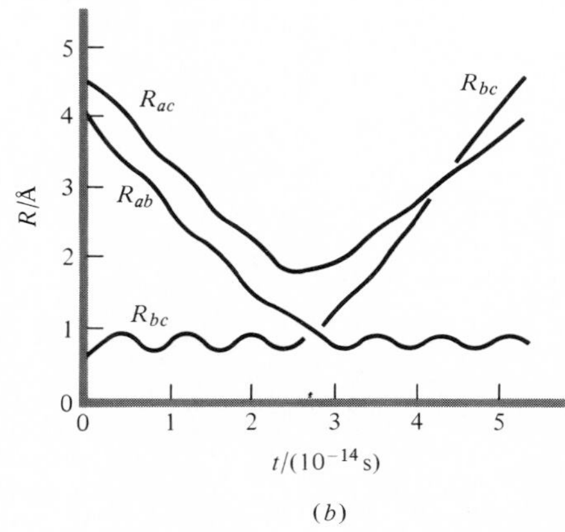

Figure 23.11 Two classical trajectories for the $H + H_2$ reaction. [M. Karplus, R. N. Porter, and R. D. Sharma, *J. Chem. Phys.*, **43**, 3259 (1965).]

that Fig. 23.12*b* is in error in two ways: (*a*) the true reaction probabilities are usually less than those of the hard-sphere theory (recall the steric factor), and (*b*) the true reaction probability reaches a maximum and then at energies much greater than ε_{thr} falls off.

Molecular beams. We now consider some experimental techniques in molecular reaction dynamics. In a crossed-molecular-beam experiment, a beam of A molecules crosses a beam of B molecules in a high-vacuum chamber. (The beams are produced by vaporization from an oven and collimation by slits, as in Fig. 15.8.) Collisions in the region of intersection of the beams can lead to the chemical reaction $A + B \rightarrow C + D$. A movable detector is used to detect the products. The speeds of the molecules in the beam can be controlled using velocity selectors (Fig. 15.8). Such experiments yield information on how the probability of reaction varies with the relative kinetic energy of the colliding molecules, on the angles at which the reaction products leave the collision site, and on the energy distribution of the products. Because of technical difficulties, most beam work has been done on ion–molecule reactions and on reactions of alkali-metal atoms with halogens and halides. (These reactions have high probabilities of occurring, and the products are readily detected.) Reactions studied in molecular beams include $D + H_2 \rightarrow HD + H$, $F + H_2 \rightarrow HF + H$, $K + Br_2 \rightarrow KBr + Br$, $K + CH_3I \rightarrow KI + CH_3$.

Infrared chemiluminescence. In this technique, a gas-phase reaction is carried out at a pressure sufficiently low for there to be a negligible probability of the product's losing vibrational and rotational energy by collisions. Instead, this energy is lost by emission of radiation (*chemiluminescence*, Sec. 21.13). For example, if the reaction $Cl + HI \rightarrow HCl + I$ is studied, measurement of the intensities of the infrared vibration–rotation emission lines from HCl tells us how the energy of the product HCl is distributed among its various vibrational–rotational states and thus tells us the relative rates of formation of HCl in these excited states. It turns out that a non-Boltzmann distribution is produced by the reaction, with a maximum population in the $v = 3$ vibrational level of HCl. (Of course, at ordinary pressures, molecular collisions would rapidly lead to the Boltzmann distribution.)

Laser techniques. Lasers can be used to give information about molecular dynamics. For example, one can use a laser to excite a large fraction of one of the reactant species to a specific vibrational level and then investigate how the reaction rate is affected by having a non-Boltzmann distribution for this reactant.

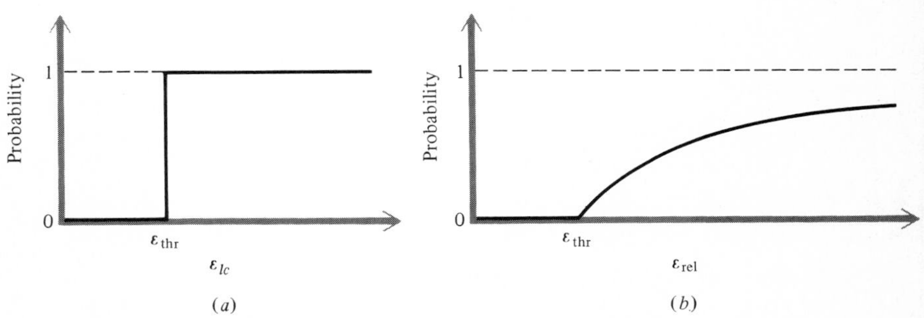

Figure 23.12 Hard-sphere collision-theory reaction probability as a function of (*a*) the line-of-centers kinetic energy and (*b*) the relative kinetic energy of the colliding molecules.

23.4 ACTIVATED-COMPLEX THEORY FOR IDEAL-GAS REACTIONS

The rigorously correct way to theoretically calculate a reaction's rate constant is (*a*) to solve the electronic (time-independent) Schrödinger equation for a very large number of configurations of the nuclei, so as to generate the complete potential-energy surface for the reaction; (*b*) if light species are not involved, use this surface to perform classical trajectory calculations for a wide variety of initial reactant states and suitably average the results to obtain k. (If light species are involved, the collisions must be dealt with using the time-dependent Schrödinger equation instead of classical mechanics.)

The very great difficulties involved in this procedure make it highly desirable to have a simpler, approximate theory of rate constants. Such a theory is the *activated-complex theory* (ACT), also called *transition-state theory*. ACT was developed in the 1930s by Pelzer and Wigner, Evans and M. Polanyi, and Eyring, and has been widely applied by Eyring and coworkers. (The theory is often called the "absolute rate theory" by its more enthusiastic practitioners, but this name is not appropriate for an approximate theory.) ACT eliminates the need for trajectory calculations and requires that the potential-energy surface be known only in the region of the reactants and the region of the transition state, rather than everywhere. This section develops ACT for ideal-gas reactions. ACT for reactions in liquid solution is discussed in Sec. 23.8.

We saw in Sec. 23.2 that the potential-energy "surface" for a reaction has a reactant region and a product region that are separated by a barrier. The activated-complex theory chooses a boundary "surface" located between the reactant and product regions and assumes that all supermolecules that cross this boundary surface become products. The boundary surface (called the *critical dividing surface*) is taken to pass through the saddle point of the potential-energy surface. For the H_3 potential-energy contour map of Fig. 23.4, the critical dividing "surface" is a straight line that starts at the origin, passes through points v, s, and w, and extends out through the H + H + H region. (Of course, most supermolecules cross the dividing line not too far from the saddle point s.) Figure 23.4 considers only collisions with $\theta = 180°$. The complete critical dividing surface for the $H + H_2$ reaction is shown in the more general figure 23.7.

ACT assumes that all supermolecules that cross the critical surface from the reactant side become products. (This is reasonable, since once a supermolecule crosses the critical surface, it's a downhill journey to products.) A second assumption of ACT is that during the reaction, the Boltzmann distribution of energy is maintained for the reactant molecules. This assumption was also used in the hard-sphere collision theory and, as noted in Sec. 23.1, is usually accurate. A third assumption is that the supermolecules crossing the critical surface from the reactant side have a Boltzmann distribution of energy corresponding to the temperature of the reacting system. These supermolecules originated from collisions of reactant molecules, and the reactant molecules have a Boltzmann distribution, so the third assumption is not unreasonable.

Let the elementary ideal-gas reaction under consideration be $A + B + \cdots \rightarrow D + E + \cdots$. The reaction may have any molecularity, so "$A + B + \cdots$" is to be interpreted as "A," "A + B," or "A + B + C," according to whether the reaction is unimolecular, bimolecular, or trimolecular. (Also, B might be the same as A.)

As noted in Sec. 23.3, not all supermolecules cross the dividing surface at precisely the saddle point of the potential-energy surface. An *activated complex* is any supermolecule whose nuclear configuration corresponds to any point on the dividing surface or to any point a short distance δ beyond the dividing surface (see below). We expect that most activated complexes will have configurations reasonably close to the saddle-point configuration. The saddle point corresponds to the "equilibrium" structure of the activated complex, and points on the dividing surface near the saddle point correspond to "vibrations" about the "equilibrium" structure.

For $H + H_2$, the saddle-point geometry ("equilibrium" activated-complex structure) is the linear symmetric structure $H \cdots H \cdots H$. Points on the dividing line through v, s, w in Fig. 23.13 and lying to either side of s correspond to configurations in which R_{ab} and R_{bc} are equal to each other but differ from the saddle-point R_{ab} distance (0.930 Å); such points correspond to the symmetric stretching vibration ν_1 in Fig. 21.15. For linear $H + H_2$ collisions, the activated complexes have configurations lying between the two parallel dashed lines in Fig. 23.13. If we start at point s and move toward products on a line perpendicular to vsw, R_{ab} is decreasing and R_{bc} is increasing; this corresponds to the asymmetric stretching vibration ν_3 in Fig. 21.15. An arbitrary point in the region between the parallel lines in Fig. 23.13 corresponds to a superposition of the vibrations ν_1 and ν_3. When all $H + H_2$ collisions (including nonlinear ones) are considered, the activated complexes have configurations lying between the dividing surface in Fig. 23.7 and a second surface (not shown) that is a distance δ beyond the dividing surface. If we move up or down in Fig. 23.7, the angle θ varies, which corresponds to the (degenerate) bending vibration ν_2 in Fig. 21.15. An arbitrary activated complex corresponds to a superposition of the vibrations ν_1, ν_2, and ν_3.

A given activated complex exists only momentarily and does not actually undergo repeated back-and-forth bending and stretching vibrations. Instead, since supermolecules cross the dividing surface at various points, any given activated complex can be considered to be in a vibrational state that corresponds to the point at which the dividing surface is crossed. ACT assumes these vibrational states to be populated in accord with the Boltzmann distribution.

The term "transition state" is often used as synonymous with "activated

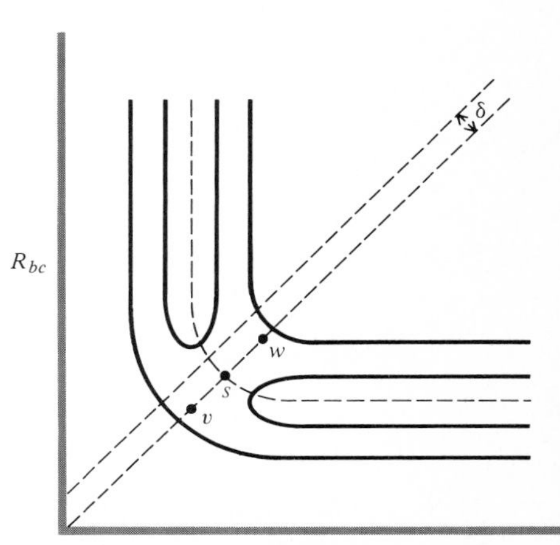

Figure 23.13
The region of existence of $\theta = 180°$ activated complexes for the $H + H_2$ reaction is between the parallel lines separated by δ.

complex," but it might be best to define the transition state as the saddle-point configuration, i.e., as the "equilibrium" configuration of the activated complex.

Denoting an activated complex by $X_f^\ddagger$, we write the elementary reaction $A + B + \cdots \to D + E + \cdots$ as

$$A + B + \cdots \to \{X_f^\ddagger\} \to D + E + \cdots \tag{23.8}$$

The braces are a reminder that an activated complex is not a stable species or a reaction intermediate (Sec. 17.6) but is simply one stage in the smooth, continuous transformation of reactants to products in the elementary reaction. The subscript f indicates that we are talking about activated complexes that are crossing the dividing surface in the forward direction from reactants to products. If any reverse reaction $D + E + \cdots \to A + B + \cdots$ is occurring in the system, there are also present activated complexes $\{X_b^\ddagger\}$ that cross the dividing surface in the reverse (back) direction; but these are of no concern to us, since we are considering only the rate of the forward reaction.

ACT postulates a Boltzmann distribution for the reactants A, B, ... and for the activated complex $X_f^\ddagger$. The arguments following Eq. (22.124) show that when the species A, B, ... and the species $X_f^\ddagger$ (which is formed from A, B, ...) are present with each species distributed among its states according to the Boltzmann distribution law, then [see Eq. (22.125)]

$$\frac{N_f^\ddagger}{N_A N_B \cdots} = \frac{z_\ddagger}{z_A z_B \cdots} e^{-\Delta\varepsilon_0{}^\ddagger/kT} \tag{23.9}$$

where $N_f^\ddagger$, N_A, ... are the numbers of molecules of $X_f^\ddagger$, A, ..., where $z_\ddagger$, z_A, ... are the molecular partition functions of $X_f^\ddagger$, A, ..., and where $\Delta\varepsilon_0^\ddagger \equiv \varepsilon_0(X_f^\ddagger) - \varepsilon_0(A) - \varepsilon_0(B) - \cdots$ is the difference between the energy of $X_f^\ddagger$ in its lowest state and the energies of the reactants A, B, ... in their lowest states. The quantity $\Delta\varepsilon_0^\ddagger$ differs somewhat from the classical barrier height ε_b, due to the zero-point vibrational energies of $X_f^\ddagger$, A, B, ...; see Fig. 23.14.

Division of each N in (23.9) by $N_0 V$ converts it into a (molar) concentration. Therefore (23.9) can be written as

$$K_f \equiv \frac{[X_f^\ddagger]}{[A][B]\cdots} = \frac{z_\ddagger/N_0 V}{(z_A/N_0 V)(z_B/N_0 V)\cdots} e^{-\Delta\varepsilon_0{}^\ddagger/kT} \tag{23.10}$$

Since (23.10) has the same form as (22.124), we defined K_f as $[X_f^\ddagger]/[A][B]\cdots$. The quantity K_f looks like an equilibrium constant, and it is often said that ACT assumes that the activated complexes are in equilibrium with reactants. This statement is misleading and is best avoided. The word "equilibrium" suggests that molecules of

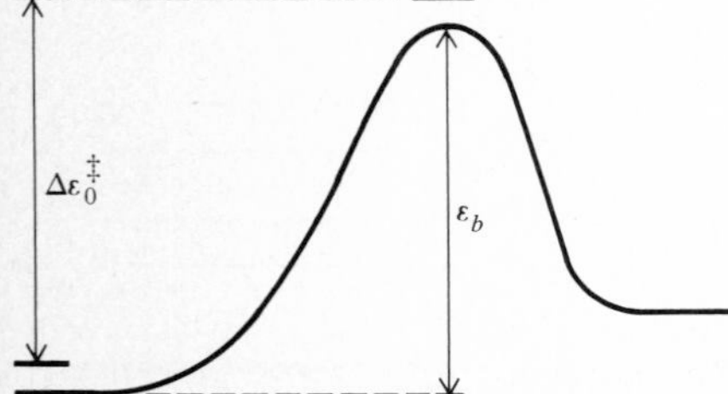

Figure 23.14
Relation between ε_b and $\Delta\varepsilon_0^\ddagger$.

$X_f^‡$ sit around for a while, and then some of them go on to form products and some go back to reactants, but this is not what happens. The symbol $X_f^‡$ denotes those supermolecules that cross the critical dividing surface from the reactant side; by hypothesis, such supermolecules always go on to form products. (It's true that supermolecules formed in low-energy collisions go only part way up the barrier, and then "roll" back down to form separated reactants, but such supermolecules are not activated complexes, since they do not reach the critical dividing surface.) The activated complexes are not in a true chemical-reaction equilibrium with the reactants. Instead the activated complexes are assumed to be in *thermal* equilibrium with the reacting system, the activated-complex states being populated according to the Boltzmann distribution appropriate to the temperature of the system.

A nonlinear activated complex with $\mathcal{N}$ atoms has 3 translational, 3 rotational, and $3\mathcal{N} - 6$ vibrational degrees of freedom. The "equilibrium" structure of the activated complex lies at a saddle point. Since it lies on the minimum-energy path, the saddle point is a point of minimum potential energy for all coordinates but one, for which it is a maximum point. The first derivative vanishes at either a minimum or a maximum, so all the first derivatives of the potential energy V for nuclear motion are zero at the saddle point and V of the activated complex can be approximated as a quadratic function of the normal vibrational coordinates, just as for an ordinary molecule (Sec. 21.7).

The one vibrational normal coordinate of the activated complex for which V at the saddle point is a maximum is called the *reaction coordinate*. The normal "vibration" corresponding to the reaction coordinate is anomalous: because the potential-energy surface slopes downhill along the reaction coordinate to either side of the saddle point, there is no restoring force for this "vibration," and a stable back-and-forth vibration along the reaction coordinate is impossible. Instead, nuclear motion along this coordinate breaks the activated complex up into products. For example, for the reaction $H_a + H_bH_c$, the anomalous vibrational mode of the H_3 activated complex is ν_3 (Fig. 21.15):

$$\overset{\longrightarrow}{H_a} \quad \overset{\longleftarrow}{H_b} \quad \overset{\longrightarrow}{H_c} \tag{23.11}$$

This mode leads to formation of the H_a—H_b bond and breaking of the H_b—H_c bond to give the products $H_aH_b + H_c$. As H_c moves away and H_a and H_b move toward each other, V decreases, so there is no restoring force to bring H_c back.

The reaction coordinate may be depicted as a line on the potential-energy contour map. In Fig. 23.13 for $H + H_2$, the reaction coordinate would be a line perpendicular to line *vsw* and going through point *s*; along this line, R_{ab} decreases and R_{bc} increases. (The reaction coordinate is defined only in the region of the saddle point.) For $H + H_2$, the direction of the reaction coordinate is tangent to the minimum-energy path at the saddle point. This is not true for reactions with unsymmetrical activated complexes (for example, $H + F_2$). The ACT critical dividing surface is defined to pass through the saddle point and to be oriented perpendicular to the reaction coordinate.

The partition function $z_‡$ of the activated complex is given by Eq. (22.58) as $z_‡ = z_{tr}^‡ z_{rot}^‡ z_{vib}^‡ z_{el}^‡$. From Eq. (22.110) $z_{vib}^‡$ is the product of partition functions for the normal vibrational modes. Singling out the reaction coordinate for special attention, we write $z_{vib}^‡ = z_{rc} z_{vib}^{‡\prime}$, where z_{rc} is the partition function for the anomalous motion along the reaction coordinate and $z_{vib}^{‡\prime}$ is the partition function for the nonanomalous vibrational modes; $z_{\text{vib}}^{\ddagger\prime}$ is a product over $3\mathcal{N} - 7$ or $3\mathcal{N} - 6$ vibrational degrees of freedom, according to whether the saddle-point geometry is nonlinear or linear. Therefore

$$z_{\ddagger} = z_{\text{rc}} z'_{\ddagger} \tag{23.12}$$

$$z'_{\ddagger} \equiv z_{\text{tr}}^{\ddagger} z_{\text{rot}}^{\ddagger} z_{\text{vib}}^{\ddagger\prime} z_{\text{el}}^{\ddagger} \tag{23.13}$$

We must decide how to treat motion along the reaction coordinate in order to calculate z_{rc}. Let Q_{rc} be the distance along the reaction coordinate. As noted above, $\partial V/\partial Q_{\text{rc}} = 0$ at the saddle point. Hence, V is nearly constant along the reaction coordinate for the short distance δ (Fig. 23.13) that defines the region of existence of activated complexes. (δ will cancel in the final expression for the rate constant, so we need not specify its value except to say that δ is short enough to prevent V from varying significantly along Q_{rc}.) With $V \approx$ const along the reaction coordinate, the force component for motion along the reaction coordinate is approximately zero: $F_{\text{rc}} = -\partial V/\partial Q_{\text{rc}} \approx 0$. The motion of the activated complexes along the reaction coordinate is therefore treated as a one-dimensional translational motion of a free particle ("free" means no forces act) confined to a region of length δ, the region of existence of activated complexes. Translation along the reaction coordinate is an internal motion of activated-complex nuclei relative to one another [see, for example, Eq. (23.11)] and is to be distinguished from the ordinary translational motion of the activated complex through the three-dimensional space of the container. The former motion corresponds to z_{rc} in (23.12); the latter motion corresponds to $z_{\text{tr}}^{\ddagger}$ in (23.13).

The partition function for a particle undergoing translation in a one-dimensional box of length a is given by the first sum in Eq. (22.78), and this sum was found to equal $(2\pi mkT)^{1/2}a/h$. Replacing a by δ and m by m_{rc}, where m_{rc} is the effective mass for motion along the reaction coordinate, we get $(2\pi m_{\text{rc}} kT)^{1/2}\delta/h$. [The expression for m_{rc} can be worked out if desired (for example, it turns out to equal one-third the mass of H_3 for the $H + H_2$ reaction; see *McClelland*, secs. 12.1 and 12.2); however, since m_{rc} will eventually cancel out, we need not worry about it.] The first sum in (22.78) assumes motion in both positive and negative directions along the coordinate axis, but we are considering only activated complexes that are moving forward along Q_{rc}. The partition function is a sum over states, and half the possible states are missing when reverse motion is excluded. We therefore must add a factor of $\frac{1}{2}$ to get z_{rc} of $X_f^{\ddagger}$; thus,

$$z_{\text{rc}} = \tfrac{1}{2}(2\pi m_{\text{rc}} kT)^{1/2}\delta/h \tag{23.14}$$

$$z_{\ddagger} = \tfrac{1}{2}(2\pi m_{\text{rc}} kT)^{1/2}\delta h^{-1} z'_{\ddagger} \tag{23.15}$$

where (23.12) was used.

Since ACT assumes that all supermolecules crossing the dividing surface become products, we need only calculate the rate at which supermolecules cross the dividing surface to find the reaction rate. Let $v_{\text{rc}} \equiv dQ_{\text{rc}}/dt$ be the velocity component of a given activated complex along the reaction coordinate. At a given time t_0 let there be $N_f^{\ddagger}$ activated complexes in the system. Let τ be the average time needed for an activated complex to move a distance δ along Q_{rc}. At time $t_0 + \tau$ all $N_f^{\ddagger}$ complexes that were present at time t_0 will have passed through the surface located a distance δ beyond the critical dividing surface and become products. The reaction rate therefore equals $N_f^{\ddagger}/\tau$. But $\tau = \delta/\langle v_{\text{rc}}\rangle$ (where $\langle v_{\text{rc}}\rangle$ is the average value of v_{rc}), so the reaction rate is $N_f^\ddagger\langle v_{rc}\rangle/\delta$. This rate is in terms of number of molecules per unit time. In kinetics, chemists use r, the rate in terms of moles per unit volume per unit time. Dividing by N_0 to convert to moles and by V to convert to rate per unit volume, we have $r = N_f^\ddagger\langle v_{rc}\rangle/N_0 V\delta = [X_f^\ddagger]\langle v_{rc}\rangle/\delta$. Use of (23.10) gives

$$r = \frac{\langle v_{rc}\rangle}{\delta}\frac{z_\ddagger/N_0 V}{(z_A/N_0 V)(z_B/N_0 V)\cdots}e^{-\Delta\varepsilon_0^\ddagger/kT}[A][B]\cdots \qquad (23.16)$$

To complete the calculation of r, we need $\langle v_{rc}\rangle$. Motion along Q_{rc} is being treated as a translation. As with the other degrees of freedom of the activated complex, we assume a Boltzmann distribution of energy for this translation. Hence the fraction of complexes with speeds along Q_{rc} in the range v_{rc} to $v_{rc} + dv_{rc}$ is $Be^{-m_{rc}v_{rc}^2/2kT}\,dv_{rc}$, where B is a constant [cf. Eq. (15.42)]. To determine B, we integrate this expression from $v_{rc} = 0$ to $v_{rc} = \infty$ (negative values of v_{rc} are excluded, as noted above) and set the total probability equal to one: $B\int_0^\infty e^{-m_{rc}v_{rc}^2/2kT}\,dv_{rc} = 1$. Use of integral 2 in Table 15.1 gives $B = 2(m_{rc}/2\pi kT)^{1/2}$. The probability density $g(v_{rc})$ for v_{rc} is therefore

$$g(v_{rc}) = 2(m_{rc}/2\pi kT)^{1/2}e^{-m_{rc}v_{rc}^2/2kT} \qquad (23.17)$$

The average value of v_{rc} is given by (15.39) as $\langle v_{rc}\rangle = \int_0^\infty v_{rc}\,g(v_{rc})\,dv_{rc}$. Substitution of (23.17) and use of integral 5 of Table 15.1 gives (Prob. 23.4)

$$\langle v_{rc}\rangle = (2kT/\pi m_{rc})^{1/2} \qquad (23.18)$$

Substitution of (23.18) and (23.15) into (23.16) gives

$$r = \frac{1}{\delta}\left(\frac{2kT}{\pi m_{rc}}\right)^{1/2}\frac{(2\pi m_{rc}kT)^{1/2}\delta}{2h}\frac{z'_\ddagger/N_0 V}{(z_A/N_0 V)(z_B/N_0 V)\cdots}e^{-\Delta\varepsilon_0^\ddagger/kT}[A][B]\cdots$$

For the elementary reaction A + B + $\cdots\rightarrow$ products, the rate per unit volume is $r = k_r[A][B]\cdots$, where k_r is the rate constant. (The subscript r avoids confusion with the Boltzmann constant k.) We get

$$k_r = \frac{kT}{h}\frac{z'_\ddagger/N_0 V}{(z_A/N_0 V)(z_B/N_0 V)\cdots}e^{-\Delta\varepsilon_0^\ddagger/kT} \qquad \text{ideal gases} \qquad (23.19)$$

which is the desired ACT expression for the rate constant of an ideal-gas elementary reaction.

Sometimes a factor κ is included on the right of (23.19). The quantity κ (which lies between 0 and 1) is called the *transmission coefficient* and allows for the possibility that the shape of the potential-energy surface might be such that some of the supermolecules crossing the critical dividing surface are reflected back to give reactants. Since there is no simple way to calculate κ, and since it is likely to lie close to 1 in most cases, it is best omitted.

To calculate $z'_\ddagger = z_{tr}^\ddagger z_{rot}^\ddagger z_{vib}^{\ddagger\prime} z_{el}^\ddagger$, we need the mass of the activated complex (to calculate $z_{tr}^\ddagger$), its equilibrium structure (to calculate the moments of inertia in $z_{rot}^\ddagger$), its vibrational frequencies, and the degeneracy of its ground electronic level. The equilibrium structure is given by the location of the saddle point. The normal-mode vibrational frequencies can be found if the potential-energy surface is known in the region of the activated complex; vibrational frequencies are related to force constants, and force constants are the second derivatives of V with respect to the normal vibrational coordinates [cf. (21.23)]. To calculate $\Delta\varepsilon_0^\ddagger$ we need the barrier height (and the vibrational frequencies, to correct for zero-point vibrations).

The quantity kT/h equals $0.6 \times 10^{13}\ s^{-1}$ at 300 K and $2 \times 10^{13}\ s^{-1}$ at 1000 K. At temperatures of 300 to 400 K, z_{tr} is typically $10^{24}\ V/cm^3$ to $10^{27}\ V/cm^3$, z_{rot} is typically 10 to 1000, and each normal vibrational mode typically contributes a factor 1 to 3 to z_{vib}.

The equation preceding (22.81), which was used to find z_{rc}, is a (semi)classical statistical-mechanical result, since it was derived by assuming the levels to be closely spaced compared with kT, so that the sum could be replaced by an integral. Moreover, $\langle v_{rc} \rangle$ was calculated using a classical-mechanical Boltzmann distribution. Thus, an additional assumption of ACT is that motion along the reaction coordinate can be treated classically. For reactions involving light species, this assumption fails, and a correction for quantum-mechanical tunneling must be applied (Sec. 23.3). The ACT tunneling-factor corrections used in calculations up to 1970 have been shown to be erroneous [D. G. Truhlar and A. Kupperman, *J. Am. Chem. Soc.*, **93**, 1840 (1971)]. Miller has derived a quantum-mechanical version of ACT that may well furnish results with a correct allowance for tunneling [W. H. Miller, *J. Chem. Phys.*, **61**, 1823 (1974)].

Example *Ab initio* calculations of the H_3 potential-energy surface give a classical barrier height of 9.8 kcal/mole and a linear transition state with bond distances $R_{ab}^{\ddagger} = R_{bc}^{\ddagger} = 0.930$ Å and vibrational wave numbers of 2048 cm^{-1} (symmetric stretching) and 972 cm^{-1} (degenerate bending). The bond length and vibrational frequency for H_2 are $R_{bc} = 0.741$ Å and 4400 cm^{-1}. Use ACT to calculate the rate constant for $H + H_2$ at 300 K.

Equation (23.19) with A = H and B = H_2 and Eq. (23.13) give

$$k_r = N_0 kTh^{-1}e^{-\Delta\varepsilon_0^{\ddagger}/kT} \frac{(z_{tr}^{\ddagger}/V)z_{rot}^{\ddagger} z_{vib}^{\ddagger\prime} z_{el}^{\ddagger}}{(z_{tr,H}/V)z_{el,H}(z_{tr,H_2}/V)z_{rot,H_2} z_{vib,H_2} z_{el,H_2}}$$

Since $\tilde{\nu} = \nu/c$, the zero-point energy (ZPE) of the activated complex is

$$\tfrac{1}{2}hc(2048\ \text{cm}^{-1} + 972\ \text{cm}^{-1} + 972\ \text{cm}^{-1}) = 3.97 \times 10^{-13}\ \text{erg}$$

The ZPE of H is zero, and that of H_2 is $\frac{1}{2}hc\tilde{\nu} = 4.37 \times 10^{-13}$ erg. The change in ZPE is -0.4×10^{-13} erg. The classical barrier height ε_b on a per molecule basis is found by dividing 9.8 kcal/mole by N_0 to give (Prob. 23.5) 6.8×10^{-13} erg. Hence, $\Delta\varepsilon_0^{\ddagger} = 6.4 \times 10^{-13}$ erg.

We saw following Eq. (22.92) that $z_{el,H} = 2$. Since H_3 also has one unpaired electron, we can expect that $z_{el}^{\ddagger} = 2$. The H_2 ground electronic level is nondegenerate, and $z_{el,H_2} = 1$.

Equations (22.85), (18.75), and (21.41) give (Prob. 23.6)

$$\frac{z_{rot}^{\ddagger}}{z_{rot,H_2}} = \frac{I^{\ddagger}}{I_{H_2}} \frac{\sigma_{H_2}}{\sigma^{\ddagger}} = \frac{m_H(R_{ab}^2 + R_{bc}^2)^{\ddagger}}{\frac{1}{2}m_H R_{bc}^2} \frac{\sigma_{H_2}}{\sigma^{\ddagger}}$$

The activated complex has a linear symmetric equilibrium geometry, and it might be thought that $\sigma^{\ddagger} = 2$. (In fact, that's what people believed until the 1960s.) However, the activated complex $X_f^{\ddagger}$ is moving only forward along the reaction coordinate; this forward motion corresponds to the unsymmetrical motion shown in Eq. (23.11). In (23.11), the two end hydrogens are not equivalent. Hence, $\sigma^{\ddagger} = 1$. Substitution of numerical values gives $z_{rot}^{\ddagger}/z_{rot,H_2} = 12.6$.

Use of $\tilde{\nu} = \nu/c$ and Eqs. (22.90) and (22.110) gives $z_{\text{vib,H}_2} = 1.000$ and $z_{\text{vib}}^{\ddagger\prime} = 1.019$ at 300 K (Prob. 23.7).

Equation (22.81) gives at 300 K

$$\frac{z_{\text{tr}}^{\ddagger}/V}{(z_{\text{tr,H}}/V)(z_{\text{tr,H}_2}/V)} = \left(\frac{m_{\text{H}_3}}{m_{\text{H}}m_{\text{H}_2}}\right)^{3/2}\left(\frac{h^2}{2\pi kT}\right)^{3/2} = 1.86 \times 10^{-24}\ \text{cm}^3$$

(Since z_{tr} is dimensionless, the above ratio must have units of volume.)

Substitution in the expression for k_r gives at 300 K

$$k_r = 1.7 \times 10^7\ \text{cm}^3\ \text{mole}^{-1}\ \text{s}^{-1} = 1.7 \times 10^4\ \text{dm}^3\ \text{mole}^{-1}\ \text{s}^{-1}$$

(This value is substantially lower than the experimental value, due to the neglect of tunneling.)

As the H_3 example shows, calculation of the symmetry number $\sigma^\ddagger$ in $z_{\text{rot}}^\ddagger$ is a bit tricky. A procedure that usually works is the following. Omit all symmetry numbers from the rotational partition functions of the reactants and the activated complex; instead, multiply k_r by $s^\ddagger$, where $s^\ddagger$ (the *statistical factor* or the *reaction-path degeneracy*) is found by putting different labels on the identical atoms in the reactants and counting the number of different activated complexes that can be formed. For example, for $H_a + H_bH_c$, we can form the two linear complexes $H_a \cdots H_b \cdots H_c$ and $H_b \cdots H_c \cdots H_a$, so $s^\ddagger = 2$. Omission of the σ's and multiplication of k_r by 2 is equivalent to taking $\sigma^\ddagger = 1$, $\sigma_{\text{H}} = 1$, $\sigma_{\text{H}_2} = 2$, as was done in the example. (For the justification of this procedure, and for further subtleties not discussed here, see *Laidler*, pp. 65–75; *Forst*, pp. 64–68; *Robinson and Holbrook*, sec. 4.9.)

Relation between ACT and hard-sphere collision theory. For the bimolecular reaction $A + B \rightarrow$ products, suppose we ignore the internal structure of the colliding molecules and treat them as hard spheres of radii r_A and r_B. Then the reactants' partition functions are $z_A = z_{\text{tr,A}}$ and $z_B = z_{\text{tr,B}}$. Equation (22.81) gives

$$z_A/V = (2\pi m_A kT/h^2)^{3/2}, \qquad z_B/V = (2\pi m_B kT/h^2)^{3/2}$$

The most reasonable choice of transition state is the two hard spheres in contact. An ordinary diatomic molecule has 1 vibrational degree of freedom, so the "diatomic" activated complex has 0 vibrational degrees of freedom, since the reaction coordinate replaces the one vibration. The spheres of masses m_A and m_B are separated by a center-to-center distance of $r_A + r_B$ in the transition state, and (18.75) gives the moment of inertia as $I = \mu(r_A + r_B)^2$, where the reduced mass is given by (18.72) as $\mu \equiv m_A m_B/(m_A + m_B)$. Equations (22.81) and (22.85) give the partition function of the activated complex as

$$\frac{z'_\ddagger}{V} = \frac{z_{\text{tr}}^\ddagger}{V} z_{\text{rot}}^\ddagger = \left[\frac{2\pi(m_A + m_B)kT}{h^2}\right]^{3/2} 8\pi^2 \frac{m_A m_B}{m_A + m_B}(r_A + r_B)^2 \frac{kT}{h^2}$$

Substitution in the ACT equation (23.19) gives

$$k_r = N_0 \pi (r_A + r_B)^2 \left[\frac{8kT}{\pi}\left(\frac{m_A + m_B}{m_A m_B}\right)\right]^{1/2} e^{-\Delta\varepsilon_0^\ddagger/kT} \tag{23.20}$$

which is identical to the hard-sphere collision-theory result (23.3) if we take $\Delta\varepsilon_0^\ddagger$ to be the threshold energy $\varepsilon_{\text{thr}} = E_{\text{thr}}/N_0$. Thus, ACT reduces to the hard-sphere collision theory when the structure of the molecules is ignored. [If A = B, then $\sigma^\ddagger = 2$, and ACT reduces to Eq. (23.4).]

Temperature dependence of the rate constant. To investigate the temperature dependence of the ACT rate constant (23.19), we must examine the temperature dependences of the partition functions. Equations (22.81), (22.85), (22.109), and (22.92) give

$$z_{\text{tr}} \propto T^{3/2}, \qquad z_{\text{rot,lin}} \propto T, \qquad z_{\text{rot,nonlin}} \propto T^{3/2}, \qquad z_{\text{el}} \propto T^0$$

(Recall the equipartition theorem.) The temperature dependence of z_{vib} is not so simple. At temperatures such that $kT \ll h\nu_s$ for all ν_s, Eq. (22.110) becomes $z_{\text{vib}} \approx 1 = T^0$. At temperatures such that $kT \gg h\nu_s$ for all ν_s, expansion of the exponentials in (22.110) gives (Prob. 23.9) $z_{\text{vib}} \propto T^{f_{\text{vib}}}$, where f_{vib} is the number of vibrational degrees of freedom of the molecule. For an intermediate temperature, $z_{\text{vib}} \propto T^b$, where b is between 0 and f_{vib}. For most vibrations, ν_s is sufficiently high for the condition $kT \gg h\nu_s$ to be reached only at quite high temperatures. For a moderate temperature, we can expect that

$$z_{\text{vib}} \propto T^a \qquad \text{where } 0 \le a \le \tfrac{1}{2} f_{\text{vib}}$$

Over a restricted temperature range, each z_{vib} in (23.19) will have its a value approximately constant, and we can write

$$k_r \approx CT^m e^{-\Delta E_0^\ddagger/RT} \tag{23.21}$$

where C and m are constants, and where $\Delta E_0^\ddagger \equiv N_0 \Delta\varepsilon_0^\ddagger$.

Using the temperature dependences of the factors in the partition functions in (23.19), one can deduce the range of values of m. This is left as an exercise (Prob. 23.10). The results are as follows: (*a*) for a bimolecular gas-phase reaction between an atom and a molecule, m usually lies between -0.5 and 0.5; (*b*) for a bimolecular gas-phase reaction between two molecules, m usually lies between -2 and 0.5.

The activation energy is defined by (17.72) as $E_a \equiv RT^2 \, d \ln k_r/dT$. Taking the log of (23.21) and differentiating, we get

$$E_a = \Delta E_0^\ddagger + mRT \tag{23.22}$$

Since m can be negative, zero, or positive, E_a can be less than, the same as, or greater than $\Delta E_0^\ddagger$. The quantity $\Delta E_0^\ddagger$ differs from the classical barrier height E_b by $\Delta\text{ZPE}^\ddagger$, the change in zero-point energy on formation of the activated complex. Since $\Delta\text{ZPE}^\ddagger$ can be negative, zero, or positive, E_a can be less than, the same as, or greater than E_b.

The Arrhenius A factor is defined by (17.73) as $A = k_r e^{E_a/RT}$. Use of (23.22), (23.19), and (23.21) gives

$$A = \frac{kTe^m}{h} \frac{z'_\ddagger/N_0 V}{(z_A/N_0 V)(z_B/N_0 V)\cdots} \approx Ce^m T^m \tag{23.23}$$

where m can be calculated if $z'_\ddagger$ is known. The accuracy of gas-kinetics data is usually too poor to allow m to be found experimentally. The T^m dependence of the A factor is overwhelmed by the exponential function $e^{-\Delta E_0^\ddagger/RT}$ in k_r, and kinetics data give only an A averaged over the temperature range of the experiments.

[For further discussion, see W. C. Gardiner, *Acc. Chem. Res.*, **10,** 326 (1977).]

Tests of ACT. Since accurate potential-energy surfaces are not known for most chemical reactions, one must usually guess at the structure of the activated complex and estimate its vibrational frequencies using approximate empirical rules that relate bond lengths and vibrational frequencies. Herschbach and coworkers used ACT to calculate the A factors of 12 bimolecular gas-phase reactions. Experimental A values can be in error by a factor of 3 or 4. One might allow a factor of perhaps 2 or 3 as the error in $z'_{\ddagger}$, since it is found by guesswork. Thus, an ACT A value that is no more than $10^{\pm 1}$ times the experimental value can be regarded as a confirmation of ACT. Ten of the 12 calculated A values were within a factor of 10 of being correct. [D. Herschbach et al., *J. Chem. Phys.*, **25**, 736 (1956); *Knox*, pp. 237–239.]

Those reactions for which the potential-energy surface has been accurately calculated involve light species, for which tunneling is important, so it is hard to test ACT in these cases. However, Miller's quantum-mechanical version of ACT has given excellent agreement with experimental k_r values for the reaction $D + H_2 \rightarrow HD + H$. [See S. Chapman, B. C. Garrett, and W. H. Miller, *J. Chem. Phys.*, **63**, 2710 (1975).]

Another kind of test is to compare the predictions of ACT with the results of trajectory calculations (Sec. 23.3). The majority of such comparisons have given pretty good agreement. However, in a few cases, trajectory calculations show that the states of the activated complex are not populated according to the Boltzmann distribution. With certain shapes of potential-energy surface, trajectory calculations show that a significant fraction of supermolecules are reflected back to reactants after crossing the critical dividing surface.

Still another way to test ACT is to make an isotopic substitution in a reactant and compare the observed effect on k_r with that calculated by ACT. Kinetic-isotope tests tend to confirm the validity of ACT. (See *Weston and Schwarz*, sec. 4.11.)

Although definitive results are not available, the weight of the evidence is that the activated-complex theory works well in most (but not all) cases.

ACT and transport properties. The applicability of ACT is wider than the above derivation implies, and the theory can be applied to such rate processes as diffusion and viscous flow in liquids; see *Hirschfelder, Curtiss, and Bird*, sec. 9.2. (To diffuse from one location to another, a molecule in a liquid must squeeze past its neighbors; the variation in potential energy for this process resembles Fig. 23.6*a*.) Problem 23.12 applies ACT to effusion of gases.

23.5 THERMODYNAMIC FORMULATION OF ACT

Comparing (23.19) with (22.124), we see that the stuff that follows kT/h in (23.19) resembles an equilibrium constant, the only difference being that $z'_{\ddagger}$ is not the complete partition function of the activated complex but omits the contribution of z_{rc} [see Eq. (23.12)]. It is therefore customary to define $K_c^{\ddagger}$ by

$$K_c^{\ddagger} \equiv \frac{z'_{\ddagger}/N_0 V}{(z_{\text{A}}/N_0 V)(z_{\text{B}}/N_0 V) \cdots} e^{-\Delta\varepsilon_0{}^{\ddagger}/kT} \tag{23.24}$$

Note from (23.10) that $K_c^{\ddagger} \neq [X_f^{\ddagger}]/[A][B] \cdots$. Instead Eqs. (23.10) and (23.15) give

$$K_c^{\ddagger} = K_f \left(\frac{2}{\pi m_{\text{rc}} kT}\right)^{1/2} \frac{h}{\delta} \tag{23.25}$$

The ACT equation (23.19) becomes

$$k_r = \frac{kT}{h} K_c^{\ddagger} \tag{23.26}$$

In thermodynamics, the standard state used for ideal gases is the state at $P = P^\circ \equiv 1$ atm and temperature T. In kinetics, rate constants are usually expressed in terms of concentrations, and it is more convenient to use a standard state having unit concentration rather than unit pressure. Equation (6.7) and $P_i = n_i RT/V = c_i RT$ give $\mu_i = \mu_i^\circ + RT \ln P_i/P^\circ = \mu_i^\circ + RT \ln c_i RT/P^\circ = \mu_i^\circ + RT \ln RTc^\circ/P^\circ + RT \ln c_i/c^\circ$, where $c^\circ \equiv 1$ mole/dm^3. When $c_i = c^\circ$, the chemical potential becomes $\mu_i^\circ + RT \ln RTc^\circ/P^\circ \equiv \mu_{c,i}^\circ$. Therefore, $\mu_i = \mu_{c,i}^\circ + RT \ln c_i/c^\circ$, where $\mu_{c,i}^\circ$ is the chemical potential of gas i at 1 mole/dm^3 concentration and temperature T. Substituting into the reaction-equilibrium condition $\sum_i \nu_i \mu_i = 0$ and following the same steps that led to (6.17), we get

$$\Delta G_c^\circ = -RT \ln K_c^\circ = -RT \ln [K_c/(c^\circ)^{\Delta|\nu|}] \tag{23.27}$$

where $\Delta G_c^\circ \equiv \sum_i \nu_i \mu_{c,i}^\circ$ and $K_c^\circ \equiv \prod_i (c_i/c^\circ)^{\nu_i} = \prod_i c_i^{\nu_i}/\prod_i (c^\circ)^{\nu_i} = K_c/(c^\circ)^{\Delta|\nu|}$, where $K_c \equiv \prod_i c_i^{\nu_i}$.

For the process $A + B + \cdots \to \{X_f^\ddagger\}$, we have $\Delta|\nu| = 1 - n$, where n is the molecularity of the reaction. By analogy to (23.27), we define $\Delta G_c^{\circ\ddagger}$, the *concentration-scale standard Gibbs free energy of activation*, as

$$\Delta G_c^{\circ\ddagger} \equiv -RT \ln [K_c^\ddagger (c^\circ)^{n-1}] \tag{23.28}$$

Use of (23.28) in (23.26) gives

$$k_r = kTh^{-1}(c^\circ)^{1-n} e^{-\Delta G_c^{\circ\ddagger}/RT} \tag{23.29}$$

This is the thermodynamic version of the ACT expression for the rate constant. The higher the value of $\Delta G_c^{\circ\ddagger}$, the slower the reaction.

In analogy with (6.35) and (6.45), we define $K_P^{\circ\ddagger}$ and $\Delta H^{\circ\ddagger}$ for gas-phase reactions as

$$K_P^{\circ\ddagger} \equiv K_c^{\circ\ddagger}(RTc^\circ/P^\circ)^{1-n} \tag{23.30}$$

$$\Delta H_c^{\circ\ddagger} = \Delta H^{\circ\ddagger} \equiv RT^2\, d \ln K_P^{\circ\ddagger}/dT \tag{23.31}$$

Since ideal-gas enthalpies depend on T only, the standard enthalpy of activation is the same whether the standard state is $P = 1$ atm or $c = 1$ mole/dm^3, that is, $\Delta H^{\circ\ddagger} = \Delta H_c^{\circ\ddagger}$.

The concentration-scale standard entropy of activation is defined by $\Delta S_c^{\circ\ddagger} \equiv (\Delta H_c^{\circ\ddagger} - \Delta G_c^{\circ\ddagger})/T$, so that

$$\Delta G_c^{\circ\ddagger} = \Delta H_c^{\circ\ddagger} - T\, \Delta S_c^{\circ\ddagger} \tag{23.32}$$

Substitution of (23.32) in (23.29) gives

$$k_r = kTh^{-1}(c^\circ)^{1-n} e^{\Delta S_c^{\circ\ddagger}/R} e^{-\Delta H_c^{\circ\ddagger}/RT} \tag{23.33}$$

The quantities $\Delta G_c^{\circ\ddagger}$, $\Delta H_c^{\circ\ddagger}$, and $\Delta S_c^{\circ\ddagger}$ are the changes in G, H, and S at temperature T when 1 mole of $X_f^\ddagger$ in its standard state (1 mole/dm^3 concentration) is formed from the pure, separated reactants in their 1 mole/dm^3 standard states, except that the contributions of motion along the reaction coordinate to the properties $\bar{G}^\ddagger$, $\bar{H}^\ddagger$, and $\bar{S}^\ddagger$ of the activated complex $X_f^\ddagger$ are omitted. In other words, in calculating $\bar{G}^\ddagger$, $\bar{H}^\ddagger$, and $\bar{S}^\ddagger$ from the statistical-mechanical equations of Chap. 22, we use $z'_\ddagger$ instead of $z_\ddagger$ (where $z'_\ddagger = z_\ddagger/z_{\text{rc}}$).

Equations (17.72) and (23.26) give for the activation energy $E_a \equiv RT^2\, d \ln k_r/dT = RT + RT^2\, d \ln K_c^{\circ\ddagger}/dT$. Equations (23.30) and (23.31) give

$d \ln K_c^{\circ\ddagger}/dT = d \ln K_P^{\circ\ddagger}/dT + (n-1)/T = \Delta H^{\circ\ddagger}/RT^2 + (n-1)/T$. Hence

$$E_a = \Delta H^{\circ\ddagger} + nRT \qquad \text{gas-phase reaction} \tag{23.34}$$

where n is the molecularity of the reaction.

Equation (17.73) gives the preexponential factor as $A \equiv k_r e^{E_a/RT}$, and use of (23.33) and (23.34) gives

$$A = (kT/h)(c^\circ)^{1-n} e^n e^{\Delta S_c^{\circ\ddagger}/R} \tag{23.35}$$

From the experimental values of A and E_a we can calculate $\Delta S_c^{\circ\ddagger}$ and $\Delta H_c^{\circ\ddagger}$ using (23.34) and (23.35). Equation (23.32) then gives $\Delta G_c^{\circ\ddagger}$.

For a bimolecular gas-phase reaction, the activated complex has fewer translational and rotational degrees of freedom and more vibrational degrees of freedom than the pair of reactant molecules. The larger spacing between vibrational levels compared with rotational and translational levels makes the contribution of a vibration to S less than that of a rotation or translation. The activation entropy $\Delta S_c^{\circ\ddagger}$ is therefore negative for a gas-phase bimolecular reaction.

Equation (5.15) gives $\Delta H^{\circ\ddagger} = \Delta U^{\circ\ddagger} + (1-n)RT$. Substitution in (23.34) gives $E_a = \Delta U^{\circ\ddagger} + RT$, where $\Delta U^{\circ\ddagger}$ equals $\bar{U}$ of $X_j^\ddagger$ at T (the contribution of motion along the reaction coordinate being omitted) minus $\bar{U}_A + \bar{U}_B + \cdots$ at T. This equation gives a simple physical interpretation to the activation energy E_a.

23.6 UNIMOLECULAR REACTIONS

In a unimolecular decomposition or isomerization of a polyatomic molecule A, the elementary chemical reaction $A^* \rightarrow$ products is preceded by the elementary physical reaction $A + M \rightarrow A^* + M$, which puts A into an excited vibrational level (Sec. 17.10).

For a gas-phase unimolecular reaction in the low-pressure falloff region, the rate of formation of vibrationally excited A molecules (symbolized by A^*) falls below that needed to maintain the Boltzmann population of A^*. Since the ACT equation (23.19) assumes an equilibrium Boltzmann population of reactant states, we cannot apply (23.19) to the overall reaction $A \rightarrow$ products in the falloff region. In the high-pressure region, the Boltzmann population of A^* is maintained, so ACT can be used to calculate $k_{\text{uni},P=\infty}$, the experimentally observed high-pressure rate constant.

For many unimolecular reactions, the potential-energy surface shows a saddle point, and we can identify the transition state with this saddle point. (The transition state in a unimolecular reaction is often called the *critical configuration*.)

For example, for the unimolecular isomerization *cis*-CHF=CHF → *trans*-CHF=CHF, the minimum-energy path between the reactant and product involves a 180° rotation of one CHF group with respect to the other. As these groups rotate, the overlap between the carbon $2p$ AOs that form the pi bond is gradually lost, becoming zero at a 90° rotation. The point of maximum potential energy on the minimum-energy path is at 90°, and this is the transition state. During the rotation from the cis to the trans compound, the bond distances will change, but these changes are small compared with the change in twist angle. Therefore, $V \approx V(\theta)$, where θ is the angle between the two CHF planes. With only one variable, the potential-energy surface becomes a line (Fig. 23.15). The carbon–carbon double- and single-bond energies are 147 and 82 kcal/mole (Table 20.1), so we expect the transition-state energy to lie about 65 kcal/mole above the reactant's energy. The observed E_a is 63 kcal/mole. The barrier is surmounted when sufficient vibrational energy flows into the vibrational normal mode that involves twisting about the CC axis.

For the unimolecular decomposition $CH_3CH_2Cl \rightarrow CH_2{=}CH_2 + HCl$, the likely transition state is

$$\begin{array}{ccc} H & \cdot\cdot & Cl \\ \vdots & & \vdots \\ H_2C & \overset{\cdot\cdot}{-} & CH_2 \end{array}$$

We can expect this transition state, with its partially broken HC and ClC bonds and partially formed CC pi and HCl bonds, to lie at a maximum on the minimum-energy path.

There are, however, many unimolecular decompositions whose potential-energy surfaces have no saddle point. Figure 23.16 shows potential-energy contour maps for the molecules HCN and CO_2 with the bond angle θ restricted to its equilibrium value of 180°. The points that are at the centers of the 2-eV contours correspond to the HCN and CO_2 equilibrium geometries; each of these points lies at the bottom of a "well."

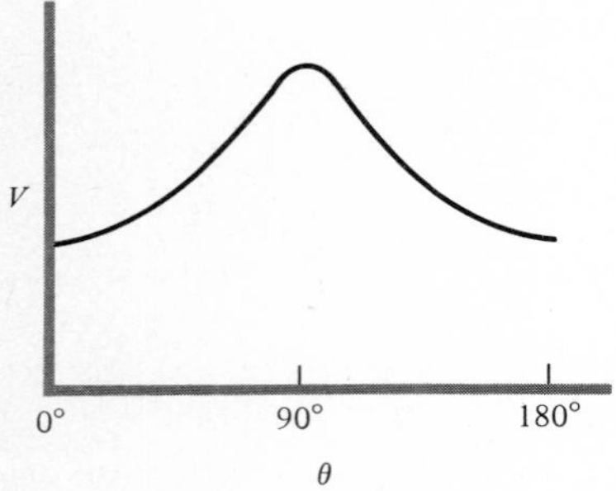

Figure 23.15
Potential energy vs. twist angle for *cis*-CHF=CHF → *trans*-CHF=CHF.

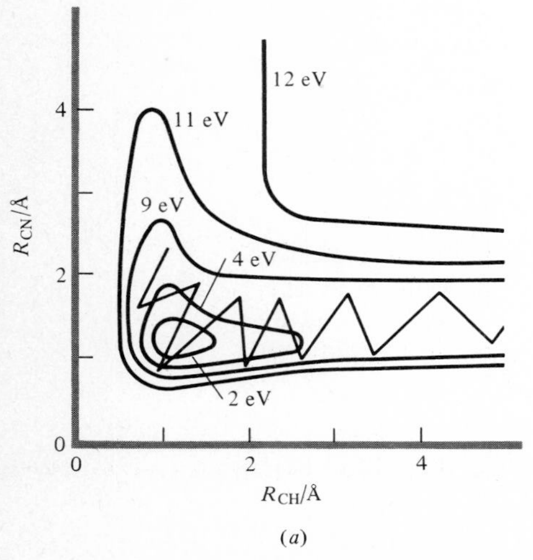

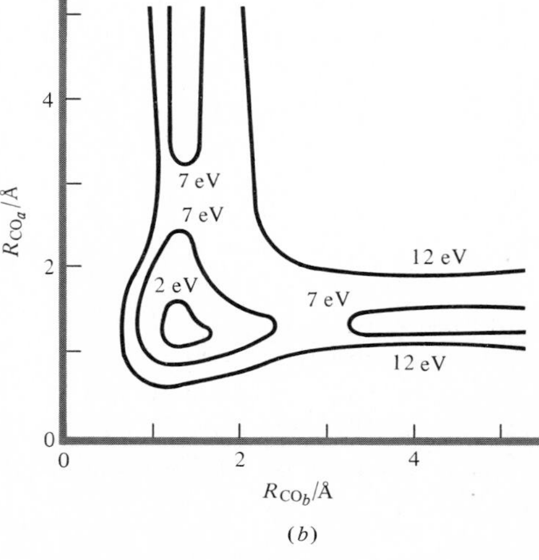

Figure 23.16
Potential-energy contour maps for linear configurations of (*a*) HCN; (*b*) CO_2.

Figure 23.16 shows that the decomposition $CO_2 \rightarrow CO + O$ has a saddle point on the minimum-energy path (which involves stretching one CO bond until it breaks, the atoms remaining collinear). However, the decomposition $HCN \rightarrow H + CN$ shows no saddle point. Figure 23.17 shows V along the minimum-energy path for these two decompositions. The drop in V at large distances along the CO_2 decomposition path is due to formation of the third bond of the triple bond in carbon monoxide. We can expect that a unimolecular decomposition where no new bonds are formed in the products will show no saddle point. (The zigzag line in Fig. 23.16 is explained later.)

To apply ACT where there is no saddle point, we must choose a dividing surface such that it is highly probable that a supermolecule crossing this surface will yield products. Clearly, the dividing surface must be much closer to the products than to the reactant molecule in this case; one speaks of a "loose" activated complex. For example, for the decomposition $C_2H_6 \rightarrow 2CH_3$, the dividing surface will correspond to a greatly elongated carbon–carbon bond; the activated complex will closely resemble two CH_3 radicals and will have more rotational degrees of freedom than C_2H_6. Two different methods have been proposed for choosing the dividing surface in a unimolecular reaction that has no saddle point (for details, see *Robinson and Holbrook*, pp. 159–160; *Forst*, chap. 11). The carbon–carbon distance in the loose $C_2H_6^{\ddagger}$ activated complex turns out to be about 5 Å, compared with 1.5 Å in the C_2H_6 molecule.

The zigzag line in Fig. 23.16 is a classical trajectory for the decomposition $HCN^* \rightarrow H + CN$. The trajectory starts well away from the equilibrium HCN geometry, corresponding to an excited vibrational level of HCN (as indicated by the star). The vibrations continue until the H—C bond has elongated sufficiently to break. (This figure is incomplete, since it omits the bending vibrations.)

For a unimolecular reaction, the ACT equation (23.19) gives the high-pressure rate constant as

$$k_{\text{uni},P=\infty} = \frac{kT}{h}\frac{z'_{\ddagger}}{z_A}e^{-\Delta\varepsilon_0{}^{\ddagger}/kT} \tag{23.36}$$

Since the activated complex $A_f^{\ddagger}$ and the reactant A have the same mass, $z_{\text{tr}}^{\ddagger} = z_{\text{tr,A}}$. Provided $A_f^{\ddagger}$ is not a loose complex, its dimensions and moments of inertia will be

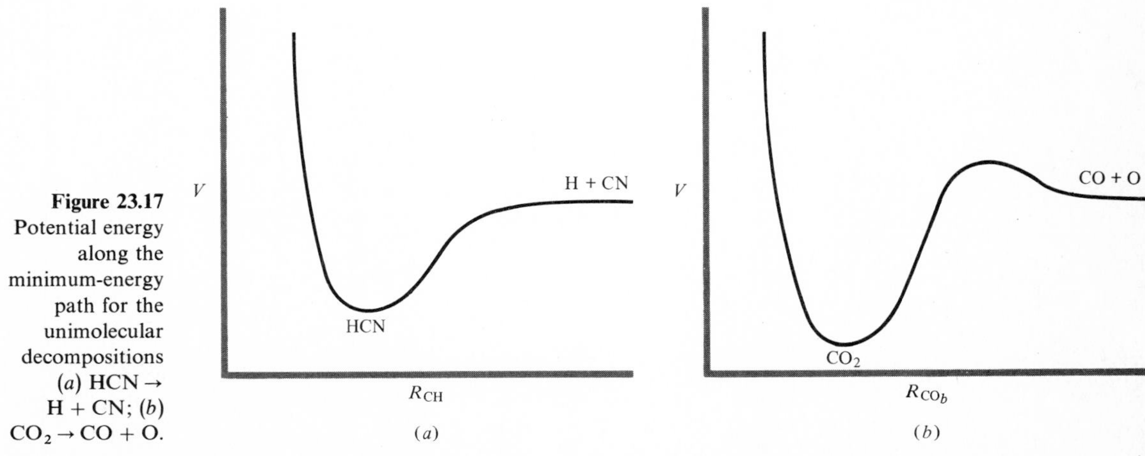

Figure 23.17 Potential energy along the minimum-energy path for the unimolecular decompositions (*a*) $HCN \rightarrow H + CN$; (*b*) $CO_2 \rightarrow CO + O$.

close to those of A, so $z_{\text{rot,A}} \approx z^{\ddagger}_{\text{rot}}$. For a rigid complex we therefore have $z'_{\ddagger}/z_{\text{A}} \approx z^{\ddagger\prime}_{\text{vib}}/z_{\text{vib,A}}$; the complex $A^{\ddagger}_{f}$ has one less vibrational mode than A, this mode having become the reaction coordinate in the complex.

What about the falloff region? A theory that allows calculation of k_{uni} in this region was developed by Marcus and Rice in 1951–1952. The Marcus–Rice theory is based in part on earlier work of Rice, Ramsperger, and Kassel and is therefore commonly called the *RRKM theory* of unimolecular reactions.

In the RRKM theory, the energization rate constant k_1 in (17.85) is calculated using statistical mechanics, making allowance for the fact that transfer of vibrational energy as well as of translational energy can raise A to an excited vibrational level, $A \rightarrow A^*$. The de-energization rate constant k_{-1} is found from the collision rate of A*. The theory recognizes that the rate constant k_2 for decomposition of an energized A* molecule will depend on how much vibrational energy A* has; the greater its vibrational energy, the faster A* decomposes. The rate constant for decomposition of an A* molecule that has a given amount of vibrational energy is calculated using ideas from ACT. For details of the RRKM theory, see *Robinson and Holbrook*, chaps. 4–6; *Gardiner*, secs. 5-2 and 5-3. Comparison with experimental data shows that the RRKM theory works very well in nearly all cases.

At high pressures, the RRKM result reduces to the ACT equation (23.36).

23.7 TRIMOLECULAR REACTIONS

The best examples of trimolecular gas-phase reactions (Sec. 17.11) are recombinations of two atoms, for which a third species (M) is needed to carry away some of the bond energy to prevent dissociation. Figure 23.4 is the potential-energy contour map for collinear configurations of the recombination reaction $H + H + H \rightarrow H_2 + H$. A trajectory that starts at point *i* and is parallel to the R_{bc} axis corresponds to H_b and H_c approaching each other while H_a remains far away. Imagine a ball rolling on the potential-energy surface along this trajectory. The ball will roll down into the $H_a + H_{bc}$ valley, roll past the $R_{bc} = 0.74$ Å equilibrium internuclear distance, roll up the side of the valley with $R_{bc} < 0.74$ Å until it reaches an altitude of 110 kcal/mole (which was the initial potential energy of the system at point *i*), and then exactly reverse its path, ending up back at point *i*. Thus, if H_a does not participate, the newly formed H_{bc} molecule will dissociate back into atoms. A trajectory that leads to formation of a stable H_{bc} molecule must involve a decrease of both R_{ab} and R_{bc}.

For the recombination reaction $A + B + M \rightarrow AB + M$ (where A and B are atoms), there is a wide variety of trajectories along which the reaction can proceed, and there is no one configuration of A, B, and M that we can pick out as *the* transition state. Thus, the activated-complex theory is not applicable to atomic recombination reactions.

Several theories of trimolecular recombination reactions have been proposed, but none is fully satisfactory. [For details, see *Gardiner*, pp. 141–147; H. O. Pritchard, *Acc. Chem. Res.*, **9**, 99 (1976).]

23.8 REACTIONS IN SOLUTION

Because of the strong intermolecular interactions in the liquid state, the theory of reactions in solution is far less developed than that of gas-phase reactions. In general, it is not currently possible to calculate reaction rates in solution from molecular properties. (An exception is diffusion-controlled reactions; see below.)

Chemical reactions in solution can be divided into (*a*) *chemically controlled reactions*, in which the rate of chemical reaction between A and B molecules in a solvent cage is much less than the rate of diffusion of A and B toward each other through the solvent; (*b*) *diffusion-controlled reactions* (Sec. 17.13), in which the rate of diffusion is much less than the rate of chemical reaction; (*c*) *mixed-control reactions*, in which the rates of diffusion and of chemical reaction have the same order of magnitude.

ACT for chemically controlled reactions in solution. For the chemically controlled elementary reaction $A + B + \cdots \rightarrow \{X_f^\ddagger\} \rightarrow C + D + \cdots$, the activated-complex theory gives r as the rate per unit volume at which supermolecules cross the critical dividing surface. From the paragraph preceding Eq. (23.16),

$$r = [X_f^\ddagger]\langle v_{rc}\rangle/\delta \tag{23.37}$$

where $[X_f^\ddagger]$ is the molar concentration of activated complexes (defined to exist for a length δ on the product side of the dividing surface). Whether the system is ideal or nonideal, it is the concentration $[X_f^\ddagger]$ that appears in the expression for r. This is because r is always defined in terms of a concentration change: $r \equiv \nu_A^{-1}\, d[A]/dt$ [Eq. (17.4)].

To get $[X_f^\ddagger]$, we modify Eq. (23.10) to allow for nonideality. Equation (23.10) expresses an apparent equilibrium between reactants and forward-moving activated complexes for an ideal-gas reaction. To allow for nonideality, the concentrations in the apparent equilibrium constant $[X_f^\ddagger]/[A][B]\cdots$ are replaced by activities, to give [cf. Eqs. (11.5) and (10.43)]

$$K_f = \frac{a_\ddagger}{a_A a_B \cdots} = \frac{\gamma_\ddagger}{\gamma_A \gamma_B \cdots} \frac{[X_f^\ddagger]/c^\circ}{([A]/c^\circ)([B]/c^\circ)\cdots} \tag{23.38}$$

where $c^\circ \equiv 1$ mole/dm^3. (The c°'s are present because the activities, the activity coefficients, and K_f are all dimensionless.) All activities and activity coefficients in this section are on the concentration scale (Sec. 10.3), so a c subscript on the a's and γ's is to be understood.

Use of (23.38) for $[X_f^\ddagger]$ and (23.18) for $\langle v_{rc}\rangle$ in (23.37) gives

$$r = \frac{\gamma_A \gamma_B \cdots}{\gamma_\ddagger}\left(\frac{2kT}{\pi m_{rc}}\right)^{1/2} \frac{1}{\delta} K_f (c^\circ)^{1-n}[A][B]\cdots \tag{23.39}$$

where n is the molecularity of the reaction. For a nonideal system, we define $K_a^\ddagger$ by analogy to (23.25) as

$$K_a^\ddagger \equiv K_f \left(\frac{2}{\pi m_{rc} kT}\right)^{1/2} \frac{h}{\delta} \tag{23.40}$$

The rate constant is $k_r = r/[A][B]\cdots$, and (23.39) and (23.40) give

$$k_r = \frac{kT}{h}\frac{\gamma_A \gamma_B \cdots}{\gamma_\ddagger}(c^\circ)^{1-n} K_a^\ddagger \tag{23.41}$$

which differs from the ideal-gas expression (23.26) by the presence of the concentration-scale activity coefficients. The $(c^\circ)^{1-n}$ is present in (23.41) [and absent from (23.45) below] because $K_a^\ddagger$ is dimensionless whereas $K_c^\ddagger$ in (23.26) has dimensions of concentration^{1-n}.

Because of the strong intermolecular interactions in the solution, we can't express K_f or $K_a^\ddagger$ in terms of individual partition functions z_A, z_B, etc. We don't have an equation like (23.19), and calculation of k_r from molecular properties is not practical for reactions in solution.

Equation (23.41) can also be applied to nonideal-gas reactions, in which case the activity coefficients are replaced by fugacity coefficients.

The apparent equilibrium constant K_f and the related constant $K_a^\ddagger$ are functions of temperature, pressure, and solvent (since the concentration-scale standard states of the species A, B, ... and $X_f^\ddagger$ depend on these things) but are independent of the solute concentrations. At infinite dilution, the γ's all become 1, and (23.41) becomes $k_r^\infty = (kT/h)(c^\circ)^{1-n}K_a^\ddagger$, where k_r^∞ is the rate constant in the limit of infinite dilution. For a fixed temperature, pressure, and solvent, the ACT equation (23.41) can therefore be written as

$$k_r = (\gamma_A \gamma_B \cdots / \gamma_\ddagger) k_r^\infty \qquad (23.42)$$

Comparison with Eq. (17.84) shows that Y in that equation equals $1/\gamma_\ddagger$. Equation (23.42) (which is the *Brønsted–Bjerrum equation*) predicts that the rate constant should vary with reactant concentrations, since the γ's change with a change in solution composition. The low accuracy of most kinetics data makes this effect not worth worrying about, except for ionic reactions. Ionic solutions are highly nonideal, even at low concentrations.

Consider the bimolecular elementary ionic reaction $A^{z_A} + B^{z_B} \rightarrow \{X_f^{\ddagger(z_A+z_B)}\} \rightarrow$ products. The log of (23.42) reads

$$\log k_r = \log k_r^\infty + \log \gamma_A + \log \gamma_B - \log \gamma_\ddagger \qquad (23.43)$$

For ionic strengths up to 0.1 mole/dm^3, the Davies equation (10.93) gives for aqueous solutions at 25°C and 1 atm

$$\log \gamma_A = -0.51 z_A^2 \left(\frac{I^{1/2}}{1 + I^{1/2}} - 0.30I \right), \qquad I \equiv I_c/c^\circ$$

where $I_c \equiv \frac{1}{2} \sum_i z_i^2 c_i$ replaces I_m since we are using concentration-scale activity coefficients. Use of the Davies equation for γ_A, γ_B, and $\gamma_\ddagger$ in (23.43) gives [since the charge factor in $\log \gamma_A + \log \gamma_B - \log \gamma_\ddagger$ is $z_A^2 + z_B^2 - (z_A + z_B)^2 = -2z_A z_B$]

$$\log k_r = \log k_r^\infty + 1.02 z_A z_B \left(\frac{I^{1/2}}{1 + I^{1/2}} - 0.30I \right) \quad \text{dil. aq. soln., 25°C} \qquad (23.44)$$

A plot of $\log k_r$ vs. $I^{1/2}/(1 + I^{1/2}) - 0.30I$ should be linear with slope $1.02 z_A z_B$, for ionic strengths up to 0.1 mole/dm^3. This has been verified for many ionic reactions in solution. (In calculating I, formation of ion pairs and complex ions must be taken into account.)

The *primary kinetic salt effect* (23.44) is large, even at modest values of ionic strength. For example, for $z_A z_B = +2$, the values of k_r/k_r^∞ at $I = 10^{-3}$, 10^{-2}, and 10^{-1} are 1.15, 1.51, and 2.7, respectively; for $z_A z_B = -2$, the corresponding values are 0.87, 0.66, and 0.37. For $z_A z_B = 4$, one finds $k_r/k_r^\infty = 7.2$ at $I = 0.1$.

If the products have different charges than the reactants, the ionic strength can change markedly during a reaction, meaning that k_r will change as the reaction proceeds. To avoid this, a large amount of inert salt is often added to keep I essentially fixed during a given run. For a reaction of unknown mechanism, one can determine $z_A z_B$ for the rate-determining step by investigating the dependence of k_r on I.

By analogy to (23.28), we define $\Delta G^{\circ\ddagger}$ for a reaction in solution by

$$\Delta G^{\circ\ddagger} \equiv -RT \ln K_a^{\ddagger} \tag{23.45}$$

Equation (23.41) becomes

$$k_r = \frac{kT}{h} \frac{\gamma_A \gamma_B \cdots}{\gamma_{\ddagger}} (c^\circ)^{1-n} e^{-\Delta G^{\circ\ddagger}/RT} \tag{23.46}$$

For nonionic reactions in dilute solutions, the γ's are reasonably close to 1 and (given the inaccuracy of most kinetics data) can be ignored:

$$k_r \approx kTh^{-1}(c^\circ)^{1-n} e^{-\Delta G^{\circ\ddagger}/RT} = kTh^{-1}(c^\circ)^{1-n} e^{\Delta S^{\circ\ddagger}/R} e^{-\Delta H^{\circ\ddagger}/RT} \tag{23.47}$$

where we used $\Delta G^{\circ\ddagger} = \Delta H^{\circ\ddagger} - T\,\Delta S^{\circ\ddagger}$. Organic chemists are especially fond of (23.47) since it can be used to rationalize observed changes in rate constants in a series of reactions in terms of changes in standard-state activation entropies and enthalpies. The solvation of the reactants and the activated complex must be taken into account in discussing $\Delta S^{\circ\ddagger}$ and $\Delta H^{\circ\ddagger}$.

Equations (17.72) and (23.41) with the γ's omitted give $E_a \equiv RT^2\, d \ln k_r/dT \approx RT^2(1/T + d \ln K_a^{\ddagger}/dT)$. By analogy with equations following Eq. (17.75), we have $d \ln K_a^{\ddagger}/dT \approx \Delta H^{\circ\ddagger}/RT^2 \approx \Delta U^{\circ\ddagger}/RT^2$. Hence

$$E_a \approx \Delta H^{\circ\ddagger} + RT \qquad \text{nonionic reaction in dil. soln.} \tag{23.48}$$

which differs from (23.34) for gases. Equation (17.73) gives $A \equiv k_r e^{E_a/RT}$, and use of (23.48) and (23.47) gives

$$A \approx kTh^{-1}e(c^\circ)^{1-n} e^{\Delta S^{\circ\ddagger}/R} \qquad \text{nonionic reaction in dil. soln.} \tag{23.49}$$

Equations (23.48) and (23.49) allow calculation of $\Delta H^{\circ\ddagger}$ and $\Delta S^{\circ\ddagger}$ (and hence $\Delta G^{\circ\ddagger}$) from experimental A and E_a values.

Diffusion-controlled and mixed-control reactions. The ACT equation (23.41) is inapplicable to diffusion-controlled and mixed-control reactions, since their rate is governed at least in part by the rate at which reactants can diffuse through the solvent to encounter each other. Let the elementary reaction be B + C → products. Very rapid reactions can be studied by producing one or both of the reactants in solution using flash photolysis or pulse radiolysis, by perturbing the equilibrium B + C = products using a T jump, or by rapid mixing of the reagents (Sec. 17.2). At the start of the reaction, species B and C are each distributed randomly in the solution. A short time after reaction begins, many pairs of B and C molecules that initially were close to each other will have encountered each other and reacted. Those B molecules which remain will most likely be B molecules for which there were no C molecules in the immediate vicinity. If diffusion is not rapid enough to restore a random distribution, the region around a given B molecule will be somewhat depleted of C molecules (and vice versa). Let R be the distance from a given B molecule. As R increases, the average concentration of C molecules will increase, reaching its value in the bulk solution at some large value of R, which we shall call infinity. There is a concentration gradient of C in the region around a given B molecule.

We shall assume that very shortly after the reaction begins a steady state is reached in which the rate of diffusion of C toward B equals the rate of reaction of C with B.

Fick's first law of diffusion, Eq. (16.33), gives the rate of diffusion of B molecules through the solvent A. To find the rate at which B and C molecules diffuse toward each other, we must use a diffusion coefficient that is the sum of the diffusion coefficients of B and C in the solvent A. We shall assume the solution is dilute, so the infinite-dilution diffusion coefficients can be used. Thus,

$$\frac{dn'_C}{dt} = (D_B^\infty + D_C^\infty)\mathscr{A}\frac{d[C]}{dR} \tag{23.50}$$

where dn'_C is the number of moles of C that in time dt cross a spherical surface of area $\mathscr{A} = 4\pi R^2$ surrounding a given B molecule and D_B^∞ and D_C^∞ are the infinite-dilution diffusion coefficients of B and C in the solvent A. The prime avoids confusion of the quantity (23.50), which refers to diffusion of C toward one particular B molecule, with the total rate of disappearance of C in the solution. We have $d[C]/dR > 0$ and we shall consider dn'_C/dt to be positive; hence a plus sign is used in (23.50).

Let us assume B and C are spherical molecules with radii r_B and r_C. For a steady state, dn'_C/dt is a constant. Integration of (23.50) between the limits $R = r_B + r_C$ (molecules B and C in contact) and $R = \infty$ (the bulk solution) gives

$$[C]_{R=\infty} - [C]_{R=r_B+r_C} = \frac{dn'_C}{dt}\frac{1}{D_B^\infty + D_C^\infty}\int_{r_B+r_C}^{\infty}\frac{1}{4\pi R^2}\,dR$$

$$[C]_{R=r_B+r_C} = [C] - \frac{dn'_C}{dt}\frac{1}{(D_B^\infty + D_C^\infty)4\pi(r_B + r_C)} \tag{23.51}$$

where $[C] = [C]_{R=\infty}$ is the concentration of C in the bulk solution.

Let r be the observed reaction rate per unit volume:

$$r = k[B][C] \tag{23.52}$$

where the observed rate constant k combines the effects of the diffusion rate and the chemical-reaction rate, and the concentrations are those in the bulk solution. The steady-state condition is that the rate at which C diffuses toward a given B molecule equal the reaction rate between C and that B molecule. Multiplication of r by the solution volume V gives the total reaction rate in the solution. The reaction rate at one particular B molecule is then rV/N_B, where N_B is the number of B molecules in the solution. The steady-state condition then gives $dn'_C/dt = rV/N_B = rV/N_0 n_B = r/N_0[B] = k[C]/N_0$, where (23.52) was used. Substituting this expression for dn'_C/dt in (23.51), we get

$$[C]_{R=r_B+r_C} = [C]\left(1 - \frac{k}{4\pi N_0(D_B^\infty + D_C^\infty)(r_B + r_C)}\right) = [C]\left(1 - \frac{k}{k_{\text{diff}}}\right) \tag{23.53}$$

$$k_{\text{diff}} \equiv 4\pi N_0(D_B^\infty + D_C^\infty)(r_B + r_C) \tag{23.54}$$

The significance of the defined quantity k_{diff} will be seen shortly.

Let k_{chem} be the rate constant for chemical reaction of B and C pairs in a solvent cage (Fig. 17.11). The steady-state hypothesis allows us to express r as

$$r = k_{\text{chem}}[\text{B}][\text{C}]_{R=r_{\text{B}}+r_{\text{C}}} \tag{23.55}$$

Equating (23.55) and (23.52), we get $k = k_{\text{chem}}[\text{C}]_{R=r_{\text{B}}+r_{\text{C}}}/[\text{C}]$. Use of (23.53) gives $k = k_{\text{chem}}(1 - k/k_{\text{diff}})$. Solving for k, we have as our final result

$$k = \frac{k_{\text{diff}}k_{\text{chem}}}{k_{\text{diff}} + k_{\text{chem}}} \qquad \text{or} \qquad \frac{1}{k} = \frac{1}{k_{\text{diff}}} + \frac{1}{k_{\text{chem}}} \tag{23.56}$$

For the limit $k_{\text{chem}} \to \infty$, the reaction is diffusion-controlled, and (23.56) becomes $1/k = 1/k_{\text{diff}}$ or $k = k_{\text{diff}}$. Thus, k_{diff} in (23.54) is the rate constant for the diffusion-controlled reaction. Equation (23.54) was given earlier as Eq. (17.100).

For the limit $k_{\text{diff}} \to \infty$ (or, more precisely, $k_{\text{diff}} \gg k_{\text{chem}}$), Eq. (23.56) becomes $k = k_{\text{chem}}$. The reaction is chemically controlled. The rate constant k_{chem} is given by the ACT equation (23.41).

When k_{diff} and k_{chem} are the same order of magnitude, the kinetics is mixed and k is given by (23.56).

The existence of a concentration gradient around each reacting molecule is a departure from the Boltzmann distribution, which would predict a uniform distribution in the absence of electric and gravitational fields.

For fast ionic reactions, the rate at which C and B diffuse toward each other is affected by the Coulombic attraction or repulsion. An extension of the above derivation (see *Weston and Schwarz*, secs. 6.2 and 6.3) gives Eq. (17.102).

PROBLEMS

23.1 (*a*) In theoretical discussions the rate law is sometimes expressed in terms of the number of molecules N_{A} (rather than moles) reacting per unit volume per unit time:

$$r_{\text{molec}} \equiv \frac{1}{\nu_{\text{A}}}\frac{d(N_{\text{A}}/V)}{dt} = k_{\text{molec}}\left(\frac{N_{\text{A}}}{V}\right)^{\alpha}\left(\frac{N_{\text{B}}}{V}\right)^{\beta}\cdots$$

Show that the molecular rate constant k_{molec} is related to the rate constant k of Eq. (17.5) by $k_{\text{molec}} = N_0^{1-n}k$, where n is the overall order. (*b*) Calculate k_{molec} in cm^3 s^{-1} if $k = 1.00$ dm^3 mole^{-1} s^{-1}. (Instead of cm^3 s^{-1}, people often write cm^3 molecule^{-1} s^{-1}.)

23.2 Give the equation that corresponds to (23.6) for the elementary reaction 2A → products.

23.3 (*a*) Use the hard-sphere collision theory to calculate the A factor for the elementary reaction $NO + O_3 \to NO_2 + O_2$; reasonable values of molecular radii (calculated from the known molecular structures) are 1.4 Å for NO and 2.0 Å for O_3. Take $T =$ 500 K. (*b*) The experimental A for this reaction is 8×10^{11} cm^3 mole^{-1} s^{-1}. Calculate the steric factor.

23.4 Verify (23.18).

23.5 Convert 9.8 kcal/mole to ergs per molecule.

23.6 (*a*) Describe the location of the principal axes of the linear $H_aH_bH_c$ transition state and express its moment of inertia I_b in terms of masses and bond distances. (*b*) Verify the expression for the ratio of rotational partition functions given in the example in Sec. 23.4 and verify the numerical value for this ratio.

23.7 (*a*) Verify the numerical values of vibrational partition functions in the example in Sec. 23.4. (*b*) Verify the numerical value of k_r in this example.

23.8 Use ACT to calculate k_r for $H + H_2$ at 600 K.

23.9 Show that in the high-T limit, z_{vib} in (22.110) is proportional to $T^{f_{\text{vib}}}$, where f_{vib} is the number of vibrational degrees of freedom.

23.10 Work out the typical range of values for m in (23.21).

23.11 Use the ACT equation (23.22) to calculate E_a for $H + H_2$ at 300 K. Assume that T is low enough for the vibrational partition functions to be neglected.

23.12 Use ACT to derive the effusion equation (15.58). Take the critical dividing surface as coinciding with the hole and use (23.16).

23.13 (*a*) For the elementary gas-phase reaction $O_3 + NO \rightarrow NO_2 + O_2$, one finds that $E_a = 2.5$ kcal/mole and $A = 6 \times 10^8$ dm^3 mole^{-1} s^{-1} for the temperature range 220 to 320 K. Calculate $\Delta G_c^{\circ\ddagger}$, $\Delta H_c^{\circ\ddagger}$, and $\Delta S_c^{\circ\ddagger}$ for the midpoint of this temperature range. (*b*) The same as (*a*) for the gas-phase elementary reaction $CO + O_2 \rightarrow CO_2 + O$; here, $E_a = 51$ kcal/mole and $A = 3.5 \times 10^9$ dm^3 mole^{-1} s^{-1} for the range 2400 to 3000 K.

23.14 (*a*) For an ionic elementary reaction between a +2 and a −3 ion, calculate k_r/k_r^∞ in water at 25°C for $I = 10^{-3}$, 10^{-2}, and 10^{-1}. (*b*) Do the same for a reaction between a −2 and a −3 ion.

23.15 Measurement of the rate constant of the reaction $S_2O_8^{2-} + 2I^- \rightarrow 2SO_4^{2-} + I_2$ as a function of $I = I_c/c^\circ$ at 25°C in water yields the following data (where $k^\circ \equiv$ 1 dm^3 mole^{-1} s^{-1}):

$10^3 I$	2.45	3.65	6.45	8.45	12.45
$10k/k^\circ$	1.05	1.12	1.18	1.26	1.40

Use a graphical method to find $z_A z_B$ for the rate-determining step.

23.16 For the elementary reaction $CH_3Br + Cl^- \rightarrow CH_3Cl + Br^-$ in acetone, one finds $A = 2 \times 10^9$ dm^3 mole^{-1} s^{-1} and $E_a = 15.7$ kcal/mole. Calculate $\Delta H^{\circ\ddagger}$, $\Delta S^{\circ\ddagger}$, and $\Delta G^{\circ\ddagger}$ for this reaction at 300 K.

23.17 Combine (23.53) and (23.56) to express the concentration of C in the immediate vicinity of B as a function of the bulk concentration of C and of k_{diff} and k_{chem}; then find the limiting values of this concentration for chemically controlled and for diffusion-controlled reactions.

23.18 Name three kinds of reactions that involve departures from the Boltzmann distribution law for the reactants.

24

SOLIDS AND LIQUIDS

24.1 SOLIDS AND LIQUIDS

A solid is classified as crystalline or amorphous. A *crystalline solid* generally shows a sharp melting point; examination by the naked eye (or by a microscope if the sample is microcrystalline) shows crystals having well-developed faces and a characteristic shape. X-ray diffraction (Sec. 24.8) shows a crystalline solid to have a regular, ordered structure composed of identical repeating units having the same orientation throughout the crystal; the repeating unit is a group of one or more atoms, molecules, or ions.

An *amorphous solid* does not have a characteristic crystal shape. When heated, it softens and melts over a wide temperature range. X-ray diffraction shows a disordered structure. Polymers often form amorphous solids; as the liquid polymer is cooled, the polymer chains become twisted and tangled together in a random, irregular way. (Some polymers do form crystalline solids; others give solids that are partly crystalline and partly amorphous.) The various kinds of glass contain chains and rings involving Si—O bonds; the structure is disordered and irregular.

In solids, the structural units (atoms, molecules, or ions) are held more or less rigidly in place. In liquids, the structural units can move about by squeezing past their neighbors. The degree of order in a liquid is much less than in a crystalline solid. Liquids have a short-range order, in that molecules in the immediate environment of a given molecule tend to adopt a preferred orientation and intermolecular distance, but liquids have no long-range order, since there is no correlation between the orientations of two widely separated molecules and no restriction on the distance between them. Amorphous solids have the rigidity of solids but resemble liquids in not having long-range order; they are often viewed as supercooled liquids.

In this chapter, the word "solid" will be understood to mean "crystalline solid."

Solid polymers are, in general, partly crystalline and partly amorphous. The degree of crystallinity depends on the polymer structure and the procedure used to prepare the solid.

At low temperatures, an amorphous polymer is hard and has a glassy appearance. When heated to a certain temperature, the amorphous solid becomes softer, rubberlike, and flexible, the polymer molecules now having sufficient energy to slide past one another; this temperature is called the *glass-transition temperature* T_g. Cooling the polymer below T_g locks the chains in fixed, random conformations to produce a hard, disordered, amorphous solid. A rubber ball cooled below T_g in liquid nitrogen will shatter when dropped on the floor.

A perfectly crystalline polymer would not show a glass transition, but would melt at some temperature to a liquid. Perfect polymer crystals do not exist. A semicrystalline polymer shows both a glass-transition temperature T_g and a melting temperature T_m that lies above T_g. We can consider T_g to be associated with the amorphous portions of the solid and T_m to be associated with the crystalline portions. For temperatures between T_g and T_m, one has tiny crystallites embedded in a rubbery matrix. For nylon 66, T_g is 60°C and T_m is 265°C. For polyethylene, T_g is -125°C and T_m is 140°C.

At the melting point of a perfect crystal, the volume undergoes a discontinuous change, $\Delta V \neq 0$. For polymers, one observes that $\Delta V = 0$ at T_g, but the slope of the V versus T curve changes; in the region of T_m, there is a very rapid change in V, but V is not discontinuous, since the polymer is not a perfect crystal.

24.2 CHEMICAL BONDING IN SOLIDS

A crystal is classified as *ionic, covalent, metallic,* or *molecular*, according to the nature of the chemical bonding and the intermolecular forces in the crystal.

Ionic crystals consist of an array of positive and negative ions, and are held together by the Coulomb attraction between oppositely charged ions. Examples are NaCl, MgO, $CaCl_2$, and KNO_3.

Metallic crystals are composed of metal atoms; some of the valence electrons are delocalized over the entire metal and hold the crystal together. Examples are Na, Cu, Fe, and various alloys.

Covalent (or *nonmetallic network*) solids consist of an "infinite" network of atoms held together by (polar or nonpolar) covalent bonds, no individual molecules being present. Examples are carbon in the forms of diamond and graphite, Si, SiO_2, and SiC. In diamond (Fig. 24.15) each carbon is bonded to four others that surround it tetrahedrally, giving a three-dimensional network that extends throughout the crystal. Silicon has the same structure. SiO_2 has a three-dimensional network in which each Si is bonded to four O's at tetrahedral angles, and each O is bonded to two Si atoms. In many covalent crystals, the covalent bonds form a two-dimensional network. Examples are graphite and mica. Graphite consists of layers of fused hexagonal rings of covalently bonded carbons, the bonds being intermediate between single and double bonds (as in benzene); weak van der Waals forces hold the layers together. The covalent solids BeH_2 and $BeCl_2$ contain one-dimensional networks (Fig. 24.1). Crystals of a long-chain polymer like polyethylene can be viewed as covalent solids with one-dimensional networks.

In ionic, metallic, and three-dimensional covalent crystals, it is not possible to pick out individual molecules. The entire crystal is a single giant molecule.

The *coordination number* of an atom or ion in a solid is the number of nearest neighbors for that atom or ion. In NaCl (Fig. 24.12) each Na^+ has six Cl^- ions as

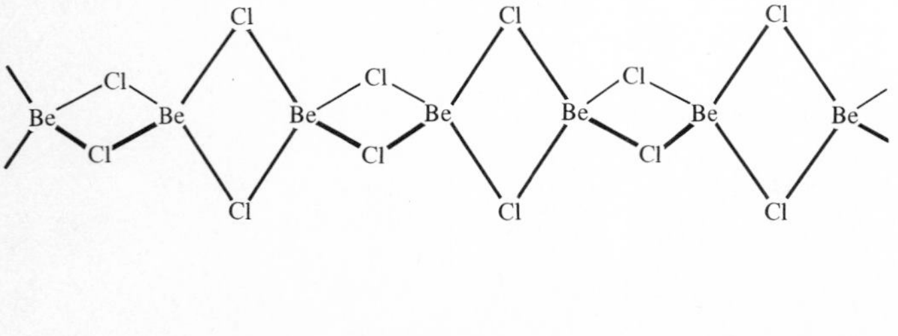

Figure 24.1
One-dimensional-network structure of solid $BeCl_2$. The four Cl's around each Be are approximately tetrahedrally disposed.

nearest neighbors, and the coordination number of Na^+ is 6. The coordination number of carbon in diamond is 4. In SiO_2, the coordination number of Si is 4 and that of O is 2. The coordination number in a metal is usually 8 or 12.

Molecular crystals are composed of individual molecules. The atoms within each molecule are held together by covalent bonds. Relatively weak intermolecular forces hold the molecules together in the crystal. Molecular crystals are subdivided into *van der Waals crystals*, in which the intermolecular attractions are dipole–dipole, dipole–induced-dipole, and dispersion forces (Sec. 22.10), and *hydrogen-bonded crystals*, in which the main intermolecular attraction is due to hydrogen bonding. Some van der Waals crystals are Ar, CO_2, CO, O_2, HI, CH_3CH_2Br, $C_6H_5NO_2$, $HgCl_2$, and $SnCl_4$. Some hydrogen-bonded crystals are H_2O, HF, NH_3, and the amino acid $H_3{}^+NCH_2COO^-$.

The distinctions between the various kinds of crystals are not always clear-cut. For example, ZnS contains a three-dimensional network of Zn and S and is often classed as a covalent solid. However, each Zn—S bond has a substantial amount of ionic character, and one might consider the structure to be an ionic one in which the S^{2-} ions are substantially polarized (distorted) to give considerable covalent bonding.

24.3 COHESIVE ENERGIES OF SOLIDS

The *cohesive energy* (or *binding energy*) E_c of a crystal is the enthalpy change $\Delta H°$ for the isothermal conversion of the crystal into isolated atoms (for metallic and covalent crystals), molecules (for molecular crystals), or ions (for ionic crystals). E_c depends on temperature, the theoretically most significant value being that at absolute zero.

Metals. The cohesive energy of the metal M is $\Delta H°$ for the process $M(c) \rightarrow M(g)$. This is the heat of sublimation of the solid to a monatomic gas at 1 atm (provided we neglect the slight difference between real-gas and ideal-gas enthalpies at 1 atm). E_c values for metals are listed in thermodynamics tables as $\Delta H_f°$ for $M(g)$. Some 25°C E_c values in kcal/mole are:

Na	K	Be	Mg	Cu	Ag	Cd	Al	Fe	W	Pt
26	21	78	35	81	68	27	78	99	203	135

The range of values is from 20 to 200 kcal/mole, which corresponds to 1 to 9 eV per atom. These energies are comparable to chemical-bond energies (Table 20.1). E_c values for transition metals tend to be high, due to covalent bonding involving d electrons. The melting points reflect the E_c values. Some melting points are: Na, 98°C; Mg, 650°C; Cu, 1083°C; Pt, 1770°C; W, 3400°C.

Covalent solids. E_c for a covalent solid can be found from $\Delta H_f°$ values of the solid and the gaseous atoms (Prob. 24.1). Some 25°C E_c values in kcal/mole are:

C(diamond)	C(graphite)	Si	SiC	SiO_2(quartz)
170.8	171.3	109	295	446

The cohesive energy of a covalent solid is due to covalent chemical bonds. The high E_c values are reflected in the hardness and the high melting points of covalent solids. Diamond sublimes (rather than melts) at 3500°C. Quartz melts at 1600°C.

Ionic solids. The cohesive energy of NaCl is ΔH° for the process $NaCl(c) \rightarrow Na^+(g) + Cl^-(g)$. Born and Haber pointed out that this process can be broken into the following isothermal steps:

$$NaCl(c) \xrightarrow{(a)} Na(c) + \tfrac{1}{2}Cl_2(g) \xrightarrow{(b)} Na(g) + Cl(g) \xrightarrow{(c)} Na^+(g) + Cl^-(g)$$

ΔH_a equals minus the enthalpy of formation ΔH_f° of NaCl(c) from its elements. ΔH_b equals ΔH_f° of Na(g) plus ΔH_f° of Cl(g). [ΔH_f° of Na(g) is the enthalpy of sublimation of solid Na. ΔH_f° of Cl(g) is closely related to the dissociation energy of Cl_2; cf. the discussion in Sec. 21.3 on H_2.] ΔH_c equals N_0 times the ionization energy I of Na minus N_0 times the electron affinity A of Cl (since excited electronic states of the atoms and ions are not populated at room temperature). Thus

$$\begin{aligned} E_c[NaCl(c)] = &-\Delta H_f^\circ[NaCl(c)] + \Delta H_f^\circ[Na(g)] \\ &+ \Delta H_f^\circ[Cl(g)] + N_0 I(Na) - N_0 A(Cl) \end{aligned} \tag{24.1}$$

Using thermodynamic data in the Appendix, $I(Na) = 5.139$ eV, $A(Cl) = 3.614$ eV, and Eq. (20.1), we get $E_c = 188$ kcal/mole at 25°C (Prob. 24.2).

Some values of E_c in kcal/mole at 25°C for ionic substances are:

NaCl	NaBr	LiF	CsCl	AgCl	$MgCl_2$	MgO	CaO
188	176	251	160	219	603	905	815

The factor-of-4 difference between NaCl and CaO is due to the ± 2 charges on the Ca^{2+} and O^{2-} ions. Some melting points are: NaCl, 801°C; NaBr, 755°C; MgO, 2800°C.

Molecular solids. E_c for a molecular solid is the enthalpy of sublimation. Some values in kcal/mole at the melting point of the crystal are:

Ar	CH_4	Kr	H_2O	O_2	Cl_2
1.8	2.3	2.5	12.2	2.0	6.4

The weakness of intermolecular forces compared with chemical bonds means that E_c for a molecular crystal is typically an order of magnitude smaller than for a covalent, ionic, or metallic crystal. For molecules of comparable size, hydrogen-bonded crystals have higher E_c values than van der Waals crystals. E_c values and melting points increase as the molecular size increases, due to the increase in dispersion energy as the number of electrons increases. Some melting points are: Ar, −189°C; CH_4, −182°C; H_2O, 0°C; Cl_2, −101°C.

24.4 THEORETICAL CALCULATION OF COHESIVE ENERGIES

Covalent solids. E_c for a covalent solid can be estimated as the sum of the bond energies for all the covalent bonds. Consider diamond, for example. In a mole, there are N_0 atoms of carbon. Each carbon atom is bonded to four others. If we were to take $4N_0$ as the number of carbon–carbon single bonds per mole, we would be counting each $C_a{-}C_b$ bond twice, once as one of the four bonds of carbon atom a and once as one of the four bonds of C_b. Hence, there are $2N_0$ bonds per mole. Table 20.1 gives the carbon–carbon single-bond energy as 82 kcal/mole. We therefore estimate E_c for diamond to be 164 kcal/mole, which is close to the actual value of 171 kcal/mole.

Ionic solids. E_c of an ionic crystal can be estimated by summing the energies of the interionic Coulomb attractions and repulsions and the Pauli repulsions arising from slight overlap of the probability densities of ions in contact (recall Prob. 20.11); this approach is due to Born and Landé.

Consider NaCl as an example. In the NaCl crystal structure (Fig. 24.12) each Na^+ ion is surrounded by six nearby Cl^- ions at an equilibrium distance (averaged over zero-point vibrations) R_0. The next nearest neighbors to an Na^+ ion are 12 other Na^+ ions at a distance $\sqrt{2}\,R_0$. Then come 8 Cl^- ions at $\sqrt{3}\,R_0$, then 6 Na^+ ions at $\sqrt{4}\,R_0$, then 24 Cl^- ions at $\sqrt{5}\,R_0$, then 24 Na^+ ions at $\sqrt{6}\,R_0$, and so on. For a general separation R between Na^+ and Cl^- nearest neighbors, Eq. (19.4) gives as the potential energy of interaction between one Na^+ ion and all the other ions in the crystal

$$\frac{e'^2}{R}\left(-\frac{6}{1}+\frac{12}{\sqrt{2}}-\frac{8}{\sqrt{3}}+\frac{6}{\sqrt{4}}-\frac{24}{\sqrt{5}}+\frac{24}{\sqrt{6}}-\cdots\right)=-\frac{e'^2}{R}M$$

where the *Madelung constant* M is found by summing the series. Because of the long-range nature of interionic forces, the series converges very slowly, and special techniques must be used to evaluate it [see the references in *Kittel*, chap. 3 and E. L. Burrows and S. F. A. Kettle, *J. Chem. Educ.*, **52**, 58 (1975)]. One finds $M = 1.74756$ for NaCl. By symmetry, the potential energy of Coulomb interaction between a Cl^- ion and the other ions of the crystal is also $-e'^2M/R$. Multiplication of $-2e'^2M/R$ by the Avogadro constant N_0 and division by 2 (to avoid counting each interionic interaction twice) gives the Coulomb contribution E_{Coul} to the NaCl cohesive energy as

$$E_{\text{Coul}} = -e'^2 M N_0 R^{-1} \tag{24.2}$$

For NaCl, x-ray diffraction data give $R_0 = 2.820$ Å at 25°C and 2.798 Å at 0 K. One finds (Prob. 24.3) $E_{\text{Coul}} = -207.4$ kcal/mol at 0 K.

The form of (24.2) and the Madelung constants for other crystal structures are discussed in D. Quane, *J. Chem. Educ.*, **47**, 396 (1970).

In addition to the Coulomb interactions between the ions considered as spheres, there is the Pauli-principle repulsion (Sec. 20.4) due to the slight overlap of neighboring ion probability densities. A crude approximation to this repulsion energy between two ions is the function A/R^n, where n is a large power and A is a constant. (Recall that Lennard-Jones took $n = 12$; Sec. 22.10.) Each ion has six nearest neighbors, so the total repulsion energy is $6AN_0/R^n \equiv B/R^n$, where $B \equiv 6AN_0$. Thus, $E_{\text{rep}} = B/R^n$.

Let $E_p = E_{\text{Coul}} + E_{\text{rep}}$ be the potential energy of interaction between the ions in 1 mole of the crystal for the arbitrary nearest-neighbor separation R. The cohesive energy E_c is defined as a positive quantity, so $E_c = -E_p$, where E_p is evaluated at $R = R_0$. We have

$$-E_p = -(E_{\text{Coul}} + E_{\text{rep}}) = e'^2 M N_0 R^{-1} - BR^{-n} \tag{24.3}$$

where B and n are as yet unknown.

To evaluate B, we use the condition that E_p is minimized at the equilibrium nearest-neighbor separation R_0. Differentiation of (24.3) gives $-dE_p/dR = -e'^2MN_0R^{-2} + nBR^{-n-1}$. Setting $dE_p/dR = 0$ at $R = R_0$ and solving for B, we get

$$B = e'^2MN_0n^{-1}R_0^{n-1} \tag{24.4}$$

Substitution of (24.4) in (24.3) gives

$$-E_p = e'^2MN_0(R^{-1} - n^{-1}R_0^{n-1}R^{-n})$$

$$-E_p = e'^2MN_0R_0^{-1}[R_0/R - n^{-1}(R_0/R)^n] \tag{24.5}$$

$$E_c = -E_p(R_0) = \frac{e'^2MN_0}{R_0}\left(1 - \frac{1}{n}\right) \tag{24.6}$$

The parameter n in the repulsive part of the potential can be evaluated from compressibility data. Equation (4.71) reads $(\partial\bar{U}/\partial\bar{V})_T = \alpha T/\kappa - P$. As $T \to 0$, Prob. 5.26 shows that $\alpha \to 0$. Also, κ is always positive. Hence, $(\partial\bar{U}/\partial\bar{V})_T = -P$ at $T = 0$. Partial differentiation with respect to $\bar{V}$ gives $(\partial^2\bar{U}/\partial\bar{V}^2)_T = -(\partial P/\partial\bar{V})_T$ at $T = 0$. Using (1.42), we have

$$(\partial^2\bar{U}/\partial\bar{V}^2)_T = 1/\bar{V}\kappa \qquad \text{at } T = 0 \tag{24.7}$$

The molar internal energy $\bar{U}$ at absolute zero equals the molar interionic interaction energy E_p plus the zero-point vibrational kinetic energy. Neglecting the variation of this kinetic energy with $\bar{V}$, we have $\partial^2\bar{U}/\partial\bar{V}^2 = \partial^2E_p/\partial\bar{V}^2$ at absolute zero.

The crystal molar volume is proportional to the cube of the nearest-neighbor distance: $\bar{V} = cR^3$, where the constant c depends on the crystal structure. Let $\bar{V}_0$ and $\bar{V}$ be the volumes corresponding to R_0 and R. Then $\bar{V}_0/\bar{V} = (R_0/R)^3$ and $R_0/R = (\bar{V}_0/\bar{V})^{1/3}$. Substitution in (24.5) gives

$$-E_p = e'^2MN_0R_0^{-1}(\bar{V}_0^{1/3}\bar{V}^{-1/3} - n^{-1}\bar{V}_0^{n/3}\bar{V}^{-n/3}) \tag{24.8}$$

Differentiating (24.8) twice with respect to $\bar{V}$ and then setting $\bar{V} = \bar{V}_0$, we get (Prob. 24.4)

$$-(\partial^2E_p/\partial\bar{V}^2)|_{R=R_0} = \tfrac{1}{9}e'^2MN_0R_0^{-1}\bar{V}_0^{-2}(1 - n) \tag{24.9}$$

Substitution of (24.9) and (24.7) into $\partial^2\bar{U}/\partial\bar{V}^2 = \partial^2E_p/\partial\bar{V}^2$ at R_0 gives

$$n = 1 + \frac{9\bar{V}_0R_0}{\kappa e'^2MN_0} \qquad \text{for } T = 0 \tag{24.10}$$

(This equation shows that the compressibility decreases with increasing Pauli repulsion, i.e., with increasing n.)

We shall see in Sec. 24.7 that the NaCl crystal is composed of cubic unit cells; each unit cell (Fig. 24.12) contains four Na^+–Cl^- ion pairs and has an edge length $2R_0$. A mole of NaCl(c) contains $N_0/4$ unit cells and has a volume $\bar{V}_0 = (N_0/4) \times (2R_0)^3 = 2N_0R_0{}^3$. Hence, for NaCl

$$n = 1 + 18R_0{}^4/\kappa e'^2M \qquad \text{for } T = 0 \tag{24.11}$$

Extrapolation of data to absolute zero gives $\kappa = 3.7 \times 10^{-12}$ cm^2 dyn^{-1} for NaCl. Using the 0-K R_0 value listed after Eq. (24.2), one finds $n = 8.4$ for NaCl (Prob. 24.5).

The theoretical NaCl cohesive energy is given by (24.6), (24.2), and the data following (24.2) as $(207.4 \text{ kcal mol}^{-1})(1 - 1/8.4) = 183$ kcal/mol at $T = 0$. The JANAF tables (Sec. 5.6) list ΔH_f° values extrapolated to 0 K, as well as $\Delta H_{f,298}^\circ$ values. Using these values in (24.1), one finds that the experimental E_c for NaCl at 0 K is 188 kcal mol^{-1}. The agreement between observed and calculated values is good.

There are two corrections that should be included. The theoretical E_c should include the potential energy of the dispersion interaction (Sec. 22.10) between the ions. One finds that this adds 5 kcal/mol to the theoretical E_c, bringing it to 188 kcal/mol. The experimental E_c is ΔH° for $\text{NaCl}(c) \rightarrow \text{Na}^+(g) + \text{Cl}^-(g)$ and is less than the theoretical E_c at $T = 0$ because of the zero-point vibrational energy of NaCl(c). This zero-point energy is $1\frac{1}{2}$ kcal/mol and reduces the theoretical E_c to $186\frac{1}{2}$ kcal/mol, compared with the experimental 0-K value 188 kcal/mol.

Metallic solids. E_c for Li and Na can be calculated quite accurately using the band theory (Sec. 24.9) of metals; for details, see *Pitzer*, sec. 10*e*. Calculations for most other metals are too difficult to give accurate results.

Molecular solids. E_c for a van der Waals crystal is found by summing the energies of the intermolecular van der Waals attractions and repulsions; for a hydrogen-bonded crystal, one also includes the energy of the hydrogen bonds.

Since the form of the intermolecular potential is not accurately known for most molecules, accurate calculation of E_c is difficult. For a simple crystal like Ar, a pretty accurate value of E_c can be calculated theoretically by using a Lennard-Jones 6-12 intermolecular potential and summing this over all pairs of Ar atoms in the crystal (see Prob. 24.8); this approach neglects three-molecule interactions.

For H_2O, spectroscopic observations on the intensity change with temperature of IR bands of the gas-phase dimer $(H_2O)_2$ show that an O···HO hydrogen bond has an energy of roughly 5 kcal/mole. Since each H atom in ice participates in one hydrogen bond, there are $2N_0$ hydrogen bonds per mole of ice. The hydrogen bonds thus account for 10 kcal/mol of the observed 12 kcal/mol cohesive energy of ice. The remaining 2 kcal/mol are due to van der Waals forces.

In summary, although accurate calculation of the cohesive energy of a crystal is not always possible, the forces that determine the cohesive energy are well understood.

24.5 INTERATOMIC DISTANCES IN CRYSTALS

Ionic crystals. Pauling used experimental nearest-neighbor distances in crystals with the NaCl-type structure (Fig. 24.12) together with arguments involving the dependence of an ion's radius on the effective nuclear charge Z_{eff} for the outer electrons to arrive at a set of *ionic radii*. Some values in Å are (L. Pauling, *The Nature of the Chemical Bond*, 3d ed., Cornell University Press, 1960, p. 514):

Li^+	Na^+	K^+	Rb^+	Cs^+	Be^{2+}	Mg^{2+}	Ca^{2+}	Al^{3+}	Fe^{2+}	Fe^{3+}
0.60	0.95	1.33	1.48	1.69	0.31	0.65	0.99	0.50	0.75	0.64

O^{2-}	S^{2-}	F^-	Cl^-	Br^-	I^-	H^-
1.40	1.84	1.36	1.81	1.95	2.16	2.08

Of course, cation radii are smaller than the corresponding atomic radii, and anion radii are larger than the corresponding atomic radii. These values can be used to estimate nearest-neighbor distances in ionic crystals.

The nearest-neighbor distance in NaCl(c) is 2.80 Å, compared with $R_e = 2.36$ Å in gas-phase NaCl. An ion in the crystal interacts with very many other ions, whereas an ion in the gas-phase molecule interacts with only one other ion. Hence, the potential energy is minimized at different values of R in the crystal and in the isolated molecule.

Molecular crystals. Because of the weakness of intermolecular van der Waals forces compared with chemical-bonding forces, one usually finds bond distances within molecules to be almost unchanged on going from the gas phase to the solid phase. For example, the Raman spectrum of gas-phase benzene gives $R_0(\text{CC}) = 1.397$ Å, x-ray diffraction of solid benzene gives $R_0(\text{CC}) = 1.39_2$ Å, and neutron diffraction of solid benzene gives $R_0(\text{CC}) = 1.39_8$ Å. (A remarkable exception is PCl_5, which exists as PCl_5 molecules in the gas phase but consists of PCl_4^+ and PCl_6^- ions in the solid.)

The distance between molecules in contact in a crystal is determined by the intermolecular van der Waals attractive and repulsive forces. From intermolecular distances in crystals, one can deduce a set of *van der Waals radii* for atoms. For example, in solid Cl_2, the shortest Cl–Cl distance is 2.02 Å, and this is the bond distance in Cl_2; the single-bond covalent radius of Cl is 1.01 Å. The Cl_2 molecules in the crystal are arranged in layers. The closest Cl–Cl distance between neighboring molecules in the same layer is 3.34 Å and between neighboring molecules in adjacent layers is 3.69 Å. The van der Waals radius of Cl thus lies between 3.34/2 and 3.69/2 Å and is usually taken as 1.8 Å.

Van der Waals radii are much larger than bond radii, since there are two pairs of valence electrons between the nonbonded atoms, compared with one pair between the bonded atoms. Van der Waals radii are found to be close to ionic radii; note that the outer part of a Cl atom in the molecule X:Ċ̤l: resembles a :Ċ̤l:⁻ ion. Some van der Waals radii in Å (Pauling, op. cit., p. 260) are:

H	He	N	O	F	Ne	P	S	Cl
1.2	0.9	1.5	1.4	1.35	1.3	1.9	1.85	1.8

Van der Waals radii are used to fix the size of atoms in space-filling molecular models. One takes each atom as a truncated sphere (Fig. 24.2).

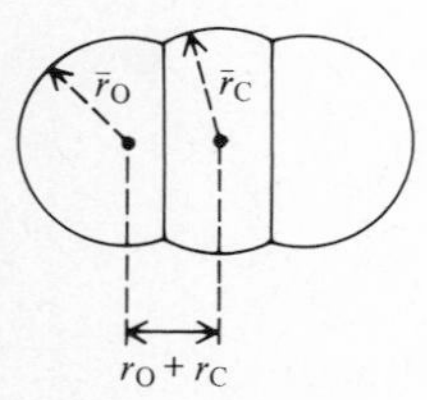

Figure 24.2 Model of CO_2. r_C and $\bar{r}_C$ are the double-bond radius and the van der Waals radius of carbon, respectively.

When hydrogen bonding exists, it may strongly influence the packing of the molecules in the crystal. Ice has a very open structure because of hydrogen bonding; when ice melts, the degree of hydrogen bonding decreases and liquid water at 0°C is denser than ice.

Covalent crystals. Interatomic distances in covalent crystals are determined by essentially the same quantum-mechanical interactions as in isolated molecules. For example, the carbon–carbon bond distance in diamond at 18°C is 1.545 Å, virtually the same as in most saturated organic compounds (1.53 to 1.54 Å).

Metallic crystals. The *metallic radius* of an atom is half the distance between adjacent atoms in a metallic crystal with coordination number 12. The metallic radius of an atom is a bit larger than its single-bond covalent radius. For example, the metallic radius of Cu is 1.28 Å, and its single-bond covalent radius is 1.17 Å.

24.6 CRYSTAL STRUCTURES

The basis. A crystal contains a structural unit, called the *basis* (or *motif*), that is repeated in three dimensions to generate the crystal structure. The environment of each repeated unit is the same throughout the crystal (neglecting surface effects). The basis may be a single atom or molecule, or it may be a small group of atoms, molecules, or ions. Each repeated basis group has the same structure and the same spatial orientation as every other one in the crystal. Of course, the basis must have the same stoichiometric composition as the crystal.

For NaCl, the basis consists of one Na^+ ion and one Cl^- ion. For Cu, the basis is a single Cu atom. For Zn, the basis consists of two Zn atoms. For diamond, the basis is two C atoms; the two atoms of the basis are each surrounded tetrahedrally by four carbons, but the four bonds at one basis atom differ in orientation from those at the other atom; see Fig. 24.15 and the accompanying discussion. For CO_2, the basis is four CO_2 molecules. For benzene, the basis is four C_6H_6 molecules.

The space lattice. If we place a single point at the same location in each repeated basis group, the set of points obtained forms the (*space*) *lattice* of the crystal. Each point of the space lattice has the same environment. The space lattice is not the same as the crystal structure. Rather, the crystal structure is generated by adding an identical structural group (the basis) to each lattice point. The space lattice is a geometrical abstraction. Figure 24.3 shows a two-dimensional lattice and a hypothetical two-dimensional crystal structure formed by associating with each lattice point a basis consisting of an M atom and a W atom.

The W atoms in Fig. 24.3 do not lie at lattice points. (Of course, the lattice points could have been chosen to coincide with the W atoms, in which case the M atoms would not lie at lattice points.) In fact, no atom need lie at a lattice point. For example, in I_2, one chooses the lattice point at the center of one I_2 molecule of the basis (which consists of two I_2 molecules); see Fig. 24.16*b*.

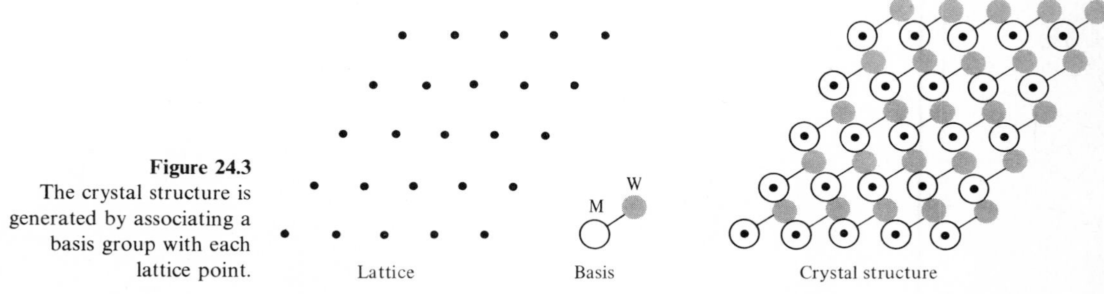

Figure 24.3
The crystal structure is generated by associating a basis group with each lattice point.

The unit cell. The space lattice of a crystal can be divided into identical parallelepipeds by joining the lattice points with straight lines. (A parallelepiped is a six-sided geometrical solid whose faces are all parallelograms.) Each such parallelepiped is called a *unit cell*. The way in which a lattice is broken up into unit cells is not unique. Figure 24.4*a* and *b* shows two different ways of forming unit cells in the two-dimensional lattice of Fig. 24.3. The same kind of choice exists for three-dimensional lattices. In crystallography, one chooses the unit cell so that it has the maximum symmetry and has the smallest volume consistent with the maximum symmetry; the maximum-symmetry requirement implies the maximum number of perpendicular unit-cell edges.

In two dimensions, a unit cell is a parallelogram with sides of length a and b and angle γ between these sides. In three dimensions, a unit cell is a parallelepiped with edges of length a, b, c and angles α, β, γ, where α is the angle between edges b and c, etc.

In 1848 Bravais showed that there are 14 different kinds of lattices in three dimensions. The unit cells of the 14 Bravais lattices are shown in Fig. 24.5. The 14 Bravais lattices are grouped into seven *crystal systems*, based on unit-cell symmetry; the relations between a, b, c and between α, β, γ for the seven systems are indicated in Fig. 24.5. (Some workers group the lattices into six systems; see *Buerger*, chap. 2.)

Those unit cells that have lattice points only at their corners are called *primitive* (or *simple*) unit cells. Seven of the Bravais lattices have primitive (P) unit cells.

A *body-centered* lattice (denoted by the letter I, from the German *innenzentrierte*) has a lattice point within the unit cell as well as at each corner of the unit cell. A *face-centered* (F) lattice has a lattice point on each of the six unit-cell faces as well as at the corners. The letter C denotes an *end-centered* lattice with a lattice point on each of the two faces bounded by edges of lengths a and b. (A and B have similar meanings.)

Each point at a unit-cell corner is shared among eight adjacent unit cells in the lattice, and so a primitive unit cell has $8/8 = 1$ lattice point and 1 basis group per unit cell. Each point on a unit-cell face is shared between two unit cells, so an F unit cell has $8/8 + 6/2 = 4$ lattice points and 4 basis groups per unit cell.

One *could* use a primitive unit cell to describe any crystal structure, but since in many cases this cell would have less symmetry than a cell of a (nonprimitive) centered lattice, it is more convenient to use the centered lattice. For example, Fig. 24.6 shows a two-dimensional lattice broken into centered unit cells and into less symmetric primitive unit cells.

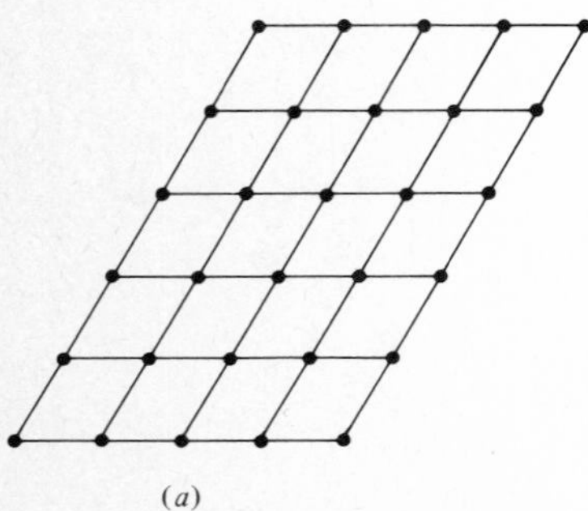
(*a*)

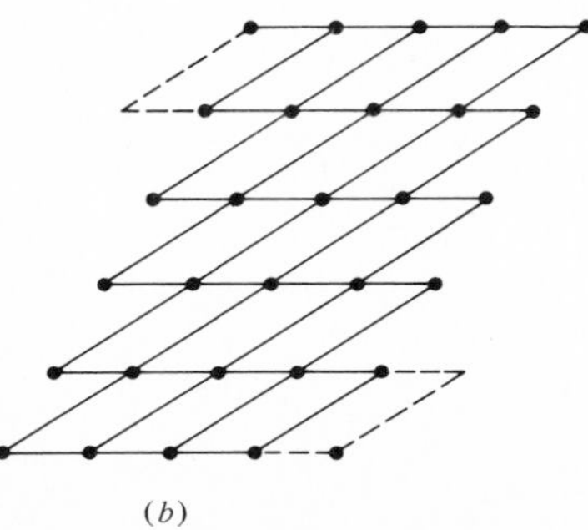
(*b*)

Figure 24.4
Two ways of breaking the two-dimensional lattice of Fig. 24.3 into unit cells.

Notation for points and planes. To designate the location of any point in the unit cell, we set up a coordinate system with origin at one corner of the unit cell and axes coinciding with the *a*, *b*, *c* edges of the cell. (Note that these axes are not necessarily mutually perpendicular.) The position of a point in the cell is specified by giving its coordinates as fractions of the unit-cell lengths *a*, *b*, and *c*. Thus, the point at the origin is 000; the interior lattice point in an I lattice is at $\frac{1}{2}\frac{1}{2}\frac{1}{2}$; the point at the center of the face bounded by the *b* and *c* axes is $0\frac{1}{2}\frac{1}{2}$.

Figure 24.5 Unit cells of the 14 Bravais lattices.

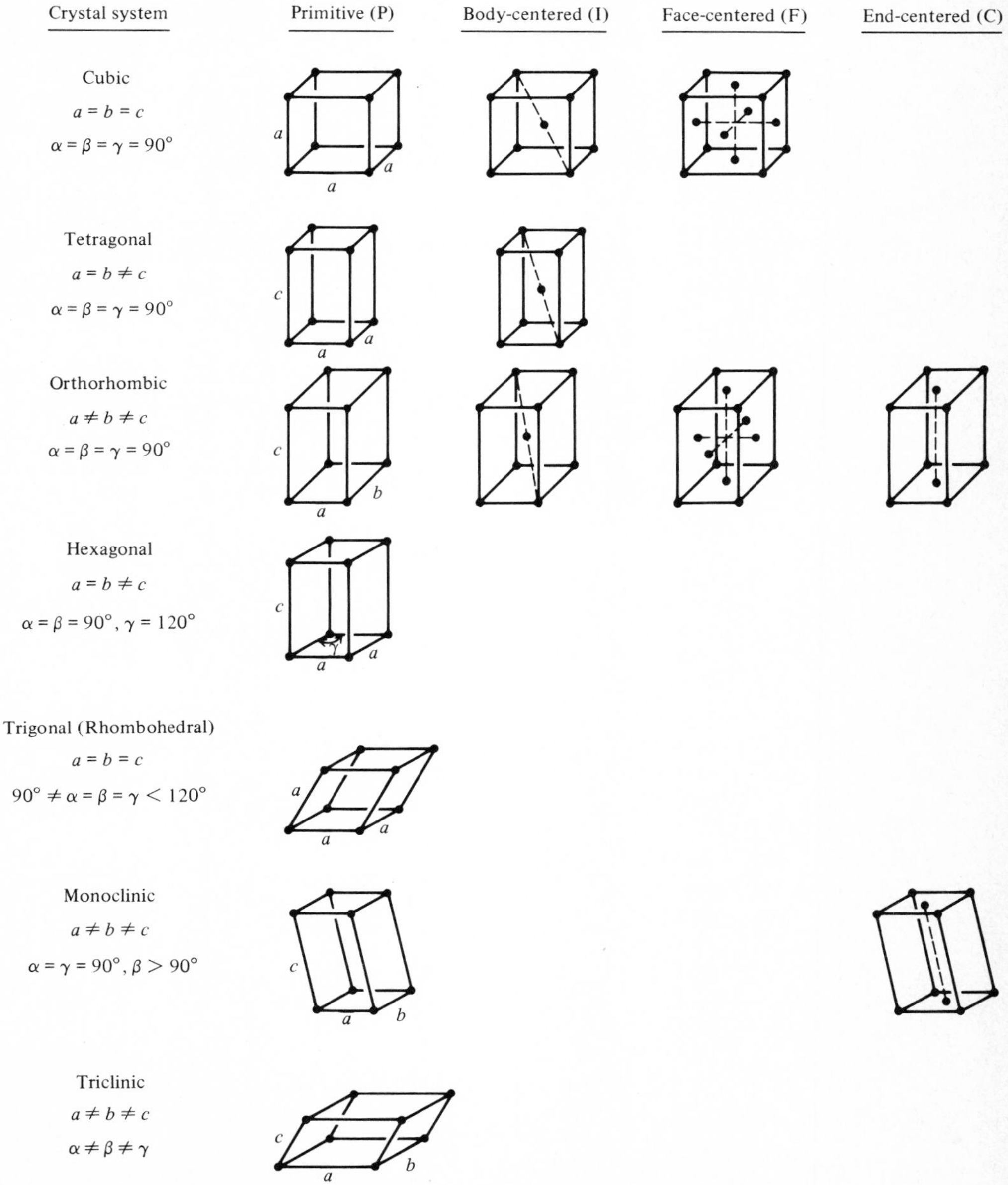

Crystal system	Primitive (P)	Body-centered (I)	Face-centered (F)	End-centered (C)
Cubic $a = b = c$ $\alpha = \beta = \gamma = 90°$				
Tetragonal $a = b \neq c$ $\alpha = \beta = \gamma = 90°$				
Orthorhombic $a \neq b \neq c$ $\alpha = \beta = \gamma = 90°$				
Hexagonal $a = b \neq c$ $\alpha = \beta = 90°, \gamma = 120°$				
Trigonal (Rhombohedral) $a = b = c$ $90° \neq \alpha = \beta = \gamma < 120°$				
Monoclinic $a \neq b \neq c$ $\alpha = \gamma = 90°, \beta > 90°$				
Triclinic $a \neq b \neq c$ $\alpha \neq \beta \neq \gamma$				

The orientation of a crystal plane is described by its *Miller indices* (hkl), which are obtained by the following steps: (1) find the intercepts of the plane on the a, b, c axes in terms of multiples of the unit-cell lengths a, b, c; (2) take the reciprocals of these numbers; (3) if fractions are obtained in step 2, multiply the three numbers by the smallest integer that will give whole numbers. (If an intercept is negative, one indicates this by a bar over the corresponding Miller index.)

One is usually interested only in planes densely populated by lattice points, since it is these planes which are important in x-ray diffraction.

As an example, the shaded plane labeled r in Fig. 24.7 intercepts the a axis at $a/2$ and the b axis at $b/2$ and lies parallel to the c axis (intercept at ∞). Step 1 gives $\frac{1}{2}$, $\frac{1}{2}$, ∞. Step 2 gives 2, 2, 0. Hence the Miller indices are (220). The plane labeled s has Miller indices (110). The plane labeled t has intercepts $\frac{3}{2}a$, $\frac{3}{2}b$, ∞; step 2 gives $\frac{2}{3}$, $\frac{2}{3}$, 0, and the Miller indices are (220). Plane u has intercepts $2a$, $2b$, ∞, so step 2 gives $\frac{1}{2}$, $\frac{1}{2}$, 0, and step 3 gives (110). Also shown are a (111) plane and (100) planes.

As well as giving the orientation of a single plane, one also uses Miller indices to denote a whole stack of parallel, equally spaced planes. The planes s, u, and an infinite number of planes parallel to them and separated by the same distance as s and u form the set of (110) planes. The set of (220) planes is considered to include the (110) planes plus the planes midway between the (110) planes. Planes r, s, t, u, ... form the (220) set. A useful rule is that the Miller index h of a stack of planes equals the number of interplanar spacings crossed in moving one lattice spacing in the direction of the a axis. Similarly for the k and l indices.

Each face of a macroscopic crystal contains a high density of lattice points. Examination of the macroscopic shape of a single crystal will generally tell which of the seven crystal systems the crystal belongs to (but will not tell what the Bravais lattice is) and will allow the a, b, and c axes to be located. One can therefore use Miller indices to specify the orientations of the macroscopic faces of the crystal (as well as to specify the orientations of planes within the crystal lattice).

Anisotropy. The repeated structural group of a crystal has a fixed orientation in space; it follows that the properties of a crystal will in general be different in different

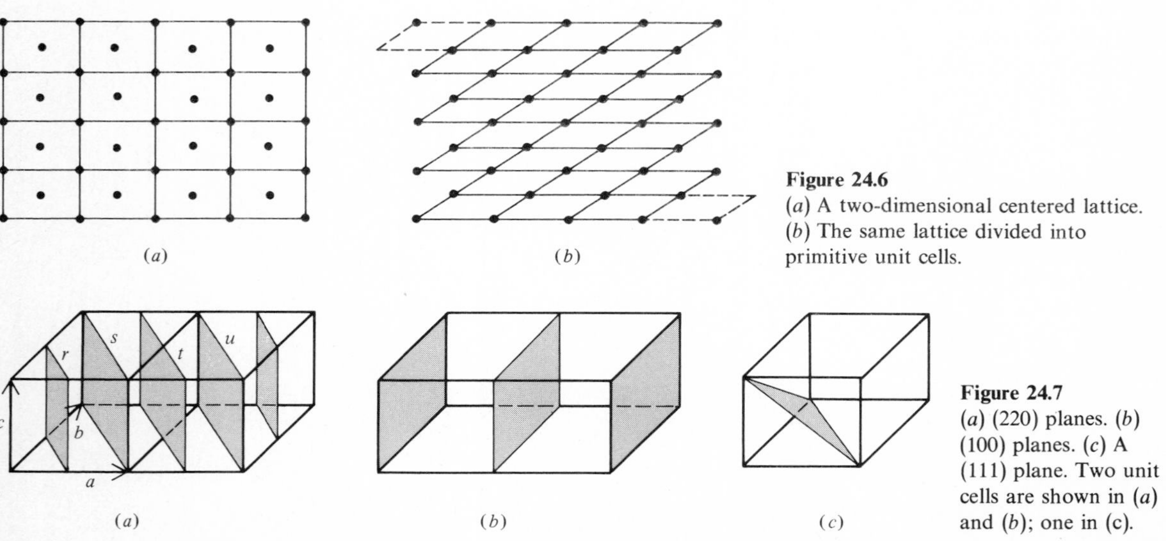

Figure 24.6
(*a*) A two-dimensional centered lattice. (*b*) The same lattice divided into primitive unit cells.

Figure 24.7
(*a*) (220) planes. (*b*) (100) planes. (*c*) A (111) plane. Two unit cells are shown in (*a*) and (*b*); one in (c).

directions. For example, we can see that the properties of the hypothetical two-dimensional crystal in Fig. 24.3 will differ in the a and b directions. A substance whose physical properties are the same in all directions is called *isotropic*; otherwise, it is *anisotropic*.

Gases and liquids are isotropic. An exception is liquid crystals. Liquid crystals flow like liquids but have much of the long-range order of solids. In a liquid crystal, the molecules can move about, and the intermolecular spacings are irregular; however, most of the molecules have the same spatial orientation (Fig. 24.8). (In certain liquid crystals, the molecules form layers; the molecules in a given layer all have the same orientation, but the orientation direction varies in a regular way from layer to layer. Such liquid crystals show bright temperature-dependent colors, and can be used to indicate temperature.) At a certain elevated temperature, a liquid crystal makes a transition to a true liquid state with random molecular orientations. Many liquid crystals are organic compounds having a long nonpolar chain and a polar group (for example, $C_6H_{13}C_6H_4COOH$). The liquid-crystal state may occur in biological-cell membranes.

A single crystal is, in general, anisotropic. A finely powdered solid is isotropic, since the random orientations of the tiny crystals produce isotropy. A single crystal has different refractive indices, coefficients of thermal expansion, electrical conductivities, speeds of sound, etc., in different directions. For example, the 20°C refractive indices of $AgNO_3$ along the a, b, and c axes are 1.73, 1.74, and 1.79 for sodium D light. The cubic crystal NaCl has the same refractive index along each axis and so is optically isotropic. However, NaCl is very anisotropic in its response to mechanical stress; e.g., when crushed, an NaCl crystal cleaves only along planes containing unit-cell faces. Different faces of a crystal may show different catalytic activities and different rates of solution.

The Avogadro constant. Let Z be the number of formula units per unit cell. The unit cell of NaCl (Fig. 24.12) has four Na^+ and four Cl^- ions, so $Z = 4$ for NaCl. The unit cell of CO_2 has four CO_2 molecules, so $Z = 4$ for $CO_2(c)$. The unit cell of diamond (Fig. 24.15) has eight carbon atoms, so $Z = 8$ for diamond.

One mole of a crystal contains N_0/Z unit cells. The volume of a right-angled unit cell is the product abc of its edges. (The volume formula for a nonrectangular cell is given in *Buerger*, p. 187.) The molar volume is therefore $\bar{V} = abcN_0/Z$. The density is $\rho = M/\bar{V}$, where M is the molar mass. Hence

$$\rho = \frac{MZ}{abcN_0} \qquad \text{for } \alpha = \beta = \gamma = 90^\circ \tag{24.12}$$

X-ray diffraction (Sec. 24.8) allows a, b, c, and Z to be determined at temperature T. An accurate measurement of ρ at T then allows the Avogadro constant N_0 to be found. This is the most accurate method for determining N_0.

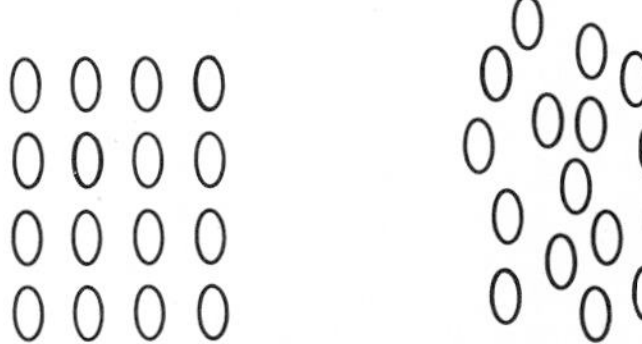

Figure 24.8
Orientation and spacing of molecules.

Silicon crystallizes in the same face-centered cubic lattice as diamond, and so $a = b = c$ and $Z = 8$ for Si. For a single crystal of very pure Si, one finds at 25°C [R. D. Deslattes et al., *Phys. Rev. Lett.*, **33**, 463 (1974)]

$$\rho = 2.328992 \text{ g/cm}^3, \qquad M = 28.08541 \text{ g/mol}, \qquad a = 5.431066 \text{ Å}$$

Substitution in (24.12) gives $N_0 = 6.022094 \times 10^{23} \text{ mol}^{-1}$.

24.7 EXAMPLES OF CRYSTAL STRUCTURES

Metallic crystals. Metal atoms are spherical, and the structures of metallic elements can be described in terms of the various ways of packing spheres.

Figure 24.9 shows a planar layer of spheres, with each sphere touching four others in the layer. (For now, ignore the lines and the shaded spheres.) Let successive layers of spheres be added with each sphere directly over a sphere in the layer beneath it. This gives a structure having a simple cubic space lattice (Fig. 24.5) with a basis of one atom at each lattice point. The coordination number (CN) is 6, since each atom touches four atoms in the same layer, one atom in the layer above it, and one atom in the layer below it. The structure is an open one, with only 52 percent of the volume occupied by the spheres. The simple cubic lattice is very rare for metals; the only known example is Po.

Instead of adding the second layer directly over the first, let us add a second layer of spheres (shown shaded in Fig. 24.9) by placing a sphere in each of the hollows formed by the first-layer spheres. A third layer can then be placed in the hollows of the second layer, with each third-layer sphere lying directly above a first-layer sphere. A fourth layer is then added, with each fourth-layer sphere directly over a second-layer sphere; etc. This produces a face-centered cubic (fcc) space lattice (Fig. 24.5), the basis being one atom at each lattice point. The square base of a unit cell is outlined in Fig. 24.9. The four spheres with dots at their centers lie at the corners of the unit-cell base, and the sphere marked with a cross lies at the center of the unit-cell base. The four shaded second-layer spheres above the sphere with a cross lie at the centers of the four side faces of the unit cell. The third-layer sphere directly over the sphere with a cross lies at the center of the top face of the unit cell. (If you have as much trouble as I do at three-dimensional visualization, construct models using coins or marbles held together with bits of clay.)

The CN of the fcc structure is 12, since each atom touches four others in the same layer, four in the layer above, and four in the layer below. The high CN makes for a very close-packed structure with 74 percent of the volume occupied. The fcc structure is often called cubic close-packed (ccp). [An alternative description of the

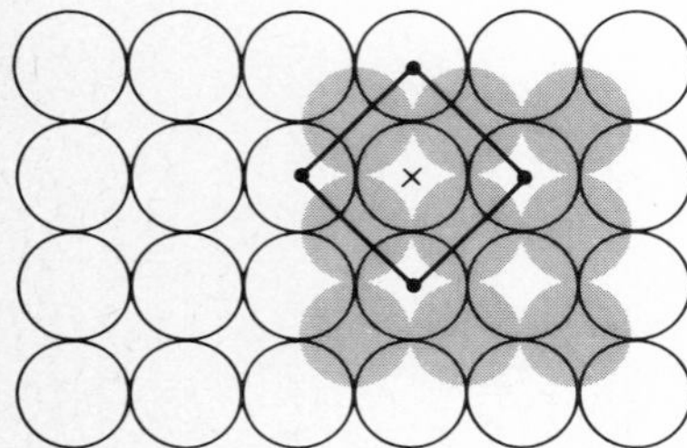

Figure 24.9

fcc structure is often used. Drawing the (111) planes in the lattice, one finds that each atom touches six others in the same (111) plane, three others in the (111) plane above it, and three others in the (111) plane below it; Prob. 24.14.]

The fcc structure is very common for metals. Examples include Al, Cu, Au, Pb, Pt, Pd, Ni, and Ca.

Suppose we start with a layer of spheres arranged as in Fig. 24.10*a* with centers separated by $2/\sqrt{3} = 1.155$ times the spheres' diameter. We place a second layer (shaded spheres) in the hollows of the first layer, and a third layer with its spheres directly over the first-layer spheres; etc. This produces a body-centered cubic (bcc) space lattice (Fig. 24.5) with a basis of one atom at each lattice point. The CN is 8, since each atom touches four atoms in the layer above it, four in the layer below it, and none in its own layer. This structure fills 68 percent of the space and is quite common for metals. Examples include Cr, Mo, W, Ta, Ba, Li, Na, K, Rb, and Cs.

Finally, we can start with a layer in which each sphere touches six others (Fig. 24.10*b*). We place spheres in the hollows marked by dots to give a second layer (shaded spheres). We then place spheres in those second-layer hollows that lie directly over the centers of the first-layer spheres to give a third layer of spheres lying directly over the first-layer spheres. (An alternative choice exists for the third-layer spheres; this turns out to give the fcc lattice discussed above.) A fourth layer is then formed with its spheres lying directly above the second-layer spheres; etc. The CN is 12, since each atom touches six atoms in its own layer, three atoms in the layer above, and three atoms in the layer below. One finds that this structure has a (primitive) hexagonal space lattice; the basis consists of two atoms associated with each lattice point (Fig. 24.11). The second atom of the basis lies at the point $\frac{2}{3}\frac{1}{3}\frac{1}{2}$, which is not a lattice point. The unit cell is drawn with heavy lines in Fig. 24.11. This structure fills 74 percent of the volume (the same as the fcc structure), and is called hexagonal close-packed (hcp). Many metals have hcp structures, including Be, Mg, Cd, Co, Zn, Ti, and Tl.

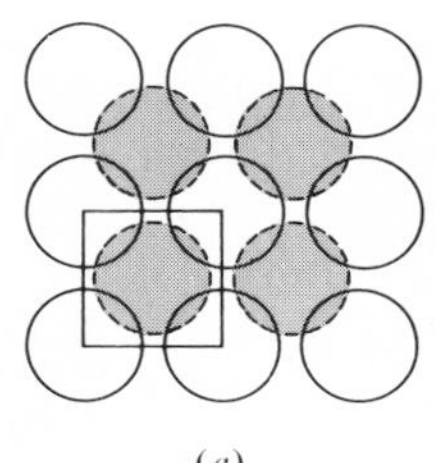

(*a*)

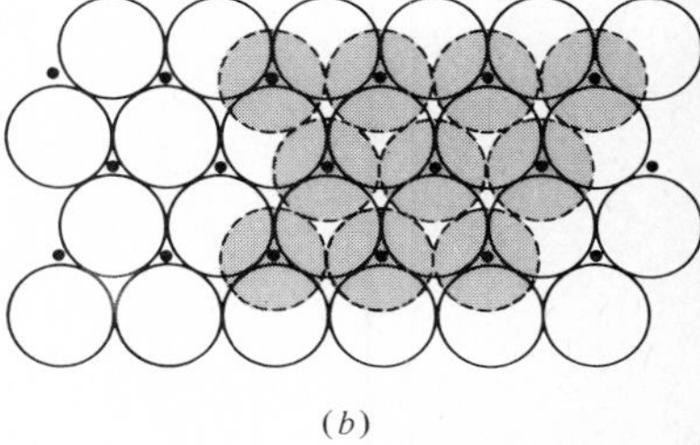

(*b*)

Figure 24.10
(*a*) Formation of a bcc lattice. (*b*) Formation of an hcp structure.

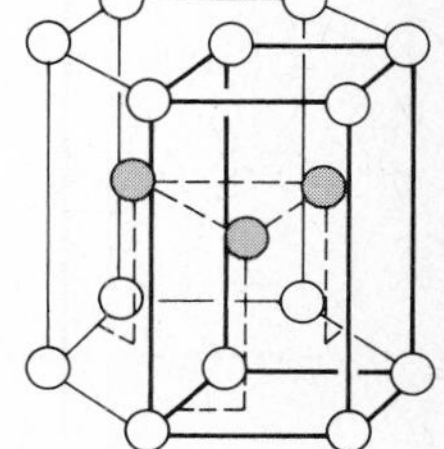

Figure 24.11
Atom positions in the hcp structure. The unit cell is rectangular and is indicated by heavy lines. With the origin at the atom at the center of the hexagonal base, the shaded atoms lie at $\frac{2}{3}\frac{1}{3}\frac{1}{2}$.

Some metals undergo changes in structure as the temperature and pressure change. For example, Fe(c) at 1 atm is ccp between 906 and 1401°C but is bcc both above and below this range. $\Delta\bar{H}$ values for the transitions between different forms of Fe(c) are small.

In summary, most metal structures are hcp (CN 12), fcc (CN 12), or bcc (CN 8).

Ionic crystals. Many M^+X^- ionic crystals have the NaCl-type structure. This is a face-centered cubic space lattice with a basis of one M^+ ion and one X^- ion associated with each lattice point. Figure 24.12a shows the NaCl unit cell. A Cl^- ion has been placed at each lattice point (the unit-cell corners and face centers), and an Na^+ ion has been placed a distance $\frac{1}{2}a$ directly above each Cl^- (where a is the cubic unit-cell edge length). The Na^+ ions do not lie at lattice points. (Of course, the lattice points could have been chosen to be at the Na^+ ions, in which case the Cl^- ions would not be at lattice points.) The four Na^+ ions on the bottom face of the unit cell are associated with lattice points in the unit cell below.

There are 8 Cl^- ions at the corners and 6 on the faces, so each unit cell has $8/8 + 6/2 = 4$ Cl^- ions. There are 12 Na^+ ions on unit-cell edges, and these are shared with three other unit cells; there is 1 Na^+ at the unit-cell center. Hence there are $12/4 + 1 = 4$ Na^+ ions per unit cell. The number of formula units per unit cell is $Z = 4$. The CN is 6.

The ions in Fig. 24.12a have been shrunk to allow the structure to be seen clearly. Figure 24.12b shows the actual arrangement of ions in a plane through the center of the unit cell. Note that oppositely charged ions are in contact. The center-to-center distance of nearest-neighbor oppositely charged ions is $\frac{1}{2}a$ [$\equiv R$ in Eq. (24.2)]; the centers of two nearest like-charged ions are $\frac{1}{2}a\sqrt{2}$ apart.

Compounds having the NaCl-type structure include many Group IA halides, hydrides, and cyanides, (for example, LiH, KF, KH, KCN, LiCl, NaBr, NaCN), many Group IIA oxides and sulfides (for example, MgO, CaO, MgS), and certain nitrides and phosphides (for example, CeN, CeP). Further examples are AgBr, MnO, PbS, and FeO.

About 15 ionic compounds have the CsCl structure, whose unit cell is shown in Fig. 24.13. The CsCl space lattice is not body-centered cubic, but is simple cubic. The basis consists of one Cs^+ ion and one Cl^- ion associated with each lattice point. The Cl^- ion at the center of the unit cell does not lie at a lattice point. The Cl^- ions

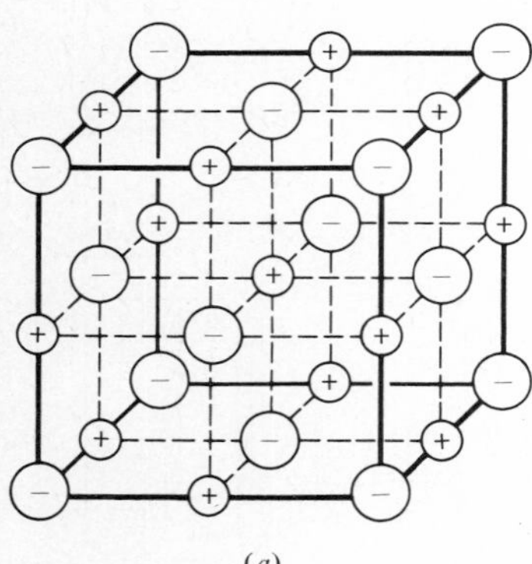

(a)

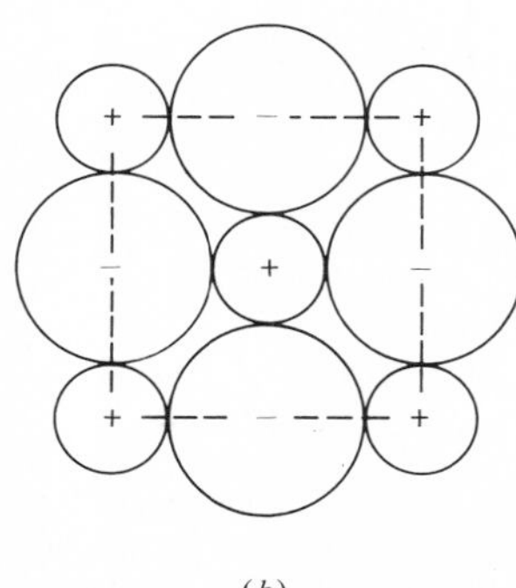

(b)

Figure 24.12
(a) The NaCl unit cell is outlined by the solid lines. (b) Packing of ions in NaCl.

associated with the other seven lattice points in Fig. 24.13 lie in adjacent unit cells. Obviously, $Z = 1$. The CN is 8. Compounds having this structure include CsCl, CsBr, CsI, TlCl, TlBr, TlI, and NH_4Cl. [Certain alloys have the CsCl structure, e.g., CuZn (β-brass) and AgZn.] (Recall the order–disorder transitions in NH_4Cl and β-brass; Sec. 7.5.)

Figure 24.14 shows the CaF_2 (fluorite) structure. The CaF_2 space lattice is face-centered cubic. The basis consists of one Ca^{2+} ion and two F^- ions, the Ca^{2+} ions lying at the lattice points. For the Ca^{2+} ion at point 000, the associated F^- ions lie at $\frac{1}{4}\frac{1}{4}\frac{1}{4}$ and $-\frac{1}{4}\,-\frac{1}{4}\,-\frac{1}{4}$. The CN of each Ca^{2+} is 8, and that of each F^- is 4. Figure 24.14*b* shows a projection of the unit-cell atom positions on the unit-cell base. The numbers in the circles give the *c* coordinate, the height above the base plane; when two numbers appear in a circle, there are two atoms directly above that position. Crystals with the CaF_2 structure include CaF_2, BaF_2, K_2O, UO_2, and Na_2S.

Several other structures occur for ionic crystals, but discussion is omitted.

Covalent crystals. The diamond space lattice is face-centered cubic; the basis consists of two C atoms. One atom of the basis occupies a lattice point, and the second atom is displaced by one-fourth the unit-cell edge in each direction. For example, there are atoms at 000 and at $\frac{1}{4}\frac{1}{4}\frac{1}{4}$ (see Fig. 24.15). Each of the two atoms of the basis is surrounded tetrahedrally by four carbons, but the bonds at each basis atom have different spatial orientations. The CN is 4. The unit cell contains $8/8 + 6/2 + 4 = 8$ carbon atoms, so $Z = 8$. Si and Ge have the diamond structure.

The ZnS space lattice is face-centered cubic, and the basis consists of one Zn atom and one S atom. The ZnS crystal structure is obtained from the diamond structure by replacing the two C atoms of each basis with one Zn atom and one S

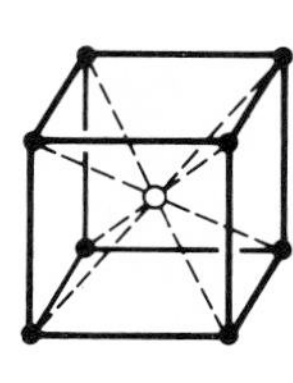

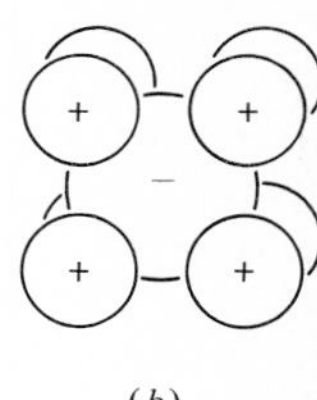

Figure 24.13
(*a*) The CsCl unit cell. (*b*) Packing of ions in CsCl.

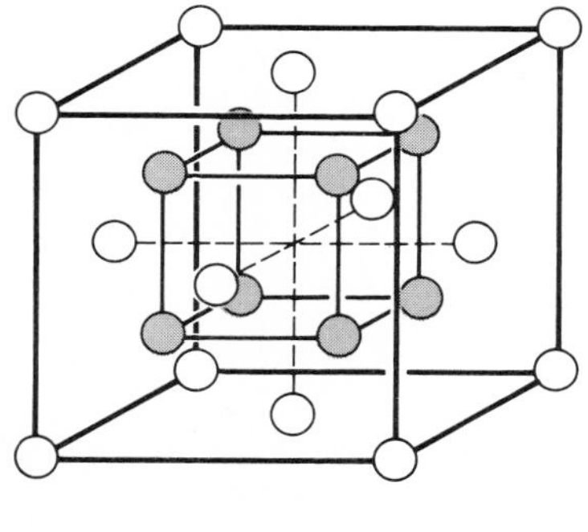

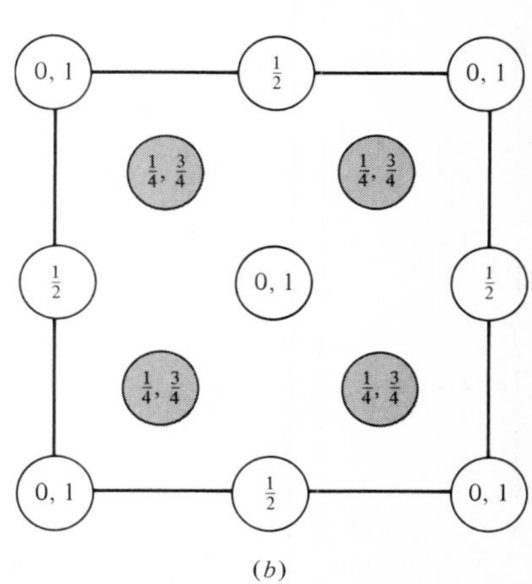

Figure 24.14
(*a*) The unit cell of CaF_2. The F atoms are shaded. (*b*) Projection of this unit cell on its base.

atom. The unit cell has four Zn atoms and four S atoms, so $Z = 4$. (Some people classify ZnS as an ionic, rather than covalent, crystal.) Crystals with the ZnS (zinc blende) structure include the following largely covalent crystals: SiC, AlP, CuCl, AgI, GaAs, ZnS, ZnSe, and CdS.

Molecular crystals. Molecular crystals show a great variety of structures, and we shall consider only a few examples.

Crystals of essentially spherical molecules often have structures determined by the packing of spheres. For example, Ne, Ar, Kr, and Xe have fcc unit cells with a basis of one atom at each lattice point. He crystallizes in the hcp structure. CH_4 crystallizes in an fcc lattice with one CH_4 molecule per lattice point.

The CO_2 space lattice is simple cubic, with a basis of four CO_2 molecules. The molecules of one basis have their centers at the corner point 0 0 0 and at the centers of the three faces that meet at this corner; each molecule is oriented with its axis parallel to one of the four cube diagonals. See Fig. 24.16*a*.

The Br_2 crystal has an end-centered orthorhombic space lattice with a basis of two Br_2 molecules associated with each lattice point; see Fig. 24.16*b*.

Crystal-structure data. Some compilations of crystal-structure data are: R. W. G. Wyckoff, *Crystal Structures*, 2d ed., vols. 1–6, Wiley-Interscience, 1963–1971; Landolt–Börnstein Tables, 6th ed., vol. 1, pt. 4, 1955 and New Series, Group III, vols. 5–7, 1971–. Detailed descriptions of many inorganic structures are given in H. D. Megaw, *Crystal Structures*, Saunders, 1973.

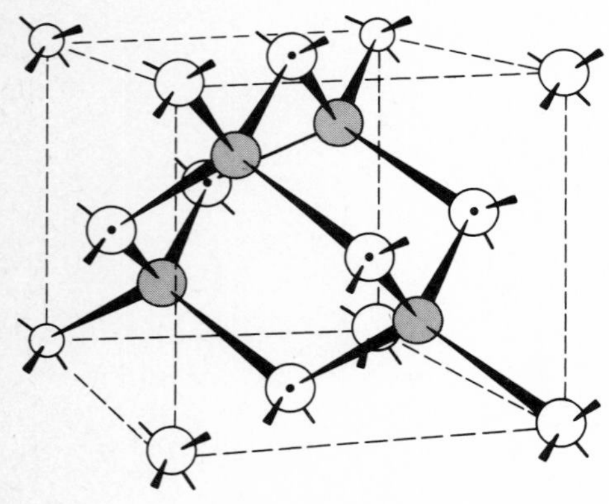

Figure 24.15
The unit cell of diamond. Shaded atoms lie within the unit cell and have c coordinate of $\frac{1}{4}$ or $\frac{3}{4}$. Dotted atoms lie on unit-cell faces.

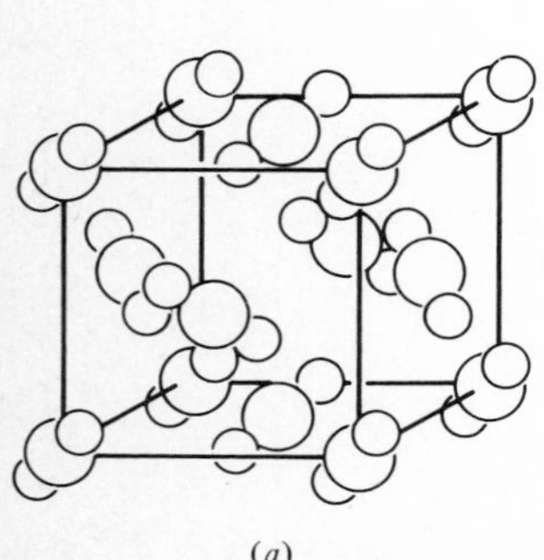

(*a*)

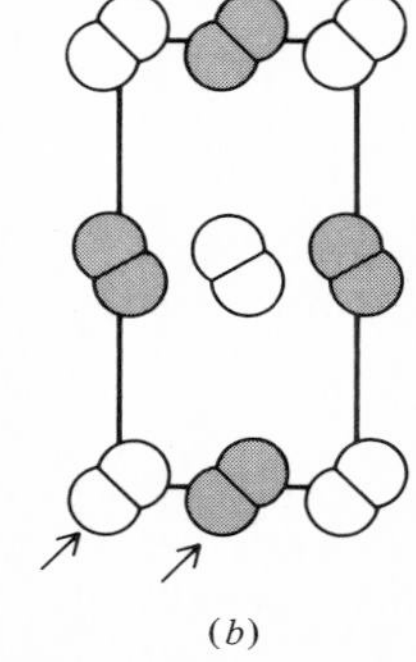

(*b*)

Figure 24.16
(*a*) The unit cell of CO_2. Each carbon atom lies at a corner or on a face of the unit cell. (*b*) The structure of $Br_2(c)$ and of $I_2(c)$. Unshaded molecules lie on the base of the unit cell. Shaded molecules have c coordinate of $\frac{1}{2}$. The top face of the unit cell contains molecules in the same positions and orientations as on the base. The arrows point to two molecules that comprise the basis. The atoms have been shrunk for clarity.

X-ray diffraction. Interatomic spacings in crystals are of the order of 1 Å. Electromagnetic radiation of 1 Å wavelength lies in the x-ray region. Hence, crystals act as diffraction gratings for x-rays. This was first realized by von Laue in 1912, and forms the basis for the determination of crystal structures.

Figure 24.17*a* shows a cross section of a primitive cubic lattice. One of the unit cells is outlined. Dashed lines have been drawn through a particular set of equally spaced parallel planes, the (210) planes. Let a beam of monochromatic x-rays of wavelength λ fall on the crystal. Although a given lattice point will in general be associated with a group of atoms, let us temporarily assume that each lattice point is associated with a single atom. Figure 24.17*b* shows the x-ray beam incident at angle θ to one of the (210) planes. Most of the x-ray photons will pass through this plane with no change in direction, but a small fraction will collide with electrons in the atoms of this plane and will be scattered, i.e., will undergo a change in direction. The x-ray photons are scattered in all directions.

Figure 24.17*b* shows waves scattered at angle β by two adjacent lattice points p and q. The observation point o for the scattered wave is essentially at infinity, so the path difference for waves scattered from p and q at angle β is $\overline{op} - (\overline{oq} + \overline{qr}) = (\overline{os} + \overline{sp}) - (\overline{oq} + \overline{qr}) = \overline{sp} - \overline{qr} = \overline{pq} \cos \beta - \overline{pq} \cos \theta = \overline{pq}(\cos \beta - \cos \theta)$. If this path difference is zero, scattered waves leaving points p and q at angle β will be in phase with each other and will give constructive interference; similarly, scattered waves leaving q and t at angle β will be in phase; etc. The zero-path-difference condition gives $0 = \overline{pq}(\cos \beta - \cos \theta)$, so $\beta = \theta$. Thus, waves scattered from a plane of lattice points at an angle equal to the angle of incidence are in phase with one another. Waves scattered at other angles will generally be out of phase with one another and will give destructive interference. The single plane of lattice points acts as a "mirror" and "reflects" a small fraction of the incident x-rays. (It is true that if the path difference between scattered waves leaving p and q equals λ, or 2λ, etc., these waves are also in phase. However, it turns out that such waves can be regarded as being "reflected" from a set of differently oriented planes in Fig. 24.17*a*, so consideration of these waves will not add anything new. See C. Kittel, *Introduction to Solid State Physics*, 2d ed., Wiley, 1956, pp. 46–48.)

So far we have considered only waves scattered by one of the (210) planes. The x-ray beam will penetrate the crystal to a depth of millions of layers of planes, so we

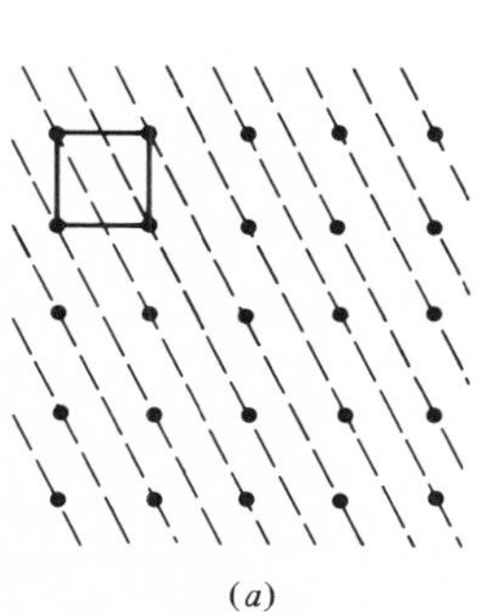

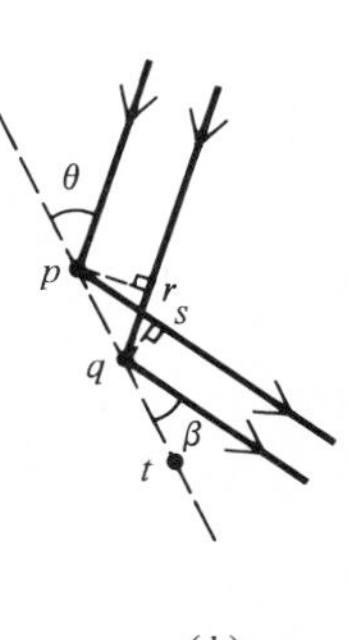

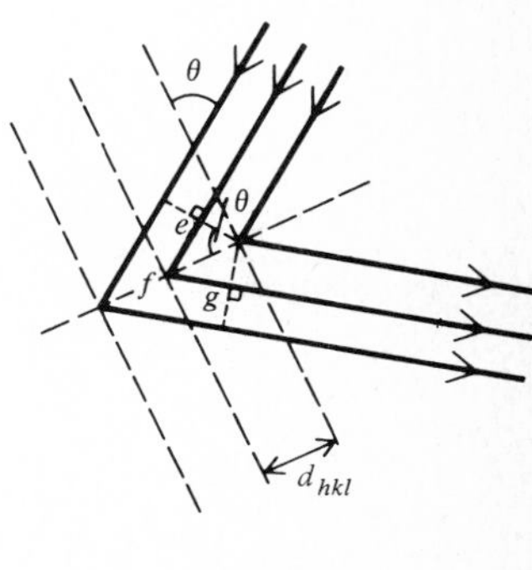

Figure 24.17
Derivation of the Bragg equation. For clarity, the scale in (*c*) differs from that in (*a*) and (*b*).

must consider the scattering from many, many (210) planes. Figure 24.17*c* shows the x-ray beam incident on three successive (210) planes. At each plane there is some scattering with constructive interference at angle θ. For constructive interference between x-rays scattered by the entire set of (210) planes, the waves constructively scattered ("reflected") by two adjacent planes must have a path difference of an integral number of x-ray wavelengths. The path difference between adjacent layers is $\overline{ef} + \overline{fg} = 2\overline{ef} = 2d_{hkl} \sin \theta$, where d_{hkl} is the spacing between adjacent planes. (In this case, *hkl* is 210.) Therefore,

$$2d_{hkl} \sin \theta = n\lambda, \qquad n = 1, 2, 3, \ldots \tag{24.13}$$

The *Bragg equation* (24.13) is the fundamental equation of x-ray crystallography. Constructive interference between waves scattered by the lattice points produces a diffracted beam of x-rays only for those angles of incidence that satisfy (24.13).

For a set of planes to give a diffracted beam of sufficient intensity to be observed, each plane of the set must have a high density of electrons; this requires a high density of atoms, so each plane must have a high density of lattice points. Because the number of sets of such planes is limited, the number of values of d_{hkl} is limited and the Bragg condition (24.13) will be met only for certain values of θ.

X-ray crystallographers prefer to write (24.13) in the form

$$2d_{nh,nk,nl} \sin \theta = \lambda \tag{24.14}$$

For example, for a primitive cubic lattice, the $n = 2$ diffracted beam from the (100) set of planes is considered to be the $n = 1$ diffracted beam from the (200) planes, whose spacing d_{200} is half that of the (100) planes. Similarly, the $n = 3$ diffracted beam for the (100) planes is considered to be the $n = 1$ diffracted beam for the (300) planes.

Example For x-rays with $\lambda = 3.00$ Å, what angles of incidence produce a diffracted beam from the (100) planes in a simple cubic lattice with $a = 5.00$ Å?

The (100) planes are spaced by a, so $\sin \theta_{100} = n\lambda/2d_{100} = n(3\text{ Å})/2(5\text{ Å}) = 0.300n = 0.300, 0.600, 0.900$. We find $\theta_{100} = 17.5°, 36.9°$, and $64.2°$. These are the 100, 200, and 300 reflections.

To apply the Bragg equation, the x-ray wavelength must be accurately known. One way to measure λ is to calculate the unit-cell dimension a for a cubic crystal from its measured density using Eq. (24.12); one then diffracts the x-rays from this crystal and uses the Bragg equation (24.13) to find λ.

For some lattices, certain reflections that satisfy the Bragg equation do not give diffracted beams. For example, consider a face-centered lattice, not necessarily cubic (Fig. 24.18). For $n = 1$ and $hkl = 100$ in Eq. (24.13), the x-rays reflected from adjacent (100) planes have a path difference of one wavelength. Hence the path difference between x-rays reflected from the (200) plane labeled v in Fig. 24.18 and x-rays reflected from either of the (100) planes labeled t and w is one-half wavelength. These waves are out of phase and cancel. The cancellation is exact, because the number of lattice points in plane v of the unit cell is the same as the number of lattice points in planes t and w of the unit cell. (There are $4/2 = 2$ lattice points in v,

Figure 24.18
Unit cell of a face-centered lattice.

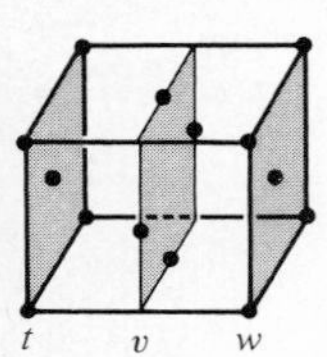

$4/8 + 1/2 = 1$ lattice point in t, and 1 lattice point in w.) For $n = 2$ in (24.13), the path difference between x-rays reflected from t and w is 2λ and between x-rays reflected from t and v is λ; hence, no cancellation occurs. Putting $n = 3, 4, \ldots$, we see that in terms of the equation (24.14), a face-centered lattice gives 200, 400, 600, ... reflections, but not 100, 300, 500, ... reflections. These missing reflections are called *extinctions* or *systematic absences.*

A more general treatment (*Buerger*, chap. 5) shows that: for a primitive lattice there are no extinctions; for a face-centered lattice, the only reflections that occur are those whose three indices in (24.14) are either all even numbers or all odd numbers; for a body-centered lattice, the only reflections that occur are those for which the sum of the indices is an even number. (The condition for end-centered lattices is omitted.)

In an x-ray crystal-structure determination, one mounts a small single crystal on a glass fiber. A monochromatic x-ray beam impinges on the crystal. The crystal is slowly rotated, thereby bringing different sets of planes into proper orientation for the Bragg equation to be satisfied. The diffracted beams are recorded on photographic film to give a pattern of spots. Alternatively, an ionization counter or a scintillation counter is used to measure the intensities of the diffracted beams. (In a scintillation counter, the x-rays fall on a material that fluoresces when illuminated with x-rays; a photoelectric cell measures the intensity of the fluorescent light.)

So far we have assumed one scatterer to be present at each lattice point. In reality, most crystals have several atoms associated with each lattice point. For simplicity, consider the crystal structure of Fig. 24.3 with two atoms, M and W, associated with each lattice point. The set of M atoms forms an array exactly like the array of lattice points, so x-ray diffraction from the M atoms occurs for those angles that satisfy the Bragg equation (24.13). Likewise, the set of W atoms forms an array exactly like the array of lattice points, so x-ray diffraction from the W atoms occurs for those angles that satisfy the Bragg equation. However, the x-rays scattered by the W atoms are in general somewhat out of phase with those scattered by the M atoms. This diminishes the intensities of the observed diffraction spots on the film. The degree to which M-scattered and W-scattered rays are out of phase depends on the M-W interatomic distance and also on which set of planes (hkl) is doing the scattering. The intensity of a given spot on the diffraction pattern therefore depends on the M-W distance; this intensity also depends on the nature of M and W, since different atoms have different x-ray scattering powers.

The same argument holds when there are more than two atoms per lattice point. We conclude that the locations of the spots in the diffraction pattern depend only on the nature of the space lattice and the lengths and angles of the unit cell and that the intensities of the spots depend on the kinds of atoms and the locations of atoms (interatomic distances) within the basis of the crystal structure.

Actually, we still have oversimplified things. Atoms are not points. Instead each atom contains a spatial distribution of electron probability density. Since it is the electrons that scatter the x-rays, the scattered-beam intensities depend on how the electron probability density varies from point to point in the basis. Analysis of the locations of the diffraction spots gives the space lattice and the lengths and angles of the unit cell. The unit-cell content Z is calculated from (24.12) using the known crystal density. Analysis of the intensities of the diffraction spots gives the electron probability density $\rho(x, y, z)$ as a function of position in the unit cell. From contours of ρ, the locations of the nuclei are obvious (see, for example, Fig. 24.19), and bond lengths and angles can be found. Bond lengths determined by x-ray diffraction are generally less accurate than those found by spectroscopic methods.

Figure 24.19 shows a cross section of contours of constant ρ for benzene(c) as determined by x-ray-diffraction data.

It should be emphasized that determination of ρ from the x-ray diffraction data is not in general a straightforward procedure. ρ could be directly determined if the relative phases of the diffracted beams were known, but measurement of intensities does not give these phases. To determine a structure, one often has to prepare derivatives with some atoms substituted by heavy atoms and examine the x-ray diffraction patterns of the derivatives. The difficulty of structure determination increases as the size of the unit cell increases.

For a crystal of a large molecule, obtaining an electron-probability-density map from the x-ray-diffraction pattern is an extremely complicated problem and requires a huge number (millions or billions) of individual calculations. Nowadays, one uses an automated computer-controlled x-ray diffractometer to obtain the x-ray-diffraction positions and intensities, and the calculations to find the molecular structure are done on a computer. (The Molecular Structure Corporation, College Station, Tex., will use x-ray diffraction to determine the structure of a crystal submitted to it; fees depend on the size of the molecule.)

The spectroscopic methods of structural determination discussed in Chap. 21 are inapplicable to large molecules. X-ray diffraction is therefore an invaluable tool in structural determinations of large molecules, including biologically important molecules such as proteins (which typically contain 10^3 nonhydrogen atoms per molecule). The three-dimensional structure of a protein whose amino acid sequence is known (from chemical analysis of fragments produced by partial hydrolysis) can frequently be determined by x-ray diffraction of a crystal of the protein. Protein crystals usually contain a substantial amount of solvent, so the environment of the protein molecule in a crystal is similar to its environment in solution, and the three-dimensional structures in the crystal and in solution are likely to be quite similar. For a tabulation of results, see B. W. Matthews, *Ann. Rev. Phys. Chem.*, **27**, 493 (1976).

X-ray diffraction has been used to study the crystal structure of many synthetic polymers. Crystalline polyethylene has a primitive orthorhombic space lattice with room-temperature unit-cell dimensions 7.39, 4.93, and 2.54 Å. A given polyethylene

Figure 24.19
Cross section of contours of constant electron-probability density in C_6H_6.

molecule in the crystal passes through many unit cells. Each unit cell contains four CH_2 groups. The basis consists of two bonded CH_2 groups at one corner of the cell plus two bonded CH_2 groups in the center of the cell. Synthetic-polymer crystal structures are tabulated and discussed in B. Wunderlich, *Macromolecular Physics*, vol. 1, Academic Press, 1973.

In the 1950s, it was found that single crystals of many polymers could be grown by cooling a solution of the polymer. These crystals are plates about 100 Å thick. The polymer chains are oriented perpendicular to the top and bottom faces of the plate and fold over at the top and bottom faces.

For a cubic crystal, one can determine the unit-cell edge length using a powdered (microcrystalline) sample. The tiny crystals in the sample have random orientations and for any given set of planes [e.g., the (220) planes], some of the tiny crystals will be oriented at the value of θ that satisfies the Bragg equation, (24.14). Hence we will get a Bragg reflection from each set of planes (except those where extinctions occur). The x-rays reflected from a plane make an angle 2θ with the direction of the incident beam (Fig. 24.20), so the diffraction angles are readily measured.

Figure 24.20

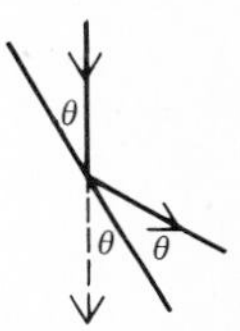

A little geometry (we omit the derivation) shows that for a cubic crystal the perpendicular distance between adjacent planes with indices (hkl) is $d_{hkl} = a/(h^2 + k^2 + l^2)^{1/2}$, where a is the unit-cell edge length. Relabeling the indices nh, nk, nl in (24.14) as hkl, this equation becomes

$$\sin\theta = (\lambda/2a)(h^2 + k^2 + l^2)^{1/2}$$

As the indices hkl of a reflection increase, θ increases.

The extinction conditions stated earlier in this section show that the first several reflections observed for primitive, face-centered, and body-centered cubic lattices are:

P 100, 110, 111, 200, 210, 211, 220, 221 and 300, 310, 311, ...

F 111, 200, 220, 311, 222, 400, 331, 420, 422, 333 and 511, ...

I 110, 200, 211, 220, 310, 222, 321, 400, 411 and 330, 420, ...

where the reflections are listed in order of increasing $h^2 + k^2 + l^2$ and hence in order of increasing θ. The ratio of two $\sin^2\theta$ values is $\sin^2\theta_1/\sin^2\theta_2 = (h_1{}^2 + k_1{}^2 + l_1{}^2)/(h_2{}^2 + k_2{}^2 + l_2{}^2)$. Comparison of the observed ratios with those expected for P, F, and I cubic lattices allows the lattice type and the hkl values of the reflections to be found; hence a can be found if λ is known.

For example, suppose that for a crystal known to be cubic (by examination of its macroscopic appearance) the first several powder diffraction angles are 17.66°, 25.40°, 31.70°, 37.35°, 42.71°, 47.98°, and 59.08°. Calculation of $\sin^2\theta$ gives 0.0920, 0.1840, 0.2761, 0.3681, 0.4601, 0.5519, and 0.7360. These numbers have the ratios 1:2:3:4:5:6:8. For P, F, and I cubic lattices, the expected $\sin^2\theta$ ratios for the first several diffraction angles are found from the above-listed reflections to be:

P $1:2:3:4:5:6:8$

F $1:1\frac{1}{3}:2\frac{2}{3}:3\frac{2}{3}:4:5\frac{1}{3}:6\frac{1}{3}$

I $1:2:3:4:5:6:7$

Hence the lattice is primitive cubic.

Except for crystals with very simple structures, the powder method is completely unsuitable for structure determination and one must use a single-crystal sample.

Neutron diffraction. A hydrogen atom contains but one electron and is therefore a very weak scatterer of x-rays. Consequently, it is very difficult to locate H-atom positions accurately by x-ray diffraction.

As noted in Chap. 18, microscopic particles like electrons, protons, and neutrons have wavelike properties and give diffraction effects. The wavelength of a neutron is given by the de Broglie relation $\lambda = h/mv$. Neutrons are scattered by atomic nuclei (not by electrons), and the neutron-scattering power of the H nucleus is comparable to that of other nuclei. This allows H atoms to be accurately located by neutron diffraction of crystals. Neutron diffraction is also being used to study structures of biological membranes; see B. P. Schoenborn, *Chem. Eng. News*., Jan. 24, 1977, p. 31.

Electron diffraction. Electrons give diffraction effects when reflected from solids or when passed through very thin sections of solids (Sec. 18.4). Because of the low penetrating power of electrons, electron diffraction of solids has been used mainly to study surface structures. A one-molecule-thick layer adsorbed on a solid can be viewed as a two-dimensional "crystal." The structure of this adsorbed layer can often be determined by electron diffraction. Since low-energy electrons are used, the technique is called *low-energy electron diffraction* (LEED). As an example, LEED shows that carbon adsorbed on Pt adopts a two-dimensional structure of graphitelike fused hexagonal rings.

Electron diffraction by gases is a valuable method in determining molecular structures. Although the gas molecules have random orientations, the orientations of atoms with respect to one another in a given molecule are fixed, so one can extract structural information (bond distances and angles) from gas-phase electron diffraction. The method is not applicable to large molecules. For details, see *Brand and Speakman*, chap. 10.

Holography. The physicist George Stroke is developing a new method for determining crystal structures. The observed x-ray diffraction spots are processed by a computer and converted into a hologram, which is printed on a photographic plate. By passing a laser beam through the hologram plate and then through a lens, one obtains an image that shows the atomic arrangement in a given crystal plane. Combining images from several planes, one obtains the three-dimensional crystal structure. The method has been successfully applied to one plane of magnesium bromide tetrahydrofuran but needs further development to be generally applicable. See *Time*, Feb. 14, 1977, p. 81.

24.9 BAND THEORY OF SOLIDS

A crystalline solid can be regarded as a single giant molecule. An approximate electronic wave function for this giant molecule can be constructed using the MO approach. The electrons are fed into "crystal" orbitals that extend over the entire crystal; each crystal orbital holds two electrons of opposite spin. Just as the MOs of a gas-phase molecule can be approximated as linear combinations of the orbitals of the atoms composing the molecule, the crystal orbitals can be approximated as linear combinations of orbitals of the species (atoms, ions, or molecules) composing the solid.

Consider Na(c) as an example. The electron configuration of an Na atom is $1s^2 2s^2 2p^6 3s$. Let the crystal contain N atoms, where N is of the order of 10^{23}. We construct crystal orbitals as linear combinations of the AOs of the N sodium atoms. Recall that when two Na atoms are brought together to form an Na_2 molecule, the two $1s$ AOs combine to give two MOs. In Na_2 the overlap between these inner-shell AOs is negligible, so their energies are nearly identical to the $1s$ AO energy of an isolated Na atom. In the Na(c) solid, which is being regarded as a single Na_N molecule, the N $1s$ AOs of the isolated atoms form N crystal orbitals with energies nearly identical to that of a sodium $1s$ AO. These N $1s$ crystal orbitals have a capacity of $2N$ electrons and hold the $2N$ $1s$ electrons of the atoms. Similarly, the inner-shell $2s$ and $2p$ AOs form crystal orbitals with energies little changed from the AO energies, and these orbitals are filled in Na_N.

The N 3s sodium AOs form N delocalized crystal orbitals. The energy difference between the lowest and highest of these N crystal orbitals is of the same order of magnitude (a couple of eV or so) as the separation between the $\sigma_g 3s$ and $\sigma_g^* 3s$ MOs in Na_2. Because we have $N \approx 10^{23}$ crystal orbitals in an energy range of only a few eV, the energy spacing between adjacent crystal orbitals is extremely small, and for all practical purposes we can consider the crystal orbitals originating from the 3s AOs to form a continuous *band* of energy levels. This 3s band contains N orbitals and so has a capacity of 2N electrons. Since there are N 3s electrons in the N sodium atoms, the 3s band is only half filled (Fig. 24.21a). Hence, there are many vacant electronic energy levels in Na(c) that lie a negligible distance above the highest occupied electronic level. This makes Na(c) an excellent conductor of electricity, since electrons are easily excited to higher energy levels by an applied electric field. (Motion of electrons through the metal involves an increase in electronic energy.)

Direct evidence for the 3s band in Na(c) is obtained from the x-ray emission spectrum (Sec. 21.10). In Na(c) and in Na(g) one finds the $2p \rightarrow 1s$ x-ray line to be narrow, indicating that the 1s and 2p inner-shell bands in Na(c) are very narrow. In Na(g), the $3s \rightarrow 2p$ line is narrow. In Na(c), the $3s \rightarrow 2p$ line is centered at 407 Å and is broad (about 30 Å wide) due to the range in energies in the 3s band. These data give an energy range of 2.3 eV for the occupied part of the 3s band in Na(c) (Prob. 24.23).

An Mg atom has the electron configuration $1s^2 2s^2 2p^6 3s^2$. The 3s band in Mg(c) is exactly filled by the 2N electrons of the N magnesium atoms. One finds that the 3p band (formed by unoccupied 3p AOs of the Mg atoms) overlaps the 3s band in Mg(c); see Fig. 24.21b. There is no energy gap between the highest occupied and lowest vacant electronic energy levels, and Mg is an excellent conductor of electricity.

Consider Ne(c). The Ne electron configuration is $1s^2 2s^2 2p^6$. The 1s and 2s electrons give bands of very narrow widths. The interaction between the 2p AOs is not great (van der Waals forces are rather weak), and the 2p band is fairly narrow. Above the 2p band lies a fairly narrow 3s band (Fig. 24.21c). Because the 2p and 3s bands are fairly narrow, and because the 2p and 3s levels in an Ne atom are very widely separated, we get a substantial energy gap (several eV) between the top of the filled 2p band in Ne(c) and the bottom of the vacant 3s band. It takes a great deal of energy to excite electrons from the 2p band to the 3s band, and Ne(c) is therefore a nonconductor of electricity, i.e., an insulator.

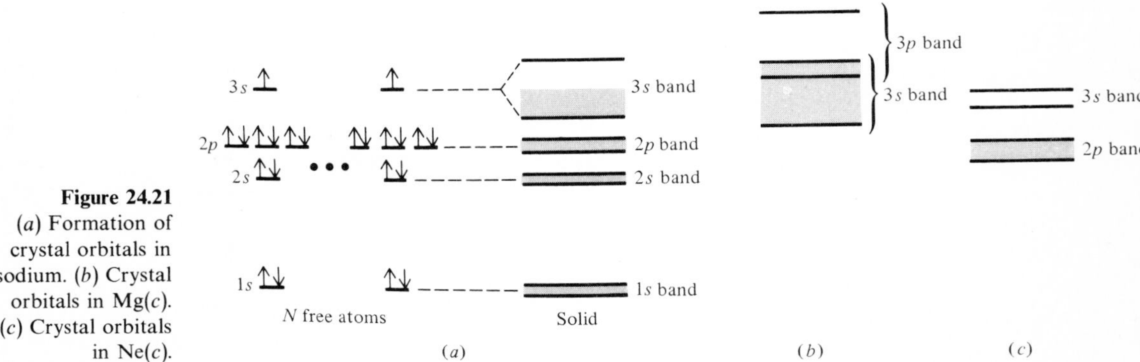

Figure 24.21 (a) Formation of crystal orbitals in sodium. (b) Crystal orbitals in Mg(c). (c) Crystal orbitals in Ne(c).

In an ionic crystal like NaCl, the bands are formed from the energy levels of the ions. The ions have rare-gas electron configurations, and arguments similar to those for Ne(*c*) show ionic crystals to be insulators. This conclusion is correct as far as conduction by electrons (electronic conduction) is concerned. However, at elevated temperatures, the ions can move fairly readily through an ionic crystal, and the crystal is an ionic conductor.

For a molecular crystal, we can construct crystal orbitals from the molecular orbitals of the molecules forming the crystal. Due to the weakness of the van der Waals interactions, the bands are rather narrow. The highest occupied band is filled, and there is a wide gap between this band and the lowest vacant band. As in Ne(*c*), the crystal is an insulator.

Consider diamond. The C atomic electron configuration is $1s^2 2s^2 2p^2$, and one might think that the band structure would involve overlapping 2*s* and 2*p* bands that have a total capacity of 8*N* electrons and are half filled with 4*N* electrons. However, a detailed quantum-mechanical calculation (see *Pitzer*, sec. 10*d*) shows that in diamond the 2*s* and 2*p* AOs combine to form two separate 2*s*-2*p* bands; the lower band holds 4*N* electrons and is completely filled; the upper band holds 4*N* electrons and is vacant. The separation between the two bands is several eV, so diamond is an insulator.

The energy gap E_g between the highest occupied band (called the *valence band*) and the lowest vacant band (the *conduction band*) in a solid can be found from the observed lowest frequency of visible or UV absorption by the solid. Results for the Group IVA elements are:

Element	C(diamond)	Si	Ge	Sn(gray)	Pb
E_g/eV	7	1.2	0.7	0.08	0

With zero E_g, lead is a metal. With a large E_g, diamond is an insulator. (For NaCl, E_g is 7 eV; this gap is between the filled Cl^- 3*p* band and the higher-lying, vacant Na^+ 3*s* band.)

Si and Ge have fairly small gaps and are classed as *intrinsic semiconductors*. Their electrical conductivity is substantially greater than that of insulators but far less than that of conductors. As the temperature is raised, the electrical conductivity of a semiconductor increases exponentially, due to excitation of electrons into the lowest vacant band. The electrons populate the bands according to the Boltzmann distribution law, with populations proportional to $e^{-\Delta E/kT}$. (The electrical conductivity of a metal decreases as T increases. The resistance of a metal arises from scattering of electrons by the positive ions. As T increases, the increase in thermal vibration of the ions increases the scattering.)

One can use light to excite electrons into the lowest vacant band of a semiconductor; semiconductors therefore exhibit *photoconductivity*, an increase in electrical conductivity when exposed to light.

24.10 STATISTICAL MECHANICS OF CRYSTALS

We shall regard the crystal as a single giant molecule with a set of allowed quantum-mechanical states having energies E_j. This description is a natural one for metallic, covalent, and ionic crystals. We shall also regard a molecular crystal as a single giant molecule containing weak "bonds" due to van der Waals forces and strong bonds due to ordinary chemical bonding. All the thermodynamic properties of the crystal are derivable from the crystal's canonical partition function $Z = \sum_j e^{-E_j/kT}$ [Eq. (22.16)].

With rare exceptions (e.g., solid H_2), molecules in crystals don't rotate. Diffusion in solids is extremely slow, so there is no need to consider translational energies of atoms, molecules, or ions. Thus, E_j is composed of vibrational and electronic energies. Most solids are insulators or semiconductors and have a significant energy gap between the highest occupied and lowest vacant electronic energy levels (Sec. 24.9); provided T is not extremely high, we can ignore electronic energy for these solids. For metals, excitation of electrons to higher levels occurs at all temperatures and must be allowed for. For now, we shall ignore the contribution of electronic energy to E_j of metals; this contribution will be added on at the end of the calculation. Thus we consider only vibrational energy.

Consider a covalent or metallic crystal or an ionic crystal composed of monatomic ions. (Molecular crystals and ionic crystals with polyatomic ions will be dealt with later.) Let there be N atoms or ions present. (For 1 mole of diamond, N is Avogadro's number. For 1 mole of NaCl, N is twice Avogadro's number.) The crystal considered as a giant molecule has $3N - 6$ vibrational degrees of freedom. (There are also 3 translational and 3 rotational degrees of freedom, which can be activated by playing catch with the crystal. Since energy of bulk motion is not included in the thermodynamic internal energy U, there is no need to consider these 6 degrees of freedom.) As we did with diatomic and polyatomic molecules in Chap. 21, we can expand the potential energy E_p for vibrational motions of the atoms or ions of the giant molecule in a Taylor series about the equilibrium configuration and neglect terms higher than quadratic. This gives the vibrational energy of the giant molecule as the sum of $3N - 6$ harmonic-oscillator energies, one for each normal mode of vibration [Eq. (21.48)]. Since N is something like 10^{23}, we omit the -6. In the harmonic-oscillator approximation, the quantum-mechanical vibrational energy levels of the giant molecule are

$$E_j = E_{p,\text{eq}} + \sum_{i=1}^{3N} (\upsilon_{i(j)} + \tfrac{1}{2})h\nu_i = U_0 + \sum_{i=1}^{3N} \upsilon_{i(j)} h\nu_i \tag{24.15}$$

$$U_0(V) \equiv E_{p,\text{eq}} + \tfrac{1}{2}h \sum_{i=1}^{3N} \nu_i \tag{24.16}$$

where $\upsilon_{i(j)}$ is the vibrational quantum number of the ith normal mode when the crystal is in the quantum-mechanical state j, and ν_i is the vibrational frequency of the ith normal mode. The quantity $E_{p,\text{eq}}$ is the potential energy when all the atoms or ions are at their equilibrium (minimum energy) separations. When the crystal's volume V is decreased by applying pressure, the equilibrium interatomic distances change and hence $E_{p,\text{eq}}$ changes. Also, the vibrational frequencies depend on V. Hence, U_0 depends on V. Because of the very large size of the giant molecule, we expect that many of the vibrational frequencies ν_i will be very low.

The sum in (24.15) is not over individual atoms (or ions). The atoms or ions in a solid interact strongly with one another, and the energy cannot be taken as the sum of energies of noninteracting atoms or ions. The sum in (24.15) is over normal modes.

In each normal mode, all the atoms or ions vibrate, and the energy of the ith normal mode is composed of kinetic and potential energies of all the atoms or ions.

The canonical partition function of the crystal is $Z = \sum_j e^{-E_j/kT}$. The crystal's Z will be like the vibrational partition function (22.110) for a polyatomic molecule. However, (22.110) takes the zero of energy at $U_0 = 0$, whereas in this chapter (to allow for the change in U_0 as V changes), we do not set $U_0 = 0$. Hence, we must add a factor $e^{-U_0/kT}$ to (22.110). Also, the product goes from 1 to $3N$. Thus, the crystal's canonical partition function is

$$Z = e^{-U_0/kT} \prod_{i=1}^{3N} \frac{1}{1 - e^{-h\nu_i/kT}} \tag{24.17}$$

$$\ln Z = -\frac{U_0}{kT} - \sum_{i=1}^{3N} \ln\,(1 - e^{-h\nu_i/kT}) \tag{24.18}$$

From ln Z, all thermodynamic properties are readily calculated using (22.37) to (22.41). The only problem is that calculation of the $3N$ normal vibrational frequencies of a giant molecule of 10^{23} atoms ain't easy.

The first application of quantum physics to the statistical mechanics of a material system was Einstein's 1907 treatment of solids. Einstein assumed (in effect) that every normal mode of the crystal has the same frequency ν_E (where E is for Einstein). Einstein knew this assumption was false, but he felt it would bring out the correct qualitative behavior of the thermodynamic properties. With each ν_i equal to ν_E, Eq. (24.18) becomes

$$\ln Z = -U_0(V)/kT - 3N \ln\,(1 - e^{-\Theta_E/T}) \qquad \text{where } \Theta_E \equiv h\nu_E/k \tag{24.19}$$

Use of (22.38) and $C_V = (\partial U/\partial T)_V$ gives for an Einstein crystal (Prob. 24.24)

$$C_V = 3Nk\left(\frac{\Theta_E}{T}\right)^2 \frac{e^{\Theta_E/T}}{(e^{\Theta_E/T} - 1)^2} \tag{24.20}$$

which may be compared with (22.102) for a gas of diatomic molecules.

In the high-temperature limit, $e^{\Theta_E/T} - 1 = 1 + \Theta_E/T + \cdots - 1 \approx \Theta_E/T$, so (24.20) approaches $3Nk$ as T goes to infinity. This result agrees with experiment. (Recall the classical equipartition theorem.) For diamond, 1 mole has N_0 molecules, so the high-T limit of $\bar{C}_V$ is $3N_0k = 3R \approx 6$ cal mol^{-1} K^{-1}. For NaCl, 1 mole has $2N_0$ ions, so the high-T limit of $\bar{C}_V$ is $6N_0k = 6R$.

The theory gives C_V, but the experimentally measured quantity is C_P. From (4.74), $\bar{C}_P = \bar{C}_V + T\bar{V}\alpha^2/\kappa$. Since $\alpha \to 0$ as $T \to 0$ (Prob. 5.26), the difference between $\bar{C}_P$ and $\bar{C}_V$ for a solid becomes negligible at low temperatures (below 100 K). At room temperature, this difference is significant for solids (10^{-1} to 10^0 cal mol^{-1} K^{-1}). At high temperatures, C_V stays essentially constant at $3Nk$, but C_P increases essentially linearly with T, due to the $TV\alpha^2/\kappa$ term.

In the low-temperature limit, $e^{\Theta_E/T} \gg 1$, so C_V in (24.20) becomes proportional to $T^{-2}e^{-\Theta_E/T}$; as $T \to 0$, the exponential dominates the T^{-2} factor and C_V goes to zero. Again, this agrees with experiment.

For intermediate temperatures, the Einstein C_V function increases as T increases, approaching the $3Nk$ limit at high T. The overall appearance of the Einstein function resembles Fig. 24.22 below for the Debye C_V function. Substitution of $T = \Theta_E$ in (24.20) shows that C_V has reached 92 percent of its high-T limit at Θ_E.

To apply the Einstein equation, one uses a single measured value of $\bar{C}_V$ to evaluate the Einstein vibrational frequency ν_E (and the Einstein characteristic temperature Θ_E). Equation (24.20) shows fairly good overall agreement with experiment. However, at very low T (below about 40 K), the agreement is poor. This is due to the very crude assumption that all normal-mode vibrations of the crystal have the same frequency. Θ_E values are 200 to 300 K for most elements, so most elements have nearly reached their high-T C_V values at room temperature. Some values of Θ_E are 67 K for Pb, 240 K for Al, and 1220 K for diamond.

In 1912, Debye presented a treatment that gives much better agreement with low-T C_V values than the Einstein theory. The large number of vibrational modes ($\approx 10^{24}$) makes it possible to treat the crystal's vibrational frequencies as continuously distributed over a range from a very low frequency (which can be taken as essentially zero) to a maximum frequency ν_m. Let $g(\nu)\,d\nu$ be the number of vibrational frequencies in the range between ν and $\nu + d\nu$. Changing the summation in (24.18) to a sum over vibrational frequencies (rather than over normal modes), we get $\ln Z = -U_0/kT - \sum_\nu \ln(1 - e^{-h\nu/kT})g(\nu)\,d\nu$. The sum over infinitesimal quantities is a definite integral, and

$$\ln Z = -\frac{U_0}{kT} - \int_0^{\nu_m} \ln(1 - e^{-h\nu/kT})g(\nu)\,d\nu \tag{24.21}$$

Debye assumed that for frequencies in the range 0 to ν_m the distribution function $g(\nu)$ is the same as the distribution function of the elastic vibrations of a continuous solid calculated ignoring the solid's atomic structure. This latter quantity can be shown to be (*Kestin and Dorfman*, sec. 9.3)

$$g(\nu) = 12\pi V v_s^{-3}\nu^2 \tag{24.22}$$

where v_s is essentially the average speed of sound waves in the solid. The total number of vibrational frequencies is $3N$, so $\int_0^{\nu_m} g(\nu)\,d\nu = 3N$. Substitution of (24.22) and integration gives

$$\nu_m = (3Nv_s^3/4\pi V)^{1/3} \tag{24.23}$$

Substitution for v_s^{-3} in (24.22) gives

$$g(\nu) = 9N\nu_m^{-3}\nu^2, \qquad 0 \le \nu \le \nu_m \tag{24.24}$$

For $\nu > \nu_m$, the Debye $g(\nu)$ is 0.

Substitution of (24.24) in (24.21) gives the Debye canonical partition function Z. Use of (22.38) and $C_V = (\partial U/\partial T)_V$ then gives (Prob. 24.27)

$$C_V = 9Nk\left(\frac{T}{\Theta_D}\right)^3 \int_0^{\Theta_D/T} \frac{x^4 e^x}{(e^x - 1)^2}\,dx \tag{24.25}$$

where the Debye characteristic temperature Θ_D is defined as

$$\Theta_D \equiv h\nu_m/k \tag{24.26}$$

(The quantity x in the definite integral is a dummy variable.) Expressions for other thermodynamic properties are readily found from (24.21) and (24.24).

The integral in (24.25) can't be expressed in terms of elementary functions but can be evaluated numerically by expanding the exponentials in infinite series and then integrating. Tabulations of the Debye $C_V/3Nk$ as a function of Θ_D/T are available. (See *McQuarrie*, app. C; *Pitzer*, app. 19.) Figure 24.22 plots the Debye C_V as a function of temperature.

At high T, the upper limit of the integral in (24.25) is close to zero, so the integration variable x is always small. Hence, $e^x \approx 1$, and $e^x - 1 = 1 + x + \cdots - 1 \approx x$. The integrand becomes x^2, and integration gives $C_V = 3Nk$, as found by the Einstein theory.

At low T, the upper limit in (24.25) can be taken as infinity. A table of definite integrals gives $\int_0^\infty x^4e^x/(e^x - 1)^2\,dx = 4\pi^4/15$, so the low-$T$ limit is

$$C_V = \frac{12\pi^4 Nk}{5}\left(\frac{T}{\Theta_D}\right)^3 \propto T^3 \qquad \text{low } T \tag{24.27}$$

The Debye T^3 law (24.27) agrees well with experimental data, and is used to extrapolate heat-capacity data to absolute zero to evaluate entropies (Chap. 5).

The Debye characteristic temperature Θ_D can be found from (24.26) and (24.23). To achieve better agreement with experiment, one usually evaluates Θ_D empirically, so as to give a good fit to the heat-capacity data. The Debye theory works quite well at all temperatures but is far from perfect. Thus, one finds that the value of Θ_D for a given solid found by fitting C_V at one temperature differs somewhat from the value obtained using C_V at another temperature. Some Θ_D values in kelvins are

Na	Cu	Ag	Be	Mg	diamond	Ge	Ar	NaCl	KF	MgO	SiO_2
160	340	225	1400	400	2200	360	93	320	340	950	470

Work subsequent to Debye's has allowed $g(\nu)$ to be calculated theoretically and measured experimentally for several solids. The Debye quadratic expression with a cutoff at ν_m turns out to be a rather crude approximation to the actual $g(\nu)$; see Fig. 24.23.

Consider the ionic crystal KNO_3 with polyatomic ions. The solid's vibrations can be divided into (*a*) "lattice" vibrations of the K^+ and NO_3^- ions, each NO_3^- being considered as a structureless unit, and (*b*) intraionic vibrations within each NO_3^- ion. The $3N$ lattice vibrations are nearly continuous and can be treated by the Debye theory; their contribution to C_V is given by (24.25). Each NO_3^- ion has six normal vibrational modes, and their vibrational frequencies will be roughly the same

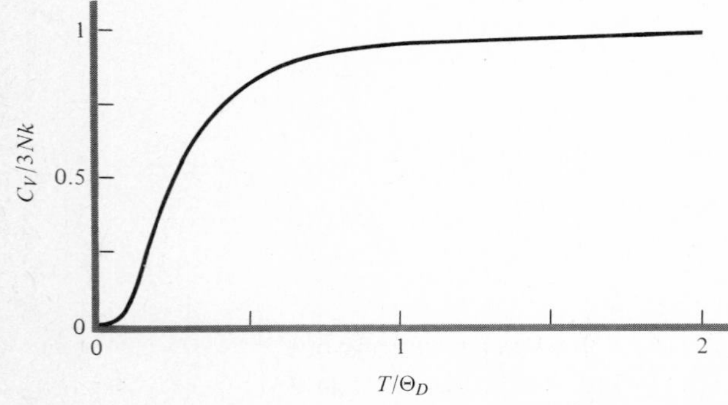

Figure 24.22
C_V of a solid according to the Debye theory.

as for an isolated NO_3^- ion. The NO_3^- frequencies are not continuous. Instead each of the six frequencies has its own characteristic temperature and gives a contribution to C_V of the Einstein form (24.20), except that N is replaced by $N/2$, the number of nitrate ions in the crystal. Since the NO_3^- frequencies are substantially higher than the lattice frequencies, at moderate temperatures the main contribution to C_V comes from the lattice vibrations. At low T, the contribution to C_V from the intraionic vibrations is entirely negligible, and the Debye T^3 law holds.

A similar treatment holds for a molecular crystal. For example, for benzene, there are $3N$ lattice vibrations (involving vibrations of the molecules considered as structureless units), whose contribution to C_V is given by the Debye expression. In addition, each C_6H_6 molecule has 30 normal vibrational modes with frequencies approximately the same as in an isolated benzene molecule, and their contributions to C_V are given by Einstein-like terms. At low T, only the lattice vibrations contribute to C_V.

For metals, excitation of electrons must be considered. An approximate treatment (*Zemansky*, sec. 11-14) shows that (provided T is not extremely high) electronic excitations contribute a term linear in T:

$$\bar{C}_{V,\text{el}} = \gamma T \tag{24.28}$$

The coefficient γ is typically 10^{-4} to 10^{-3} cal mol^{-1} K^{-2}. The room-temperature contribution of (24.28) to $\bar{C}_V$ is 0.03 to 0.3 cal mol^{-1} K^{-1}, which is far less than $3R = 6$ cal mol^{-1} K^{-1}. At moderate temperatures, the electronic contribution (24.28) is thus only a small fraction of the total C_V. At very low T (below 4 K), the electronic contribution exceeds that of the lattice vibrations, since $(T/\Theta_D)^3$ goes to zero faster than γT.

24.11 DEFECTS IN SOLIDS

So far, we have assumed crystals to be perfect. In reality, all crystals show defects. These defects affect the crystal's density, heat capacity, and entropy only slightly but profoundly alter the mechanical strength, electrical conductivity, diffusion rate, and catalytic activity. Imperfections in solids are classified as point, line, or plane defects.

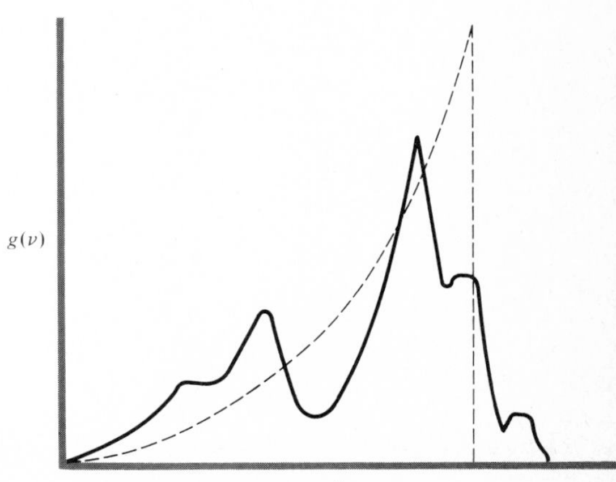

Figure 24.23
Typical vibrational distribution function in a solid. The dashed line is the Debye expression (24.22).

Point defects. A *vacancy* is the absence of an atom, ion, or molecule from a site that would be occupied in a perfect crystal. A *substitutional impurity* is an impurity atom, molecule, or ion located at a place that is occupied by some other species in a perfect crystal; an *interstitial impurity* is located at a place (void) that would be unoccupied in a perfect crystal. A *self interstitial* is a nonimpurity atom, molecule, or ion located at a void. As the crystal's temperature is increased, the number of atoms, molecules, or ions having enough vibrational energy to break away from their perfect-crystal sites increases, thereby increasing the numbers of vacancies and self interstitials.

Catalytic sites on the surfaces of metal oxides are often due to anion or cation vacancies. Diffusion in solids and ionic conduction in solid salts involve vacancies and interstitials. Semiconductors used in transistors are generally *extrinsic* (or *impurity*) semiconductors (as contrasted to intrinsic semiconductors; Sec. 24.9), in which the electrical conductivity is due mainly to defects. For example, addition of a small amount of P as a substitutional impurity to Si greatly enhances the semiconductivity of the Si; the P atoms have five valence electrons compared with four for Si, and this produces extra electronic energy levels lying only slightly below the conduction band, allowing electrons to be more easily excited into the conduction band than in pure Si.

Line defects. An *edge dislocation* is an extra plane of atoms that extends only part way through the crystal, thereby distorting its structure in the nearby planes (Fig. 24.24) and making the crystal mechanically weak. A more complicated kind of dislocation is a *screw dislocation*; see *Kittel*, chap. 18.

Plane defects. One kind of plane defect is a stacking error. For example, a hexagonal close-packed crystal might contain a few planes in which the packing is cubic close-packed.

Most crystalline solids do not consist of a single crystal but are composed of many tiny crystals held together. Neighboring crystals have random orientations, and the boundaries between the faces of neighboring crystals are plane defects.

24.12 LIQUIDS

Liquids have neither the long-range structural order of solids nor the small intermolecular interaction energy of gases and so are much harder to deal with theoretically than solids or gases.

Liquids do have a short-range order that shows up in diffraction effects when x-rays are scattered by liquids. The scattered x-ray intensity as a function of angle

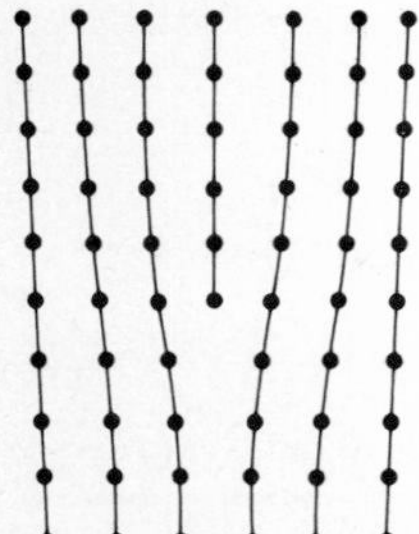

Figure 24.24
An edge dislocation.

shows broad maxima, in contrast to the sharp maxima obtained from solids. Analysis of these x-ray diffraction patterns allows one to obtain the *radial distribution function* (or *pair-correlation function*) $g(r)$ for the liquid; this function shows the variation in the average density of molecules with distance r from a given molecule. More explicitly, $g(r) = \rho(r)/\rho_{\text{bulk}}$, where ρ_{bulk} is the bulk molecular density ($\rho_{\text{bulk}} = N/V$, where N is the total number of molecules and V the total volume) and $\rho(r)$ is the local molecular density in the thin spherical shell from r to $r + dr$ around a given molecule. For nonspherical molecules ρ/ρ_{bulk} depends on the directional angles θ and ϕ from the central molecule as well as on r; one obtains a radial pair-correlation function by averaging over the angles.

Figure 24.25 shows $g(r)$ calculated theoretically for a liquid of spherical molecules with a Lennard-Jones 6-12 intermolecular potential (22.131). The maxima at $r = \sigma$, 2σ, and 3σ indicate three "shells" of molecules surrounding the central molecule. As r increases, the oscillations in $g(r)$ die out and $g(r)$ equals 1 for large r, indicating the lack of long-range order. For $r < \sigma$, $g(r)$ goes rapidly to zero, due to the intermolecular Pauli repulsion.

The complexity of the intermolecular potential and the large number of degrees of freedom make evaluation of the configuration integral (22.143) extraordinarily difficult for a liquid. Many approximate statistical-mechanical theories of liquids have been developed. For a simple liquid like Ar, some of these theories give rather accurate values of thermodynamic and transport properties, but the theoretical understanding of complex liquids is far from complete. Statistical-mechanical theories of liquids are discussed in *McQuarrie*, chap. 12; D. A. McQuarrie, *Statistical Mechanics*, Harper & Row, 1976, chaps. 13–15; M. S. Jhon and H. Eyring, *Ann. Rev. Phys. Chem.*, **27**, 45 (1976); R. O. Watts and I. J. McGee, *Liquid State Chemical Physics*, Wiley, 1976.

A valuable approach to the theoretical study of liquids is *molecular-dynamics calculations*. Here, one takes a system of several hundred molecules, confines the molecules to a box of size suitable to give a liquid density, and gives the molecules initial energies consistent with some desired temperature; one assumes some form of intermolecular potential and uses a computer to solve for the classical-mechanical motions of the molecules over a period of time. Molecular-dynamics calculations allow the pair-correlation function to be found and give information on structural

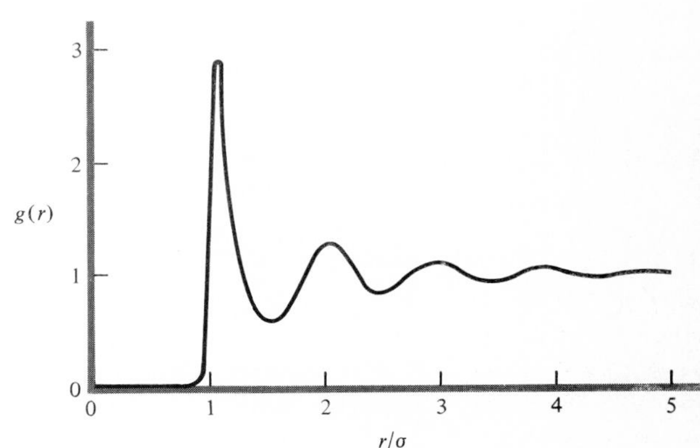

Figure 24.25
Radial distribution function for a liquid obeying the Lennard-Jones potential plotted at a typical temperature and density. [Data from the molecular-dynamics calculations of L. Verlet, *Phys. Rev.*, **165**, 201 (1968).]

details not accessible to experiment. For example, the formation and breaking of hydrogen bonds in liquid water has been studied by Rahman and Stillinger [*J. Chem. Phys.*, **55**, 3336 (1971); **57**, 1281 (1972); *J. Am. Chem. Soc.*, **95**, 7943 (1973)].

One of the most intensively studied liquids is, of course, water. The many competing theories are discussed in the Jhon and Eyring reference cited above. The state of water in biological tissues is the subject of much controversy; some workers argue that such water has, at least in part, an ordered structure resembling that of a liquid crystal.

PROBLEMS

24.1 Use thermodynamic data in the Appendix to find E_c at 25°C for (*a*) graphite; (*b*) SiC; (*c*) SiO_2.

24.2 Use (24.1) to calculate E_c of NaCl.

24.3 Use (24.2) to calculate E_{Coul} for NaCl at 0 K.

24.4 Verify Eq. (24.9).

24.5 Use (24.11) to calculate n for NaCl at 0 K.

24.6 Instead of $E_{rep} = B/R^n$, quantum-mechanical calculations indicate that the Pauli repulsion is more accurately represented by $E_{rep} = ae^{-cR}$, where a and c are constants. (*a*) Use procedures like those which gave (24.4) and (24.10) to find expressions for a and c. (*b*) Evaluate E_c for NaCl(c) using this form of E_{rep}.

24.7 Evaluate the Madelung constant for a hypothetical one-dimensional crystal consisting of alternating +1 and −1 ions with equal spacing between ions. *Hint:* Use (8.41).

24.8 Using the Lennard-Jones intermolecular potential (22.131) for the interactions in Ar(c) and summing over the interactions between all pairs of atoms, one finds (*Kittel*, chap. 3) $-E_c = N_0\varepsilon[24.264(\sigma/R)^{12} - 28.908(\sigma/R)^6]$, where R is the nearest-neighbor distance. (*a*) Use the gas-phase Ar Lennard-Jones parameters $\varepsilon/k = 118$ K (where k is Boltzmann's constant) and $\sigma = 3.50$ Å and the experimental nearest-neighbor separation $R_0 = 3.75$ Å at 0 K to calculate E_c of Ar. (The experimental E_c is 1.85 kcal/mole at 0 K.) (*b*) Show that the expression for E_c in (*a*) predicts that $R_0/\sigma = 1.09$. Compare with the experimental value of this ratio.

24.9 How many lattice points are there in a unit cell of (*a*) a body-centered lattice; (*b*) an end-centered lattice?

24.10 How many basis groups are there in a unit cell of a crystal with a face-centered lattice?

24.11 How many basis groups are there in a unit cell of a crystal with a primitive lattice?

24.12 One form of $CaCO_3(c)$ is the mineral aragonite. Its room-temperature density is 2.93 g/cm³. Its lattice is orthorhombic with $a = 4.94$ Å, $b = 7.94$ Å, and $c = 5.72$ Å at room temperature. Calculate the number of Ca^{2+} ions per unit cell of aragonite.

24.13 One form of TiO_2 is the mineral rutile, which has a tetragonal lattice with $a = 4.594$ Å and $c = 2.959$ Å at 25°C. There are two formula units per unit cell. Calculate the density of rutile at 25°C.

24.14 Sketch a (111) plane of a metallic fcc structure and indicate which atoms touch an atom in the (111) plane.

24.15 The solid KF has the NaCl-type structure. The density of KF at 20°C is 2.48 g/cm³. Calculate the unit-cell length and the nearest-neighbor distance in KF at 20°C.

24.16 CsBr has the CsCl-type structure. The density of CsBr at 20°C is 4.44 g/cm³. Calculate the unit-cell a value and the nearest-neighbor distance in CsBr.

24.17 The density of CaF_2 at 20°C is 3.18 g/cm³. Calculate the unit-cell length for CaF_2 at 20°C.

24.18 The density of diamond is 3.51 g/cm³ at 25°C. Calculate the carbon–carbon bond distance in diamond.

24.19 Ar crystallizes in an fcc lattice with one atom at each lattice point. (*a*) The unit-cell length is 5.311 Å at 0 K. Calculate the nearest-neighbor distance in Ar(c) at 0 K. (*b*) Assume a Lennard-Jones intermolecular potential and write down an expression for the cohesive energy of Ar(c) considering only nearest-neighbor interactions. Compare this expression with the more accurate one given in Prob. 24.8.

24.20 The lattice of crystalline $COCl_2$ is body-centered tetragonal with 16 formula units per unit cell. How many molecules does the basis consist of?

24.21 A crystal has a simple cubic lattice with a unit-cell length of 4.70 Å. For x-rays with $\lambda = 1.54$ Å, calculate the diffraction angles from (*a*) the (100) planes; (*b*) the (110) planes.

24.22 Visual examination of crystals of Ag shows Ag(c) to belong to the cubic system. Using x-rays with

$\lambda = 1.542$ Å, one finds the first few diffraction angles for a powder of Ag(c) to be 19.08°, 22.17°, 32.26°, 38.74°, 40.82°, 49.00°, and 55.35°. (a) Is the lattice P, F, or I? (b) Assign each of these diffraction angles to a set of planes. (c) Calculate the unit-cell edge length from each of these angles.

24.23 From data in Sec. 24.9 calculate the energy range for the occupied part of the 3s band in Na(c).

24.24 (a) Derive the expression for U of an Einstein crystal. (b) Verify (24.20) for C_V of an Einstein crystal.

24.25 Find the expression for A of an Einstein crystal.

24.26 Find the expression for S of an Einstein crystal.

24.27 (a) From (24.24), (24.21), and (22.38), derive the expression for U of a Debye crystal. (To simplify the final result, multiply numerator and denominator of the integrand in U by $e^{h\nu/kT}$.) (b) From the expression for U, find the Debye expression for C_V. Verify that the substitution $x \equiv h\nu/kT$ in the integrand gives (24.25).

24.28 The 1816 law of Dulong and Petit states that the product of the specific heat and the atomic weight of a metallic element is approximately 6 cal/(g °C). What is the statistical-mechanical basis for this law?

24.29 The Einstein temperature is 240 K for Al. What value of $\bar{C}_V$ for Al does the Einstein model predict at (a) 50 K; (b) 100 K; (c) 240 K; (d) 400 K?

24.30 For I_2 one finds $\bar{C}_V = 0.96$ cal mol^{-1} K^{-1} at 10 K. (a) Calculate Θ_D of I_2. (b) Calculate $\bar{C}_V$ of I_2 at 12 K.

24.31 (a) Show that for a metal, a plot of $\bar{C}_V/T$ vs. T^2 at low T should give a straight line whose slope and intercept allow Θ_D and γ to be calculated. (b) Use the following data for Ag to find Θ_D and γ from a graph:

T/K	1.35	2.00	3.00	4.00
$10^3\bar{C}_V$/(cal mol^{-1} K^{-1})	0.254	0.626	1.57	3.03

BIBLIOGRAPHY

Adamson, A. W.: *The Physical Chemistry of Surfaces*, 3d ed., Wiley-Interscience, 1976.

Andrews, F. C.: *Equilibrium Statistical Mechanics*, 2d ed., Wiley, 1975.

Aveyard, R., and D. A. Haydon: *An Introduction to the Principles of Surface Chemistry*, Cambridge University Press, 1973.

Bamford, C. H., and C. F. H. Tipper (eds.): *Comprehensive Chemical Kinetics*, Elsevier, 1969–.

Bates, R. G.: *Determination of* pH, 2d ed., Wiley-Interscience, 1973.

Bird, R. B., W. E. Stewart, and E. N. Lightfoot: *Transport Phenomena*, Wiley, 1960.

Blinder, S. M.: *Advanced Physical Chemistry*, Macmillan, 1969.

Bockris, J. O.'M., and D. M. Drazic: *Electro-chemical Science*, Taylor and Francis, 1972.

———, and A. K. Reddy: *Modern Electrochemistry*, Plenum, 1970.

Brand, J. C. D., J. C. Speakman, and J. K. Tyler: *Molecular Structure*, 2d ed., Halsted, 1975.

Buerger, M. J.: *Contemporary Crystallography*, McGraw-Hill, 1970.

Chang, R.: *Basic Principles of Spectroscopy*, McGraw-Hill, 1971.

Davidson, N. R.: *Statistical Mechanics*, McGraw-Hill, 1962.

Davies, C. W.: *Ion Association*, Butterworth, 1962.

Defay, R., I. Prigogine, A. Bellemans, and D. H. Everett: *Surface Tension and Adsorption*, Wiley, 1966.

Denbigh, K.: *The Principles of Chemical Equilibrium*, 3d ed., Cambridge University Press, 1971.

Eyring, H., D. Henderson, and W. Jost (eds.): *Physical Chemistry: An Advanced Treatise*, Academic, 1967–1975.

Forst, W.: *Theory of Unimolecular Reactions*, Academic, 1973.

Frost, A. A., and R. G. Pearson: *Kinetics and Mechanism*, 2d ed., Wiley, 1961.

Gardiner, W. C.: *Rates and Mechanisms of Chemical Reactions*, Benjamin, 1969.

Gatz, C. R.: *Introduction to Quantum Chemistry*, Merrill, 1971.

Gray, H.: *Chemical Bonds*, Benjamin, 1973.

Guggenheim, E. A.: *Thermodynamics*, 5th ed., North-Holland, 1967.

Hague, D. N.: *Fast Reactions*, Wiley-Interscience, 1971.

Halliday, D., and R. Resnick: *Physics*, Wiley, 1966.

Hammes, G. G. (ed.): *Investigation of Rates and Mechanisms of Reactions*, 3d ed. [vol. VI, pt. II of A. Weissberger (ed.), *Techniques of Chemistry*], Wiley, 1974.

Hill, T. L.: *An Introduction to Statistical Thermodynamics*, Addison-Wesley, 1960.

Hirschfelder, J. O., C. F. Curtiss, and R. B. Bird: *The Molecular Theory of Gases and Liquids*, Wiley, 1954.

Ives, D. J. G., and G. J. Janz (eds.): *Reference Electrodes*, Academic, 1961.

Jackson, E. A.: *Equilibrium Statistical Mechanics*, Prentice-Hall, 1968.

Johnston, H. S.: *Gas Phase Reaction Rate Theory*, Ronald, 1966.

Karplus, M., and R. N. Porter: *Atoms and Molecules*, Benjamin, 1970.

Kauzmann, W.: *Kinetic Theory of Gases*, Benjamin, 1966.

Kennard, E. H.: *Kinetic Theory of Gases*, McGraw-Hill, 1938.

Kestin, J., and J. R. Dorfman: *A Course in Statistical Thermodynamics*, Academic, 1971.
Kirkwood, J. G., and I. Oppenheim: *Chemical Thermodynamics*, McGraw-Hill, 1961.
Kittel, C.: *Introduction to Solid State Physics*, 5th ed., Wiley, 1976.
Knox, J. H.: *Molecular Thermodynamics*, Wiley-Interscience, 1971.
Laidler, K. L.: *Theories of Chemical Reaction Rates*, McGraw-Hill, 1969.
Levine, I. N.: *Molecular Spectroscopy*, Wiley-Interscience, 1975 (referred to as *M. S.*).
———: *Quantum Chemistry*, 2d ed., Allyn & Bacon, 1974 (referred to as *Q. C.*).
Levine, R. D., and R. B. Bernstein: *Molecular Reaction Dynamics*, Oxford University Press, 1974.
Lewis, E. S. (ed.): *Investigation of Rates and Mechanisms of Reactions*, 3d ed. [vol. VI, pt. II of A. Weissberger (ed.), *Techniques of Chemistry*], Wiley, 1974.
McClelland, B. J.: *Statistical Thermodynamics*, Chapman & Hall, 1973.
McQuarrie, D. A.: *Statistical Thermodynamics*, Harper & Row, 1973.
Mandl, F.: *Statistical Physics*, Wiley, 1971.
Münster, A.: *Classical Thermodynamics*, Wiley-Interscience, 1970.
Pitzer, K. S.: *Quantum Chemistry*, Prentice-Hall, 1953.
Present, R. D.: *Kinetic Theory of Gases*, McGraw-Hill, 1958.
Reed, T. R., and K. E. Gubbins: *Applied Statistical Mechanics*, McGraw-Hill, 1973.
Ricci, J. E.: *The Phase Rule and Heterogeneous Equilibrium*, Dover, 1966.
Robinson, P. J., and K. A. Holbrook: *Unimolecular Reactions*, Wiley-Interscience, 1972.
Robinson, R. A., and R. H. Stokes: *Electrolyte Solutions*, 2d ed., Butterworth, 1959.
Sands, D. E.: *Introduction to Crystallography*, Benjamin, 1969.
Shoemaker, D. P., G. W. Garland, and J. I. Steinfeld: *Experiments in Physical Chemistry*, 3d ed., McGraw-Hill, 1974.
Sokolnikoff, I. S., and R. M. Redheffer: *Mathematics of Physics and Modern Engineering*, 2d ed., McGraw-Hill, 1966.
Sparnaay, M. J.: *The Electrical Double Layer*, Pergamon, 1972.
Thomas, G. B.: *Calculus and Analytic Geometry*, alternate ed., Addison-Wesley, 1972.
Vold, M. J., and R. D. Vold: *Colloid Chemistry*, Reinhold, 1964.
Weston, R., and H. Schwarz: *Chemical Kinetics*, Prentice-Hall, 1972.
Zemansky, M.: *Heat and Thermodynamics*, 5th ed., McGraw-Hill, 1968.

APPENDIX

Standard-state thermodynamic properties at 25°C[a]

Substance	$\Delta H^\circ_{f,298}$ / kcal mol^{-1}	$\Delta G^\circ_{f,298}$ / kcal mol^{-1}	$\bar{S}^\circ_{298}$ / cal mol^{-1} K^{-1}	$\bar{C}^\circ_{P,298}$ / cal mol^{-1} K^{-1}
$Ag^+(aq)$	25.23	18.43	17.4	
$Br(g)$	26.74	19.70	41.80	4.97
$Br^-(aq)$	−29.05	−24.85	19.7	−33.9
$Br_2(l)$	0	0	36.38	18.09
$Br_2(g)$	7.39	0.75	58.64	8.61
C(graphite)	0	0	1.37	2.04
C(diamond)	0.453	0.693	0.568	1.462
$C(g)$	171.29	160.44	37.76	4.98
$CF_4(g)$	−221	−210	62.50	14.60
$CH_4(g)$	−17.88	−12.13	44.49	8.44
$CO(g)$	−26.416	−32.78	47.22	6.96
$CO_2(g)$	−94.051	−94.25	51.06	8.87
$CO_3^{2-}(aq)$	−161.84	−126.17	−13.6	
$COF_2(g)$	−151.7	−148.0	61.78	11.19
$C_2H_2(g)$	54.19	50.00	48.00	10.50
$C_2H_4(g)$	12.49	16.28	52.45	10.41
$C_2H_6(g)$	−20.24	−7.86	54.85	12.58
$C_2H_5OH(l)$	−66.37	−41.80	38.4	26.64
$(CH_3)_2O(g)$	−43.99	−26.93	63.64	15.39
$C_3H_8(g)$	−24.82	−5.61	64.51	17.57
$C_6H_6(g)$	19.82	30.99	64.34	19.52
Cyclohexene(g) (C_6H_{10})	−1.28	25.54	74.27	25.10
α-D-Glucose(c) ($C_6H_{12}O_6$)	−304.6	−217.6	50.7	52.3
Sucrose(c) ($C_{12}H_{22}O_{11}$)	−531.0	−369.1	86.1	101.7
$CH_3(CH_2)_{14}COOH(c)$	−212.9	−75.3	108.8	110.1
$CaCO_3$(calcite)	−288.46	−269.80	22.2	19.57
$CaCO_3$(aragonite)	−288.51	−269.55	21.2	19.42
$CaO(c)$	−151.79	−144.37	9.50	10.23
$Cl(g)$	29.08	25.26	39.46	5.22

Standard-state thermodynamic properties at 25°C[a] (*continued*)

Substance	$\Delta H^\circ_{f,298}$ / kcal mol^{-1}	$\Delta G^\circ_{f,298}$ / kcal mol^{-1}	$\bar{S}^\circ_{298}$ / cal mol^{-1} K^{-1}	$\bar{C}^\circ_{P,298}$ / cal mol^{-1} K^{-1}
$Cl^-(aq)$	−39.95	−31.37	13.5	−32.6
$Cl_2(g)$	0	0	53.29	8.10
Cu(*c*)	0	0	7.92	5.84
$Cu^{2+}(aq)$	15.48	15.66	−23.8	
$F_2(g)$	0	0	48.44	7.48
Fe(*c*)	0	0	6.52	6.00
$Fe^{3+}(aq)$	−11.6	−1.1	−75.5	
H(*g*)	52.10	48.58	27.39	4.968
$H^+(aq)$	0	0	0	0
$H_2(g)$	0	0	31.208	6.889
HD(*g*)	0.08	−0.35	34.343	6.978
$D_2(g)$	0	0	34.620	6.978
HBr(*g*)	−8.70	−12.77	47.46	6.96
HCl(*g*)	−22.06	−22.78	44.65	6.96
HF(*g*)	−64.8	−65.3	41.51	6.96
$HN_3(g)$	70.3	78.4	57.09	10.44
$H_2O(l)$	−68.315	−56.687	16.71	18.00
$H_2O(g)$	−57.796	−54.634	45.10	8.02
$H_2O_2(l)$	−44.88	−28.78	26.2	21.3
$H_2S(g)$	−4.93	−8.02	49.16	8.18
$K^+(aq)$	−60.3	−67.5	24.2	5
KCl(*c*)	−104.37	−97.70	19.73	12.26
Mg(*c*)	0	0	7.81	5.95
Mg(*g*)	35.30	27.04	35.50	4.97
MgO(*c*)	−143.8	−136.1	6.4	8.9
N(*g*)	112.98	108.88	36.62	4.97
$N_2(g)$	0	0	45.77	6.96
$NH_3(g)$	−11.02	−3.94	45.97	8.38
NO(*g*)	21.57	20.69	50.35	7.13
$NO_2(g)$	7.93	12.26	57.35	8.89
$NO_3^-(aq)$	−49.56	−26.61	35.0	−20.7
$N_2O_4(g)$	2.19	23.38	72.70	18.47
Na(*g*)	25.76	18.48	36.71	4.97
$Na^+(aq)$	−57.4	−62.6	14.0	11
NaCl(*c*)	−98.26	−91.79	17.24	12.07
O(*g*)	59.55	55.39	38.47	5.24
$O_2(g)$	0	0	49.003	7.02
$OH^-(aq)$	−54.97	−37.59	−2.57	−35.5
$PCl_3(g)$	−68.6	−64.0	74.49	17.17
$PCl_5(g)$	−89.6	−72.9	87.11	26.96
$SO_2(g)$	−70.94	−71.75	59.30	9.53
Si(*g*)	108.9	98.3	40.12	5.32
SiC(*β*, cubic)	−15.6	−15.0	3.97	6.42
SiO_2(quartz)	−217.7	−204.8	10.00	10.62

[a] Data mainly from D. D. Wagman et al., Selected Values of Chemical Thermodynamic Properties, *Natl. Bur. Stand. Tech. Note* 270-3, 270-4, 270-5, and 270-6, Washington, 1968–1971. The molality-scale standard state is used for solutes in aqueous solutions.

Properties of some isotopes[a]

Isotope	Abundance, %	Atomic mass	I	g_N
^{1}H	99.985	1.007825	$\frac{1}{2}$	5.5856
^{2}H	0.015	2.01410	1	0.8574
^{11}B	81	11.00930	$\frac{3}{2}$	1.792
^{12}C	98.892	12.000...	0	—
^{13}C	1.108	13.00335	$\frac{1}{2}$	1.4048
^{14}N	99.635	14.00307	1	0.4036
^{15}N	0.365	15.0001	$\frac{1}{2}$	−0.566
^{16}O	99.76	15.99491	0	—
^{19}F	100	18.99840	$\frac{1}{2}$	5.257
^{23}Na	100	22.98977	$\frac{3}{2}$	1.478
^{32}S	95.0	31.97207	0	—
^{35}Cl	75.5	34.96885	$\frac{3}{2}$	0.5479
^{37}Cl	24.5	36.96590	$\frac{3}{2}$	0.4560
^{39}K	93.2	38.9637	$\frac{3}{2}$	0.2609
^{79}Br	50.6	78.91833	$\frac{3}{2}$	1.404
^{81}Br	49.4	80.91629	$\frac{3}{2}$	1.513
^{127}I	100	126.90447	$\frac{5}{2}$	1.124

[a] Abundances are for the earth's crust. Atomic masses are the relative masses of the neutral atoms on the ^{12}C scale. I and g_N are the nuclear spin and g factor (Sec. 21.11).

ANSWERS TO SELECTED PROBLEMS

1.1 (*a*) 33.9 ft; (*b*) 0.995 atm. **1.4** 8.66 atm. **1.6** 31.0_6. **1.7** 0.133 mol N_2, 0.400 mol H_2, 1.33 mol NH_3. **1.10** (*a*) 1.99×10^{-23} g; (*b*) 2.99×10^{-23} g. **1.11** (*a*) 17.5 atm; (*b*) 0.857 for H_2 and 0.143 for CH_4. **1.12** 32.3 cm^3. **1.16** (*a*) 0.60 mol and 0.40 mol; (*b*) 0.60 g and 0.40 g. **1.21** 2.6×10^{-4} K^{-1}, 4.9×10^{-5} atm^{-1}, 5.3 atm/K. **1.23** 717.8 K. **1.25** 23 atm. **1.26** (*a*) 2.5×10^{19}; (*b*) 3.2×10^{10}; (*c*) 3.2×10^{5}. **1.27** 0.0361 g and 0.619. **1.28** 5.3×10^{21} g. **1.29** (*a*) 6.4×10^{-5} atm; (*b*) 460 K. **1.30** 0.24.

2.4 (*d*) 11,200 m/s, 25,000 mi/hr. **2.7** (*a*) 1 cal = 5.5 J; (*b*) 1 cal = 4.15 J; (*c*) 1 cal = 3.58 J. **2.12** 0.14°C. **2.17** (*a*) 146 J/m^3. **2.22** (*a*) ∞; (*b*) $-\infty$; (*c*) ∞; (*d*) 0. **2.24** (*a*) 1436 cal, 0.039 cal, 1436 cal, 1436 cal; (*b*) 1801 cal, −0.019 cal, 1801 cal, 1801 cal; (*c*) 9717 cal, −741 cal, 8976 cal, 9717 cal. **2.27** 3270 cal, −3270 cal, 0, 0. **2.29** 0, 98.8 J, 98.8 J, 138 J. **2.31** 15°C. **2.32** 18.001°C.

3.1 (*a*) 74.6%; (*b*) 746 J, 254 J. **3.6** (*a*) 373.2°M; (*b*) 199.99°M. **3.9** 4.16 cal/K. **3.11** (*a*) 6.66 J/K; (*b*) 14.1 J/K; (*c*) 8.06 J/K. **3.14** (*a*) 20.2 J/K; (*b*) −14.4 J/K; (*c*) −5.76 J/K. Zero for cycle. **3.15** −2.73 cal/K. **3.17** (*a*) 32.0°C; (*b*) −1.59 cal/K; (*c*) 1.87 cal/K; (*d*) 0.28 cal/K. **3.25** (*b*) 7. **3.30** -2.03×10^{-5} cal/K.

4.4 −3220 J, −3220 J. **4.6** (*a*) 0; (*b*) −28 cal. **4.7** 2040 atm. **4.8** (*a*) 40 cal mol^{-1} K^{-1}; (*b*) −0.4 cal mol^{-1} atm^{-1}; (*c*) 70 cal cm^{-3}; (*d*) 0.13 cal mol^{-1} K^{-2}; (*e*) −0.0012 cal mol^{-1} K^{-1} atm^{-1}; (*f*) 36.4 cal mol^{-1} K^{-1}. **4.10** −0.0221 K/atm. **4.13** (*a*) -1.66×10^{-6} cal mol^{-1} K^{-1} atm^{-1}; (*b*) 6.64 cal mol^{-1} K^{-1}. **4.15** 72.1 cal. **4.27** 4960 ft/s.

5.1 (*a*) 0; (*b*) 689 cal/mol; (*c*) 0. **5.3** (*b*) −1.09 cal/mol. **5.5** (*a*) 18 kcal/mol; (*b*) 21 kcal/mol. **5.7** (*a*) −152 kcal/mol; (*b*) −304 kcal/mol; (*c*) 76 kcal/mol. **5.9** (*a*) −268.65 kcal/mol; (*b*) −247.61 kcal/mol; (*c*) −228.6 kcal/mol. **5.11** 33.99 kcal/mol. **5.13** (*a*) 16.71 cal mol^{-1} K^{-1}. **5.15** (*a*) −93.31 cal mol^{-1} K^{-1}; (*b*) −36.53 cal mol^{-1} K^{-1}; (*c*) −5.6 cal mol^{-1} K^{-1}. **5.16** (*a*) −240.83 kcal/mol, −236.72 kcal/mol, −226.9 kcal/mol; (*b*) −240.83 kcal/mol, −236.73 kcal/mol, −227.0 kcal/mol. **5.22** (*a*) 10.5 kcal/mol; (*b*) 28.3 cal mol^{-1} K^{-1}; (*c*) 2.0_6 kcal/mol; (*d*) 2.0_6 kcal/mol; (*e*) 2.05 kcal/mol. **5.24** 10 kcal/mol. **5.27** 0.067 cal mol^{-1} K^{-1}.

6.1 −24.32 kcal/mol. **6.2** −2060 cal/mol. **6.9** 12.5. **6.10** 0.633 mol, 2.633 mol, 2.367 mol. **6.15** 1.50 mol CO_2, 0.50 mol CF_4, 8×10^{-4} mol COF_2. **6.20** 22.5. **6.24** 4.83 kcal/mol, 22.10 kcal/mol, 34.5 cal mol^{-1} K^{-1}, 0.

7.5 (*a*) 0.130 g liquid and 0.230 g vapor; (*b*) 0.360 g vapor. **7.7** 10,200 cal mol^{-1}. **7.8** (*a*) 1481 torr; (*b*) 85°C. **7.12** (*a*) −38.4°C; (*b*) −34.7°C. **7.15** 14,900 atm. **7.18** 2052 cal/mol, 2053 cal/mol.

8.3 (*a*) 317 atm; (*b*) 804 atm; (*c*) 172 atm. **8.7** (*a*) 3.61×10^6 cm^6 atm mol^{-2}, 42.9 cm^3 mol^{-1}; (*b*) 2.5 cal/mol; (*c*) 0.012 cal mol^{-1} K^{-1} and 6.1 cal/mol. **8.12** 0.01745. **8.18** 49 atm, 565 K, 260 cm^3/mol.

9.2 0.00493, 0.0283, 0.283 mol/kg, 1.62 mol/kg. **9.7** 36.9 cm^3/mol, 18.1 cm^3/mol. **9.10** 25.0 cm^3/mol. **9.13** (*a*) 40.4 torr, 10.2 torr; (*b*) 0.798, 0.202. **9.15** 2.755 cal/K. **9.18** (*a*) $6.8_2 \times 10^4$ atm; (*b*) 1.64 mg. **9.20** (*b*) 0.0164. **9.33** 6.5×10^{-3}, 5.6×10^{-5}. **9.34** 0.211, 0.789. **9.40** 173 torr, 60.8 torr.

10.1 (*a*) 1.11, 2.04; (*b*) −168 cal; (*c*) −306 cal. **10.3** (*b*) 0.843, 0.0337, 1.0036, 0.9635. **10.13** (*b*) 2.85, 3.15. **10.14** (*a*) 0.457; (*b*) 0.288. **10.17** 0.988, 0.961. **10.21** (*a*) −19.10 kcal/mol; −13.34 kcal/mol; 19.28 cal mol^{-1} K^{-1}; (*b*) −24.77 kcal/mol, −0.53 kcal/mol; 81.4 cal mol^{-1} K^{-1}. **10.24** (*a*) 3749 cal; (*b*) 4080 cal.

11.1 1.34×10^{-7} mol/kg. **11.2** 1.55×10^{-7} mol/kg. **11.3** 1.05×10^{-7} mol/kg. **11.4** 4.28×10^{-4} mol/kg. **11.6** 19.10 kcal/mol, −18.0 cal mol^{-1} K^{-1}, 13.73 kcal/mol. **11.15** (*a*) $7._2$; (*b*) 0.56. **11.16** 1.4×10^{-23} atm. **11.17** 2.72 mol Fe_3O_4, 3.72 mol CO, 1.84 mol FeO, 4.28 mol CO_2. **11.18** 2.51×10^{-5} mol^2/kg^2. **11.23** (*c*) 1.35×10^{-5} mol/kg. **11.25** (*a*) 0.039 mol $CaCO_3$, 0.0109 mol CaO, 0.0109 mol CO_2; (*b*) 0 mol $CaCO_3$, 0.0050 mol CaO, 0.0050 mol CO_2.

12.1 1488 torr. **12.4** 339. **12.6** (*a*) 180; (*b*) 10.8 kcal/mol. **12.7** 6.89 kcal/mol. **12.10** 0.027 mol naphthalene, 0.014 mol anthracene. **12.13** 736 cm. **12.17** (*b*) 1.95 atm. **12.32** (*a*) 0.306; (*b*) 0.306; (*c*) 0.0360. **12.40** 29.0.

13.1 (*a*) 4.8 cm^2; (*b*) 1.0×10^6 cm^2. **13.2** 220 ergs. **13.3** 20.2 mN/m. **13.4** 6.0%. **13.6** 4.9×10^{-6} atm. **13.7** 22.6 dyn/cm. **13.8** 3.22 cm. **13.11** 53.0 dyn/cm. **13.16** 25 Å. **13.18** -3.0×10^{-10} mol/cm^2. **13.21** (*b*) 17.726 torr.

14.1 4.6×10^{-8} N. **14.2** (*a*) 3.6×10^{10} V/m; (*b*) 0.90×10^{10} V/m. **14.3** −3.60 V. **14.4** 2.75×10^{-20} C, 0.172. **14.7** 10,000 J/mol. **14.8** (*a*) 5.79×10^5 C; (*b*) -5.79×10^4 C. **14.10** (*a*) 2; (*b*) 1; (*c*) 2; (*d*) 6; (*e*) 2. **14.12** (*a*) 0.355 V; (*b*) 0.38 V. **14.15** (*b*) 1.00057 atm; (*c*) 7.3×10^{-6} V. **14.16** (*b*) 0.046 V; (*c*) 0.046 V; (*d*) −8900 J/mol; 10,500 J/mol; 65.2 J mol^{-1} K^{-1}. **14.19** (*a*) 2.1×10^{41}; (*b*) 5.0×10^{-27}. **14.21** (*b*) −0.16 V. **14.22** (*b*) 0.36 V; (*c*) 0.27_5 V. **14.24** 1.3. **14.26** 4.7×10^{-50}. **14.28** (*a*) −56,700 J/mol, 8.8×10^9; (*b*) −28,400 J/mol; 9.4×10^4. **14.30** −0.152 V. **14.32** −31.34 kcal/mol, −31.34 kcal/mol. **14.37** (*b*) 1.75×10^{-5}. **14.39** −0.0205 V. **14.40** −1.599 V. **14.43** 4.68. **14.44** −0.0104 V. **14.46** 1.25×10^{-39} C^2 N^{-1} m, 1.12×10^{-29} m^3.

15.1 (*a*) 3720 J. **15.2** (*a*) 1.18×10^{-20} J. **15.3** 1.366. **15.4** 5/3. **15.6** 4.50×10^4. **15.8** (*a*) 5.32×10^4 cm/s; (*b*) 4.90×10^4 cm/s; (*c*) 4.35×10^4 cm/s. **15.23** 0.0152 torr. **15.25** 8.9×10^{17}, 5.8×10^{-4} g. **15.26** (*a*) 3.5×10^9 s^{-1}; (*b*) 4.3×10^{28} s^{-1} cm^{-3}. **15.28** (*a*) 510 Å; (*b*) 770,000 Å. **15.31** 25.8 cal mol^{-1} K^{-1}. **15.34** 3.5, 15.17.

16.1 (*a*) 288 J; (*b*) 0.161 J/K. **16.2** (*b*) 1.4×10^{-3} J K^{-1} cm^{-1} s^{-1}. **16.4** 6.05 mJ K^{-1} cm^{-1} s^{-1}. **16.8** (*a*) 5.66 cP; (*b*) 187. **16.10** 4.17×10^4 cm/s. **16.12** 6.04×10^{-5} P. **16.14** (*a*) 0.025 cm; (*b*) 0.19 cm; (*c*) 0.95 cm. **16.18** −0.083 μm, 2.5 μm. **16.19** (*a*) 0.16 cm^2 s^{-1}; (*b*) 0.016 cm^2 s^{-1}. **16.21** 2.4×10^{-5} cm^2 s^{-1}. **16.23** (*b*) 0.40 cm^2 s^{-1}. **16.26** 6.2×10^{18}. **16.27** 0.0104 Ω, 95.8 Ω^{-1}. **16.28** 0.25 A. **16.29** 1.19 g. **16.31** (*a*) 248.4 Ω^{-1} cm^2 mol^{-1}; (*b*) 124.2 Ω^{-1} cm^2 equiv^{-1}. **16.32** 4.668×10^{-4} cm^2 V^{-1} s^{-1}, 0.3894. **16.35** (*a*) 100.4 Ω^{-1} cm^2 mol^{-1}; (*b*) 391 Ω^{-1} cm^2 mol^{-1}. **16.37** (*a*) 6.99×10^{-4} cm^2 V^{-1} s^{-1}; (*b*) 0.017 cm/s; (*c*) 1.37 Å. **16.38** (*a*) 145.0 Ω^{-1} cm^2 mol^{-1}; (*b*) 307 Ω^{-1} cm^2 mol^{-1}. **16.39** (*b*) 0.426, 0.574. **16.42** $9.9_7 \times 10^{-15}$ mol^2/dm^6. **16.48** 89.9 Ω^{-1} cm^2 mol^{-1}. **16.50** (*a*) 400,000 g/mol, 500,000 g/mol; (*b*) 300,000 g/mol; 400,000 g/mol. **16.52** 63,000.

17.1 (*a*) 7.1×10^{-8} mol dm^{-3} s^{-1}, 8.5×10^{-7} mol s^{-1}; (*b*) -1.4×10^{-7} mol dm^{-3} s^{-1}; (*c*) 1.0×10^{18}; (*d*) 3.46×10^{-5} s^{-1}, 14×10^{-8} mol dm^{-3} s^{-1}, 17×10^{-7} mol s^{-1}. **17.2** (*c*) 6.09×10^{-6} atm^{-1} s^{-1}. **17.5** (*a*) 0.00066 s^{-1}, 0.00133 s^{-1}; (*b*) 905 s, 1731 s. **17.6** (*a*) 2.00 ×

10^4 s; (*b*) 5.0×10^{-4} mol dm^{-3}. **17.8** 0.030 mol NO_2, 2.01_5 mol F_2, 1.97 mol NO_2F. **17.15** (*a*) 8.0×10^6 dm^3 mol^{-1} s^{-1}; (*b*) 2.2×10^{-6} s. **17.21** 0.036_1 dm^6 mol^{-2} s^{-1}. **17.33** 0.018 dm^3 mole^{-1} s^{-1}. **17.35** 45.5 kcal mol^{-1}, 1.9×10^{11} dm^3 mole^{-1} s^{-1}. **17.39** (*b*) 3.87×10^{-7} s^{-1}; (*c*) 2.36×10^{10} s, 8.95×10^5 s, 795 s. **17.41** 81.8 kcal mol^{-1}. **17.55** 1.3×10^{10} dm^3 mole^{-1}s^{-1}. **17.58** 1. **17.62** -3.154×10^{11} J/mol. **17.63** 2.41×10^5 yr. **17.65** (*a*) $1.0_8 \times 10^{-10}$ %; (*b*) 0.0286 min^{-1} (g C)$^{-1}$; (*c*) 4800 yr. **17.69** 5.9×10^9 yr. **17.70** 1.4×10^{14} g/cm^3.

18.1 (*c*) 1.76×10^{13} s^{-1}, 1.76×10^{14} s^{-1}; (*d*) 5800 K. **18.2** (*b*) $9._9 \times 10^{25}$ J/s; (*c*) $3._5 \times 10^{16}$ kg. **18.3** (*a*) 5.3×10^{14} s^{-1}, $1.2_1 \times 10^{15}$ s^{-1}, 5.7×10^{-5} cm, 2.5×10^{-5} cm. **18.4** 2.8×10^{-12} erg. **18.7** (*a*) 6.6×10^{-10} cm, 1.1×10^{-30} cm. **18.8** (*a*) 2.2×10^{-33} g; (*b*) 2.2×10^{-32} g. **18.10** 2.9×10^{-6} cm. **18.11** (*b*) 0.0908, 0.2500, 0.3031. **18.12** (*a*) 3.45×10^{-5}, 9.05×10^{-5}. **18.23** (*a*) 1.02×10^4 dyn/cm; (*b*) 5.1×10^{30}.

19.1 (*a*) 7.7×10^{-19} J, 7.7×10^{-12} erg, 4.8 eV. **19.2** 7.5×10^{-26} g. **19.4** 1.23 Å. **19.9** (*a*) 54.4 V; (*b*) 122.4 V. **19.12** 6.8 V. **19.14** 45°, 90°, 135°. **19.27** 54.7°, 125.3°. **19.34** 3, 3/2. **19.37** 4406 V.

20.1 (*a*) 1.07 Å, 1.43 Å, 0.96 Å; (*b*) 1.07 Å, 1.15 Å. **20.4** 1.4_5 D, 1.2 D. **20.6** 4.3 D. **20.7** (*a*) 0.5; (*b*) 1.1; (*c*) 0.6. **20.10** (*a*) 5.70 eV; (*b*) 10.4 D. **20.21** (*a*) 58 kcal/mol; (*b*) -1 kcal/mol.

21.1 2.42×10^{14} s^{-1}, 1.24×10^{-4} cm, 8.06×10^3 cm^{-1}. **21.2** 2.25×10^{10} cm/s, 4430 Å. **21.5** (*a*) 4.2×10^{-6}; (*b*) 1.6×10^{-54}. **21.7** $1.1_3 \times 10^3$ dm^3 mol^{-1} cm^{-1}. **21.9** 20.6%. **21.11** (*a*) 9.757 eV; (*b*) 11.092 eV. **21.13** 1.1308_9 Å; (*b*) 1.1308_3 Å. **21.16** 1.176×10^6 dyn/cm. **21.18** The ratios to the $J = 0$ population are 1, 2.714, 3.703, 3.839, 3.307, 2.450, 1.588. **21.22** (*a*) 0, 3.046×10^9 s^{-1}, 7.249×10^9 s^{-1}; (*b*) 3046 MHz, 6093 MHz. **21.27** 364.7 nm. **21.29** 13.50 nm. **21.31** 5.0×10^{-20} J/T. **21.33** (*a*) 20.49 MHz; (*b*) 27.32 MHz. **21.34** (*a*) 1.41 T; (*b*) 7.05 T. **21.40** 70.0 GHz. **21.42** 13.8.

22.5 (*a*) 3,628,810; (*b*) $9.332621569 \times 10^{157}$; (*c*) $4.0238726002 \times 10^{2567}$. **22.7** 1.8×10^{-9}. **22.10** (*a*) 1.3×10^{-5}; (*b*) 0.044; (*c*) 0.36. **22.12** (*a*) 0.999955; (*b*) 0.950; (*c*) 0.865; (*d*) 0.632; (*e*) 0.095. **22.14** (*b*) 480 K; (*c*) 0.0393. **22.15** (*a*) 3.935; (*b*) 1.53×10^{23}, 2.78×10^{23}, 1.71×10^{23}; (*c*) 0.669×10^{23}; 2.01×10^{23}, 3.35×10^{23}. **22.23** 30.12, 34.94, 36.98, 39.19, and 40.53 cal mol^{-1} K^{-1}. **22.25** 889 cal mol^{-1}. **22.27** $10^{4.48 \times 10^{24}}$, 6.0×10^{30}. **22.28** (*a*) 41.50 cal mol^{-1} K^{-1}; (*b*) 4.967 cal mol^{-1} K^{-1}; (*c*) 6.954 cal mol^{-1} K^{-1}. **22.34** 3.51×10^{30}, 108, 1.0000322, 1. **22.37** 49.14 cal mol^{-1} K^{-1}. **22.46** (*a*) 7.1×10^{-8} dyn; (*b*) 1.2×10^{-6} dyn; (*c*) -5.0×10^{-4} dyn. **22.49** (*a*) 4.86×10^{-15} erg; (*b*) 3.42×10^{-14} erg.

23.1 (*b*) 1.66×10^{-21} cm^3 s^{-1}. **23.3** (*a*) 3×10^{14} cm^3 mole^{-1} s^{-1}; (*b*) 0.003. **23.11** 8.9 kcal mol^{-1}. **23.14** (*a*) 0.65, 0.29, 0.052; (*b*) 1.53, 3.45, 19.4. **23.16** 15.1 kcal mol^{-1}, -4.2 cal mol^{-1} K^{-1}, 16.4 kcal mol^{-1}.

24.1 (*a*) 171.3 kcal/mol; (*b*) 295.8 kcal/mol; (*c*) 445.7 kcal/mol. **24.7** 1.38629. **24.8** (*a*) 1.99 kcal mol^{-1}. **24.13** 4.249 g/cm^3. **24.15** 5.38 Å, 2.69 Å. **24.18** 1.54_5 Å. **24.21** (*a*) 9.4°, 19.1°, 29.4°, 40.9°, 55.0°, 79.4°; (*b*) 13.4°; 27.6°; 44.0°, 67.9°. **24.29** (*a*) 1.15 cal mol^{-1} K^{-1}; (*b*) 3.77 cal mol^{-1} K^{-1}; (*c*) 5.49 cal mol^{-1} K^{-1}; (*d*) 5.79 cal mol^{-1} K^{-1}. **24.30** (*a*) 78 K; (*b*) 1.66 cal mol^{-1} K^{-1}.

INDEX

INDEX

Fundamental constants[a]

Constant	Symbol	SI value	Gaussian value
Avogadro constant	N_0	6.0220×10^{23} mol^{-1}	6.0220×10^{23} mol^{-1}
Gas constant	R	8.314 J mol^{-1} K^{-1}	8.314×10^7 erg mol^{-1} K^{-1} 82.06 cm^3 atm mol^{-1} K^{-1}
Proton charge	e	1.60219×10^{-19} C	
	e'		4.80324×10^{-10} statC
Planck constant	h	6.6262×10^{-34} J s	6.6262×10^{-27} erg s
Electron rest mass	m	9.1095×10^{-31} kg	9.1095×10^{-28} g
Proton rest mass	m_p	1.67265×10^{-27} kg	1.67265×10^{-24} g
Faraday constant	$\mathscr{F}$	96,485 C mol^{-1}	
Boltzmann constant	k	1.3807×10^{-23} J K^{-1}	1.3807×10^{-16} erg K^{-1}
Speed of light in vacuum	c	2.99792×10^8 m s^{-1}	2.99792×10^{10} cm s^{-1}
Permittivity of vacuum	ε_0	8.85419×10^{-12} C^2 N^{-1} m^{-2}	
Permeability of vacuum	μ_0	$4\pi \times 10^{-7}$ N C^{-2} s^2	
Gravitational constant	G	6.67×10^{-11} m^3 s^{-2} kg^{-1}	6.67×10^{-8} cm^3 s^{-2} g^{-1}

[a] Adapted from E. R. Cohen and B. N. Taylor, *J. Phys. Chem. Ref. Data*, **2,** 663 (1973).

Greek alphabet

Alpha	A	α	Iota	I	ι	Rho	P	ρ
Beta	B	β	Kappa	K	κ	Sigma	Σ	σ
Gamma	Γ	γ	Lambda	Λ	λ	Tau	T	τ
Delta	Δ	δ	Mu	M	μ	Upsilon	Υ	υ
Epsilon	E	ε	Nu	N	ν	Phi	Φ	ϕ
Zeta	Z	ζ	Xi	Ξ	ξ	Chi	X	χ
Eta	H	η	Omicron	O	o	Psi	Ψ	ψ
Theta	Θ	θ	Pi	Π	π	Omega	Ω	ω